HANDBOOK OF
HETEROCYCLIC CHEMISTRY

SECOND EDITION 2000

Related Titles of Interest

BOOKS

Tetrahedron Organic Chemistry Series
CARRUTHERS: Cycloaddition Reactions in Organic Synthesis
CLARIDGE: High-Resolution NMR Techniques in Organic Chemistry
DEROME: Modern NMR Techniques for Chemistry Research
FINET: Ligand Coupling Reactions with Heteroaromatic Compounds
GAWLEY & AUBÉ: Principles of Asymmetric Synthesis
HASSNER & STUMER: Organic Syntheses based on Named Reactions and Unnamed Reactions
LEVY & TANG: The Chemistry of *C*-Glycosides
LI & GRIBBLE: Palladium in Heteroaryl Synthesis
McKILLOP: Advanced Problems in Organic Reaction Mechanisms
OBRECHT: Solid Supported Combinatorial & Parallel Synthesis of Small Molecular-Weight Compound Libraries
PERLMUTTER: Conjugate Addition Reactions in Organic Synthesis
SESSLER & WEGHORN: Expanded Contracted and Isomeric Porphyrins
WONG & WHITESIDES: Enzymes in Synthetic Organic Chemistry

Major Reference Works
BARTON, NAKANISHI, METH-COHN: Comprehensive Natural Products Chemistry
BARTON & OLLIS: Comprehensive Organic Chemistry
KATRITZKY & REES: Comprehensive Heterocyclic Chemistry I CD-Rom
KATRITZKY, REES & SCRIVEN: Comprehensive Heterocyclic Chemistry II
KATRITZKY, METH-COHN & REES: Comprehensive Organic Functional Group Transformations
SAINSBURY: Rodd's Chemistry of Carbon Compounds, 2nd edition
TROST, FLEMING: Comprehensive Organic Synthesis

Also of Interest
GRIBBLE & GILCHRIST: Progress in Heterocyclic Chemistry
MATHEY: Phosphorus Heterocycles
PELLETIER: Alkaloids: Chemical and Biological Perspectives

Journals
BIOORGANIC & MEDICINAL CHEMISTRY
BIOORGANIC & MEDICINAL CHEMISTRY LETTERS
CARBOHYDRATE RESEARCH
HETEROCYCLES (distributed by Elsevier)
IL FARMACO
JOURNAL OF FLUORINE CHEMISTRY
PHYTOCHEMISTRY
TETRAHEDRON
TETRAHEDRON: ASYMMETRY
TETRAHEDRON LETTERS

Full details of all Elsevier Science publications, and a free specimen copy of any Elsevier Science journal, are available on request from your nearest Elsevier Science office or from the Elsevier Science web site www.elsevier.nl

HANDBOOK OF HETEROCYCLIC CHEMISTRY

SECOND EDITION 2000

ALAN R. KATRITZKY

University of Florida, USA

ALEXANDER F. POZHARSKII

University of Rostov, Russia

PERGAMON
AMSTERDAM – BOSTON – LONDON – NEW YORK – OXFORD – PARIS – SAN DIEGO
SAN FRANCISCO – SINGAPORE – SYDNEY – TOKYO

ELSEVIER SCIENCE Ltd
The Boulevard, Langford Lane
Kidlington, Oxford OX5 1GB, UK

First edition 2000
Second impression 2003

Library of Congress Cataloging in Publication Data catalog record number is 00-020752.

British Library Cataloguing in Publication Data
A catalogue record from the British Library has been applied for.

ISBN: 0-08-042988-2 (HB)
ISBN: 0-08-042989-0 (PB)

♾ The paper used in this publication meets the requirements of ANSI/NISO Z39.48-1992 (Permanence of Paper).
Printed in The Netherlands.

Outline Contents

Foreword to Second Edition of the Handbook   vii

Detailed Contents   ix

1   Preliminaries   1
    1.1   Foreword to First Edition of the Handbook   3
    1.2   Explanation of the Reference System   5
    1.3   Topics Covered and Arrangement of Handbook-II   7

2   Structure of Heterocycles   9
    2.1   Overview   11
    2.2   Structure of Six-membered Rings   15
    2.3   Structure of Five-membered Rings with One Heteroatom   55
    2.4   Structure of Five-membered Rings with Two or More Heteroatoms   91
    2.5   Structure of Small and Large Rings   141

3   Reactivity of Heterocycles   165
    3.1   Overview   167
    3.2   Reactivity of Six-membered Rings   169
    3.3   Reactivity of Five-membered Rings with One Heteroatom   297
    3.4   Reactivity of Five-membered Rings with Two or More Heteroatoms   367
    3.5   Reactivity of Small and Large Rings   475

4   Synthesis of Heterocycles   499
    4.1   Overview   501
    4.2   Synthesis of Monocyclic Rings with One Heteroatom   511
    4.3   Synthesis of Monocyclic Rings with Two or More Heteroatoms   551
    4.4   Synthesis of Bicyclic Ring Systems without Ring Junction Heteroatoms   605
    4.5   Synthesis of Tri- and Poly-cyclic Ring Systems without Ring Junction Heteroatoms   657
    4.6   Synthesis of Fused Ring Systems with Ring Junction Heteroatoms   667

5   Appendixes   685
    Appendix A: Introduction to "Comprehensive Heterocyclic Chemistry - II"   687
    Appendix B: Short Contents of "Comprehensive Heterocyclic Chemistry-II"   695
    Appendix C: Short Contents of "Comprehensive Heterocyclic Chemistry"   701
    Appendix D: Miscellaneous (MI) References   705

Subject Index   709

Foreword to Second Edition of "Handbook of Heterocyclic Chemistry"

In the fourteen years since the first edition of this Handbook (Handbook-I) was published much has happened. In particular, "Comprehensive Heterocyclic Chemistry-II" (CHEC-II) has been published, updating "Comprehensive Heterocyclic Chemistry" (CHEC), on which the first edition of the Handbook was based. In view of the very large volume of heterocyclic chemistry which has appeared, it was essential to thoroughly revise and update the Handbook for its second edition.

In this endeavor the original author of Handbook-I has been joined by Alexander Pozharskii of the University of Rostov, Russia. We have adopted for Handbook-II (for the most part) the arrangement and classification of Handbook-I. However, a large amount of new material has been added, most of which has been taken from the 10 volumes of CHEC-II. In many instances we have utilized pieces of actual text and structural schemes directly from CHEC-II. Because of the close relationship to CHEC and to CHEC-II, the Foreword, Introduction to CHEC-II and Short Contents of both CHEC and CHEC-II are reprinted as part of the "Part I: Preliminaries" of Handbook-II. Just as for Handbook-I, the organization of Handbook-II follows that of the CHEC and CHEC-II rather closely except in the case of ring synthesis, and Chapter 1.3 of the "Handbook" describes these divergencies.

The authors of the Handbook-II have many acknowledgments to make for help. We particularly thank Charles Rees, Eric Scriven and John Zoltewicz who have read the work in its entirety. We are most grateful for the technical assistance from Ms. Hong Wu, Dr. Olga V. Denisko, Dr. O. V. Dyablo, Dr. A. V. Gulevskaya, Mr. S. Kakasyev and Ms. V. N. Stepanova and to Dr. Linde Katritzky for her constant support.

The updating required to produce Handbook-II has not been an easy one, and we are quite aware that there are numerous deficiencies. Many hard decisions had to be made to keep the overall length within reasonable bounds. The choice of what to put in and what to leave out was often difficult, and we have probably overlooked many important contributions that should have been included. Despite our best efforts, undoubtedly errors remain and we, the authors, take full responsibility for them.

ALAN R. KATRITZKY
Gainesville, Florida USA

ALEXANDER F. POZHARSKII
Rostov, Russia

Detailed Contents

PART 1: PRELIMINARIES

1.1 Foreword to the First Edition of the Handbook   3

1.2 Explanation of the Reference System   5

1.3 Topics Covered and Arrangement of the Handbook-II   7

PART 2: STRUCTURE OF HETEROCYCLES

2.1 Overview   11

2.1.1 RELATIONSHIP OF HETEROCYCLIC AND CARBOCYCLIC AROMATIC COMPOUNDS   11

2.1.2 ARRANGEMENT OF STRUCTURE CHAPTERS   12

2.1.3 NOMENCLATURE   12

2.2 Structure of Six-membered Rings   15

2.2.1 SURVEY OF POSSIBLE STRUCTURES: NOMENCLATURE   15

   2.2.1.1 Aromatic Nitrogen Systems without Exocyclic Conjugation   15
   2.2.1.2 Aromatic Systems with Exocyclic Conjugation   16
   2.2.1.3 Ring Systems Containing One Oxygen or Sulfur   18
   2.2.1.4 Rings Containing Nitrogen with Oxygen and/or Sulfur   19
   2.2.1.5 Fully Conjugated but Non-aromatic Compounds   19

2.2.2 THEORETICAL METHODS: CALCULATIONS   20

   2.2.2.1 The σ,π-MO Approximation (Hückel theory)   20
   2.2.2.2 More Sophisticated Semi-Empirical Methods   21
   2.2.2.3 Ab Initio Calculations   21
   2.2.2.4 Molecular Mechanics   22
   2.2.2.5 Some Results of Calculations of Six-membered Heterocyclic Molecules   22

2.2.3 STRUCTURAL METHODS   24

   2.2.3.1 X-Ray Diffraction   24
   2.2.3.2 Microwave Spectroscopy   26
   2.2.3.3 ^1H NMR Spectra   26
     2.2.3.3.1 Chemical shifts   26
     2.2.3.3.2 Coupling constants   27
   2.2.3.4 ^{13}C NMR Spectra   27
     2.2.3.4.1 Aromatic systems: chemical shifts   27
     2.2.3.4.2 Aromatic systems: coupling constants   30
     2.2.3.4.3 Saturated systems   32
   2.2.3.5 Nitrogen and Oxygen NMR Spectra   33
   2.2.3.6 UV and Related Spectra   35
     2.2.3.6.1 Features of UV spectra   35
     2.2.3.6.2 Applications of UV spectroscopy   36
   2.2.3.7 IR and Raman Spectra   37
   2.2.3.8 Mass Spectrometry   38
   2.2.3.9 Photoelectron Spectroscopy   41

2.2.4 THERMODYNAMIC ASPECTS   41

   2.2.4.1 Intermolecular Forces   41
     2.2.4.1.1 Melting and boiling points   41
     2.2.4.1.2 Solubility   41
     2.2.4.1.3 Gas-liquid chromatography   43
   2.2.4.2 Fully Conjugated Rings: Aromaticity   43
     2.2.4.2.1 Background   43

	2.2.4.2.2 *Energetical criteria*	44
	2.2.4.2.3 *Structural criteria*	45
	2.2.4.2.4 *Magnetic criteria*	46
	2.2.4.3 *Partially and Fully Reduced Rings*	47
2.2.5	TAUTOMERISM	47
	2.2.5.1 *Prototropic Tautomerism of Substituent Groups*	47
	2.2.5.1.1 *Pyridones and Hydroxypyridines*	47
	2.2.5.1.2 *Other substituted azines*	51
	2.2.5.2 *Tautomerism of Dihydro and Tetrahydro Compounds*	52
	2.2.5.3 *Tautomerism of Other Substituents (Non-prototropic)*	52
	2.2.5.4 *Ring-Chain and Valence Bond Tautomerism*	52
2.2.6	SUPRAMOLECULAR STRUCTURES	53

2.3	**Structure of Five-membered Rings with One Heteroatom**	**55**
2.3.1	SURVEY OF POSSIBLE STRUCTURES	55
	2.3.1.1 *Monocyclic Compounds*	55
	2.3.1.2 *Benzo Derivatives*	56
2.3.2	THEORETICAL METHODS	58
2.3.3	STRUCTURAL METHODS	60
	2.3.3.1 *X-Ray Diffraction*	60
	2.3.3.2 *Microwave Spectroscopy*	61
	2.3.3.2.1 *Molecular geometry*	61
	2.3.3.2.2 *Partially and fully saturated compounds*	62
	2.3.3.3 *^{1}H NMR Spectroscopy*	62
	2.3.3.3.1 *Parent aromatic compounds*	62
	2.3.3.3.2 *Substituted aromatic compounds*	63
	2.3.3.3.3 *Saturated and partially saturated compounds*	64
	2.3.3.4 *^{13}C NMR Spectroscopy*	65
	2.3.3.5 *Heteroatom NMR Spectroscopy*	66
	2.3.3.6 *UV Spectroscopy*	68
	2.3.3.7 *IR Spectroscopy*	69
	2.3.3.7.1 *Ring vibrations*	69
	2.3.3.7.2 *Substituent vibrations*	72
	2.3.3.8 *Mass Spectrometry*	73
	2.3.3.8.1 *Parent monocycles*	73
	2.3.3.8.2 *Substituted monocycles*	74
	2.3.3.8.3 *Benzo derivatives*	75
	2.3.3.8.4 *Saturated compounds*	75
	2.3.3.9 *Photoelectron Spectroscopy*	76
	2.3.3.9.1 *Parent monocycles*	76
	2.3.3.9.2 *Substituted compounds*	77
	2.3.3.9.3 *Reduced compounds*	77
	2.3.3.9.4 *Core ionization energies*	78
2.3.4	THERMODYNAMIC ASPECTS	79
	2.3.4.1 *Intermolecular Forces*	79
	2.3.4.1.1 *Melting and boiling points*	79
	2.3.4.1.2 *Solubility*	79
	2.3.4.1.3 *Gas chromatography*	79
	2.3.4.2 *Stability and Stabilization*	79
	2.3.4.2.1 *Thermochemistry and conformation of saturated heterocycles*	79
	2.3.4.2.2 *Aromaticity*	79
	2.3.4.3 *Conformation*	80
	2.3.4.3.1 *Aromatic compounds*	80
	2.3.4.3.2 *Reduced ring compounds*	85
2.3.5	TAUTOMERISM	85
	2.3.5.1 *Annular Tautomerism*	85
	2.3.5.2 *Compounds with a Potential Hydroxy Group*	86
	2.3.5.3 *Compounds with Two Potential Hydroxy Groups*	88
	2.3.5.4 *Compounds with Potential Mercapto Groups*	89
	2.3.5.5 *Compounds with Potential Amino Groups*	89

2.4	**Structure of Five-membered Rings with Two or More Heteroatoms**	**91**
2.4.1	SURVEY OF POSSIBLE STRUCTURES	91
	2.4.1.1 *Aromatic Systems without Exocyclic Conjugation*	91
	2.4.1.2 *Aromatic Systems with Exocyclic Conjugation*	92
	2.4.1.3 *Non-aromatic Systems*	93

2.4.2	THEORETICAL METHODS	95
2.4.2.1	*Electron Densities and Frontier Orbital Energies*	95
2.4.2.2	*Other Applications of Theory*	97
2.4.3	STRUCTURAL METHODS	99
2.4.3.1	*X-Ray Diffraction*	99
2.4.3.2	*Microwave Spectroscopy*	99
2.4.3.2.1	*Molecular geometry*	99
2.4.3.2.2	*Partially and fully saturated ring systems*	103
2.4.3.3	*^{1}H NMR Spectroscopy*	103
2.4.3.4	*^{13}C NMR Spectroscopy*	108
2.4.3.5	*Nitrogen and Oxygen NMR Spectroscopy*	112
2.4.3.6	*UV Spectroscopy*	114
2.4.3.6.1	*Parent compounds*	114
2.4.3.6.2	*Benzo derivatives*	114
2.4.3.6.3	*Effect of substituents*	115
2.4.3.7	*IR Spectroscopy*	117
2.4.3.7.1	*Aromatic rings without carbonyl groups*	117
2.4.3.7.2	*Azole rings containing carbonyl groups*	118
2.4.3.7.3	*Substituent vibrations*	119
2.4.3.8	*Mass Spectrometry*	121
2.4.3.9	*Photoelectron Spectroscopy*	121
2.4.4	THERMODYNAMIC ASPECTS	124
2.4.4.1	*Intermolecular Forces*	124
2.4.4.1.1	*Melting and boiling points*	124
2.4.4.1.2	*Solubility of heterocyclic compounds*	124
2.4.4.1.3	*Gas-liquid chromatography*	125
2.4.4.2	*Stability and Stabilization*	125
2.4.4.2.1	*Thermochemistry and conformation of saturated heterocycles*	125
2.4.4.2.2	*Aromaticity*	125
2.4.4.2.3	*Stable carbenes, silylenes, germylenes*	128
2.4.4.3	*Conformation*	129
2.4.5	TAUTOMERISM	131
2.4.5.1	*Annular Tautomerism*	131
2.4.5.1.1	*Prototropy*	131
2.4.5.1.2	*Annular Elementotropy*	134
2.4.5.2	*Substituent Taulomerism*	135
2.4.5.2.1	*Azoles with heteroatoms in the 1,2-positions*	135
2.4.5.2.2	*Azoles with heteroatoms in the 1,3-positions*	136
2.4.5.3	*Ring-chain Tautomerism*	138
2.4.5.3.1	*Without direct involvement of substituents*	138
2.4.5.3.2	*With involvement of substituents*	138
2.5	**Structure of Small and Large Rings**	**141**
2.5.1	SURVEY OF POSSIBLE STRUCTURES	141
2.5.1.1	*Small Rings*	141
2.5.1.2	*Large Rings*	141
2.5.2	THEORETICAL METHODS	141
2.5.3	STRUCTURAL METHODS	148
2.5.3.1	*X-Ray Diffraction*	148
2.5.3.2	*Microwave Spectroscopy*	150
2.5.3.3	*^{1}H NMR Spectroscopy*	150
2.5.3.4	*Heteronuclear NMR Spectroscopy*	152
2.5.3.5	*UV Spectroscopy*	153
2.5.3.5.1	*Electronic spectra of small-ring heterocyclic compounds*	153
2.5.3.5.2	*Electronic spectra of large-ring heterocyclic compounds*	153
2.5.3.6	*IR Spectroscopy*	153
2.5.3.7	*Mass Spectrometry*	155
2.5.3.8	*Photoelectron Spectroscopy*	156
2.5.4	THERMODYNAMIC ASPECTS	157
2.5.4.1	*Stability and Stabilization*	157
2.5.4.1.1	*Ring strain*	157
2.5.4.1.2	*Aromaticity and antiaromaticity*	157
2.5.4.2	*Conformation*	159
2.5.4.2.1	*Small rings*	159
2.5.4.2.2	*Large rings*	160

2.5.5   TAUTOMERISM   161
    2.5.5.1 Valence Bond Tautomerism   161
    2.5.5.2 Annular Prototropy   162

PART 3: REACTIVITY OF HETEROCYCLES

3.1   Overview   167

3.1.1   REACTION TYPES   167

3.1.2   HETEROAROMATIC REACTIVITY   167

3.1.3   ARRANGEMENT OF THE REACTIVITY SECTIONS   167

3.2   Reactivity of Six-membered Rings   169

3.2.1   REACTIVITY OF AROMATIC RINGS   169

    3.2.1.1   General Survey of Reactivity   169
      3.2.1.1.1   Pyridines   169
      3.2.1.1.2   Azines   170
      3.2.1.1.3   Cationic rings   170
      3.2.1.1.4   Pyridones, N-oxides and related compounds: betainoid rings   170
      3.2.1.1.5   Anionic rings   171
      3.2.1.1.6   Aromaticity and reversion to type   171
    3.2.1.2   Intramolecular Thermal and Photochemical Reactions   172
      3.2.1.2.1   Fragmentation   172
      3.2.1.2.2   Rearrangement to or elimination via Dewar heterobenzenes   173
      3.2.1.2.3   Rearrangement to or via hetero-prismanes and -benzvalenes   174
      3.2.1.2.4   Rearrangement to or via 1,3-bridged heterocycles   175
      3.2.1.2.5   Ring opening   176
    3.2.1.3   Electrophilic Attack at Nitrogen   176
      3.2.1.3.1   Introduction   176
      3.2.1.3.2   Effect of substituents   176
      3.2.1.3.3   Orientation of reaction of azines   177
      3.2.1.3.4   Proton acids   177
      3.2.1.3.5   Metal ions   179
      3.2.1.3.6   Alkyl and aryl halides and related compounds   180
      3.2.1.3.7   Acyl halides and related compounds and Michael-type reactions   181
      3.2.1.3.8   Halogens   182
      3.2.1.3.9   Peracids   183
      3.2.1.3.10   Aminating agents   184
      3.2.1.3.11   Other Lewis acids   184
    3.2.1.4   Electrophilic Attack at Carbon   184
      3.2.1.4.1   Species undergoing reaction and reaction mechanism   184
      3.2.1.4.2   Reactivity and effect of substituents   185
      3.2.1.4.3   Orientation   185
      3.2.1.4.4   Nitration   186
      3.2.1.4.5   Sulfonation   188
      3.2.1.4.6   Acid-catalyzed hydrogen exchange   188
      3.2.1.4.7   Halogenation   189
      3.2.1.4.8   Acylation and alkylation   192
      3.2.1.4.9   Mercuration   192
      3.2.1.4.10   Nitrosation, diazo coupling, Mannich reaction, Kolbe reaction and reaction with aldehydes   192
      3.2.1.4.11   Oxidation   193
    3.2.1.5   Attack at Ring Sulfur Atoms   194
      3.2.1.5.1   Reactions with electrophiles   194
      3.2.1.5.2   Reactions with nucleophiles   195
    3.2.1.6   Nucleophilic Attack at Carbon   195
      3.2.1.6.1   Ease of reaction   195
      3.2.1.6.2   Effect of substituents   197
      3.2.1.6.3   Hydroxide ion   197
      3.2.1.6.4   Amines and amide ions   204
      3.2.1.6.5   Sulfur nucleophiles   208
      3.2.1.6.6   Phosphorus nucleophiles   209
      3.2.1.6.7   Halide ions   209
      3.2.1.6.8   Carbon nucleophiles   210
      3.2.1.6.9   Chemical reduction   217
    3.2.1.7   Nucleophilic Attack at Ring Nitrogen   220
    3.2.1.8   Nucleophilic Attack at Hydrogen attached to Ring Carbon or Ring Nitrogen   220
      3.2.1.8.1   Metallation at a ring carbon atom   220
      3.2.1.8.2   Hydrogen exchange at ring carbon in neutral azines, N-oxides and azinones   221
      3.2.1.8.3   Hydrogen exchange at ring carbon in azinium cations   222
      3.2.1.8.4   Proton loss from a ring nitrogen atom   222
    3.2.1.9   Reactions with Radicals and with Electron-deficient Species; Reactions at Surfaces   223

3.2.1.9.1   Carbenes and nitrenes 223
3.2.1.9.2   Free radical attack at ring carbon atoms 223
3.2.1.9.3   Electrochemical reactions and reactions with free electrons ... 225
3.2.1.9.4   Other reactions at surfaces 226
3.2.1.10   Intermolecular Reactions with Cyclic Transition States 227
3.2.1.10.1   Introduction 227
3.2.1.10.2   Heterocycles as inner dienes in [2 + 4] cycloaddition 227
3.2.1.10.3   Heterocycles as inner dienes in [1 + 4] cycloaddition 232
3.2.1.10.4   Heterocycles as 1,3-dipoles 232
3.2.1.10.5   Heterocycles as dienophiles 234
3.2.1.10.6   [2 + 2] Cycloaddition 235
3.2.1.10.7   Heterocycles as 4π component in [4 + 4] cycloaddition 235

3.2.2   REACTIONS OF NON-AROMATIC COMPOUNDS 235

3.2.2.1   Eight π-electron Systems: 1,2- and 1,4-Dioxins, -Oxathiins and -Dithiins ... 236
3.2.2.1.1   Intramolecular thermolysis and photolysis reactions 236
3.2.2.1.2   Reactions with electrophiles 236
3.2.2.1.3   Reactions with nucleophiles 237
3.2.2.2   Thiabenzenes and Related Compounds 237
3.2.2.3   Dihydro Compounds 238
3.2.2.3.1   Introduction 238
3.2.2.3.2   Tautomerism 239
3.2.2.3.3   Aromatization 239
3.2.2.3.4   Electron loss to form radicals 242
3.2.2.3.5   Electrocyclic ring opening 243
3.2.2.3.6   Proton loss to an eight π-electron conjugated system 244
3.2.2.3.7   Electrophilic substitution 244
3.2.2.3.8   Cycloaddition reactions 245
3.2.2.3.9   Other reactions 246
3.2.2.4   Tetra- and Hexa-hydro Compounds 246
3.2.2.4.1   Tautomeric equilibria 246
3.2.2.4.2   Aromatization 246
3.2.2.4.3   Ring fission 246
3.2.2.4.4   Other reactions 247
3.2.2.4.5   Stereochemistry 248

3.2.3   REACTIONS OF SUBSTITUENTS 248

3.2.3.1   General Survey of Reactivity of Substituents on Ring Carbon Atoms ... 248
3.2.3.1.1   The carbonyl analogy 248
3.2.3.1.2   Effect of number, type and orientation of heteroatoms 249
3.2.3.1.3   The effect of one substituent on the reactivity of another ... 251
3.2.3.1.4   Reactions of substituents not directly attached to the heterocyclic ring ... 251
3.2.3.2   Benzenoid Rings 252
3.2.3.2.1   Fused benzene rings: unsubstituted 252
3.2.3.2.2   Fused benzene rings: substituted 254
3.2.3.3   Alkyl Groups 256
3.2.3.3.1   Reactions similar to those of toluene 256
3.2.3.3.2   Alkyl groups: reactions via proton loss 256
3.2.3.3.3   Alkylazines: reactions involving essentially complete anion formation ... 256
3.2.3.3.4   Alkylazines: reactions involving traces of reactive anions or traces of methylene bases ... 257
3.2.3.3.5   Alkyl-azonium and -pyrylium compounds 259
3.2.3.3.6   Tautomerism of alkyl derivatives 261
3.2.3.4   Further Carbon Functional Groups 261
3.2.3.4.1   Aryl groups 261
3.2.3.4.2   Carboxylic acids and derivatives 262
3.2.3.4.3   Aldehydes and ketones 264
3.2.3.4.4   Other substituted alkyl groups 264
3.2.3.4.5   Vinyl groups 265
3.2.3.5   Amino and Imino Groups 265
3.2.3.5.1   Orientation of reactions of amino-pyridines and -azines with electrophiles ... 265
3.2.3.5.2   Reaction of aminoazines with electrophiles at the amino group ... 266
3.2.3.5.3   Diazotization of amino compounds 266
3.2.3.5.4   Reactions of amino compounds with nucleophiles 267
3.2.3.5.5   Amino-imino tautomerism 269
3.2.3.6   Other N-Linked Substituents 269
3.2.3.6.1   Nitro groups 269
3.2.3.6.2   Nitramino compounds 270
3.2.3.6.3   Hydrazino groups 270
3.2.3.6.4   Azides 271
3.2.3.6.5   Nitroso groups 272
3.2.3.7   Hydroxy and Oxo Groups 272
3.2.3.7.1   Hydroxy groups and hydroxy-oxo tautomeric equilibria 272
3.2.3.7.2   Pyridones, pyrones, thiinones, azinones, etc.: general pattern of reactivity ... 272

3.2.3.7.3   *Pyridones, pyrones and azinones: electrophilic attack at carbonyl oxygen*   274
3.2.3.7.4   *Pyridones, pyrones and azinones: nucleophilic displacement of carbonyl oxygen*   274
3.2.3.7.5   *Heterocyclic quinones*   276
3.2.3.8   *Other O-Linked Substituents*   276
3.2.3.8.1   *Alkoxy and aryloxy groups*   276
3.2.3.8.2   *Acyloxy groups*   278
3.2.3.9   *S-Linked Substituents*   278
3.2.3.9.1   *Mercapto-thione tautomerism*   278
3.2.3.9.2   *Thiones*   278
3.2.3.9.3   *Alkylthio, alkylsulfinyl and alkylsulfonyl groups*   279
3.2.3.9.4   *Sulfonic acid groups*   280
3.2.3.10   *Halogen Atoms*   280
3.2.3.10.1   *Pattern of reactivity*   280
3.2.3.10.2   *Replacement of halogen by hydrogen or a metal (including transmetallation), or by coupling*   280
3.2.3.10.3   *Reactions via hetarynes*   282
3.2.3.10.4   *The S_{RN} mechanistic pathway*   283
3.2.3.10.5   *ANRORC reactions*   283
3.2.3.10.6   *Nucleophilic displacement by classical S_{AE} mechanism*   284
3.2.3.10.7   *Nucleophilic displacement with transition metal catalysis*   286
3.2.3.11   *Metals and Metalloid Derivatives*   287
3.2.3.12   *Substituents Attached to Ring Nitrogen Atoms*   288
3.2.3.12.1   *Introduction*   288
3.2.3.12.2   *Alkyl groups*   289
3.2.3.12.3   *Other C-linked substituents*   291
3.2.3.12.4   *N-Linked substituents*   291
3.2.3.12.5   *O-Linked substituents*   293
3.2.3.12.6   *Other substituents attached to nitrogen*   295
3.2.3.13   *Substituents Attached to Ring Sulfur Atoms*   296

3.3   Reactivity of Five-membered Rings with One Heteroatom   **297**

3.3.1   REACTIONS AT HETEROAROMATIC RINGS   297
3.3.1.1   *General Survey of Reactivity*   297
3.3.1.1.1   *Comparison with aliphatic series*   297
3.3.1.1.2   *Effect of aromaticity*   298
3.3.1.2   *Thermal and Photochemical Reactions Involving No Other Species*   298
3.3.1.3   *Electrophilic Attack on Ring Heteroatoms*   299
3.3.1.3.1   *Pyrrole anions*   299
3.3.1.3.2   *Thiophenes, selenophenes and tellurophenes*   301
3.3.1.4   *Electrophilic Attack on Carbon: General Considerations*   302
3.3.1.4.1   *Relative reactivities of heterocycles*   302
3.3.1.4.2   *Directing effects of the ring heteroatom*   303
3.3.1.4.3   *Directing effects of substituents in monocyclic compounds*   304
3.3.1.4.4   *Directing effects of fused benzene rings*   305
3.3.1.4.5   *Range of substitution reactions*   305
3.3.1.5   *Electrophilic Attack on Carbon: Specific Reactions*   305
3.3.1.5.1   *Proton acids*   305
3.3.1.5.2   *Nitration*   307
3.3.1.5.3   *Sulfonation*   308
3.3.1.5.4   *Halogenation*   308
3.3.1.5.5   *Acylation*   310
3.3.1.5.6   *Alkylation*   312
3.3.1.5.7   *Reactions with aldehydes and ketones*   314
3.3.1.5.8   *Mercuration*   316
3.3.1.5.9   *Diazo coupling*   316
3.3.1.5.10   *Nitrosation*   317
3.3.1.5.11   *Electrophilic oxidation*   318
3.3.1.6   *Reactions with Nucleophiles*   320
3.3.1.6.1   *Deprotonation at nitrogen*   320
3.3.1.6.2   *Deprotonation at carbon*   320
3.3.1.6.3   *Reactions of cationic species with nucleophiles*   321
3.3.1.6.4   *Vicarious nucleophilic substitution and related reactions*   322
3.3.1.6.5   *Nucleophilic attack on sulfur*   323
3.3.1.7   *Reactions with Radicals and Electron-deficient Species: Reactions at Surfaces*   324
3.3.1.7.1   *Carbenes and nitrenes*   324
3.3.1.7.2   *Free radical attack*   325
3.3.1.7.3   *Electrochemical reactions*   326
3.3.1.7.4   *Reactions with free electrons*   326
3.3.1.7.5   *Catalytic hydrogenation*   327
3.3.1.7.6   *Reduction by dissolving metals*   327
3.3.1.7.7   *Desulfurization*   328
3.3.1.8   *Reactions with Cyclic Transition States*   328
3.3.1.8.1   *Heterocycles as inner ring dienes*   328
3.3.1.8.2   *Heterocycles as dienophiles*   331

3.3.1.8.3   *[2 + 2] Cycloaddition reactions*   332
3.3.1.8.4   *Other cycloaddition reactions*   332

3.3.2   REACTIVITY OF NON-AROMATIC COMPOUNDS   333

3.3.2.1   *Pyrrolenines and Indolenines*   333
3.3.2.2   *Thiophene Sulfones and Sulfoxides*   334
3.3.2.3   *Dihydro Derivatives*   335
3.3.2.3.1   *Aromatization of dihydro compounds*   335
3.3.2.3.2   *Behavior analogous to aliphatic analogues*   336
3.3.2.3.3   *Other reactions*   336
3.3.2.4   *Tetrahydro Derivatives*   337
3.3.2.5   *Ring Carbonyl Compounds and their Hydroxy Tautomers*   338
3.3.2.5.1   *Survey of structures*   338
3.3.2.5.2   *Interconversion and reactivity of tautomeric forms*   338
3.3.2.5.3   *Reactions of hydroxy compounds with electrophiles*   339
3.3.2.5.4   *Reactions of anions with electrophiles*   340
3.3.2.5.5   *Reactions of carbonyl compounds with nucleophiles*   341
3.3.2.5.6   *Reductions of carbonyl and hydroxy compounds*   342

3.3.3   REACTIVITY OF SUBSTITUENTS   342

3.3.3.1   *General Survey of Reactivity*   342
3.3.3.1.1   *Reaction types*   342
3.3.3.1.2   *Nucleophilic substitution of substituents*   342
3.3.3.2   *Fused Benzene Rings*   343
3.3.3.2.1   *Electrophilic attack*   343
3.3.3.2.2   *Nucleophilic attack*   344
3.3.3.2.3   *Reactions with electrons*   345
3.3.3.2.4 *Reactions of substituents on benzene rings*   345
3.3.3.3   *Other C-Linked Substituents*   345
3.3.3.3.1   *Alkyl groups*   345
3.3.3.3.2   *Vinyl groups*   346
3.3.3.3.3   *Substituted alkyl groups: general*   347
3.3.3.3.4   *Halomethyl*   349
3.3.3.3.5   *Hydroxymethyl*   349
3.3.3.3.6   *Aminomethyl*   350
3.3.3.3.7   *Carboxylic acids, esters and anhydrides*   350
3.3.3.3.8   *Acyl groups*   352
3.3.3.4   *N-Linked Substituents*   353
3.3.3.4.1   *Nitro*   353
3.3.3.4.2   *Amino*   354
3.3.3.4.3   *Azides*   355
3.3.3.5   *O-Linked Substituents*   355
3.3.3.6   *S-Linked Substituents*   355
3.3.3.7   *Halo Groups*   356
3.3.3.7.1   *Nucleophilic displacement*   356
3.3.3.7.2   *Reductive dehalogenation*   357
3.3.3.7.3   *Rearrangement*   357
3.3.3.7.4   *Formation of Grignard reagents*   357
3.3.3.8   *Metallo Groups*   357
3.3.3.8.1   *General*   357
3.3.3.8.2   *Formation of C-C bonds*   359
3.3.3.8.3   *Formation of C-O bonds*   360
3.3.3.8.4   *Formation of C-S bonds*   360
3.3.3.8.5   *Formation of C-N bonds*   361
3.3.3.8.6   *Formation of C-halogen bonds*   361
3.3.3.8.7   *Ring-opening reactions*   361
3.3.3.8.8   *Palladium- and nickel-catalyzed cross-coupling reactions*   362
3.3.3.8.9   *Mercury derivatives*   365
3.3.3.9   *Substituents Attached to the Pyrrole Nitrogen Atom*   365
3.3.3.10   *Substituents Attached to the Thiophene Sulfur Atom*   366

3.4   Reactivity of Five-membered Rings with Two or More Heteroatoms   367

3.4.1   REACTIONS AT HETEROAROMATIC RINGS   367

3.4.1.1   *General Survey of Reactivity*   367
3.4.1.1.1   *Reactivity of neutral azoles*   367
3.4.1.1.2   *Azolium salts*   368
3.4.1.1.3   *Azole   anions*   368
3.4.1.1.4   *Azolinones, azolinethiones, azolinimines*   368
3.4.1.1.5   N-*Oxides,* N-*imides,* N-*ylides of azoles*   369
3.4.1.2   *Thermal and Photochemical Reactions Formally Involving No Other Species*   369
3.4.1.2.1   *Thermal fragmentation*   370
3.4.1.2.2   *Photochemical fragmentation*   372

3.4.1.2.3 Equilibria with open-chain compounds	373
3.4.1.2.4 Rearrangement to other heterocyclic species	374
3.4.1.2.5 Polymerization	375
3.4.1.3 Electrophilic Attack at Nitrogen	375
3.4.1.3.1 Introduction	375
3.4.1.3.2 Reaction sequence	376
3.4.1.3.3 Orientation in azole rings containing three or four heteroatoms	376
3.4.1.3.4 Effect of azole ring structure and of substituents	377
3.4.1.3.5 Proton acids on neutral azoles: basicity of azoles	377
3.4.1.3.6 Proton acids on azole anions: acidity of azoles	379
3.4.1.3.7 Basicity and acidity in gas phase	379
3.4.1.3.8 Metal ions	380
3.4.1.3.9 Alkyl halides and related compounds: azoles without a free NH group	381
3.4.1.3.10 Alkyl halides and related compounds: compounds with a free NH group	383
3.4.1.3.11 Acyl halides and related compounds	385
3.4.1.3.12 Halogens	386
3.4.1.3.13 Peracids	386
3.4.1.3.14 Aminating agents	387
3.4.1.3.15 Other Lewis acids	388
3.4.1.4 Electrophilic Attack at Carbon	388
3.4.1.4.1 Reactivity and orientation	388
3.4.1.4.2 Nitration	389
3.4.1.4.3 Sulfonation	390
3.4.1.4.4 Acid-catalyzed hydrogen exchange	390
3.4.1.4.5 Halogenation	391
3.4.1.4.6 Acylation, formylation and alkylation	392
3.4.1.4.7 Mercuration	392
3.4.1.4.8 Diazo coupling	393
3.4.1.4.9 Nitrosation	394
3.4.1.4.10 Reactions with aldehydes and ketones	394
3.4.1.4.11 Oxidation	395
3.4.1.5 Attack at Sulfur	396
3.4.1.5.1 Electrophilic attack	396
3.4.1.5.2 Nucleophilic attack	397
3.4.1.6 Nucleophilic Attack at Carbon	397
3.4.1.6.1 Hydroxide ion and other O-nucleophiles	398
3.4.1.6.2 Amines and amide ions	401
3.4.1.6.3 S-Nucleophiles	403
3.4.1.6.4 Halide ions	403
3.4.1.6.5 Carbanions	404
3.4.1.6.6 Reduction by complex hydrides	406
3.4.1.6.7 Phosphorus nucleophiles	407
3.4.1.7 Nucleophilic Attack at Nitrogen Heteroatom	407
3.4.1.8 Nucleophilic Attack at Hydrogen Attached to Ring Carbon or Ring Nitrogen	408
3.4.1.8.1 Metallation at a ring carbon atom	408
3.4.1.8.2 Hydrogen exchange at ring carbon in neutral azoles	409
3.4.1.8.3 Hydrogen exchange at ring carbon in azolium ions and dimerization	410
3.4.1.8.4 C-Substitution via electrophilic attack at N, deprotonation and rearrangement	411
3.4.1.8.5 Formation and reactions of stable carbenes	412
3.4.1.8.6 Ring cleavage via C-deprotonation	412
3.4.1.8.7 Proton loss from a ring nitrogen atom	413
3.4.1.9 Reactions with Radicals and Electron-deficient Species; Reactions at Surfaces	413
3.4.1.9.1 Carbenes and nitrenes	413
3.4.1.9.2 Free radical attack at the ring carbon atoms	414
3.4.1.9.3 Thiation	415
3.4.1.9.4 Electrochemical reactions and reactions with free electrons	415
3.4.1.9.5 Other reactions at surfaces (catalytic hydrogenation and reduction by dissolving metals)	417
3.4.1.10 Reactions with Cyclic Transition States	418
3.4.1.10.1 Heterocycles as inner-ring dienes	419
3.4.1.10.2 Heterocycles derivatives as inner-outer ring dienes	422
3.4.1.10.3 Heterocycles derivatives as outer-ring dienes	422
3.4.1.10.4 Heterocycles as dienophiles	423
3.4.1.10.5 [2 + 2] Cycloaddition reactions	424
3.4.1.10.6 Other cycloaddition reactions	424
3.4.2 REACTIONS OF NON-AROMATIC COMPOUNDS	425
3.4.2.1 Isomers of Aromatic Derivatives	425
3.4.2.1.1 Compounds not in tautomeric equilibrium with aromatic derivatives	425
3.4.2.1.2 Compounds in tautomeric equilibrium with aromatic derivatives	426
3.4.2.2 Dihydro Compounds	427
3.4.2.2.1 Tautomerism	427
3.4.2.2.2 Aromatization	427
3.4.2.2.3 Ring contraction	428

3.4.2.2.4 Other reactions		429
3.4.2.3 Tetrahydro Compounds		431
3.4.2.3.1 Aromatization		431
3.4.2.3.2 Ring fission		431
3.4.2.3.3 Other reactions		432

3.4.3 REACTIONS OF SUBSTITUENTS — 433

3.4.3.1 General Survey of Substituents on Carbon		433
3.4.3.1.1 Substituent environment		433
3.4.3.1.2 The carbonyl analogy		434
3.4.3.1.3 Two heteroatoms in the 1,3-positions		435
3.4.3.1.4 Two heteroatoms in the 1,2-positions		435
3.4.3.1.5 Three heteroatoms		435
3.4.3.1.6 Four heteroatoms		435
3.4.3.1.7 The effect of one substituent on the reactivity of another		435
3.4.3.1.8 Reactions of substituents not directly attached to the heterocyclic ring		435
3.4.3.1.9 Reactions of substituents involving ring transformations		435
3.4.3.2 Fused Benzene Rings		437
3.4.3.2.1 Electrophilic substitution		437
3.4.3.2.2 Oxidative degradation		438
3.4.3.2.3 Nucleophilic attack		438
3.4.3.2.4 Rearrangements		439
3.4.3.3 Alkyl Groups		440
3.4.3.3.1 Reactions similar to those of toluene		440
3.4.3.3.2 Alkylazoles: reactions involving essentially complete anion formation		440
3.4.3.3.3 Reactions of alkylazoles involving traces of reactive anions		441
3.4.3.3.4 C-Alkyl-azoliums, -dithiolyliums, etc.		442
3.4.3.4 Other C-Linked Substituents		444
3.4.3.4.1 Aryl groups: electrophilic substitution		444
3.4.3.4.2 Aryl groups: other reactions		444
3.4.3.4.3 Carboxylic acids		445
3.4.3.4.4 Aldehydes and ketones		446
3.4.3.4.5 Vinyl and ethynyl groups		447
3.4.3.4.6 Ring fission		448
3.4.3.5 Aminoazoles		448
3.4.3.5.1 Dimroth rearrangement		448
3.4.3.5.2 Reactions with electrophiles (except nitrous acid)		450
3.4.3.5.3 Reaction with nitrous acid. Diazotization		450
3.4.3.5.4 Deprotonation of aminoazoles		452
3.4.3.5.5 Aminoazolium ions/neutral imines		453
3.4.3.5.6 Oxidation of aminoazoles		453
3.4.3.6 Other N-Linked Substituents		453
3.4.3.6.1 Nitro groups		453
3.4.3.6.2 Azidoazoles		454
3.4.3.7 O-Linked Substituents		454
3.4.3.7.1 Tautomeric forms: interconversion and modes of reaction		454
3.4.3.7.2 2-Hydroxyazoles, heteroatoms-1,3		455
3.4.3.7.3 3-Hydroxyazoles, heteroatoms-1,2		456
3.4.3.7.4 5-Hydroxyazoles, heteroatoms-1,2		456
3.4.3.7.5 4- and 5-Hydroxyazoles, heteroatoms-1,3 and 4-hydroxyazoles, heteroatoms-1,2		456
3.4.3.7.6 Hydroxy derivatives with three heteroatoms		457
3.4.3.7.7 Alkoxy and aryloxy groups		458
3.4.3.8 S-Linked Substituents		458
3.4.3.8.1 Mercapto compounds: tautomerism		458
3.4.3.8.2 Thiones		458
3.4.3.8.3 Alkylthio groups		460
3.4.3.8.4 Sulfonic acid groups		460
3.4.3.9 Halogen Atoms		460
3.4.3.9.1 Nucleophilic displacements: neutral azoles		460
3.4.3.9.2 Nucleophilic displacements: haloazoliums		462
3.4.3.9.3 Other reactions		462
3.4.3.10 Metals and Metalloid-linked Substituents		463
3.4.3.11 Fused Heterocyclic Rings		464
3.4.3.12 Substituents Attached to Ring Nitrogen Atoms		465
3.4.3.12.1 N-Linked azole as a substituent		465
3.4.3.12.2 Aryl groups		465
3.4.3.12.3 Alkyl and alkenyl groups		466
3.4.3.12.4 Acyl and carboxy groups		468
3.4.3.12.5 N-Amino group		469
3.4.3.12.6 N-Nitro group		471
3.4.3.12.7 N-Hydroxy groups and N-oxides		472
3.4.3.12.8 N-Halo groups		472
3.4.3.12.9 N-Silicon, phosphorus, sulfur and related groups		473

3.5   Reactivity of Small and Large Rings   475

   3.5.1   GENERAL SURVEY   475

      3.5.1.1   Neutral Molecules   475
      3.5.1.2   Cations   475
      3.5.1.3   Anions   475
      3.5.1.4   Radicals   476

   3.5.2   THERMAL AND PHOTOCHEMICAL REACTIONS, NOT FORMALLY INVOLVING OTHER SPECIES   476

      3.5.2.1   Fragmentation Reactions   476
      3.5.2.2   Rearrangements   478

   3.5.3   ELECTROPHILIC ATTACK ON RING HETEROATOMS   481

      3.5.3.1   Protonation   481
      3.5.3.2   Complex formation   483
      3.5.3.3   Alkylation and Acylation   483

   3.5.4   NUCLEOPHILIC ATTACK ON RING HETEROATOMS   484

   3.5.5   NUCLEOPHILIC ATTACK ON RING CARBON ATOMS   485

      3.5.5.1   Reactions of Three-membered Rings   485
      3.5.5.2   Reactions of Four-membered Rings   486
      3.5.5.3   Reactions of Carbonyl Derivatives of Four-membered Rings   487
      3.5.5.4   Large Rings   487

   3.5.6   NUCLEOPHILIC ATTACK ON PROTONS ATTACHED TO RING ATOMS   487

   3.5.7   ATTACK BY RADICALS OR ELECTRON-DEFICIENT SPECIES. OXIDATION AND REDUCTION   489

      3.5.7.1   Reactions with Radicals and Carbenes   489
      3.5.7.2   Oxidation   490
      3.5.7.3   Reduction   491

   3.5.8   REACTIONS WITH CYCLIC TRANSITION STATES   491

      3.5.8.1   [2 + 4] Cycloadditions   492
         3.5.8.1.1   Heterocycles as dienophiles   492
         3.5.8.1.2   Heterocycles as dienes   492
      3.5.8.2   1,3-Dipolar Cycloadditions   494

   3.5.9   REACTIVITY OF TRANSITION METAL COMPLEXES   495

   3.5.10   REACTIVITY OF SUBSTITUENTS ATTACHED TO HETEROATOM OR RING CARBON ATOMS   496

      3.5.10.1   C-Linked Substituents   496
      3.5.10.2   N-Linked Substituents   498

PART 4: SYNTHESIS OF HETEROCYCLES

4.1   Overview   501

   4.1.1   AIMS AND ORGANIZATION   501

   4.1.2   RING FORMATION FROM TWO COMPONENTS   502

      4.1.2.1   By Reaction Between Electrophilic and Nucleophilic Carbons   502
      4.1.2.2   Ring Formation via Cycloaddition   502
         4.1.2.2.1   [2 + 2] Cycloadditions   502
         4.1.2.2.2   1,3-Dipolar cycloadditions   503
         4.1.2.2.3   Diels-Alder reactions   506

   4.1.3   RING CLOSURE OF A SINGLE COMPONENT   506

      4.1.3.1   By Reaction between Electrophilic and Nucleophilic Centers   506
      4.1.3.2   Electrocyclic Reactions   507
      4.1.3.3   By Radical, Carbene or Nitrene Intermediates   507
      4.1.3.4   By Intramolecular Cycloadditions   508

   4.1.4   MODIFICATION OF AN EXISTING RING   508

      4.1.4.1   Ring Atom Interchange   508
      4.1.4.2   Incorporation of New Ring Atoms: No Change in Ring Size   508
      4.1.4.3   Ring Expansions   509
      4.1.4.4   Ring Contractions   509
      4.1.4.5   Ring Closure with Simultaneous Ring Opening   510

4.2 Synthesis of Monocyclic Rings with One Heteroatom 511

4.2.1 RINGS CONTAINING NO ENDOCYCLIC DOUBLE BONDS 511

   4.2.1.1 *From Acyclic Compounds by Concerted Formation of Two Bonds* 511
      4.2.1.1.1 *Three-membered rings* 511
      4.2.1.1.2 *Four-membered rings* 514
      4.2.1.1.3 *Five-membered rings* 515
   4.2.1.2 *From Acyclic Compounds by Formation of One or Two C-Z Bonds* 515
      4.2.1.2.1 *Three-membered rings* 515
      4.2.1.2.2 *Four-membered rings* 518
      4.2.1.2.3 *Five-membered rings* 519
      4.2.1.2.4 *Six-membered rings* 521
      4.2.1.2.5 *Larger rings* 522
   4.2.1.3 *From Acyclic Compounds by Formation of One C-C Bond* 522
   4.2.1.4 *From Carbocyclic Compounds* 522
   4.2.1.5 *From Other Heterocyclic Compounds* 523
      4.2.1.5.1 *Reactions involving ring expansion* 523
      4.2.1.5.2 *Reactions without change in ring size* 524
      4.2.1.5.3 *Ring contraction* 524

4.2.2 RINGS CONTAINING ONE ENDOCYCLIC DOUBLE BOND 525

   4.2.2.1 *From Acyclic Compounds by Concerted Formation of Two Bonds* 525
   4.2.2.2 *From Acyclic Compounds by Formation of One or Consecutive Formation of Two C-Z Bond(s)* 526
      4.2.2.2.1 *Z atom component acting as nucleophile* 526
      4.2.2.2.2 *Z atom component acting as electrophile* 527
   4.2.2.3 *From Carbocycles* 528
   4.2.2.4 *From Heterocycles* 528

4.2.3 RINGS CONTAINING TWO ENDOCYCLIC DOUBLE BONDS 529

   4.2.3.1 *Overview* 529
   4.2.3.2 *Synthesis of Pyrroles, Furans and Thiophenes by Substituent Introduction or Modification* 529
   4.2.3.3 *Synthesis of Pyrroles, Furans and Thiophenes from Acyclic Precursors* 529
      4.2.3.3.1 *From C_4Z or C_4 units* 529
      4.2.3.3.2 *From C_3ZC or C_3 and CZ units* 534
      4.2.3.3.3 *From C_2 and ZCC units* 534
      4.2.3.3.4 *From C_2 and CZC units* 537
      4.2.3.3.5 *From two C2 and Z units* 538
   4.2.3.4 *Synthesis of Pyrans, Dihydropyridines and their Thio and Oxo Derivatives from Acyclic Precursors* 538
      4.2.3.4.1 *From C_5 units* 538
      4.2.3.4.2 *With C-C bond formation* 539
   4.2.3.5 *Synthesis of Four-, Five- and Six-membered Rings from Carbocyclic or Heterocyclic Precursors* 540
      4.2.3.5.1 *With ring expansion* 540
      4.2.3.5.2 *No change in ring size* 542
      4.2.3.5.3 *With ring contraction* 542
   4.2.3.6 *Synthesis of Seven- and Eight-membered Rings* 544

4.2.4 RINGS CONTAINING THREE ENDOCYCLIC DOUBLE BONDS 545

   4.2.4.1 *Synthetic Methods for Substituted Pyridines* 545
   4.2.4.2 *Synthesis of Six-membered Rings from Acyclic Compounds* 546
      4.2.4.2.1 *From or via pentane-1,5-diones* 546
      4.2.4.2.2 *From pent-2-ene-1,5-diones* 546
      4.2.4.2.3 *Other methods* 547
   4.2.4.3 *Synthesis of Six-membered Rings from Other Heterocycles* 548
      4.2.4.3.1 *From five-membered rings* 548
      4.2.4.3.2 *From other six-membered rings* 548
   4.2.4.4 *Synthesis of Seven-membered and Larger Rings* 549

4.3 Synthesis of Monocyclic Rings with Two or More Heteroatoms 551

4.3.1 SUBSTITUENT INTRODUCTION AND MODIFICATION 551

   4.3.1.1 *Overview* 551
   4.3.1.2 *Substituent Introduction and Modification in Azoles* 551
   4.3.1.3 *Substituent Introduction and Modification in Azines* 551

4.3.2 TWO HETEROATOMS IN THE 1,2-POSITIONS 552

   4.3.2.1 *Three-membered Rings* 552
   4.3.2.2 *Four-membered Rings* 552
      4.3.2.2.1 *1,2-Diazetidines* 552
      4.3.2.2.2 *1,2-Oxazetidines* 553
      4.3.2.2.3 *1,2-Thiazetidines* 554
      4.3.2.2.4 *1,2-Dioxetanes* 554
      4.3.2.2.5 *1,2-Oxathietanes* 555

4.3.2.2.6	*1,2-Dithietanes*	555
4.3.2.3	*Five-membered Rings: Pyrazoles, Isoxazoles, Isothiazoles, etc.*	556
4.3.2.3.1	*Synthesis from hydrazine, hydroxylamine and hydrogen disulfide*	556
4.3.2.3.2	*Synthesis by Z-Z bond formation*	558
4.3.2.3.3	*Other methods from acyclic precursors*	559
4.3.2.3.4	*From other heterocycles*	560
4.3.2.4	*Six-membered Rings: Pyridazines, 1,2-Oxazines, etc.*	561
4.3.2.4.1	*Synthesis from hydrazine or hydroxylamine derivatives*	561
4.3.2.4.2	*By cycloaddition reactions*	562
4.3.2.4.3	*Other methods from acyclic precursors*	563
4.3.2.4.4	*From other heterocycles*	563
4.3.2.5	*Seven-membered Rings*	564
4.3.2.5.1	*1,2-Diazepines*	564
4.3.2.5.2	*1,2-Oxazepines and 1,2-thiazepines*	566
4.3.2.5.3	*1,2-Dioxepins and 1,2-dithiepins*	566
4.3.3	**TWO HETEROATOMS IN THE 1,3-POSITIONS**	566
4.3.3.1	*Four-membered Rings*	566
4.3.3.1.1	*1,3-Diazetidines*	566
4.3.3.1.2	*1,3-Oxazetidines*	566
4.3.3.1.3	*1,3-Thiazetidines*	567
4.3.3.1.4	*1,3-Dithietanes*	567
4.3.3.2	*Five-membered Rings: Imidazoles, Oxazoles, Thiazoles, Dithiolium Salts and Derivatives*	568
4.3.3.2.1	*Overview*	569
4.3.3.2.2	*Synthesis from C_2 + ZCZ' components*	569
4.3.3.2.3	*Synthesis of imidazoles, oxazoles and thiazoles from acylamino ketones*	569
4.3.3.2.4	*Other syntheses of imidazoles, oxazoles, thiazoles, dithiolyliums and oxathiolyliums by cyclization of C_2ZCZ' components*	570
4.3.3.2.5	*Synthesis of imidazoles, oxazoles and thiazoles by C-C bond formation or 1,3-dipolar addition*	570
4.3.3.2.6	*Synthesis of azolinones and reduced rings from acyclic precursors*	572
4.3.3.2.7	*Synthesis from heterocycles*	573
4.3.3.3	*Six-membered Rings*	574
4.3.3.3.1	*C_3 + ZCZ type*	576
4.3.3.3.2	*ZC_3Z + C(5 + 1) and (6 + 0) cyclizations*	576
4.3.3.3.3	*[4 + 2] Cyclizations*	577
4.3.3.3.4	*Syntheses from heterocycles*	577
4.3.3.4	*Seven-membered Rings*	578
4.3.3.4.1	*1,3-Diazepines*	579
4.3.3.4.2	*1,3-Oxazepines and 1,3-thiazepines*	579
4.3.3.4.3	*1,3-Dioxepins and 1,3-dithiepins*	580
4.3.4	**TWO HETEROATOMS IN THE 1,4-POSITIONS**	581
4.3.4.1	*Six-membered Rings*	581
4.3.4.1.1	*Pyrazines from acyclic compounds*	581
4.3.4.1.2	*1,4-Dioxins, 1,4-dithiins, 1,4-oxazines and 1,4-thiazines*	581
4.3.4.1.3	*Non-aromatic rings from acyclic compounds*	582
4.3.4.1.4	*From heterocyclic precursors*	584
4.3.4.2	*Seven-membered Rings*	584
4.3.4.2.1	*1,4-Diazepines*	585
4.3.4.2.2	*1,4-Oxazepines and 1,4-thiazepines*	585
4.3.4.2.3	*1,4-Dioxepins and 1,4-dithiepins*	586
4.3.5	**THREE HETEROATOMS IN THE 1,2,3-POSITIONS**	587
4.3.5.1	*Three- and Four-membered Rings*	588
4.3.5.2	*Five-membered Rings*	588
4.3.5.2.1	*Formation of a bond between two of the heteroatoms*	588
4.3.5.2.2	*Other methods*	588
4.3.5.3	*Six-membered Rings*	590
4.3.6	**THREE HETEROATOMS IN THE 1,2,4-POSITIONS**	591
4.3.6.1	*Five-membered Rings*	591
4.3.6.1.1	*From acyclic intermediates containing the preformed Z-Z' bond*	591
4.3.6.1.2	*From acyclic intermediates by formation of the Z-Z' bond*	591
4.3.6.1.3	*From heterocycles*	593
4.3.6.2	*Six-membered Rings*	596
4.3.6.2.1	*1,2,4-Triazines*	596
4.3.6.2.2	*Rings containing O or S atoms*	597
4.3.6.3	*Seven-membered Rings*	598
4.3.6.3.1	*Heteroatoms in the 1,2,4-positions*	598
4.3.6.3.2	*Seven-membered rings with heteroatoms in the 1,2,5-positions*	599
4.3.7	**THREE HETEROATOMS IN THE 1,3,5-POSITIONS**	599
4.3.7.1	*s-Triazines*	599

4.3.7.2	*Compounds Containing O or S Atoms*	600
4.3.7.3	*Synthesis from Heterocyclic Precursors*	601
4.3.7.4	*Seven-membered Rings*	602

4.3.8   FOUR OR MORE HETEROATOMS   602

| *4.3.8.1* | *Five-membered Rings* | 602 |
| *4.3.8.2* | *Six-membered Rings* | 603 |

4.4   Synthesis of Bicyclic Ring Systems without Ring Junction Heteroatoms   605

4.4.1   SYNTHESIS BY SUBSTITUENT INTRODUCTION AND MODIFICATION   605

| *4.4.1.1* | *In the Heterocyclic Ring* | 605 |
| *4.4.1.2* | *In the Benzene Ring* | 605 |

4.4.2   ONE HETEROATOM ADJACENT TO RING JUNCTION   605

4.4.2.1	*Three- and Four-membered Rings*	605
4.4.2.1.1	*Three-membered rings*	605
4.4.2.1.2	*Four-membered rings*	606
4.4.2.2	*Five-membered Rings*	607
4.4.2.2.1	*Survey of syntheses for indoles, benzofurans and benzothiophenes*	607
4.4.2.2.2	*Ring closure by formation of Z-C(2) bond*	607
4.4.2.2.3	*Ring closure by formation of ring-C bond*	610
4.4.2.2.4	*Ring closure by formation of C(2)-C(3) bond*	614
4.4.2.2.5	*Ring closure by formation of ring-Z bond*	615
4.4.2.2.6	*From other heterocycles*	615
4.4.2.3	*Six-membered Rings*	616
4.4.2.3.1	*Survey of synthetic methods for quinolines, benzo[b]pyrans and their derivatives*	616
4.4.2.3.2	*Ring closure of o-substituted anilines or phenols*	616
4.4.2.3.3	*Formation of a C-C bond by reaction of a multiple bond with a benzene ring*	618
4.4.2.3.4	*Synthesis via cycloaddition reactions*	621
4.4.2.3.5	*Synthesis from heterocycles*	621
4.4.2.4	*Seven-membered and Larger Rings*	622

4.4.3   ONE HETEROATOM NOT ADJACENT TO RING JUNCTION   623

4.4.3.1	*Five-membered Rings: Isoindoles and Related Compounds*	623
4.4.3.2	*Six-membered Rings*	625
4.4.3.2.1	*Overview of ring syntheses of isoquinolines, benzo[c]pyrans and their derivatives*	625
4.4.3.2.2	*Ring closure of an o-disubstituted benzene*	625
4.4.3.2.3	*From a β-phenethylamine*	626
4.4.3.2.4	*From a benzylimine*	627
4.4.3.3	*Seven-membered and Larger Rings*	627

4.4.4   TWO HETEROATOMS 1,2 TO RING JUNCTION   627

4.4.4.1	*Four-membered Rings*	627
4.4.4.2	*Five-membered Rings*	628
4.4.4.2.1	*Indazoles*	628
4.4.4.2.2	*Anthranils, benzisothiazoles and saccharins*	629
4.4.4.3	*Six-membered Rings*	629
4.4.4.3.1	*Cinnolines*	629
4.4.4.3.2	*Rings containing O or S atoms*	631
4.4.4.4	*Seven-membered Rings*	631

4.4.5   TWO HETEROATOMS 1,3 TO RING JUNCTION   632

4.4.5.1	*Five-membered Rings*	632
4.4.5.1.1	*Ring closure of o-disubstituted benzene or hetarene*	632
4.4.5.1.2	*Other methods*	633
4.4.5.2	*Six-membered Rings*	634
4.4.5.2.1	*Quinazolines and azinopyrimidines by cyclization procedures*	634
4.4.5.2.2	*Rings containing O or S atoms*	635
4.4.5.2.3	*From other heterocycles*	636
4.4.5.3	*Seven-membered Rings*	637
4.4.5.3.1	*Seven-membered rings with heteroatoms 1,3 to ring junction*	637
4.4.5.3.2	*Seven-membered rings with heteroatoms 2,4 to ring junction*	638

4.4.6   TWO HETEROATOMS 1,4 TO RING JUNCTION   638

4.4.6.1	*Quinoxalines and Azinopyrazines*	638
4.4.6.2	*1,4-Benzoxazines and 1,4-Benzothiazines*	639
4.4.6.3	*Rings Containing Oxygen and/or Sulfur Atoms*	640
4.4.6.4	*Synthesis from Heterocyclic Precursors*	641
4.4.6.5	*Seven-membered Rings with Two Heteroatoms 1,4 to the Ring Junction*	641
4.4.6.5.1	*1,4-Benzodiazepines*	641
4.4.6.5.2	*1,4- and 4,1-Benzoxazepines, 1,4- and 1,5-benzothiazepines, and 1,4-benzodioxepins*	642

4.4.6.6   *Seven-membered Rings with Two Heteroatoms 1,5 to the Ring Junction*   643

4.4.7   TWO HETEROATOMS 2,3 TO RING JUNCTION   644

4.4.7.1   *Six-membered Rings*   644
4.4.7.2   *Seven-membered Rings*   646

4.4.8   THREE OR MORE HETEROATOMS   647

4.4.8.1   *Five-membered Heterocyclic Rings*   647
4.4.8.2   *Six-membered Heterocyclic Rings*   647
4.4.8.2.1   *Three heteroatoms in the 1,2,3-positions*   647
4.4.8.2.2   *Three heteroatoms in the 1,2,4- or 1,3,4-positions*   649
4.4.8.2.3   *Four heteroatoms*   651
4.4.8.3   *Seven-membered and Larger Rings*   652
4.4.8.3.1   *Heteroatoms 1,2,4 to ring junction*   652
4.4.8.3.2   *Heteroatoms 1,2,5 to ring junction*   653
4.4.8.3.3   *Heteroatoms 1,3,4 to ring junction*   653
4.4.8.3.4   *Heteroatoms 1,3,5 to ring junction*   654
4.4.8.3.5   *Four or more heteroatoms*   655

4.5   Synthesis of Tri- and Poly-cyclic Ring Systems without Ring Junction Heteroatoms   657

4.5.1   TWO ADJACENT FUSED RINGS, ONE HETEROATOM   657

4.5.1.1   *Five-membered Heterocyclic Rings*   657
4.5.1.1.1   *Overview of synthetic methods for carbazoles, dibenzofurans and dibenzothiophenes* 657
4.5.1.1.2   *Formation of C-C bond*   657
4.5.1.1.3   *Formation of C-Z bond*   659
4.5.1.1.4   *Miscellaneous methods*   660
4.5.1.2   *Six-membered Rings*   660

4.5.2   TWO ADJACENT FUSED RINGS, TWO HETEROATOMS   661

4.5.3   TWO NON-ADJACENT FUSED RINGS, ONE HETEROATOM   662

4.5.4   TWO NON-ADJACENT FUSED RINGS, TWO HETEROATOMS   663

4.5.4.1   *Phenazines*   663
4.5.4.2   *Phenoxazines and Phenothiazines*   663
4.5.4.3   *Dibenzo[1,4]dioxin, Phenoxathiin and Thianthrene*   665
4.5.4.4   *Dibenzoxepins and Dibenzothiepins*   665

4.5.5   PERI-ANNULATED HETEROCYCLIC SYSTEMS   665

4.5.6   THREE FUSED RINGS   666

4.6   Synthesis of Fused Ring Systems with Ring Junction Heteroatoms   667

4.6.1   FORMATION OF THREE- OR FOUR-MEMBERED RINGS WITH ONE N ATOM AT A RING JUNCTION   667

4.6.2   FORMATION OF A FIVE-MEMBERED RING WITH ONE N ATOM AT A RING JUNCTION   668

4.6.2.1   *No Other Heteroatoms*   668
4.6.2.1.1   *5-5 Systems*   668
4.6.2.1.2   *5-6 Systems*   669
4.6.2.2   *One Additional Heteroatom*   670
4.6.2.2.1   *Pyrazolo-fused systems*   670
4.6.2.2.2   *Imidazo-fused systems*   671
4.6.2.2.3   *Thiazolo-fused systems*   672
4.6.2.2.4   *Oxazolo- and isoxazolo-fused systems*   673
4.6.2.3   *Two Other Heteroatoms*   674
4.6.2.3.1   *1,2,4-Triazolo[b]-, 1,2,4-thiadiazolo[b]- and 1,3,4-thiadiazolo[b]-fused systems* 674
4.6.2.3.2   *1,2,4-Triazolo[c]- and 1,2,4-thiadiazolo[c]-fused systems*   675
4.6.2.3.3   *1,2,3-Triazolo[c]-fused systems*   676
4.6.2.4   *Three Other Heteroatoms: Fused Tetrazoles*   676

4.6.3   FORMATION OF A SIX-MEMBERED RING WITH ONE N ATOM AT A RING JUNCTION   676

4.6.3.1   *Ring Formation Using a Three-atom Fragment*   676
4.6.3.2   *Ring Formation Using a Two-atom Fragment*   678
4.6.3.3   *Ring Formation Using a One-atom Fragment*   678
4.6.3.4   *Cycloaddition and Ring-Transformation Reactions*   679
4.6.3.5   *Other Methods*   681

4.6.4   FORMATION OF A SEVEN-MEMBERED RING WITH ONE N ATOM AT A RING JUNCTION   681

4.6.5   TWO NITROGEN ATOMS AT A RING JUNCTION   682

4.6.5.1   *Five-membered Rings*   682
4.6.5.2   *Six-membered Rings*   683

4.6.6   SULFUR AT A RING JUNCTION   683

PART 5: APPENDIXES

Appendix A: Introduction to "Comprehensive Heterocyclic Chemistry - II"   687

  1   SCOPE, SIGNIFICANCE, AND AIMS   687

    1.1   Scope   687
    1.2   Significance   687
    1.3   Aims of CHEC and of the Present Work   687

  2   ARRANGEMENT OF THE WORK IN VOLUMES   688

    2.1   Relationship of CHEC-II to CHEC   688
    2.2   Arrangement of CHEC-II in Volumes   688
    2.3   Arrangement of CHEC in Volumes   689

  3   RATIONALE FOR ARRANGEMENT OF MATERIAL IN EACH VOLUME   689

    3.1   Major Division of Carbocyclic and Heterocyclic Chemistry   689
    3.2   Saturated Heterocyclic Compounds   690
    3.3   Partially Unsaturated Heterocyclic Compounds   690
    3.4   Heteroaromatic Compounds   690
    3.5   Characteristics of Heteroatoms in Rings   690
    3.6   The General Chapters   691

  4   ORGANIZATION OF INDIVIDUAL MONOGRAPH CHAPTERS   691

    4.1   Introduction   691
    4.2   Theoretical Methods   692
    4.3   Experimental Structural Methods   692
    4.4   Thermodynamic Aspects   692
    4.5   Reactivity of Fully Conjugated Rings   692
    4.6   Reactivity of Nonconjugated Rings   692
    4.7   Reactivity of Substituents Attached to Ring Carbon Atoms   693
    4.8   Reactivity of Substituents Attached to Ring Heteroatoms   693
    4.9   Ring Syntheses Classified by Number of Ring Atoms in Each Component   693
    4.10   Ring Synthesis by Transformation of Another Ring   693
    4.11   Synthesis of Particular Classes of Compounds and Critical Comparison of the Various Routes Available   693
    4.12   Important Compounds and Applications   693

  5   THE REFERENCE SYSTEM   693

  6   THE INDEXES   694

    6.1   Author Index   694
    6.2   Ring Index   694
    6.3   Subject Index   694

Appendix B: Short Contents of "Comprehensive Heterocyclic Chemistry - II"   695

Appendix C: Short Contents of "Comprehensive Heterocyclic Chemistry"   701

Appendix D: Miscellaneous (MI) References   705

Subject Index   709

Part 1
Preliminaries

1.1

Foreword to First Edition of the Handbook

This Handbook is designed with two purposes in mind.

(a) It is the sequel to the earlier textbooks by Katritzky and Lagowski: 'Heterocyclic Chemistry', published in 1960, and 'The Principles of Heterocyclic Chemistry', published in 1967. It is thus designed to provide a one-volume overall picture of the subject suitable for the graduate or advanced undergraduate student, as well as for research workers, be they specialists in the field or engaged in another discipline and requiring knowledge of heterocyclic chemistry.

(b) It represents Volume 9 of 'Comprehensive Heterocyclic Chemistry' (CHEC) and is based on the general chapters which appear scattered throughout the first eight volumes of the compendium. The reader is referred for more detailed factual matter to these volumes; specific references are denoted 'CHEC'.

The two purposes are symbiotic: the growth of the subject, particularly in the last two decades, renders impossible the inclusion of much detail in a one-volume work. However, the highly systematic nature of the subject and that of the treatment adopted in 'Comprehensive Heterocyclic Chemistry' are ideally suited for the treatment of the principles in one volume. The user can conveniently refer to the other eight volumes for further examples and more detailed explanations.

Because of the close relationship to 'Comprehensive Heterocyclic Chemistry' the Foreword, Contents and Introduction to the complete work, which appear in Volume I of the set, have been reprinted in the Handbook. The Foreword outlines the scope, significance and aim of the whole work, whereas the Introduction describes the plan of the work and the novel features it incorporates.

Finally, in Chapter 1.3 of the Handbook, I describe the divergencies between the treatment of heterocyclic chemistry in the Handbook and that found in the main work.

The author has many acknowledgments to make for help given toward the present book. As stated on the title page, it was a collaborative effort and I wish to thank the Volume Editors of 'Comprehensive Heterocyclic Chemistry' whose names are listed there for their considerable efforts. I also acknowledge my debt to the authors of all the monograph chapters from which material has been abstracted for this Handbook. Especially I thank Charles Rees for his constant encouragement and perceptive criticism. It was initially intended that this book would be written with Jeanne Lagowski; although pressure of work prevented this, her joint authorship of two vital chapters was crucial. The work would not have been finished without the expert assistance of Christa Schwarz and Cecilia Bell to whom I also owe my gratitude. The publishers, Pergamon Press, and their Senior Managing Editor, Dr. Colin Drayton, have been most supportive of this exciting venture. Finally, I thank my wife for all that she has contributed.

ALAN R. KATRITZKY
Florida

1.2

Explanation of the Reference System

As in CHEC and CHEC-II references are designated by a number-letter coding of which the first two numbers denote tens and units of the year of publication, the next one to three letters denote the journal, and the final numbers denote the page. This code appears in the text each time a reference is quoted; the advantages of this system are outlined in Appendix A (Section 5). The system is based on that previously used in the following two monographs: (a) A. R. Katritzky and J. M. Lagowski, 'Chemistry of the Heterocyclic N-Oxides', Academic Press, New York, 1971; (b) J. Elguero, C. Marzin, A. R. Katritzky and P. Linda, 'The Tautomerism of Heterocycles', in 'Advances in Heterocyclic Chemistry', Supplement 1, Academic Press, New York, 1976.

A list of journal codes is given in alphabetical order together with the journals to which they refer on the end papers of each volume, including the present Handbook. However, unlike CHEC and CHEC-II where the references are given in full at the end of the relevant volume, no reference lists are included in the Handbook for reasons of space. The full reference can usually be ascertained from the relevant volume of CHEC or CHEC-II. The list of miscellaneous (MI) references is given in Appendix D.

For non-twentieth century references the year is given in full in the code. For journals which are published in separate parts, the part letter or number is given (when necessary) in parentheses immediately after the journal code letters. Journal volume numbers are *not* included in the code numbers unless more than one volume was published in the year in question, in which case the volume number is included in parentheses immediately after the journal code letters. Patents are assigned appropriate three letter codes.

1.3

Topics Covered and Arrangement of Handbook-II

The individual chapters of Handbook-II are based on the general chapters which appeared in CHEC (but not in CHEC-II). However, considerable modifications were made as the general chapters for CHEC were written by five different groups of authors. Although each started out from the same master plan, individual initiatives very properly led to somewhat different interpretations. For both Handbooks-I and -II, it was considered important where possible to present the material in a strictly standard order.

For the 'Structure' and 'Reactivity' chapters, the Handbooks-I and -II usually follow the order of material used in the corresponding general chapters for the five-membered ring with several heteroatoms. Certain portions of the text have been deleted and additional material has been incorporated, including such new and fast developing fields of heterocyclic chemistry as palladium-catalyzed cross-coupling reactions, intramolecular cycloadditions, stable azole carbenes and others. It is hoped that this gives the Handbook-II a more appropriate balance. Reference lists are not included in Handbooks-I and -II (except the miscellaneous reference list, which is given in Appendix D). Under the reference system employed, a reference can usually be identified by its citation (see Appendix A, Section 5); complete details are given in the reference list in the appropriate volume of CHEC or CHEC-II.

The synthesis section, while retaining many concepts from the original, has been largely rewritten with the aim of achieving an integrated approach.

Part 2
Structure of Heterocycles

2.1
Overview

2.1.1 RELATIONSHIP OF HETEROCYCLIC AND CARBOCYCLIC AROMATIC COMPOUNDS

Heterocyclic compounds (like carbocyclic compounds) can be divided into heteroaromatic and heteroalicyclic types. The chemistry of the heteroalicyclic compounds is in general similar to that of their aliphatic analogues, but that of heteroaromatic compounds involves additional principles.

Aromatic compounds possess rings in which each of the ring atoms is in the same plane and has a p-orbital perpendicular to the ring plane, and in which $(4n+2)$ π-electrons are associated with each ring.

For a better understanding of the genesis and electronic nature of basic heteroaromatic systems it is convenient to take as a starting point their carbocyclic precursors. The latter can be divided into three main groups: neutral (*e.g.* benzene), anionic (*e.g.* cyclopentadienyl anion) and cationic (*e.g.* tropylium ion). Each of these carbocyclic systems is parent to a large number of isoelectronic heteroaromatic compounds. Six-membered aromatic heterocycles are derived from benzene by replacing CH groups with N, O^+, S^+ or BH^-: these are isoelectronic with the CH group. One CH group can be replaced to give pyridine (**1**), the pyrylium ion (**2**) and the thiinium (thiopyrylium) ion (**3**) or $1H$-boratobenzene anion (**4**) ⟨95JA8480⟩. The heteroatom in all these molecules is in a double-bonded state (Kekule structure) and formally contributes one p-electron to the aromatic π-ensemble. Such a heteroatom is called "pyridine-like". Replacement of two or more CH groups in such a manner is possible with retention of aromaticity.

The five-membered aromatic heterocycles pyrrole (**5**), furan (**6**) and thiophene (**7**) are formally derived from cyclopentadienyl anion by replacement of one CH^- group with NH, O or S, each of which can contribute two p-electrons to the aromatic π-electron sextet. Heteroatoms of this type have in classical structures only single bonds and are called "pyrrole-like". Other five-membered aromatic heterocycles are derived from compounds (**5**), (**6**) and (**7**) by further replacement of CH groups with N, O^+ or S^+.

Transition from the tropylium ion to its neutral heteroaromatic counterparts is possible by replacement of a CH^+ group by a heteroatom with a vacant p-orbital. The latter effectively accepts π-electrons, thus providing ring electron delocalization. A typical example is the boron atom in a seven-membered borepine (**8**) ⟨92AG1267⟩. Correspondingly, this type of heteroatom can be referred to as "borepine-like". Other little-known representatives of this family are alumopine (**8**, Z = AlH) or gallepine (**8**, Z = GaH).

These three fundamental types of heteroatoms are also found in small and large heterocycles.

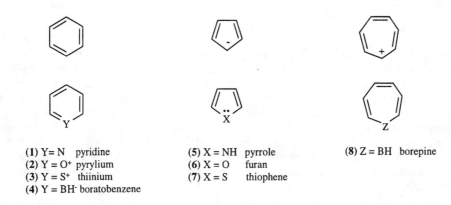

(**1**) Y = N   pyridine
(**2**) Y = O^+ pyrylium
(**3**) Y = S^+ thiinium
(**4**) Y = BH^- boratobenzene

(**5**) X = NH pyrrole
(**6**) X = O   furan
(**7**) X = S   thiophene

(**8**) Z = BH   borepine

Scheme 1 Three possible types of heteroatoms and the relationship between carbocyclic and heterocyclic aromatic systems

2.1.2 ARRANGEMENT OF STRUCTURE CHAPTERS

Each of the chapters on structure commences with a survey of the possible heterocyclic structures with the size of the ring in question. Structures are generally subdivided into those in which the ring atoms are all in conjugation with each other *(aromatic* or *antiaromatic)* and those in which at least one sp^3-hybridized ring atom interrupts the conjugation. The first class is further subdivided into those possessing exocyclic conjugation, and those without.

Theoretical methods are surveyed, followed by data on molecular dimensions obtained from X-ray diffraction or microwave spectroscopy. The various types of NMR spectroscopic characteristics are then surveyed, including ^{1}H, ^{13}C and nitrogen NMR spectroscopy. This is followed by a discussion of UV and visible and then IR, mass and photoelectron spectroscopy. Each of the spectroscopic sections deals with both the various parent rings and the effect of substituents.

The next main section deals with thermodynamic aspects. It starts by consideration of the intramolecular forces between heterocyclic molecules which influence melting and boiling points, solubility and chromatographic characteristics. This is followed by a section on stability and stabilization, including thermochemistry and conformation of the saturated ring systems, and then a discussion of aromaticity.

The last major section deals with tautomerism, including angular tautomerism where applicable and then substituent tautomerism.

2.1.3 NOMENCLATURE*

Some of the rules of systematic nomenclature used in *Chemical Abstracts* and approved by the International Union of Pure and Applied Chemistry are collected here. Important trivial names are listed at the beginning of individual chapters.

The types of heteroatom present in a ring are indicated by prefixes: 'oxa', 'thia' and 'aza' denote oxygen, sulfur and nitrogen, respectively (the final 'a' is elided before a vowel). Two or more identical heteroatoms are indicated by 'dioxa', 'triaza', *etc.*, and different heteroatoms by combining the above prefixes in order of preference, *i.e.* O, S and N.

Ring size and the number of double bonds are indicated by the suffixes shown in Table 1. Maximum unsaturation is defined as the largest possible number of non-cumulative double bonds (O, S and N having valencies of 2, 2 and 3, respectively). Partially saturated rings are indicated by the prefixes 'dihydro', 'tetrahydro', *etc.*

Table 1 Stem Suffixes for Hantzsch-Widman Names

Ring size	Rings with Nitrogen			Rings without Nitrogen		
	Maximum unsaturation	One double bond	Saturated	Maximum unsaturation	One double bond	Saturated
3	-irine	—	-iridine	-irene	—	-irane
4	-ete	-etine	-etidine	-ete	-etene	-etane
5	-ole	-oline	-olidine	-ole	-olene	-olane
6	-ine	—	—	-in	—	-ane
7	-epine	—	—	-epin	—	-epane
8	-ocine	—	—	-ocin	—	-ocane
9	-onine	—	—	-onin	—	-onane
10	-ecine	—	—	-ecin	—	-ecane

Numbering starts at an oxygen, sulfur or nitrogen atom (in decreasing order of preference) and continues in such a way that the heteroatoms are assigned the lowest possible numbers. Other things being equal, numbering starts at a substituted rather than at a multiply bonded nitrogen atom. In compounds with maximum unsaturation, if the double bonds can be arranged in more than one way, their positions are defined by indicating the nitrogen or carbon atoms which are not multiply bonded, and consequently carry an 'extra' hydrogen atom, by '1*H*-', '2*H*-', *etc.* In partially saturated compounds, the positions of the hydrogen atoms can be indicated by '1,2-dihydro', *etc.* (together with the 1*H*-type notation if necessary); alternatively, the positions of the double bonds can be specified; for example 'Δ^3-' indicates that a double bond is between atoms 3 and 4. A positively charged ring is denoted by the suffix '-ium'.

* For a detailed discussion, see CHEC 1.02.

The presence of a ring carbonyl group is indicated by the suffix '-one' and its position by a numeral, *e.g.* '1-one', '2-one', *etc.;* the numeral indicating the position of the carbonyl group is placed immediately before the name of the parent compound unless numerals are used to designate the position of heteroatoms, when it is placed immediately before the suffix. Compounds containing groups (**10**) or (**13**) are frequently named either as derivatives of (**9**) and (**12**) or of (**11**) and (**14**).

$$
\begin{array}{cccccc}
\diagdown CH_2 & \diagdown CO & \diagdown CH & \diagdown CH_2 & \diagdown CO & \diagdown CH \\
| & | & || & | & | & || \\
\diagup CH_2 & \diagup CH_2 & \diagup CH & \diagup NH & \diagup NH & \diagup N \\
(\mathbf{9}) & (\mathbf{10}) & (\mathbf{11}) & (\mathbf{12}) & (\mathbf{13}) & (\mathbf{14})
\end{array}
$$

Ring $C=S$ and $C=NH$ groups are denoted by the suffixes '-thione' and '-imine'; *cf.* '-one' for the $C=O$ group.

2.2
Structure of Six-membered Rings

2.2.1 SURVEY OF POSSIBLE STRUCTURES: NOMENCLATURE

2.2.1.1 Aromatic Nitrogen Systems without Exocyclic Conjugation

Since N^+ and C are isoelectronic, the simplest and most direct hetero-analogue of benzene (**1**) is the pyridinium ion (**2**). Further 'azonia substitution' of this kind gives the disubstituted species (**3**)–(**5**).

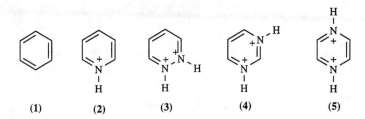

(1)   (2)   (3)   (4)   (5)

Deprotonation from the azonium group leaves a lone pair of electrons on the nitrogen atom, and a neutral aza substituent. The known parent monocyclic azines (see Scheme 1a) include all the possible diazines and triazines, but only one tetrazine, the 1,2,4,5-isomer. Some 1,2,3,5-tetrazines have been reported, but only when heavily substituted or fused. Some aromatic bicyclic 1,2,3,4-tetrazines have been prepared (see Section 4.4.8.2.3) as well as reduced 1,2,3,4-tetrazines (see CHEC 2.21). No pentazines are known. All attempts to prepare hexazine also failed though several claims about fixation of the latter in a matrix have appeared.

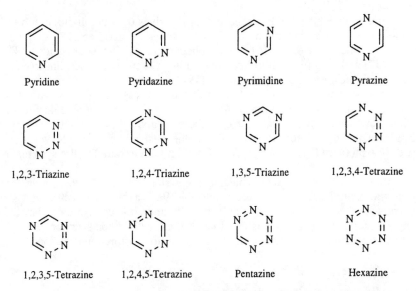

Pyridine   Pyridazine   Pyrimidine   Pyrazine

1,2,3-Triazine   1,2,4-Triazine   1,3,5-Triazine   1,2,3,4-Tetrazine

1,2,3,5-Tetrazine   1,2,4,5-Tetrazine   Pentazine   Hexazine

Scheme 1a   Monocyclic heteroaromatic nitrogen systems with six-membered rings

Pyrimidine natural products are particularly important (Scheme 1b). The nucleic acids contain pyrimidine and purine bases; ribonucleic acids (RNA) contain D-ribose and uracil, deoxyribonucleic acids (DNA) contain 2-deoxy-D-ribose and thymine and both types contain phosphate residues, cytosine, adenine and guanine. Nucleosides are pyrimidine or purine glycosides (*e.g.* uridine, cytidine).

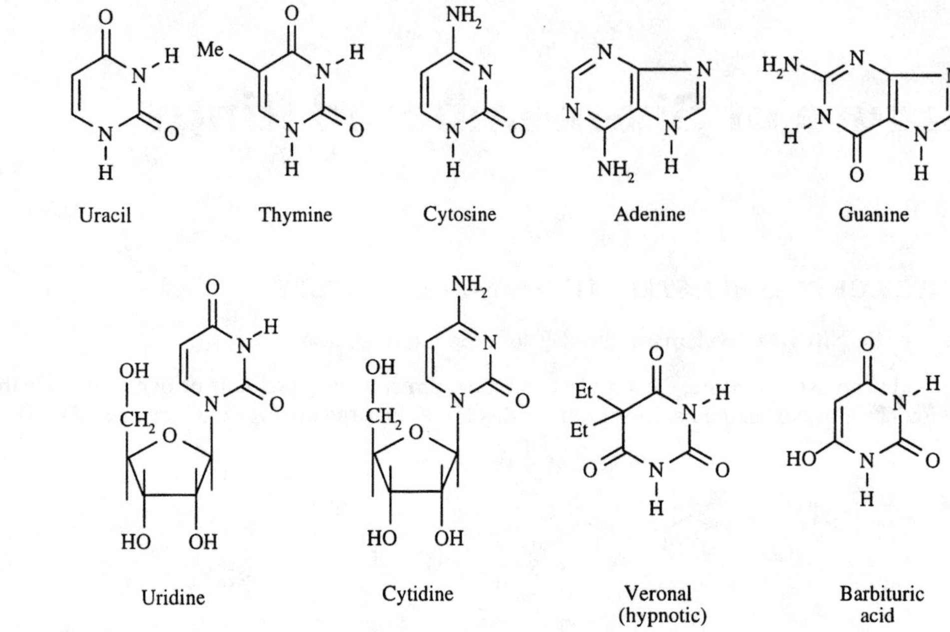

Scheme 1b Important derivatives

Synthetic derivatives (*e.g.* veronal) of barbituric acid are hypnotics.

When two fused six-membered rings (naphthalene analogues) are considered, possibilities become very numerous. There are two monoazanaphthalenes, quinoline (**6**) and isoquinoline (**7**), four benzodiazines [cinnoline (**8**), phthalazine (**9**), quinazoline (**10**) and quinoxaline (**11**)], with the two nitrogen atoms in the same ring, and six naphthyridines [*e.g.* (**12**), named and numbered in a systematic way] with the nitrogens in different rings. Of the higher polyazanaphthalenes which have been prepared (examples with up to six nitrogen atoms are known), the important pteridine system (**13**) should be noted. Both the benzotriazines are known, but 1,2,3,4-benzotetrazine is not. Other fused diazines include purine and alloxazine.

Azonia substitution at a naphthalene bridgehead position gives the quinolizinium ion (**14**). The two best known monoaza systems with three aromatic fused rings are acridine (**15**), derived structurally from anthracene, and phenanthridine (**16**), an azaphenanthrene. The better-known diaza systems include phenazine (**17**) and 1,10-phenanthroline (**18**), while systems with three linearly fused pyridine rings are called anthyridines, *e.g.* the 1,9, 10-isomer (**19**).

Although most ring systems are numbered according to a fairly straightforward set of rules (see CHEC 1.02), there are several exceptions. Thus numbering in purine molecule follows historical tradition. Acridine now has its 'meso' positions numbered 9 and 10 (see structure **15**). (At least two other numbering systems have been widely used in the past.) Phenazine, however, is numbered systematically, as in structure (**17**). Phenanthridine (**16**) is now numbered in the systematic fashion as indicated.

Heterocycles structurally based on the phenalene ring system frequently possess distinctive colors. With nitrogen as the central atom we have the unstable 9a-azaphenalene (**20**), ⟨76JCS(P1)341⟩. The cyclazine nomenclature is commonly applied to this and related compounds: thus, (**20**) is (3.3.3)cyclazine. Further aza substitution is possible, *e.g.* as in the heptaazaphenalene (**21**). Another well-known and important representative of this family is yellow colored perimidine.

2.2.1.2 Aromatic Systems with Exocyclic Conjugation

Important compounds of three classes result when a positively charged ring carries a negatively charged substituent.

(i) In certain relative orientations the charges may formally cancel. The 2- and 4-pyridone structures (**22**, **23**; X = NH) possess complete π-conjugation and 'aromatic' stabilization.

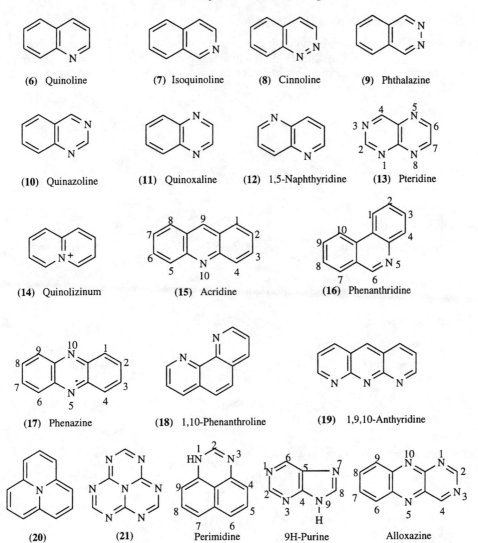

(6) Quinoline   **(7)** Isoquinoline   **(8)** Cinnoline   **(9)** Phthalazine

(10) Quinazoline   **(11)** Quinoxaline   **(12)** 1,5-Naphthyridine   **(13)** Pteridine

(14) Quinolizinum   **(15)** Acridine   **(16)** Phenanthridine

(17) Phenazine   **(18)** 1,10-Phenanthroline   **(19)** 1,9,10-Anthyridine

(20)   **(21)**   Perimidine   9H-Purine   Alloxazine

Scheme 2 Polycyclic aromatic nitrogen systems

(22)   **(23)**   **(24)**

(ii) The isomeric structure (**24**; X = NH, O, S) contains the elements of azomethine or carbonyl ylides, which are 1,3-dipoles. More complex combinations can lead to '1,4-dipoles', for instance the pyrimidine derivative (**25**). The 'cross-conjugated ylide' (**26**) is a higher order dipole.

(iii) Charge-separated dipoles are formed by attachment of an anionic group to an azonia nitrogen: pyridine 1-oxide (**27**) and the ylid (**28**) possess 1,3-dipoles canonical forms.

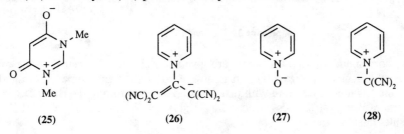

(25)   **(26)**   **(27)**   **(28)**

2.2.1.3   Ring Systems Containing One Oxygen or Sulfur

Replacement of CH in benzene by an oxonia group (O$^+$) gives the pyrylium cation, but no neutral oxygen analogue of pyridine is possible. Both 2*H*- and 4*H*-pyran contain *sp³*-hybridized carbon atoms. Many trivial names exist for oxygen heterocycles and the more important of these are shown in Scheme 3.

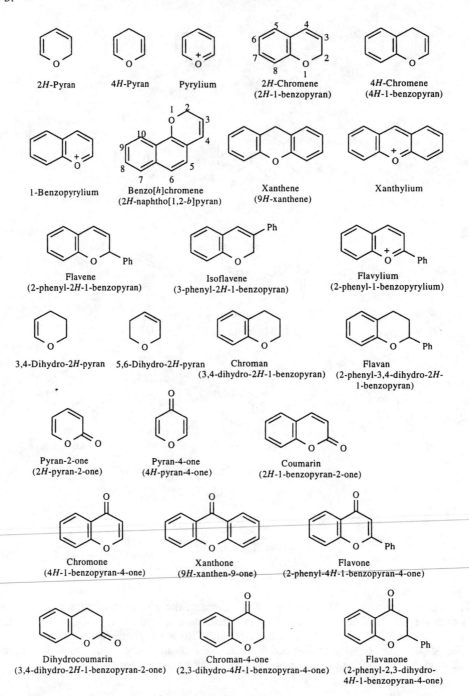

Scheme 3   Oxygen heterocycles

A few of the corresponding sulfur systems are given in Scheme 4; because of the availability of sulfur *d*-orbitals, the neutral species 1*H*-thiin is now a possibility and some substituted derivatives exist. These are the 'thiabenzenes', and with just the single heteroatom in the ring (C$_5$S rings) they are very unstable, non-planar structures which are largely ylidic in character (CHEC 2.25). Aza substitution results in stabilization.

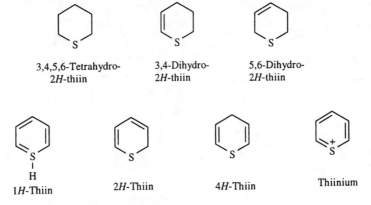

Scheme 4 Monocyclic thiins: structure and nomenclature

2.2.1.4 Rings Containing Nitrogen with Oxygen and/or Sulfur

Compounds with two heteroatoms are illustrated in Scheme 5. The oxazines and thiazines contain a saturated carbon atom; the corresponding aromatic cations are oxazinium and thiazinium.

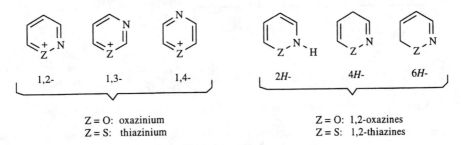

Scheme 5 Oxazines and thiazines

The range of possible ring systems with three, four or five heteroatoms is considerable: some of the more common systems are delineated in Scheme 6.

2.2.1.5 Fully Conjugated but Non-aromatic Compounds

Cyclic π-conjugation is also present in six-membered rings containing two double bonds and two atoms (NH, O, S) each carrying a lone pair of electrons. The eight π-electrons imply 'antiaromaticity'. Many such systems are known. Particularly when the heteroatoms are oxygen or sulfur, the rings are often folded (heteroatoms 1,4-) or puckered (1,2-). The best-known examples are the dibenzo-fused systems phenothiazine (**29**; X = S), phenoxazine (**29**; X = O), thianthrene (**30**; X = Y = S), phenoxathiin, or phenoxathionine (**30**; X = S, Y = O) and dibenzo-*p*-dioxin (**30**; X = Y = O). An aromatic sextet of electrons can be formulated for the sulfur-containing rings by invoking *d*-orbital participation for one of the S atoms (acceptor) and assuming that the other S atom behaves as a *p*-donor. Although most physical methods indicate at least non-aromaticity for these compounds, chemical reactivity does show some evidence of preference for substitution rather than addition reactions, particularly for 1,4-dithiin derivatives (see Section 3.2.2.1.2).

(29)   (30)

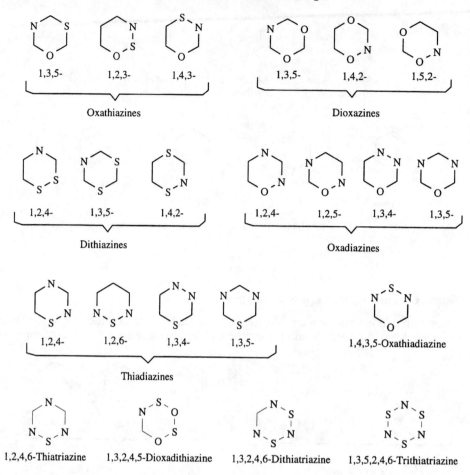

Scheme 6   Six-membered monocyclic systems containing nitrogen with oxygen and/or sulfur (the name printed corresponds to the ring with maximum unsaturation)

2.2.2   THEORETICAL METHODS: CALCULATIONS

The main objective of quantum chemical calculation is to solve the Schrödinger equation, *i.e.* to obtain (i) the so called eigen-functions which are the composition of the MO's and (ii) the eigen values of the energies MO's. One major group of physical molecular properties depends on the molecular orbital composition including molecular geometry, electron density, net atomic charges, bond orders, frontier MO electron densities, free valences, electrostatic potential maps, and dipole moments; these depend on the energies of the MO's. A second major group includes ionization potentials, electron affinities, delocalization and localization energies. Several levels of MO approximations are applied to achieve these goals: from simple Hückel theory to *ab initio* calculations.

2.2.2.1   The σ,π-MO Approximation (Hückel theory)

In the simpler MO approximations the π-electrons are assumed to move independently in MOs that can be represented as linear combinations of the atomic p-orbitals. The energies of MOs and the distribution of the π-electrons in each depend on the values of certain integrals. The Coulomb integrals are characteristic of individual atomic p-orbital in a given molecular environment and are a measure of the effective electronegativity of that atom towards π-electrons. The resonance integrals pertain to bonds between two p-atomic orbitals and are a measure of the stability that a localized π-bond would possess when formed between them. These calculations are based on certain simplifying assumptions concerning the relative values of the different Coulomb and resonance integrals.

For unsubstituted aromatic hydrocarbons all the carbon atoms are assigned the same Coulomb integral (α^o) and all C—C bonds are assigned the same resonance integral (β^o). In heteroaromatic molecules the approximate Coulomb integral for heteroatom α_x is expressed as in Eq. (1) in terms of α^o and β^o and the electronegativity parameter h.

$$\alpha_X = \alpha^O + h\beta^O \tag{1}$$

$$\beta_{XY} = k_{XY}\beta^O \tag{2}$$

Resonance integrals of bonds between atoms X and Y, β_{XY}, are expressed as defined in Eq. (2), where k_{XY} depends on the bond length. There has been considerable variation in the values taken for the Coulomb and resonance integrals for heterocyclic molecules. One of the best available set of parameters is still that originally suggested by A. Streitwieser (Molecular orbital theory. J. Wiley & Sons, Inc., N.Y.-L., 1961):

$$h_{\dot{N}} = 0.5; \ h_{\ddot{N}} = 1.5; \ h_{\overset{+}{N}} = 2.0; \ h_{\dot{O}} = 1.0; \ h_{\ddot{O}} = 2.0; \ h_{\overset{+}{O}} = 2.5$$

$$k_{C-\dot{N}} = k_{C-\dot{O}} = 1.0; \ k_{C-\ddot{N}} = k_{C-\ddot{O}} = 0.8$$

In this notation, heteroatoms which contribute one and two π-electrons to the aromatic system are designed accordingly.

A semi-empirical π-electron theory that takes electron repulsion into account is the Pariser-Parr-Pople (PPP) method (R.G. Parr. Quantum Theory of Molecular Electronic Structure, Benjamin, N.Y., 1963).

2.2.2.2 More Sophisticated Semi-Empirical Methods

The simple, or Hückel based, molecular orbital theory (HMO and PPP methods) frequently provides useful qualitative insights but cannot be used reliably in a quantitative manner. For this purpose it is necessary to use a method which takes account of all the electrons as well as their mutual repulsions. A major bottleneck in such calculations is in the computation and storage of the enormous number of electron-repulsion integrals involved. Early efforts to reduce this problem led Hoffmann to the EH approximation (I.N. Levine, Quantum Chemistry, 4-th ed., 1991, Prentice-Hall, Inc., Ch. 16, 17), and Pople and co-workers to the CNDO, INDO and NDDO-approximations ⟨B-70MI 40100⟩.

Dewar has implemented modifications of the latter approximations which proved to be very successful, known as the MINDO/3 ⟨75JA1285⟩ and MNDO ⟨77JA4899⟩ methods. The MNDO method was further developed into the AM1 and PM3 methods. Since their inception these methods have been somewhat controversial since they were parameterized to reproduce not the results of the *ab initio* methods from which they are derived (as was done in CNDO/2 and INDO), but experimental geometries and heats of formation. These semi-empirical MO procedures therefore appear to involve several theoretical inconsistencies which have yet to be fully resolved ⟨B-77MI 40100⟩. Nevertheless, judged on purely empirical criteria, they seem to work well. Indeed they are frequently comparable in accuracy (relative to experiment) to *ab initio* calculations using moderately sized basis sets and the semi-empirical methods are much better than *ab* initio calculations using minimal basis sets ⟨79JA5558⟩. Semi-empirical procedures of this kind have been used to study the reactions of various five- and six-membered heterocyclic systems ⟨77JCS(P2)724⟩.

MNDO gives substantially improved results as compared to MINDO/3. In 1985 an improved version of MNDO, called AM1, was published ⟨85JA3902, 88TH1⟩. Later MNDO was reparameterized to give MNDO-PM3 ⟨89JCC10, 90JCC543⟩.

The computer program package MOPAC contains – MINDO/3, MNDO, AM1 and PM3 [J.J.P. Stewart, MOPAC – 7.0: A semi-empirical Molecular Orbital Program, Program No 455, Quantum Chemistry Program Exchange (QCPE), Indiana University, Bloomington, IN 47405 USA].

2.2.2.3 *Ab Initio* Calculation

Non-empirical or *ab initio* calculations, in contrast to semi-empirical treatment, do not use experimental data other than the values of the fundamental physical constants. A formalism of *ab initio* calculations was developed independently by Roothaan ⟨51MI 40100⟩ and Hall ⟨51MI 40101⟩. Most practical applications of *ab initio* MO theory now utilize one of the carefully optimized expansions of Gaussian functions introduced by Pople and his co-workers ⟨74MI 40100⟩. The simplest (and least accurate) is designated STO-3G, while one of the most complex in common use is designated 6-31G*. Fortunately, even the simplest basis sets seem to give excellent descriptions of the electron distribution, especially when expressed in terms of integrated spatial populations ⟨80MI 40101⟩. Palmer, in particular, has presented electron distribution data in many heterocyclic systems ⟨78JST(43)33⟩. Unfortunately, apart from those special cases in which species having the same numbers of each type of bond are compared, the quantitative estimation of relative energies in *ab initio* theory usually requires a basis set of at least 3-21G quality.

Successive increases in basis-set size (STO-3G → 3-21G → 3-21G* → 6-31G*) give improved bond-length accuracy. Moreover, for reliable results the geometries of the species being compared must be calculated at the *ab initio* level. The expense inherent in the use of the more complex basis set, as well as that geometry optimization, originally limited the most detailed studies to rather small heterocyclic ring systems ⟨77JA7806, 78JA3674, 83JA309⟩.

2.2.2.4   Molecular Mechanics

The molecular mechanics (MM) or force field method is an empirical method based on classical mechanics and adjustable parameters. It has the disadvantage of being limited in its application to certain kinds of compounds for which the required parameters have been determined (experimentally or by theoretical calculations). Its advantage is a considerably shorter computation time in comparison with other procedures having the same purpose. This method has been shown to be very reliable and efficient in determing molecular geometries, energies, and other properties for a wide variety of compounds.

In the analysis of the structural properties of heterocyclic compounds the most frequently used force field among several available seems to be Allinger's MM force field. The earlier version was developed for application to conjugated systems by including a p-system molecular orbital treatment in calculations ⟨87JCC581, 88JA2050⟩ and with parameters extended to include furan and related compounds ⟨85TL2403, 88JOC5471, 89JCC635⟩.

2.2.2.5   Some Results of Calculations of Six-membered Heterocyclic Molecules

Heteroatoms cause an irregular distribution of electron density in heterocyclic rings and thus strongly influence reactivity as well as almost all physico-chemical properties. In Table 1 effective atomic π-charges in key azine molecules, calculated by various methods, are given. One can see that a major contradiction exists between the *ab initio* method, on the one hand, and SCF and HMO, on the other hand, concerning their estimates of the relative π-deficiency for positions 2(6) and 4 in pyridine, for positions 2 and 4(6) in pyrimidine, and again for positions 3(6) and 4(5) in pyridazine. Though there is no unequivocal way to check these calculations by experiment, the HMO and SCF

Table 1   The Values of Atomic π-Charges in Azine Molecules Calculated by Several Quantum Chemical Methods

Compound	Position	*ab initio*[a]	SCF MO[b]	HMO[c]
			Method	
Pyridine	1	− 0.11	− 0.100	− 0.195
	2,6	+ 0.05	+ 0.050	+ 0.077
	3,5	− 0.04	− 0.010	− 0.004
	4	+ 0.07	+ 0.020	+ 0.050
Pyridazine	1,2	− 0.06	− 0.054	− 0.124
	3,6	+ 0.03	+ 0.042	+ 0.077
	4,5	+ 0.04	+ 0.014	+ 0.047
Pyrimidine	1,3	− 0.14	− 0.112	− 0.199
	2	+ 0.11	+ 0.100	+ 0.155
	4,6	+ 0.12	+ 0.074	+ 0.126
	5	− 0.07	− 0.026	− 0.009
Pyrazine	1,4	− 0.05	− 0.080	− 0.147
	2,3,5,6	+ 0.02	+ 0.040	+ 0.074
1,3,5-Triazine	1,3,5	− 0.17	− 0.118	− 0.203
	2,4,6	+ 0.17	+ 0.118	+ 0.203
1,2,4,5-Tetrazine	1,2,4,5	− 0.03	—	− 0.074
	3,6	+ 0.05	—	+ 0.147

[a] J. Almlof, B. Ross, U. Wahlgren, H.J. Johanson, J. Electron Spectrosc. Related Phenomenon, 1973, vol. 2, p. 51.
[b] ⟨60JCS1946⟩.
[c] ⟨79KGS1155⟩.

ordering seem to be the more realistic. They are in better agreement with larger deshielding of nuclei ^{1}H and ^{13}C at positions 2 of pyridine and pyrimidine and position 3 of pyridazine in NMR spectra (see Table 5 and 11) (for a more detailed discussion see A.F. Pozharskii, Theoretical Principles of Heterocyclic Chemistry (in Russian), "Khimia", Moscow, 1985, p. 58).

The relative π-deficiency sequence of azines does not change significantly on going from neutral molecules to their cations, though π-deficiency itself of course strongly increases. Most data indicate higher positive charge in positions 2 and 6 of pyridinium and pyrylium cations than in position 4. In the pyrylium cation only 30–35% of positive charge is localized on the oxygen atom.

The π-deficiency of azines is determined by the number of heteroatoms and their mutual disposition. According to calculations, the total π-deficiency (the total positive charge on all carbon atoms) changes in the sequence: s-triazine > pyrimidine > s-tetrazine ~ pyrazine > pyridazine > pyridine. Relative local π-deficiency (the largest positive charge on any one carbon atom in a molecule) is somewhat different: s-triazine > pyrimidine > s-tetrazine > pyridazine ~ pyridine > pyrazine. Thus, though s-tetrazine has the largest number of nitrogen atoms among stable azines its π-deficiency is less than that of s-triazine and even pyrimidine. Evidently the π-acceptor action of heteroatoms in azines is most effective when they are *meta* to each other (s-triazine, pyrimidine). The *ortho-para* disposition as in the case of pyrazine subjects each carbon atom to two contradictory forces: the strong electron acceptor influence of an *ortho-* or *para*-nitrogen and the weak electron-donor influence (due to reorganization of π-cloud) of a *meta*-nitrogen. As a result some decrease of π-deficiency can occur which is different for different systems.

On transition from pyridine to quinoline (**31**) or isoquinoline, both the total and the local π-deficiencies increase somewhat, and the benzene ring is also slightly π-deficient. A remarkable exception among six-membered heterocycles is perimidine (**32**). Its molecule has 14π-electrons but in the hetero-ring there are formally seven π-electrons. The expulsion of the "superfluous" extra-Hückel electron from the hetero-ring to the naphthalene moiety leads to considerable polarization of the π-cloud. As a result the naphthalene moiety becomes strongly π-excessive whereas the μ-carbon atom in the hetero-ring becomes strongly π-deficient.

(31)   (32)

Other fundamental characteristics of heteroaromatic systems are their electron donor and electron acceptor properties. The energies of the highest occupied and the lowest empty molecular orbitals (frontier orbitals) can serve as measure of such properties. Pyridine-like heteroatoms, as rather strong electron acceptors, lower the energies of all the molecular orbitals. This means that compounds containing heteroatoms of this type should show more π-acceptor and less π-donor character. In accordance with expectation, π-acceptor properties of azines are increased in the sequence: s-tetrazine > pyrazine > pyridazine > pyrimidine > pyridine (Table 2). This agrees with the relative ease of polarographic reduction of the same heterocycles. However, there is no strict dependence between π-deficiency and π-acceptor strength of the compounds. π-Acceptor properties of azines are also increased on transition to benzoderivatives, *e.g.* acridine > quinoline > pyridine.

Table 2 HMO Energies of Frontier Orbitals of Some Six-membered Heterocycles (in β-units)[a]

Compounds	HOMO	LUMO	Compounds	HOMO	LUMO
Pyridine	1.000	− 0.841	1,2,4,5-Tetrazine	1.186	− 0.500
Pyridazine	1.101	− 0.727	Quinoline	0.703	− 0.527
Pyrimidine	1.077	− 0.781	Isoquinoline	0.646	− 0.576
Pyrazine	1.000	− 0.686	Acridine	0.514	− 0.325
1,3,5-Triazine	1.281	− 0.781	Perimidine	0.328	− 0.674

[a] Note that β is negative value

Monocyclic azines are very weak π-donors and behave mostly as n-donors on interaction with electrophiles. However, π-donor character is significantly increased in their benzo-derivatives. For instance, acridine forms with chloranil a highly colored 1 : 1 molecular complex. Perimidine is one of the strongest heterocyclic π-donors which gives deeply colored molecular complexes with a variety of organic electron acceptors. On the other hand, the π-acceptor ability of perimidine is moderate.

2.2.3 STRUCTURAL METHODS

2.2.3.1 X-Ray Diffraction

An early compilation of X-ray data for heterocyclic compounds ⟨70PMH(5)1⟩ contains many examples of six-membered rings (see Table 3). More recent data are contained in the Cambridge crystallographic database.

Table 3 X-Ray Structures of Compounds with Six-membered Rings and One or More Heteroatoms[a]

	Ring position						
Ring	1	2	3	4	5	6	*Examples of compounds studied*
C_5N	N	—	—	—	—	—	Pyridines, pyridones, pyridine *N*-oxides, piperidines
C_4N_2	N	—	N	—	—	—	Pyrimidines, thymine, barbituric acid
	N	—	—	N	—	—	Pyrazines, piperazines
C_3N_3	N	—	N	—	N	—	*s*-Triazines, cyanuric acid
C_2N_4	N	N	—	N	N	—	1,2,4,5-Tetrazines, dihydro derivatives[b]
C_5O	O	—	—	—	—	—	α-D-Glucose, α-L-sorbose, kanamycine monosulfate monohydrate
C_4O_2	O	—	—	O	—	—	1,4-Dioxanes
C_3O_3	O	—	O	—	O	—	Trioxane
C_5S	S	—	—	—	—	—	Thiins, 6-benzoyl-3-hydroxy-1-methyl-5-phenylthia-Benzene 1-oxide
C_4S_2	S	S	—	—	—	—	1,2-Dithiane-3,6-dicarboxylic acid
	S	—	S	—	—	—	2-Phenyl-1,3-dithiane
	S	—	—	S	—	—	1,4-Dithianes, 1,4-dithiins
C_3S_3	S	—	S	—	S	—	1,3,5-Trithiane
C_4NO	O	—	—	N	—	—	Morpholine
C_4NS	S	—	—	N	—	—	Cycloalliin hydrochloride monohydrate, 4-methyl-thio-morpholine-1,1-dioxide
C_4OS	O	—	—	N	—	—	*Trans*-2,3-Dichloro-1,4-oxathiane
C_3O_2S	S	O	—	—	—	O	Trimethylene sulfite

[a] For further details see ⟨70PMH(5)1⟩. [b] See CHEC 2.21.3.2.

In Scheme 7 the bond lengths and internal bond angles are given for some of the simple azines, based on X-ray diffraction, gas-phase electron diffraction or microwave spectroscopy.

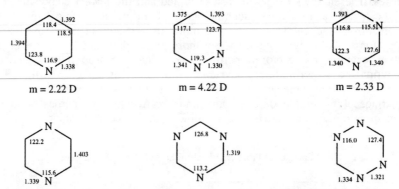

Scheme 7 Molecular dimensions and dipole moments of some simple azines

The C—N bond length is usually some 4% shorter than C—C. To accommodate this with minimal disturbance of the other bond angles a small displacement of the N atoms toward the center of the ring, with consequent opening of the CNC bond angle from 120°, would be required. This is more or less what is observed in pyridinium salts (where NH^+ replaces one CH in benzene). However, the internal angle at the aza nitrogen in the free bases (where N: replaces CH) is generally found to be slightly less than 120°; the nitrogen nuclei are slightly further from the center of gravity than the carbons. These are comparatively minor deviations from a completely regular hexagon. Available data on pyrylium or thiinium salts are generally of heavily substituted ions. In the oxygen rings the trend outlined above with the pyridines is apparently not continued; in the studied examples the COC angle (about 124°) is slightly larger than that for a regular hexagon, and the interior angles in the ring at the carbons next to the oxygen are smaller (*ca.* 118°). Thiinium salts are characterized by long C—S bonds (*ca.* 1.72 Å) and a small CSC angle of *ca.* 104°.

X-ray analysis has been useful in structural investigation of intermediates arising in important reactions. For example, the catalytic effect of pyridines in acylation reactions is supposed to arise from N-acylpyridinium formation (See Section 3.2.1.3.7). Thus, the tetrafluoroborate structure (**33**) is involved in the transfer of Boc groups (Boc = *t*-butoxycarbonyl). X-ray crystallography shows that the carbonyl, pyridyl, and amino groups are virtually planar, explained by the through conjugation shown in structures (**33a**) and (**33b**), while bond lengths show this to be muchenhanced over 4-dimethylami-nopyridine ⟨91AG(E)1656⟩. The positioning of the anion over the carbonyl carbon atom as in (**33b**) indicates the trajectory of nucleophilic approach, furnishing a further example of crystal structure as a guide to reaction trajectories ⟨83ACR153⟩.

(33)   (33a)   (33b)

With two O, S or NR groups and two C=C double bonds in a ring, the systems are not aromatic, and usually not planar. Thianthrene (**34**; X = Y = S) and its various sulfoxides and sulfones have a fairly sharp dihedral angle between the planes of the benzene rings (θ = 128° in thianthrene ⟨63AX310⟩). Phenoxathiin (**34**; X = S, Y = O) and phenothiazine (**34**; X = S, Y = NH) are also non-planar (θ = 136° and *ca.* 158°, respectively) ⟨66AX429, 68CC833⟩, but dibenzodioxin (**34**; X = Y = O) is apparently planar (θ = 180°) ⟨78AX(B)2956⟩. Phenoxazine (**34**; X = O, Y = NH) probably has coplanar benzene rings, according to an early study ⟨40MI20100⟩, with a pyramidal nitrogen atom. When the central ring is rendered aromatic by oxidation (in the phenoxazinium and phenothiazinium ions (**35**; X = O and S), respectively), planarity is found as expected. The same effect is seen with the alloxazines (**36**), which are planar, and the dihydroalloxazines (**37**), which are not.

Thus, 6-ethyl-3-phenyl-1,6-dihydro-1,2,4,5-tetrazine (**38**), a possible homoaromatic system ⟨82JOC2856⟩ exists in a boat conformation, with C(6) and C(3) pointing upward with dihedral angles of 49.3° and 26.7° respectively. N(1) was *sp*² hybridized, and the bond lengths N(1) to N(5) of the ring were between single and double bond lengths, in agreement with the expected electron delocalization.

Many X-ray studies have also been carried out on saturated and partially saturated rings, *e.g.* piperidines and piperidones.

(34)   (35)

(36)   (37)   (38)

2.2.3.2 Microwave Spectroscopy ⟨74PMH(6)53⟩

Microwave spectra allow the determination of precise bond lengths and angles and some of the data in Scheme 7 (Section 2.2.3.1) were obtained in this way. In Table 4 details of molecular geometry are deduced from microwave spectra.

Table 4 Geometry of Six-membered Rings from Microwave Spectra[a]

Shape	Molecules
Planar	4-Pyrone,[b] 2 -pyrone, 4-pyranthione, 4-thiopyrone
Twist-chair	2,3-Dihydropyran
Chair	Piperidine,[c] morpholine, 1-methylmorpholine, 1,3,5-trioxane

[a] Abstracted from ⟨74PMH(6)53⟩. [b] Conjugation shown by bond lengths: short C—O and C—C, long C=C and C=O. [c] Both *ax*-NH and *eq*-NH forms found; the latter predominates.

Microwave spectra allow the determination of conformations. For example, 2-formylpyridine is shown to be planar with the carbonyl group *trans* to the nitrogen atom.

Microwave spectra also provide precise values for dipole moments and values for pyridine and many substituted pyridines are available (Table 3 of CHEC 2.04).

2.2.3.3 ¹H NMR Spectra

2.2.3.3.1 *Chemical shifts*

The protons on the benzene ring experience a deshielding effect due to the aromatic ring current, which brings the chemical shift of the benzene protons to δ 7.24 ppm. The same ring current persists in the polyazabenzenes as in benzene itself. Attempts to use estimates of ring current for calculating aromaticity from comparisons of proton chemical shifts suffer from the difficulty of finding suitable models (see Section 2.2.5.4).

In the aza derivatives of the aromatic hydrocarbons the nitrogen atoms exert a strong deshielding influence on the α-hydrogen atoms, and a similar but smaller effect on the γ-hydrogens. The protons at the β positions in pyridine are in fact shifted slightly upfield of the benzene resonance. Further aza substitution produces similar effects, but strict additivity is not observed. For instance, two adjacent nitrogen atoms, as in pyridazine, exert a much larger deshielding effect on the α-protons than the sum of the α- and β-effects of a single nitrogen atom. Conversion of pyridine into the pyridinium cation causes a downfield shift of all the hydrogens, especially the β and γ.

Strongly electron-releasing groups at the 2-, 4- and 6-positions of the pyrylium salts (**39**) exert a profound effect on the chemical shift of H(3) and H(5), which are moved upfield by 2–3 ppm. It is proposed that such compounds are better described as bridged pentamethine cyanines (**40**) ⟨88MRC707⟩.

(39)   (40)

Tables 5 and 6 summarize the chemical shifts of the protons in various aza heterocycles. ¹H NMR data for the corresponding heteroaromatic cations are given in Table 7.

Table 5 ^{1}H NMR Chemical Shifts of the Simple Monocyclic Azines (*cf* benzene, δ 7.24)[a]

| | δ (^{1}H) (ppm, TMS) (Position of N atoms indicated) | | | | | | | |
Position	Pyridine	Pyridazine	Pyrimidine	Pyrazine	1,2,3-Triazine	1,2,4-Triazine	1,3,5-Triazine	1,2,4,5-Tetrazine
1	N	N	N	N	N	N	N	N
2	8.52	N	9.26[b]	8.6	N	N	9.18	N
3	7.16	9.17	N	8.6	N	9.63	N	(10.48)[c]
4	7.55	7.52	8.78	N	9.06	N	9.18	N
5	7.16	7.52	7.36	8.6	7.45	8.53	N	N
6	8.52	9.17	8.78	8.6	9.06	9.24	9.18	(10.48)

[a] ⟨B-73NMR⟩. [b] Measured in CDCl$_3$. [c] Calculated value; shifts for some monosubstituted tetrazine derivatives lie in the range 10.26—10.45 ppm in CD$_3$OD, 10.11—10.25 ppm in CDCl$_3$ ⟨81JOC5102⟩.

Table 6 ^{1}H NMR Chemical Shifts of Protons on the Heterocyclic Rings of Simple Benzazines (*cf.* naphthalene, column 1)

Position	δ *(^{1}H)* (ppm, TMS) (Position of N atoms indicated)							
1	7.72	N	9.15	N	N	N	9.44	N
2	7.33	8.81	N	N	9.23	8.74	N	N
3	7.33	7.27	8.45	9.15	N	8.74	N	N
4	7.72	8.00	7.50	7.75	9.29	N	9.44	9.85

Table 7 ^{1}H NMR Spectra Data for Six-membered Heteroaromatic Cations[a]

| | | δ (^{1}H) (ppm) | | | J (Hz) | | | |
Compound	Anion/Solvent	2	3	4	2:3	2:4	2:6	3:4
Pyridinium	DSO$_4$$^-$/D$_2$O	8.78	8.09	8.62	5.8	1.7	0.7	7.5
1-Methylpyridinium	I$^-$/D$_2$O	8.77	8.04	8.53	—	—	—	—
Pyrylium	ClO$_4$$^-$/MeCN	9.59	8.40	9.20	3.5	2.4	1.5	8.0
2,6-Dimethylpyrylium	SbCl$_6$$^-$/SO$_2$	3.04[b]	7.98	8.86	—	—	—	—
2,4,6-Trimethylpyrylium	ClO$_4$$^-$/SO$_2$	2.90[b]	7.77	2.74[b]	—	—	—	—
Thiinium	ClO$_4$$^-$/MeCN	10.13	8.97	9.00	7.7	1.7	1.6	6.1

[a] From ⟨B-73NMR⟩ which see for original references. [b] Methyl shifts.

The spectra of protonated polyaza heterocycles are frequently complicated by the occurrence of covalent hydration. This is more common with polycyclic systems, *e.g.* pteridine.

The extensive delocalization and aromatic character of pyridones, pyrones, *etc.* are shown by their chemical shift and coupling constant values (Table 8). By contrast, pyrans and thiins show chemical shifts characteristic of alkenic systems (Table 9). For these and for rings containing only a single endocyclic bond (Table 10), ^{1}H NMR spectroscopy offers a most useful tool for structure determination.

Solvent-induced and lanthanide-induced shifts are of great value in structure assignments in heterocyclic compounds, because cyclic nitrogen atoms, carbonyl groups, *etc.* undergo specific hydrogen bonding or coordination resulting in differential shifts of groups in different positions.

2.2.3.3.2 *Coupling constants*

The normal pattern of coupling constants for aromatic six-membered rings is found in the heterocyclic aza systems, except that the *ortho* coupling to a proton α to a heterocyclic nitrogen is reduced from 7–8 Hz to 4.5–6 Hz. The $J_{2,3}$ of pyrylium salts is still lower (*ca.* 3.5 Hz), but in pyridinium salts and pyridine *N*-oxide it is of intermediate value (*ca.* 6.5 Hz) (see Table 7).

The 'direct' coupling constants (D_{ij}) of pyridine, obtained from a spectrum of the molecule in a nematic liquid crystal solvent ⟨B-73NMR, p. 10⟩, provide information about the geometry.

2.2.3.4 ^{13}C NMR Spectra
2.2.3.4.1 *Aromatic systems: chemical shifts*

Chemical shift data for a number of monocyclic, unsubstituted six-membered heteroaromatic compounds are given in Table 11.

Table 8   ¹H NMR Spectral Data for Six-membered Heteroaromatic Rings with Exocyclic Carbonyl or Thione Groups[a]

A γ-Compounds

| Compound | | Neutral species (D₂O, CCl₄ or CDCl₃) | | | | Cation (TFA or D₂SO₄) | | | |
| | | δ(¹H) (ppm) | | J (Hz) | | δ(¹H) (ppm) | | J (Hz) | |
Z	X	2	3	2:3	2:5	2	3	2.3	2:5
NH	O	7.98	6.63	7.532	0.260	8.5	7.4	7.6	—
NMe	O	7.81	6.49	8.4	—	8.4	7.4	7.6	—
NH	S	8.37	7.87	7.0	—	—	—	—	—
O	O	7.92	6.39	5.9	0.4	8.49	7.06	—	—
O	S	7.51	7.15	5.7	0.5	8.87	8.14	~5.7	~0.5
S	O	7.89	7.09	10.4	0.5	9.40	8.19	~10.1	~0.5
S	S	7.90	7.58	10.1	0.5	—	—	—	—

B α-Compounds

| Compound | | | δ(¹H) (ppm) | | | |
Z	X	Solvent	3	4	5	6
NH	O	CDCl₃	6.60	7.3	6.20	7.23
NH	O	10–20% D₂SO₄	7.3	8.2	7.3	8.2
O	O	CDCl₃	6.38	7.56	6.43	7.77

| | | | J (Hz) | | | |
			3:4	4:5	4:6	5:6
NH	O	DMSO	10	7	2	7
O	O	CDCl₃	9.4	6.3	2.4	5.0

[a] Data abstracted from ⟨B-73NMR⟩ and ⟨71PMH(4)121⟩ which see for original references.

Table 9   ¹H NMR Data or Six-membered Heterocyclic Rings with Two Endocyclic Double Bonds[a]

A 4H-Systems

| Z | Solvent | δ(¹H) (ppm) | | | J (Hz) | | | |
		2	3	4	2:3	2:4	2:6	3:4
NH	C₆D₆	5.73	4.42	3.15	—	—	—	—
NPh	CCl₄	6.27	4.53	2.92	9.0	1.6	—	3.9
O	—	6.12	4.63	2.66	7.0	1.7	1.5	3.4
S	—	5.97	5.54	2.84	10.0	1.1	2.9	3.9

B 2H-Systems

| Z | Solvent | δ (¹H) (ppm) | | | | |
		2	3	4	5	6
NPh	CCl₄	4.26	5.21	5.88	4.94	6.41
S[b]	CCl₄	3.16	4.99	—	6.08	6.08

[a] Data abstracted from ⟨B-73NMR⟩ and ⟨71PMH(4)121⟩ which see for original references.
[b] Data refer to the 4-methyl derivative.

Table 10 ¹H NMR Data for Six-membered Heterocyclic Rings with One Endocyclic Double Bond[a]

A Δ²-Systems

Compound			δ (¹H) (ppm)				
Z	X	Solvent	2	3	4	5	6
NH	COMe	CDCl₃	7.12	7.48	2.28	1.77	3.23
O	H	CCl₄	6.22	4.54	1.93	1.93	—

B Δ³-Systems

		δ (¹H) (ppm)				
Z	Solvent	2	3	4	5	6
NH	CDCl₃	3.33	5.75	5.75	2.07	2.95
NMe	CCl₄	2.76	5.58	5.58	—	—
O	—	4.03	5.76	5.76	2.10	—

[a] Data abstracted from ⟨B-73NMR⟩ and ⟨71PMH(4)121⟩ which see for original references.

Ring carbon atoms α to a heteroatom are most heavily deshielded, those γ to a heteroatom are also deshielded relative to benzene, while those in a β-position are more benzene-like. Introduction of a second nitrogen atom α or γ to a ring carbon atom results in further deshielding by approximately 10 and 3 ppm, respectively, whereas the effect on a β carbon atom is a shielding of approximately 3 ppm. Substituent effects follow the same general trend as in substituted benzenes, *i.e.* the chemical shifts of ring carbon atoms which either carry the substituent or are *para* to it differ in a predictable way relative to the unsubstituted heterocycle, whereas those of ring carbon atoms *meta* to the substituent are little affected by it. These effects are most conveniently exemplified in the pyridine series; typical data for a variety of monosubstituted pyridines are listed in Table 12. Fusion of an aromatic or heteroaromatic ring to an azine changes the electronic distribution and hence the chemical shifts of remaining ring carbon atoms in the azine portion of the molecule, although the difference from those in the parent azine is usually less than 10 ppm. Shift data for a number of common fused azine systems are given in Scheme 8.

Protonation of azines results in shielding of the α carbon atoms and deshielding of the β and γ carbon atoms (Table 13), particularly the latter, and these effects have been accounted for in terms of additivity parameters. The upfield protonation parameter for the α carbon atom has been assigned to changes in the C-N bond order, while the β and γ parameters have been assigned to charge polarization effects. The parameters are highly reproducible for monoprotonation but deviate significantly from additivity for diprotonated heterocycles. A related effect is observed on quaternization, but in this case the operation of a β-substituent effect results in the overall change at the α carbon atom normally being small (Table 13). A further important general trend in the azines arises on *N*-oxidation, which results in shielding of the α and γ carbon atoms, especially the latter, and clearly indicates the high electron

Table 11 ¹³C NMR Chemical Shifts of the Simple Monocyclic Azines (*cf.* benzene, δ 128.5 ppm)

	δ (¹³C) (ppm, TMS) (Position of N atoms indicated)							
Position	Pyridine	Pyridazine	Pyrimidine	Pyrazine	1,2,3-Triazine	1,2,4-Triazine	1,3,5-Triazine	1,2,4,5-Tetrazine
1	N	N	N	N	N	N	N	N
2	149.5	N	158.4	145.9	N	N	166.1	N
3	125.6	153.0	N	145.9	N	158.1	N	161.9
4	138.7	130.3	156.9	N	149.7	N	166.1	N
5	125.6	130.3	121.9	145.9	117.9	149.6	N	N
6	149.5	153.0	156.9	145.9	149.7	150.8	166.1	161.9

Table 12 ^{13}C NMR Chemical Shifts of Monosubstituted Pyridines[a]

Substituent position	Substituent	δ (^{13}C) (ppm)				
		C-2	C-3	C-4	C-5	C-6
	H	150.6	124.5	136.4	124.5	150.6
2	Br	*142.9*	129.0	139.5	123.7	151.0
2	CHO	*153.1*	121.6	137.5	128.3	150.3
2	CN	*133.8*	129.2	137.9	127.8	151.5
2	COMe	*153.9*	121.4	136.9	127.5	149.3
2	Me	*158.7*	123.5	136.1	120.8	149.5
2	NH$_2$ [b]	*160.9*	109.5	138.5	113.6	148.7
2	OH [b,c]	*162.3*	119.8	140.8	104.8	135.2
2	OMe [b]	*163.1*	110.5	138.7	116.7	146.6
3	Br	151.7	*121.6*	139.1	125.4	148.7
3	CHO	152.0	*132.1*	136.2	124.8	155.0
3	CN	153.2	*110.5*	140.6	124.8	153.8
3	COMe	150.1	*123.9*	132.5	121.5	153.8
3	Me	150.9	*133.1*	136.4	123.4	147.3
3	NH$_2$ [b]	137.7	*145.7*	122.0	125.1	138.8
3	OH [b]	137.8	*153.5*	121.4	123.8	140.0
3	OMe [b]	137.3	*155.2*	120.0	123.8	141.4
4	Br	152.6	127.6	*133.2*	127.6	152.6
4	CHO	151.3	123.6	*141.7*	123.6	151.3
4	CN	151.7	126.4	*120.5*	126.4	151.7
4	COMe	151.2	121.6	*143.0*	121.6	151.2
4	Me	*150.1*	125.0	*147.0*	125.0	150.1
4	NH$_2$ [b]	148.5	110.4	*155.8*	110.4	148.5
4	OH [b,c]	139.8	115.9	*175.7*	115.9	139.8
4	OMe [b]	150.7	109.8	*164.9*	109.8	150.7

[a] Neat liquids, unless otherwise specified; values for *ipso* position
(carrying substituent) italicized; data from ⟨72CPB429, 73OMR(5)551⟩.
[b] In DMSO-d$_6$.
[c] Compound in NH form.

density at these positions in the ring (Scheme 9). The corresponding conjugate acids of the *N*-oxides have chemical shifts very similar to those of the protonated parent heterocycles.

The pyrones and thiinones show general ^{13}C NMR spectral characteristics similar to the pyridones which reflect charge distributions in the heterocyclic rings. Thus, carbon atoms α or γ to the heteroatom are deshielded relative to benzene, while those β are shielded. Substituent effects are in general as expected, although fewer detailed studies have been carried out in this area with the oxygen and sulfur heterocycles than with the azines. Chemical shift data for representative compounds are given in Scheme 10.

2.2.3.4.2 *Aromatic systems: coupling constants*

Single-bond ^{13}C-^{1}H coupling constants for six-membered heteroaromatic compounds lie in the approximate range 150–220 Hz, the magnitude varying with substituent electronegativity. Data for

Table 13 ^{13}C NMR Chemical Shifts (δ, ppm from TMS) and One-bond ^{13}C-^{1}H Coupling Constants (Hz) of
some Simple Heterocyclic Cations (*cf.* pyridine, column 2)

Position			Chemical shift (Coupling constant)			
1	N	O$^+$	NH$^+$	NMe$^+$	NPh$^+$	S$^+$
2	150.3 (178)	169.3 (218)	148.3 (192)	145.8	145.3 (191)	158.8
3	124.3 (162)	127.7 (180)	128.6 (173)	128.5	129.5 (178)	138.3
4	136.4 (162)	161.2 (180)	142.2 (173)	145.8	47.8 (174)	150.8
Anion	—	ClO$_4$$^-$	CF$_3$CO$_2$$^-$	I$^-$	Cl$^-$	
Solvent	DMSO-d$_6$	CD$_3$CN	D$_2$O	DMSO-d$_6$	D$_2$O	
Ref.	70MI20100	73OMR(5)251	80CCC2766	76OMR(8)21	80CCC2766	87PS(29)187
	76OMR(8)21	77OMR(9)16	70MI20100			

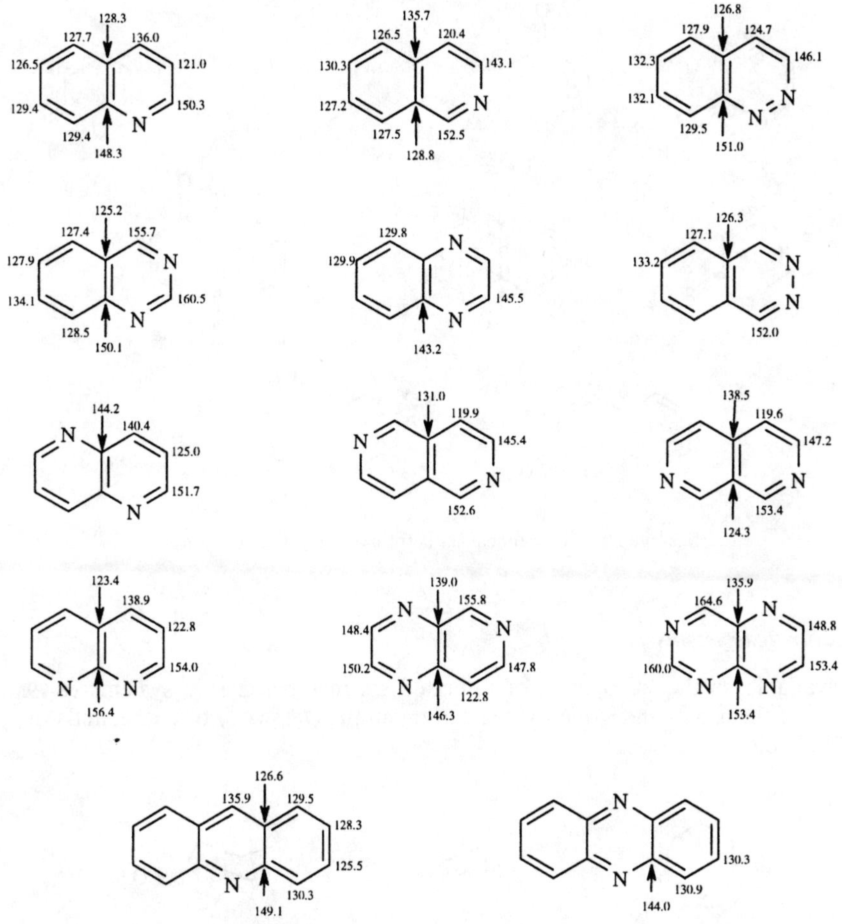

Scheme 8 ^{13}C chemical shifts for bicyclic systems

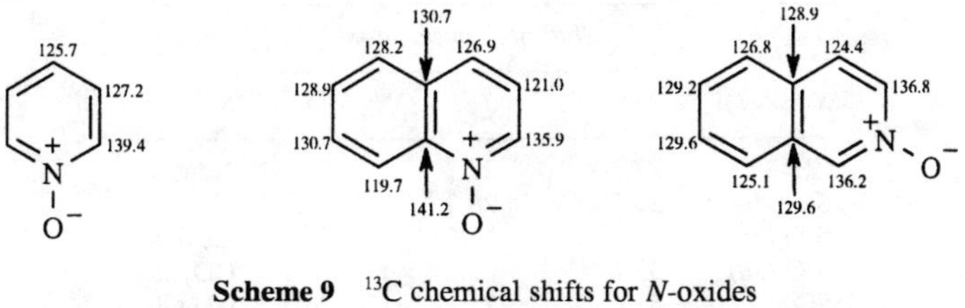

Scheme 9 ^{13}C chemical shifts for *N*-oxides

simple azines are summarized in Table 14. Longer range couplings are much smaller (up to *ca.* 12 Hz), and the values are difficult to predict.

Table 14 One-bond ^{13}C-^{1}H Coupling Constants (Hz) in the Simple Monocyclic Aromatic Azines (*cf.* 159 Hz for benzene)

| Position | *J* (Hz) (Position of N atoms indicated) | | | | | | |
	Pyridine	Pyridazine	Pyrimidine	Pyrazine	1,2,4-Triazine	1,3,5-Triazine	1,2,4,5-Tetrazine
1	N	N	N	N	N	N	N
2	178	N	211	184	N	206	N
3	162	186	N	184	207	N	214
4	162	174	182	N	N	206	N
5	162	174	171	184	188	N	N
6	178	186	182	184	188	206	214

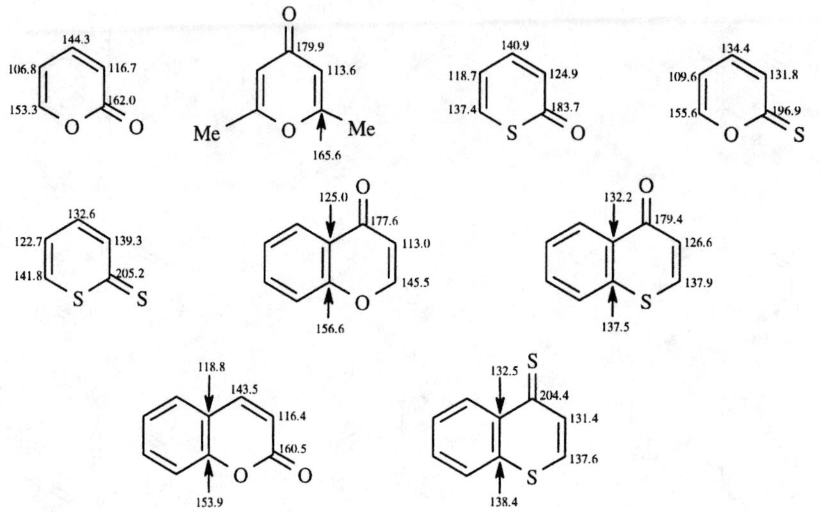

Scheme 10 ^{13}C chemical shifts for oxygen and sulfur systems

2.2.3.4.3   Saturated systems

Data are available on the ^{13}C spectra of saturated six-membered ring systems ⟨B-79MI20101⟩. The chemical shifts of the α-, β- and γ-methylene carbon atoms ⟨76JA3778⟩ of compounds of type (**41**) are summarized in Table 15.

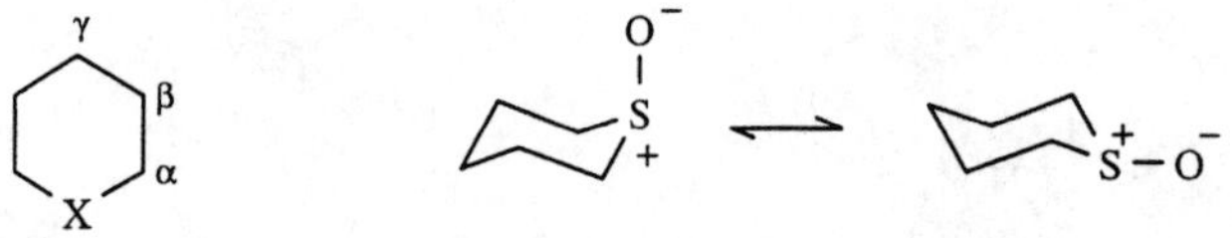

Table 15 ^{13}C NMR Chemical Shifts (δ, ppm from TMS) in Saturated Six-membered Rings (**41**)[a]

	Shift of C at position			
X in (41)	*2*	*3*	*4*	*Solvent*
CH$_2$	27.7	27.7	27.7	None
S	29.3	28.2	26.9	None
S-O *(ax)*	45.1	15.5	24.7	CD$_2$Cl$_2$
S-O *(eq)*	52.1	23.3	24.7	CD$_2$Cl$_2$
SO$_2$	52.6	25.1	24.3	CD$_2$Cl$_2$
NH	47.5	27.2	25.5	None
NMe	56.7	26.3	24.3	None
NH$_2^+$ I$^-$	45.6	23.2	22.5	H$_2$O
NHMe$^+$I$^-$	55.6	23.9	21.8	H$_2$O
NMe$_2^+$I$^-$	63.5	20.6	21.0	H$_2$O
NCOPh[b]	42.8, 48.5	25.5, 26.3	24.4	CDCl$_3$
N-NO[c]	39.0, 50.8	25.5, 27.2	24.7	None
O	68.0	26.6	23.6	None

[a] From ⟨76JA3778⟩, unless otherwise indicated.
[b] ⟨75JOC3547⟩.
[c] ⟨74JCS(P2)1381⟩.

A sulfur (S^{II}) atom exerts only a very small effect on the chemical shifts of the carbon atoms in the ring: cyclohexane absorbs at δ 27.7, and the signals of the α, β and γ carbons of tetrahydrothiin (**41**; X = S) are all very close to this value. The corresponding sulfone (X = SO$_2$) shows a rather large shift of the α carbon (to δ 52.6), while the β and γ protons are moved slightly upfield. The sulfoxide group is variable in its effect, depending on whether the oxygen atom is axial (**42**) (the preferred conformation) or equatorial (43): an equatorial oxygen has considerably the more deshielded α and β carbons; in the axial conformer the β carbon absorbs at a field over 11 ppm higher than cyclohexane.

A nitrogen atom at X results in a variable downfield shift of the α carbons, depending in its extent on what else is attached to the nitrogen. In piperidine (41; X = NH) the α carbon signal is shifted by about 20 ppm, to *ca.* δ 47.7, while in *N*-methylpiperidine (41; X = NMe) it appears at δ 56.7. Quaternization at nitrogen produces further effects similar to replacement of NH by *N*-alkyl, but simple protonation has only a small effect. *N*-Acylpiperidines (41, X = NCOR) show two distinct α carbon atoms, because of restricted rotation about the amide bond. The chemical shift separation is about 6 ppm, and the mean shift is close to that of the unsubstituted amine (41; X = NH). The nitroso compound (41; X = N-NO) is similar, but the shift separation of the two α carbons is somewhat greater (*ca.* 12 ppm). The β and γ carbon atoms of piperidines, *N*-acylpiperidines and piperidinium salts are all upfield of the cyclohexane resonance, by 0–7 ppm.

The ether oxygen of tetrahydropyran (41; X = O) induces a large downfield shift of the α carbons, while the β and γ carbons move slightly upfield, the γ more noticeably.

When two heteroatoms are present in a saturated six-membered ring their effects are approximately additive to within 5 ppm. Observed shifts for a few representative examples are shown in structures (44)-(48).

Many studies have been devoted to substituent effects in saturated heterocyclic six-membered rings ⟨B-79MI20101⟩. The so-called γ-gauche effect' induces an upfield (screening) shift of the ^{13}C signal from a ring carbon *meta* to an axial substituent (49). The shielding experienced by the γ carbon atom caused by an electronegative heteroatom *para* to it (50) is also a manifestation of the γ-gauche effect.

The one-bond coupling ($^1J_{CH}$) for cyclohexane (an average of couplings to axial and equatorial protons) is 123 Hz, and is increased by substitution adjacent to the carbon by an electronegative element, as with the aromatic systems discussed above.

2.2.3.5 Nitrogen and Oxygen NMR Spectra ⟨B-73MI20100, B-79MI20102⟩

Some nitrogen chemical shifts relating to azines are given in Table 16.
The following trends are observed:

(1) A pyridine-type nitrogen absorbs at comparatively low field (+ 63 ppm for pyridine itself, without solvent).

(2) Substituent effects are often considerable, particularly when strongly electron-donating effects (to the aza nitrogen) are present, when upfield shifts of up to 60 ppm (2-NH$_2$) may be observed.

(3) Further aza substitution *ortho* or *para* in the same ring deshields the nitrogen; the effect is moderate for a *para-*, and large for an *ortho*-nitrogen. The latter is probably a special 'azo effect', since the nitrogens of a simple azo group absorb at still lower field (− 130 ppm, in ether).

(4) Hydrogen bonding to the nitrogen lone pair leads to an upfield shift, the extent of which depends on the proton-donor ability of the solvent, and the acceptor ability of the base: shifts of some 20 ppm are commonly found.

(5) When the lone electron pair is protonated, the nitrogen chemical shift moves by *ca.* 100 ppm to higher field. Large upfield shifts are also found when a compound exists in a tautomeric form with a proton on the nitrogen. The nitrogen NMR spectrum is often of considerable value in studies of tautomerism of this type.

Table 16 ^{14}N and ^{15}N NMR Chemical Shifts of Typical Azines and their Derivatives[a]

Compound	Solvent	N-shift[b]	[14]N bandwidth at half height (Hz)[c]
Pyridine	CCl₄	57	170
Pyridine	MeOH	83	
Pyridinium	HCl/H₂O	181	20
Pyridine 1-oxide	Acetone	85	66
1-Hydroxypyridinium	HCl/H₂O	133	570
1-Methylpyridinium	H₂O	174	(Narrow)
4-Nitropyridine	Acetone	35	490
4-Bromopyridine	None	56	1100
4-Methylpyridine	Acetone	74	480
4-Methoxypyridine	Acetone	90	1100
4-Aminopyridine	Acetone	105	680
4(1*H*)-Pyridone	Acetone	201	390
Pyridazine	CHCl₃	20	410
Pyridazine	DMSO	20.3	
Pyridazine 1-oxide	DMSO	55.1 (N-1)	33.6 (N-2)
Pyrimidine	DMSO	84.8	
Pyrimidine 1-oxide	DMSO	90 (N-1)	80.3 (N-3)
Pyrazine	DMSO	46.3	
Pyrazine 1-oxide	DMSO	75.7 (N-1)	70.4 (N-4)
1,2,3-Triazine[d]		13.7 (N-1)	80.8 (N-2)
1,2,4-Triazine[e]	DMSO	40.0 (N-1)	2.0 (N-2), 62.0 (N-4)
Quinoline	None	72	650
Isoquinoline	None	68	680
Cinnoline	DMSO	− 44.6 (N-1)	− 41.3 (N-2)
Phthalazine	Dioxane	11	800
Quinoxaline	Dioxane	46	950

[a] [14]N data from ⟨B-73MI20100⟩; ^{15}N data from ⟨80HCA504⟩.
[b] Screening constants, ppm upfield from MeNO₂.
[c] No entry in this column means data are from ^{15}N spectra.
[d] ⟨85LA1732, 86JCS(P1)1249, 88JCS(P1)1509⟩.
[e] ⟨84SA(A)637⟩.

NMR longitudinal relaxation times can be used for determination of the site of protonation in polyfunctional acids and bases ⟨93JCS(P2)283⟩. Thus, the ^{14}N NMR spectrum of 4-aminopyridine shows clearly from the sharpening of the signal for the ring nitrogen that protonation has occurred here. This procedure is an important innovation in the elucidation of heterocyclic tautomeric structures, especially for the cases of fast exchange.

Observation of the one-bond [13]C-^{15}N coupling in quaternized heterocycles containing specific labeling with ^{15}N has been used to identify the site of quaternization ⟨76JOC3051⟩.

(6) N-Oxidation of an azine nitrogen usually shifts the signal upfield insignificantly (10–30 ppm). [In five-membered rings, however, downfield shifts have been claimed ⟨78JOC2542⟩.]

^{17}O NMR spectroscopy directly measures properties of the oxygen atom and it allows examination of the relative back donation of various diazine N-oxides. The order of [17]O chemical shifts is the N-oxides of pyridazine (412 ppm) > pyrazine > pyridine (349 ppm) > pyrimidine (338 ppm), with the larger value corresponding to the greater double-bond character for the nitrogen-oxygen bond in the N-oxide group. This is in agreement with the ^{15}N chemical shifts of the parent compounds as well as the differences in chemical shifts between the parent and the oxidized nitrogens. Additionally, ^{15}N chemical shifts of nonoxidized nitrogen and the difference from the parent nitrogen are also in same sequence. These facts strongly suggest that the ring nitrogen *para* and particularly *ortho* to the N-oxide function affects the back donation by attracting the ring electrons on the N-oxide oxygen atom. In the case of pyrazine N-oxides, the resonance structure (**52**) is predicted to be preferred by far over others, such as (**51**) ⟨85JHC981⟩.

(**51**)   (**52**)

2.2.3.6 UV and Related Spectra ⟨71PMH(3)67⟩

2.2.3.6.1 Features of UV spectra

The spectra of saturated heterocycles show amine n→σ* absorptions and transitions associated with sulfur. Saturated ethers are usually transparent down to 210 nm.

The six-membered aza-aromatic compounds possess the basic π-electron systems of benzene and its homologues, and in addition there are non-bonding lone pairs of electrons on the nitrogen atoms. These lone pairs are responsible for weak transitions, denoted n→π*, at the long-wavelength end of the spectrum. These absorptions are usually weak, in comparison to the transitions (π→π*) of the π-electrons, and are frequently difficult to locate, except when two aza nitrogen atoms are adjacent. In these cases the two filled non-bonding orbitals interact (see Figure 1). The n→π* bands are more obvious features of the spectra of compounds such as pyridazine, cinnoline and 1,2,4,5-tetrazine. In the last compound there are two n→π* bands, a weak system near 320 nm, and a stronger one, giving a band at *ca.* 550 nm (ε 830), which is responsible for the red color. Transannular interaction in (53) leads to a hypsochromic shift of the second n→π* absorption ⟨86TL5367⟩. Table 17 gives the positions of the n→π* absorptions of a number of the simpler aza-aromatics.

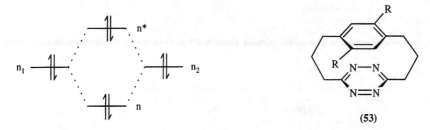

Figure 1 The splitting of the adjacent non-bonding orbitals in pyridazine

Table 17 UV Absorption Bands of the Simple Monocyclic Azines[a]

Compound	π→π* bands	n→π* bands
Benzene	203.5 (3.87), 254 (2.04)	—
Pyridine	251 (3.30)	*ca.* 270 (sh)
Pyridazine	246 (3.11)	340 (2.50)
Pyrimidine	243 (3.31)	298 (2.51)
Pyrazine	260 (3.75)	328 (3.02)
1,2,4-Triazine[b]	264 (3.71)	384 (2.72)
1,3,5-Triazine	222 (2.18)	272 (2.95)
1,2,4,5-Tetrazine	252 (3.33)	*ca.* 320 (w), 542 (2.92)

[a] Solvent cyclohexane; the positions of the peak maxima (nm) are given, with $\log_{10} \varepsilon$ values in parentheses; mainly from Mason ⟨59JCS1240, 59JCS1247⟩. [b] Trimethyl derivative.

The transition energies from bonding to antibonding π-orbitals (π→π*) correspond fairly well in their levels to those of the isosteric hydrocarbons, although the band intensities are often very different. Thus, the spectra of naphthalene, quinoline, isoquinoline and the quinolizinium ion bear a remarkable similarity, except in the intensities of the long-wavelength bands. Cinnoline is slightly different, with long-wavelength peaks at 309 nm (log ε 3.29), 317 (3.25) and 322 (3.32) and the n→π* band at 390 nm (log ε 2.43). The differences are probably a result of the pronounced asymmetry of the molecule, compared with naphthalene. Table 17 lists the principal bands in the spectra of some representative monocyclic systems, with those of benzene added for comparison. In the aza derivatives of the higher polyarenes, UV spectral comparisons have frequently been used as an indication of structural correspondence ⟨58HC(12)551⟩. For example, the spectra of the benzo, dibenzo, *etc.* derivatives of the quinolizinium ion bear similar qualitative relationships to those of the polycyclic hydrocarbons as does quinolizinium ion itself to naphthalene, *i.e.* a bathochromic shift accompanied by a pronounced intensification (see Table 18).

The effects of substituents on UV spectral maxima are illustrated in Table 19. The effects are greatest when conjugation is significantly increased, *e.g.* for strong electron donor substituents in the pyridinium cation.

The non-aromatic but fully conjugated 1,4-dioxin and its sulfur analogues show absorption at quite long wavelengths.

Table 18   UV Spectral Maxima (nm) for Benzazines[a] [λ_{max} (log ε)]

Position of nitrogen						
atom(s)	Neutral	Monocation	Neutral	Monocation	Neutral	Monocation
—[b]	275 (4.0) 310 (2.81)	—	292 (4.30) 330 (2.54)	—	375 (4.88)	—
1	312 (3.52)	313 (3.79)	346 (3.24)	365 (3.56)	354 (4.02)	402 (3.48)
2	319 (3.47)	332 (3.63)	—	—	—	—
1,2	321 (3.44)	353 (3.40)	370 (3.2)	?	—	—
1,3	305 (3.38)	260 (3.91)	—	—	—	—
1,4	316 (3.79)	331 (3.93)	—	—	365 (4.2)	430 (3.2)
2,3	305 (3.11)	314 (3.45)	—	—	—	—

[a] Abstracted from ⟨71PMH(3)67⟩ which see for original references.
[b] Values in this row taken from E. S. Stern and C. J. Timmons, 'Electronic Absorption Spectroscopy in Organic Chemistry', Arnold, London, 1970, apply to the hydrocarbon species.

Table 19   UV Spectral Maxima (nm) of Substituted Azines[a]

Substituents	Position	Pyridine (neutral)	Pyridine (cation)	Pyridine 1-oxide	Pyridazine	Pyrimidine	Pyrazine
—	—	257	256	265	300	243	261
Me	2	262	262	—	—	249	271
	3	263	262	254	310	252[b]	—
	4	255	252	256	292	245	—
Cl	2	264	271	—	—	—	268
	3	267	270	—	308	—	—
	4	258	257	265	—	—	—
OMe	2	269	279	249	—	264	292
	3	276	284	262	265	—	—
	4	245	236	261	254	248	—
NH$_2$	2	229	229	239	—	292	316
	3	231	250	—	—	298[b]	—
	4	241	263	215	249	268	—
NO$_2$	2	269	—	—	—	—	—
	3	241	—	—	—	237[b]	—
	4	286	—	328	—	—	—
CO$_2$H	2	264	265	259	—	246	—
	3	262	260	260	—	247[b]	—
	4	271	275	281	—	253	—
Ph	2	241	294	240	—	250	—
	3	244	—	249	254	256[b]	—
	4	256	286	293	—	275	—

[a] Abstracted from ⟨71PMH(3)67⟩ which see for original references. [b] Refers to 5-substituted compound.

2.2.3.6.2   Applications of UV spectroscopy

(i) UV spectroscopy has been much used in determining ionization constants for both proton addition and proton loss. Conversely, it is important that the pH of a solution is known when a UV spectrum of a potentially basic or acidic compound is obtained.

(ii) UV spectroscopy has been particularly useful in studies of tautomeric compounds. Thus 2-pyridone has a spectrum very similar to 1-methyl-2-pyridone, and quite different from 2-methoxypyridine; this is illustrated in Figure 14 of CHEC 2.04.

(iii) UV absorption spectra are also very useful in the investigation of covalent hydration, important in polyaza six-membered heteroaromatics, especially when bicyclic (see Section 3.2.1.6.3).

(iv) UV-visible spectra demonstrate charge transfer complex formation, e.g. between polycyclic quinolizinium ions and polycyclic aromatic hydrocarbons.

2.2.3.7 IR and Raman Spectra ⟨63PMH(2)161, 71PMH(4)265⟩

Systematic studies of the IR spectra of substituted pyridines establish: (1) that substituents vibrate largely independently of the rings; (2) that the vibrational modes of the ring skeleton are related, and approximate in position, to those found in benzene derivatives; (3) that the bending modes of the ring hydrogen atoms are similar to those of the corresponding arrangements of adjacent hydrogen atoms on a benzene ring. Although it is reasonable to assume that these generalizations are applicable to all azines, other systems have not been examined in such detail as the pyridines.

Four ring-stretching modes for pyridines and pyrimidines are listed in Table 20, together with the corresponding bands of a monosubstituted benzene. Quinolines and isoquinolines show seven or eight bands in the region 1650–1350 cm^{-1}.

In six-membered aromatic rings the intensity of the ring-stretching mode near 1600 cm^{-1} is related to the square of the $\sigma_R°$ parameter for the substituent and/or ring atoms. Such intensity measurements on 4-substituted pyridine *N*-oxides confirm the ability of the *N*-oxide group to both donate and withdraw electrons according to the nature of the 4-substituent.

Six-membered heteroaromatic rings show bands characteristic of in-plane CH bending in the region 1300–1000 cm^{-1} (Table 21), and of out-of-plane CH bending below 1000 cm^{-1} (Table 22).

Table 20 Approximate Positions of Ring-stretching Modes for Pyridines, Pyrimidines and Benzenes (cm^{-1})

| Compounds | Mode (see structures A-D) | | | |
	A	B	C	D
Monosubstuted benzene[a]	1610–1600	1590–1580	1520–1470	1460–1440
Pyridine[b]	1580	1572	1482	1439
4-Substituted pyridines[a]	1610–1595	1570–1550	1520–1480	1420–1410
Pyridine 1-oxides[a]	1645–1590	1585–1560	1540–1470	1440–1410
Pyrimidines[c]	1600–1545	1575–1540	1510–1410	1470–1330

[a] In CHCl$_3$. [b] Vapor phase. [c] Various media.

  A *B* *C* *D*

Table 21 Azines: Characteristic IR Bands (cm^{-1}) in the 1300–1000 cm^{-1} Region[a]

Compounds	β CH modes				Ring modes
2-Substituted pyridines	1293–1265	1150–1143	1097–1089	1053–1043	998–990
3-Substituted pyridines	1202–1182	1129–1119	1108–1098	1045–1031	1027–1023
4-Substituted pyridines	1232–1208	—	1070–1064	—	995–991
Pyridine 1-oxides	1158–1142	—	1130–1112	—	—
Pyrimidines	1280–1200	1210–1130	—	—	—
Pyridazines	—	1150–1100	—	—	1065–935

[a] Data taken from ⟨71PMH(4)265⟩ which see for references to the original literature.

Table 22 Azines: Characteristic IR Bands (cm^{-1}) below 1000 cm^{-1} [a]

Compounds	Characteristic IR Bands (cm^{-1})				
2-Substituted pyridines	794–781[b]	752–746[b]	—	780–740[c]	—
3-Substituted pyridines	810–789[b]	715–712[b]	920–880[c]	820–770[c]	730–690[c]
4-Substituted pyridines	820–794[b]	775–709[b]	—	850–790[c]	ca. 725[c]
Pyridine 1-oxides	—	—	886–858	825–817	—
Pyrimidines	1010–960	—	850–780	—	—
Pyridazines	—	—	860–830	—	—

[a] Based on ⟨71PMH(4)265⟩ which see for references to the original literature. [b] Alkyl substituents only.
[c] Other substituents.

In the IR spectra of six-membered heterocycles containing one or more carbonyl groups in the ring (pyridin-2- and -4-ones, the pyrones and pyrimidones) one of the higher-frequency bands in the 1700–1500 cm^{-1} region can usually be assigned to the carbonyl stretching vibration. However, this is by no means always the highest frequency band; solvent and isotopic substitution effects on the band positions have shown that there is a considerable degree of mixing of the ring and carbonyl modes in the pyridones. The assignments in Table 23 refer to the principal motion in these non-localized vibrations.

The position of the carbonyl group in the IR spectra of various pyranones and pyridones, *etc.,* is indicative of the C-O bond order, and therefore of the importance of the betaine structure. However, this criterion must be used with caution because of the non-localized nature of the vibrations just mentioned.

Assignments have been suggested for the absorptions of many of the saturated heterocyclic six-membered rings ⟨63PMH(2)240, 71PMH(4)339⟩. As expected, force constants between the atoms of the ring are much lower than in the aromatic rings, and the absorptions which are due to skeletal modes are generally found below *ca.* 1200 cm^{-1}; bands in the region 1500–1200 cm^{-1} arise from the various deformation modes of the CH bands. The so-called Bohlmann bands at 2750 cm^{-1} are diagnostic of *trans*-fused quinolizidine structures.

Table 23 $\nu(C=O)$ Frequencies (cm$^-$) for some Azinones[a]

Other structural features	*Bond*	α-*Series*		γ-*Series*	
		Z = NR	*Z = O*	*Z = NR*	*Z = O*
—	$\nu(C=O)$	1666–1655	1736–1730	1577–1550	1634
	ring	1619–1570	1647–1612	1643–1624	1660
5,6-Benzo	$\nu(C=O)$	1667–1633	1710–1700	1647–1620	1650
6-N	$\nu(C=O)$	1681	—	1662	—
3-N	$\nu(C=O)$	1670	—	1653	—

[a] Abstracted from ⟨63PMH(2)161⟩ and ⟨71PMH(4)265⟩ which see for references to the original literature.

2.2.3.8   Mass Spectrometry ⟨B-67MI20100, 66AHC(7)301, 71PMH(3)223⟩

The behavior of the simple azines and their benzo analogues on electron impact under mass spectrometric conditions is complex. Extensive randomization of ring hydrogen atoms, which increases with increasing lifetime of the ions, takes place prior to fragmentation, as does independent scrambling of ring carbon atoms. Skeletal and Dimroth-type rearrangements (see Section 3.2.3.5.4.iii) are also common.

The mass spectra of the aromatic six-membered heterocycles and their benzo analogues reflect the stability of the ring systems, with the molecular ion in many cases also being the base peak. Fragmentation of the azines by loss of HCN (*M*-27) is the common pathway and for pyridine the *M*-27 ion is the only fragment of any significance in the spectrum apart from the molecular ion. Fragmentation by successive losses of molecules of HCN is common in polyaza systems. Pyrimidine, for example, loses two molecules of HCN in succession to give the radical cation of acetylene, and pteridine fragments similarly to the dehydropyrazine radical cation. Loss of nitrogen from systems containing an -N=N- unit is also a common feature although the ease with which this occurs can vary substantially and, to some extent at least, predictably. It is, for example, very common with cinnolines and 1,2,3-benzotriazines but is of much less importance with phthalazines.

The general fragmentation pattern of monocyclic 1,2,3-triazines shows peaks for [M$^+$–N$_2$], for an acetylene, and for a nitrile, in accordance with the results of thermolysis and photolysis. The mass spectrum of the parent triazine showed peaks as follows: 81 (M$^+$, 47%), 53 (M$^+$–N$_2$, 69%), 27 (HCN,

13%), and 26 (C$_2$H$_2$, 100%) ⟨81CC1174⟩. The mass spectrum of the unsubstituted 1,2,4-triazine shows peaks at 81, 53, 52, 51, 40, 39, 38 28, 27, 26, and 25 Da, *cf.* Scheme 11 ⟨67JHC224⟩. [4-^{15}N]-3-Methyl-1,2,4-triazine shows a similar fragmentation pattern ⟨72LA(760)102⟩. From these data it follows that the fragmentation starts with loss of nitrogen.

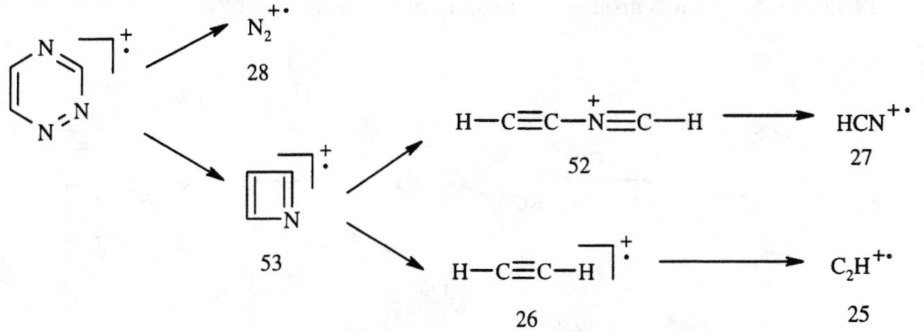

N$_2^{+\bullet}$
28

H—C≡C—$\overset{+}{N}$≡C—H   ⟶   HCN$^{+\bullet}$
52   27

53

H—C≡C—H$^{+\bullet}$ ⟶   C$_2$H$^{+\bullet}$
26   25

Scheme 11 Mass-spectral fragmentation pattern for 1,2,4-triazine

There has been relatively little detailed study of the mass spectrometry of pyrylium and thiinium salts, due partly no doubt to the involatility of the compounds; elimination of CO or CS is the major fragmentation pathway. *N*-Oxides generally show an abundant *M*-16 peak which is sometimes the base peak and di-*N*-oxides show successive elimination of two atoms of oxygen. The intensity of the *M*-16 peak is often very substantially reduced in compounds containing a substituent group α to the ring *N*-oxide which has at least one C-H bond, due to abstraction of a hydrogen atom by the oxygen of the *N*-oxide followed by elimination of a hydroxyl radical. The *M*-17 peak is then a correspondingly significant fragment. Methyl group substitution α to a ring nitrogen atom usually results in fragmentation *via* elimination of MeCN rather than HCN. In monomethyl polyazines both processes are observed, although which fragmentation occurs first appears to be determined by the position of the methyl group. 2-Methylazines also give rise to fragments formed by loss of both a hydrogen atom and a methyl radical. Unlike the general situation pertaining with benzenoid aromatics, where β-cleavage is the preferred fragmentation pathway, the decomposition mode of azines substituted with alkyl groups larger than methyl depends both on the nature of the substituent and on its position relative to the ring nitrogen atom. β-Cleavage occurs with all of the alkylpyridines, but the extent varies in the order 3 > 4 > 2, reflecting the relative electron densities at these positions. The resulting azabenzylic ions rearrange to the isomeric azatropylium ions, and these in turn fragment with loss of HCN or MeCN. γ-Cleavage of a carbon-hydrogen bond can also be important for 2-alkylazines and may even give rise to the base peak, but this form of fragmentation is generally much less important than β-cleavage with the 3- and 4-isomers. McLafferty-type rearrangements are usually pronounced with 2-substituted azines, *e.g.* (**54**), and this is a general process.

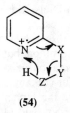

(**54**)

2-Pyridone undergoes fragmentation by loss of CO and formation of the pyrrole radical cation. 3-Hydroxypyridine, on the other hand, loses HCN to give the furan radical cation while 4-pyridone shows both modes of cleavage. The loss of CO from azinones is highly characteristic, *e.g.* (**55**) → (**56**) and (**57**) → (**58**), but with compounds such as uracils and their benzo analogues, the 1-substituted 4-quinazolinones and quinazolinediones, the retro-Diels-Alder fragmentation is favored, *e.g.* (**59**), followed by loss of CO, then HCN or a hydrogen atom, or both. Rather complex fragmentation is often observed with alkoxy-substituted azines, especially with quinolines and isoquinolines, and intramolecular transfer of a hydrogen atom from the ether alkyl group to a ring carbon atom appears to be common. Loss of the alkyl radical and formation of quinonoid intermediates, followed by loss of CO, also appear to be common. Amino groups are usually eliminated as HCN.

The mass spectrum of 2-pyrone shows an abundant molecular ion and a very prominent ion due to loss of CO and formation of the furan radical cation. Loss of CO from 4-pyrone, on the other hand, is almost negligible, and the retro-Diels-Alder fragmentation pathway dominates. In alkyl-substituted 2-pyrones, loss of CO is followed by loss of a hydrogen atom from the alkyl substituent and ring expansion of the resultant cation to the very stable pyrylium cation. Similar trends are observed with the benzo analogues of the pyrones, although in some cases both modes of fragmentation are observed. Thus, coumarins, chromones, flavones and xanthones, for example, all show significant (*i.e.* > 20% relative abundance) or dominant fragmentation by loss of CO, while the retro-Diels-Alder pathway is dominant or significant in the fragmentation of 4-hydroxycoumarins, isocoumarins, chromones and flavones. Dihydrocoumarins fragment with loss of CO while the retro-Diels-Alder mode is important in the fragmentation of dihydrocoumarins, dihydrochromones, chromans, flavanones, isoflavanones and rotenoids. Not unexpectedly, the retro-Diels-Alder pathway also tends to dominate in the fragmentation of monocyclic dihydroheterocycles.

In the fully saturated heterocyclic systems, loss of hydrogen atoms from α carbons, (**60**) → (**61**), and β-cleavage, (**62**) → (**63**) and (**64**) → (**65**), are common, but it is often impossible to discern clear trends in the later stages of fragmentation. Fragmentation of sulfur-containing heterocycles almost always proceeds with loss of sulfur atoms or of sulfur-containing fragments. Thianthrene (**66**) fragments by two successive losses of sulfur atoms to give the biphenylene radical cation, and two of the major fragments from 1,3-dithiane (**67**) are $(M-CH_2S)^+$ and $(M-C_3H_6S)^{+\cdot}$. Transfer of hydrogen atoms to sulfur is common: both thiane and 1,2-dithiane, for example, lose $H_2S^\cdot$, in the latter case as a major fragment, while a third major fragment from 1,3-dithiane arises by loss of CH_3S, requiring a hydrogen migration. In other cases, loss of sulfur-containing fragments clearly involves initial rearrangement; the pyrimidothiazinone (**68**), for instance, fragments with loss of COS. In cyclic sulfones loss of SO_2 is a major fragmentation process.

(60) (61) (62) (63)

(64) (65) (66)

(67) (68)

2.2.3.9 Photoelectron Spectroscopy

In the ultraviolet photoelectron spectrum, the most readily ionized level of pyridine is the non-bonding orbital (with contributions from the σ-framework). The three diazines show two lone-pair levels, with the greatest splitting in the case of pyridazine but considerable also in pyrimidine and pyrazine. These long-distance splittings are attributed to both through-space and through-bond interactions, particularly the latter.

The energy levels corresponding to the second and subsequent ionization potentials of pyridine have been correlated with those in benzene.

PE spectroscopy offers a method of investigating the tautomeric structure of pyridones and other related compounds in the gas phase by comparison of ionization potentials of the potentially tautomeric compound with those of fixed models. This method is discussed in CHEC 2.04.

2.2.4 THERMODYNAMIC ASPECTS

2.2.4.1 Intermolecular Forces

2.2.4.1.1 *Melting and boiling points* (Table 24)

The introduction of a pyridine-like nitrogen into a benzene ring *tends* to make a derivative more crystalline and less volatile; this effect is greater for the diazines, especially pyridazine and pyrazine. When a hydrogen-bond donor substituent is also carried, the difference from the benzenoid compound becomes even more marked.

Examination of the effects of substituents on the melting and boiling points of the parent compounds is instructive.

(i) Methyl and ethyl groups attached to ring carbon atoms usually increase the boiling point by *ca.* 20–30 and *ca.* 50–60°C, respectively.

(ii) Acids and amides are solids, as are amino and 'hydroxy' compounds. The latter generally exist in the tautomeric oxo structures. Strong hydrogen bonding is possible for all these classes of derivatives.

(iii) Methoxy, methylthio and dimethylamino derivatives are often liquids.

(iv) Chloro compounds usually have boiling points similar to those of the corresponding ethyl compounds. Bromo compounds boil approximately 25°C higher than their chloro analogues.

2.2.4.1.2 *Solubility* (Table 25)

The solubility in water is much enhanced by the presence of a pyridine nitrogen atom due to the possibility of hydrogen bonding. However, if this possibility is increased sufficiently, then the

Table 24   Melting and Boiling Points[a,b,c]

Ring system	H	Me	Et	COMe	CO₂H	CO₂Et	CONH₂	CN	NH₂	OH	OMe	SH	SMe	Cl	Br
Benzene	80	**111**	136	202	**122**	211	**130**	190	184	**43**	37	168	187	131	155
Pyridine-2	115	128	148	192	**137**	243	**107**	222	**57**	107	252	128	197	171	193
Pyridine-3	115	144	163	220	**235**	223	**129**	**50**	**65**	125	179[d]	79	—	150	173
Pyridine-4	115	145	171	211	**306**	219	**156**	**79**	157	148	93	186	**44**	147	174
Pyridazine-3	208	**215**	—	**90**	**200**	**68**	**182**	**44**	169	103	219	170	**38**	**35**	**73**
Pyridazine-4	208	**225**	—	—	**240**	255	**191**	**80**	130	250	**44**	210[d]	**45**	**76**	—
Pyrimidine-2	123	138	—	—	**270**	—	—	—	127	320	—	230[d]	218	**65**	—
Pyrimidine-4	123	141	—	—	**240**[d]	—	—	—	151	164	—	187	—	**37**	—
Pyrimidine-5	123	153	—	—	**270**	**38**	**212**	—	170	210[d]	—	215	—	**37**	**75**
Pyrazine-2	**57**	135	153	77	**229**[d]	50	189	205	120	119	187	215	**46**	160	180

[a] Melting points above 30°C are given in bold; melting points below 30°C are not included.

[b] Boiling points are given at atmospheric pressure to facilitate comparison; those reported at other than atmospheric pressure were converted using a nomogram (*Ind. Eng. Chem.*, 1957, **49**, 125).

[c] A dash indicates that the compound is unstable, unknown, or the data are not readily available.

[d] With decomposition

compound will prefer intermolecular hydrogen bonding with itself, and this can decrease water solubility.

Introduction of amino groups into pteridine lowers the solubility in all solvents despite the fact that the amino group almost invariably increases the solubility in water of aliphatics and aromatics. The reduced solubility of aminopteridines is due to intermolecular hydrogen bonding.

Table 25 Solubilities in Water (parts soluble in 1 part of water) at 20°C[a]

Parent monocyclic compounds		Polycyclic derivatives		OH-Substituted derivatives		NH₂-Substituted derivatives	
Benzene	0.0015	Naphthalene	0.00002	Phenol	0.07	Aniline	0.03
Pyridine	misc.[b]	Quinoline	0.007	2-Pyridone	1	2-Aminopyridine	1
Pyridazine	misc.	Isoquinoline	0.004	3-Hydroxypyridine	0.03	3-Aminopyridine	> 1
Pyrimidine	> 1	Quinoxaline	0.7	4-Pyridone	1	2,6-Diaminopyridine	0.1
Pyrazine	1.7	Pteridine	*0.15*	2-Pyrimidinone	*0.5*	2-Aminopteridine	0.0007

[a] Data abstracted from ⟨63PMH(1)177⟩

[b] Misc. = miscible.

2.2.4.1.3 *Gas-liquid chromatography*

Typical operating conditions for the GLC separation of azines are shown in Table 26.

Table 26 Operating Conditions for the GLC Separation of Azines[a]

Compounds	Conditions
Pyridines	Diphenyl phthalate on 'Tide'
Quinolines	Silicone E-301
Pyrimidines	15% Hallcomid M-18 on firebrick
Pyrazines	Apiezon M and N or Reoplex 400
Piperazines	Flexol 8N8 on firebrick
Phenazines	SE-30 on Chromosorb W
s-Triazines	Ethylene glycol adibatic on glass beads

[a] Data abstracted from ⟨7PMH(3)297⟩ which see for original references and further information.

Chromatographic techniques in combination with other physical methods have solved difficult chemical problems. Sensitive compounds can be detected and characterized by HPLC as indicated during the oxidation of the (6R)- and (6S)-diastereoisomers of 5,6,7,8-tetrahydrobiopterin to the (6R)- and (6S)-quinonoid 6,7,8-dihydrobiopterins which rearrange with a half-life of about five minutes to form the stable 7,8-dihydrobiopterin ⟨83JC361, 84MI 718-06⟩. Separation of regioisomers such as 6,8- and 7,8-dimethylpterin can be achieved under special conditions using strong cation exchange columns ⟨92MI 718-11⟩. The identification and determination of the stereoconfiguration of pterins and lumazines with chiral side chains was successfully performed by ligand exchange chromatography using a reverse-phase column with a chiral mobile phase ⟨92MI 718-05, B-93MI 718-06, 93MI 718-08, 94MI 718-02⟩. Cation-exchange, reverse-phase, and ion-pair reverse-phase chromatography were evaluated for the separation of various pterins ⟨84JLC2561⟩. HPLC methods have also been developed for the determination of pteridines in biological samples ⟨83MI 718-09⟩ and in blood cells and plasma ⟨83MI 718-04⟩.

2.2.4.2 Fully Conjugated Rings: Aromaticity

2.2.4.2.1 *Background*

Aromaticity has been long recognized as one of the most useful theoretical concepts in organic chemistry. It is essential in understanding the reactivity, structure and many physico-chemical characteristics of heterocyclic compounds. Aromaticity can be defined as a measure of the basic state of cyclic conjugated π-electron systems, which is manifested in increased thermodynamic stability, planar geometry with non-localized cyclic bonds, and the ability to sustain an induced ring current. In contrast to aromatic compounds there exist nonaromatic and antiaromatic systems. Thus, pyrazine (**69**)

is an aromatic molecule. Its 1,2,3,4-tetrahydro derivative (**70**) is certainly nonaromatic, since only one
C—C bond participates in the conjugation with the two nitrogen atoms. The 1,4-dihydropyrazine
derivative (**71**) is a typical representative of antiaromatic compounds. In this molecule there exist cyclic
conjugation but it leads to destabilization, since two extra π-electrons above the Hückel limit are
forced to occupy antibonding orbitals.

| (69) | (70) | (71) |

Though it is usually not difficult to classify a given compound as aromatic, nonaromatic or
antiaromatic from a qualitative point of view, much more complex problems arise in attempting to
describe the aromaticity in quantitative terms. Until now, three main groups of quantitative criteria of
aromaticity have been elaborated: energetical, structural and magnetic.

2.2.4.2.2 *Energetical criteria*

The very different thermodynamic stability of aromatic, nonaromatic and antiaromatic systems
provides a good basis for the development of energetical criteria of aromaticity. In theoretical organic
chemistry, resonance energy is accepted as a measure of such stability. This is a difference between
electron energies of a real (conjugated) molecule and a hypothetical Kekule-type structure with
localized bonds. Since the latter do not exist in reality, the choice of an adequate model of localized
structure creates a problem. There are two main approaches to the calculation of resonance energy –
purely theoretical and semi-empirical. In the first case, both energy values are calculated quantum
mechanically. In the second case, the energy of a real molecule is determined experimentally (usually
through the measurement of heats of combustion or hydrogenation), whereas for a localized structure
the energy is calculated by summing tabulated values of the energies of isolated bonds. Energies
determined by a semi-empirical approach are called empirical resonance energies (ERE). Relatively
few ERE values exist for six-membered heterocyclic compounds, some of which are listed in Table 27.
As easily seen, ERE values are quite divergent.

The Dewar approach is considered to be the most successful. Dewar resonance energies (DRE) are
derived from heats of atomization, ΔH_a. For real molecules, ΔH_a values are taken from the
experimental heats of combustion. If the experimental data are not available, the ΔH_a value is obtained
from a quantum mechanical calculation. A major achievment by Dewar was his choice of acyclic
polyenes or heteropolyenes as reference structures. Unlike classical Kekule-type structures, their
formally single bonds contain a π-component. To compare the aromaticities of compounds with
different numbers of π-electrons, Dewar introduced REPE indices which denote "resonance energy per
electron". They are obtained by dividing the full resonance energy by the number of ring π-electrons.
Hess and Shaad had successfully adapted this approach to an HMO method (HSRE). Another type of
resonance energy is the topological resonance energy (TRE). In Table 27, REPE indices of some
simple six-membered heterocycles are given. One can see that REPE values of the DRE, HSRE and
TRE show in most cases the same trend. Aromaticity changes in the following sequences: (i) benzene
> pyridine > pyrimidine > pyrazine > pyridazine; and (ii) per ring, pyridine > quinoline $\geq$
isoquinoline > acridine. The calculated energies for the addition of one equivalent of hydrogen
suggested that resonance energies of benzene (151 kJ mol^{-1}), pyridine (142), pyrimidine (138) and
pyrazine (134) are quite close; while pyridazine (109), 1,3,5-triazine (105) and 1,2,4,5-tetrazine (77)
are somewhat less $\langle$96CHEC-II(6),98,903$\rangle$.

Heats of combustion can give useful comparative data on the thermodynamic stabilities of
heterocyclic compounds $\langle$74PMH(6)199$\rangle$. The heats of formation of the isomeric diazines pyridazine,
pyrimidine and pyrazine are respectively 4397.8, 4480.2 and 4480.6 kJ mol^{-1} $\langle$62ACS916$\rangle$; pyridazine
is almost 83 kJ mol^{-1} less stable than the other two.

The concept of aromaticity has been linked to those of tautomerism and equilibrium by using K^T, or
an equilibrium constant as a measure of the binding energy difference between the pyridinoid and
pyridonoid forms, and comparing this to the corresponding quantities for saturated derivatives (Scheme
12).

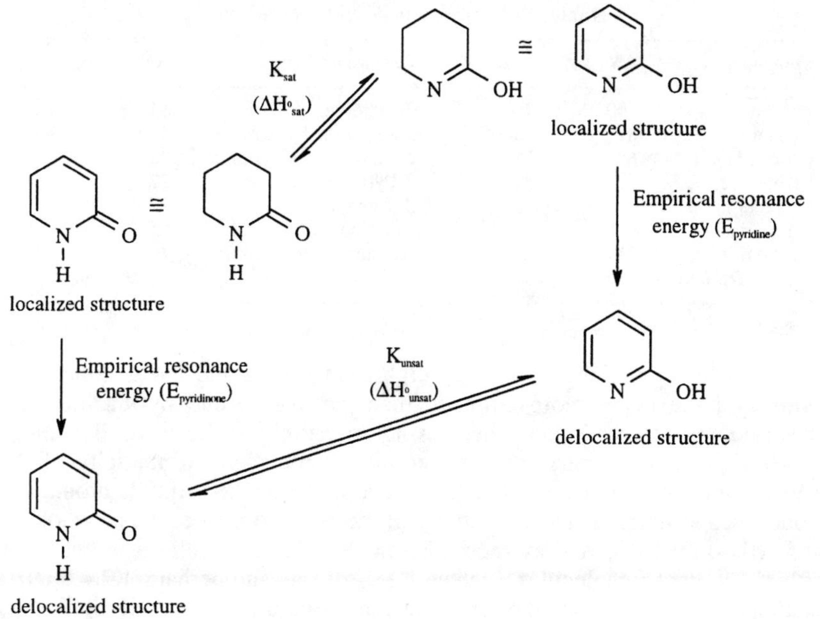

Scheme 12

In this way, pyridin-2-one and pyridin-4-one are calculated to be *ca.* -30 kJ mol^{-1} less aromatic than pyridine. In the bicyclic quinolone, the difference in aromaticity between the two forms is less. The precise degree of aromatic character possessed by 2- and 4-pyranone is not settled; various methods of estimation give divergent values.

Table 27 REPE Indices Estimated by Various Methods

Compounds	ERE (kJ mol^{-1})	DRE (kJ mol^{-1})[c]	HSRE (β)[d]	TRE (β)[e]
Benzene	28.5[a]	14.0	0.065	0.046
Pyridine	17.0[a]	16.1	0.058	0.038
Pyridazine	8.7[a]			
Pyrimidine	5.7[a]	14.1	0.049	0.032
Pyrazine	5.7[a]	12.0	0.049	0.022
Quinoline	19.8[b]	14.3	0.052	0.036
Isoquinoline	—	14.3	0.051	0.033
Acridine	—	12.3	—	—

[a] From bonding data given in G.W. Wheland, "Resonance in Organic Chemistry", Wiley, New York, 1955, p. 275.
[b] Derived from heat of combustion ⟨49CB358, 51CB916⟩.
[c] ⟨70TCA235⟩.
[d] ⟨75T295⟩.
[e] ⟨93AHC(56)303⟩.

2.2.4.2.3 *Structural criteria*

Cyclic π-electron conjugation tends to equalize the lengths of the ring bonds. This is why molecules of benzene, tropylium cation, and cyclopentadienyl anion are flat rectilinear polyhedrons with sides of length 1.40 Å (the standard length of an aromatic C—C bond). On the substitution of a ring carbon atom by a heteroatom, the distribution of π-electron density becomes less even, conjugation is partially disturbed and bond lengths now deviate from the standard values. The extent of such deviations can be treated statistically to give a measure of the aromaticity of heterocycles. Two closely related structural indices of this sort have been suggested: $\Delta \bar{N}$ and I_6. Both reflect mean value differences of bond orders of all ring bonds. Experimental bond lengths are taken for the calculation of these indices. Since bonds of different type (*e.g.* C—C and C—N), at equal length have non-equal bond orders, bond distances are converted by calculation into bond orders. $\Delta \bar{N}$ and I_6 indices can be conveniently expressed as percentage of the aromaticity of benzene, taken as 100%.

Table 28 Structural Indices of Aromaticity

Compound	$\Delta\bar{N}$ (%)[a]	I (%)[b]	Compound	$\Delta\bar{N}$ (%)[a]	I (%)[b]
Benzene	100	100	Pyridinium	54	66.7
Pyridine	82	85.7	2,4,6-Trimethylpyrylium	43	—
Pyridazine	65	78.9	2,4,6-Triphenylpyrylium	28	—
Pyrimidine	67	84.3	2-Pyridone	22	—
Pyrazine	75	88.8	2-Thiopyridone	22	—
1,2,3-Triazine	65	76.9	Acridine (all molecule)	45	—
1,2,4-Triazine	71	86.1	Acridine (hetero-ring)	20	—
1,2,4,5-Tetrazine	88	97.8			

[a] ⟨85KGS867⟩ [b] ⟨86T89⟩

Table 28 indicates that pyridine shows very high aromaticity, and the diazines somewhat lower aromaticity with pyrazine being more aromatic than pyrimidine and pyridazine. In keeping with expectation the structural indices indicate decreasing aromaticity of heterocyclic cations (pyridine > pyridinium > pyrylium) in comparison with neutral heterocycles. Aromaticity of 2-pyridone and 2-thiopyridones is even lower. Structural indices give a possibility to estimate aromaticity of separate rings in polycondensed systems. Thus, aromaticity of the acridine molecule as a whole is 45%, but for the central ring it is only 20%. A drawback inherent in structural indices is overestimation of the aromaticity of high-symmetry molecules, such as 1,3,5-triazine and perhaps 1,2,4,5-tetrazine.

A linear relationship has been shown to exist between structural aromaticity indices and resonance energies ⟨92T335⟩. From this so-called unified aromaticity index, I_A, has been proposed. It makes for a more appropriate comparison the aromaticity of heterocycles of different size.

2.2.4.2.4 *Magnetic criteria*

Magnetic properties of conjugated cyclic systems are very important for understanding their electron distribution and physico-chemical characteristics. All magnetic criteria of aromaticity are based on the ring currents that are induced in conjugated cyclic molecules by external magnetic fields. In particular, ring currents are manifested in the anisotropy and the exaltation of diamagnetic susceptibility and in the deshielding of protons outside the conjugated system and stronger shielding of protons within the cyclic conjugated system.

Magnetic susceptibility anisotropy has been used for the estimation of relative aromaticity of some azines in comparison with benzene ⟨77JCS(P2)897⟩. If the extent of π-electron delocalization for benzene is taken as 1.0, the corresponding values for azines are: pyridine 0.7, pyridazine 0.7, pyrimidine 0.5, and 1,3,5-triazine 0.3.

Another quantitative magnetic index is the exaltation of the total magnetic susceptibility Λ. This parameter is given by the difference between χ_m and $\chi_{m'}$ which represent, respectively, the experimentally measured molar susceptibility and the molar susceptibility calculated by an additive scheme for hypothetical reference structures with localized double bonds in which the ring current vanishes. All aromatic systems reveal large Λ values, whereas for non-aromatic ones Λ is close to zero. It is assumed that aromaticity increases with Λ. Values of Λ seem to be a good diagnostic test for aromaticity, but as a quantitative index they are not practical. Thus, Λ for benzene, pyridine, pyrazine, quinoline, isoquinoline and 1-ethyl-2-pyridone are (in units of -10^6): 17.9, 18.3, 12.7, 36.3, 34.2 and 13.0 cm^3 mol^{-1}, respectively. Though aromaticity of 1-ethyl-2-pyridone should be certainly the lowest in this series, the relative aromaticity of the other compounds looks doubtful. The same is true for proton chemical shifts. Thus, in the ^{1}H NMR spectra of pyrazine protons resonate at δ 8.6 ppm, whereas in its nonaromatic (**70**) and antiaromatic (**71**) derivatives, the protons at the double bonds show signals at 5.38 and 4.64 ppm respectively ⟨83AG(E)171⟩. Obviously, this is qualitatively in agreement with the general idea of ring current shielding-deshielding influence. However, attempts to deduce from proton chemical shifts the relative aromaticity of different heteroaromatic systems usually fail. This is understandable because the shielding constants of a nucleus A is determined not only by the contribution from the ring current ($\sigma^A_{ring\ curr}$) but also by contributions σ^{AA}_d, σ^{AA}_p and $\Sigma\ \sigma^{AB}$ B($\neq$A), respectively, stemming from the diamagnetic and paramagnetic currents induced in the atom A itself and diamagnetic and paramagnetic currents induced in the atom B adjacent to atom A as expressed in Eq. (3).

$$\sigma = \sigma^{AA}_d + \sigma^{AA}_p + \sum_{B(\neq A)} \sigma^{AB} + \sigma^A_{ring\ curr} \tag{3}$$

One of the most obvious examples is strong deshielding of the α-protons in the series: pyridine (δ 8.29 ppm), phosphabenzene (8.61), arsabenzene (9.68), stibabenzene (10.94), and bismabenzene (13.25), although other data unambiguously point to a falling off of the aromaticity in this sequence. Here the contribution by $\sigma^A_{ring\,curr}$ is mostly obscured by local effects connecting with nonuniform distribution of the electron density and by the anisotropy of the heavier heteroatoms.

Finally one can conclude that despite still remaining uncertainties there exist some firmly established aromaticity regularities in series of six-membered heterocycles. They are: (1) Increase in the electronegativity of the heteroatom leads to decrease of aromaticity; (2) Aromaticity falls with increasing the number of pyridine-like nitrogens in the ring; (3) Transition from simple azines to their benzoderivatives is also accompanied by a decrease of aromaticity.

For more detailed discussion on problems of heteroaromaticity see ⟨85KGS867⟩, ⟨91H127⟩, ⟨93AHC(56)303⟩, ⟨98JOC5228⟩.

2.2.4.3 Partially and Fully Reduced Rings

The fully saturated rings share with cyclohexane the property of being able to adopt one or more conformations which are virtually free of torsion or bond angle strain.

Hetero-substituted cyclohexanes, with one or more CH_2 groups being replaced by O or NR, almost invariably exist predominantly in chair forms. Inclusion of a sulfur atom changes the geometry more significantly, because of the different bond lengths and angles, but again the overall shapes of the molecules are generally chair-like.

Exceptionally the tetrathiane (72) prefers the twist form in the solid phase. In CS_2 solution at 0°C chair and twist forms coexist with a free energy difference of *ca.* 3.5 kJ between them ⟨67JA5978, 68JA2450⟩. Strain energies for several thianes have been derived from heats of formation.

(72)

Besides the shapes adopted by the rings, considerable attention has been paid to the conformational preferences of substituents, both on carbon and on the heteroatoms (nitrogen and sulfur):

(1) The sulfoxide group tends to occupy an axial rather than an equatorial position in the thiane *S*-oxides.

(2) An electronegative substituent adjacent to a ring oxygen atom also shows a preference for an axial orientation. This is known as the 'anomeric effect', and is particularly significant in the conformations of carbohydrates ⟨B-71MI20100, B-83MI20100⟩.

(3) Since a nitrogen freely undergoes 'umbrella' inversion, a substituent on that atom can exchange between axial and equatorial positions without interchange of the rest of the ring atoms. The inversion at nitrogen is usually faster than the ring inversion, unless an electronegative element is attached to it.

(4) The nitrogen lone pair is sterically undemanding and so usually occupies an axial site predominantly.

Hetero-substituted cyclohexenes, *e.g.* dihydropyrans, exist in half-chair conformations.

2.2.5 TAUTOMERISM ⟨76AHC(SI)71⟩

2.2.5.1 Prototropic Tautomerism of Substituent Groups

2.2.5.1.1 Pyridones and Hydroxypyridines

In the consideration of the tautomerism of pyridines and azines there is usually little need to consider the transfer of hydrogen to a ring carbon atom. In other words, in the set of formulae (73)-(75) the non-aromatic form (75) is not significantly populated. However, there are exceptions in, for example, polyhydroxy-pyridines and -azines.

Thus, the major area of consideration is tautomerism of the type between structures (73) and (74). Table 29 summarizes the main results on the tautomerism of mono-hydroxy-, -mercapto-, -amino- and -methyl-azines and their benzo derivatives for dilute solutions in water at *ca.* 20°C.

Table 29 Tautomeric Equilibria of some Monofunctional Azines and Benzazines[a]

X =	O	S	NH	CH$_2$
Parent	B(3.0)	B(4.7)	A(ca.6)	A(13.3)
Benzo[a]	B(3.9)	B(5.1)	A(4.3)	A(9.4)
Benzo[b]	B(low)	B(3.0)	—	—
Benzo[c]	B(4.8)	B(5.8)	A(3.8)	A(9.5)
Dibenzo[a,c]	B(3.9)	—	A(2.8)	A(high)
Parent	B(0.1)	B(2.2)	A	A
Benzo[a]	A(ca.1)	B(1.5)	A	A
Benzo[b]	B(ca. 0.5)	—	A	A
Parent	B(3.3)	B(4.6)	A(8.7)	A(13.4)
Benzo[a]	B(4.2)	B(5.0)	A(3.2)	A
Dibenzo[a,b]	B(7.0)	B(high)	A	A
Parent	B(4.3)	B	A	—
Benzo[a]	B(2.6)	—	A(8.3 vs. B)	—
Benzo[b]	B	B	A	—

Table 29 Continued

X =	O	S	NH	CH$_2$
Parent	B(2.6)	B	A	—
Benzo[a]	B(3.6)	B	A(4.1 vs. B)	—
Parent	B ≡ C	B ≡ C	A(ca. 6 vs. B)	—
Benzo[a]	B	—	—	—
Parent	B(0.4 vs. C)	B	A(ca. 6 vs. B/C)	—
Benzo[a]	B(0.85 vs. C)	B	A	
Parent	B	B	A	—
Benzo[a]	B(>1.6 vs. A)	B	A	—

[a] From ⟨76AHC(S1)206⟩; results are expressed as log ([major form]/[minor form]); when a form is simply indicated, no quantitative data are available on the equilibrium.

(73)   (74)   (75)

Of the important canonical forms of 2-hydroxypyridine (**76**) and 2-pyridone (**77**), the fact that positive charge prefers nitrogen and negative prefers oxygen indicates that charge-separated canonicals (**77c**) are more important than (**76c**). This partly explains the importance of the 2-pyridone tautomeric form. When the carbonyl oxygen of (**77**) is replaced by less electronegative atoms, as in the imine tautomers of amino heterocycles, or the methylene tautomers of methyl derivatives, the tendency toward polarization in forms corresponding to (**77b**) and (**77c**) is considerably less, and the amino and methyl tautomers are therefore favored in most instances.

(76c)   (76b)   (76a)   (77a)   (77b)   (77c)

Polar solvents stabilize polar forms. In the vapor phase at equilibrium both 2- and 4-hydroxypyridines exist as such, rather than as the pyridones. 3-Hydroxypyridine, which in water is an approximate 1:1 mixture of OH and NH forms, also exists as the OH form in the vapor phase. However, 2- and 4-quinolinones remain dominantly in the NH (oxo) forms, even in the vapor phase. Hydrocarbon or other solvents of very low polarity would be expected to give results similar to those in the vapor phase, but intermolecular association by hydrogen bonding often leads to a considerably greater proportion of polar tautomers being present than would otherwise have been predicted.

Aqueous solvation is of particular importance and complexity. Hydrogen bonding in 2-pyridone—H_2O and 2-pyridone—$2H_2O$ has been investigated by rotationally resolved fluorescence excitation spectra in both the S_0 and S_1 electronic states ⟨93JA9708⟩, using hydrogen- and oxygen-labelled species to enable interpretation. The monohydrate shows an increased importance of zwitterionic canonical form over the anhydrous species, although its contribution to the resonance hybrid decreases on excitation. The monohydrate (**78**) contains two nonlinear hydrogen bonds with the amine hydrogen and the carbonyl oxygen, which increase in length and thus decrease in strength on excitation. The dihydrated species (**79**) shows stronger hydrogen bonding.

(78)   (79)   (80)

Experimental assessments of the concentration of the minor hydroxy tautomer of 2-pyridone and substituted derivatives in cyclohexane and acetonitrile solution may be carried out by the use of fluorescence spectroscopy ⟨85JCS(P2)1423⟩. For the parent compound, the pyridinol component in cyclohexane is estimated to be 4% and in acetonitrile 1.2%; this preference for the hydroxy form in the former over the latter solvent is maintained over a fair range of variously substituted pyridones. *Ab initio* calculations ⟨85JA7569⟩ on 4-hydroxypyridine, the minor tautomer in aqueous solution, include 92 water molecules in the estimations, and thus give a very detailed picture of the solvated molecule, while the experimental technique of microwave spectroscopy not only gives an accurate estimation of the 2-hydroxypyridine / 2-pyridone ratio of 3:1 in the gas phase but also reveals that the former isomer is predominantly in the (Z)-form (**80**) and that both isomers are planar ⟨93JPC46⟩.

2.2.5.1.2 *Other Substituted Azines*

Thiol-thione tautomerism in pyridines and azines in general parallels that of the corresponding hydroxy-oxo systems. Although a thiol group is in general more acidic than a hydroxy, the thioamide resonance (**81a**↔**81b**) is strongly polarized toward the form (**81b**), on account of the generally weak C=C π-bond in (**81b**). This tends to increase the acidity of the proton on the nitrogen atom, and as a result the thiol-thione equilibrium constants in heterocyclic systems carrying these functions favor the thione form over the thiol (**82**) to an extent only slightly greater than that found in the hydroxy-oxo equilibria. In the vapor phase, 2- and 4-pyridinethiols predominate over the thiones ⟨76JA171⟩.

In amine-imine systems (**83**⇌**84**) the mobile proton can in principle be located at either of the two basic nitrogen sites in the anion (**85**). Since the canonical form with aromatic (benzenoid) structure is polar in the imine (**84b**) and non-polar in the amine (**83a**), the amine structure should be, and is favored, particularly in non-polar solvents.

The tautomerism of a methyl group α or γ to a ring nitrogen (**86**⇌**87**) is still less favorable than that of the amine; simple valence and electronegativity considerations of the type employed above suggest much reduced aromaticity associated with the methylene tautomer (**87**). These tautomers are therefore present to only a very small extent at equilibrium.

(**81a**)   (**81b**)   (**82a**)   (**82b**)

(**83a**)   (**83b**)   (**84a**)   (**84b**)

(**85**)   (**86**)   (**87a**)   (**87b**)

The effects of other substituents on tautomeric equilibria can usually be predicted from general chemical principles. Thus, an electron-withdrawing group adjacent to a ring nitrogen atom tends to decrease its basicity, and so a tautomer with a proton at that nitrogen atom would be correspondingly destabilized, and the equilibrium displaced toward alternative forms. Substituents may also favor one tautomer or another by intramolecular hydrogen bonding. The effect of substituents on a group involved in the tautomerism (for instance, on an amino or methyl group) can be very profound, as, for instance, carbanion-stabilizing groups on the methyl group equilibrium (in structure **87b**).

Benzo-fusion to a heterocyclic ring involved in tautomerism has the effect of steering the equilibrium in directions which tend to retain the full aromaticity of the benzene ring. Thus, while the 3-hydroxyisoquinoline *versus* 3-isoquinolinone equilibrium (**88**⇌**89**) and that of the cinnoline derivatives (**90**⇌**91**) favor the oxo forms (**89**, **91**), the proportion of hydroxy tautomers is considerably greater than in the corresponding unfused systems. By contrast the benzo-fusion in 2- and 4-quinolinone and in 1-isoquinolinone has the effect of reducing the aromaticity of the heterocyclic ring, and consequently of lowering the proportion of the hydroxy tautomers.

(**88**) X = CH   (**89**) X = CH
(**90**) X = N   (**91**) X = N

Extra nitrogen heteroatoms in the ring provide alternative sites for the tautomeric proton. 4-Hydroxypyrimidine, for instance, can exist as such (**92**), or in the 1*H*- (**93**) or 3*H*- (**94**) pyrimidone forms. In the hydroxypyrone (**95**), the 2-hydroxy form (**96**) is possible, and, since aromaticity is not very strong in the pyrones, the diketo structure (**97**) may also be important. The general tendency in compounds of these types is for the forms with amide or ester groups (**94**, **95**) to be preferred over the vinylogous amides or esters (**93**, **96**). In compounds with several potential hydroxy groups, *e.g.* barbituric acid (**98**), non-aromatic structures, with interrupted cyclic conjugation, are commonly favored. In polar media the dipolar 4,6-pyrimidinedione molecule (**99**), with two CO-NH groups, is the predominant tautomer. Cyanuric acid was shown to exist as the trioxo structure by UV and X-ray data and trithiocyanuric acid similarly exists mainly in the trithione tautomeric structure.

(92)   (93)   (94)   (95)   (96)   (97)

(98)   (99)   (100)   (101)

2.2.5.2 Tautomerism of Dihydro and Tetrahydro Compounds

There are manifold possibilities for tautomerism in partially saturated derivatives. As an example, dihydro-1,2,4,5-tetrazines have been formulated in the 1,2-, 1,4-, 1,6- and 3,6-dihydro structures, but the 1,4-dihydro structure is probably the most stable (see also Section 3.2.2.3.2).

2.2.5.3 Tautomerism of Other Substituents (Non-prototropic)

Like protons, acyl groups can occupy alternative positions, on a ring atom or a substituent (*e.g.* **100** ⇌ **101**), and their migration from one position to another is sometimes rapid enough for the system to be considered tautomeric.

2.2.5.4 Ring-Chain and Valence Bond Tautomerism

Ring-chain tautomerism can occur when a substituent group can interact with an NH group or nitrogen atom of a heterocyclic ring.

An example of the first case is provided by the 1,2,4-triazinone side-chain aldehydes (**102a**) which exist in equilibrium with the cyclized forms (**102b**).

The second possibility is illustrated by α-azidoazines which exist in equilibrium with tetrazole fused systems: thus, 3-azido-1,2,4-triazines (**103a**) exist predominantly as the tetrazolotriazines (**103b**).

(102a)   (102b)   (103a)   (103b)

A ^{13}C NMR study on (**104**) under strongly protonating conditions ⟨86AQ(C)61⟩ has determined the equilibrium between the protonated azide (**105**) and the protonated tetrazolopyridinium salt (**106**). Comparison of the ^{13}C NMR values of (**104**) in trifluoroacetic acid (*i.e.*, those of (**105**) or (**106**)) with those of 2-azido-1-ethylpyridinium salt (**107**) and 1-methyltetrazolo[1,5-a]pyridinium salt (**108**) revealed unambiguously that the bicyclic cation (**106**) is present under these conditions.

(104)   (105)   (106)   (107)   (108)

The 2*H*-pyran (**109**) exists in valence bond tautomeric equilibrium with *cis*-β-ionone (**110**). The equilibria are strongly dependent on both the solvent and temperature. In tetrachloroethene, only 9% of the open chain form (**110**) is present at about 30°C, but this rises to 40% at 113°C ⟨81T1571⟩. The nature of R also plays a significant role ⟨85HCA192⟩.

Whilst the penta-substituted 2*H*-pyran (**111**; R = Me) contains only traces of its acyclic tautomer (**112**; R = Me), ^{13}C NMR indicates that 5-acetyl-2,2,6-trimethyl-2*H*-pyran (**111**; R = H) contains about 35% of the valence isomer (**112**; R = H) at 30°C. It appears that the 3-methyl group sterically destabilizes the dienone ⟨80IZV1011⟩.

The valence isomerization of the 2*H*-pyran system is the basis for the commercially valuable photochromic properties exhibited by the benzologs.

(109)   (110)   (111)   (112)

2.2.6   SUPRAMOLECULAR STRUCTURES

The chemistry of supramolecules forms an area of ever-increasing interest and application, arising from the extreme importance of such molecules, whether viewed in an academic or an applied perspective. Supramolecules are molecular assemblages, aggregates, or associations, and pyridine derivatives, particularly the bipyridines ⟨84AHC281⟩ and acridine, frequently play a key role in their construction and thus in their subsequent structure and behavior ⟨95CRV2725⟩.

In the realm of new materials and molecular devices, nonlinear optics deals with the interaction of applied electromagnetic fields with chemical compounds, resulting in a change of frequency or phase. Molecules are required of the form in which an acceptor functionality is linked *via* a conjugate system to a donor function. In 1994 the electromagnetic field susceptibilities of the charged chromophores (**113**) and (**114**), which can be used as building blocks for self-assembled superlattices, were calculated ⟨94CRV195⟩. It was shown that this property depends strongly on the conjugated chain length, but only slightly on its shape. Thus, the inclusion of an additional ethylene unit (**113**, n = 2 compared to **113**, n = 1) nearly doubles the second-order susceptibility while the rod-like alkyne (**114**) demonstrates almost the same response as its alkenic analogue (**113**, n = 1).

(113) n = 1, 2   (114)

Assembling [2] rotaxanes (wheel and axle) can involve three basic processes (Scheme 13). One of these ⟨93CC1269⟩ involves slippage, in which the axle (**115**) is linked by a 4,4'-bipyridine, and the wheel (**116**) is a bisparaphenylene-34-crown-10 ether. Heating the two components in acetonitrile at 60°C yields the rotaxane, which can be characterized by FABMS and ^{1}H and ^{13}C NMR, but extrusion of the wheel occurs at 100°C. In other developments ⟨94NAT(369)133⟩, the pyridine component may be incorporated in the wheel, as in structure (**117**), where two bipyridinium units are connected by *p*-xylyl groups, and here the rotaxane acts as molecular switch. At room temperature in acetonitrile the wheel moves at random along the axle, but favors the benzidine station at 229K. In acid solution, the amino groups on the axle are protonated, repelling the positively charged wheel, and this is subsequently restricted to the biphenol area.

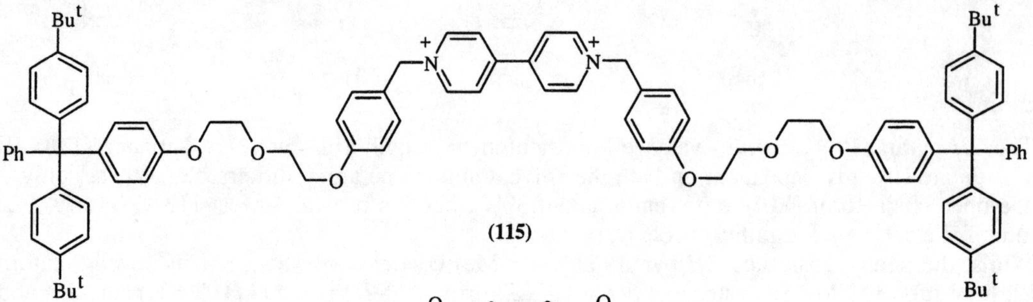

(115)

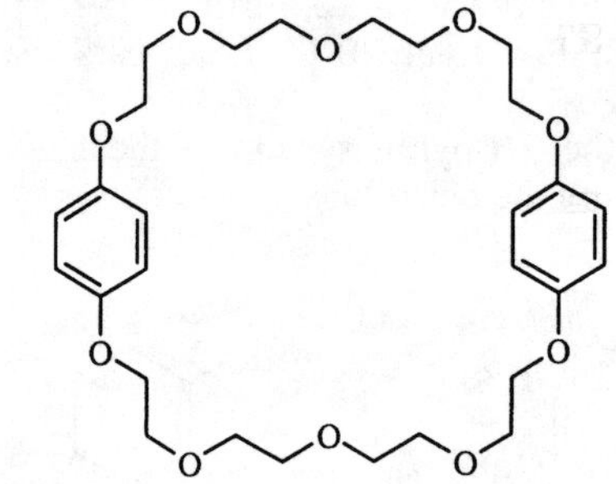

(116)

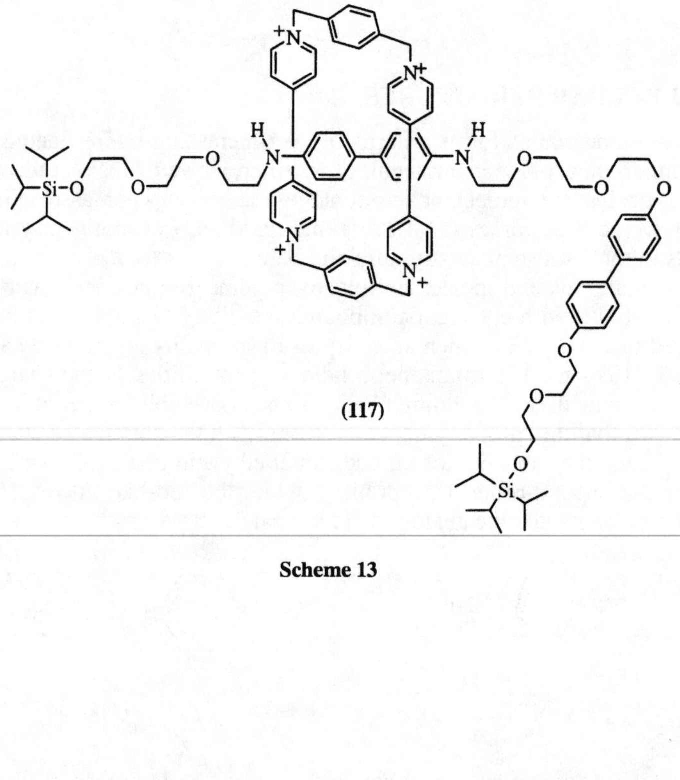

(117)

Scheme 13

2.3

Structure of Five-membered Rings with One Heteroatom

2.3.1 SURVEY OF POSSIBLE STRUCTURES

2.3.1.1 Monocyclic Compounds

Aromatic species include the neutral molecules pyrrole, furan and thiophene (**1**; Z = NH, O, S) and the pyrrole anion (**2**). The radicals derived from these rings are named pyrryl, furyl and thienyl. The 2-furylmethyl radical is called furfuryl. Compounds in which two pyrrole nuclei are joined by a CH_2 group are called 'dipyrromethanes'; when the linkage is by a CH group, they are named 'dipyrromethenes'. The 2- (**3**) and 3-pyrrolenines (**4**) are isomeric with the pyrroles, but are nonaromatic as the ring conjugation is broken by an sp^3-hybridized carbon atom.

Three types of cationic species exist (**5–7**); all are non-aromatic. The 2-oxo derivative exists in two tautomeric forms (**8** and **9**); the 3-oxo derivatives have a unique structure (**10**).

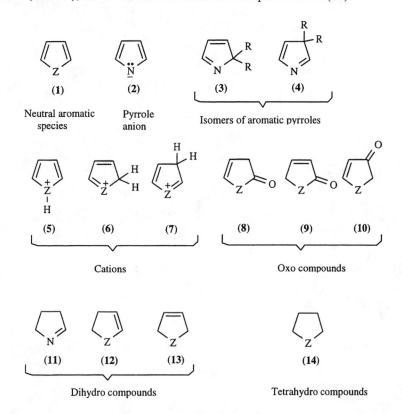

Three types of dihydropyrrole (**11–13**) and two types of dihydro-furan and -thiophene exist (**12–13**), but only a single class of tetrahydro compounds (**14**).

Reduced thiophenes and furans are named systematically as 2,3-dihydro (**12**), 2,5-dihydro (**13**) and 2,3,4,5-tetrahydro compounds (**14**). Alternatively, *delta* (Δ) can be used to indicate the position of the remaining double bond. Thus, (**12**) and (**13**) are named as Δ^2- and Δ^3-dihydro compounds, respectively; tetrahydrothiophene is also called thiophane.

Reduced pyrroles have trivial names; the dihydro derivatives, of which there are three types, are designated as Δ^1- (**11**), Δ^2 - and Δ^3-pyrrolines, and the tetrahydropyrroles are called pyrrolidines.

Reduced furan rings occur in many important anhydrides, lactones, hemiacetals and ethers. Maleic anhydride is frequently used as a dienophile in Diels-Alder reactions and it is a component of alkyd resins. Several unsaturated γ-lactones are natural products, while the furanose sugars are cyclic hemiacetals.

Thiophene and its homologues occur in coal-tar benzene, shale oil and crude petroleum. They show the indophenine test (Section 3.3.1.5.7.ii), and the discovery of thiophene followed the observation that pure benzene did not give this test.

Furfural (furan-2-carbaldehyde; Scheme 1) arises from the decomposition of sugars and is a commercially important raw material used in furfural-phenol resins and as a synthetic intermediate (see CHEC 1.15).

The bile pigments are metabolic products having chains of four pyrrole rings. Their precursors are the porphyrins, which comprise the blood pigments, the chlorophylls and vitamin B_{12} and consist of four pyrrole units joined in a macro ring. The phthalocyanines are important synthetic pigments (see Scheme 1).

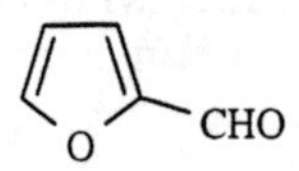

Furfural

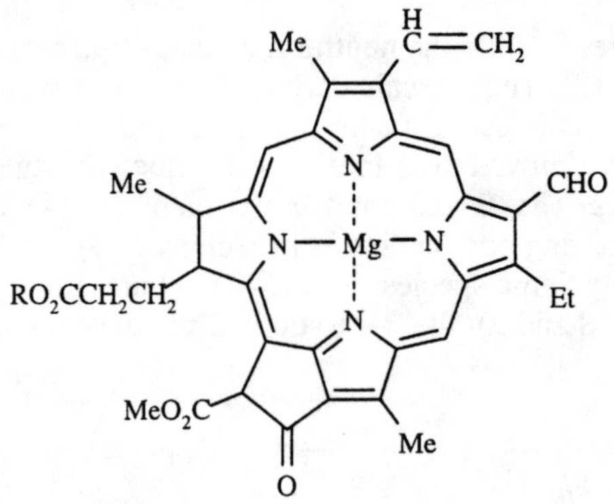

Chlorophyll-*b*
R = phytyl, $C_{20}H_{34}$

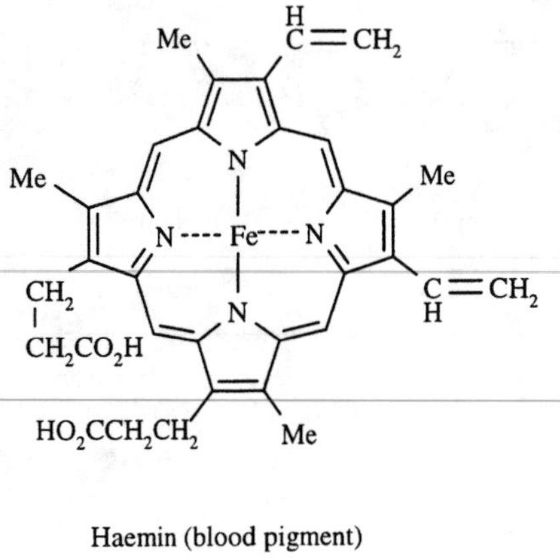

Haemin (blood pigment)

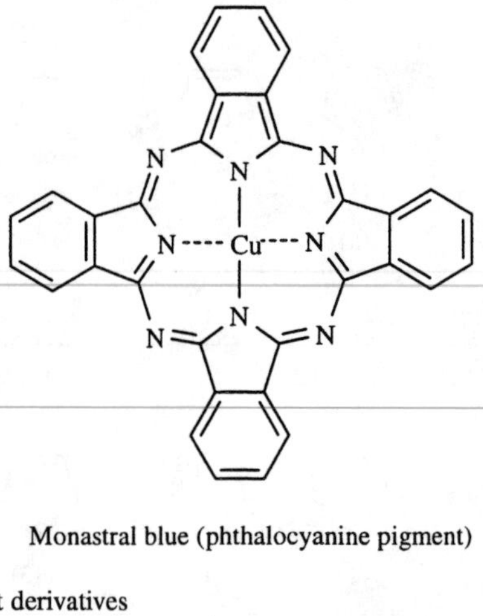

Monastral blue (phthalocyanine pigment)

Scheme 1   Important derivatives

2.3.1.2   Benzo Derivatives

The benzo[*a*] (**19**), benzo[*b*] (**15**) and benzo[*c*] (**16**) fused heterocycles are heterocyclic analogues of naphthalene, with the dibenzo heterocycles (**17**) bearing a similar electronic relationship to phenanthrene. Some of these compounds are still known by their trivial names: indole (**15**; Z = NH), isoindole (**16**; Z = NH), carbazole (**18**) and indolizine (**19**). The names thianaphthene and pyrrocoline for (**15**; Z = S) and (**19**) respectively are now little used. Particular confusion can arise in consulting the literature on indolizine (**19**), where differing numbering systems have been used. Note that carbazole (**18**) is an exception to the IUPAC rules for numbering the other dibenzo heterocycles.

The non-aromatic isomers of indoles (**15**; Z = NH) are indolenines (**20**), which are stable only if R is not equal to H; however, isoindoles often exist as (**21**). Some of the carbonyl derivatives have trivial names: oxindole (**22**; Z = NH), coumarone (**22**; Z = O), indoxyl (**23**; Z = NH) and phthalide (**24**; Z = O). The common names indoline (**25**; Z = NH) and coumaran (**25**; Z = O) are used for the 2,3-dihydro derivatives of indole and benzofuran, respectively. Similarly, isoindoline is (**26**, Z = NH).

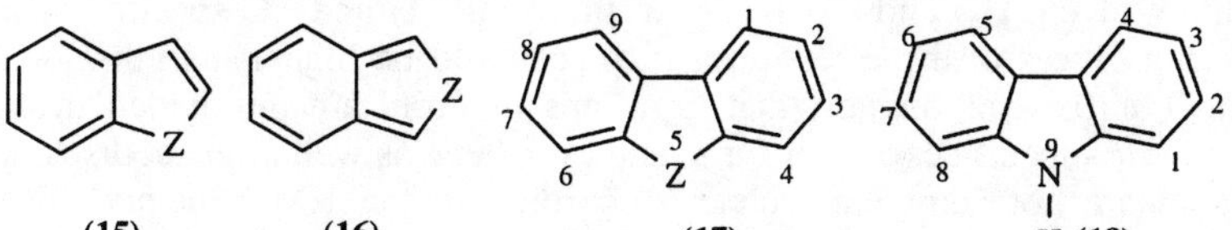

(**15**)   (**16**)   (**17**)   H (**18**)

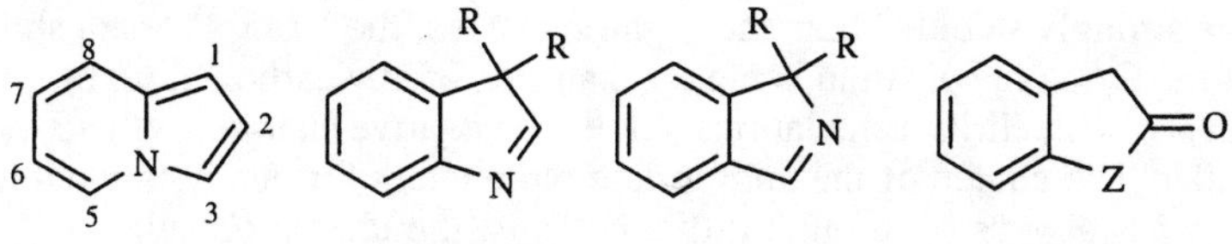

(**19**)   (**20**)   (**21**)   (**22**)

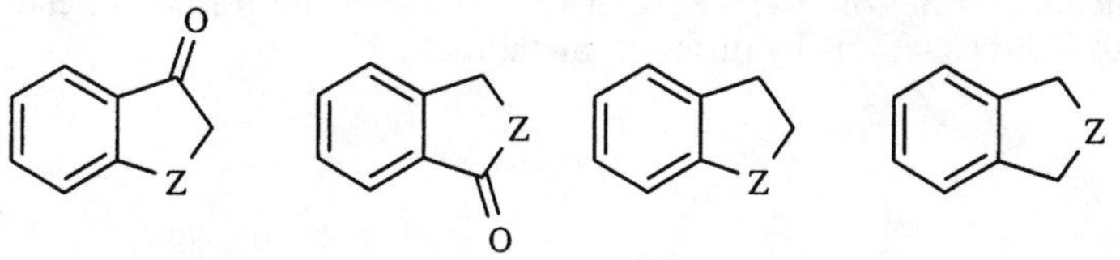

(**23**)   (**24**)   (**25**)   (**26**)

Important indole derivatives (see Scheme 2) include: (i) indigo, a vat dye known and widely used since antiquity, and originally obtained from indican, a β-glucoside of indoxyl which occurs in some plants. Indigo is now prepared synthetically. Tyrian purple, a natural dye used since classical times, is 6,6′-dibromoindigo; (ii) the numerous indole alkaloids, with complex derivatives such as yohimbine and strychnine; (iii) tryptophan, an essential amino acid found in most proteins. Its metabolites include skatole and tryptamine; and (iv) 3-indoleacetic acid, which is important as a plant growth hormone.

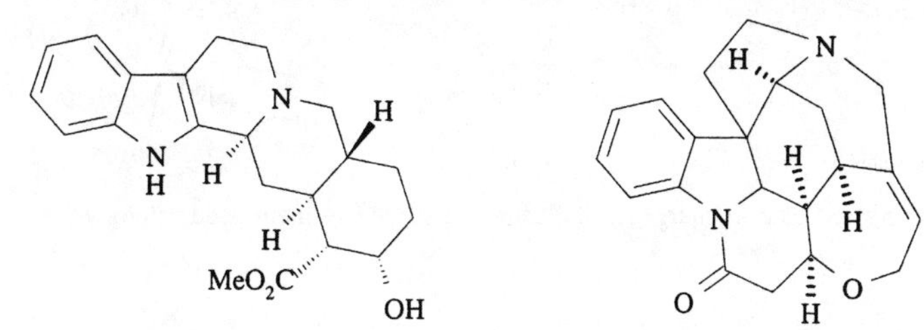

Yohimbine

Strychnine

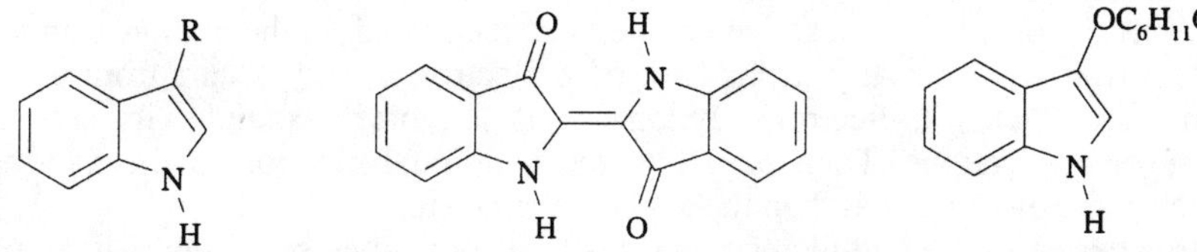

Tryptophan   R = $CH_2CH(NH_2)CO_2H$
Skatole   R = Me
Tryptamine   R = $CH_2CH_2NH_2$
3-Indoleacetic acid   R = CH_2CO_2H

Indigo

Indican

Scheme 2

2.3.2. THEORETICAL METHODS

The heteroatom in these five-membered heterocycles provides two π-electrons for the aromatic sextet and thus acquires considerable positive charge. Correspondingly, the ring carbon atoms together gain the same amount of negative charge (Scheme 3). In accordance with HMO calculations, the total π-electron excess of the three main representatives of this series is in the sequence: pyrrole > thiophene > furan. Pyrrole takes first place also by the criterion of local π-electron excess, the β-carbon atoms carry larger negative π-changes than α-carbons. This distribution of π-electron density is in agreement with larger shielding of the H_β- and C_β- nuclei in the NMR ^1H and ^{13}C spectra (Tables 4 and 9). In furan, the π-electron excess of the β-positions also seems to be higher than that of the α-positions. NMR ^{13}C chemical shifts even suggest that C_α-atoms in furan are deshielded in comparison with benzene. The relative π-electron excess of furan and thiophene as well as exact distribution of electron density in the latter are not completely clear. According to the HMO method, the distribution in thiophene is similar to that of furan. However, judged from the NMR ^{13}C chemical shifts, the C_α-atoms in thiophene are more strongly shielded than the C_β-atoms. Also, the NMR ^{13}C data show that the local π-electron excess of the C_β-atoms of furan is higher than that of any carbon atom of thiophene. On the other hand in accordance with HMO calculations values of negative charges in the β-positions of both heterocycles are similar. The choice of the appropriate parameters for quantum chemical calculations on sulfur heterocycles has always been quite difficult. Thus, the charge on sulfur in thiophene varies from 0.12 to 0.75 by HMO calculations while from 0.009 to 0.36 in the PPP method, and between 0.0 to 0.49 by *ab initio* methods ⟨CHEC-II(2)443⟩. Similar discrepancies are formed in estimates of relative π-electron excess of C_α- and C_β-atoms by different methods.

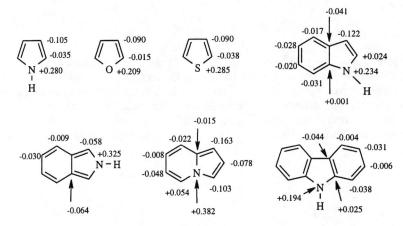

Scheme 3 π-Electron charges in molecules of some five-membered heterocycles

The total π-electron excess of benzo-derivatives of pyrrole varies in the order: indolizine > isoindole > pyrrole > indole > carbazole. Local π-electron excess shows a somewhat different order: indolizine (C-1) > indole (C-3) > pyrrole (C-3) > isoindole (C-1) > carbazole (C-1). Benzene rings thus gain considerable negative charge. For example, in indole and carbazole molecules, more than 40 and 80%, respectively, of the total negative charge is concentrated on the carbon atoms which are not part of the hetero-ring. An especially strong shift of π-electron density occurs from those hetero-rings which contain extra-Hückel π-electrons. Indolizine is a typical example: its six-membered ring formally has seven π-electrons. Transfer out of the additional electron to the five-membered ring renders the latter more electron-rich than the six-membered ring.

The π-electron excess of the five-membered rings is accompanied by a high π-donor character. The best measure of π-donation is the value of first ionization potential, IP_1, which for all aromatic heterocycles with one heteroatom of pyrrole type reflects the energy of highest occupied π-orbital. IP_1 values decrease in the sequence: pyrrole > indole > carbazole; furan > benzo[*b*]furan > dibenzofuran; thiophene > benzo[*b*]thiophene (Section 2.3.3.9, Tables 21 and 23). Thus, the more extensive the π-system, the stronger is its electron donor ability. Furan and thiophene possess almost equal π-donation, which is considerably lower than that of pyrrole.

Inclusion of sulfur *3d*-orbitals in *ab initio* calculations on thiophene makes little difference to the total energy ⟨70CC319, 72TCA(27)171⟩. Their principal role is to act as polarization functions rather than as an extra valence orbital. Thus, the population of the *3d*-orbital is very small but its introduction into the basis set causes considerable changes in the population of *3s*- and *3p*-orbitals so that electron density on sulfur is increased mainly at the expense of the flanking carbon atoms. This conclusion has received experimental support ⟨76JCP(64)3021⟩ from the analysis of fluorescent sulfur $L_{II,III}$ emission spectra, which supplies electronic populations of the *3s* and *3d* valence bands. A consequence of the inclusion of *d*-orbitals is that the calculated dipole moment is lowered from 0.96 to 0.62 D (*cf.* experimental value of 0.54 D). A similar effect has been observed upon the inclusion of selenium *4d*-orbitals in calculations on selenophene ⟨75JCS(F2)1397⟩. These and other *ab initio* calculations ⟨74JCS(P2)420, 77JCP(67)5738⟩ have provided, *inter alia,* satisfactory dipole moments, second moments and ionization potentials for pyrrole, furan, thiophene and selenophene.

Extension of these calculations to thiophene *S,S*-dioxide has led to the deduction of an OSO bond angle of *ca.* 118° ⟨75JCS(P2)1223⟩, in close agreement with the angle of 120° found for dibenzothiophene *S,S*-dioxide, and an out-of-plane angle of 45° has been predicted for thiophene *S*-oxide with an estimated inversion barrier of 44 kJ mol^{-1}, which is of the same order of magnitude as that observed, 62 kJ mol^{-1}, for the 2,5-bis(2,2-dimethylhexyl) derivative. Out-of-plane angles of 28.4° and 75° have been computed for *O*-protonated furan and *S*-protonated thiophene respectively, with corresponding inversion barriers of 5.48 and 138 kJ mol^{-1} ⟨77G55⟩. These species may serve as models for *O*- and *S*-alkyl cations. Out-of-plane angles of 68.9° and 68.2° have been reported for the *S*-methyl cations of dibenzothiophene and naphtho[2,3-*b*]thiophene ⟨81JA289⟩ and a similar geometry has been found for the 1,2,3,5-tetramethylbenzo[*b*]thiophenium cation ⟨81JCS(P2)266⟩. The conformational preferences of 2- and 3-monosubstituted furans, pyrroles and thiophenes also have been the subject of *ab initio* calculations ⟨77JCS(P2)1601, 78JA3981, 79JA311⟩ and will be discussed in Section 2.3.4.3.1.

The MINDO/3 semi-empirical MO method provides ⟨75JA1302, 75JA1311, 79JOC374⟩ satisfactory geometries, heats of formation and ionization potentials for furan, thiophene and pyrrole, with somewhat less satisfactory dipole moments. Calculated vibrational frequencies for these heterocycles also agree well with experimental values, any errors tending to be systematic for a given vibration type ⟨77JA1685⟩.

The relative importance of through-bond (hyperconjugative) and through-space (homoconjugative) interactions between the heteroatom and the double bond in 2,5-dihydro-furan, -thiophene and -pyrrole has been the subject of a CNDO/S study ⟨76ZN(A)215⟩. This analysis concluded that the proportion of through-space interaction increased from 19% in the dihydrofuran and 20% for dihydrothiophene to 83% for the dihydropyrrole (*cf.* Section 2.3.3.9).

The less sophisticated PPP approximation has been shown to reproduce well the electronic spectral features not only of the monocyclic furan, pyrrole, thiophene, selenophene and tellurophene but also many of the benzo-fused derivatives as well ⟨79MI30101, 68JPC3975, 68TCA(12)247⟩.

A useful predicative application of the relatively crude Hückel method to illustrate quantitatively the effect of benzenoid annulation on the resonance energies of furan and thiophene is summarized in Scheme 4. As expected, thiophenes are more stable than the corresponding furans and 3,4-fusion results in less stable compounds than 2,3-fusion ⟨77CR(C)(285)421⟩.

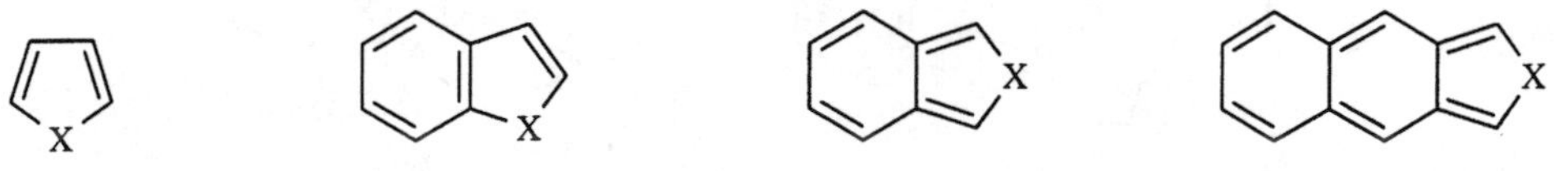

X = O 75 kJ mol^{-1}	188	138	201
X = S 121	222	184	251

Scheme 4

The *ab initio* STO-3G basis set has been used (after MNDO) to calculate electron densities, Mulliken π-overlap populations, and ionization potentials for pyrrole ⟨85JCS(P2)97⟩. As expected, the nitrogen atom has the highest electron density at 7.31, with the C-3 at 6.10 only slightly higher than the C-2 at 5.98.

AM1 studies for pyrrole, indole, and carbazole, and their *N*-deprotonated analogues have been reported ⟨90JCS(P2)65, 90JCS(P2)1881⟩. Of note are the geometrical changes brought about by anion formation. In all three systems, deprotonation brings about a decrease in C—N—C bond angle of the pyrrole ring of about 3°.

The *ab initio* techniques have also been employed to estimate other molecular properties such as electronic spectra and ionization potentials. These studies have been performed on furan ⟨84CPH(90)231, 85JCP(83)723, 89JCS(P2)263, 93JA6184⟩, in comparison with other five-membered rings and ⟨83JST(105)375⟩ with benzoheterocyclic derivatives. These *ab initio* calculations provide values for molecular properties in accordance with experimental trends, but it is necessary to consider the effects of electron correlation for the calculations to quantitatively reproduce experimental values.

Molecular mechanics has been applied extensively to the conformation analysis of hydrofurans. For 2,3-dihydrofuran and tetrahydrofuran, the MM2 method accurately identifies the preferred equilibrium conformation and predicts reasonably well in the barriers to planarity and pseudorotation ⟨87JSP(124)369, 90JPC1830⟩. Application of this methodology to substituted tetrahydrofurans is efficient for the analysis of conformational preferences, both for substituted ⟨86JMC1917, 86JST(142)387, 88JCC188, 88T429, 91T297, 91T1291, 92JST(265)225⟩ and condensed derivatives ⟨86JST(142)383, 91CJC521, 91JOC6180, 92CAR(229)245, 92JOC5271, 93CAR(244)49⟩. MM2 has also been employed in reactivity studies, as in the nitrile oxide cycloadditions to unsaturated sugars ⟨89JOC793⟩, and the intramolecular Diels-Alder reaction ⟨92JA10738⟩, with encouraging results.

The molecular mechanics force field MM3 is better parametrized for carbohydrates ⟨89JA8551, 90JA8551, 90JA8293, 92CAR(244)49⟩.

2.3.3 STRUCTURAL METHODS

2.3.3.1 *X-Ray Diffraction*

Bond lengths and angles from X-ray structures of heterocyclic compounds through 1970 have been listed ⟨78PMH(5)1⟩. This compilation contains many examples, particularly of furans, thiophenes and pyrroles and their benzo derivatives; further examples are contained in the monograph chapters of CHEC. The dimensions in Table 1 for the 2-carboxylic acid derivatives of furan, thiophene, selenophene and tellurophene are generally in good accord with those obtained for the parent heterocycles (see Table 2). Further bond lengths are tabulated and discussed in Section 2.3.4.3.1 dealing with conformations of aromatic compounds.

Solid-state structures of *N*-lithio- (**27**) and *N*-potassio- (**28**) indole-TMEDA complexes ⟨90OM1485⟩ show they aggregate as dimers. The lithioindole crystallizes as a *syn*-dimer, while the potassioindole forms the *anti*-dimer.

(27)   (28)

The solid-state structures of carbazole anions are also largely determined by the counterion ⟨89AG(E)1224⟩. The *N*-caesiocarbazole-PMTDA complex deviates from the expected simple dimer; the caesium atoms show normal σ-bonding to the nitrogen atom, but also have significant π interactions with all five atoms comprising the pyrrole ring. The *N*-potassiocarbazole-PMTDA complex shows a much smaller interaction between the metal ion and π cloud, interacting only with carbons 8a and 9a, whereas the *N*-lithioindole shows interaction only between the lithium and the C-2 ⟨90OM1485⟩ . Similarly, the *N*-lithiocarbazole-THF dimer has two different orientations of the cation: one σ interaction between the lithium and the in-plane nitrogen lone pair, and the other a perpendicular interaction between a second lithium and the N-π orbital ⟨87CB1533⟩.

Table 1 Bond Lengths and Angles of Heterocycle 2-Carboxylic Acids

Bond length (Å)	X = O	X = S	X = Se	X = Te
X-C(5)	1.312	1.701	1.850	2.047
X-C(2)	1.368	1.693	1.872	2.057
C(5)-C(4)	1.446	1.363	1.355	1.357
C(2)-C(3)	1.288	1.362	1.356	1.384
C(3)-C(4)	1.351	1.414	1.421	1.412
C(2)-C(6)	1.414	1.481	1.438	1.423
Bond angle (degrees)				
C(2)XC(5)	109	92.0	87.1	81.5
XC(5)C(4)	109	111.8	112.25	111.7
XC(2)C(3)	109	111.8	110.7	111.7
C(5)C(4)C(3)	105	111.9	114.2	118.8
C(4)C(3)C(2)	105	112.4	115.7	116.3
XC(2)C(6)	120	122.2	121.0	123.4
C(3)C(2)C(6)	131	125.9	128.3	124.8
Ref.	62AX919	62AX737	62AX737	72CSC273

2.3.3.2 Microwave Spectroscopy

For some general remarks on microwave spectra see Section 2.4.3.2.

2.3.3.2.1 *Molecular geometry*

High accuracy molecular dimensions for the planar parent heterocycles in the gas phase have been obtained by microwave spectroscopy and are recorded in Table 2. Increasing size of the heteroatom results in lengthening of the X-C(2) bond and decrease of the C(2)XC(5) bond angle, so that the shape of the molecule is progressively elongated. Delocalization in the case of pyrrole brings the imino hydrogen into the plane of the other atoms with an N-H bond length of 0.996 Å.

The distortions caused by substitution are usually of small magnitude. Thus in 2-cyanopyrrole ⟨80JPC1767⟩ microwave spectroscopy shows that the N-C(2) and C(2)-C(3) bond lengths are only shortened by 0.009 Å relative to those observed for pyrrole with consequent increases in the ring bond angles at N and C(3) of 0.1°. Distortions of the same magnitude have also been detected for the 2-cyano derivatives of furan and thiophene ⟨80JPC1767⟩, and for 3-cyanothiophene ⟨80ZN(A)770⟩.

Table 2 Comparison of Bond Lengths and Angles of Monoheterocycles

Bond length (Å)	X = NH	X = O	X = S	X = Se	X = Te
X-C(2)	1.370	1.362	1.714	1.855	2.055
C(2)-C(3)	1.382	1.361	1.370	1.369	1.375
C(3)-C(4)	1.417	1.430	1.423	1.433	1.423
C(2)-H	1.076	1.075	1.078	1.070	1.078
C(3)-H	1.077	1.077	1.081	1.079	1.081
Bond angle (degrees)					
C(2)XC(5)	109.8	106.5	92.17	87.76	82.53
XC(2)C(3)	107.7	110.65	111.47	111.56	110.81
C(2)C(3)C(4)	107.4	106.07	112.45	114.55	117.93
XC(2)H	121.5	115.98	119.85	121.73	124.59
C(2)C(3)H	125.5	127.83	123.28	122.59	121.04
Ref.	69JST(3)491	78JST(48)157	61JSP(7)58	69DOK(185)384	73CPH(1)217

2.3.3.2.2   *Partially and fully saturated compounds*

The determination of the bond lengths of the fully saturated heterocycles has been complicated by their conformational mobility, which is considered in Section 2.3.4.3.2. The data which have been provided by electron diffraction are listed in Table 3 and show the expected trends consonant with increasing size of heteroatom.

Table 3   Bond Lengths and Angles for Saturated Monoheterocycles

$$H_2C - CH_2$$
$$H_2C \diagdown_X{}^{\diagup} CH_2$$

Bond lengths (Å)	X = O	X = S	X = Se
X-C	1.428	1.839	1.975
C-C	1.535	1.536	1.538
Bond angles (degrees)			
CXC	106.4–110.6	93.4	89.1
XCC	103.7–107.5	106.1	105.8
CCC	100.3–104.4	105.0	106.0
Ref.	69ACS2748, 69T3045	69ACS3534	70ACS1903

2.3.3.3   ¹H NMR Spectroscopy

2.3.3.3.1   *Parent aromatic compounds*

The ¹H NMR spectra of the parent heterocycles (Table 4) each consist of two multiplets, of which the one at lower field is assigned to the α-hydrogens.

Table 4   ¹H NMR Spectral Data for Monoheterocycles (in CDCl₃)[b]

	Pyrrole[a]	Furan	Thiophene	Selenophene	Tellurophene	Cyclopentadiene
H-2	6.68	7.29	7.18	7.88	8.87	6.28
H-3	6.22	6.24	6.99	7.22	7.78	6.43
$J_{2,3}$	2.70	1.75	4.90	5.40	6.70	5.05
$J_{2,4}$	1.44	0.85	1.04	1.46	1.30	1.09
$J_{2,5}$	1.87	1.40	2.84	2.34	2.60	1.93
$J_{3,4}$	3.35	3.30	3.50	3.74	4.00	1.93
Ref.	b	65SA85	65SA85	65SA85	72JCS(P1)199	70JCP(53)2343

[a] For pyrrole signal of NH proton appears at 7–12 ppm. [b] Varian Catalog. [c] Units are ppm and Hz.

Apart from pyrrole, the chemical shifts for the β-protons increase with decreasing electronegativity of the heteroatom. In contrast, the chemical shifts of the α-protons do not display any obvious regularity, probably due to paramagnetic shielding contributions which will become more important with increasing availability of d-orbitals. The chemical shift of the pyrrole N-H is solvent dependent.

In the case of pyrrole, the ring protons are also coupled to the N-H with $J_{1,2} = J_{1,5} = 2.58$ Hz and $J_{1,3} = J_{1,4} = 2.46$ Hz, while satellites due to spin-spin coupling between the α-protons and the ring heteroatom are observed for selenophene, $J(^{77}Se, H) = 47.5$ Hz, and tellurophene, $J(^{125}Te, H) = 100.4$ Hz ⟨74ACS(B)175⟩.

The magnitudes of the vicinal vinylic proton coupling constants are much smaller than the 7.6 Hz observed for benzene and reflect in part the greater separation of the protons when attached to five-membered rings; however, factors such as electronegativity and double-bond character also affect coupling sizes: $J_{2,3}$ and $J_{3,4}$ increase systematically, *i.e.* O < NH < S < Se < Te. Comparison with the corresponding coupling constants for cyclopentadiene indicates the important role of heteroatom electronegativity in determining the magnitude of $J_{2,3}$. The much larger $J_{3,4}$, observed for all of the heterocycles, emphasizes the greatly increased conjugative interaction between carbons 3 and 4 relative to that in cyclopentadiene. The longer range coupling constant $J_{2,4}$ is lower for all of the heterocycles than corresponding benzenoid *meta* coupling constants (*ca.* 2–3 Hz), whereas the $J_{2,5}$

couplings, which increase in the sequence $O < NH < CH_2 < Se < Te < S$, are larger than benzenoid *para* coupling constants (*ca.* 0–1 Hz).

The ring proton chemical shifts of 2H-pyrroles and 3H-pyrroles are as expected for conjugated imines ⟨B-90MI 201-02⟩.

2.3.3.3.2 *Substituted aromatic compounds*

As in benzenoid analogues, electron-withdrawing substituents deshield and electron-releasing groups shield the ring protons. Quantitative correlations between the ^{1}H NMR spectral properties of monosubstituted furans ⟨75CS(7)211⟩, thiophenes ⟨75CS(7)76⟩, selenophenes ⟨75CS(7)111⟩ and tellurophenes ⟨76ACS(B)605⟩ have been extensively explored and rationalizations offered in terms of inductive and mesomeric effects. Linear correlations exist (a) between the relative shifts of H-2 and H-3 in 3-substituted furans, thiophenes and selenophenes and (b) between heteroatom electronegativity and the shifts of H-5 in 2- and 3-substituted, and H-2 in 3-substituted, heterocycles.

Long-range couplings occur between ring protons and hydrogens attached to substituent atoms. For example, in heterocyclic 2-aldehydes couplings of the order of 1 Hz have been observed.

Annulation of a benzene ring on to the [*b*] face of the heterocyclic ring does not have any pronounced effect upon the chemical shifts of the heterocyclic protons (*cf.* Table 5). The chemical shift of the indole N-H is solvent dependent and as in pyrrole it is also coupled to the ring protons with $J_{1,2} = 2.4$ Hz and $J_{1,3} = 2.1$ Hz. The benzenoid protons H-4 and H-7 appear downfield from H-5 and H-6. Long-range coupling between H-3 and H-7 (*cf.* **29**) is of considerable diagnostic value in establishing the orientation of a substituent on the heterocyclic ring.

Table 5 ^{1}H NMR Spectral Data for Benzo[*b*] Heterocycles (in CCl$_4$)[b]

	Indole	Benzofuran	Benzothiophene	Benzoselenophene	Benzotellurophene
H-2	6.52 (7.27)[a]	7.52 (7.78)[a]	7.33	7.90	8.65
H-3	6.29 (6.45)	6.66 (6.76)	7.22	7.50	7.91
H-4	(7.55)	7.49 (7.63)	7.72	7.76	7.79
H-5	(7.00)	7.13 (7.23)	7.26	7.19–7.29	7.08–7.30
H-6	(7.08)	7.19 (7.30)	7.24	7.19–7.29	7.08–7.30
H-7	(7.40)	7.42 (7.51)	7.79	7.86	7.90
$J_{2,3}$	3.1	2.19	5.5	6.0	7.1
$J_{2,6}$	—	—	0.5	0.3	—
$J_{3,7}$	0.7	0.87	0.75	0.67	—
$J_{4,5}$	7.8	7.89	8.5	—	—
$J_{4,6}$	1.2	1.28	1.14	—	—
$J_{4,7}$	0.9	0.80	0.7	—	—
$J_{5,6}$	7.0	7.27	7.0–7.5	—	—
$J_{5,7}$	1.2	0.92	0.5–1.0	—	—
$J_{6,7}$	8.0	8.43	8.0–7.5	—	—
Ref.	64JCS981, 65AJC353	65AJC353	66BCJ2316	72BSF3193	72BSF3193

[a] Chemical shifts in parentheses are for acetone solutions. [b] Units are ppm and Hz.

^{1}H NMR spectroscopy has provided useful structural information on the unstable benzo[*c*]-fused heterocycles (Table 6). The protons H-1 and H-3, adjacent to the heteroatom, are observed at lower field than are the corresponding ones in the parent heterocycles or their benzo[*b*] isomers.

(29)   **(30)**

However, the same sequence of downfield shifts, namely $NR < S < O < Se$, is observed. Thus, at least the heterocyclic moiety appears to be aromatic. The ratio of the vicinal coupling constants $J_{5,6} / J_{4,5}$ (0.70–0.74) is significantly less than for compounds of the Kekulé series (*ca.* 0.9), much higher than observed for a polyenic π-system (cyclohexa-1,3-diene displays a value of 0.52), and close to the predicted value of 0.71 for the corresponding ratio for $J_{4,5}$ / $J_{3,4}$ in octatetra-1,3,5,7-ene with conformation (**30**). NMR data for dibenzo heterocycles are given in Table 7.

Table 6 ^{1}H NMR Spectral Data for Benzo[*c*] Heterocycles[a]

	H-1, *H*-3	*H*-4, *H*-7	*H*- 5, *H*-6	$J_{1,4}$	$J_{4,5}$	$J_{4,6}$	$J_{4,7}$	$J_{5,6}$	*Solvent*	*Ref.*
Isoindole	6.28	7.5	6.8	—	8.49	—	—	6.29	(CD$_3$)$_2$CO	73JCS(P1)1432
N-Methylisoindole	7.05	7.51	6.92	0.46	8.69	0.90	0.79	6.46	CDCl$_3$	76JCS(P2)81
Benzo[*c*]furan	7.99	7.38	6.84	0.64	8.52	1.01	0.57	6.22	CDCl$_3$	76JCS(P2)81
Benzo[*c*]thiophene	7.63	7.59	7.04	0.42	8.64	1.03	0.79	6.36	CDCl$_3$	76JCS(P2)81
Benzo[*c*]selenophene	8.40	7.33–7.54	6.77–7.02	—	9.16	—	—	6.79	CDCl$_3$	77JA8248, 76JA867

[a] Units are ppm and Hz.

Table 7 ^{1}H NMR Spectral Data for Dibenzo
Heterocycles (in Acetone)[b]

	Carbazole[a]	*Dibenzofuran*	*Dibenzothiophene*
H-1	7.49	7.61	7.95
H-2	7.36	7.51	7.50
H-3	7.16	7.37	7.49
H-4	8.08	8.07	8.29
$J_{1,2}$	8.21	8.42	8.07
$J_{1,3}$	0.89	0.72	1.11
$J_{1,4}$	0.67	0.57	0.70
$J_{2,3}$	7.17	7.56	7.23
$J_{2,4}$	1.18	1.36	1.17
$J_{3,4}$	7.80	7.87	8.06
Ref.	65AJC353	65AJC353	71AJC2293

[a] This is not the normal numbering system for carbazole.
[b] Units are ppm and Hz.

2.3.3.3.3 *Saturated and partially saturated compounds*

The α-hydrogen resonances for the fully saturated monocycles (Table 8) occur at lower fields than the β-hydrogen resonances. The chemical shifts of the β-hydrogens vary systematically with the electronegativity of the heteroatom, but the chemical shifts of the α-hydrogens do not.

Table 8 ^{1}H NMR Spectral Data (ppm) for
Tetrahydro Heterocycles (in CCl$_4$)

Heterocycle	*H-2*	*H-3*	*Ref.*
Pyrrolidine	2.77	1.63	72JA8854
Tetrahydrofuran	3.62	1.79	72JA8854
Tetrahydrothiophene	2.75	1.92	72JA8854
Tetrahydroselenophene	2.79	1.96	74JHC827
Tetrahydrotellurophene	3.10	2.03	74JHC827

The spectra of 2,5-dihydrofuran and 2,5-dihydrothiophene are simple, showing two singlets at 4.51 and 5.85 ppm for the dihydrofuran, and 3.66 and 5.79 ppm for the sulfur analogue; the lack of coupling is in accord with an almost planar conformation ⟨73JMR(12)244⟩.

The ^{1}H NMR spectra of pyrrolizines has been reviewed ⟨84AHC(37)1⟩, with particular attention to substituent effects.

2.3.3.4 ^{13}C NMR Spectroscopy

The ^{13}C NMR spectral properties of the parent heterocycles are summarized in Table 9. The signal for the pyrrole α-carbon is broadened as a result of coupling with the adjacent nitrogen-14 atom (*cf.* Section 2.3.3.5). While the frequencies observed for the β-carbon atoms show a fairly systematic upfield shift with increasing electronegativity of the heteroatom, the shifts for the α-carbon atoms vary irregularly. All the shifts are in the region of that for benzene, δ 128.7.

Table 9 ^{13}C NMR Spectral Data for Monoheterocycles (in Acetone)[a]

	Pyrrole	*Furan*	*Thiophene*	*Selenophene*	*Tellurophene*
C-2	118.2	143.6	125.6	131.0	127.3
C-3	107.2	110.4	127.3	129.8	138.0
$J_{C-2,H-2}$	184	201	185	189	183
$J_{C-3,H-3}$	170	175	168	166	159
$J_{C-2,H-3}$	(7.6)	14	7.35	7.0	—
$J_{C-2,H-4}$	(7.6)	5.8	10.0	10.0	—
$J_{C-2,H-5}$	(7.6)	7.0	5.15	3.5	—
$J_{C-3,H-2}$	7.8	7.0	4.7	4.5	—
$J_{C-3,H-4}$	4.6	4.0	5.9	6.0	—
$J_{C-3,H-5}$	7.8	10.8	9.5	10.4	—
Ref.	68JA3543	68JA3543, 65JA5333	68JA3543, 65JA5333	68JA3543, 74ACS(B)175	74ACS(B)175

[a] Units are ppm and Hz.

The direct $J_{C-3,H-3}$ coupling constants decrease regularly along the series O > NH > S > Se > Te. The values for $J_{C-2,H-2}$, are appreciably larger than the 159 Hz observed for benzene and the 170 Hz for the alkenic protons of cyclopentadiene, while the $J_{C-3,H-3}$ coupling constants span this range.

Similar correlations have been observed ⟨75CS(7)211⟩ between the ^{13}C NMR spectra of mono-substituted furans, thiophenes, selenophenes and tellurophenes as with their ^{1}H NMR spectra (*cf.* Section 2.3.3.3). Thus, for the 2-substituted compounds the Δ(C-3)/Δ(C-5) ratios decrease systematically in the series furan (2.58) > thiophene (1.34) = selenophene (1.34) > tellurophene (0.91). Extensive quantitative correlations have been established between the shifts of the corresponding carbon atoms in the different heterocycles ⟨75CS(7)211⟩. In most cases $^{1}J_{(C,Hβ)}$ in both 2- and 3-substituted heterocycles can be linearly correlated with the electronegativity of the heteroatom, with the couplings being largest for the furans.

The signal for C-2 of indole, benzo[*b*]furan and benzo[*b*]thiophene (Table 10) is shifted to lower field than C-3. However, the shifts for C-2 (O, 144.8; Se, 128.8; S, 126. 1; NH, 124.7; Te, 120.8) and C-7a (O, 155.0; Se, 141.3; S, 139.6; NH, 135.7; Te, 133.0) in the benzo[*b*] heterocycles vary irregularly ⟨80OMR(13)319⟩, and the sequence is different to that observed for C-2 in the parent heterocycles, namely O > Se > Te > S > NH. Also noteworthy is the upfield position of C-7, especially in indole and benzofuran, relative to the other benzenoid carbons at positions 4, 5 and 6. Thus for indole, the effect of the heteroatom is strongest at the β-carbons C-3, C-3a, and C-7 with upfield shifts of 29.6, 16.7, and 12.4 ppm, respectively. Only moderate effects (0.1–9.5 ppm) are seen at the other carbons ⟨87MRC377⟩.

Table 10 ^{13}C NMR Spectral Data (ppm) for Benzo[*b*] Heterocycles

	Indole	*Benzofuran*	*Benzothiophene*
C-2	124.67	145.1	126.21
C-3	102.14	106.9	123.79
C-4	120.76	121.6	123.57
C-5	121.81	123.2	124.10
C-6	119.76	124.6	124.17
C-7	111.35	111.8	122.44
C-7a	135.65	*155.5*	139.71
C-3a	128.26	127.9	139.57
Solvent	dioxane	CS$_2$	CDCl$_3$
Ref.	70JOC996, 75JOC3720	74BCJ1263	76OMR(8)252

The ^{13}C shifts for indoles have almost no solvent dependence, allowing facile comparison between spectra obtained in different solvents. Substitution on the nitrogen has varying effects on the ring carbons. N-Deprotonation causes large downfield shifts of the C-2 and C-3a, with slight upfield shifts of C-5, C-6, and C-7a. On the other hand, replacement of the N-proton with a phenyl group does not noticeably affect any of the ring carbons. Alkyl substitution seems to only affect the chemical shift of the C-2, moving its signal downfield by about 5 ppm.

In the dibenzo heterocycles (Table 11), C-1 and C-8 are shifted upfield in carbazole and dibenzofuran relative to the corresponding carbons in dibenzothiophene and fluorene, and similar, though smaller, shifts can be discerned for C-3 and C-6 in the former compounds.

Table 11 ^{13}C NMR Spectral Data (ppm) for Dibenzo Heterocycles
(in DMSO-d_6)[a]

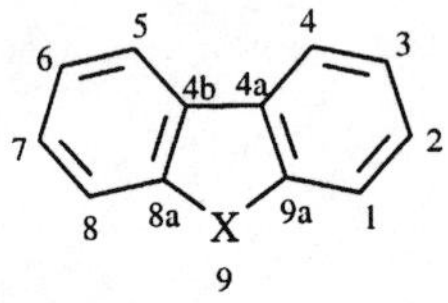

	Carbazole	Dibenzofuran[b]	Dibenzothiophene[b]	Fluorene
C-1,8	110.8	111.6	122.9	125.0
C-2,7	125.4	127.5	127.0	126.7
C-3,6	118.4	123.0	124.6	126.7
C-4,5	120.0	121.1	121.9	119.9
C-4a,4b	122.6	123.5	134.9	141.0
C-8a,9a	139.6	155.4	138.5	142.8

[a] ⟨79OMR(12)647⟩. [b] This is not the normal numbering system for these molecules.

Some information on partially and fully saturated heterocycles is summarized in Scheme 5. As would be expected, the downfield shift of the α-carbon atom decreases with decreasing electronegativity of the heteroatom in the sequence O < NH < S < CH_2.

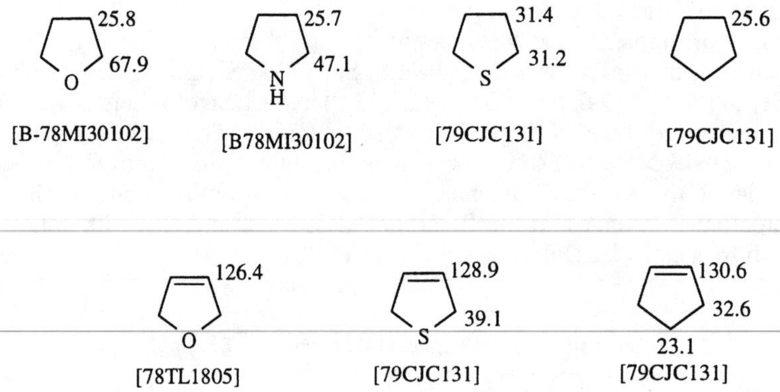

Scheme 5

2.3.3.5 Heteroatom NMR Spectroscopy

All of the common heteroatoms possess at least one naturally occurring isotope with a magnetic moment (Table 12). The electric quadrupole of ^{14}N, ^{17}O and ^{33}S broadens the NMR signals so that line widths may be 50–1000 Hz or even wider. To some extent this problem is offset by the more extensive chemical shifts that are observed. The low natural abundances and/or sensitivities have necessitated the use of accumulation techniques for all of these heteroatoms. ^{14}N and ^{15}N chemical shifts are identical.

Table 12 Magnetic Properties of Heteronuclei

Isotope	Natural abundance (%)	Nuclear spin	Electric quadrupole moment[a]	NMR frequency for a 23.5 kG field (MHz)	Relative sensitivity
^{1}H	99.98	1/2	—	100	1
^{13}C	1.11	1/2	—	25.19	0.016
^{14}N	99.635	1	2.0	7.224	0.00101
^{15}N	0.365	1/2	—	10.133	0.00104
^{17}O	0.037	5/2	−0.4	13.56	0.029
^{33}S	0.74	3/2	−6.4	7.67	0.0023
^{77}Se	7.58	1/2	—	19.135	0.00693
^{125}Te	7	1/2	—	30.6	0.032

[a] $e \times 10^{-26}$ cm^2.

Despite its unfavorable NMR properties, the ^{17}O nucleus has attracted considerable interest, since its chemical shifts represent a discriminating probe for structural and molecular properties. In a study of some 5-membered heterocycles (furan and isoxazole methyl derivatives) ⟨84OMR(22)55⟩ it was found that the ^{17}O chemical shifts are mainly determined by the *p*-electron density on the oxygen atom. A ^{17}O downfield shift of 222 ppm is observed on the formal aromatization of tetrahydrofuran to furan ⟨61HCA865⟩.

Examination of ^{17}O chemical shifts of several 2-substituted and 2,5-disubstituted furans ⟨85MRC985⟩ has demonstrated the additivity of substituent effects by plotting the ^{17}O chemical shifts of 2,5-disubstituted furans *versus* those of the 2-substituted analogues. The electronic character of substituents alone does not determine ^{17}O chemical shifts, because they do not correlate with Hammett constants.

Downfield shifts have been observed for ^{33}S (tetrahydrothiophene → thiophene 309 ppm) ⟨72JA6579⟩ and ^{14}N (pyrrolidine → pyrrole 106 ppm) ⟨71TL1653⟩, and a very much larger one can be anticipated for tetrahydroselenophene to selenophene ⟨76OMR(8)354, 72JCS(D)1397⟩. Benzoannulation of pyrrole causes *ca.* 10–20 ppm upfield shift per benzene ring. Much larger upfield shifts are observed in proceeding from selenophene → benzoselenophene (79 ppm) → dibenzoselenophene (75 ppm) ⟨76OMR(8)354⟩. Similar upfield shifts of 65 ppm and 56 ppm, respectively, are observed on annulation with a thiophene ring or selenophene ring to give the selenolothiophene (**31**) ⟨76OMR(8)354⟩ and the selenoloselenophene (**32**) ⟨74CS(5)236⟩.

(**31**)   (**32**)

The effects of 2- and 3-substituents on the ^{77}Se NMR chemical shifts in selenophene and benzo[*b*]selenophene have been investigated in detail ⟨75CS(8)8, 81OMR(16)14⟩. The ^{125}Te shifts in 2-substituted tellurophenes exactly parallel the ^{77}Se shifts in similarly substituted selenophenes ⟨76CS(10)139⟩.

The relatively modest shifts observed for substituted pyrroles are more suitably probed by ^{15}N rather than ^{14}N NMR and seem to parallel the ^{77}Se and ^{125}Te behaviors. Electron withdrawing groups such as 2-nitro, 2-formyl and 2-acetyl cause downfield ^{15}N shifts ⟨76OMR(8)208⟩. Perhaps unexpected is the much larger downfield shift found for 3-nitropyrrole than its 2-nitro isomer. Similar disparities with electron accepting groups also arise with selenophenes, and may be accounted for in terms of an important contribution to the hybrid from the resonance structure (**33**). Similar considerations apply to the relatively large ^{15}N downfield shifts observed when such groups are present at positions 3 or 5 of the indole nucleus, the relevant resonance structures being (**34**) and (**35**), respectively ⟨76OMR(8)117⟩.

(**33**)   (**34**)   (**35**)

In a study of protonation effects ⟨93MRC791⟩, ^{15}N NMR spectra were taken for carbazole, *N*-methylcarbazole, and *N*-methylindole in both DMSO-d$_6$ and acidic media (CF$_3$CO$_2$H or H$_2$SO$_4$) relative to external CH$_3$NO$_2$. The shifts (35–45 ppm) for carbazoles are much larger than for anilines, probably owing to disruption of the carbazole aromaticity. Indole shifts approximately 94 ppm downfield upon protonation, this combined with the couplings, suggests the formation of a 3*H*-indole cation rather than the *N*-protonated species, and agrees well with AM1 calculations ⟨90JCS(P2)65⟩ (see also Section 3.3.1.3).

2.3.3.6.   UV Spectroscopy

The parent heterocycles (Table 13) display a strong band near 220 nm with one additional band at longer wavelengths for thiophene and selenophene, and two for tellurophene. Analogous weak bands reported in the older literature for furan and pyrrole are now generally accepted as arising from autoxidation products.

The cyclopentadiene anion is a carbocyclic analogue of these heterocycles. Simple theoretical treatment shows that, in contrast to the iso-π-electronic molecule benzene, there is no breaking of the degeneracy of the low-energy forbidden transitions or the high-energy allowed ones as a result of the lower symmetry of the anion. The absorption spectrum of the cyclopentadienyl anion and its hetero analogues is thus expected to consist of a moderate intensity band followed at shorter wavelengths by a high intensity band. Bands arising from promotion of an electron from the lone pair orbital of a heteroatom to a π-orbital of the heterocyclic ring ($n \rightarrow \pi^*$) would not be expected for pyrrole, where the nitrogen lone pair is involved in the π-bonding. For the other heterocycles where additional heteroatom lone pairs are available they have yet to be identified, even by electron impact studies ⟨76JCP(64)1315, 76CPL(41)535⟩. These did, however, show singlet → triplet transitions at 3.99 eV (311 nm) and 5.22 eV (238 nm) in furan, 3.75 eV (331 nm) and 4.62 eV (268 nm) in thiophene and 4.21 eV (294 nm) in pyrrole. The positions and energy splitting are analogous to the lowest π → π* transitions in benzene and very different to those anticipated for conjugated dienes.

The marked progressive shift of absorption to longer wavelengths in the sequence furan < pyrrole < thiophene < selenophene < tellurophene is also observed with their 2-substituted derivatives. Increasing conjugating powers of 2-substituents result in displacement of absorption bands to longer wavelength in the expected sequence CO$_2$H < COMe < CHO < NO$_2$. Smaller bathochromic shifts occur when conjugating substituents are introduced into position 3, but now pyrrole displays larger shifts than thiophene.

Table 13   UV Spectra (nm) of Monosubstituted Heterocycles

Substituent	O	NH	λ_{max}(log ε) S	Se	Te
None	208 (3.90)	210 (4.20)	215 (3.8)	232 (3.56)	209 (3.57)
			231 (3.87)	249 (3.75)	241 (3.36)
					279 (3.93)
2-CO$_2$H	214sh (3.58)	222sh (3.65)	246 (3.96)	258 (3.94)	—
	243 (4.03)	258 (4.10)	260 (3.84)	282 (3.70)	
2-CO$_2$Me	252 (4.13)	238sh (3.63)	248 (3.97)	260 (4.0)	—
		263 (4.14)	268 (3.86)	284 (3.80)	
2-COMe	226 (3.38)	250sh (3.70)	260 (4.01)	271 (4.02)	211 (3.93)
	270 (4.15)	287 (4.21)	283 (3.87)	302 (3.70)	282 (3.87)
					346 (3.58)
2-CHO	227 (3.48)	251 (3.49)	260 (4.04)	271 (4.07)	—
	272 (4.12)	287 (4.12)	286 (3.86)	304 (3.71)	
2-NO$_2$	225 (3.53)	231 (3.61)	270 (3.80)	—	—
	315 (3.91)	335 (4.23)	296 (3.78)		
3-CO$_2$H	200 (3.85)	223 (3.89)	241 (3.92)	—	—
	235 (3.39)	245 (3.71)			
3-CO$_2$Me	238 (3.40)	224 (3.90)	241 (3.95)	—	—
		247 (3.73)			
3-COMe	—	243 (3.97)	250 (4.08)	—	—
		270sh (3.66)			
3-CHO	—	243 (3.97)	251 (4.12)	—	—
		270 (3.66)			
Solvent	EtOH	EtOH, MeOH	EtOH	EtOH	*n*-hexane
Ref.	71PMH(3)79	71PMH(3)79, 71T245	58AK(13)23, 58SA350	58G453	72JCS(P1)199

Annulation increases the complexity of the spectra just as it does in the carbocyclic series, and the spectra are not unlike those of the aromatic carbocycles obtained by formally replacing the heteroatom by two aromatic carbon atoms (—CH=CH—). Although quantitatively less marked, the same trend for the longest wavelength band to undergo a bathochromic shift in the heteroatom sequence O < NH < S < Se < Te is discernible in the spectra of the benzo[*b*] heterocycles (Table 14). As might perhaps have been anticipated, the effect of the fusion of a second benzenoid ring on to these heterocycles is to reduce further the differences in their spectroscopic properties (*cf.* Table 15). The absorption of the benzo[*c*]heterocycles (Table 16) at longer wavelengths than their benzo[*b*] counterparts is a reflection of the lower aromaticity of the former compounds and the consequential differences in the energies of the highest occupied molecular orbitals.

Furan, pyrrole or thiophene do not fluoresce or phosphoresce; however, indole and carbazole both fluoresce and phosphoresce strongly (CHEC 3.04.4.4) ⟨74PMH(6)166⟩. The characteristic fluorescence of indoles has found extensive application in the detection and estimation of naturally occurring derivatives. The fluorescence maximum of indole observed at 297 nm in cyclohexane undergoes dramatic shifts in hydrogen-bonding solvents, possibly due to strong interactions between the solvent and the lowest excited singlet state of the indole. The wavelength of phosphorescence of benzothiophene (λ_p 416 nm) is somewhat longer than that for indole (λ_p 404 nm), reflecting the effect of the higher energy $3p_\pi$ orbital of sulfur on the energy of the highest occupied MO as opposed to the nitrogen $2p_\pi$ one. The effect is less marked in the case of dibenzothiophene (λ_p 411 nm), relative to dibenzofuran (λ_p 408 nm) and carbazole (λ_p 407 nm).

Table 14 UV Spectra of Benzo[b] Heterocycles (in Heptane)

Compound		λ_{max} (nm) (log ε)					
Benzofuran[a]	—	239.5 (4.03)	240.5 (4.03)	244.5 (4.04)	250.5 (3.91)	269 (3.23) 275 (3.40)	271 (3.25) 281 (3.49)
Indole[a]	215 (4.38)	261 (3.69)	266.5 (3.70)	277sh (3.58)	279 (3.62)	287 (3.51)	—
Benzothiophene[a]	228 (4.45)	263sh (3.71)	258 (3.76)	281 (3.19)	288.5 (3.33)	290.5 (3.33)	297 (3.52)
Benzoselenophene[a]	236 (4.45)	260 (3.70)	270sh (3.52)	296 (3.56)	298 (3.54)	305 (3.79)	—
Benzotellurophene[b]	214 (4.43)	251 (4.43)	—	312sh (4.15)	318 (4.2)	—	—

[a] 57ACH(11)365; [b] 71BSB521.

Table 15 UV Spectra of Dibenzo Hetcrocycles (in EtOH)[a]

Compound	Zone A	Zone B	Zone C
Dibenzofuran	218 (4.51), 227sh (4.31), 241 (4.04), 244 (4.04), 249 (4.26)	275sh (4.09), 280 (4.22), 286 (4.19)	296 (3.95), 300sh (3.65)
Carbazole	211 (4.43), 227sh (4.53), 233 (4.57), 243 (4.38), 253 (4.25)	282sh (3.99), 291 (4.14)	322 (3.52), 334 (3.44)
Dibenzothiophene	237 (4.57), 258 (4.12), 264 (3.99)	280sh (3.75), 289 (4.05)	317 (3.28), 328 (3.39)
Dibenzoselenophene	238 (4.77), 260sh (4.14)	278 (3.84), 286 (3.98)	316 (3.37), 326 (3.46)
Dibenzotellurophene	212 (4.39), 232sh (4.93), 235 (4.93), 255 (4.32), 263 (4.20)	280sh (3.99), 286 (4.24)	303 (3.45), 312 (3.53), 325 (3.63)

[a] ⟨58AC(R)738⟩.

Table 16 UV Spectra of Benzo[c] Heterocycles (in Hexane)

Compound	λ_{max} (nm) (log ε)
Benzo[c]indole[a]	263.5, 268.5, 275, 286.5, 294sh, 300, 306.5, 312.5, 320, 326.5, 335
Benzo[c]furan[b]	215 (4.17), 244 (3.4), 249 (3.37), 254 (3.35), 261 (3.12), 292sh (3.35), 299sh (3.47), 305sh (3.56), 313 (3.7), 319 (3.7), 327 (3.87), 334 (3.66), 343 (3.79)
Benzo[c]thiophene[c] (in MeOH)	215 (4.84) 257sh (4.04), 272 (3.94), 278 (3.94), 283sh (3.93), 290 (3.97), 295 (3.96), 298sh (3.94), 305 (4.06), 313 (3.89), 318 (3.90), 322 (3.88), 328 (3.91), 333 (3.90), 343 (3.79)
Benzo[c]selenophene[d]	273, 286, 291, 298, 302sh, 305sh, 312, 323, 328, 336sh, 340, 344sh, 353, 357, 362sh

[a] ⟨73JCS(P1)1432⟩; [b] ⟨71TL2337⟩; [c] ⟨63JPR(20)244⟩; [d] ⟨76JA867⟩.

2.3.3.7 IR Spectroscopy

2.3.3.7.1 *Ring vibrations*

The literature concerning the IR spectra of these heterocycles has been extensively surveyed ⟨63PMH(2)165, 71PMH(4) 265⟩.

The vibrational assignments for the parent heterocycles are summarized in Table 17. These have been derived from IR and Raman spectra of both the parent heterocycles and deuterated derivatives. In the case of the vibrations of A_1 symmetry, these have been further supported by photoelectron spectroscopic studies ⟨71SA(A)2525⟩. The variety of factors responsible for the observed differences precludes a complete rationalization as they may variously operate in concert or in opposition. Pertinent factors include the mass and electronegativity of the heteroatom, π-electron delocalization, ring geometry and vibrational coupling of normal modes. The ring modes permit the best qualitative comparison, especially as the interaction of the C-H and N-H in-plane deformation modes of pyrrole precludes direct comparisons of its β-CH vibrations with those of the other heterocycles. The increase in frequency of the ν_5 mode, which is associated with the symmetric stretching of the double bonds, occurs in the sequence thiophene < selenophene < tellurophene < pyrrole < furan and may be related to increasing localization of double bonds attendant upon decreasing aromaticity. The large decreases in frequency of the ν_3 and ν_{17}, modes, which depend upon symmetric and anti-symmetric C-X-C stretching, respectively, and the ring deformation modes ν_8, ν_{18}, and ν_{21} have been attributed to mass and geometry effects.

Table 17   Fundamental Vibrational Frequencies (cm^{-1}) of Parent Heterocycles

Vibration [a]	Approximate description[a]	C_{2v}	Pyrrole[b]	Furan	Thiophene	Selenophene	Tellurophene
ν_1	C-H stretch	A_1	3133	3159	3110	3110	3084
ν_2	C-H stretch		3108	3128	3086	3063	3045
ν_5	Ring stretch		1466	1483	1408	1419	1432
ν_4	Ring stretch		1384	1380	1360	1341	1316
ν_6	C-H def. i.p.		—	1140	1081	1080	1079
ν_7	C-H def. i.p.		1076	1061	1033	1010	984
ν_3	Ring stretch		—	986	833	758	687
ν_8	Ring def. i.p.		—	873	606	456	380
ν_9	C-H def. o.o.p.	A_2	869	863	900	905	912
ν_{10}	C-H def. o.o.p.		—	728	686	685	690
ν_{11}	Ring def. o.o.p.		—	613	565	541	507
ν_{12}	C-H stretch	B_1	3133	3148	3110	3100	3084
ν_{13}	C-H stretch		3108	3120	3073	3054	3030
ν_{14}	Ring stretch		1531	1556	1506	1515	1516
ν_{15}	C-H def. i.p.		1047	1270	1250	1243	1246
ν_{16}	C-H def. i.p.		1015	1171	1081	1080	1079
ν_{17}	Ring (def. + stretch)			1040	871	820	797
ν_{18}	Ring def. i.p.		652	873	750	623	552
ν_{20}	C-H def. o.o.p.	B_2	1047	839	864	870	884
ν_{19}	C-H def. o.o.p.		768	745	712	700	674
ν_{21}	Ring def. o.o.p.		649	601	453	394	354
	Ref.		B-77MI30100	67JSP(24)133	65SA689	70AHC(12)1	76SA(A)1089

[a] Numbering and approximate description from ⟨67JSP(24)133⟩.
[b] Also 3410 N-H (A_1) stretch, 1146 NH in-plane def. (B_1), 561 NH out-of-plane def. (B_2), 1418 ring stretch (B_1).

In the case of 2- and 3-substituted heterocycles, bands attributed to three ring stretching modes are generally observed (*cf.* Table 18) with frequencies decreasing in the order furan > pyrrole > selenophene ≥ thiophene > tellurophene. The intensities of these bands are frequently increased by increasing electron withdrawing capabilities of substituents. The effect of this increased electron withdrawal will be to cause tighter conjugation, steeper charge gradients, and hence larger dipole moment changes during the vibrations. As a consequence of the occurrence of hydrogen bonding, either between molecules in concentrated solution or between molecules and solvent, the stretching frequency of the pyrrole N-H is subject to considerable variation ⟨70SA(A)269⟩. In the absence of association the NH stretching vibration is also very susceptible to the electronic character of ring substituents. This is particularly noticeable with electron-withdrawing groups such as ethoxycarbonyl, acetyl, benzoyl and cyano, and the effect of an α-substituent is always greater than that of a β-analogue ⟨65SA295, 66AJC1O7, 70MI30100⟩.

Three β-CH modes corresponding to in-plane C-H deformations are also observed (Table 19) and are probably best depicted as in (**36**), (**37**) and (**38**), although those for pyrrole will be modified as a result of interaction with the in-plane N-H deformation. The skeletal ring breathing mode (**39**) observed at *ca.* 1137 cm^{-1} for 2-substituted pyrroles and at 1015 cm^{-1} for 2-substituted furans is displaced to the 800 cm^{-1} region for thiophenes and presumably to even lower wavelengths for selenophene and tellurophene derivatives. The NH deformation mode of substituted pyrroles is responsible for a band at *ca.* 1120 cm^{-1} ⟨63AJC93⟩.

The γ-CH modes arising from out-of-plane CH deformations characterize the substitution pattern, and the observed frequencies are summarized in Table 20. For 2-substituted compounds these may be assigned as (**40**), (**41**) and (**42**). Additional characteristic bands for 2-substituted thiophenes are observed at 870–840m and 740–690s cm^{-1} ⟨67RTC37⟩.

Table 18 Monosubstituted Heterocycles: Ring Stretching Bands in the 1600–1300 cm^{-1} Region

Heterocycle	Substit	Phase	β-CH modes		Ring breathing		Ref.
			(cm^{-1}) (ε_A)	(cm^{-1}) (ε_A)	(cm^{-1}) (ε_A)	(cm^{-1}) (ε_A)	
Pyrrole	1-	CHCl$_3$	1549 + 3 (15–25)	1477 ± 7 (25–185)	—	1394 ± 10 (15–35)	66AJC289
Furan	2-	CHCl$_3$	1585 ± 26 (10–145)	1498 ± 28 (20–290)	1391 ± 14 (5–115)	—	59JCS657
Thiophene	2-	CHCl$_3$	1523 ± 9 (3–110)	1422 ± 12 (20–290)	1354 ± 7 (15–150)	1231 ± 10	59JCS3500, 70SA(A)1651
Selenophene	2-	CHCl$_3$	1532 ± 28 (5–213)	1432 ± 28 (13–290)	1332 ± 28 (6–65)	—	75JST(27)195
Tellurophene	2-D	liquid	1505	1423	1300	1224	76SA(A)1089
Pyrrole	2-	CHCl$_3$	1558 ± 9 (20–80)	1471 ± 3 (20–70)	1415 ± 8 (20–325)	—	63AJC93
Furan	3-	liquid	*ca.* 1562	*ca.* 1512	—	—	59G913, 58T(4)68
Thiophene	3-	CHCl$_3$	1512 ± 17 (90–240)	1413 ± 15 (65–135)	1365 ± 11 (5–45)	1210 ± 12	63JCS3881
Selenophene	3-	CHCl$_3$	1532 ± 28 (6–307)	1432 ± 28 (12–139)	1332 ± 28 (9–360)	—	75JST(27)195
Pyrrole	3-	CHCl$_3$	1549 ± 7 (20–180)	1491 ± 9 (55–195)	1427 ± 4 (60–160)		71PMH(4)265

Table 19 Monosubstituted Heterocycles: Characteristic Bands in the 1300–1000 cm⁻¹ Region

Heterocycle	Substit.	Phase	β-CH (36) (cm^{-1})	β-CH (37) $(cm^{-1})(\varepsilon_A)$	β-CH (38) $(cm^{-1})(\varepsilon_A)$	Ring breathing (39) $(cm^{-1})(\varepsilon_A)$	Ref.
Pyrrole	1-	CHCl₃	—	1069 ± 6 (100–240)	1027 ± 9 (35–135)	—	66AJC289
Furan	2-	CHCl₃	1220 ± 20	1158 ± 7 (70–120)	1076 ± 3 (25–65)	1015 ± 4 (60–280)	59JCS657
Thiophene	2-	CHCl₃	—	1081 ± 3 (5–15)	1043 ± 7 (15–95)	—	59JCS3500, 70SA(A)1651
Pyrrole	2-	CHCl₃	—	1088 ± 18 (25–450)	1033 ± 13 (40–400)	1137 ± 8 (25–130)	63AJC93
Selenophene	2-D	liquid	—	1027	1083	—	70AHC(12)1
Tellurophene	2-D	liquid	—	1079	1021	—	76SA(A)1089
Furan	3-	liquid	—	ca. 1156	—	—	59G913
Selenophene	3-D	liquid	—	1013	1076	—	70AHC(12)1
Pyrrole	3-	CHCl₃	—	1077 ± 3 (60–100)	1041 ± 4 (30–150)	—	B-71MI30100

Table 20 Monosubstituted Heterocycles: Characteristic Bands in the 1000–600 cm⁻¹ Region

Heterocycle	Subst.	Phase	γ-CH or β-ring modes (cm^{-1}) (ε_A)					Ref.
Pyrrole	1-	CHCl₃	—	926 ± 4 (20–140)	722 ± 2 (375–475)	—	—	66AJC289
Furan	2-	CHCl₃	—	925 ± 9	884 ± 2	808 ± 28	—	59JCS657
Thiophene	2-	CHCl₃	—	925 ± 8 (5–15)	853 ± 7 (50–100)	823 ± 20 (30–70)	752 ± 16	59JCS3500, 70SA(A)1651
Selenophene	2-D	liquid	—	873	785	845	612	70AHC(12)1
Tellurophene	2-D	liquid	—	910	856	855	748	76SA(A)1089
Pyrrole	2-	—	961 ± 8 (15–45)	929 ± 3 (10–75)	882 ± 4 (10–120)	—	—	63AJC93
Furan	3-	liquid	—	—	878 ± 8	—	741	59G913, 58T(4)68
Thiophene	3-	liquid	—	—	—	—	755s	57AK(12)239
Selenophene	3-D	liquid	—	852	810	794	609	70AHC(12)1
Pyrrole	3-	—	953 ± 4 (30–110)	—	886 ± 2 (10–20)	—	—	B-77MI30100

2.3.3.7.2 Substituent vibrations

In most cases the frequencies of substituent groups attached to these heterocycles differ little from those observed for their benzenoid counterparts. The only notable exception is the spectral behavior of carbonyl groups attached to position 2. These have attracted much attention as they frequently give rise to doublets, and occasionally multiplets. In the case of (**43**), (**44**) ⟨76JCS(P2)1⟩ and (**45**) ⟨76JCS(P2)597⟩ the doublets arise from the presence of two conformers (*cf.* Section 2.3.4.3.1), whereas for the aldehydes (**46**) the doublets are attributed to Fermi resonance ⟨75JCS(P2)604⟩. This phenomenon has also been found to occur with several compounds of types (**43**), (**44**) and (**45**), causing the observance of multiple peaks. Fermi resonance is also responsible for the splitting of the -C≡N stretching band of some cyanopyrroles ⟨70JCS(B)79⟩. In concentrated solutions 2-acylpyrroles and pyrrole-2-carboxylic esters exist as NH—O=C bonded dimers with a consequent lowering of the carbonyl stretching frequency. In dilute solution the spectra of the monomeric species show a carbonyl frequency for the 2-substituted pyrroles some 20–30 cm⁻¹ lower than displayed by the 3-isomers ⟨66AJC107, 65SA1011, 65T2197⟩. The carbonyl frequencies of 1-acyl- and 1-alkoxycarbonyl-pyrroles are some 70 cm⁻¹ higher than those of the corresponding 2- or 3-substituted compounds ⟨65T2197, 66AJC289⟩.

(**43**) Z = F, Cl   (**44**) Z = F, Cl   (**45**)   (**46**)

IR spectroscopy has been particularly helpful in detecting the presence of keto tautomers of the hydroxy heterocycles discussed in CHEC 3.01.6. Some typical frequencies for such compounds are indicated in Scheme 6. Here again the doublets observed for some of the carbonyl stretching frequencies have been ascribed to Fermi resonance.

	X = NH	O	S
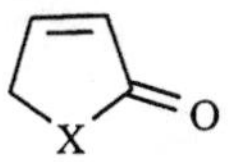 (structure 1)	1690, 1660 <57LA(604)178> (1638)	1785, 1755 <64CRV353> (1630-1620)	1678-1670 <60AK(15)499> (1607)
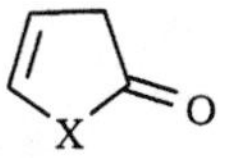 (structure 2)	–	1800 <64CRV353>	1715 <60AK(15)499> (1639)
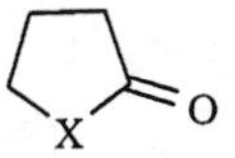 (structure 3)	1695 <57CB975>	1770 <57CB975>	–
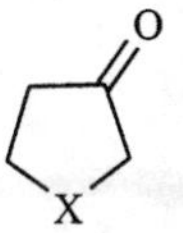 (structure 4)	1757 <58BSF350>	1770 <64T1763>	–
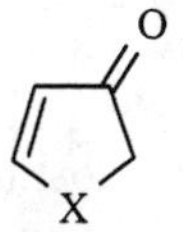 (structure 5)	1675 <63CJC625>	1712 <63HCA1259>	1680 <64CRV353>

Scheme 6 IR frequencies (cm^{-1}) for keto heterocycles (carbonyl stretching frequencies; bracketed frequencies are for C=C stretches)

2.3.3.8 Mass Spectrometry

In contrast to the other heteroatoms under consideration here, the uneven valence and the even atomic weight of the principal isotope, ^{14}N, of nitrogen ensure that pyrroles always display a molecular ion of uneven mass unless bearing a nitrogen-containing substituent. Like nitrogen, oxygen has only one principal naturally occurring isotope but sulfur with a natural isotope distribution ^{32}S/^{34}S of 25:1 ensures that thiophenes have two molecular ions, two mass units apart, of appropriate intensity ratio. A far more complex situation arises with selenium and tellurium, of which each has a number of naturally occurring isotopes, namely ^{76}Se (9.1%), ^{71}Se (7.5%), ^{78}Se (23.6%), ^{80}Se (50%), ^{82}Se (8.8%) and ^{122}Te (2.5%), ^{123}Te (0.9%), ^{124}Te (4.7%), ^{125}Te (7%), ^{126}Te (18.7%), ^{128}Te (31.8%), ^{130}Te (34.8%).

2.3.3.8.1 Parent monocycles

The principal fragmentation pathways encountered for pyrrole ⟨64JCS1949⟩, furan ⟨60BSB449⟩, thiophene ⟨59CCC1602⟩, selenophene ⟨78JHC137⟩ and tellurophene ⟨78JHC137⟩ are indicated in Scheme 7. The molecular ions are the base peaks in their respective spectra, for all except furan where the molecular ion is the strongest peak (70%) after the cyclopropenyl cation. The cyclopropenyl ion is also an important feature of the spectrum of pyrrole but much less important in the fragmentation of thiophene and selenophene, and apparently not observed for tellurophene. Another important ion in the spectra of pyrrole, thiophene and selenophene, but of low abundance for tellurophene and absent from the spectrum of furan, is (b) formed by loss of acetylene from the molecular ion. The ion (c) is much less abundant with furan and selenophene than for pyrrole or thiophene and only just detectable for tellurophene. In keeping with the much weaker nature of the carbon-tellurium and carbon-selenium bonds the spectrum of tellurophene contains ions corresponding to *M*-Te and *M*-HTe, and that of selenophene less abundant *M*-Se and *M*-HSe ions. The spectrum of tellurophene contains a very intense Te$^+$ ion and the one for selenophene a weaker Se$^+$ ion.

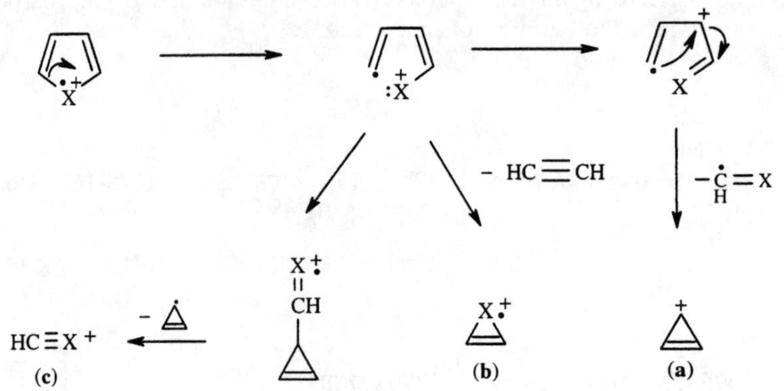

Scheme 7

2.3.3.8.2   *Substituted monocycles*

Similar mass spectra are obtained for 2- and 3-alkyl derivatives of furan ⟨66T2223⟩, thiophene ⟨59CCC1602⟩ or pyrrole ⟨64JCS1949⟩. Apart from modest contributions from ions corresponding to (a), (b) and (c) above, a major fragmentation pathway is initiated by β-cleavage of the alkyl substituent (Scheme 8). The resulting ions (d) and (e) are believed to rearrange to the common ion (f) which is generally the base peak.

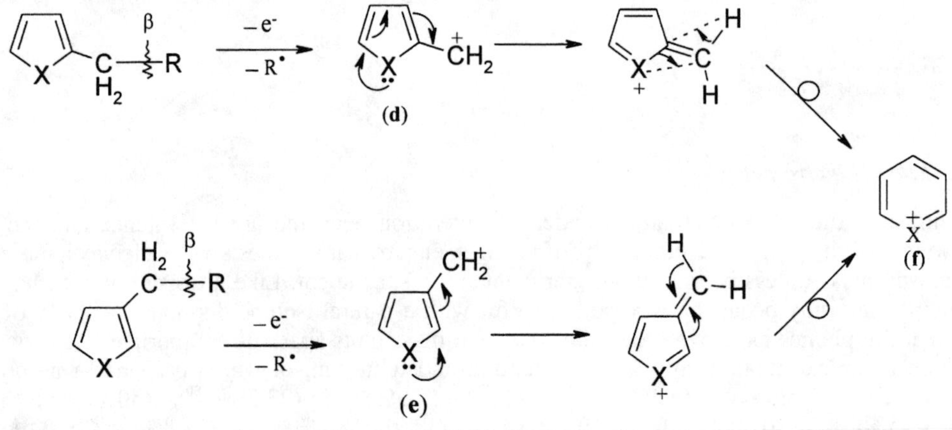

Scheme 8

N-Alkylpyrroles ⟨64JCS1949⟩ fragment in a somewhat different fashion, typified by *N*-pentylpyrrole. Apart from β-cleavage of the alkyl groups yielding the ion (g) (Scheme 9), which is believed to rearrange to the pyridinium ion (h) as it subsequently undergoes the characteristic fragmentation of elimination of HCN, the molecular ion also generates the *N*-methylpyrrole cation (i) which is the base peak. Deuterium labeling experiments ⟨65JA805⟩ indicate that the hydrogen transferred in the formation of (i) comes mainly (78%) from C-3 of the alkyl chain with the remainder supplied by C-4. A lesser amount of the pyrrole cation (j) is also formed with the N-H being mainly derived from C-2 and C-4.

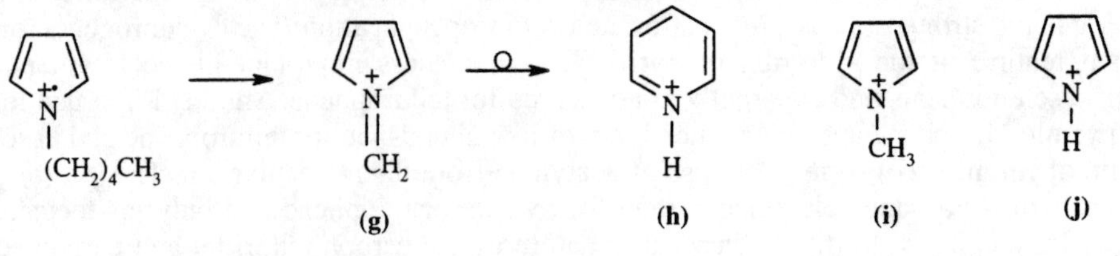

Scheme 9

2.3.3.8.3 *Benzo derivatives*

The fragmentation pathways displayed by the benzo[*b*]-fused analogues of these heterocycles parallel those of the monocyclic compounds (X = NH ⟨68AJC997⟩, O ⟨64AJC975⟩, S ⟨67AJC103⟩, Se ⟨69JCS(B)971⟩, Te ⟨70JHC219⟩) (Scheme 10). As befits aromatic molecules, the molecular ion is also the base peak except for benzotellurophene. The fragment **k** (or **k′**) has only been noted as a minor ion in the spectra of benzo[*b*]thiophene and benzo[*b*]tellurophene. The elimination of the heteroatom resulting in the formation of the benzocyclobutadiene cation radical (**l**) is most prevalent with benzotellurophene where it provides the base peak, less important with benzoselenophene and only just detectable for benzothiophene. Finally, the ions (**n**) and (**p**) are prominent in the spectra of all of the benzo[*b*] heterocycles apart from benzotellurophene. As with the monocyclic series, 2- and/or 3-alkyl derivatives undergo β-fission of the alkyl group and rearrangement of the initial resulting radical ion to the cation radical (**q**) (Scheme 11). A similar cleavage occurs when the alkyl group is attached to the benzene ring but the resulting ion is probably the isomeric tropylium one (**r**). The fragmentation of 1-methylindole proceeds *via* (**s**).

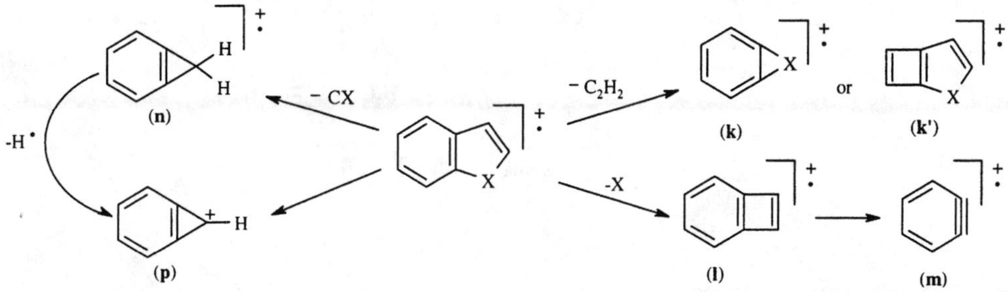

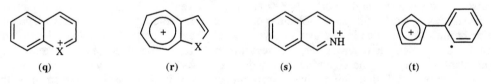

Scheme 10

The only notable fragments derived from the very stable molecular ions of carbazole ⟨64JA3729⟩, dibenzofuran ⟨64AJC975⟩ and dibenzothiophene ⟨68T3255⟩ are at *m/e* 140 ($C_{11}H_8$) and 139 ($C_{11}H_7$),, corresponding to loss of CX and HCX respectively; structure (**t**) has been suggested.

$$(q) \qquad (r) \qquad (s) \qquad (t)$$

Scheme 11

2.3.3.8.4 *Saturated compounds*

The mass spectral fragmentations of the fully saturated parent heterocycles ⟨65JA2920⟩ are indicated in Scheme 12. All exhibit appreciable molecular ions. The *M*-1 ions (**v**) from pyrrolidine, tetrahydrofuran and tetrahydrothiophene are formed by the predominant loss of an α-hydrogen atom (94%, 70% and 65%, respectively), while tetrahydroselenophene loses only a β-hydrogen. The percentage total ionization of the *M*-1 species decreases in the order NH > O > S > Se > Te. Ions (**x, x′**) corresponding to loss of C_2H_4, were generally abundant with a notable exception in the case of tetrahydrofuran.

The ions (**w**) resulting from loss of cyclopropane from the molecular ions were only observed for the sulfur, selenium and tellurium analogues. The alternative mode of fragmentation in which the hydrocarbon fragment (**y**) carries the charge provides the base peak for the tetrahydrofuran spectrum, but is only a minor feature of the spectra of the selenium and tellurium analogues. The hydrocarbon ion $C_4H_7^+$ (**z**) is a minor feature of the tetrahydrothiophene spectrum but provides the base peak of the spectra of the selenium and tellurium analogues.

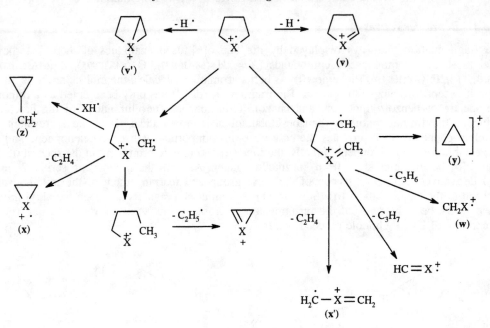

Scheme 12

2.3.3.9 Photoelectron Spectroscopy

2.3.3.9.1 Parent monocycles

The He(Iα) photoelectron spectra of the parent heterocycles are compared in Table 21. The assignments are based upon comparisons with the spectra of the reduced heterocycles, the effect of ring substituents and comparisons with results of MO calculations. In the case of pyrrole, furan and thiophene the assignments have been further supported by measurements of photoelectron angular distribution which permit evaluation of the asymmetry factors, thereby providing a useful criterion for distinguishing between π- and σ-orbitals ⟨79CPL(61)355⟩. Even so, the third π-orbital cannot be unambiguously assigned. The three π-molecular orbitals can be depicted as in Scheme 13. The energy of the π_3-orbital which extends exclusively over the carbocyclic part of the molecules is almost constant for the chalcogen heterocycles, whereas the π_2-orbital energy depends markedly on the heteroatom and increases as the electronegativity decreases. As expected, the values for π_3 are close to the ionization potentials determined by electron impact. There is also excellent agreement with the energies of the π_3- and π_2-orbitals obtained from the charge transfer spectra of these heterocycles with tetracyanoethylene ⟨75JCS(F1)2045⟩.

Table 21 Vertical Ionization Energies (eV) of π-Molecular Orbitals of Parent Heterocycles

Molecular orbital	Heterocycle				
	Pyrrole	Furan	Thiophene	Selenophene	Tellurophene
π_3	8.2	8.89	8.87	8.92	8.88
π_2	9.2	10.32	9.49	9.18	8.40
π_1		14.4	12.1	12.0	11.8
Ref.	79CPL(61)355	73CPL(22)132	73CPL(22)132	73CPL(22)132	73CPL(22)132

Scheme 13

2.3.3.9.2 Substituted compounds

Vertical ionization energies are available ⟨76JCS(P2)276⟩ for a range of α-substituted heterocycles (Table 22). Excellent linear correlations of unitary slope are obtained between corresponding π-orbital energies of the different α-substituted heterocycles, such as π_3 for thiophenes *versus* π_3 for tellurophenes. α-Methyl substitution increases the separation of π_2 and π_3 in furan, thiophene and selenophene, in agreement with the expectation that electron-releasing substituents in the α-position would exert a more pronounced destabilizing effect on the energy of the π_3-orbital than on the π_2 one.

Table 22 Vertical Ionization Energies (eV) of π_3 and π_2 Molecular Orbitals of 2-Substituted Heterocycles [a]

Substituent	Furan π_3	π_2	Thiophene π_3	π_2	Selenophene π_3	π_2	Tellurophene π_3	π_2
Me	8.37	10.13	8.43	9.23	8.40	8.96	8.20	8.43
H	8.89	10.31	8.87	9.49	8.92	9.18	8.40	8.88
CONMe$_2$	8.86	10.41	8.84	9.40	8.85	9.10	8.39	8.89
Cl	—	—	8.89	9.63	8.83	9.34	8.68	8.89
Br	—	—	8.82	9.58	—	—	8.59	8.84
I	—	—	8.52	9.47	—	—	8.34	8.52
CO$_2$H	9.16	10.72	9.14	9.73	9.19	9.45	8.62	9.15
CO$_2$Me	9.00	10.56	8.98	9.61	9.05	9.26	8.51	9.00
NO$_2$	9.75	11.13	9.73	10.21	9.64	9.88	—	—
CH$_2$Cl	—	—	8.89	9.49	—	—	—	—
CHO	—	—	9.37	9.87	—	—	—	—
CN	9.47	10.99	—	—	—	—	—	—
SMe	8.58	10.32	8.63	9.37	—	—	—	—

[a] ⟨76JCS(P2)276⟩

The spectra of the halogen-containing compounds show two bands due to the halogen lone pairs which are non-equivalent due to the different interaction of the p_x- and p_y-orbitals with the aromatic π-electron system. The peak at lower ionization energy has been assigned to the electrons occupying the p_y-orbital coplanar with the ring. The band at higher energy is thus due to electrons in the p_y-orbital which is perpendicular to the plane of the ring and overlaps the ring π-orbitals. Linear relationships are observed between the energies of the lone pair orbitals of Cl, Br and I and their corresponding Pauling electronegativities for 2-halo-thiophenes and -tellurophenes ⟨76JCS(P2)276⟩. The ionization energies of the halogen lone pairs vary with the ring decreasing in the series furan > thiophene > selenophene > tellurophene.

The energy separation between the lone pairs of a particular halogen is constant (Cl 0.38, Br 0.53, I 0.68 eV) and independent of the heteroatom. This is consistent with the p_y-orbital interacting with the π_3-orbital whose energy is constant for these chalcogen heterocycles. These energy separations are similar to those observed for halobenzenes, namely Cl 0.34, Br 0.55, I 0.84 eV. The effect of benzenoid annulation is to lower ionization potentials (Table 23). Perhaps the most noteworthy feature is that the ionization potentials of the benzo[c] heterocycles are lower than those of the benzo[b] isomers.

2.3.3.9.3 Reduced compounds

Ionization energies for fully reduced heterocycles are recorded in Table 24. Interpretation is based on a local C_{2v} symmetry of the -CH$_2$-X-CH$_2$-fragment. Mixing of the non-bonding electrons of the heteroatoms (O, S, Se, Te) with the σ-system is highest for tetrahydrofuran and gradually decreases down to tetrahydrotellurophene. Transannular interaction between the heteroatom and the double bond in 2,5-dihydro-furans and -thiophenes is indicated as 'throughbond' rather than 'through-space' ⟨78H(11)443, 74CB725, 73TL1437⟩.

Table 23 The Lowest Vertical Ionization
Potentials (eV) of Benzo Annulated Heterocycles

	Benzo[b]	Benzo[c]	Dibenzo
Pyrrole	7.76[a]	6.93[c,e]	7.60[d]
	8.38	8.60	7.99
	9.78	9.61	9.06
Furan	8.37[a]	7.63[c]	8.09[d]
	8.99	9.81	8.48
	10.40	10.30	9.35
Thiophene	8.13[a]	7.50[c]	7.93[d]
	8.73	8.95	8.34
	10.02	10.20	9.26
Selenophene	8.03[b]	—	—
	8.64		
	9.86		
Tellurophene	7.76[b]	—	—
	8.52		
	9.54		

[a] ⟨76ZN(A)1051⟩; [b] ⟨75HCA2646⟩;
[c] ⟨76JCS(P2)81⟩; [d] ⟨78ZN(A)1006⟩;
[e] These values are for *N*-methylisoindole.

Table 24 Vertical Ionization Potentials (eV) of Tetrahydro Heterocycles

Heterocycle	n_π (b_1)	C_2X (a_1)	C_2X (b_2)	Ref.
Pyrrolidine	8.82	—	—	78MI30100
Tetrahydrofuran	9.53	11.4	13.0 ± 0.5	74CPL(26)356
	9.65	—	—	—
Tetrahydrothiophene	8.42	10.9	≥ 11.9	74CPL(26)356
Tetrahydroselenophene	8.14	10.5	≥ 11.9	74CPL(26)356
Tetrahydrotellurophene	7.73	10.0	10.7	74CPL(26)356

2.3.3.9.4 Core ionization energies

Available core ionization energies for the parent heterocycles and some of their tetrahydro derivatives are listed in Table 25. The main C_{1s} band of these compounds consists of two overlapping signals about 1 eV apart ⟨77JCP(67)2596, 72TCA(26)357⟩ due to the non-equivalent carbon atoms. The core ionization energy of the heteroatom in the aromatic compounds is higher than that for the corresponding tetrahydro derivatives. Conversely the averaged C_{1s} ionization energies of the aromatic compounds are lower than for the corresponding tetrahydro derivatives. These trends are due to the net drift of charge from the heteroatoms toward the carbon atoms which occurs on going from the tetrahydro derivative to the aromatic compound.

Table 25 Core Ionization and Shake-up Energies (eV) for Some Heterocycles [a]

| | Ionization energy | | Shake up energy | |
	Heteroatom	Carbon (1s) (average)	Heteroatom	Carbon (1s)
Tetrahydrofuran	533.1 (O_{1s})	285.9	—	—
Furan	534.3 (O_{1s})	285.0	8.6 (O_{1s})	7.7
Pyrrolidine	399.7 (N_{1s})	285.7	—	—
Pyrrole	400.4 (N_{1s})	284.8	8.0 (N_{1s})	7.2
Tetrahydrothiophene	163.2 ($S_{2p}^{3/2}$)	285.1	—	—
Thiophene	163.8 ($S_{2p}^{3/2}$)	284.3	7.5 ± 0.5 (S_{2p})	5.7

[a] ⟨73CPL(22)352⟩.

2.3.4 THERMODYNAMIC ASPECTS

2.3.4.1 Intermolecular Forces

2.3.4.1.1 *Melting and boiling points*

A selection of these physical constants for pyrroles, furans and thiophenes is included in Table 32 of Chapter 2.4 and trends are discussed there (Section 2.4.4.1.1), together with data for five-membered rings containing two or more heteroatoms. A computer-assisted QSPR method has been proposed for predicting the normal boiling point for new furans, tetrahydrofurans, and thiophenes ⟨91JCI301, 98JCI28⟩.

2.3.4.1.2 *Solubility* ⟨63PMH(1)177⟩

Pyrrole, furan and thiophene have only limited solubility in water, decreasing in the order cited (6, 3 and 0.1%). The hydrogen donor property of the NH site of pyrrole and the hydrogen acceptor property of the oxygen atom of furan probably account for the much greater solubility of the first two over that of thiophene.

2.3.4.1.3 *Gas chromatography* ⟨71PMH(3)297⟩

Hydrogen bonding with polar phases (tristearine, tween, polyethylene glycol 1000) by pyrrole lengthens its retention time. Thus, *N*-methylpyrrole on these stationary phases has a shorter retention time. Ethyl or larger alkyl groups at the 2-position sterically hinder such bonding and also shorten the time. 3-Alkylpyrroles therefore have longer retention times than the 2-isomers.

Furans have been separated on columns using tricresyl phosphate, triethylene glycol and dinonyl phthalate stationary phases, often with Chromosorb as a support.

Stationary phases used for thiophenes include: pentaerythritol benzoate, polyethylene glycol adipate, tricresyl phosphate and benzyldiphenyl. Celite 545 is a useful support.

2.3.4.2 Stability and Stabilization

2.3.4.2.1 *Thermochemistry and conformation of saturated heterocycles*

Strain energies of 23.5, 24.8 and 8.3 kJ mol^{-1} were estimated for tetrahydrofuran, pyrrolidine and tetrahydrothiophene, respectively ⟨74PMH(6)199⟩. The larger sulfur covalent radius of 1.04 Å lowers angular strain.

The effect on strain energy of introducing unsaturation into these rings has been evaluated in the cases of 2,3- and 2,5-dihydrothiophene, where the additional values are 18 and 15.8 kJ mol^{-1}.

2.3.4.2.2 *Aromaticity*

All of the parent heterocycles possess some degree of aromaticity, as based upon chemical behavior such as their proclivity to undergo substitution reactions with electrophilic reagents. Quantification of the relative aromaticities of these heterocycles is less easily achieved. The wide range of potential criteria available for this purpose has been surveyed in Section 2.2.4.2.

Aromaticity indices for parent heterocycles based upon energetical and structural properties are summarized in Tables 26 and 27. Empirical and Dewar resonance energies indicated a decrease in aromaticity in the sequence benzene > thiophene > pyrrole > furan. However, the Hess-Shaad index as well as the TRE index place pyrrole before thiophene. The latter sequence is obviously doubtful and could be caused by incorrect parametrization of the sulfur atom. Similar sequence (S > N > O) are also found for the benzo[*b*] and dibenzo analogues. A somewhat different sequence is found for the benzo[*c*]fused heterocycles with isoindole > benzo[*c*]thiophene > benzo[*c*]furan. As would be anticipated, the resonance energies for the benzo[*c*] heterocycles are substantially lower than those for their benzo[*b*] isomers. A peculiarity of five-membered rings with one heteroatom, unlike six-membered rings, is that benzoannulation generally stabilizes an aromatic system, *e.g.* (Dewar REPE index): carbazole > indole > isoindole > pyrrole > indolizine. The bridge-head indolizine, with an aromaticity the lowest among benzo derivatives according to all estimates, is the only exception.

Table 26   REPE Indices of Some Five-membered Heterocycles, Estimated by
Various Methods

Compound	ERE (kJ mol^{-1})[a]	DRE (kJ mol^{-1})[b]	HSRE (β)[c]	TRE (β)[d]
Benzene	25.0	14.0	0.065	0.046
Pyrrole	15.1	3.7	0.039	0.040
Indole	21.8	12.6	0.047	—
Isoindole	—	7.4	0.029	—
Carbazole	28.4	14.3	0.051	—
Indolizine	—	2.9	0.027	—
Furan	11.3	3.0	0.007	0.007
Thiophene	20.3	4.5	0.032	0.033

[a] ⟨74AHC(17)255⟩ [b] ⟨70TCA235, 70T4505⟩ [c] ⟨72T3657, 73JA3907⟩
[d] ⟨93AHC(56)303⟩

According to structural indices $\Delta\bar{N}$ and I, the aromaticity of non-condensed heterocycles varies in the sequence thiophene > pyrrole ~ selenophene > tellurophene > furan.

Table 27   Structural Indices of Aromaticity of Some Five-membered
Heterocycles

Compound	$\Delta\bar{N}$ (%)[a]	I (%)[b]	Compound	$\Delta\bar{N}$ (%)[a]	I (%)[b]
Benzene	100	100	Selenophene	37	59
Pyrrole	37	59	Tellurophene	20	48
Furan	12	43	Phosphole	8	36[c]
Thiophene	45	66			

[a] ⟨85KGS867⟩ [b] ⟨85T1409⟩ [c] 1-Benzylderivative.

NMR has been widely invoked in assessing aromaticity. Comparison of the chemical shifts of furan, H-2 7.46 and H-3 6.41, with those observed for 4,5-dihydrofuran, H-2 6.31 and H-3 4.95 ⟨66JCS(B)127⟩, indicates there is *ca.* 1–1.5 ppm downfield shift attributable to the presence of an aromatic ring current in furan. The same effect is observed for thiophene, H-2 7.35 and H-3 7.13, and 4,5-dihydrothiophene, H-2 6.17 and H-3 5.63 ppm. The similar range of chemical shifts observed for all of the parent heterocycles may be compared with that for benzene, 7.27 δ, and further attests to their possessing appreciable ring currents.

The validity of using chemical shifts as a quantitative measure of ring currents has frequently been questioned, *e.g.* ⟨66JCS(B)127⟩. As with other approaches to a quantitative assessment of aromaticity, a major difficulty is the selection of appropriate non-aromatic models. However, the order of decreasing aromaticity arrived at in the present case, namely benzene 1, thiophene 0.75, pyrrole 0.59 and furan 0.46 ⟨65CC160, 65T515⟩ is in keeping with that derived by other means.

Several methods based on NMR spectroscopy have been devised which attempt to assess the relative magnetic susceptibilities of aromatic molecules, parallel and perpendicular to the plane of the ring. Values of the dilution shift parameter ($\delta\Delta V_m 2/3$) ⟨74JCS(P2)332⟩ gave a linear correlation with the Pauling resonance energies ⟨33JCP(1)606⟩ for benzene, thiophene and furan and permitted the estimation of resonance energies for selenophene and tellurophene of 121.3 and 104.6 kJ mol^{-1}, respectively.

In summary, most of the presently available criteria point to an order of decreasing aromaticity of benzene > thiophene > selenophene ≈ pyrrole > tellurophene > furan.

2.3.4.3   Conformation

2.3.4.3.1   *Aromatic compounds*

Comparison of the data in Table 28 with that in Table 2 shows that the internal bond angles of the heterocyclic ring do not change appreciably on annulation. However, the bond lengths are increased and this is particularly noticeable in the case of the C-X bond (*cf.* Table 29). It may be noted, however, that in no case does the length of the C(6)-C(6′) bond reach the 1.497 Å length ⟨61NAT(191)593⟩ of the interannular bond of biphenyl, implying that the central heterocyclic ring retains some modicum of aromaticity. A most intriguing feature of these molecules is that they adopt a slightly bow-shaped configuration with small dihedral angles between the planes of the five-membered and benzenoid rings. The observed values are carbazole 1.0°, dibenzofuran 1.12°, dibenzothiophene 0.4–1.2°, and dibenzoselenophene 0.5–1.2°.

Table 28 Comparison of Bond Lengths and Angles of Dibenzo Heterocycles

Bond Length (Å)	X = NH	X = O	X = S	X = Se	X = Te
C(1)-C(2)	1.403	1.385	1.384	1.395	1.397
C(2)-C(3)	1.372	1.388	1.384	1.371	1.381
C(3)-C(4)	1.393	1.385	1.385	1.377	1.386
C(4)-C(5)	1.392	1.389	1.370	1.380	1.375
C(5)-C(6)	1.391	1.384	1.392	1.395	1.403
C(6)-C(6′)	1.477	1.481	1.441	1.453	1.460
C(1)-C(6)	1.408	1.393	1.409	1.398	1.394
C(1)-X	1.393	1.404	1.740	1.899	2.087
X-H	1.02	—	—	—	—
Bond angle (degrees)					
C(1)XC(1′)	108.3	104.1	91.5	86.6	81.7
XC(1)C(6)	109.7	112.3	112.3	112.4	112.1
C(1)C(2)C(3)	115.6	116.7	117.8	118.7	119.1
C(2)C(3)C(4)	123.9	120.9	121.6	121.1	120.6
C(3)C(4)C(5)	120.1	121.9	120.5	120.6	120.1
C(4)C(5)C(6)	117.9	117.9	120.0	120.3	120.9
C(5)C(6)C(1)	120.6	119.6	118.7	118.1	118.0
C(6)C(1)C(2)	121.9	123.0	121.6	121.6	121.3
C(1)C(6)C(6′)	106.1	105.3	111.9	114.3	117.1
Ref.	69BCJ2174	72AX(B)1002	70JCS(A)1561	70AX(B)628	75IC2639

Table 29 Comparison of C-X Bond Lengths for Parent Heterocycles and their Dibenzo Derivatives

X	Parent heterocycle: C-X (Å)	Dibenzo derivative: C-X (Å)
NH	1.383	1.414
O	1.362	1.404
S	1.714	1.740
Se	1.855	1.899
Te	2.055	2.087

In benzo[b]thiophene itself, both rings are coplanar, but the introduction of a substituent on the thiophene ring usually causes the two rings to be inclined to each other at about 1.0° ⟨74AX(B)2058, 84AX(A)C277⟩.

For benzo[b]thiophene-1,1-dioxide, its 2-bromo, and 2,3-dimethyl derivatives ⟨87AX(C)2421⟩ as well as for the 2- and 3-methyl analogues ⟨88AX(C)498⟩, the bond lengths and angles are approximately similar to their non-oxidized precursors, the largest difference being in the C(2)=C(3) bond which is much smaller in the dioxide. As before, the two rings are tilted towards each other at about 1°. The geometries are also in good agreement with 1,2,3,5-tetramethylbenzo[b]thiophenium tetrafluoroborate (**47**) ⟨81JCS(P2)266⟩. The bonds about the sulfur atom are pyramidal in nature with the methyl group being out of the plane of the molecule. In comparison to benzo[b]thiophene derivatives, the salt has longer C-S and C(3)-C(3a) bonds but a shorter C(2)=C(3) bond indicating more bond alternation here due to the lack of delocalization.

Large distortions of the heterocyclic ring are encountered in [2,2]heterophanes. In the furanonaphthalenophane (**48**) the furan ring is completely planar but the extra annular bonds are directed about 7° out of the plane of the furan ring ⟨78TI641⟩. While the non-bridged portion of the naphthalenoid ring is planar, the portion which is bridged to the furanoid ring through its 1 and 4 carbon atoms is puckered and boat shaped. A similar situation is observed in the furanopyridophane (**49**) ⟨75JHC433⟩. In 2,5-furanophane (**50**) ⟨84BCJ3552⟩ the furan ring is slightly puckered. The furan ring takes an envelope form with the oxygen atom away from the benzene ring: the dihedral angle formed between the planes by the O, C-2, C-5 atoms and C-2, C-3, C-4, and C-5 is 3.3°.

(47)   (48)   (49)   (50)

X-Ray crystallographic studies show that the three isomeric bithienyls are planar in the solid state ⟨68AX(B)467⟩. However, in the vapor phase the principal conformation of 2,2′-bithienyl is indicated by electron diffraction to be a non-planar one with an angle of twist of 34° ⟨58ACS1671⟩ relative to the planar *transoid* conformation (51) adopted in the solid state. However, 2-(2-furyl)pyrrole and 2-(2-thienyl)pyrrole preferentially adopt a *cis* planar conformation in solution ⟨81JCS(P2)127⟩. Bulky substituents in the positions adjacent to the interannular bond permit the separation of the resulting stereoisomers. Typical examples are provided by (52) and (53) ⟨75CS(7)173⟩, whose X-ray structures ⟨75CS(7)204, 76CS(9)66⟩ show that *cis* skew conformations are adopted in the solid state. Circular dichroism spectra of these compounds in solution and the solid state are very similar and comparable with those of the necessarily *cisoid* dithienothiepins (54) ⟨76CS(10)120⟩. Analogous behavior has been observed for 3,3′-biselenenyls where both (55) and its thiophene analogue have been resolved ⟨75CS(7)131⟩.

(51)   (52)   (53)

(54)   (55)

Comparable examples of restricted rotation involving pyrroles were encountered much earlier and resolutions effected, amongst others, of compounds (56) ⟨53JOC1413⟩, (57) ⟨31JA2353⟩ and (58) ⟨31JA3519⟩. A consequence of the greater separation of adjacent substituents on these five-membered rings relative to six-membered ones is a lower barrier to rotation (see Scheme 14). Thus, the rotational barrier for pentaarylpyrroles is about 75 kJ mol^{-1} lower than for the corresponding hexaarylbenzenes ⟨81JOC1499⟩.

(56)   (57)   (58)

The conformational preferences of 2- and 3-monosubstituted derivatives of furans, thiophenes and pyrroles have been the subject of *ab initio* MO studies ⟨77JCS(P2)1601, 78JA3981, 79JA311⟩. Of the systems considered, experimental information from microwave spectroscopy is available for 2- and 3-methyl derivatives of furan ⟨69JCP(51)403, 70ZN(A)570, 71BCJ2344⟩ and thiophene ⟨70MI30100, 74JSP(42)38⟩ and agrees with prediction. The preferred conformations are (**59**), (**61**), (**60**) and (**62**) with barriers to rotation of 5.0, 4.6, 2.3 and 3.1 kJ mol^{-1} respectively, which are somewhat higher than those calculated.

Gas electron diffraction of 2,5-dimethylthiophene ⟨93JST(301)107⟩ show one of the C–H bonds of the methyl substituent to be *cis* with respect to the S–C bond of thiophene. The C(2)–Me distance is smaller than a normal C–C single bond and quite close to the C–C bond length in non-conjugated alkenes. The C–S bond length here is larger than in thiophene as is also the C–S–C angle. In general, STO-3G* calculations reproduce most of the experimental differences between 2,5-dimethylthiophene and thiophene.

Early work on the experimentally established conformational preferences in solution for a variety of other 2-substituted heterocycles is summarized in Table 30. Most of these conclusions have been deduced either from dipole moment measurements in benzene or by the use of lanthanide induced shifts for chloroform solutions. The aforementioned MO studies correctly predict the preferred conformations, (**63**, R = H) or (**64**, R = H), of pyrrole-2-carbaldehyde, thiophene-2-carbaldehyde and furfural in the gas phase.

In furfural the *anti-(E)*-conformer is the most stable in the gas phase, while the *syn-(Z)*-conformer is the one preferred in the liquid phase and is the only rotamer present in the solid state. This behavior has been attributed to the difference in polarity [(*E*) is the less polar conformer] and to the small difference in stability between the conformers. The calculated preference of 4.1 kJ mol^{-1} for the *anti* isomer and barrier to rotation of *ca.* 24.7 kJ mol^{-1} compare favorably with the values that have been estimated from microwave spectra, 3.1 and 36 kJ mol^{-1} ⟨66ZN(A)1633⟩, and from far-IR spectroscopy, 8.5 and 25 kJ mol^{-1}, respectively ⟨67SA(A)891⟩. These conformational effects have also been rationalized by MINDO/3 calculations, which also took solvent effects into account ⟨81JHC1055⟩. The presence of other ring substituents would be expected to modify the *syn-anti* preference *inter alia* by modifying the dipole moment of the ring, and this aspect has been extensively studied for pyrrole-2-carbaldehyde ⟨75JCS(P2)333, 75JCS(P2)337⟩.

(**59**) X=O

(**60**) X=S

(**61**) X=O

(**62**) X=S

(**63**) *anti* (E)

(**64**) *syn* (Z)

Table 30 Conformational Preference of 2-Substituted Five-membered Heterocycles

Substituent	Pyrrole	Furan	Percent of syn form (**64**) Thiophene	Selenophene	Tellurophene	Phase
CHO	Mainly[a]	Minor[c]	Mainly[f]	—	—	Vapor
	100[b]	83[d]	100[d]	80[d]	72[d]	C$_6$H$_6$
		70–75[e,i]	99[e,i]	98[e]	96[e]	CDCl$_3$
COMe	100[b]	51[d]	92[d]	70[d]	47[d]	C$_6$H$_6$
		53[e]	79[e]	87[e]	90[e]	CDCl$_3$
CONMe$_2$	—	11[d]	56[d]	33[d]	11[d]	C$_6$H$_6$
		5[e]	2[e]	5[e]	—	CDCl$_3$
CO$_2$Me	Predominates[g]	55[d,h]	33[d]	59[d]	42[d]	C$_6$H$_6$
SMe	—	48[d,h]	51[d]	59[d]	65[d]	C$_6$H$_6$

[a] ⟨74JST(23)93⟩. [b] ⟨74JCS(P2)1318⟩. [c] ⟨66ZN(A)1633⟩. [d] ⟨77JCS(P2)775⟩
[e] ⟨74T4129⟩. [f] ⟨73JST(17)161⟩. [g] ⟨80JCS(P2)1631⟩. [h] ⟨80SA(A)633⟩. [i] ⟨82T1485⟩.

Although theoretical studies indicate very small gas-phase energy differences between the *syn* and *anti* conformers of the 3-carbaldehydes of furan, thiophene and pyrrole with a slight preference for the *syn* conformer (**65a**, R = H), in chloroform solution the furan- and thiophene-3-carbaldehydes adopt the *anti* conformers (**65b**, R = H) to the extent of 100 and 80%, respectively ⟨82T3245⟩. However, *N*-substituted 3-(trifluoroacetyl)pyrroles (**65**, R = CF$_3$) exist in solution as mixtures of rotational isomers ⟨80JCR(S)42⟩.

(65a) *syn* (Z)   **(65b)** *anti* (E)   **(66)**

The situation in 2-acetylfuran (**63**, R = Me) resembles qualitatively that of the 2-formyl (**63**, R = H). Accurate NMR measurements ⟨85JCS(P2)1839⟩ at low temperature and in dimethyl ether performed on 2-acyl derivatives of furan have enabled direct detection of the two conformers. At 173K the (*E*)-conformer population for 2-acetylfuran amounts to 53%.

In 2,4-dinitrophenylhydrazone of 2-acetylfuran, the (*E*)-conformer (**66**) is predominant (70%) and its stability has been explained by the possible formation of an intramolecular hydrogen bond between the N-H proton and the furan oxygen atom ⟨92MI 205-02⟩. Such an intramolecular hydrogen bond is supported by the crystal structure of ethyl α-(*p*-tolylhydrazono)-2-furopropionate ⟨88AX(C)1252⟩.

In the 2-carboxylic acids the oxygen of the OH group faces the heteroatom in all cases except furoic acid. Progressive shortening of the C(2)-C(CO₂H) bond, presumably indicating an increase in conjugative interaction, occurs in the sequence thiophene > selenophene > tellurophene > furan, which corresponds to the order of decreasing heterocyclic aromaticity.

NMR-evaluated free energies of activation for rotation about the C(O)-N bond in furan-, pyrrole- and thiophene-2- and -3-*N*,*N*-dimethylcarboxamides ⟨76JOC3591, 77TI337⟩ are summarized in Table 31. The barrier to rotation arises from resonance interaction between the lone pair of electrons on the nitrogen atom and the electronegative oxygen of the carbonyl group. An electron donor attached to this carbonyl group will reduce this interaction and hence lower the rotational barrier. As expected, the values show that the order of electron donation is pyrrolyl > thienyl > furyl, and that donation is greater from the 2- than the 3-position. All are electron donating compared with benzene.

The rotational isomerism in carbonyl-containing five-membered heterocycles has been extensively reviewed ⟨77JCS(P2)1601, 81RCR336, 84KGS579, 87AHC(41)75⟩. Generally, in carbonyl containing heterocyclic compounds the extent of π-conjugation between the carbonyl group and the ring is the main factor in determining the energy of rotation.

Table 31 Free Energies of Activation
(kJ mol⁻¹) for Rotation of Five-membered
Heterocyclic *N*,*N*-Dimethylcarboxamides (*cf.* for
N,*N*-dimethylbenzamide E = 66.5 kJ mol⁻¹)ᵃ

Heterocyclic Ring	Substituent position	
	2	3
Furan	63.1	63.5
Thiophene	60.6	—
Pyrrole	60.2	61.0

ᵃ ⟨76JOC3591⟩

Inspection of Scheme 14 shows that, in contrast to benzene, *ortho* substituents should experience less steric interference in these five-membered heterocycles and for any individual ring system this will be smaller for 2,3- rather than 3,4-disubstitution. The tetrasubstituted pyrrole (**67**) has a planar ring but most substituent atoms deviate significantly from this plane ⟨72JCS(P2)902⟩. The acetyl group is twisted about 15° out of the ring plane whereas the ester group is only twisted 1°. Distortions of the ring are also obviously present in 3,4-di-*t*-butylthiophene (**68**) ⟨80CC922⟩, where in particular the C(3)-C(4) bond length is 1.667 Å in contrast to 1.423 Å in the parent heterocycle, and the Buᵗ-C-C-Buᵗ bond angles are increased to 133° from their preferred value of 124°. No appreciable distortions of the thiophene ring are observed for (**69**) ⟨78CSC703⟩, the overcrowding being relieved by the nitro groups being twisted out of the plane of the ring by 37° and 44° and the C-N bonds being displaced out of the plane of the ring by 5.3° and 6.7° in opposite directions.

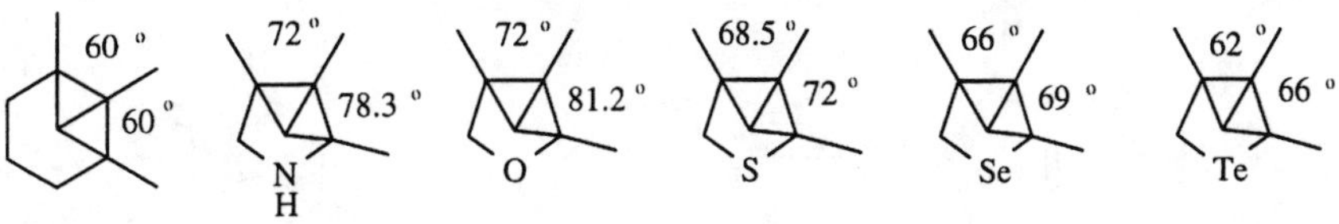

Scheme 14

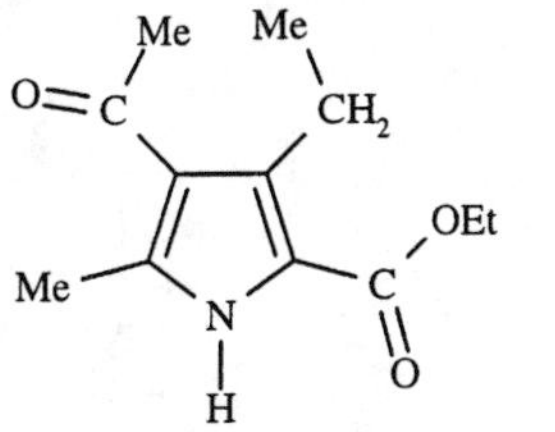

(67) 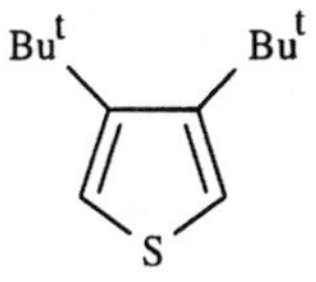   (68) 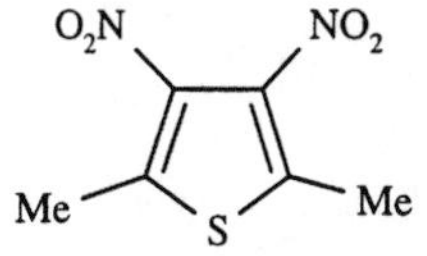   (69)

2.3.4.3.2 *Reduced ring compounds*

Some reduced heterocyclic rings are non-planar. In 2,3-dihydrofuran and 2,3-dihydrothiophene, the C-2 methylene group is out of the plane of the other ring atoms with barriers to ring inversion of about 1 and 4 kJ mol^{-1} respectively ⟨72JCP(56)5692, 73JCP(59)2249, 86JST(147)255⟩. The higher barrier for the sulfur compound is presumably due to a decrease in the ring strain forces relative to those in the dihydrofuran. Torsional forces are comparable for both molecules and tend to overcome the lower ring strain forces of the dihydrothiophene and pucker the ring to a larger degree. 2,5-Dihydrofuran ⟨67JCP(47)4042, 84JA20, 93JSP(160)158⟩ and 2,5-dihydrothiophene ⟨69SA(A)723⟩ are planar, but 2,5-dihydropyrrole is non-planar with an asymmetric double minimum potential on account of the axial-equatorial conversion of the N-H bond by ring inversion ⟨72CPL(12)499⟩. The equatorial form is favored energetically.

The fully reduced ring systems are non-planar. For tetrahydrofuran, analysis of IR ⟨69JCP(50)124⟩, dipole moment, microwave ⟨69JCP(50)2446⟩ and ^{1}H NMR ⟨74JMR(16)136⟩ data, together with *ab initio* MO calculations ⟨75JAI358⟩, indicates a freely pseudorotating system with 10 twist and 10 envelope conformations. In the envelope conformation, one of the ring atoms is out of the plane of the other four atoms. Intermediate between these envelope conformations are half chair twist conformations (only three adjacent ring atoms coplanar), slightly more stable than the envelope and with a barrier to pseudorotation of about 0.7 kJ mol^{-1}. Similarly, pyrrolidine is a free or only slightly restricted pseudorotator ⟨58JCP(29)966⟩, but the appreciably increased size of the heteroatom in tetrahydrothiophene ⟨79MI30102, 80CJC2340, 80OMR(13)282⟩ and tetrahydroselenophene ⟨78JCP(69)3714, 79JMR(36)113⟩ results in higher barriers to pseudorotation and these molecules preferentially adopt twisted conformations.

2.3.5 TAUTOMERISM

All tautomeric equilibria of these heterocycles involve one or more non-aromatic tautomers. An important factor in determining the extent to which such non-aromatic tautomers are involved is the magnitude of the potential loss of resonance energy.

2.3.5.1 Annular Tautomerism

Tautomerism not involving a functional group can only occur with pyrroles. There is no authenticated case of a non-annulated pyrrolenine tautomeric form (70a) or (70b) predominating, but the pyrrolopyrimidinedione (71) adopts the tautomeric form shown ⟨77JOC1919⟩, possibly because this minimizes steric repulsions between the adjacent *t*-butyl and acetate groups. The potential loss of resonance energy is much less for tautomerism of an indole to the corresponding indolenine (73). Even so, most indoles exist in the indole form but the introduction of a strong electron donating group at C-2 can tip the balance. Thus, the equilibrium is progressively shifted from (72) to (73) as the substituent is changed from -SEt to -OEt to -NC$_5$H$_{10}$, due to increasing electron donating ability as implied by the zwitterionic canonical form (74) ⟨70T4491, 71T775⟩.

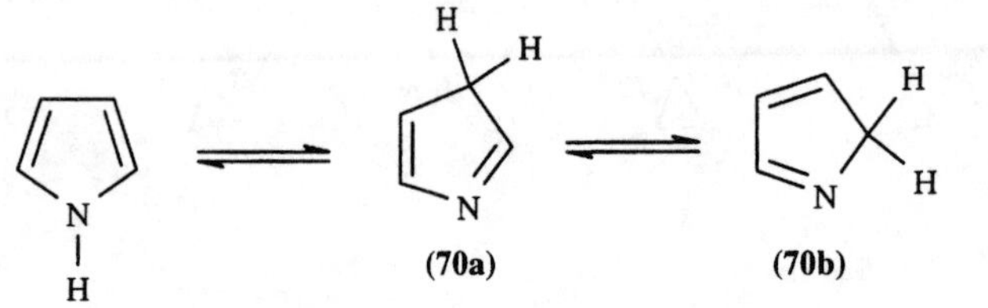

(70a)   (70b) 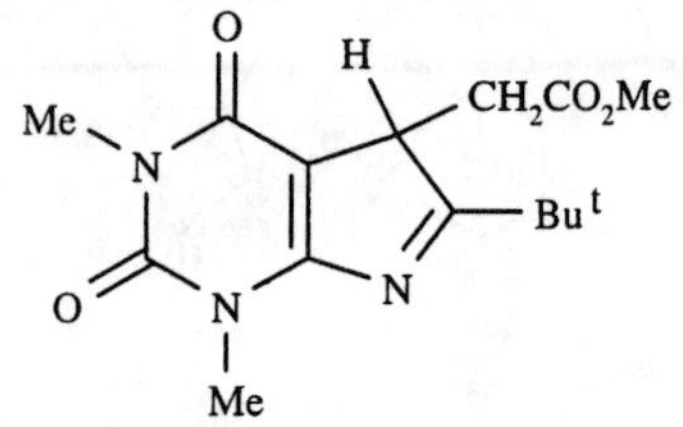   (71)

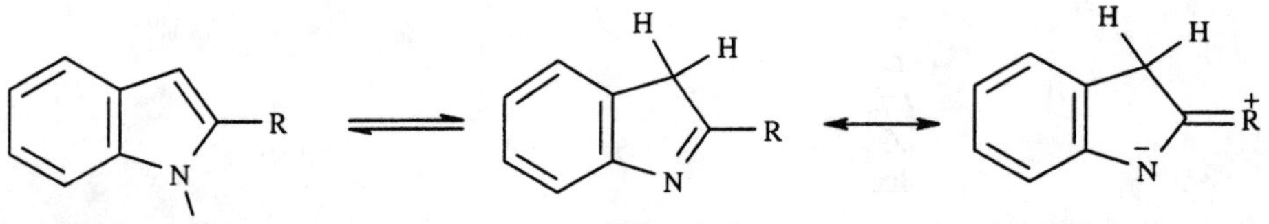

(72)   (73)   (74)

Although isoindole itself exists preferentially as such ⟨73JCS(P1)1432⟩, in accordance with predictions of MO calculations ⟨64JA4152, 67TL3669⟩, the lower resonance energy of isoindole often results in the isoindole–isoindolenine equilibrium favoring the latter species in substituted derivatives. Thus, the presence of an aryl group at C-1 capable of conjugating with C=N results in increasing proportions of the isoindolenine tautomer (75) as the *p*-substituent (R) is changed from hydrogen (9%) to methoxy (31%) to dimethylamino (50%) in CDCl₃, ⟨64JA4152⟩.

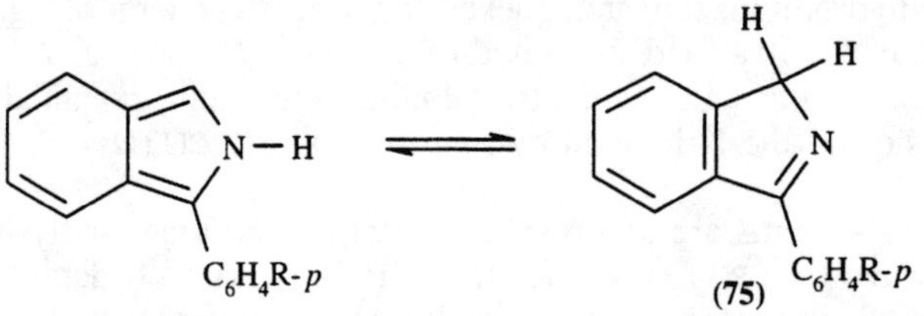

2.3.5.2 Compounds with a Potential Hydroxy Group

Although *N*-hydroxypyrroles possess in principle several tautomeric forms, *e.g.* (76), (77) and (78), only the *N*-hydroxy form (76) has been observed for 1-hydroxy-2-cyanopyrrole ⟨73JOC173⟩. In the case of 1-hydroxyindoles, where the potential loss of aromatic resonance energy will be much less, both tautomers (79) and (80) coexist in solution with the relative proportions being dependent on the solvent ⟨67BSF1296⟩.

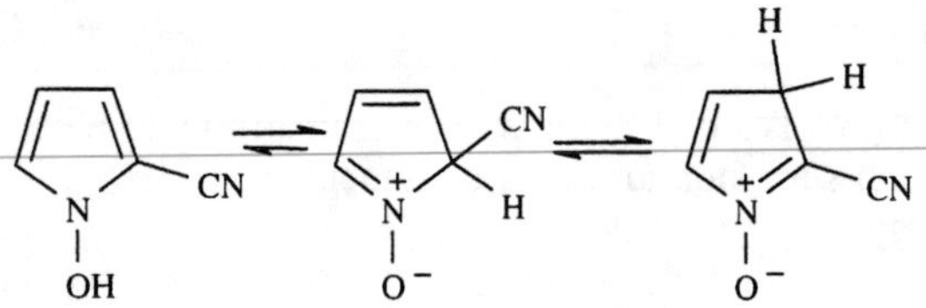

(76)   (77)   (78) 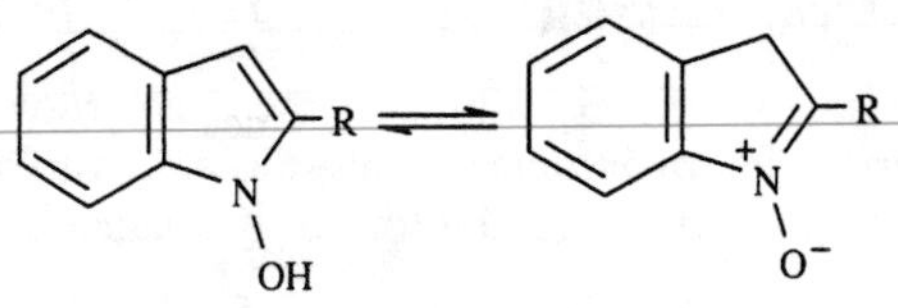   (79)   (80)

Potential *C*-hydroxy compounds usually exist as the oxo tautomers, unless the hydroxy tautomer is appreciably stabilized by electron withdrawing or chelating substituents. The tendency for enolic hydroxy compounds to revert to the oxo form is easily comprehended by reference to simple aliphatic ketones where the keto–enol tautomeric equilibrium constants are of the order of 10^8. In the heterocycles under consideration this tendency will be in opposition to the attendant loss of aromatic conjugation energy which will increase in the order furan << thiophene ≤ pyrrole. For the 2-hydroxy compounds (81) some extra stabilization of the oxo tautomers (82) and (83) will come from the resonance energy of the -X-C(=O)- group, which by analogy with open-chain groups should increase in the sequence thiolester, ester << amide.

Variation in tautomeric behavior is observed with change of heteroatom. Thus, 2-hydroxyfurans exist predominantly as the tautomer (**83**) although the energy of activation for the conversion of (**82**) to (**83**) is sufficiently high to permit the isolation of (**82**) ⟨64CRV353⟩. 2-Hydroxypyrroles behave similarly but equilibration between (**82**) and the more stable (**83**) occurs in polar solvents at room temperature ⟨65JOC3824, 71OMR(3)7, 80HCAI21⟩. For thiophenes the ratio of (**82**) to (**83**) depends upon the substitution at C-5 ⟨63T1867, 64AK(22)211, 67T3737, 69AK(29)427⟩. Whereas (**83**) predominates in the parent molecule, the introduction of a 5-aryl group causes the equilibrium to shift largely toward (**82**) in the solid state or carbon tetrachloride solution, and in methanol (**82**) is accompanied by some 25–30% of the hydroxy tautomer (**81**). Tautomer (**83**) also predominates in the case of 2-hydroxyselenophene ⟨71BSF3547⟩.

In the benzo[*b*] (**84**) and benzo[*c*] (**86**) heterocycles, where the loss of heterocycle ring resonance energy on tautomerism to (**85**) and (**87**) will be much less than for the non-annulated heterocycle, the latter oxo tautomers are preferred. The hydroxy form of 2-hydroxybenzo[*b*]thiophenes (**84**; X = S) was detected by ^{1}H NMR spectroscopy when its trimethylsilyl precursor was hydrolyzed (CD$_3$COCD$_3$/D$_2$O/DCl). At 25°C, the signal at 6.45 ppm which is assigned to the alkenic proton at the 3-position disappears within 15 min and is replaced by a broad singlet at 4.2 ppm, indicating rapid conversion to the keto form (**85**; X = S) ⟨89JA5346⟩.

(**81**)   (**82**)   (**83**)

X = O ⟨69JOC4164⟩
X = S ⟨70JCS(C)1926⟩
X = Se ⟨76JCS(P1)2452⟩
X = NH ⟨68T6093⟩

(**84**)   (**85**)

X = O, S, Se ⟨70JA4447⟩
X = NH ⟨71PMH(4)265, p.289⟩

(**86**)   (**87**)

Most 3-hydroxyfurans ⟨65HCA1322, 76CS(10)126⟩ and 3-hydroxybenzofuran exist exclusively in the keto form at equilibrium (**89**, **91**; X = O), but the enolic forms (**88**, **90**; X = O) were generated transiently in solution by hydrolysis of their trimethylsilyl derivatives ⟨89JA5346⟩. Most 3-hydroxy-pyrroles ⟨70LA(736)1⟩ and selenophenes ⟨76CS(10)126, 71BSF3547⟩ also exist preferentially as the oxo tautomer (**89**), but introduction of an acyl function into position 2 of furan ⟨65HCA1322⟩ or pyrrole ⟨67AJC935⟩ causes the equilibrium to favor the hydroxy tautomer (**88**), a consequence of intramolecular hydrogen bonding between the hydroxy and acyl groups and of their mesomeric interaction. The importance of the latter factor is indicated by the existence of the potential 3-hydroxy-4-acyl analogues in the oxo form.

(**88**)   (**89**)   (**90**)   (**91**)

The 3-hydroxy form is considerably more favored in the more aromatic thiophene system. Thus, 2-methyl-3-hydroxythiophene exists at equilibrium as a mixture of the keto form (**92**; R^1 = Me, R^2 = H) (20%) and enol form (**93**; R^1 = Me, R^2 = H) (80%) ⟨86HC(44/3)1⟩. For the *t*-butyl derivative (R^1 = Bu^t, R^2 = H) it was 45% of (**92**) and 55% of (**93**). For the 2,5-dimethyl derivative (R^1 = R^2 = Me), the keto–enol equilibrium constant in cyclohexane was 2.61, while in methanol it was 1.21. 3-Hydroxythiophene was prepared from the trimethylsilyl derivative and was sufficiently stable to obtain its NMR in a variety of solvents ⟨89JA5346⟩. In CCl$_4$, both tautomers were observed, contrary to an earlier report where only the keto form was observed ⟨86TL5155⟩. Ketonization of 3-hydroxyfuran in water or DMSO affords over 99% of the keto form at equilibrium ⟨87PAC1577⟩. Electron withdrawing groups such as alkoxycarbonyl generally cause the hydroxy form to predominate ⟨65T3331⟩.

(92)   (93)

A closely similar pattern of behavior is also found for the corresponding benzo-annulated derivatives. Thus, 3-hydroxybenzofuran ⟨66CB3076⟩ and 3-hydroxyindoles ⟨58JCS1217⟩ adopt the oxo form (91), but enolize to (90) when an acetyl group is present at position 2 ⟨65T3331, 60JA1187⟩.

Equilibrium and rate constants for the keto-enol tautomerization of 3-hydroxy-indoles and -pyrroles are collected in Table 32 ⟨86TL3275⟩. The pyrroles ketonize substantially (10^3–10^4 times) faster than their sulfur or oxygen analogues, and faster still than the benzo-fused systems, indole, benzofuran, and benzothiophene. The rate of ketonization of the hydroxy-thiophenes and -benzothiophenes in acetonitrile–water (9:1) is as follows: 2-hydroxybenzo[*b*]thiophene > 2,5-dihydroxythiophene > 2-hydroxythiophene > 3-hydroxybenzo[*b*]thiophene > 3-hydroxythiophene. 3-Hydroxythiophene does not ketonize readily in the above solvent system, but in 1:1 acetonitrile–water, it ketonizes 6.5 times slower than 2-hydroxythiophene ⟨87PAC1577⟩.

The solvent has a significant effect on the equilibrium constants ⟨89JA5346⟩. Generally speaking, increasing solvent polarity favors the hydroxy tautomer (90) which becomes the almost exclusive species in 2-acetyl ⟨65T3331⟩ and 2-aryl ⟨76CS(9)216⟩ derivatives even in non-polar media.

Table 32 Rate and equilibrium constants for ketonization of hydroxy-heterocyclic compounds at 25°C[a]

Compound	k_{H+} (mol^{-1} sec^{-1})	K_{enol} (H_2O)	K_{enol} (DMSO)
3-Hydroxypyrrole	2.38×10^4	0.13	$> 10^2$
1-Me-3-hydroxypyrrole	9.65×10^3	0.18	$> 10^2$
3-Hydroxyindole	3.44	0.086	28.3
1-Me-3-hydroxyindole	5.82	0.303	11.4
1-Hydroxyindene	9.03×10^2	1.85×10^{-8}	$< 2 \times 10^{-2}$

[a] ⟨86TL3275⟩

2.3.5.3 Compounds with Two Potential Hydroxy Groups

2,3-Dihydroxy-furan ⟨71T3839⟩, -pyrrole ⟨53JOC382, 69JOC3279⟩ and -thiophene ⟨71T3839⟩ all adopt the tautomeric structure (95) rather than (94). 2,3-Dihydroxyindole adopts the structure (96) whereas the behavior of 2,3-dihydroxybenzofuran is strongly temperature dependent. At 20°C it exists solely as the tautomer (97) but as the temperature is raised the equilibrium is shifted progressively toward (96) whose proportion reaches 95% at 100°C ⟨68M2223⟩.

(94)   (95)   (96)   (97)

The potentially 2,4-dihydroxy derivatives of furan and thiophene exist in the solid state and in polar solvents as the monoenols (98) ⟨71T3839⟩. However, in non-polar solvents the furan derivatives exist predominantly in the dioxo form (99). The 2,5-dioxo structure (100) is well established for X = O, NR,

S and Se ⟨71BSF3547⟩ and there is no evidence for intervention of any enolic species. The formal tautomer (101) of succinimide has been prepared and is reasonably stable ⟨62CI(L)1576⟩.

Examples of the potential 3,4-dihydroxy heterocycles are presently restricted to furan and thiophene. Although the parent 3,4-dihydroxyfuran apparently exists as the dioxo tautomer (102), derivatives bearing 2-alkyl or 2,5-dialkyl substituents prefer the keto-enol structure (103) ⟨71T3839, 73HCA1882⟩. The thiophene analogues also prefer the tautomeric structure (103), except in the case of the 2,5-diethoxycarbonyl derivative which has the fully aromatic structure (104) ⟨71T3839⟩.

2.3.5.4 Compounds with Potential Mercapto Groups

The mercapto form is much more strongly favored than is the hydroxy form for the corresponding oxygen compounds. A pertinent comparison in this respect is the greatly reduced inclination of enethiols to tautomerize to the corresponding thiocarbonyl compounds, in contrast to the facile ketonization of vinyl alcohols.

2-Mercapto derivatives of furan, thiophene, selenophene ⟨77ACS(B)198⟩ and pyrrole ⟨72AJC985⟩ all exist predominantly in the thiol form. 2-Mercaptobenzothiophene is also a thiol ⟨70JCS(C)2431⟩ whereas 2-mercaptoindole is mainly indoline-2-thione (105) ⟨69CPB550⟩. This is not due solely to the greater resonance energy associated with thioamides since *N*-alkylation (which disallows hydrogen bonding) shifts the tautomeric equilibrium back to the thiol form.

The known 3-mercapto derivatives of furan, thiophene, selenophene ⟨77ACS(B)198⟩, benzothiophene ⟨70JCS(C)243t⟩ and indole ⟨69TL4465⟩ all exist as the 3-thiol tautomers.

2.3.5.5 Compounds with Potential Amino Groups

MO calculations predict that 2- and 3-amino derivatives of furan and pyrrole will preferentially exist as such rather than adopt tautomeric imino forms ⟨70JA2929⟩. These conclusions appear to be borne out for 2-aminofurans ⟨66CB1002⟩, as well as 2-amino- ⟨54HCA1256⟩ and 3-amino-pyrroles ⟨64UP30100, 70JCS(C)1658⟩. 3-Aminofuran still eludes isolation but 3-acylaminofurans are known ⟨82T2783⟩. The preference for the amino form is also observed for 2-amino- ⟨69JHC147, 71T5873⟩ and 3-amino- ⟨73JHC1067⟩ thiophenes, 2-amino-3-ethoxycarbonyl- ⟨71T5873⟩ and 3-amino-benzofurans ⟨73JPR779⟩, 2-aminobenzo[*b*]thiophene ⟨65JOC4074⟩, 3-aminoindoles ⟨69BSF2004⟩ and 1-aminoindolizines ⟨65JCS2948⟩. Among the few well-established exceptions is the aminobenzo[*c*]thiophene (106) which preferentially exists as the imine (107) ⟨64JOC607⟩. The existence of 2-aminoindole (108) as the tautomer (109) ⟨56HCA116, 71T775⟩ has been mentioned earlier and is a consequence of the appreciable resonance energy of the amidine group and the low resonance energy of the indole pyrrole ring. In the corresponding *N*-methyl compound (110) the equilibrium is displaced toward (111) and this displacement increases with increasing solvent polarity; however, replacement of the 2-amino group by an *N*-alkylamino causes (112) to predominate over (113).

(106)   (107)   (108)   (109)

(110)   (111)   (112)   (113)

2.4

Structure of Five-membered Rings with Two or More Heteroatoms

2.4.1 SURVEY OF POSSIBLE STRUCTURES

We classify compounds as aromatic [(4n + 2) π-electron systems] or antiaromatic (2n π-electron systems), if there is continuous conjugation around the ring, and as non-aromatic. Aromatic compounds are further subdivided into those without exocyclic double bonds and those in which canonical forms containing exocyclic double bonds contribute.

2.4.1.1 Aromatic Systems without Exocyclic Conjugation

The neutral aromatic azole systems (without exocyclic conjugation) are shown in Scheme 1; throughout, Z is O, S or NR. There are, thus, 24 possible systems; however, for NR = NH, tautomerism renders (3) $\equiv$ (5), (4) $\equiv$ (6), and (7) $\equiv$ (8). Ring-fused derivatives without a bridgehead nitrogen atom are possible for systems (1), (2), (3) and (5). Ring-fused derivatives with a bridgehead nitrogen atom can be derived from all.

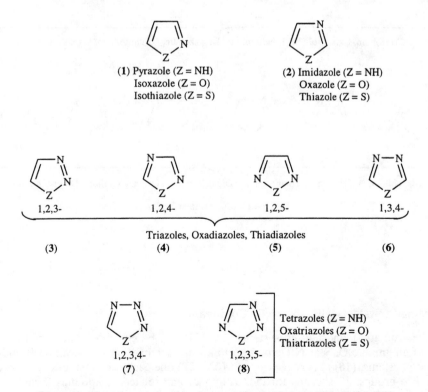

(1) Pyrazole (Z = NH)
Isoxazole (Z = O)
Isothiazole (Z = S)

(2) Imidazole (Z = NH)
Oxazole (Z = O)
Thiazole (Z = S)

1,2,3-   1,2,4-   1,2,5-   1,3,4-

Triazoles, Oxadiazoles, Thiadiazoles

(3)   (4)   (5)   (6)

Tetrazoles (Z = NH)
Oxatriazoles (Z = O)
Thiatriazoles (Z = S)

1,2,3,4-   1,2,3,5-

(7)   (8)

Scheme 1   Neutral aromatic azoles (no exocyclic double bonds) (Z = O, S or NR)

The five possible azole monoanions are shown (one canonical form only) in Scheme 2; all heteroatoms are now nitrogens.

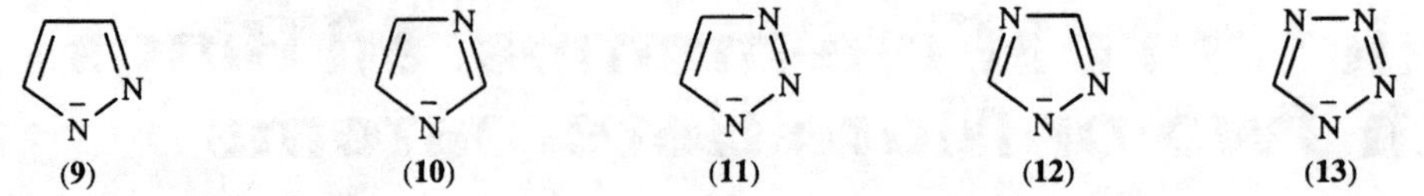

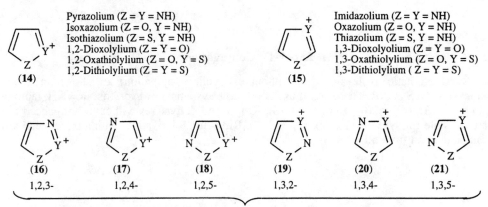

Scheme 2   Monoanionic aromatic azoles

The aromatic azole monocations are given in Scheme 3; here Z and Y are both O, S or NR; there are therefore three mixed sets. If Z = Y, then (**16**) ≡ (**18**), (**20**) ≡ (**21**), (**22**) ≡ (**24**), and (**25**) ≡ (**26**). Hence there are (3 × 14) + (3 × 10) = 72 possible systems.

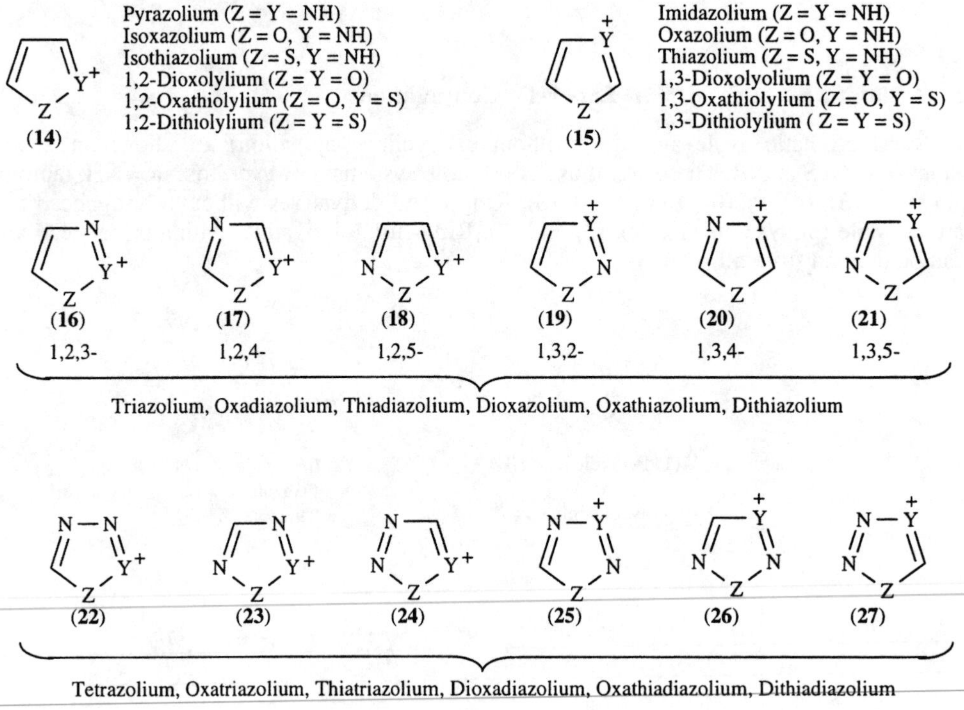

Pyrazolium (Z = Y = NH)
Isoxazolium (Z = O, Y = NH)
Isothiazolium (Z = S, Y = NH)
1,2-Dioxolylium (Z = Y = O)
1,2-Oxathiolylium (Z = O, Y = S)
1,2-Dithiolylium (Z = Y = S)
(**14**)

Imidazolium (Z = Y = NH)
Oxazolium (Z = O, Y = NH)
Thiazolium (Z = S, Y = NH)
1,3-Dioxolyolium (Z = Y = O)
1,3-Oxatholyolium (Z = O, Y = S)
1,3-Dithiolylium (Z = Y = S)
(**15**)

(**16**)   (**17**)   (**18**)   (**19**)   (**20**)   (**21**)
1,2,3-   1,2,4-   1,2,5-   1,3,2-   1,3,4-   1,3,5-

Triazolium, Oxadiazolium, Thiadiazolium, Dioxazolium, Oxathiazolium, Dithiazolium

(**22**)   (**23**)   (**24**)   (**25**)   (**26**)   (**27**)

Tetrazolium, Oxatriazolium, Thiatriazolium, Dioxadiazolium, Oxathiadiazolium, Dithiadiazolium

Scheme 3   Monocationic azoles

2.4.1.2   Aromatic Systems with Exocyclic Conjugation

Each of the aromatic monocationic systems (**14**)-(**27**) can be converted into a neutral system by substitution of an anionic O, S or NR group on to a ring carbon atom. However, (**14**) and (**15**) each give three such systems, (**16**)-(**21**) two each, and (**22**)-(**27**) one each. The resulting 24 systems can be divided into two groups: 12 systems for the azolinones and related compounds (Scheme 4) and 12 systems for the mesoionic (betaine) compounds (Scheme 5).

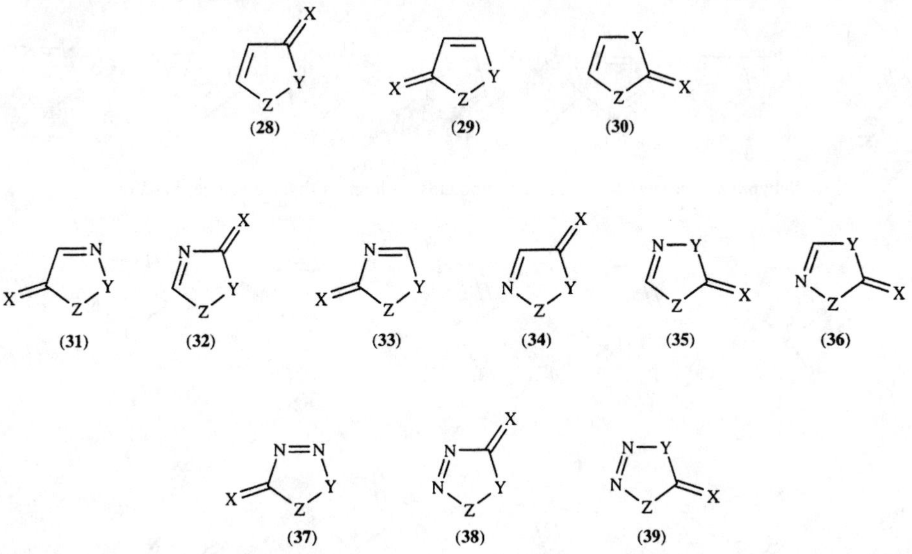

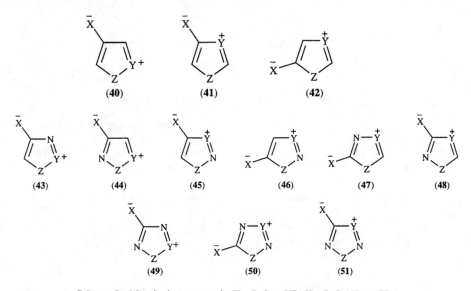

Scheme 4 Azolinones and related compounds (X = O, azolinones; X = S, azolinethiones; X = NR, azolinimines)

Of the mesoionic systems, (**40**) and its aza derivatives (**43**), (**44**) and (**49**) have been designated as Class B by Ollis ⟨76AHC(19)1⟩, including compounds with X = CRR'; there are 88 total systems. Class A mesoionic compounds include (**41**), (**42**) and their aza derivatives (**45**)-(**48**), (**50**) and (**51**), giving a total of 144 systems. Members of the latter group contain 1,3-dipoles, often reflected in their pronounced ability to undergo cycloaddition reactions.

Scheme 5 Mesoionic compounds (Z = O, S or NR; X = O, S, NR or CR$_2$)

2.4.1.3 Non-aromatic Systems

These are subdivided into: (a) compounds isomeric with aromatic compounds in which the ring contains two double bonds but also an sp^3-hybridized carbon (7 systems; Scheme 6) or a quaternary nitrogen atom (9 systems; Scheme 7). The second type is rarely encountered.

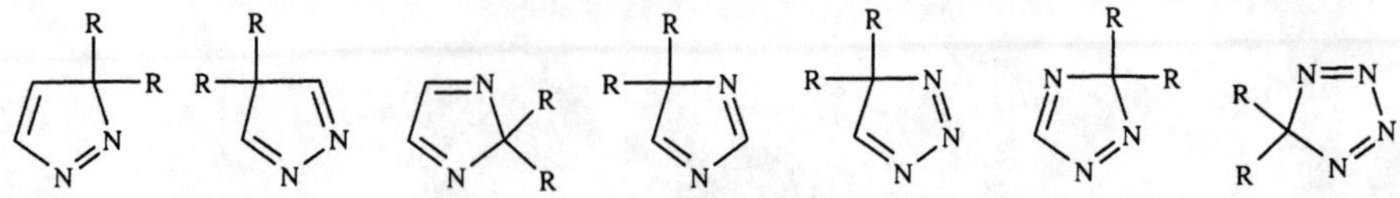

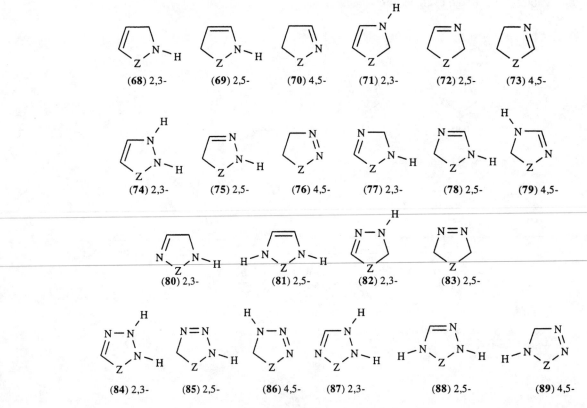

Scheme 6 Isomers of aromatic compounds with an *sp³*-hybridized carbon atom

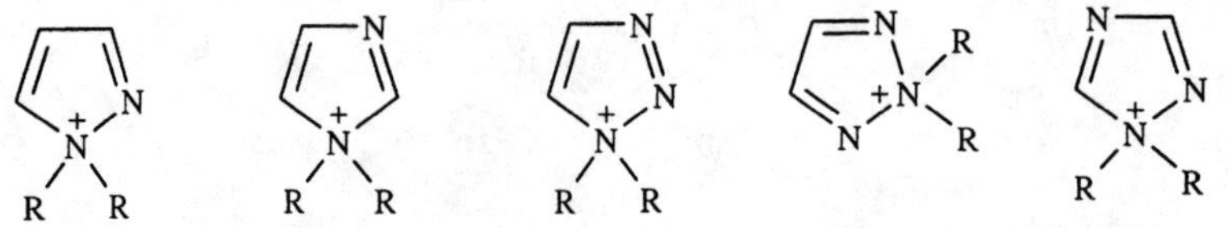

(52)   (53)   (54)   (55)   (56)   (57)   (58)

(59)   (60)   (61)   (62)   (63)

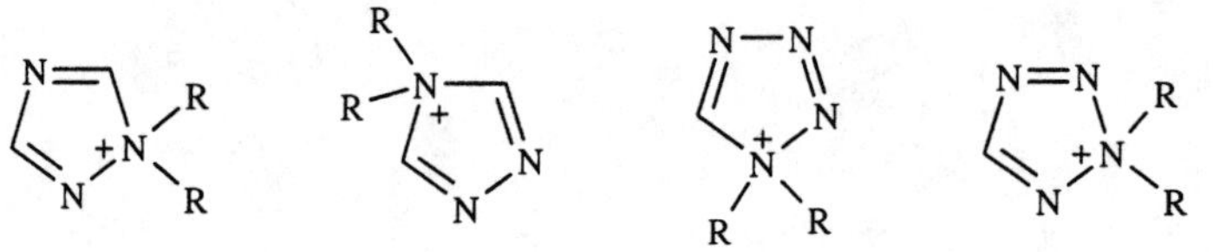

(64)   (65)   (66)   (67)

Scheme 7 Isomers of aromatic compounds with a quaternary nitrogen atom

(b) Dihydro compounds in which the ring contains one double bond (66 systems; Scheme 8).

(68) 2,3-   (69) 2,5-   (70) 4,5-   (71) 2,3-   (72) 2,5-   (73) 4,5-

(74) 2,3-   (75) 2,5-   (76) 4,5-   (77) 2,3-   (78) 2,5-   (79) 4,5-

(80) 2,3-   (81) 2,5-   (82) 2,3-   (83) 2,5-

(84) 2,3-   (85) 2,5-   (86) 4,5-   (87) 2,3-   (88) 2,5-   (89) 4,5-

Scheme 8 Dihydroazoles (Z = O, S, or NR)

(c) Tetrahydro compounds in which the ring contains no double bonds (24 systems; Scheme 9): derivatives with three (**92**, **94**) or four contiguous heteroatoms (**96**, **97**) are rare or unknown.

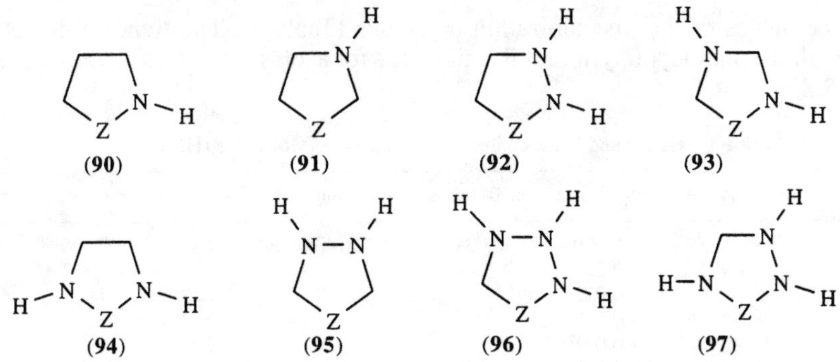

(90) (91) (92) (93)

(94) (95) (96) (97)

Scheme 9 Tetrahydro compounds (Z = O, S or NR)

2.4.2 THEORETICAL METHODS

2.4.2.1 Electron Densities and Frontier Orbital Energies

The presence in azoles of both pyrrole-like and pyridine-like heteroatoms leads to a highly perturbed π-electron distribution. As a result, these molecules often display along with π-excessive centers, atoms with a rather high π-deficiency, sometimes even higher than in typical azines. We first consider non-fused azoles (Table 1).

Pyrazole, imidazole and 1*H*-1,2,3-triazole, according to the total π-charge (π-balance) on their carbon atoms, can be classified as weakly π-excessive systems. On the other hand, 2*H*-1,2,3-triazole, 1*H*- and 4*H*-1,2,4-triazole, and 1*H*-, 2*H*-tetrazole are clearly π-deficient heterocycles. Pyrazole probably possesses the largest local π-excess among simple azoles. The negative charge in the position 4 is somewhat higher than even that at the β-carbons of pyrrole. This is also shown by a larger shielding of the C-4 carbon in the ^{13}C NMR spectrum of pyrazole in comparison with the position of the C-3(4) atoms of pyrrole. 1,2,4-Triazole and tetrazole have the largest local π-deficiency. Attention is drawn to a good correlation between the π-electron densities on the corresponding carbon atoms and the chemical shifts of ^{13}C ⟨68CC1337, 68JA697, 68JA3543⟩ and ^{1}H nuclei ⟨66TL2627⟩ of azoles. As a rule for positions with negative π-charge, the corresponding signals move towards higher field with respect to benzene, whereas shifts towards lower field are observed for nuclei carrying positive charge.

The electronegativity of the cyclic oxygen atom in isoxazole and oxazole has a considerable influence both on their general π-balance and π-electron distribution. This influence explains the π-deficiency of positions 2 and 5 and π-excess of position 4 in both these heterocycles. According to the sign of their π-balance, isoxazole and oxazole ought to be classified as weakly π-deficient systems. Isothiazole and thiazole molecules are also characterized by an irregular π-electron distribution, but both they have almost zero sum charge on the carbon atoms.

Peculiarities of π-electron distribution in fused azoles may be conveniently analyzed by using imidazole as a typical example. Benzoannulation dramatically influences both the π-electron distribution and the reactivity of azoles. Thus, benzimidazole and the isomeric naphthoimidazoles possess moderate general π-deficiency unlike imidazole. The existance of a highly π-deficient ring carbon atom is an especially important feature of all condensed 1,3-diazoles (Figure 1). The value of effective positive charge on this atom exceeds that in most azines (Sec. 2.2.2.5). This explains why 1-*R*-benzimidazoles and 1-*R* or 3-*R*-naphthoimidazoles undergo reactions of nucleophilic substitution (*e.g.* sodamide amination) with great facility (Sec. 3.4.1.6-2).

On first sight it seems surprising that the pyrrole heteroatom in a fused imidazole does not prevent the appearance of a large positive charge at the neighbouring ring carbon atom. The reason for this is that the pyrrole nitrogen displays its π-donor effect almost exclusively in the direction of the fused nucleus.

Purine is also notable for its high π-deficiency, 7*H*-purine being more π-deficient than the 9*H*-tautomer. Possibly this very circumstance is responsible for the natural occurrence of the less π-deficient and therefore chemically more stable derivatives of 9*H*-purine.

Transition from a neutral azole to an azolium cation strongly increases the π-deficiency. In azines, the HOMO is normally occupied by the nonbonded electron pair of a nitrogen atom. By contrast, in azoles the HOMO is of the π-type. This is supported by the parallel behavior of calculated HOMO energies and the values of the first ionization potentials (Table 2). The trend to decreasing electron donor ability with the number of pyridine-like nitrogens for azoles is seen, as already noted in the azine series (Sec. 2.2.2.5).

Table 1 Calculated Values of π-Charges in Azole Molecules (HMO method)

Compounds	Atom	π-Charge	π-Balance	Compounds	Atom	π-Charge	π-Balance
Pyrazole	N-1	+0.300	−0.031	2H-Tetrazole	N-1	−0.190	+0.066
	N-2	−0.269			N-2	+0.362	
	C-3	+0.025			N-3	−0.065	
	C-4	−0.107			N-4	−0.174	
	C-5	+0.051			C-5	+0.066	
Imidazole	N-1	+0.298	−0.011	Isoxazole	O	+0.200	+0.079
	N-3	−0.287			N	−0.279	
	C-2	+0.094			C-3	+0.075	
	C-4	−0.068			C-4	−0.097	
	C-5	−0.037			C-5	+0.101	
1H-1,2,3-Triazole	N-1	+0.332	−0.015	Oxazole	O	+0.222	+0.051
	N-2	−0.150			N	−0.273	
	N-3	−0.167			C-2	+0.126	
	C-4	−0.059			C-4	−0.055	
	C-5	+0.044			C-5	−0.02	
2H-1,2,3-Triazole	N-1(3)	−0.184	+0.036	Isothiazole	S	+0.280	+0.004
	N-2	+0.332			N	−0.283	
	C-4(5)	+0.018			C-3	+0.047	
1H-1,2,4-Triazole	N-1	+0.315	+0.252		C-4	−0.095	
	N-2	−0.267			C-5	+0.052	
	N-4	−0.299					
	C-3	+0.075		Thiazole	S	+0.278	−0.002
	C-5	+0.177			N	−0.275	
4H-1,2,4-Triazole	N-1(2)	−0.257	+0.202		C-2	+0.098	
	N-4	+0.312			C-4	−0.056	
	C-3(5)	+0.101			C-5	−0.044	
1H-Tetrazole	N-1	+0.341	+0.163				
	N-2	−0.135					
	N-3	−0.124					
	N-4	−0.245					
	C-5	+0.163					

Table 2 HMO Energies of Frontier Orbitals and Values of First Ionization Potentials (IP-1) of Some Azoles

Compounds	HOMO (β)	LUMO (β)	IP-1 (eV)
Imidazole	0.674	−0.959	8.78
Pyrazole	0.811	−0.846	9.15
1H-1,2,3-Triazole	0.830	−0.747	—
2H-1,2,3-Triazole	1.000	−0.681	10.06
1H-1,2,4-Triazole	0.885	−0.824	10.0
4H-1,2,4-Triazole	0.781	−0.936	—
1H-Tetrazole	0.952	−0.745	—
2H-Tetrazole*	1.028	−0.621	11.3

[a] *Tautomer dominant in the gas phase

Neutral non-fused azoles, with the exception of oxa-, thia- and seleno-diazoles, generally display low electron acceptor ability and are not polarographically reduced. Their radical-anions have been observed in an argon matrix at 4K ⟨73JA27, 73JA4801⟩. Accordingly to their ESR spectra the radical-anions of pyrazole and imidazole possess structures (**98**) and (**99**), whereas the opened-ring structures (**100**) and (**101**) are more stable for isoxazole, oxazole and their benzo derivatives (a single possible resonance structure is given in each instance).

(**98**)   (**99**)   (**100**)   (**101**)

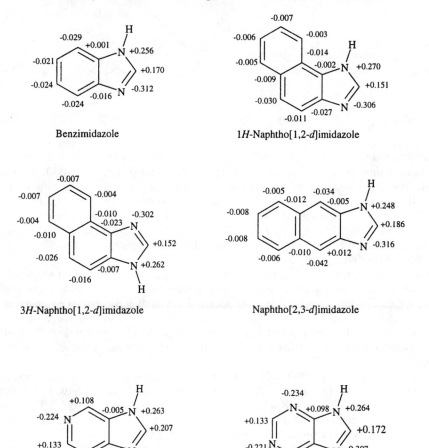

Benzimidazole

1*H*-Naphtho[1,2-*d*]imidazole

3*H*-Naphtho[1,2-*d*]imidazole

Naphtho[2,3-*d*]imidazole

7*H*-Purine

9*H*-Purine

Figure 1 The effective π-electron charges in condensed imidazole systems (HMO method).

2.4.2.2 Other Applications of Theory

Applications of MO methods to such diverse problems as aromaticity, tautomeric structure, dipole moments, and UV, NMR and PE spectroscopy are discussed in various monograph chapters of CHEC and CHEC-II. Below some typical examples are given.

The reliability of semi-empirical methods (AM1, PM3, and MNDO) for the treatment of tautomeric equilibria has been tested for a series of five-membered nitrogen heterocycles, including 1,2,3-triazole and benzotriazole. The known tendency of MNDO to overestimate the stability of heterocycles with two or more adjacent pyridine-like lone pairs is also present in AM1 and to a somewhat lesser extent in PM3. Tautomers with a different number of adjacent pyridine-like nitrogens cannot be adequately treated by these semi-empirical methods. Both AM1 and PM3 represent major improvements over MNDO in the case of lactam-lactim tautomerism. The stability of N-oxides as compared to N-hydroxy tautomers is overestimated by PM3 method. All three methods give reliable ionization potentials and dipole moments ⟨90ZN(A)1328⟩.

Fully optimized geometries have been calculated for the two tautomeric forms of 1,2,3-triazole at the Hartree-Fock SCF level employing a basis set of double zeta quality. The calculations show the 2H form to be the more stable, the energy difference being $\Delta E_{1H-2H} = -14.7$ kJ mol^{-1} ⟨88ACS(A)500⟩.

Ab initio calculations on the equilibrium between (**102**) and (**103**) are carried out with the 3-21G basis set. The plot of the calculated activation energy E_a *vs.* the reaction energy ΔE_r for the cyclization reactions is linear and that provides a striking confirmation of Hammond's postulate.

(102)	**(103)**	**(104a)**	**(105)**	**(104b)**

The structures and relative stabilities of furoxan (**104**, R = R′ = H) and its open-chain tautomers have been calculated using *ab initio* Hartree-Fock (HF/3-21G, HF/3-21G*) and Moller-Plesset (MP4/6-31G*, MP2/6-31G*) procedures 〈92JCO177〉; the results support a mechanism of isomerization (**104a**)⇌(**104b**) involving the *cis*-1,2-dinitrosoethene (**105**) as an intermediate / transition state with an energy *ca.* 120 kJ mol^{-1} above that for furoxan. The presence of the *vicinal* nitroso groups strengthens the C-N bonds and weakens the C=C bonds 〈88JPC5094〉. It is reported that semiempirical MINDO/3 calculations are capable of predicting the relative stabilities of isomeric pairs of various monocyclic furoxans (**104a**) and (**104b**) 〈86KGS264〉.

A graph-theoretical approach based on the notion of ring-bond redistribution graphs (RBR graphs) has been used for the classification of ring transformations in heterocycles, such as the Dimroth rearrangement (see Section 3.4.3.5.1). The vertices and edges of RBR graphs exclusively correspond to those atoms and skeleton bonds which are present in both the final and initial heterocyclic rings. Structurally similar reactions of the same level have the same RBR graphs and differ only by their labeling. This approach has been utilized in a computer program for the systematization of heterocyclic recyclization 〈93JA2416〉.

Hartree-Fock calculations with the 3-21G* and 6-31G* basis sets have been performed to study the structure and energetics of Na$^+$, K$^+$ and Al$^+$ -azole complexes. Structures have been fully optimized at the 3-21G* level. Calculated X$^+$ (X = H, Li, Na, K, Al) binding energies of 1,2,3-triazole show that cation association energies follow the sequence Li$^+$ > Al$^+$ > Na$^+$ > K$^+$, and all of them are much smaller than the corresponding protonation energies 〈92JPC3022〉.

There is an excellent correlation between the experimental microwave dipole moments of a variety of azoles and those calculated by the *ab initio* method at the 6-31G*//6-31G level: $\mu_{exp} = 0.942\mu_{cal} + 0.008$ 〈86JPC5597〉. With this equation, the experimental values for 1H- and 2H-1,2,3-triazole are predicted as 4.35 D and 0.32 D, respectively 〈89JCO(10)426〉. Thus, quantum chemical calculations are a valuable aid in estimating tautomeric equilibria by dipole moment studies 〈90ZN(A)1328〉.

Molecular orbital calculations at various levels of approximation have been applied to both furazans and furoxans. *Ab initio* procedures using minimal (STO-3G) and split valence (3-21G) basis sets have been used to determine bond orders, total energies, ionization potentials, and dipole moments for the parent furazan and furoxan, and several derivatives 〈88JCS(P2)661〉; the calculated molecular geometries (3-21G) are compared in Table 3 with those obtained experimentally.

Table 3　Observed and calculated geometries for furazan and furoxan

	Bond lengths (Å)						Bond angles (deg.)						
	a	*b*	*c*	*d*	*e*	*f*	*ab*	*bc*	*cd*	*de*	*ea*	*af*	*bf*
Furazan													
observed[a]	1.373	1.303	1.421	1.303	1.373		105.5	108.9	108.9	105.5	111.2		
calculated[b]	1.427	1.279	1.444	1.279	1.427		105.5	110.1	110.1	105.5	108.8		
Furoxan													
observed[c]	1.441	1.302	1.401	1.292	1.379	1.240	107.2	107.2	111.9	106.6	107.1	116.4	136.4
calculated[b]	1.430	1.295	1.424	1.284	1.442	1.289	108.0	107.7	112.3	105.3	106.7	118.8	133.2

[a] 〈88ZN(A)597〉. [b] 〈88JCS(P2)661〉. [c] 〈94MI405-01〉.

Radicals (**106**) and (**107**) are calculated by GAUSSIAN-82 to be stable. They should be easily derived (activation energy 1.3–2.0 kcal mol^{-1}) from the parent NH compounds by elimination of H atoms and ought to be highly reluctant (activation energy > 33 kcal mol^{-1}) to N-N dimerize ⟨88CCC2128⟩. These expectations were largely confirmed for the radical (**106**) synthesized at almost the same time ⟨89IZV1819⟩. The cation derived from (**106**) by removal of the unpaired electron (calculated ionization energy 8.86 eV) should be more stable by ∼ 130 kcal mol^{-1} ⟨88CCC2128⟩. On the energy surface calculated by the GAUSSIAN-80 and -83 methods for reaction between ethylene and NO$_2$$^+$ in gaseous phase, the cation in question occupied an energy minimum as a transition state ⟨84JA1074⟩.

(**106**)   (**107**)

2.4.3   STRUCTURAL METHODS

2.4.3.1   X-Ray Diffraction

Details of bond lengths and bond angles for the X-ray structures of heterocyclic compounds through 1970 are listed in 'Physical Methods in Heterocyclic Chemistry', volume 5. This compilation contains many examples for five-membered rings containing two heteroatoms, particularly pyrazoles, imidazoles, isoxazoles, oxazoles, isothiazoles, thiazoles, 1,2-dithioles and 1,3-dithioles. Further examples of more recent measurements on these heterocyclic compounds can be found in the monograph chapters of CHEC and CHEC-II.

For compounds with three or four heteroatoms in the ring the number of measurements is less; some of these are summarized in Table 4.

2.4.3.2   Microwave Spectroscopy

Microwave spectra provide a rich source of minute details of molecular structures. They tell us about the molecular geometry because the spectra are primarily analyzed in terms of the accurate average values of the reciprocals of the three moments of inertia. This generally gives at once the general molecular conformation and some precise structural features may emerge. To obtain a complete structure it is necessary to measure the changes in moments of inertia which accompany the isotopic replacements of each atom in turn ⟨74PMH(6)53⟩.

From accurate measurements of the Stark effect when electrostatic fields are applied, information regarding the electron distribution is obtained. Further information on this point is obtained from nuclear quadrupole coupling effects and Zeeman effects ⟨74PMH(6)53⟩.

Microwave studies also provide important information regarding molecular force fields, particularly with reference to low frequency vibrational modes in cyclic structures ⟨74PMH(6)53⟩.

2.4.3.2.1   Molecular geometry

Structural parameters in aromatic five-membered rings are shown in Table 5. All the C-H distances are near 107.5 pm, close to the C-H link in ethylene. With heteroatoms at adjacent ring positions, the C-H groups are displaced from the bisector of the ring angles toward the adjacent heteroatom ⟨74PMH(6)53⟩.

The N-H bond lengths in pyrazole and imidazole (99.8 pm) are a little shorter than those found in dimethylamine. Delocalization in pyrazole, imidazole, 1,2,3-triazole and 1,2,4-triazole is sufficient to bring the hydrogen attached to nitrogen into the plane of the other atoms. The N-H bond in pyrazole does not lie in the bisector of the ring angle (as is required by symmetry in pyrrole), but is displaced by around 5° toward the second nitrogen ⟨74PMH(6)53⟩.

Table 4   X-Ray Structures of Compounds with Five-membered Rings and Two, Three or Four Heteroatoms[a]

Ring	Ring position 1	2	3	4	5	Examples of compounds studied
C₃N₂	N	N	—	—	—	Pyrazole;[b] substituted pyrazoles; Δ¹- and Δ²-pyrazolines; pyrazolinones; pyrazolidines; pyrazolidinones
	N	—	N	—	—	Imidazole;[b] 4,5-di-*t*-butylimidazole; histamine dihydrochloride; 2-thiohydantoin
C₂N₃	N	N	N	—	—	1,3-Dimethyl-4-(1,2,3-triazolyl)sulfide; 3-methyl-2-phenyl-1,2,3-triazol-1-ine-4-thione
	N	N	—	N	—	1,2,4-Triazole[b]
CN₄	N	N	N	N	—	5-Amino-2-methyltetrazole;[b] 5-aminotetrazole monohydrate; sodium tetrazolate monohydrate[b]
C₃NO	O	N	—	—	—	5,5′-Bis(isoxazole);[b] 3-hydroxy-5-phenylisoxazole;[b] 3,3′-bi-2-isoxazoline;[b] 3-hydroxy-5-phenylisoxazole;[b] 3-phenylisoxazolin-5-one[b]
C₃NS	S	N	—	—	—	Methyl 3-hydroxy-4-phenylisothiazole-5-sulfonate; dehydromethionine
C₃NO	O	—	N	—	—	2,2′-*p*-Phenylenebis(5- phenyloxazole); 2-(4-pyridyl)oxazole;[b] 2,4-dimethyl-5-(*p*-nitrophenyl)oxazole;[b] 2-oxazolidinone[b]
C₃NS	S	—	N	—	—	Thiamine hydrochloride monohydrate;[b] rhodanine;[b] 2-imino-5-phenyl-4-thiazolidinone[b]
C₂N₂O	O	N	N	—	—	N-(*p*-Bromophenyl)sydnone;[b] 4,4′-dichloro-3,3′-ethylenebis(sydnone)
	O	N	—	N	—	3-(2-Aminopyridyl)-5-methyl-1,2,4-oxadiazole
	O	N	—	—	N	3-(*p*-Bromophenyl)-4-methyl-1,2,5-oxadiazole 2-oxide;[b] 3-(*p*-bromophenyl)-4-methyl-1,2,5-oxadiazole 5-oxide;[b] 3,4-diphenyl-1,2,5-oxadiazole
	O	—	N	N	—	Monoaryl-1,3,4-oxadiazoles[c]
C₂N₂S	S	N	N	—	—	5-Acylamino-3-methyl-1,2,3-thiadiazole; 5-phenyl-1,2,3-thiadiazole 3-oxide
	S	N	—	N	—	5-Imino-4-phenyl-3-phenylamino-4*H*-1,2,4-thiadiazoline[d]
	S	—	—	—	N	3,4-Diphenyl-1,2,5-thiadiazole; 1,2,5-thiadiazole-3,4-dicarboxamide
	S	—	N	N	—	1,3,4-Thiadiazole; 2,5-diphenyl-1,3,4-thiadiazole[b]
CN₃O	O	N	N	N	—	Mesoionic 3-phenyl-1,2,3,4-oxatriazole-5-phenylimine; mesoionic 3-phenyl-1,2,3,4-oxatriazol-5-one
CN₃S	S	N	N	N	—	5-Phenyl-1,2,3,4-thiatriazole; 5-amino-1,2,3,4-thiatriazole; 5-phenyl-1,2,3,4-thiatriazole 3-oxide
CN₃S	S	N	N	—	N	2-Acetyl-5-chloro-2*H*-1,2,3,5-thiatriazolo[4,5-α]isoquinoline 5-oxide
C₃O₂	O	—	O	—	—	Bis(dioxolane); ethylene carbonate; *cis*-2-*t*-butyl-5-carboxymethyl-1,3-dioxolan-4-one; 2-methyl-1,3-dioxolan-2-ylium perchlorate
C₃OS	O	S	—	—	—	3,3-Diphenyl-1,2-oxathiolane 2,2-dioxide; 5*H*-1,2-benzoxathiol 2,2-dioxide
	O	—	S	—	—	Cholestan-4-one-3-spiro(2,5-oxathiolane)[b]
C₃S₂	S	S	—	—	—	1,2-Dithiolane-4-carboxylic acid ;[b] 3-phenyl-1,2-dithiolylium iodide;[b] 4 methyl-1,2-dithiole-3-thione[b]
	S	—	S	—	—	Bis-1,3-dithiol-2-yl;[b] 4,5-dioxo-2-thioxo-1,3-dithiolane;[b] 1,3-dithiolane-2-thione 5-oxide;[b] tetrathiafulvalene[e]
C₂O₃	O	O	—	O	—	*trans*-5-Anisyl-3-methoxycarbonyl-1,2,4-trioxolane[f]
C₂S₃	S	S	—	S	—	1,2,4-Trithiolane-3,5-dione diphenylhydrazone[g]
C₂O₂S	O	S	O	—	—	1,3,2-Dioxathiolane 2,2-dioxide;[h] 1,3,2-dioxathiol 2,2-dioxide[h]
C₂NS₂	S	S	N	—	—	2,4,6-Tri-*t*-butyl-7,8,9-dithiazabicyclo[4.3.0]nona-1(9),2,4-triene[i]
C₂NS₂	S	S	—	N	—	5-Amino-1,2,4-dithiazolin-3-one ('Rhodan hydrate');[j] 5-amino-1,2,4-dithiazoline-3-thione ('Xanthane hydride');[k] 3,5-diamino-1,2,4-dithiazolium chloride ('Thiuret hydrochloride')[l]
C₂NOS	O	S	N	—	—	2,4-Dioxo-2-(4-methylphenyl)-1,2,3-oxathiazoline;[m] 1,2,3-oxathiazolo-[5,4-*d*][1,2,3]oxathiazole 2,2,5,5-tetraoxide[n]
	O	S	—	N	—	6,10b-Dihydro-3-(2,2,6,6-tetramethylcyclohexyliden)-1,2,4-oxathiazolo-[5,4-α]isoquinoline[o]
	O	N	S	—	—	4-Phenyl-1,3,2-oxathiazolin-5-one[p]
	O	—	S	N	—	2-Trichloromethyl-5-phenyl-Δ⁴-1,3,4-oxathiazoline[q]

[a]Unless otherwise indicated data are taken from the appropriate chapter of 'Comprehensive Heterocyclic Chemistry'. Data taken from ⟨72PMH(5)1⟩, which gives references to the original literature. [c]X-Ray powder data. *cf*. Chapter 4.23 [d]⟨78CC652⟩. [e]⟨71CC889⟩. [f]⟨70ACS2137⟩. [g]⟨71JCS(B)415⟩. [h]⟨68JA2970⟩. [i]⟨80AX(B)1466⟩. [j]⟨66ACS754⟩. [k]⟨63ACS2575, 63AX1157⟩. [l]⟨66ACS1907⟩. [m]⟨71TL4243⟩. [n]⟨80CPL(72)489⟩. [o]⟨78AG(E)455⟩. [p]⟨72G23⟩. [q]⟨81JCS(P1)2991⟩.

The ring angles at C and N are usually $108 \pm 5°$, but are nearer to 90° at sulfur and selenium.

Minor variations in the bond lengths reflect variations in the double bond character and, for example, suggest larger delocalization in 1,3,4-thiadiazole than in 1,3,4-oxadiazole ⟨74PMH(6)53⟩. In thiazole the geometry of the SCN part of the ring resembles the corresponding part of 1,3,4-thiadiazole, while the remaining part of the ring resembles the corresponding moiety in thiophene.

Table 5   Structure Parameters in Five-membered Rings from Microwave Spectra

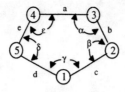

| Compd | Bond length (pm)[a,b] | | | | | Angle (°)[b] | | | | Dipole moment |
	a	b	c	d	e	α	β	γ	δ	ε (10⁻³⁰ C m)[a]
[c]	141.6	133.1	134.9	135.9	137.3	111.9	104.1	113.1	106.4	104.5 7.37
[d]	(137.8)	(132.6)	(134.9)	(136.9)	(135.8)	(105.4)	(111.3)	(107.2)	(106.3)	(109.8) 12.8
[d]	—	—	—	—	—	—	—	—	—	— 5.97
[e]	135.9	132.3	135.9	133.1	132.4	114.6	102.1	110.2	110.1	103.0 9.07
[e]	135.1	128.4	132.4	133.4	131.0	—	—	—	—	— 7.31
[f]	(134.5)	(128.3)	(134.7)	(135.1)	(129.0)	(112.2)	(106.9)	(105.3)	—	— 17.05
[g]	—	—	—	—	—	—	—	—	—	— 9.21
[h]	—	—	—	—	—	—	—	—	—	— 8.67
[e]	139.5	129.3	135.7	137.0	135.3	103.9	115.0	103.9	108.1	109.1 5.00
[e]	137.2	130.4	172.4	171.3	136.7	110.1	115.2	89.3	109.6	115.8 5.37
[i]	(133.4)	(131.3)	(138.9)	—	—	(111.3)	(103.8)	(114.4)	—	21.12
[j]	(138.0)	(130.3)	(141.8)	—	—	(106.1)	(103.2)	(114.2)	—	— 5.44
[c]	142.1	130.0	138.0	138.0	130.0	109.0	105.8	110.4	105.8	109.0 11.28
[e]	141.7	132.7	163.0	163.0	132.7	113.8	106.5	99.4	106.5	113.8 5.27

Table 5 Continued

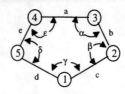

| Compd | Bond length (pm)[a,b] | | | | | Angle (°)[b] | | | | Dipole moment |
	a	b	c	d	e	α	β	γ	δ	ε (10⁻³⁰ C m)[a]
(N—N ring) c	139.9	129.7	134.8	134.8	129.7	105.6	113.4	102.0	113.4	105.6	10.14
(N—N ring) k	136.6	129.0	169.2	168.9	136.9	114.0	111.2	92.9	107.8	114.2	11.98
(N ring) l	136.6	131.7	164.9	170.7	131.3	120.1	107.1	92.8	112.3	107.7	1.5
(N ring) c	142.0	132.8	163.1	163.1	132.8	113.8	106.4	99.6	106.4	113.8	5.24
(N—N ring) c	137.1	130.2	172.1	172.1	130.2	112.2	114.6	86.4	114.6	112.2	10.94
(O ring) e	—	—	—	—	—	—	—	—	—	—	3.97
(O ring) m	156.3	142.8	—	142.8	156.3	105.1	—	—	105.1	101.7	—
(O ring) n	143.6	139.5	147.0	139.5	143.6	106.2	99.2	99.2	106.2	99.2	3.64
(N—S ring) o	169.0	176.6	140.2	135.6	128.6	93.8	106.3	110.8	121.1	107.9	—

[a] 1 Å = 100 pm; 1 D = 3.336 × 10⁻³⁰ C m. [b] X-Ray diffraction data are enclosed in parentheses. [c] Data taken from ⟨74PMH(6)53⟩ which see for references to the original literature. [d] Dipole moment is concentration-dependent (*cf.* CHEC 4.06.3.2). [e] Data taken from appropriate chapter, 'Comprehensive Heterocyclic Chemistry'. [f] No bond lengths or angles given in ⟨74JSP(49)423⟩; calculated values taken from CHEC 4.13. [i] Bond lengths and angles for X-ray data for 4,4'-dichloro-3,3'-ethylenebis(sydnone) ⟨67JA5977⟩; calculated dipole moment for 3-methylsydone ⟨71JST(9)321⟩. [j] Bond lengths and angles from X-ray data for 3-(2-aminopyridyl)-1,2,4-oxadiazole ⟨79AX(B)2256⟩; dipole moment for 3-methyl-5-phenyl-1,2,4-oxadiazole ⟨35G152⟩. [k] Dipole moment from ⟨76CPH(13)73⟩. [l]⟨74TH40100⟩. [m] Values calculated from experimentally derived rotational constants; *cf.* CHEC 4.30.1.3.2. [n]⟨72JA6337⟩. [o] Values for 1,3,4-oxathiazolin-2-one determined from electron diffraction measurements and refined using rotational parameters from microwave spectra; *cf.* CHEC 4.34.2.3.2.

The microwave spectrum of the parent 1,2,3-triazole was interpreted in terms of the 1*H*-form ⟨70SA(A)825⟩. The microwave spectrum of a highly enriched sample of *N*-deuterio-1,2,3-triazole was also apparently assigned unambiguously to the 1-deuterio form ⟨74CC605⟩. However, later analysis of the microwave spectra of the parent molecule, the highly enriched ¹⁵N₃ species, and the *N*-deuterated derivative, reveals 1,2,3-triazole to exist as a mixture of two planar tautomers: a 1*H* form with C_s symmetry and a 2*H* form with C_{2v} symmetry. This is the first recorded microwave spectrum of the deuteriated 2*H* tautomer. The population ratio of the two tautomers, estimated from the analysis of the microwave spectrum of the triple ¹⁵N sample, is 1*H* : 2*H* ≈ 1 : 1000 at room temperature. The dipole moments of the tautomers are measured as μ_{1H} = 4.38 D and μ_{2H} = 0.218 D for the 1*H* and 2*H* forms of the ¹⁵N₃ triazole, respectively ⟨88ACS(A)500⟩.

1,2,4-Triazole exists in the 1*H*-form. In tetrazole both the 1*H*- and 2*H*-forms are detected ⟨74PMH(6)53⟩.

Dipole moments can also be obtained from the microwave spectral data ⟨74PMH(6)53⟩ and available values are given in Table 5.

2.4.3.2.2 *Partially and fully saturated ring systems*

Relatively few such heterocyclic systems have been studied by microwave spectroscopy; some data are included in Table 5. In 1,3-dioxolane the bent form is more stable than the twisted, and pseudorotation occurs. In 1,2,4-trioxocyclopentane the equilibrium conformation is twisted, and there is a barrier of 6.3 kJ mol $^{-1}$ opposing pseudorotation ⟨74PMH(6)53⟩.

2.4.3.3 ¹H NMR Spectroscopy

Proton chemical shifts and spin coupling constants for ring CH of fully aromatic neutral azoles are recorded in Tables 6–9. Vicinal CH-CH coupling constants are small; where they have been measured they are 1–2 Hz.

For the NH azoles (Table 6), the two tautomeric forms are usually rapidly equilibrating on the NMR timescale (except for triazole in HMPT). The *N*-methylazoles (Table 7) are 'fixed'; chemical shifts are shifted downfield by adjacent nitrogen atoms, but more by a 'pyridine-like' nitrogen than by a 'pyrrole-like' *N*-methyl group.

A significant development in the ¹H NMR spectroscopy of NH-pyrazoles, is the use of low temperature to block proton transfer and, thus, determine tautomeric equilibrium constants by simple integration of the signals. In this way, the K_T values of 3(5)-phenylpyrazole (**108**, R = H) ⟨91G477⟩ and 3(5)-phenyl-5(3)-methylpyrazole (**108**, R = Me) ⟨92JCS(P2)1737⟩ were obtained. In the latter case and using the [¹⁵N₂] derivative, the ¹H—¹⁵N coupling constants of each tautomer were measured.

Compound (**109**) presents an unusual long range $^6J(^1H—^1H)$ coupling constant between pyrazole H-3 and azomethine =CH protons ⟨87AP(320)115⟩. A systematic study of *ortho*-benzylic coupling constants $^4J_{Me-C=C-H}$ has been published ⟨92JHC935⟩: 1,3-dimethylpyrazole (− 0.55 Hz), 1,4-dimethylpyrazole (− 0.68 Hz with H_3, and − 0.92 Hz with H_5), 1,5-dimethylpyrazole (− 0.83 Hz), the coupling values being linearly related to bond orders.

(**108a**)   (**108b**)   (**109**) (R = 5-nitrofuryl-2)

Differentiation between 1,4- and 1,5-disubstituted imidazoles has frequently proved to be difficult; however, by using cross-ring coupling constants proton NMR can distinguish them when sufficient appropriate ring protons are present. Usually, $J_{2,5}$ is somewhat larger than $J_{2,4}$ (1.1–1.5 Hz; 0.9–1.0 Hz, respectively) ⟨82JHC253, 89CPB1481, 90CHE407⟩.

Nuclear Overhauser effect (NOE) difference spectroscopy has been used for the assignment of pyrazoles ⟨90M837, 93JHC865⟩, for the unambiguous discrimination between H(4) and H(5) signals in 1-substituted 1*H*-1,2,3-triazoles ⟨91T9783, 92JHC1203⟩ and in the differentiation between regioisomeric 1,4-, 1,5- and 2,4-disubstituted 1,2,3-triazoles. The NOE experiment also supports the literature assignments with δ(H(5)) < δ(H(4)) in CDCl₃ and δ(H(5)) > δ(H(4)) in DMSO-d₆ for 1-substituted 1,2,3-triazoles ⟨91T9783⟩. The NOE method for discriminating between isomeric disubstituted 1,2,3-triazoles is illustrated by compounds (**110**)-(**112**). Simple NOE experiments allow the identification of (**110**); with this isomer only, an NOE enhancement of the triazole-H singlet is observed upon irradiation of the benzyl CH₂ group. Compounds (**111**) can be distinguished from (**112**) by a moderate NOE enhancement between the NCH₂ and the ester group.

(**110**)   (**111**)   (**112**)

Table 6 ^{1}H NMR Spectral Data for Ring Hydrogens of Nitrogenous Azoles: (a) NH Derivatives

	[1]H Chemical shifts (δ, ppm)				Coupling constants,		
Compound	*H-2*	*H-3*	*H-4*	*H-5*	*J (Hz)*	*Solvent*	*Ref.*
Pyrazole	—	7.61	7.31	7.61	2.1	—	71PMH(4)121, B-73NMR
Imidazole	7.86	—	7.25	7.25	1.0	$CDCl_3$	71PMH(4)121, B-73NMR
1,2,3-Triazole	—	—	7.75	7.75	—	$CDCl_3$	[a]
1,2,4-Triazole	—	7.92	—	8.85	—	HMPT	71PMH(4)121, B-73NMR
Tetrazole	—	—	—	9.5	—	D_2O	

[a]Data taken from appropriate chapter of 'Comprehensive Heterocyclic Chemistry'.

Table 7 ^{1}H NMR Spectral Data for Ring Hydrogens of Nitrogenous Azoles: (b) *N* (1)-Methyl Derivatives

	[1]H Chemical shifts[a] (δ, ppm)			
Compound	*H-2*	*H-3*	*H-4*	*H-5*
Pyrazole[b,c]	—	7.49	6.22	7.35
Imidazole[b]	7.47	—	7.08	6.88
1,2,3-Triazole[b]	—	—	7.74	7.59
1,2,5-Triazole[b]	—	7.75	7.75	—
1,2,4-Triazole[b]	—	7.94	—	8.09
1,3,4-Triazole[b]	8.23	—	—	8.23
1,2,3,4-Tetrazole[d]	—	—	—	8.98
1,2,3,5-Tetrazole[d]	—	—	8.60	—

[a]Spectra measured in $CDCl_3$. [b]Data taken from ⟨B-73NMR⟩ which contains references to the original literature. [c]Coupling constants are $J_{3,4} = 2.0$ Hz; $J_{3,5} = 0.7$ Hz; $J_{4,5} = 2.3$ Hz. [d]Value for *N*-methyl derivative; *cf.* CHEC 4.13.

Comparison of the relevant data shows that an adjacent oxygen (Table 8) and especially a sulfur atom (Table 9) induce lower field shifts than either type of nitrogen atom.

Table 8 ^{1}H NMR Spectral Data for Ring Hydrogens of Azoles Containing Oxygen

	[1]H Chemical shifts (δ, ppm)				
Compound	*H-2*	*H-3*	*H-4*	*H-5*	*Solvent*
Isoxazole[a]	—	8.14	6.28	8.39	CS_2
Oxazole[b]	7.95	—	7.09	7.69	CCl_4
1,2,4-Oxadiazole[c]	—	8.2	—	8.7	C_6H_6
1,2,5-Oxadiazole[d]	—	8.19	8.19	—	$CHCl_3$
1,3,4-Oxadiazole[e]	8.73	—	—	8.73	$CDCl_3$

[a]Coupling constants: $J_{3,4} = 1.78$ Hz; $J_{3,5} = 0.27$ Hz; $J_{4,5} = 1.69$ Hz ⟨74CJC833⟩.
[b]Coupling constants: $J_{2,4} = 0$ Hz; $J_{2,5} = 0.8$ Hz; $J_{4,5} = 0.8$ Hz.
[c]⟨76AHC(20)65, 64HCA942⟩.
[d]CHEC 4.22.2.3.1. [e]CHEC 4.23.2.2.1.

Table 9 ^{1}H NMR Spectral Data for Ring Hydrogens of Azoles Containing Sulfur

	[1]H Chemical shifts (δ, ppm)				
Compound	H-2	H-3	H-4	H-5	Solvent
Isothiazole[a]	—	8.54	7.26	8.72	CCl$_4$
Thiazole[b,c]	8.88	—	7.98	7.41	CDCl$_3$
1,2,3-Thiadiazole[g]	—	—	(AB multiplet centered at 8.80)		CCl$_4$
1,2,4-Thiadiazole	—	8.66[d]	—	9.90[e]	
1,2,5-Thiadiazole[h]	—	8.70	8.70	—	CCl$_4$
1,3,4-Thiadiazole[f]	7.55	—	—	7.55	CDCl$_3$

[a]Coupling constants: $J_{3,4} = 11.66$ Hz; $J_{3,5} = 0.15$ Hz; $J_{4,5} = 4.66$ Hz. [b]Coupling constants: $J_{2,4} = 0$ Hz; $J_{2,5} = 1.95$ Hz; $J_{4,5} = 3.15$ Hz. [c]⟨79HC(34-1)67, 79HC(34-1)73⟩. [d]Value given for 5-phenyl derivative ⟨80JOC3750⟩. [e]Value given for 3-phenyl derivative ⟨74JOC962⟩.
[f]⟨78BAP291⟩. [g]⟨78JOC2487⟩. [h]⟨64DIS2690⟩.

The effects of anion and cation formation on ^{1}H chemical shifts can be assessed from data in Tables 10 and 11. Anion formation always results in shifts to higher field; however, the effect is relatively modest except for the 4-position of pyrazole because in all other cases the adjacent nitrogen lone pair partially cancels the shift. Conversely, in the cations (Table 11), the downfield shift is especially large for the CH groups next to nitrogen. The coupling constants appear to be significantly greater in the cations.

Table 10 ^{1}H NMR Spectral Data for Ring Hydrogens of Azole Anions

	[1]H Chemical shifts (δ, ppm)					
Compound	H-2	H-3	H-4	H-5	Solvent	Ref.
	—	7.35	6.05	7.35	KOD/D$_2$O	68JA4232
	7.80	—	7.21	7.21	—	71PMH(4)121
	—	—	7.86	7.86	NaOD/D$_2$O	[a]
	—	8.19	—	8.19	NaOD/D$_2$O	71PMH(4)121
	—	—	—	8.73	—	71PMH(4)121

[a] Data taken from CHEC 4.11.

Table 11 ^{1}H NMR Spectral Data for Ring Hydrogens of Azole and Related Cations: Two Heteroatoms

Compound	^{1}H Chemical shifts (δ, ppm)				Solvent	Coupling constants, J (Hz)				
	H-2	H-3	H-4	H-5		2,4	2,5	3,4	3,5	4,5
Pyrazolium[a,b]	—	8.57	6.87	8.57	DMSO-d_6	—	—	2.9	—	2.9
Imidazolium[c]	8.6	—	7.5	7.5	H$_2$SO$_4$	1.4	1.4	—	—	2.4
Isoxazolium[i]	—	9.18	7.26	9.01	D$_2$SO$_4$	—	—	2.8	—	1.9
Isothiazolium[a]	—	9.1	7.9	9.6	H$_2$SO$_4$	—	—	2.7	0.6	5.6
Thiazolium[d]	9.55	—	8.23	7.93	—	0.70	1.55	—	—	3.10
1,2-Oxathiolylium[e]	—	—	7.64	—	—	—	—	—	—	—
1,2-Dithiolylium[c]	—	6.71	1.44	−0.26	—	—	—	—	4.9	4.9
1,3-Dioxolylium[f]	10.4	—	—	—	—	—	—	—	—	—
1,3-Oxathiolylium[g]	—	—	8.12	—	—	—	—	—	—	—
1,3-Dithiolylium[h]	11.65	—	9.67	9.67	—	2.0	2.0	—	—	—

[a]Data taken from appropriate chapter of 'Comprehensive Heterocyclic Chemistry'. [b]Values given for 1,2-dimethylpyrazolium. [c]Data taken from ⟨71PMH(4)12⟩ and ⟨B-73NMR1⟩ which see for references to the original literature. [d]⟨66BSF3524⟩. [e]Value given for 5-methyl-3-(2-oxo-1-propyl)-1,2-oxathiolylium perchlorate. [f]Value given for 1,3-benzodioxolylium fluorosulfonate. [g]Value given for 2,5-diphenyl-1,3-oxathiolylium perchlorate. [h]⟨74JOC3608⟩. [i]⟨83PC40100⟩.

Relatively few data are available on the ^{1}H NMR spectra of azolinones and related thiones and imines (Table 12).

Table 12 ^{1}H NMR Spectral Data for Ring C-H of Azolinones

Compound	^{1}H Chemical shifts (δ, ppm)				Solvent
	H-2	H-3	H-4	H-5	
Pyrazolin-3-one[b]	[a]	—	5.25	7.22	CDCl$_3$
Imidazolin-2-one[c]	—	—	6.50	6.50	—
Pyrazoline-3-thione[b]	—	—	6.23	[d]	—
1,2,4-Triazoline-3-thione	—	—	—	8.20	DMSO
Pyrazolin-3-imine[f]	—	—	5.46	[e]	CDCl$_3$
1,2,4-Triazolin-5-imine	—	2.05	—	—	CDCl$_3$
Δ^2-1,3,4-Oxadiazoline-5-thione[g]	8.88	—	—	—	DMSO-d_6
Isothiazolin-3-one[h,i]	—	—	6.05	7.98	—
1,3-Thiazolin-2-one[j,k]	—	−1.14	3.21	3.7	DMSO—d_6
Isothiazoline-3-thione[l,m]	—	—	6.90	8.25	—
1,3-Thiazoline-2-thione[k,n]	—	6.68	2.7	3.05	C$_3$D$_6$O
1,3-Thiazolin-2-imine[o,p]	—	—	3.03	3.37	CDCl$_3$

[a]Values given for 1,2-dimethylpyrazolin-3-one; $J_{4,5} = 3.5$ Hz. [b]⟨76AHC(S1)1⟩. [c]Data taken from ⟨B-73NMR⟩ which contains references to the original literature. [d]Values given for 1,5-dimethyl-2-phenylpyrazoline-3-thione. [e]Values given for 1,5-dimethyl-2-phenylpyrazolin-3-imine. [f]⟨72BSF2807⟩. [g]CHEC 4.23.2.2.1. [h]Values given for 2-methylisothiazolin-3-one; $J_{4,5} = 6.0$ Hz. [i]⟨71JHC571⟩. [j]Coupling constants: $J_{3,4} = 2.5$ Hz; $J_{3,5} = 1.1$ Hz; $J_{4,5} = 5.3$ Hz. [k]⟨79HC(34-2)385⟩. [l]Values given for 2-methylisothiazoline-3-thione; $J_{4,5} = 6.0$ Hz. [m]⟨80CPB487⟩. [n]$J_{4,5} = 4.6$ Hz. [o]Values given for 2-ethoxycarbonylimino-3-ethyl-Δ^4-1,3-thiazoline. [p]⟨79HC(34-2)26⟩.

Some available data on ^{1}H NMR spectra of non-aromatic azoles containing two ring-double bonds are given in Table 13. Here there is no ring current effect and the chemical shifts are consequently more upfield.

Tables 14 and 15 give some available chemical shifts for azolines and azolidines, respectively. Unfortunately, data for many of the parent compounds are lacking, sometimes because the compounds themselves are unknown.

The ^{1}H NMR spectra of 128 Δ^2-pyrazolines have been reported ⟨90JCR(S)200⟩. Additive contributions of substituents were calculated which allowed the prediction of the chemical shifts of unknown pyrazolines. The J_{gem}, J_{cis}, and J_{trans} coupling constants of the protons at positions 4 and 5, are quite sensitive to the nature of the substituent on the nitrogen atom. A series of papers ⟨89JCS(P2)319, 93T863⟩ deals with the conformational analysis of fused- and spiro-pyrazolines using ^{1}H, ^{13}C, and two-dimensional methods. ^{1}H NMR spectroscopy (chemical shifts and ^{1}H − ^{1}H coupling constants) can determine the protonation site of Δ^2-pyrazolines (**113**, R^1, R^2 = H, Me, Ph) ⟨87CS283⟩.

The stable carbene (**114**; R = H) has the ring proton signal at 6.92 ppm. This chemical shift very much resembles that of 1-methylimidazole (Table 7). The 1,3,4,5-tetramethyl carbene (**114**; R = Me) has methyl signals at 2.01 (C-methyl) and 3.48 ppm (N-methyl), cf. methyl shifts for 4-methyl- (2.23 ppm), and 1-methylimidazole (3.70 ppm) ⟨92JA5530, 92JA9724⟩ (see also Section 2.4.4.2.3).

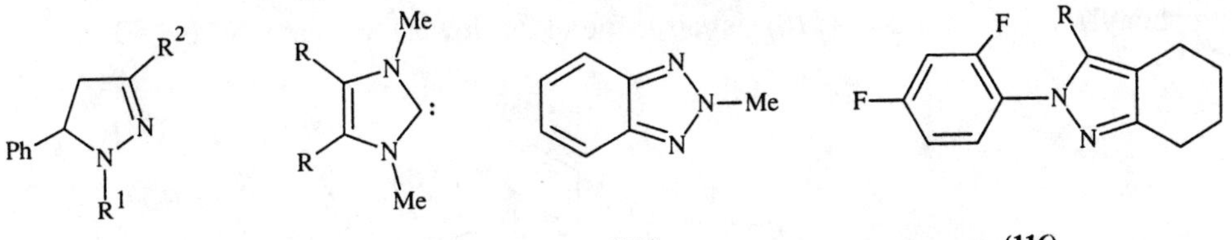

(113)   **(114)**   **(115)**   **(116)**

Table 13 ^{1}H NMR Spectral Data (δ, ppm) for Ring Hydrogens of Non-aromatic Azoles with Two Ring Double Bonds

3*H*-Pyrazoles[a]

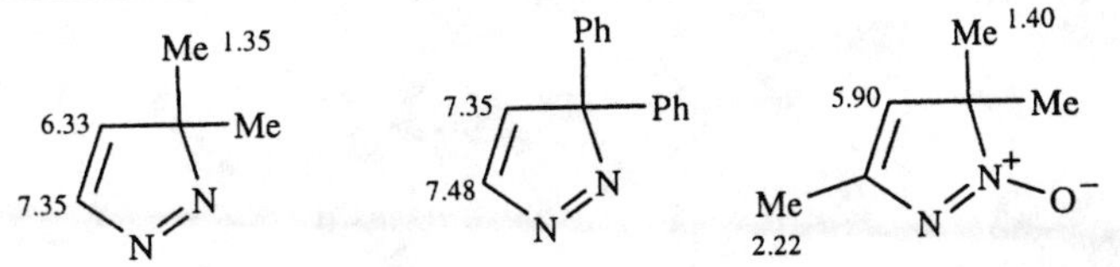

Me 1.35; 6.33; Me; 7.35; N—N
Ph; 7.35; Ph; 7.48; N—N
Me 1.40; 5.90; Me; Me; 2.22; N—N$^+$—O$^-$

4*H*-Pyrazoles[b]

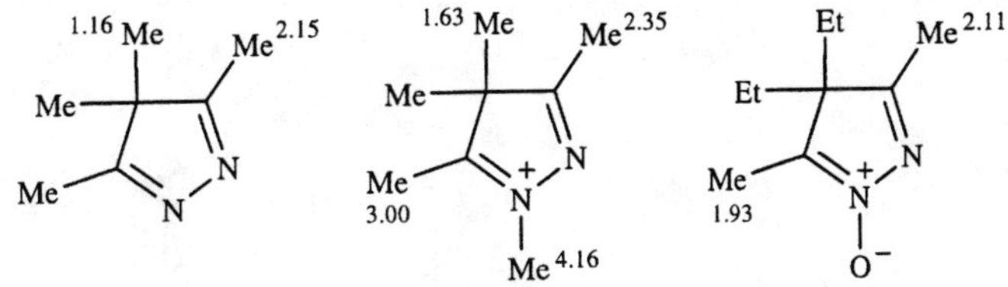

1.16 Me; Me 2.15; Me; Me; N—N
1.63 Me; Me 2.35; Me; Me; 3.00; N$^+$—N; Me 4.16
Et; Me 2.11; Et; Me; 1.93; N$^+$—N; O$^-$

2*H*-Imidazoles[c]

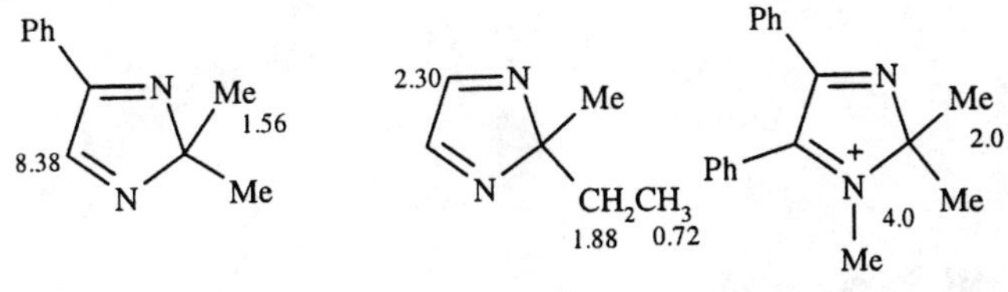

Ph; N; Me 1.56; 8.38; N; Me
2.30; N; Me; N; CH$_2$CH$_3$; 1.88 0.72
Ph; N; Me 2.0; Ph; N$^+$; 4.0; Me; Me

4*H*-Imidazoles[d]

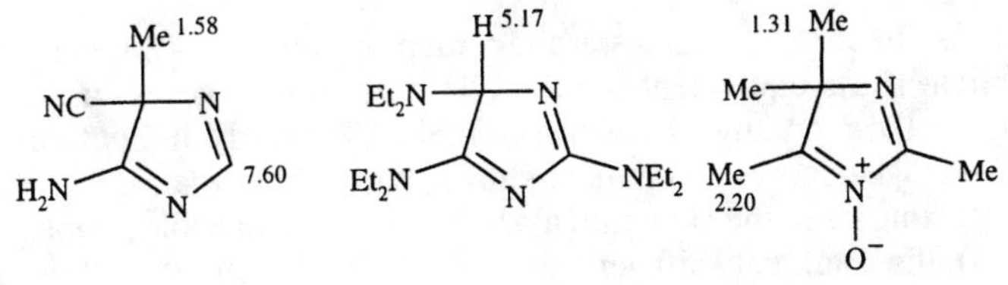

Me 1.58; NC; N; H$_2$N; N 7.60
H 5.17; Et$_2$N; N; Et$_2$N; N; NEt$_2$
1.31 Me; Me; N; Me; 2.20; N$^+$; Me; O$^-$

[a] ⟨83AHC(34)2⟩. [b] ⟨83AHC(34)54⟩. [c] ⟨84AHC(35)376⟩. [d] ⟨84AHC(35)414⟩.

Proton-proton coupling constants of benzo rings of benzazoles can illuminate the bonding in such compounds. Thus, comparison of the *J* values for naphthalene with those for benzotriazoles of different types (Table 16) shows evidence of bond fixation, particularly in the 2-methyl derivative (**115**) ⟨71PMH(4)121⟩.

Long range "through—space" spin—spin couplings ($^6J(^1H—^{19}F)$ and $^6J(^{19}F—^{19}F)$) are found in pyrazoles (**116**) between the *ortho* fluorine atom and R (R = CH$_3$, CHF$_2$, and CF$_3$) ⟨93MRC323⟩.

Variable-temperature NMR is useful for studying rotational barriers and tautomeric equilibria. Thus a degenerate rearrangement involving rapid exchange of aryl rings in the N-aryl triazole *N*-arylimides (**117a⇌117b**) is studied between 120 and −83°C. A single exchange-narrowed AA′BB′ spectrum for aromatic rings and an averaged signal for H$_1$H$_1'$ and H$_2$H$_2'$ is observed. As the temperature is lowered the signals are broadened. The coalescence temperature is 40°C and the rate of exchange at this point is 20 s^{-1} in DMSO. By −46°C the exchange is frozen out and the spectrum is that of form (**118**) with two separate AA′BB′ systems and two separate signals for H$_1$H$_1'$ and H$_2$H$_2'$. Restricted rotation of the

N-(p-nitrophenyl) bond in form (118) is also found at lower temperature (-50 to $-83°C$) ⟨87JCR(S)332⟩.

(117a)　　　　　　　　　　　　　　　　**(117b)**

(118)

2.4.3.4　^{13}C NMR Spectroscopy

Chemical shifts for aromatic azoles are recorded in Tables 17–20. Fast tautomerism renders two of the C-13 chemical shifts equivalent for the NH derivatives just as in the proton spectra (Table 17). However, data for the N-methyl derivatives (Table 18) clearly indicate that the carbon adjacent to a 'pyridine-like' nitrogen shows a chemical shift at lower field than that adjacent to a 'pyrrole-like' N-methyl group (in contrast to the H chemical shift behavior). In azoles containing oxygen (Table 19) and sulfur (Table 20), the chemical shifts are generally at lower field than those for the wholly nitrogenous analogues, but the precise positions vary.

The ^{13}C NMR parameters of 169 azoles with no other substitution in the ring show the effect of N-substituents ⟨88MRC134⟩.

N-Substituents have little effect on the ^{13}C chemical shifts of azoles with a few exceptions, just as for the ^{1}H chemical shifts. N-SnR$_3$ Substituents exchange readily between the different nitrogens of azoles. There is a parallelism with annular metallotropy (see Section 2.4.5.1.2); thus the chemical shifts of 1-[tri-(n-butyl)stannyl]benzotriazole are quite different from others ⟨88MRC134⟩. 1-Tri-fluoromethylsulfonyl-1,2,3-triazole does not exist as named but as its open-chain isomer, the diazoimine. Thus, the chemical shifts of C(4) and C(5) are 64.6 ppm and 172.6 ppm, respectively ⟨88MRC134⟩. This is the normal behavior of 1-substituted 1,2,3-triazoles with strong electron-withdrawing groups, such as 1-cyano-1,2,3-triazole ⟨81BSB615⟩ (see Sections 2.4.5.3.1 and 3.4.1.2.3) The C(4) and C(5) ^{13}C chemical shifts of 1-alkyl-1,2,3-triazolines appear between 41–48 ppm and 63–66 ppm, respectively ⟨93JOC2097⟩.

A comparison between the ^{13}C NMR chemical shifts of thiadiazole, oxadiazole, and isosydnone derivatives indicates that the effect of substituents is more pronounced for thiadiazoles than for oxadiazoles ⟨82OMR(18)159⟩.

Azolinone derivatives and the corresponding thiones and imines are listed in Table 21; only substituted derivatives have been measured frequently. The ^{13}C chemical shifts of non-aromatic azole derivatives are given in Tables 22–24; relatively few data are available and these are generally for substituted derivatives rather than for the parent compounds.

^{13}C NMR spectroscopy is a powerful technique for solving different structural problems. Thus, the protonation site of Δ^2-pyrazolines has been determined as N-1 using ^{13}C NMR spectroscopy. This general rule is not followed for 1-phenyl-3-aminopyrazolines which are protonated on N-2 〈95JCS(P2)1875〉. ^{13}C NMR spectroscopy easily differentiates the 1,1- (**119**) and the 1,2-disubstituted (**120**) pyrazolinium cations 〈86MRC551〉. For instance, the *N*-methyl groups appear at 54.1 ppm for (**119**), and at 42.1 ppm (position 1) and 36.0 ppm (position 2) for (**120**).

(**119**)   (**120**)

Long range ^{1}H—^{13}C coupling constants in pyrazoles and $^1J(^{13}$C—^{13}C) (C$_3$, C$_4$ = 51.5 Hz, C$_4$, C$_5$ = 64.6 Hz) coupling constants in 1-methylpyrazole have been used for confident assignment of isomeric 1,3- and 1,5-disubstituted pyrazoles 〈87T4663, 94MRC62〉.

^{13}C NMR studies, especially in the solid state 〈83H(20)1713〉, are of value in studies of tautomerism 〈83H(20)1713, 86ZC378〉. Solid state studies on imidazole (and pyrazole) show there are three distinct signals for the annular carbon atoms (imidazole: C-2, 136.3; C-4, 126.8; C-5, 115.3 ppm). Proton exchange does not occur in the solid, hence the compounds resemble their crystal structures. Comparison with the corresponding chemical shifts for 1-methylimidazole (137.6, 129.3, 119.7 ppm) implies that the tautomerism has been frozen in the solid state 〈81CC1207〉. Solid-state examination of 2,2′-bis-1H-imidazole also reveals "frozen" tautomerism.

Table 14 ^{1}H NMR Spectral Data for Ring Hydrogens of Azolines (Non-aromatic Azoles with One Ring Double Bond)

Class and Sub-Class	Substituents	^{1}H Chemical shifts (δ, ppm),						Coupling constants
		H-1	H-2	H-3	H-4	H-5	Solvent	J (Hz)
		Pyrazole						
2,3-Dihydro- [a]	1,2,3-trimethyl-	2.55	2.60	1.71	4.60	3.65	CDCl$_3$	3,5 = 1.8; 4,5 = 1.8
4,5-Dihydro-3*H*- [a]	—	—	—	4.27	1.46	4.27	Neat	—
4,5-Dihydro-[a]	—	5.33	—	6.88	2.65	3.31	CDCl$_3$	3,4 = 1.4; 4,5 = 9.8
		1,2,3-Triazole						
4,5-Dihydro- [b,d]	—	—	—	—	5.35	4.77	—	4,5 = 2.0
4,5-Dihydro- [c,d]	—	—	—	—	5.73	4.60	—	4,5 = 7.5
		1,2,4-Oxadiazole						
2,5-Dihydro- [e]	2,5-dimethyl-3-phenyl-	—	—	—	—	6.0	—	—
4,5-Dihydro- [f]	5-ethyl-3-phenyl-	—	—	—	5.38	5.64	—	4,5 = 4.5
		1,3,4-Oxadiazole						
2,3-Dihydro-[a]	3-benzoyl-5-phenyl-	—	5.9	—	—	—	—	—
		Thiazole						
2,5-Dihydro- [i]	—	—	4.79[g]	3.03[g]	6.12[g]	—	—	—
4,5-Dihydro- [j]	—	—	2.26	—	5.83	6.84	—	2,4 = 2.2; 4,5 = 8.6
		1,3,4-Thiadiazole						
2,3-Dihydro-[k]	2,3,5-triphenyl-	—	—	—	—	6.38	—	—
		1,2,4-Dioxazole						
2,3-Dihydro- [a]	3-carboxymethyl-5-phenyl-	—	—	6.39	—	—[l]	CDCl$_3$	—
		1,3,4-Dioxazole						
2,3-Dihydro-	2-benzyl-5-phenyl-	—	6.20	—	—	—[l]	CDCl$_3$	—
		1,3,4-Oxathiazole						
2,3-Dihydro-[a]	5-methyl-2-trichloromethyl-	—	6.30	—	—	—[l]	CDCl$_3$	—
		1,2,4-Dithiazole						
2,3-Dihydro-	5-phenyl-	—	—	5.70	—	—[l]	CCl$_4$	—

[a]Data taken from appropriate chapter, 'Comprehensive Heterocyclic Chemistry'. [b]Values given for *trans*-5-propyloxy-4-methyl-1-(4-nitrophenyl) derivative. [c]Values given for *cis* isomer of compound named in footnote b. [d]〈65CB1153〉. [e]〈73BSF2996〉. [f]〈77JOC1555〉. [g]Value taken from 5,5-dimethyl-Δ^3-thiazoline. [h]Value taken from 2,2,4-trimethyl-Δ^3-thiazoline. [i]〈66BSF3524〉. [j]〈64MI40100〉. [k]〈81JCS(P1)360〉. [l]Substituent in this position.

Table 15 ^{1}H NMR Spectral Data for Ring Hydrogens of Azolidines (Non-aromatic Azoles without Ring Double Bonds)

		[1]H Chemical shifts (δ, ppm)				
Compound	Substituents	H-2	H-3	H-4	H-5	Solvent
Tetrahydro						
-pyrazole[a]	1,2-dimethyl-3-phenyl-	—	6.51	7.85	6.76	CDCl$_3$
–thiazole[b,c]	—	5.9	8.2	6.8	7.2	—
-1,2,4-oxadiazole[d]	2-t-butyl-3,4-diphenyl-5-thioxo-	—	5.93	—	—	—
-1,3,4-oxadiazole[e]	3,4-dimethyl-	4.27	—	4.27	—	—
-1,2,4-dioxazole[e]	3,5-di-n-propyl-	—	4.63	—	4.63	—
-1,3,2-dioxazole[e]	N-alkyl-	—[f]	—	3.97	3.97	—
-1,3,4-dioxazole[e]	2,5-di-t-butyl-4-phenyl-	4.96	—	—[f]	4.58	—
-1,2,4-dithiazole[e]	3,4-dialkyl-3-phenyl-	—	—[f]	—[f]	4.72	—
-1,3,2-dithiazole[e]	2-methyl-	—[f]	—	3.58	3.58	—

[a]⟨71T133⟩. [b]$J_{4,5}$ values are $J_{A,B'} = J_{A',B} = 7.61$ Hz; $J_{A,B} = J_{A',B'} = 4.71$ Hz; $J_{5,5} = 13.7$ Hz; $J_{4,4} = -7.2$ Hz. [c]⟨74JA1465⟩. [d]⟨74JOC957⟩. [e]Data taken from appropriate chapter of 'Comprehensive Heterocyclic Chemistry'. [f]Substituent in this position.

Table 16 Proton-Proton Coupling Constants (Hz) in Benzotriazole[a]

Compound	$J_{4,5}$	$J_{5,6}$	$J_{4,6}$
Benzotriazole[b]	8.3	6.7	1.4
2-Methylbenzotriazole	9.4	3.6	0.5
Naphthalene[c]	8.6	6.0	1.4

[a]Data taken from ⟨71PMH(4)141⟩ which contains references to the original literature. [b]Rapid tautomerism between 1- and 3-positions occurs. [c]J values for naphthalene refer to $J_{1,2}$, $J_{2,3}$ and $J_{1,3}$, respectively.

Table 17 ^{13}C NMR Chemical Shifts for Nitrogenous Azoles: (a) NH Derivatives

	[13]C Chemical shifts (ppm)[a]				
Compound	C-2	C-3	C-4	C-5	Solvent
Pyrazole[b]	—	134.6	105.8	134.6	CH$_2$Cl$_2$
Imidazole[c]	135.9	—	122.0	122.0	—
1,2,3-Triazole[b]	—	—	130.4	130.4	Me$_2$CO
1,2,4-Triazole[c]	—	147.8	—	147.8	—
Tetrazole[c]	—	—	—	144.2	—

[a]All chemical shifts expressed in ppm from TMS (original values converted where necessary).
[b]Data taken from appropriate chapter, 'Comprehensive Heterocyclic Chemistry'. For pyrazoles ^{13}C NMR data see also ⟨84OMR(22)603, 93MRC107⟩.
[c]Data taken from ⟨71PMH(4)121⟩, which contains references to the original literature.

Table 18 ^{13}C NMR Chemical Shifts for Nitrogenous Azoles: (b) N(1)-Methyl Derivatives

	[13]C Chemical shifts (ppm)[a]				
Compound	C-2	C-3	C-4	C-5	Solvent	Ref.
Pyrazole	—	138.7	105.1	129.3	CDCl$_3$	[b]
Imidazole	138.3	—	129.6	120.3	—	74JOC357
1,2,3-Triazole	—	—	134.3	125.5	DMSO-d_6	[b]
1,2,5-Triazole	—	133.8	133.8	—	—	CHEC-II(4), 14
1,2,3,4-Tetrazole	—	—	—	142.1	DMSO-d_6	74JOC357
1,2,3,5-Tetrazole	—	—	151.9	—	DMSO-d_6	74JOC357

[a]All chemical shifts expressed in ppm from TMS (original values converted where necessary).
[b]Data taken from appropriate chapter, 'Comprehensive Heterocyclic Chemistry'.

Table 19 ^{13}C NMR Chemical Shifts for Azoles Containing Oxygen[a]

Compound	C-2	C-3	C-4	C-5	Solvent
		[13]C Chemical shifts (ppm)[b]			
Oxazole	150.6	—	125.4	138.1	CDCl$_3$
1,3,4-Oxadiazole[c]	159.5	—	—	166.3	—
1,2,5-Oxadiazoles[d]	—	139–160	—	—	—

[a]Data taken from appropriate chapter, 'Comprehensive Heterocyclic Chemistry'. [b]All chemical shifts expressed in ppm from TMS (original values converted where necessary). [c]Values given for 5-methoxy-2-methyl derivative. [d]3-Substituted 4-phenyl derivatives.

Table 20 ^{13}C NMR Chemical Shifts for Azoles Containing Sulfur

Compound	C-2	C-3	C-4	C-5	Solvent	Ref.
		[13]C Chemical shifts (ppm)[a]				
Isothiazole	—	157.0	123.4	147.8	CDCl$_3$	[b]
Thiazole	153.4	—	143.7	119.7	—	79HC(34-1)76
1,2,3-Thiadiazole	—	—	147.3	135.8	CDCl$_3$	[b]
1,2,4-Thiadiazole	—	170.0	—	187.1	—	81JPR279
1,3,4-Thiadiazole	152.7	—	—	152.7	CDCl$_3$	78BAP291
1,2,3,4-Thiatriazole						
5-phenyl-	—	—	—	178.5	CDCl$_3$	[b]

[a]All chemical shifts expressed in ppm from TMS (original values converted where necessary).
[b]Data taken from appropriate chapter, 'Comprehensive Heterocyclic Chemistry'.

Table 21 ^{13}C NMR Chemical Shifts for Azolinones

Class and Compound	C-2	C-3	C-4	C-5	Solvent	Ref.
		[13]C Chemical shifts (ppm)[a]				
2-Pyrazolin-5-one						
3-methyl-1-phenyl-	—	156.2	43.0	170.6	—	[b]
3-Pyrazolin-5-one						
2,3-dimethyl-1-phenyl-	—	156.0	98.1	165.7	—	[b]
Imidazolin-2-one	183.8	—	44.4	44.4	DMSO-d_6	80CS(15)193
Imidazoline-2-thione	164.8	—	40.3	40.3	DMSO-d_6	80CS(15)193
Tetrazolin-5-one						
1-phenyl-	—	—	—	148.7	—	[b]
Tetrazoline-5-thione	—	—	—	162.9	—	[b]
Δ[4]-Thiazoline-2-thione	—	118.4	128.9	114.0	CHCl$_3$	79HC(34-1)388
1,2,3-Oxadiazolin-5-imine						
3-methyl-	—	—	97.3	170.4	—	80RCR28
Δ[2]-1,3,4-Oxadiazolin-5-one	145.7	—	—	155.7	—	82OMR(18)159
Δ[2]-1,3,4-Thiadiazoline-						
5-thione	157.2	—	—	186.4	—	77JOC3725
1,3,4-Oxathiazolin-2-one						
5-methyl-	174.2	—	—	158.7	—	[b]
1,2,4-Dithiazoline-3-thione						
5-anilino-	—	209.3	—	179.1	—	[b]

[a] All chemical shifts expressed in ppm from TMS (original values converted where necessary).
[b] Data taken from appropriate chapter, 'Comprehensive Heterocyclic Chemistry'.

Table 22 ^{13}C NMR Chemical Shifts for Non-aromatic Azoles with Two Ring Double Bonds

Compound	C-2	C-3	C-4	C-5
		[13]C Chemical shifts (ppm)		
4*H*-Pyrazoles[a]	—	178–182	63–68	178–182
3*H*-Pyrazoles[b]	—	93–110	125–150	130–170
2*H*-Imidazoles[c]	101–119	—	158–165	158–165
4*H*-Imidazoles[d]	163–181	—	66–115	180–190

[a]⟨83AHC(34)54⟩. [b] ⟨83AHC(34)2⟩. [c] ⟨84AHC(35)376⟩. [d] ⟨84AHC(35)414⟩.

Table 23 ^{13}C NMR Chemical Shifts for Azolines (Non-aromatic Azoles with One Ring Double Bond)

Class and Compound	[13]C Chemical shifts (ppm)[a]				Ref.
	C-2	C-3	C-4	C-5	
Δ^2-**Pyrazoline**	—	142.9	33.2	45.4	80TH40100
Δ^2-**Imidazoline**					
1-methyl-2-methylthio-	165.3	—	[b]	54.3	76TL3313
Δ^4-**Imidazoline-2-thione**					
1-methyl-	161.3	—	[b]	119.8	76TL3313
Δ^2-**Thiazoline**	229.4	—	321.7	354.7	70CR(C)(270)1688
1,3,4-Thiadiazoline					
2-benzylamino-5-phenyl-5-tosyl-	151.5	—	—	149.5	[c]
Δ^2-**1,3,4-Oxathiazoline**					
2-phenyl-5-trichloromethyl-	95.9	—	—	157.6	[c]
1,2,4-Dithiazoline	—	146.36[d]	—	156.19[e]	[c]

[a] All chemical shifts expressed in ppm from TMS (original values converted where necessary).
[b] No C-4 shift reported. [c] Data taken from appropriate chapter, 'Comprehensive Heterocyclic Chemistry'.

Table 24 ^{13}C NMR Chemical Shifts for Azolidines (Non-aromatic Azoles without Ring Double Bonds)

Class and Compound	[13]C Chemical shifts (ppm)[a]				Ref.
	C-2	C-3	C-4	C-5	
Imidazolidine-2-thione, 1-methyl-	183.2	—	[b]	50.4	76TL3313
Oxazolidine, cis-4-methyl-5-phenyl-	85.4	—	60.3	80.3	79MI40100
Thiazole, tetrahydro-	248.4	—	245.9	226.9	74CR(C)(279)717
1,3,4-Thiadiazolidine, 2,2-dimethyl- 4-phenyl-5-phenylimino-	79.2	—	—	176.3	80JCS(P1)574
1,3,2-Dioxazole, tetrahydro-*N*-alkyl-	—	—	67.6	67.6	[c]
1,2,3-Oxathiazole, dihydro-*N*-phenyl, *S*-oxide	—	—	45.9	70.9	[c]

[a] All chemical shifts expressed in ppm from TMS (original values converted where necessary).
[b] C-4 shift not reported. [c] Data taken from appropriate chapter, 'Comprehensive Heterocyclic Chemistry'.

2.4.3.5 Nitrogen and Oxygen NMR Spectroscopy

A comprehensive ^{15}N study of azoles has been reported ⟨84OMR(22)215⟩. The main trends can be seen from a consideration of the corresponding data for 1,2,3-triazole and benzotriazole and some of their N-substituted derivatives (Scheme 10). The ^{15}N chemical shifts are determined using CDCl[3] and DMSO with respect to external nitromethane and are corrected to external ammonia by the addition of 380.2 ppm. The shielding of a pyrrole-type nitrogen is greater than that of a pyridine-type nitrogen. The pyridine-type nitrogen atom is deshielded when connected to the pyrrole-type nitrogen atom. Thus the order of shielding in 1-methyl-1,2,3-triazole and 1-methylbenzotriazole is N(1) >> N(3) > N(2), and in the 2-methyl isomers, N(2) > N(1), N(3). The large downfield shift of the pyridine-type nitrogen atoms is attributed to a large paramagnetic shielding term associated with the nonbonding electron pair. Fusion of benzene ring leads to shielding of adjacent nitrogen atoms, both pyridine- and pyrrole-type. The non-adjacent N(2) atom on benzoannulation is deshielded in 1-methylbenzotriazole and is shielded in 2-methylbenzotriazole. Solvent effects on chemical shifts are significant, particularly for unsubstituted triazole and benzotriazole.

Some other data on ^{15}N chemical shifts, referred to internal MeNO[2], are recorded in Table 25. Comparison of pyrrole with imidazole demonstrates that the deshielding of the pyrrole-type nitrogen is about 15 ppm. Increased nitrogen shielding is observed for 2,1,3-benzothiadiazole (+49 ppm) and the parent 1,2,5-thiadiazole (+34 ppm). In contrast, the corresponding oxadiazoles were deshielded, as in 2,1,3-benzooxadiazole (−36 ppm) and 1,2,5-oxadiazole (−34 ppm).

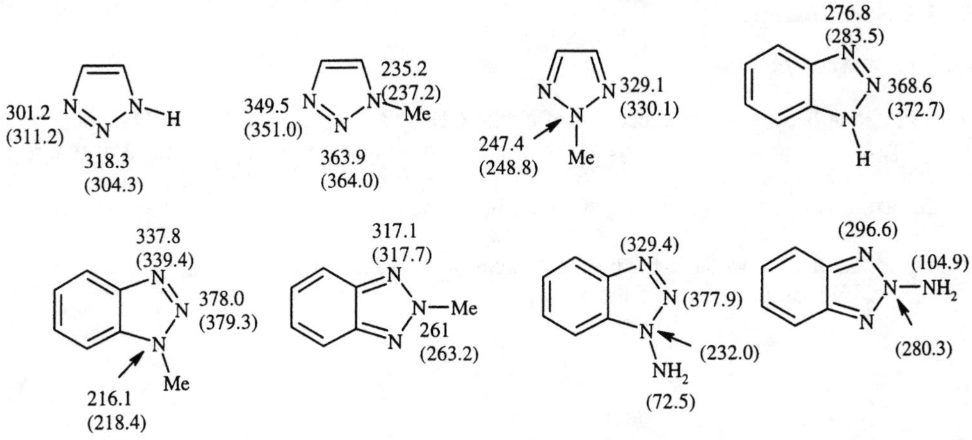

Scheme 10 The ^{15}N chemical shifts of 1,2,3-triazoles and benzotriazoles (in CDCl$_3$ or DMSO-d$_6$ (in brackets))

There is a growing interest in the use of ^{15}N NMR spectroscopy for elucidation of various structural problems of azole chemistry, especially tautomerism. For example, the mole fractions of the prototropic tautomers were obtained from the ^{15}N chemical shifts of the NH tautomers and the corresponding N-methyl derivatives. By this method, the average mole fraction for the 2-NH tautomer of benzotriazole is 0.02 in both CDCl$_3$ and DMSO, and that of 1,2,3-triazole is 0.34 in CDCl$_3$ and 0.55 in DMSO ⟨82JOC5132⟩.

Low temperatures studies, for instance in THF at 175K, can elucidate annular tautomerism ⟨92JCS(P2)1737⟩; a large collection of ^{15}N chemical shifts of NH-pyrazoles with tautomerism frozen has been published ⟨94MRC699⟩. The tautomerism is also frozen in the solid state. Thus two ^{15}N shifts at 150 ppm (NH) and 222 ppm (=N-) are observed for solid imidazole; the average is close to the single signal at 186 ppm found in solution.

The discovery, by the combined use of crystallography and CPMAS NMR, of the dynamic properties of 3,5-dimethylpyrazole (an H-bonded trimer, *cf.* **122**) in the solid state opened a new field of research ⟨85JA5290, 92JA9657⟩. Other cyclic structures, dimers like (**121**) and tetramers like (**123**) were discovered ⟨94JHC695⟩. ^{15}N CPMAS (with labeled derivatives) proved to be superior to ^{13}C CPMAS.

Some other examples of application of ^{15}N NMR spectroscopy include establishing the protonation site of Δ^2-pyrazolines ⟨87MI301-01⟩ and assigning the structures to isomeric N-7 and N-9 substituted purines ⟨86T5073⟩.

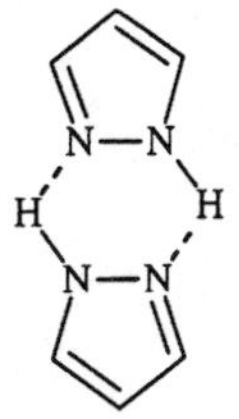

(**121**)   (**122**)   (**123**)

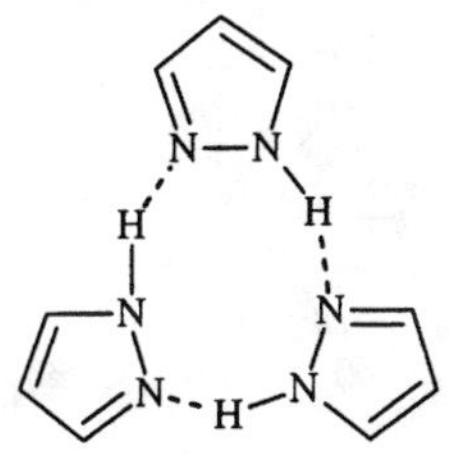

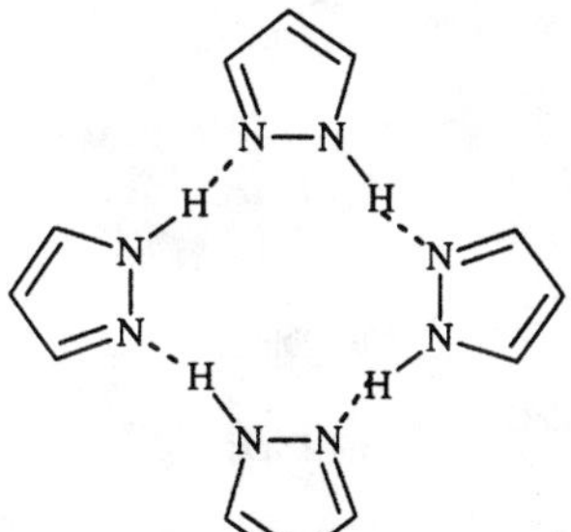

Data on ^{17}O NMR spectra of five-membered oxygen-containing heterocycles are scarce. ^{17}O chemical shifts, δ_0 (with respect to H$_2$$^{17}O$), near 350 ppm have been observed for isoxazoles and 280 and 303 ppm for 1,2-dioxolane (**124**) and its bridged analogue (**125**) ⟨85JOC4484⟩.

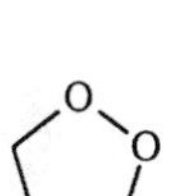

(**124**)   (**125**)

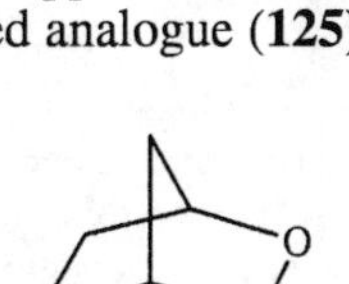

2.4.3.6 UV Spectroscopy

2.4.3.6.1 *Parent compounds*

In general, aza substitution (replacement of cyclic CH by N) has little effect on UV spectra (Table 26). Typically, aza analogues of pyrrole show λ_{max} 217 nm or lower with log ε *ca.* 3.5, whereas aza analogues of thiophene show λ_{max} 230–260 nm with log ε *ca.* 3.7 (Table 26); insufficient data are available for aza analogues of furan to generalize but these compounds appear to have maxima below 220 nm.

Relatively few data are available for protonated cationic species, it appears that protonation has little effect on the position and intensity of the absorption.

2.4.3.6.2 *Benzo derivatives*

Benzo derivatives show at least two, and up to seven, maxima in the range 200–320 nm (Table 27). The longest wavelength maximum occurs at 275–315 nm and generally at rather longer wavelengths for the sulfur derivatives than for their N or O analogues, and also for the benzo[*c*] compared with the benzo[*b*] derivatives.

Table 25 ^{15}N NMR Chemical Shifts for Nitrogenous Azoles[a,b]

Nitrogen position(s)	Positions of ring nitrogen	^{15}N Chemical shifts with heteroatom in 1-position (ppm)			
		NH	NMe	O	S
2-Aza	1	−135	−181	—	—
	2	−135	−76	+3	−82
3-Aza	1	−173	−221	—	—
	3	−173	−125	−124	−57
2,3-Diaza	1	−128	−144	—	—
	2	−60	−15	−35[c]	—
	3	−60	−28	−106[d]	—
2,4-Diaza	1	−174	−173	—	−106
	2	—	−84	−20	−186[e]
	4	−132	−130	−140	−70
2,5-Diaza	1	−128	−132	—	—
	2,5	−60	−53	+33	−35
3,4-Diaza	1	—	−222	—	—
	3,4	—	−82	−82	−10
2,3,4-Triaza	1	−106	−159	—	—
	2	−15	−14	—	—
	3	−25	+9	—	—
	4	−106	−54	—	—
2,3,5-Triaza	1	—	−99	—	—
	2	—	−4	—	—
	3	—	−43	—	—
	5	—	−69	—	—

[a]Chemical shifts expressed in ppm referred to internal MeNO$_2$.
[b]Data taken from ⟨B-73MI40100⟩ or ⟨81MI40100⟩, which contain references to the original literature. [c]Value for 3-methylsydnone ⟨80OMR(13)274⟩. [d]Value for 5-acetyl-3-methylsydnonimine ⟨80OMR(13)274⟩. [e]Value for 2-amino-5-methyl-1,3,4-thiadiazole ⟨78H(11)121⟩.

Table 26 UV Absorption Maxima for Azoles[a]

Additional nitrogen atoms	Z = NH λ_{max} (nm) (log ε)	Z = O λ_{max} (nm) (log ε)	Z = S λ_{max} (nm) (log ε)
2	210 (3.53)	211 (3.60)	244 (3.72)
3	207–208 (3.07) end absorption	240	207.5 (3.41), 233 (3.57)[b]
2,3-di	210 (3.64)	—	211 (3.64), 249 (3.16), 294 (2.29)[c]
2,4-di	216.5 (3.66)	—	229 (3.73)
2,5-di	—	220	250 (3.86), 253 (3.87), 257 (2.83), 260 (3.68)[d]
3,4-di	—	200[e]	220
2,3,4-tri	205	—	280 (4.03)[f]

[a] Unless otherwise indicated data taken from ⟨71PMH(3)67⟩ which contains references to the original literature. [b] Measured in ethanol. [c] Measured in cyclohexane. [d] Measured in isooctane; *cf.* CHEC 4.26.2.4. [e] 2-Methyl-1,3,4-oxadiazole exhibits a maximum at 206 nm (log ε 2.62) in MeOH; *cf.* CHEC 4.23.2.2.2. [f] 5-Phenyl-1,2,3,4-thiatriazole.

Table 27 UV Absorption Maxima for Benzazoles[a]

X	Y	Z = NH λ_{max} (nm) (log ε)	Z = O λ_{max} (nm) (log ε)	Z = S λ_{max} (nm) (log ε)
N	CH	250 (3.65), 284 (3.63), 296 (3.52)	235 (4.00), 243 (3.91), 280 (3.46)	205 (4.20), 222 (4.37), 252 (3.56), 261 (3.39), 297 (3.58), 302 (3.57), 3.08 (3.56)[b]
CH	N	242 (3.72), 265 (3.58), 271 (3.70), 277 (3.69)	231 (3.90), 263 (3.38), 270 (3.53), 276 (3.51)	217(4.27), 251(3.74), 285 (3.23), 295 (3.13)
N	N	259 (3.75), 275 (3.71)	[c]	213.5 (4.20), 266 (3.72), 312.5 (3.40)
S[+]	N	—	—	238 (3.91), 350 (4.35), 425 (3.29)[d]

X	Y	Z = NH λ_{max} (nm) (log ε)	Z = O λ_{max} (nm) (log ε)	Z = S λ_{max} (nm) (log ε)
N	CH	275 (3.80), 292 (3.79), 295 (3.78)	—	203 (4.16), 221 (4.21), 288s (3.88), 298 (3.46), 315s (3.60)[e]
N	N	274 (3.96), 280 (3.98), 285 (3.97)	310 (3.5), 275 (3.6)	221–222 (4.16), 304 (4.14), 310 (4.14), 330s (3.39)

[a] Unless otherwise indicated data taken from ⟨71PMH(3)67⟩ which contains references to the original literature. [b] ⟨73JHC267⟩. [c] Unknown; see Scheme 2 in CHEC 4.2.1.1. [d] In concentrated H_2SO_4 ⟨69ZOR153⟩. [e] Data taken from CHEC 4.17.3.6.

2.4.3.6.3 Effect of substituents

Although UV spectra have been measured for a large number of substituted azoles, there has been little systematic attempt to explain substituent effects on such spectral maxima. Readily available data are summarized in Table 28, and some major trends are apparent. However, detailed interpretation is

Table 28   Effects of Substituents on UV Absorption Maxima (nm)[a,b]

Ring numbering: Z = position 1; positions 2, 3, 4, 5 around the five-membered ring.

Substituent	Aza positions	Z = NH Substituent position				Z = O Substituent position				Z = S Substituent position			
		2	3	4	5	2	3	4	5	2	3	4	5
Me	2	210	213	220	(≡3)	—	217	221	213	—	246.5	251[c]	243[c]
	3	—	—	215	(≡4)	235	—	—	(≡2)	235	—	241	239
	2,4	—	—	216	(≡4)	—	—	—	—	—	—	213	217
	2,5	—	<220	—	(≡3)	—	—	—	—	—	—	—	—
	3,4	—	—	—	—	—	—	—	—	—	—	—	—
Ph	2	271	248	250.5	(≡3)	206	239	243	260	287	257[m]	266[d]	266[d]
	2,3	—	—	260	(≡4)	263	310[e]	—	267	—	270, 291	252	275
	2,4	—	—	245	(≡4)	—	238[e]	—	—	—	—	—	—
	2,5	—	241.5	—	(≡3)	—	—	—	250[e]	—	—	—	(≡2)
	3,4	—	—	—	241	—	—	—	(≡2)	—	—	—	—
Ph	2,3,4	—	—	—	(≡3)	247.5	—	(≡3)	(≡2)	268	280[m]	256[c]	241[i]
Br (Cl)	3	220*	—	221	(≡4)	245	220	—	—	245	(244)[f]	246	247
	2,3	—	—	217	(≡3)	239	—	—	—	239	255[g]	—	—
OMe (OH)	3	—	<220	234*	(≡4)	—	—	(≡3)	—	—	273	(≡3)	(258)
	2,3	—	—	—	(≡4)	—	—	—	—	—	—	—	—
	2,3,4	—	—	—	213*	—	228.5	—	245.5	254.4*	278[h]	286[e]	247[i]
NH₂ (NMe₂)	2	—	228	230*	(≡3)	—	—	—	(≡2)	251	—	—	(≡2)
	3	—	—	228	(≡4)	—	—	—	—	—	—	—	—
	2,4	—	<220	—	(≡4)	—	—	—	—	—	—	—	—
	3,4	—	—	—	(≡3)	—	—	—	244	—	—	—	263[i]
CO₂Me (CO₂H, CHO)	2	285	217	256.5	218	243[k]	—	216[j]	(≡2)	—	256	(250)[i]	263[i]
NO₂	3	325	261	275	(≡4)	—	—	(≡3)	(≡2)	—	263	230	(≡2)
	2,5	—	—	298	(≡3)	—	—	(≡3)	—	—	—	(≡3)	—
	3,4	—	215, 245[l]	—	(≡4)	—	239	239	—	—	—	272	—
	2,4	—	—	—	(≡4)	—	—	—	—	—	—	—	—
	2,5	—	—	—	(≡3)	—	—	(≡3)	—	—	—	(≡3)	—

[a] Unless otherwise indicated, data taken from ⟨71PMH(3)67⟩, which contains references to the original literature. A dash indicates that a substituent cannot be in that position; a blank means that the value has not been reported. [b] Asterisk (*) indicates, exceptionally, acidified or basified solvent; for details, see references quoted in ⟨71PMH(3)67⟩. [c] ⟨64CS446⟩. [d] In cyclohexane ⟨66T2119⟩. [e] ⟨64HCA942⟩. [f] Value based on 'average increment' observed for a large number of compounds; cf. CHEC 4.17. [g] ⟨70T2497⟩. [h] Perchlorate salt in MeOH ⟨71HC657⟩. [i] ⟨54CB57⟩. [j] 2-Benzyl-5-methyl derivative. [k] 2-Benzyl-5-methyl derivative. [l] In water. [m] cf. CHEC 4.26.2.4.

hindered by the fact that different solvents have been used and that in aqueous media it is not always clear whether a neutral, cationic or anionic species is being measured. Furthermore, values below 220 nm are of doubtful quantitative significance.

π-Excessive five-membered heterocycles with strong π-acceptor substituents rather often display phenomenon of intramolecular charge transfer which can be of practical significance. Thus, intramolecular charge transfer in 1,2-dithioles of the type (**126**) allows a photoinduced color response, of use in optical filters ⟨88GEP3636157⟩.

(126)

2.4.3.7 IR Spectroscopy ⟨71PMH(4)265, 63PMH(2)161⟩

Most fundamental work on the vibrational spectra of azoles appeared in the period 1960–1980. Examples of more recent work include: (i) a complete assignment of the gas-phase IR spectrum of indazole ⟨93JCS(F1)4005⟩; (ii) IR spectral data were used to determine the enthalpies of O—H . . . N and N—H . . . O bonds in complexes of formic acid and 3,5-dimethylpyrazole ⟨87MI301-01⟩; (iii) the vibrational assignment of the Raman spectrum of polycrystalline pyrazole ⟨92MI301-01⟩ based on 3-21G calculations.

2.4.3.7.1 Aromatic rings without carbonyl groups

For many of the parent compounds, complete assignments have been made ⟨71PMH(4)265⟩. For substituted derivatives, group frequencies have been derived. In the ring systems under discussion, IR bands may be placed in the following categories: (a) CH stretching modes near 3000 cm^{-1} which are of little diagnostic utility; (b) ring stretching modes at 1650–1300 cm^{-1} (Table 29), four or five bands generally being found which lie in well-defined regions. The intensities of these bands vary according to the nature and orientation of substituents; (c) CH and ring deformation modes at 1300–1000 cm^{-1} (Table 30) and at 1000–800 cm^{-1} (Table 31). The former are largely in-plane CH and the latter out-of-plane CH and in-plane ring deformation modes; (d) substituent vibrations: these are discussed later.

Table 29 Azoles: IR Ring Stretching Modes in the 1650–1300 cm^{-1} Region[a]

Compound	Stretching modes (cm^{-1})				
Isoxazoles	1650–1610	1580–1520	1510–1470	1460–1430	1430–1370
Isothiazoles	—	—	1488	1392	1342
Pyrazoles	—	1600–1570	1540–1510	1490–1470	1380–1370
Oxazoles	1650–1610	1580–1550	1510–1470	1485	1380–1290
Thiazoles	1625–1550	—	1550–1470	1440–1380	1340–1290
Imidazoles	1605	1550–1520	1500–1480	1470–1450	1380–1320
1,2,4-Oxadiazoles	—	1590–1560	—	1470–1430	1390–1360
1,2,5-Oxadiazoles	—	1630–1560	1530–1515	1475–1410	1395–1370
1,3,4-Oxadiazoles	1680–1650	1630–1610	1600–1580	1430–1410	—
1,2,4-Thiadiazoles	—	1590–1560	1540–1490	—	—
1,2,3-Thiadiazoles	1650–1590	—	1560–1420	1350–1325	1260–1180
1,2,5-Thiadiazoles[b]	—	—	—	1461	1350
1,2,3,4-Thiatriazoles	1720–1690	1610–1530	—	—	1300–1260
1,2,3-Triazoles	1650–1615w	—	1530–1485	1440w	1420–1400w
1,2,4-Triazoles	—	—	1545–1535	1470–1460	1365–1335
Tetrazoles	1640–1615	—	1450–1410	1400–1335	1300–1260

[a]Data taken from ⟨71PMH(4)265⟩, which contains references to the original literature; w = weak.
[b]Data taken from CHEC 4.26.2.5.

Table 30 Azoles: Characteristic IR Bands in the 1300–1000 cm^{-1} Region[a]

Compound	β (CH) modes (cm^{-1})			Ring breathing (cm^{-1})
Isoxazoles	1218m	1155–1130	1088s	1028–1000
Isothiazoles	—	1070m	1060m	980s
Pyrazoles	1310–1130	1160–1090	1090–990	1040–975
Thiazoles	1240–1230	1160–1075	1105–1055	1040
Imidazoles	1285–1260	1140	1100	1060
1,2,4-Oxadiazoles	—	—	1070–1050	—
1,2,5-Oxadiazoles	1360–1175	1190–1150	1160–1150	1035–1000
1,2,4-Thiadiazoles	1270–1215	1185–1170	1160–1080	1050
1,2,3-Thiadiazoles	—	—	1150–950	—
1,2,5-Thiadiazoles[b]	1251–1227	—	1041	—
1,3,4-Thiadiazoles	1230–1165	1190–1120	1075–1045	1040
1,2,3,4-Thiatriazoles	1235–1210	1120–1090	1060–1030	—
1,2,3-Triazoles	1300–1275	1150–1070	1095–1045	1005–970
Tetrazoles	1210–1110	1170–1035	1060	995–900

[a] Data taken from ⟨71PMH(4)265⟩, which contains references to the original literature; s = strong, m = medium, w = weak. [b] Data taken from CHEC 4.26.2.5.

Table 31 Azoles: Characteristic IR Bands below 1000 cm^{-1} [a]

Compound	CH modes (?) (cm^{-1})		β-Ring (?) (cm^{-1})	CH modes (?) (cm^{-1})
Isoxazoles	970–920	899	945–845	774
Isothiazoles	915w	—	810s	740vs
Pyrazoles	960–930	860–855	805–790	765–750
Thiazoles	980–880	890–785	800–700	745–715
Imidazoles	970–930	895	840	760
1,2,4-Oxadiazoles	—	—	915–885	750–710
1,2,5-Oxadiazoles	980–900	—	890–825	715–700
1,2,3-Thiadiazoles	—	—	910–890	—
1,2,4-Thiadiazoles	1030–935	890	860–795	750–740
1,2,5-Thiadiazoles	—	—	860–800	780[b]
1,3,4-Thiadiazoles	975–905	905–875	850	775–750
1,2,3,4-Thiatriazoles	960–930	—	910–890	—
1,2,3-Triazoles	—	855–825	970–700	—
1,2,4-Triazoles	—	—	865–855(?)	—
Tetrazoles	960	—	—	810–775

[a] Data taken from ⟨71PMH(4)265⟩ which contains references to the original literature; vs = very strong, s = strong, w = weak. [b] Data taken from CHEC 4.26.2.5.

2.4.3.7.2 Azole rings containing carbonyl groups

The carbonyl group is found in a wide variety of situations in five-membered rings accommodating diverse heteroatoms. Carbonyl frequencies to be discussed range from about 1800 to below 1600 cm^{-1}; some such frequencies are so low that they often fail to be recognized as carbonyl absorptions at all, *e.g.* carbonyl joined to a heterocyclic ring *via* an exocyclic double bond as in structure (**127**). Low frequency $v(C=O)$ bands are widespread because five-membered heterocyclic rings are inherently π-electron donors. The transfer of electrons into the exocyclic double bond [$v(C=Z)$ of **127**] is likely to be extensive when both X and Y are electron donor atoms.

(127)

(128)

In Table 32 the $v(C=O)$ and other characteristic bands are given for some saturated five-membered heterocycles, and compared with the corresponding absorption frequencies for cyclopentanone. Adjacent NH groups and sulfur atoms have the expected bathochromic effect on $v(C=O)$, whereas an adjacent oxygen atom acts in the reverse direction. The CH_2 vibrations of cyclopentanone are repeated to a considerable extent in the heterocyclic analogues.

Table 33 reports $v(C=O)$ for a variety of azolinones containing ring double bonds. The hypsochromic effect of an oxygen atom or CR_2 group *versus* the bathochromic effect of NR, S or $C=C$ can readily be traced.

IR spectroscopy can be applied to study molecular association with participation of a carbonyl group. Thus, examination of the hydrogen bonding of saccharin, thiosaccharin, and their salts ⟨86JST(142)275, 92JST(267)197⟩ shows that saccharin exists in the solid state as a symmetric dimer (128), which gives three bands in the N—H stretching region (3090 cm^{-1}, 2970 cm^{-1}, and 2700 cm^{-1}), one of which is an overtone.

2.4.3.7.3 Substituent vibrations

In general, substituent frequencies in azoles are consistent with those characteristic of the same substituents in other classes of compounds. Some characteristic trends are found, and these have been used to measure electronic effects. Thus, for example, the frequencies of $v(C=O)$ in 3-, 4- and 5-alkoxycarbonylisoxazoles (*cf.* 129) are respectively 9–12, 2–8 and 17–18 cm^{-1} higher than those of the corresponding alkyl benzoates, indicating the following order of electron donor power: phenyl > 4- > 3- > 5-position of isoxazole. Similar work has been reported for other ring systems and substituents ⟨63PMH(2)161⟩.

(129) (130) (131) (132)

Table 32 IR Absorption Assignments for 2,5-Dihetero Derivatives of Cyclopentanone[a]

C=X:	C=O	C=O	C=S	C=O	C=O	C=S	C=S	C=O	C=S	C=O	C=S
2-Position:	CH_2	CH_2	CH_2	O	S	S	O	O	S	NH	NH
5-Position:	O	N	N	O	S	S	NH	NH	NH	NH	NH
Vib type	(cm^{-1})	(cm^{-1})	(cm^{-1})	(cm^{-1})	(cm^{-1})	(cm^{-1})	(cm^{-1})	(cm^{-1})	(cm^{-1})	(cm^{-1})	(cm^{-1})
$v(C=X)$	1770	1695	1109	1795	1638	1058	1171	1724	1047	1661	1208
CH_2 scissors	1490	1494	1478	1483	1434	1416	1464	1485	1458	1488	1459
	1466	1440	1458	1422	1422	1370	1402	1412	1432	1449	1368
	1426	1426	1420	1394					1380		
	1379	1378									
CH_2 wag	1284	1270	1309	1226	1275	1275	1319	1333	1250	1200	—
	1242	1230	1215		1254	1243	1230	1230	1203		
CH_2 twist	1192	1169	1168	1175	1158	1148	1203	—	1160	—	—
CH_2 rock	990	995	970	1005	983	983	969	967	998	988	980
v ring	1166	1285	1290	1140	888	882	1290	1250	1294	1270	1273
	1037	1068	1059	1071	826	831	1035	1077	1080	1103	1042
	890	915	915	971	677	670	942	1021	650	1037	1003
	929	887	880	894	939	946	914	918	927	933	919
γ ring	800	805	493	773	—	457	—	770	—	768	—

[a] Data taken from ⟨63PMH(2)161⟩, which contains references to the original literature.

Otting ⟨56CB1940, 57MI40100⟩ has shown that the $v(C=O)$ for acetylazoles (*cf.* **130**) increases with the number of cyclic nitrogen atoms (Table 34). Additional nitrogen atoms in the ring act as powerful electron-withdrawing substituents and decrease the importance of forms such as (**131**). Staab *et al.* ⟨57MI40100⟩ give data for benzo analogues (Table 34) and show that $v(C=O)$ values for substituted benzoyl-imidazoles (**132**) and -triazoles follow the Hammett equation. For data on *N*-acylpyrazoles, see reference ⟨59MI40100⟩.

Table 33 $v(C=O)$ Frequencies for Some Azolinones[a]

Ring system substituent(s) R = H or Me	Z = NH (cm^{-1})	Z = O (cm^{-1})	Z = S (cm^{-1})
	1705–1695[b]	—	—
	1680–1630	—	—
	—	—	1660[c]
	1684–1677[d]	1780–1750	1640[e]
	—	1770–1740	1780–1700
	—	*ca.*1820	1725[f]

[a]Unless otherwise indicated, data taken from appropriate chapter, 'Comprehensive Heterocyclic Chemistry'. [b]Z = NR', R' = alkyl, aryl. [c]R = Me ⟨64TL1477⟩. [d]Data taken from ⟨63PMH(2)161⟩ which contains references to the original literature. [e]⟨79HC(34-2)421⟩. [f]⟨79HC(34-2)430⟩.

Table 34 Carbonyl Frequencies for N-Acetylazoles[a]

N-Acetylazole	$v(C=O)$ (cm^{-1})	N-Acetylazole	$v(C=O)$ (cm^{-1})
Pyrrole	1732	Indole	1711
Imidazole	1747	Benzimidazole	1729
1,2,4-Triazole	1765	Benzotriazole	1735
Tetrazole	1779		

[a]Data taken from ⟨56CB1940⟩ and ⟨57MI40100⟩ which contain references to the original literature.

Substituent vibrations in IR spectra have been extensively used to determine tautomeric structure, particularly C=O, NH, OH and SH stretching modes. For example, 3-hydroxy-isoxazoles show broad $v(OH)$ at 2700 cm^{-1} (dimers). Boulton and Katritzky developed a technique for determining the structure of potentially tautomeric amino compounds by partial deuterium exchange. If the compound under investigation contains an amino group, the change from NH$_2$ to NHD produces a new single $v(NH)$ at a frequency between those of the original doublet for asymmetrical and symmetrical $v(NH_2)$.

If the compound does not contain an amino group and the two bands are derived from two separate NH groups, there should be no new band between the original two in the partially deuterated derivative. The method was applied to aminoisoxazoles ⟨61T(12)51⟩.

2.4.3.8 Mass Spectrometry ⟨66AHC(7)301, B-71MS⟩

Among the most important fragmentation pathways of the molecular ions of azoles are the following.

(a) Loss of RCN or HCN; this occurs particularly readily for systems containing (133), *i.e.* imidazoles and thiazoles. It does not occur so readily for oxazoles, but is found again in the 1,2,4-oxadiazoles (134) and also for pyrazoles (135).

(b) Loss of RCO⁺; this is important for systems (136) in oxazoles and in 1,3,4-oxadiazoles. It is also found in isoxazoles, where it probably occurs after skeletal rearrangement of the isoxazole *via* (137) into an isomeric oxazole.

(c) Loss of NO⁺ and/or NO· occurs for furazans (138) and sydnones (139).

(d) Loss of N₂ from triazoles and tetrazoles (but *not* pyrazoles).

(e) Whenever the structural element (140) occurs, a McLafferty rearrangement can take place (140→141).

(f) Isothiazoles are rather stable and show intense molecular ions with some HCN loss, probably *via* rearrangements analogous to (137) to give thiazoles.

(133) X = S or NR¹   (134)   (135)   (136)

(137)   (138)   (139)   (140)   (141)

There are correlations between mass spectral fragmentations and thermal and photochemical fragmentations and rearrangements; see Sections 3.4.1.2.1 and 3.4.1.2.2.

2.4.3.9 *Photoelectron Spectroscopy* ⟨74PMH(6)1⟩

In this method, photons of an energy much above of the ionization potential are directed onto a molecule. The photoelectron spectrum which results allows assessment of the energies of filled orbitals in the molecule, and thus provides a characterization of a molecule. Comparisons between photoelectron spectra of related compounds give structural information, for example, on the tautomeric structure of a compound by comparison of its spectrum with those of models of each of the fixed forms.

Photoelectron spectroscopy has demonstrated there is no linear correlation between the ionization energies and pK_a values of azoles ⟨84JHC269⟩.

Photoelectron spectra have been discussed and assigned in the following series: pyrazole (CHEC 4.04.1.4.9), 1,2,3-triazole (CHEC 4.11.3.2.9), isothiazole (CHEC 4.17.2.2), 1,3,4-oxadiazole (CHEC 4.23.2.2.5), 1,2,5-thiadiazole (CHEC 4.26.2.2), 1,3,4-thiadiazole (CHEC 4.27.2.3.10), 1,3-dioxolane (CHEC 4.30.1.4.5), 1,2-dithiole (CHEC 4.31.1.4), and 1,2,4-trioxolane and 1,2,4-trithiolane (CHEC 4.33.2.2.5).

The photoelectron spectra of pyrazole has been assigned of using *ab initio* CI calculations ⟨87CPH249⟩.

Table 35 Melting and Boiling Points[a,b,c]

Ring system	H	Me	Et	COMe	CO₂H	CO₂Et	CONH₂	CN	NH₂	OH	OMe	SH	SMe	Cl	Br
Benzene	80	111	136	212	122	211	130	190	184	43	154	168	187	131	155
Pyrrole-1	130	114	129	180	95d	180	166	—	—	83[g]	—	—	—	—	—
Pyrrole-2	130	148	181	90	205d	39	174	—	—	—	—	—	208–14	—	—
Pyrrole-3	130	158	179	115	148	78d	152	—	—	—	—	—	—	—	—
Furan-2	31	64	92	31	133	34	142	147	68	80	110	—	—	78	103
Furan-3	31	65	—	54	122	179	168	—	—	58	—	—	—	80	103
Thiophene-2	84	113	133	214	129	218	180	196	214	217	156	166	—	128	150
Thiophene-3	84	115	135	57	138	208	178	179	—	—	—	171	—	136	157
Pyrazole-1	70	127	137	234	103	213	141	37	—	—	—	—	—	—	—
Pyrazole-3	70	205	209	101	214d	160	148	150	40	166	—	—	>370d	40	70
Pyrazole-4	70	207	244	114	278	79	—	92	81	118	62	—	—	77	97
Isoxazole-3	95	118	139	16	149d	—	134	168	—	—	—	—	—	—	—
Isoxazole-4	95	127	—	—	—	—	—	—	—	—	—	—	—	—	130
Isoxazole-5	95	121	—	52	149	218	174	—	—	—	—	—	—	—	—
Imidazole-1	90	199	226	102	—	—	—	—	—	—	—	—	—	—	—
Imidazole-2	90	141	80	80	164d	157	—	—	—	250d	—	227	139	165	207
Imidazole-4	90	56	—	—	275d	—	215	—	—	—	—	—	—	—	130
Oxazole-2	69	87	—	—	—	—	—	—	97	—	—	—	—	—	—
Oxazole-4	69	—	—	—	142	48	—	—	—	—	—	—	—	—	—
Thiazole-2	118	128	158	226	102d	48	118	31	92	—	—	79	230	145	147
Thiazole-4	118	133	—	56	196	52	150	60	—	—	164	—	230	165	190
Thiazole-5	118	141	142	—	218	217	186	53	83	—	176	—	222	140	192

np > 370

np79.5–80.5

Table 35 continued

Ring system	H	Me	Et	COMe	CO_2H	CO_2Et	$CONH_2$	CN	NH_2	OH	OMe	SH	SMe	Cl	Br
1,2,3-Triazole-1	206	228	238	40-2	—	—	—	—	51	—	—	—	—	—	—
1,2,4-Triazole-1	121	20	199	—	—	230	138	—	—	—	—	—	—	81-2d	136-8d
1,2,4-Triazole-3	121	95	65-6	—	137	178	312	187	159	234	oil	216	105	167	189
1,2,4-Triazole-4	121	90	oil	—	—	—	—	—	76-7	—	—	—	—	—	—
Tetrazole-5	156	148	—	—	—	86	234	99	203	260	154	205	151	73	148
Tetrazole-1	156	39	265	—	—	—	—	—	>370	—	—	—	—	—	—
Tetrazole-2	156	147	163	—	—	—	—	—	>370	—	147	—	—	—	—
Isothiazole-3	113	134	—	32	135	290	154	60	33	74	—	—	—	160	—
Isothiazole-4	113	146	—	—	161	174	192	94	45	—	—	—	—	143	34
Isothiazole-5	113	142	340	250	201d	—	172	47	112	—	—	—	—	149	150
1,2,3-Oxadiazole (sydnone)	—	36	—	—	—	—	—	—	—	—	—	—	—	—	—
1,2,4-Oxadiazole-3	87	105	—	—	—	—	—	—	—	—	—	—	—	—	—
1,2,4-Oxadiazole-5	87	104	—	—	—	—	—	—	—	—	—	—	—	—	—
1,3,4-Oxadiazole-2	150	164	174.5	140	—	255	—	—	—	120	—	89-91	—	170	—
1,2,3-Thiadiazole-4	160	87-9	—	—	227-8	86	220-2	62-3	156	—	—	—	—	—	—
1,2,3-Thiadiazole-5	160	—	—	—	104-6	222	—	—	44-6	—	—	—	—	—	—
1,2,4-Thiadiazole-3	121	132	—	—	—	—	—	—	145-7	—	—	—	—	—	—
1,2,4-Thiadiazole-5	121	—	—	—	—	—	—	—	—	—	—	—	32	122	—
1,2,3,4-Thiatriazole-5	—	—	—	—	—	—	—	—	119	120	44-5d[e]	93	220	expl.[f]	—
1,3,4-Thiadiazole	43	201d	—	—	—	—	221	—	128-30	—	—	50-65d	34	33	—
1,3-Dioxole-4	51	76	—	—	—	—	—	—	193	—	—	143	—	—	—
1,2-Dithiol-3-one-4	>370	>370	>370	—	—	—	—	—	79.5-80.5	—	—	—	—	62-3	73
1,2-Dithiol-3-one-5	>370	—	>370	—	214-6	>370	223	—	—	—	—	—	—	—	—
1,2,4-Trioxolane-3,3-di	116	88	—	—	—	42	—	—	—	—	—	—	—	—	—
1,2,4-Trioxolane-3,5-di	116	90	140	—	—	—	—	—	—	—	—	—	—	—	—
1,2,4-Trithiolane-3,5-di	78	>200	>230	—	—	—	—	—	—	—	—	—	—	—	—

[a]Melting points above 30°C are given in bold; melting points below 30°C are not included.
[b]Boiling points are given at atmospheric pressure; those reported at other than atmospheric pressure were converted using a nomogram ⟨57MI40101⟩. [c]A dash indicates that the compound is unstable, unknown, or the data are not readily available. [d]Value given for EtO derivative. [e]Value given for EtS derivative. [f]Explodes. [g]Value given for the monohydrate.

2.4.4 THERMODYNAMIC ASPECTS

2.4.4.1 Intermolecular Forces

2.4.4.1.1 Melting and boiling points

In the parent unsubstituted ring systems (*cf.* first column of Table 35) replacement of a -CH=CH- group with a sulfur atom has little effect, while replacement of a -CH=CH- group with an oxygen atom lowers the boiling point by *ca.* 40°C.

Introduction of nitrogen atoms into the ring is accompanied by less regular changes. Substitution of (i) a -CH=CH- group by an NH group and (ii) of a =CH- group by a nitrogen atom both increases the boiling point. When both of these changes are made simultaneously the boiling point is increased by an especially large amount due to the possibilities of association by hydrogen bonding.

The effect of substituents on melting and boiling points can be summarized as follows.

(a) Methyl and ethyl groups attached to ring carbon atoms usually increase the boiling point by *ca.* 20–30 and *ca.* 40–60°C, respectively. However, conversion of an NH group into an NR group results in large decreases in the boiling point (*e.g.* pyrazole to 1-methylpyrazole) because of decreased association.

(b) The acids and amides are all solids. Many of the amides melt in the range 130–180°C; the melting points of the acids vary widely.

(c) Compounds containing a hydroxy, mercapto or amino group are usually relatively high-melting solids. For many hydroxy and mercapto compounds this can be attributed to their tautomerism with hydrogen-bonded 'one' and 'thione' forms. However, hydrogen bonding can evidently also occur in amino compounds.

(d) Methoxy, methylthio and dimethylamino derivatives are often liquids.

(e) Chloro compounds are usually liquids which have boiling points similar to those of the corresponding ethyl compounds. Bromo compounds boil approximately 25° C higher than their chloro analogues.

2.4.4.1.2 Solubility of heterocyclic compounds ⟨63PMH(1)177⟩

In general, the solubility of heterocyclic compounds in water (Table 36) is enhanced by the possibility of hydrogen bonding. 'Pyridine-like' nitrogen atoms facilitate this (compare benzene and pyridine). In the same way, oxazole is miscible with water, and isoxazole is very soluble, more so than furan.

The effect of amino, hydroxy or mercapto substituents is to increase hydrogen bonding properties. However, if stable hydrogen bonds are formed in the crystal, then this can decrease their solubility in water ⟨63PMH(1)177⟩, e.g. indazole and benzimidazole are less soluble than benzoxazole.

Other solvents can be divided into several classes. In hydrogen bond-breaking solvents (dipolar aprotics), the simple amino, hydroxy and mercapto heterocycles all dissolve. In hydrophobic solvents, hydrogen bonding substituents greatly decrease the solubility. Ethanol and other alcohols take up a position intermediate between water and the hydrophobic solvents ⟨63PMH(1)177⟩.

Table 36 Solubilities of Some Five-membered Heterocycles in Water at 20°C[a]

Compound	Parts soluble in 1 part of water	Compound	Parts soluble in 1 part of water
Furan	0.03	Pyrrole	0.06
Isoxazole	0.02	Pyrazole	0.40
Oxazole	Misc.[b]	2-amino-4-hydroxy	0.009
Benzoxazole	0.008	Imidazole	1.8
Sydnone	—	2,5-dihydroxy-	0.02
3-methyl-[c]	Misc.	1-methyl-	Misc.
1,2,4-Oxadiazole[d]	Misc.	1,2,4-Triazole	1
Thiophene	0.001	3-amino-	0.3
Isothiazole	0.03	3-hydroxy-	0.05
Thiazole	—	Tetrazole	1
2-methyl-	Misc.	Benzimidazole	0.002
2-amino-	0.05	Indazole	0.0008
1,2,3-Thiadiazole	0.03	3-hydroxy-	0.002
1,3,4-Thiadiazole	Misc.	Purine	0.5

[a]Unless otherwise indicated, data taken from ⟨63PMH(1)177⟩ or from appropriate chapter, 'Comprehensive Heterocyclic Chemistry', which contain references to the original literature.
[b]Misc. = miscible. [c]At 40°C ⟨80JCS(P2)553⟩. [d]⟨65MI40100⟩.

2.4.4.1.3 *Gas-liquid chromatography* ⟨71PMH(3)297⟩

Gas-liquid chromatography has been widely used for the identification of reaction mixtures and for the separation of heterocycles. Some typical conditions are shown in Table 37.

Table 37 Operating Conditions for GLC Separation of Five-membered Heterocycles with More Than One Heteroatom[a]

Compound	Conditions
Pyrazolines	10% Cyanethylated mannite on Celite 545.
Imidazoles	5% OV-t7 on Chromosorb W, AW-DMCS (H.P.).[b]
Thiazoles	Carbowax 4000, dioleate on firebrick, 190°C.
Oxadiazoles	Silicone grease on Chromosorb P.
1-Phenylpyrazoles	Apiezon L on firebrick C-22, 220°C.
Purines	15% Hallcomid M-18 on firebrick.
Dioxolanes	Carbowax 20M on Gas-Chrom P.
Benzotriazole	Ethyleneglycol succinate on Diatoport-S.

[a]Data taken from ⟨71PMH(3)297⟩, which contains references to the original literature.
[b]Simple alkyl- and aryl-imidazoles. *N*-Unsubstituted compounds are *N*-acylated prior to injection.

2.4.4.2 Stability and Stabilization

2.4.4.2.1 *Thermochemistry and conformation of saturated heterocycles* ⟨74PMH(6)199⟩

Calculation of group increments for oxygen, sulfur and nitrogen compounds has allowed the estimation of conventional ring-strain energies (CRSE) for saturated heterocycles from enthalpies of formation. For 1,3-dioxolane, CRSE is about 20 kJ mol^{-1}. In 2,4-dialkyl-1,3-dioxolanes the *cis* form is always thermodynamically the more stable by approximately 1 kJ mol^{-1}.

For 1,3-dithiolanes the ring is flexible and only small energy differences are observed between the diastereoisomeric 2,4-dialkyl derivatives. The 1,3-oxathiolane ring is less mobile and pseudoaxial 2- or 5-alkyl groups possess conformational energy differences (*cf.* **142**⇌**143**); see also the discussion of conformational behavior in Section 2.4.4.3.

(**142**) 13% (*a*)   (**143**) 87% (*e*)

2.4.4.2.2 *Aromaticity*

This subject has been dealt with in ⟨85KGS867, 91AHC(56)303⟩, which should be consulted for further details and references to the original literature.

Data on structural and energetic indices of aromaticity are contained in Tables 38–40. Azoles occupy an average position between five-membered heterocycles with one heteroatom and six-membered heterocycles. Thus, Bird's structural index I decreases in the sequence pyridine (86) > pyrazole (73) > pyrrole (59). This corresponds with the general rule according to which a pyridine-like heteroatom provides more effective cyclic π-conjugation than a pyrrole-like heteroatom. Interestingly, this tendency is manifested even inside a single condensed heteroaromatic system. For instance, structural index $\Delta \bar{N}$ shows better levelling of the righ bonds in the pyrimidine ring of purine ($\Delta \bar{N} = 80\%$) in comparison with its imidazole ring ($\Delta \bar{N} = 59\%$).

Table 38　Structural Indices of Aromaticity (percentage to benzene molecule, taken as 100%)

Compound	I (%) [a]	$\Delta\bar{N}$ (%) [b]	Compound	I (%) [a]	$\Delta\bar{N}$ (%) [b]
Pyrazole	73	61	Isoxazole	47	—
Imidazole	64	45	Oxazole	38	—
1H-1,2,3-Triazole	73	—	Isothiazole	59	—
2H-1,2,3-Triazole	88	—	Thiazole	64	45
1H-1,2,4-Triazole	81	72	1,2-Dithiolium	62	—
1H-Tetrazole [c]	72	80	Sydnone [d]	—	17
Purine (all molecule)	—	69			
Purine (imidazole ring)	—	59			
Purine (pyrimidine ring)	—	80			

[a] ⟨85T1409⟩. [b] ⟨85KGS867⟩. [c] Sodium derivative. [d] 3-*p*-Bromophenyl derivative.

According to the indices, pyrazole is more aromatic than imidazole. The stability of azoles generally increases with an increasing number of aza-groups, though some exceptions are known. The relative aromaticities of triazoles and tetrazole are questionable. 2*H*-1,2,3-Triazole ($I = 88\%$) which is the more stable in the gas phase reveals more bond levelling than 1*H*-1,2,3-triazole ($I = 73\%$).

π-Electron delocalization in isoxazole seems to be more effective than in oxazole; however, isothiazole is less aromatic than thiazole; thus it is not a general rule that 1,2-diazoles possess higher aromaticity in comparison with 1,3-diazoles. Oxygen-containing heterocycles are always less aromatic than their sulfur and nitrogen counterparts, e.g. thiazole > imidazole >> oxazole. At the same time, the relative aromaticity of S- and N-containing heterocycles can interchange (pyrazole > isothiazole > isoxazole).

The structural indices of aromaticity, I, of oxadiazoles (**145–148**), thiadiazoles (**150–153**) and selenadiazoles (**155, 156**) are compared with that of the parent furan (**144**), thiophene (**149**) and selenophene (**154**) (Scheme 11). 1,2,3-Oxadiazole (**145**) is the least stable among them since all attempts to synthesize this compound were unsuccessful, most likely because of its easy isomerization to the acyclic isomer. At the same time its sulfur analogue (**150**) possesses good stability and has been synthesized. Its 2,4-diaza- (**151**), 3,4-diaza- (**152**) and 2,5-diaza-(**153**) isomers demonstrate even more the extent of π-electron delocalization. There exists a well-known tendency of decreasing aromaticity depending on the type of pyrrole-like heteroatom: S > Se > O. However, there is no uniformity in the change in aromaticity in the horizontal rows, *i. e.,* dependence on heteroatom disposition.

Oxidation at the sulfur atom of thiadiazole to the mono-oxide (**157**) causes a loss in aromatic character which can be seen by X-ray analysis and by the 33 kcal mol^{-1} inversion barrier of the pyramidal sulfur which agrees with the calculated barrier (31.9 kcal mol^{-1}) ⟨82JA1375⟩.

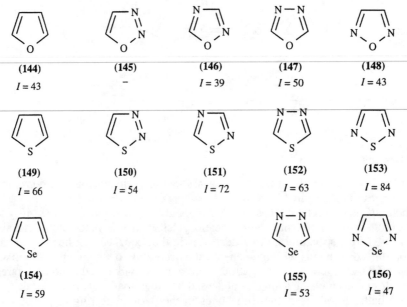

(144)	**(145)**	**(146)**	**(147)**	**(148)**
$I = 43$	–	$I = 39$	$I = 50$	$I = 43$
(149)	**(150)**	**(151)**	**(152)**	**(153)**
$I = 66$	$I = 54$	$I = 72$	$I = 63$	$I = 84$
(154)			**(155)**	**(156)**
$I = 59$			$I = 53$	$I = 47$

Scheme 11　Structural indices of aromaticity, I

The 1,2-dithiolium (**158**) and 1,3-dithiolium ions (**159**) are iso-π-electronic with the tropylium ion, from which they may be formally derived by replacing two pairs of double bonds by sulfur atoms. Structural data and calculations demonstrate that the rings are substantially stabilized by π-electron delocalization.

(**157**)   (**158a**)   (**158b**)   (**159a**)   (**159b**)

The sydnones may be represented by structures (**160a-d**), of which the zwitterionic structures (**160a,b**) most clearly imply an aromatic sextet. *p*-Bromophenylsydnone (**160**; R = H, R′ = *p*-bromophenyl) is essentially planar. However, the O—N bond and O(1)—C(5) bond lengths are not very different from normal single bond distances. Contributions of structures (**160c,d**) to the resonance hybrid are important, and the structural index of sydnone reveals its a rather low aromaticity ($\Delta \bar{N} = 17\%$).

(**160a**)   (**160b**)   (**160c**)   (**160d**)

The influence of benzoannulation on the aromaticity of azoles is not unambiguous. Thus, benzimidazole according to energy indices is more stable than imidazole (Table 39). On the other hand, transition from pyrazole to indazole is accompanied by a fall in stability. As a result, the extent of π-electron delocalization of benzimidazole and indazole is comparable. The further transition from the benzo derivatives to their naphtho analogues decreases aromaticity, with angular isomers being more stable than linear ones, *e.g.* benzimidazole > 1*H*-naphtho [1,2-*d*]imidazole > naphtho[2,3-*d*]imidazole. This tendency agrees with a similar well-known sequence among arenes. The aromaticity of polynuclear bridge-head heterosystems, as it is exemplified by isomeric imidazo[1,2-*a*]benzimidazole (**161**) and (**162**), are also lower than in the case of their benzo predecessors (Table 39).

(**161**)   (**162**)

Empirical resonance energies, being characterized by large discrepancies of values, nevertheless suggest the same tendencies in changing of aromaticity (Table 40).

Table 39   Hess-Shaad REPE indices for some azoles [a]

Compounds	REPE (β)	Compounds	REPE (β)
Pyrazole	0.055	1H-Naphtho[1,2-*d*]imidazole	0.048
Imidazole	0.042	Naphtho[2,3-*d*]imidazole	0.046
Indazole	0.050	9*H*-Imidazo[1,2-*a*]benzimidazole (**161**)	0.046
Benzimidazole	0.050	1*H*-Imidazo[1,2-*a*]benzimidazole (**162**)	0.039

[a] ⟨75T295, 85KGS867⟩

Table 40   Empirical Resonance Energy Data (kJ mol^{-1}) for Azoles[a,b]

Compound	Method A[c]	Method B[d]	Method C[e]
Pyrazole	173.6	112.1	136.8
Imidazole	134.3	53.1	74.1
1,2,4-Triazole	205.8	83.7	151.5
Tetrazole	—	231.0	264.0
Indazole	309.2	246.9	248.9
Benzimidazole	287.9	203.8	204.2
Benzotriazole	—	346.9	312.5

[a] Adapted from ⟨74AHC(17)255⟩ which contains further details and references to the original literature.
[b] 1×10^{-7} J = 2.3901 $\times$ 10^{-8} cal. [c] Results obtained using Pauling's bond energy terms. [d] Using the bond energy terms reported by Cottrell. [e] Using Coates and Sutton's bond energy terms.

The experimental heat of formation of tetrazole of 28.11 eV, compared to values for pyrazole (39.391 eV), imidazole (39.790 eV) and 1,2,4-triazole (34.377 eV), suggests that tetrazole is less stable than the lower azoles ⟨66JCP759⟩. Dewar has calculated resonance energies of 26.0 kcal mol^{-1} for 1*H*-tetrazole and 29.25 kcal mol^{-1} for 2*H*-tetrazole which compare with values for pyrazole (35.5), imidazole (22.9) and 1*H*-1,2,3-triazole (27.9 kcal mol^{-1}) ⟨66JCP759⟩.

Magnetic criteria have received wide application mainly as a qualitative test for aromaticity and antiaromaticity. The values of the exaltation of diamagnetic susceptibility (in 10^{-6}Λ cm^{-3} mol^{-1}), and therefore aromaticity, decrease in the sequence: thiazole (17.0) > pyrazole (15.5) > sydnone (14.1). The relative aromaticity of heterocycles with a similar type of heteroatom can be judged from values of the chemical shifts of ring protons. The latter reveals paramagnetic shifts when π-electron delocalization is weakened. For example, in the series of isomeric naphthoimidazoles aromaticity decreases in the sequence: naphtho[1,2-*d*]imidazole (δ = 7.7–8.7 ppm) > naphtho[2,3-*d*]imidazole (δ = 7.5–8.2 ppm) > perimidine (δ = 6.1–7.2 ppm). This sequence agrees with other estimates, in particular with energetic criteria.

2.4.4.2.3   Stable carbenes, silylenes, germylenes

The nitrogen heteroatoms in imidazole and some closely related heterocycles can stabilize a carbene center at the 2-position ⟨97AG(E)2162⟩. Thus, 1,3-disubstituted imidazole-2-ylidenes (**163**)–(**170**), 1,3-dimesitylimidazoline-2-ylidene (**171**), 1,3,4-triphenyl-1*H*-1,2,4-triazole-5-ylidene (**172**), and their silylene (**173**) and germylene (**174**) analogues are stable (in the absence of oxygen and moisture) solids with definite melting points, which can be recrystallized from appropriate hydrocarbon solvents. The exception is carbene (**163**) which is an unstable liquid; however, it is stable in solution.

(**163**) R = CH$_3$
(**164**) R = C(CH$_3$)$_3$
(**165**) R = 1-Adamantyl
(**166**) R = Mesityl
(**167**) R = 4-CH$_3$C$_6$H$_4$
(**168**) R = 4-ClC$_6$H$_4$

(**169**) R = CH$_3$
(**170**) R = *i*-C$_3$H$_7$

(**171**)

(**172**)

(**173**) X = Si
(**174**) X = Ge

Many crystallographic structures of these carbenes and carbenoids has been solved, indicating: (1) they are true carbenes with little ylidic character; thus the contribution of ionic structures (**175b**) is relatively small in comparison with (**175a**); (2) small N-C-N angles at the carbene center (101–102°) in comparison with typical values (108.5–109.7°) for the corresponding angle in imidazolium salts;

(3) π-delocalization in carbenes is diminished relatively to imidazolium cations. The last conclusion is supported by the upfield shift of the imidazole ring protons in carbenes (from near 7.9 → 6.9 ppm). Other NMR chemical shifts are given in Table 41.

(175a)   (175b)

Table 41   Carbenes: Selected NMR Chemical Shifts (ppm)[a]

Carbene	^{13}C-2	^{13}C-4(5)	^1H-4(5)	^{15}N-1(3)	^{14}N-1(3)
			Nucleus		
163	215.2	120.5	6.92	−197.3	−197.5
165	211.4 [b]	113.9 [b]	7.02 (6.91 [b])	−160.5 [b]	−161 [b]
166	219.7	121.3	7.04 (6.48 [b])	−178.9	−180
167	215.8	118.8	7.64 (6.96 [b])	—	−171
168	216.3	119.2	7.76 (6.68 [b])	—	−174
169	213.7	123.1	—	−198.5	−198
171	244.5	51.36	3.71	−236.7	−234
172	214.6 [b,c]	152.2 [b,d]	—	—	—

[a] In THF-d_8 solution. References are tetramethylsilane or $NH_4^+NO_3^-$.
[b] In benzene-d_6 solution.
[c] For C-5 atom. [d] For C-3 atom.

The stability of azole carbenes can be attributed to electronic factors which operate in both the π- and σ-frameworks ⟨92JA5530⟩. In the π-framework, electron donation into the carbene out-of-plane p-orbital by the electron-rich system moderates the typical electrophilic reactivity of carbenes. In the σ-framework, additional stability for the carbene electron pair may be gained from the σ-electron-withdrawal effects on the carbene center by the more electronegative nitrogens, which moderates the carbene nucleophilic reactivity. The combination of these σ- and π-effects serves to increase the singlet-triplet gap and stabilize the singlet carbene over the more reactive triplet state. For carbenes with bulky substituents (tert-butyl, 1-adamantyl, *etc.*) steric effects provide additional stabilization.

2.4.4.3   Conformation

Saturated five-membered heterocyclic compounds are non-planar, existing in half-chair or envelope conformations. The far-IR spectra of THF and 1,3-dioxolane (**176**) show both to have barriers of *ca.* 0.42 kJ mol^{-1}.

(176)   (177)   (178)   (179)

1,3-Dioxolane also pseudorotates essentially freely in the vapor phase. 2,2′-Bi-1,3-dioxolane (**177**) has been shown by X-ray crystallography to have a conformation midway between the half-chair and envelope forms. The related compound 2-oxo-1,3-dioxolane (**178**) shows a half-chair conformation. This result is confirmed by microwave spectroscopy and by ^{13}C NMR data. Analysis of the AA′BB′ NMR spectra of the ring hydrogen atoms in some 1,3-dioxolane derivatives is in agreement with a puckered ring. Some 2-alkoxy-1,3-dioxolanes (**179**) display anti and gauche forms about the exocyclic C(2)-O bond.

The 1,2,4-trioxolane ring prefers a half-chair conformation (**180**); the C-O-C portion of the ring forms the reference plane, and alkyl substituents prefer the equatorial positions.

(**180**)   (**181**)   (**182**)   (**183**)

1,3-Dithiolane (**181**) derivatives also possess non-planar skeletons; the most important conformation is probably of symmetry C_2 (half-chair). The dithiolane ring may be quite flexible and a minimum energy conformation is only well defined if there is a bulky substituent at the 2-position.

Fully saturated isothiazolidines normally exist in an envelope conformation. The X-ray of dehydromethionine (**182**) show C-3 to be out of the plane of the other four ring atoms, and on the same face of the ring as the S—Me group; solution NMR suggests rapid equilibrium between the conformers ⟨82JCS(P1)1283⟩. A combination of ¹H and ¹³C NMR, X-ray studies, and theoretical calculations show that dibenzo[*b,g*]thiazocinium salts exist as ammoniosulfuranes, with fairly rigid conformations as shown in (**183**), with the sulfur atom having a distorted trigonal bipyramidal geometry ⟨86JA6320⟩.

Proton NMR (sometimes low temperature NMR or NOE) is the major reliable method for conformational analysis of isoxazolidines which undergo pyramidal inversion at the ring nitrogen atom ⟨82JOC4397, 84OMR(22)399⟩.

Variable temperature ¹H and ¹³C NMR show that for 1,3-diacylimidazolidines (**184**, R = Ph), (**184b**) is favored over (**184a**) by 6.3 kJ mol⁻¹ and over (**184c**) by 3.2 kJ mol⁻¹. When R = butyl at 193K (**184b**) is favored over (**184a**) by 1.5 kJ mol⁻¹ and over (**184c**) by 1.2 kJ mol⁻¹ ⟨87JCS(P2)1701⟩. Dynamic NMR studies, allied with molecular mechanics calculations, have assisted conformational analysis of 1,3-disubstituted imidazoline-2-thiones ⟨87JOC5177⟩.

(**184a**)   (**184b**)   (**184c**)

The conformation of N,N-linked biazoles has been calculated by the MNDO method. The energy minimum for 1,1'-bipyrazole (**185**) is for a dihedral angle of 108° with N-atoms being in opposite positions ⟨84CJC687⟩. The two triazole rings in 1,1'- and 2,2'-bi-1,2,3-triazole as well as in 4,4'-bi-1,2,4-triazole are also approximately orthogonal ⟨84CJC687⟩.

(**185**)   (**186**)   (**187**)   (**188**)

In N,C-linked biheteroaryls, *e.g.* (**186**)–(**188**), two rings are virtually coplanar, the conformation in such molecules is controlled by weak hydrogen bonding between an acidic α-hydrogen of one ring and the lone pair of a pyridine type nitrogen in another ring.

A conformational study of bis-, tris-, and tetrakis-pyrazolyl-1-methanes was carried out using crystallography, NMR and MM2 and MNDO calculations ⟨89T7805⟩.

Studies of absorption and fluorescence spectra of 2,2'-bis(benzimidazole) have demonstrated that its conformation in solution is determined by hydrogen bonding as shown in (**189**), as well as by considerable internuclear conjugation ⟨89CJC1200⟩. In *C*-arylazoles, the dihedral angles between the rings are normally less than in biaryls. Thus, in 1-methyl-4-phenylimidazole (**190a**), according X-ray measurements, this angle is 7.3°. However, steric hindrance can increase this angle considerably, *e.g.* in 1-methyl-2-phenylimidazole (**190b**) to 32.3° ⟨94JHC899⟩.

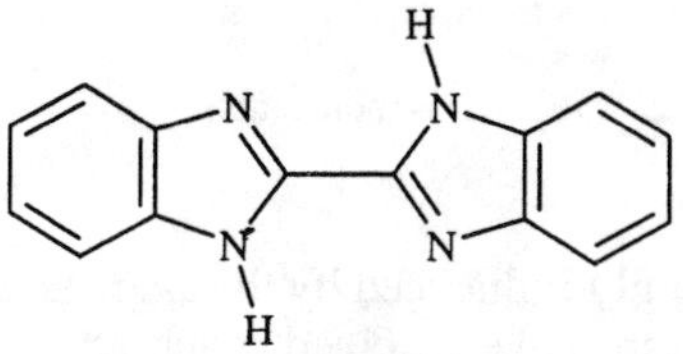

(**189**) 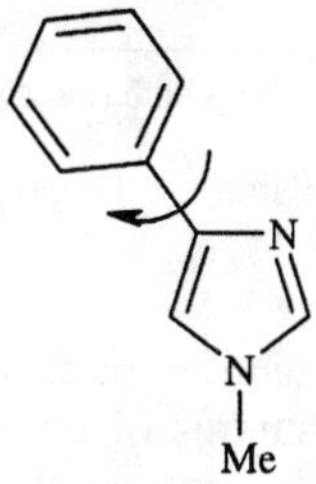   (**190a**) 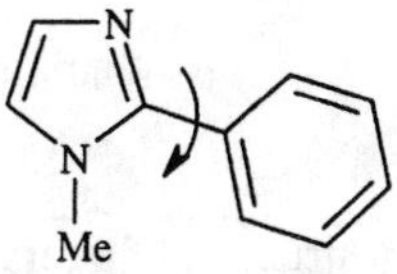   (**190b**)

2.4.5   TAUTOMERISM ⟨76AHC(S1)1⟩

2.4.5.1   Annular Tautomerism

2.4.5.1.1   *Prototropy*

(i) General view

Annular tautomerism (*e.g.* **191**⇌**192**) involves the movement of a proton between two annular nitrogen atoms. For unsubstituted imidazole (**191**; R = H) and pyrazole (**193**; R = H) the two tautomers are identical, but this does not apply to substituted derivatives. For triazoles and tetrazoles, even the unsubstituted parent compounds show two distinct tautomers. However, interconversion occurs readily and such tautomers cannot be separated. Sometimes one tautomeric form predominates. Thus the mesomerism of the benzene ring is greater in (**195**) than in (**196**), and UV spectral comparisons show that indazole exists predominantly as (**195**).

Table 42 summarizes the known annular tautomerism data for azoles. The tautomeric preferences of substituted pyrazoles and imidazoles can be rationalized in terms of the differential substituent effect on the acidity of the two NH groups in the conjugate acid, *e.g.* in (**197**; EWG = electron-withdrawing group) the 2-NH is more acidic than 1-NH and hence, for the neutral compound, the 3-substituted pyrazole is the more stable.

The situation is more complex for triazoles and tetrazoles where other effects such as lone-pair repulsions intervene; see discussion in ⟨76AHC(S1)296⟩.

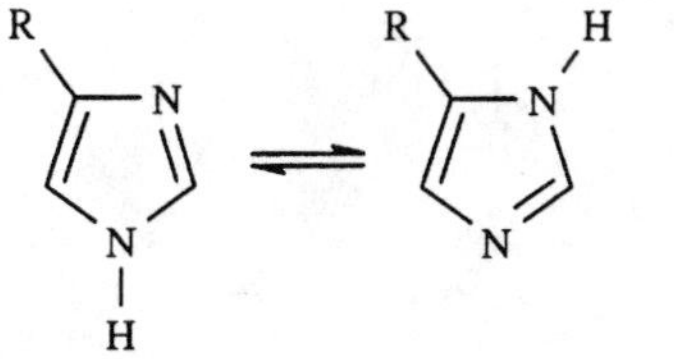

(**191**)   (**192**) 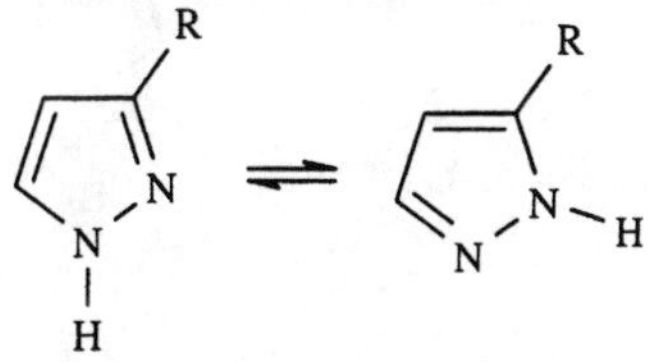   (**193**)   (**194**)

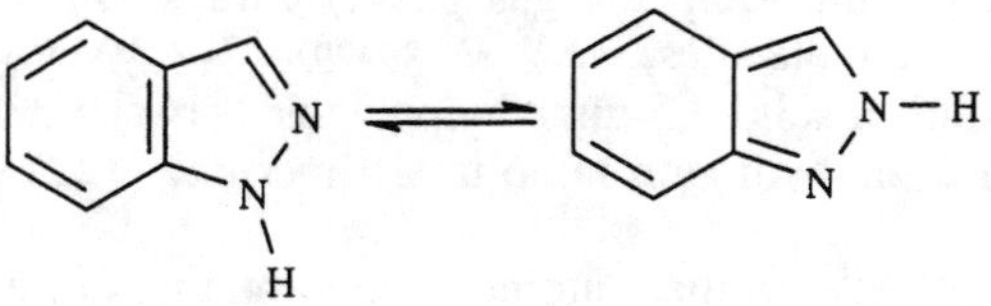

(**195**)   (**196**) 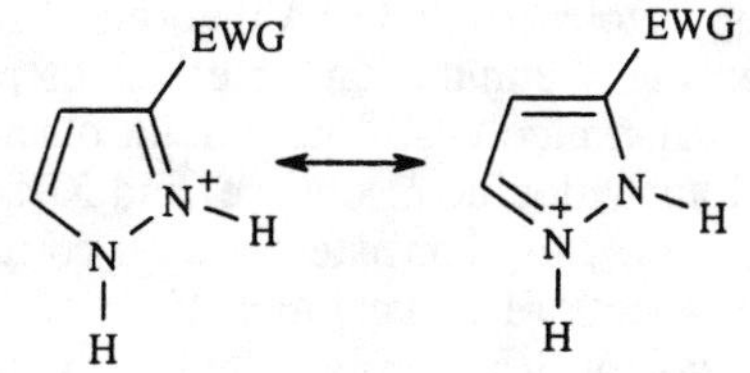   (**197**)

Table 42 Annular Tautomerism of Azoles[a]

Azole	Parent	Substituted compounds
Pyrazole	Equivalent	Electron-withdrawing group prefers 3-position
Imidazole	Equivalent	Electron-withdrawing group prefers 4-position
v-Triazole	1,2,5 > 1,2,3	—
s-Triazole	1,2,4 > 1,3,4	—
Tetrazole	1,2,3,4 > 1,2,3,5 [b]	—

[a] Data taken from ⟨76AHC(S1)296⟩ which contains references to the original
literature.
[b] In the solid state and in solution; in the gas phase 2*H*-tautomer is more stable.

Apart from structural factors annular tautomerism is strongly influenced by the aggregate state of the compound. Gas-phase and solid state studies of tautomerism have received much attention recently because of the development of methods such as X-ray crystallography, CPMAS NMR, microwave spectroscopy, *etc*. Low temperature ^{1}H NMR is the simplest and most straightforward method to determine K_T; it only requires that proton transfer be slow enough to observe separate signals for both tautomers, and that the equilibrium is not shifted too much towards one of the tautomers (about 5% of the minor tautomers seems the limit). More detailed information on tautomerism of key azole systems is given below.

(ii) Pyrazoles
A systematic study of annular tautomerism of N*H*-pyrazoles in the solid state has been published ⟨88CJC1141⟩.

^{13}C CPMAS NMR studies of N*H*-pyrazoles and indazoles ⟨93CJC678⟩ concluded (i) the tautomer present in the solid is also the major tautomer in solution; (ii) 3(5)-ferrocenylpyrazole is a 50:50 mixture of both tautomers in the solid state; (iii) certain pyrazole *C*-substituents (CF$_3$, Br, Ar, Het) prefer position 3 (that is, they are 3-substituted pyrazoles (**193**)), others (Bu^t, Pr^i, CH$_3$) prefer position 5 (that is, they are 5-substituted pyrazoles (**194**)).

Low temperature ^{13}C NMR spectroscopy of 3(5)-phenylpyrazole shows that at $-20°$C it exists a mixture of 80% of the 3-phenyl tautomer (**193**, R = Ph) and 20% of the 5-phenyl tautomer (**194**, R = Ph) ⟨91G477⟩.

A rare example of desmotropy (two tautomers that crystallize in two different crystals) in azoles, found by X-ray crystallography, concerns 3-methyl-4-nitropyrazole (**198**) and its tautomer, 5-methyl-4-nitropyrazole (**199**) ⟨94CC1143⟩.

(**198**)   (**199**)

3,5-Dimethylpyrazole shows prototropic tautomerism in the solid state (crystallography and CPMAS NMR) ⟨85JA5290⟩ as do other N*H*-pyrazoles, *e.g.* 3,5-diphenyl- and 3,5-diphenyl-4-bromopyrazole ⟨92JA9657⟩. A theoretical study (INDO, STO-3G) concludes that 1,2-proton shifts are forbidden, thus, annular tautomerism of pyrazoles (including the gas phase) always implies other entities (water, alcohols, formic acid, other pyrazole molecules, *etc*.) ⟨86BSF429⟩. ^{15}N CPMAS NMR, ^{14}N quadrupole double resonance, and X-ray studies of solid 3,5-dimethylpyrazole between 270K and 350K ⟨89JA7304⟩ are consistent with a correlated triple hydrogen jump in a trimer like (**122**) with an activation energy of 11 kcal mol^{-1}.

In the case of 3(5),4-polymethylenepyrazoles ⟨91JHC647⟩, the tautomeric equilibrium is directed by the Mills–Nixon effect; thus, the 3,4-trimethylene tautomer (**201**) is more stable than the 4,5-one (**200**) (ΔG = 1.3 kcal mol^{-1}) ⟨94NJC269⟩.

(200)   (201)   (202)

Tautomerism of 3,5-bis(4-methylpyrazol-1-yl)-4-methylpyrazole (**202**) also involves the simultaneous rotation of the two lateral rings and has a value of 12 kcal mol^{-1} (in methanol), almost the same as 3,5-dimethyl-4-chloropyrazole and 3,5-dimethyl-4-nitropyrazole ⟨93CJC1443⟩.

In the gas phase 1*H*-indazole (**195**) is 4.7 kcal mol^{-1} more stable than 2*H*-indazole (**196**), while in water the difference is reduced to 2.2 kcal mol^{-1}. Thus, water more effectively stabilizes the 2*H*-tautomer because of its higher dipole moment ⟨88JA4105, 94JPC10606⟩ (see ⟨92JST(155)1⟩ for the existence of (**195**) in the gas phase). 1*H*-Indazole (**195**) is also more stable in the excited state (S_1) ⟨94JPC10606⟩ although the difference in energy is lower in the excited (1.6 kcal mol^{-1}) than in the ground state.

(iii) Imidazoles

Crystalline 2-methylimidazole exhibits different ^{13}C (CPMAS) chemical shifts for C-4 and C-5 (125.0, 115.7 ppm). The average (120.3 ppm) is close to that reported for imidazole in deuterated DMSO (121.2 ppm). These results imply that solid state chemical shifts can be used instead of *N*-methyl models in tautomerism studies ⟨87H(26)333⟩. For imidazole the solid state ^{13}C shifts are 137.6 (C-2), 129.3 (C-4), and 119.7 (C-5) ⟨81JA6011⟩. No proton exchange occurs in the solid, and the data support a structure resembling the crystal structure. Cooling imidazole solutions has not yet allowed the detection of individual tautomers, but by symmetry the compound exists in equal tautomeric forms, as does pyrazole ⟨81CC1207⟩.

The two signal ^{15}N NMR spectrum of solid imidazole indicates that tautomerism is slow in the solid state. The average (186 ppm) is the same as the shift observed in solution. As the solution pH decreases the ^{15}N signal moves progressively upfield until both nitrogens are protonated ⟨82JA1192⟩.

(iv) Triazoles

Tautomerism in 1,2,3-triazole has been extensively investigated using various spectroscopic methods. Microwave and photoelectron spectroscopic studies conclude that in the gas phase the 2*H*-tautomer strongly dominates ⟨81ZN(A)34, 81ZN(A)1246⟩. In solution, both the 2*H* and the 1*H* tautomers are present with their relative proportion depending upon temperature and concentration ⟨84JCS(P2)1025⟩. 1*H*-1,2,3-Triazole is more stable (66%) in CDCl$_3$, but in DMSO-d$_6$ the 2*H*-1,2,3-triazole is favored (55%) ⟨82JOC5132⟩. In aqueous solution, however, the 2*H*-tautomer is favored over the 1*H*-tautomer by a factor of about two based on arguments from basicity and partitioning ⟨89JCS(P2)1903⟩.

The tautomeric equilibrium of unsubstituted benzotriazole has been studied extensively as summarized in reviews ⟨63AHC(2)27, 76AHCS295⟩. In the crystalline state the sole existence of 1*H*-benzotriazole is demonstrated by X-ray ⟨74AX(B)1490⟩, by ^{13}C NMR ⟨83H(20)1713⟩, and by microwave spectra ⟨93JSP(161)136⟩. The 1*H*-form also predominates strongly under most other conditions.

However, measurements of gas-phase UV spectra of benzotriazole and its 1-methyl- and 2-methyl derivatives have shown that the percentage of 2*H*-tautomer is 45% at 30°C, 35% at 50°C and 25% at 80°C ⟨93JOC5276⟩. Calculations suggest that the 2*H*-tautomer is enthalpically more stable than 1*H*-tautomer by about 3.8 kcal mol^{-1}.

Theoretical studies favour the 1*H*-structure (**203**) for 1,2,4-triazole rather than the more symmetrical 4*H*-structure (**204**). Calculations suggest that the destabilization of (**204**) is due to the two contiguous heteroatoms of the same hydridization type ⟨86JA3237, 89JOC3553, 90JPC5499⟩. In the gas phase the energy difference between (**203**) and (**204**) was −26.3 kJ mol^{-1}.

(203)   (204)

(v) Tetrazoles

The 2*H*-form (**206**) of tetrazole dominates in the gas phase while in solution the 1*H*-form (**205**) is favoured ⟨85JCS(F2)1555, 93JA2465⟩. *Ab initio* calculations using RHF/6-31G suggested the 2*H*-form (**206**) to be 1.7 kcal mol^{-1} more stable than the 1*H*-form (**205**) in the gas phase ⟨85JPC460⟩. High level *ab initio* calculations have also been applied to the 5*H*-tautomer (**207**) which was determined to lie 19.6 kcal mol^{-1} above (**205**) ⟨93JA2465⟩.

(**205**)   (**206**)   (**207**)   (**208**)

A calculated transition state for the symmetry allowed 1,2-H shift from (**205**) to (**206**) involves the migrating hydrogen atom moving 59° out of the ring plane with a transfer angle at the hydrogen atom of 71° as in structure (**208**) (R = H). The activation barrier is 207 kJ mol^{-1} in the gas phase and is expected to be higher in a polar solvent since (**206**) is less polar than (**205**) ⟨93JA2465⟩. Since the tautomerism is a rapid process in solution and individual tautomers cannot be detected even at low temperatures, an intermolecular mechanism with a lower activation barrier must operate in solution.

^{13}C CPMAS NMR measurements with solid tetrazole gave a C-5 shift of 144.0 ppm which is similar to that for 1-methyltetrazole (143.4) in (CD$_3$)$_2$SO and suggests that the 1-NH form is the major tautomer in the solid state ⟨83H(20)1713⟩.

2.4.5.1.2 Annular Elementotropy

There are an increasing number of examples of the isomerism involving reverse migration of organic and inorganic groups heavier than proton. This is called elementotropy. The annular elementotropy involves alkylotropy, acylotropy, sililotropy, metallotropy, *etc.*

At room temperature, solutions of 1-trimethylsilylbenzimidazole show an averaged ^{13}C NMR spectrum, but once the tautomerism is frozen the molecule loses its symmetry ⟨83H(20)1713⟩. Migrations of acyl, trimethylsilyl, trialkylstannyl and *p*-methoxybenzyl groups between the nitrogens in 1,2,3-triazoles are also well-known ⟨66CB2512, 70TL5225, 72JOM(44)117, 82JCR(S)292⟩.

N-Dialkylaminomethylbenzotriazoles usually exist in the crystalline state solely as the N(1) isomers, but in solution they form equilibrium mixtures of the N(1) (**209, 210**) and N(2) (**211**) isomers ⟨75JCS(P1)1181, 87JCS(P1)2673⟩. The N(1) and N(2) isomers are of nearly equal stability in nonpolar solvents and in the gas phase (2:1 ratio on statistical grounds). Polar solvents favor the 1- and 3-substituted forms over the 2-substituted, and conversely substituents at positions four and seven favor the 2-substituted form. The interconversion of these N(1) and N(2) isomers proceeds intermolecularly, as demonstrated by cross-over experiments, by a dissociation-recombination mechanism involving the formation of intermediate iminium cations and the benzotriazole anion.

(**209**)   (**210**)   (**211**)

Under flash vacuum pyrolysis conditions 1- and 2-(1-adamantyl)indazoles as well as 1-, 2- and 3-tritylindazoles are mutually interconverted ⟨91BSF592, 89BSB349⟩.

Enthalpies of gas-phase thermal isomerisations of 1-methyl- to 2-methyltetrazoles have been reported to be small (2.1–3.9 kJ mol^{-1}) ⟨90ZPH656⟩.

2.4.5.2 Substituent Tautomerism

2.4.5.2.1 *Azoles with heteroatoms in the 1,2-positions*

3-Substituted isoxazoles, pyrazoles and isothiazoles can exist in two tautomeric forms (**212**, **213**; Z = O, NR or S; Table 43). Amino compounds exist as such, as expected, and so do the hydroxy compounds under most conditions. The stability of the OH forms of these 3-hydroxy-1,2-azoles is explained by the weakened basicity of the ring nitrogen atom in the 2-position due to the adjacent heteroatom at the 1-position and the oxygen substituent at the 3-position. This concentration of electron-withdrawing groups near the basic nitrogen atom causes these compounds to exist mainly in the OH form.

(212)   (213)

Table 43   Tautomerism of 3-Substituted Azoles with Heteroatoms-1,2[a]

Substituent	Ring	Phase (s)	Conclusions
OH	1-Substituted pyrazole	C_6H_{12}, $CHCl_3$, H_2O, xtal.	OH except in H_2O where OH/NH coexist
	Isoxazole	C_6H_{12}, $CHCl_3$, H_2O, xtal.	Mainly OH in all media
	Isothiazole	MeOH, xtal.	OH
NH_2	1-Substituted pyrazole	MeOH, KBr	NH_2
	Isoxazole	$CHCl_3$, H_2O	NH_2
	Isothiazole	CCl_4	NH_2
SH	Isothiazole	CCl_4	SH

[a]For further details and original references see ⟨70C134⟩ and ⟨76AHC(S1)1⟩.

The 4-substituted analogues can exist in two uncharged tautomeric forms (**214**) and (**215**) and, in addition, in the zwitterionic form (**216**), but all the evidence shows that the compounds all exist predominantly in the NH_2 or OH form (**214**).

(214)   (215)   (216)

For 5-substituted isoxazoles, pyrazoles and isothiazoles, three uncharged tautomeric forms are possible: (**217**), (**218**) and (**219**). Some generalizations are recorded in Table 44. Again, the amino derivatives exist as such. In the case of the hydroxy compounds, the hydroxy form is of little importance (except in special cases where a suitable substituent in the 4-position can form a hydrogen bond with the 5-hydroxy group). The relative occurrence of the 4*H*-oxo form (**218**) and 2*H*-oxo tautomer (**219**) depends on the substitution pattern and on the solvent. Tautomer (**219**) is considerably more polar than (**218**), with a large contribution from the charge-separated canonical structure (**220**). Hence, it is not unexpected that the 2*H*-oxo tautomer (**219**) is strongly favored by polar media. A substituent at the 4-position also tends to favor form (**219**) over (**218**) because of conjugation or hyperconjugation of the 4-substituent with the 3,4-double bond.

(217)   (218)   (219)   (220)

Table 44   Tautomerism of 5-Substituted Azoles with Heteroatoms-1,2[a]

Substituent	Ring	Phase(s)	Conclusions
OH	1-Substituted pyrazole	C_6H_{12}, CHCl$_3$, EtOH, H$_2$O, xtal.	CH in non-polar;
	Isoxazole	C_6H_{12}, CHCl$_3$, EtOH, H$_2$O, xtal.	Increasing NH in polar media; NH favored by 4-substituent
NH$_2$	1-Substituted pyrazole	CCl$_4$	NH$_2$
	Isoxazole	CHCl$_3$, H$_2$O	NH$_2$
	Isothiazole	CCl$_4$	NH$_2$
SH	Isoxazole	C_6H_{12}, CCl$_4$, MeOH, xtal.	SH

[a] For further details and original references see ⟨70C134⟩ and ⟨76AHC(S1)1⟩.

Complex tautomerism for azoles with heteroatoms in the 1,2-positions occurs for pyrazoles which are not substituted on nitrogen. Scheme 12 shows the four important tautomeric structures (**221**)-(**224**) for 3-methylpyrazolin-5-one, and (**225**) and (**226**) as examples of other possible structures. A detailed investigation of this system disclosed that in aqueous solution (polar medium) the importance of the tautomers is (**222**) > (**224**) >> (**223**) or (**221**), whereas in cyclohexane solution (non-polar medium) (**224**) > (**221**) >> (**222**) or (**223**).

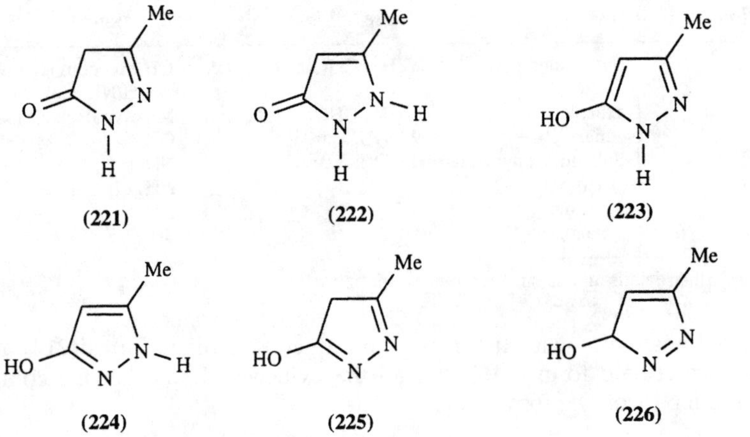

Scheme 12

Indazolinone exists as such (**227**) in the solid state but only as a minor isomer (15%) in DMSO solution, where the 3-hydroxy-1*H*-indazole tautomer (**228**) predominates (85%) ⟨86JCS(P2)1677⟩. No evidence for the existence of 3-hydroxy-2*H*-indazole (**229**) has been found.

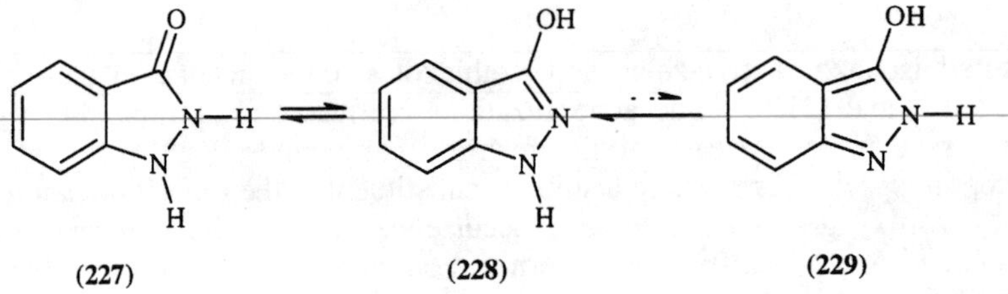

2.4.5.2.2   Azoles with heteroatoms in the 1,3-positions

The tautomerism of 2-substituted 1,3-azoles (**230** ⇌ **231**) is summarized in Table 45. Whereas amino compounds occur invariably as such, all the potential hydroxy derivatives exist in the oxo form, and in this series the sulfur compounds resemble their oxygen analogues. There is a close analogy between the tautomerism for all these derivatives with the corresponding 2-substituted pyridines.

(230)　　　　　　　　(231)

For 2-aminoazoles the tautomeric equilibrium can be expected to shift towards the imino form when electron-withdrawing substituents (*e.g.* acyl groups) are bonded to an exocyclic nitrogen ⟨66ZOR917, 78KGS113⟩. The tautomeric constants for the amine-imine equilibrium were determined by measuring pK_a values ⟨66ZOR917, 71KGS807, 82JCS(P2)535⟩. The effect of the solvent and/or additives can be important. In fact, the tautomeric equilibrium in 2-aminothiazoles in toluene or carbon tetrachloride is shifted towards the imino form by adding small amounts of tetrabutylammonium bromide ⟨92JHC1461⟩ or dimethyl sulfoxide ⟨94JCS(P2)615⟩.

Table 45 Tautomerism of 2-Substituted Azoles with Heteroatoms-1,3[a]

Substituent	Ring	Phase(s)	Conclusions
OH	1-Substituted imidazole	KBr disc	NH, C=O
	Oxazole	MeOH, CCl₄	NH, C=O
	Thiazole	Alcohol, CS₂	NH, C=O
NH₂	1-Substituted imidazole	0.1N aq.KCl, EtOH/aq. KCl	NH₂
	Oxazole	MeOH	NH₂
	Thiazole	EtOH, CDCl₃	NH₂
SH	1-Substituted imidazole	Liquid film, MeOH, EtOH	NH, C=S
	Oxazole	CCl₄, MeOH	NH, C=S
	Thiazole	CCl₄, EtOH	NH, C=S

[a]For further details and original references see ⟨70C134⟩ and ⟨76AHC(S1)1⟩.

4-Substituted 1,3-azoles exist in two non-charged tautomeric forms (**232**) and (**233**) together with the zwitterionic form (**234**). 5-Substituted 1,3-azoles also exist in forms (**235**) and (**236**) together with the zwitterionic forms (**237**). Some results are summarized in Table 46; for the potential hydroxy forms, the non-aromatic tautomers of types (**233**) and (**236**) clearly can be of importance.

(232)　　　(233)　　　(234)　　　(235)　　　(236)　　　(237)

Table 46 Tautomerism of 4- and 5-Substituted Azoles with Heteroatoms-1,3[a]

Substituent	Ring	Phase (s)	Conclusions
4-OH	Oxazole	EtOH, xtal., Me₂SO	CH
4-OH	Thiazole	Me₂SO, Me₂CO	OH and CH
5-OH	Oxazole	Xtal.	CH
5-NH₂	Oxazole	CHCl₃, xtal.	NH₂

[a]For further details and original references, see ⟨70C134⟩ and ⟨76AHC(S1)1⟩.

Time-resolved electron paramagnetic resonance served to detected the short-lived triplet state of the keto tautomer (**238**) of 2-(2-hydroxyphenyl)benzothiazole (**239**) generated by excited state intra-molecular proton transfer ⟨92CC641⟩.

(238)   (239)

2.4.5.3   Ring-chain Tautomerism

All known examples of ring-chain tautomerism can be conveniently divided into two large groups: (i) those without direct involvement of any substituent in the heterocyclic ring and (ii) isomerizations with participation of certain substituents.

2.4.5.3.1   *Without direct involvement of substituents*

Oxazolidines (**240**, X = O) are subject to ring-chain tautomerism. The process can be considered as a reversible intramolecular nucleophilic addition to the C=N bond. A variety of substituted oxazolidines in the solid state exist in the open-chain form (**241**, X = O) ⟨85T5919, 92T4979⟩. In solution, the two forms are in equilibrium, the position of which depends on the solvent and the substituent. In case of imidazolidines (**240**, X = NR), 1,3-thiazolidines (**240**, X = S) as well as benzimidazolines and benzothiazolines the system highly prefers the ring form (probably because of the higher nucleophilicity of an NH_2 or SH group in **241** in comparison with OH group) and only in some instances can the open-chain tautomer of type (**241**) be detected by NMR studies ⟨90T6545⟩ or can it be proved indirectly by several ways such as the formation of metal chelates ⟨87RRC151⟩, hydrolysis ⟨79JA420⟩, or C-2 epimerization ⟨87JHC1629, 89JHC589⟩.

(240)   (241)

1-Nitrobenzotriazole (**242**) is in equilibrium with the isomeric α-diazo-1-nitroimine (**243**). In the presence of nucleophiles, such as piperidine, this open-chain isomer can be trapped by attack at the diazo group. A second molecule of the amine acts as a base to afford the piperidinium salt (**244**) ⟨84JOC2197⟩. Similar equilibria of 1,2,3-triazoles with strong electron-withdrawing groups at a nitrogen heteroatom. such as CN or OSO_2CF_3. are also known.

(242)   (243)   (244)

With a few exceptions simple 1,2,3-oxadiazole (**245**, X = O) are not isolable because they are less stable than their open tautomers (**246**, X = O). The geometry and energy of 1,2,3-oxadiazole, calculated by an *ab initio* method ⟨85AG(E)713⟩ indicate that the heterocycle is unlikely to be isolated as a discrete species, even in a matrix at low temperature, because there is a low energy barrier to ring opening to diazoacetaldehyde.

The sterically protected 1,2,3-oxadiazole (**247**) is the only known oxadiazole bearing alkyl substituents. It exists in the cyclic form in the crystalline state but as the diazoketone in chloroform solution ⟨80TH403-01⟩. 1,2,3-Benzoxadiazole has been detected in a matrix at 15K ⟨84AG(E)509⟩; this and some substituted 1,2,3-benzoxadiazoles have been shown to exist in equilibrium with their open-chain tautomers ⟨91JST(247)135⟩. Naphtho[2,3-*d*]-1,2,3-oxadiazole (**248**) is stable in the solid state at 215 °C in the absence of light ⟨91AG(E)1476⟩.

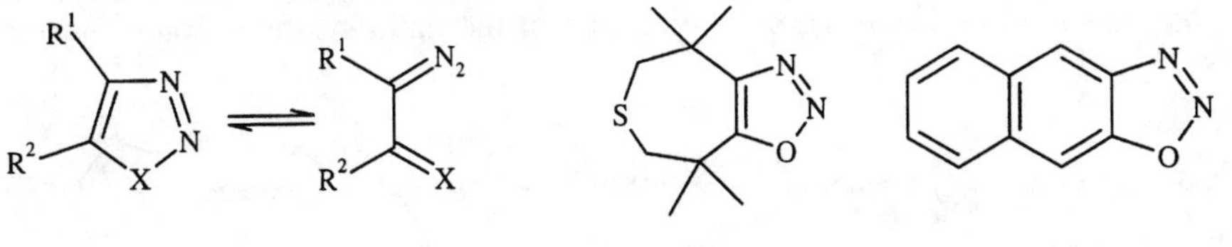

(245)   (246)   (247)   (248)

The 1,2,3-thiadiazoles (**245**, X = S) exist as a remarkably stable neutral aromatic compound. It is isomeric with the ring opened α-diazoketones (**246**, X = S) and there is evidence that it can react through (**246**) as an intermediate.

For mesoionic 1,2-dithioles, IR studies indicate that the cyclic structure (**249**) is favored over acyclic structure (**250**) except where R is an amino substituent ⟨87PS(31)109⟩.

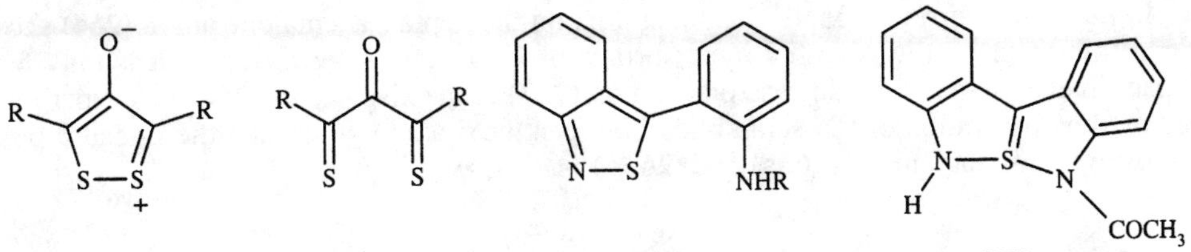

(249)   (250)   (251)   (252)

3-(*o*-Aminophenyl)benzo[*c*]isothiazole (**251**; R = H) exists as such, but when R = CH₃CO the tetracyclic tautomer (**252**) which has a hypervalent sulfur atom is preferred ⟨88JHC1095⟩.

Ab initio calculation of energy changes accompanying distortions of the dioxathiaazapentalene cation showed structure (**254**) to be a transition state between the two valence bond structures (**253a**) and (**253b**).

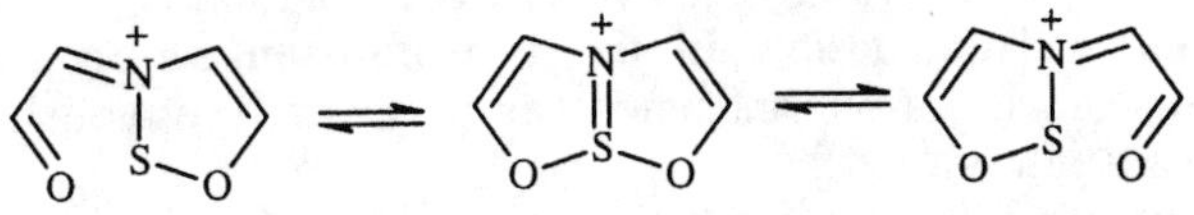

(253a)   (254)   (253b)

2.4.5.3.2 *With involvement of substituents*

The most well-known tautomeric systems of this type involve α-azidoazoles. Unlike their azine counterparts normally existing as condensed tetrazoles (Sec. 2.2.5.4), they exist as a rule in an open-chain structure, *e.g.* (**255**). However, in alkaline media N-anion (**256**) has a strong tendency to cyclize into tetrazole (**257**) which often becomes predominant in an equilibrium mixture.

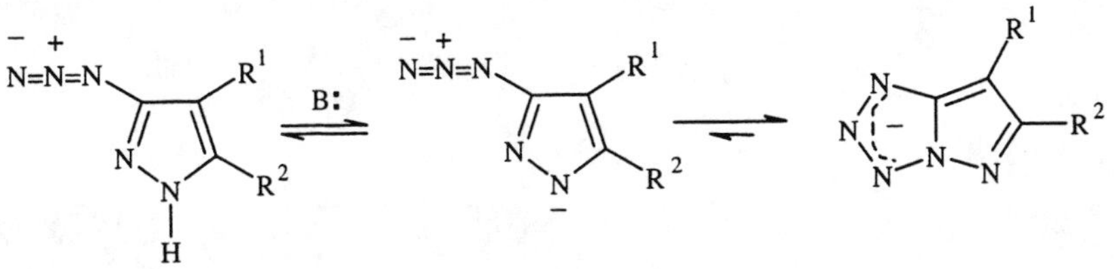

(255)   (256)   (257)

Azido-tetrazole tautomerism is also observed for partially hydrogenated azoles and even for monocyclic tetrazoles. The latter (**258**, X = NR¹) may exist in equilibrium with an acyclic imidoyl azide

(**259**, X = NR[1]). This is also the case for systems (**258**) with X = O and S and their behavior provides a guide to some of the factors influencing the tetrazole case. When with (**258**, X = O) the azido form (**259**) dominates and in general when strongly electron-withdrawing groups are present on the imidoyl azide, cyclization to tetrazole is not observed ⟨84CHEC-(5)791⟩. In (**258**, X = S) the system exists in the cyclic form (**258**). When electron-donating substituents are present, the imidoyl azide (**259**) rapidly ring closes to tetrazole (**258**, X = NH), this being one of the main synthetic routes to tetrazoles (see Section 4.3.8.1).

(**258**)   (**259**)   (**260**)   (**261**)   (**262**)

Higher temperatures favor azido form (**261**) in equilibrium with (**260**) and this ring opening is responsible for the Dimroth rearrangement of substituted 5-aminotetrazoles ⟨82JHC943, 84JHC627⟩ where 5-alkylamino-1-aryltetrazoles (**260**) rearrange into 5-arylamino-1-alkyltetrazoles (**262**) on heating at 180–200°C. Both forms are present in the melt (see Section 3.4.3.5.1).

4-31G *Ab initio* and corrected MNDO calculations on the heptaazapentalene anion (**264**) showed it to be a true minimum on the potential hypersurface of CN_7^- and suggested that it is only 6.6 kcal mol^{-1} higher in energy than 5-azidotetrazolate anion (**263**). The existence of (**264**) was experimentally supported by the observation of ^{15}N scrambling between the 5-azido group and the 2- and 3-positions of the tetrazole ring in the anions (**263a**) and (**263b**) ⟨86CC959⟩.

(**263a**)   (**264**)   (**263b**)

Another example of ring-chain tautomerism with participation of a substituent is the interconversion of furoxan-2- and furoxan-5-oxides (**104a** ⇌ **104b**). The equilibration is presumed to proceed *via* an intermediate *cis*-1,2-dinitrosoalkene (**105**). In the benzofuroxan series the involvement of the corresponding 1,2-dinitrosoarenes as intermediates has been firmly established by matrix isolation experiments ⟨91CC1178, 91JOC5216, 92CL57⟩.

For ring-chain tautomerism of azolo pseudobases see Section 3.4.1.6.1, ii.

2.5

Structure of Small and Large Rings

The division of compounds of these classes into aromatic and non-aromatic (also antiaromatic) types is much less clear-cut than for the five- and six-membered rings. Heterocyclic aromaticity requires a planar or nearly planar conjugated system containing $(4n + 2)$ π-electrons. The subject has been reviewed ⟨74AHC(17)339, 85KGS867, 93AHC(56)303⟩. Planarity often cannot be achieved because of increased strain in going from puckered to planar geometries. The Hückel condition often can be met, utilizing the system's π-bonds and n electron pairs, and sometimes by forming cations or anions to adjust the number of participating electrons. Larger heteroaromatic systems are known with 7 to 21 ring members. Still, a great many heterocycles having the right number of electrons are polyenic rather than aromatic, due to the excessive energy required to achieve near planarity. The aromatic systems are diatropic as seen in their ^{1}H NMR spectra, and show double bond delocalization in their electronic spectra and bond length equalization in their X-ray or electron diffraction structures. Cases intermediate between polyenic and aromatic are also found (CHEC 5.20.2.2.1). Well-known aromatic systems include 1,3-dithiepin anions (CHEC 5.18.4.2; CHEC-II, 9.11.3.3) 1,4-dihydro-1,4-diazocines (CHEC 5.19.4.4; CHEC-II, 9.23.3), azonine (CHEC 5.20.2.2.2; CHEC-II, 9.27), 2,7-methanoaza-[10]annulene (CHEC 5.20.2.3) ⟨78AG(E)853⟩ and *trans,trans,trans*-aza[13]annulene (CHEC 5.20.2.4).

2.5.1 SURVEY OF POSSIBLE STRUCTURES

2.5.1.1 Small Rings

Tables 1 and 2 list the three- and four-membered rings, respectively, which have been made or for which there is strong evidence (such as from isotope or stereochemical studies). Triaziridine, a long sought molecule, turns out to be relatively stable.

2.5.1.2 Large Rings

The number of possible heterocyclic rings of seven or more members is enormous. Many are known. Some of the most important ring systems are listed in Scheme 1, with reference to the CHEC and CHEC-II chapters where they are discussed.

2.5.2 THEORETICAL METHODS

Elaborate *ab initio* calculations employing large basis set correlated wave functions have been conducted on three-membered rings, for example, investigations on the antiaromatic analogues of the cyclopropenium ion ⟨82JOC1869, 83CJC2596, 83JA5541, 83JA396⟩.

Molecular orbital calculations, including HF, SCF, MP2, and MP3 *ab initio* methods using a wide variety of basis sets, provide data on the geometry, energy, and bond order of aziridine (**1**) ⟨85JA3800, 87JA3224, 87JPC6484, 88JST(165)99, 89JCC468, 93JA11074, 93MI 101-02⟩. Calculated proton affinities (*ca.* 218.2–223.3 kcal mol^{-1}) ⟨93JA11074⟩ agree with experimental values (215.7 kcal mol^{-1}) ⟨81JA486⟩. *Ab initio* modeling of the pyramidal inversion of the nitrogen of aziridine and simple substituted aziridines, provided inversion barriers (12–22.3 kcal mol^{-1}, depending on the method), transition structures, and calculated rate constants for inversion ⟨87JA3224, 87JA6290, 89JCC468, 93CPL(204)175⟩.

The MNDO SCF method gives results which compare well with experimental values, including the high barriers of *N*-halo- and *N*-amino-aziridines, and the low ones for *N*-trimethylsilyl- and *N*-phosphino-aziridines ⟨80JCS(P2)1512⟩.

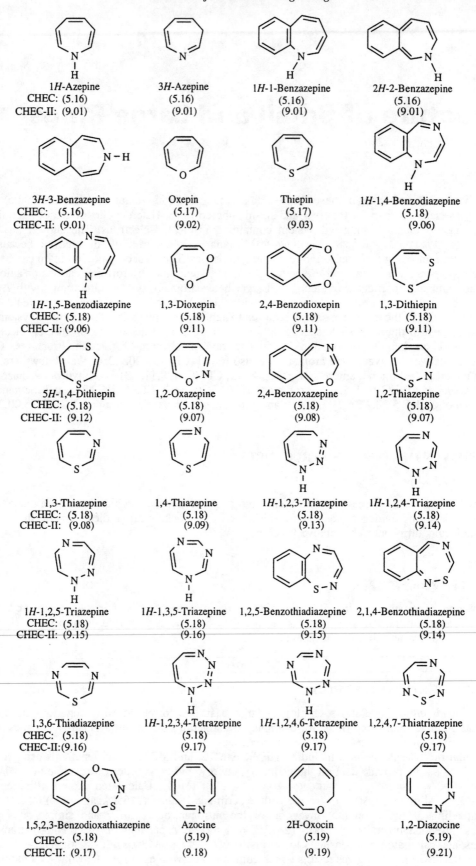

1*H*-Azepine
CHEC: (5.16)
CHEC-II: (9.01)

3*H*-Azepine
(5.16)
(9.01)

1*H*-1-Benzazepine
(5.16)
(9.01)

2*H*-2-Benzazepine
(5.16)
(9.01)

3*H*-3-Benzazepine
CHEC: (5.16)
CHEC-II: (9.01)

Oxepin
(5.17)
(9.02)

Thiepin
(5.17)
(9.03)

1*H*-1,4-Benzodiazepine
(5.18)
(9.06)

1*H*-1,5-Benzodiazepine
CHEC: (5.18)
CHEC-II: (9.06)

1,3-Dioxepin
(5.18)
(9.11)

2,4-Benzodioxepin
(5.18)
(9.11)

1,3-Dithiepin
(5.18)
(9.11)

5*H*-1,4-Dithiepin
CHEC: (5.18)
CHEC-II: (9.12)

1,2-Oxazepine
(5.18)
(9.07)

2,4-Benzoxazepine
(5.18)
(9.08)

1,2-Thiazepine
(5.18)
(9.07)

1,3-Thiazepine
CHEC: (5.18)
CHEC-II: (9.08)

1,4-Thiazepine
(5.18)
(9.09)

1*H*-1,2,3-Triazepine
(5.18)
(9.13)

1*H*-1,2,4-Triazepine
(5.18)
(9.14)

1*H*-1,2,5-Triazepine
CHEC: (5.18)
CHEC-II: (9.15)

1*H*-1,3,5-Triazepine
(5.18)
(9.16)

1,2,5-Benzothiadiazepine
(5.18)
(9.15)

2,1,4-Benzothiadiazepine
(5.18)
(9.14)

1,3,6-Thiadiazepine
CHEC: (5.18)
CHEC-II:(9.16)

1*H*-1,2,3,4-Tetrazepine
(5.18)
(9.17)

1*H*-1,2,4,6-Tetrazepine
(5.18)
(9.17)

1,2,4,7-Thiatriazepine
(5.18)
(9.17)

1,5,2,3-Benzodioxathiazepine
CHEC: (5.18)
CHEC-II: (9.17)

Azocine
(5.19)
(9.18)

2H-Oxocin
(5.19)
(9.19)

1,2-Diazocine
(5.19)
(9.21)

Scheme 1   Examples of seven-membered and larger rings with references to the relevant CHEC and CHEC-II chapters in parentheses

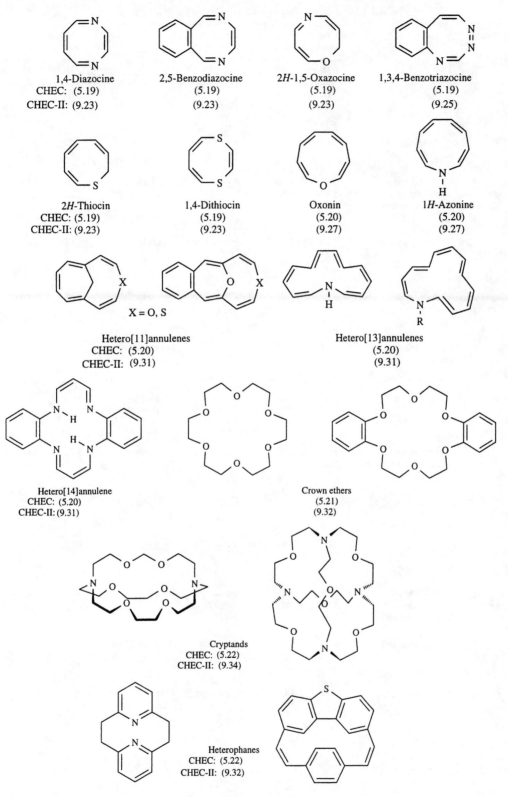

1,4-Diazocine
CHEC: (5.19)
CHEC-II: (9.23)

2,5-Benzodiazocine
(5.19)
(9.23)

2*H*-1,5-Oxazocine
(5.19)
(9.23)

1,3,4-Benzotriazocine
(5.19)
(9.25)

2*H*-Thiocin
CHEC: (5.19)
CHEC-II: (9.23)

1,4-Dithiocin
(5.19)
(9.23)

Oxonin
(5.20)
(9.27)

1*H*-Azonine
(5.20)
(9.27)

X = O, S

Hetero[11]annulenes
CHEC: (5.20)
CHEC-II: (9.31)

Hetero[13]annulenes
(5.20)
(9.31)

Hetero[14]annulene
CHEC: (5.20)
CHEC-II: (9.31)

Crown ethers
(5.21)
(9.32)

Cryptands
CHEC: (5.22)
CHEC-II: (9.34)

Heterophanes
CHEC: (5.22)
CHEC-II: (9.32)

Scheme 1 (Continued)

Table 1   Structures of Three-membered Heterocyclic Compounds

Skeleton	Name	CHEC	CHEC-II	Skeleton	Name	CHEC	CHEC-II
(NH)	Aziridine	5.04	1.01	(NH, N–H)	Diaziridine	5.08	1.11
(N)	1-Azirine	5.04	1.01	(N, N)	1-Diazirine	5.08	1.11
(X; N–H)	Aziridinone Alkylideneaziridine Aziridineimine	5.04 5.04 5.04	1.01 1.01 1.01	(X; NH, N–H)	Diaziridinone Diaziridinimine	5.08 5.08	1.11 1.11
(X X; N–H)	Aziridinedione	5.04	1.01	(NH, O)	Oxaziridine	5.08	1.12
(X; N)	Azirinimine	5.04	1.01	(O, O)	Dioxirane	5.08	1.14
(N–H)	2-Azirine	5.04	1.01	(X; O, O)	Dioxiranone	—	1.14
(O)	Oxirane	5.05	1.03	(NH, S, O₂)	Thiaziridine 1,1-dioxide	5.08	1.13
(X; O)	Oxiranone Alkylideneoxirane Oxiranimine	5.05 5.05 5.05	1.03 1.03 1.03	(S, O)	Oxathiirane	5.08	1.14
(O)	Oxirene	5.05	1.03	(X; NH, S)	Thiaziridineimine	5.08	1.13
(S)	Thiirane	5.06	1.05	(HN–NH, N–H)	Triaziridine	5.08	—
(X; S)	Thiiranone Thiiranimine Alkylidenethiirane	5.06 5.06 5.06	1.05 1.05 1.05	(HN–N, N)	Triazirine	—	—
(X X; S)	Thiiranediimine	5.06	1.05	(HN–NH, O)	Oxadiaziridine	—	1.01
(S)	Thiirene	5.06	1.05	(HN–NH, S, O₂)	Thiadiaziridine 1,1-dioxide	—	1.01

Table 2 Structures of Four-membered Heterocyclic Compounds

Skeleton	Name	CHEC chapter or Section number		Skeleton	Name	CHEC chapter or Section number	
		CHEC	CHEC-II			CHEC	CHEC-II
	Azetidine	5.09	1.18		Thietan-2-one (benzo[b]-)	5.14.2.5	1.24
	1-Azetine	5.09.4.1	1.18		Thietane-2,4-diimine	5.14.2	1.24
	2-Azetine	5.09.4.3	1.18		Thietane-2,3,4-triimine	5.14	1.24
	Azetidin-2-one	5.09	1.18		1,2-Diazetidine	5.15.1.2	1.30
	Azetidin-3-one	5.09	1.18		3H,4H-Diazetine	5.15	1.30
	Azete	5.09.5	1.18		1H,2H-Diazetine	5.15	1.30
	2-Azetin-4-one	5.09.4.3	1.18		1,2-Diazetidine-3-one	5.15	1.30
	Azetidine-2,4-dione 4-Thioxoazetidin-2-one	5.09	1.18 1.18		1,2-Diazetidine-3,4-dione	5.15	1.30
	X = N, Y = O X = CR2, Y = O	5.09 5.09	1.18 1.18		1,2-Oxazetidine	5.15.1.2	1.31
	X = CR2, Y = Z = O X = CR2, Y = O, Z = N	5.09 5.09	1.18 1.18		4H-1,2-Oxazetine	5.15.1.2	1.31
	Oxetane	5.13	1.22		1,2-Thiazetine(1,1-dioxide, benzo[b]-)	5.15.1.2	1.32
	Oxete	5.13	1.22		1,3-Diazetidin-2-one 1,3-Diazetidin-2-imine	5,15.1.2.4 5.15.1.2.4	1.30 1.30
	Oxetan-2-one Oxetan-2-imine	5.13 5.13	1.22 1.22		1,3-Diazetidine-2,4-dione	5.15.1.2.4	1.30
	Oxetan-3-one Oxetan-3-imine 3-Alkylidenoxetane	5.13 5.13 5.13	1.22 1.22 1.22		1,3-Oxazetidin-2-one	5.15.1.2.5	1.31

Table 2 (Continued)

Skeleton	Name	CHEC chapter or Section number		Skeleton	Name	CHEC chapter or Section number	
		CHEC	CHEC-II			CHEC	CHEC-II
	Thietane	5.14	1.25		1,3-Thiazetidin-2-imine	5.15.1.2.6	1.32
	Thiete	5.14	1.25		1,2-Dioxetane	5.15.1.2.2	1.33
	Thiacyclobutadiene	5.14.2.6	—		1,2-Dioxetanone	5.15.1.2.2	1.33
	1,2-Dioxetanedione	5.15.1.2.2	1.33		1,2- oxathietane 1,1 -dioxide	5.15.1.2.8	1.34
	1,2-Oxathietane	5.15.1.2.8	1.34		1,2-Dithietane	5.15.1.2.9	1.35
	1,2-Oxathiete	5.15.1.2	1.34		1,3-Dithietane	5.15.1.2.10	1.35

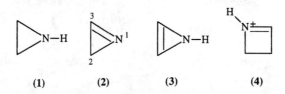

(1)   (2)   (3)   (4)

The structures of 1-azirine (**2**) and its complexes with H^+ and Li^+ ⟨93JA11074⟩ have been calculated by *ab initio* methods. An unsuccessful study aimed at matrix isolation of 2-azirine (**3**) calculated its infrared spectrum with *ab initio* methods and predicted that 2-azirine is $32.7–33.2$ kcal mol^{-1} higher in energy than 1-azirine ⟨93CB2337⟩. The relative basicities of (**1**)-(**3**) have been calculated: 2-azirine is more basic than aziridine, which is more basic than 1-azirine. The antiaromaticity of 2-azirine is relieved upon protonation, thus accounting for its higher basicity. Whereas 2-azirine is considerably higher in energy than 1-azirine (see above), protonated 2-azirine is only 4.7 kcal mol^{-1} higher in energy than protonated 1-azirine. The structures of the potentially aromatic azirinyl cation $C_2H_2N^+$ and the azirinyl radical cation ⟨91JA3689⟩ have also been studied by semiempirical and *ab initio* MO methods.

Structures and energies of the cyclic $C_3H_6N^+$ cations have been examined by *ab initio* molecular orbital calculations with the 3-21-G basis set. The iminium ion (**4**) is the most stable of the four-membered ring isomers ⟨89JA5560⟩.

MNDO/3 has been applied to thiirene (**5**) and the isomeric carbene (**5a**), zwitterion (**5b**), heterocumulene (**5c**), and cyclic carbene (**5d**). Structure (**5**) is calculated to lie in a local energy minimum.

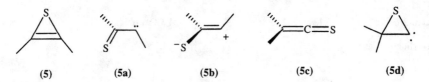

(5)   (5a)   (5b)   (5c)   (5d)

Studies on larger systems have tended to be limited to semi-empirical or small basis set *ab initio* methods.

Restricted Hartree-Fock (HF) calculations have been carried out on 1*H*-azepine and the results compared with those obtained by semi-empirical MNDO to test whether the latter were sufficiently rigorous to apply to larger cyclopolyenes for which *ab initio* calculations may be prohibitive ⟨84JST(19)277⟩. According to these results, 1*H*-azepine possesses a boat configuration with 22% chair character. This permits π-delocalization of the cyclic double bond system to a level comparable to that of a linear polyene. The data further suggest that the ring puckering is a consequence of ring strain rather than a destabilizing antiaromatic 8π-interaction. Overall ring shape is better predicted by the HF calculation; alternating bond distances are more accurately represented, however, by MNDO calculations. The MOMM calculations also show that the π-electrons of 3*H*-azepine are highly localized and that the molecule prefers to exist in a boat conformation, which is predicted to be 4.7 kcal mol^{-1} more stable than the planar form and 17 kcal mol^{-1} more stable than 1*H*-azepine ⟨88JCO(9)905⟩.

Using both semi-empirical and *ab initio* calculations the study of oxepin (**6**), benzene oxide (**7**), and their equilibrium (Scheme 2) has been conducted. The fully optimized geometry ⟨90MI 902-01⟩ agrees with that experimentally found for several substituted oxepines. The carbon skeleton of benzene oxide is practically planar while the angle between the epoxide ring and the adjacent plane is *ca.* 106°. The oxepin molecule is boat-shaped with a fold angles between C_2—C_7 and C_3—C_6 of *ca.* 137 and 159°, respectively.

Calculated resonance energies of some seven-membered heterocycles are given in Table 7.

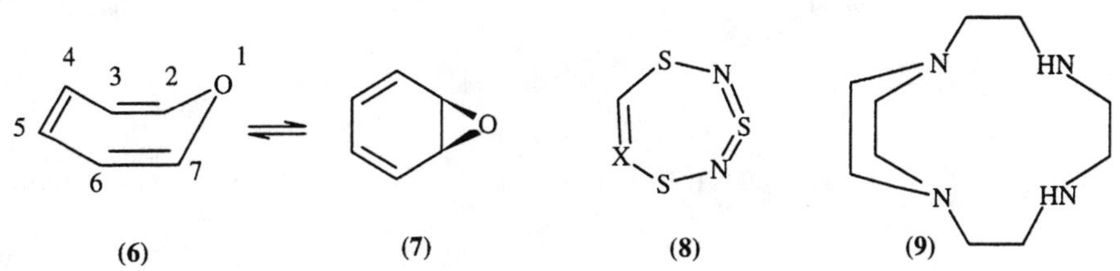

Scheme 2

A possible role in commercially useful conducting polymers has stimulated significant interest in the calculated properties of large rings incorporating sulfur and nitrogen heteroatoms ⟨84JA312, 82JA2691, 80JA6687⟩. The delocalized 10π aromatic nature of the trithiadiazepine (**8**, X = CH) and trithiatriazepine (**8**, X = N) rings was supported by MNDO and *ab initio* MO calculations. The results are consistent with the photoelectron spectrum of (**8**) and confirm the fundamental aromatic nature of these systems, in accord with their stability, chemical reactivity, and other spectroscopic properties ⟨85CC398, 87CS(P1)203⟩.

SINDO1 calculations, which successfully reproduced both the geometry and the aromaticity of the cyclooctatetraene dianion, predict a high degree of bond localization and nonplanarity for 1,4-dioxocin and its derivatives ⟨84JOC4475⟩.

Molecular mechanics have become more and more important in the study of the correlation between cavity size and ionic radius of coordinated metal ions ⟨84CCR1, 93CCR177⟩. The different aspects of metal ion reactivity and ligand design for specific and selective binding can be modeled and understood in detail.

The concept of matching between macrocyclic cavity size and size of the complexed metal ion is supported by several lines of evidence ⟨78ACR49⟩. The high selectivity of crown ethers or cryptands towards alkali metal ions has been shown to stem from such correlations and has resulted in the so-called "peak selectivity", typical for cyclic compounds. However, work on tetraazamacrocycles ⟨85IC3378⟩ has shown that for these ligands, the actual factors inducing complexation selectivity for transition-metal ions are the sizes of the chelate rings formed by coordination of the metal ions. Thus, when chelate ring size is changed in the process of varying the macrocyclic ring size, it is found that the higher the number of six-membered chelate rings, the lower the stability of the complexes with larger metal ions ⟨89MI 928-02⟩. Exactly the same effect is observed in open-chain ligands indicating that the chelate ring and bite size and not the cyclic structure are important. If a more rigid structure is built into the macrocycle as for (**9**), the ring cannot fold and adapt to the geometrical requirements of the metal ion.

2.5.3 STRUCTURAL METHODS

2.5.3.1 X-Ray Diffraction

The geometries of many three- and four-membered rings have been determined by X-ray diffraction on crystalline materials ⟨72PMH(5)1, p.12⟩. Three-membered heterocycles generally have shorter C-C bonds than does cyclopropane, an exception being thiirane dioxides. The C-X bonds are longer than in CH_3-X-CH_3. The CXC angles in oxirane and aziridine are close to 60°, and the peripheral HCH bond angles are near 118°. Table 3 gives representative data.

The geometry of four-membered rings is far more complex; the rings are usually not planar (see Table 4). Thus, the X-ray crystal structure of the parent oxetane revealed exact C_s ring symmetry puckered with an angle of 10.7° at 90K and 8.7° at 140K. Interestingly, the carbon-oxygen bond length (1.460 Å) was unusually large compared with tetrahydrofuran (1.429 Å) and dioxane (1.433 Å).

Heterocyclics with seven and more ring members display an enormous variety of shapes. Bond lengths are often close to those of open chain counterparts, but bond angles can be greatly different. 3,7-Di-*t*-butyl-5-phenyl-2-(*p*-tolyl)oxepin ⟨82CB385⟩ data show the substituted oxepin in a boat conformation with dihedral angles 117.6° between C_2OC_7 and $C_2C_3C_6C_7$ planes and 150.7° between

Table 3 Bond Lengths and Bond Angles of Three-membered Heterocyclic Compounds

| Structure | Bond lengths (Å) | | | Bond angles (°) | | | CRSE (kJ mol^{-1}) | CHEC section number/ref.[a] |
	a	b	c	ca	ab	bc		
(aziridine, N-H)	1.482	1.491	1.482	—	—	—	113	5.04.2.2
(aziridine, N-CH(OH)CCl₃)	1.49	1.49	1.50	59.9	60.4	59.8	—	74PMH(6)1, p.8
(N-adamantyl, =O)	1.33	1.45	1.51	60.9	—	53.4	—	5.04.2.2
(N, Me)	1.256	1.463	1.598	60.3	—	48.2	—	5.04.2.2
(oxirane, O)	1.44	1.47	1.44	61.24	59.18	59.18	114	5.05.2.2
(thiirane, S)	1.815	1.484	1.815	48.3	—	—	—	5.06.2.1
(S, =O)	1.726	1.460	1.916	49.6	73.4	59.7	—	5.06.2
(S)	1.79–1.98	1.25–1.29	—	—	—	—	—	5.06.2.1
(Me, N-H, N-H)	1.468	1.479	1.479	59.5	59.5	61	—	5.08.2.2
(N, N)	1.228	1.428	1.428	64.5	64.5	50.9	—	5.08.2.2
(p-O₂NC₆H₄, O, N-Pr^i)	1.50	1.405	1.434	57.2	59.0	63.8	—	5.08.2.2

Table 4 Bond Lengths and Bond Angles of Four-membered Heterocyclic Compounds

Structure	Bond lengths (Å)				Bond angles (°)				Pucker angles(°) and Ring inversion barrier (kJ mol^{-1})	CHEC section Number/ref.
	a	b	c	d	da	ab	bc	cd		
azetidine (NH; b, a, c, d)	1.477	1.560	1.560	1.477	88	—	—	—	33 (5.3)	5.09.2.1, 80CRV231
(CO$_2$H; NH)	1.51	1.54	1.53	1.52	88	90	88	90	—	5.09.2.1
(Me$_2$; N(CH$_2$Ph)$_2$)	1.51	1.53	1.51	1.51	84	96	84	97	—	B-79MI50100
(O; N-COC$_6$H$_4$Br-o)	1.38	1.51	1.57	1.50	94.0	93.4	86.3	86.3	—	5.09.3.1
(PhN, C$_6$H$_4$F-p, (CF$_3$)$_2$, N)	1.307	1.467	1.575	1.504	92.4	97.9	83.6	86.1	Flat	5.09.4.1
oxetane (O)	1.443	1.517	1.517	1.443	90.5	91.89	85.0	92.61		84JA7118
	1.449	1.549	1.549	1.449	91.89	91.73	84.55	91.74	10	80CRV231, 74MI150100
				CRSE = 106 kJ mol^{-1}						
(H$_2$C; O, O)	1.39	1.51	1.54	1.47	95.8	83	91.3	90	Flat	80CRV231, 72PMH(5)1, p12
(S)	1.847	1.549	1.549	1.847	76.8	90.6	95.6	90.6	26(3.3)	5.14.2.1, 80CRV231, 74MI50100
(O=; S)	1.826	1.528	1.528	1.826	—	—	100.5	—	Flat	5.14.2.1
(Me$_2$, NAr; Ph$_2$—S)	1.777	1.524	1.591	1.868	77	95.8	—	90.1	20	5.14.2.1
(SO$_2$)	1.79	1.43	1.39	1.77	80.5	—	—	104.5	Flat	5.14.2.1
(NH; NH)	1.427	1.481	1.537	1.471	—	—	—	—	24.3	5.15.1.2.1
(ade, O, O, ade)	1.48	1.549	1.475	1.549	—	—	—	—	21.3	5.15.1.2.2
ade = adamantylidene (S, S)	2.146	1.835	1.564	1.835	99.1	—	80.9	—		5.15.1.2
(Cl, Cl, Cl, Cl; S, S)	1.801	—	1.77	—	83.9	—	96.1	—	Flat	5.15.1.2.6

$C_2C_3C_6C_7$ and $C_3C_4C_5C_6$ planes. Double and single bonds in the molecule show some degree of equalizing: O—C_2, 1.424, C_2—C_3, 1.393, C_3—C_4 1.426, C_4—C_5 1.367, C_5—C_6 1.441, C_6—C_7 1.323, and C_7—O 1.404 Å. Similar geometry has also been demonstrated for 2,7-diphenyloxepin ⟨86JOC2784⟩ and 7-*t*-butoxycarbonyloxepin ⟨86CC970⟩.

The crystal structures of some thiepines are nonplanar.

The X-ray crystal structure of the 1-tosyl-1*H*-1,2-diazepine (**10**) shows a boat conformation with N(1) at the prow (Figure 1) ⟨72T581⟩. Double bonds are clearly localized at N(2)—C(3) 1.255 Å, C(4)—C(5) 1.326 Å, and C(6)—C(7) 1.333 Å, and the imine bond is isolated, whereas bond delocalization is present in the butadiene-like part C(4)—C(5)—C(6)—C(7). In contrast, when the diazepine is complexed with Fe(CO)₃, for example (**11**), the iron atom is bound to the butadiene moiety C(4) to C(7), as also occurs in the 1*H*-azepine and 1-methoxycarbonyl-1*H*-azepine complexes; the N(1) atom adopts a planar sp^2 configuration and the seven-membered ring consists of the two planar parts bent along the C(4) . . . C(7) line by 140° ⟨70AG(E)958⟩.

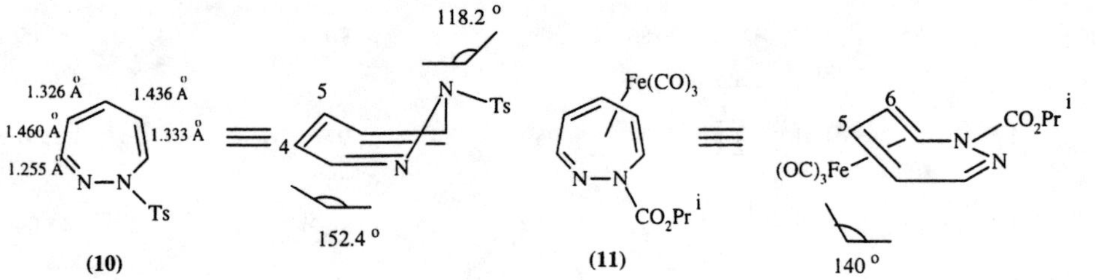

Figure 1   Selected geometrical parameters of some 1,2-diazepines

The bond lengths of fully saturated seven-membered rings are the same as those in the corresponding open chain compounds, while the bond angles tend to be larger.

2.5.3.2   Microwave Spectroscopy

Microwave spectra have been used to determine molecular dimensions for, *inter alia*, aziridines and 1-azirines (CHEC 2.04.2.2), diazirine and substituted derivatives (CHEC 5.08.2.2), oxetanes (CHEC 5.13.2.2), thietanes and thietes (CHEC 5.14.2.1).

2.5.3.3   ¹H NMR Spectroscopy

The NMR spectra of three- and four-membered heterocyclics display regularities of great value to structure determination. For protons on adjacent carbons the coupling constants J_{cis} seem to be always greater than J_{trans}. In three-membered rings J_{gem}, is almost always smaller than J_{cis} and J_{trans}. Extensive tables are found in ⟨B-73NMR138⟩. The average values for 64 aziridines are $J_{gem} = 1.4$, $J_{trans} = 3.3$ and $J_{cis} = 6.4$ Hz ⟨71PMH(4)121, p.126⟩. The size of J_{gem} and of the vicinal C-H coupling constants seems to depend more on the number of non-bonding electron pairs at the heteroatom than on its electronegativity. Each electron pair contributes $+5.5$ Hz to J_{gem}, -2.5 Hz to J_{cis} and -2.7 Hz to J_{trans} ⟨80OMR(13)45⟩. Table 5 gives some examples, the data being taken from the monograph chapters of CHEC and from ⟨B-73NMR, 71PMH(4)121, 80OMR(13)45⟩. Data on some three-membered heterocycles with an *exo* methylene group are found in Table 6 ⟨78RTC214⟩.

Table 5   Ranges of NMR Data of Three-membered Heterocyclic Systems

Skeleton	$\delta(C\text{-}H)$(ppm)	J_{gem} (Hz)	J_{cis} (Hz)	J_{trans} (Hz)	$J(^{13}C\text{-}H)$ (Hz)	$\delta(^{13}C)$(ppm)
Cyclopropane	0.22	-3 to -1	6–12	-4 to 8	164	-2.2
Aziridine [a]	1.48	0.9–4	5–9	2–7	168	18–22
Oxirane	2.54	5–7	2–5	1–3	176	39.7
Thiirane	2.27	-14 to 1	6–7	5–6	170	18
1-Azirine C-2	10	—	—	—	—	160–170
C-3	0.2–2.5					19–45
Diaziridine	1.2	—	—	—	—	56
1-Diazirine	0.4	—	—	—	—	—
Oxaziridine	4.5–5	—	—	—	—	56

[a] $\delta(^{15}N) = -8.5$ (vs. NH₃)

Table 6 Ranges of NMR Data for Three-membered Heterocyclic Systems with Exocyclic Unsaturation

Skeleton	$\delta(C_2)$ (ppm)	$\delta(C_3)$ (ppm)	$\delta(C_{ex})$ (ppm)	$J(C_3\text{-}H)$ (ppm)	$J(C_{ex}\text{-}H)$ (ppm)
Cyclopropane	−2.6	− 2.6	—	160.5	—
Methylenecyclopropane	131.0	3.0	103.5	161.5	160.8
N-t-Butyl-2-methyleneaziridine	134.0	23.8	80.6	170.7	165
2-Methylene-3-*t*-butyloxirane	144.3	68	70.5	—	—
2-Methylenethiirane	130.1	18.5	99.5	174	166

In azetidine derivatives the proton-proton coupling constants J_{gem} on the carbons adjacent to N are 5–7.5 Hz, and J_{cis} is larger than J_{trans}. Long range coupling between ring protons is common. Scheme 3 gives some examples ⟨B-73NMR142, 71PMH(4)121, p.144⟩.

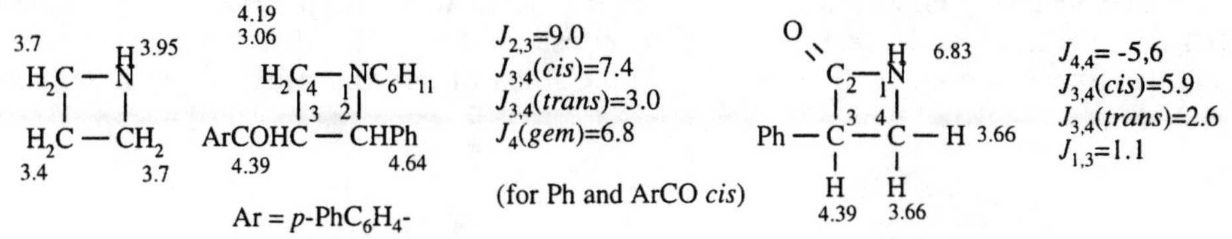
3.7
3.95
3.4
3.7
4.19
3.06
4.39
4.64
Ar = p-PhC6H4-
J2,3=9.0
J3,4(cis)=7.4
J3,4(trans)=3.0
J4(gem)=6.8
(for Ph and ArCO cis)
6.83
3.66
4.39
3.66
J4,4= -5,6
J3,4(cis)=5.9
J3,4(trans)=2.6
J1,3=1.1

Scheme 3 ^{1}H NMR shifts and coupling constants of azetidine derivatives

The NMR spectra of heterocyclic compounds with seven or more ring members are as diverse as the shape, size and degree of unsaturation of the compounds. Proton-proton coupling constants provide a wealth of data on the shape of the molecules, while chemical shift data, heteroatom-proton coupling constants and heteronuclear spectra give information of the electronic structure. Some data on seven-membered rings are included in Table 7. Several additional examples of NMR spectroscopy for large heterocycles are discussed below.

Table 7 Structure and ^{1}H NMR Data of Seven-membered Heterocyclic Compounds

		Bond lengths (Å)				*Resonance*	$\delta(^1H)$(ppm)						*Longest Wave-length UV absorption*	
X	Y	a	b	c	d	(kJ mol^{-1})	H-2	H-3	H-4	H-5	H-6	H-7	(nm)	log ε
C	C	1.45	1.37	1.48	1.35	—	—	—	—	—	—	—	—	—
O	C	1.39	1.35	1.46	1.35	+0.5	5.7	5.7	6.3	—	—	—	305	2.95
HN	C	1.42	1.35	1.46	1.35	−0.75	—	—	—	—	—	—	—	—
EtOCON	C	—	—	—	—	—	5.95	5.51	6.15	—	—	—	318	2.83
EtOCON	N	—	—	—	—	—	—	6.23	5.75	6.55	6.25	7.4	355	2.38
S	C	1.79	1.35	1.46	1.35	−6.1	—	—	—	—	—	—	—	—

Ring numbering is 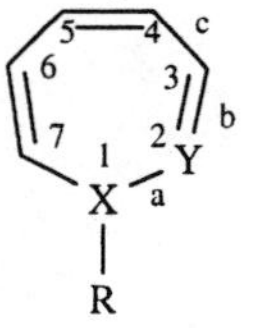
d
c
5
4
6
3
b
7
1
2
Y
X
a
R

Accidental degeneracy can lead to extraordinarily simple ^{1}H NMR spectra. Compound (**12**; R = Ph) displays a singlet (at 6.49 ppm in $CDCl_3$) for the four ring protons ⟨82CL1579⟩, while parent 1,2-diazocine (**12**; R = H) shows a broad singlet at 6.93 ppm (2H) and a sharp singlet at 6.03 ppm (4H). An europium shift reagent splits the last resonance into an AB pair of doublets, ruling out isomeric bicyclic structures for (12), indicating that this compound exists as the bond-localized 2,4,6,8-tetraene (12) rather than the 1,3,5,7-tetraene (12′) ⟨79JOC1264⟩ which, by virtue of the nonplanarity of the ring, must be considered a distinct isomer of (12) rather than as a resonance structure.

(12)   (12′)   (13)   (14)

The 4H-thiepinium ion (14), generated from thiepine (13) in $FSO_3H/SO_2/CD_2Cl_2$ (−70°C) has been shown to possess the homothiopyrylium ion structure ⟨84CC604⟩. The ^1H NMR spectrum of (14; $R^2 = H$) indicates charge delocalization, with a chemical shift difference ($\Delta\delta = 2.5$ ppm) between the methylene protons, (4o)-H and (4i)-H, and a geminal coupling constant of 11.6 Hz.

NMR spectroscopy has been used extensively in the study of the conformations of nine-membered rings.

2.5.3.4 Heteronuclear NMR Spectroscopy

The ^{13}C–H coupling constants (often obtained from ^{13}C satellites in ^1H NMR spectra) of small heterocycles have been listed ⟨B-79MI50101⟩.

Nitrogen-15 NMR spectra of aziridines and azetidines have been measured ⟨80JOC1277⟩. Relative to anhydrous ammonia, the aziridine nitrogen absorbs at −8.5 ppm and N-alkylation moves this shift downfield. For N-Me the signal is at 0.7 ppm, for N-CHMe$_2$ at 30.2 ppm, and for N-CMe$_3$, at 33.5 ppm. Substitution on the β-carbon shifts the ^{15}N resonance downfield relative to unsubstituted aziridine. This effect decreases with increasing bulk of the substituent, so that 2-methylaziridine has the nitrogen signal at 10.5 ppm, and 2-t-butylaziridine has it at 3.4 ppm. Further substitution on one or both β-carbons causes more downfield shift, the effect being only poorly reproduced by assuming group contributions to be additive.

Aziridine ^{15}N shifts parallel the ^{13}C shifts; in a plot of ^{13}C vs. ^{15}N shifts of 13 aziridines, the correlation coefficient was 0.953 and the slope 2.1 ppm N/ppm C ⟨80JOC1277⟩.

Azetidine ^{15}N shifts are similar to those of the aziridines. Unsubstituted azetidine has its ^{15}N resonance (relative to anhydrous ammonia) at 25.3 ppm, and N-t-butylazetidine shows the signal at 52 ppm ⟨80JOC1277⟩.

The ^{17}O NMR shift data distinguish different molecular configurations for isomeric compounds. Oxygen-17 NMR data of some substituted oxiranes are depicted in Table 8 ⟨83OMR(21)403⟩, and discussed in ⟨B-91MI 103-01⟩. The ^{17}O chemical shifts of mono- and disubstituted oxiranes can be calculated using additivity parameters ⟨86MRC15⟩. Discrepancies between experimental and calculated ^{17}O shift values fall in the range 0 ± 14 ppm. Oxygen-17 NMR shifts are reported for 2-oxetanone: (carbonyl O, 347 ppm; ether O, 241 ppm) and oxetane (13 ppm). The ^{17}O NMR chemical shift data for 30 lactones, including some oxetanones, have been compiled ⟨89H(29)301⟩.

Table 8   Oxygen-17 NMR shift values of some oxiranes

R	H	Me	Et	CH$_2$=CH	Ph
δ(^{17}O)(ppm)	−49	−16	−18	−9.5	−8

Chemical shifts of ^{33}S NMR (in ppm relative to CS_2) in thiirane, thiirane 1-oxide, and thiirane 1,1-dioxide were observed at −240, 120 and 245 ppm, respectively ⟨87JOC3857⟩.

2.5.3.5 UV Spectroscopy

2.5.3.5.1 *Electronic spectra of small-ring heterocyclic compounds*

Saturated three- and four-membered heterocyclics absorb little in the readily accessible regions of the UV spectrum. Sulfur-containing rings are an exception, as can be seen in Table 9. Despite the lack of absorption of most parent compounds, there is a wealth of photochemistry of small heterocyclics. Light absorption by substituents, and energy transfer from photoexcited molecules present in the photoreactive system make photoconversion of the heterocycles practical. On the other hand, the lack of substantial absorption of their own can be exploited in the preparation of small heterocycles, by designing the system to be unsuitable for destructive energy transfer.

The introduction of a second heteroatom (other than sulfur) does not drastically change the absorption characteristics of small heterocycles. Oxaziridine and diaziridine are still 'transparent' to light of wavelengths above 220 nm (CHEC 5.08.2.3.2).

2.5.3.5.2 *Electronic spectra of large-ring heterocyclic compounds*

Electronic spectra, so very important in the characterization of five- and six-membered heterocycles, have played a lesser role in the study of large heterocyclic rings, and fewer data are available for comparison. Table 7 which gives data on bond lengths and calculated resonance energies, includes the longest wavelength electronic absorption of some seven-membered heterocycles ⟨70T4269, 70JAI453, 81H(15)1569⟩. Aromaticity and some other peculiarities of electronic structure of large heterocycles can be detected by their electronic absorption.

Many 1,2-diazocines, including the parent (**12**) ⟨79JOC1264⟩ and its 3,8-diaryl-4,7-dichloro derivatives, are yellow ⟨85CJC1829⟩. However, the color is quite sensitive to substituents, as replacing one or both chlorines in the 4 and 7 positions by nitrogen or sulfur substituents, leads to colorless compounds ⟨87BCJ731⟩.

The parent 1,4-dioxocin (**15**) is colorless, with its major absorption band well into the ultraviolet [238 nm (3.86)] and a weak shoulder at 285 nm (2.50), a spectrum similar to that of 1,3,6-cyclooctatriene ⟨72AG(E)935⟩.

(**15**)   (**16**)   (**17**)

The ultraviolet spectra of 1,5-dithia-2,4,6,8-tetrazocines (**16**; R = Ph) (λ_{max} 306.5 nm) and (**16**; R = NMe$_2$) (λ_{max} 229nm) reflect the unusual structural dependence on substitution that exists for these compounds ⟨81JA1540⟩. Whereas the former compound has perfectly planar eight-membered ring, the 3,7-bis(dimethylamino) derivative was folded about an axis drawn through the two sulfur atoms as shown in (**17**) (see also Sec. 2.5.4.2.2).

2.5.3.6 IR Spectroscopy

IR spectroscopy can give a great deal of information on small ring heterocyclics, because of the effects of ring strain on the frequencies of vibration of substituents attached to the ring, and because the ring vibrations fall into a readily accessible region of the IR spectrum. A wealth of data has been gathered and can be found in the monograph chapters of CHEC and in the following reviews ⟨71PMH(4)265, 63PMH(2)161, B-75MI50100, B-75MI50101⟩. This section concentrates on vibrations of general diagnostic value. A treatment emphasizing the theoretical foundations is available ⟨63PMH(2)161, 71PMH(4)265⟩.

Small rings show high C-H absorption frequencies for the ring C-H bonds (between 3080 and 3000 cm^{-1}). The asymmetric C-H stretching frequency decreases with increasing ring size, from 3047 cm^{-1} for aziridine to 2966 cm^{-1} for azetidine and 2950 cm^{-1} for pyrrolidine. Analogous changes are found in saturated oxygen heterocycles (3052, 2978, 2958 cm^{-1}) and their sulfur analogues (3047, 2968, 2959 cm^{-1}) ⟨71PMH(4)265, p. 278⟩. The stretching frequencies for exocyclic C = X bonds follow a similar sequence, with the smallest rings having the highest frequencies, as seen in Table 10. Four-membered rings have somewhat lower C=X frequencies; the carbonyl frequency of azetidin-2-one is 1786 cm^{-1} and that of oxetan-2-one is 1832 cm^{-1}.

Table 9 Electronic Absorption Spectra of Small Heterocyclic Systems

Skeleton	λ_{max} (nm)	Absorption coefficient	CHEC section number/ref.
	179 145 118	4200 6100 6300	69JCP(51)52, 72BCJ3026
	171.3 158 143	5600	5.05.2.5, 76JCP(64)2062
	260 205	40 4000	5.06.2.4
	191.7		76JCP(64)2062
	187 174 161 153	2000 2750	76JCP(64)2062
	275 218	30 600	5.14.2.4
	340 238	80 7440	5.15.1.2.9

Table 10 Stretching Frequencies for Exocyclic Double Bonds on Small Rings[a]

Skeleton	Stretching frequencies (cm^{-1})		
	X=C	X=N	X=O
	1770 (R=R'=H, R"=Et, X=CH$_2$)	1805 (R=H, R'=Bu^t, R"=Me, X=NMe)	1837 (R=R'=Me, R"=Bu^t, X=O)
	1780 (R=R'=Bu^t, X=CHBu^t)	—	1890 (R=R'=Bu^t) 1990, 1945[b] (R=R'=CF$_3$)
	1738 (R=R'=Me, X=CMe$_2$)	1700, 1630[c] (R=R'=Ph, X=NTs)	1785 (RR'=CH$_2$=)
	1690–1650[d] (R=Ar, R'=SO$_2$Me, X=CMe$_2$)	1790 (R=R'=trans-Bu^t, X=NBu^t)	1882[e] (R=R'=Me)
	—	—	2045

[a]Taken from ⟨80AG(E)276⟩ unless otherwise stated. [b]⟨82CC362⟩. [c]⟨78AG(E)195⟩. [d]⟨77AG(E)475⟩. [e]⟨75AG(E)428⟩.

Table 11 Ring Breathing and C-H IR Absorptions of Small Heterocycles [a]

Structure	Symmetry	Ring breathing	C-H stretch	CH_2 scissoring	CH_2 wagging	Ref.
 (aziridine, N–H)	C_s	1268, 1210	3078, 3012	1475, 1455	1128, 1131, 1088, 998	b c
 (oxirane, O)	C_{2v}	1266	3079, 3063, 3016, 3005	1490, 1470	1153, 1120	b c
 (thiirane, S)	C_{2v}	1112	3080, 3000	1446, 1427	1051, 1025	b c
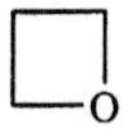 (oxetane, O)	—	980–970, 900	—	—	—	b c
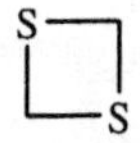 (dithietane, S, S)	—	738	2980, 2942	1438	1187	d

[a] cm^{-1}.
[b] ⟨63PMH(2)161⟩
[c] ⟨71PMH(4)265⟩, p. 277
[d] ⟨79MI50102⟩

Ring breathing frequencies are shown in Table 11, together with some C-H IR absorptions. As expected, the ring breathing frequencies are lower for four- than for three-membered rings ⟨63PMH(2)161, 71PMH(4)265⟩.

IR spectra are less useful for the structure determination of large heterocyclic rings than for small ones. The ring breathing vibrations fall into a range well below that commonly used in the laboratory, and the absorptions are often broad and ill defined. Owing to the almost infinite variety of special effects, bond angle deformation and the consequent effect on the absorption frequencies of ring C-H, C=O and C=X bands are not easily used in the diagnosis of an unknown structure.

The IR spectra of, for example, oxepanes (C-O-C, C-H stretching, and CH_2 deformation) are similar to analogous non-cyclic compounds.

2.5.3.7 Mass Spectrometry

The mass spectral fragmentation patterns of three- and four-membered heterocycles consist of (i) cleavages typical for substituents: (ii) of those due to the formation of particularly stable and accessible fragments (such as N_2), and (iii) the more characteristic patterns attributable to fragmentations promoted by ring strain and by stereochemical factors. Thus, small rings usually open after ionization. In aziridines this can be accomplished by loss of the substituent on the nitrogen, *i.e.* of H·, R·, *etc.*, to give ions of the type $R_2C=N^+=CR_2$ ⟨B-71MS296⟩. More generally, three-membered heterocyclics cleave into a radical and a cation, either of which can contain one or two of the original ring atoms (Scheme 4) (CHEC 5.04.2.8, 5.05.2.4 and 5.06.2.3). Especially in thiiranes, this may involve rearrangements, such as path (c) in Scheme 4. α-Cleavage, particularly important in oxiranes and thiiranes, may give a substituent radical and a cyclic ion (Scheme 5). β-Cleavage, more important in aziridines, gives a radical and an ion (Scheme 5). Longer side chains permit rearrangements, such as that in Scheme 6 ⟨B-71MS11⟩.

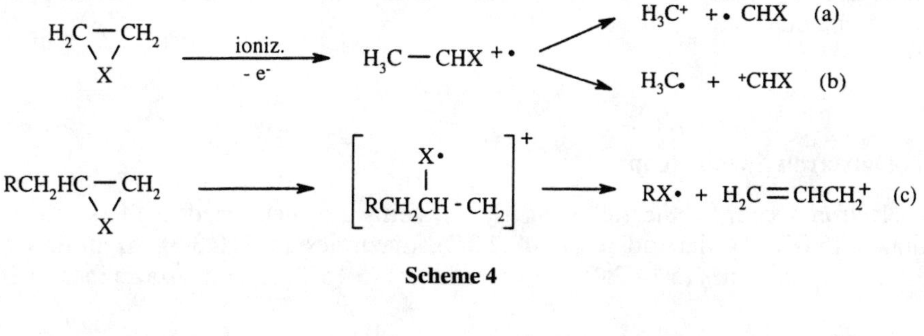

Scheme 4

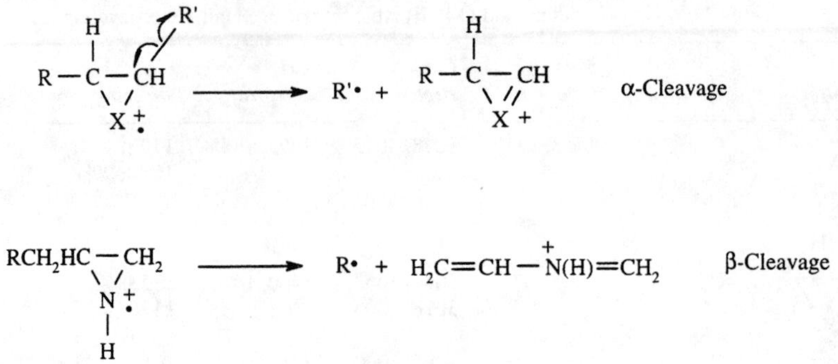

α-Cleavage

β-Cleavage

Scheme 5

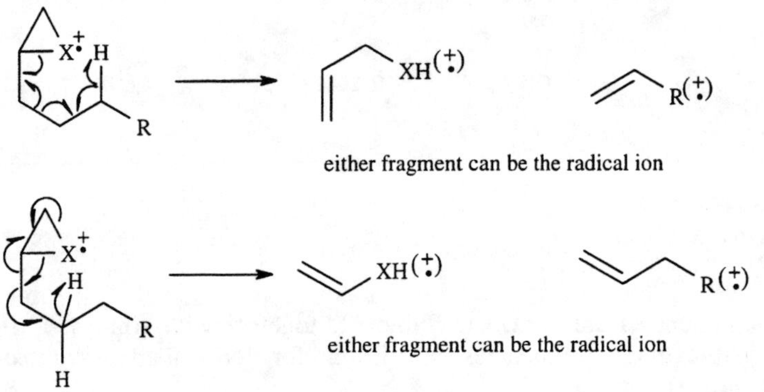

either fragment can be the radical ion

either fragment can be the radical ion

Scheme 6

Four-membered heterocycles prefer to cleave, upon ionization, into two fragments, each containing two of the ring atoms. Further cleavages commence from these initial fragments (Scheme 7). Specific details can be found as follows: azetidines ⟨B-71MS296⟩, oxetanes ⟨B-71MS34⟩, thietanes (CHEC 5.14.2.3) ⟨B-71MS229⟩. The cleavage to two sets of two ring-atom fragments is illustrated by the formation of fragments with the masses of ethylene, methyleneimine from azetidine (HC≡CNH is also formed) and that of those with the masses of RNCO, ketenes and imines from azetidin-2-ones ⟨B-71MS300⟩.

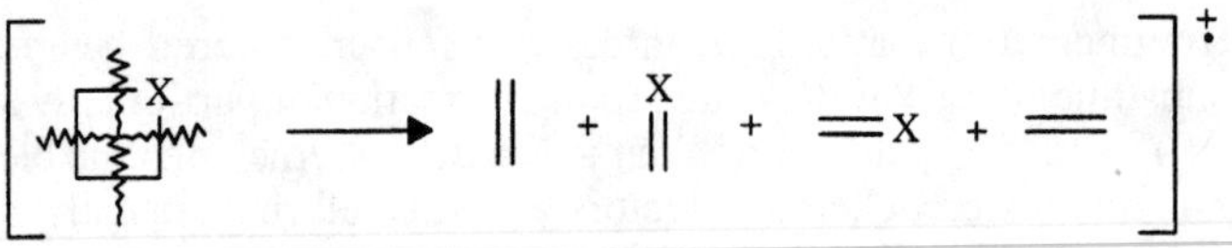

Scheme 7

As example of general trends of fragmentation of large heterocycles one can mention 1,2-diazocine (**12**; R = H). It displays the anticipated parent ion upon electron-impact mass spectral analysis, as well as significant fragments of *m/e* 79 (base peak) and 80 (15%) which have been assigned to pyridine and pyridazine, respectively ⟨79JOC1264⟩. The mass spectral fragmentation more-or-less parallels the thermal decomposition of (**12**); both afford pyridine as the major product. However, thermolysis also affords benzene and no pyridazine, while mass spectral analysis does not suggest formation of benzene as a major fragmentation path.

2.5.3.8 Photoelectron Spectroscopy

The photoelectron spectra of the following ring systems are mentioned in the section of CHEC quoted: thiiranes (5.06.2.3), diaziridines (5.08.2.3.2), diazirines (5.08.2.3.3), azetidines (5.09.2.1), oxetanes (5.13.2.3.4), thietanes (5.14.2.4), 1,2-diazetidines (5.15.1.2.1), dibenzazepines (5.16.2.6).

The exceptionally low ionization energies of the bridged diazonines (**18**) indicate through-space and through-bond coupling effects ⟨81JA6137⟩.

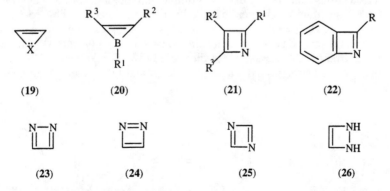

(**18**; n = 1,2)

2.5.4 THERMODYNAMIC ASPECTS

2.5.4.1 Stability and Stabilization

2.5.4.1.1 Ring strain

The strain in three- and four-membered rings is mostly due to bond angle deformation. Some conventional ring strain energies (CRSE$_s$) ⟨74PMH(6)199, p. 228⟩ are given in Tables 3 and 4. The ring strain in three- and four-membered rings is roughly the same magnitude, depending more on the nature of the heteroatom(s) than on the ring size. For comparison, the CRSE is 115 kJ mol^{-1} for cyclopropane and 111 kJ mol^{-1} for cyclobutane. As long as non-bonding interactions are avoided, alkyl substituents stabilize small rings by a few kJ mol^{-1}. For example, 2-methyloxirane is more stable than oxirane by 4 kJ mol^{-1} ⟨74PMH(6)199, p. 229⟩.

Exocyclic unsaturation can stabilize small ring heterocycles. In three-membered rings it is difficult to separate the contributions from increased angle strain and from electronic interactions between the unsaturation and the heteroatom. In four-membered rings such separation has been done ⟨74PMH(6)199, p. 235⟩. The CRSEs change from oxetane (106 kJ mol^{-1}) by –11 kJ mol^{-1} to oxetan-2-one (95 kJ mol^{-1}) (corrected for electronic effects) and 4-methyleneoxetan-2-one (95 kJ mol^{-1}). In contrast, an increase of 10 kJ mol^{-1} over the value for cyclobutane (111 kJ mol^{-1}) is observed on going to both methylenecyclobutane and 1,3-bis(methylene)cyclobutane.

Strain enormously influences the tendency for ring formation to give azetidines. Within the homologous series of azaheterocycles, the tendency for cyclization is smallest for the nitrogen-containing four-membered ring (5 > 3 > 6 > 7 ≈ 4) ⟨85JCS(P2)1345⟩.

2.5.4.1.2 Aromaticity and antiaromaticity

(i) Small rings

1H-Azirine, oxirene and thiirene (**19**; X = NH, O, S) have never been observed. Bond lengths have been calculated, however, for these antiaromatic 4π-systems ⟨80PAC1623⟩. In comparison with the corresponding saturated heterocycles, the C—X bond lengths are increased by 0.05 to 0.17 Å and the C—C bond length is decreased by 0.2 Å.

(**19**) (**20**) (**21**) (**22**)

(**23**) (**24**) (**25**) (**26**)

In borirene (**20**; R^1 = R^2 = R^3 = 2,3,4,5-tetramethylphenyl), the three-atom ring formed an equilateral triangle with all bond distances at 1.42 Å and bond angles at 60° ⟨87JA2526⟩, strongly supporting the completely delocalization of the two π-electrons in the borirene ring. This agrees with the result of molecular orbital calculations of the parent molecule (**20**; R^1 = R^2 = R^3 = H) ⟨81JA2589, 84JOC4475⟩.

From an aromaticity–antiaromaticity point of view, an intriguing representative of the four-membered heterocycles is undoubtedly the hitherto unknown azete (**21**; R^1 = R^2 = R^3 = H). Theory predicts that it possesses a negative resonance energy of 15 kcal mol^{-1}, slightly more stable than cyclobutadiene ⟨CHEC-II(1B)584⟩. There exists two general approaches to stabilize the azete system. The first one involves obtaining the thermodynamically stabilized compounds such as tris(dimethylamino)azete (**21**; R^1 = R^2 = R^3 = NMe$_2$) ⟨73AG(E)847⟩ and benzo fused azetes (**22**). The second approach consists in the synthesis of kinetically stabilized compounds containing bulky substituents. A remarkable example is 2,3,4-tri-*t*-butylazete (**21**; R^1 = R^2 = R^3 = Bu^t), reddish-brown needles stable for several days at 100°C ⟨86AG(E)842⟩.

The existence of ring strain, nitrogen lone-pair interactions, or both seems to be crucial in the absence of aromaticity of 1,2-Δ^3-diazetine (**26**).

Bonding parameters and energies at the MCSCF level of the parent 1,2-diazete (**23**) and (**24**) and 1,3-diazete (**25**) ⟨93AG(E)617⟩ suggest that the (**25**) can be considered as a "push-pull" cyclobutadiene; its singlet ground state is stabilized relative to that of cyclobutadiene. The 1,2-diazetes are higher in energy than the 1,3-isomer (**25**): the 1,2-diazete with two C = N bonds (**23**) is higher in energy by 50.7 kJ mol^{-1}, and that with one N = N and C = C bonds (**24**) by 94.2 kJ mol^{-1}.

(ii) Large Rings

For large rings, aromaticity is possible where the conditions of planarity and Hückel's rule are met, but the majority of fully unsaturated large heterocycles are not aromatic.

The heteropines (**27**; X = NH, O, S) are formally isoelectronic with 8π-electron cyclooctatetraene and cycloheptatrienide anion and, if planar, should be antiaromatic by the Hückel rule. Resonance energy obtained by simple (HMO) and advanced calculations (SCF-MO) for thiepine predict planar thiepine to be antiaromatic with a negative resonance energy of − 29.7 kJ mol^{-1}. In reality all heteropines possess a boat conformation that permits π-delocalization of the cyclic double bond to a level comparable to that of a linear polyene. Indeed, the structural index of aromaticity, $\Delta\bar{N}$, is 6%, for 1-*p*-bromobenzenesulphonylazepine (**27**; X = NSO$_2$C$_6$H$_4$Br-*p*) and reflects strong bond alternation. The non-aromatic nature of 1-R-azepines, as well as 1,2- and 1,4-diazepines, is also evident in their NMR spectra, where the ring protons are found in the typical range for alkenic protons, *i.e.*, δ 5.0–7.0 ppm.

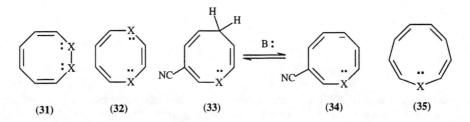

| (27) | (28) | (29) | (30) |

Unlike the tub-shaped parent azocines which are antiaromatic, their 10π-electron dianions (*e.g.* **30**) are planar and aromatic in nature ⟨71JA161⟩. The dianiones are formed by two-electron reduction of azocines. An intermediate radical-anion (**29**) was obtained from 3,8-dimethyl-2-methoxyazocine (**28**) ⟨83JA6078⟩ which has a strong tendency to disproportionate into dianion (**30**) and neutral azocine.

Semiempirical AM1 quantum mechanical calculations suggest that 1,2-dihydro-1,2-diazocine (**31**; X = NH) and 1,2-dioxocin (**31**; X = O) are antiaromatic systems, representing not a stable structure but rather a saddle point on the reaction coordinate leading to ring opening ⟨91RRC333⟩. Similar calculations for 1,4-dihydro-1,4-diazocine (**32**; X = NH), 1,4-dioxocine (**32**; X = O), and 1,4-oxazocine predicted their planarity, but suggested strong bond length alternation and little delocalization ⟨84JOC4475⟩, (see also Sec. 2.5.4.2.2).

| (31) | (32) | (33) | (34) | (35) |

1,4-Dioxocine (**32**; X = O) behaves chemically as an alkene rather than as an aromatic compound. Thus, it is readily hydrogenated to 1,4-dioxocane and polymerizes readily upon standing ⟨72AG(E)935⟩.

Proton and ^{13}C NMR analyses allowed assessment of both the conformation and potential aromaticity of 1,4-dihydro-1,4-diazocines as a function of the N-substituents ⟨79AG(E)964, 79AG(E)967⟩. Derivatives (**32**; X = NH, NMe, NTMS, NCONMe$_2$) display downfield chemical shifts for the ring protons relative to those predicted for nonplanar, localized bond structures. More dramatically, the ^{13}C shifts of these derivatives appear substantially upfield from those in (**32**; X = NSO$_2$Me, NCO$_2$Me). In addition, ^3JHH coupling constants for the diene portion of the latter compounds are unequal ($^3J_{56}$ *ca.* 9 Hz, $^3J_{67}$ *ca.* 5–7 Hz) whereas they are nearly equal in the former derivatives (*ca.* 10.5 Hz), suggesting that the former are planar, with substantially delocalized bonding, while the latter are nonplanar, bond-localized structures.

The NMR ^{1}H spectra of 1*H*-aza and oxaheterocinyl-anions (**34**) (X = N-^{t}Bu, N-Tos, O), formed by deprotonation of (**33**), show that (**34**) are planar-diatropic 10π systems. Evidently, the availability of heteroatom lone-pair electrons strongly influences aromaticity of fully unsaturated large heterocycles ⟨87TL2517⟩.

Results of MNDO calculation of 1*H*-azonine (**35**; X = NH) are in agreement with experimental evidence that this is a planar, aromatic molecule. The calculated geometry of oxonin (**35**; X = O), as a buckled, unsymmetrical polyenic heterocycle, is also in agreement with its known properties. The MNDO calculations on thionin (**35**; X = S) indicate that this molecule is planar, which should allow effective π delocalization, and at least some aromatic character ⟨86MI927-01⟩. The topological resonance energy model also predicts 1*H*-azonine and thionin to be aromatic, and oxonin nonaromatic ⟨84JHC273⟩.

2.5.4.2   Conformation

2.5.4.2.1   *Small rings*

(i) Rings

Three-membered rings are necessarily planar. Four-membered heterocycles are often puckered rather than planar (Table 4). As expected *exo*- and *endo*-unsaturation tend to make these systems planar. Substituted rings have ring inversion conformers of different energies. Moreover, inversion of substituents on heteroatoms may multiply the number of conformers of different energies. The ring inversion barriers of saturated four-membered systems are often very low. From IR and microwave data the barriers are 5.27 kJ mol^{-1} for azetidine, nearly zero for oxetane, and 3.14 kJ mol^{-1} for thietane ⟨74MI50100, p. 273⟩. Table 4 gives bond lengths and angles for some four-membered heterocycles.

(ii) Inversion at ring nitrogen

Variable temperature studies on aziridines and diaziridines show a remarkable range of nitrogen inversion rates. Electron-delocalizing substituents on small-ring nitrogen lower the inversion barrier by lowering the energy of the transitional, 'flat' geometry in which three substituents on the nitrogen are all in the same plane ⟨67JA352⟩. Substituents bearing unshared electron pairs raise the inversion barrier to levels at which enantiomers can be isolated ⟨68JA508⟩. Pure invertomers of *N*-haloaziridines ⟨B-73NMR137⟩ and *N*-alkoxyaziridines have been obtained ⟨70JA1079, 73TL619⟩. Some absolute configurations of the nitrogen of aziridines are known, such as that of diethyl 1-methoxyaziridine-2,2-dicarboxylate ⟨79DOK(246)1150⟩. Dynamic NMR methods and theoretical methods used to study nitrogen inversion barriers have been reviewed ⟨B-92MI 101-01⟩, with data tabulated on the inversion barriers of many types of aziridines.

For oxaziridines the *N*-inversion barrier is considerably higher than that for similar aziridines. *N*-Alkyl-3,3-dialkyloxaziridines are resolvable and absolute configurations have been determined (CHEC 5.08.2.3.1).

Diaziridines also show slow nitrogen inversion, and carbon-substituted compounds can be resolved into enantiomers, which typically racemize slowly at room temperature (when *N*-substituted with alkyl and/or hydrogen). For example, 1-methyl-3-benzyl-3-methyldiaziridine in tetrachloroethylene showed a half-life at 70°C of 431 min ⟨69AG(E)212⟩. Preparative resolution has been done both by classical methods, using chiral partners in salts ⟨77DOK(232)1081⟩, and by chromatography on triacetyl cellulose (CHEC 5.08.2.3.1). *N*-Inversion in azetidine and azetidin-2-one is rapid, even at −77° and −40°C, respectively ⟨B-73NMR144⟩. Again, halo substituents on nitrogen drastically slow the inversion rate, so that *N*-chloro-2-methylazetidine can be separated into two diastereomers ⟨B-77SH(1)54⟩. Substituent effects on *N*-inversion are much the same as in the aziridines: *N*-aryl and N-acyl compounds undergo *N*-inversion faster, whereas *N*-halo, *N*-amino and *N*-nitroso compounds are slower ⟨B-77SH(1)56⟩. By and large, the *N*-inversion barrier of azetidines is 38 kJ mol^{-1} lower than that of similarly substituted aziridines ⟨B-77SH(1)55, 85JA4335⟩. In 1,2-diazetidines one finds the *N*-inversion rate lowered, and coalescence temperatures and free energies of activation have been reported for a number of 1,2-diaryldiazetidin-3-ones ⟨B-77SH(1)61⟩.

The rotational barriers of *N*-nitroso-, *N*-formyl and *N*-(*N*,*N*-dimethylcarbamoyl)-azetidines, compared with those of analogous acyclic amides, suggest that amide conjugation is weaker when the nitrogen is part of an azetidine ring ⟨87KGS912⟩.

2.5.4.2.2 Large rings

Fully unsaturated seven-membered heterocyclics have alternating bond lengths and are normally in boat conformations. Ring inversion barriers are 42.7 kJ mol^{-1} for 3-methyl-3*H*-azepine and 35.6 kJ mol^{-1} for 3*H*-azepin-2-one (CHEC 5.16.2.3). The barriers for oxepin and thiepin are somewhat lower. Annulation can introduce large conformational barriers, to the extent of making possible the resolution into enantiomers of a tribenzoxepin ⟨71CB2923⟩.

Fully saturated seven-membered heterocycles with one or two heteroatoms are normally in mobile twist-chair conformations (CHEC 5.17.1.1, CHEC 5.18) ⟨B-77SH(2)123⟩. Annulation and the introduction of exocyclic double bonds can have profound effects; oxepan-2-one, for example, is in a near chair conformation ⟨67JA5646⟩.

(36) (37) (38)

The ring in the trithiadiazepine (**36**; $R^1 = R^2 = CO_2Me$) is planar within 0.019 Å ⟨84CC55⟩. The S—N bond lengths (1.54–1.60 Å) are all similar and are nearer to double (1.55 Å) than to single (1.67 Å) bonds. The ring C—C and C—S bonds are very similar in length to the corresponding bonds in thiophene.

Minimal interaction between the amine lone pair of electrons and the delocalized π system of the ring is confirmed by X-ray crystal structures for 6-amino-, 6-dimethylamino-, and 6-morpholino-1,3,5,2,4-trithiadiazepine (**36**; $R^1 = H$, $R^2 = NH_2$, NMe_2, morpholino) ⟨90CC1315⟩. In each case, the amino group is tetrahedral and rotated out-of-the-plane of the ring such that the nitrogen lone pair is approximately orthogonal to the π-system.

X-ray diffraction studies on azocine (**37**) revealed normal dimensions and a conformation with N(1), C(2), C(5), C(6) in one plane and C(3), C(4), C(7), C(8) in another, both within 0.02 Å. The two planes are 0.81 Å apart and approximately parallel (angle of intersection 2.2(1)°) ⟨85CC85⟩.

Unsubstituted 1,4-dihydro-1,4-diazocine (**32**; X = NH) and its *N*,*N*′-bis(trimethylsilyl) derivative both display planar diazocine rings; some π-electron delocalization is suggested by partial bond length equalization. In contrast, monocyclic derivatives bearing at both nitrogens electron-withdrawing substituents ($MeSO_2$, MeO_2C) display markedly nonplanar diazocine rings and completely localized bonding ⟨79AG(E)964, 79AG(E)967⟩ (see Section 2.5.4.1.2). Oxazocine (**38**; R = *p*-MeC$_6$H$_4$SO$_2$) is markedly nonplanar, with pyramidal nitrogen. By contrast, **38** (R = 3,4,5-(MeO)$_3$C$_6$H$_2$CH$_2$) is planar, with an average deviation from the ring plane of only 0.021 Å.

The 1,5-dithia-2,4,6,8-tetrazocine system prefers to adopt a planar 10π-electron monocyclic structure (**16**). However substituents at the carbon atoms with π-donor capacity (*e.g.* NMe_2) can induce pseudo-Jahn-Teller distortion leading to the bent, bicyclic 8π system of type (**17**) with an S—S transannular partial bond. X-Ray crystallographic analysis showed (**17**) was folded about an axis drawn through the two sulfur atoms. The S—S transannular distance in (**16**; R = Ph) is 3.79 Å and 2.428 Å in (**17**), which is longer than a disulfide bond (*ca.* 2.06 Å) but shorter than the sum of two sulfur van der Waals radii (3.6 Å) suggesting a partial bond ⟨81JA1540⟩. One dimethylamino substituent is sufficient to destroy the planarity and aromaticity of dithiatetrazocines ⟨89CC1137⟩.

Ab initio calculations on (**39**) estimate activation energies of amine rotation and ring inversion (**39a**) and (**39b**) to be 17.1 and 17.3 kcal mol^{-1}, respectively, which is in good agreement with values derived from NMR experiments ⟨94JA5167⟩.

(39a) (39b)

Transannular interaction is a unique feature of medium-sized heterocycles. The chemistry of eight-membered rings with two functional groups in appropriate positions may be dominated by transannular interactions. Isoelectronic aminoketone (**40**) and aminoalkene (**41**) represent typical examples. The photoelectron spectra were used to study the transannular interactions and conformations in their molecules. In total, the results indicate that considerable lone pair/π interactions occur in eight-membered rings. This is consistent with IR results (C=O frequency is 20–30 cm^{-1} lower than that in a cyclic ketone) indicating that the partial single bond character of the C=O is due to the transannular amide resonance. In aminoalkene, the C=C stretching vibration is located at 1625 cm^{-1} indicating a weaker C=C bond due to the transannular interaction with the amino group. This transannular interaction may be due to the preferred conformation (**41**) ⟨86JOC592⟩.

(40) (41) (42)

Transannular hydrogen bonding also can considerably influence the conformation of large rings. Thus in *N,N'*-dimethyl-1,6-diazacyclodeca-3,8-diyne (**42**) both methyl groups are in axial position of the chair conformation so that the interaction between the lone pairs of the nitrogen atoms and the triple bond is minimal ⟨91TL2887⟩. However, in the case of monoprotonated form (**42**) as its iodide salt, the structure resembles that of *cis*-decalin, with the central C—C bond replaced by a symmetrical hydrogen bond, with an N—N distance of 2.600 Å and a N—H—N angle of 169° ⟨88CC1528⟩.

The structures of nine-membered heterocycles as determined by X-ray crystallography seldom show strikingly unusual bond lengths or angles compared with acyclic analogues, other than a general increase in the magnitude of the endocyclic bond angles. Rings containing *trans*-C=C bonds, ester bonds or amide bonds, however, often exhibit significant deviations from planarity of these bonds.

2.5.5 TAUTOMERISM

There are two important types of tautomerism found for small and large rings: valence bond tautomerism and (to a somewhat lesser extent) annular prototropy.

2.5.5.1 Valence Bond Tautomerism

Study of kinetically stable azetes (*e.g.* **43**, R = Bu^t, Mes) reveals their capability to exist in the form of two valence tautomers differentiated by the position of ring double bonds. Their interconversion can be followed by temperature-dependent NMR spectroscopy ⟨86AG(E)842, 88AG(E)1559⟩. Interestingly, for 2,3-di-*tert*-butyl-4-mesitylazete form (**43a**) is thermodynamically more stable.

(43a) (43b) (44a) (44b)

The heterepine-heteronorcaradiene equilibrium exemplifed for oxepin in Scheme 2 represents perhaps the best known type of valence tautomerism met in seven-membered and larger heterocyclics.

The equilibrium position is strongly affected by substitution (see CHEC 5.17.1.2) and the nature of heteroatom. For instance, in the case of thiepine its potential valence isomer, benzene episulfide, is too unstable to be detected.

Thermal cycloreversion of 6-bromo-1,4-dioxocin (**44a**; X = Y = O, R = Br) to its benzene dioxide valence tautomer (**44b**; X = Y = O, R = Br) displays an activation energy of 28.5 kcal mol^{-1}, with an A factor of 3.8×10^{13} in CCl$_4$. Cycloreversion back to the dioxocin has a barrier of 25.6 kcal mol^{-1}, with an A factor of 3.2×10^{12}, giving a 65 : 35 mixture of dioxocin and benzene dioxide at 77°C ⟨72AG(E)935⟩.

The parent 1,4-dioxocin shows no evidence of tautomerizing in CCl$_4$ at 100°C, but gives a 95:5 ratio of dioxocin and benzene dioxide at 60°C in benzene ⟨72AG(E)937⟩. In contrast to dioxocines, the thermal rearrangement of diaza- (**44b**; X = Y = NR, R = H) and oxoaza-benzene tautomers (**44b**; X = O, Y = NR, R = H) to the corresponding 1,4-dihydro-1,4-diazocines or 1,4-dihydro-1-oxa-4-azaazocines (**44a**) appears to be irreversible ⟨79AG(E)967, 83CB2492⟩.

Diazabicyclo[4.2.0]octatrienes (*e.g.* **45a**) appear not to exist in equilibrium with (**45b**) (Scheme 8) in solution at room temperature. The structure (**45b**) was calculated to be more stable than (**45c**) by about 33 kJ mol^{-1}, and the enthalpy changes accompanying the isomerizations of (**45b**) to the bicyclo[4.2.0]octatrienes (**45a, d-g**) are 0–13 kJ mol^{-1} more endothermic than the corresponding isomerizations of the carbocyclic analogue ⟨79JOC1264⟩. However, the thermal and photochemical reactions of (**45**), as well as its fragmentation upon mass spectral analysis, may occur through the intermediacy of these valence tautomers, as has also been suggested for related 1,2-diazocines ⟨86BCJ1087, 87BCJ731⟩.

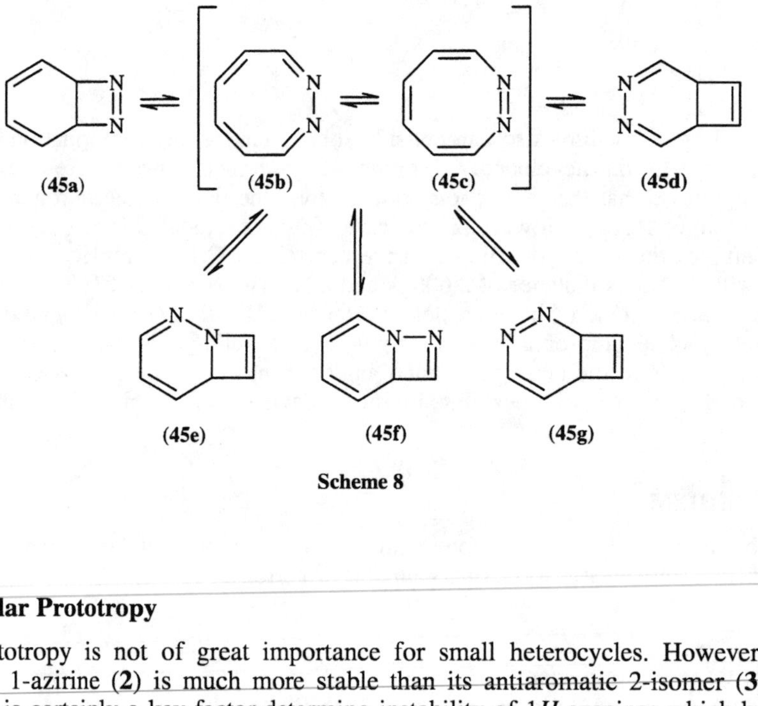

Scheme 8

2.5.5.2 Annular Prototropy

Annular prototropy is not of great importance for small heterocycles. However, it should be mentioned that 1-azirine (**2**) is much more stable than its antiaromatic 2-isomer (**3**). By analogy, antiaromaticity is certainly a key factor determing instability of 1*H*-azepines which have never been observed. Thus, demethoxycarbonylation of methyl-3,6-di-*t*-butylazepine-1-carboxylate (**46**) by DBU gives a mixture of the corresponding 2*H*-, 3*H*-, and 4*H*-azepines in the approximate ratio 13:56:1 (Scheme 9) ⟨94JCS(P1)1753⟩. The distribution of the azepine isomers is proportional to their relative thermal stabilities as they interconvert *via* allowed 1,5-hydrogen shifts.

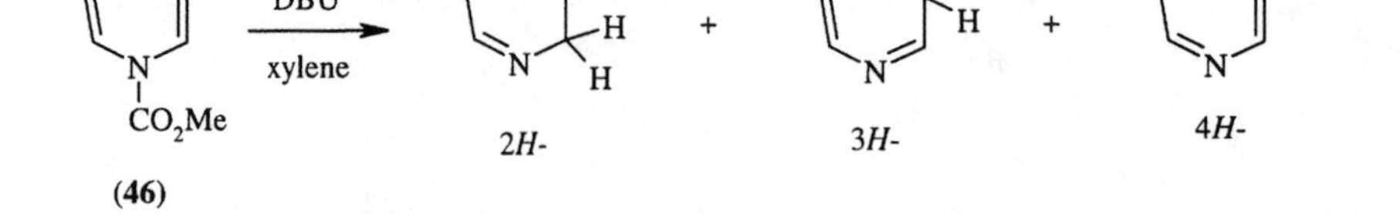

Scheme 9

Annular tautomerism between one or more NH-forms and various CH forms of 1,2-diazepines is possible with the favored species dependent upon the nature and position of substituents ⟨69ACS3125, 70T739⟩. Interestingly, unlike monocyclic azepines and tautomeric 1,2-diazepines, 5H-1,2-diazepines (**47**) exist preferentially in their bicyclic diazanorcaradiene form (**48**) ⟨72JA2770⟩.

(**47**)   (**48**)

For monocyclic fully unsaturated 1,4-diazepines, the NH-forms are unstable. Azonine (**35**; X = NH) exists exclusively as NH-tautomer that is due to its 10π electron aromatic character ⟨70TL825⟩.

Part 3
Reactivity of Heterocycles

3.1

Overview

3.1.1 REACTION TYPES

All reactions can be broken down into a succession of individual steps in each of which bonds are broken and/or formed. A chemical bond can be formed (or broken) in three ways.

(i) In a generalized ionic reaction step, one of the atoms contributes both electrons to the bond either from a lone pair (**1**) or from another bond, often a multiple bond (**2**). The atom, molecule or ion which contributes the electron pair is a nucleophile and that which accepts is an electrophile.

(ii) In a free radical step each atom contributes one electron to the bond (**3**; the single-headed arrows represent the movement of single electrons). At least one of the reactants or products must contain an unpaired electron.

(iii) In a cyclic transition state the bond is formed or broken by the electrons moving in a ring.

$$A: \overset{\frown}{} B \longrightarrow A^+ \!\!-\! B^- \qquad D\!\!=\!\!A\overset{\frown}{} B \longrightarrow \overset{+}{D}\overset{\frown\!\!A}{}\!\!\overset{\frown}{}B^- \qquad A\cdot\overset{\frown}{}\overset{\frown}{}\cdot B \longrightarrow A\!\!-\!\!B$$

$$\text{(1)} \qquad\qquad\qquad\qquad \text{(2)} \qquad\qquad\qquad\qquad\qquad \text{(3)}$$

3.1.2 HETEROAROMATIC REACTIVITY

The basic principles governing the degree and type of reactivity shown by heteroaromatic compounds are familiar from aliphatic and benzenoid chemistry. Three are very important:

(i) Oxygen, nitrogen or sulfur multiply-bonded to carbon can accept the whole of a shared pair of π-electrons (**4**) and thus allow a nucleophilic reagent to attack the carbon atom, as in many common reactions of carbonyl compounds. The attack by a nucleophilic reagent is easier when the heteroatom carries a positive charge (**5**).

$$\text{(4)} \qquad\qquad \text{(5)} \qquad\qquad \text{(6)} \qquad\qquad \text{(7)}$$

(ii) A shared pair of electrons on oxygen, nitrogen or sulfur adjacent to an unsaturated system can be made available for reaction through that system (**6**). This can also happen when the heteroatom carries a negative charge (**7**); the alkylation of the acetoacetate anion on carbon is an analogous reaction in aliphatic chemistry.

(iii) Aromatic compounds tend to 'revert to type', *i.e.* to return to their initial delocalized system, if disturbed.

These basic principles give much insight into the reactions of aromatic heterocyclic compounds.

3.1.3 ARRANGEMENT OF THE REACTIVITY SECTIONS

Within each of the main groups of ring systems, the reactivity chapters are arranged in the same way, as is described in Part 5 (Appendix "A", Section 4).

3.2

Reactivity of Six-membered Rings

3.2.1 REACTIVITY OF AROMATIC RINGS

3.2.1.1 General Survey of Reactivity

3.2.1.1.1 Pyridines

(i) Most pyridines are thermally and photochemically stable, but, just as in benzenes, polysubstitution can lead to susceptibility to such reaction modes.

(ii) As a first approximation, the reactions of pyridines with electrophiles may be compared with those of trimethylamine and benzene. Thus, pyridine reacts easily at the nitrogen atom with reagents such as proton acids, Lewis acids, metal ions and reactive halides to form salts, coordination compounds, complexes and quaternary salts, respectively. Under much more vigorous conditions it reacts at ring carbons to form substitution products in nitration, sulfonation and halogenation reactions.

Pyridine is a weaker base (pK_a 5.2) than trimethylamine (pK_a 9.8): the sp^2-hybridized lone pair of the pyridine nitrogen atom is less available than the sp^3 lone pair. The conditions required for nitration or sulfonation of pyridine are far harsher than those needed for benzene. Substitution of a nitrogen atom for a CH group in benzene is equivalent to introducing an electron-withdrawing group (nitrogen is more electronegative than carbon); thus, pyridine itself should be substituted in the 3-position (about as readily as nitrobenzene). However, electrophilic reagents react at the pyridine nitrogen atom very readily, and in the strongly acid media used for nitration and sulfonation conversion to cation is essentially complete. Thus, the CH in benzene is replaced by NH^+ and the positively charged nitrogen reduces the reactivity toward electrophilic substitution very much. It is for this reason that most pyridines are nitrated and sulfonated only with difficulty, and at high temperatures. Halogenation of pyridines is easier: *N*-halogenation is incomplete and *C*-halogenation can occur on the free base. Dihalogenation occurs since a halogen atom causes relatively little additional deactivation of the ring.

(iii) The electron pull toward the nitrogen atom allows nucleophilic reagents to attack pyridines. Such attack may occur at α or γ-ring carbon atoms or at the hydrogen of ring CH groups.

Nucleophilic attack at ring carbon occurs in benzenes only when electron-withdrawing substituents are present. Even with pyridine, only the strongest nucleophiles react. This is because the formation of the initial adduct (**2**) involves de-aromatization of the pyridine ring and, once formed, many such adducts tend to re-aromatize by dissociation (**1⇌2**). Benzo fusion decreases the loss in aromaticity for the formation of the adduct and thus quinoline (**3**) and especially acridine (**4**) react more readily with nucleophiles.

(1) (2) (3) (4)

Reaction with nucleophiles by deprotonation at a CH bond occurs in pyridine much more readily than in benzene. The reactivity order is γ > β > α rather than γ, α > β because of lone pair lone pair repulsion in the α-deprotonated species (*cf.* discussion in Section 3.2.1.8).

(iv) Pyridines undergo a variety of reactions with free radical reagents, and at surfaces: many of these parallel the corresponding reactions of benzenes. Electron uptake from a metal to form a radical anion occurs readily.

(v) Propensity toward cyclic transition state reactions again shows a parallel with benzenes: generally it is small for pyridines, but increases with suitable polysubstitution.

3.2.1.1.2 *Azines*

Extrapolation from benzene through pyridine to the diazines and then to the triazines and tetrazines delineates the main trends of azine chemistry.

(i) Reactions with electrophilic reagents become successively more difficult than those with pyridine, both at nitrogen (weakened basicity) and on ring carbon atoms (no reaction at all without activation, even in diazines).

(ii) Conversely, nucleophilic attack is increasingly easier than in pyridine. Nucleophiles which react only with quaternized pyridines will sometimes react with the parent diazines. Triazines and tetrazines are even attacked by weak nucleophiles.

(iii) Successive introduction of nitrogen atoms into benzene causes a gradual reduction in aromatic stabilization. The diazines still show typical aromatic behavior in that in most of their reactions they revert to type. However, with the triazines and tetrazines decreasing aromaticity increases the ease both of thermal and photochemical fragmentations and rearrangements, and of cyclic transition state reactions with other reagents.

(iv) Successive replacement of carbon by nitrogen lowers the energy of the LUMO. As a consequence, the ease of reduction and the stability of radical-anion are increased from benzene to pyridine, diazines, triazines and tetrazines.

3.2.1.1.3 *Cationic rings*

(i) In the pyridinium, pyrylium and thiinium cations, there is no available nitrogen lone pair, and electrophilic attack at ring carbon is severely discouraged by the positive charge, although it can occur if sufficiently activating substituents are present.

Diazinium, oxazinium and thiazinium cations possess a 'pyridine-like' nitrogen atom, but it is of very weak basicity and nucleophilicity. However, pyridazines do form diquaternary salts with very strong alkylating agents such as oxonium compounds. Electrophilic attack at ring carbon in these compounds is practically unknown. These trends are emphasized in cationic rings with an increased number of nitrogen atoms.

(ii) A positive charge facilitates attack by nucleophilic reagents at positions α or γ to the heteroatom. Amines, hydroxide, alkoxide, sulfide, cyanide and borohydride ions, certain carbanions, and in some cases chloride ions react with pyridinium, pyrylium and thiinium cations under mild conditions to give initial adducts of types (5) and (6). These adducts undergo a wide variety of further transformations. Such reactions are further encouraged by the additional nitrogen atoms in diazinium, triazinium, *etc.*, cations.

(iii) A positive charge perturbs the electron distribution and thus reduces the aromaticity of a six-membered cationic ring. As expected, reaction with free radicals and reactions *via* cyclic transition states (both intra- and inter-molecular) are facilitated. The uptake of an electron to form a neutral radical is especially easy.

3.2.1.1.4 *Pyridones, N-oxides and related compounds: betainoid rings*

Three types of compound can be considered (see Section 2.2.1.2) to be derived from a cationic ring carrying a negatively charged substituent (7):

(i) If Y is α or γ to Z, then alternative uncharged canonicals (**8**) and (**9**) exist as in pyridones, pyridinethiones, *etc.* (see Section 3.2.3.7.2 for an overall survey of the reactivity of this type of compounds).

(**7**)   (**8**)   (**9**)   (**10**)   (**11**)

(ii) If Y is in the β-position, the compounds are true zwitterions (**10**) as in 3-oxidopyridinium, *etc.*

(iii) If Z is nitrogen, then the substituent Y can be directly attached to it, to give pyridine *N*-oxides, *N*-imides, *etc.*, (**11**) (see Section 3.2.3.12.5 for an overall survey of their reactivity).

These compounds all contain both an electron source (Y$^-$) and an electron sink (Z$^+$). Furthermore, their aromaticity is significantly reduced by the non-uniform electron distribution. Hence they are highly reactive. The orientations of the reactions of these compounds with electrophilic and nucleophilic reagents are deducible from their canonical forms (see structures **12-14**).

(**12**)   (**13**)   (**14**)

(i) Electrophiles readily attack at the Y (not at the Z) atom and at C atoms *ortho* and *para* to Y. As a generalization, electrophilic attack at Y is relatively easy and relatively easily reversible while that at C is more difficult, but less easily reversible.

(ii) Nucleophiles readily attack at C atoms α and γ to Z and at hydrogens attached to C atoms α to Z. If Z is NH, then nucleophiles can remove the NH proton to give a pyridone (*etc.*) anion.

(iii) Compounds of this type undergo a wide variety of thermal and photochemical rearrangements, and cycloaddition reactions *via* cyclic transition states.

(iv) Additional ring nitrogen atoms, as occur in, *e.g.* diazinones and thiazinones, alter but little the reactivity patterns of these compounds.

3.2.1.1.5   *Anionic rings*

Stable anions can be formed by the loss of a proton from *N*-unsubstituted pyridones or hydroxypyridines. They are the pyridine analogues of phenolate anions and react very readily with electrophilic agents at N, O or ring carbon (see Section 3.2.1.8.4).

3.2.1.1.6   *Aromaticity and reversion to type*

The aromaticity of six-membered rings is discussed in Section 2.2.4.2. In general most of these compounds tend to react typically by substitution rather than addition, *i.e.* they tend to revert to type. However, ring oxygen atoms, an increasing number of ring heteroatoms, benzannulation and ring carbonyl groups all reduce the aromaticity. Thus, phenoxazonium and phenothiazonium salts, oxazones and thiazones show increasing tendency to addition reactions.

3.2.1.2 Intramolecular Thermal and Photochemical Reactions

The fundamental types of thermally and photochemically induced intramolecular transformations are summarized in Scheme 1. All reactions of this class involve intermediates in which aromaticity is lost; hence they are most common in the classes of less aromaticity, *i.e.* polyhetero rings, cationic rings, rings containing carbonyl groups. However, polysubstitution, especially by bulky groups, can also induce reactions by strain relief in transition states. Most of the reactions known are photochemical.

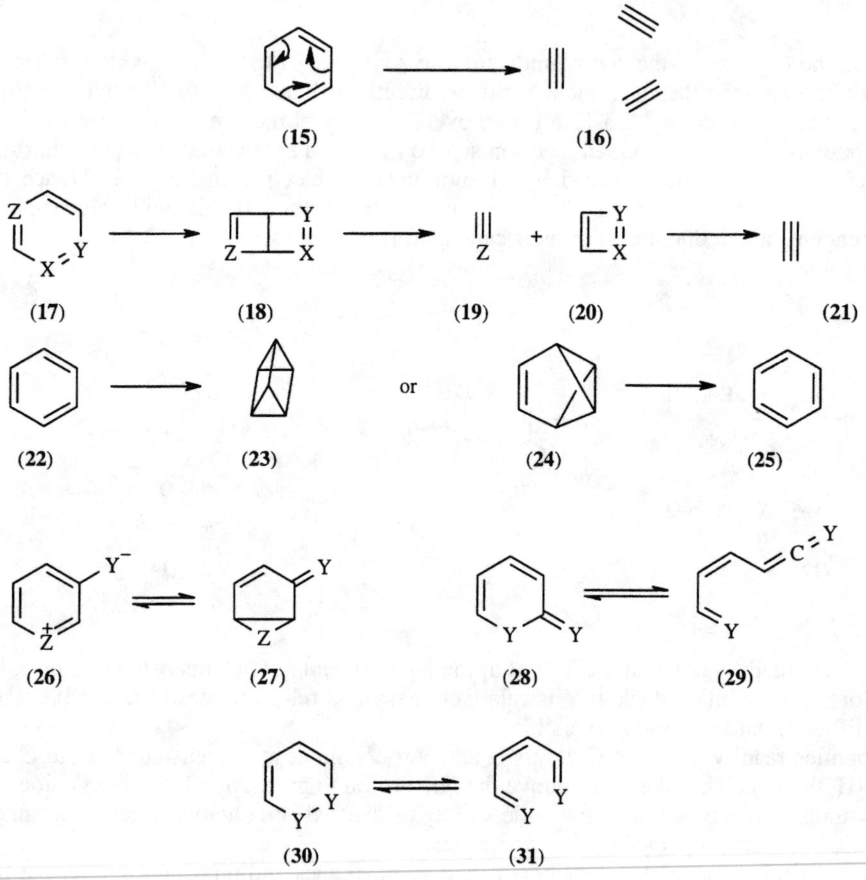

Scheme 1

3.2.1.2.1 *Fragmentation (15 → 16)*

Direct fragmentation (as opposed to those *via* rearrangement; see next section) most often occurs in polyhetero rings. It is implicit in Scheme 1 that the presence of contiguous nitrogen atoms tends to labilize the rings, as well as provide extra stability to the fragmentation products (by increasing the possibilities for generating N_2 molecules). Thus, 1,2,4,5-tetrazines are thermolyzed to nitrogen and nitriles. (Photolysis affords the same products, but this may involve an intermediate of type (18)).

Under flash vacuum thermolysis (FVT) the 1,2,3-triazine (32) was readily thermolyzed to an alkyne, a nitrile, and nitrogen in high yields. For unsymmetrically substituted 1,2,3-triazines as FVT substrates, fragmentation proceeded selectively: a bulky substituent at C-4(6) made the adjacent C—N bond break more easily than the opposite C—N bond. The FVT method was applied to the synthesis of the fluorinated alkynes, perfluoro-3-methyl-1-butyne and difluoroethyne [(33), R = $(CF_3)_2CF$, F] (Equation (1)) ⟨89CC1657, 91CC456⟩. It has been claimed, however, that tris(dimethylamino)-1,2,3-triazine forms the monocyclic azete.

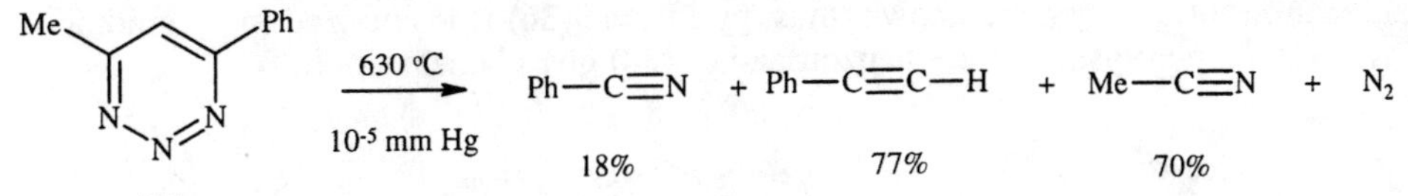

$$Ph-C{\equiv}N \quad + \quad Ph-C{\equiv}C-H \quad + \quad Me-C{\equiv}N \quad + \quad N_2$$

18%   77%   70%

(32)

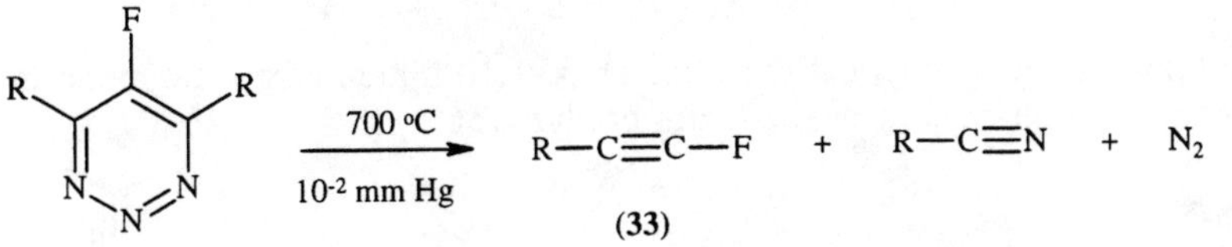

$$R-C{\equiv}C-F \quad + \quad R-C{\equiv}N \quad + \quad N_2 \tag{1}$$

(33)

Thermally above 600°C, and also photochemically, 1,3,5-triazine decomposes to three molecules of hydrogen cyanide. Most 1,2,4-triazines are thermally very stable.

3.2.1.2.2 *Rearrangement to or elimination via Dewar heterobenzenes* (17) → (18) → (19), (20) → (21)

Certain polysubstituted pyridines yield isolable Dewar pyridines, as illustrated in equation (2) (see CHEC 2.05).

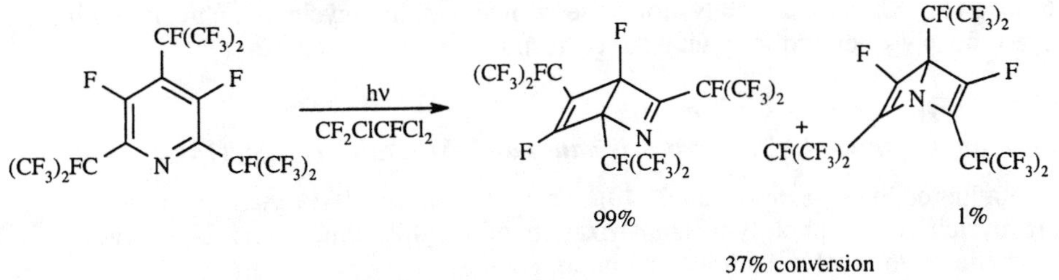

$$\tag{2}$$

99%   1%

37% conversion

Irradiation of pyridine itself gives Dewar pyridine, observable spectroscopically, which in water is hydrolytically ring-opened to form $H_2N(CH{=}CH)_2CHO$, but in a matrix fragments to cyclobutadiene and HCN.

3,4,5,6-Tetra-*t*-butylpyridazine (**34**) is converted into its Dewar isomer (**35**) when irradiated in pentane with UV light of wavelength > 300 nm. Irradiation of this product at shorter wavelengths, or thermolysis, gives rise to further reaction ⟨91TL57⟩. Irradiation of 4-amino-2,6-dimethylpyrimidine gives the acyclic amino imine *via* the Dewar pyrimidine as shown in Scheme 2a. The photoisomerization of perfluoropyridazines to pyrazines is considered also to involve Dewar diazine intermediates.

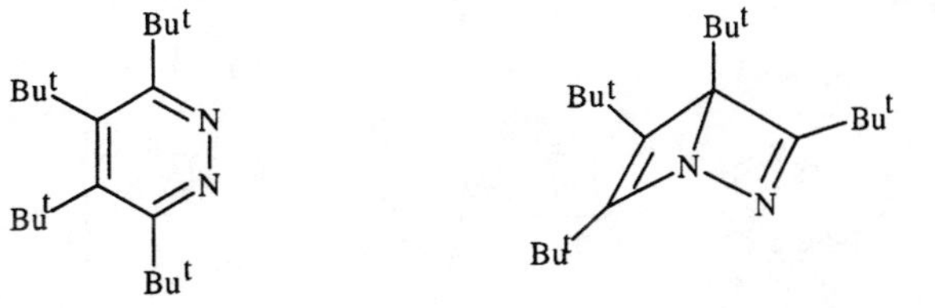

(34)   (35)

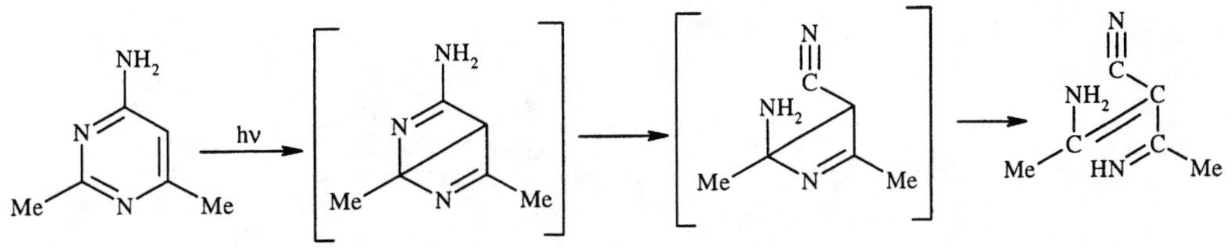

Scheme 2a

Thermal reactions of this type are known, thus, pyridazine (36) is isomerized to pyrimidine (37) at 300°C. Flash vacuum pyrolysis of 1,2,3-benzotriazines (38) gives benzazetes (39).

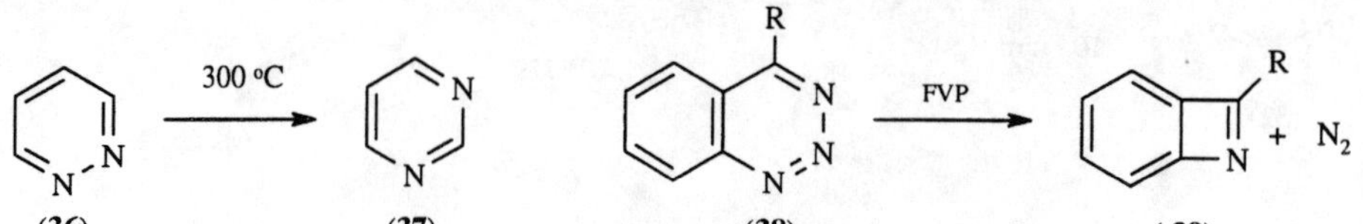

Quite often such reactions can be synthetically useful. Thus, 1,10-dimethylbenzo[c]cinnoline gave a practical route to 1,8-dimethylbiphenylene as the major product (56%) (Scheme 2b).

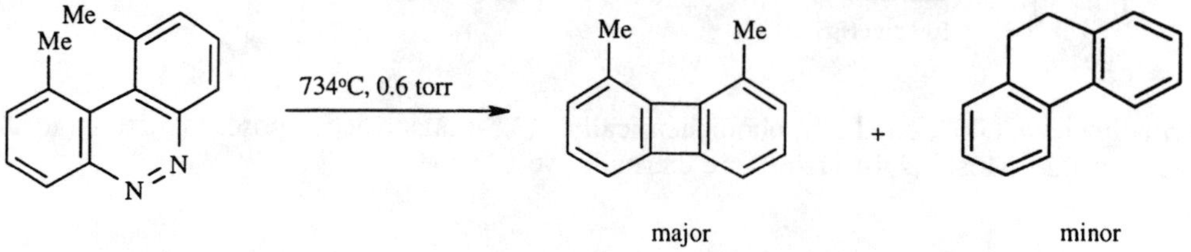

Scheme 2b

This type of isomerization is much more common in carbonyl-containing rings. A well known example is the generation of cyclobutadiene by photolysis of pyran-2-one with loss of the CO_2 molecule. 1-Methyl-2-pyridone (17; XY = NMeCO, Z = CH) gives (18; XY = NMeCO, Z = CH); 1,3-oxazin-6-ones (17; XY = OCO, Z = N) form the corresponding bicycle, which can eliminate CO_2, and 1,2,3-benzotriazin-4-ones similarly give the corresponding benzazetones.

3.2.1.2.3 *Rearrangement to or* via *hetero-prismanes and -benzvalenes* (23), (24)

Pentakis(pentafluoroethyl)-1-azaprismane (40) can be isolated in 91% yield by irradiation of the corresponding pyridine. The photolytic isomerization of alkylpyridines (*e.g.* 2-picoline to 3- and 4-picolines) is believed to involve azaprismane intermediates.

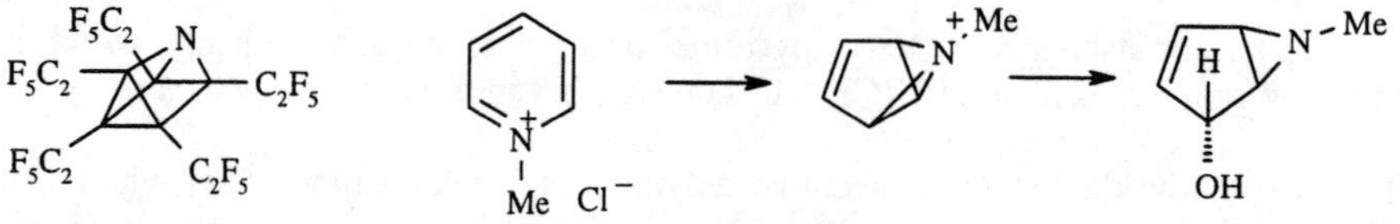

Photolysis of 1-methylpyridinium chloride (41 → 43) is considered to involve the azoniabenzvalene (42) as an intermediate, and similar behavior has been found in certain pyrylium cations (Schemes 3 and 4). Diazabenzvalenes are implicated in the rearrangement at 300°C of certain perfluoropyridazines to pyrimidines and pyrazines (Scheme 5).

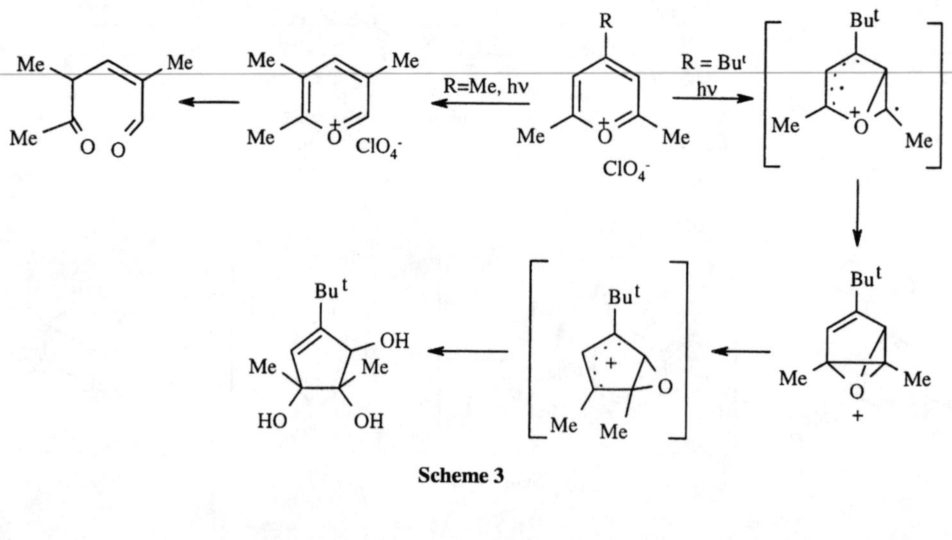

Scheme 3

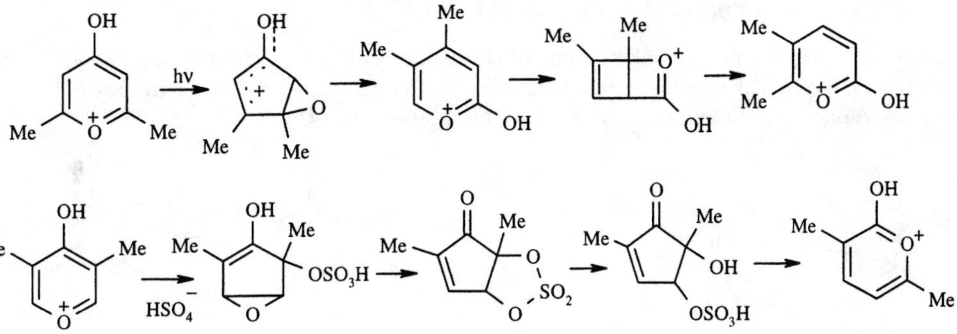

Scheme 4

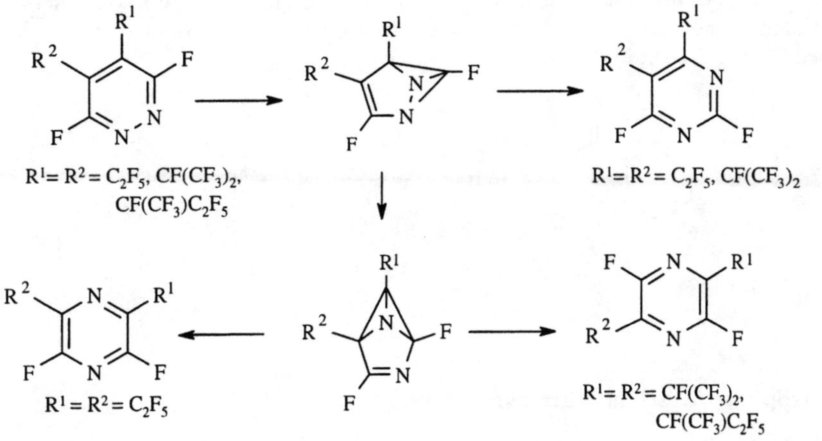

$R^1 = R^2 = C_2F_5, CF(CF_3)_2,$
$CF(CF_3)C_2F_5$

$R^1 = R^2 = C_2F_5, CF(CF_3)_2$

$R^1 = R^2 = C_2F_5$

$R^1 = R^2 = CF(CF_3)_2,$
$CF(CF_3)C_2F_5$

Scheme 5

3.2.1.2.4 *Rearrangement to or via 1,3-bridged heterocycles* (26) → (27)

3-Oxidopyridiniums (**26**; Z = NR, Y = O) are converted photochemically into the bicycle (**27**); corresponding 3-oxidopyryliums and especially 4-oxidoisochromyliums isomerize more easily (*cf.* **44** → **45**). 5-Oxidopyridazinium betaines are isomerized photochemically into corresponding pyrimidin-4-ones by a similar path (Scheme 6).

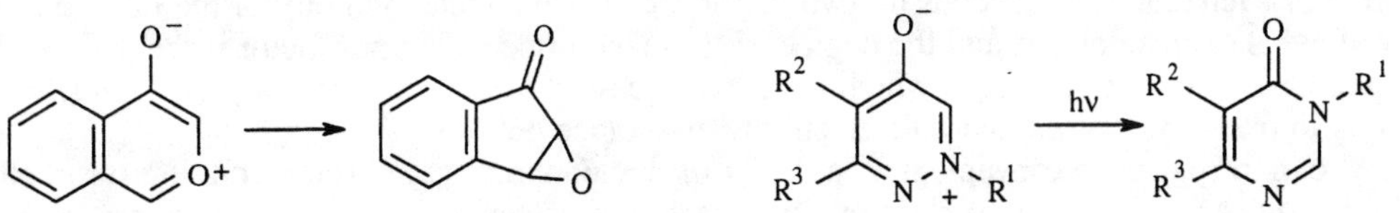

Scheme 6

Photoisomerization of pyran-4-one and substituted derivatives to pyran-2-ones involves a zwitterionic intermediate of similar type (Scheme 7).

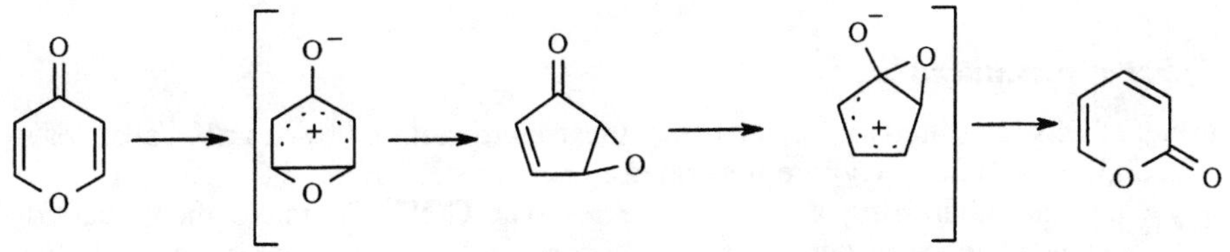

Scheme 7

3.2.1.2.5   Ring opening (28) → (29), (30) → (31)

Irradiation of pyran-2-one gives the ketene (46) reversibly. Similar reactions are known for aza and diaza analogues. Thus, 1,3-oxazin-6-ones isomerize photochemically to ketene imines (47), and flash vacuum pyrolysis converts the oxadiazinone (48) reversibly into (49).

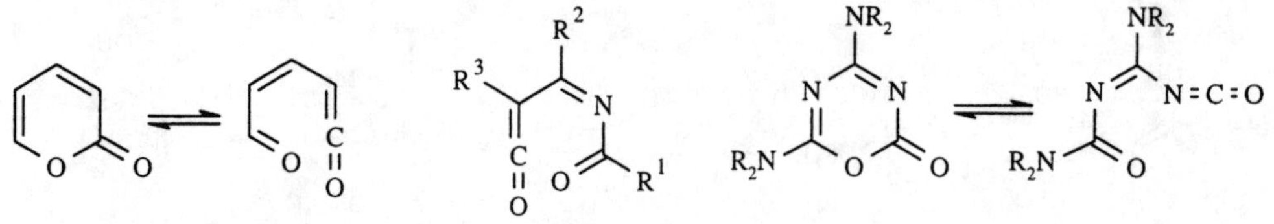

(46)   (47)   (48)   (49)

2H-1,2-Oxazines and thiazines are unstable with respect to ring-opened isomers (*cf.* 30→31). 1,2,3-Benzotriazin-4-ones on protonation undergo ring-chain tautomerism to yield diazonium ions (Scheme 8; see CHEC 2.18).

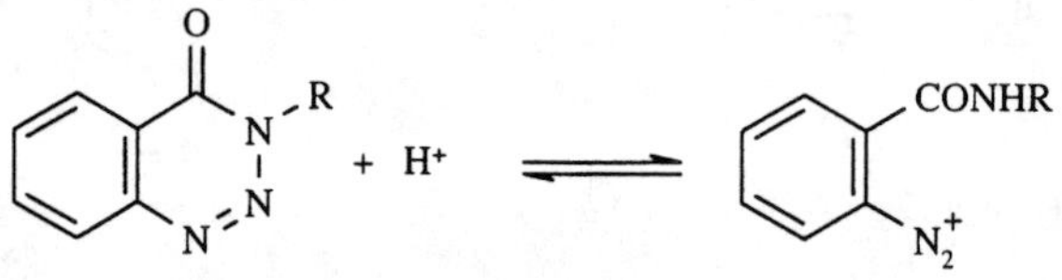

Scheme 8

3.2.1.3   Electrophilic Attack at Nitrogen

3.2.1.3.1   Introduction

Pyridines and azines behave as tertiary amines in their reactions with a wide range of electrophiles:
  (i)   proton acids give salts;
  (ii)   Lewis acids form coordination compounds;
  (iii)   transition metal ions form complex ions;
  (iv)   reactive halogen compounds give quaternary salts;
  (v)   activated alkenes (and alkynes) give quaternary salts (by Michael addition);
  (vi)   halogens form adducts;
  (vii)   certain oxidizing agents yield amine oxides;
  (viii) electrophilic amination reagents form *N*-aminoazinium salts.
The ease of such reactions depends on two major factors: the nucleophilicity of the nitrogen atom, dominated by its charge density, and the degree of steric hindrance. A minor factor is the juxtaposition of nitrogen lone pairs (the α-effect), which increases the reactivity at nitrogen in pyridazines, but not sufficiently to overcome the unfavorable electronic effect (see below).

The pK_a of a nitrogen is a convenient measure of its nucleophilicity: in proton addition steric effects are unimportant. All other types of electrophilic attack at nitrogen are sensitive in varying degrees to steric effects from α-substituents. (Exception: certain ring formation reactions as in metal chelation.)

Additional aza substitution decreases nitrogen charge density considerably, and the azines are all less nucleophilic than pyridine. Pyridine-like nitrogen atoms in cationic rings, *e.g.* diazinonium, oxazinium and thiazinium, are still less nucleophilic and few reactions with electrophiles are known, although diazines can be converted by reactive alkylating agents into diquaternary salts.

3.2.1.3.2   Effect of substituents

The electronic effects are summarized in (i)-(iii): these are quantified by the pK_a values of pyridines in Section 3.2.1.3.3. Steric effects (iv) are illustrated in Sections 3.2.1.3.4–3.2.1.3.11.
  (i) Strongly electron-withdrawing substituents, *e.g.* NO_2, COR, Cl, make these reactions more difficult by decreasing the electron density on the nitrogen atom; the effect is largely inductive and therefore is particularly strong from the α-position.

(ii) Strongly electron-donating substituents, *e.g.* NH$_2$, OR, facilitate electrophilic attack by increasing the electron density on the nitrogen. This operates by the mesomeric effect and is strongest from the γ-position. From the α-position opposing inductive effects possessed by these same substituents can partially or wholly cancel the increase in reactivity caused by α-NH$_2$ or α-OR. The especially powerful donor O$^-$ is formed in azinone anions, which can react with electrophiles at O (see Section 3.2.3.7), at C (see Section 3.2.1.4) or N (considered in this section).

(iii) Fused benzene rings, aryl and alkyl groups, and other groups with relatively weak electronic effects have little influence.

(iv) Reactions other than proton addition are hindered by all types of α-group. The shape of the substituent is important: thus, Me, Et and Pr^i generally show rather similar effects (as Et and Pr^i can rotate) whereas Bu^t shows a much larger steric effect. However, buttressing and the 'gear effect' ⟨76JA2847⟩ can alter this situation. A fused five-membered ring is generally less hindering than a fused six-membered ring.

3.2.1.3.3 Orientation of reaction of azines

The position of attack in azines containing more than one ring nitrogen atom is determined by the substituents according to the above guidelines. Thus, in 3-substituted pyridazines protonation will occur at position 2 only for strong electron donor substituents (effectively NR$_2$, O$^-$). All other protonations, and all other reactions with electrophiles occur predominantly at N-1. In 4-substituted pyridazines, where steric effects are unimportant and inductive effects of the substituent less important, reaction will occur at N-1 for electron donor and N-2 for electron acceptor substituents. In 3,6-disubstituted pyridazines, the less bulky and most activating substituent will direct the substitution α to it. In monocyclic 1,2,3-triazines, the 2-nitrogen atom is usually the preferred site of electrophilic attack (*e.g.*, see Section 3.2.1.3.9).

3.2.1.3.4 Proton acids

(i) Pyridine

Pyridines form stable salts with strong acids. Yellow ionic picrates were used for characterization in the past. Pyridine itself is often used to neutralize acid formed in a reaction and as a basic solvent. The basicity of pyridine (as measured by the dissociation constant of its conjugate acid, pK_a 5.2) is less than that of aliphatic amines (*cf.* NH$_3$, pKa 9.5; NMe$_3$, pK_a 9.8). This reduced basicity is probably due to the changed hybridization of the nitrogen atom: in ammonia the lone electron pair is in an sp^3-orbital, but in pyridine it is in an sp^2-orbital. The higher the *s* character of an orbital, the more it is concentrated near the nucleus, and the less available for bond formation. Nitriles, where the lone electron pair is in an *sp*-orbital, are of even lower basicity.

(ii) Azines

The basicity of the diazines is sharply reduced from that of pyridine (pK_a 5.2): the pK_a of pyrazine is 0.4, pyrimidine is 1.1 and pyridazine is 2.1. The significantly higher basicity of pyridazine as compared to pyrazine, unexpected for mesomeric and inductive effects, is attributed to the lone pair-lone pair repulsion which is removed in the cation.

The basicities of triazines and tetrazines are also low (e.g., for 1,3,5-triazine pK_a 1.0), but few quantitative data are available.

A fused benzene ring has little effect on the pK_a values in the cases of quinoxaline (*ca.* 0.6) and cinnoline (2.6). Quinazoline has an apparent pK_a of 3.3 which makes it a much stronger base than pyrimidine, but this is due to covalent hydration of the quinazolinium cation (see Section 3.2.1.6.3); the true anhydrous pK_a for equilibrium between the anhydrous cation and anhydrous neutral species of quinazoline is 1.95 ⟨76AHC(20)128⟩.

Quino[7,8-*h*]quinoline (**50**) and benzo[1,2-*h* : 4,3-*h*']diquinoline (**52**) belong to the so-called "proton sponges" and possess abnormally high basicities, pK_a = 12.8 and 10.3, respectively ⟨89AG(E)84, 89AG(E)86⟩. This is mostly due to strong destabilization of both bases because of electrostatic repulsion of their nitrogen electron pairs. Such repulsion is canceled on transition to the protonated forms (e.g. (51)), showing a strong intramolecular hydrogen bond. Both compounds give only monocations.

(50)   (51)   (52)

(iii) Effects of substituents on basicity of pyridine

The ΔpK_a values of representative substituted pyridines as compared with pyridine itself are shown in Table 1. Substituent effects are in line with the discussion in Section 3.2.1.3.2.

(a) Methyl groups are weakly base strengthening due to hyperconjugative and inductive effects. The increase in pK_a is somewhat greater for α- and γ- than for β-methyl groups.

(b) Phenyl groups are weakly electron-withdrawing by the inductive effect but can release electrons by the resonance effect. The resonance effect does not operate for the meta-position, and 3-phenylpyridine has a reduced basicity. The inductive effect for the 4-position is weak leading to an increased basicity, whereas the two effects cancel in 2-phenylpyridine.

(c) Amino groups are strong resonance electron donors and hence base strengthening. The base strength is 4-amino > 2-amino (increased importance of opposing inductive effect) > 3-amino (small influence of resonance effect).

(d) Methoxy groups are resonance donors but inductive acceptors. The inductive effect is dominant for the 2-position, the resonance effect for the 4-position.

(e) Halogen atoms are strong inductive acceptors and weak resonance donors: they cause a marked decrease in basicity, especially from α-positions.

(f) The nitro group, strongly electron-withdrawing by both inductive and resonance effects, causes an especially large drop in basicity.

Table 1 ΔpK_a Values for Monosubstituted Pyridines
(in H_2O)[a]

	Me	Ph	NH₂	OMe	Cl	NO₂
2-Position	0.8	0.1	1.7	−1.9	−4.5	−7.8
3-Position	0.5	−0.4	0.9	−0.3	−2.4	−4.4
4-Position	0.8	0.3	4.0	1.4	−1.4	−3.6

[a] cf. pyridine, pK_a 5.2.

(g) Fused benzene rings usually have little effect; cf. pK_a values: quinoline, 4.85; isoquinoline, 5.14; acridine, 5.6. Substituents on them usually have little effect on the basicity; however, those which can lead to significant charge delocalization in the conjugate acid by a p-quinoid canonical form are base strengthening (cf. pK_a, values: 7-aminoquinoline, 6.5; 3-aminoacridine (53), 8.04).

(53)   (54)

(h) Intramolecular hydrogen bond formation with the pyridine nitrogen atom is base weakening; cf. pK_a values: 8-aminoquinoline (54), 3.93; 4-aminoacridine, 4.40.

(i) Steric effects are usually unimportant; however, in extreme cases as in 2,6-di-t-butylpyridine (pK_a 3.6) the pK_a, does fall significantly below that of pyridine. This is attributed to entropy rather than enthalpy effects as it is the entropy change for the transfer of the protonated 2,6-di-t-butylpyridine from the gas phase to the aqueous phase which is abnormal. Aqueous protonated 2,6-di-*t*-butylpyridine is hydrogen bonded to a water molecule *via* the N—H bond and the observed loss of entropy is due to the substantial restrictions in the internal rotations in the solution complex of protonated 2,6-di-*t*-butylpyridine ⟨84JA4341⟩.

Much work has been done on the quantitative correlation of the basicity of pyridines with Hammett substituent constants. The best single parameter correlation for 4-substituents is with σ_p ⟨78AHC(22)81⟩.

(iv) Proton acids and azinone anions: acidity of azinones

Some pK_a values are collected in Table 2. Thiones are *ca.* 2 p*K* units more acidic than the corresponding azinones. Fused benzene rings have little effect except in the 3-substituted isoquinoline series where partial bond fixation lowers the acidity. Additional aza substitution increases the acidity significantly.

Table 2 Acidity of Azinones and Azinethiones[a]

Additional Structure	α-Series		γ-Series	
Features	X = O	X = S	X = O	X = S
—	11.7	9.97	11.09	8.83
3,4-Benzo	>11	10.82	—	—
4,5-Benzo	9.62	8.58	—	—
5,6-Benzo	>11	10.21	11.25	8.83
2N	—	—	8.68	6.54
3N	9.17	7.14	8.59	6.90
4N	8.23	6.32	—	—
5N	8.59	6.90	—	—
6N	10.46	8.30	—	—

[a] pK_a values in aqueous solution taken from ⟨63PMH(1)1⟩.

3.2.1.3.5 Metal ions

Six-membered heteroaromatic compounds can act as both σ- and π-ligands (for review see ⟨99AHC(72)1⟩.

(i) Simple complexes

Many transition and B-subgroup metals form complex ions with pyridines in aqueous solution, *e.g.* $Ni^{2+} \rightarrow Ni(C_5H_5N)_4^{2+}$; $Ag^+ \rightarrow Ag(C_5H_5N)_2^+$; if certain anions are also present, uncharged complexes can result, *e.g.* $Cu^{2+} + 2OCN^- + 2C_5H_5N \rightarrow Cu(OCN)_2 \cdot (C_5H_5N)_2$, soluble in H_2O and $CHCl_3$; Ni^{2+}, Cd^{2+} and Zn^{2+} react similarly.

The diazines also form metal complexes. Thus, pyrazine forms tetrahedral and octahedral complexes with Co[II] and other transition metals: it functions as a monodentate and also as a bidentate bridging ligand to give polymeric complexes. The stability of these complexes is increased by back-bonding.

Pyridine and pyrazine can also replace up to three of the carbonyl groups in Group VI metal carbonyls to form compounds of type $Cr(CO)_3py_3$. The reaction of 4,5,6-triphenyl-1,2,3-triazine with nonacarbonyldiiron afforded 3,4,5-triphenylpyrazole in 80% yield, probably through the initial n-donor complexation of the triazine to iron carbonyl ⟨87BCJ3062⟩.

(ii) Chelate complexes

Chelate rings can be formed by pyridines containing α-substituents such as carboxyl or CH=NR. Important bicyclic chelating agents are 2,2'-bipyridyl (**55**; Y = H), *o*-phenanthroline (**56**) and

8-hydroxyquinoline (**57**), which all form bis- and tris-complexes with many metals ⟨94CRV327⟩. This type of complex formation has many analytical applications. Overlap between the *d*-orbitals of the metal atom and the pyridine σ-orbitals is believed to increase the stability of many of these complexes. Steric effects can hinder complex formation as in (**55**;Y = Me).

(**55**)   (**56**)   (**57**)   (**58**)

Suitably substituted diazines also form chelate complexes, *e.g.* 2,3,5,6-tetrakis(α-pyridyl)pyrazine yields red tridentate complexes with Fe^{II}. In the fast development of metallosupramolecular chemistry, many other polydentate ligands, based on 2,2′-bipyridyl units, have been obtained and studied ⟨94CI(L)56⟩. Thus, two molecules of oligopyridine (**58**) interact with various metal ions (Fe^{2+}, Co^{2+}, Cu^{2+} etc.) to form a double-helical [M$_2$L$_2$]$^{4+}$ complex in which each metal is bonded to a tridentate region from each ligand.

(iii) π-Complexes

Azine π-complexes, *e.g.* (**59**)-(**61**) also formed from their components and their X-ray structure and reactivity have been studied ⟨88CB1983, 91JCS(P1)757⟩.

(**59**)   (**60**)   (**61**)

3.2.1.3.6 *Alkyl and aryl halides and related compounds*

(i) N-Alkylation of pyridines

Pyridines displace halide, sulfate, toluene-*p*-sulfonate and other ions from the corresponding alkyl compounds to form alkylpyridinium salts. These reactions are of the S$_N$2 type and are sensitive to steric changes in the pyridine or alkyl derivatives. Pyridine reacts exothermically with methyl iodide or dimethyl sulfate. Reactions involving pyridines with α-substituents of any type, electron-withdrawing β- or γ-substituents, or alkyl halides other than methyl, are slower and are often carried out by heating in a solvent of suitable high dielectric constant, such as acetonitrile, to promote ion formation. Alternatively a highly active alkylating agent can be used, such as an alkyl triflate.

The quantitative effects of α-substituents in decreasing the rates of these reactions are not additive and also depend considerably on solvent and alkylating agent. They are low in liquid sulfur dioxide as a solvent where solvation effects are small and the high dielectric constant increases the bond breaking in the transition state. For 3- and 4-substituted pyridines a Bronsted correlation exists between the rates of quaternization and the pK_a values ⟨78AHC(22)86⟩.

With tertiary halides, bimolecular elimination usually occurs; if isomeric alkenes can result, the proportions formed depend on the steric requirements of the pyridine because formation of the more substituted alkene (Saytsev Rule) is more sensitive to steric hindrance than formation of the less substituted alkene (Hofmann Rule). Pyridine and *t*-amyl bromide give 25% of 2-methylbut-1-ene (less substituted alkene), but 2,6-lutidine gives it in 45% yield (*cf.* **62**). However, pyridine and *t*-butyl bromide in the presence of AgBF$_4$ yield the 1-*t*-butylpyridinium ion.

(**62**)

(ii) Diazines and triazines

Alkyl halides react with diazines less readily than with pyridines. All the diazines are, nevertheless, more reactive toward methyl iodide than predicted by their pK_a values and the Bronsted relationship. The significant although modest rate enhancements found are considered to arise from interactions between the two lone pairs on the nitrogen atoms; this interaction is largest in pyridazine. Use of oxonium ions can convert the diazines into diquaternary salts. Quinoxalines and phenazines similarly yield diquaternary salts under forcing conditions.

The alkylation of pyridones and azinones is considered in Section 3.2.1.8.4.

(iii) N-Arylation

Only highly activated aryl halides react with pyridines. Thus, 2,4-dinitrochlorobenzene with pyridine forms 1-(2,4-dinitrophenyl)pyridinium chloride; active heteroaryl halides such as 2-chloropyrimidine react similarly. To phenylate pyridine, diphenyliodonium ions are needed: Ph$_2$I$^+$BF$_4^-$ + pyridine → 1-phenylpyridinium BF$_4^-$ + PhI. This reaction may involve initial electron transfer.

3.2.1.3.7 Acyl halides and related compounds and Michael-type reactions

Acyl and sulfonyl halides and anhydrides react instantaneously with pyridine to form quaternary salts which are excellent acylating and sulfonylating agents. The familiar use of pyridine as a solvent in such reactions reflects this. While usually used *in situ*, stable *N*-acylpyridinium and *N*-acylisoquinolinium salts, *e.g.* (**63**), have been isolated and their structures confirmed by single crystal X-ray analysis ⟨92JOC5136, 94AX(B)25⟩ (see also Section 2.2.3.1).

(**63**) (**64**) (**65**)

4-Dimethylaminopyridine (DMAP) is very much more effective than pyridine in catalyzing acylation and related reactions ⟨78AG602⟩, and in this case many of the highly reactive intermediates can be isolated.

Bromocyanogen and pyridine give 1-cyanopyridinium bromide, important for ring-opening reactions (see *e.g.* Section 3.2.1.6.3.iv).

Pyridines add to quinones in Michael-type reactions to give phenolbetaines (**64**). Many other Michael acceptors behave similarly, *e.g.* acrylate esters and acrylamides in the presence of acid yield quaternary ions py$^+$CH$_2$CH$_2$COY. Pyridazine at room temperature with maleic anhydride gives the 2:1 adduct (**65**).

Such reactions occur readily with alkynic esters, but the products isolated are often complex. Thus, the initial Michael adduct of type (**68**) from pyridine with acetylenedicarboxylic ester reacts with more alkynic ester to yield (**66**), (**67**) and other products. Quinoline and isoquinoline react similarly.

(66)   (67)   (68)

Ylide (**68**) can also be trapped by carbon dioxide to give (**69**). In the presence of water the reaction can take a different course to give the indolizine (**70**) ⟨78AHC(23)350⟩.

Acridine and dimethyl acetylenedicarboxylate in the presence of methanol yield the addition compound (**71**) ⟨63AHC(1)159⟩.

(69)   (70)   (71)   (72)

E = CO$_2$Me

1,2,3-Triazines react with tetracyanoethylene oxide to afford the stable triazinium 2-dicyanomethylides (**72**).

3.2.1.3.8 Halogens

At room temperature pyridines react reversibly with halogens and interhalogens, *e.g.* ICl, to give unstable adducts, which behave as mild halogenating agents. X-Ray diffraction studies of the pyridine-iodine complex have given its structure (**73**).

(73)   (74)   (75)

The molar enthalpies of complexation of pyridine with iodine are -36 kJ mol^{-1} (CCl$_4$) and -34 kJ mol^{-1} (C$_6$H$_{12}$) by calorimetry, with corresponding values of -33.1 and -34.6, respectively, using the UV absorbance of the pyridine/iodine complex ⟨84JCS(P2)731, 87JCS(P2)1713⟩. For 2,6-dimethylpyridine the complex formation constant is reduced considerably from that for pyridine itself (from 106 to 46 in cyclohexane), but the enthalpy of formation remains sensibly constant, and thus independent of steric considerations.

Study of the complexes of pyridine, methyl nicotinate, and methyl isonicotinate with chlorine ⟨87JA7204⟩ suggest that the pyridine/chlorine atom complex has a long three-electron σ bond between nitrogen and chlorine. Pyridine iodo compounds can be prepared by treating TiI$_3$[AsF$_6$] with pyridines, from which the pyridinium salt [C$_5$H$_5$NI]$^+$[AsF$_6$]$^-$ has been isolated and characterized ⟨90ZAAC(586)93⟩. *N*-Fluoropyridinium salts, of growing interest as fluorinating agents ⟨91MI 502-01⟩, are synthesized with either FONO$_2$ or CF$_3$SO$_3$Na/F$_2$.

3.2.1.3.9 *Peracids*

Pyridine *N*-oxides are formed by the treatment of pyridines with peracids (74)→(75). Typical conditions are $MeCO_2H/H_2O_2$ at 100°C or m-$ClC_6H_4CO_3H/CHCl_3$ at 0°C. The pyridine nitrogen atom reacts less readily with peracids than do aliphatic tertiary amines, as expected. Large α-substituents and any electron-withdrawing substituents slow the reaction; thus the *N*-oxidation of 2,6-diphenylpyridine proceeds in poor yield, and efficient conversion of pentachloropyridine to the *N*-oxide requires a powerful oxidant such as peroxytrifluoroacetic acid.

The formation of *N*-oxides by peracid oxidation of azines proceeds less readily than in the pyridine series. The orientation of *N*-oxide formation in diazines follows the rules outlined in Section 3.2.1.3.2. Thus, 3-aminopyridazines give mainly 2-oxides; other 3-substituted pyridazines form 1-oxides. However, the orientation of *N*-oxide formation in pyrazines can depend on the conditions: 2-chloropyrazine normally forms the 4-oxide as expected, but in strongly acidic conditions the 4-nitrogen atom is protonated, and *N*-oxide formation takes place at the 1-position. Pyrimidines and methylpyrimidines are susceptible to decomposition, ring-carbon oxidation and ring-opening reactions on direct N-oxidation, resulting in low yields of *N*-oxides. Activating substituents are required ⟨84CJC1176⟩. In strong acid low yields of pyrimidinones may result ⟨85JOC3073⟩. With *m*-chloroperbenzoic acid in chloroform, pyrimidine *N*-oxides result: 48% from pyrimidine and 55% from 2-methylpyrimidine ⟨81H(16)573⟩. Unsymmetrical pyrimidines are oxidized preferentially at sites *para* to strong electron donors. Bulky groups and electron-withdrawing substituents decrease oxidation, and direct the attack to the more remote nitrogen ⟨84CJC1176⟩.

Monocyclic 1,2,3-triazines on treatment with MCPBA or $AcOH/H_2O_2$, give 1- and/or 2-oxides (76) and/or (77). 1,2,3-Triazines with bulky aryl groups on C-4 and C-6 give 2-oxides predominantly, whereas 4- and/or 6-alkyl groups allow 1-oxides to form. Equation (3) summarizes results in this area. The synthesis of benzo-1,2,3-triazine 2-oxides was also reported using MCPBA ⟨88JCS(P1)1509⟩.

	(76)	(77)
$R^1 = R^3 = Ph, R^2 = H$	0	81%
$R^1 = Me, R^2 = H, R^3 = Ph$	29%	60%
$R^1 = R^3 = Me, R^2 = H$	20%	48%
$R^1 = Me, R^2 = R^3 = H$	15%	32%

Only pyrazine and its benzo derivatives are easily converted into di-*N*-oxides, although di-*N*-oxides have been reported, for example, in the pyridazine, pyrimidine and cinnoline series. Oxidation of 3-amino-1,2,4-triazine 2-oxide with H_2O_2 in polyphosphoric acid at 24°C affords 3-amino-1,2,4-triazine 2,4-dioxide (78). However, 3-amino-1,2,4-triazine with peracetic or peroxytrifluoroacetic acid at 60°C formed only 3-amino-1,2,4-triazin-5(2H)-one (79) ⟨86H(24)951⟩.

(78)   (79)   (80)   (81)

With hydrogen peroxide in acetic acid 1,2,4-benzotriazine gave 1,2,4-benzotriazine 1-oxide and benzotriazole, but oxidation with MCPBA afforded a mixture of the 1-oxide and the 2-oxide ⟨82T1793⟩.

3.2.1.3.10 Aminating agents

Hydroxylamine *O*-sulfonic acid converts pyridine into the 1-aminopyridinium cation. Pyridazines undergo *N*-amination readily. 1,2,3-Triazine *N*-imines (**80**) were obtained from the reaction with *O*-mesitylenesulfonylhydroxylamine (MSH), followed by a neutralization with potassium carbonate.

3-Methylthio-1,2,4-triazin-5(2H)-ones are aminated regioselectively with *O*-(2,4-dinitrophenyl)-hydroxylamine to give 2-amino-3-methylthio-1,2,4-triazin-5(2H)-ones (**81**) ⟨82JHC1583, 83JHC1671⟩.

Treatment of 1*H*-pyrrolo[2,3-*b*]pyridine (**82**) with MSH resulted in *N*-amination of the pyridine nitrogen in high yields. Upon addition of a base such as potassium hydroxide the ylide (**83**), which formed rapidly, underwent rearrangement to give 7-amino-7*H*-pyrrolo[2,3-*b*]pyridine (**84**) in nearly quantitative yield. ⟨79CPB2183⟩.

(82)   (83)   (84)

Reactions with nitrenes, which also give *N*-amination, are considered in Section 3.2.1.9.1.

3.2.1.3.11 Other Lewis acids

Pyridine readily forms stable coordination compounds. Thus, boron, aluminum and gallium trihalides react at 0°C in an inert solvent to give 1:1 adducts (*cf.* **85**). Steric factors are important, and α-substituents decrease the ease of reaction. This is illustrated by the heats of reaction of pyridine, 2-methylpyridine and 2,6-dimethylpyridine with boron trifluoride which are 101.3, 94.1 and 73.2 kJ mol^{-1}, respectively. The marked decrease in exothermicity here should be contrasted with the small steric requirement of the proton as shown by the pK_a values of substituted pyridines (see Section 3.2.1.3.4).

Alkyl-substituted pyridines have been complexed with a wide variety of other boron Lewis acids, BH_3, $B(OH)_3$, and BX_3, where X is a selection of alkyl, aryl, hydroxy, alkoxy, and aryloxy groups ⟨87JCS(P2)771⟩.

Sulfur trioxide gives the expected adduct, a sulfonating agent (see also Section 3.2.1.4.5); similarly *N*-nitropyridinium tetrafluoroborate is formed with $NO_2^+ BF_4^-$.

(85)

Van der Waals complexes of C_{2v} symmetry between 1,2,4,5-tetrazine and a number of light gases (He, Ar, H_2) have been observed and characterized by laser spectroscopic studies of free supersonic jet expansion of the tetrazine in the carrier gas ⟨84CHEC-(3)531⟩. In these complexes, one equivalent of noble gas sits on top of the aromatic π-system of the heterocycle. 1,2,4,5-Tetrazine, its 3-methyl, and 3,6-dimethyl derivative as well as aminotetrazine have all been used as heterocycles with noble gases, water, HCl, benzene, and acetylene, playing the role of the second partner.

3.2.1.4 Electrophilic Attack at Carbon

3.2.1.4.1 Species undergoing reaction and reaction mechanism

The intrinsic difficulty of electrophilic substitution of pyridines and azines is exacerbated because most of these reactions are carried out in acidic conditions where the pyridine nitrogen atom has become protonated. However, although electrophilic reagents react at the nitrogen atoms very readily, these reactions are often reversible, and even in strongly acidic solution there is a small proportion of the free base present. Thus, *a priori,* reaction is possible either on the conjugate acid majority species or on the minority free base species. In fact, considerable work has shown that some reactions occur on the pyridine or diazine free base, other reactions on conjugate acids.

The weaker the basicity of the pyridine nitrogen, the more likely it is that the reaction could occur on the free base. α-Halogen atoms are particularly effective, in that they sharply reduce basicity but not very much the susceptibility toward electrophilic substitution.

Halogenation of pyridines is easier than nitration or sulfonation because it can be carried out in non-acidic media and the pyridine-halogen adducts are appreciably dissociated. Dihalogenation can occur since one halogen atom causes little additional deactivation of the ring. The mercuration of pyridines (Section 3.2.1.4.9) probably involves coordination of the pyridine nitrogen to the mercury atom, and such coordination causes less ring deactivation than protonation.

In some instances, especially with the oxygen and sulfur heterocycles, the overall reaction leading to a substituted product does not involve an S_EAr mechanism but proceeds by an addition followed by elimination sequence, as outlined for the bromination of coumarin in equation (4). The choice of experimental conditions can affect the outcome of the reaction, as illustrated in the formation of (**114**) and (**115**) in Section 3.2.1.4.7.

3.2.1.4.2. Reactivity and effect of substituents

The reactivity of six-membered rings toward electrophilic substitution reactions can be summarized as follows:

(i) Triazines, diazines *without* strongly activating substituents (NH_2, OR) and pyridines *with* strongly deactivating substituents (NO_2, SO_3H, COR, *etc.*) do not react.

(ii) Pyridines without strong activation, diazines with a single strongly activating substituent and diazinones undergo nitration and sulfonation with difficulty (reactivity *ca.* that of *m*-dinitrobenzene) and halogenation somewhat more readily.

(iii) Pyridones, aminopyridines, and diazines with two strongly activating substituents readily undergo nitration, sulfonation and halogenation (reactivity *ca.* that of benzene).

(iv) Pyridines with two, and diazines with three strongly activating substituents are very reactive toward electrophilic substitution.

(v) Pyridines, pyridones and pyrones containing an amino or hydroxy group also undergo diazo coupling, nitrosation and Mannich reactions, as do their benzenoid analogues, phenol or aniline. Such reactions take place under conditions of relatively low acidity where less of the compounds is in the form of unreactive cations.

(vi) Alkyl groups and halogen atoms behave normally as weakly activating and deactivating substituents, respectively.

(vii) Fused benzene rings do not much affect the intrinsic reactivity, but electrophilic substitution frequently occurs in the benzene ring (see next section).

(viii) It follows from the above that the influence of substituent groups on the ease of electrophilic attack on ring carbon atoms can be largely predicted from a knowledge of benzene chemistry.

3.2.1.4.3. Orientation

Substituents exert their normal directive effects, and aza substitution directs *meta*. If there is conflict, then a strongly *para* directing substituent dominates. Thus, 3-hydroxypyridine reacts first at the 2- and then at the 6-position. Pyridine 1-oxide is nitrated at position 4 as the free base but sulfonated at position 3 as the conjugate acid.

meta-Disubstituted benzenes containing one strongly *o*, *p*-directing and one strongly *m*-directing group are often further substituted between the two groups, and this may be compared with the orientation observed in (**86**). In the quinolizinium ion, amino or hydroxy groups in position 1 direct to the 2-position and those in position 2 direct to the 1-position (Scheme 9).

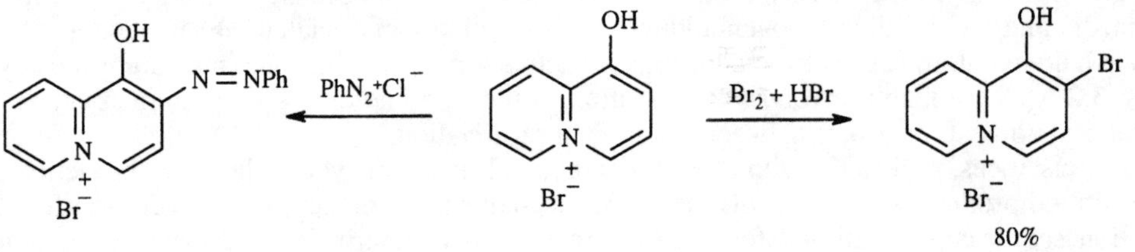

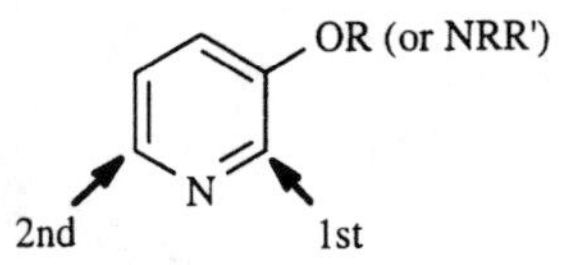

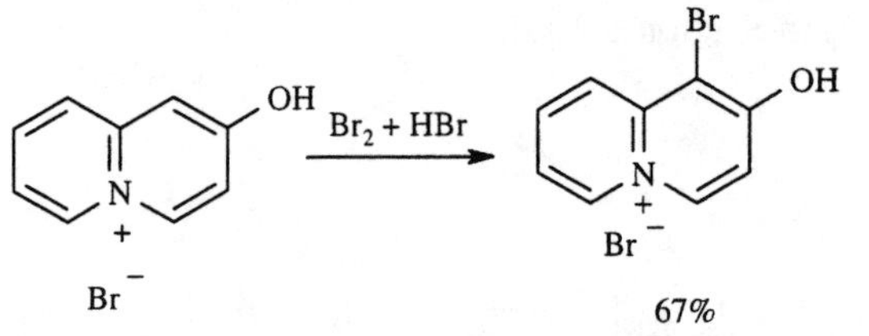

Scheme 9

In benzo- and phenyl-pyridines and in phenylpyridine 1-oxides, electrophilic substitution usually takes place in the benzene ring. In benzo-pyridones, -pyrones and -pyridine *N*-oxides, electrophilic substitution often occurs in either the benzene or the heterocyclic ring depending on the conditions; sometimes mixtures are formed (see Section 3.2.3.2.1).

3.2.1.4.4   *Nitration*

(i) Pyridines

Pyridine itself requires vigorous conditions for nitration (H_2SO_4-SO_3-KNO_3 at 300°C); 3-nitropyridine is obtained only in poor yield. A single methyl group is insufficient activation; on attempted nitration, the picolines are extensively oxidized. However, 2,6-lutidine and 2,4,6-collidine afford the corresponding 3-nitro derivative in fair yield in milder conditions (H_2SO_4-SO_3-HNO_3 at 100°C).

As expected, an amino group facilitates nitration strongly. 2-, 3- and 4-Aminopyridines are nitrated smoothly (H_2SO_4-HNO_3 at 40–70°C to form mono- (5-, 2- and 3-, respectively) and dinitro- (3,5-, 2,6- and 3,5-, respectively) derivatives (Section 3.2.1.3.2). Alkylamino-, alkoxy- and 3-hydroxy-pyridines react analogously to the corresponding amino compounds.

Most nitrations of pyridines take place on the *N*-protonated species, and this includes the conversion of 2,6-dimethoxypyridine to the 3-nitro derivative. However, the further nitration of 2,6-dimethoxy-3-nitropyridine to the 3,5-dinitro derivative occurs on the free base. The introduction of the first nitro group reduces the basicity such that sufficient free base is now present for the reaction to take place through this minority species. 2,6-Dihalopyridines also undergo nitration as free base.

In general pyridines with $pK_a > 1$ nitrate as conjugate acids at the α- or β-position depending on the orientating effect of the attached substituents, while derivatives with $pK_a < -2.5$ nitrate as free bases. Those pyridines with intermediate pK_a values often show a mechanistic changeover, with change in pH (H_o).

(ii) Azines

Two or more electron-releasing substituents make 5-nitration in pyrimidines relatively easy: 2,4-diamino-6-chloropyrimidine yields the 5-nitro compound *via* a nitroamine intermediate. Similarly, 4-amino-3,6-dimethoxypyridazine undergoes easy nitration to the corresponding 5-nitro compound. The less activated 3-methoxy-5-methylpyridazine requires more vigorous conditions, yielding 4-, 6- and 4,6-di-nitro derivatives.

(iii) Cationic rings

Few examples are known, but 1,2,4,6-tetramethylpyridinium can be nitrated to yield the 3-nitro derivative.

(iv) Pyridones, pyrones and azinones

2- and 4- Pyridone and their 1-alkyl derivatives are readily nitrated to form first the 3- or 5-mono-(H_2SO_4/HNO_3, 30°C) and then the 3,5-di-nitro derivatives. These nitrations involve reactions of the neutral pyridone species. The proportions of 3- and 5-nitration in 2-pyridone vary with the conditions. 3-Nitration is favored at low acidity and high temperature and 5-nitration by the reverse.

Quinolin-2- and -4-ones can be nitrated in the 3-position (HNO_3, 100°C); under conditions of higher acidity reaction occurs on the protonated species in the benzene ring (see Section 3.2.3.2.1).

Nitration of 6-phenyl-2-pyrone (**87**) depends on the conditions. The free base reacts at the 3-position, the conjugate acid at higher acidities in the *p*-position of the phenyl group.

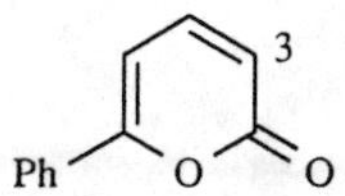

(**87**) 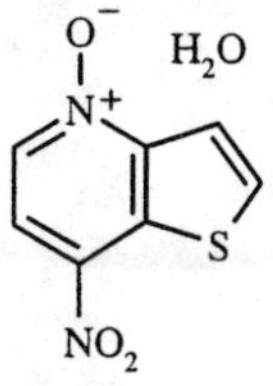   (**88**)

Nitration of 4,5-dichloro-2-methylpyridazin-3-one occurs at position 6. Pyrimidin-2-one is nitrated under vigorous conditions to give the 5-nitro derivative, whereas 1-methylpyrimidine-2,4-dione yields the 5-nitro derivative at 25°C.

(v) Azine N-oxides

Pyridine 1-oxide is nitrated (H_2SO_4/HNO_3, 100°C) to give the 4-nitro derivative in good yield. Substituted pyridine oxides such as the 2- and 3-methyl, -halo and -methoxy derivatives also give 4-nitro compounds in high yield. Quinoline 1-oxides are selectively nitrated in the 4-position at temperatures above *ca.* 80°C whereas at lower temperatures nitration occurs in the benzo ring (Section 3.2.3.2.1).

Nitrations of pyridine 1-oxides in the 4-position take place on the neutral free base species. 2,6-Dimethoxypyridine 1-oxide is nitrated as the conjugate acid to yield the 3-nitro derivative; a second nitration to give the 3,5-dinitro analogue takes place on the free base. Thieno[3,2-*b*]pyridine *N*-oxide is nitrated by HNO_3-H_2SO_4 to produce (**88**) with 63% yield ⟨96CHEC-II(7)199⟩.

Pyridazine 1-oxide and many of its substituted derivatives undergo nitration with nitric and sulfuric acids to form the corresponding 4-nitropyridazine 1-oxides. If the 4-position is occupied nitration can occur at the 6-position.

Pyrimidine *N*-oxides cannot be nitrated unless they possess electron-donating substituents, *e.g.*, 5-nitration in 2,6-diamino-4-ethylaminopyrimidine *N*-oxide ⟨79AJC2049⟩.

Reaction of pyridine 1-oxide with benzoyl nitrate leads to the 3-nitro derivative: the postulated mechanism is shown in Scheme 10.

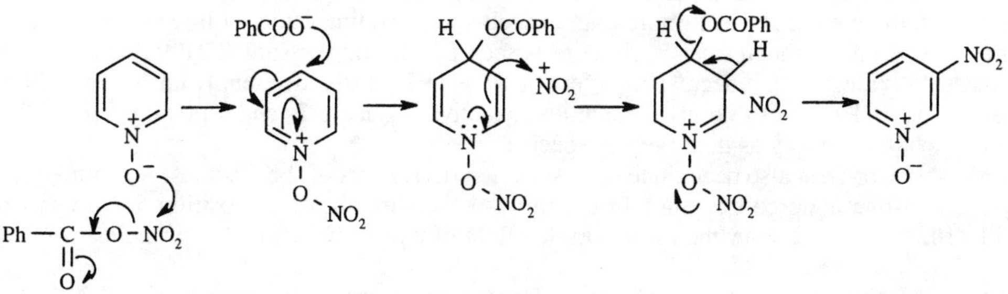

Scheme 10

Nitration of pyridazine *N*-oxides with acyl nitrates prepared from acyl chlorides and silver nitrate also occurs at the β-position relative to the *N*-oxide group. Thus, pyridazine 1-oxide yields 3-nitropyridazine 1-oxide.

Pyridine-*N*-(2,4,6-trinitrophenyl)imine (**89**) can be nitrated in the 4-position (equation 5).

$$(5)$$

(**89**)

3.2.1.4.5 Sulfonation

Sulfonation of pyridine affords the 3-sulfonic acid in 70% yield, but vigorous conditions (H_2SO_4-SO_3, 230°C) and $HgSO_4$ catalyst are required ⟨90AHC(47)310⟩. The picolines form β-sulfonic acids similarly. Sulfonation of pyridine at 360°C gives a considerable amount of the 4-sulfonic acid. Heating the 3-sulfonic acid at this temperature produces a similar result; presumably, thermodynamic control takes over.

2-Aminopyridine and 1-methyl-2-pyridone are sulfonated under milder conditions (H_2SO_4-SO_3, 140°C) in the 5-position. 2,6-Di-*t*-butylpyridine is converted into the 3-sulfonic acid under mild conditions (SO_2-SO_3, 0°C) because reaction of SO_3 at the nitrogen atom is prevented sterically: thus, reaction occurs on the free base, under conditions where this is the majority species.

Sulfonation of pyridine 1-oxide requires vigorous conditions (H_2SO_4-SO_3-Hg^{2+}, 230°C) and gives the 3-sulfonic acid (*cf.* Section 3.2.1.4.1).

3.2.1.4.6 Acid-catalyzed hydrogen exchange

Acid-catalyzed hydrogen exchange can be detected by isotopic labeling. Deuteration (followed by NMR) and tritiation (followed by radioactivity) at various ring positions at different acidities and temperatures have been investigated. By extrapolating all measurements to 100°C and pH = 0, standardized rates of hydrogen exchange have been established for a large number of heterocycles, some of which are given in Scheme 11.

Compared to the rate for one position in benzene ($\log k_0 = -11$), the large effect of the N^+ pole is apparent: from the *meta*-position it more than cancels the activating effect of three *o/p* methyl groups. A neutral nitrogen has a much smaller effect as is seen by the comparisons (**92**)/(**93**) and (**94**)/(**95**). In a protonated *N*-oxide, the rate-decreasing effect of $N^+ - OH$ is little different from N^+H [*cf.* (**91**)/(**99**); however, the neutral $N^+ - O^-$ group is much less deactivating at the 2- and 4-positions [*cf.* (**90**)/(**98**)]. Pyridones exchange rapidly as their neutral species.

Considerable work has also been done with activated derivatives of the diazines. 4-Aminopyridazine and pyridazin-4-one undergo exchange in the form of their free bases at position 5. Comparisons of (**95**) with (**102**) and (**103**) show the considerable effect of a *para*-N, while that of the *meta*-N is much less.

In the quinolinium cation, there is little difference in reactivity between position 3 and all the positions of the benzene ring (**105**).

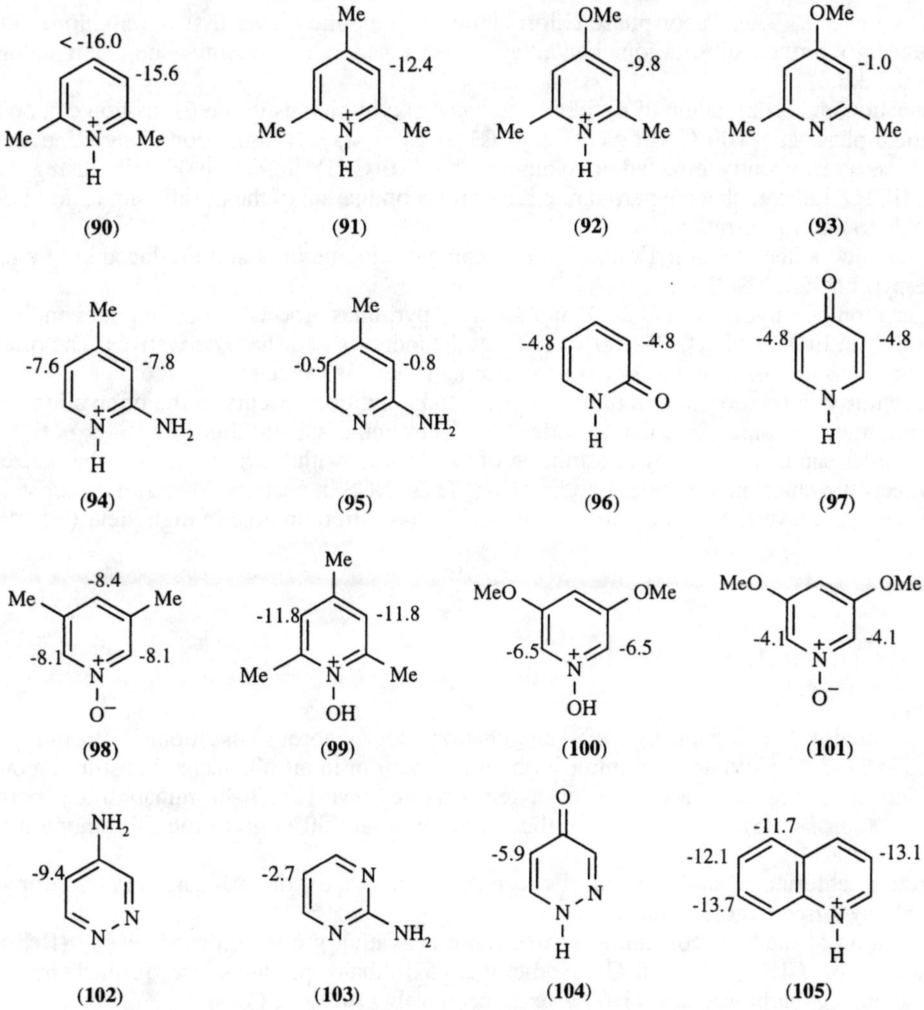

(90) (91) (92) (93)

(94) (95) (96) (97)

(98) (99) (100) (101)

(102) (103) (104) (105)

Scheme 11 Rats of hydrogen exchange standardized to 100°C and pH 0 (log k_o)

Pyrimidin-2-one exchanges its 5-hydrogen much faster than pyridin-2-one. However, this is due to the existence of a small proportion of the covalent hydrate (106) which undergoes rapid exchange.

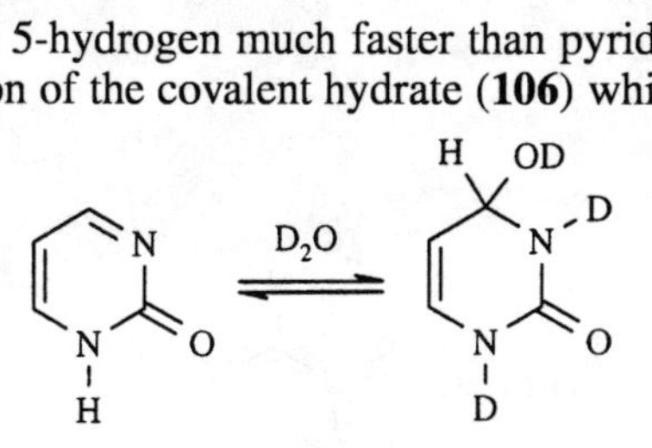

(106)

3.2.1.4.7 Halogenation

(i) Pyridines

Several review articles have appeared on the halogenation of pyridines ⟨84AHC(35)281, 88AHC(44)199, 90AHC(47)303, 93AHC(58)272, 94AHC(59)286⟩. Electrophilic halogenation of pyridine at carbon follows the orientation order 3 > 4 ≥ 2, whereas free radical halogenation occur at the 2-position (see Section 3.2.1.9.2.).

Pyridine gives perfluoropiperidine with CoF_3-F_2; conversion of pyridine to mainly 2-fluoropyridine occurs with xenon difluoride.

Vapor phase chlorination at 150–200°C and bromination at 300°C of pyridine give fair yields of the 3-mono- and 3,5-di-halo derivatives. As the temperature is raised increasing amounts of α-substitution occur (Section 3.2.1.9.2). Vapor phase chlorination of quinoline yields first the 3-chloro derivatives which undergo further substitution, but alkylpyridines generally undergo side-chain halogenation (Section 3.2.3.3.3).

Chlorination and bromination of pyridine and some alkylpyridines in the β-position can be effected in the liquid phase at ~ 100°C using excess $AlCl_3$ as catalyst. β-Bromination of pyridine and 2- and 4-picoline is conveniently effected in oleum at 80–120°C. Bromination kinetics using HOBr in aqueous $HClO_4$ indicate that the partial rate factor for bromination of the pyridinium cation is ~ 10^{-13}, comparable to that for nitration.

Electrophilic iodinations of pyridines are less common: iodine in oleum produced only a low yield of 3-iodopyridine ⟨57JCS387⟩.

Halogenation of 3-hydroxy- and 2-, 3- and 4-amino-pyridines proceeds under milder conditions (*e.g.* Cl_2, Br_2 or I_2 in EtOH or H_2O, 20–100°C) to form the mono- and di-halo derivatives. The orientations of these products are *ortho* and *para* to the activating group, as expected.

Electrophilic chlorination of quinoline under neutral conditions occurs in the orientation order 3 > 6 > 8. Hammett σ^+ values predict an order for electrophilic substitution of 5 > 8 ≈ 6 > 3. The reactivity order can be affected by substitution of an electron-withdrawing group in the benzene ring, which directs the chlorination to the pyridine ring. Thus, NCS in acetic acid or sulfuryl chloride in *o*-dichlorobenzene converts 8-nitroquinoline into 3-chloro-8-nitroquinoline in high yield ⟨91M935⟩.

(ii) Azines

Unsubstituted pyrimidine undergoes 5-halogenation under vigorous conditions. 5-Bromopyrimidine is formed in 71–88% yield using bromine without a solvent or in nitrobenzene. Substitution on carbon is preceded by a vigorous reaction at lower temperatures involving *N*-bromination and perbromide formation ⟨90AHC(47)325⟩. Bromination in the vapor phase at 230°C gives the 5-bromopyrimidine in 62% yield ⟨73JHC153⟩.

Pyrazine is chlorinated at 400°C to give a mixture of mono-, di-, tri- and tetra-chloropyrazines, presumably by a free radical mechanism.

Halogenation of diazines containing one or more activating groups proceeds easily (Br₂ or Cl₂ in H_2O, AcOH, or $CHCl_3$, 20–100°C). Sometimes 5,5-dihalo products are formed by ring de-aromatization, *e.g.* barbituric acid (**107**) gives successively (**108**) and (**109**).

(**107**)   (**108**)   (**109**)

Electrophiles can not attack ring carbons of 1,2,3-triazine because of the intense π-electron deficiency of the ring system. The reaction of halogenating reagents with 4,6-dimethyltriazine (**110**) to afford 5-halotriazines (**111**) and (**112**) is probably not an electrophilic substitution. The use of interhalogen reagents affords 5-halotriazines derived completely or mainly from the more electronegative halogen, which suggests that the reaction is initiated by the quaternization of (**110**) with cationic halogen, followed by nucleophilic addition of halide ion and elimination of hydrogen halide (Scheme 12) ⟨86CPB4432⟩.

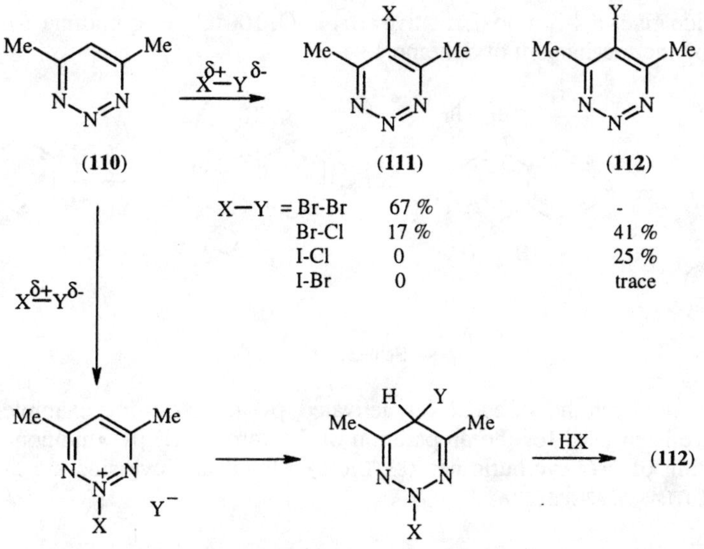

Scheme 12

(iii) Pyridones, pyrones, azinones and N-oxides

2- and 4-Pyridones and 2- and 4-pyrones readily give their 3-mono- and 3,5-di-halo derivatives; even chelidamic acid (113) reacts in this way. Bromination of pyran-2-one gives the substitution product (115) or the addition product (114) depending on the conditions.

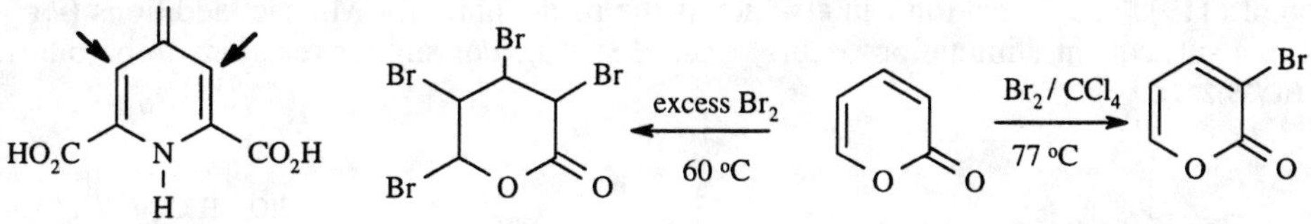

Quinolin-4-one forms a 3-bromo derivative, but coumarin gives the addition compound (116) which is easily re-aromatized (116 → 117). 4-Thiopyrones are halogenated in position 3. Pyridazinones and cinnolinones are also readily halogenated in the expected position.

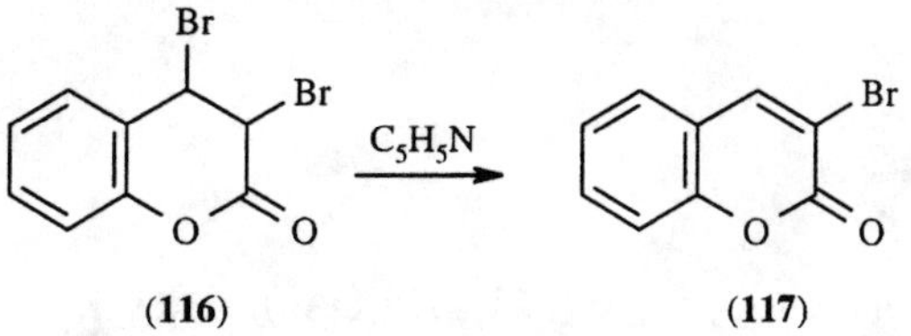

Halogenation of pyridine 1-oxide is not easy. Bromination in oleum gives the 3-bromo derivative, presumably *via* the conjugate acid, but only small yields of 4-bromopyridine 1-oxide have been obtained under less acidic conditions. The electrophilic bromination of substituted pyridine 1-oxides has been more widely studied than the corresponding chlorination. Strong electron-releasing groups on the pyridine 1-oxide favor nuclear bromination. In the case of 4-methoxypyridine 1-oxide, bromination occurred at C-3 along with ether hydrolysis. Bromination of 2-dimethylaminopyridine 1-oxide gave the 5-bromo-derivative as the major product (Equation (6)) ⟨83JOC1064⟩.

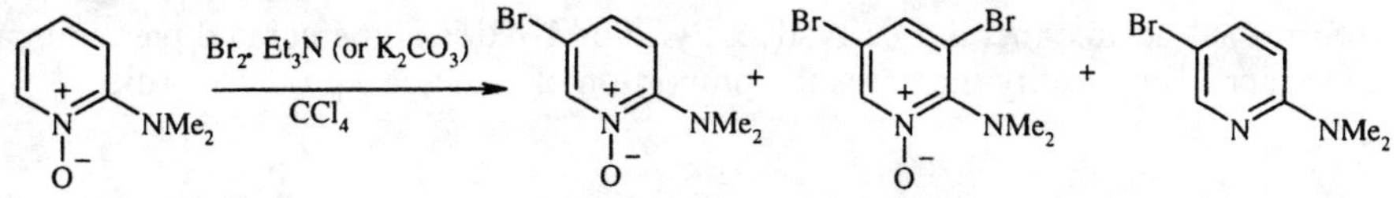

Electrophilic chlorination of pyridine 1-oxides is more difficult than for the analogous pyridines. An example is chlorination of 2-acetamidopyridine 1-oxide with hydrogen chloride and hydrogen peroxide, which gives a mixture of 5-chloro and 3,5-dichloro derivatives.

Quinoline 1-oxide gives a 4-bromo derivative (Br$_2$-H$_2$O, 100°C). Isoquinoline 2-oxide is brominated in the 4-position by the mechanism of Scheme 13.

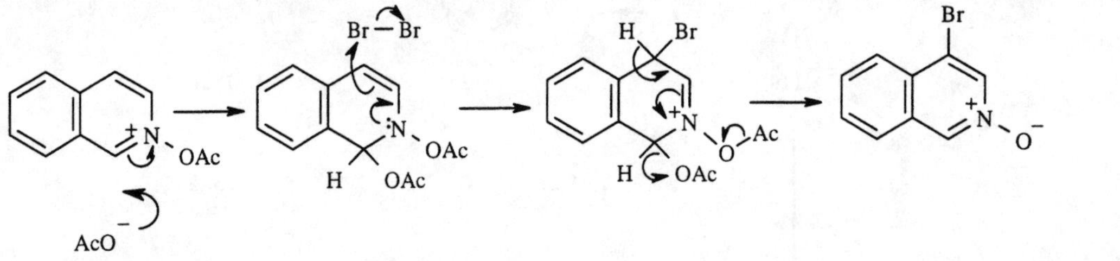

Scheme 13

Direct 5-fluorination can be effected on activated pyrimidines; for example, F$_2$ in HOAc or anhydrous HF have been used for the preparation of 5-fluoro-2(1H)-pyrimidinones ⟨77CCC2694⟩, and for 5,5-difluorination of 6-*O*-cyclouridines ⟨83TL1055⟩. Uracil and cytosine are 5-fluorinated by the same procedure ⟨80TL4605, 82CPB887⟩.

3.2.1.4.8 *Acylation and alkylation*

Friedel-Crafts reactions are almost unknown in pyridine and azine chemistry. Direct electrophilic alkylation in the pyrimidine 5-position can be carried out on pyrimidines with at least two strongly donating groups, and more readily with three such groups. Thus, α-haloketones and α-bromocarboxylic esters can be used for direct alkylation of 6-aminouracils (**118**), for example in the formation of (**119**). The 5-position can also act as the nucleophile for Michael additions (*e.g.* **118** → **120**, where a subsequent elimination occurs) ⟨92AHC(55)129⟩. For similar reactions in barbituric acids see ⟨85AHC(38)229⟩.

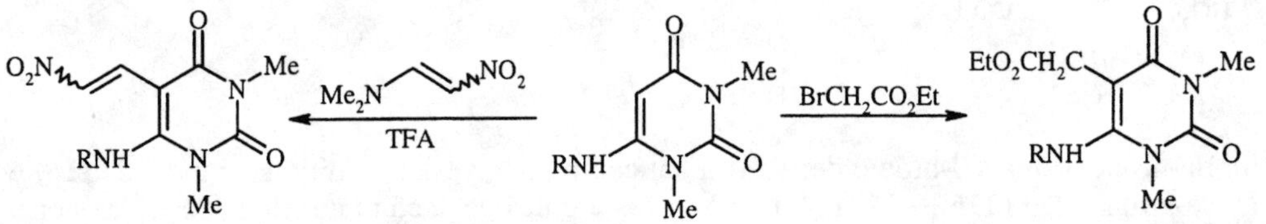

(**120**)   (**118**)   (**119**)

3.2.1.4.9 *Mercuration*

Pyridine forms 3-acetoxymercuripyridine (**121**) (Hg(OAc)$_2$, -H$_2$O, 155°C); 2-amino- and 4-methyl-pyridine give the 5- and 3-acetoxymercuri compounds, respectively, at somewhat lower temperatures. Mercuric acetate mercurates 4-pyridone in the 3-position and 3-hydroxypyridine in the 2-position.

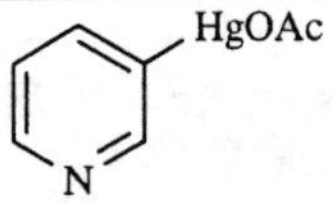

(**121**)

Isoquinoline forms a 4-acetoxymercuri derivative. Pyridine 1-oxide is mercurated predominantly in the 2-position, but increasing acidity increases the proportion of 3-mercuri product formed.

3.2.1.4.10 *Nitrosation, diazo coupling, Mannich reaction, Kolbe reaction and reaction with aldehydes*

In general, only those pyridines containing hydroxy or amino groups, diazines with multiple amino or hydroxy substitution, and pyridones will undergo these reactions. Some examples are now given.

(i) Nitrosation

2,6-Diaminopyridine with nitrous acid forms the 3-nitroso derivatives. Nitrosation of pyrimidines occurs readily in the presence of three electron-releasing substituents; pyrimidine-4,6-diamine is also nitrosated to the blue 5-nitroso derivative.

(ii) Diazo coupling and Mannich reaction

4-Quinolone (**122**), 2-pyridone (**123**), kojic acid (**124**) and 4-hydroxycoumarin (**125**) couple with diazonium salts (to form azo compounds, *e.g.* **126**) and undergo Mannich reactions (*e.g.* with $HCHO + HNMe_2$ to form $-CH_2NMe_2$ derivatives) at the positions indicated. Chromones undergo the Mannich reaction to give, for example, (**127**).

Mannich reactions and diazo coupling proceed readily in pyrimidines provided activating groups are present in each of the 2-, 4- and 6-positions.

(iii) Kolbe and Reimer-Tiemann reactions

3-Hydroxypyridine undergoes the Kolbe reaction (with carbon dioxide to give the carboxylic acid); the Na salt reacts mainly at the 2-, and the K salt at the 6-position. Uracil undergoes the Reimer-Tiemann reaction with sodium hydroxide/chloroform to give 5-formyluracil.

(iv) Reactions with aldehydes

3-Hydroxypyridine and formaldehyde give 2-hydroxymethyl-3-hydroxypyridine. Pyrones can be chloromethylated, *e.g.* (**128**) → (**129**). Hydroxymethylation of the 5-position is possible in pyrimidines with at least two electron-releasing substituents. Under Vilsmeier-Haack conditions pyrimidones are converted into 5-formylchloropyrimidines; for example (**130**) gives (**131**).

3.2.1.4.11   Oxidation

It is convenient to discuss oxidative attack at ring carbon here although this can involve radical as well as electrophilic oxidizing agents. Pyridine rings are generally very resistant to oxidation. CrO_3, dissolved in pyridine is used as a reagent to oxidize hydroxy groups, particularly in steroids. Many pyridine substituent groups can be selectively oxidized by $KMnO_4$ or $K_2Cr_2O_7$, especially under acidic conditions. In alkaline media, some oxidative degradation of pyridine rings occurs; thus, isoquinoline gives both cinchomeronic (**132**) and phthalic acids ($KMnO_4$, $NaOH-H_2O$) (Section 3.2.3.2.1). Ozone reacts with pyridines, although less readily than with benzenes; products corresponding to both Kekule forms can be isolated, *e.g.* (134) → (133) + (135).

(132)   (133)   (134)   (135)

Acridine is less stable toward oxidizing agents and yields acridone (with $Na_2Cr_2O_7$-HOAc). Some diazine derivatives are apparently oxidized directly to diazones: quinoxaline gives (136) (with $K_2S_2O_8$-H_2O) and quinazolin-4-one yields (137) (with $KMnO_4$, CrO_3). These reactions probably involve nucleophilic attack of water followed by oxidation of the adduct (see Section 3.2.1.6.3).

(136)   (137)

Highly activated rings are hydroxylated by $K_2S_2O_8$-$FeSO_4$: 2-pyridone and 3-hydroxypyridine are both hydroxylated *para* to the substituent; thus, each gives the same compound (5-hydroxy-2-pyridone). 2-Pyrimidinone affords the 5-hydroxy derivative. Addition of hydrogen peroxide to the 5,6-double bond of pyrimidines gives 5,6-dihydroxy adducts. When 1,3-dimethyluracil was oxidized by dimethyldioxirane, which allows mild reaction conditions, a mixture of 1,3-dimethyl-5,6-epoxy-5,6-dihydrouracil (10%) (138) and 5,6-dihydroxy-5,6-dihydrouracils (139) as the *cis* (50%) and the *trans* (25%) isomers was formed. Under conditions where most of the water was removed from the reagent, the oxirane (138) was obtained in 50% yield, and in water only the diols were obtained. 5- and 6-Methyl-*N*-alkyluracils gave higher yields of the epoxide ⟨93TL6313⟩.

(138)   (139)

3.2.1.5   Attack at Ring Sulfur Atoms

Reactions of this type are rare. Heterocyclic sulfur is usually electron-deficient, which hinders the normal attack of electrophiles, and the normal position for nucleophilic attack is at a ring carbon or hydrogen. However, examples of both types of reactions are known.

3.2.1.5.1   *Reactions with electrophiles*

Reactions take place in two circumstances:
(i)   in formally 8π-electron compounds (see Section 3.2.2.1.2.iii);
(ii) if the S atom is already in a higher oxidation state (see Section 3.2.2.2).

3.2.1.5.2 Reactions with nucleophiles

Thiinium cations, *e.g.* (**140**), react with lithium aryls to give complex mixtures containing some 1,2,4,6-tetrasubstituted thiabenzenes (**141**) (see CHEC 2.02.5.4.5).

(**140**)   (**141**)   + Other products

3.2.1.6 Nucleophilic Attack at Carbon

Before discussing nucleophilic attack specifically at ring carbon we enumerate five general pathways which have been distinguished for the attack of nucleophiles on heteroaromatic six-membered rings:

Path A: Nucleophilic attack at a hydrogen atom of a substituent with subsequent elimination (discussed under the relevant substituent in Section 3.2.3.1).

Path B: Attack at α- or γ-ring carbon, with subsequent reaction not involving ring opening (discussed in this section).

Path C: Nucleophilic attack at a substituent atom other than hydrogen (discussed under the relevant substituent in Section 3.2.3.1).

Path D: Attack at α-ring carbon followed by ring opening (discussed in this section).

Path E: Ylide formation by removal of a ring hydrogen atom followed by *(i)* addition of an electrophile (discussed in Section 3.2.1.8), or *(ii)* dimerization.

3.2.1.6.1 Ease of reaction

(i) Pyridines

The electron displacement toward the nitrogen atom allows nucleophilic reagents to attack pyridines at the α-position (**142**), a type of reactivity shown in benzenoid chemistry only by derivatives with electron-withdrawing substituents. However, formation of the initial adduct (**143**) in an appreciable amount is difficult because this involves de-aromatization of the pyridine ring and, once formed, the adduct tends to re-aromatize by dissociation (**143 → 142**). Only very strong nucleophilic reagents (*e.g.* NH_2^-, LiR, $LiAlH_4$, Na/NH_3, and, at high temperatures, OH^-) react. The tiny proportion of adducts of type (**143**) formed by the addition of amide or hydroxide ions can also re-aromatize by hydride ion loss, thus gradually causing complete reaction (**143 → 144**). The adducts formed by the addition of hydride ions (from $LiAlH_4$) or carbanions (from LiR) are more stable; at low temperatures they are converted to dihydropyridines (**145**) by proton addition, but at higher temperatures re-aromatization occurs by hydride ion loss.

(**142**)   (**143**)   (**144**)   (**145**)

(ii) Azines

Diazines are considerably more reactive toward nucleophiles than pyridines and as the number of ring nitrogens increases the propensity for nucleophilic addition reactions increases still more. Many 1,2,4-triazines give addition products with various nucleophiles which are formally dihydrotriazines.

The limit is reached with 1,3,5-triazine. This reacts very easily even with weak nucleophiles, and ring cleavage nearly always follows. Thus, it behaves as a formylating agent toward amines and other active hydrogen compounds.

Pteridines also undergo nucleophilic addition reactions particularly easily, including covalent hydration, addition of bisulfite and others.

(iii) Cationic rings

In a pyridinium ring the positive charge facilitates attack by nucleophilic reagents at positions α or γ to the heteroatom. Hydroxide, alkoxide, sulfide, cyanide and borohydride ions, certain carbanions and amines react, usually at the α-position, under mild conditions, to give initial adducts of type (**147**). These non-aromatic adducts can be isolated in certain cases but undergo further reactions with alacrity. The most important of these include variations (a)-(d) listed below. If the group Z^+ of (**146**) is a nitrogen with a leaving group, usually N^+OX, then the further possibilities (e)-(f) exist.

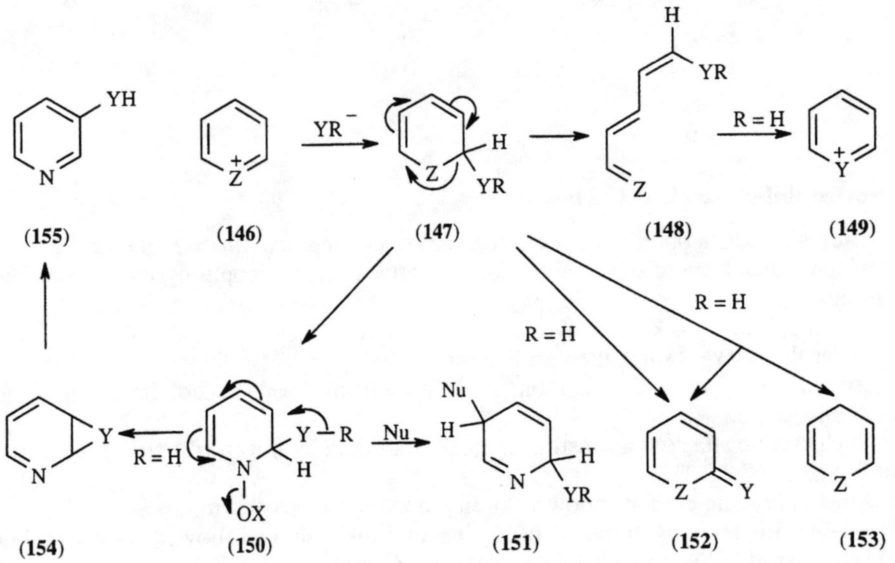

(a) Oxidation: *e.g.* (**147**; YR = OH) → pyridones (**152**); (**147**; YR = CH$_2$ — Heterocycle) → cyanine dyes.

(b) Disproportionation: *e.g.* (**147**; YR = OH) → pyridone (**152**) and dihydropyridine (**153**).

(c) Ring opening with subsequent closure (**146** → **149**): *e.g.* reaction of pyrylium salts with RNH$_2$ or S^{2-}.

(d) Ring opening without subsequent closure (**146** → **148**): *e.g.* reactions with OH$^-$ salts carrying electron-withdrawing groups on nitrogen and pyrylium salts.

(e) Rearrangement of attacking group to the 3-position (**150** → **154** → **155**).

(f) Addition of a second equivalent of nucleophile (**150** → **151**).

Pyrylium and thiinium salts are very easily attacked by nucleophiles, particularly if there is an unsubstituted α- or γ-position, and 1,3,5-oxadiazinium and -thiadiazinium salts are still more susceptible (see Scheme 14).

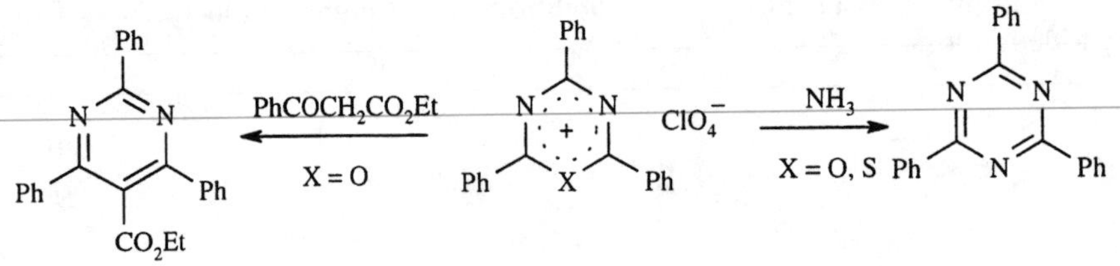

Scheme 14

(iv) Pyridones and azinones

(a) In both α- and γ-pyridones the carbon atom of the carbonyl group can be attacked by a powerful nucleophile (as in **156**). The reaction then proceeds by complete loss of the carbonyl oxygen atom and aromatization. These reactions, which also occur in α- and γ-pyrones, are all considered as substituent reactions in Section 3.2.3.7.2.

(b) Adducts (**156**) formed by reaction of α-pyrones at the carbonyl carbon atom can react further by ring opening as in the reactions with hydroxide ion, ammonia and amines (see Sections 3.2.1.6.3 and 3.2.1.6.4).

(156)

(v) N-Oxides

The intrinsic reactivity of pyridine 1-oxides toward nucleophiles is little greater than that of pyridine: the strongest nucleophiles react. However, after initial reaction with an electrophile at the *N*-oxide oxygen, subsequent attack by nucleophiles is easy: see above discussion under 'cationic rings'.

3.2.1.6.2 Effect of substituents

Nucleophilic attack on the ring carbon atoms of pyridines should be, and is, facilitated by electron-attracting substituents and hindered by electron-donating substituents. In pyridinium salts the effect of strongly electron-withdrawing substituents attached to the nitrogen atom, *e.g.* -C_6H_3(NO_2)_2 or -CN, is particularly marked and facilitates ring opening (Section 3.2.1.6) which is otherwise unusual.

Fused benzene rings aid nucleophilic attack on pyridines, pyridinium and pyrylium ions, and pyrones; the loss of aromaticity involved in the formation of the initial adduct is less in monobenzo derivatives and still less in linear dibenzo derivatives than in monocyclic compounds. For the same reason, the tendency for this initial adduct to re-aromatize is less for benzopyridines. Fused benzene rings also influence the point of attack by nucleophilic reagents; attack rarely occurs on a carbon atom shared with a benzene ring. Thus, in linear dibenzo derivatives, nucleophilic attack is at the γ-position (**157**).

Similarly, reclosure to a new heterocyclic system after ring opening is possible in benzo[*c*] derivatives (**158**); in these last compounds initial attack is always in the α-position adjacent to the benzene ring because of partial double bond fixation. By contrast, ring opening of an αβ-benzo derivative gives a phenol or aniline (**159**) in which the Z—C bond is not easily broken.

(157)   (158)   (159)

However, in the quinolizinium ion (**160**) the fused ring increases the stability and makes it more difficult for nucleophilic attack, because now the aromaticity of both rings is lost in the intermediate addition product (**161**). Conversely, once the addition product is formed, ring opening is particularly easy (**162**).

(160)   (161)   (162)

3.2.1.6.3 Hydroxide ion

(i) Pyridine and its benzoderivatives

Uncharged pyridines are resistant to hydroxide ion at usual temperatures. Pyridine itself reacts with hydroxide ions under extreme conditions (KOH-air, 300°C) to give 2-pyridone, the stable tautomer of 2-hydroxypyridine, which is formed by oxidation of the initial adduct. As is expected, this reaction is facilitated by electron-withdrawing groups and fused benzene rings; quinoline and isoquinoline form

2-quinolone and 1-isoquinolone, respectively, rather more readily. However, γ-hydroxylation is more difficult. Thus, acridine is hydroxylated with KOH at 300°C to give 9-acridone in 28% yield ⟨72KGS1673⟩.

(ii) Other azines

1-*R*-Perimidines are hydroxylated and oxidized (KOH, 180°C) to form 1-*R*-perimidones (**163**) in excellent yield ⟨81RCR1559⟩.

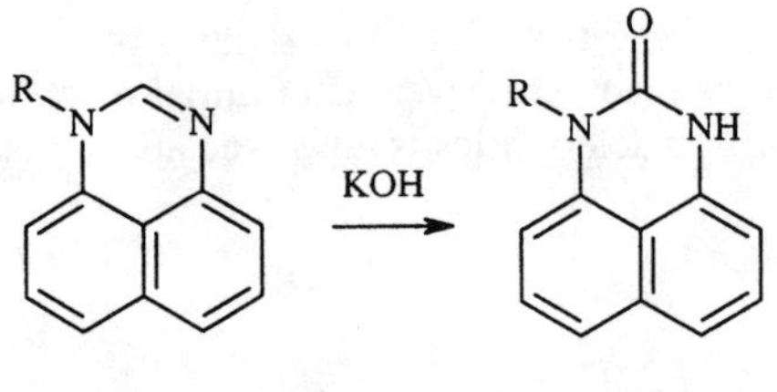

(163)

Increasing numbers of nitrogen atoms increase not only the kinetic susceptibility toward attack but also the thermodynamic stability of the adducts. Reversible covalent hydration of C = N bonds has been observed in a number of heterocyclic compounds ⟨76AHC(20)117⟩. Pyrimidines with electron-withdrawing groups and most quinazolines show this phenomenon of 'covalent hydration'. Thus, in aqueous solution the cation of 5-nitropyrimidine exists as (**164**) and quinazoline cation largely as (**165**). These cations possess amidinium cation resonance. The neutral pteridine molecule is covalently hydrated in aqueous solution. Solvent isotope effects on the equilibria of mono- (**166**) and dihydration (**167**) of neutral pteridine as followed by NMR are near unity ⟨83JOC2280⟩. The cation of 1,4,5,8-tetraazanaphthalene exists as a bis-covalent hydrate (**168**).

The following factors help stabilize covalent hydrates ⟨65AHC(4)33⟩:

(1) amidine-type resonance, particularly in cationic (*e.g.* **169**) but also in neutral species (*e.g.* **170**);
(2) guanidinium-type resonance (**171**);
(3) urea-type resonance (*e.g.* **172**);
(4) 4-aminopyridinium-type resonance (**173**).

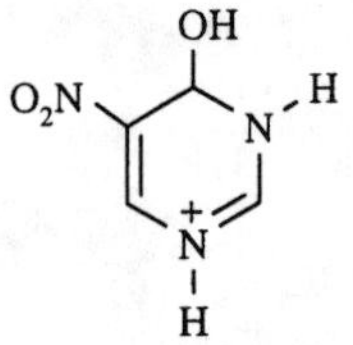

| (164) | (165) | (166) | (167) |

| (168) | (169) | (170) |

| (171) | (172) | (173) |

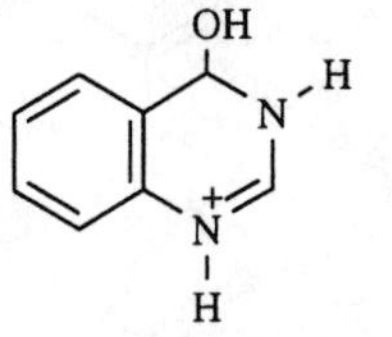

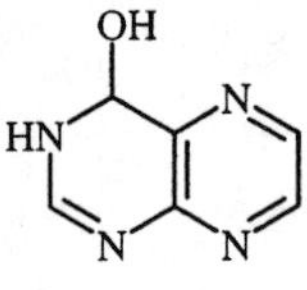

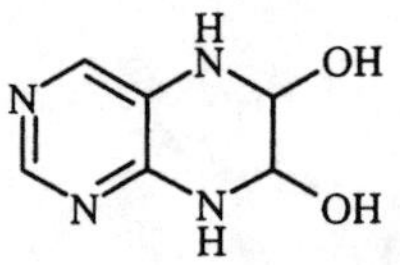

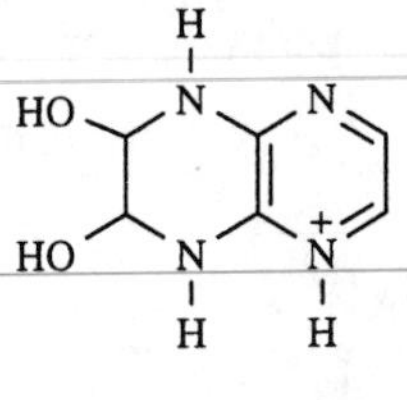

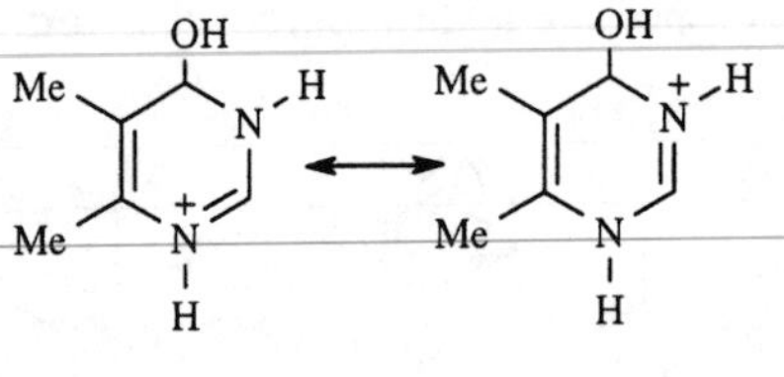

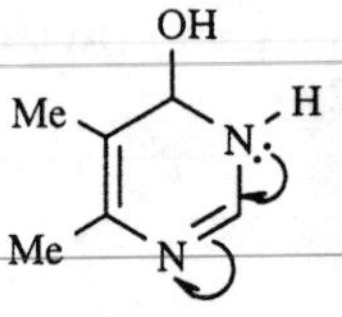

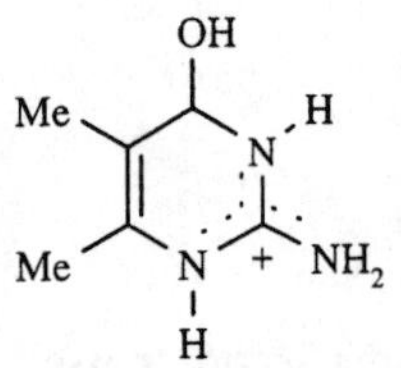

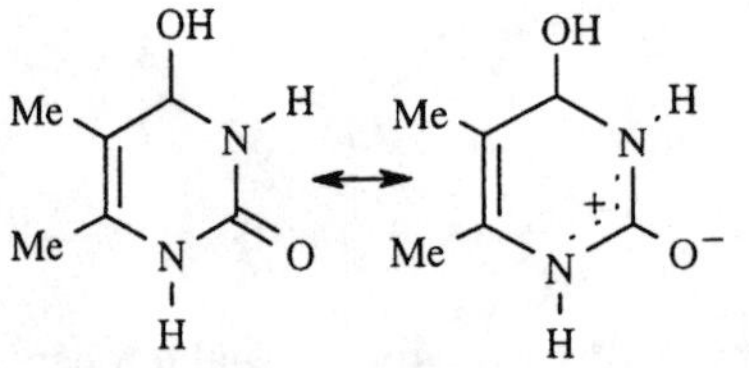

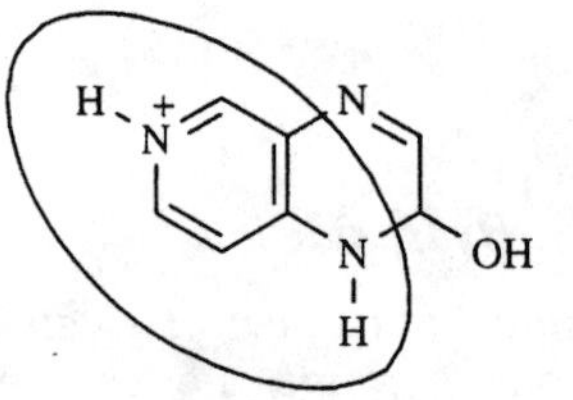

On irradiation uracil and related pyrimidines undergo photohydration across the 5,6-bond.

Many 1,2,4-triazine derivatives undergo photochemical hydration reactions, *e.g.* (174) → (175). Such reactions with 1,2,4-triazine 2-oxides are followed by loss of hydroxide ion to give overall substitution (176) → (177).

Insertion of a methyl group at the site where nucleophilic attack occurs during hydration considerably hinders the reaction and lowers the percentage of covalently hydrated species at equilibrium. Covalent hydrates are converted by mild oxidation into oxo compounds.

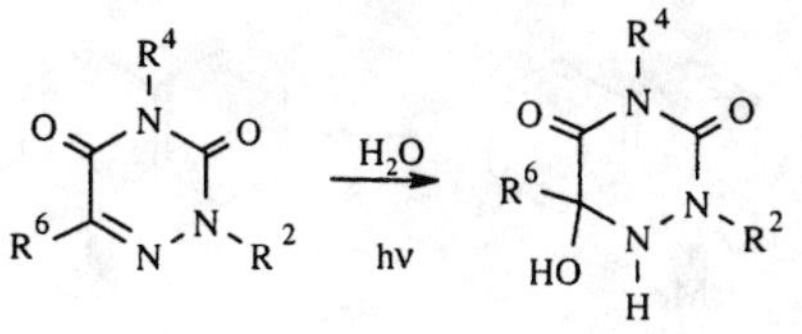

(174)   (175)

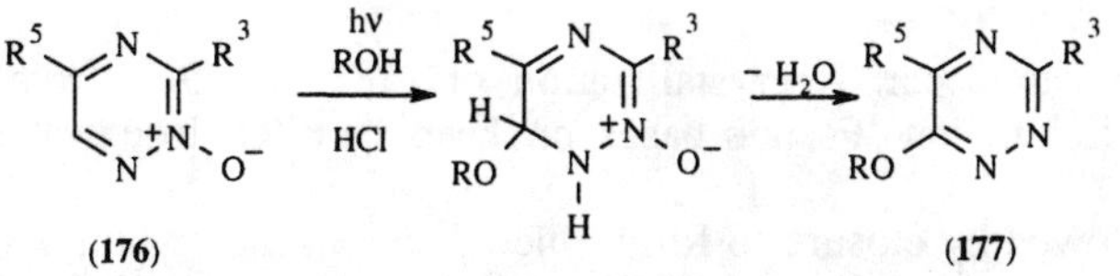

(176)   (177)

Covalent hydrates can undergo ring opening especially in acidic media, for example the triazanaphthalene (178 → 179).

Polyaza rings suffer complete hydrolytic ring cleavage. Monocyclic 1,2,3-triazines are hydrolyzed by acid to yield 1,3-dicarbonyl compounds (Scheme 15). 1,2,3-Benzotriazines are easily converted into derivatives of 2-aminobenzaldehyde. 1,2,4,5-Tetrazines (180) are hydrolyzed with a base to give aldehyde hydrazones, RCH=NNHCOR.

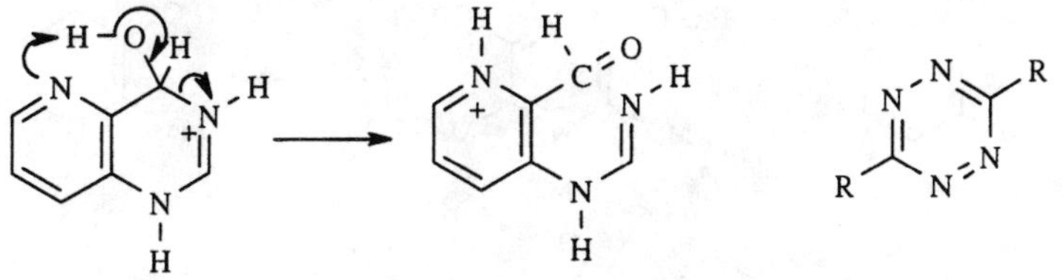

(178)   (179)   (180)

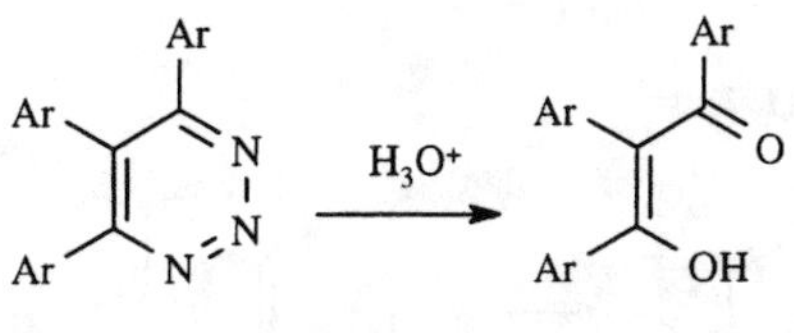

Scheme 15

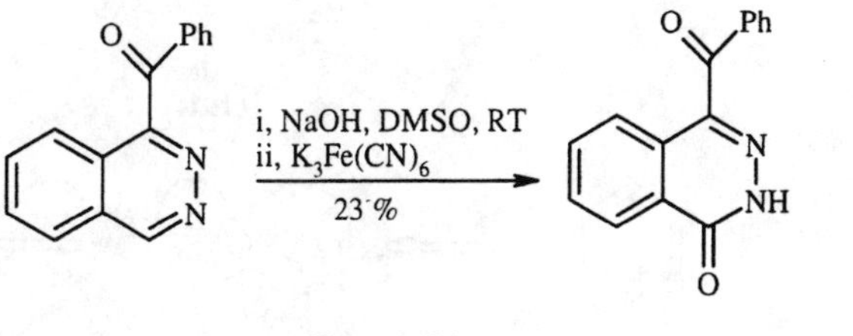

Scheme 16

1-Benzoylphthalazine with sodium hydroxide in dimethyl sulfoxide hydroxylates the phthalazine ring to give, after oxidation, 4-benzoyl-1(2H)phthalazinone (Scheme 16) ⟨85CPB4193⟩.

(iii) Alkylpyridinium cations

1-Methylpyridinium ions (181) react reversibly with hydroxide to form a small proportion of the pseudo-base (182). The term 'pseudo' is used to designate bases that react with acids measurably slowly, not instantaneously as for normal acid–base reactions. Fused benzene rings reduce the loss of resonance energy when the hetero ring loses its aromaticity and hence pseudo-bases are formed somewhat more readily by 1-methylquinolinium, 2-methylisoquinolinium and 10-methylphenanthridinium, and much more readily by 10-methylacridinium ions. Pseudo-bases carrying the hydroxy group in the α-position are usually formed preferentially, but acridinium ions react at the γ-position.

Pseudo-bases can undergo a number of further reactions:

(a) Oxidation. 1-Alkylpyridinium ions in alkaline solution are oxidized by $K_3Fe(CN)_6$, to give 2-pyridones (*e.g.* 183). 2-Quinolones, 1-isoquinolones, 9-phenanthridones and 9-acridones can be prepared similarly.

(b) Many pseudo-bases disproportionate on standing to dihydropyridines and pyridones, *e.g* (185) → (186) + (187). The mechanism shown, which resembles that for the Cannizzaro reaction, has been supported by kinetic studies ⟨78JOC3662⟩.

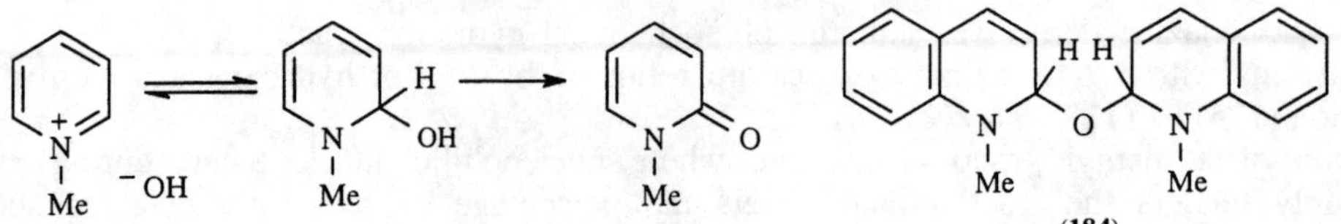

(181)　　(182)　　(183)　　(184)

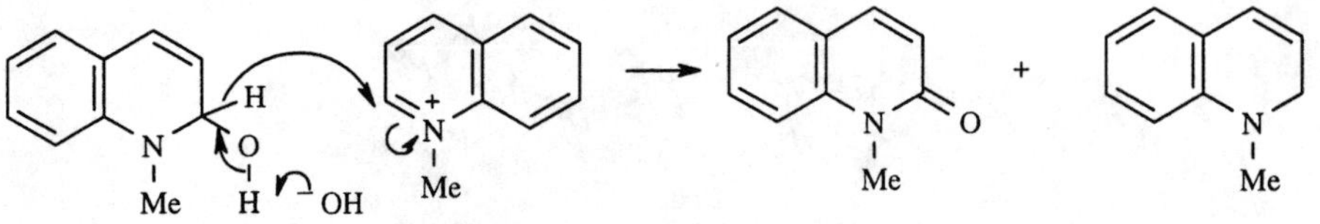

(185)　　　　　　(186)　　　　　　(187)

(c) Ether formation can occur. Recrystallization of (188; Y = OH) from ethanol forms (188; Y = OEt) *via* the acridinium ion. Pseudo-bases on keeping often lose water to give bimolecular products (*e.g.* 184).

(d) Ring fission followed by closure to form a new heterocyclic or homocyclic ring can occur in pyridinium ions carrying suitable substituents. Examples are (189) + KOH + $H_2O \rightarrow$ (190) + EtOH, and also, under vigorous conditions (200°C), (191) + NaOH → 10% of (192).

(e) Ring fission can be reversible as in for example, 1-methoxypyridinium (193) giving the glutaconic aldehyde derivative (194); this is followed by irreversible scission of the N-O bond (see Section 3.2.3.12.5).

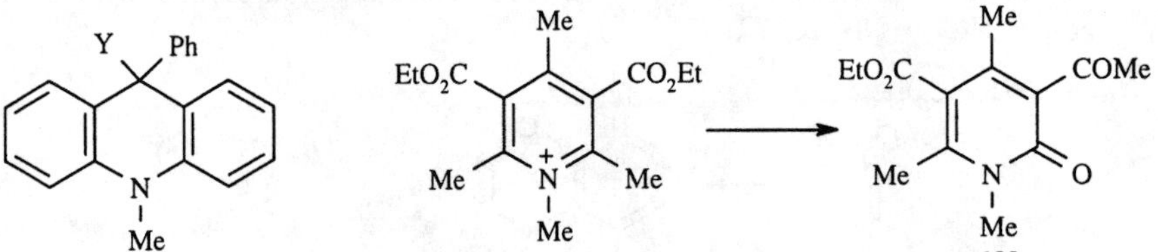

(188)　　　(189)　　　(190)

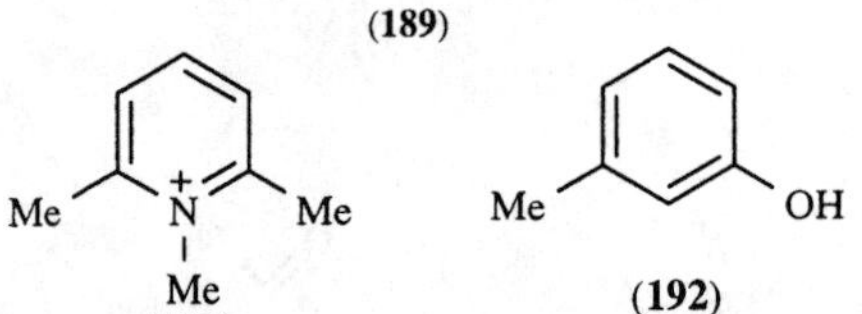

(191)　　　　　(192)

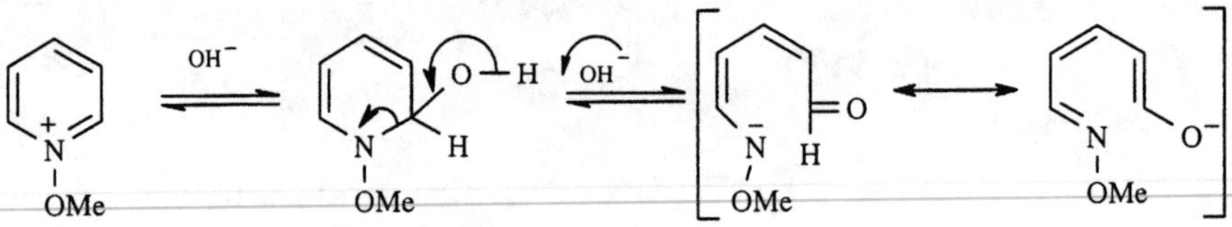

(193)　　　　　　　　　　　　(194)

(iv) Other pyridinium ions

Pseudo-bases derived from pyridinium ions carrying a strongly electron-withdrawing substituent on the nitrogen atom are unstable and undergo ring fission by hydroxide ions under mild conditions (NaOH-H_2O, 20°C). Pyridine-sulfur trioxide (195) and 1-cyano- and 1-(4-pyridyl)-pyridinium ions all give glutaconic aldehyde (196 → 197); the other products are sulfamic acid, cyanamide (→ $NH_3 + CO_2$) and 4-aminopyridine, respectively. Similarly, isoquinoline-sulfur trioxide gives homophthalaldehyde (198).

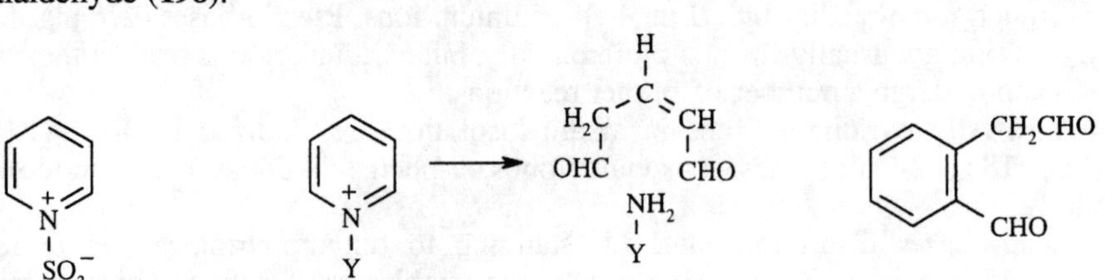

(195)　　　(196)　　　(197)　　　(198)

(v) Other cationic rings

Diazinium salts resemble pyridinium salts in their behavior. They form pseudo-bases with hydroxide ions which can disproportionate (*e.g.* 2-methylphthalazinium ion (**199**) → 2-methylphthalaz-1-one + 2-methyl-1,2-dihydrophthalazine) or undergo ring fission (*e.g.* 3-methylquinazolinium ion → (**200**). Aqueous acid converts (**201**) into (**202**), presumably by attack of a water molecule on a protonated species with subsequent intramolecular oxidative-reductive rearrangement of an intermediate carbinol base (**201a**) as shown.

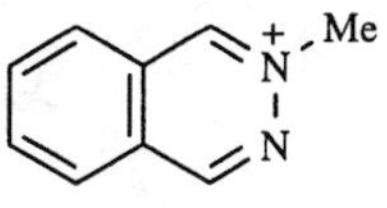

(**199**) 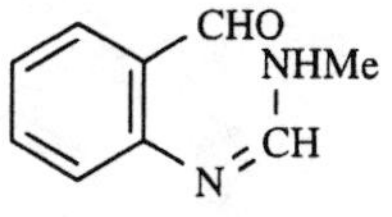   (**200**)

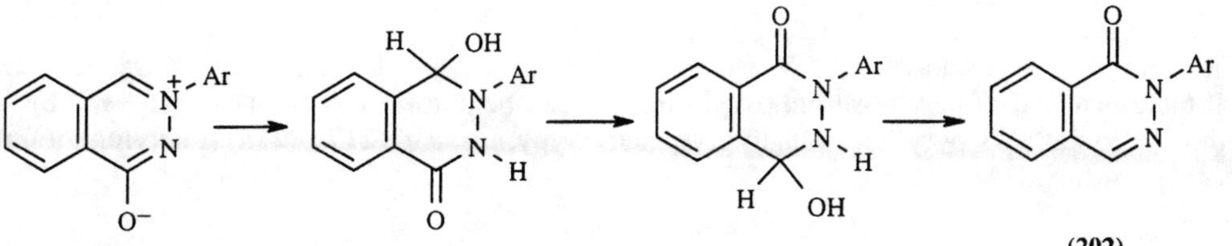

(**201**)   (**201a**)   (**202**)

Pseudo-bases derived from 2-unsubstituted 1,3-dialkylperimidinium salts disproportionate into 1,3-dialkylperimidones and 1,3-dialkyl-2,3-dihydroperimidines, the former always predominate. By contrast, 1-aroyl-3-alkylperimidinium salts (**203**), formed *in situ* from 1-*R*-perimidines and aroyl chlorides, produce pseudobases (**204**) which exist exclusively in the acyclic form (**205**). Interestingly, heating with alkali cleanly converted (**205**) into the corresponding 1,2-disubstituted perimidines (**206**). This reaction is of preparative significance for the introduction of aryl and heteroaryl groups in position 2 of perimidines ⟨81RCR1559⟩.

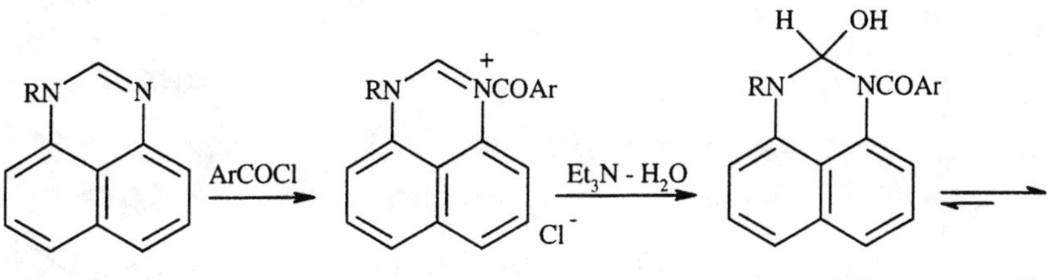

(**203**)   (**204**)

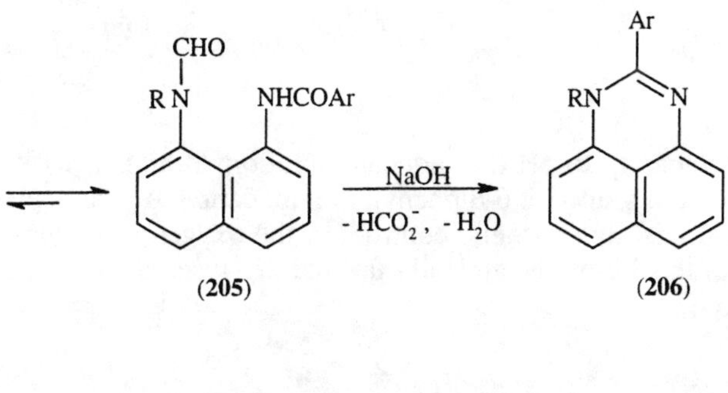

(**205**)   (**206**)

Pyrylium salts react with hydroxide ions in a complex series of equilibria involving the pseudo-base (**207**) and ring-opened forms (**208**) and (**209**).

Some pseudo-bases do not ring open; the xanthylium ion (**210**) gives xanthydrol (**211**) which can be isolated or oxidized with dilute nitric acid to xanthone (**212**).

Hydrogen peroxide reacts with 2,4,6-trisubstituted pyrylium salts to cause ring contraction (**213** → **214**). If there is a free α- or γ-position, pyrylium salts can be oxidized to pyrones (**215** → **216**).

Methoxide ion attacks pyrylium salts to give methoxypyrans, *e.g.* (**217**), (**218**). Flavylium ion (**220**) gives (**221**) (with NaOAc-EtOH).

Unsubstituted thiinium cation is stable up to pH 6 in aqueous solution; at higher pH it ring opens to the aldehyde (**219**). Methoxide adds to 4-alkyl-2,6-diphenylthiinium cation to give a mixture of the 4-methoxy-4*H*- and 2-methoxy-2*H*-thiins under kinetic control (*cf.* the oxygen analogues **217**, **218**). The mixture then equilibrates toward the thermodynamically favored 2*H*-system.

(vi) Pyridones, pyrones, azinones, etc.

Although pyridones are usually resistant to alkali, pyrone rings are often easily opened. Pyran-2-ones are reversibly ring-opened by aqueous alkali to acid anions (**222**). Hydroxide ions convert coumarins (**223**) reversibly into salts of coumarinic acids (**224**) which can be converted into the *trans* isomers (**225**), and chromones (**226**) into β-dicarbonyl compounds (**227**).

3-Bromo-2-pyrones and 3-bromocoumarins give furan- and benzofuran-2-carboxylic acids by ring fission and subsequent closure, *e.g.* (228) → (229); (116) or (117) → (230). Pyrone rings are opened by aqueous acid in some cases, probably by successive protonation and attack of a water molecule, *e.g.* dehydroacetic acid (231) gives (233) which immediately forms (232) or (234) with HCl or H_2SO_4, respectively.

Oxazinone rings, *e.g.* (235) and (236), and oxazinedione rings, *e.g.* isatoic anhydride (238), are rapidly cleaved by alkali as expected. Thus, (236) yields (237).

(vii) N-*Oxides*

N-Oxides are normally resistant to hydroxide attack. However, acetic anhydride converts the N-oxide group into N-acetoxy cations, and such compounds can be attacked by acetate anions. Thus, the reaction of acetic anhydride with pyridine N-oxide gives 2-pyridone (Scheme 17). This is a very general reaction; thus, for example, pyridazine 1-oxides unsubstituted at position 6 rearrange in the presence of acetic anhydride to the pyridazin-6-ones.

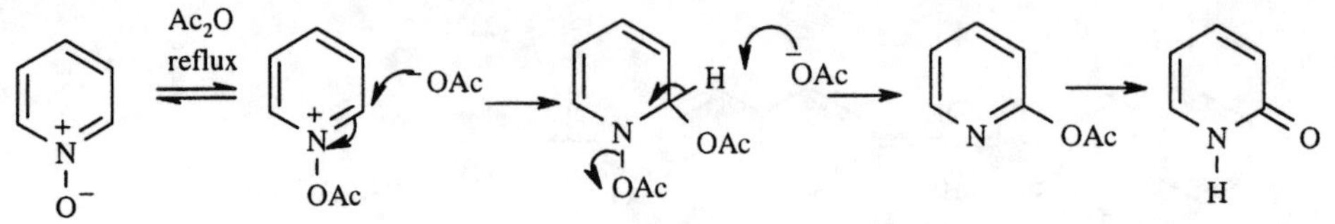

Scheme 17

3.2.1.6.4 *Amines and amide ions*

(i) *Pyridines and azines*

Amines are insufficiently nucleophilic to react with pyridines, and the stronger nucleophile NH_2^- is required. Pyridine reacts with sodium amide (toluene, 110°C) giving 2-aminopyridine (75%) and a small amount of 4-aminopyridine. At 180°C 2,6-diaminopyridine is produced in good yield along with a small amount of 2,4,6-triaminopyridine. This, the Chichibabin reaction, is widely used for direct introduction of an amino group into electron-deficient positions of many azines and some azoles (Section 3.4.1.7.2) ⟨78RCR1042, 83AHC95, 87KGS1011, 88AHC2⟩ Quinoline is aminated at atoms C-2 and C-4, isoquinoline at C-1, acridine at C-9, phenanthridine at C-6, quinazoline at C-4, 1-*R*-perimidines at C-2.

There are two general procedures for conducting the Chichibabin reaction. According to the classical procedure, the reaction is conducted at realtively high temperatures, in a solvent inert towards sodium amide (arenes, *N*,*N*-dialkylanilines, mineral oil, *etc.*) or without any solvent. In this case, the reaction proceeds under heterogeneous conditions on the surface of the sodium amide particles. Under such conditions, heterocycles with pKa values of 5–6 are aminated smoothly, but for heterocycles of lower basicity, the ease of amination is markedly decreased. Dependence of the reaction on basicity suggests that formation of an adsorption complex of type (**239**) with a weak coordination bond between the cyclic nitrogen and a sodium ion may be important. Obviously such coordination increases positive charge on the ring α-carbon atom and thus favors the amination.

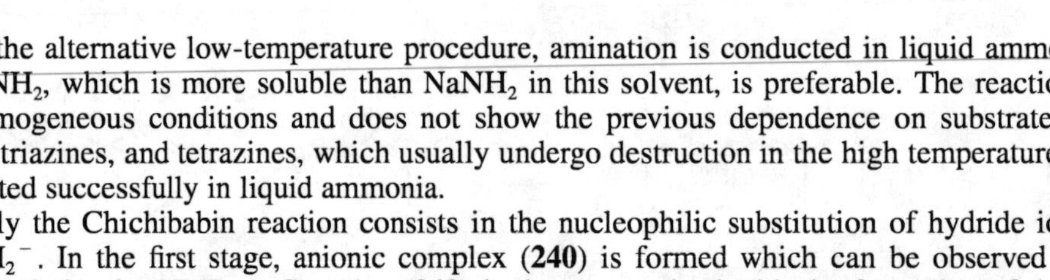

(239)   (240)   (241)

Under the alternative low-temperature procedure, amination is conducted in liquid ammonia. The use of KNH_2, which is more soluble than $NaNH_2$ in this solvent, is preferable. The reaction occurs under homogeneous conditions and does not show the previous dependence on substrate basicity. Diazines, triazines, and tetrazines, which usually undergo destruction in the high temperature process, are aminated successfully in liquid ammonia.

Formally the Chichibabin reaction consists in the nucleophilic substitution of hydride ion by the group NH_2^-. In the first stage, anionic complex (**240**) is formed which can be observed in liquid ammonia solution by NMR. σ-Complex (**240**) is then aromatized with the formation of the sodium derivative of amine (**241**). From the latter, the free amine is obtained by the addition of water or NH_4Cl. In the high-temperature method, hydride ion is eliminated as hydrogen gas (the second hydrogen atom in H_2 comes as a proton from an amino group). It is convenient to follow the reaction by the volume of hydrogen gas evolved. There is indirect evidence that under heterogeneous conditions one-electron transfer from the NH_2^- nucleophile to substrate may play a significant role. Thus, the formation of dimers or hydrogenated dimers of the starting heterocycle is often observed. For example, the yield of 4,4'-dipyridyl in the sodamide amination of pyridine in mixtures of xylene-hexamethyl phosphoro-triamide reaches 20%.

In liquid ammonia, hydride ion, which is a very poor leaving group, cannot eliminate spontaneously. An oxidant (KNO$_3$ or better KMnO$_4$) is usually added to aromatize the σ-complex ⟨B-94MI 502-06⟩.

In the amination of quinoline in liquid ammonia, the initially formed σ-complex (**242**) on heating is fully isomerized to the more stable σ-complex (**243**). This transformation is used in preparative work. Adding potassium permanganate to the σ-complex formed under kinetic or under thermodynamic control allows either 2-amino- or 4-amino-quinoline to be obtained in good yield.

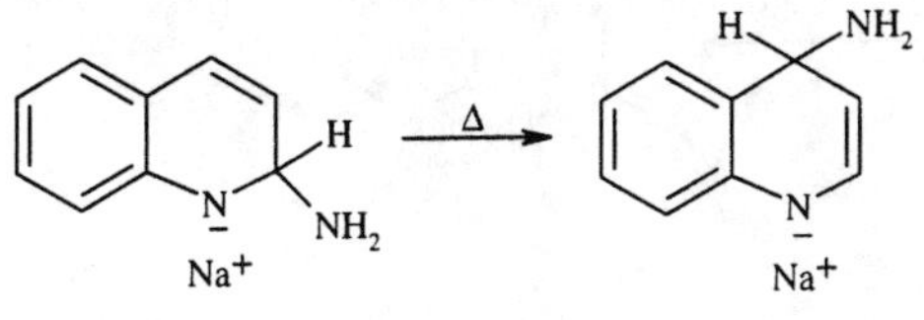

(**242**)   (**243**)

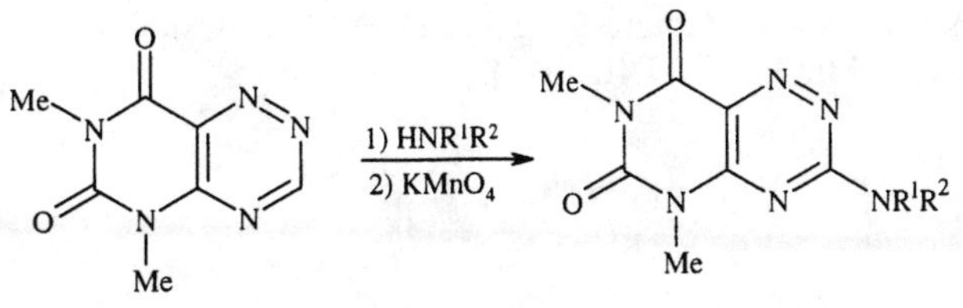

(**244**)   (**245**)

Of all the aza-heterocycles, pyridine possesses the least electron deficiency. Because of this, pyridine itself does not form a σ-complex in liquid ammonia and cannot be aminated under these conditions. By the contrast, highly π-deficient polyaza-heterocycles (diazines, triazines, tetrazines, pteridines, *etc.*) undergo oxidative amination, sometimes even by liquid ammonia itself. Sodamide converts 4-methylpyrimidine successively into the 2-mono- and 2,6-di-amino derivatives, and pyrazine gives 2-aminopyrazine. Nitro groups, the activating ability of which is even stronger than that of an aza-group, facilitate the amination. For instance, 4-nitroquinoline on treatment with liquid ammonia and KMnO$_4$ at $-33°C$ gives 3-amino-4-nitroquinoline in 86% yield.

Oxidative alkylamination of strongly π-deficient azines by neutral alkylamines is also possible. Thus, 5,7-dimethylpyrimido[4,5-*e*]-1,2,4-triazine-6,8-dione (**244**) can be converted in this way into its 3-alkylamino derivatives (**245**) with excellent yield.

Pteridine adds ammonia at low temperature to form 4-amino-3,4-dihydropteridine (**246**) which is transformed in a slower reaction into 6,7-diamino-5,6,7,8-tetrahydropteridine (**247**) (*cf.* similar adducts with water, **166** and **167**).

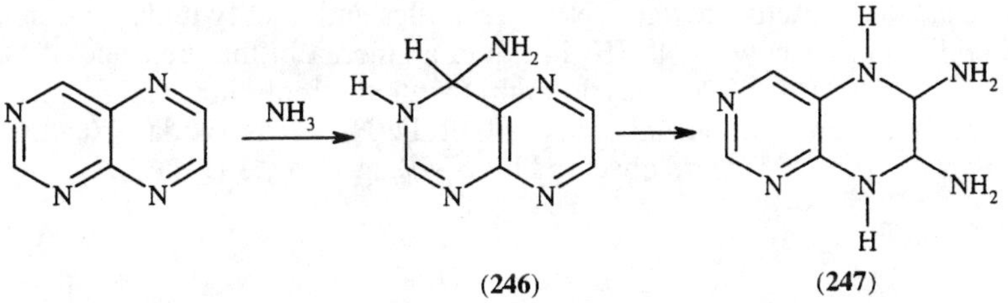

(**246**)   (**247**)

Alkylsodamides (AlkNH$^-$ Na$^+$) (but not dialkylsodamides), under heterogeneous conditions convert pyridine into its 2-alkylamino derivatives. An intramolecular version of the reaction has also been described (Scheme 18).

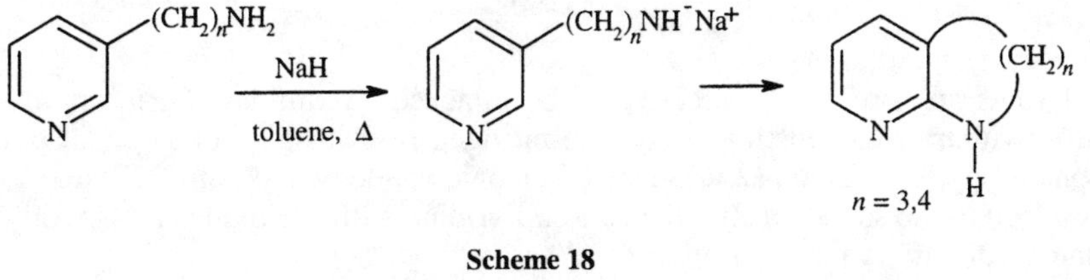

$n = 3,4$

Scheme 18

During the amination or alkylamination of halogeno azines, aminodehalogenation and aminodehydrogenation reactions often compete. Thus, 6-chloro-1,3-dimethyllumazine (**248**) reacts with 1,2-diaminoethane in the presence of $[Ag(C_5H_5N)_2]MnO_4$ (this oxidant is more soluble in alkylamines than $KMnO_4$) producing compounds (**249**) (30%) and (**250**) (38%) ⟨92KGS1202⟩.

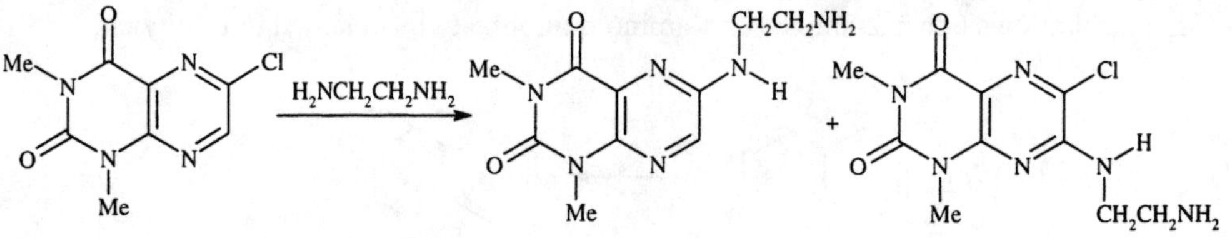

(248)   (249)   (250)

Some azine N-oxides can also be aminated, *e.g.* (**251**) → (**252**). Usually the reaction proceeds better in presence of an oxidant ($KMnO_4$, *i*-PrONO, *etc.*).

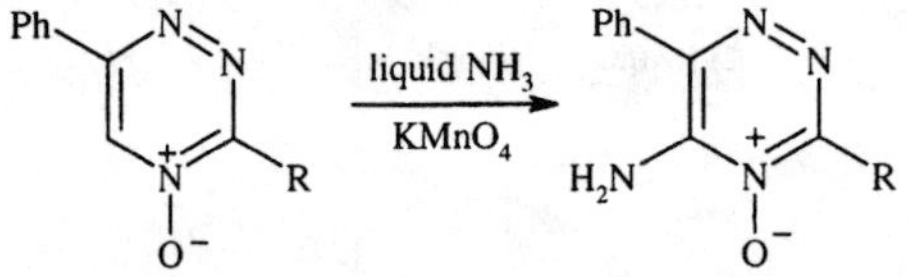

(251)   (252)

R = H, SMe, OMe, Ph

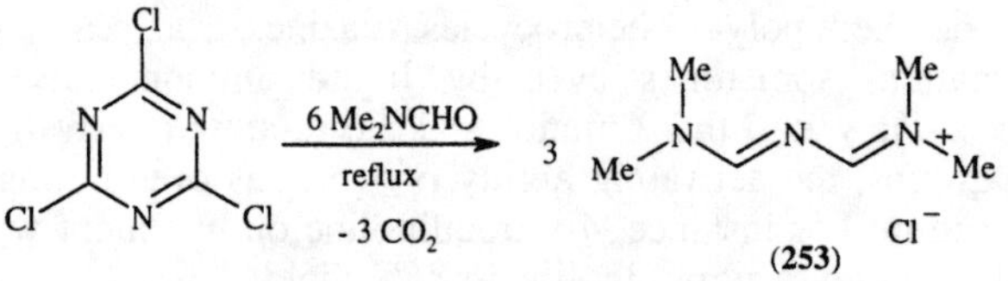

(253)

Nitrogen nucleophiles can ring-open azine rings in some conditions. Thus, cyanuryl chloride is ring cleaved with DMF to form [3-(dimethylamino)-2-azaprop-2-en-1-ylidene]dimethyl ammonium chloride (Gold's reagent) (**253**), a general β-dimethylaminomethylenating agent, in 95% yield ⟨86OS(64)85⟩.

(ii) Pyridinium ions

Azinium cations are very reactive towards N-nucleophiles and readily undergo oxidative imination. Thus, *N*-methylacridinium reacts with $NaNH_2$ in liquid ammonia in the presence of $Fe(NO_3)_3$ to give 9-acridonimine (**255**) ⟨76KGS356⟩. *N*-Alkylpyridinium, *N*-alkylquinolinium and *N*-alkylisoquinolinium salts also enter this reaction with $KMnO_4$ instead of $Fe(NO_3)_3$ as oxidant ⟨85JHC765, 87JHC1377⟩. Intermediate adducts of type (**254**) were observed by NMR spectra ⟨73JOC1949⟩.

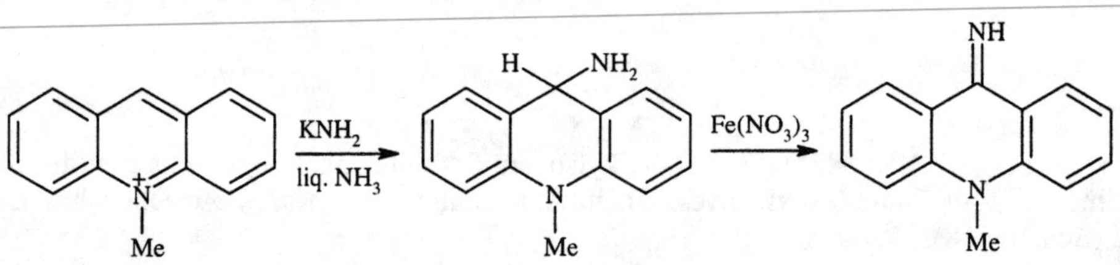

(254)   (255)

The charged rings are sufficiently reactive to be attacked by amines. Pyridinium ions carrying strongly electron-withdrawing substituents on the nitrogen react to give open-chain products. Thus, 1-(2,4-dinitrophenyl)pyridinium ion (**256**) gives glutaconic dialdehyde dianil (**257**) and 2,4-dinitroaniline ($PhNH_2$, 100°C) in the so-called Zinke reaction. Pyridine-sulfur trioxide, 1-(4-pyridyl)pyridinium ion and 1-cyanopyridinium ion react similarly.

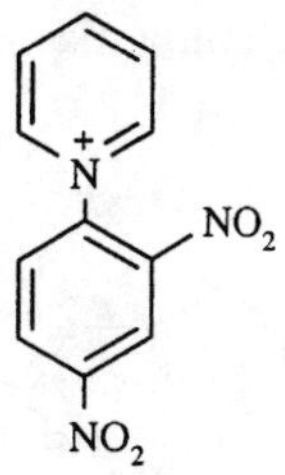

(256) 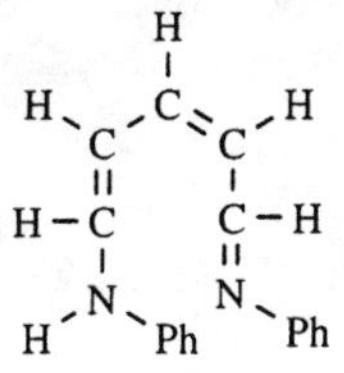   (257) 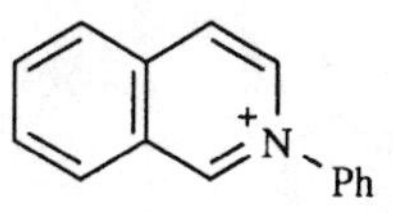   (258)

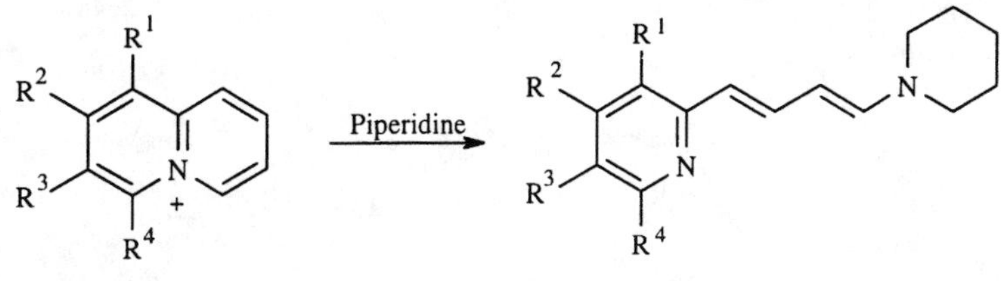

Piperidine

(259)   (260)

Quinolizinium ions (**259**) react with piperidine to give the ring-opened products (**260**). Subsequent closure of the ring can also occur, *e.g.* 2-(2,4-dinitrophenyl)isoquinolinium ion with PhNH$_2$ at 190°C forms (**258**).

(iii) Other cationic rings

Pyrylium cations form pyridines with ammonia and pyridinium salts with primary amines ⟨B-82MI 505-02⟩. For example, 2,4,6-triphenylpyrylium cation (**261**; Z = O) yields 2,4,6-triphenylpyridine with ammonia, the corresponding 1-methylpyridinium salt with methylamine, and pyridine 1-arylimines with phenylhydrazine. Xanthylium ions (**210**), where ring opening cannot readily occur, form adducts (**262**) with ammonia, amines, amides, ureas, sulfonamides and imides. Similar adducts (*e.g.* **263**) are formed by benzo[*b*]pyrylium ions.

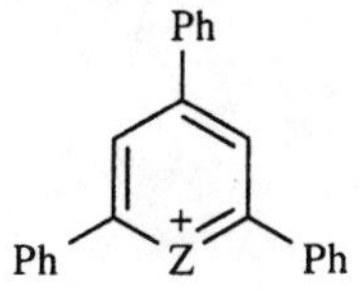

(261) 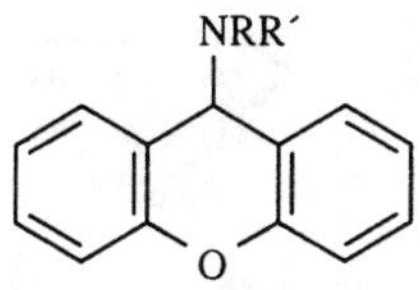   (262) 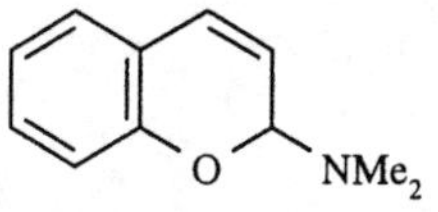   (263)

Amines also react with thiinium salts to give pyridinium salts, but reaction goes less easily than with the pyrylium analogues. 1,3-Oxazinium and 1,3-thiazinium cations react with ammonia and primary amines to give pyrimidines and pyrimidinium cations, respectively.

(iv) Pyridones, pyrones and azinones

2- and 4-Pyridones both undergo the Chichibabin reaction to give the 6- and 2-amino derivatives, respectively. Pyran-4-ones react with ammonia and amines to give ring-opened products, which reclose to yield pyridones. Pyran-2-ones are similarly converted by ammonia or amines into 2-pyridones. Isocoumarins form isoquinolones on treatment with ammonia or primary amines.

However, chromones react differently, because the phenolic hydroxy group in the ring-opened intermediate is unreactive. Thus, isoxazoles (**264**) result from the reaction with hydroxylamine, and a pyrazole is formed with hydrazine. γ-Pyrones also give pyrazoles with hydrazine.

Oxazones in which the heteroatoms are not adjacent react with ammonia and amines to give diazones, *e.g.* (**265, 266**; Z = O) yield (**265, 266**; Z = NR).

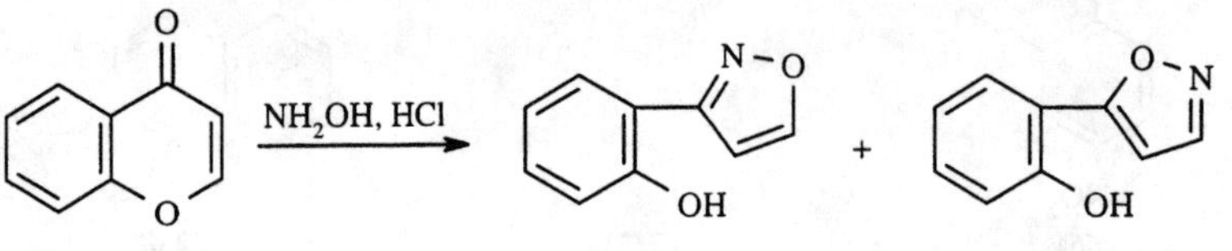

(264a)   +   (264b)

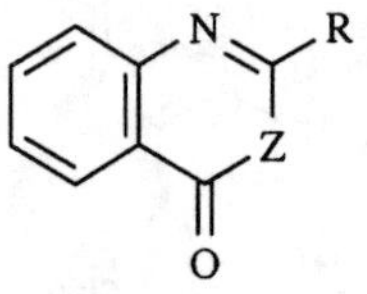

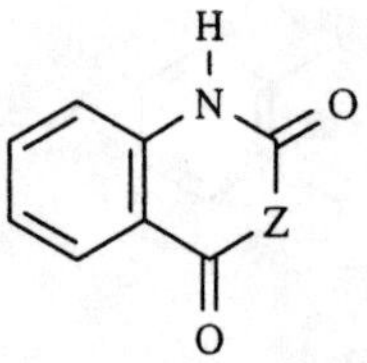

(265)   (266)

3.2.1.6.5 *Sulfur nucleophiles*

Neutral pyridines do not normally react with sulfur nucleophiles. Such reactions are known, however, for *N*-oxides and cationic rings.

Pyridine 1-oxide and methyl analogues undergo thioalkylation at the α- and γ-positions with alkanethiols in acetic anhydride (Scheme 19). Some β-substitution occurs under these conditions probably *via* episulfonium intermediates, evidence for which comes from the isolation of tetrahydropyridines, *e.g.* (**267**).

Pyrylium salts are converted by sodium sulfide into thiinium salts, *e.g.* (**261**; Z = O) → (**261**; Z = S).

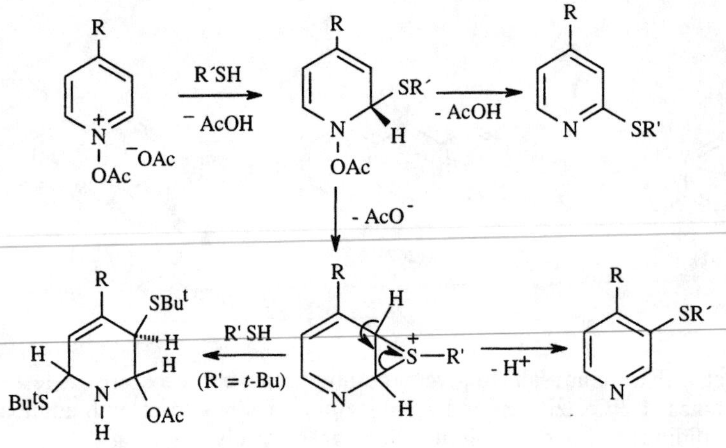

Scheme 19

The 5,6-double bond in uracil, 5-fluorouracil, *N*-alkyluracils, thiouracils, and uridines adds sodium sulfite or bisulfite to give the corresponding 5,6-dihydro-6-sulfonic acid salts. Bisulfite addition to cytosines and cytidine may be succeeded by a second reaction involving nucleophilic replacement of the amino group, for example, by water.

Perimidine and its N-substituted derivatives on heating with sulfur undergo thiation to form perimidine-2-thiones in good yields ⟨81RCR1559⟩. Thiopyrimidines are prepared by the thiolysis of halopyrimidines. using sodium hydrogen sulfide, or thioureas.

3.2.1.6.6 *Phosphorus nucleophiles*

1-Ethoxycarbonylpyridinium cations are attacked by phosphites (Scheme 20). The intermediates can be further reacted to give 4-alkylpyridines as shown 〈81TL4093〉. Pyrylium salts with nucleophilic phosphines yield phosphorines (*e.g.* **268**) 〈67AG(E)458〉.

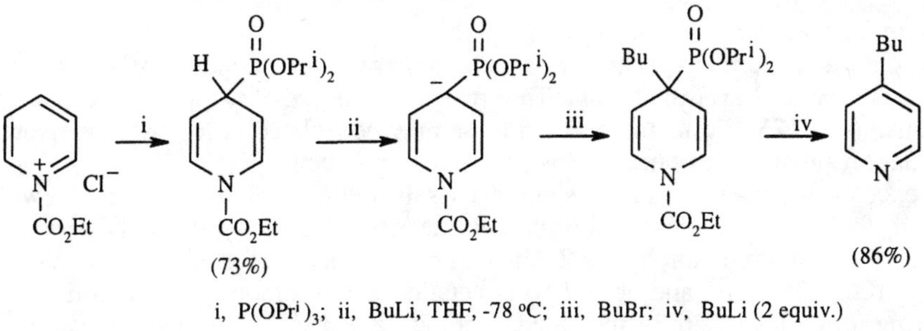

i, P(OPr^i)$_3$; ii, BuLi, THF, -78 °C; iii, BuBr; iv, BuLi (2 equiv.)

Scheme 20

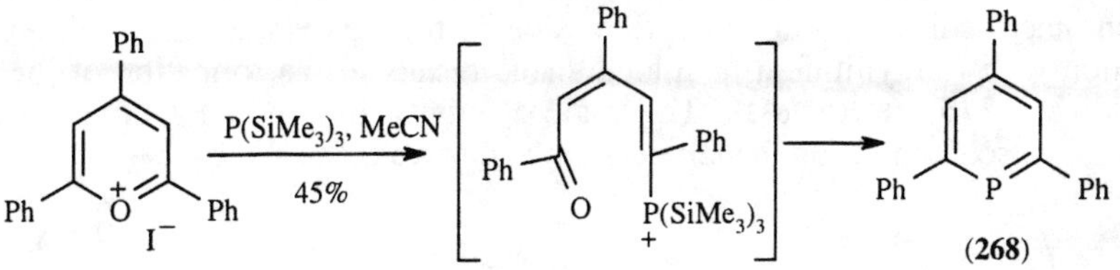

3.2.1.6.7 *Halide ions*

Chloride ions are comparatively weak nucleophiles, and do not react with pyridines. In general, there is also no interaction with pyridinium and pyrylium compounds, but xanthylium chloride is in equilibrium with an appreciable amount of (**269**), this being a particularly favorable case with little loss of resonance energy on adduct formation.

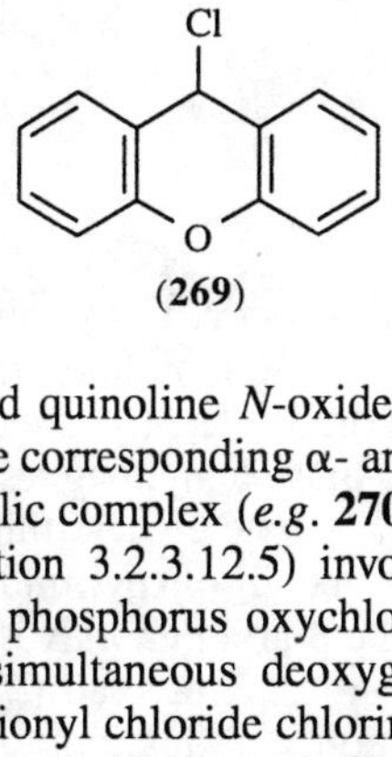

(**269**) 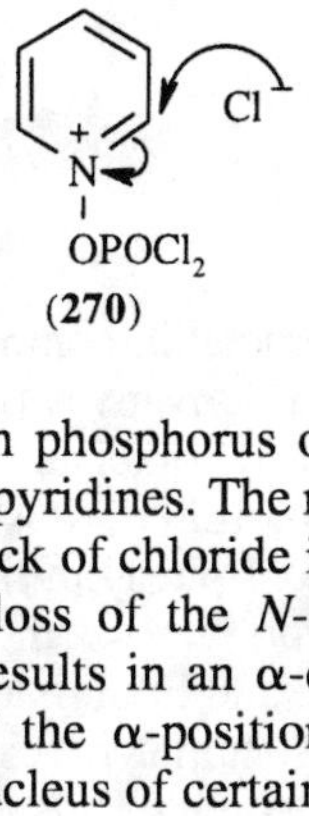 (**270**) 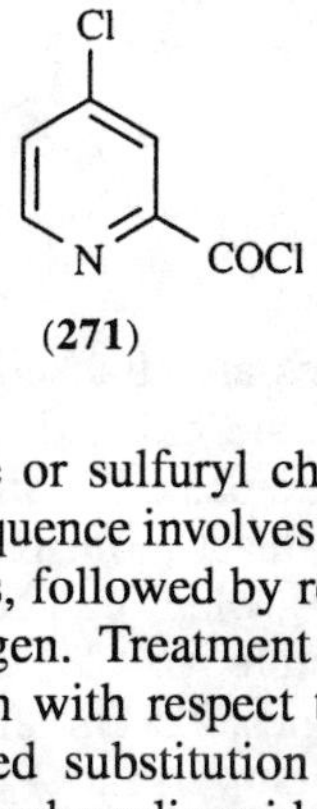 (**271**)

Pyridine and quinoline *N*-oxides react with phosphorus oxychloride or sulfuryl chloride to form mixtures of the corresponding α- and γ-chloropyridines. The reaction sequence involves first formation of a nucleophilic complex (*e.g.* **270**), then attack of chloride ions on this, followed by rearomatization (see also Section 3.2.3.12.5) involving the loss of the *N*-oxide oxygen. Treatment of pyridazine 1-oxides with phosphorus oxychloride also results in an α-chlorination with respect to the *N*-oxide groups with simultaneous deoxygenation. If the α-position is blocked substitution occurs at the γ-position. Thionyl chloride chlorinates the nucleus of certain pyridine carboxylic acids, *e.g.* picolinic acid → (**271**), probably by a similar mechanism.

Regioselectivity in the chlorination of pyrazine *N*-oxides with phosphoryl chloride depends upon the substituent on the C-3 carbon. Thus, 3-aminopyrazine 1-oxide is chlorinated with loss of the *N*-oxide oxygen to give 2-amino-3-chloropyrazine as the sole product whereas 3-methoxy and 3-chloropyrazine 1-oxides form approximately equal amounts of 3-chloro- and 6-chloro-2-substituted pyrazines along with a trace of the 5-chloro derivatives.

Ring fluorination of pyridine and its benzo derivatives is suggested to occur through nucleophilic attack of fluoride ion on an initial pyridine fluorine complex 〈87TL255, 91BCJ1081, 93AHC(58)291〉. In electron-deficient pyridines and their benzo derivatives, fluorination on the pyridine ring is kinetically competitive with annular and side chain fluorination.

Xenon difluoride and xenon hexafluoride fluorinate pyridine to give a mixture of 2- and 3-fluoropyridines along with 2,6-difluoropyridine 〈76JFC(7)179〉. Cesium fluoroxysulfate at room temperature in ether or chloroform converted pyridine to 2-fluoropyridine in 61 and 47% yield, respectively, along with 2-fluorosulfoxypyridine ⟨90TL775⟩.

3.2.1.6.8   Carbon nucleophiles

(i) Organometallic compounds

Pyridine reacts with lithium alkyls and aryls under rather vigorous conditions (*e.g.* xylene at 100°C) to afford 2-alkyl- and 2-aryl-pyridines. The reaction proceeds by way of the corresponding dihydropyridines (*e.g.* **275** or a tautomer), and these may be isolated at lower temperatures. The less reactive Grignard reagents give poorer yields of the same products.

In the presence of free magnesium, considerable 4-substitution is also observed, and when pyridine is reacted with, for example, *n*-butyl chloride and magnesium directly, 4-*n*-butylpyridine is formed without appreciable contamination by the 2-isomer. Possible reasons for this change in orientation are discussed in CHEC 2.05. Allyl and benzyl type metalloorganic compounds also add to 4-position of the pyridine ring ⟨69JCS(B)901⟩. In 3-substituted pyridines, attack at C-2 is favored over C-6 unless the C-3 substituent or the attacking alkyl group is very large. Thus, a 4,4-dimethyloxazolinyl-2 group in the 3-position of a pyridine ring is often used to direct addition of Grignard reactions and alkyllithiums to the 4-position, intermediate 1,4-dihydropyridines being further oxidized (*e.g.* **272** → **273** → **274**). The same reaction with *t*-butyllithium or lithiodithiane results in the formation of the 2-substituted 1,2-dihydropyridines (**275**) ⟨82JOC2633⟩. The regioselectivity of the organolithium reaction with 3-pyridyloxazolines also depends on the temperature and solvent.

Benzopyridines are attacked by organometallic compounds at a position α to the nitrogen unless both α-positions are blocked. The dihydro derivatives of quinoline and isoquinoline are more stable and less easily aromatized than those from pyridine, and are hence more frequently isolated.

Diazines also react more readily than pyridine. Thus, pyrimidine and phenylmagnesium bromide give adduct (**276**), which can be oxidized to 4-phenylpyrimidine. Aryl- and heteroaryl-lithium reagents at low temperature ⟨79AG1, 80RTC234⟩ add across the 3,4-double bond of pyrimidines to give dihydropyrimidines. 2,5-Dimethylpyrazine and lithium aryls afford the 3-aryl derivatives.

Grignard reagents add to 1,2,4-triazines. Initial attack at the 5-position is favored (**277** → **278** → **279**); if this position is substituted the nucleophile adds to the 6-position, and finally to the 3-position. Starting from the parent 1,2,4-triazine, 3,5,6-triaryl-1,2,4-triazines (**280**) have been prepared by successive addition of Grignard reagents to the ring and oxidation of the dihydro-1,2,4-triazine so formed.

Cationic rings react readily with organometallic compounds: Grignard reagents with *N*-alkylpyridinium salts generally give 1,2-dihydropyridines. If the size of the *N*-substituent is increased, then the orientation can be directed to the 4-position. If the *N*-substituent is also a leaving group, then re-aromatization can occur (Scheme 21).

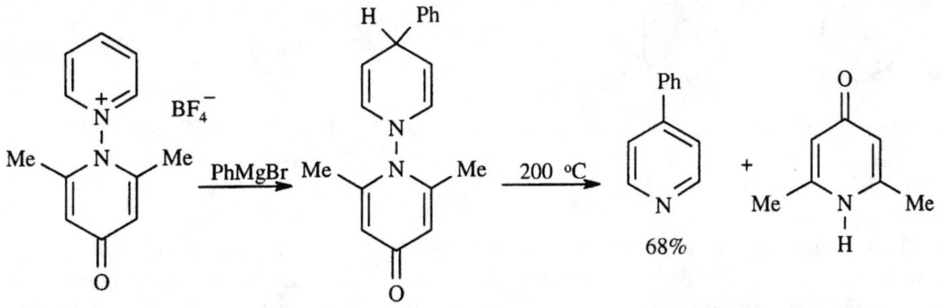

Scheme 21

N-Acylpyridinium salts (usually generated *in situ*) react with organometallic reagents efficiently and in many cases regioselectively ⟨88AHC(44)199⟩. With alkyl Grignards reagents *N*-acylpyridinium salts (**281**) give mixtures of 1,2- (**282**) and 1,4-dihydropyridines (**283**), whereas aryl, vinyl, and alkynyl Grignard reagents yield mainly the 1,2-dihydropyridine products (**282**) ⟨88AHC(44)199⟩.

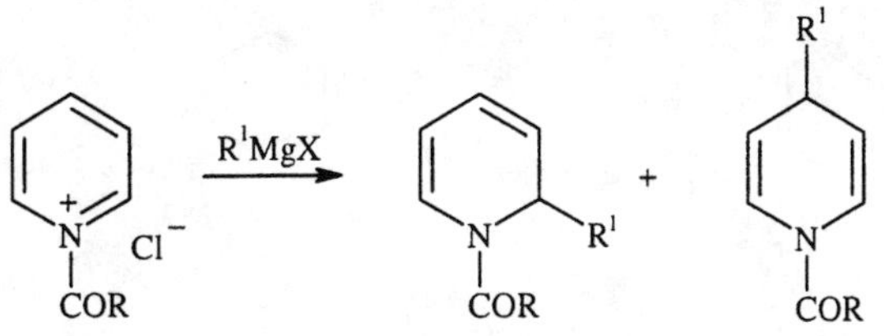

The regioselectivity of Grignard addition to *N*-acylpyridinium salts can often be controlled by changing the conditions. Thus, *e.g.*, the regioselectivity towards C-4 addition products can be enhanced by the presence of catalytic amounts of copper salts. Direct reaction of salt (**281**, R = OMe) with lithium dialkyl cuprate gives also almost exclusively the 1,4-dihydro product (**283**, R = OMe, R^1 = Me).

An asymmetric synthesis of 1,2-dihydropyridines has been achieved by the addition of Grignard reagents to the pyridinium salt generated from 3-(triisopropylstannyl)pyridine and the chloroformate of 8-arylmenthyl based chiral auxiliaries (Scheme 22) ⟨91JOC7167⟩.

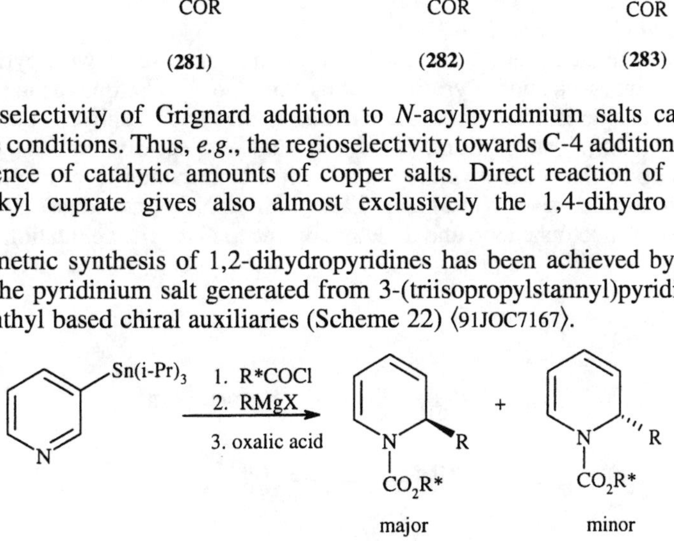

R	R*= (-)-8-phenylmenthyl de	R*= (-)-8-(4-phenoxyphenyl)menthyl de
n-Pr	78	82
c-C_6H_{11}		91
Bn		76
vinyl		90
Ph	84	89
4-MeC$_6$H$_4$		92

(de ≡ diastereomeric excess)

Scheme 22

With a Grignard reagent, 1-methylquinolinium ions give products of type (**284**). A notable reaction of this class is that between the pyridine-sulfur trioxide complex and sodium cyclopentadienide (**286**) which forms azulene (**287**) by a sequence involving opening of the pyridinium ring and subsequent closure to the seven-membered carbocyclic ring.

(284)   (285)   (286)   (287)

Corresponding reactions with *N*-oxides give α-substituted aromatic products by the loss of hydroxide ions from intermediates of type (**285**). However, yields are poor, both because the *N*-oxide also acts as an oxidizing agent toward the organometallic compound and because ring opening can occur. Moreover, monocyclic 1,2,3-triazine 2-oxides are stable and unreactive toward Grignard reagents, alkyl- and aryl-lithiums, and other organometallics.

Pyran-4-ones with Grignard reagents give pseudo-bases which form the pyrylium salts with acid, *e.g.* (**288**) → (**289**).

(288)   (289)

(ii) Activated methyl and methylene carbanions

The mesomeric anions of activated methyl and methylene compounds react with pyridinium and pyrylium cations, generally at the γ-position. Pyridinium ions combine with ketones as in (**290**) to give products of type (**291**) which can be isolated or oxidized *in situ* to mesomeric anhydro-bases (**292**) [*cf*. (**647**) → (**648**), Section 3.2.3.3.5]. Quinolinium, isoquinolinium and acridinium ions give similar adducts of stability increasing in the order given. Aliphatic nitro compounds react analogously, *e.g.* 1-methylquinolinium ion gives successively (**293**) and (**294**) (with MeNO$_2$-piperidine). 1-Benzoylpyridinium ions react similarly with acetophenone and dimethylaniline to give, after oxidation, aromatized products (**295**) and (**296**).

(290)   (291)   (292)

(293)   (294)   (295)   (296)

The application of *N—N*-linked pyridinium salts to induce reaction with active hydrogen compounds in the 4-position is illustrated in Scheme 23.

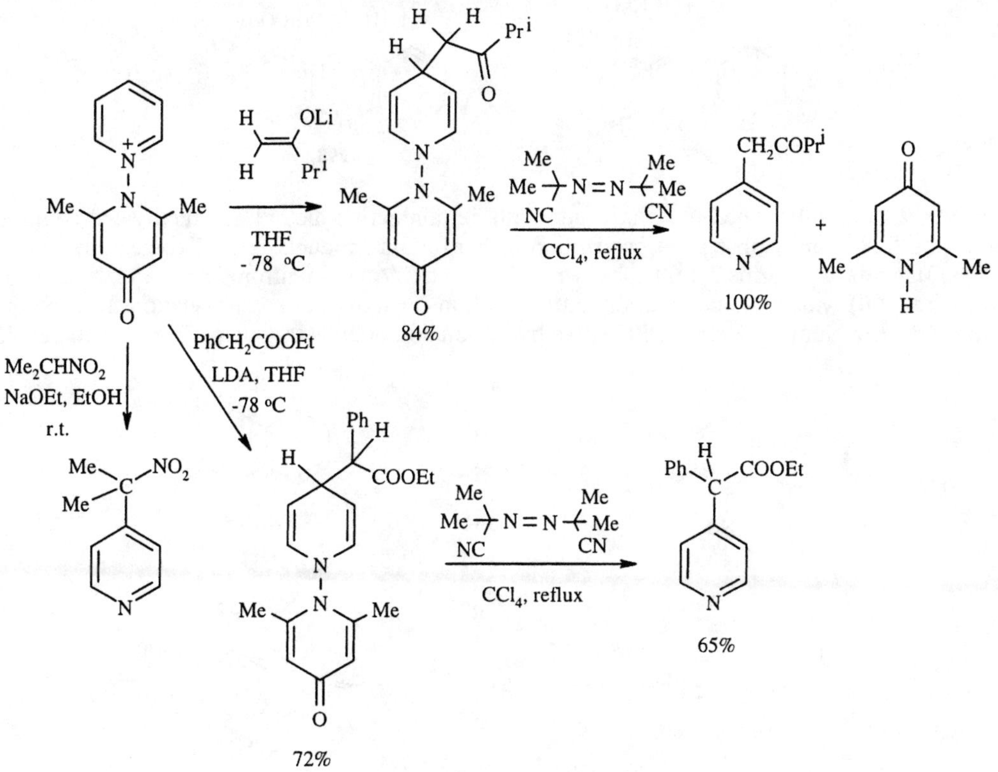

Scheme 23

Zinc enolates have been shown to react with chiral *N*-acylpyridinium salts with high diastereo-selectivity (Scheme 24) ⟨93JOC5035⟩. Other metallo enolates gave products with lower diastereoselectivity. Titanium enolates react with *N*-acylpyridinium salts to yield 1,4-dihydropyridines, which on subsequent oxidation (aromatization) give 4-(2-oxoalkyl)pyridines ⟨84TL3297⟩.

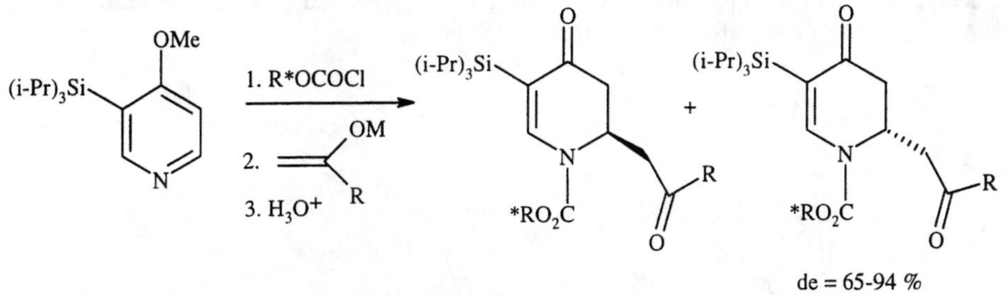

R = n-Pr, Ph, Me, 1-butenyl
M = ZnCl, Ti(i-PrO)$_3$, TiCl$_3$, SnCl$_3$, Li, MgBr
R* = (-)-8-phenylmenthyl, (-)-*trans*-(α-cumyl)cyclohexyl

Scheme 24

1-Trimethylsilylpyrimidinium triflate, derived from pyrimidine and trimethylsilyl triflate, adds silylated enol ethers to form 1,4-dihydropyrimidines. *N*-Acylpyrimidinium tetrafluoroborates undergo analogous reactions ⟨85H(23)207⟩.

Pyrylium salts with a free α- or γ-position react in a similar way without ring fission, *e.g.* flavylium ions (**220**) add dimethylaniline and the product aromatizes to give (**297**); xanthylium ions (**210**) form adducts at the 9-position with β-diketones, β-keto esters and malonic esters (*e.g.* **298**).

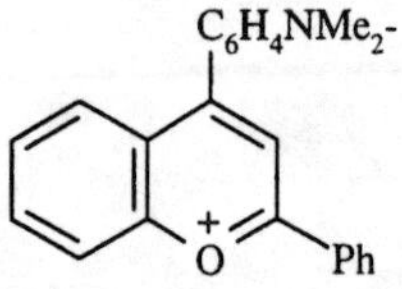

(297) 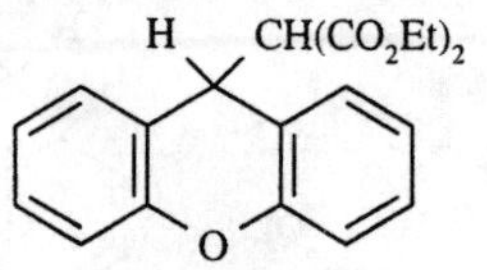   (298)

However, 2,4,6-trisubstituted pyrylium salts with certain active methyl and methylene compounds undergo ring fission and subsequent cyclization to benzenoid products. 2,4,6-Triphenylpyrylium ion (261; Z = O) in this way forms 2,4,6-triphenylnitrobenzene (299) with nitromethane and the substituted benzoic acid (300) with malonic acid, the latter reaction involving a decarboxylation. In reactions of this type, 1,3-oxazinium salts react with active hydrogen compounds to give pyridines (Scheme 25).

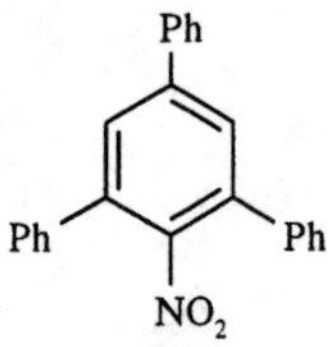

(299) 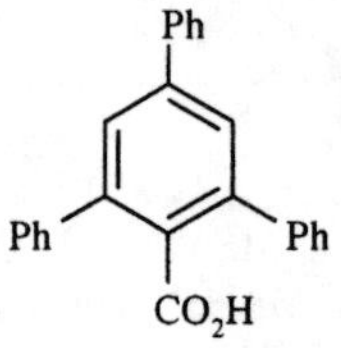   (300) 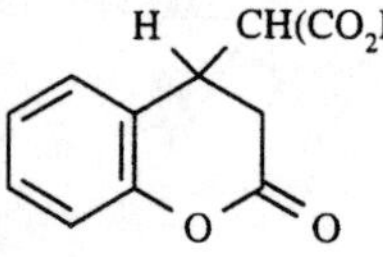   (301)

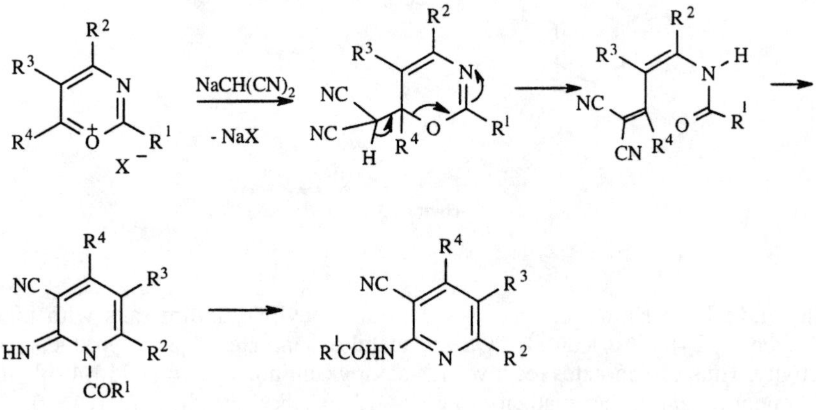

Scheme 25

Sodium acetylide reacts with pyridine 1-oxide to give ring-opened products (Scheme 26). However, in general, *N*-oxides react only after quaternization (Scheme 27).

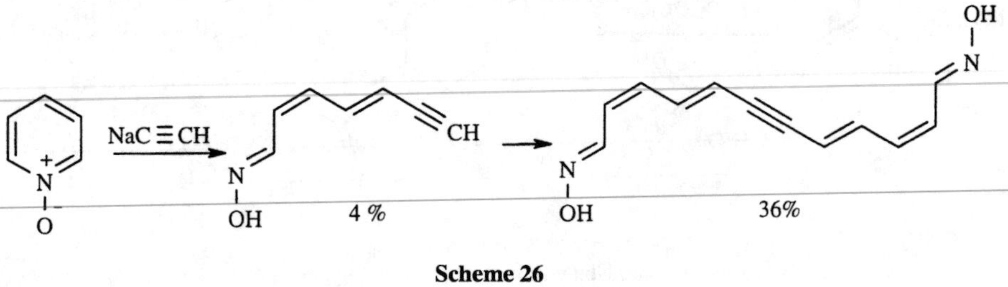

Scheme 26

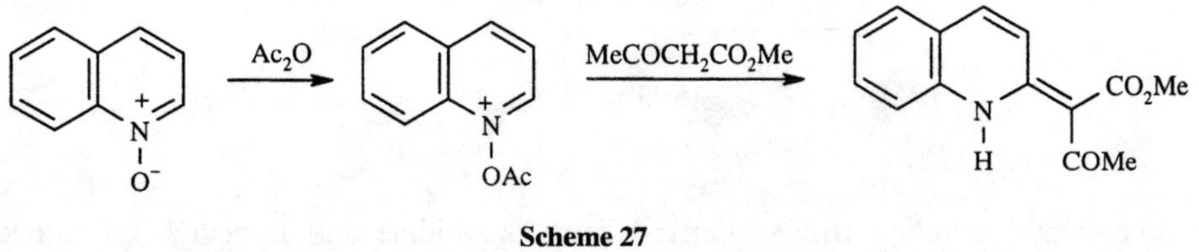

Scheme 27

Certain pyrones react with active hydrogen compounds in a quite different way: they display their alkenic character by a Michael reaction. Thus, coumarin yields (301) with malonic ester.

Bifunctional C-nucleophiles react with azines to cause *meta*-bridging cyclization in two steps ⟨88AHC(43)301⟩. Transformation of 5-nitropyrimidin-2-one into nitrophenols (303) by β-dicarbonyl compounds proceeds *via* the bicyclic *meta*-bridged intermediate (**302**) ⟨82JOC1018⟩, During the course of the reaction the N1-C-N3 part of the uracil is replaced by the C-C-C part of the 1,3-dinucleophilic reagent to form (**303**).

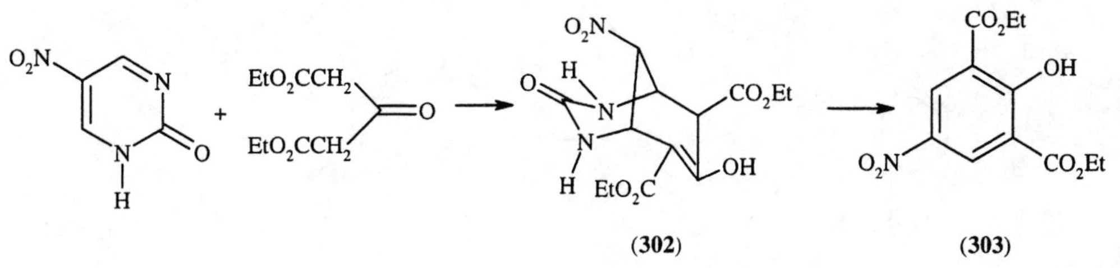

(302)   (303)

Nucleophilic attack at the fully conjugated 1,3,5-triazine usually results in ring cleavage. Thus, active methylene compounds react with 1,3,5-triazine in aminomethylenation reactions. A three-component reaction of cyclopentadiene with 1,3,5-triazine and a secondary amine leads to *N,N*-disubstituted pentafulven-6-amines (**304**) and N^2-(6-pentafulvenyl)formamidines (**305**) ⟨86LA374⟩.

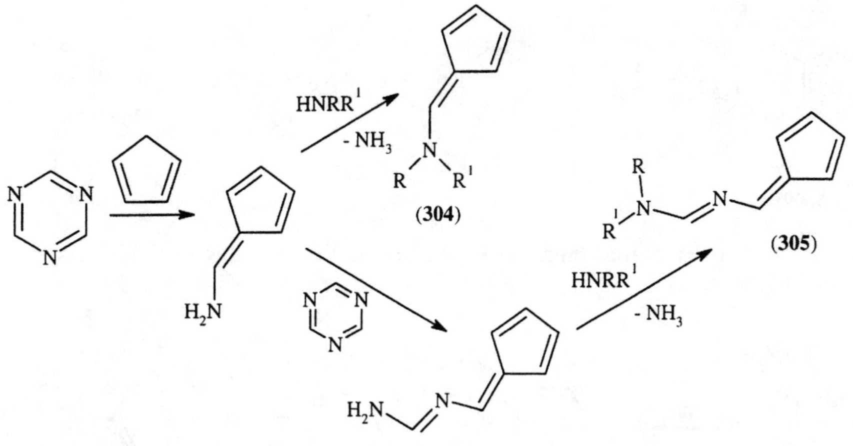

(304)   (305)

(iii) Anhydro-bases and enamines

Anhydro-bases with cationic rings give adducts (*e.g.* **306** → **307**) which are spontaneously oxidized to cyanine dyes (**308**).

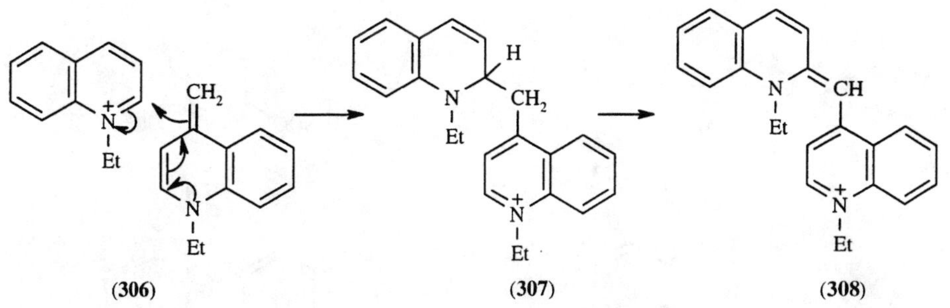

(306)   (307)   (308)

Pyridine *N*-oxides in the presence of acyl halides react with enamines and indoles as shown (Scheme 28).

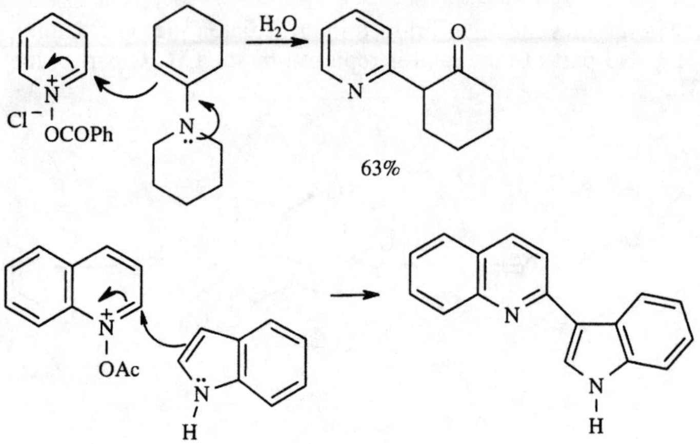

Scheme 28

(iv) Cyanide ions

So-called 'pseudocyanides', analogous to pseudo-bases, are formed by reaction of cyanide anions with benzopyridinium cations. Quite often, different isomeric pseudocyanides are formed depending on the temperature. Thus, at − 70 to − 30°C (kinetic control), 1-methylquinolinium ion gives only the 2-cyanoadduct (**309**), whereas at 20°C (thermodynamic control) exclusively the 4-isomer (**310**) is observed in the NMR spectrum ⟨80ZOR671⟩.

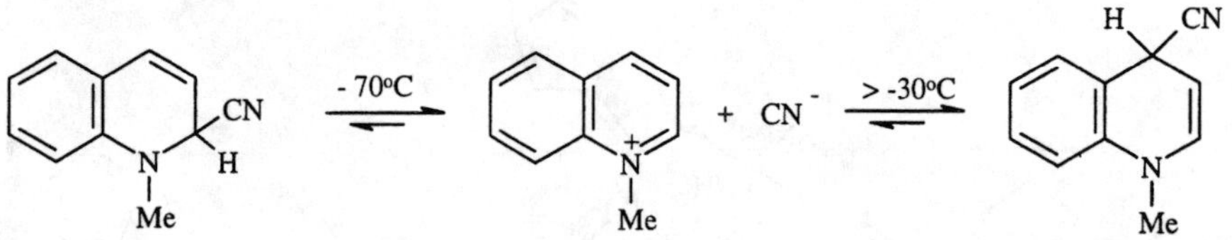

(309) **(310)**

Pseudocyanides are important intermediates in the conversion of 1-alkylpyridines into 4,4'-bipyridyl diquaternary salts (Scheme 29).

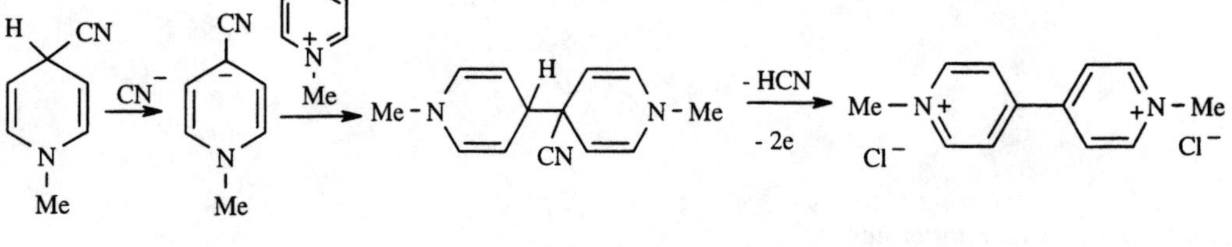

Scheme 29

Analogously, 5-unsubstituted 1,2,4-triazines with cyanide ions afford bi-1,2,4-triazin-5,5′-yls (**311**) and 1,2,4-triazine-5-carboxamides (**312**) ⟨84CHEC-(3)369, 87CPB1378⟩.

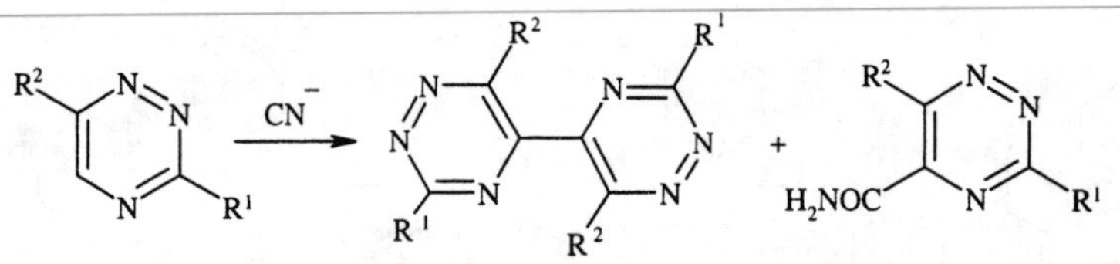

(311) **(312)**

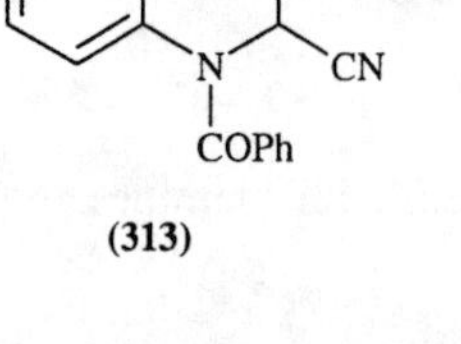

(313)

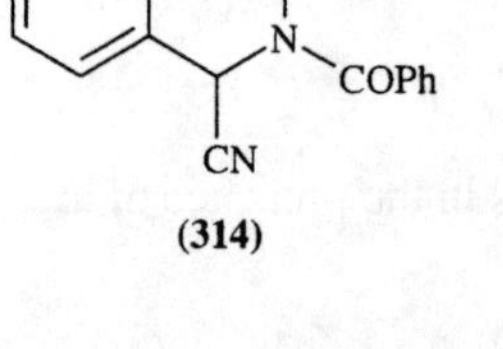

(314)

In the Reissert reaction, 1-benzoylquinolinium ions (formed *in situ* from quinolines and PhCOCl) and cyanide ions give 'Reissert compounds'; thus, quinoline itself forms (313). These Reissert compounds are hydrolyzed by dilute alkali to, for example, quinoline-2-carboxylic acids and benzaldehyde. Isoquinolines also form Reissert compounds (*e.g.* 314).

In the Reissert-Henze reaction, quinoline 1-oxide reacts with benzoyl chloride and potassium cyanide to give 2-cyanoquinoline in good yield (Scheme 30). Pyridine 1-oxides undergo the Reissert-Henze reaction readily when the reaction is carried out in non-aqueous medium using PhCOCl-Me$_3$SiCN (Scheme 31). Pyrimidine *N*-oxides and pyrazine *N*-oxides also undergo Reissert-Henze reactions.

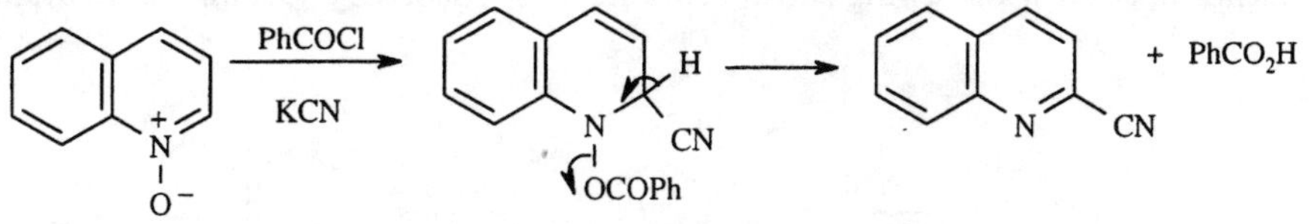

Scheme 30

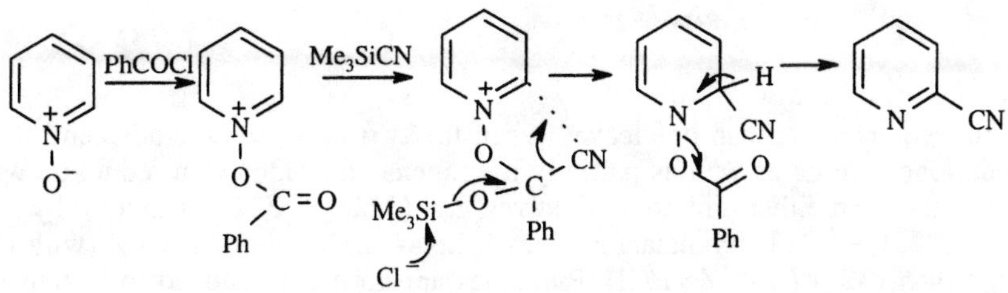

Scheme 31

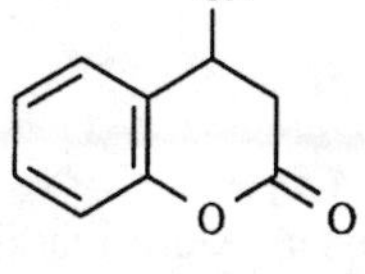

1-Alkoxypyridinium and 1-methoxypyridazinium salts yield cyanopyridines and cyanopyridazines, respectively, on treatment with potassium cyanide; the cyano group enters the α-position with respect to the *N*-oxide of the starting material.

Coumarin gives compound (315) by Michael addition of hydrogen cyanide.

3.2.1.6.9 *Chemical reduction*

(i) Pyridines

Pyridines are more susceptible to reduction than benzenes. Sodium in ethanol or in liquid ammonia evidently reduces pyridine to 1,4-dihydropyridine (or a tautomer) because hydrolysis of the reaction mixture affords glutaric dialdehyde (318 → 317 → 316). Reduction of pyridines with sodium and ethanol can proceed past the dihydro stages to Δ^3-tetrahydropyridines and piperidines (318 → 319 and 320).

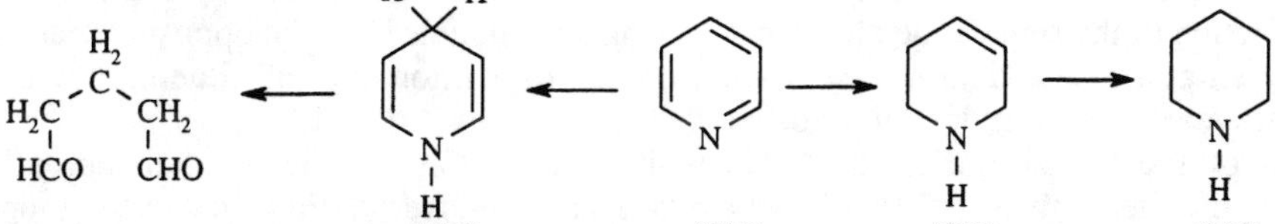

Comparison of the experimental results with theoretical calculations (MNDO/3MO) of the isomeric anions formed by addition of hydride to pyridines and azines suggests that the thermodynamically most stable products are obtained with mild reducing agents. By monitoring reduction of pyridine with strong reducing agents (*i.e.*, MgH$_2$ and AlH$_3$), it was shown by ^{1}H NMR that initially there is a mixture of 1,2- and 1,4-dihydro-1-pyridyl anions, but the final product is the 1,4-dihydro-1-pyridyl anion. The proposed mechanism for this selective reduction is a hydride exchange between reduced and unreduced substrate molecules coordinated to the same metal ion ⟨83JCS(P2)989⟩. Reduction of pyridines with diisobutylaluminum hydride ⟨85JOC2443⟩, lithium 9-borabicyclo[3.3.1]nonane ⟨84JOC3091⟩, 9-*sec*-amyl-9-borabicyclo[3.3.1]nonane ⟨89MI 502-01⟩, aluminum hydride-triethylamine complex $\langle$93JOC3974$\rangle$, bis(diethylamino)aluminum hydride $\langle$94MI 502-01$\rangle$, and sodium diethyldihy-droaluminate $\langle$92MI 502-03$\rangle$ have been reported to occur slowly, whereas lithium triethylborohydride (super hydride) reduces isoquinolines, quinoline, and pyridines effectively to give 1,2,3,4-tetra-hydroisoquinolines, 1,2,3,4-tetrahydroquinoline, and piperidines. The mechanism of this reduction has been explored using lithium triethylborodeuteride $\langle$93TL7239$\rangle$.

Samarium diiodide rapidly reduces pyridine to piperidine in the presence of water at room temperature in excellent yield $\langle$93H(36)2383$\rangle$, and various substituted pyridines have been reduced similarly.

Reduction of quinoline with lithium aluminum hydride gives 1,2-dihydroquinoline. Neutral pyridines bearing electron-withdrawing substituents are also reduced by sodium borohydride (Scheme 32).

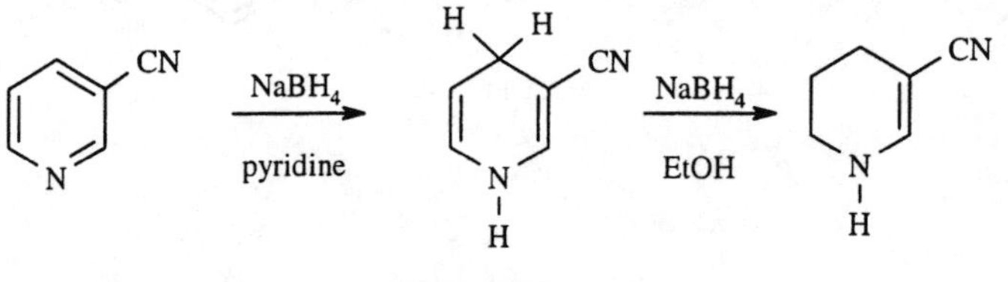

Scheme 32

(ii) Azines

Diazines are readily reduced. The ring can be cleaved when the two nitrogens are adjacent: thus, pyridazine gives tetramethylenediamine as well as partially hydrogenated products on reduction with sodium and ethanol. Cinnolines form either dihydro derivatives, *e.g.* (**322**) → (**321**), or indoles by ring opening and reclosure, *e.g.* (**322**) → (**323**). Phthalazine gives 1,2,3,4-tetrahydrophthalazine (with Na/Hg) or the ring-opened product (**324**) (with Zn-HCl). Pyrazines are normally reduced to hexahydro derivatives (*e.g.* with nickel-aluminium alloy in potassium hydroxide $\langle$87JOC1043$\rangle$), whereas quinoxalines usually give 1,2,3,4-tetrahydro derivatives. Thus, treatment of quinoxalines with sodium borohydride and benzyl chloroformate in methanol at $-78°C$ forms tetrahydro derivatives (**325**; $R = CO_2Bn$) $\langle$92JCS(P1)1245$\rangle$.

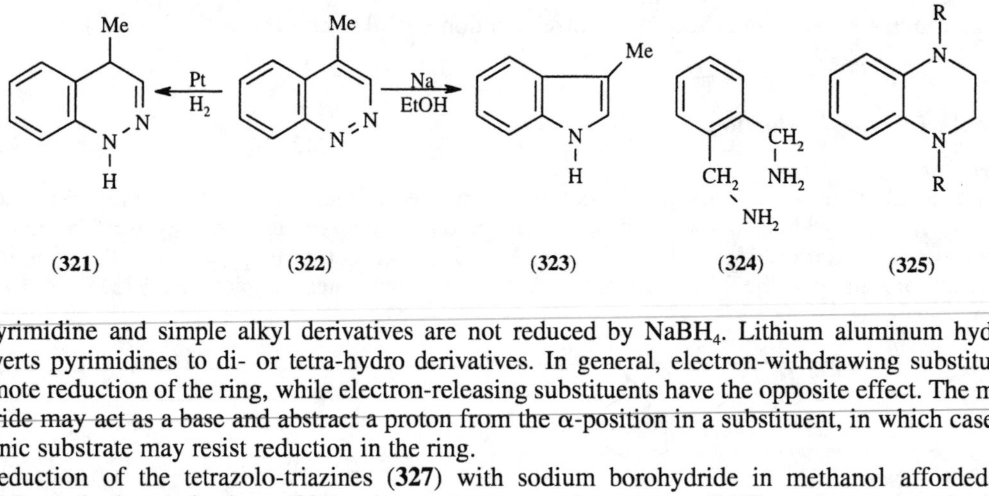

| (321) | (322) | (323) | (324) | (325) |

Pyrimidine and simple alkyl derivatives are not reduced by $NaBH_4$. Lithium aluminum hydride converts pyrimidines to di- or tetra-hydro derivatives. In general, electron-withdrawing substituents promote reduction of the ring, while electron-releasing substituents have the opposite effect. The metal hydride may act as a base and abstract a proton from the α-position in a substituent, in which case the anionic substrate may resist reduction in the ring.

Reduction of the tetrazolo-triazines (**327**) with sodium borohydride in methanol afforded the 5,6,7,8-tetrahydro derivatives (**328**) whereas hydrogenation over a Pd/C catalyst stopped at the 7,8-dihydro-stage (**326**) $\langle$88JOC5371$\rangle$.

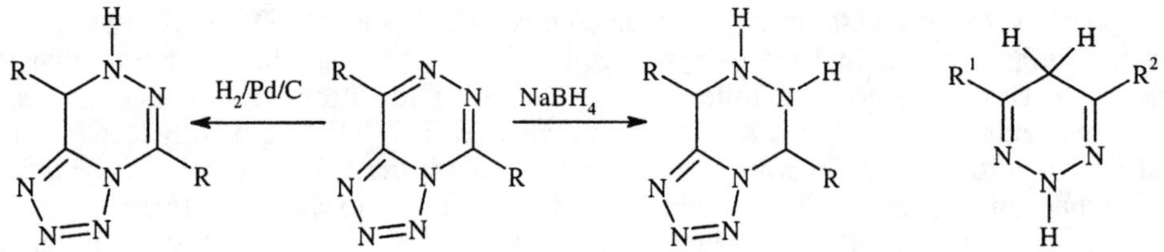

| (326) | (327) | (328) | (329) |

Sodium borohydride reduction of 4,6-disubstituted 1,2,3-triazines in methanol afforded 2,5-dihydro-1,2,3-triazines (**329**) in good yields. 1,2,4,5-Tetrazines with mild reducing agents give dihydro derivatives.

(iii) Cationic rings

Cationic rings are readily reduced under relatively mild conditions. 1-Methylpyridinium ion with sodium borohydride (in H_2O, 15°C) gives the 1,2-dihydro derivative (**330**) at pH > 7 and the 1,2,3,6-tetrahydro derivative (**331**) at pH 2–5. The tetrahydro compound is probably formed *via* (**332**) which results from proton addition to (**330**). Pyridine cations are also reduced to 1,2-dihydropyridines by dissolving metals, *e.g.* Na/Hg.

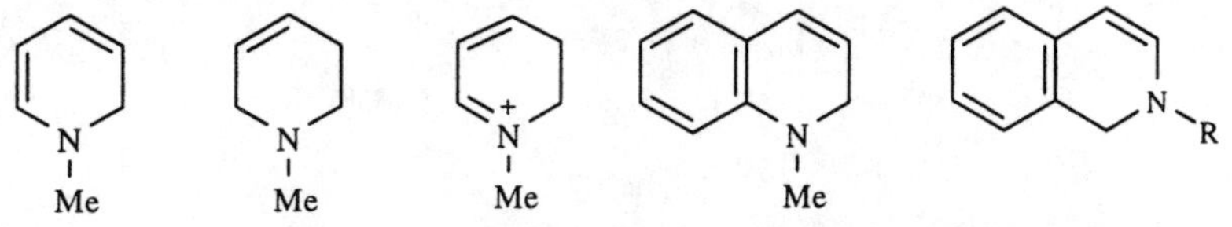

(**330**)   (**331**)   (**332**)   (**333**)   (**334**)

Complex hydride reduction (NaBH$_4$ or LiAlH$_4$) of 1-methylquinolinium ions proceeds analogously to 1,2-dihydro compounds (*e.g.* **333**). 1-Methyl- and 1-acylisoquinolinium ions (the latter with Bu$_3$SnH ⟨88CL913⟩) give the corresponding 1,2-dihydro compounds (**334**). Borohydride reduces pyrylium salts to mixtures of 2*H*- and 4*H*-pyrans; the former immediately ring opens to form the dienone (Scheme 33).

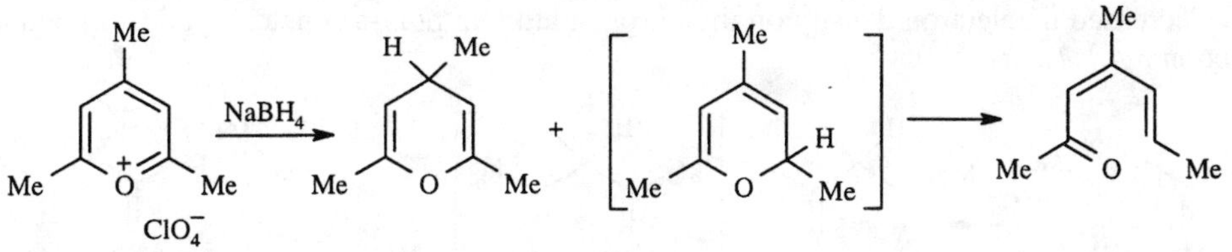

Scheme 33

Reduction of pyridinium ions with sodium dithionite (Na$_2$S$_2$O$_4$-H$_2$O-Na$_2$CO$_3$) gives 1,4-dihydro products ⟨80JA1092⟩. The mechanism involves initial formation of a sodium sulfinate intermediate which is stable in alkaline solution, but which decomposes as shown in Scheme 34 in acid or neutral solution.

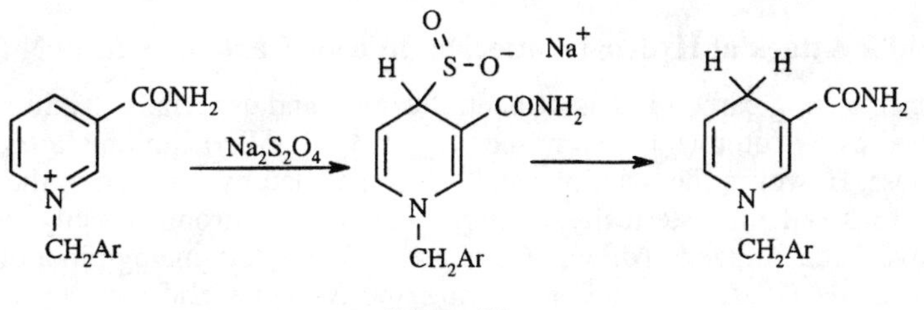

Scheme 34

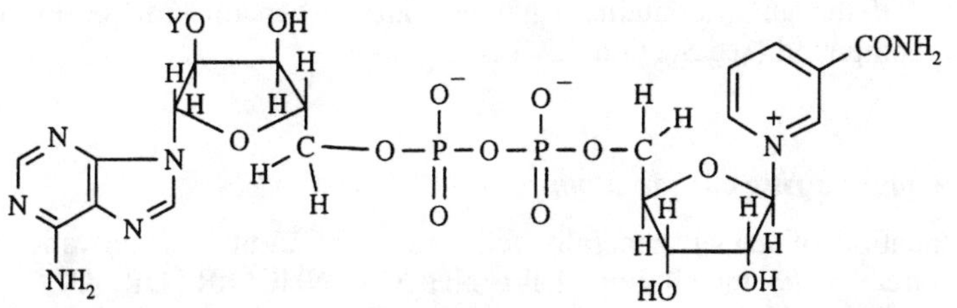

(**335**)

Vigorous chemical reduction (*e.g.* Sn-HCl or Zn-HCl) effects complete reduction of the heterocyclic ring, *e.g.* 1-methylquinolinium ion yields 1-methyl-1,2,3,4-tetrahydroquinoline. Reversible reduction of the pyridinium ring of coenzymes I and II (**335**; Y = H and PO$_3$H$_2$, respectively) is important physiologically ⟨B-97MI 502-07⟩.

Chemical reduction of azine *N*-oxides, depending on substrate structure, reductant and reaction conditions can proceed both with or without deoxygenation of the N-atom. Thus, 1,2,3-triazine 1-oxide (**337**) with NaBH$_4$ gives 2,5-dihydro-1,2,3-triazine (**336**), suggesting that the N-oxide moiety back-donates electrons to the triazine ring. On the other hand, on reduction of the isomeric 2-oxide leading to tetrahydro derivatives (**338**) and (**339**) the N-oxide function is not touched ⟨82H(17)317⟩.

(**336**)   (**337**)   (**338**)   (**339**)

3.2.1.7 Nucleophilic Attack at Ring Nitrogen

Some azines are capable to add a carbanion from an organometallic compound at a nitrogen heteroatom ⟨96CHEC-II(6)491⟩. Thus, 4,6-dimethyltriazine reacts with alkyl and aryl Grignard reagents to form N-2-addition compounds (**340**) and products of deeper transformation (**341**) and (**342**) along with "normal" C-4 addition product (**343**). Such "azaphilic addition" has been also shown for 1,2,3-benzotriazines and 1,2,4,5-tetrazines. The driving force of "azaphilic addition" is supposed to be strongly decreased by electron density on the nitrogen atom in poly-aza heterocyclic compounds (see also Section 3.4.1.7).

(**340**) 17-37 %   (**341**) 6-8 %   (**342**) 10 %   (**343**)   6-15 %

3.2.1.8 Nucleophilic Attack at Hydrogen attached to Ring Carbon or Ring Nitrogen

Hydrogen attached to ring carbon atoms of neutral azines, and especially azinium cations, is acidic and can be removed as a proton (for review see ⟨74KGS1587⟩). Alkyl lithiums are ordinarily used as bases for this purpose. However, the reaction can be accompanied by addition of the alkyl anion to the ring C=N bond. To avoid this, sterically hindered bases with strong basicity but extremely low nucleophilicity have been utilized. Among them are: lithium tetramethylpiperidide (LiTMP) and lithium diisopropylamide (LDA). The anions from azine *N*-oxides and some neutral azines can be stabilized as lithium derivatives. In other cases, the anion again adds a proton and hydrogen isotopic exchange can result. If the anion contains a halogen atom, then this can be eliminated to form a pyridyne or similar compound (see Section 3.2.3.10.1).

3.2.1.8.1 *Metallation at a ring carbon atom*

Direct ring metallation of azines generally requires some additional activating substituent. The following groups direct *ortho* metallation: halo, NHCOR, NHCO$_2$R, OR, OCONR$_2$, 2-oxazolino, CONHR, CONR$_2$, and masked RCHO, RCOR, and SO$_2$NR$_2$ groups. The directing groups have lone-pair electrons on heteroatoms which enable formation of coordination complexes with metal ions such as lithium, resulting in metallation at sites adjacent to the substituent. For example, 2-(4-pyridyl)-oxazoline (**344**) is metallated at C-3 with methyllithium while the 3-isomer (**272**) reacts at C-4 with lithium tetramethylpiperidide ⟨82JOC2633⟩. Treatment of 3-ethoxy- or 3-butoxy-pyridine with *n*-butyllithium in the presence of TMEDA results in apparently exclusive metallation at the 2-position

⟨82S235⟩. Examples of halogen directed metallation include the lithiation of 5-bromopyrimidine (**345**) or chloropyrazine (**346**) by LDA. 3-Halopyridines are also lithiated regiospecifically at C-4, but pyridyne formation is then rapid.

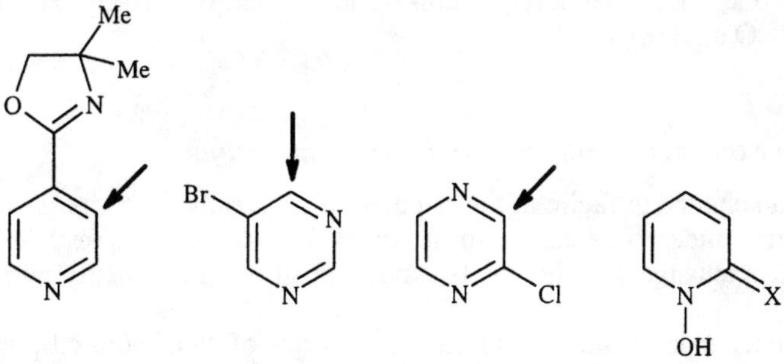

(**344**)   (**345**)   (**346**)   (**347**)

The carbanion can usually be masked or trapped as a more metalloid derivative and be released in the presence of an electrophile. Trialkylsilane and trialkylstannane derivatives are most often used ⟨81T4069, 88G211⟩, for example in *ipso* substitution with acyl halides.

Ring metallation generally succeeds with *N*-oxides. α-Lithio derivatives (**348**) can be generated in non-protic conditions by treating pyridine 1-oxides with *n*-butyllithium. These may be intercepted by various electrophiles such as cyclohexanone (Scheme 35). Reaction of substituted lithio *N*-oxides with carbon dioxide gives carboxylic acids; with elementary O_2 and S_8 (**347**, X = O, S) are produced.

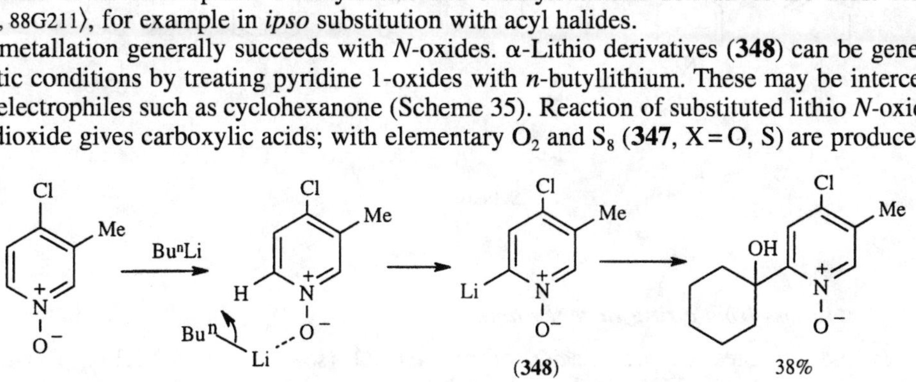

(**348**)   38%

Scheme 35

3.2.1.8.2 *Hydrogen exchange at ring carbon in neutral azines, N-oxides and azinones*

Pyridine undergoes base-catalyzed hydrogen-deuterium exchange much more readily than benzene, resulting in eventual replacement of all hydrogen atoms by deprotonation followed by rapid deuteration of the intermediate negatively charged species in a sequential manner. The reactivity order γ > β > α is found for exchange in NaOMe-MeOD at 160°C, NaOD-D_2O at 200°C and NaND$_2$-ND$_3$ at −25°C. The low reactivity of the 2-position reflects the unfavorable lone pair-lone pair interaction in the intermediate carbanion (**349**). Pyridine *N*-oxides exchange all protons, but now the C-2 protons react the most readily *via* the anion (**350**). In aqueous solution at low pH, base-catalyzed hydrogen deuterium exchange of pyridine can involve the pyridinium ion (**351**) and ylide (**352**) ⟨74AHC(16)1⟩. The exchange is facilitated by electron-withdrawing groups; in particular an electron-withdrawing group at the 3-position of pyridine accelerates exchange at the 4-position and *vice versa*. Quinoline undergoes hydrogen-deuterium exchange in NaOMe-MeOD, at 190.6°C at the 2-, 3-, 4- and 8-positions (**353**) ⟨73JA3928⟩.

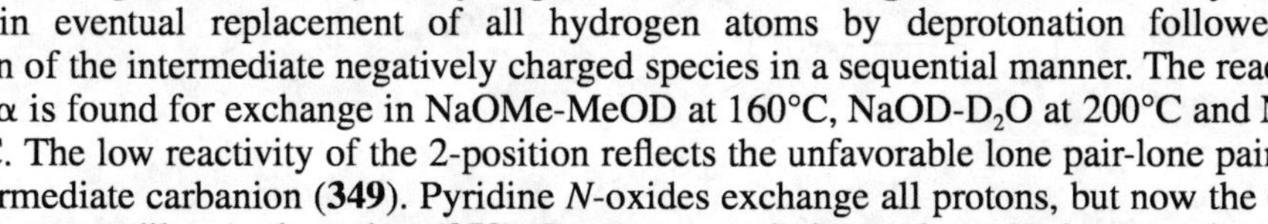

(**349**)   (**350**)   (**351**)   (**352**)   (**353**)

In pyridazines, base-catalyzed hydrogen-deuterium exchange takes place at positions 4 and 5 more easily than at positions 3 and 6. Pyridazine 1-oxide reacts first at positions 5 and 6 and then at positions

3 and 4. Pyrimidine exchanges as expected most readily at the 5-position, next at the 4-, and least readily at the 2-position. In pyrimidine 1-oxide, the reactivity order is 2 > 6 > 4 >> 5. 1,2,4-Triazines easily undergo base-catalyzed hydrogen exchange in the 3-position.

1-Methylpyridin-4-one (and 1-methylpyridin-2-one) undergoes H-D exchange at the 2- and 6-positions in basic D_2O at 100°C.

3.2.1.8.3 *Hydrogen exchange at ring carbon in azinium cations*

Hydrogen isotope exchange is facile at the α-position in pyridinium salts. 3-Methyl- and 3-cyano-pyridinium methiodides undergo exchange in the order 2 > 6 >> 4, 5 in 0.01 M NaOD-D_2O. The relative rates of H-D exchange for the α-, β- and γ-positions in 1-methylpyridinium chloride are 3400:3:1.

The rates of H-D exchange at the α-positions for a series of *N*-substituted pyridinium cations and pyridine 1-oxide derivatives in D_2O at 75°C (Scheme 36) ⟨70JA7547⟩ correlate well with the Taft inductive parameter σ_I (ρ_I = 15). A positively charged nitrogen in a ring is estimated to activate the α-position toward deprotonation and ylide formation by a factor of 10^{15}.

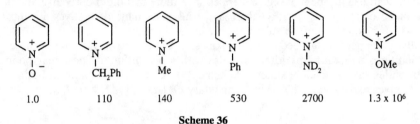

Scheme 36

3.2.1.8.4 *Proton loss from a ring nitrogen atom*

Pyridones and azinones are weak acids of pK_a *ca.* 11 (see Section 3.2.1.3.4.iv). They form mesomeric anions (*cf.* **354** → **355**) which react very readily with electrophilic reagents at the nitrogen, oxygen or carbon atom, depending on the circumstances (see Section 3.2.3.7.2). The anion (**355**) from 2-pyridone is alkylated and aminated mainly on nitrogen (**355** → **356, 357**), acylated on oxygen (**355** → **358**) and reacts at a ring carbon atom in the Kolbe reaction (**355** → **359**). Attack on the pyridone anion (*cf.* **355**) is probably involved in certain other electrophilic substitutions, *e.g.* the diazo coupling of 4-quinolone (see Section 3.2.1.4.10).

The position of alkylation can depend on counter ion, solvent and reagent. Thus, Ag salts tend to give *O*-alkylation, whereas Na or K salts predominantly undergo *N*-alkylation (see CHEC 2.05.2.5).

Reaction of pyridones with diazoalkanes involves deprotonation as the first step, forming an alkyldiazonium cation which then rapidly alkylates the pyridone anion; 2-pyridone gives mainly *O*-methyl derivatives, but 4-pyridone gives mixed *O*- and *N*-methyl derivatives.

The reactivity pattern of azinones containing an NH group is similar to that of the pyridones just discussed. Thus, for example, the alkylation of pyridazinones under basic conditions also gives mixtures of *N*- and *O*-substitution products. At the same time, uracil, unlike 2-pyridone, is benzoylated at a nitrogen, not an oxygen, atom. Depending on the quantity of benzoyl chloride, 1-benzoyl- or 1,3-dibenzoyl derivatives can be obtained. Selective removal of the 1-benzoyl group can be effected under mild basic conditions to furnish the 3-benzoyl derivative ⟨84T681⟩.

Perimidines with a free NH-group (**360**) in basic media form readily oxidizable anions (**361**). The latter can be *N*-alkylated in inert atmosphere, the reaction being sensitive to steric interference from 2- and 4(9)-substituents ⟨81RCR1559⟩.

(**360**)   (**361**)   (**362**)

3.2.1.9 Reactions with Radicals and with Electron-deficient Species; Reactions at Surfaces

3.2.1.9.1 *Carbenes and nitrenes*

Attack generally occurs at a pyridine nitrogen atom for both carbenes and nitrenes.

The reaction of triplet diphenylcarbene with pyridine has been well studied, and a mechanism proposed from kinetic data ⟨90TL953⟩. The carbenes generated from laser flash photolysis of alkylbromo- and alkylfluoro-diazirines were trapped by pyridine to form the pyridinium ylides ⟨94JPO24⟩.

Ylide (**364**) has been obtained by heating (**363**) in pyridine. Isoquinoline yields a stable ylide upon reaction with ethoxycarbonylcarbene ⟨70TL941⟩ and the relative rates of reaction of ethoxy-carbonylcarbene with pyridine, quinoline, and acridine have been studied ⟨88JOC4374⟩. However, the isoquinolone (**365**) undergoes attack at the isoquinoline double bond to give (**366**).

(**363**)   (**364**)   (**365**)   (**366**)   (**367**)

Sulfonyl azides react with pyridine to give pyridine 1-sulfonylimides (*e.g.* **367**). However, the analogous reaction with 2,4,6-trimethylpyridine gives some 3-(phenylsulfonylamino) derivative together with the 1-sulfonylimide. Nitrenes derived from photolysis of acyl azides also add to the nitrogen atom to form the corresponding pyridine *N*-imines ⟨74AHC(17)220⟩.

3.2.1.9.2 *Radical attack at ring carbon atoms*

(i) *Halogen atoms*

Regiospecific chlorination of pyridine at the α-position occurs rapidly during irradiation in the presence of chlorine dissolved in carbon tetrachloride ⟨67JHC375, 93AHC(58)274⟩. Both ring and side-chain chlorination of 4-picoline occurs under γ-irradiation conditions ⟨80JGU354⟩.

Vapor phase halogenation of pyridine at high temperatures gives mixtures of 2- and 2,6-dibromopyridine (Br$_2$, 500°C or CuBr-Br$_2$, 350°C) and 2- and 2,6-di-chloropyridine (Cl$_2$, 270°C).

2-Fluoropyridines have been prepared by direct reaction of fluorine diluted in an inert gas and dissolved in a polyhalogenated solvent ⟨91BCJ1081⟩. Presumably these reactions involve attack by free halogen atoms as distinct from the ionic halogenation at lower temperatures which gives β-orientation (*cf.* Section 3.2.1.4.7). Under similar conditions (Br$_2$, 450°C) quinoline gives 2-bromoquinoline.

(ii) Alkyl and α-hydroxyalkyl radicals

Alkyl radicals, prepared *in situ*, react with pyridine to form mainly 2-alkyl derivatives and this regioselectivity is shown in the free radical reactions of alkylmercurial compounds with pyridines (Scheme 37).

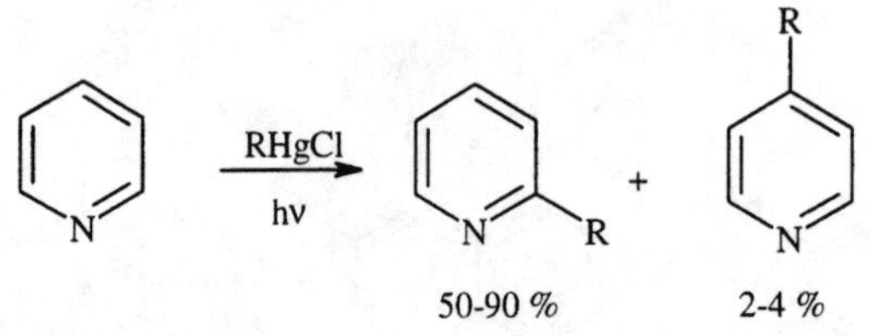

50-90 %   2-4 %

Scheme 37

These and other homolytic alkylations of neutral heteroaromatics usually proceed in poor yields, but if protonated heteroaromatic bases are used, many of the side reactions are minimized and selectivity is high and yields are good. Selectivity is increased because the alkyl radicals are nucleophilic in character and thus selectively attack the α-position.

Alkyl radicals for such reactions are available from many sources such as acyl peroxides, alkyl hydroperoxides, particularly by the oxidative decarboxylation of carboxylic acids using peroxydisulfate catalyzed by silver. Pyridine and various substituted pyridines have been alkylated in the 2-position in high yield by these methods. Quinoline similarly reacts in the 2-, isoquinoline in the 1-, and acridine in the 9-position. Pyrazine and quinoxaline also give high yields of 2-substituted alkyl derivatives ⟨74AHC(16)123⟩.

3,6-Dichloropyridazine can be *tert*-butylated to give 4-*t*-butyl-3,6-dichloropyridazine (**368**) in high yield ⟨88OPP117⟩, and hydroxy-*t*-butyl radicals, generated *in situ* from 2,2-dimethyl-1,3-propanediol, give the 4-hydroxy-*t*-butyl derivative (**369**) (55%) along with some of the fused dihydrofuropyridazine by-product (**370**) ⟨88JOC5704⟩.

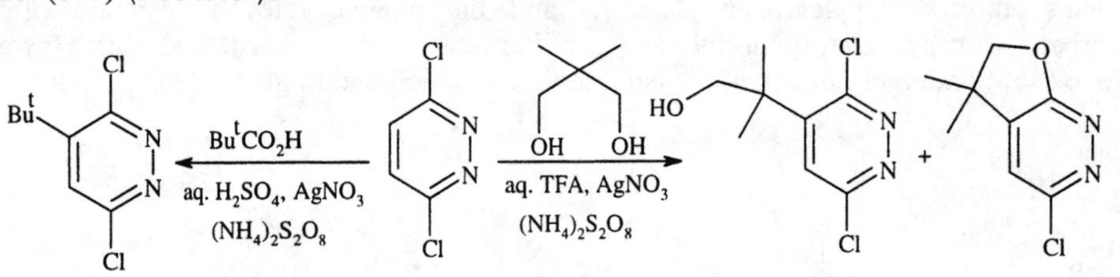

(368)   (369)   (370)

N-ω-Iodoalkylpyridinium salts (**371**, R = H, Me, n = 0–2) undergo intramolecular radical cyclization upon treatment with tributyltin hydride and AIBN to give the [6,5], [6,6], and [6,7] fused pyridinium salts (**372**) in good yields ⟨90TL1625⟩.

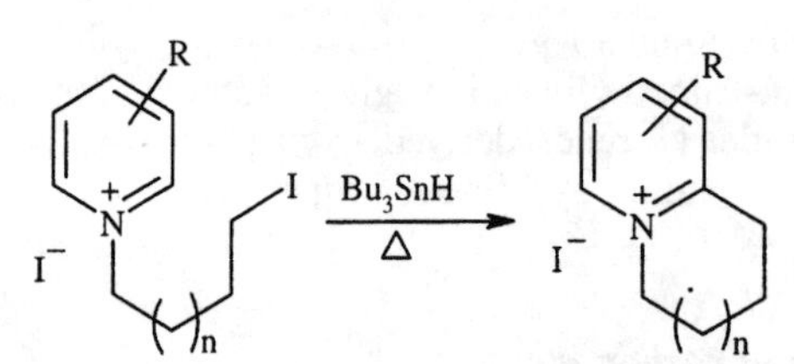

(371)   (372)

Hydrogen abstraction from a position α to the oxygen of alcohols and ethers gives α-oxyalkyl radicals which add readily to electron-deficient heterocycles (Minisci reaction): for example, pyridine is hydroxymethylated at C-2 and C-4 using methanol and ammonium persulfate; 4-methylquinoline yields the 2-CH$_2$OH derivative with NH$_2$OSO$_3$H + MeOH + FeCl$_3$ ⟨83CC916⟩.

(iii) Acyl radicals

Acyl radicals obtained by the oxidation of aldehydes or the oxidative decarboxylation of α-keto acids react selectively at the α- or γ-position of the protonated heterocyclic nitrogen. Pyridines, quinolines, pyrazines and quinoxalines all react as expected; yields are typically 40 to 70%. Similarly, pyridines can be carbamoylated in acid media at C-2 (Scheme 38).

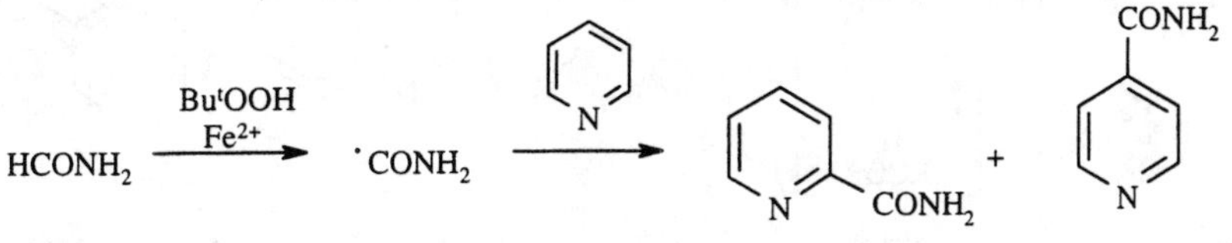

Scheme 38

This reaction also succeeds with diazines. Homolytic acylations of methoxy and chloro substituted pyrazines are directed *ortho*, giving the corresponding 2,3-disubstituted pyrazines. Acetyl, alkoxy-carbonyl, carboxy, cyano and carbamoyl groups direct radical substitution *para*, leading to the corresponding 2,5-disubstituted pyrazines. These selectivities result from a combination of the inductive and resonance effects of the substituents ⟨88S119, 89JOC640, 91S581, 92JHC1685⟩. The Minisci-type alkoxycarbonylation reaction produced an 89% yield of ethyl 2-pyrazinecarboxylate by redox treatment of pyrazine with ethyl pyruvate in a water-dichloromethane two-phase system ⟨86T5973⟩.

(iv) Aryl radicals

In sharp contrast, homolytic arylation is unselective and gives low yields. Phenyl radicals attack pyridine unselectively to form a mixture of 2-, 3- and 4-phenylpyridines in proportions of *ca.* 53, 33 and 14%, respectively. The phenyl radicals may be prepared from the normal precursors: PhN(NO)COMe, Pb(OCOPh)$_4$, (PhCO$_2$)$_2$ or PhI(OCOPh)$_2$. Substituted phenyl radicals react similarly.

Photolysis of aryl or pyridyl oxime esters in pyridine provides α-phenylpyridines as the major products together with bipyridyls ⟨84TL3887⟩. Rate constants for the addition of phenyl radical to protonated and non-protonated 4-substituted pyridines have been determined by studing the competition between phenyl radical addition and chlorine abstraction from carbon. The 4-arylpyridines were the major products, and no 3-substituted pyridines were observed. Among the solvents studied (MeCN, DMF, DMSO, and HMPA), MeCN gave the highest yields and selectivity ⟨91OPP438⟩.

(v) Hydrogen atoms

As for any other organic molecules containing CH bonds, heteroaromatics can be labeled with tritium by reaction with energetic tritium atoms from neutron irradiation.

3.2.1.9.3 *Electrochemical reactions and reactions with free electrons* (See also Section 3.2.1.6.9 – Chemical reduction)

(i) Neutral species: reactions with metals

Certain metals (*e.g.* Na, Zn) add one electron to pyridine to form a radical anion (**373** ↔ **374**) which can dimerize by reaction at the α- or γ-position; these dimers form bipyridyls by hydride ion loss. On treatment with sodium at 20°C, pyridine forms mixtures of 2,2'-, 2,3'-, 2,4'- and 4,4'-bipyridyl, probably by aromatization of intermediate dihydro compounds (**375** → **376**). The reaction can be directed to give largely 4,4'-bipyridyl, a product of commercial importance.

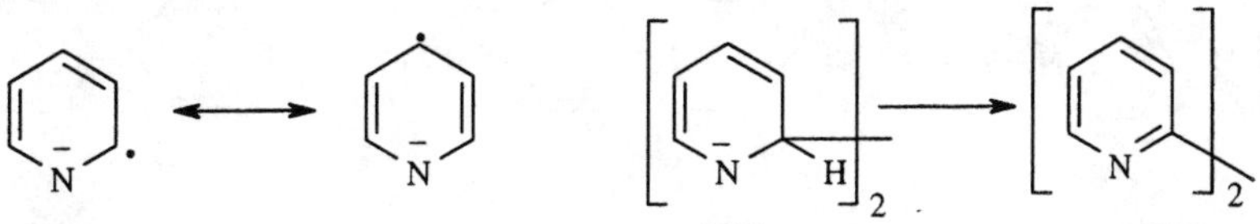

Pyridine is converted by a modified Raney nickel catalyst into 2,2'-bipyridyl, and the reaction has been extended to many substituted pyridines and quinolines. 2-Substituted pyridines give the 6,6'-bipyridyls. 3-Methylpyridine gives 5,5'-dimethyl-2,2'-bipyridyl, but none of the other isomers.

These dimerizations are analogous to those of the radical anions R$_2$C$^{\bullet}$ – O$^-$ which are intermediates in the reduction of ketones to pinacols. Indeed, in the presence of magnesium amalgam, pyridine condenses with a ketone to give an alcohol (**377**) by oxidation of the intermediate dihydropyridine. In a similar reaction type, pyridine with zinc and acetic anhydride or ethyl chloroformate yields (**378**; R = Me and OEt, respectively).

(**377**)   (**378**)   (**379**)   (**380**)

Treatment of quinoline with sodium gives mainly 2,3′-biquinolyl, the formation of which can possibly be explained as initial reduction followed by reaction of the dihydroquinoline anion with another molecule of quinoline (**379**). Quinoxaline derivatives in an ether solvent are reduced by sodium to the corresponding anti-aromatic doubly charged anions, *e.g.* (**380**) ⟨85JA1501⟩.

(ii) Neutral species: electrochemical reduction

Electrochemical reduction of pyridines to piperidines can be achieved using various methods. Piperidines can be obtained in high yield by the electrochemical reduction of pyridine on a lead cathode in the presence of carbon dioxide and Pd − Ni or Cu − Ni catalysts ⟨89KFZ1120⟩. In the absence of catalyst, 4,4′-bipyridine was produced as the major product.

Electron-attracting substituents facilitate electrochemical reduction. The reduction potentials for the polarographic reduction of quinoline and isoquinoline derivatives are much less negative than those for the pyridine analogues. Diazines are reduced electrochemically stepwise, usually as far as tetrahydro derivatives ⟨70AHC(12)262⟩.

The behavior of pyrimidine during polarographic reduction depends on the pH of the aqueous solution. In acidic solution two one-electron waves are observed, while in neutral solution two two-electron waves result. In alkaline solution four-electron reduction is effected *via* 1,6-dihydropyrimidine to give tetrahydropyrimidine ⟨84AHC(36)235⟩.

(iii) Cationic rings

Pyridinium cations are reduced electrochemically or by metals to neutral radicals of considerable stability, especially when merostabilization by an α- or γ-substituent occurs; thus (**381**) has been isolated. Bispyridinium compounds are particularly readily reduced to radical cations, such as (**383**). Radical (**383**; R = Me) is the active species of the herbicide paraquat. Pyridyl radicals without such stabilization dimerize and form bispyridinium compounds by oxidation.

One-electron reduction of pyrylium salts, with dissolving metals or electrochemically, gives dimers (*e.g.* **382**) *via* pyranyl radicals ⟨80AHC(27)46⟩.

(**381**)   (**382**)

(**383**)

3.2.1.9.4 Other reactions at surfaces

(i) Catalytic hydrogenation

Hydrogenation of pyridines and their benzoderivatives can be carried out with various catalysts. Pyridines are readily hydrogenated to piperidines over Raney nickel at 120°C. The optimal pressure for hydrogenation of pyridine to piperidine over platinum is reported to be 2.0–2.5 mpa, and at 20–8.1 mpa the activation energy is 35.5–42.2 kJ mol^{-1} ⟨85ZPK322⟩.

Reductions with noble metal catalysts proceed smoothly (at 20°C) when the bases are in the form of hydrochlorides; the free bases tend to poison the catalyst. A pyridine ring is reduced more easily than a benzene ring; thus, 2-phenylpyridine gives 2-phenylpiperidine (**384**), quinoline gives 1,2,3,4-tetrahydroquinoline (**385**) and acridine gives 9,10-dihydroacridine (**386**).

Pyridinium and pyrylium cations, pyridones and pyrones are all readily hydrogenated; *e.g.* flavylium ion (**220**) and coumarin yield (**387**) and (**388**), respectively.

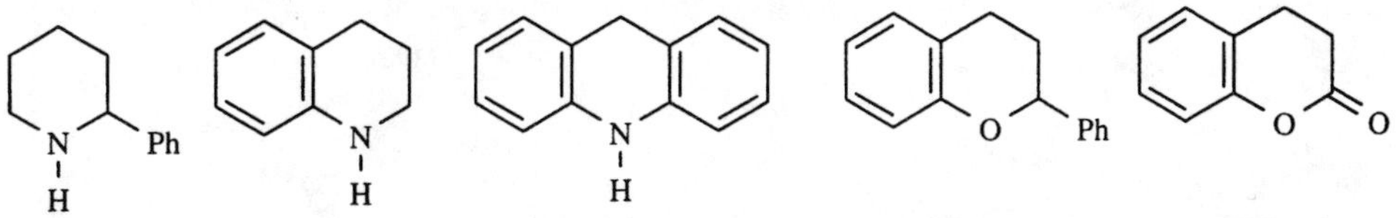

Palladium on charcoal (Pd/C) is commonly used in the catalytic hydrogenation of pyrimidines in acidic media to form 1,2,4,5-tetrahydro derivatives which are stabilized as amidinium salts ⟨62JOC2170, 65JCS1406⟩. Platinum effects hydrogenation of the 5,6-double bond of uracils, for example, in the addition of deuterium to produce [5,6-^{2}H$_2$]5,6-dihydrouracil. The use of rhodium-on-charcoal and Raney nickel also gives good results. The addition of hydrogen to the 5,6-bond of thymidine and other 5-substituted uridines is stereospecific with rhodium-on-alumina as catalyst.

(ii) Isotopic hydrogen exchange

Transition metals also catalyze isotopic exchange reactions. Platinum is the most active catalyst for most heterocycles. The mechanism may involve metallation, addition, σ-addition and π-complex formation. α-Hydrogen exchange in pyridine is favored over β- and γ-positions, particularly by a cobalt catalyst whereas platinum is much less selective. In isoquinoline both the 1- and 3-position protons are exchanged at almost the same rates with very little exchange at any other position. In 3-substituted pyridines exchange is preferred at the 6-position, the more so as the size of the 3-substituent increases ⟨73AHC(15)140⟩.

3.2.1.10 Reactions with Cyclic Transition States

3.2.1.10.1 Introduction

Reactions of this type are characteristic of compounds with low aromaticity. While rare in pyridine, they are favored by the following structural modifications which lower the aromatic stabilization energy:

(i) Benzo ring fusion, especially three or more rings linearly fused.
(ii) Polyhetero rings and especially two adjacent ring nitrogens or a ring oxygen.
(iii) Enhanced π-deficiency especially caused by presence of positively charged heteroatom.
(iv) Exocyclic carbonyl groups and especially betaine structures.

We classify these reactions according to the type of cycloaddition reaction and type of heterocyclic systems that participate in it.

3.2.1.10.2 Heterocycles as inner dienes in [2+4] cycloaddition

Because of the strong π-deficiency of most six-membered heteroaromatic compounds, cyclo-additions of this type belong to Diels-Alder reactions with inverse electron demands; in other words, they are LUMO$_{diene}$-HOMO$_{phil}$ controlled reactions (for review see ⟨B-87MI 502-08⟩). Acceptor substituents in the heterocyclic diene and donor substituents in the dienophile accelerate the reaction, as shown by kinetic data ⟨83TL1481, 84TL2541, 90TL6851⟩.

(i) Heterocyclic systems with one heteroatom

Condensed heteroaromatic cations are reactive in [2+4] cycloaddition reactions with inverse electron demand. For instance, 2-benzopyrylium salts (**389**) react with vinyl ethyl ether to afford naphthalene derivatives (**392**) in good yield ⟨90KGS315⟩. Similar transformations (Bradsher reaction) are also known for isoquinolinium salts ⟨84CC761⟩.

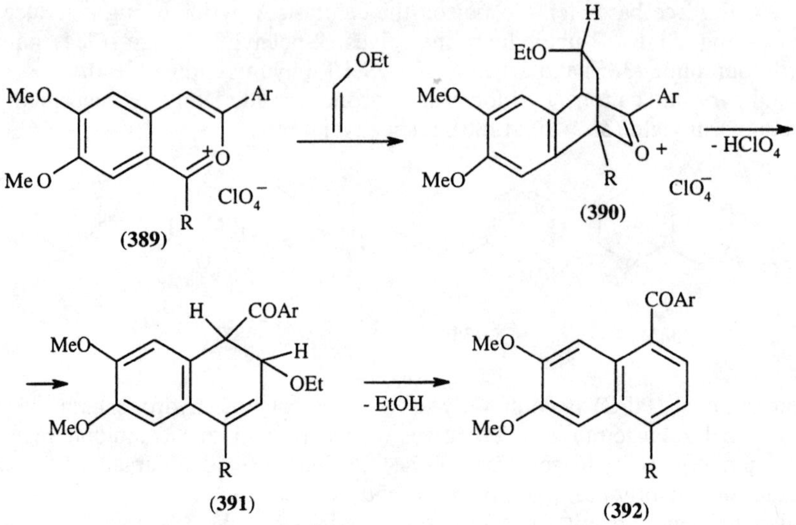

(**389**)   (**390**)

(**391**)   (**392**)

The acridizinium ion adds to various dienophiles to give products of the type (**393**; Scheme 39).

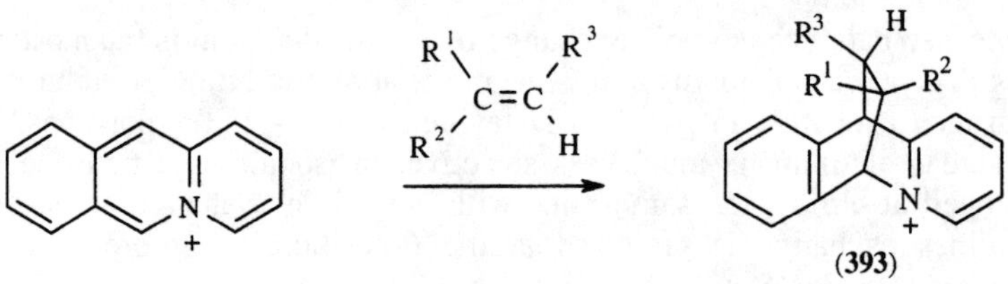

(**393**)

Scheme 39

α-Pyranones undergo Diels–Alder reactions: with maleic anhydride; adducts of type (**394**) are formed which can lose carbon dioxide and react with more anhydride to give (**395**). 2-Pyranones also react with singlet oxygen to give endo-peroxides (**396**). Reaction of α-pyranone with nitrosobenzene leads first to the anti-aromatic transient 2-phenyl-1,2-oxazine which appears to undergo further cycloaddition immediately (Scheme 40). Benzyne reacts with 1-methyl-2-pyridone to give (**397**).

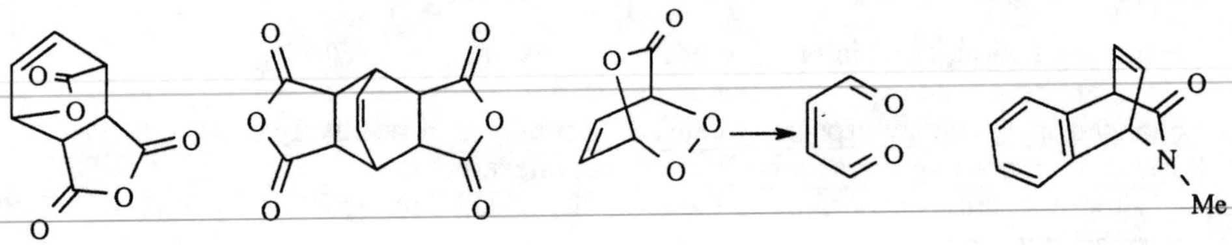

(**394**)   (**395**)   (**396**)   (**397**) R = Me

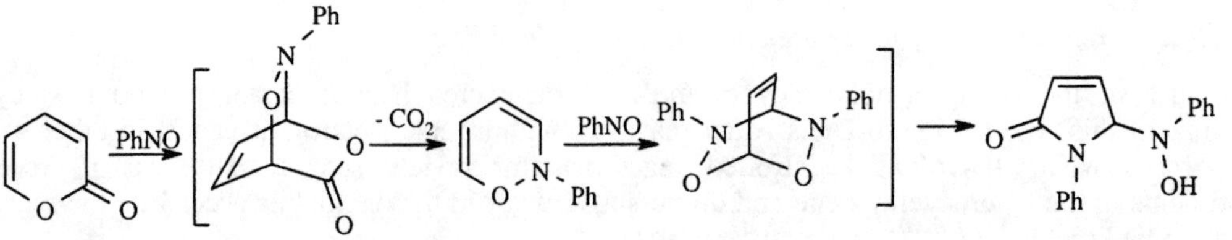

Scheme 40

(ii) Diazines

Pyridazinecarboxylic esters undergo cycloaddition reactions as exemplified in Scheme 41.

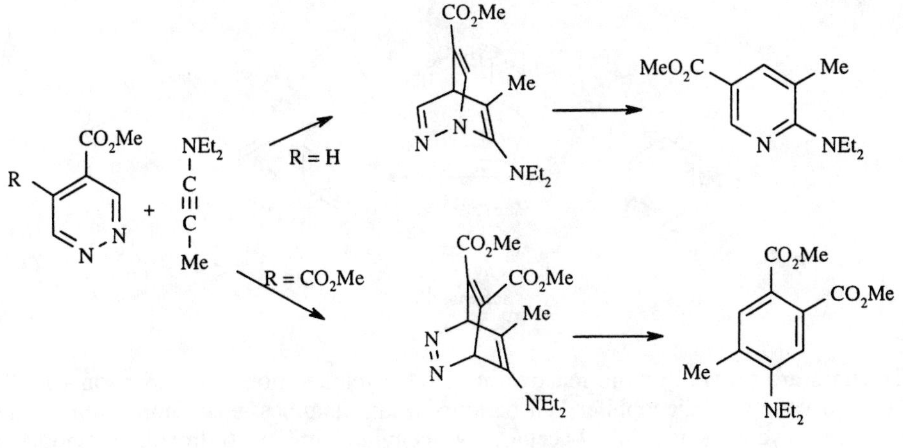

Scheme 41

Phthalazines undergo [2+4] cycloaddition with enamines to give naphthalene derivatives. Pyrimidines also react as azadienes in reactions with enamines, *e.g.* (**398** → **399**).

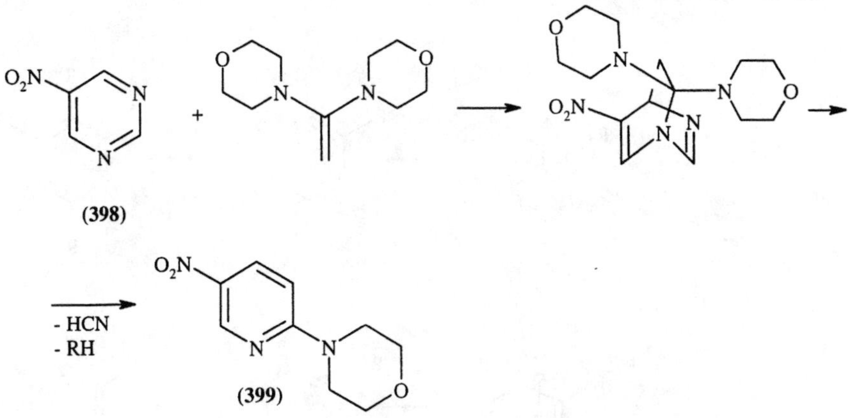

2,5-Dimethoxy-3-alkenyl substituted pyrazines (**400**) undergo facile addition of singlet oxygen to form relatively stable endo-peroxides (**401**), which decompose when heated ⟨89JCS(P1)453⟩. The peroxides having a benzyl or isobutyl substituent at C-3 decompose mainly by loss of oxygen, regenerating the pyrazines in high yields. Deoxygenation with triphenylphosphine induces skeletal rearrangement, the pyrazine undergoing ring contraction to a substituted imidazole (**402**).

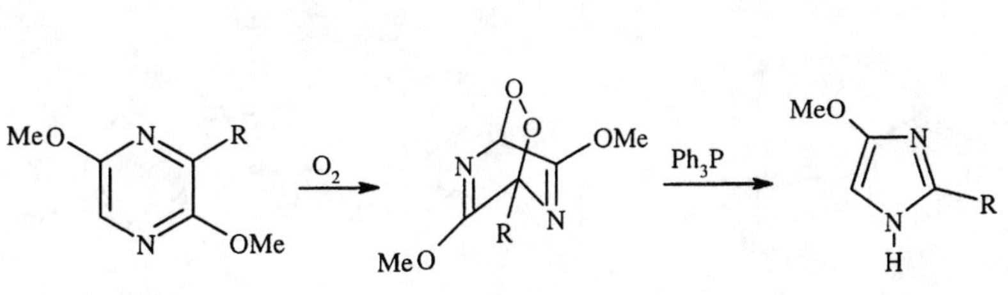

(iii) Triazines

1,2,3-Triazines react with various kinds of enamines to afford pyridine derivatives (see Scheme 42). This approach was applied to the total syntheses of several pyridine alkaloids.

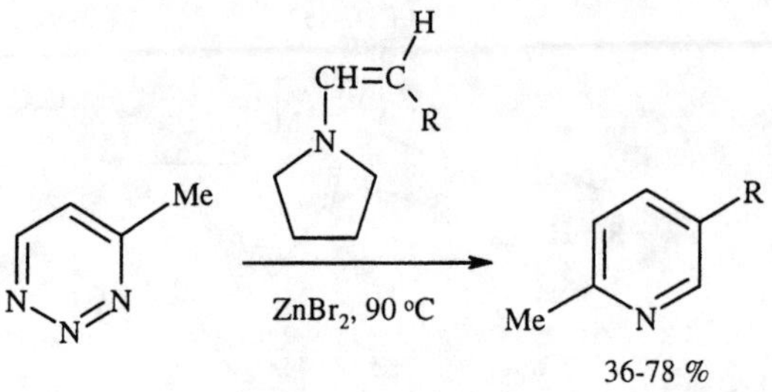
H
CH=C
R
N
Me
R
N
N
N
ZnBr2, 90 °C
Me
N
36-78 %

Scheme 42

1,2,4-Triazines are generally more reactive in [2+4] cycloaddition in comparison with 1,2,3-triazines. The wide variety of dienophiles can be employed: enamines, enaminones, vinyl silyl ethers, vinyl thioethers, cyclic ketene *N,O*-acetals, *N*-phenylmaleimide, 6-dimethylaminopentafulvene, 2-alkylidene-imidazolidines (cyclic ketene aminals), cyclic vinyl ethers, arynes, benzocyclopropene, acetylenes, and alkenes like ethylene, (*Z*)-but-2-ene, cyclopentene, cyclooctene and bicyclo[2.2.1]hept-2-ene, hexa-1,5-diene, cycloocta-1,5-diene, diallyl ether, cyclododeca-1,5,9-triene, di-(3-methylcyclopropen-3-yl), and norbornadiene.

In most Diels-Alder reactions, dienophile addition occurs across the 3- and 6-positions of the 1,2,4-triazine ring as shown in Scheme 43, but acetylenes can also add across the 2- and 5-positions (Scheme 44). The orientations can be explained using the frontier-orbital method, or through secondary orbital interactions.

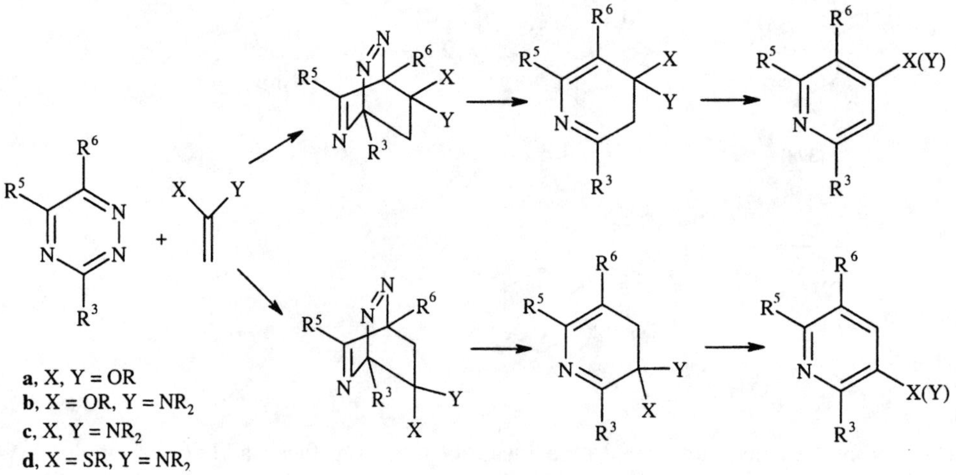
a, X, Y = OR
b, X = OR, Y = NR2
c, X, Y = NR2
d, X = SR, Y = NR2

Scheme 43

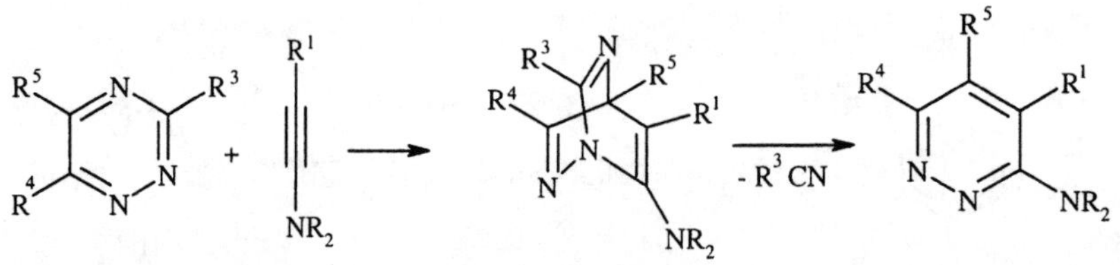
- R3CN

Scheme 44

1,3,5-Triazines usually react with dienophiles in Diels-Alder reactions to afford pyrimidine derivatives (Equation 7).

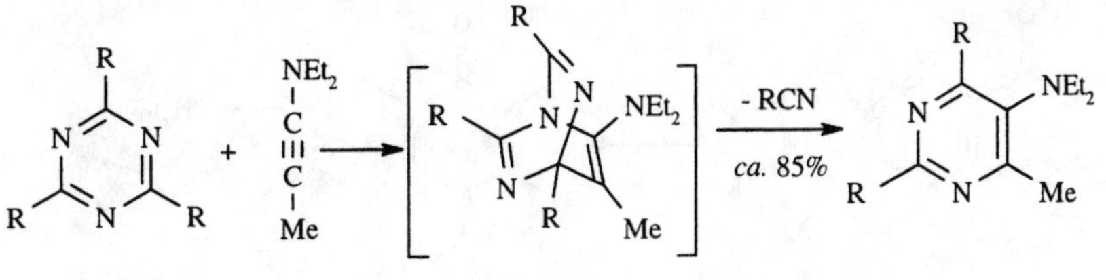

$$R = H, Me, CO_2Et, \textit{etc.}$$

(7)

(iv) Tetrazines

1,2,4,5-Tetrazines react with alkenes to give bicycles (**403**) which lose nitrogen to give the 4,5-dihydropyridazine (**404**). This can either tautomerize to a 1,4-dihydropyridazine, be oxidized to the aromatic pyridazine, or undergo a second Diels-Alder reaction to give (**405**). Many heterocycles can act as the dienophiles in such reactions; for example thiophene gives (**406**). The reaction is also used to trap unstable compounds, for example, 2-phenylbenzazete (**407**) as compound (**408**).

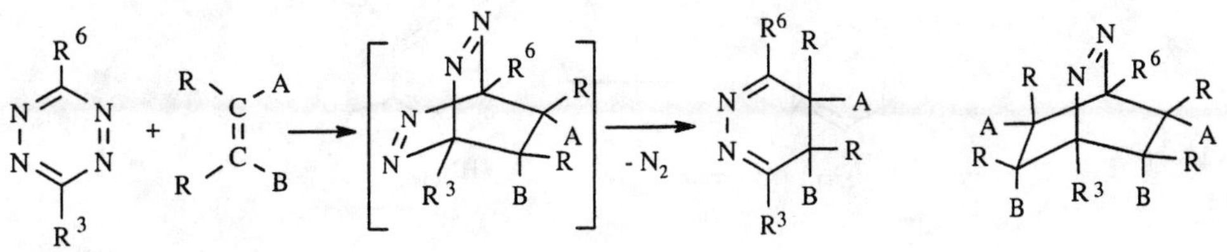

(403)   (404)   (405)

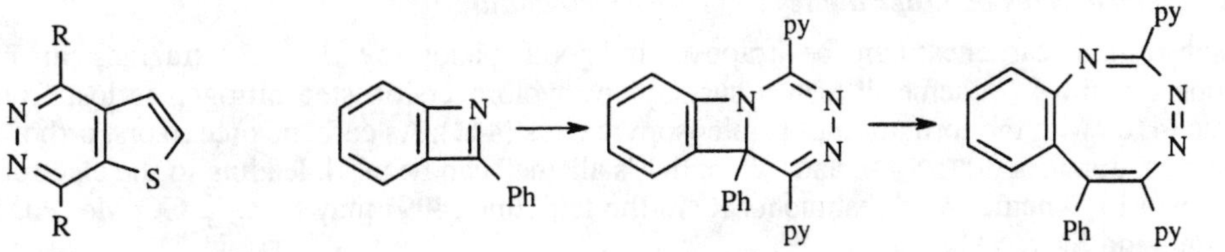

(406)   (407)   (408)

The reactivity of tetrazines with electron-attracting substituents (*e.g.* $R^3 = R^6 = CO_2Me$) is particularly high, so it was proposed to use them as titrating agents to determine the purity of liquid alkenes ⟨62CB2248⟩.

(v) Intramolecular cycloadditions

Intramolecular Diels-Alder reactions with inverse electron demands are widely used for the synthesis of condensed pyridines, pyrimidines and pyrazines. Typical examples are shown on Schemes 45–48. The heterocyclic substrate should possess an appropriate dienophilic side-chain with terminal alkyne or nitrile group. The yields as a rule are high though the reactions usually demand prolonged heating in high boiling solvents ($C_6H_5NO_2$, 1,2-$Cl_2C_6H_4$, *etc.*). Reactivity in intramolecular [2 + 4] cycloadditions is strongly related to the conformational properties of the side-chain. Rate reduction has been found for molecules that are able to form stable conformations in which the azadiene and dienophile side-chain are positioned in such a way that their interactive approach becomes more difficult ⟨92JOC3000⟩. Alkynylsulfinyl compounds are particularly reactive ⟨85TL2419⟩.

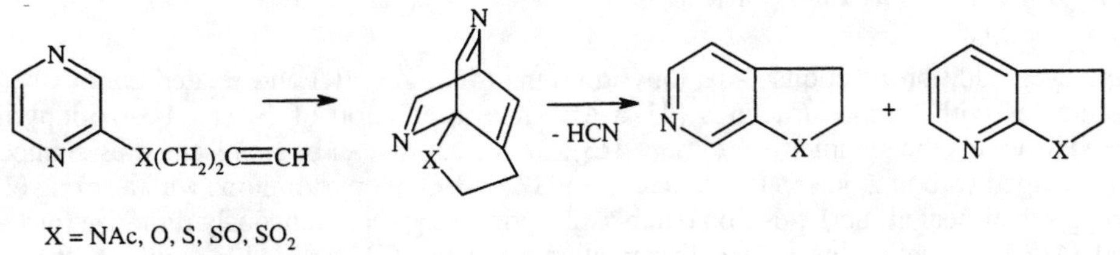

$$X = NAc, O, S, SO, SO_2$$

Scheme 45

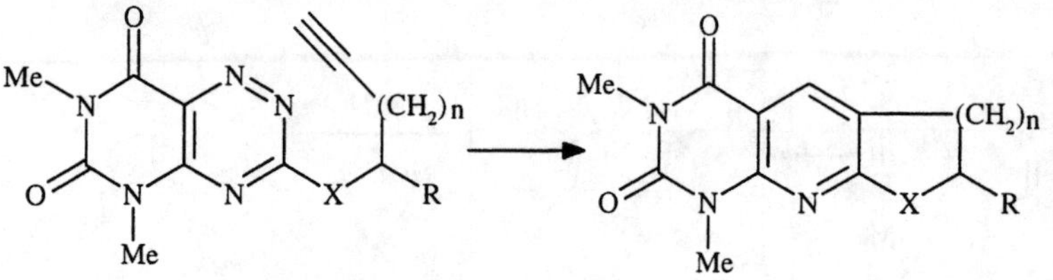

Scheme 46

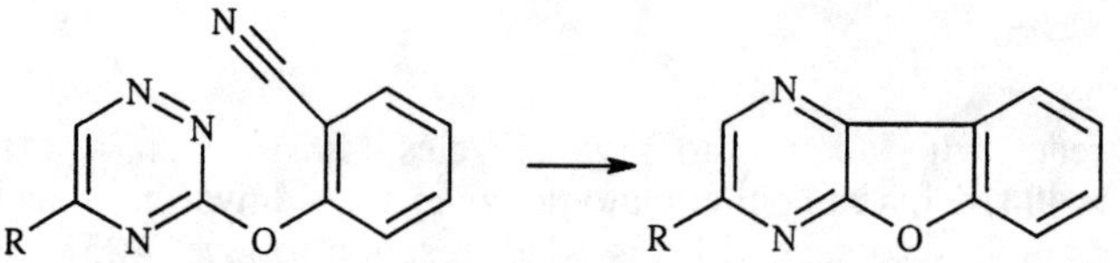

Scheme 47

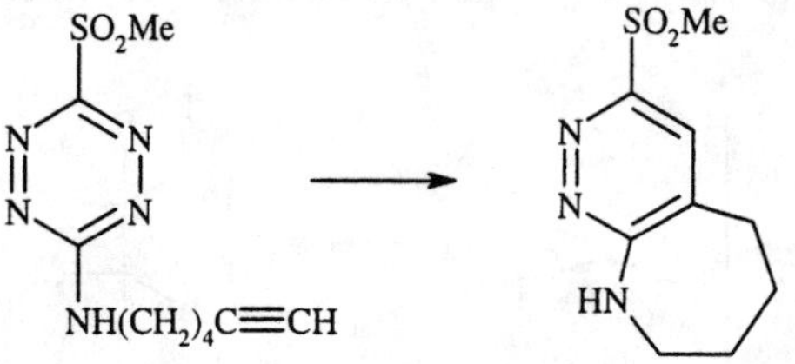

Scheme 48

3.2.1.10.3 *Heterocycles as inner dienes in [1 + 4] cycloaddition*

Donor-substituted carbenes can be trapped in good yields by 1,2,4,5-tetrazines in [1+4] cycloaddition reactions (Scheme 49); in a subsequent cycloreversion step nitrogen is lost from the intermediate (**410**) with the formation of stable isopyrazoles (**411**). As carbene precursors, orthoformic acid derivatives, diazines ⟨91TL2743⟩, and "Wanzlick's alkene" can be used, leading to the electron-rich carbenes shown in Scheme 49. Substituents R^1 in the tetrazine (**409**) may be CF_3, CO_2Me, Ph, SMe, SO_2Me/NMe_2, and SMe/NMe_2.

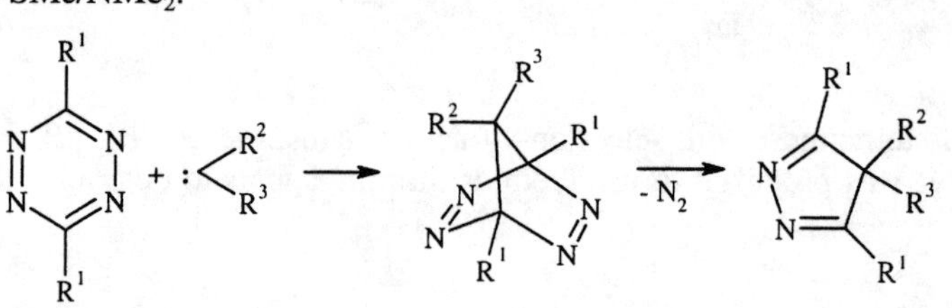

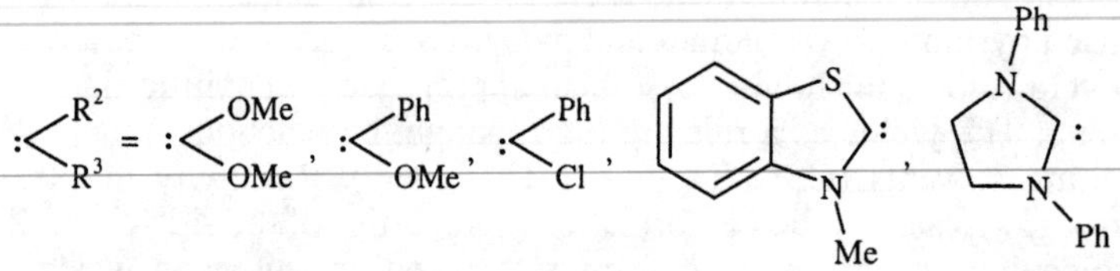

Scheme 49

3.2.1.10.4 *Heterocycles as 1,3-dipoles*

(i) *Inner 1,3-dipoles*

In many cycloaddition reactions 3-oxidopyridiniums (**412**, $Z = NR$) and related cations behave as inner 1,3-dipoles with formal structures (**412a-d**). Thus, irradiation of 1-phenyl-3-oxidopyridinium (**412**, $Z = NPh$) yields the symmetrical dimer (**413**). Its formation can be represented as connection of oppositely charged carbon atoms in two molecules (**412b**). 3-Oxidopyridiniums with a strong electron-withdrawing substituent at the 1-position (such as 2-pyrimidinyl) spontaneously dimerize into isomers (**414**) and (**415**). Formally this is the dimerization of types (**412b**) + (**412c**) and (**412b**) + (**412d**). 3-Oxidopyryliums and 3-oxidothiiniums behave similarly. These cyclodimerizations are reversible.

(412a)   (412b)   (412c)   (412d)   *etc.*

(413)   (414)   (415)   (416)

(417)   (418)   (419)   (420)

3-Oxidopyridiniums and 3-oxidopyryliums react with a variety of monoenes, dienes, and trienes according to the Woodward-Hoffmann rules. 3-Oxido-pyridiniums and -pyryliums react readily with dienophiles to yield cycloadducts of type (416). For Z = NMe, an electron-withdrawing X group is required in the dienophile, but with Z = O or *N*-(2-pyridyl) even unactivated alkenes react. 1-Phenyl-3-oxidopyridinium and benzyne give (417); dienes give adducts of type (418). Fulvenes behave as trienes to give adducts across the 2,6-positions (419). 4-Oxido-2-benzothiinium (420) gives a thermal dimer across the 1,3-positions.

1-Methyl-3-oxidopyrazinium salt (421) undergoes 1,3-dipolar cycloaddition with a variety of dipolarophiles to give the bicyclic compounds (422) existing in enamide tautomeric form ⟨87TL2187⟩.

(421)   (422)

(ii) Inner-outer 1,3-dipoles

Azine *N*-oxides and azine *N*-imides undergo a deep ring transformation on interaction with various dipolarophiles. Thus, fervenulin-4-oxides (423) react with DMAD to afford derivatives of pyrrolo[3,2-*d*]pyrimidines (425). The first step of the reaction is assumed to involve the formation of cycloaddition products (424) which transform by subsequent multi-stage process into (425) ⟨79JOC3830⟩.

(423)   (424)

(425) R = H, Me, Et, Ph

Similarly, 1,2,3-triazine *N*-imines react with DMAD or *N,N*-diethylamino-1-propyne to give the products (426) and (426a), respectively ⟨90CPB2108⟩.

(426)   (426a)

See also Sections 3.2.3.12.4 and 3.2.3.12.5.

3.2.1.10.5 Heterocycles as dienophiles

The 5,6-double bond in activated pyrimidines such as 2-acylamino-6-acetyl-4(1H)-pyrimidinones (427) undergo Diels-Alder reactions to yield on heating with 1-acetoxy-3-methylbutadiene (428) hydroquinazolines (429). The yields are modest, but the regiochemistry is clean. With 1,3-butadiene and with isoprene the cycloadducts were formed in low yields ⟨83JOC3627⟩.

(427)   (428)   (429)

cis   AcO 10-23 %
trans AcO 6-10 %

Pyran-2-ones can give unsymmetrical dimers (430) and (431) where one molecule behaves as diene and another as dienophile.

(430)   (431)

3.2.1.10.6 [2 + 2]Cycloaddition

Uracil and thymine, their 5- and/or 6-substituted derivatives, *N*-alkyl derivatives and their nucleosides and nucleotides, dimerize in solution when irradiated by UV light in the range 200–300 nm to yield cyclobutane dimers (Scheme 50). This reaction also occurs in living tissue between adjacent thymine residues in a polynucleotide to form cyclobutane-type dimers, which are lethal in a variety of organisms. Photolyses repair damaged DNA by utilizing the energy of near-UV or visible light (300–500 nm) to cleave the cyclobutane ring of the dimer ⟨94JA3115⟩. In the dimerization of uracils two regioisomers are formed in which the pyrimidine rings can be regarded as parallel or anti-parallel, and which are referred to as *syn* (**432**) or *anti* (**433**), respectively. Each regioisomer can have a *cis* or *trans* form in which the pyrimidine rings are on the same or opposite side of the cyclobutane plane.

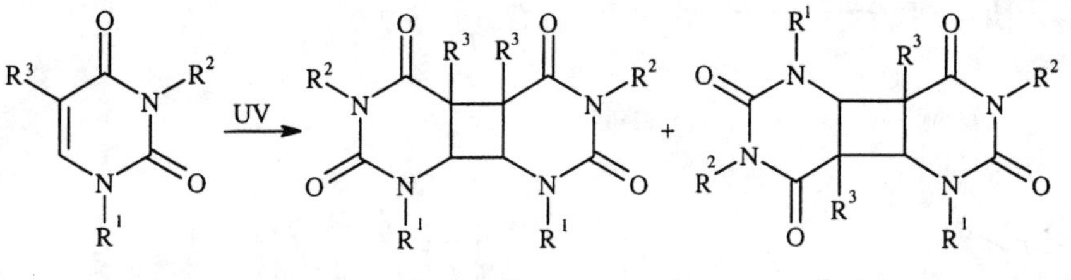

syn: cis, trans   anti: cis, trans

(**432**)   (**433**)

Scheme 50

2-Quinolone undergoes photochemical addition of tetramethylethylene to give (**434**) ⟨70AHC(11)50⟩, 1,3-oxazin-4-ones photocycloadd ketene acetals to give (**435**), and irradiation of 2,6-dimethylpyran-4-one yields the cage dimer (**436**). 2-Pyranones form [2 + 2] photodimers whose structure is similar to that of uracil dimers (**432** or **433**).

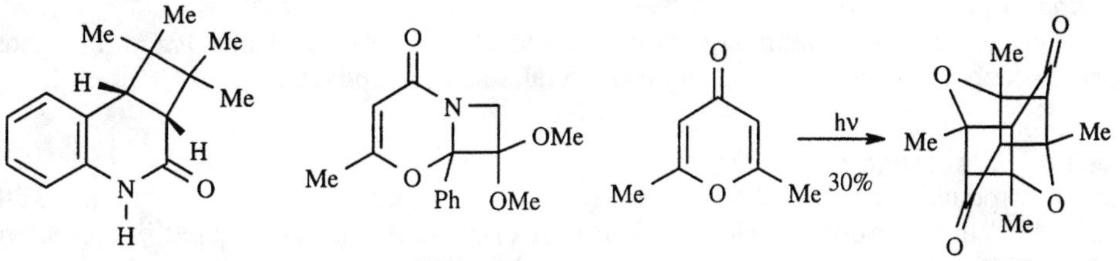

(**434**)   (**435**)   (**436**)

3.2.1.10.7 *Heterocycles as 4π component in [4 + 4] cycloaddition*

2-Pyridone on irradiation in concentrated solution gives the dimer (**437**); 2-aminopyridine behaves similarly. The acridizinium ion (see Scheme 39) like anthracene undergoes [4 + 4] photocycloaddition to yield (**438**).

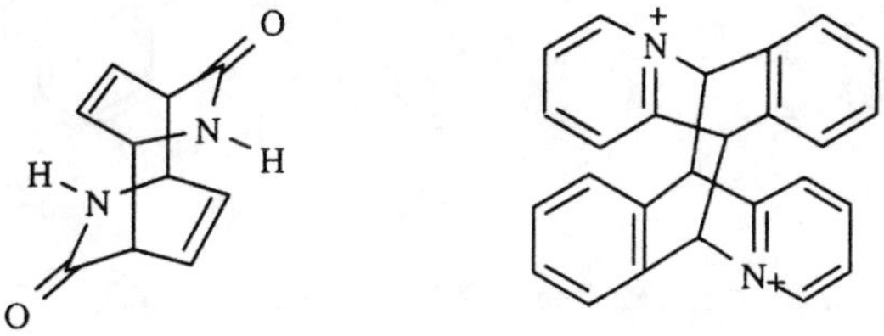

(**437**)   (**438**)

3.2.2 REACTIONS OF NON-AROMATIC COMPOUNDS

We classify these compounds according to their degree of unsaturation. There is a class of eight π-electron systems containing two O, S or NR atoms in the ring which possess significant stability. We then consider the thiabenzenes, which behave as cyclic sulfonium betaines, and related compounds.

Among the hydrogenated derivatives, we distinguish dihydro from the tetrahydro/hexahydro class, as the former bear an intimate relationship to their aromatic analogues.

3.2.2.1   Eight-π-electron Systems: 1,2- and 1,4-Dioxins, -Oxathiins and -Dithiins

3.2.2.1.1   *Intramolecular thermolysis and photolysis reactions*

Substituted 1,4-dioxins thermolyze and photolyze to complex mixtures (see CHEC 2.26.3.1.2). By contrast, 1,4-dithiins and the corresponding sulfoxides generally extrude sulfur or sulfur monoxide to give the corresponding thiophene [(**439**) → (**440**); Y = S]. 1,4,2,6-Dithiadiazines (*e.g.* **441**) similarly extrude sulfur to give 1,2,5-thiadiazoles. 2-Nitro-1,4-benzodithiin undergoes photochemical dimerization (Equation 8) 〈70AHC(11)63〉. The dibenzo-fused systems are rather stable.

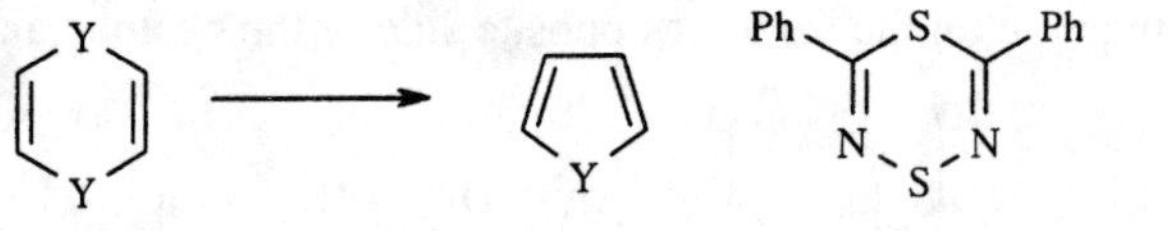

(439)   (440)   (441)

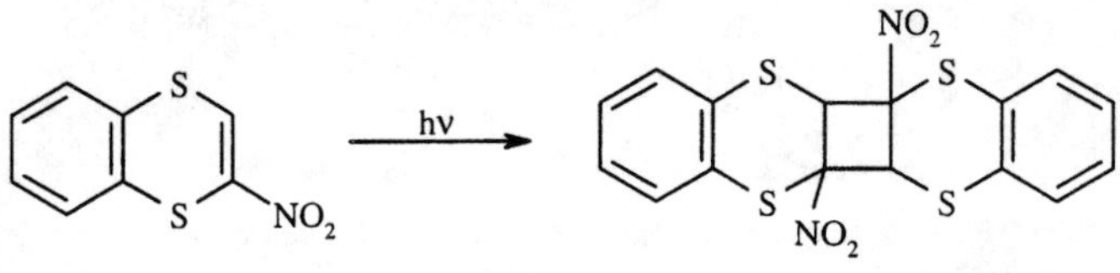

(8)

3.2.2.1.2   *Reaction with electrophiles*

(i) *By addition to C = C double bonds*

1,4-Dioxin and 1,4-dithiin both undergo easy electrophilic addition reactions, *e.g.* of halogens to the double bonds. Alcohols under acid catalysis form ketal addition products.

(ii) *By electrophilic substitution*

Substitution products can be obtained from some 1,4-dithiins. Thus, 2,5-diphenyldithiin is formylated under Vilsmeier conditions, and mono- or di-nitrated and brominated in the heterocyclic ring. 1,4-Benzodithiin shows similar properties (see Scheme 51).

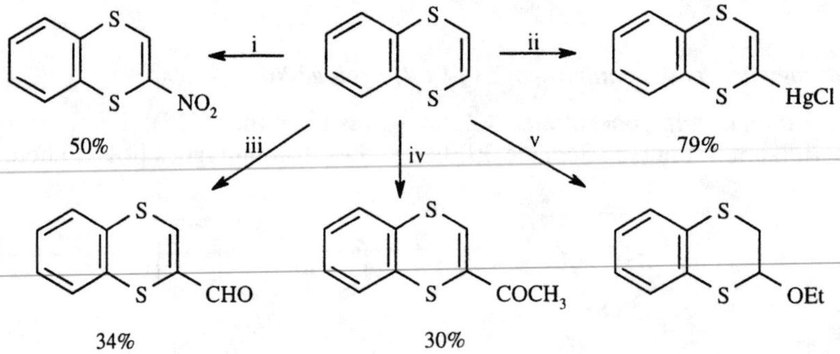

i, HNO₃, AcOH; ii, HgCl₂; iii, PhNMeCHO, POCl₃; iv, Ac₂O, H₃PO₄; v, EtOH, H⁺

Scheme 51

(iii) *By reaction at sulfur*

1,4-Dithiins react readily at sulfur with peracids, alkyl halides and hydroxylamine *O*-sulfonic acid to give sulfoxides, thiinium salts and sulfilimines, respectively. Similar reactions are known for 1,4-benzodithiins.

N-Alkylphenothiazines are oxidized to *S*-monoxides and *S,S*-dioxides with hydrogen peroxide in acetic acid. Phenothiazine can also be *S*-phenylated to form a phenothiazinium cation (**442**), which loses HBF₄ to give (**443**).

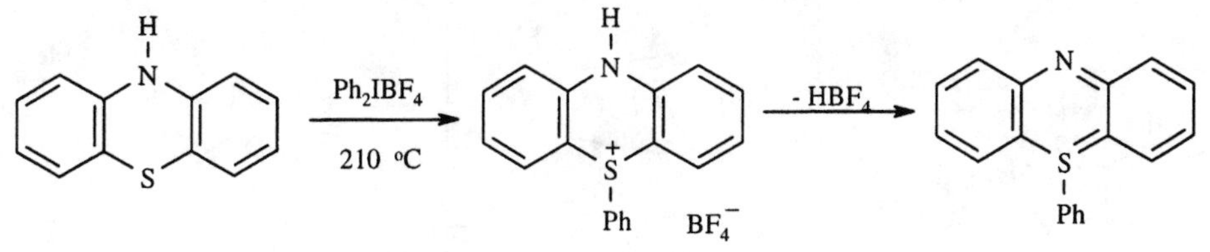

(442) (443)

(iv) By electron loss

These compounds undergo one-electron oxidations to give radical cations and further loss of a second electron to give a six-π-electron dication (see Scheme 52). These reactions can be carried out electrochemically or chemically. Dication salts such as (444) and (445) have been isolated.

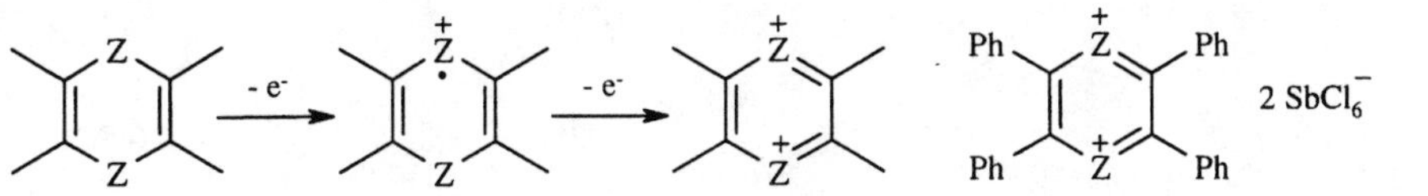

(444) Z = O
(445) Z = S

Scheme 52

Phenoxazines and phenothiazines (446; X = O, S) may be oxidized to phenoxazonium and phenothiazinium salts (447; X = O, S). Radical cations are intermediates; these lose H⁺ to form a neutral radical followed by another electron to form the six-π-electron system (Scheme 53). On careful oxidation, radical cations (446a) can be isolated as deep-colored crystalline salts, stable enough for X-ray analysis ⟨80JHC1053, 81JCS(P2)852, 88CB2059⟩.

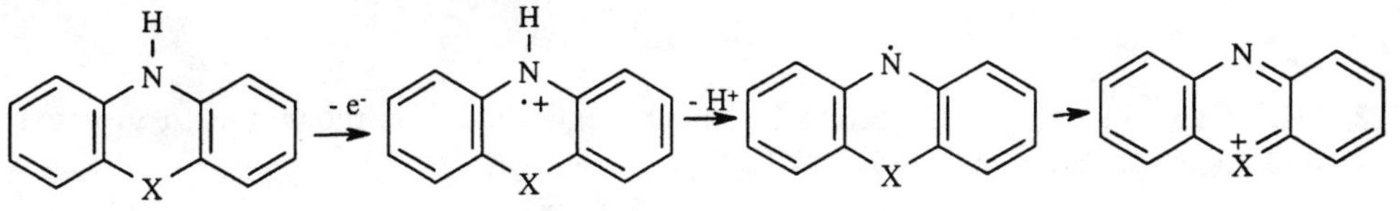

(446) (446a) (447)

Scheme 53

3.2.2.1.3 Reactions with nucleophiles

These electron-rich systems usually show little tendency to react with nucleophiles, but 1,2-dithiins suffer nucleophilic attack at sulfur followed by ring cleavage.

1,4-Dithiin is readily metallated at the 2-position by *n*-butyllithium at −110°C, and (448) can be trapped at this temperature. At −60°C ring opening occurs to give (449).

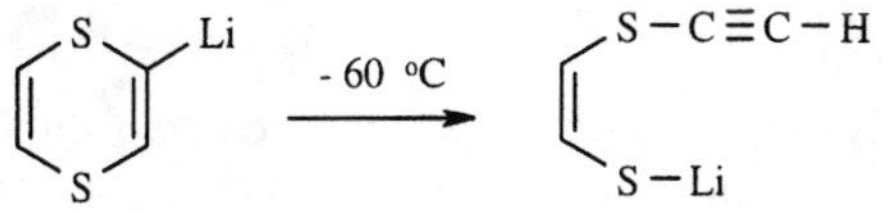

(448) (449)

3.2.2.2 Thiabenzenes and Related Compounds

Thiabenzenes should be considered as sulfonium betaines. They react readily with acids to give mixtures of 2*H*- and 4*H*-thiinium salts, behave as dienes with dienophiles (Equation 9), and can be oxidized to sulfoxides. The sulfimide (450) is an aza analogue of a thiabenzene and it is oxidized by KMnO₄ to the corresponding sulfoximide.

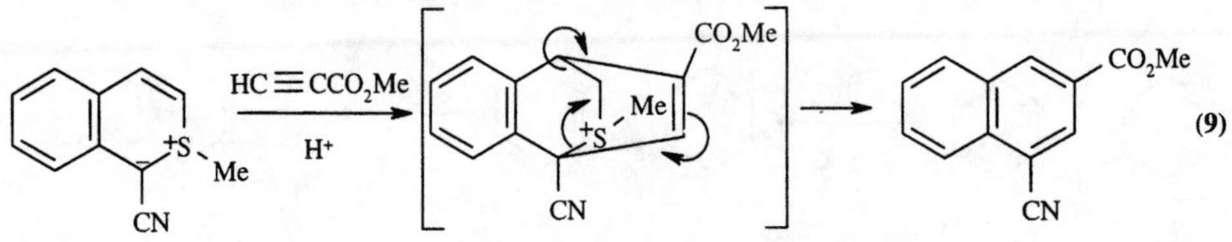

$$(9)$$

Thiabenzene sulfoxides can be nitrated and brominated; thus, (451) gives (452) and (453).

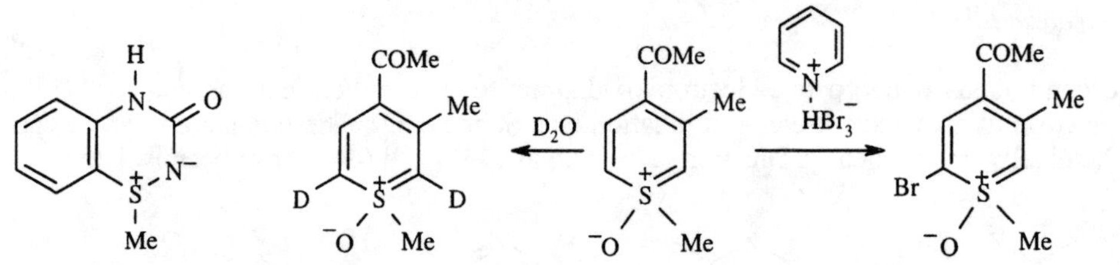

(450)   (453)   (451)   (452)

2H-Thiin dioxides form anions which appear to have some homoaromatic stability. The anions react with aldehydes (Scheme 54).

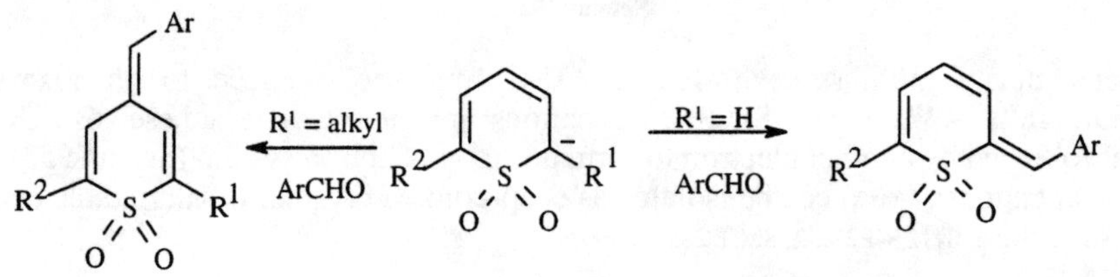

Scheme 54

2H-1,2,6-Thiadiazine 1-oxides on heating in toluene extrude sulfur monoxide to give pyrazole derivatives in high yields (Scheme 55) ⟨83JCS(P1)2273⟩.

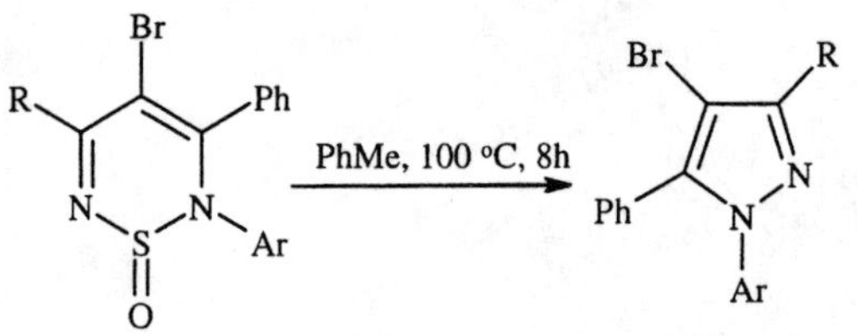

Scheme 55

3.2.2.3 Dihydro Compounds

3.2.2.3.1 Introduction

We consider as 'dihydro derivatives' those rings which contain either one or two sp^3-hybridized carbon atoms. According to this definition, all reactions of the aromatic compounds with electrophiles, nucleophiles or free radicals involve dihydro intermediates. Such reactions with electrophiles afford Wheland intermediates which usually easily lose H^+ to re-aromatize. However, nucleophilic substitution (in the absence of a leaving group such as halogen) gives an intermediate which must lose H^- and such intermediates often possess considerable stability. Radical attack at ring carbon affords another radical which usually reacts further rapidly. In this section we consider the reactions of isolable dihydro compounds; it is obvious that much of the discussion on the aromatic heterocycles is concerned with dihydro derivatives as intermediates.

The reactions of dihydro compounds are of two main classes. The first class comprises reactions to regain aromaticity which depend intrinsically on the 'dihydro six-membered heterocyclic' structure and these can in turn be subdivided into four groups, of which the first is by far the most important:

(i) Loss of a group attached to an sp^3-hybridized ring carbon with its electrons to regain aromaticity.

(ii) Electrocyclic ring opening.

(iii) Loss of a group attached to an sp^3-hybridized ring carbon without its electron to form an eight-π-electron conjugated ring.

(iv) Loss of an electron or H· from a radical cation or neutral radical.

The other class of reactions includes those which are common to alicyclic analogues: reactions with electrophiles and nucleophiles and through cyclic transition states.

We discuss these in turn, but first consider tautomeric structures.

3.2.2.3.2 Tautomerism

N-Unsubstituted dihydropyridines can exist in at least five tautomeric forms (Section 2.2.5.2). At least for *N*-substituted compounds 1,4-dihydropyridines (*cf.* **453**) are generally more stable, by *ca.* 9 kJ mol^{-1}, than the 3,4-dihydro and the 1,2-dihydro isomers (*cf.* **454**). By contrast 2*H*-pyrans appear to be thermodynamically more stable than 4*H*-pyrans. All three types of 1,3-oxazine are known.

(453)   (454)

There are five dihydropyrimidines (**455**)-(**459**). Most of those known have either the 1,2- or the tautomeric 1,4- or 1,6-dihydro structures. Gaussian 70 *ab initio* calculations of the energy of unsubstituted dihydropyrimidines yielded the following order of stability: (**457**) > (**456**) > (**455**) > (**458**) > (**459**). The results agree with the experimentally observed behavior of these compounds ⟨85AHC(38)1⟩.

(455)   (456)   (457)   (458)   (459)

3.2.2.3.3 Aromatization

(i) By tautomerism

Compounds of type (**460**) can aromatize by isomerization [(**460**) → (**461**), Y = CHR, NR]. In a few cases such tautomerism is reversible: barbituric acid (**462**) exists mainly in the trioxo form whereas its anion is aromatic.

(460)   (461)   (462)

1-Benzoyl-2-cyano-1,2-dihydroquinolines and the corresponding isoquinolines (Reissert compounds) (*cf.* Section 3.2.1.6.8.iv) are cleaved by acid into aldehydes plus quinoline-2- or isoquinoline-1-carboxamides. The mechanism of this reaction involves the sequence (**463**) → (**466**); note the aromatization step (**464**) → (**465**).

(463)   (464)   (465)   (466)

(ii) By loss of attached leaving groups with bonding electrons

Those dihydro compounds which carry a leaving group attached to their single sp^3-hybridized carbon atom exist in equilibrium with the corresponding aromatic compounds (*e.g.* the pseudobases **467**⇌**468**; see Section 3.2.1.6.3.iv). Another similar example is that of the covalently hydrated cations of neutral azines (see Section 3.2.1.6.3). A somewhat less obvious example is the acid-catalyzed cleavage (**469**) → (**470**) with the loss of $MeCO_2H$.

(467)   (468)   (469)   (470)

(iii) By loss of attached group without bonding electrons

This includes expulsion of a carbonium ion (*e.g.* **471** → **472** + $PhCH_2OH$) or nucleophilic removal of a positive halogen atom (**473** → **474** + Br^+).

(471)   (472)   (473)   (474)

(iv) By disproportionation

Dihydro compounds often disproportionate (*e.g.* **475** → 2 moles of the corresponding pyridine + 1 mole of the piperidine). Some dihydrodiazines disproportionate: the dihydrocinnoline (**476**) on treatment with hydrochloric acid gives 4-phenylcinnoline and 4-phenyl-1,2,3,4-tetrahydrocinnoline.

(475)   (476)   (477)   (478)

2*H*-(*e.g.* **477**) and 4*H*-thiins on treatment with acid disproportionate into thiinium salts (**478**) and the corresponding tetrahydrothiin.

(v) By oxidation or dehydrogenation

Dihydropyridines, 1,2-dihydro-quinolines and -isoquinolines, pyrans and chromenes are very easily oxidized. The oxidation mechanism, which can involve either hydride transfer or single electron steps, has been extensively studied. Thus, oxidative aromatization of 10-methyl-9-R-9,10-dihydroacridines (**479**) with $Fe(ClO_4)_3$ in MeCN occurs with cleavage of either the C_9-H or C_9-R bond to yield acridinium ions (**480**) or (**481**) or both. Fission of the C-C bond is facilitated by a substituent R which is capable of forming a stable carbenium ion. This gives strong support to a single electron transfer mechanism of the reaction and participation of radical-cations as key intermediate. Indeed, the later have been observed by ESR studies ⟨93JA8960⟩.

R	Yield, %	
	(**480**)	(**481**)
Me	100	0
Ph	100	0
PhCH$_2$	45	55
i-Pr	40	60
t-Bu	0	100

Similar studies have been carried out in the 1,4-dihydropyridine series. Nevertheless, the detailed mechanism remains questionable and evidence in support of both possible reaction pathways, direct hydride transfer and electron-proton-electron transfer, were presented.

1-Substituted dihydropyridines can be aromatized in various ways, *e.g.* with nitrous fumes (NO-N$_2$O$_4$). Compound (**482**) is converted into 4-acetylpyridine by sulfur, or to 4-ethylpyridine by Zn–HOAc.

Pyrans and thiins are also easily aromatized, *e.g.* (**483**) + S$_2$Cl$_2$ → 1-benzothiinium ion. 2*H*-Thiins are aromatized by hydride acceptors such as triphenylmethyl cations to give thiinium salts, and similar conversions produce pyrylium salts from pyrans.

3,4-Dihydroisoquinolines (*e.g.* **484**), 3,4-dihydrocoumarins (*e.g.* **485**) and 2,3-dihydrochromones (*e.g.* **486**) are aromatized by either oxidation or dehydrogenation with S or Se at 300°C or Pd at 200°C. 9,10-Dihydroacridines (*e.g.* **487**) and 5,6-dihydrophenanthridines (*e.g.* **488**) with NH groups are oxidized to the fully aromatic compounds on exposure to air or by other oxidizing agents such as chromic oxide.

Oxidation agents for dihydrodiazines include KMnO$_4$, K$_3$Fe(CN)$_6$, MnO$_2$ and the widely used DDQ. 2,3-Dihydroperimidines (**489**) with NH groups can be effectively aromatized with Na$_2$S$_2$O$_5$ ⟨80KGS96⟩. Some dihydrodiazines are oxidized directly to diazinones (*e.g.* **490** → **491** with KMnO$_4$–OH⁻). Dihydro-triazines and -tetrazines also readily yield the corresponding aromatic azine (*e.g.* **492** + Br$_2$ or O$_2$ → **493**).

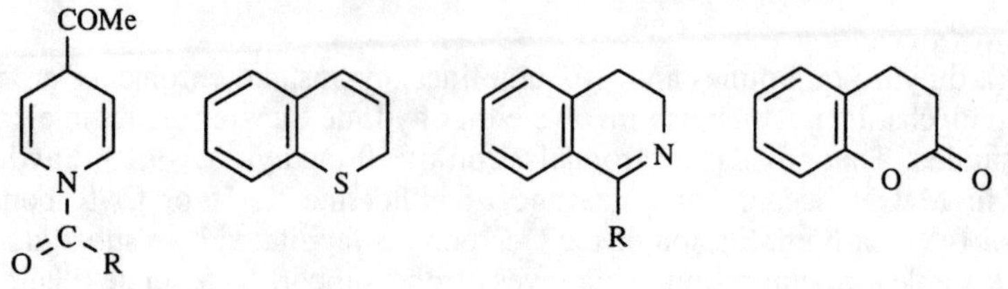

(482)   (483)   (484)   (485)

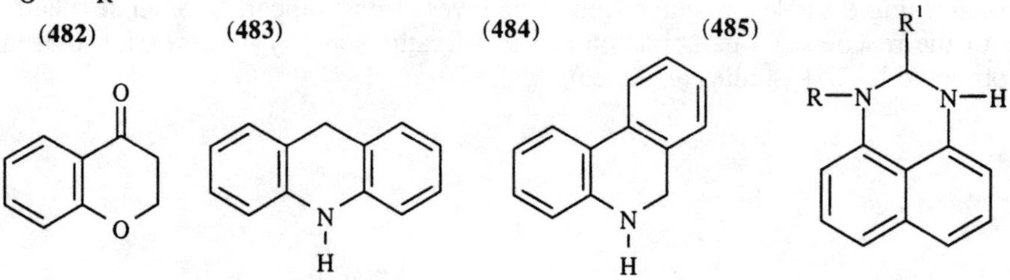

(486)   (487)   (488)   (489)

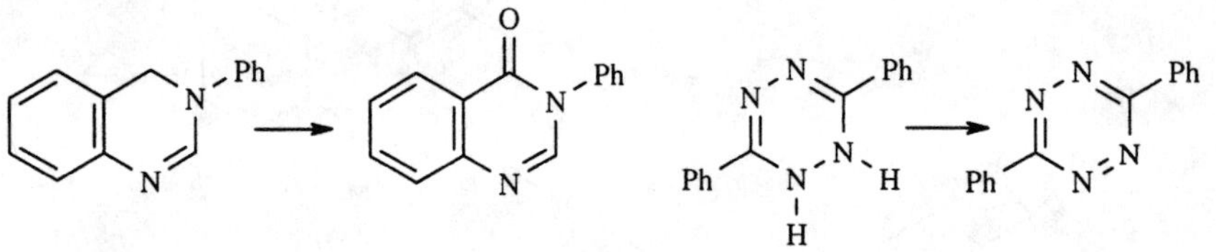

(490)   (491)   (492)   (493)

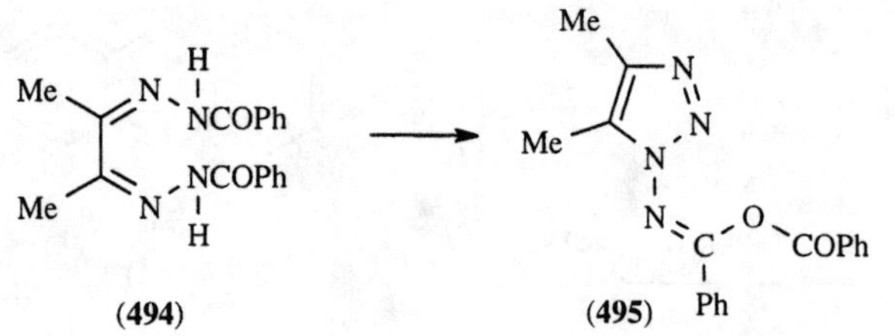

(494)   (495)

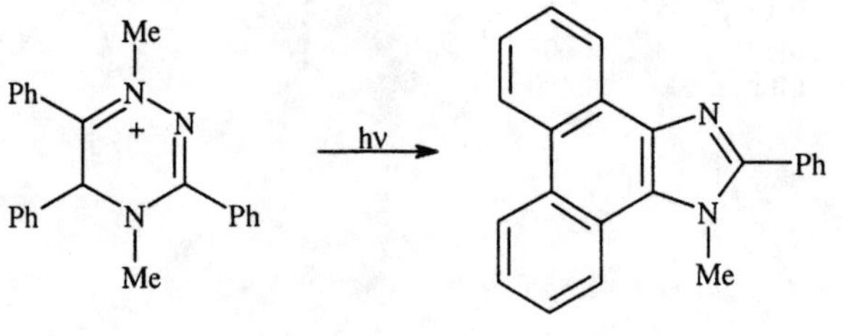

(496)   (497)

(vi) By ring contraction

Oxidation of 1,2-bishydrazones (**494**) is now known to give triazoles (**495**) and not dihydro-1,2,3,4-tetrazines. Possibly the latter are intermediates but are aromatized by ring contraction.

Photolysis of 1,4-dimethyl-3,5,6-triphenyl-4,5-dihydro-1,2,4-triazinium iodide (**496**) gave 1-methyl-2-phenylphenanthro[9,10-*d*]imidazole (**497**) in 90% yield ⟨96CHEC-II(6)539⟩.

3.2.2.3.4 *Electron loss to form radicals*

The formation of a radical cation by electron loss, or a neutral radical by successive loss of e⁻ and H⁺ is probably an important pathway in many of the oxidative and other reactions of these dihydro compounds. In most cases, the radicals is merely a transient intermediate, but 1,4-dihydro-

1,2,4,5-tetrazines are electrochemically oxidized to stable radical cations (**498**) or to neutral verdazyls (**499**). In turn, verdazyls are easily reduced to 1,2,3,4-tetrahydrotetrazines (**500**).

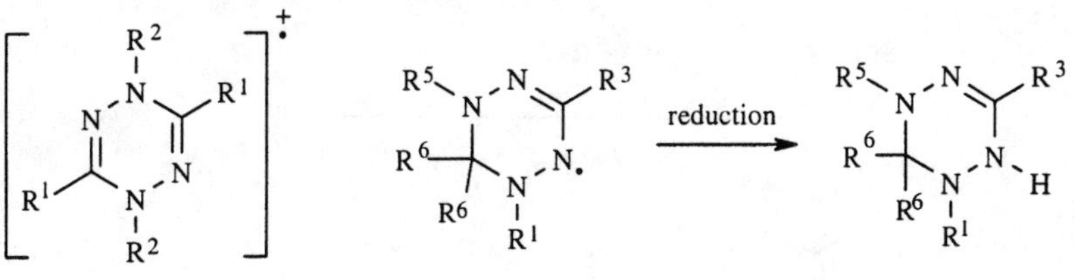

(498)   (499)   (500)

3.2.2.3.5 *Electrocyclic ring opening*

The concept of electrocyclic ring opening of a 1,2-dihydro six-membered heterocycles is familiar from the numerous examples found after nucleophilic attack, especially on cationic rings. Similar reactions occur with isolated 1,2-dihydro derivatives. Dihydropyridines can undergo isomerization by electrocyclic ring opening (see Equation 10). 1-Vinyl-1,2-dihydropyridines in a somewhat similar sequence yield pyridines *via* azacyclooctatrienes (Equation 11).

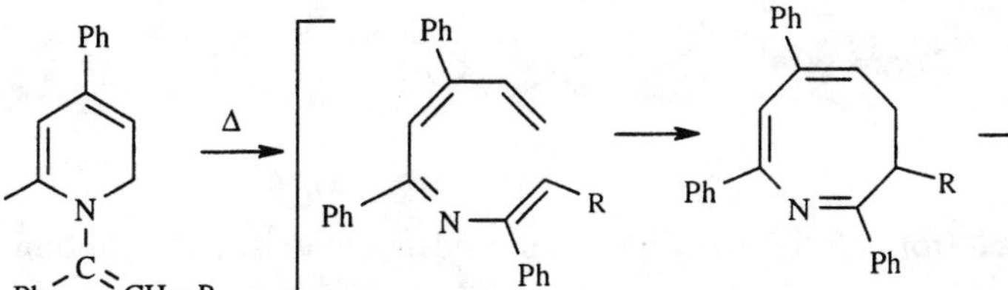

(10)

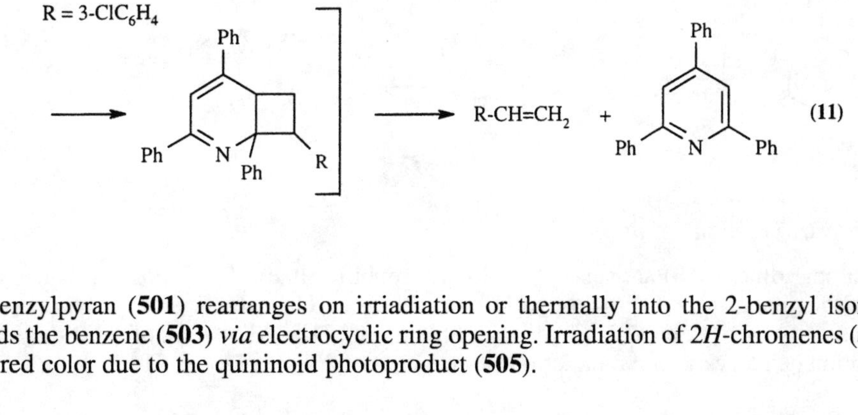

$R = 3\text{-}ClC_6H_4$

(11)

The 4-benzylpyran (**501**) rearranges on irriadiation or thermally into the 2-benzyl isomer (**502**) which yields the benzene (**503**) *via* electrocyclic ring opening. Irradiation of 2*H*-chromenes (**504**) gives an intense red color due to the quininoid photoproduct (**505**).

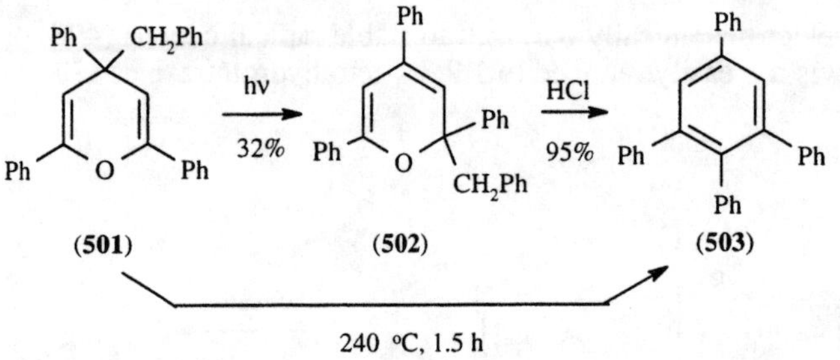

(501)   (502)   (503)

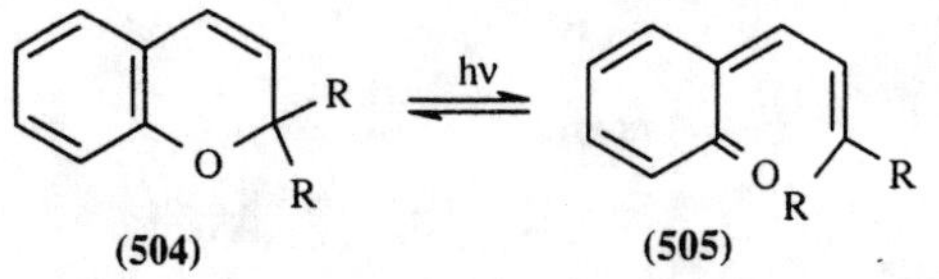

240 °C, 1.5 h

(504)   (505)

Several types of 1,2-oxazines undergo thermal pericyclic reactions in which the N—O bond is cleaved. Thus (506, R = Me, Ph) undergo a thermal retro Diels-Alder reaction on heating to give the corresponding nitrile and *o*-benzoquinone methide, which can be intercepted by alkenes (Scheme 56) ⟨90JA5341, 94TL7273⟩.

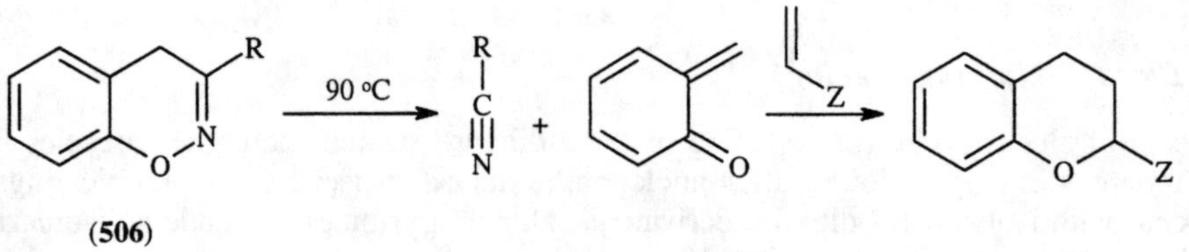

(506)

Scheme 56

3.2.2.3.6 *Proton loss to an eight-π-electron conjugated system*

Very strong bases can extract a proton from the 1,2- or 1,4-dihydropyridine ring giving a fully conjugated eight-π-electron antiaromatic system, which can be trapped by electrophiles.

Reissert compounds (*cf.* Section 3.2.1.6.8.iv) can be deprotonated (NaH/HCONMe$_2$) to give anions (*e.g.* **507**) which react with electrophiles to give intermediates (**508**) which can be hydrolyzed to substituted heterocycles (**509**). Electrophiles utilized include alkyl and reactive aryl halides and carbonyl compounds.

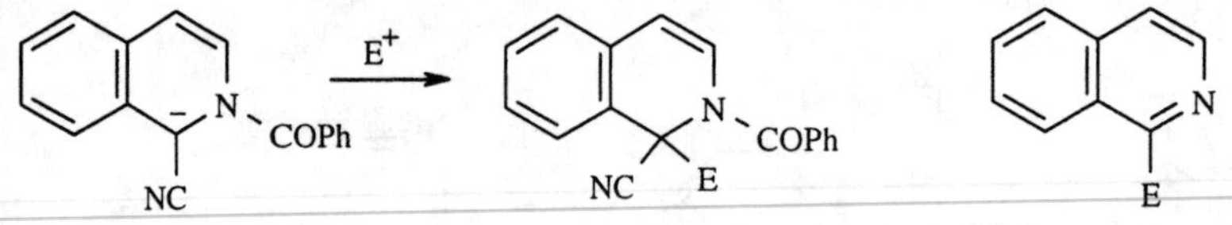

(507)   (508)   (509)

Oxazinyl anions obtained from 4*H*-1,3-oxazines are considered to exist in equilibrium with valence bond tautomers (Scheme 57).

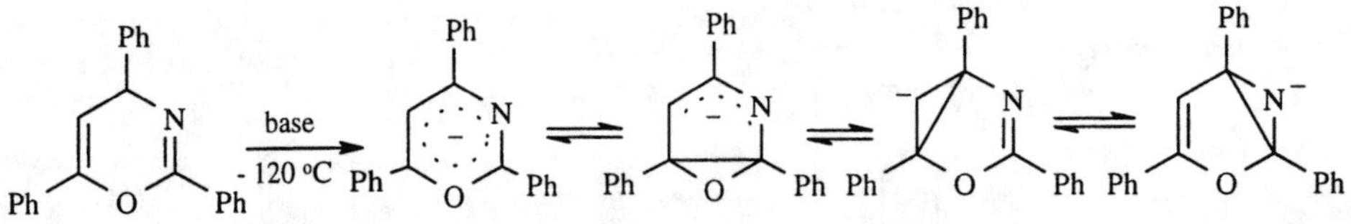

Scheme 57

3.2.2.3.7 *Electrophilic substitution*

1,2-Dihydropyridines (**510**) are susceptible to electrophilic attack. The 5-position is the kinetic site of protonation giving a 2,5-dihydropyridinium cation (**511**) which slowly rearranges to the thermodynamically more stable 2,3-dihydropyridinium ion (**512**). Alkylation of a 1,2-dihydropyridine at the 5-position can be carried out under phase-transfer conditions (**513** → **514**).

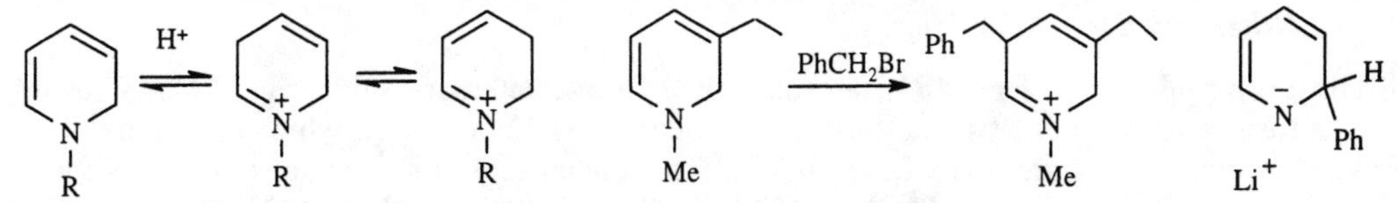

(510)   (511)   (512)   (513)   (514)   (515)

The *N*-lithio-2-phenyl-1,2-dihydro adduct (515) (Section 3.2.1.6.8.i) is a useful synthetic intermediate that reacts with alkyl halides, bromine ⟨70CC478⟩, carbon dioxide ⟨70TL3371⟩ and benzophenone ⟨70CC921⟩ to give 2,5-disubstituted derivatives.

3.2.2.3.8   Cycloaddition reactions

1,2-Dihydropyridines are reactive 1,3-dienes, but their reactions with dienophiles are strongly dependent on the nature of the latter and the reaction conditions. Thus, 1-methyl-1,2-dihydropyridine (516) reacts with methylacrylate at − 10°C to give product of [2 + 2] cycloaddition (517). On the other hand, at + 80°C compound (516) behaves as a 1,3-diene forming adducts (518) and (519) in a 3.2:1 ratio ⟨80JOC1657⟩. Interestingly, adduct (517) on heating is isomerized to the thermodynamically more stable (518) and (519). In this reaction 1,2-dihydropyridine reacts not as simple diene, but rather as an enamine; the regioselectivity can be explained *via* participation of the bipolar structure (516c) in cycloaddition.

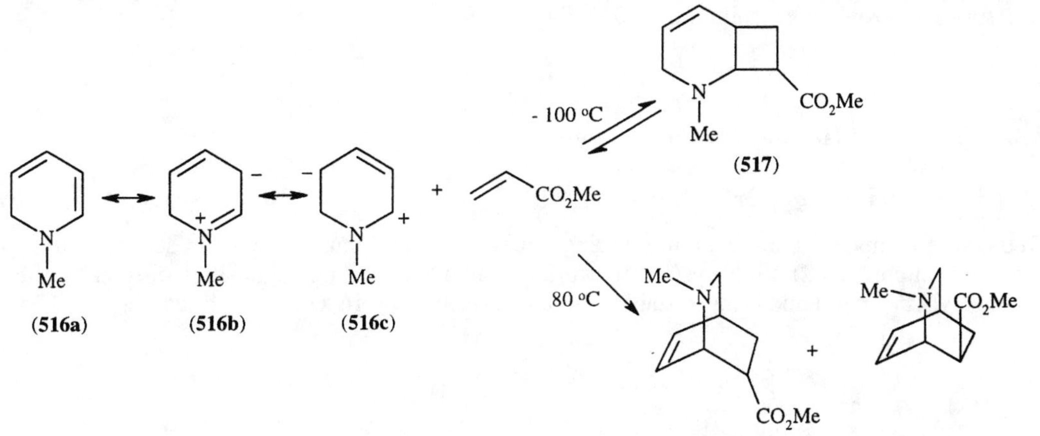

(516a)   (516b)   (516c)

(517)

(518)   (519)

The 1,2-dihydropyridine ring can also undergo [2 + 2] cycloaddition with alkynes (Scheme 58).

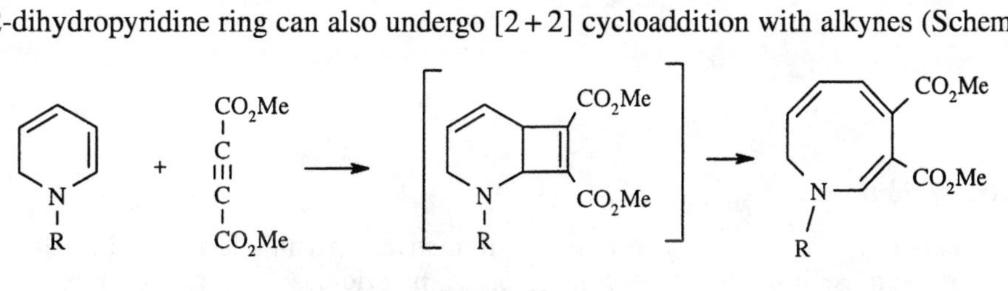

Scheme 58

1,2-Dihydropyridines are isomerized photochemically to 2-azabicyclohexenes (520). 2*H*-Thiins behave as dienes in Diels-Alder reactions (Scheme 59).

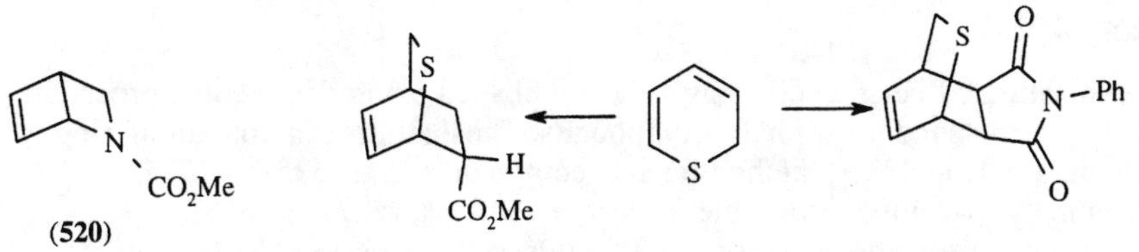

(520)

Scheme 59

3.2.2.3.9 Other reactions

3,4-Dihydroisoquinolines, *e.g.* (**484**), are basic and form quaternary salts, *e.g.* (**521**). With alkali these salts form carbinolamine pseudo-bases, *e.g.* cotarnine (**522**; Y = OH), which can be oxidized to lactams or which disproportionate on standing. The quaternary ions can also react with other nucleophilic reagents, *e.g.* (**521**) + RMgBr → (**522**; Y = R); (**521**) + MeCOMe → (**522**; Y = CH$_2$COMe); (**521**) + CN$^-$ → (**522**; Y = CN); (**521**) + RNH$_2$ → (**522**; Y = NHR). The pseudo-bases are in equilibrium with open-chain compounds since aldehyde derivatives can be prepared.

Reduction of dihydro compounds to the tetra- or hexa-hydro derivatives is usually possible. For example, dihydroisoquinolines of type (**484**) form the corresponding tetrahydroisoquinolines with H$_2$/Pd or with Na/Hg-EtOH.

(521)	(522)	(523)	(524)

C-Styrylpyridines undergo photocyclization to give azaphenanthrenes, and *N*-styrylpyridinium cation forms an azoniaophenanthrene (**523** → **524**).

3.2.2.4 Tetra- and Hexa-hydro Compounds

3.2.2.4.1 Tautomeric equilibria

Tetrahydro compounds still contain one ring double bond and thus can exist in several tautomeric forms (*cf.* Section 2.2.5.2). Little systematic work is available regarding the position of such equilibria, but 1-methyl-Δ^2-piperideine is more stable than the Δ^3-isomer by 16 kJ mol^{-1} (Equation 12).

$$ (12) $$

3.2.2.4.2 Aromatization

Tetra- and hexa-hydro compounds can often be aromatized, but this is more difficult than in the corresponding dihydro series. Thus, the conversion of piperidines to pyridines typically requires dehydrogenation with Pd at 250°C.

3.2.2.4.3 Ring fission

Cleavage of a saturated heterocyclic ring is accomplished using degradative procedures which are also applicable to corresponding aliphatic compounds. Thus, a nitrogen-containing ring is opened by:
(i)   the von Braun amide and PCl$_5$ method for NH compounds, *e.g.* (**525**) → (**526**);
(ii) the von Braun cyano-ammonium route for tertiary amines, (**527**) → (**528**);
(iii) Hofmann exhaustive methylation for tertiary amines, *e.g.* (**530**) → (**529**);
(iv) the Emde reaction, specifically for tetrahydroisoquinolines, *e.g.* (**530**; Y = Me) → (**531**; Y = Me).

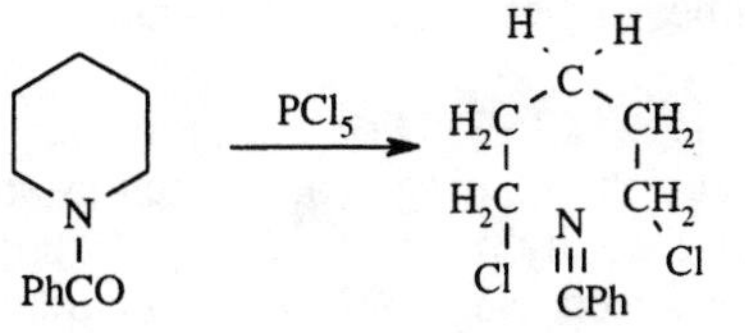

(525) (526) 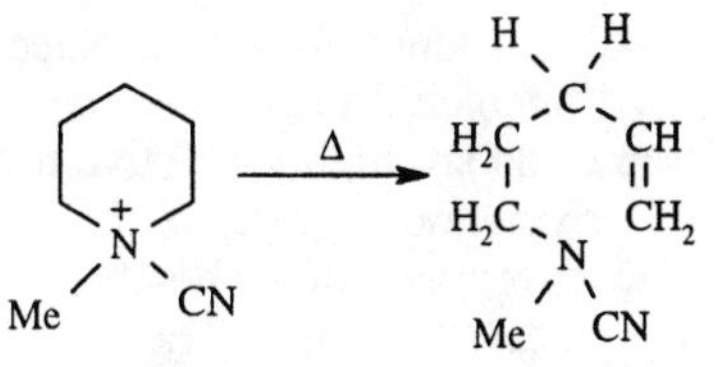 (527) (528)

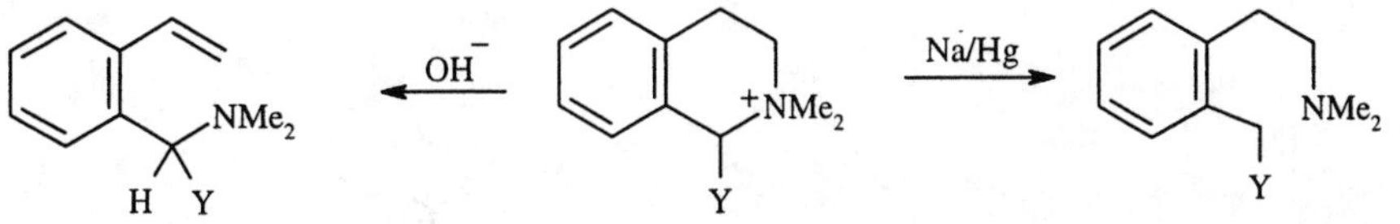

(529) (530) (531)

The rings of cyclic ethers are opened more readily than in the acyclic series; *e.g.* tetrahydropyran with aqueous hydrochloric acid at 100°C gives $Cl(CH_2)_5Cl$.

1,2,3,4-Tetrahydroisoquinoline undergoes a thermal retro-Diels-Alder reaction to give *o*-quinodimethane (Equation 13).

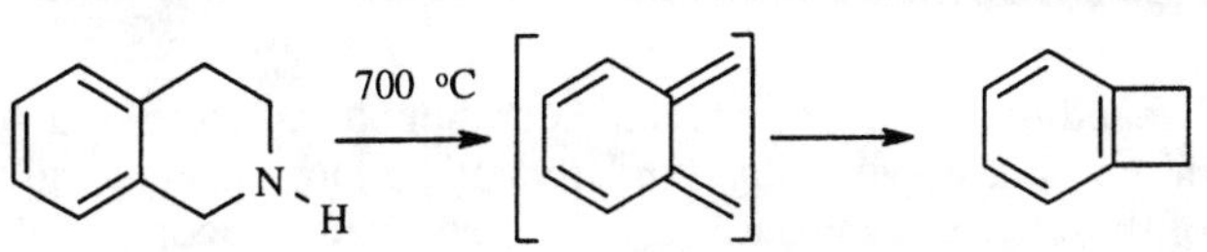

$$ \text{(13)} $$

3.2.2.4.4 Other reactions

These compounds usually show other reactions typical of their aliphatic analogues. 1,2,3,4-Tetrahydroquinoline (**532**; Z = NH) is thus an *N*-alkylaniline; chroman (**532**, X = O) is an alkyl aryl ether.

3-Piperidone (**533**) behaves as an amino ketone, although on Clemmensen reduction it gives 2-methylpyrrolidine (**534**). 2-Piperidone (**535**) is a lactam.

Δ^2-Dihydropyran (**536**) is an enol ether and as such adds hydroxy compounds (hydroxy groups are thus 'protected') to give adducts (**537**) which dissociate on heating.

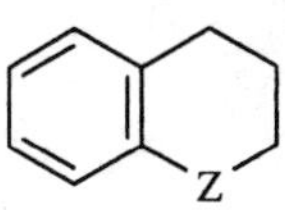

(532) 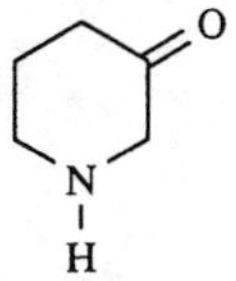 (533) 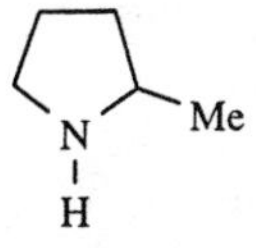 (534) 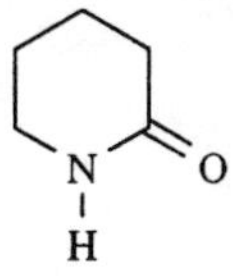 (535) 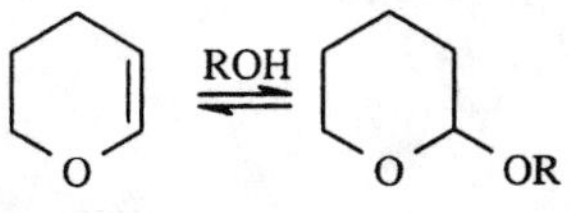 (536) (537)

N-Methylpiperidine is a tertiary amine, and as such it is converted by mercuric acetate into 1-methyl-Δ^2-tetrahydropyridine.

1,3-Dioxane behaves as an acetal, 1,4-dioxane as a bis-ether, and 2,5-dioxopiperazine (**538**) as a bis-lactam.

Piperazine (**539**; Z = NH) and morpholine (**539**; Z = O) show typical aliphatic secondary amine properties, but their pK_a values, 9.8 and 8.4, respectively (*cf.* piperidine pK_a 11.2), reflect the inductive effect of the second heteroatom.

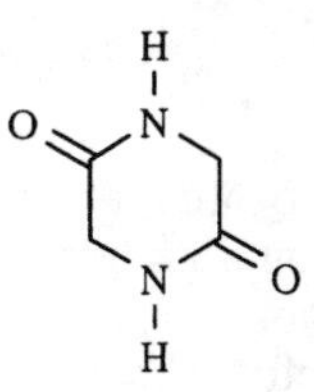

(538) 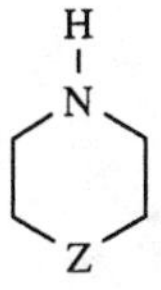 (539) 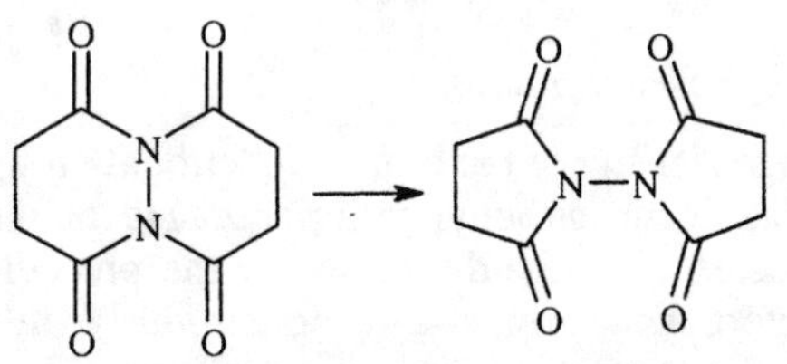 (540) (541)

Tetrone (**540**) rearranges to *N,N'*-bissuccinimide (**541**) either on heating or on reaction with succinyl chloride or hydrogen chloride in dioxane ⟨63JA3052, 67JA4875⟩.

Tetrahydropterins are highly reactive towards oxidation (*e.g.* **542** → **544**); even molecular oxygen can cause hydroxylation. The autoxidation is due to the electron donating groups such as amino and hydroxy, whereas removal of such substituents enhances the stability of the reduced pteridine nucleus tremendously ⟨96CHEC-II(7)701⟩. The reaction appears to proceed *via* single electron transfer. The radical cation (**543**) can be observed by cyclic voltammetry.

(542)   (543)   (544)

3.2.2.4.5 Stereochemistry

Steric effects can alter the reactivity of a heterocyclic compound as compared to its aliphatic analogue. For example, piperidine is less sterically hindered and more strongly nucleophilic than diethylamine. Conformations of these compounds are discussed elsewhere (Section 2.2.4.3).

The quaternization of simple quinolizidine and related alkaloids has attracted much attention, directed primarily at the stereochemistry of the reaction. Methylation of *trans*-quinolizidine (**545**) affords the *cis*-fused 5-methylquinolizidinium salt (**546**); its isomer (**548**) is available only by the cyclization of piperidine derivative (**547**) ⟨51JA3681⟩.

(545)   (546)   (547)   (548)

3.2.3 REACTIONS OF SUBSTITUENTS

3.2.3.1 General Survey of Reactivity of Substituents on Ring Carbon Atoms

We consider here substituents attached to carbon (see Section 3.2.3.12.1 for a general discussion of substituents attached to nitrogen).

The difference in the reactivities of the same substituents on heteroaromatic nuclei and on benzene rings are a measure of the influence of the heteroatom(s). For six-membered heteroaromatic rings, the typical effect of the heteroatom(s) is to attract electrons away from the carbon atoms of the ring. This influence is relatively small when the heteroatom is β to the substituent, but large for the α- and γ-orientations.

3.2.3.1.1 The carbonyl analogy

The reactions of many of the typical functional groups of organic chemistry are influenced to a large extent by an adjacent carbonyl group because of the conjugation electron-withdrawing effect. As would be expected from the discussion in the preceding section, the reactions of substituents in six-membered heterocyclic rings can be similarly influenced. It is therefore helpful to consider systematically the familiar effects on substituents attached to carbonyl groups in aliphatic chemistry. These can be classified into six groups:

(i) Groups which can form anions are readily displaced by nucleophilic reagents (**549**).

(ii) α-Hydrogen atoms are easily lost as protons (**550**).

(iii) As a consequence of (ii) tautomerism is possible (**551⇌552**).

(iv) Carbon dioxide is lost very easily from a carboxymethyl group (**553**) and readily from a carboxyl group (**554**).

(v) These effects are transferred through a vinyl group, and nucleophilic reagents will add to vinyl and ethynyl groups (**555**) (Michael reaction).

(vi) Electrons are withdrawn from aryl groups; thus, electrophiles attack the *meta* position (**556**).

In Table 3 the consequences of effects (i)-(vi) are listed systematically for heterocycles, and compared with the similar effect found in the corresponding aliphatic carbonyl compound.

Table 3   The Carbonyl Analogy for Reactions of Azine Substituents

Reaction type	Group	Behavior of substituent in azine	Section	Compare with
Nucleophilic displacement	Nitro	Readily displaced	3.2.3.6.1	—
	Halogen	Displaced	3.2.3.10.1	Acid chloride
	Alkoxy	Displaced when additionally activated	3.2.3.8.1	Ester
	Amino		3.2.3.5.4	Amide
	Cyano		3.2.3.4.2.iv	Acyl cyanide
Proton loss	Hydroxy	Acidity raised	3.2.3.7.1	Carboxylic acid
	Amino	Basicity lowered	3.2.3.5.4	Amide
	Alkyl	'Active'	3.2.3.3.2	Ketone
Tautomerism	Hydroxy	Exist in oxo form	3.2.3.7.1	Carboxylic acid (two equivalent structures)
	Amino	Exist as amine	3.2.3.5.5	Amide
	Mercapto	Exist in thione form	3.2.3.9.1	Thiocarboxylic acid
Decarboxylation	Carboxyl	Decarboxylate	3.2.3.4.2	α-Keto acids
	Carboxymethyl	Decarboxylate easily	3.2.3.4.2	β-Keto acids
Michael reactions	Vinyl	Undergo Michael addition	3.2.3.4.5	α,β-Unsaturated ketones
	β-Hydroxyethyl	Undergo reverse Michael reaction (loss of H_2O)	3.2.3.4.4	β-Hydroxy ketone
Electrophilic attack on phenyl group	Phenyl	Undergo electrophilic substitution in the *meta* and *para* positions	3.2.3.4.1	Phenyl ketones

3.2.3.1.2 *Effect of number, type and orientation of heteroatoms*

(i) Pyridines and azines

An α-substituent in pyridine (**557**) is in an electronic environment approaching that of a substituent in the imino compound (**558**). Since the reactions of the carbonyl compounds (**559**) are better known than those of the imino compounds (**558**), the reactions of α-substituted pyridines are compared with those of the analogous carbonyl compounds (see preceding section). However, the electron pull is much greater in carbonyl compounds than in pyridine; α-substituents on pyridine accordingly show reactivities intermediate between those of substituents on benzene and substituents attached to carbonyl groups.

The electron-withdrawal to the cyclic nitrogen atom can be transmitted to the γ-position of pyridine (**560**) (illustrating the principle of vinylogy). Hence, γ-substituents have properties similar to those of α-substituents. β-Substituents in pyridine are not directly conjugated with the heteroatom; usually the reactivity is intermediate between that of the same substituent attached to a benzene ring, and that of an α- or γ-substituted pyridine.

(557)   (558)   (559)   (560)

In the diazines, triazines and tetrazines, the effects of the additional nitrogen atom(s) are roughly additive. In Table 4 the positions of substituents in the common azine ring systems are listed in order of increasing reactivity. The limit is reached in 2-, 4- or 6-substituted 1,3,5-triazines for which the reactivity approximates to that in the corresponding carbonyl compound (**559**).

Table 4 Substituent Environments in Azines Listed in Order of Increasing Reactivity

Position of substituent	Ring system	Number of α- or γ-N	Number of β-N
Any	Benzene	—	—
3 or 5	Pyridine	—	1
5	Pyrimidine	—	2
2, 4 or 6	Pyridine	1	—
3, 4, 5 or 6	Pyridazine ⎫		
		1	1
2, 3, 5 or 6	Pyrazine ⎭		
6	1,2,4-Triazine	1	2
2, 4 or 6	Pyrimidine	2	—
3 or 5	1,2,4-Triazine	2	—
3 or 6	1 2,4,5-Tetrazine	2	2
2, 4 or 6	1,3,5-Triazine	3	—

The influence of additional nitrogen atoms in the azines sometimes allows new reactions. An example of this is that of nucleophilic displacement of a cyano group, as in (**561**) → (**562**); this does not normally occur in the pyridine series, but is analogous to a reaction of acyl cyanides (RCOCN).

Substituents in the 5-position of pyrimidines (**563**) are the only substituents on diazines which are not α or β to a ring nitrogen atom, and these behave similarly to the substituents in the 3-position of pyridines.

(561)   (562)   (563)   (564)

The above considerations apply to the reactivity of *neutral species*. If a proton or other electrophile adds to a pyridine nitrogen atom, this is now transformed into a positive pole, with a far greater influence, as outlined in the next section.

(ii) Cationic rings

In cationic rings the electron pull of the positively charged heteroatom is much greater than that of an uncharged nitrogen atom. The effect of a single positively charged N, O or S atom at the α- or γ-position is somewhat stronger than that of *three* α, γ-nitrogen atoms and significantly stronger than that of a carbonyl group. Hence substituents attached to the α- or γ-positions of pyridinium, pyrylium and thiinium ions (**564**) correspondingly show reactivity greater than that of the analogous carbonyl derivative (**559**). Additional nitrogen atoms, and especially a second positively charged atom, enhance the reactivity still further.

(iii) Rings with exocyclic conjugation

For pyridones, pyrones, azinones (*cf.* **565**) and also for *N*-oxides (*cf.* **566**) and β-oxidocationic rings (*cf.* **567**) the situation is more complex. The combined effect of the heteroatoms in such compounds is to act either as an electron source or as an electron sink depending on the requirements of the reaction (see Section 3.2.1.1.4). In practice, in reactions involving neutral species, substituents α or γ to the heteroatom in 2- and 4-pyridones, 2- and 4-pyrones, and 2- and 4-thiinones (*e.g.* **565**) and pyridine *N*-oxides (**566**) are usually activated by electron withdrawal almost as much as they are in pyridine itself. Additional nitrogen atoms increase the reactivity as expected.

However, an important consideration again here is the species undergoing reaction. The reactivity of X in the cation (**569**) will be much more, and in the anion (**570**) much less, affected by electron withdrawal than that of X in (**568**).

(565)   (566)   (567)

(568)   (569)   (570)   (571)

3.2.3.1.3   The effect of one substituent on the reactivity of another

This effect is generally similar to that observed in polysubstituted benzenes. Thus, groups such as NO$_2$ and CN reinforce the electron withdrawal from the substituent which is caused by the heteroatom(s).

As in naphthalene, a fused benzene ring induces bond fixation. Hence, whereas substituents in the 1-position of isoquinoline (**571**; note numbering) behave like substituents in the 2-position of the pyridine nucleus, substituents in the 3-position of isoquinoline show reactivity less than that of true α-substituents and about midway between those of 2- and 3-substituents on pyridine ⟨90AHC(47)390⟩.

3.2.3.1.4   Reactions of substituents not directly attached to the heterocyclic ring

In general, substituents removed from the ring by two or more saturated carbon atoms undergo normal aliphatic reactions. A notable exception is the reverse Michael reaction which is undergone by β′-substituted ethyl compounds such as 2-(β-hydroxyethyl)pyridine (see Section 3.2.3.4.4).

Substituents directly attached to fused benzene rings or aryl groups mostly undergo the reactions of those on normal benzenoid rings. Naturally a substituent on the benzenoid ring in quinoline or isoquinoline should be compared with that on a naphthalene rather than with a benzene nucleus; *e.g.*

such hydroxy derivatives undergo the Bucherer reaction, $ArOH + (NH_4)_2SO_3 \rightarrow ArNH_2$, typical for naphthols (see also Section 3.2.3.2.2).

3.2.3.2 Benzenoid Rings

3.2.3.2.1 Fused benzene rings: unsubstituted

(i) Electrophilic substitution of benzazines

In azines with fused benzene rings, electrophilic substitution on carbon usually occurs in the benzenoid ring in preference to the heterocyclic ring. For quinoline and isoquinoline the only common exception is mercuration which in both occurs in the pyridine ring. However, a strong electron donor substituent such as NH_2 in the pyridine ring, or one or two strong electron acceptor substituents such as NO_2, in the benzenoid ring can also direct attack toward the pyridine ring.

Frequently the orientation of substitution in benzazines parallels that of naphthalene. Nitration (H_2SO_4-HNO_3, 0°C) of quinoline and isoquinoline proceeds in positions corresponding to α-substitution in naphthalene as shown in diagrams (**572**) and (**573**). Sulfonation of isoquinoline gives the 5- and 8-sulfonic acids, the former predominating below 180°C. Sulfonation of quinoline (H_2SO_4-SO_3) is also temperature dependent (100–300°C) yielding 5-, 7- and 8-quinolinesulfonic acids. Heating at 170°C gives a mixture of the 8- and 5-isomers with the former predominating. At higher temperatures (300°C) the main product is the thermodynamically favored 6-isomer and, as expected, both the 5- and 8-isomers undergo rearrangement to the 6-isomer under the appropriate conditions. Acridine (**574**), phenanthridine (**575**), cinnoline (**576**) and quinazoline (**577**) are nitrated as shown. Halogenation follows a somewhat similar pattern.

(572)   (573)   (574)

(575)   (576)   (577)

These substitution reactions probably all occur on the conjugate acid species as supported by the fact that the corresponding cations, such as *N*-methylquinolinium, undergo nitration at approximately the same rate.

The relative reactivity of different positions toward electrophilic substitution is conveniently studied by acid-catalyzed deuterium exchange; reaction rates can be followed by NMR and introduction of deuterium hardly affects the reactivity of the remaining positions. In D_2SO_4 both quinoline and quinoline 1-oxide react as the conjugate acid at positions 8 > 5, 6 > 7 > 3.

(ii) Electrophilic substitution of benzopyridones and related compounds

If the hetero ring is in the form of a pyridone, pyrone or *N*-oxide, or contains a strongly electron-donating substituent (OR or NR_2), electrophilic substitution into the hetero ring can compete with substitution into a fused benzene ring. In some such compounds, substitution occurs entirely in the heterocyclic ring; *e.g.* nitration of 2-quinolone (fuming HNO_3, no H_2SO_4) gives a 3-substituted product (see Section 3.2.1.4.4).

In other compounds, reaction can occur in both rings. Under such circumstances the orientation can depend on the conditions; frequently reaction in the benzene ring involves the cationic species, whereas that in the pyridine ring involves the free base. Thus, the temperature dependent nitration of quinoline 1-oxide (**578**) reflects the decrease in intrinsic acidity as the temperature rises, which in turn increases the available amount of free base species.

Coumarins undergo nitration readily in the 6-position while bromination results in substitution at the 3-position as a consequence of addition-elimination.

(578)   (579)

As outlined above, the orientation of substitution into bicyclic benzazines frequently occurs preferentially at the 5- and/or 8-positions. However, when the heterocyclic ring contains a carbonyl group, the orientation of substitution into a fused benzene ring frequently occurs in the 6-position. For 2-quinolone (**579**; Z = NH) (nitration, H_2SO_4-HNO_3, 20°C) and for coumarin [nitration (H_2SO_4-HNO_3) and sulfonation (H_2SO_4)] can be compared with the *para*-substitution of acetanilide and phenyl acetate.

(iii) Oxidation

Vigorous oxidation (*e.g.* $KMnO_4$) usually degrades fused benzene rings in preference to pyridine rings, especially under acid conditions. Quinoline and isoquinoline yield the dicarboxylic acids (**580**) and (**581**), respectively; phthalazine gives (**582**) and phenazine yields (**583**) and (**584**). Oxidation of such a fused benzene ring is facilitated when it carries electron-donating groups and is hindered by electron-withdrawing groups.

(580)   (581)   (582)   (583)   (584)

Ozonolysis of quinoline gives glyoxal and pyridine-2,3-dicarboxaldehyde.

(iv) Radical reactions

Reactions with radicals are often unselective and form complex mixtures; thus, phenylation of quinoline with benzoyl peroxide gives all seven phenylquinolines (*cf.* Section 3.2.1.9.2).

(v) Nucleophilic attack

Nucleophiles normally attack the heterocyclic ring (Section 3.2.1.6). However, if all positions on the heterocyclic ring are blocked, and if it is highly electron deficient, reactivity toward nucleophilic attack is found in a fused benzene ring. Such conditions apply in phenazinium, phenoxazinylium and phenothiazinylium ions (**585**; Z = NR, O, S). Hydroxide ions give pseudo-base intermediates (**586**) which are easily oxidized (with air, Br_2, *etc.*) to the pyridone analogues (**587**).

Similarly, ammonia and amines (*e.g.* $PhNH_2$, Me_2NH) give initial adducts of type (**589**), which are then oxidized (with air, Br_2, *etc.*) to new onium salts (**590**). The adduct with cyanide ion tautomerizes to (**588**); phenothiazonium chloride forms (**591**) *via* a similar addition and tautomerism (HCl-H_2O, 100°C). Repeated reaction with amines gives products of type (**592**).

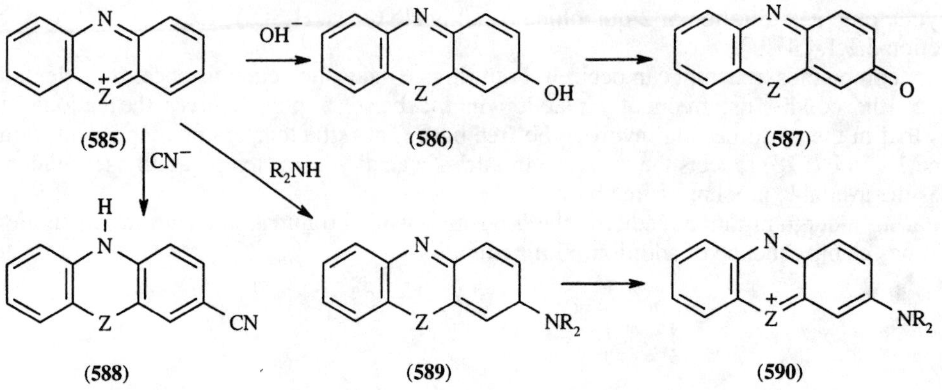

(585)   (586)   (587)

(588)   (589)   (590)

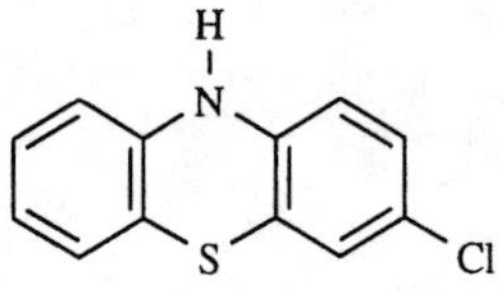

(591) 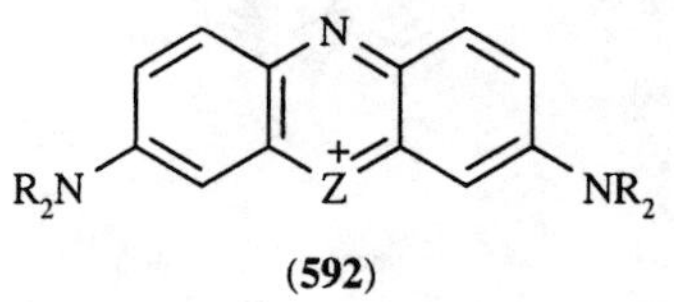   (592)

An important analogue of these conversions is radical-cation electrophilic substitution in such π-excessive six-membered heterocycles as phenothiazines, perimidines and 1,6-diazaphenalenes. These compounds can be quite effectively nitrated with nitrous acid ($NaNO_2$-AcOH) ⟨81H(16)1453, 81RCR816⟩. The process probably occurs in accordance with Scheme 59a. In the first step, a radical-cation is generated *in situ*; it is then attacked by nitrite ion at the positions of the benzene ring carrying the largest positive charge (spin density). These radical-cation salts can be isolated and subsequently treated with various nucleophiles, *e.g.* Hal⁻, SCN⁻, to produce the corresponding substitution products. Only heterocycles with extended π-systems, which can form relatively stable radical-cations, display this type of reactivity.

$$HetH \quad + \quad NO^+ \longrightarrow HetH^{+\cdot} \quad + \quad NO$$

$$HetH^{+\cdot} \quad + \quad NO_2^- \longrightarrow HetH(NO_2)^\cdot$$

$$HetH(NO_2)^\cdot \xrightarrow{-H^\cdot} Het\text{-}NO_2$$

Scheme 59a

3.2.3.2.2   *Fused benzene rings: substituted*

(i) Electrophilic substitution

Substituents on the benzene rings exert their usual influence on the orientation and ease of electrophilic substitution reactions. For example, further nitration (HNO_3-H_2SO_4-SO_3) of nitroquinolines occurs *meta* to the nitro group as shown in diagrams (**593**) and (**594**). Friedel-Crafts acylation of 8-methoxyquinoline succeeds (*cf.* **595**) although this reaction fails with quinoline itself.

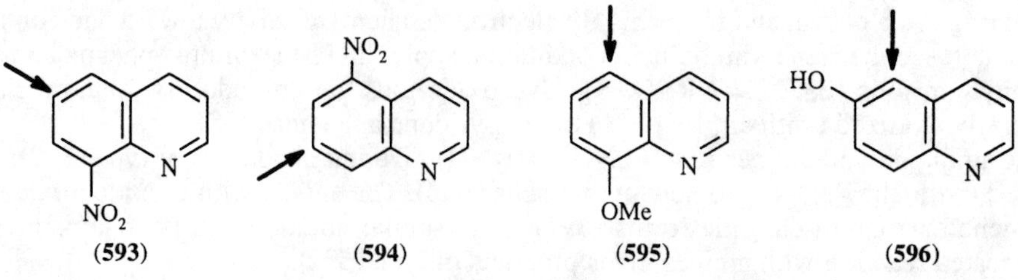

(593)   (594)   (595)   (596)

A heterocyclic ring induces partial double-bond fixation in a fused benzene ring. Hence diazo coupling occurs at the 5-position of 6-hydroxyquinoline (**596**), and not at the 7-position.

(ii) Amino groups

Amino groups on fused benzene rings in benzopyridines show basicity lower than aniline (initial proton addition occurs mainly on the hetero nitrogen atom) but are diazotized normally. Displacements of a diazonium group often occur under Gattermann but not under Sandmeyer conditions, probably because complexes are formed with Cu^{2+}.

However, a strongly electron-deficient heterocyclic ring can induce unusual reactivity as occurs for the 3-amino groups in phenazonium, phenoxazonium and phenothiazonium salts (**597**; Z = NR, O or S) which are important in dye chemistry. Thus, phenosafranine (**592**; Z = NPh, R = H) is converted by alkali into the imine (**598**; Y = NH) or, on more vigorous treatment, into the phenazone (**598**; Y = O). Methylene blue (**592**; Z = S, R = Me) on oxidation ($K_2Cr_2O_7$-HCl) gives the imine (**599**; Y = NMe), and on treatment with alkali the oxo derivative (**599**; Y = O).

(**597**) (**598**) (**599**)

(iii) Hydroxy groups

In general, hydroxy groups on fused benzene rings undergo the expected reactions. *O*-Methylation is effected by diazomethane, methyl iodide or dimethyl sulfate. *O*-Alkylation is reversed by aluminum trichloride or tribromide in benzene or nitrobenzene.

However, the reactivity of phenolic hydroxy groups can be modified by a fused heterocyclic ring. Thus, hydroxy groups *peri* to a carbonyl group, *e.g.* (**600**), are hydrogen bonded; they do not react with diazomethane, and are difficult to acylate. This allows selective reactions in polyhydroxychromones.

Hydroxy group acidity is increased, sometimes quite dramatically. Thus, hydroxy groups on benzene rings fused to pyrylium or pyridinium rings can lose a proton if the resulting anhydrobases are stabilized by mesomerism with a non-charged *p*-quinonoid canonical form, *e.g.* (**601**) → (**602**); the corresponding *o*-quinonoid anhydro-bases are less stable. If two suitably oriented hydroxy groups are present, a further proton can be lost to give a mesomeric anion (*e.g.* **604**).

In favorable cases, a phenolic hydroxy group can show a reactivity usually associated only with substituents on the heterocyclic ring and be converted into a chloro group (**601** → **603**).

(**600**) (**601**) (**602**)

(**605**) (**603**) (**604**)

(iv) Halogen atoms

In extreme cases, halogen atoms on fused benzene rings are labile, *e.g.* (**603**) + $PhNH_2$ → (**605**).

3.2.3.3 Alkyl Groups

3.2.3.3.1 *Reactions similar to those of toluene*

The typical reactions of alkyl groups attached to benzenoid rings involve benzyl-type radical intermediates. An azine ring can stabilize a methyl radical just as can a phenyl ring, and thus most alkylpyridines and azines show these reactions.

(i) Oxidation in solution ($KMnO_4$, CrO_3, *etc.*) gives the carboxylic acid, *e.g.* 3-picoline → nicotinic acid (**606**), or ketone, *e.g.* 2-benzylpyridine → 2-benzoylpyridine (**607**).

(ii) Controlled catalytic vapor phase oxidation converts, for example, 2-, 3- and 4-picolines into 2-, 3-and 4-pyridine carboxaldehydes.

(iii) Free radical bromination with *N*-bromosuccinimide succeeds, *e.g.* 2,6-dimethyl-4-pyrone → (**608**).

(iv) So-called am-ox vapor phase conversion by O_2—NH_3 of CH_3 into CN.

(606)   (607)   (608)   (609)

3.2.3.3.2 *Alkyl groups: reactions via proton loss*

Alkyl groups α or γ to a pyridine nitrogen show additional reactions because of the possibility of losing a proton from the carbon atom of the alkyl group which is adjacent to the ring. The ease of proton loss depends on the number, orientation and nature of the heteroatoms in the ring carrying the alkyl groups as discussed in Section 3.2.3.1. Reactions of this type consist of two essential steps: loss of the proton and then subsequent reaction with an electrophile. For the neutral alkylazines, we distinguish between:

(i) use of a strong base which removes the proton completely before addition of electrophile;

(ii) use of a weaker base which sets up a pre-equilibrium giving traces of reactive anion which reacts with the electrophile and is then replenished by the equilibrium; and

(iii) use of an *acid* catalyst which affords small amounts of the tautomeric methylene form (*e.g.* **609**) which reacts.

A similar distinction between three analogous mechanisms of reaction with electrophiles can be made in the chemistry of ketones.

3.2.3.3.3 *Alkylazines: reactions involving essentially complete anion formation*

The strongest bases such as sodamide ($NaNH_2$-NH_3, − 40°C), lithium diisopropylamide or organometallic compounds (PhLi-Et_2O, 40°C) convert 2- and 4-alkylpyridines, 2- and 4-methylpyrimidines and 3- and 4-methylpyridazines, *etc.*, essentially completely into the corresponding anions (*e.g.* **610**). These anions react readily (as **611**) with even mild electrophilic reagents; thus, the original alkyl groups can be substituted in the following ways:

(i) Alkylation, *e.g.* 2-picoline (**612**) → 2-*n*-propylpyridine (**613**).

(ii) Acylation, *e.g.* lepidine (**614**) → 4-phenacylquinoline (**615**).

(iii) Carboxylation, *e.g.* 2-picoline (**612**) → 2-pyridylacetic acid (**616**) which is esterified before isolation (see Section 3.2.3.4.2).

(iv) Reaction with carbonyl compounds to form alcohols, *e.g.* 2-picoline (**612**) → the tertiary alcohol (**617**).

(v) An amyl nitrite and sodamide in liquid ammonia give an oxime (Scheme 60).

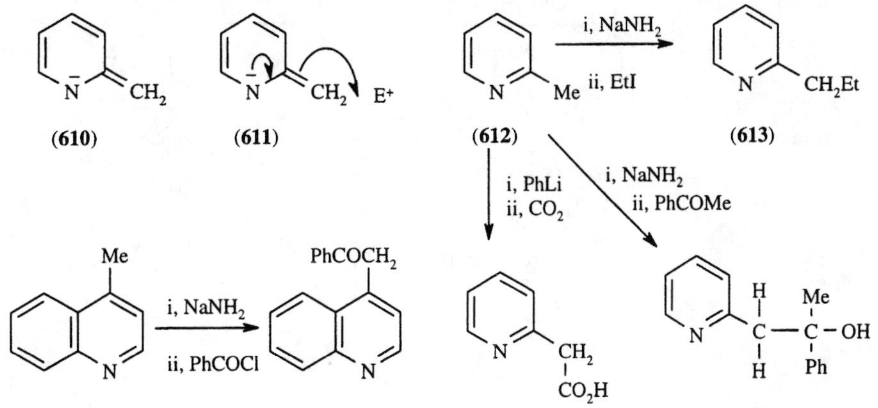

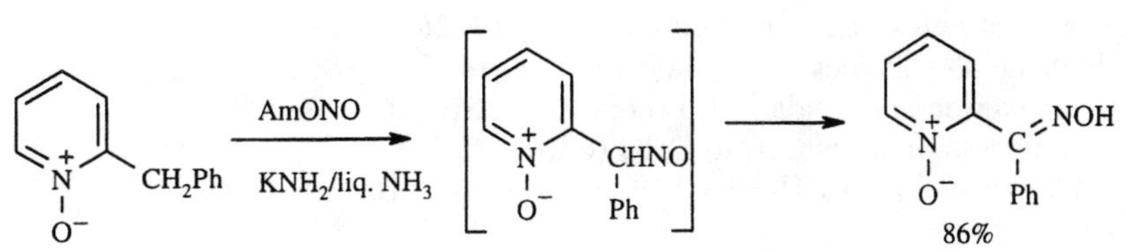

Scheme 60

The methyl group of 3-picoline is sufficiently reactive to be alkylated and acylated in this way, although the yields are poor.

With 2,4-dimethyl-pyridine and -quinoline, selective alkylation or acylation may be achieved at either position. *n*-Butyllithium promotes ionization at the 2-methyl groups, whereas lithium diisopropylamide reacts at the 4-methyl group.

3.2.3.3.4 Alkylazines: reactions involving traces of reactive anions or traces of methylene bases

In aqueous or alcoholic solution, activated alkyl groups in heterocyclic rings react with bases to give traces of anions of type (**610**). In such reactions, alkoxide or hydroxide ions, aliphatic amines (*e.g.* NEt$_3$, piperidine) or the alkylpyridine itself can act as the base. With suitable electrophilic reagents the anions (*cf.* **610**) undergo reasonably rapid and essentially non-reversible reaction, gradually converting the whole of the heterocyclic compound. An obvious prerequisite for reaction under these conditions is that the base used does not react irreversibly with the electrophile.

In acidic media, loss of a proton can give traces of methylene forms of type (**609**). Alternatively, a Lewis acid catalyst such as acetic anhydride may be used which involves formation of complexes of type (**618**) from which proton loss is facile. Such methylene bases can also react with electrophiles, gradually causing complete conversion of the heterocycle.

Reactions of these types will be illustrated using 4-picoline (**620**) and quinaldine (**626**) as typical substrates:

(i) Formaldehyde gives poly-alcohols (**620 → 623**).
(ii) Other aliphatic aldehydes form mono-alcohols (**620 → 622**).
(iii) Aromatic aldehydes give styryl derivatives (**620 → 621**) by spontaneous dehydration of the intermediate alcohol (*cf.* Section 3.2.3.4.4).
(iv) Nitroso compounds form Schiffs bases (**620 → 619**).

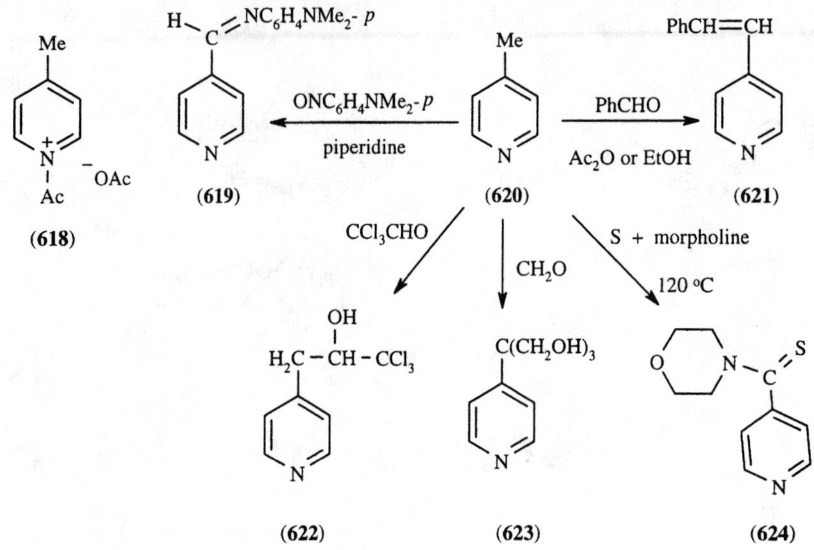

(618)   (619)   (620)   (621)

(622)   (623)   (624)

(v)   Halogens substitute all adjacent hydrogen atoms (626 → 625).
(vi)   Formaldehyde plus amines yield Mannich bases (626 → 628).
(vii)   Phthalic anhydride gives indane-1,3-dione derivatives (626 → 629).
(viii)   Selenium dioxide oxidizes CH_3 to CHO (626 → 627).
(ix)   Willgerodt conversion of CH_3 → $CSNR_2$ with S/R_2NH (620 → 624).

Although the last two examples have radical intermediates, they probably involve electron transfer from the type (610) anion to the reagent in the rate-limiting step.

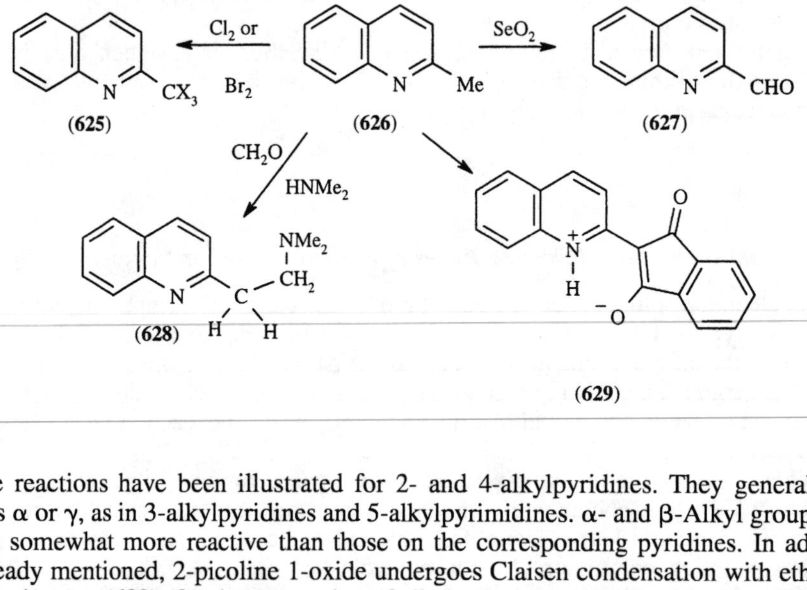

(625)   (626)   (627)

(628)

(629)

The above reactions have been illustrated for 2- and 4-alkylpyridines. They generally fail if no heteroatom is α or γ, as in 3-alkylpyridines and 5-alkylpyrimidines. α- and β-Alkyl groups in pyridine *N*-oxides are somewhat more reactive than those on the corresponding pyridines. In addition to the reactions already mentioned, 2-picoline 1-oxide undergoes Claisen condensation with ethyl oxalate to yield the pyruvic ester (630) (for the conversion of alkyl substituents in *N*-oxides into CH_2OAc groups see Section 3.2.3.12.5.iv).

Methyl groups in the 2-, 4- or 6-position of pyrimidine are also more reactive. In addition to typical reactions such as condensation with benzaldehyde, selenium dioxide oxidation and halogenation, they can be converted into oximino groups by nitrous acid, and undergo Claisen condensation with $(CO_2Et)_2$. In the reaction of 2,5-dimethylpyrimidine with benzaldehyde, only the electrophilic 2-methyl group reacts preferentially to yield the 2-styryl derivative (631). In quinazolines partial double bond fixation makes a methyl group in the 4-position more reactive than that in the 2-position.

(630)　　　　(631)　　　　(632)　　　　(633)

α- and β-Alkyl groups in pyrones and pyridones also undergo many reactions of these types. For example, with benzaldehyde, 2,6-dimethylpyrone gives the styryl derivative (632) and 1,4- and 1,6-dimethylpyrid-2-ones condense with ethyl oxalate. 2-Methyl groups in pyran-4-ones and chromones condense with benzaldehyde and can be halogenated. However, reactions sometimes fail; (633) will not condense with benzaldehyde.

If an α- or γ-alkyl group itself carries an electron-withdrawing substituent, proton loss is facilitated, and additional reactions can occur (*e.g.* 634 → 635 + CO_2) (*cf.* the Japp-Klingemann reaction: $MeCOCH_2CO_2H + PhN_2^+ → MeCOCH=NNHPh$).

(634)　　　　　　　　(635)

3.2.3.3.5 Alkyl-azonium and -pyrylium compounds

(i) Formation of stable anhydro-bases

Compounds containing methylene groups activated by both a cationic ring and another electron-withdrawing group easily form stable anhydro-bases, *e.g.* (636) → (637), (638) → (639). Stabilization is also achieved by utilization of the aromatic character of the cyclopentadiene anion or the pyrrole anion; compounds of type (640; Z = NR, O, S) and (643) readily lose protons to give the mesomeric anhydro-bases (as 641 ↔ 642) which are called pseudoazulenes.

(636)　　　　(637)　　　　(638)　　　　(639)

(640)　　　　(641)　　　　(642)　　　　(643)

(ii) Anhydro-bases as intermediates

Proton loss from α- and γ-alkyl groups on a cationic (pyridinium, pyrylium or thiinium) ring is comparatively easy. The resulting unstable and highly reactive neutral anhydro-bases or 'pyridone methides' (*cf.* 645) can be isolated by using 10M sodium hydroxide, but are generally used directly.

These anhydro-bases are heterocyclic equivalents of enamines and enol ethers and react readily with electrophilic reagents to give products which can often lose a proton to give a new resonance-stabilized anhydro-base. Thus, anhydro-1,2-dimethylpyridinium hydroxide (645) reacts with phenyl isocyanate to give an adduct (646) which is converted to the stabilized product (647 ↔ 648). A similar sequence with carbon disulfide yields the dithio acid (644).

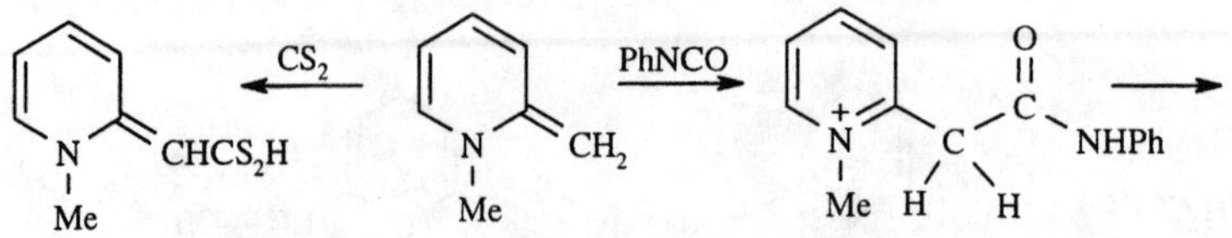

(644)   (645)   (646)

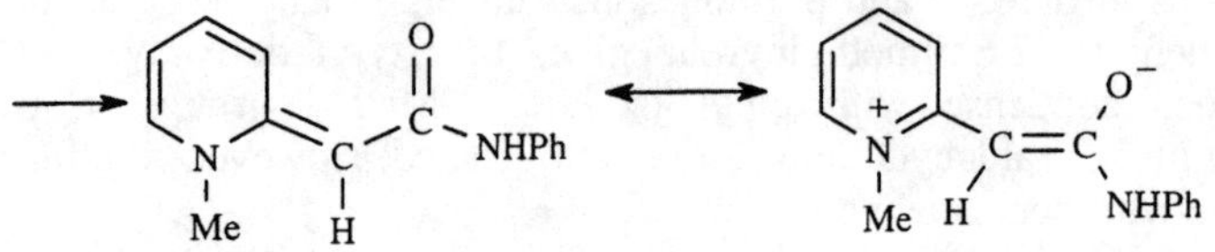

(647)   (648)

α- and γ-Alkyl cationic heterocycles, analogous to the 2- and 4-alkylpyridines, can also react with electrophilic reagents without initial complete deprotonation. They undergo the same types of reaction as the alkylpyridines under milder conditions; such reactions are often catalyzed by piperidine. For example, 1,2-dimethylpyridinium cation (650) with PhCHO and with p-Me$_2$NC$_6$H$_4$NO yields (649) and (651), respectively, and 2-methylchromylium (652) gives (653).

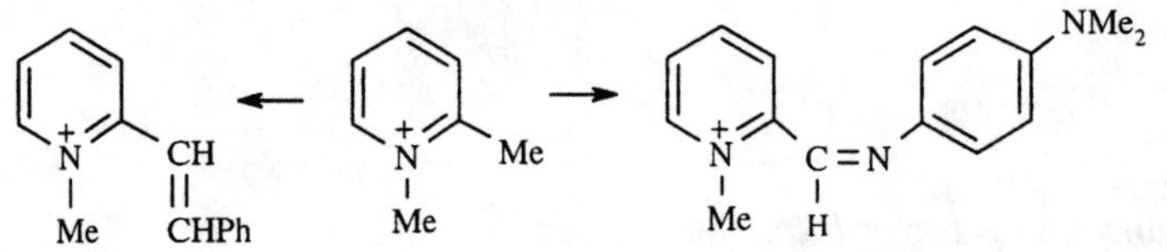

(649)   (650)   (651)

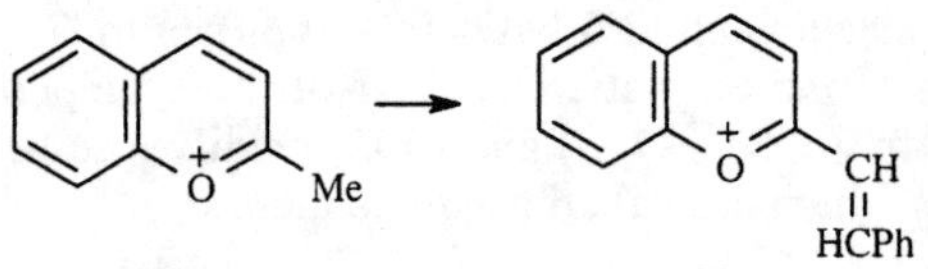

(652)   (653)

Some weak electrophilic reagents, which are usually inert toward neutral pyridines and azines, also react

(a)   Diazonium salts yield phenylhydrazones (*e.g.* 654 → 655; Z = NMe, O) in a reaction analogous to the Japp–Klingemann transformation of β-keto esters to phenylhydrazones.

(b)   Monomethine cyanines are formed by reaction with an iodoquaternary salt (*e.g.* 656 + 657 → 658; $n = 0$).

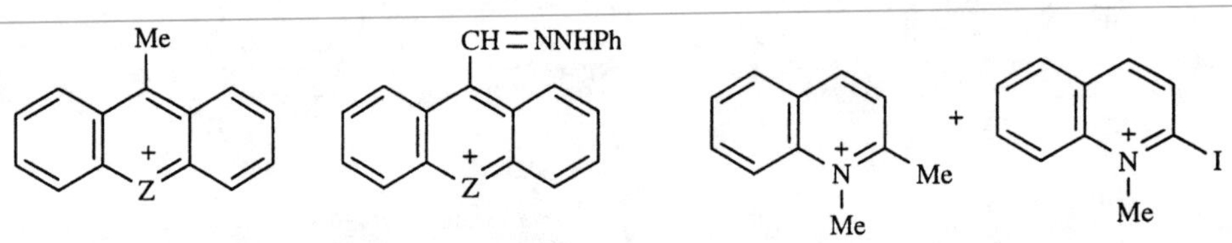

(654)   (655)   (656)   (657)

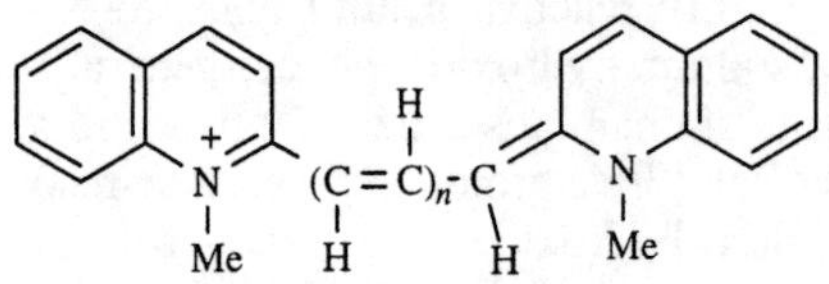

(658)

EtO(CH=CH)$_n$CH(OEt)$_2$
(659)

PhNH(CH=CH)$_n$CH=NPh
(660)

(c) Tri-, penta- and hepta-carbocyanines (*e.g.* **658**; $n = 1, 2$ and 3, respectively) are obtained by the reaction of two molecules of a quaternary salt with one molecule of ethyl orthoformate (**659**; $n = 0$), β-ethoxyacrolein acetal (**659**; $n = 1$) or glutacondialdehyde dianil (**660**; $n = 2$), respectively.

(d) With the anils (**660**; $n = 0, 1$ or 2), it is possible to isolate intermediates of type (**661**; $n = 1, 2$ or 3) which react with another molecule of the same or a different quaternary base to give symmetrical or unsymmetrical tri-, penta- and hepta-carbocyanines. A similar reaction sequence in the pyrylium series is shown by (**662**) → (**663**) → (**664**).

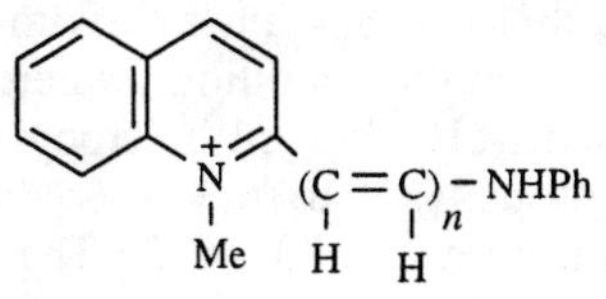

(**661**)

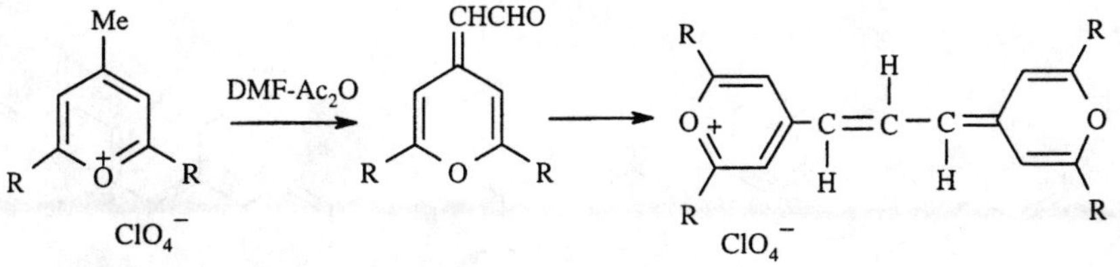

(**662**)   (**663**)   (**664**)

(e) α- and γ-Alkyl groups of pyrylium salts condense with pyrones to yield trinuclear cyanine dyes, *e.g.* (**665**) + (**666**) → (**667**).

(f) γ-Alkylpyridines with benzenesulfonyl chloride yield products of type (**668**), probably *via* intermediates such as (**669**).

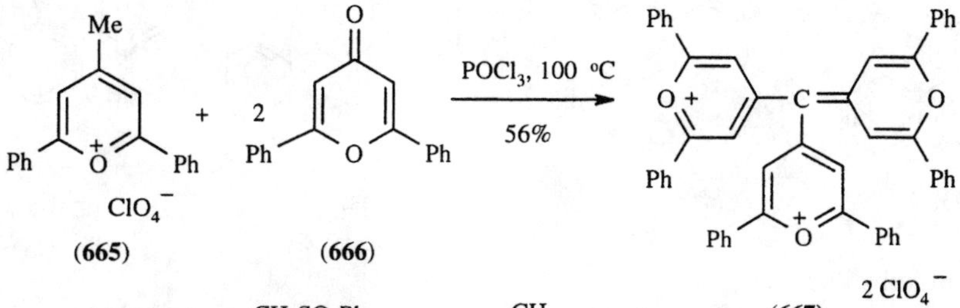

(**665**)   (**666**)   (**667**)

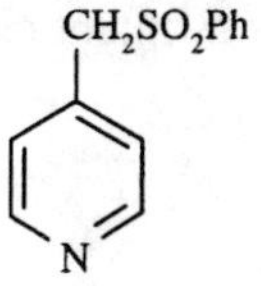

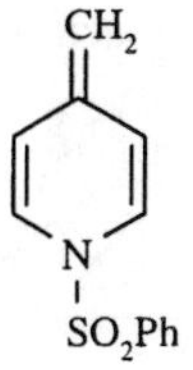

(**668**)   (**669**)

3.2.3.3.6   *Tautomerism of alkyl derivatives*

Analogous to the tautomerism of the hydroxy- and amino-pyridines (Sections 3.2.3.7.1 and 3.2.3.5.5), there are alternative tautomeric alkylidene forms of the 2- and 4-alkylpyridines (*e.g.* **609** for 2-picoline). Although the proportion of alkylidene form at equilibrium is very small (as is discussed in Section 2.2.5.1.2), it can be important as a reactive intermediate (see above).

3.2.3.4   Further Carbon Functional Groups
3.2.3.4.1   *Aryl groups*

(i) Electrophilic substitution

Electrophilic substitution usually occurs preferentially in the aryl group. In compounds containing both an aryl group and a fused benzene ring, electrophiles usually attack the aryl group exclusively.

α- and γ-Phenylpyridines are nitrated to form mixtures of the α- and γ-(*o-*, *m-* and *p*-nitrophenyl)pyridines (*cf.* **670**); the proportion of the isomers formed does not greatly vary with the acidity of the reaction mixture. Likewise, nitration (H₂SO₄-MeOH-HNO₃, 0°C) of flavone also gives a mixture of the 2'-, 3'- and 4'-nitro derivatives (**671**). This represents reactivity midway between the corresponding carbonyl and benzenoid derivatives: acetophenone is nitrated exclusively *meta*; biphenyl, exclusively *ortho* and *para*.

The tendency to be nitrated *meta* increases with the electron deficiency of the parent ring, and presumably depends on the species that reacts. 4-Phenylpyrimidine is nitrated in the phenyl group at all positions in proportions depending on the conditions, whereas the pyrazine derivative (**672**) reacts at the *meta*-positions as shown. Positively charged heterocyclic rings direct the substitution to the *meta*-position of α- or γ-phenyl groups but to the *ortho, para*-positions of β-phenyl groups as exemplified by the orientation for nitration in (**673**)-(**675**). The activating and *para*-directing influence of an *N*-oxide group toward electrophilic substitution (*cf.* Section 3.2.1.4.4.v) does not extend to phenyl substituents, *e.g.* 2-phenylpyridine 1-oxide is nitrated as shown (**676**).

(**670**)   (**671**)   (**672**)

(**673**)   (**674**)   (**675**)   (**676**)

(ii) Reactions of substituents

An example of a significant effect on the reactivity of a substituent on a phenyl ring is found in the easy proton loss in (**677**) → (**678**).

(**677**)   (**678**)   (**679**)

3.2.3.4.2 *Carboxylic acids and derivatives*

(i) General

Pyridine- and azine-carboxylic acids, as amino acids, exist partly as betaines (*e.g.* **679**) in aqueous solution, but very dominantly as neutral molecules in ethanol which has a lower dielectric constant.

In most of their reactions, the pyridine- and azine-carboxylic acids and their derivatives behave as expected (*cf.* Scheme 61). However, some acid chlorides can be obtained only as hydrochlorides, and we must also consider decarboxylation.

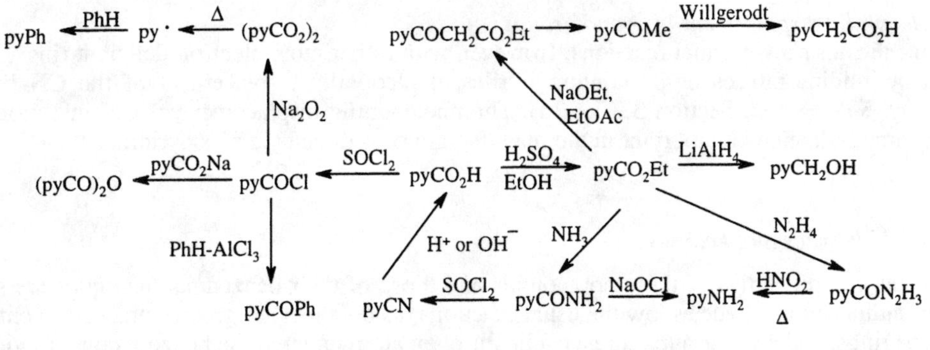

(py represents 2-, 3- or 4-pyridyl)

Scheme 61

(ii) Decarboxylation of carboxy groups directly attached to ring

Azinecarboxylic acids lose CO_2 significantly more easily than benzoic acid. Pyridinecarboxylic acids decarboxylate on heating with increasing ease in the order $\beta << \gamma < \alpha$. 2-Pyridazinecarboxylic acid gives pyrazine at 200°C, and 4,5-pyrimidinedicarboxylic acid forms the 5-mono-acid on vacuum distillation. Pyrone- and pyridone-carboxylic acids also decarboxylate relatively easily; thus, chelidonic acid (**680**; Z=O) at 160°C over copper powder and chelidamic acid (**680**; Z=NH) at 260°C give (**681**; Z=O, NH).

The relatively easy decarboxylation of α- (**682**) and γ-carboxylic acids is a result of inductive stabilization of intermediate ylides of type (**683**) (*cf.* Section 3.2.1.8.2). By carrying out the decarboxylation in the presence of aldehydes or ketones, products of type (**684**) are formed (Hammick Reaction).

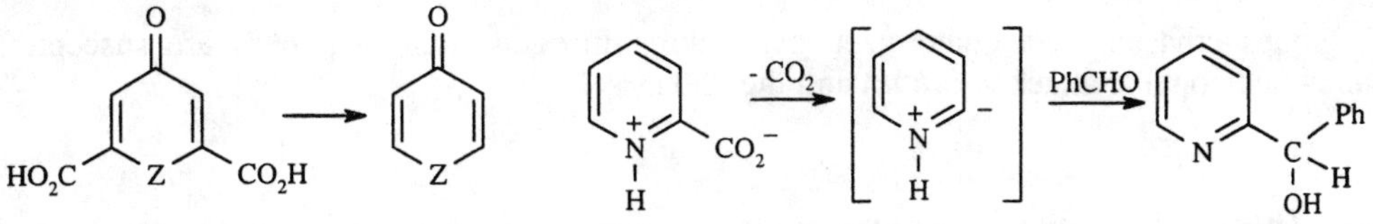

(680) **(681)** **(682)** **(683)** **(684)**

(iii) Decarboxylation of carboxymethyl groups

Pyridines with an α- or γ-carboxymethyl group (*e.g.* **685**) undergo facile decarboxylation by a zwitterion mechanism (**685** → **688**) somewhat similar to that for the decarboxylation of β-keto acids (*cf.* Section 3.2.3.1.1). Carboxymethylpyridines often decarboxylate spontaneously on formation; thus, hydrolysis of (**689**) gives (**690**). The corresponding 2- and 4-pyridone and 2- and 4-pyrone acids are somewhat more stable, *e.g.* (**691**) decarboxylates at 170°C. 3-Pyridineacetic acid shows no pronounced tendency to decarboxylate.

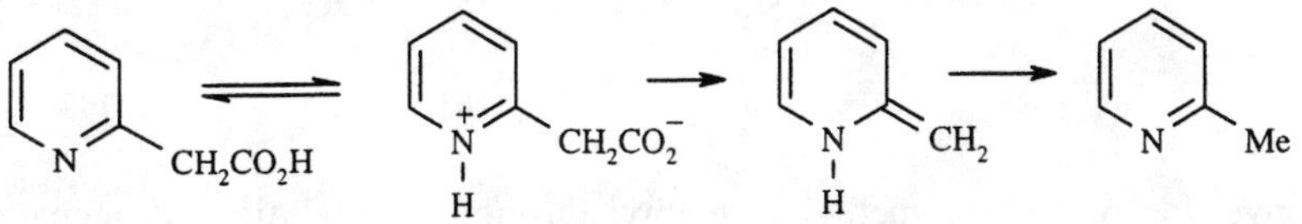

(685) **(686)** **(687)** **(688)**

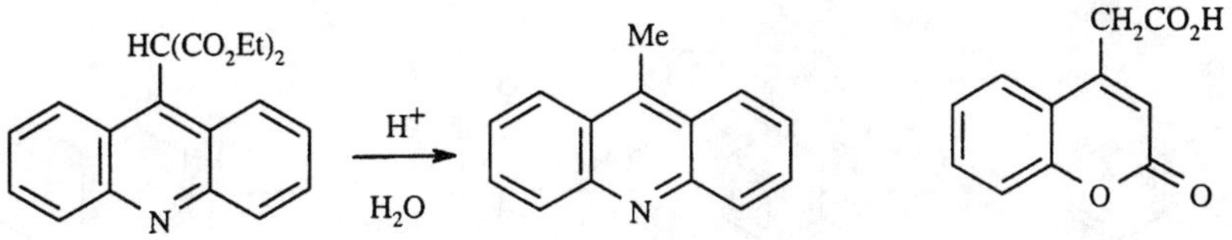

(689) **(690)** **(691)**

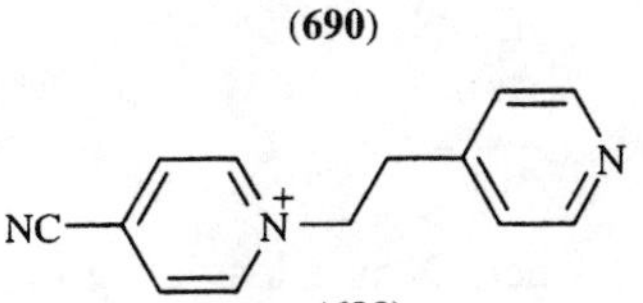

(692)

(iv) Nucleophilic displacement of cyano groups

Pyridine nitriles show normal reactions. However, with rather more electron-deficient rings, such as those in pyrimidine nitriles or pyridinium nitriles, nucleophilic displacement of the CN becomes possible (*cf.* **561** → **562**, Section 3.2.3.1.2.i). The nucleophilic displacement of CN in cationic ring (**692**) by dimethylamine is important in the manufacture of 4-dimethylaminopyridine.

3.2.3.4.3 Aldehydes and ketones

In general, the properties of these compounds and those of their benzenoid analogues are similar: thus, pyrimidinecarbaldehydes show the usual reactions, as do aldehyde groups attached to chromone and pyrone rings. Aldehyde groups α to a cyclic nitrogen atom undergo the benzoin condensation very readily because the end products are stabilized as hydrogen-bonded ene-diols (*e.g.* **693**).

2,4,6-Tris(dimorpholinomethyl)-1,3,5-triazine (**694**, R = morpholino) can behave as a synthetic equivalent of the still unknown 1,3,5-triazinetricarbaldehyde. Trisphenylhydrazones, trioximes or trisemicarbazones of this tri-carbaldehyde have been obtained from (**694**) ⟨87JHC793⟩.

| (693) | (694) | (695) | (696) | (697) |

Acyl groups adjacent to a quaternized pyridinium nitrogen atom (*e.g.* **695**) are susceptible to removal by nucleophilic attack *via* (**696**) and the ylide (**697**).

3.2.3.4.4 Other substituted alkyl groups

(i) Examples of normal reactivity

These include halogen atoms in side chains: (**698**) + H_2SO_4 + H_2O → pyridine-2-carboxaldehyde; (**699**; Y = Cl) + KCN → (**699**; Y = CN); (**622**) + KOH → (**700**).

| (698) | (699) | (700) | (701) | (702) |

Reductive conversion of hydroxymethyl to methyl during the catalytic hydrogenation of 6- and 7-hydroxymethylpterins (*e.g.* **703** → **706**) probably involves H_2O elimination from the intermediate 5,8-dihydro species (**704**) and subsequent reduction of the C=C and C=N bonds in (**705**) ⟨B-89MI 718-02⟩.

| (703) | (704) | (705) | (706) |

(ii) Reverse Michael reactions

Compounds of type Het-CH_2CH_2X where X is a leaving group (*e.g.* halogen, OR, NR_2, SR, *etc.*) undergo reverse Michael reaction *via*

$$\text{Het-}\bar{\text{C}}\text{HCH}_2\text{X} \rightarrow \text{Het-CH}=\text{CH}_2 + \text{X}^-$$

In this way we find ready dehydration of α- or γ-(2-hydroxyethyl) groups, *e.g.* (**701**; X = OH) → (**702**).

(iii) Diazoalkyl groups and related carbenes

2-(Diazomethyl)pyridine (**708**) which normally exists in the ring-closed form (**707**) thermolyzes to 2-pyridylcarbene (**709**) which interconverts in the gas phase with phenylnitrene (**711**). Photolysis of 2-(diazomethyl)pyridine in an argon matrix allows identification of 1-aza-1,2,4,6-cycloheptatetraene (**710**). 3- and 4-Pyridylcarbenes also interconvert with phenylnitrene ⟨81AHC(28)279⟩.

(**707**)   (**708**)   (**709**)   (**710**)   (**711**)

(iv) Nucleophilic displacements

Trihalomethyl groups can be replaced in nucleophilic substitution reactions in sufficiently activated systems, as for example in *s*-triazines. 2,4,6-Tris(trichloromethyl)-*s*-triazine is converted into 2,4,6-triamino-*s*-triazine by ammonia.

(v) Formation of zwitterion intermediate

2-(α-Hydroxyalkyl) groups are removed by nucleophiles in 'retro-aldol' type reactions. The heterocycle is eliminated as the zwitterion species, or 'nucleophilic carbene' (**712**).

3.2.3.4.5 Vinyl groups

Vinyl groups α or γ to the pyridine nitrogen atom readily undergo Michael additions. Water, alcohols, ammonia, amines and hydrogen cyanide are among the nucleophiles which may be added. For example, 2-vinylpyridine and dimethylamine give the adduct (**701**; X = NMe₂).

The usual alkenic reactions are also shown by *C*-vinyl heterocycles, including ready free radical or nucleophilic polymerization.

(**712**)   (**713**)

3.2.3.5 Amino and Imino Groups

3.2.3.5.1 Orientation of reactions of amino-pyridines and -azines with electrophiles

These compounds contain three types of site for electrophilic attack: ring nitrogen, amino nitrogen and ring carbon. In 2- and 4-aminopyridines, canonical forms of type (**713**) increase the nucleophilicity of the hetero nitrogen atom and the α- and γ-carbon atoms but decrease that of the amino group. Consequently, all electrophiles would be expected to attack the ring nitrogen preferentially. Indeed, as is discussed in Section 3.2.1.3, protons, alkylating agents, metal ions and percarboxylic acids do react at the hetero nitrogen atom.

However, certain other electrophilic reagents form products derived by reaction at the amino group or at a ring carbon atom. There are four different sets of circumstances under which this behavior is found:

(i) The initial reversible reaction at the pyridine nitrogen forms an unstable product which dissociates to regenerate the reactants, or undergoes inter- or intra-molecular rearrangement [see examples (i), (iv)-(vi) and (viii) in Section 3.2.3.5.2].

(ii) In acid media the pyridine nitrogen is protonated, and reaction on this species now occurs on the amino nitrogen [see examples (ii), (iii) and (vii) in Section 3.2.3.5.2].

(iii) Reaction at the ring nitrogen can be sterically hindered. Whereas 4-dimethylaminopyridine undergoes methylation and *N*-oxide formation at the ring nitrogen as expected, in 2-dimethylaminopyridine it is the dimethylamino nitrogen which is both quaternized and *N*-oxidized, because the ring nitrogen is shielded.

(iv) If the reaction of the electrophile at the amino nitrogen is also reversible, then subsequent slower but irreversible reaction at the ring carbon can slowly go to completion. In this way, the electrophiles responsible for nitration, sulfonation and halogenation react at the ring carbon atoms, as discussed in Section 3.2.1.4.

3.2.3.5.2 *Reaction of aminoazines with electrophiles at the amino group*

These reactions are illustrated (**714–725**) for 2-aminopyridine.

(i)   Carboxylic and sulfonic acid chlorides and anhydrides give acylamino- and sulfonamido-pyridines (**716**) and (**717**). Evidence exists that initial products of the reaction, the corresponding N_1-acyl-2-aminopyridinium salts, are unstable and rapidly rearranges into the thermodynamically more stable 2-acylaminoderivatives.

(ii) Nitric acid-sulfuric acid gives nitramino compounds (**715**) which are easily rearranged to *C*-nitro derivatives (**714**) (*cf.* Section 3.2.1.4.4).

(iii) Oxidation by permonosulfuric acid yields nitropyridines (**723**).

(iv) Nitrosobenzene yields phenylazopyridines (**724**).

(v)   Sodium hypochlorite gives symmetrical azopyridines.

(vi) With dimethyl sulfide and *N*-chlorosuccinimide the sulfilimine (**725**) is formed, which can be oxidized to the corresponding nitroso heterocycle.

(vii) Nitrous acid gives highly unstable diazonium salts (**720**) (*cf.* next section).

(viii) Quaternary heterocyclic iodides give products (*e.g.* **718** from 1-methylquinolinium iodide) which can be converted into azacyanine dyes (*e.g.* **722**).

3.2.3.5.3 *Diazotization of amino compounds*

The stabilities of pyridine-2- and -4-diazonium ions resemble those of aliphatic rather than benzenoid diazonium cations. Benzenediazonium ions are stabilized by mesomerism (**726**) which involves electron donation from the ring and such electron donation is unfavorable in 2- and 4-substituted pyridines. On formation they normally immediately react with the aqueous solvent to form pyridones. However, by carrying out the diazotization in concentrated HCl or HBr, useful yields of chloro- and bromo-pyridines may be obtained. Iodinated pyridines can be obtained in good yield using the Sandmeyer reaction. Amino-pyridazines and -pyrazines, 2- and 4-aminopyrimidines and amino-1,2,4-triazines behave similarly. Nucleophilic fluorination *via* the Balz–Schiemann reaction of diazonium fluoroborates is reported to yield fluoropyridines. 2-Fluoropyridines can be prepared from diazonium fluorides, even with weak nucleophiles such as fluoride ion in hydrofluoric acid ⟨81JOC4567⟩. Fluoroborates can also be converted to fluoro compounds by ultraviolet irradiation.

The reactions of 3-amino groups in pyridines and 5-amino groups in pyrimidines, by contrast, are close to those of the amino group in aniline. The diazonium salts are reasonably stable and undergo coupling and replacement reactions and can be reduced to hydrazines. Amino groups in the 3-position of isoquinolines (727) are subject to bond fixation and react in a manner intermediate between those of α- and β-amino groups; they can be diazotized under the conditions normally employed.

Aminopyridine *N*-oxides can be diazotized, and the diazonium salts undergo coupling, *etc.* These diazonium salts are stabilized by mesomerism (728), *cf.* (726). Amino groups in pyridazine *N*-oxides can also be diazotized and the diazonium group further replaced by other functionality. Nitrosation of 3-amino-1,2,4-triazine 2-oxides and subsequent thermolysis of the diazonium tetrafluoroborate salts afforded 3-fluoro-1,2,4-triazine 2-oxides ⟨85H(23)1969⟩.

β-Aminopyridones form diazo anhydrides (*e.g.* 729) (*cf.* aminophenols) which on irradiation give pyrrolecarboxylic acids (731) *via* (730).

(726)   (727)   (728)

(729)   (730)   (731)

3.2.3.5.4 *Reactions of amino compounds with nucleophiles*

Three modes of reaction are possible: proton loss, nucleophilic displacement and Dimroth rearrangement.

(i) *Proton loss*

Just as they decrease the susceptibility of 2- and 4-amino groups to electrophilic attack, canonical forms of type (713) facilitate proton loss from amino groups. Aminoazines are thus weak acids; their anions, *e.g.* (732) ↔ (733), react with electrophilic reagents preferentially at the amino nitrogen. 2-, 3- and 4-Aminopyridines are converted by NaNH$_2$ – MeI in liquid ammonia into the corresponding methylamino or dimethylamino derivatives (*e.g.* 735 → 736, 737). However, this reaction does not stop at the monoalkylation stage regardless of the amine : alkyl halide ratio. Even when an excess of an amine is used, the dialkylamino derivative predominates with only a small amount of monoalkylamine formed ⟨72CI(L)256⟩. To prepare the monoalkylated amine, an aprotic non-polar solvent should be used (*cf.* Section 3.4.3.5.4). With EtONO–NaOEt, the reaction (735) → (734) gives a sodium diazotate which couples with phenols.

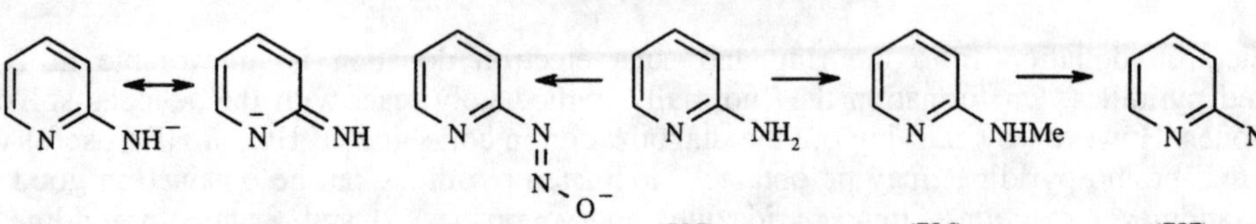

(732)   (733)   (734)   (735)   (736)   (737)

(ii) Nucleophilic displacement

Nucleophilic reagents can also react with 2- and 4-aminopyridines at the carbon atom which carries the amino group in a replacement reaction (*e.g.* **738 → 739**) similar to, but far less facile than, that undergone by chloro and alkoxy compounds, *etc.* In this way aminopyrimidines can be converted into pyrimidinones by direct acidic or alkaline hydrolysis under rather vigorous conditions.

Reactions of this type are easy if the amino group is quaternized as in, for example, 1-(4′-pyridyl)pyridinium chloride (**740**), which gives pyridine and 4-substituted pyridines [**741**; Nu = Cl, Br (with PX$_5$), Nu = SH, SR (with SH$^-$, SR$^-$), Nu = NH$_2$, NHR (with NH$_3$, NH$_2$R)]. Similarly, NMe$_3^+$ groups in pyrimidines undergo nucleophilic displacement.

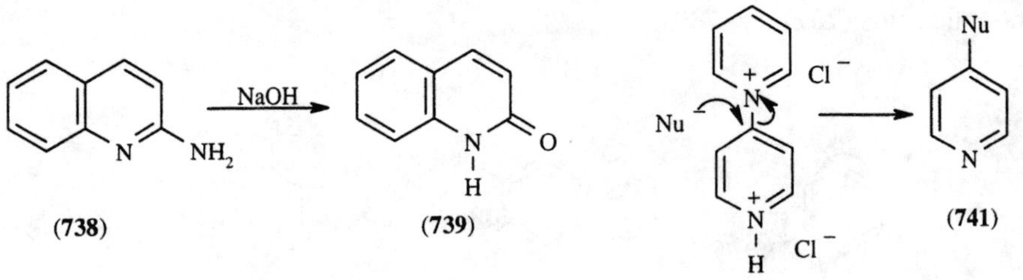

(738)   (739)   (741)

(740)

(iii) Dimroth rearrangement

Cationic α-amino derivatives with base undergo a rearrangement in which the two nitrogen atoms change places. Thus, 1-methyl-2-aminopyrimidinium ion (**742**) yields 2-methylaminopyrimidine. This, the Dimroth rearrangement, involves nucleophilic (usually OH$^-$) addition at the α-position (**743**), followed by electrocyclic ring opening to (**744**) and reclosure to (**745**). 2-Aminopyrones and 2-aminopyrylium salts similarly rearrange readily into substituted pyridones (**746 → 747**).

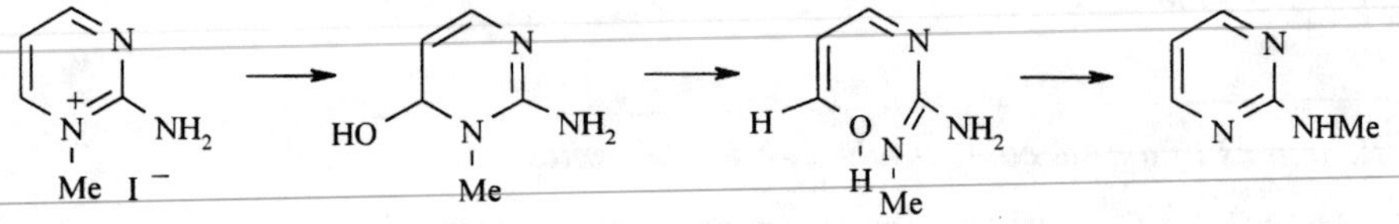

(742)   (743)   (744)   (745)

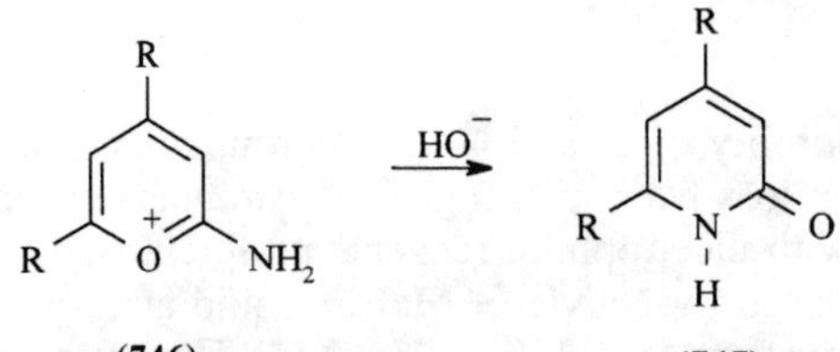

(746)   (747)

Dimroth rearrangement of neutral heterocyclic bases is also possible. For example, 1-ethyladenine heated in NaOH produces N(6)-ethyladenine (91%), hypoxanthine (2%) and 1-ethylhypoxanthine

(2%). The minor products are formed by hydrolysis of the first formed imidazole carboxamidines and recyclization (Scheme 62).

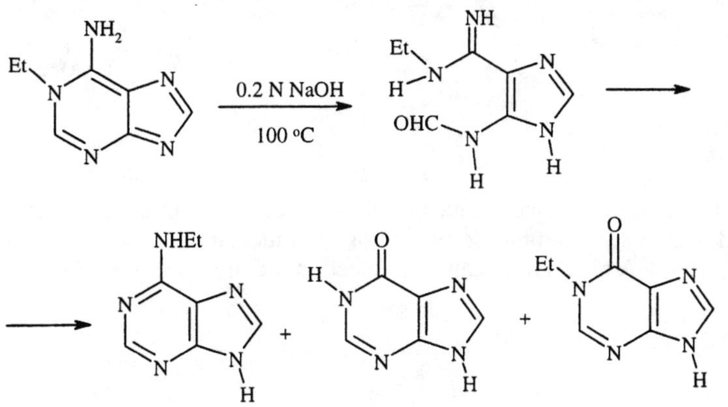

Scheme 62

3.2.3.5.5 *Amino-imino tautomerism*

2- and 4-Aminopyridines (*e.g.* **748**) can also exist in tautomeric pyridonimine forms (*e.g.* **750**). However, the pyridonimine forms are unimportant (Section 2.2.5.1) in direct contrast to the 2- and 4-hydroxypyridines which exist largely as pyridones. This difference can be rationalized by consideration of the mesomerism of the alternative forms. Resonance stabilization of aminopyridines (**748** ↔ **749**) is greater than that of hydroxypyridines, while resonance stabilization of pyridonimines (**750** ↔ **751**) is less than that of pyridones.

α- and γ-Amino *N*-oxides also exist predominantly in the amino form, *e.g.* as (**752**) rather than (**753**).

3.2.3.6 Other N-Linked Substituents

3.2.3.6.1 *Nitro groups*

2-and 4-Nitro groups on pyridines and pyridine *N*-oxides are smoothly displaced by nucleophilic reagents, indeed, more readily than the corresponding halogen compounds. Thus, 4-nitropyridine is converted by sodium ethoxide at 80°C into 4-ethoxypyridine. Such reactions are of particular importance in *N*-oxides where the nitro derivatives are readily available by direct nitration and are exemplified by the transformations (**755**) → (**754**) and (**755**) → (**756**; X = Cl, Br). A good method for preparing bromopyridines is the nucleophilic displacement of nitro groups using HBr, either on the substituted pyridine or the corresponding *N*-oxide ⟨81JAP(K)115776, 83JAP(K)5818360, 87MI 502-06⟩. The reactions involving hydrogen bromide and chloride are acid catalyzed, while those with acetyl chloride probably proceed *via* intermediates of type (**757**). 4-Nitropyridine gives (**758**) and other products on keeping; *cf.* polymerization of 4-halopyridines (Section 3.2.3.10.6). Nitro groups at all the positions of pyridazine 1-oxide are easily substituted by halogen or other nucleophiles.

(754)   (755)   (756)   (757)

Nitro compounds are easily reduced, catalytically or chemically, to amino compounds. Incomplete reduction can lead to a hydroxylamino derivative or to binuclear azo, azoxy and hydrazo compounds, *e.g.* (760) → (759), (761). A nitro group can be reduced in the presence of an *N*-oxide group, *e.g.* (755) → (752).

(758)   (759)   (760)   (761)

3.2.3.6.2 Nitramino compounds

These compounds can be rearranged (*cf.* Section 3.2.3.5.1.iv), reduced to hydrazino derivatives or hydrolyzed to pyridones.

3.2.3.6.3 Hydrazino groups

These form derivatives with carbonyl compounds and can be acylated, sulfonylated, eliminated by mild oxidation [*e.g.* (762; Y = NHNH$_2$) + CuSO$_4$ + AcOH → (762; Y = H)]. They are converted by nitrous acid into azides as in benzenoid chemistry.

Hydrazino groups attached to cationic rings (as 763) undergo oxidative coupling reactions with amines, phenols (to give *e.g.* 764) and reactive methylene compounds, *e.g.* (763) + CH$_2$(CN)$_2$ → (765).

α- or γ-Phenylsulfonylhydrazino groups are eliminated by alkali (*e.g.* 766 → acridine + N$_2$ + PhSO$_2$H); *cf.* the McFadyen and Stevens reaction (RCONHNHSO$_2$Ph → RCHO + N$_2$ + PhSO$_2$H).

(762)   (763)   (764)   (765)   (766)

Thermolysis of hydrazine (767) in toluene-methanol catalyzed by TFA led to a translocative rearrangement to give the 2-aminopyridine derivative (768) ⟨88JOC5309⟩.

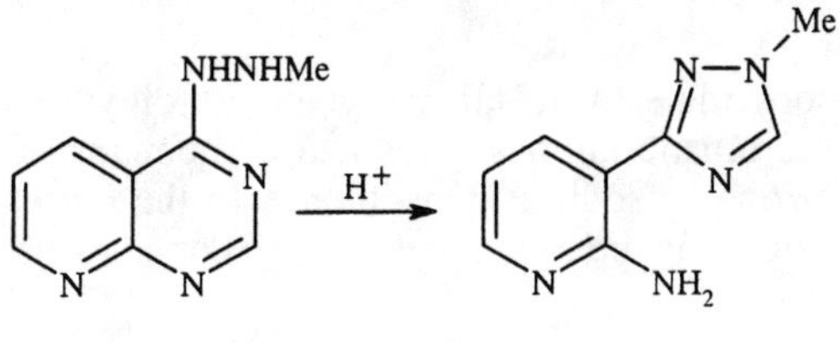

(767)   (768)

3.2.3.6.4 Azides

2-Azidopyridines exist largely, and 3-azidopyridazines completely, in the bicyclic form (*cf.* **769** ⇌ **770**) (see also Section 2.2.5.4). A synthetically useful reaction of azidoazines is their thermal fragmentation with loss of a nitrogen molecule. This leads to formation of a *C*-nitrene which undergoes subsequent deep rearrangements leading to ring contraction or ring enlargement products. Thus, pyrolysis of 2-azidopyrazines or their irradiation in ethanol gives 1-cyanoimidazoles in excellent yields (Scheme 63) ⟨83JHC1277⟩.

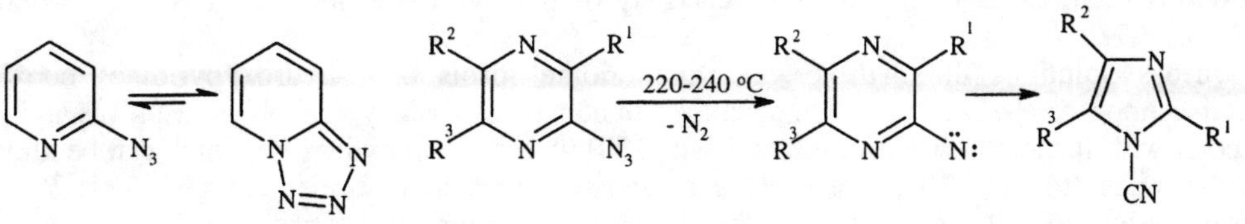

(769)   (770)

Scheme 63

The photolysis of 4-azidopyridazines gives significant yields of the ring expansion products, unsaturated 1,2,5-triazepines. Thus, 3-methoxy derivative (**771**) reacts, *via* formation of a nitrene and ring opening of an intermediate fused azirine, with methoxide or diethylamine, to give 4-methoxy (or 4-diethylamino)-1,2,5-triazepines (**772**, X = OMe, NEt$_2$) (Scheme 64).

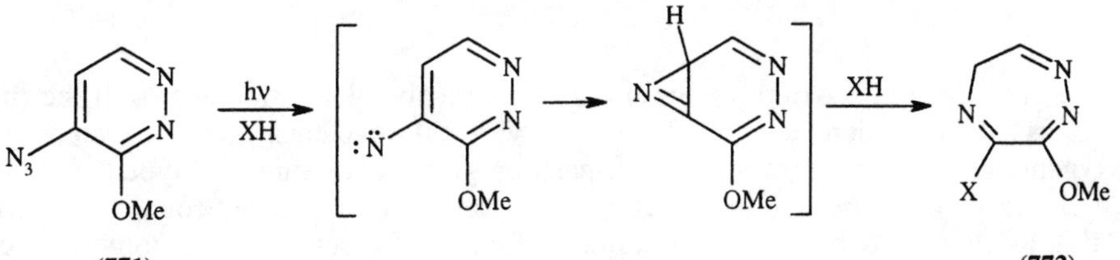

(771)   (772)

Scheme 64

Nitrotriazines are reported to be formed by photolysis of azido-1,3,5-triazines in CHCl$_3$ or MeCN in the presence of air (Scheme 65) ⟨87IZV1196⟩.

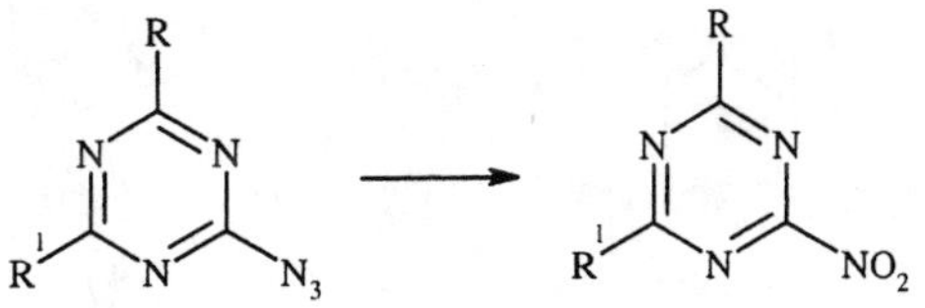

R, R^1 = Me$_2$N, MeO

Scheme 65

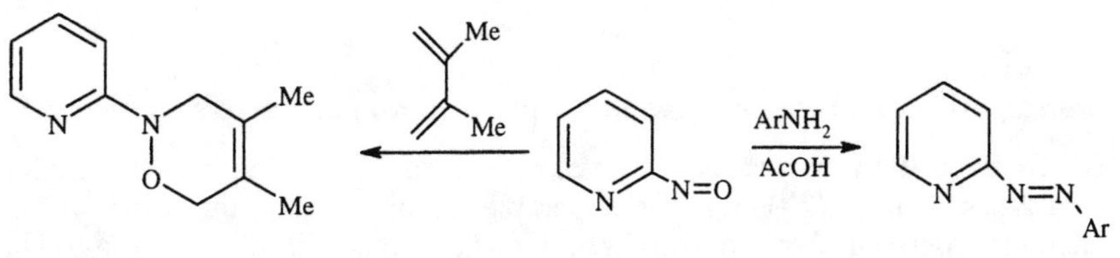

(773)   (774)   (775)

3.2.3.6.5 Nitroso groups

Ready addition of 2-nitrosopyridine to 1,3-dienes gives 3,6-dihydro-1,2-oxazines, *e.g.* (774) →
(773), and condensation with aromatic amines gives azo compounds, *e.g.* (774) → (775). Nitroso
compounds are oxidized by ozone or sodium hypochlorite to the corresponding nitro compounds.
5-Nitrosopyrimidines can be reduced to the 5-amino derivatives or condensed with activated methylene
groups.

3.2.3.7 Hydroxy and Oxo Groups

3.2.3.7.1 Hydroxy groups and hydroxy-oxo tautomeric equilibria

Hydroxypyridines (776) are both weak acids and bases and can therefore exist as zwitterions (777)
(see Section 2.2.5.1). The zwitterions of 2- and 4-hydroxypyridines are known as 2- and 4-pyridones
because of their uncharged canonical forms, *e.g.* (778) and (779). α- and γ-Hydroxypyridines exist in
aqueous solution very predominantly as the oxo or pyridone form. For α- and γ-hydroxy-
benzopyridines and -benzazines, the equilibrium favors the benzopyridone form still more, with the
exception of 3-hydroxyisoquinoline. The reactivity of the pyridones and azinones is considered in
Sections 3.2.3.7.2–4.

In aqueous solutions the hydroxy and zwitterionic forms of β-hydroxypyridines coexist in
comparable amounts. 3-Hydroxypyridine behaves in many ways as a typical phenol. It gives an intense
violet color with ferric chloride and forms a salt (780) with sodium hydroxide which can be alkylated
by alkyl halides (to give 781; Y = alkyl) and acylated by acid chlorides (to give 781; Y = acyl).
5-Hydroxypyrimidines also exist as such, they behave as phenols and are easily acylated.

(776)　　(777)　　(778)　　(779)　　(780)　　(781)

Cycloaddition reactions of 3-oxidopyridinium betaines involving addition at two of the ring atoms
have been discussed in Section 3.2.1.10. However, with chloroketenes reaction occurs across the
exocyclic oxygen atom and either the 4- or the 2-position giving compounds of type (782).

Hydroxypyridine 1-oxides are also tautomeric; the 4-isomer exists in about equal amounts of forms
(783) and (784). 4-Hydroxy-pyrones and -pyridones exhibit a different type of tautomerism: the α-one
(*e.g.* 785) structure is favored relative to the γ-one structure. β-Hydroxy-4-pyrones such as kojic acid
(786) show phenolic properties (CHEC 3.28, Scheme 30). α- and β-Hydroxy cations (*e.g.* 787) are the
conjugate acids of pyridones and pyrones and are considered in the next section.

(782)　　(783)　　(784)　　(785)　　(786)　　(787)

3.2.3.7.2 Pyridones, pyrones, thiinones, azinones, etc.: general pattern of reactivity

These compounds are usually written in the uncharged form (788, 789; Z = NH, NR, O, S), but
canonical forms of types (790) or (791) are of comparable importance, *i.e.* the compounds can also be
considered as betaines derived from pyridinium, pyrylium and thiinium cations. They possess
considerable stability and aromaticity in that in many of their reactions they 'revert to type'.

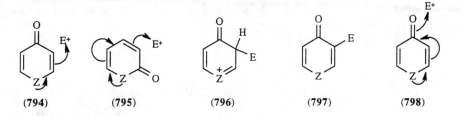

(788)   (789)   (790)   (791)   (792)   (793)

The reactivity pattern of these compounds considered briefly in Section 3.2.1.1.4 will now be summarized. The system of heteroatoms in these molecules can act either as an electron source or an electron sink. This, together with the possibility of readily forming cationic (**792**) and anionic (**793**) species, increases considerably the possibilities for reaction in these compounds.

(i) Reactions with electrophiles

These can attack ring carbon atoms β to the cyclic heteroatom as shown in (**794**) and (**795**); the intermediates (*e.g.* **796**) usually revert to type by proton loss (**796** → **797**). These electrophilic substitution reactions are considered in Section 3.2.1.4.

Electrophilic reagents can also attack the carbonyl oxygen atom (*e.g.* **798**); reactions of this type are considered in Section 3.2.3.7.3.

(794)   (795)   (796)   (797)   (798)

(ii) Reactions with nucleophiles

Four modes of reaction exist:

(a) Attack at a ring carbon atom, other than that of the carbonyl group, can be followed by proton addition, *i.e.* overall Michael-type reaction. An example of this rather rare reaction type, which involves loss of aromaticity, is given in Section 3.2.1.6.8.

(b) Nucleophilic reagents can attack the carbon atom of the carbonyl group (as in **799**). The reaction sequence proceeds by complete loss of the carbonyl oxygen and subsequent rearomatization. These reactions are considered in Section 3.2.3.7.4.

(c) Nucleophiles can attack α- and γ-pyrones and oxazinones, *etc.*, at a carbon α or γ to the ring oxygen to give initial adducts (**799, 800**) which undergo ring opening (*e.g.* with OH⁻) which can be followed by reclosure in suitable cases (*e.g.* with NH₃, RNH₂). These reactions are considered in Section 3.2.1.6.

(d) A hydrogen atom on the heterocyclic nitrogen atom of pyridones can be removed as a proton by nucleophilic reagents, *e.g.* (**801**) → (**802**). The resulting mesomeric anion, *e.g.* (**802**) ↔ (**803**), reacts readily with electrophilic reagents at the nitrogen, β-carbon or oxygen atoms as discussed in Section 3.2.1.8.4.

(799)   (800)   (801)   (802)   (803)

(iii) Reactions with radicals and electron-deficient species

These reactions are discussed in Section 3.2.1.9. Pyridones and pyrones are easily reduced catalytically.

(iv) Intra- and inter-molecular reactions with cyclic transition states

Reactions of these types are discussed in Sections 3.2.1.2 and 3.2.1.10 respectively; due to the reduced aromaticity and polarizability, reactions of these types are of considerable importance.

3.2.3.7.3 *Pyridones, pyrones and azinones: electrophilic attack at carbonyl oxygen*

Pyridones and pyrones are weak bases: 4- and 2-pyridone have pK_a values of 3.3 and 0.8, respectively, for proton addition to the carbonyl oxygen atom, *e.g.* (804) ↔ (805) → (806).

(804)   (805)   (806)

O-Alkylation of pyridones can be effected with diazomethane: 2-pyridone forms 2-methoxypyridine. Frequently *O*- and *N*-alkylation occur together: 4-pyridone with CH_2N_2 yields 4-methoxypyridine and 1-methyl-4-pyridone. Et_3O^+ and similar active alkylating agents also alkylate the carbonyl oxygen of pyridones and pyrones.

Silylation is important for hydroxyl group protection during synthetic transformations (*e.g.* **807** → **808** → **809**). The trimethylsilyl group is frequently used. Treatment of a pyrimidinone with chlorotrimethylsilane in the presence of a tertiary base, or often more conveniently by heating the pyrimidinone with hexamethyldisilazane, gives the corresponding *O*-trimethylsilane. The silyl ethers of hydroxypyrimidines are very sensitive to hydrolysis, more so than alkyl silyl ethers. The stability increases with the size of the silyl substituents. The dimethylthexylsilyl group is very useful when a relatively resistant silyl group is desirable.

(807)   (808)   (809)

Alkylation of pyridones *via* the anion is discussed in Section 3.2.1.8.4.

3.2.3.7.4 *Pyridones, pyrones and azinones: nucleophilic displacement of carbonyl oxygen*

Nucleophilic attack on the carbon atom of the carbonyl group, in reactions which lead to substitution rather than to ring opening, is discussed in this section.

Pyridones and pyrones behave as cyclic amides and esters and, predictably, do not normally react with nucleophilic 'ketonic reagents' such as HCN, RNH_2, $NaHSO_3$, NH_2OH, N_2H_4, PhN_2H_3 and $NH_2CON_2H_3$. Strong nucleophiles of the type that attack amides generally do react with pyridones and pyrones, as described in (i)-(v) below. However, in all these reactions, the nucleophilic attack is preceded by electrophilic attack at the carbonyl oxygen and it is this that allows the nucleophilic attack that is being discussed.

(i) Pyridones are converted into chloropyridines with $POCl_3$, PCl_5, $SOCl_2$ or $COCl_2$ in DMF, *e.g.* 2-methyl-4-quinolinone (810) gives (811). Bromopyridines may be prepared using $POBr_3$ or PBr_5. Azinones react similarly; thus, pyrimidinones yield chloropyrimidines, and uric acid (812) yields 2,6,8-trichloropurine (813).

(810)   (811)   (812)   (813)

Alkyl substituents on the pyridone nitrogen atom are usually lost in reactions of this type, but the quaternary salts from *N*-substituted acridones can be isolated. Pyrones (with PCl_5 or $POCl_3$) form

highly reactive chloropyrylium ions which are used *in situ* as reaction intermediates, *e.g.* (**814**) →
(**815**) → (**816**).

(**814**)   (**815**)   (**816**)   (**817**)

(ii) Phosphorus pentasulfide by heating in an inert solvent, in pyridine or another tertiary amine
converts carbonyl groups into thiocarbonyl groups, *e.g.* (**810**) → (**817**). Pyridazinethiones,
pyrimidinethiones and triazinethiones are similarly prepared. Direct thiation of pyrimidinones can also
be effected by the Lawesson's reagent (2,4-bis-*p*-methoxyphenyl-1,3,2,4-dithia-diphosphetane-
2,4-disulfide) and has been found useful in the conversion of oxo to thioxo groups in nucleosides.
4(6)-Oxo substituents undergo thiation more readily than 2-oxo substituents, which allows for
4-monothiation in 2,4-pyrimidinediones.

(iii) Pyrones react with active methylene compounds with Ac$_2$O as a catalyst, *e.g.* 2,6-dimethyl-
4-pyrone and malononitrile give (**818**).

(iv) Lithium aluminum hydride and Grignard reagents react with chromones, coumarins and
xanthones to give the pseudo-bases, *e.g.* (**819**) → (**820**; R = alkyl, H) of pyrylium salts (*e.g.* **821**). Some
pyridones react analogously (*e.g.* **822** + LiAlH$_4$ → phenantridine). LiAlH$_4$ will reduce in pyrimidinones
a C-4(C-6) carbonyl group to CH$_2$ before reducing a C-2 carbonyl group ⟨83OMR(21)334⟩.

(**818**)   (**819**)   (**820**)   (**821**)

(**822**)   (**823**)

(v) Reduction of xanthones (**825**) in acid solution gives bimolecular reduction products (**824, 826**).
Similarly, treatment of 2,5-diphenyl-1,3,5-thiadiazin-4-one with Lawesson's reagent, yields (instead of
the 4-thione) the 4,4-bis(1,3,5-thiadiazinylidene) (**823**) which, with iodine, forms an electrical
conducting charge-transfer complex ⟨84SUL127⟩.

(**824**)   (**825**)   (**826**)

(vi) Amines can be formed directly from tautomeric pyrimidinones by heating with an amide of
phosphoric or phosphorous acid, but the vigorous reaction conditions limit the application of this
method ⟨B-94MI 602-01⟩.

Pyrimidinones are efficiently aminated in a one-pot procedure *via* silylation. The silylating agent converts the hydroxy groups into silyl ethers which are activated for nucleophilic substitution, and react *in situ* with ammonia, primary, or secondary amines to form the corresponding amines. With Lewis acid catalysis, the reactions usually proceed in high yield if the trimethylsilanol leaving group is converted *in situ* into hexamethyldisiloxane by an excess of the silylating agent; for example, 2(1*H*)-pyrimidinone is aminated and uracil diaminated in amination reactions with HMDS ⟨84CB1523⟩. This reaction has proven to be a sufficiently mild amination method for nucleosides.

3.2.3.7.5 *Heterocyclic quinones*

Pyridine quinones are little known, but the diazine analogues of benzoquinones include alloxan (**827**) in which the carbonyl group in the 5-position shows ketonic properties. Alloxan undergoes the benzylic acid rearrangement (Na$_2$CO$_3$ → alloxanic acid **828**) and can be reduced to a dimeric product (H$_2$S → alloxantoin **829**), or to the hydroquinone analogue (SnCl$_2$ → dialuric acid **830**).

(**827**)   (**828**)   (**829**)   (**830**)

3.2.3.8 Other *O*-Linked Substituents

3.2.3.8.1 *Alkoxy and aryloxy groups*

2- and 4-Alkoxy groups in pyridines undergo nucleophilic replacement when some additional activation is present as is the case for 3-nitro-4-methoxypyridine (**831** → **832**). Such reactions are facilitated in the pyrimidine series for 2-, 4- and 6-alkoxy groups which are α and γ to *two* nitrogen atoms; they are often used to prepare aminopyrimidines. Alkyl cyanurates (**833**) behave as esters being hydrolyzed to cyanuric acid and the alcohol and readily undergoing transesterification. Nucleophilic displacement of alkoxy groups on cationic rings occurs exceedingly readily as is illustrated for 4-methoxy-2,6-dimethylpyrylium cation (**837** → **838**, Nu = EtO, NMe$_2$).

Pyridines and benzopyridines with alkoxy groups in the α- or γ-position rearrange to *N*-alkyl-2-pyridones and *N*-alkyl-4-pyridones on heating. 2-Methoxypyridine gives (**839**) at 300°C whereas 2-methoxyquinoline forms (**840**) at 100°C. One molecule of alkoxy compound acts as an alkylating agent for another in these intermolecular reactions. Similarly, 2-alkoxypyridine *N*-oxides rearrange thermally to *N*-alkoxy-2-pyridones (**841** → **842**). The thermal rearrangement of alkoxypyrimidines to *N*-alkylpyrimidinones is known as the Hilbert-Johnson reaction. Alkyl cyanurates (**833**) on heating isomerize to isocyanurates (**836**) in a step-wise manner *via* (**834**) and (**835**).

(**831**)   (**832**)

(**833**)   (**834**)   (**835**)   (**836**)

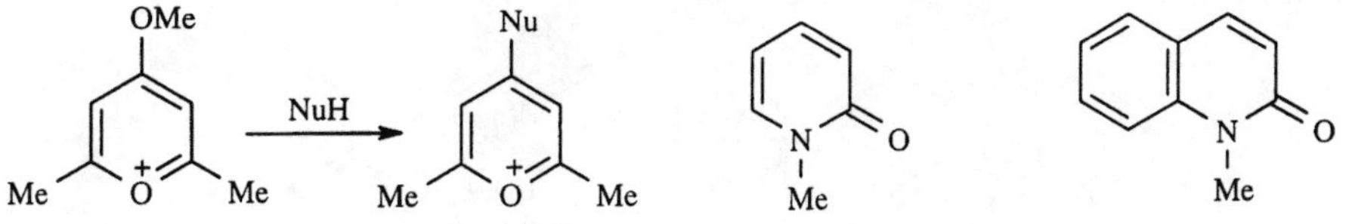

(837)   (838)   (839)   (840)

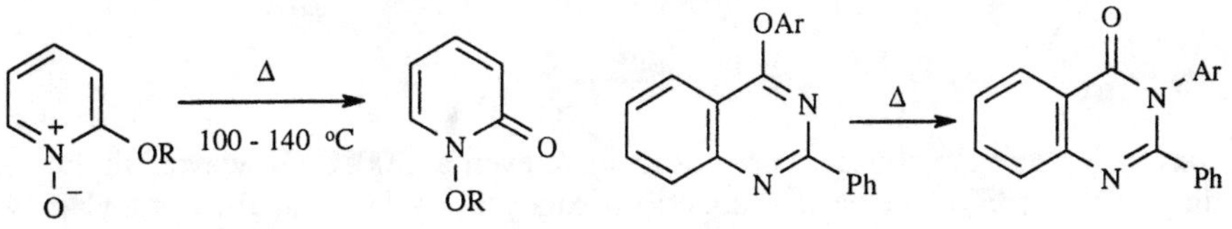

(841)   (842)   (843)   (844)

Aryloxy groups are more resistant to rearrangement (unless nitro-activation of the aryl group is present); however, under forcing conditions the reaction (**843** → **844**) forms the basis for a synthetic technique for converting a phenol into an amine.

Allyl ethers undergo true intramolecular rearrangement from oxygen to nitrogen or ring carbon. Thus, 2-(1-methylallyloxy)pyridine (**845**) gives approximately equal amounts of 1- (**846**) and 3- (**847**) crotyloxy-2-pyridones. 4-Allyloxyquinolines rearrange to 3-allyl-4-quinolones. If the 3-position is occupied by a methyl group (*cf.* **848a**) the allyl group rearranges to the benzene ring to give (**848b**) of uncertain orientation ⟨67AHC(8)147⟩.

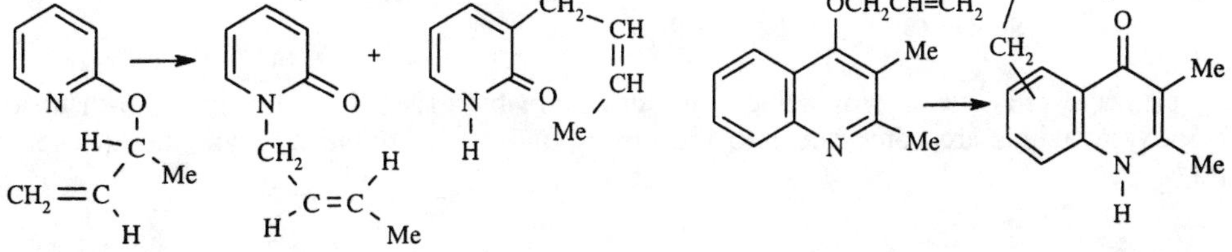

(845)   (846)   (847)   (848a)   (848b)

Thermally promoted Claisen rearrangement of simple allyloxypyrimidines is difficult to effect. 4-Allyloxypyrimidines can be rearranged at elevated temperatures; a mixture of the C5- (**849**) and the N3-allyl derivative (**850**) was formed from 4-allyloxy-2-phenylpyrimidine on heating in aniline. 4-Allyloxyquinazoline is rearranged at 190–200°C to form 3-allylquinazolin-4(3H)-one in 75% yield ⟨86T4873⟩. Rearrangement of 5-allyloxy-1,3-dimethyluracil proceeds with quantitative conversion to the 6-allyl derivative (Scheme 66). 6-Allyloxy-1,3-dimethyluracil behaves similarly to the 5-isomer; heating in dioxane gives the Claisen product (**851**) ⟨84H(22)2217⟩.

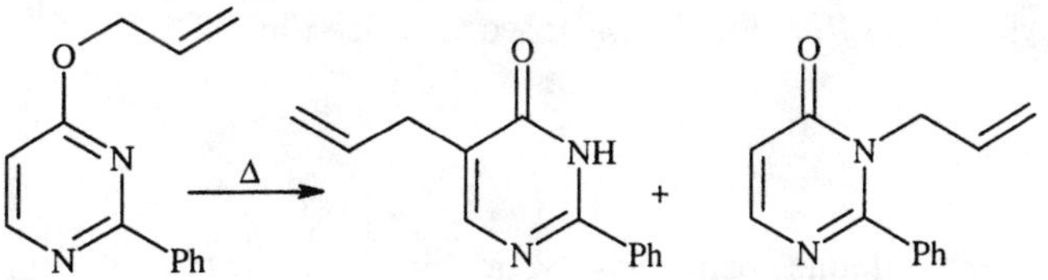

(849)   (850)

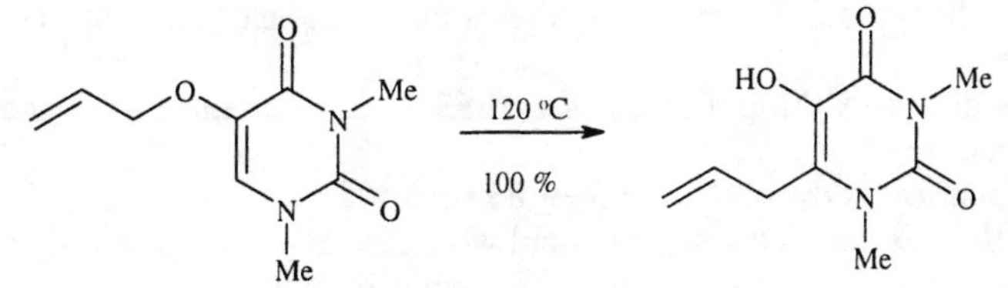

Scheme 66

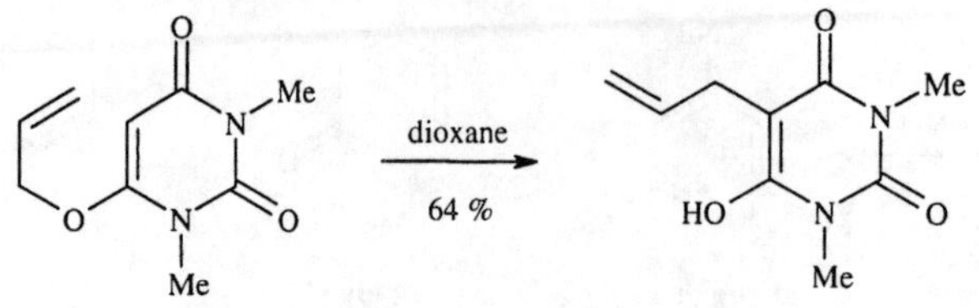

(851)

Scheme 67

2-Allyloxypyrimidine largely resists rearrangement even at 200°C. However, allylic cyclopentenyl 2-pyrimidinyl ethers (852) can be rearranged on catalysis by tetrakis(tri-isopropyl phosphite)palladium, a Pd(0)-complex (Scheme 68).

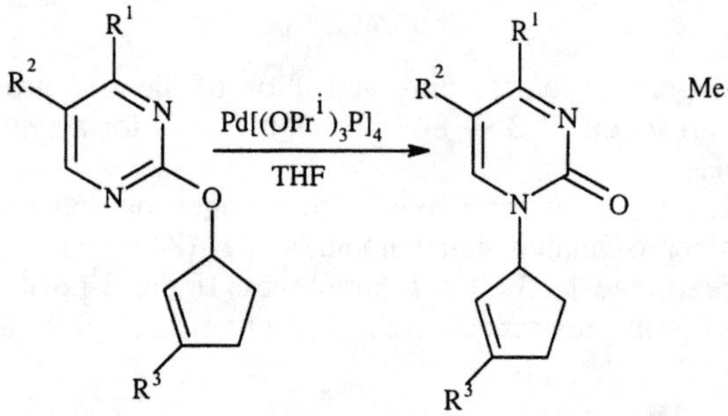

(852)

Scheme 68

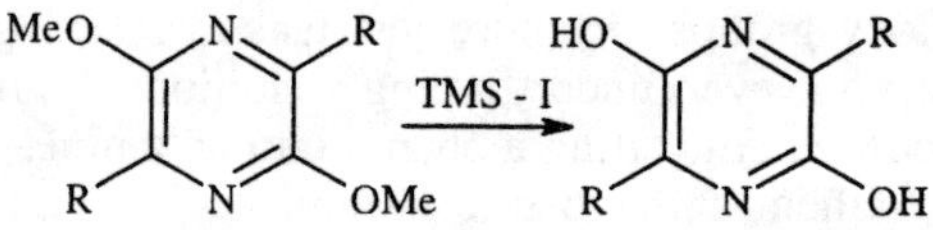

Scheme 69

Ether cleavage in methoxypyrazines is effected by TMS-I, whereby 3,6-disubstituted 2,5-dimethoxypyrazines are converted into the corresponding 2,5-dihydroxypyrazines (Scheme 69) ⟨86JHC1677⟩.

3.2.3.8.2   *Acyloxy groups*

2- and 4-Acyloxypyridines are so easily hydrolyzed that they are difficult to isolate; the reactions of 3-acyloxypyridines parallel those of phenyl acetate

3.2.3.9   **S-Linked Substituents**

3.2.3.9.1   *Mercapto-thione tautomerism*

Pyridines and azines with α- or γ-mercapto groups exist predominantly in the pyridinethione, *etc.*, forms, *e.g.* as (855) rather than in the mercapto form (854). This behavior is analogous to that of the corresponding hydroxypyridines (Section 3.2.3.7); see also Section 2.2.5.1.

3.2.3.9.2   *Thiones*

Pyridine-, pyran- and azine-thiones behave as cyclic thioamides or thioesters and show their typical reactions. Thus, they react with electrophiles at the sulfur atom [as exemplified in (i)-(iv)], and with nucleophiles including the typical 'ketonic' reagents at the thione carbon atom [as exemplified in (vi)-(viii)].

    (i)   Alkyl halides give alkylthiopyridines, *e.g.* (855 → 856), azines, or in the absence of an NH group, alkylthio cationic rings.

    (ii)   Iodine oxidation forms disulfides (855 → 853).

    (iii)   Oxidation with H_2O_2 forms the sulfinic acid which usually spontaneously loses SO_2 to give the CH derivative. Mercapto groups in any position of the pyrimidine ring can be replaced by hydrogen in this way.

(iv) Strong oxidation forms a sulfonic acid (**855 → 857**).

(v) *S*-Amination is possible. 3,3-Pentamethyleneoxaziridine (**859**) ⟨91S327⟩ or hydroxylamine-*O*-sulfonic acid reacts with 1-acylamino-, with 1-phenylureido-, or with 1-arylaminopyrimidine-2-thiones (**858**) to form *S*-aminated derivatives (**860**).

(vi) Phenylhydrazine forms (**862**; Y = NHPh) from thiocoumarin (**861**).

(vii) PCl$_5$ gives chloro compounds, *e.g.* chloropyrimidines from pyrimidinethiones (*e.g.* **863**).

(viii) Hydrolysis with HCl-H$_2$O converts compounds such as pyrimidinethiones into pyrimidinones (**863 → 864**).

(ix) Pyran-4-thiones and diazomethane are converted into unusual dimeric products (**865**).

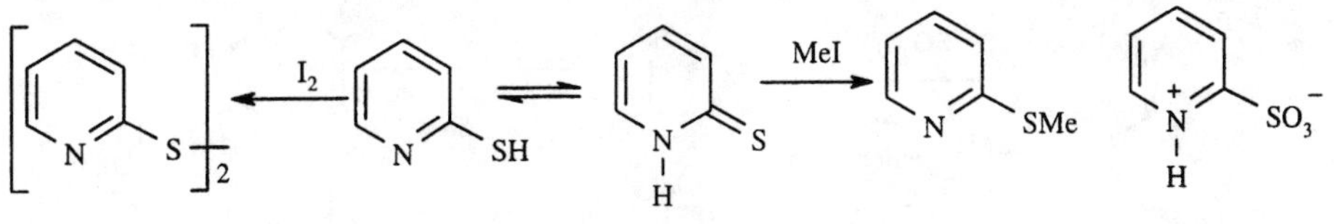

(**853**)   (**854**)   (**855**)   (**856**)   (**857**)

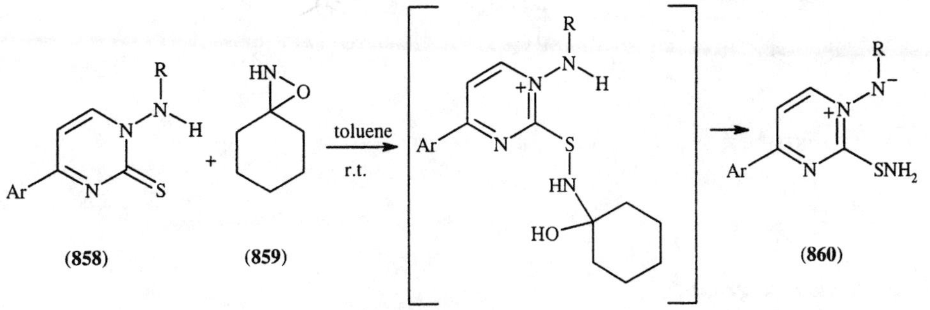

(**858**)   (**859**)   (**860**)

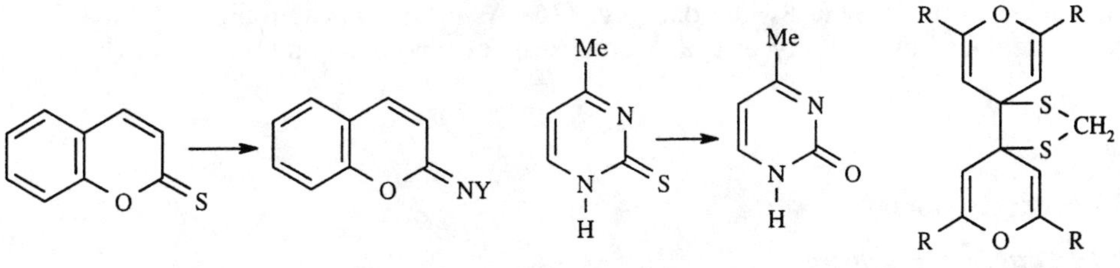

(**861**)   (**862**)   (**863**)   (**864**)   (**865**)

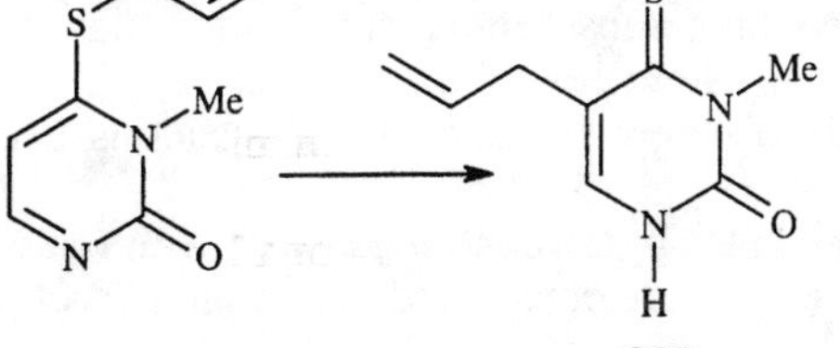

(**14**)

(**866**)

(x) Claisen type rearrangements are known. 4-Allylthio-3-methyl-2(3*H*)-pyrimidinone thus yields the 5-allyl-4-thiouracil (**866**) quantitatively (Equation (14)), although the 1-methyl isomer resists rearrangement on heating in benzene.

3.2.3.9.3 Alkylthio, alkylsulfinyl and alkylsulfonyl groups

The SR substituent can be displaced nucleophilically by amines and hydroxide (**867 → 868**) and removed reductively by dissolving metals, *e.g.* from (**869**) with Zn/H$^+$. Oxidation gives the corresponding sulfoxide and sulfone, in which nucleophilic displacement is easier. Thus, 2- and 4-(phenylsulfonyl)pyrimidines give the corresponding replacement products with various nitrogen and oxygen nucleophiles. Direct halogenation of SMe group in methylthioazines is possible, *e.g.* **870** → **871** → **872**).

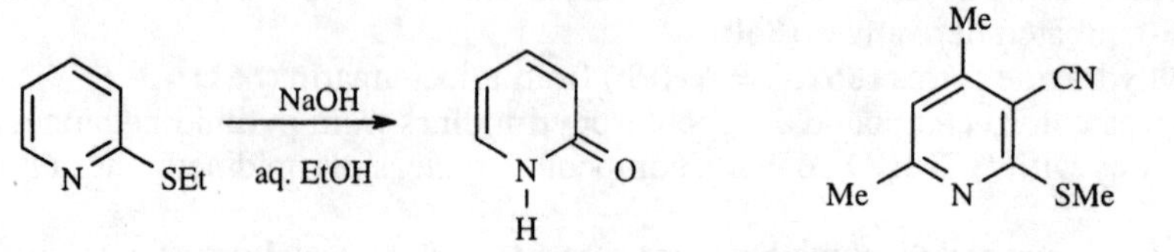

(867)   (868)   (869)

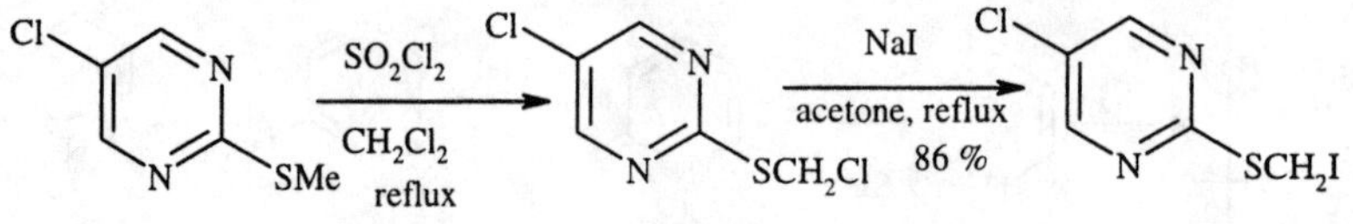

(870)   (871)   (872)

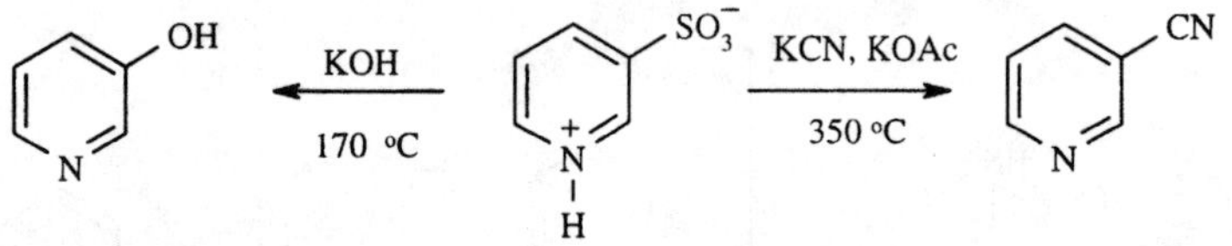

(873)   (874)   (875)

3.2.3.9.4 Sulfonic acid groups

Pyridinesulfonic acids exist as zwitterions (*e.g.* **874**). As for benzenesulfonic acid, the sulfonic acid group can be replaced by hydroxy or cyano groups under vigorous conditions, *e.g.* (**874**) → (**873**), (**875**).

3.2.3.10 Halogen Atoms

3.2.3.10.1 Pattern of reactivity

Halogen atoms attached to ring carbon of heteroaromatic six-membered rings show both reactions typical of aryl halides and their own characteristic reactions.

Just as in phenyl halides, the halogen can be replaced by hydrogen, by a metal, or be coupled. Two of the four mechanisms of such nucleophilic substitutions are also familiar from benzene chemistry: *via* arynes and by the $S_{RN}1$ mechanism. However, of the two further mechanisms of nucleophilic replacement, the ANRORC is unique to heterocycles, and S_{AE} reactions occur only with strongly activated benzenoid systems.

Recently, substitution of heteroaryl halides by carbon nucleophile with transition metal catalyst has become of great importance. These reactions are considered in a separate section.

3.2.3.10.2 Replacement of halogen by hydrogen or a metal (including transmetallation), or by coupling

Heterocyclic nuclear halogen atoms undergo the following reactions which are typical of aryl halides:

(i)   They can be replaced with hydrogen atoms by catalytic (Pd, Ni, *etc.*) or chemical reduction (HI or Zn/H_2SO_4). Reductive removal of halogen atoms is accompanied by reduction of the ring in compounds of relatively low aromaticity (*e.g.* quinazolines).

(ii) They can be converted into Grignard reagents which show the normal reactions. However, in the preparation of such Grignard reagents ethyl bromide usually has to be added to activate the magnesium ('entrainment method').

(iii) Pyridyllithium reagents can be formed by halogen-metal exchange using *n*-butyllithium at −60°C. The products react with electrophiles in the manner typical of aryllithiums: $CO_2 \rightarrow$ acids, RCHO → alcohols, esters → ketones, *etc.* Attempts to lithiate 2-chloro-, 2-bromo-, or 2-iodo-pyrimidine met with little success; adduct formation by addition of lithiated species to electrophilic pyrimidine positions is a major side-reaction. 2-Lithiopyrimidine (**878**), however, is formed selectively by metal exchange, *i.e.*, 2-stannylpyrimidines (**877**) react with butyllithium by metal-metal exchange to form (**878**); the transfer is confirmed by addition reactions to carbonyl compounds (**878** → **879**) ⟨94T275⟩.

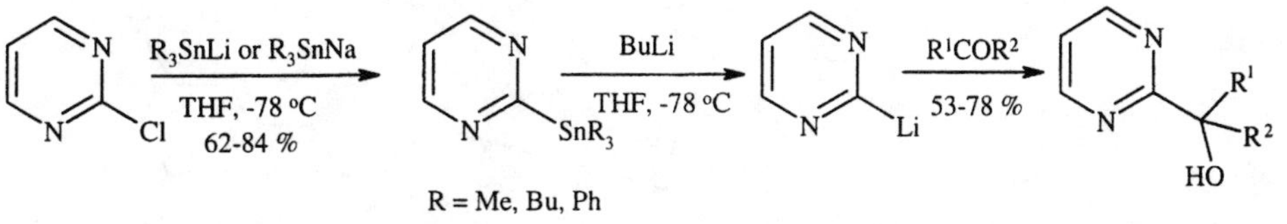

R = Me, Bu, Ph

(**876**) (**877**) (**878**) (**879**)

(iv) Organotin reagents can be prepared. 2-Chloropyrimidine is stannylated by tributyl-, trimethyl- or triphenyl-stannyllithium (**876** → **877**) ⟨94T275⟩. Direct stannylation in the 4-position has not been successful. 4-Iodo-2-methylthiopyrimidine has been stannylated by way of lithiation followed by quenching with a stannyl chloride.

(v) Organometalics by transmetallation: in general, zinc reagents are prepared by a transmetallation reaction of aryllithium or arylmagnesium halides with zinc halides at low temperature. The same approach can be used for heteroarenes in the absence of incompatible substituents, but has been little applied to pyrimidines.

Pyrimidines lithiated in the 5-position can be boronated (Scheme 70). Cupration is well established in pyrimidines ⟨86T3981, 87CS519⟩.

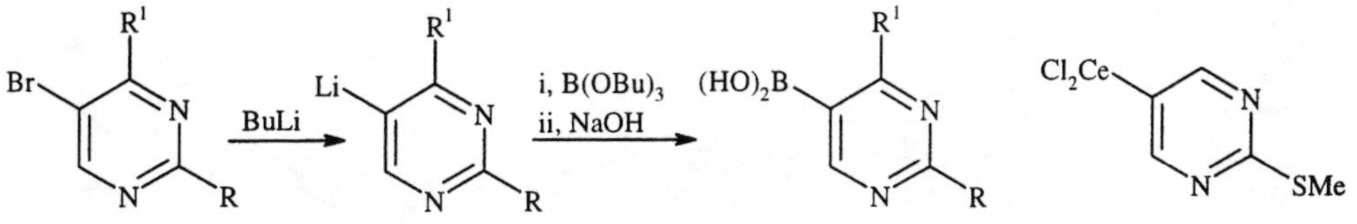

(**880**) (**881**)

Scheme 70

Organocerium derivatives posses low basicity. 5-Pyrimidylcerium dichlorides (*e.g.* **881**) are available by low-temperature metal-metal exchange using cerium trichloride and 5-lithiated pyrimidines (*e.g.* **880**, R = SMe, R^1 = H). The 5-pyrimidinylcerium dichloride was superior to its 5-lithio analogue in reactions with aldehydes and ketones, especially in reactions with enolizable aldehydes and ketones ⟨89ACS816⟩.

(vi) Heck reactions. In alkenylation reactions mediated by Pd-catalysis, the active species is a palladized heterocycle formed by oxidative insertion of the metal either into a halogen-carbon or oxygen-carbon bond, or by electrophilic substitution using a Pd(II)-species, or by a metal-metal exchange reaction. The subsequent addition of the metal complex to the alkene in the formation of the carbon-carbon bond is followed by metal elimination.

The alkenylation protocol between styrene, ethyl acrylate or acrylonitrile, and 5-bromo- and 5-iodo-pyrimidine gives the coupling product (**882**) ⟨79CPB193, 81H(16)965⟩. Formation of 4,4′-bipyrimidines (**884**) is a major pathway from 4-iodopyrimidines (**883**) under the relatively vigorous conditions required to form the cross-coupled product (**885**). In the absence of alkenes homo-coupling (160°C) is almost quantitative ⟨79CPB193⟩.

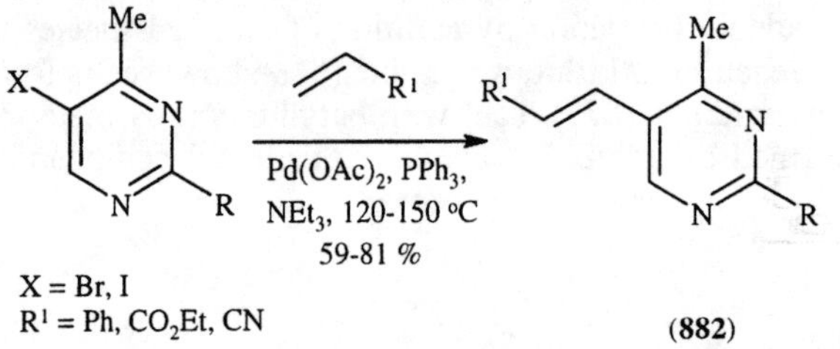

X = Br, I
R^1 = Ph, CO_2Et, CN

(**882**)

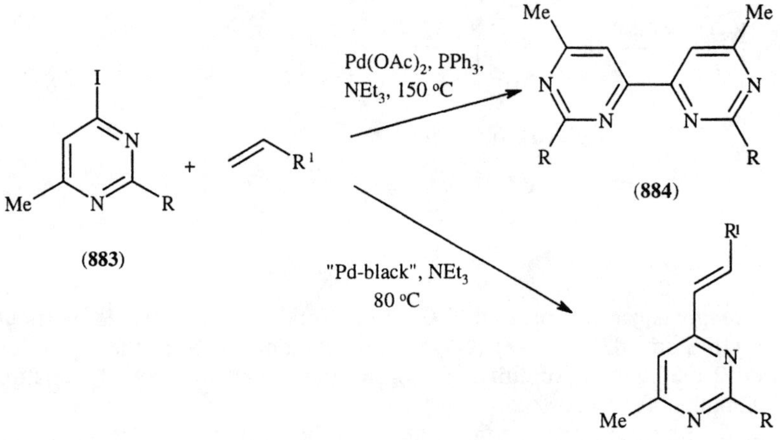

(**883**)

(**884**)

(**885**)

2-Iodo- and 4-iodo-pyrimidines were readily alkynylated, *e.g.* (**886** → **887**). 4,6-Diiodo-2-methylpyrimidine was dialkynylated to (**888**) ⟨78H(9)271⟩.

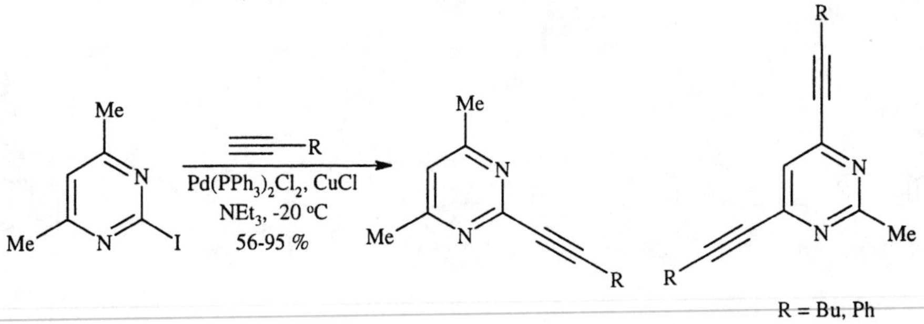

R = Bu, Ph

(**886**)   (**887**)   (**888**)

The Heck reaction of iodo-1,2,4-triazines with acetylenes has been also reported ⟨84H(22)2245, 93LA583⟩.

(vii) Ullmann reactions succeed; *e.g.* 2-bromopyridine yields 2,2'-bipyridyl (with Cu).

3.2.3.10.3 *Reactions via hetarynes* ⟨65AHC(4)127⟩

Very strong bases such as $NaNH_2$ convert unactivated aryl halides into benzyne intermediates which react rapidly with nucleophiles to form the products of an apparently simple nucleophilic substitution. It is now clear that hetarynes are frequent intermediates in reactions of not too highly activated heteroaromatic halides.

Thus, reaction of 3-chloro-, 3-bromo- and 3-iodo-pyridine with potassium amide in liquid ammonia give in each case the same mixture of 3- and 4-aminopyridine, showing the intermediacy of 3,4-pyridyne. Under these conditions the corresponding 4-halopyridines react entirely through 3,4-pyridyne (**890**).

Pyridynes are also formed from *ortho*-dihalides and alkali metals. Thus, reaction of 3-bromo-4-chloropyridine (**889**) with lithium amalgam and furan gives product (**891**) by trapping of the 3,4-pyridyne (**890**). Although 2,3-pyridyne is not formed from 3-halopyridines, because of the weaker acidity of the 2- as compared to the 4-hydrogen atom (see Section 3.2.1.8.2), it can be trapped by furan in small yield from the reaction of 3-bromo-2-chloropyridine with lithium amalgam.

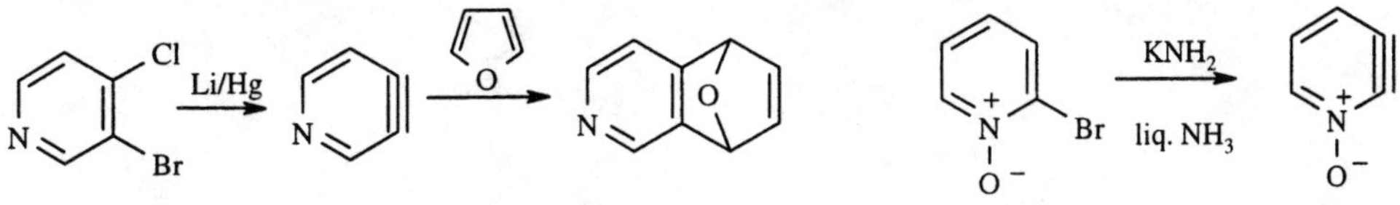

2,3-Pyridyne 1-oxide (**893**) is obtained by the action of potassium amide on 2-bromopyridine 1-oxide (**892**), as shown by the formation of a mixture of the 2- and 3-aminopyridine oxides. Reaction of 5-bromopyrimidine with sodium amide in liquid ammonia involves 4,5-pyrimidyne as an intermediate.

3.2.3.10.4   The S$_{RN}$ mechanistic pathway

Unactivated aryl halides also undergo nucleophilic displacement *via* electron transfer in the initial step: the so-called S$_{RN}$1 mechanism. It is now clear that in the case of heteroaromatic compounds, nucleophilic substitution by the S$_{RN}$ process often competes with the addition-elimination pathway. The S$_{RN}$ reactions are radical chain processes, and are usually photochemically promoted. For example, ketone (**895**) is formed by the S$_{RN}$1 pathway from 2-chloroquinoxaline (**894**) ⟨82JOC1036⟩.

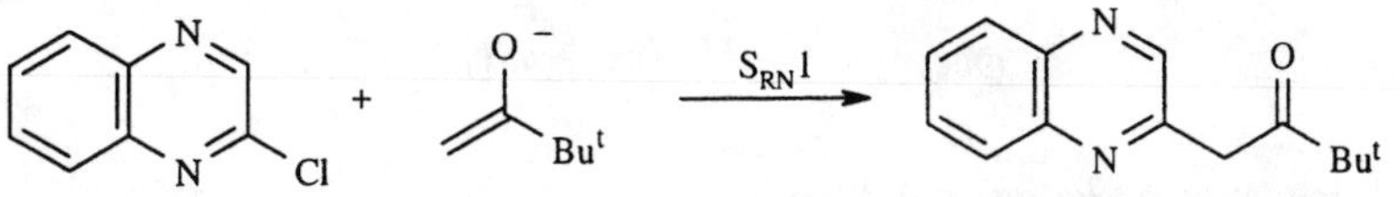

3.2.3.10.5   ANRORC reactions

The ANRORC (*A*ddition of *N*ucleophile, *R*ing *O*pening, *R*ing *C*losure) reaction involves the initial addition of a nucleophile to a ring carbon atom *not* carrying the halogen followed by electrocyclic ring opening ⟨78ACR462⟩. The sequence, exemplified by Scheme 71, is similar to that of the Dimroth rearrangement (Section 3.2.3.5.4.iii) but in the ANRORC reaction the initial ring opening involves elimination of the halogen atom. The ring formed can be the same or different to that started with.

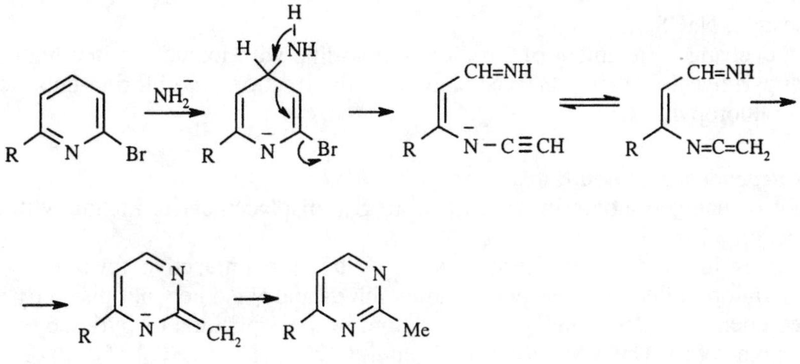

Scheme 71

Further examples of ANRORC reactions are the conversions of quinoline (**896**) into quinazoline (**897**) and of pyrimidine (**898**) into triazine (**899**).

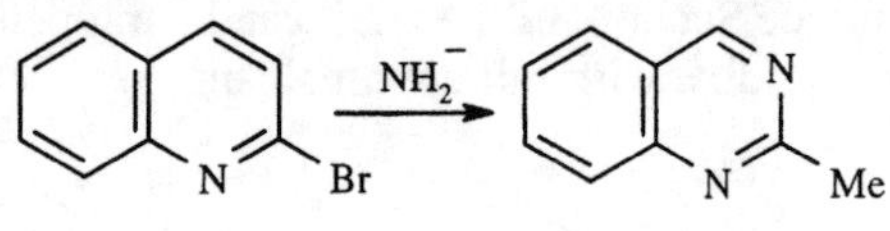

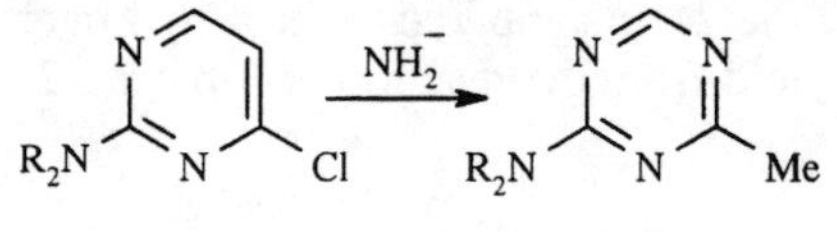

(896)   (897)   (898)   (899)

3.2.3.10.6   Nucleophilic displacement by classical S_{AE} mechanism

(i) Dichotomy of mechanisms

Nucleophilic displacement of an α- or γ-halogen atom by the classical S_{AE} mechanism of nucleophilic displacement *via* a Meisenheimer intermediate (*e.g.* **900**) is facilitated by mesomeric stabilization of the transition state. However, the mechanistic balance is fine: whereas 4-bromopyridine 1-oxide reacts with potassium amide by the A-E mechanism, the 2-bromo analogue prefers the E-A route. Again, nucleophilic displacement in chloropyrazines can involve the ANRORC mechanism.

Some of the reactions of α- and γ-halopyridines are acid catalyzed, *i.e.* ions of type (**901**) are formed and react with nucleophilic reagents (*e.g.* 2-chloro-5-nitropyridine + PhNH$_2$ $\rightarrow$ 2-anilino-5-nitropyridine, catalyzed by H$_2$O − HCl).

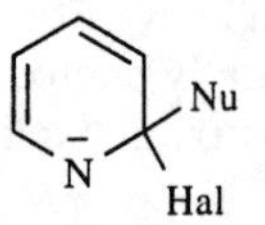

(900) 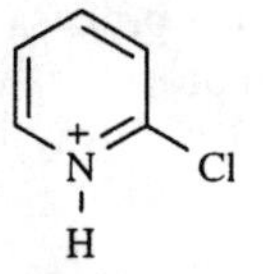   (901)

(ii) Reactivity dependence on halogen and nucleophile

The relative reactivities with respect to nucleophilic S_{AE} displacement increase in the order Cl $\leq$ Br $\leq$ I $\leq$ F. The relative reactivities of nucleophiles are illustrated by the reactions of 2-bromopyridine: replacement by the following groups occurs under the conditions given (an * indicates that the product spontaneously tautomerizes):

(a)   Hydroxy*, by NaOH-H$_2$O, 150°C.
(b)   Alkoxy, *e.g.* methoxy by NaOMe-MeOH, 65°C.
(c)   Phenoxy, PhONa-EtOH.
(d)   Mercapto*, KSH-propylene glycol.
(e)   Methylmercapto, NaSMe-MeOH, 65°C.
(f)   Amino (NH$_3$-H$_2$O, 200°C), dimethylamino (NHMe$_2$, 150°C) or hydrazino (N$_2$H$_4$, 100°C).
(g)   Cyano, by distillation with CuCN.
(h)   Di(ethoxycarbonyl)methyl by the sodio derivative of malonic ester.
(i)   Isothiouronium, (NH$_2$)$_2$CS-EtOH, reflux; the product is converted by alkali into urea and pyridine-2-thione.
(j)   Sulfonic acid, NaHSO$_3$.
(k)   Halogen exchange. Treatment of 2,6-dichloropyridine with iodide ion in a high concentration gives 2,6-diiodopyridine ⟨80JOM(186)147⟩. Nucleophilic fluorination *via* chlorine-fluorine exchange is reported to yield fluoropyridines.

(iii) Reactivity dependence on nature of heterocyclic rings

The reactivity of halogen atoms in azines toward S_{AE} displacement is in line with the sequence discussed in Section 3.2.3.1.1.

β-Halogen atoms in pyrones and pyridones (*e.g.* **902**) are unreactive toward S_{AE} nucleophilic displacement. β-Halopyridines are less reactive than the α- and γ-isomers but distinctly more reactive than unactivated phenyl halides. Thus, a bromine atom in the 3-position of pyridine or quinoline can be replaced by methoxy (NaOMe-MeOH, 150°C), amino (NH$_3$-H$_2$O-CuSO$_4$, 160°C) or cyano (CuCN, 165°C). 5-Halogens in pyrimidines are also relatively unreactive.

(902)   (903)   (904)

(905)   (906)   (907)

The reactions of the 4-halopyridines parallel those of the corresponding 2-isomers, with the exception that 4-halopyridines polymerize much more readily (*e.g.* to **903**) because the pyridine nitrogen atom is not sterically hindered and is more basic (*cf.* Section 3.2.1.3.4). As expected, the chlorine atom in the 1-position of 1,3-dichloroisoquinoline is more reactive than that in the 3-position, thus, mild treatment with sodium ethoxide gives (**904**). Halogens in the 9-position of acridine are more reactive, *e.g.* (**906**) → (**905**), (**907**).

α-and γ-Halogen atoms on benzo-pyridines, -pyridones, -pyrones (*e.g.* **908**) and -pyridine *N*-oxides (*e.g.* **909**) are about as reactive as those in the corresponding monocyclic compounds. A 2-chlorine atom in chromone is readily displaced by nucleophiles; a 3-halogen atom is less reactive, but can still undergo nucleophilic displacement.

(908)   (909)   (910)   (911)   (912)

Reactivity increases in the diazines as compared with pyridines. 3-Chloropyridazine (**910**) and 2-chloropyrazine, for example, undergo the usual nucleophilic replacements (*cf.* Section 3.2.3.10.6.ii) rather more readily than does 2-chloropyridine. 2-, 4- and 6-Halogen atoms in pyrimidines are easily displaced. The reactivity of halogens in pyridazine 1-oxides toward nucleophilic substitution is in the sequence 5 > 3 > 6 > 4.

Amides aminate 2-fluoropyrazine smoothly at 40–50°C in excellent yield ⟨84H(22)1105⟩. The amination of 2-chloropyrazines 1-oxide needs higher temperatures (100–110°C).

Halogen atoms in the α- and γ-positions of cationic nuclei are very reactive, as illustrated by the hydrolysis to 2,6-dibromo-1-methylpyridinium ion at 20°C (**911** → **912**). Halogen groups in the α- or γ-positions of thiinium cations are also highly reactive.

(iv) Reactivity in polyhalo compounds

Regioselective substitutions in di- and tri-halogenopyrimidine derivatives can in many cases be achieved. Chloro, bromo, and iodo substituents undergo aminolysis at approximately the same rates whereas a fluoro substituent reacts 60–200 times faster. 4(6)-Halo substituents react up to 10 times faster than 2-halo substituents. Electron-donating substituents (*e.g.*, Me, Ph, OMe, NMe₂) decrease the rate of aminolysis whereas electron-withdrawing substituents (Cl, CF₃, NO₂) have the opposite effect ⟨B-94MI 602-01⟩.

Hence, in polyhalo compounds such as 2,4,6-trichloropyrimidine, each successive chlorine atom is replaced more slowly than the last because the groups introduced (*e.g.* NH₂) partially cancel the activating effect of the annular nitrogen atoms. Similarly, the chlorine atoms in cyanuric chloride (2,4,6-trichloro-1,3,5-triazine) are replaced sequentially: the reactivity of the first chlorine is very high. Chlorine has been replaced by carbon, hydrogen, fluorine, nitrogen, oxygen, phosphorus, and sulfur nucleophiles ⟨88AHC(44)243⟩.

A halogen atom in the 4-position of quinazoline is more reactive than one in the 2-position because of partial double bond fixation. Thus, in 2,4-dichloroquinazoline (**913**), replacement occurs almost exclusively in the 4-position, whereas 2,4-dichloropyrimidine (**914**) yields approximately equal amounts of the 2- and 4-monoreplacement products. Similarly, in 2,6,8-trichloropurine (**915**) the order of replacement of the halogen atoms is 6, 2 and 8, successively. Only the 4-chlorine atom in 3,4-dichlorocinnoline is readily replaced. In tetrachloropteridine, the 6- and 7-chlorine atoms are the most reactive followed by the 2-chlorine.

Just as in benzene chemistry, all types of halogen atom are activated toward nucleophilic displacement by the presence of other electron-withdrawing substituents. This is illustrated by the conversion of 5-nitro-2 chloropyridine (**916**) to the 2-hydrazine derivative (N_2H_4, 20°C) and the 2-thione (**917**) under relatively mild conditions.

3.2.3.10.7 Nucleophilic displacement with transition metal catalysis

Transition metal-catalyzed cross-coupling reactions between heteroaryl halides and organotin compounds have become important because of ease of preparation and handling of the organotins and their high reactivity in the presence of Pd catalysts ⟨86AG504, B-87MI 602-01, 92S803⟩. Activation to the reactive species by the transition metals is a result of transmetallation; the organotin compound is converted into a new organometallic derivative of the catalyst metal which is the actual reacting species in carbon-carbon bond formation. The regio and chemoselectivity in these reactions is demonstrated by the stepwise introduction of three different carbosubstituents into 5-bromo-2,4-dichloropyrimidine. Initial styrylation occurs at the 4-position (**918**), subsequent phenylation is at the 5-position (**919**), and finally thienylation is at the 2-position (**920**) ⟨89ACS62⟩.

sp^3-Hybridized carbon attached directly to a metal is less reactive than sp^2- or sp-hybridized carbon in Pd-catalyzed reactions. Tetramethyl- or tetrabutyl-stannane can be used for the preparation of methyl and butyl derivatives, but these reactions require heating in DMF for a long time. Reactivity is enhanced when the sp^3-hybridized carbon carries an electronegative group. Therefore, in the reaction between benzyltributylstannane and 4-iodopyrimidine, it is the benzyl group which is transferred to the pyrimidine. With allyltributylstannane the product is the 4-*trans*-propenylpyrimidine ⟨87ACS(B)712⟩.

Cross-coupling reactions between organoboranes and heteroaryl halides are effectively catalyzed by Pd(0) in the presence of a base. Couplings in simple pyrimidines are illustrated by the reaction between 2-chloropyrimidine and 2- or 3-thiophene- and selenophene-boronic acids which give the corresponding 2-substituted pyrimidines (**921**) (Scheme 72). In 2,4-dichloro- or 2,4-dibromo-pyrimidine it is the 4-halo substituent which is the more reactive.

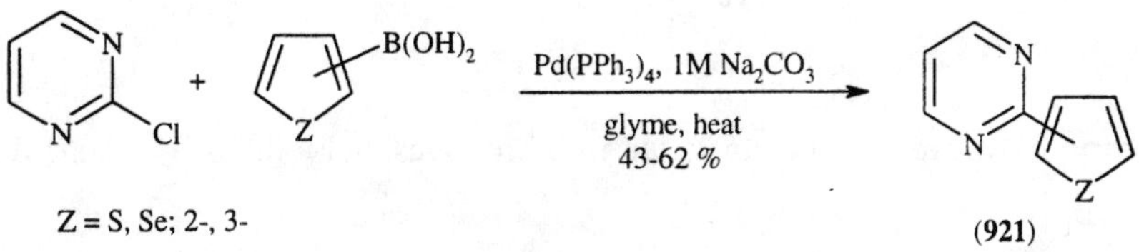

Scheme 72

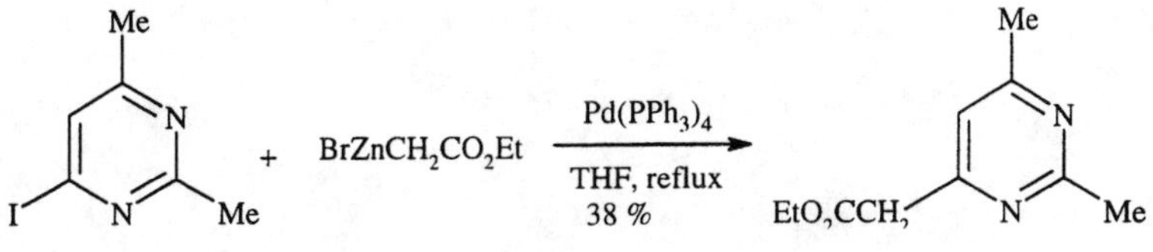

Scheme 73

Cross-coupling with organozinc reagents using transition metal catalysis proceeds with high chemo-, regio-, and stereoselectivity ⟨92T9577⟩. Organozinc reagents are more useful for the transfer of alkyl groups, than organotin analogues. This is exemplified by the introduction of an acetic acid unit through substitution of a 4-iodopyrimidine (Scheme 73).

Alkylation and arylation of chloropyrazine or its *N*-oxides is effected by cross-coupling with organoaluminum ⟨85H(23)133, 87H(26)2449⟩, organoboron ⟨87H(26)2449, 89H(29)939⟩, and organotin compounds ⟨86H(24)785, 89H(29)123⟩. Trimethylaluminum, triethylboron, and tetraaryltins give the corresponding alkyl or aryl substituted pyrazines in good yields. Areneboronic acids are also effective for arylations, notably improved by use of the binuclear catalyst Pd(dppf)(OAc)$_2$ ⟨88JOC2052⟩. Classical alkylation by a cross-coupling reaction between a halogenopyrazine and a Grignard reagent ⟨78JOC3367⟩ is promoted by a nickel catalyst ⟨92JA5269⟩. Thus, 2,6-diethylpyrazine is formed spontaneously at room temperature in the reaction of diethylzinc with 2,6-dichloropyrazine employing catalytic bis-1,3-(diphenylphosphino)propanenickel(II) dichloride.

3.2.3.11 Metals and Metalloid Derivatives

Azine Grignard reagents and sodium and lithium compounds in which the metal atom is attached directly to the ring are considered under the corresponding halogen compounds (Section 3.2.3.10) because they are generally prepared from their halogen derivatives by metal-halogen exchange, and are seldom isolated.

Lithium derivatives of azine *N*-oxides are likewise considered in Section 3.2.1.8.1 because they are generally prepared by direct lithiation of the *N*-oxide, and again are seldom isolated.

Finally, those organometallic derivatives of azines in which the metal is separated from the ring by one carbon atom are considered under the corresponding alkyl compound (Section 3.2.3.3).

As exemplified in the sections indicated, these compounds show most of the typical reactions of Grignard reagents and alkyllithiums. Thus, pyridyllithiums and their benzo analogues allow the introduction of other metals, and non-metals, on to the ring, such as mercury, boron, phosphorus, tin and arsenic (Scheme 74) (see also Section 3.2.3.10.2.v).

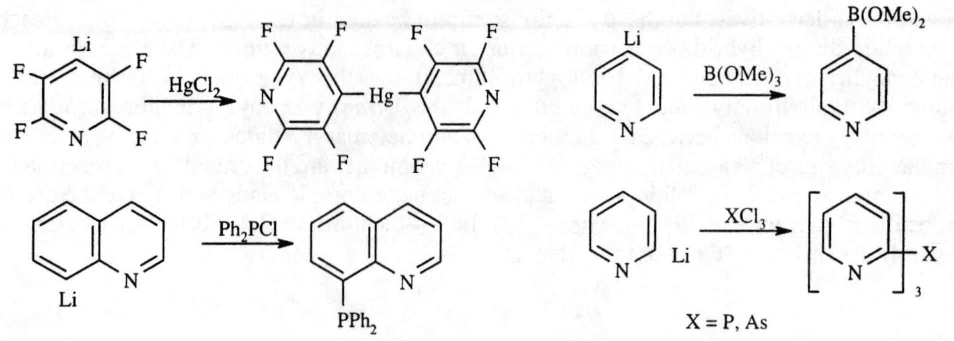

Scheme 74

Organomercury derivatives can be converted into bromides and iodides by standard methods, *e.g.* Scheme 75.

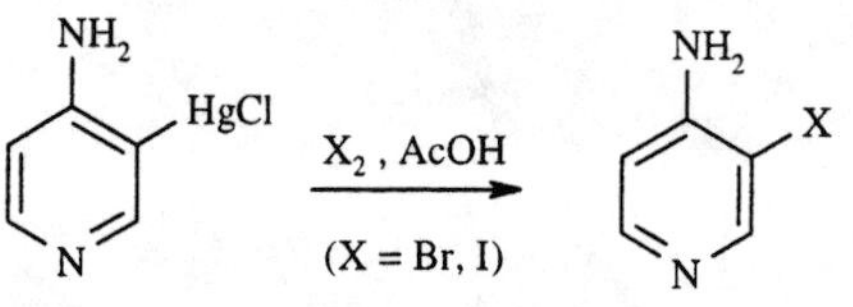

Scheme 75

3.2.3.12   Substituents Attached to Ring Nitrogen Atoms

3.2.3.12.1   *Introduction*

(i) Types

Substituents attached to a nitrogen atom in a six-membered heteroaromatic ring are to be found in compounds of the following types:

(a)   cations, *e.g.* (**922**);
(b)   ylides, including *C*-ylides (**923**), *N*-ylides (**924**) and *N*-oxides (**925**);
(c)   zwitterions, *e.g.* (**926**), (**927**);
(d)   compounds with exocyclic conjugation, *e.g.* (**928**).

Significantly, types (b), (c) and (d) can all be derived from a compound of type (a) by deprotonation.

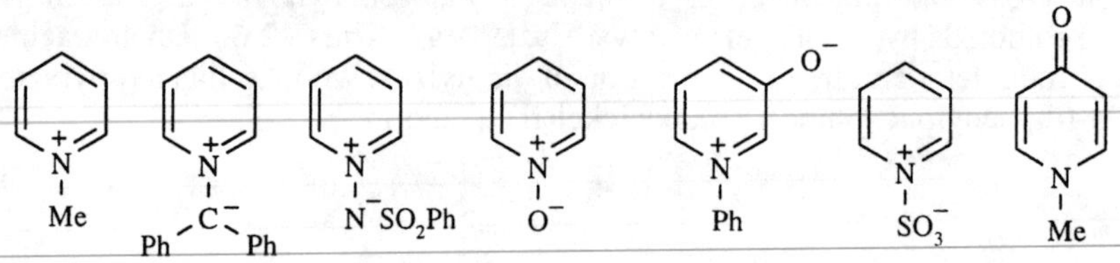

(**922**)   (**923**)   (**924**)   (**925**)   (**926**)   (**927**)   (**928**)

(ii) Overall survey of reactivity

We will survey the reactions of N-linked substituents classified by the atom attached to the cyclic nitrogen. Unlike heterocyclic C-substituents, where the benzene prototype and the carbonyl analogy link much of the typical chemical behavior to familiar compounds, no simple model exists for N-substituents. However, certain trends are clear. The existence of the positive pole in cations of type (**922**) ensures that nucleophilic attacks (a)-(d) are the most important of the following reaction types, many of which occur for several of the different classes of N-substituents.

(a)   The N-substituent can be completely removed (**929**). This is the reverse of the reaction of an electrophile at the pyridine nitrogen atom (Section 3.2.1.3).

(b)   An α-proton in the substituent can be removed (**930 → 931**); this gives an ylide (*cf.* **923**, **924** above). Such ylides can revert to their precursors by protonation (**931 → 930**), or react with other electrophiles (**931 → 932**).

(c) An αβ bond can cleave in a substituent by nucleophilic attack or spontaneously, in the reverse (**932** → **931**) of the reaction just mentioned.

(d) An elimination reaction (**933**) can occur.

(e) A nucleophile can add to an αβ-multiple bond in the substituent (**934**) to give an ylide.

(f) Rearrangements can occur of *an N*-substituent into the ring. A variety of thermal and photochemical rearrangements are known for *C-, N-* and *O*-linked substituents.

(g) Electrocyclic addition involving N-linked substituent and an α-position of the ring.

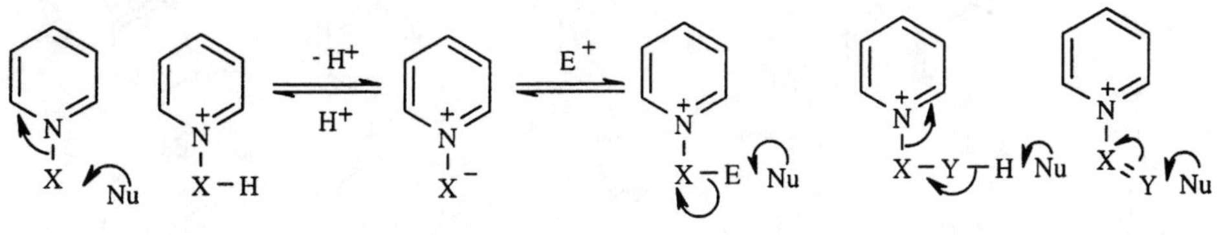

(929)   (930)   (931)   (932)   (933)   (934)

3.2.3.12.2 Alkyl groups

(i) Loss of alkyl groups

1-Alkylpyridinium halides dissociate reversibly into the alkyl halide and pyridine on vacuum distillation. Reactions of type (**935** → **936**) + RNu are accelerated by bulky α-substituents R′ and also by electron-withdrawing groups in the ring. Reactions occur with a wide range of halogen, oxygen, sulfur, nitrogen, phosphorus and carbon nucleophiles. If R can form a stabilized carbon cation (*p*-methoxybenzyl, *s*-alkyl, *etc.*), the N-C cleavage can occur by the S_N1 as well as the S_N2 mechanism.

Transfer of an *N*-alkyl group to certain nucleophiles can also occur by a radicaloid non-chain mechanism, *e.g.* (**935**; R = CH$_2$Ph, R′ = Ph) + $^-$CMe$_2$NO$_2$ → (**937**) → (**938**) + (**939**).

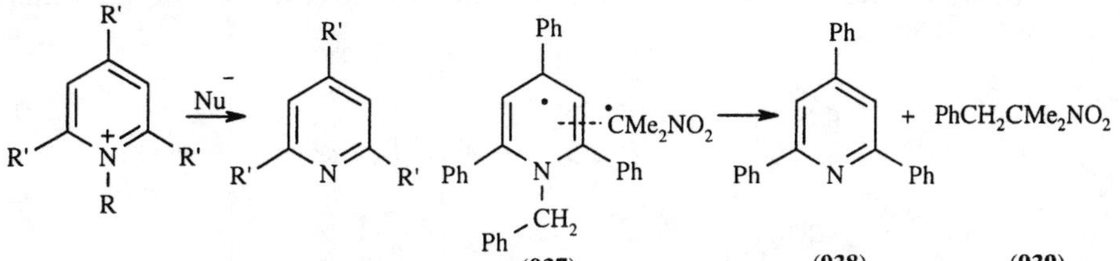

(935)   (936)   (937)   (938)   (939)

Demethylation of toxoflavins (**941**) to reumycin (**942**) is effected by dimethylformamide or diethylamine and also proceeds *via* radical-anions ⟨87JHC1373⟩. On treatment with DMF and methyl iodide, (**941**, R = H) undergoes a concomitant demethylation and methylation to give the fervenulin (**943**) ⟨74JHC271, 75BCJ2884⟩.

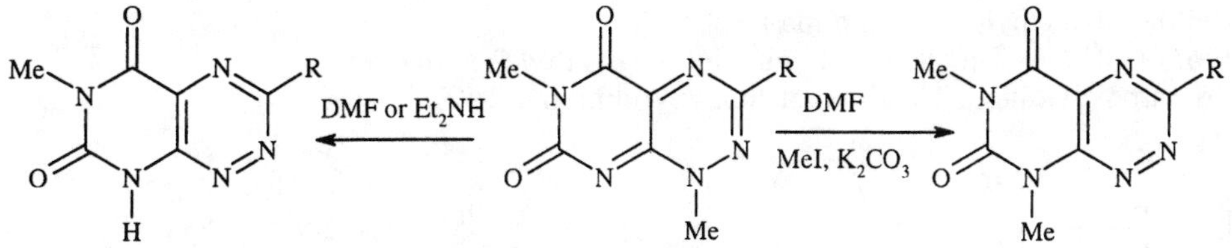

(942)   (941)   (943)

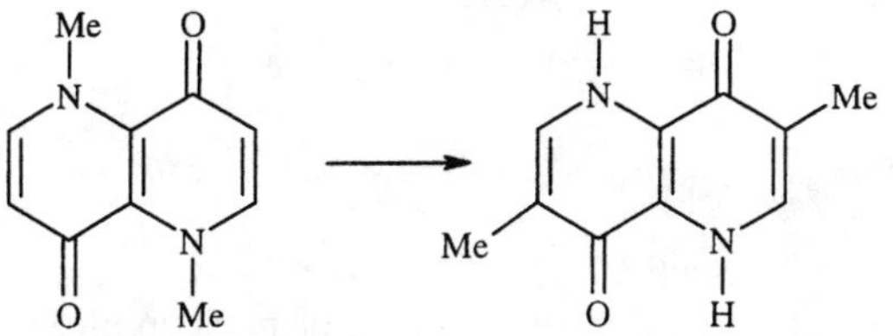

(944)   (945)

Compound (**944**) on heating with methyl iodide gives (**945**) (53% yield) by a *C*-methylation and *N*-demethylation ⟨77CC87⟩.

(ii) Proton loss from a carbon atom

The ease with which this occurs is determined by the other groups attached to the carbon (*cf.* **946**). The resulting ylide can be isolated only in special cases (*e.g.* **947-950**); ylide stability increases with increasing possibility for spreading the negative charge (*cf.* **948** ↔ **949**; **950**).

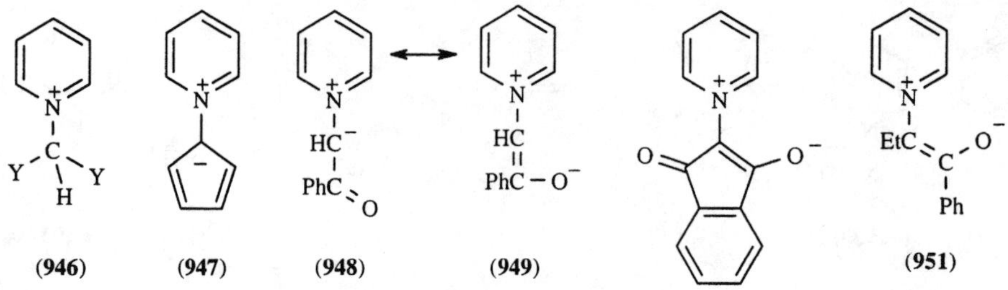

(**946**)   (**947**)   (**948**)   (**949**)   (**950**)   (**951**)

Stabilized ylides show the following properties:

(a) they form salts (**946**) with proton acids;

(b) they can be alkylated, *e.g.* (**948**) + EtBr + OH⁻ → (**951**);

(c) they can be halogenated: (**948**) + Br₂ + H₂O → PhCOCHO + C₅H₅N at 20°C probably *via* (**952**).

Ylids of type (**953**) are important as intermediates in the following reactions:

(a) the formation of β-hydroxyalkyl derivatives with aldehydes and subsequent dehydration, *e.g.* 1,3-dimethylpyridinium + PhCHO + NaOH → (**954**);

(b) the Kröhnke reaction of benzylpyridines (**955**) with nitroso compounds to give nitrones (**957**) *via* (**956**).

(c) Ylide thermolysis and photolysis can cause N-C cleavage to give carbene intermediates, *e.g.* (**958**) → (**959**) + pyridine.

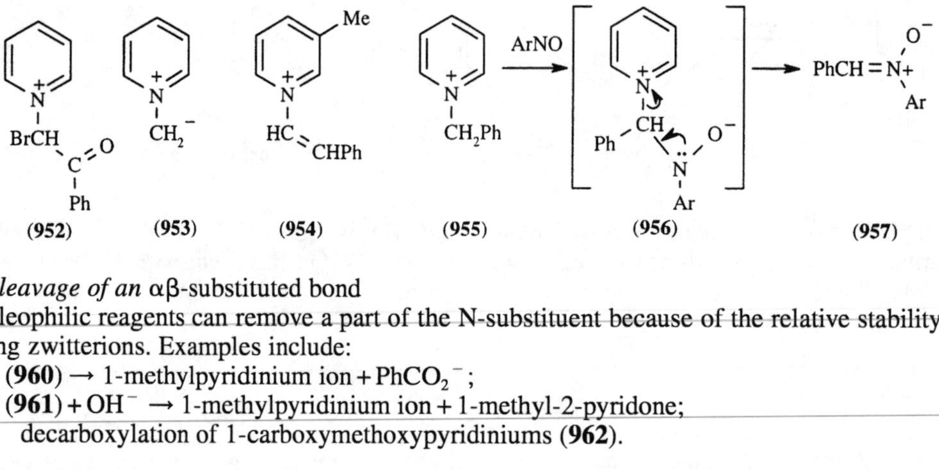

(**952**)   (**953**)   (**954**)   (**955**)   (**956**)   (**957**)

(iii) Cleavage of an αβ-substituted bond

Nucleophilic reagents can remove a part of the N-substituent because of the relative stability of the resulting zwitterions. Examples include:

(a) (**960**) → 1-methylpyridinium ion + PhCO₂⁻;

(b) (**961**) + OH⁻ → 1-methylpyridinium ion + 1-methyl-2-pyridone;

(c) decarboxylation of 1-carboxymethoxypyridiniums (**962**).

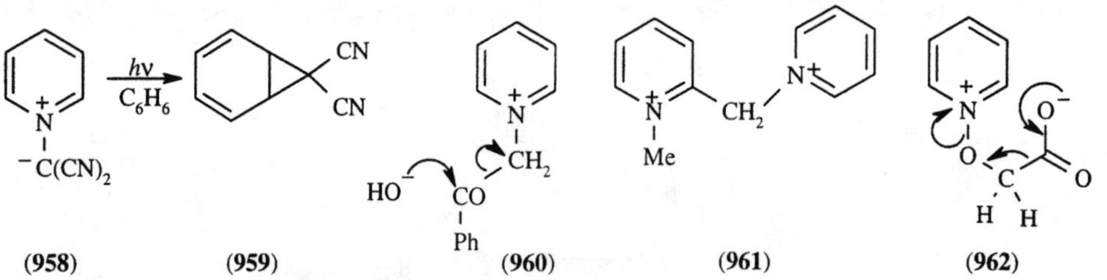

(**958**)   (**959**)   (**960**)   (**961**)   (**962**)

*(iv) Elimination reactions (cf. **933**, Section 3.2.3.12.1.ii)*

If compounds of type (**963**) are heated in the absence of nucleophile, E1 elimination occurs: thus, the pentacyclic triflates (**963**; R = primary alkyl) decompose at 150°C. The *s*-alkyl analogues form alkenes already at 20°C.

(v) Rearrangements

1-Alkylpyridinium halides give mixtures of alkylpyridines on heating, *e.g.* (**964**) gives 2- and 4-picoline, with other minor products. This reaction is known as the Ladenburg rearrangement, and involves *N*-alkyl bond homolysis.

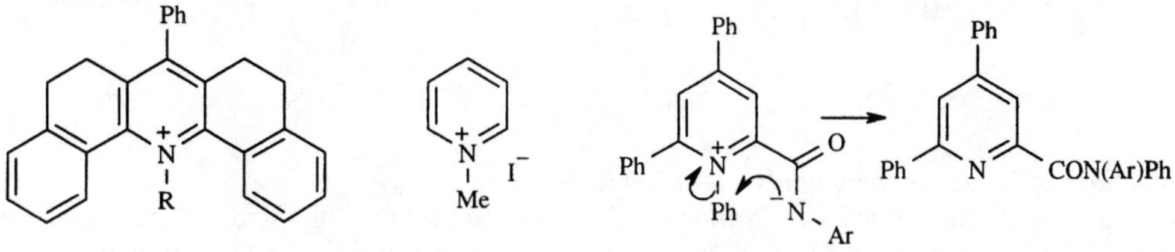

(963)   (964)   (965)   (966)

3.2.3.12.3 Other C-linked substituents

(i) N-Aryl groups

Typical reactions include:

(a) Electrophilic substitution. Nitration and sulfonation of the 1-phenylpyridinium cation occurs at the *meta* position to the 1-substituent.

(b) C-N Bond cleavage. Normally this is very difficult, but it can be achieved intramolecularly, *e.g.* (**965**) → (**966**).

(ii) N-Acyl and related groups

1-Acylpyridinium ions are very susceptible to attack by nucleophilic reagents and are good acylating agents. They are generally encountered only as intermediates (Section 3.2.1.3.7). *N*-Cyano and *N*-imidoyl groups are also easily transferred to nucleophiles; the former render the α ring position particularly susceptible to nucleophilic attack (see Section 3.2.1.6).

Derivatives of this type formed from 4-dimethylaminopyridine possess considerably more stability and can be isolated readily.

(iii) N-Vinyl groups

The typical reaction is Michael addition of nucleophiles: thus, 1-vinylpyridinium ion (**967**) adds *N*-, *S*- and *C*-nucleophiles to give products of type (**968**). *N*-Vinyl groups can also undergo polymerization.

(iv) Other substituted alkyl groups

N-Allyl compounds can be rearranged to *N*-propenyl derivatives. Similarly, *N*-propargyl compounds (**969**) give *N*-allenyl derivatives (**970**).

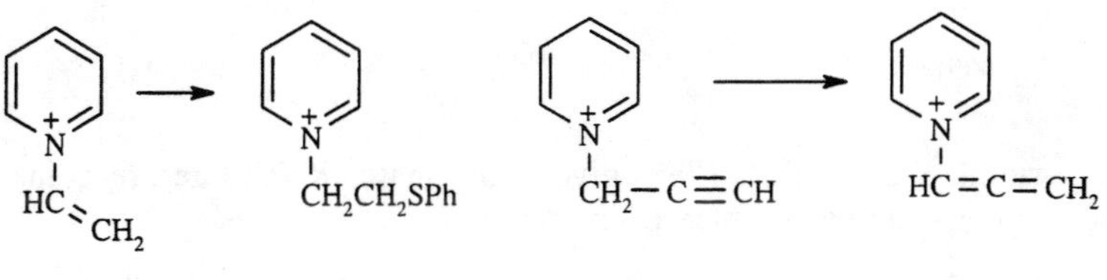

(967)   (968)   (969)   (970)

3.2.3.12.4 N-Linked substituents

(i) Prototropic equilibria

N-(Monosubstituted amino)pyridiniums (**972**) are in prototropic equilibrium with *N*-imides (**973**) and dications (**971**). For R = H or allyl, the *N*-imides (**973**) are very strong bases and cannot usually be isolated; if R is an electron-withdrawing group (*e.g.* acyl, sulfonyl, nitro), then the imide (**973**) is less basic and more stable. The cations (**971**) are only obtained in very strongly acid media.

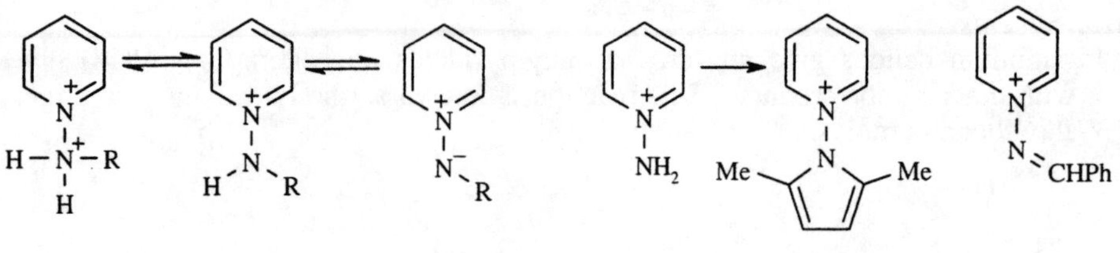

(971)   (972)   (973)   (974)   (975)   (976)

(ii) Reactions of N-amino compounds with electrophiles

N-Aminopyridinium cations can be acylated or sulfonylated (with acid halides) and nitrated (H_2SO_4 − HNO_3) to give the corresponding N-(substituted amino)pyridines (972), often isolated as the imides (973; R = COR', SO_2R' or NO_2).

Both the NH_2 protons in (974) can be replaced by electrophiles in this way; e.g. with hexane-2,5-dione the N-(pyrrol-1-yl)pyridinium (975) is formed.

Aldehydes readily yield imines, cf. (976), and aryldiazonium salts form aryl azides and pyridine, presumably via ArN = N − N⁻ − Py⁺. Nitrous acid and N-aminopyridinium cations yield pyridine and N_2O.

(iii) Other reactions of N-amino compounds

N-Aminopyridones can be oxidized to nitrenes. Thus, 3-amino-1,2,3-benzotriazin-4-ones (977) with lead tetraacetate lead to an intermediate nitrene (978), which can lose one or two molecules of nitrogen.

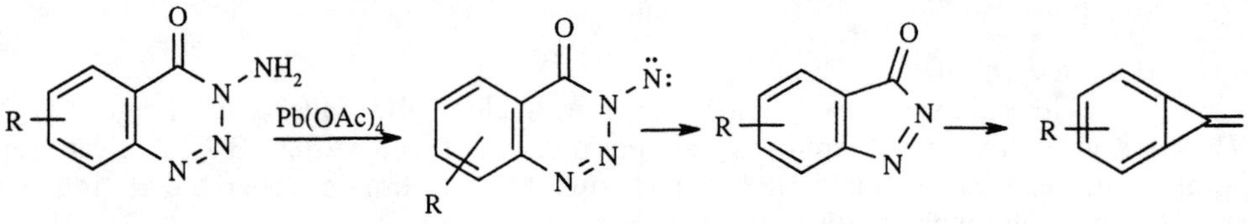

(977)   (978)

N-Aminopyridiniums and N-imines yield pyridines and amines upon reduction, e.g. (979) → (980) + $PhNH_2$ (H_2/Pt or Zn/NaOH).

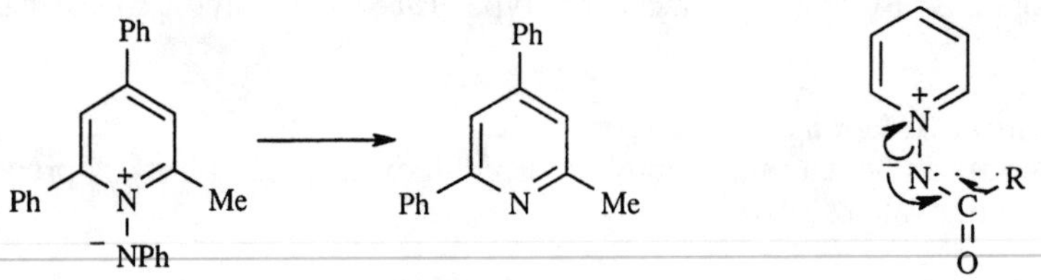

(979)   (980)   (981)

Thermolysis of pyridine N-acylimides (981) gives isocyanates RNCO and pyridine. Photochemical rearrangements of N-imides can be complex (Scheme 76).

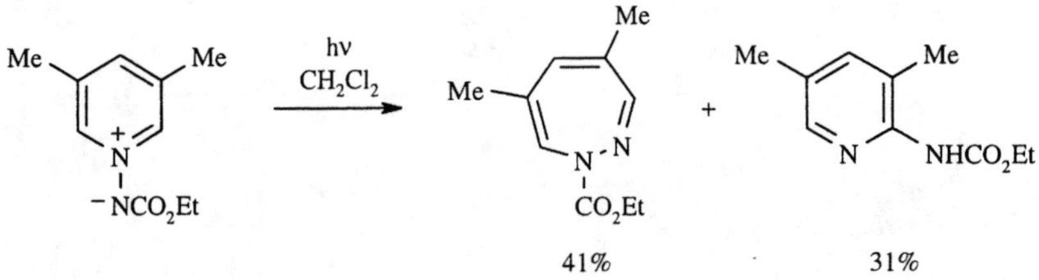

41%   31%

Scheme 76

Heteroaromatic N-imines undergo 1,3-dipolar cycloaddition reactions across the N-imine nitrogen atom and the α-position of the ring. With nitriles, triazolopyridines are formed by dehydrogenation of intermediate adducts (**982**). The more stable the *N*-imine, the more reactive the dipolarophile has to be. *N*-Arylimines react with acrylonitrile to give similar adducts (**983**) ⟨74AHC(17)246⟩.

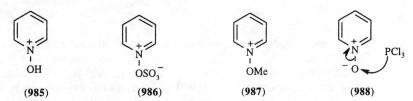

(**982**)   (**983**)   (**984**)

(iv) Other N-linked substituents

N-Nitropyridiniums (**984**) are mild and selective nitrating agents ⟨68JA4742⟩.

3.2.3.12.5 O-Linked substituents

(i) Reactivity pattern of N-*oxides*

1-Hydroxypyridinium ions (**985**) readily lose a proton to give *N*-oxides; *N*-oxides are themselves weak bases which form 1-hydroxypyridinium ions by proton addition.

Pyridine 1-oxides possess a unique pattern of reactivity:

(a) Electrophiles can attack *either* at ring carbon (usually γ to the *N*-oxide on the free base, β-position on the conjugate acid) in electrophilic substitution reactions (see Section 3.2.1.4), *or* at the oxygen atom as discussed below.

(b) Nucleophiles can attack at *either* α *or* γ ring carbon atoms (strong nucleophiles on free bases, weaker nucleophiles on intermediates first formed by electrophile addition to the *N*-oxide oxygen — see Section 3.2.1.6). Certain nucleophiles also lead to *N*-oxygen removal as discussed below.

(c) *N*-Oxides undergo a variety of electrocyclic additions and rearrangements as discussed below.

(d) *N*-Oxides being reactive 1,3-dipoles can enter various cycloaddition reactions (*cf.* Section 3.2.1.10.4).

(ii) Reactions of N-oxides with electrophiles at the N-oxide oxygen

These include the following:

(a) Proton acids give 1-hydroxypyridinium salts (**985**).

(b) Lewis acids give complexes, *e.g.* $SO_3 \rightarrow$ (**986**).

(c) Alkyl halides form 1-alkoxypyridinium salts (**987**).

(d) Pyridine 1-oxides with arenediazonium salts yield *N*-aryloxypyridinium salts.

(e) Metal ions give coordination compounds.

(**985**)   (**986**)   (**987**)   (**988**)

(iii) Loss of the N-*oxide group*

Reduction of *N*-oxides affords the parent heterocycle and can be achieved by:

(a) Catalytic hydrogenation, *e.g.* over Pd.

(b) Chemical reduction, *e.g.* with Fe-HOAc.

(c) Electrochemical means.

(d) Deoxygenation with a trivalent phosphorus compound, *e.g.* PCl_3, PPh_3, $P(OEt)_3$. This can be formulated as nucleophilic attack on oxygen (**988**), but probably initial electron donation of the *N*-oxide lone pair into the vacant phosphorus *d*-orbital is involved. In this way, for example, pyrimidine *N*-oxides are converted to pyrimidines by phosphorus trichloride in chloroform.

(iv) Rearrangement reactions

The most important rearrangements include:

(a) Photochemical. Photolysis of pyridine 1-oxide in benzene solution yields, *via* the excited triplet state, phenol and pyridine. Photolysis to the excited triplet state in a polar solvent like water takes a different course *via* the oxaziridine (**990**) to 2-quinolinone (**989**; R = H). Methyl migration can occur in such reactions: (**990**) → (**989**), R = Me. The primary oxaziridine can isomerize and undergo valence bond tautomerism to an oxazepin, *e.g.* (**990**) → (**991**) → (**992**). Hydrolytic ring opening of the oxazepin followed by ring closure can give a pyrrole or indole (**992** → **993**).

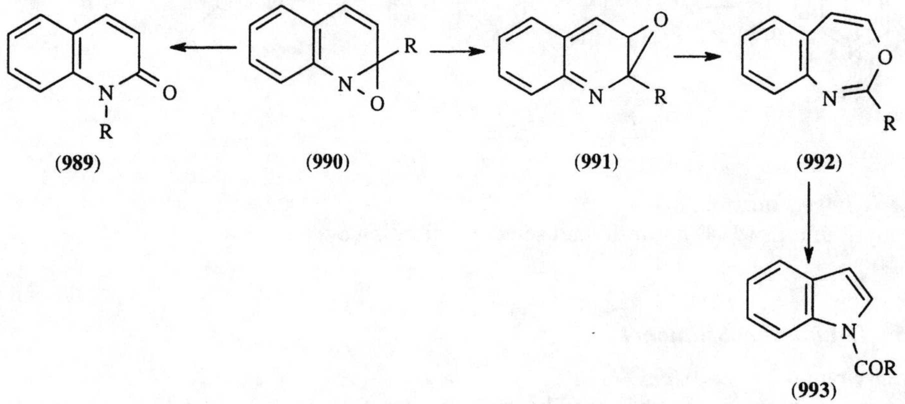

(**989**)   (**990**)   (**991**)   (**992**)

(**993**)

Certain pyridazine *N*-oxides are isomerized into the corresponding diazoketones, *e.g.* (**994**) → (**995**); ring contraction to the corresponding furan (**996**) can then occur.

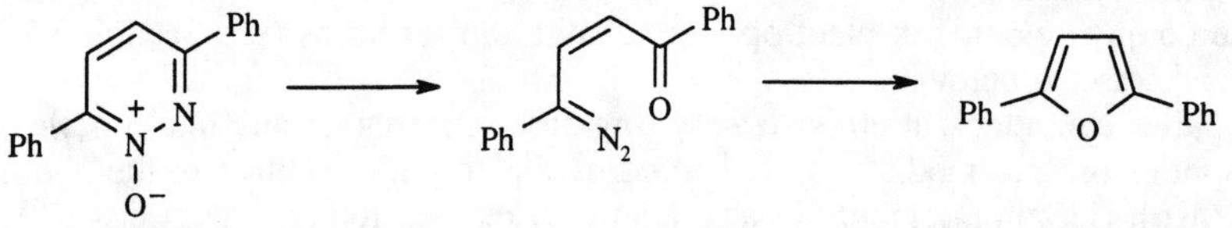

(**994**)   (**995**)   (**996**)

Phenazine 9,10-dioxide (**997**) is converted by irradiation in benzene to the 1,6-oxidoannulene (**999**) *via* (**998**) and an oxygen walk ⟨82AHC(30)253⟩.

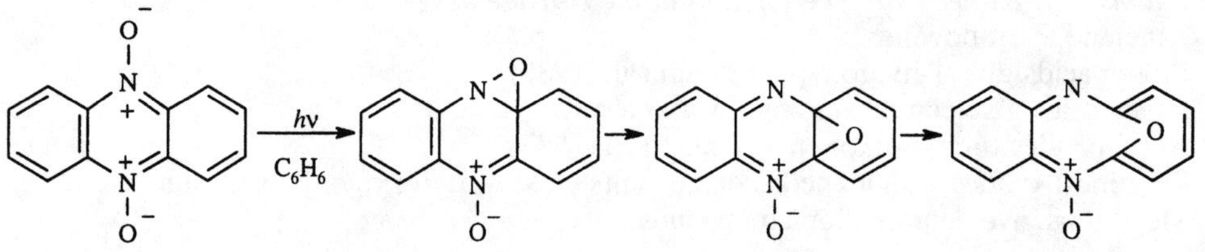

(**997**)   (**998**)   (**999**)

(b) Acid anhydrides. Pyridine 1-oxides heated with acid anhydrides are converted in good yield into pyridones (Scheme 77) (see Section 3.2.1.6.3.vii) unless the *N*-oxide contains an α- or γ-alkyl group. In the latter case an alternative reaction occurs with acetic anhydride to form an α- or γ-(acetoxyalkyl)pyridine (*e.g.* **1001**). For attack on α- (**1000** → **1001**) and γ-alkyl groups these reactions appear to be similar to the *ortho*- and *para*-Claisen rearrangement, respectively, of allyl phenol ethers; however, the formation of (**1001**) proceeds *via* a radical cage. In these reactions, β-acetoxy compounds are formed as by-products, *e.g.* 3-acetoxy-2-methylpyridine from (**1000**).

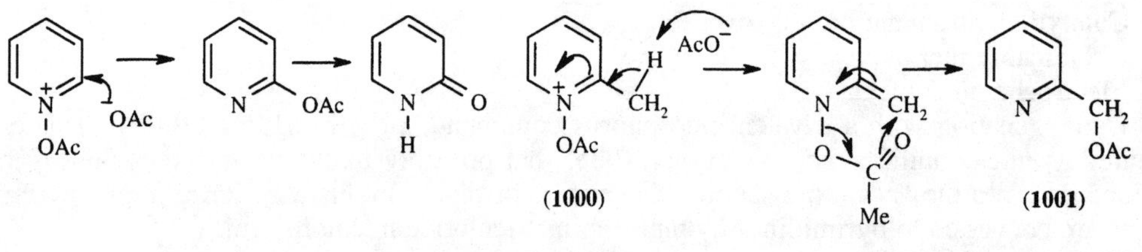

(**1000**)   (**1001**)

Scheme 77

(c) Imidoyl chlorides. The treatment of *N*-oxides with imidoyl chlorides leads to intramolecular amination *via* the intermediates as shown. When the 2- and 6-positions are both substituted, reaction goes further to give a 3-hydroxypyridine (Scheme 78).

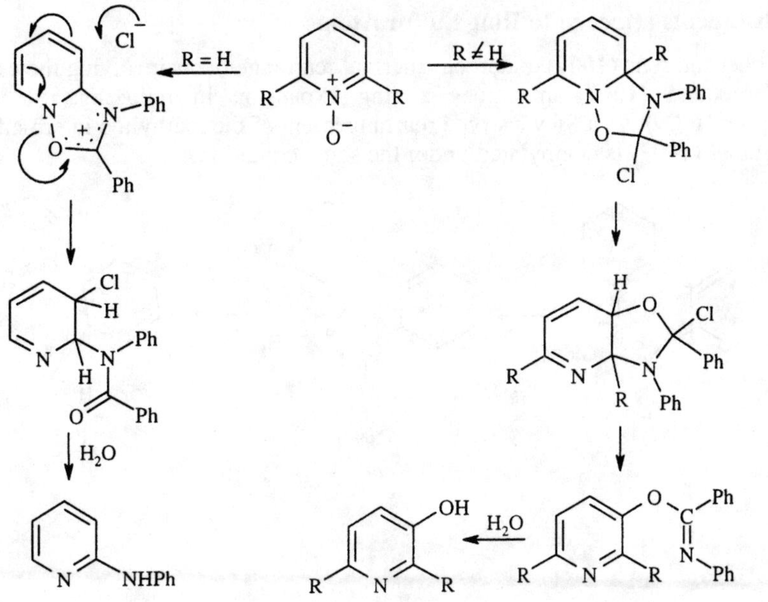

Scheme 78

(d) Benzenesulfonyl chlorides. The reaction with *N*-oxides gives 3-benzenesulfonyloxy compounds, *e.g.* (**1004**) *via* (**1002**) and (**1003**).

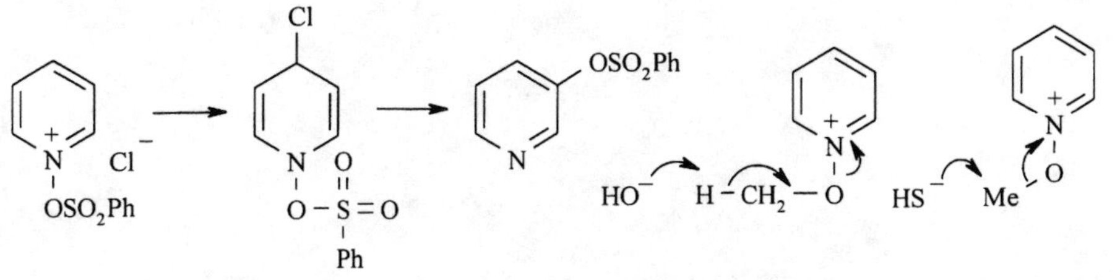

(**1002**)   (**1003**)   (**1004**)   (**1005**)   (**1006**)

(v) Reactions of N-alkoxy-pyridines and -azines

Two distinct types of reaction are common:

(a) 1-Alkoxypyridinium compounds react with hydroxide ions to give aldehydes and pyridines in an elimination reaction (*cf.* **1005**).

(b) Soft nucleophiles can remove the alkyl group, *e.g.* (**1006**) → pyridine 1-oxide + MeSH.

3.2.3.12.6 Other substituents attached to nitrogen

(i) S-Linked

Pyridine *N*-sulfides are known only in the form of their derivatives. Thus, 1-arylthiopyridinium cations (from pyridine and sulfenyl chloride) react with KCN to form ArSCN and pyridine ⟨81CC703⟩.

Pyridine–sulfur trioxide is a mild sulfonating reagent, used for sulfonation of furan and pyrrole.

(ii) Halogens

Pyridine–halogen complexes (**73**) dissociate on heating; halogen is lost so readily that these compounds act as mild halogenating agents toward phenol or aniline, for example (see Section 3.2.1.3.8).

(iii) Metalloids

The complexes formed with boron trihalides are decomposed to pyridine by boiling water. Complexes with other Lewis acids behave similarly.

3.2.3.13   Substituents attached to Ring Sulfur Atoms

The azathiaphenanthrene (**1007**) undergoes thermal rearrangements involving the carbon substituent on sulfur. Compound (**1007**) undergoes a ring expansion in refluxing xylene to yield the dibenzothiazepine (**1009**) *via* a Stevens-type rearrangement of the methylide intermediate (**1008**). The S-ethyl analogue of (**1007**) is deethylated under the same conditions.

(**1007**)   (**1008**)   (**1009**)

3.3

Reactivity of Five-membered Rings with One Heteroatom

3.3.1 REACTIONS AT HETEROAROMATIC RINGS

3.3.1.1 General Survey of Reactivity

We first consider the different types of reactivity of which five-membered heteroaromatic rings with one heteroatom are capable. We compare and contrast the effects of the different heteroatoms and compare these compounds with analogous aliphatic and benzenoid derivatives. All these reactions are considered later in this chapter in more detail.

Electrophilic attack at ring heteroatoms is rare for the neutral compounds, although examples are known for thiophenes and selenophenes. However, pyrrole anions undergo easy reaction with electrophiles at both C and N atoms.

The most important reactions involve electrophilic attack on ring carbon atoms, a wide variety of which are known for pyrroles, furans and thiophenes. Most frequently, such electrophilic attack is followed by proton loss, resulting in overall substitution.

Nucleophilic attack on neutral pyrroles, furans and thiophenes without strongly electron-withdrawing substituents is restricted to deprotonation at N or C atoms. However, the cations formed by electrophilic attack on pyrrole, furan and thiophene rings react readily with weak nucleophiles, resulting in overall addition or ring-opening reactions.

The neutral rings react readily with radicals and other electron-deficient species, and a variety of reactions at surfaces are known.

Several types of reaction involving cyclic transition states are known.

The five-membered ring heterocycles possess Diels-Alder reactivity of varying degree. This is most pronounced in the case of furan and benzo[c]-fused heterocycles such as isoindole. In this capacity they are functioning as heterocyclic analogues of cyclopentadiene, and high Diels-Alder reactivity can be correlated with low aromaticity.

(1)   (2)   (3)   (4)

3.3.1.1.1 Comparison with aliphatic series

Many common reactions of aliphatic amines, ethers and sulfides (1) involve initial attack by an electrophilic reagent at a lone pair of electrons on the heteroatom; salts, quaternary salts, coordination compounds, amine oxides, sulfoxides and sulfones are formed in this way. Corresponding reactions are very rare (cf. Section 3.3.1.3) with pyrroles, furans and thiophenes. These heterocycles react with electrophilic reagents at the carbon atoms (2–3) rather than at the heteroatom. Vinyl ethers and amines (4) show intermediate behavior reacting frequently at the β-carbon but sometimes at the heteroatom.

(5)   (6)   (7)

The heteroatoms of pyrrole, furan and thiophene carry partial positive charges in the ground state which hinder reaction with electrophilic reagents. Conversely, the carbon atoms of these compounds are partially negatively charged, which aids reaction with electrophilic reagents at the ring carbons. This charge distribution follows from the valence bond theory as a consequence of contributions to the resonance hybrids of canonical forms (5) ↔ (6). Molecular orbital theory leads to similar predictions, the heteroatom contributing two electrons to the π-molecular orbitals and the carbon atoms only one each (7).

3.3.1.1.2 *Effect of aromaticity*

Vinyl ethers and amines disclose little tendency to 'revert to type'; thus, the intermediate formed by reaction with an electrophilic reagent reacts further by adding a nucleophilic species to yield an addition compound; *cf.* the sequence (8) → (11). Thiophene and pyrrole have a high degree of aromatic character; consequently the initial product formed by reaction of thiophene or pyrrole with an electrophilic species subsequently loses a proton to give a substituted compound; *cf.* the reaction sequence (12) → (15). Furan has less aromatic character and often reacts by overall addition as well as by substitution. In electrophilic addition, the first step is the same as for substitution, *i.e.* the formation of a σ-complex (*e.g.* 13), but instead of losing a proton this now adds a nucleophile.

The Kekulé resonance of the benzene ring is impaired in the 3,4-benzo derivatives (16), and these compounds are unstable and usually react by overall addition. 2,3-Benzo derivatives (17) have appreciable resonance energies and usually 'revert to type'.

3.3.1.2 Thermal and Photochemical Reactions Involving No Other Species

Pyrrole, furan and thiophene rings are thermally very stable.

Photochemical scrambling of ring atoms can involve a 'ring-walk' or a cyclopropene mechanism.

2-Cyanopyrrole undergoes photochemical rearrangement to 3-cyanopyrrole. Analogous rearrangement of *N*-methyl-2-cyanopyrrole (18) in methanol gives in addition to the 3-cyanopyrrole (20) the bicyclic intermediate (19) trapped as the methanol adduct (21) ⟨78CC131⟩.

The light-induced rearrangement of 2-phenyl- to 3-phenyl-thiophene may involve an equilibrium between the bicyclic intermediate (22) and the cyclopropenylthioaldehyde (23) (Scheme 1). The formation of *N*-substituted pyrroles on irradiation of either furans or thiophenes in the presence of a primary amine supports this suggestion (Scheme 2).

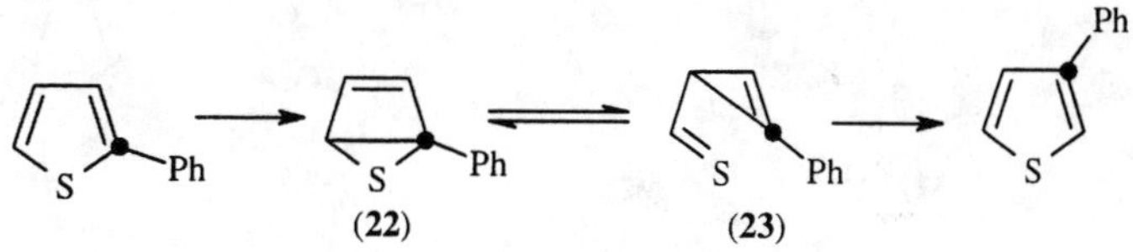

Scheme 1

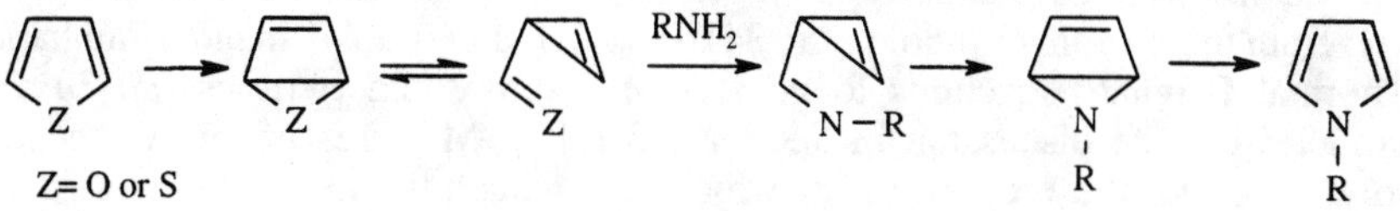

Z= O or S

Scheme 2

The parent Dewar thiophene (**24**) has been generated and trapped ⟨85JA723⟩. When a solution of thiophene in furan (mole ratio, 1:10) was irradiated at 229 nm at 25°C, the two 1:1 adducts (**25**) and (**26**) were formed.

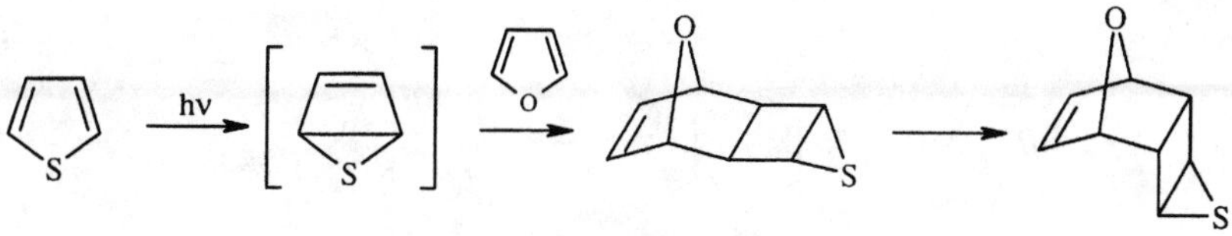

2,3-Bis(trifluoromethyl)thiophene, on photolysis, gives an equilibrium mixture of the two Dewar thiophenes (**27**) and (**28**) (8:1), along with the rearranged 2,5-, 2,4-, and 3,4-bis(trifluoromethyl)-thiophene. Interestingly, the Dewar isomer (**28**) seems to be more reactive than (**27**) towards dienes; thus, on reaction with 2,5-dimethylfuran, the mixture of (**27**) and (**28**) gives exclusively the adduct (**29**) derived from (**28**). Irradiation of 2,5-bis(trifluoromethyl)thiophene gives no Dewar isomers; only the rearranged 2,4-disubstituted isomer is formed, probably *via* a cyclopropenylthioketone or a tricyclic isomer. Photolysis of 3,4-bis(trifluoromethyl)thiophene also does not lead to any Dewar thiophene.

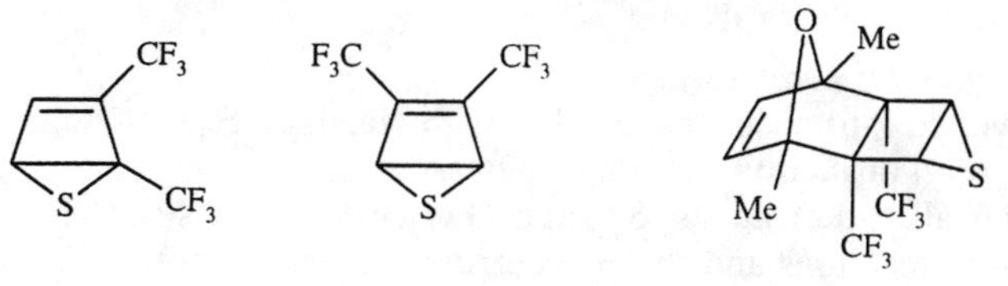

3.3.1.3   Electrophilic Attack on Ring Heteroatoms

Reactions of this type are rare for reasons discussed in Section 3.3.1.4. No examples are known of electrophilic attack at the furan oxygen atom.

Neutral pyrroles and indoles also are not susceptible to attack at the cyclic nitrogen. However, the pK_a of carbazole and 9-methylcarbazole equal to -6.0 and -8.2, respectively, (in H_2SO_4–EtOH, 4:1) refer to N-protonation ⟨71JA5102⟩. Pyrrole anions undergo easy reaction with various electrophiles.

Increasing tendency towards electrophilic attack at the ring heteroatom is shown in thiophenes, selenophenes and tellurophenes.

3.3.1.3.1   *Pyrrole anions*

As discussed in Section 3.3.1.6.1, pyrroles are weak acids. The resulting ions react exceedingly readily, even with weak electrophilic reagents at either carbon (**30**) or nitrogen (**31**); this behavior is similar to that of the ambident anion from acetoacetic ester which shows alternative reactions (**32**, **33**).

(30)    (31) 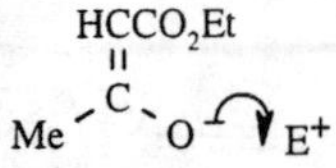   (32) 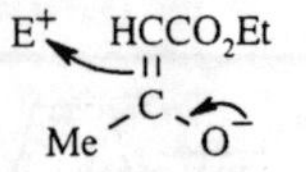   (33) 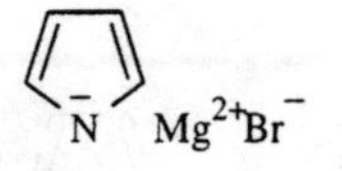   (34)

(i) Pyrrole Grignard reagents

Pyrroles and indoles with Grignard reagents give hydrocarbons and new, largely ionic Grignard reagents derived from the pyrrole or indole (*e.g.* **34**). Pyrrolyl- and indolyl-magnesium halides undergo many of the normal Grignard reactions to give 1- or 2-substituted pyrroles (*cf.* **30**, **31**) or 1- or 3-substituted indoles (*cf.* the discussion in Section 3.3.1.4.2). Mixtures of the *N*- and *C*-substituted products are often formed, the proportions of which are frequently altered by changing the solvent, temperature or reagent.

A classical method for preparing *C*-acylated pyrroles involves the acylation of pyrrylmagnesium bromide. In general, tightly coordinated *N*-pyrryl and indolyl salts, exemplified by their Grignard derivatives, undergo preferential *C*-acylation and *C*-alkylation (Scheme 3) ⟨81TL4647⟩.

Ketones and esters usually react further with Grignard reagents; however, both ketones and esters of type (**35**) and pyrrolyl Grignard reagents are stabilized by mesomerism, and are therefore less reactive.

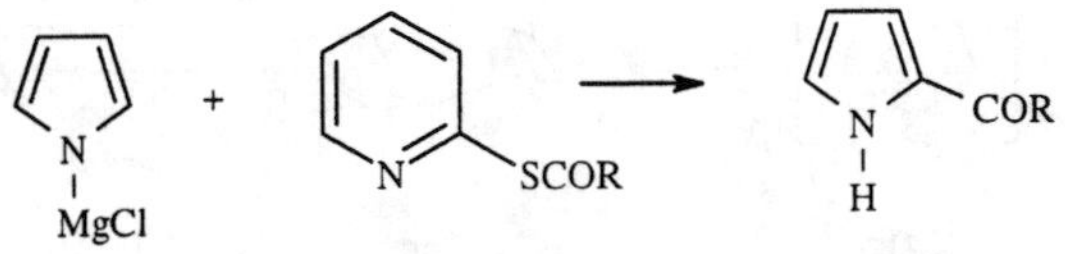

(35)

Scheme 3

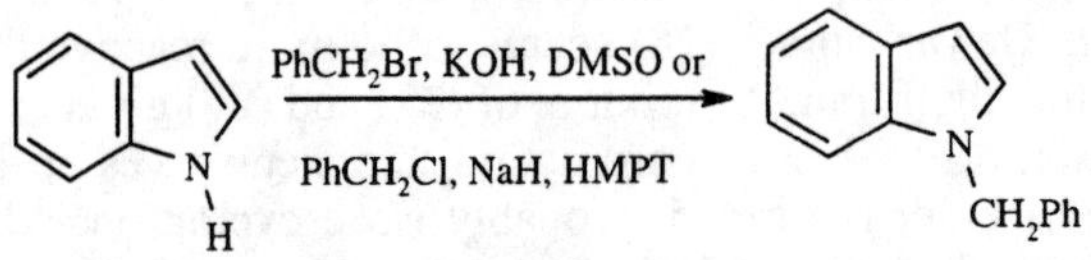

Scheme 4

(ii) Further pyrrole anion intermediates

Pyrroles can be converted into alkali metal salts (with $NaNH_2/NH_3$ or K/toluene) (see also Section 2.3.3.1). The use of pyrrolyl or indolyl sodium or potassium salts under ionizing conditions favors the formation of *N*-acyl or *N*-alkyl derivatives (Scheme 4) ⟨74OS(54)58, 74OS(54)60⟩.

In addition to Grignard reactions and those occurring through complete transformation into alkali metal salts, there are reactions which take place under conditions of only partial conversion into anions. In some cases, 1-substituted compounds are formed, *e.g.* pyrrole is benzoylated in the presence of $NaOH-H_2O$ to yield 1-benzoylpyrrole.

More often, 2-substituted pyrroles and 3-substituted indoles result; the following exemplify reactions of this type:

(a) Boiling aqueous potassium carbonate converts pyrrole into its 2-carboxylic acid; the anion reacts with carbon dioxide as in (**30**).

(b) 1,2,3,4-Tetramethylpyrrole with MgO-MeI gives pentamethylpyrrolenine (**36**).

(c) Alkyl nitrites or nitrates with sodium ethoxide convert indole into 3-nitroso- (*e.g.* **37** → **38**) or 3-nitro-indoles, respectively.

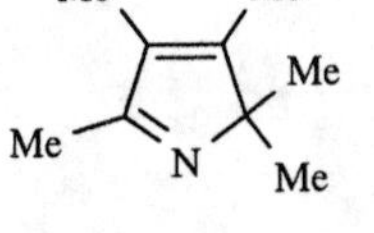

(36) 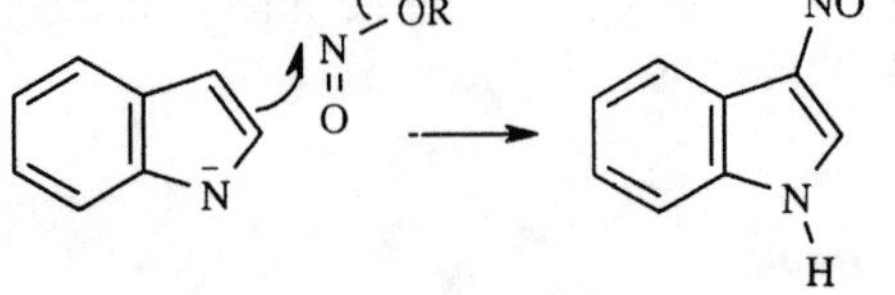   (37)   (38)

Magnesium indolates and allyl alcohols react in the presence of bis(triphenylphosphine)nickel chloride to give 3-allylindoles in moderate yields ⟨86JOC2343⟩.

A mild and effective method for obtaining *N*-acyl- and *N*-alkyl-pyrroles and -indoles is to carry out these reactions under phase-transfer conditions ⟨80JOC3172⟩. For example, *N*-benzenesulfonylpyrrole is best prepared from pyrrole under phase-transfer conditions rather than by intermediate generation of the potassium salt ⟨81TL4901⟩. In this case the softer nature of the tetraalkylammonium cation facilitates reaction on nitrogen. The thallium salts of indoles prepared by reaction with thallium(I) ethoxide, a benzene-soluble liquid, also undergo *N*-alkylation and *N*-acylation ⟨81S389⟩.

N-Alkylation is favored by sodium and potassium cations, whereas *C*-alkylation is promoted by the more strongly coordinating lithium and magnesium cations ⟨90JCS(P1)111⟩. *N*-Alkylation is also favored by polar solvents and the use of tosylates rather than iodides. Under phase transfer conditions, the preference for *N*-alkylation in indoles ranges from 1.5:1 for benzyl bromide to 10:1 for benzyl chloride and *n*-alkyl bromides, over 3-alkylation ⟨90H(31)447⟩.

Relatively electron-deficient indoles such as (**39**) can be alkylated on nitrogen using Mitsunobu reaction conditions (Equation 1) ⟨94TL1847⟩.

$$\text{(39)} \qquad + \quad ROH \quad \xrightarrow{\text{DEAD, Ph}_3\text{P, CH}_2\text{Cl}_2} \qquad \qquad \text{(1)}$$

N-Chlorination of pyrrole occurs when a solution of pyrrole in carbon tetrachloride is stirred at 0°C with an aqueous solution of sodium hypochlorite.

3.3.1.3.2 *Thiophenes, selenophenes and tellurophenes*

(i) *Alkylation and halogenation*

Alkylating agents capable of forming thiophenium salts include trimethyloxonium tetrafluoroborate ($Me_3O^+\ BF_4^-$) and alkyl fluorosulfonates ($ROSO_2F$). The salts (*e.g.* **40**) are conveniently isolated as hexafluorophosphates (**41**).

Halogens attack the ring heteroatom in selenophene and tellurophene. Thus, the 1-bromoselenophene bromide (**42**) is among the bromination products of benzo[*b*]selenophene. Tellurophene reacts with halogens to give 1,1-dihalo derivatives (*e.g.* **43**).

(ii) *Ylid Formation*

Thiophenium bis(alkoxycarbonyl)methylides (**44**) are obtained in high yield by rhodium(II) carboxylate-catalyzed reaction of diazomalonate esters with thiophene derivatives ⟨88JCS(P1)1023⟩. Likewise, ylides from benzo[*b*]thiophene and dibenzothiophene (*e.g.* **45**) have also been reported by *trans*-ylidation using phenyliodonium bis(phenylsulfonyl)methylide ⟨88JHC1599⟩.

It was formerly considered that nitrenes attack only the carbons atom of thiophene ⟨84CHEC-(4)741⟩. Since 1984, several S,N-ylides formed by the attack of a nitrene on the ring sulfur atom of thiophene have been prepared ⟨89AHC(45)151⟩. Thus, ethoxycarbonyl nitrene with polyhalogenothiophenes forms the S,N-ylide (**46**) in 44% yield ⟨84CC190⟩. The X-ray crystal structure reveals that the sulfur in such ylides is pyramidal.

Thiophene S,N-ylides are comparable with the 1-mono- and 1,1-dioxides in that they exhibit diene rather than aromatic properties ⟨86JCS(P1)233⟩. The S,C-ylides appear to be more sluggish in cycloaddition reactions and are therefore considered more "aromatic".

(44)   (45)   (46)

(iii) Oxidation

Oxidation of thiophene with peracid under carefully controlled conditions gives a mixture of thiophene 1-oxide and 2-hydroxythiophene 1-oxide (**48**) (for a recent review on thiophene 1-oxide see ⟨97SR(19)349⟩). These compounds are trapped by addition to benzoquinone to give ultimately naphthoquinone (**47**) and its 5-hydro derivative (**49**) ⟨71ACS(B)353⟩. Diels-Alder reaction also gives (**50**).

(47)   (48)   (49)   (50)

Stable sulfones have been obtained from oxidation of 2,5-dimethylthiophene and benzo[*b*]thiophene. Dimethyldioxirane is a general reagent for oxidizing thiophenes to thiophene 1,1-dioxides ⟨90TL5955⟩. By contrast, singlet oxygen oxidation of 2,5-dimethylthiophene results in the formation of a cyclic peroxide which subsequently ring opens (*e.g.* Section 3.3.1.8.1) Very vigorous oxidation of the thiophene ring results in breakdown to maleic and oxalic acids and ring sulfur is oxidized to sulfuric acid.

3.3.1.4 Electrophilic Attack on Carbon: General Considerations

3.3.1.4.1 *Relative reactivities of heterocycles*

Electrophilic substitution is much easier than in benzene. Thiophene reacts about as readily as mesitylene; pyrrole and furan react as readily as phenol or even resorcinol. Electrophilic substitution of heterocycles has been reviewed ⟨86HC(44/2)1⟩, ⟨90AHC(47)87⟩, and the following conclusions were reached:

(i) Aromaticity and relative reactivity: The ground-state stability and aromaticity decrease in the order benzene > thiophene > pyrrole > selenophene > tellurophene > furan. Hence, all these heterocycles are more reactive than benzene. However, among them, the reactivity is governed by the polarizability of the heteroatom. Thus, sulfur is more polarizable than oxygen, but thiophene has greater ground-state stability than furan. A single σ^+ parameter will therefore not accurately describe the quantitative reactivities of these heterocycles under all conditions (for definition of σ^+ see ⟨84CHEC-(4)54⟩).

(ii) Relative rates: The order of reactivities at position 2 is pyrrole > furan > tellurophene > selenophene > thiophene. Where data are available for both 2- and 3-positions, the following order is seen: 2-furan > 2-thiophene > 3-furan > 3-furan > 3-thiophene. Reactivity parameters (σ^+) for the 2- and 3-positions of thiophene for reactions of varying ρ-values (ranging from -0.66 to -12.0) have been given.

The reactivity of five-membered rings with one heteroatom to electrophilic reagents has been quantitatively compared. Table 1 shows that the rates of substitution for (a) formylation by phosgene and *N,N*-dimethylformamide, (b) acetylation by acetic anhydride and tin(IV) chloride, and (c) trifluoroacetylation with trifluoroacetic anhydride ⟨71AHC(13)235⟩ are all in the sequence furan > tellurophene > selenophene > thiophene. Pyrrole is still more reactive as shown by the rate for trifluoroacetylation, by the relative rates of bromination of the 2-methoxycarbonyl derivatives (pyrrole > furan > selenophene > thiophene), and by the rate data on the reaction of the iron tricarbonyl-complexed carbocation $[C_6H_7Fe(CO)_3]^+$ (Scheme 5) (2-methylindole $\cong$ *N*-methylindole > indole > pyrrole > furan > thiophene ⟨73CC540⟩).

Table 1 Relative Rates of Reaction of Thiophene, Selenophene, Tellurophene and
Furan in Selected Electrophilic Substitution Reactions

Heterocycle	Acetylation (25°C)	Trifluoroacetylation (75°C)	Formylation (30°C)
Thiophene	1	1	1
Selenophene	2.28	7.33	3.64
Tellurophene	7.75	46.4	36.8
Furan	11.9	140.0	107.0

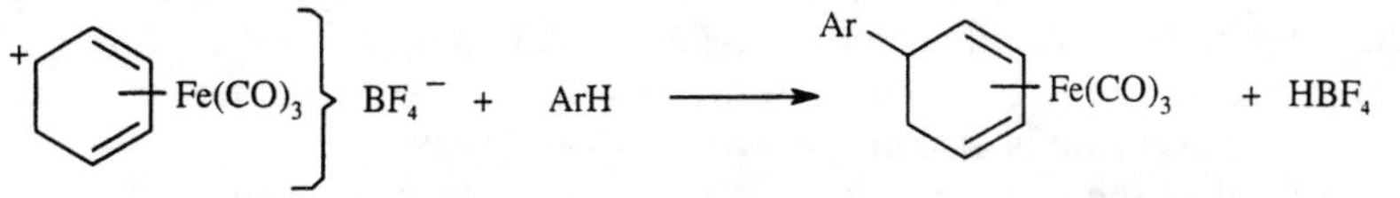

Scheme 5

The electrophilic substitution of thiophene is much easier than that of benzene; thus, thiophene is protonated in aqueous sulfuric acid about 10^3 times more rapidly than benzene, and it is brominated by molecular bromine in acetic acid about 10^9 times more rapidly than benzene. Benzene in turn is between 10^3 and 10^7 times more reactive than an uncharged pyridine ring to electrophilic substitution.

(iii) Sensitivity to substituent effect: conjugative effects are strongly transmitted in the above five-membered heterocycles. Obviously, the four-carbon chain is important in this transmission. The greater the degree of bond fixation in the heterocycle, the more difficult will it be to place a double bond across C-3 and C-4. The ease of transmission will therefore parallel the aromaticity, *viz.* benzene > thiophene > pyrrole > selenophene > tellurophene > furan. Thiophene is less effective at transmitting substituent effects than benzene (except in protiodesilylation). Hammett ρ factors for benzene and thiophene have been compared. The greater ability of substituents on thiophene to attain coplanarity with the heteroaromatic ring, when compared with the situation with benzene, may result in a magnification of the resonance effect (+ M or − M). For the same reason, for reactions involving side-chain carbocations, the heterocycle may appear to be a better transmitter than benzene.

The effect of substituents on the reactivity of heterocyclic nuclei is broadly similar to that on benzene. Thus, *meta*-directing groups such as methoxycarbonyl and nitro are deactivating. The effects of strongly activating groups such as amino and hydroxy are difficult to assess since simple amino compounds are unstable and hydroxy compounds exist in an alternative tautomeric form. Comparison of the rates of formylation and trifluoroacetylation of the parent heterocycle and its 2-methyl derivative indicates the following order of sensitivity to substituent effects: furan > tellurophene > selenophene ≈ thiophene ⟨77AHC(21)119⟩.

(iv) The effect on electrophilic substitution reactions of the fusion of a benzene ring to the '*b*' face of a furan or thiophene ring is to decrease reactivity; this decrease is much more pronounced in the case of fusion to a furan than to a thiophene ring. As a consequence the overall reactivities of benzo[*b*]furan and benzo[*b*]thiophene are approximately equal ⟨71AHC(13)235⟩.

Partial rate factors and σ⁺ values for detritiation have been calculated from rates of deuteration for indole, 1-methylindole, 2-methylindole, 1,2-dimethylindole, and *N*-methylisoindole and confirm the known reactivity patterns. Isoindole is the most reactive, being approximately 10^4 times more reactive than indole ⟨87JCS(P2)591⟩.

3.3.1.4.2 *Directing effects of the ring heteroatom*

Monocyclic five-membered heteroaromatics with one heteroatom all undergo preferential α rather than β electrophilic substitution. This is rationalized in terms of the more effective delocalization of charge in the intermediate (**51**) leading to α substitution than in the intermediate (**52**) leading to β substitution.

(51)   (52)

Z = NH, O, S, Se or Te

However, these considerations apply to reactions in solution. In the dilute gas phase, where the intrinsic orientating properties of pyrrole can be examined without the complication of variable phenomena such as solvation, ion-pairing and catalyst which are attendant on electrophilic substitution reactions in solution, preferential β-attack occurs on pyrrole. In gas phase *t*-butylation, the relative order of reactivity at β-carbon, α-carbon and nitrogen is 10.3 : 3.0 : 1.0 ⟨81CC1177⟩.

The α directing effect of the heteroatom in furan >> thiophene ≈ selenophene >> pyrrole. The α-directing effect in tellurophene is also pronounced ⟨77AHC(21)119⟩.

Possible reasons for the high regioselectivity of furan in electrophilic substitution reactions include complex formation between substrates and reagents and the ability of heteroatoms to assist in the stabilization of cationic intermediates ⟨80CHE1195⟩.

The observed ratio of α- to β-substitution products may also be influenced by reaction temperature. For example, in the acylation of thiophene a higher proportion of β-substituted product is obtained by reaction at higher temperatures ⟨71AHC(13)235⟩. The isolated product ratios may reflect thermodynamic rather than kinetic control because of acid-catalyzed rearrangements. 2- or 3-Substituted pyrroles undergo acid-mediated rearrangement, in some cases under extremely mild conditions. Migration of bromo, chloro, acyl, sulfinyl, sulfonyl and sulfenyl groups have all been observed ⟨82JOC3668⟩. Acid-mediated rearrangements have also been documented in the thiophene series.

Positional selectivity in electrophilic substitution reactions, including methods of orientation control have been reviewed ⟨94H(37)2029⟩.

3.3.1.4.3 *Directing effects of substituents in monocyclic compounds*

Large 1-alkyl substituents increase β-substitution in the Vilsmeier formylation of pyrroles ⟨70JCS(C)2573⟩. A similar trend occurs in trifluoroacetylation of *N*-alkylpyrroles ⟨80JCR(S)42⟩. The trifluoroacetylation, formylation and bromination of 1-tritylpyrrole occur regioselectively at the 3-position in high yield ⟨83JCS(P1)93⟩.

Electron donor 2-substituents orient substitution in furan, thiophene and selenophene into the 5-position. In pyrrole, although the ratio of α to β reactivity is much smaller than in the other five-membered rings, 5-substituted 2-alkylpyrroles still appear to be the major products.

Electron donor 3-substituents supplement the α-directing effect of the heteroatom and direct the incoming substituent into the 2-position. However, steric effects can result in an increased proportion of 5-substitution.

For pyrroles with electron acceptor substituents in the 1-position electrophilic substitution with soft electrophiles can be frontier orbital controlled and occur at the 2-position, whereas electrophilic substitution with hard electrophiles can be charge controlled and occur at the 3-position.

Nitration and Friedel-Crafts acylation of 1-benzenesulfonylpyrrole occur at the 3-position, whereas the softer electrophiles generated in the Mannich reaction ($R_2N^+=CH_2$), in formylation under Vilsmeier conditions ($R_2N^+=CHCl$) or in formylation with dichloromethyl methyl ether and aluminum chloride ($MeO^+=CHCl$) effect substitution mainly in the 2-position ⟨81TL4899, 81TL4901⟩.

For electron acceptor substituents such as NO_2, CN and COR in the 2-position, position 4 is least deactivated by the substituent, but position 5 is most activated by the ring heteroatoms. In fact, such 2-substituted furans give exclusively 5-substitution, whereas for analogous thiophenes and especially pyrroles increasing amounts of 4-substitution occur. The harder the electrophile, the greater the tendency to 4-substitution.

Formylation of 2-methoxycarbonyl-1-methylpyrrole with dichloromethyl methyl ether and aluminum chloride occurs in the 4-position, while under Vilsmeier conditions the main product is the 5-formyl derivative ⟨78JOC4849⟩.

With electron-withdrawing substituents in the 3-position mutually reinforcing directing effects combine to direct substitution into the 5-position.

Electrophilic substitution of 2,5-disubstituted compounds normally occurs at that β-position expected from normal rules, but in some cases displacement of an α-substituent such as carboxyl, acyl or halogen takes place (Scheme 6).

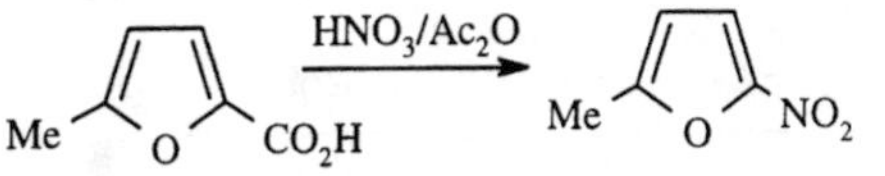

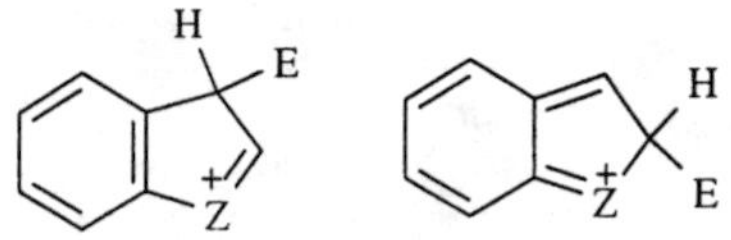

Scheme 6

3.3.1.4.4 Directing effects of fused benzene rings

A [b]-fused benzene ring would be expected to favor β-substitution in the heterocyclic ring (53) over α-substitution (54) based on the expected σ-complex stability.

In fact, benzo[b]furan undergoes mainly α-substitution, benzo[b]thiophene undergoes mainly β-substitution and indole undergoes almost exclusive β-substitution. This again illustrates the very strong directing effect of the oxygen atom to the α-position.

Electrophilic reactions of benzothiophenes have been analyzed ⟨90AHC(47)181⟩. There are no quantitative reactivity data for benzo[c]thiophenes because of stability problems. In benzo[b]thiophene, position 3 is usually more reactive than position 2; this can be easily understood in terms of the number of canonical structures having an aromatic benzene ring for the transition state for substitution. Thus, when conjugative effects are relatively unimportant, the reactivity order is 3 > 2. But for reactions with transition states nearer to the Wheland intermediate, position 2 might become as reactive as position 3. Turning to the quantitative aspect, the σ^+ values are not constant; they generally increase with increasing demand for resonance stabilization of the transition state, such as in acetylation and molecular halogenation.

3.3.1.4.5 Range of substitution reactions

The range of preparatively useful electrophilic substitution reactions is often limited by the acid sensitivity of the substrates. Whereas thiophene can be successfully sulfonated in 95% sulfuric acid at room temperature, such strongly acidic conditions cannot be used for the sulfonation of furan or pyrrole. Attempts to nitrate thiophene, furan or pyrrole under conditions used to nitrate benzene and its derivatives invariably result in failure. In the case of sulfonation and nitration milder reagents can be employed, *i.e.* the pyridine-sulfur trioxide complex and acetyl nitrate, respectively. Attempts to carry out the Friedel-Crafts alkylation of furan are often unsuccessful because the catalysts required cause polymerization.

The higher relative rate of reaction of pyrrole with electophilic reagents, compared with the other five-membered rings with one heteroatom, is paralleled by the greater range of reactions it undergoes. Thus pyrrole, unlike furan and thiophene, can be C-nitrosated and can undergo diazo coupling reactions. In a similar fashion, N,N-dimethylaniline and enamines show enhanced reactivity over anisole and enol ethers, respectively.

The benzo[b] heterocycles are generally less reactive than their monocyclic counterparts. Thus, benzo[b]thiophene unlike thiophene does not undergo Vilsmeier formylation or the Mannich reaction.

3.3.1.5 Electrophilic Attack on Carbon: Specific Reactions

3.3.1.5.1 Proton acids

The σ-complexes formed by proton addition to a carbon atom can sometimes be isolated to give, for example, pyrrole salts. Reversal by proton loss leads to acid-catalyzed hydrogen exchange of the ring hydrogen atoms. The σ-complex can undergo rearrangement before proton loss. The σ-complex can itself react as an electrophile, with other molecules of heterocycle, leading to oligomerization or polymerization, or with other nucleophiles; such reactions are discussed in Section 3.3.1.6.3.

(i) Base strengths

The pK_a values of pyrroles and benzopyrroles are given in Table 2. These basicities are lower than those of enamines in consequence of the loss of aromaticity which accompanies protonation on the ring nitrogen or on carbon 2 or carbon 3 of the ring.

Table 2 pK_a Values of Some Pyrroles and Benzopyrroles

Parent heterocycle	Substituent	pK_a	Position of protonation	Ref.
Pyrrole	–	− 3.8	2	63JA2763
	1-methyl	− 2.9	2	63JA2763
	2-methyl	− 0.2	2	63JA2763
	3-methyl	− 1.0	2	63JA2763
Indole		− 3.6	3	64JA3796
	1-methyl	− 2.3	3	64JA3796
	2-methyl	− 0.3	3	64JA3796
	3-methyl	− 4.6	3	64JA3796
Isoindole			1 or 3	
	2,5-dimethyl-1,3 -diphenyl	+ 2.05	1 or 3	76TI767
Indolizine		+ 3.9	3	76TI767
Carbazole		− 6.0	9	76TI767

The pK_a for protonation of pyrrole at C-2 to the thermodynamically stable 2*H*-pyrrolium ion is − 3.8; the corresponding pK_a values for protonation at C-3 and at nitrogen are − 5.9 and *ca.* − 10 (Scheme 7).

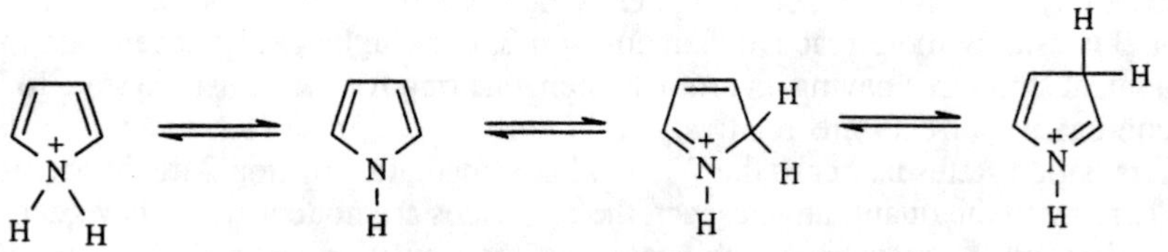

Scheme 7

For pyrrole, furan and thiophene the pK_a values of their 2,5-di-*t*-butyl derivatives are − 1.01, − 10.01 and − 10.16, respectively. In each case protonation occurs at position 2. The base-strengthening effect of alkyl substitution is clearly apparent by comparison of pyrrole and its alkyl derivatives, *e.g.* *N*-methylpyrrole has a pK_a for α-protonation of − 2.9 and 2,3,4,5-tetramethylpyrrole has a pK_a of + 3.7. In general, protonation of α-alkylpyrroles occurs at the α-position whereas β-alkylpyrroles are protonated at the adjacent α-position. As expected, electron-withdrawing groups are base-weakening; thus *N*-phenylpyrrole is reported to have a pK_a of − 5.8.

The pK_a for the protonation of indole at position 3 is − 3.6 and the pK_a values of 2-methylindole, 2-methylbenzo[*b*]furan and 2-methylbenzo[*b*]thiophene for β-protonation are − 0.3, − 13.3 and − 10.4, respectively (Scheme 8).

Z = NH, O or S 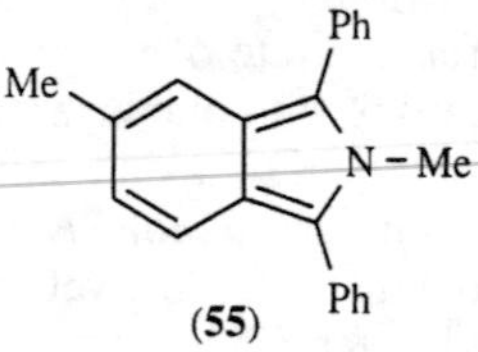   (55)

Scheme 8

Isoindoles are more basic than indoles or pyrroles. For example, 2,5-dimethyl-1,3-diphenyl-isoindole (**55**) has a pK_a of + 2.05; protonation of isoindoles occurs at positions 1 or 3. The pK_a for protonation of indolizine (**56**) at position 3 is + 3.94 and that for carbazole for protonation on nitrogen (**57** → **58**) is estimated as − 6.0.

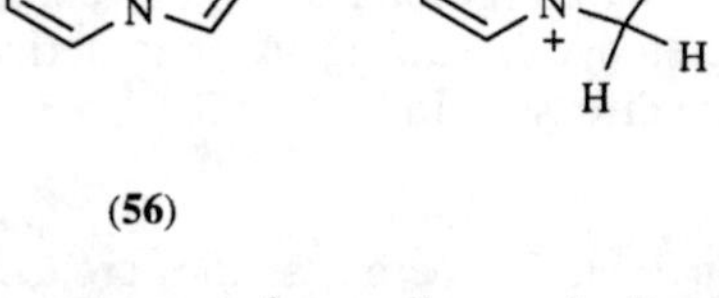

(56) 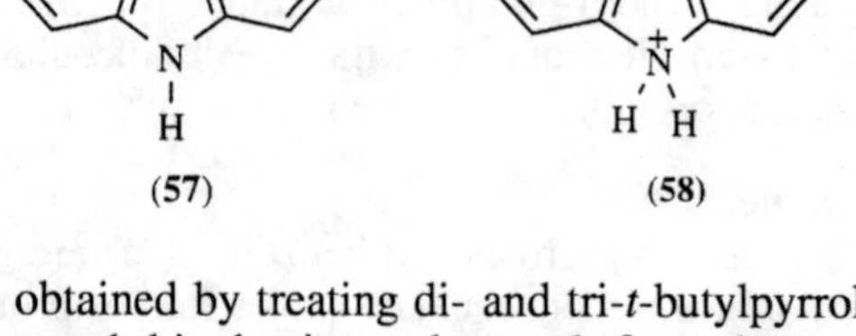   (57)   (58)

Stable α-protonated pyrrolium salts have been obtained by treating di- and tri-*t*-butylpyrroles with tetrafluoroboric acid ⟨81LA789⟩ and stable α-protonated thiophenium salts result from the reaction of thiophenes with hydrogen chloride and aluminum trichloride in an inert solvent (*e.g.* methylene chloride) ⟨75ZOR424⟩.

The gas phase basicities at both the α- and β-positions in five-membered heterocycles have been studied by ion cyclotron resonance equilibrium and bracketing experiments on deuteriated substrates. α-Protonation is preferred by 2.8–4.6 kcal mol^{-1} for both furan and thiophene as compared to 0–2.9 kcal mol^{-1} for pyrrole ⟨81NJC505⟩ and heteroatom protonation is much less favored than α-protonation. Semiempirical (MNDO) molecular orbital calculations have provided quantitative confirmation of the above conclusions.

(ii) Acid-catalyzed hydrogen exchange

The ring hydrogen atoms of pyrrole, furan and thiophene exchange in acid. Relative rates are in the order 2-pyrrole > 3-pyrrole > 2-thiophene > 3-thiophene ≈ 2-furan. No exchange of the 3-position in furan could be found; ring-opening intervened. For pyrrole the exchange rate is *ca.* 10^{15} times that for benzene. The 1-position protons of pyrroles exchange in neutral solution, presumably *via* the anion (see Section 3.3.1.3.1).

Acid-catalyzed hydrogen exchange in thiophenes has been reviewed ⟨86HC(44/2)1, 90AHC(47)87⟩. A conceptually new development in this area is the recognition that thiophene (as well as furan and pyrrole) can participate in hydrogen-bond formation in certain acidic solvents, thereby reducing the reactivity. Evidence for this small effect came initially from the determination of rate coefficients for detritiation of 2-tritio- and 3-tritiothiophene in 100% TFA and TFA - HOAc (35:65). The relative rate for thiophene in these two systems is 2420, compared with 5230 for mesitylene. The lower value for thiophene is ascribed to hydrogen bonding of the solvent to sulfur, thereby reducing the availability of the lone-pair for resonance with the π-electrons of the ring. The extent of hydrogen bonding decreases with increasing proportion of HOAc in the mixture.

(iii) Rearrangement

The acid-catalyzed rearrangements of substituted pyrroles and thiophenes consequent on *ipso* protonation have been referred to previously (Section 3.3.1.4.2). There is some evidence that these two rearrangements are intramolecular in nature since in the case of acid-induced rearrangement of 2-acylpyrroles to 3-acylpyrroles no intermolecular acylation of suitable substrates could be demonstrated (Scheme 9) ⟨81JOC839⟩.

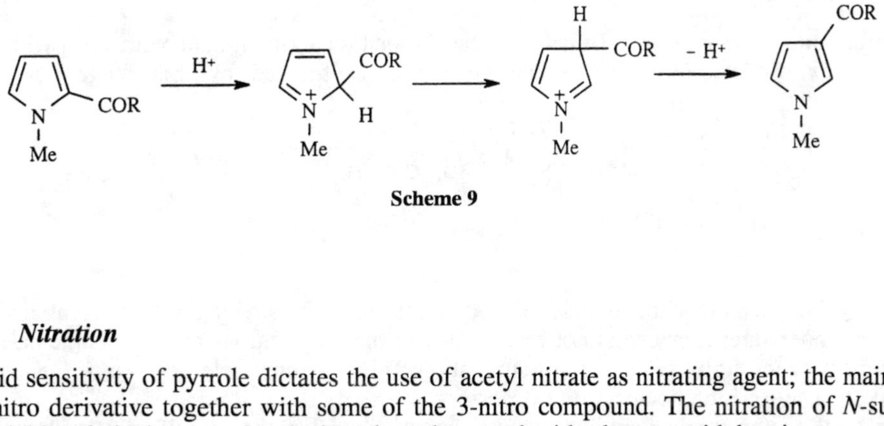

Scheme 9

3.3.1.5.2 Nitration

The acid sensitivity of pyrrole dictates the use of acetyl nitrate as nitrating agent; the main product is the 2-nitro derivative together with some of the 3-nitro compound. The nitration of *N*-substituted pyrroles yields relatively more β-substituted product, and with electron-withdrawing groups such as acetyl or ethoxycarbonyl in the 2-position comparable amounts of 4- and 5-nitro derivatives are obtained. 3-Nitropyrrole is efficiently prepared by nitration of 1-benzenesulfonylpyrrole followed by hydrolysis of the blocking group ⟨81TL4899⟩.

Furan and acetyl nitrate give an addition product which is converted by pyridine into 2-nitrofuran (Scheme 10). The positions in which substituted furans undergo nitration with acetyl nitrate are shown in diagrams (**59**)–(**63**) and illustrate the rules of orientation.

The nitration of furfural in the presence of acetic anhydride gave 5-nitrofurfural diacetate (74%).

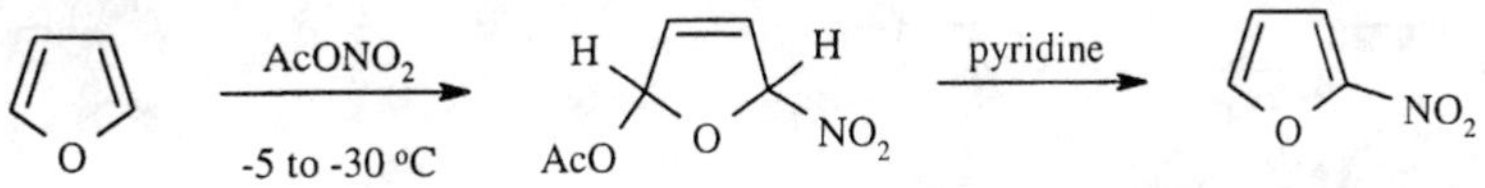

Scheme 10

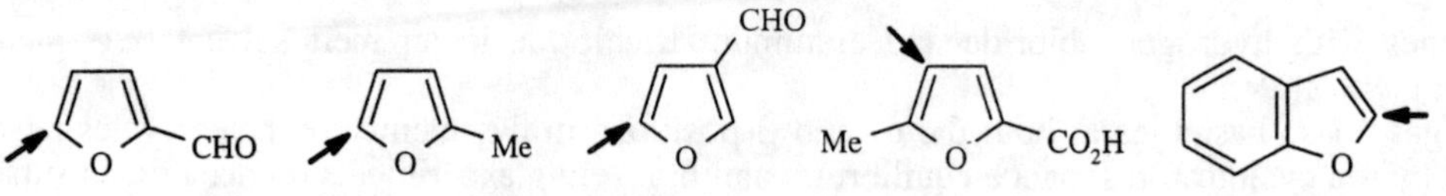

(59)   (60)   (61)   (62)   (63)

Thiophene is nitrated by mild nitrating agents such as acetyl or benzoyl nitrate, mainly in the 2-position. The α selectivity decreases with increasing vigor of the reagent and up to 15% of the 3-isomer has been obtained. 2-Cyanothiophene is nitrated predominantly at position 4.

Indole with benzoyl nitrate at low temperatures gives 3-nitroindole. More vigorous conditions can be used for the nitration of 2-methylindole because of its resistance to acid-catalyzed polymerization. In nitric acid alone it is converted into the 3-nitro derivative, but in a mixture of concentrated nitric and sulfuric acids 2-methyl-5-nitroindole is formed, by conjugate acid nitration (see section 3.3.3.2.1).

There are examples of *ipso* attack during the nitration of pyrroles, furans and thiophenes and in the corresponding benzo-fused systems. Reactions resulting in nitro-dealkylation, nitrodeacylation, nitro-decarboxylation and nitro-dehalogenation are to be found in the monograph reactivity chapters of CHEC. Treatment of the 3-azophenylindole (64) with nitric acid in acetic acid at room temperature gives 80% of the 3-nitroindole (65) ⟨81JCS(P2)628⟩.

Nitration of benzo[b]thiophene (HNO₃/AcOH) yields mainly the 3-nitro derivative. Under these conditions the β to α ratio of substitution is approximately 5:1, which is significantly less than for indole itself which undergoes almost exclusive β substitution. Indolizines are readily nitrated, *e.g.* brief treatment of 2-methylindolizine with a mixture of concentrated nitric and sulfuric acids gives 2-methyl-1-nitroindolizine (66).

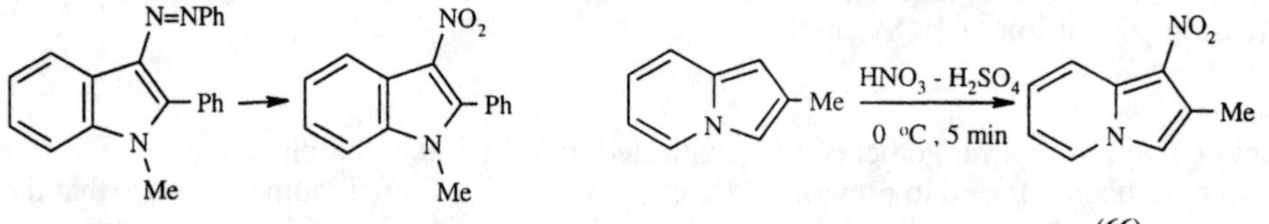

(64)   (65)   (66)

3.3.1.5.3 Sulfonation

Both pyrrole and furan can be sulfonated in the 2-position by treatment with the pyridine-sulfur trioxide complex (Scheme 11). Furan can be further sulfonated by this reagent to give the 2,5-disulfonate.

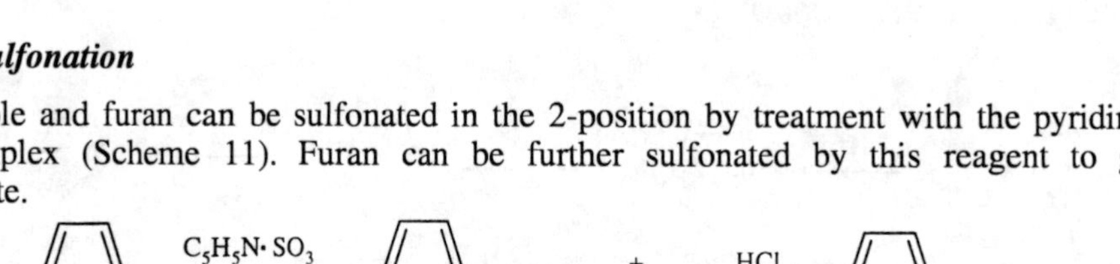

(Z = NH or O)

Scheme 11

Thiophene, which is more stable to acid, is readily sulfonated by shaking with concentrated sulfuric acid at room temperature. Benzene is not reactive under these conditions and this is the basis for the purification of benzene from thiophene contamination. With all three heterocycles, if the α-positions are blocked, then sulfonation occurs at the β-position.

Indole is sulfonated under similar conditions to pyrrole; the 3-sulfonic acid is formed. Benzo[b]thiophene is also sulfonated in the 3-position ⟨71AHC(13)235⟩.

3.3.1.5.4 Halogenation

Pyrrole is readily halogenated. Chlorination with one equivalent of sulfuryl chloride in ether at 0°C gives 2-chloropyrrole; further chlorination with this reagent yields di-, tri- and tetra-chloro derivatives and ultimately pentachloropyrrolenine (67) ⟨62JOC2585⟩.

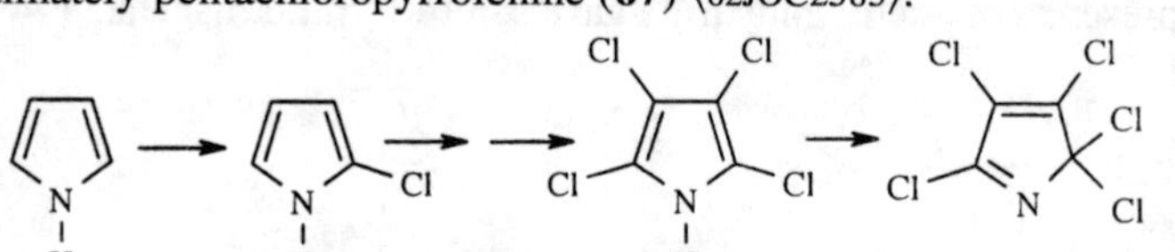

(67) 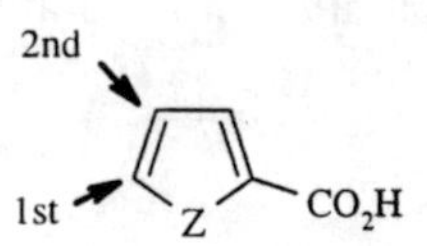   (68)

Treatment of pyrrole, 1-methyl-, 1-benzyl- and 1-phenyl-pyrrole with one mole of *N*-bromosuccinimide in THF results in the regiospecific formation of the corresponding 2-bromopyrroles. Pyrroles with electron-withdrawing groups such as acyl or ethoxycarbonyl at C-3, undergo bromination using copper(II) bromide. If the 2- and 5- positions are blocked, reaction occurs at C-4 ⟨90JHC1209, 90JHC1277⟩. Pyrrole with bromine in acetic acid gives 2,3,4,5-tetrabromopyrrole and iodine in aqueous potassium iodide yields the corresponding tetraiodo compound.

Chlorination with *N*-chlorosuccinimide is less selective ⟨81JOC2221⟩. 1-Methylpyrrole reacts with *N*-chloroimides to give (69) in which the imidyl group is attached to C-2; the 2-chloro compound is proposed as an intermediate (Equation 2) ⟨89JOC5347, 93CC1757⟩.

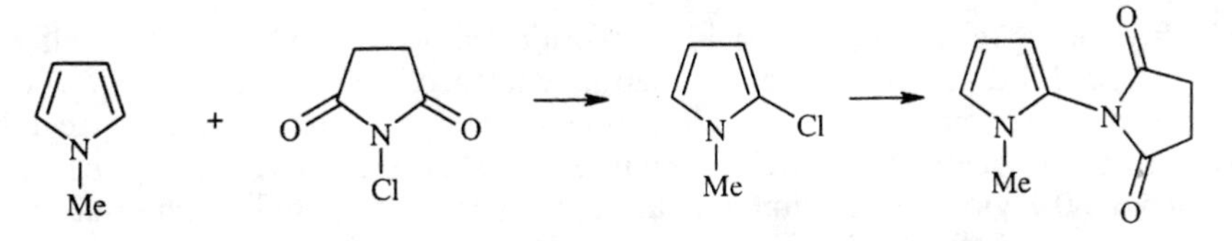

(69)

Pyrroles containing an electron-withdrawing group at C-2 undergo fluorination at C-5 in 25–54% yield using xenon difluoride in acetonitrile; only α-fluorination is observed ⟨94TL3679, 94T6181⟩. The reaction fails if the pyrroles contain no electron-withdrawing groups or two electron-withdrawing groups. Replacement of the solvent acetonitrile with dichloromethane results in chlorination instead of fluorination.

Furan reacts vigorously with chlorine and bromine at room temperature to give polyhalogenated products. Low temperature ($-40°C$) reaction of furan with chlorine in dichloromethane yields mainly 2-chlorofuran and reaction of furan with dioxane dibromide at 0°C affords 2-bromofuran in good yield. Furans stabilized by electron-withdrawing groups are halogenated more smoothly; thus, 2-furoic acid is brominated to form successively the 5-monobromo and 4,5-dibromo derivatives (*cf*. 68; Z = O). The bromination of furan when carried out in DMF using one or two equivalents of bromine gives 2-bromo- or 2,5-dibromo-furan ⟨90SC3371⟩ and a synthesis of the marine natural product bromobeckelleride has been reported starting with the bromination at the 4-position of 5-methylfuran-2-carboxaldehyde ⟨89TL1237⟩. 2-Iodofuran is obtained by treatment of 2-furoic acid with iodine and potassium iodide in aqueous sodium hydroxide.

Chlorine and bromine react with thiophene to give successively the halogenation products shown (70–73). The bromination can be interrupted at the intermediate stages; monochloro and dichloro derivatives have been obtained preparatively by chlorination with MeCONHCl. Addition products are also formed during chlorination; prolonged action (with Cl₂-I₂) gives the dihydrothiophene derivative (74; Z = S). Iodination (I₂-HgO) results in mono- and di-iodothiophenes (70) and (71) (X = I) only. Substituted compounds are halogenated as expected, *e.g.* (75).

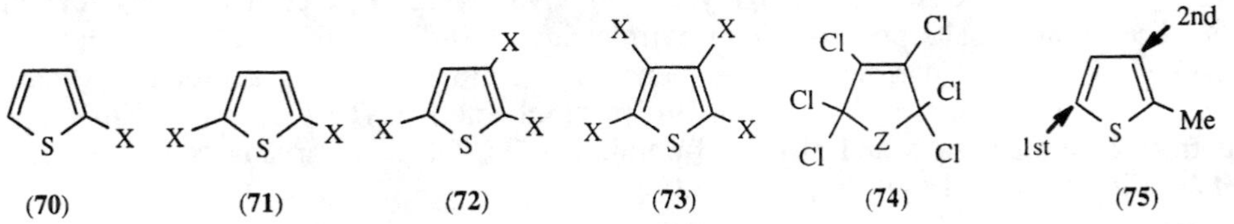

(70)   (71)   (72)   (73)   (74)   (75)

3-Chloroindole has been prepared from indole and sulfuryl chloride ⟨66JOC2627⟩ and 3-bromo- and 3-iodo-indole have been obtained by direct halogenation in DMF ⟨82S1096⟩. 2-Methylindole reacts with sodium hypochlorite in carbon tetrachloride to give a 2:1 mixture of 1,3- and 3,3- dichloro derivatives ⟨81JOC2054⟩. 3-Substituted indoles are halogenated to yield 3-haloindolenium ions which react in a variety of ways, as illustrated by the reaction of 3-methylindole with NBS in aqueous acetic acid (Scheme 12).

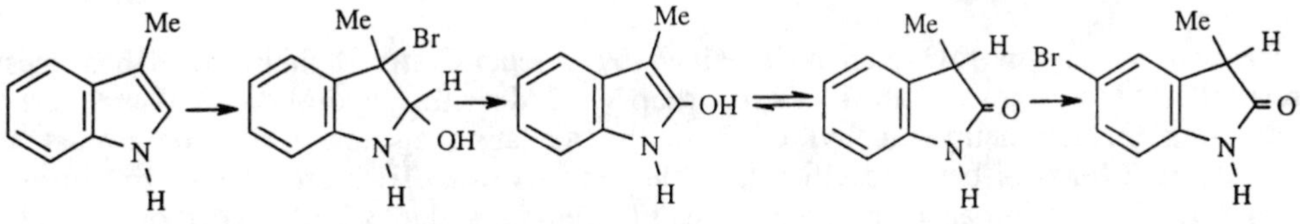

Scheme 12

Halogens react with benzo[*b*]furan by an addition-elimination mechanism to give 2- and 3-substituted products ⟨76JCS(P2)266⟩. Treatment of benzo[*b*]thiophene with chlorine or bromine in acetic acid gives predominantly 3-substituted products ⟨71JCS(B)79⟩. 2,2,3,3,4,5,6,7-Octachloro-2,3-dihydrobenzothiophene is obtained when benzo[*b*]thiophene is treated with chlorine in the presence of 1 equivalent of iodine ⟨80JOC2151⟩.

3.3.1.5.5 *Acylation*

(i) *Pyrroles*

Pyrrole and alkylpyrroles are acylated by acid anhydrides above 100°C. Pyrrole itself gives a mixture of 2-acetyl- and 2,5-diacetyl-pyrrole on heating with acetic anhydride at 150–200°C. *N*-Acylpyrroles are obtained by reaction of the alkali metal salts of pyrrole with an acyl halide. *N*-Acetylimidazole efficiently acetylates pyrrole on nitrogen ⟨65CI(L)1426⟩. Pyrrole-2-carbaldehyde is acetylated on nitrogen in 80% yield by reaction with acetic anhydride in methylene chloride and in the presence of triethylamine and 4-dimethylaminopyridine ⟨80CB2036⟩.

The most useful general method for the *C*-acylation of pyrroles is the Vilsmeier-Haack procedure with the phosphoryl chloride complex (**76a,b**) of an *N,N*-dialkylamide. The intermediate imine salt (**77**) is hydrolyzed subsequently under mildly alkaline conditions to give the acylated pyrrole (**78**). On treatment of the imminium salt (**77**; R′ = H) with hydroxylamine hydrochloride and one equivalent of pyridine and heating in DMF, 2-cyanopyrrole is formed ⟨80CJC409⟩.

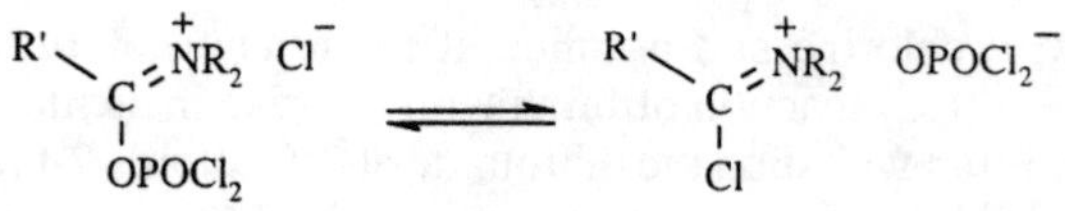

(**76a**)   (**76b**)

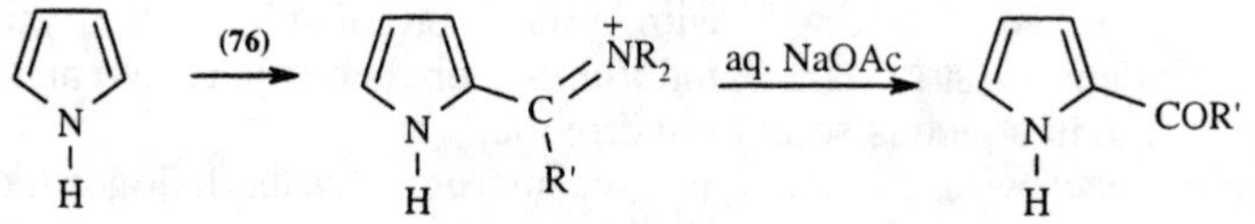

(**77**)   (**78**)

N-Methylnitrilium fluoroborates react readily at low temperatures with indoles and pyrroles to give iminium salts, which can be isolated in high yields and hydrolysed to the corresponding acyl products ⟨85TL4649⟩. The iminium salts prepared from pyrroles and *N,N*-dimethylchloroformiminium chloride can be converted into acyl pyrroles by a sequence of reactions involving addition of cyanide ion, alkylation of the resulting intermediate, and hydrolysis (Scheme 13). These transformations apply both to the normal 2-iminium salts and also to 3-iminium salts, the latter arising from reaction of the hindered *N*-(triisopropylsilyl)pyrrole ⟨88JOC6115⟩.

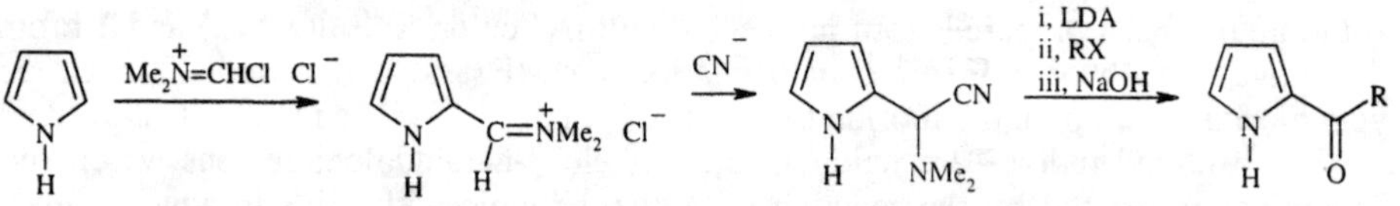

Scheme 13

3-Aroyl- and 3-acetyl-pyrroles can be obtained by Friedel-Crafts acylation of 1-benzenesulfonylpyrrole followed by removal of the blocking group by mild alkaline hydrolysis ⟨81TL4899, 81TL4901⟩. However, using dichloromethyl methyl ether as the acylating agent, substitution occurs at the 2-position to give 2-formyl-1-benzenesulfonylpyrrole, exclusively. 2-Trichloroacetylpyrrole, formed on treating pyrrole with trichloroacetyl chloride, is readily transformed into 2-ethoxycarbonylpyrrole (**79**) ⟨71OS(51)100⟩.

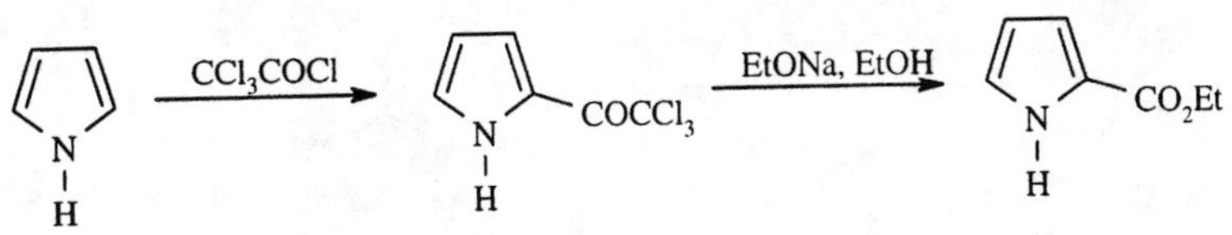

(79)

Although a 2-acylpyrrole undergoes a second acylation at C-4, yields are variable. A more consistent route to 2,4-diacylpyrroles is by the acylation of a 3-acylpyrrole, which can in turn be obtained through the acylation of a 1-benzenesulfonylpyrrole ⟨93JCS(P1)273⟩. An indirect method for preparing pyrrole-2,5-dicarbaldehydes is outlined in Scheme 14 ⟨78S295, 82CJC383⟩.

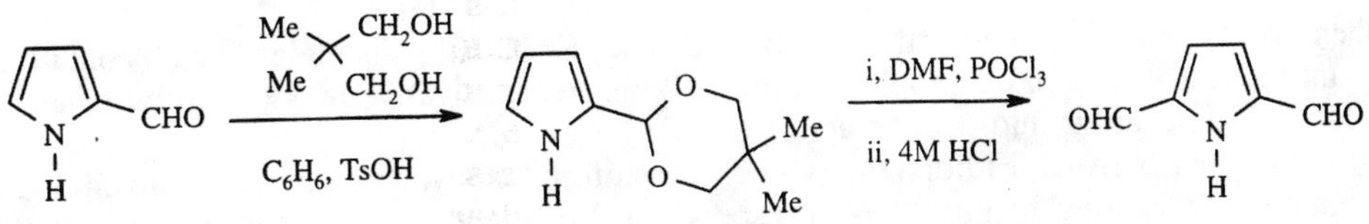

Scheme 14

Indole with acetic anhydride at 140°C gives 1,3-diacetylindole *via* the 3-acetyl compound. In the presence of sodium acetate, acetic anhydride gives exclusively 1-acetylindole. The Vilsmeier-Haack reaction is an efficient route to 3-acylindoles. The usual difference in orientation of substitution is observed in the acetylation (Ac$_2$O/SnCl$_4$/dichloroethane) of benzo[*b*]furan and benzo[*b*]thiophene. Thus, the oxygen heterocycle yields mainly the 2-acetyl derivative and the sulfur heterocycle yields mainly the 3-acetyl derivative ⟨71AHC(13)235⟩. 1-Phenylisoindole readily undergoes acetylation to give a 3-acetyl derivative (Ac$_2$O/pyridine, room temperature) and indolizine gives a 3-acetyl derivative (**80**) on heating with acetic anhydride and acetic acid; it also undergoes Vilsmeier-Haack formylation.

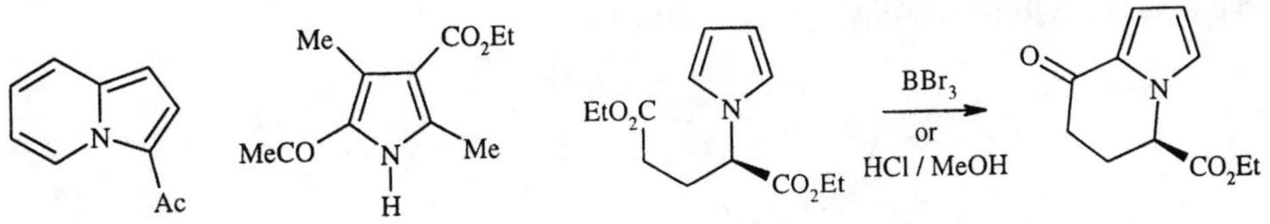

(80)　　　　　(81)　　　　　(82)　　　　　(83)

Intramolecular acylations of pyrroles and indoles are useful synthetically, for example *N*-glutamyl substituted pyrrole (**82**) gives regioselectively (**83**), with retention of configuration, on brief treatment with methanolic hydrogen chloride ⟨93JCS(P1)2241⟩ or boron tribromide ⟨94TL3905, 94TL4759⟩.

(ii) Furans

Furan can also be acylated by the Vilsmeier-Haack method or with acid anhydrides and acyl halides in the presence of Friedel-Crafts catalysts (BF$_3$·Et$_2$O, SnCl$_4$ or H$_3$PO$_4$). Reactive anhydride such as trifluoroacetic anhydride, however, requires no catalyst. Acetylation with acetyl *p*-toluenesulfonate gives high yields.

Pyrroles and furans also undergo the Gattermann aldehyde synthesis: with HCl and HCN, furan gives furfuraldehyde and 2-methylindole gives 2-methylindole-3-carboxaldehyde. The Houben-Hoesch ketone synthesis is also applicable to the preparation of acyl derivatives of furans and pyrroles, *e.g.* ethyl 2,4-dimethylpyrrole-3-carboxylate with MeCN and HCl yields (**81**).

(iii) Thiophenes

The acylation of thiophenes has been reviewed ⟨86HC(44/3)309⟩. The best catalyst for the Friedel-Crafts acylation of thiophene, using free carboxylic acids, appears to be 2-(trifluoromethylsulfonyloxy)pyridine (**84**) in conjunction with TFA. 2-Acylthiophene is formed in quantitative yield, without any evidence of the thiophene dimerizing. The reaction might be mediated by the mixed anhydride (**85**) (Scheme 15) ⟨88BCJ455⟩ (*cf.* ⟨84CHEC-(4)741⟩).

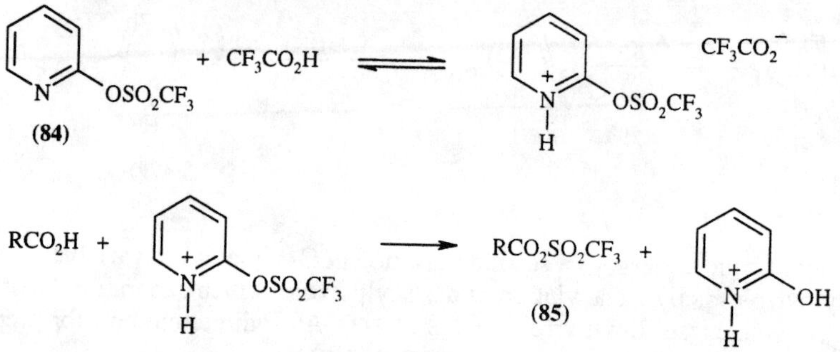

Scheme 15

Thiophene is also readily 2-acylated under both Friedel-Crafts and Vilsmeier-Haack conditions. An almost quantitative conversion of thiophene into its 2-benzoyl derivative is obtained by reaction with 2-benzoyloxypyridine and trifluoroacetic acid ⟨80S139⟩.

Acylation of either pyrrole or thiophene with chlorosulfonyl isocyanate gives a 2-substituted amide which fragments on heating to give the corresponding 2-cyano derivative (Scheme 16) ⟨81CJC2673, 70OS(50)52⟩. An alternative procedure for obtaining 2-cyanopyrrole and 3-cyanoindole involves treatment of the parent heterocycle with thiocyanogen and triphenylphosphine ⟨80JCS(P1)1132⟩.

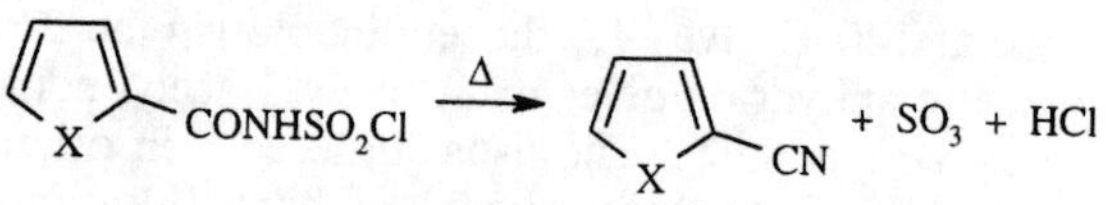

Scheme 16

A muscone synthesis involves selective intramolecular acylation at a vacant α-position (Scheme 17) ⟨80JOC1906⟩. In attempts to prepare 5,5-fused system *via* intramolecular acylation reactions on to a β-position of a thiophene or a pyrrole, *ipso* substitution occurs in some cases with the result that rearranged products are formed (Scheme 18) ⟨82TH30200⟩.

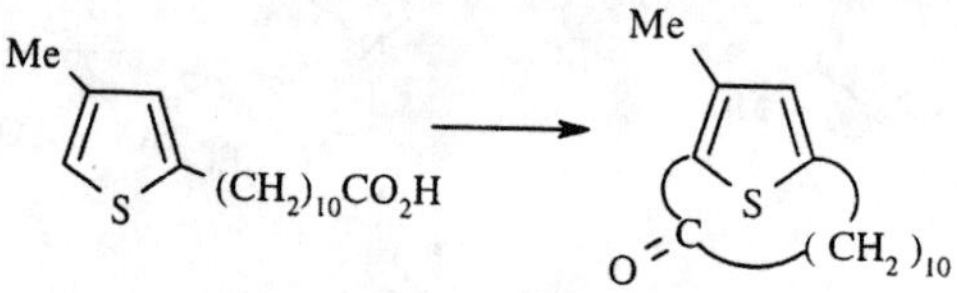

Scheme 17

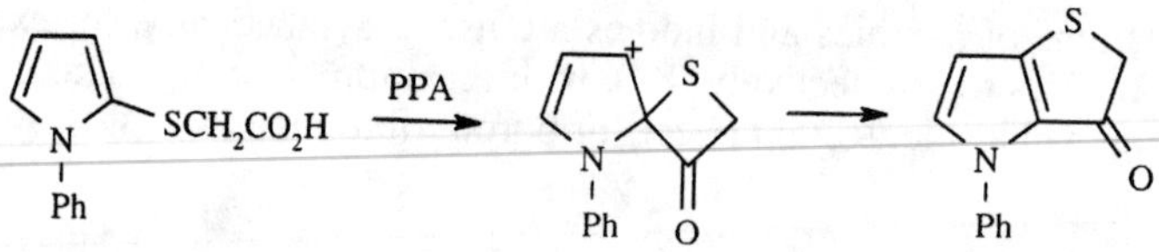

Scheme 18

3.3.1.5.6 Alkylation

Pyrroles give polyalkylated products on reaction with methyl iodide at elevated temperatures and the more reactive allyl and benzyl halides under milder conditions in the presence of weak bases. Alkylation of pyrrole Grignard reagents gives mainly 2-alkylated pyrroles whereas N-alkylated pyrroles are obtained by alkylation of pyrrole alkali metal salts in ionizing solvents (see Section 3.3.1.3.1). Pyrrolenines are formed by the C-alkylation of the Grignard derivatives of polyalkylated pyrroles. Alkylation occurs at positions 2 and 3, although the former predominates.

Indoles can be alkylated by alkyl alcohols in the presence of lithium perchlorate and acetic acid: for example, products such as the precursor (86) to the anti-fertility alkaloid yuehchukene can be produced (Equation 3) ⟨93CC510⟩.

(3)

(86)

Pyrroles and indoles are alkylated by electrophilic alkenes in reactions catalyzed by acidic clays, which are often more effective than Lewis acids ⟨88TL2577⟩. Indoles are alkylated as usual at C-3, and the alkylation of 3-substituted indoles has also been shown to take place at C-3 with subsequent migration occurring to C-2.

C-Alkylation of pyrroles and indoles can be mediated by alumina in association with the appropriate halide ⟨92JCS(P1)2821⟩. 3-Indolyl sugar derivatives have been obtained by alkylation reactions of indolylmagnesium bromide ⟨91JOC5466⟩. This type of alkylation has been extended to ring-fused indoles (87) and yields 3*H*-indole derivatives (88) in good yield ⟨91JHC1293⟩.

i, MeMgI
ii, MeI

n = 3-6

(87)   (88)

Indole with excess methyl iodide at 110°C gives a tetramethyl derivative (91). The intermediate 2,3-dimethylindole (90) is thought to arise by rearrangement of the 3,3-dimethyl-3*H*-indolium cation (89).

(89)   (90)   (91)

Pyrroles can be α-alkenylated by acetylenedicarboxylic ester, as discussed in Section 3.3.1.8.1.

Alkylation of furan and thiophene has been effected with alkenes and catalysts such as phosphoric acid and boron trifluoride. In general, Friedel-Crafts alkylation of furans or thiophenes is not preparatively useful, partly because of polymerization by the catalyst and partly because of polyalkylation.

The competition between Michael addition of α,β-unsaturated ketones and Diels-Alder reactions involving furan and 2-methylfuran is affected by the catalyst used. Methyl vinyl ketone gives the alkylation product with furan and 2-methylfuran in the presence of silica gel ⟨88TL175⟩. Bis(alkylated) products have also been obtained in reactions of 2-methylene-1,3-dicarbonyl compounds ⟨90H(31)1699⟩. An intramolecular proton catalyzed alkylation reaction of an α,β-unsaturated ketone provided a straightforward synthesis of norpinguisone ⟨90TL4343⟩ and in the example shown in Equation (4) the cyclization reaction involved an α,β-γ,δ-dienone ⟨94TL4887⟩.

Et₂O · BF₃

Et₃N, - 78 °C

(4)

A successful procedure for the formation of 2,5-di-*t*-butylfuran involves reaction of the parent heterocycle with *t*-butyl chloride in the presence of iron(III) chloride and iron(III) oxide ⟨82CI(L)603⟩.

Ipso intramolecular alkylation to (93) occurs in the acid-promoted cyclization of the amino alcohols (92) with trifluoroacetic acid. Polyphosphoric acid gave the non-rearranged thieno-pyridine (94) ⟨82CC793⟩.

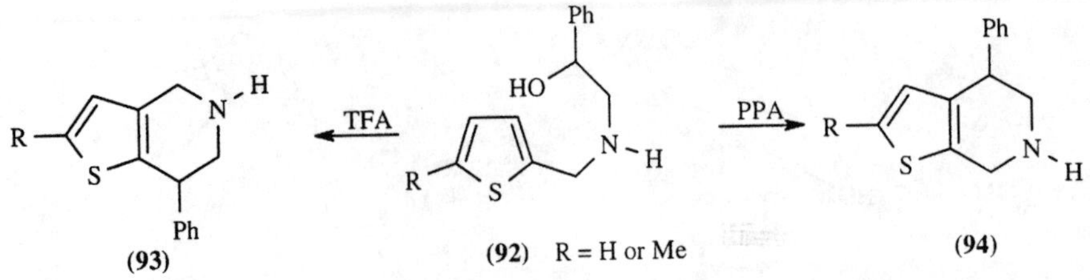

(93)   (92)   R = H or Me   (94)

Silica gel is an effective catalyst for the *t*-butylation of thiophene and benzo[*b*]thiophene using *t*-butyl bromide. 2,5-Di-*t*-butylthiophene and 3-*t*-butylbenzo[*b*]thiophene can be prepared easily by this procedure ⟨84JOC4161⟩. Alkylation of thiophene with *t*-butyl chloride, isopropyl chloride or ethyl chloride at $-70°C$ in the presence of $AlCl_3$ produced σ-complexes under kinetic control. On thermal equilibration, migration of alkyl from position 3 to position 2, as well as disproportionation to dialkyl- and trialkyl thiophenes can occur ⟨86T759⟩.

3.3.1.5.7   *Reactions with aldehydes and ketones*

(i) *Formation of carbinols or carbonium ions*

Thiophenes, pyrroles and furans react with the conjugate acids of aldehydes and ketones to give carbinols (*e.g.* **95**) which cannot normally be isolated but which undergo proton-catalyzed loss of water to give reactive electrophiles (*e.g.* **96**).

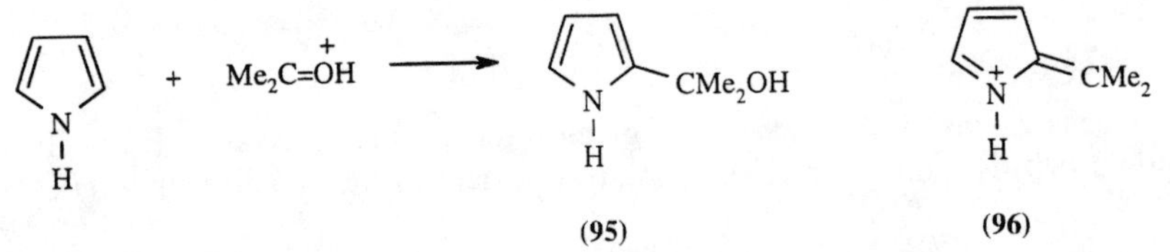

(95)   (96)

By using an aromatic aldehyde carrying an electron-releasing group the intermediate cation can be stabilized. This is the basis of the classical Ehrlich color reaction for pyrroles, indoles and furans which have a free reactive nuclear position (Scheme 19). As expected, pyrroles react preferentially in the α-position and indoles in the β-position, but if these positions are filled, reaction can occur at other sites.

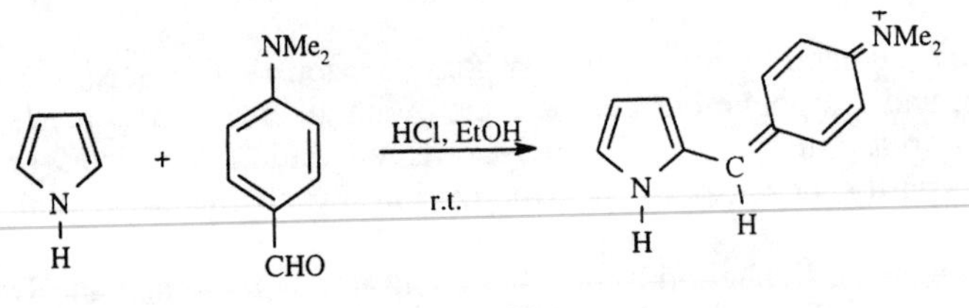

Scheme 19

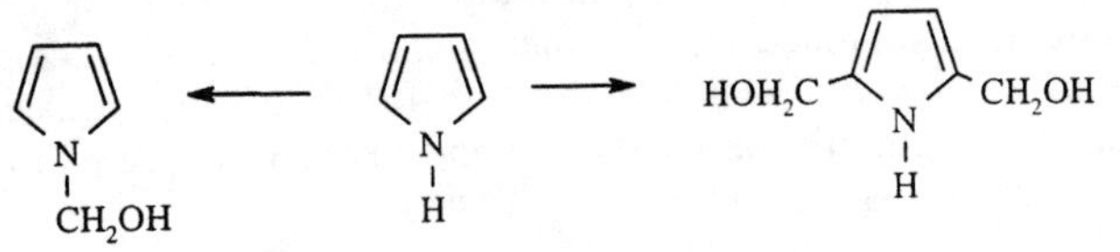

Scheme 20

Pyrrole has been condensed under alkaline conditions with formaldehyde to give products of either *N*- or *C*-hydroxymethylation (Scheme 20).

(ii) *Further reactions of carbonium ions*

With *N*-unsubstituted pyrroles, ions of types (**97**; Z = NH) can lose a proton from the ring nitrogen, *e.g.* indole with Ph_2CO or PhCHO in HCl/EtOH gives products of type (**98**). Indole with HCO_2H or PhCOCl gives 'rosindoles' (**99**; R = H, Ph), involving the formation of ketones (*cf.* discussion in Section 3.3.1.5.5.i) and a subsequent reaction of this type with a second molecule of indole.

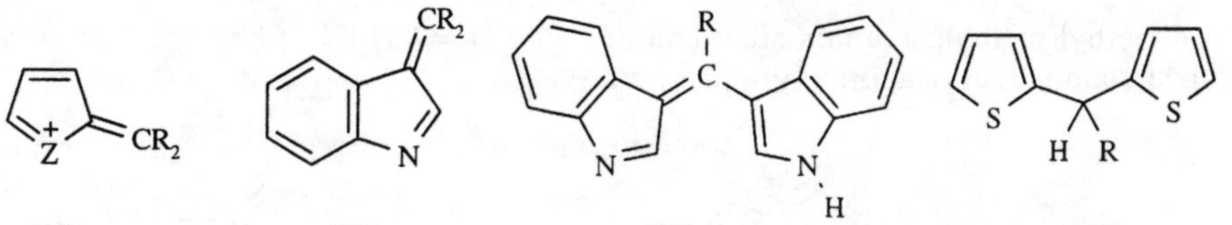

(97)   (98)   (99)   (100)

The ion (97), acting as an electrophilic reagent, can also attack another molecule of the heterocyclic compound. Thiophene with benzaldehyde or chloral gives the dinuclear product (100; R = Ph, CC1$_3$). Pyrrole and furan react with acetone to form tetranuclear derivatives of type (101; Z = NH, O). Pyrroles with a single free position react analogously to thiophene; *e.g.* two molecules of 3-ethoxycarbonyl-2,4-dimethylpyrrole with formaldehyde afford the dipyrromethane (102).

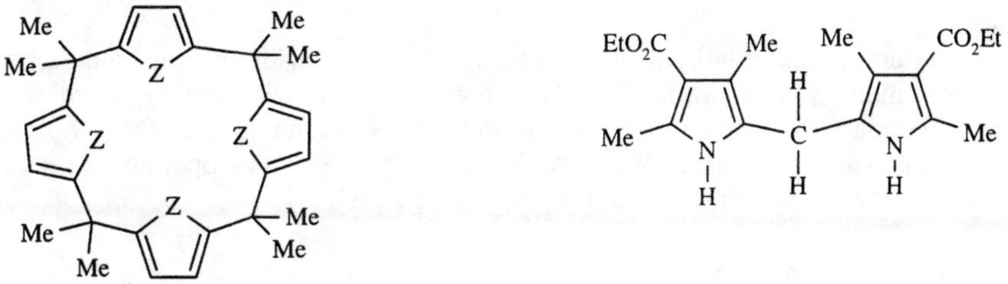

(101)   (102)

More rarely, ions of type (97) form dimeric products (possibly by initial loss of nuclear protons); thus, thiophenes with two free α-positions, or free adjacent α- and β-positions, give indophenines (*e.g.* 103) with isatin (104). This reaction is used as a test for thiophene, the so-called 'indophenine test'.

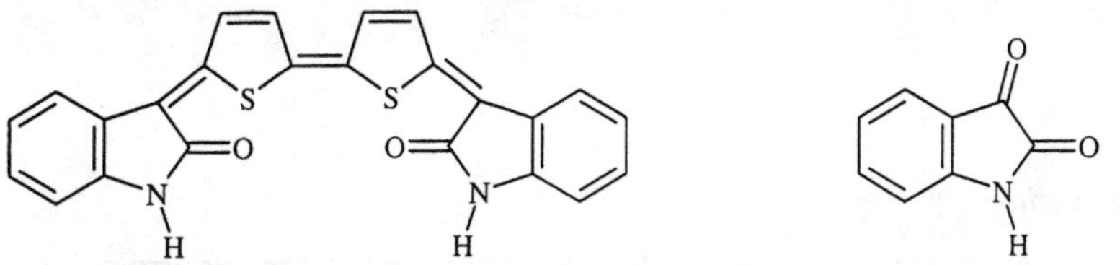

(103)   (104)

Although acid-catalyzed hydroxymethylation is not a practical possibility, by the addition of a reducing agent to the reaction mixture overall reductive alkylation can be achieved (Scheme 21).

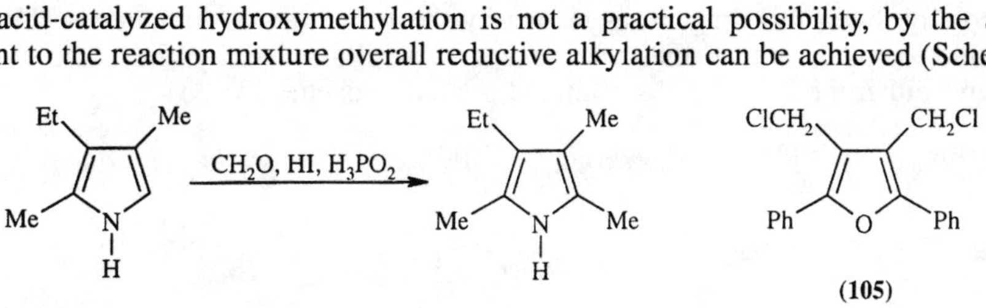

(105)

Scheme 21

(iii) Chloromethylation

Thiophene and selenophene can be chloromethylated by treatment with formaldehyde and hydrochloric acid. Depending on the conditions, 2-chloromethyl or 2,5-bis(chloromethyl) derivatives are obtained. The chloromethylation of benzo[*b*]thiophene gives the 3-chloromethyl derivative and that of benzo[*b*]furan the 2-chloromethyl compound 〈71AHC(13)235〉. Furan is destroyed by this treatment, but 2,5-diphenylfuran, for example, gives the 3,4-bischloromethyl derivative (105).

(iv) Mannich reaction

Pyrrole is aminomethylated to give products of type (106). The intermediate immonium ion generated from formaldehyde, dimethylamine and acetic acid is not sufficiently reactive to aminomethylate furan, but it will form substitution products with alkylfurans. The Mannich reaction appears to be still more limited in its application to thiophene chemistry, although 2-aminomethyl-thiophene has been prepared by reaction of thiophene with formaldehyde and ammonium chloride. The use of *N,N*-dimethyl(methylene)ammonium chloride ($Me_2N^+=CH_2Cl^-$) has been recommended for the *N,N*-dimethylaminomethylation of thiophenes ⟨83S73⟩.

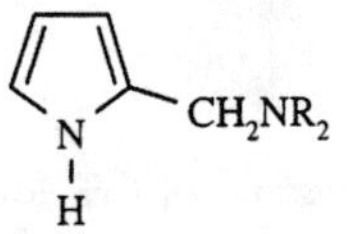

An important application of the Mannich reaction is the synthesis of 3-dialkylaminoindoles. Intramolecular versions of this reaction are also possible, as illustrated by the formation of the α-carboline (**107**).

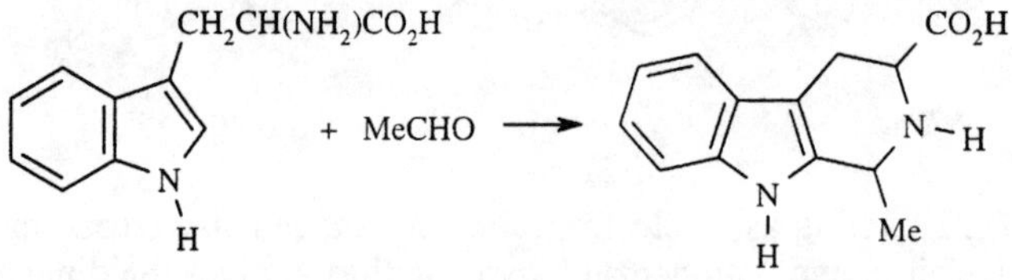

Secondary and tertiary aminoalkylbenzotriazoles (**108**) react with pyrrole, indole, and their *N*-methyl analogues under mild conditions in the presence of a Lewis acid to afford selectively the corresponding secondary or tertiary aminoalkyl derivatives (Scheme 22) ⟨92T4971⟩. *N,N*-Bis(alkoxymethyl)amines can also be used to give secondary alkylamines, which operate *via* reactive iminium salts formed by treatment with trimethylsilyl chloride ⟨90TL4229⟩.

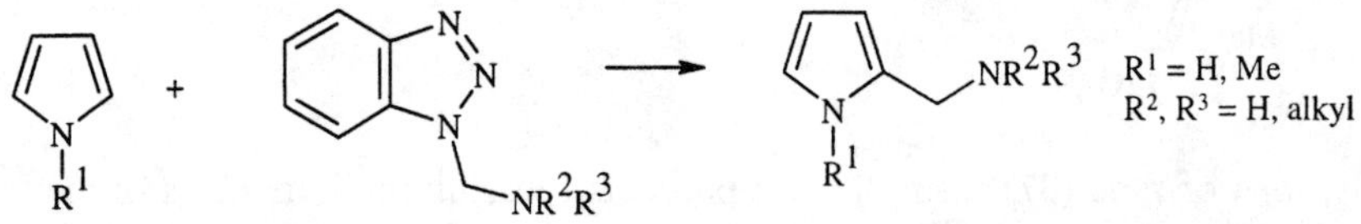

R^1 = H, Me
R^2, R^3 = H, alkyl

Scheme 22

3.3.1.5.8 Mercuration

Mercury(II) acetate tends to mercurate all the free nuclear positions in pyrrole, furan and thiophene to give derivatives of type (**109**). The acetoxymercuration of thiophene proceeds *ca.* 10^5 times faster than that of benzene. Mercuration of rings with deactivating substituents such as ethoxycarbonyl and nitro is still possible with this reagent, as shown by the formation of compounds (**110**) and (**111**). Mercury(II) chloride is a milder mercurating agent, as illustrated by the chloromercuration of thiophene to give either the 2- or 2,5-disubstituted product (Scheme 23).

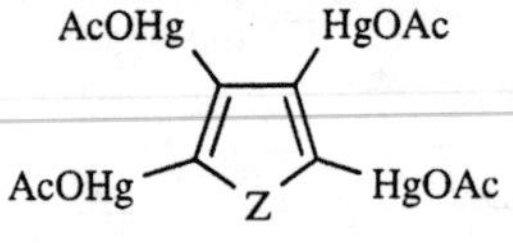

(**109**) Z = NH, O or S 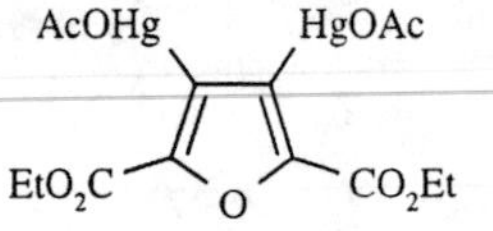   (**110**) 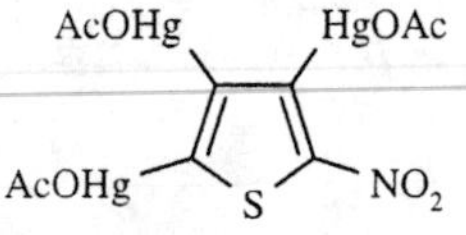   (**111**)

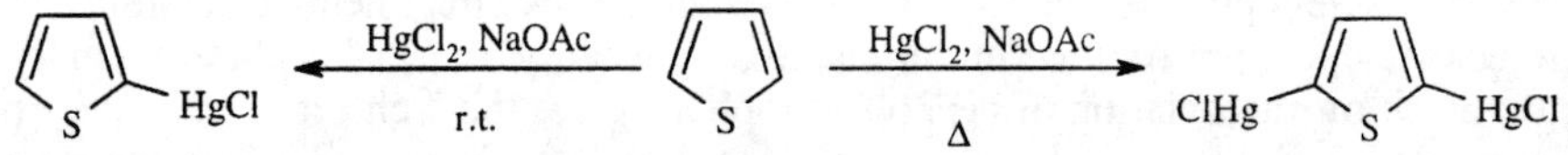

Scheme 23

3.3.1.5.9 Diazo coupling

Diazo coupling occurs readily between pyrroles and indoles and benzenediazonium salts. Reaction is much more rapid in alkaline solution when the species undergoing reaction is the *N*-deprotonated heterocycle. Depending on the conditions, pyrrole yields either 2-azo or 2,5-bis(azo) derivatives, *e.g.* (**112**) or (**113**), and indole gives a 3-substituted product (**114**).

An α-demethylated product (**116**) is formed unexpectedly when the tetrasubstituted pyrrole (**115**) is reacted with *p*-nitrobenzenediazonium chloride ⟨82JOC1750⟩. *N-t*-Butylisoindole (**117**) couples with *p*-nitrobenzenediazonium fluoroborate to give the hydrazone (**118**) ⟨80AG(E)320⟩.

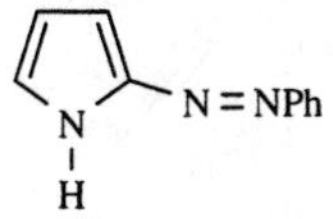

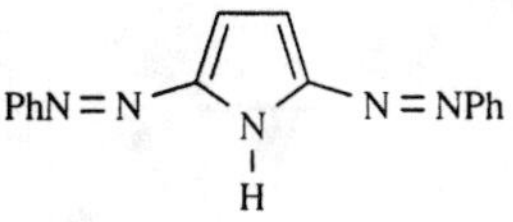

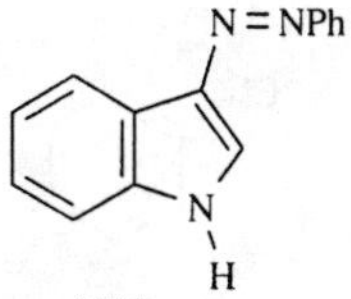

Furan undergoes phenylation rather than diazo coupling on reaction with benzenediazonium salts, and thiophene similarly yields 2- or 2,5-diaryl derivatives rather than coupled products (see Section 3.3.1.7.2). However, 2,5-dimethylfuran and 2-*t*-butylfuran give normal coupled products with 2,4-dinitrobenzenediazonium ion.

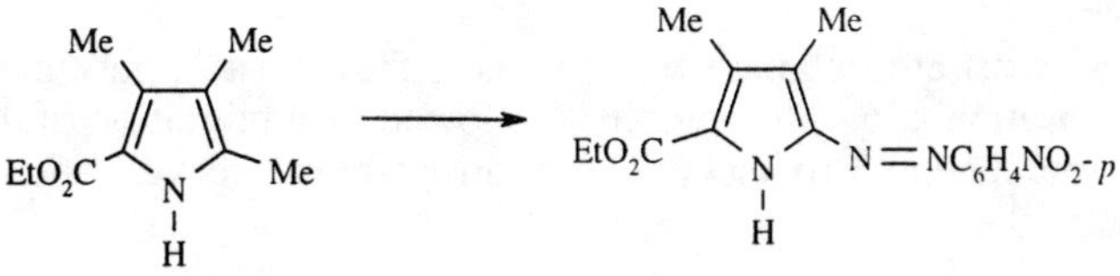

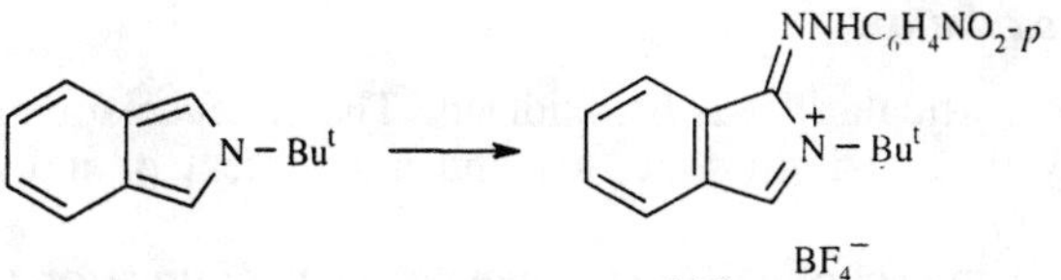

Intramolecular diazo coupling can be also achieved ⟨85JCS(P1)131⟩. Starting from suitable (*o*-aminophenyl)thiophenes various isomeric thieno[*c*]cinnolines have been obtained (Scheme 24). The reaction can proceed either at an α- or at β-position of the thiophene ring.

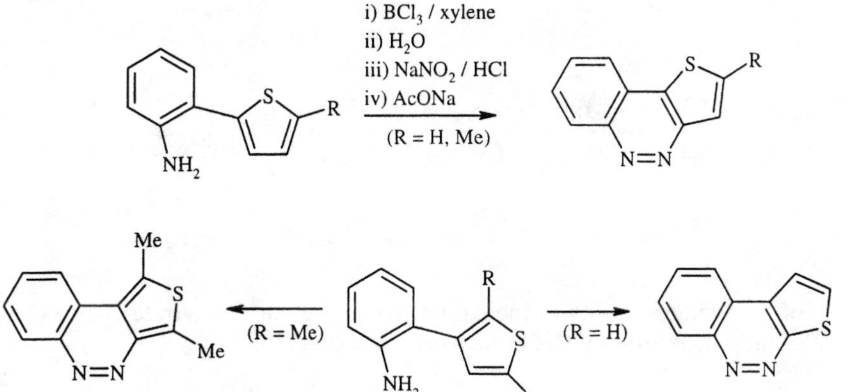

Scheme 24

3.3.1.5.10 Nitrosation

Nitrosation of pyrrole or alkylpyrroles may result in ring opening or oxidation of the ring and removal of the alkyl groups. This is illustrated by the formation of the maleimide (**119**) from 2,3,4-trimethylpyrrole.

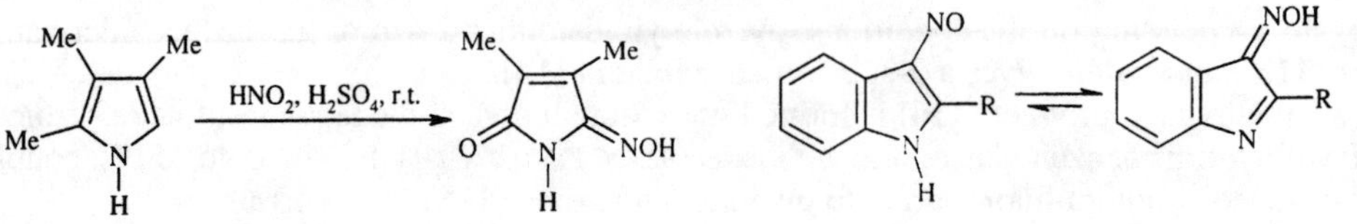

(119)   (120)   (121)

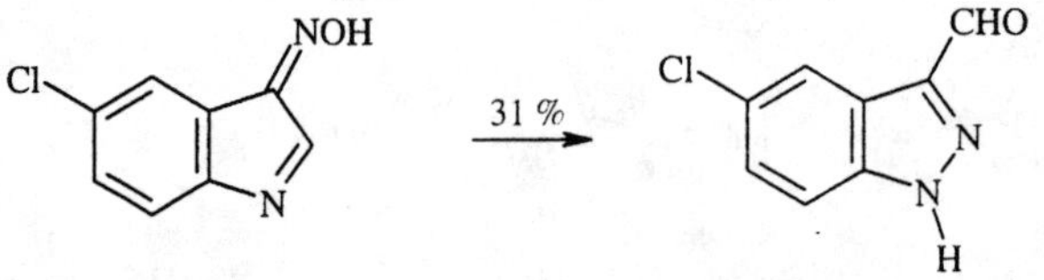

(122)   31 %   (123)

3-Nitroso derivatives (120) are obtained from indoles; they exist largely in oximino forms (121) ⟨80IJC(B)767⟩. The *N*-nitrosation of 5-chloroindole is followed by a migration of the nitroso group from N to C-3, to give an indolenine-3-oxime (122): hydrolysis and recyclization leads to a indazole carbaldehyde (123) ⟨86JA4115⟩.

3.3.1.5.11 Electrophilic oxidation

Pyrroles and furans are particularly easily oxidized. The mechanism of primary attack can be electrophilic, radical or cyclic transition state, and the assignment of individual reactions to these classes is sometimes arbitrary.

Simple pyrroles frequently give complex breakdown products. With strong oxidizing agents such as chromium trioxide in aqueous sulfuric acid, alkylpyrroles are converted into maleimides, *e.g.* (124). This oxidative technique played an important part in the classical determination of porphyrin structure. Milder oxidizing agents, such as hydrogen peroxide, convert pyrroles to pyrrolinones, *e.g.* oxidation of the parent heterocycle gives a tautomeric mixture of pyrrolin-2-ones (125) and (126).

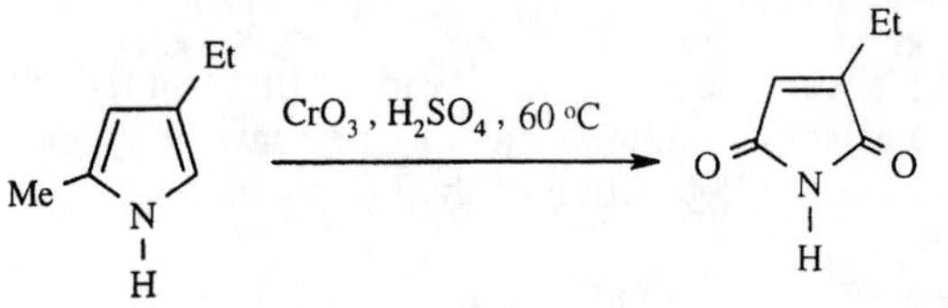

(124)

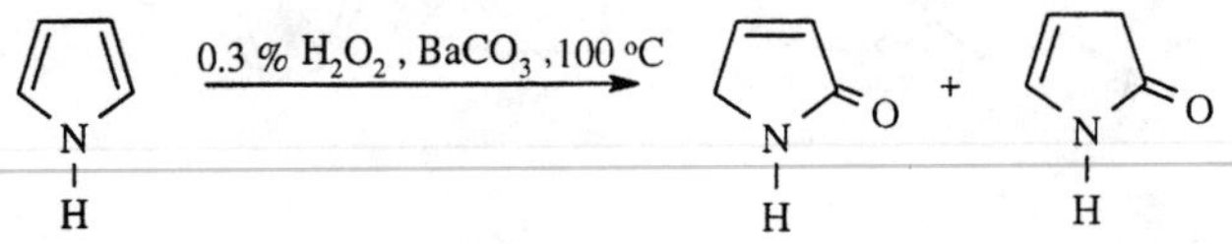

(125)   (126)

An example of the synthetic utility of the oxidative cleavage of the pyrrole C-2 to C-3 bond is the ring expansion of bicyclic pyrroles (127) to lactams (128) ⟨83S390⟩.

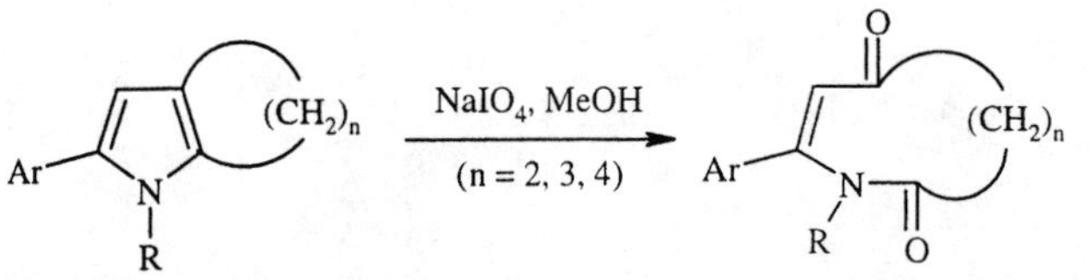

(127) R = H, Me   (128)

Bromine or electrolytic oxidation of furan in alcoholic solution gives the corresponding 2,5-dialkoxy-2,5-dihydrofuran (129, R = alkyl). Lead tetraacetate in acetic acid oxidation yields 2,5-diacetoxy-2,5-dihydrofuran (129, R = Ac).

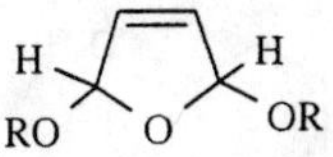

(129) *cis* and *trans*

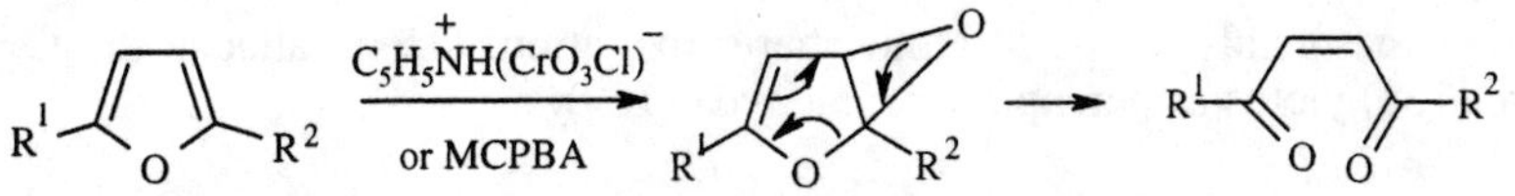

Scheme 25

Oxidation of 2,5-dialkylfurans with pyridinium chlorochromate results in high yields of α,β-unsaturated γ-dicarbonyl compounds (Scheme 25) ⟨82S245, 80T661⟩. Similar results are obtained by peracid oxidation of furans (see review ⟨90MI 206–01⟩). The acid most frequently used is MCPBA. It is assumed that the first step involves epoxidation as shown in Scheme 25 ⟨83CL1771⟩.

Oxidants and electrophilic reagents attack pyrroles and furans at positions 2 and 5; in the case of indoles the common point of attack is position 3. Thus, autoxidation of indoles (*e.g.* 130) gives 3-hydroperoxy-3*H*-indoles (*e.g.* 131). Lead tetraacetate similarly reacts at the 3-position to give a 3-acetoxy-3*H*-indole. Ozone and other oxidants have been used to cleave the 2,3-bond in indoles (132 → 133) ⟨81BCJ2369⟩.

The oxidation of 2,3-dimethylindole with peroxodisulfate and peroxomonosulfate yields 3-methylindole-2-carbaldehyde (136) ⟨88JCS(P2)1065⟩. The reaction proceeds in three steps, namely electrophilic attack of the peroxide at C-3 to give an indolenine intermediate (134), then peroxide attack on the enamine tautomer (135), and hydrolysis (Scheme 26) ⟨93JOC7469⟩.

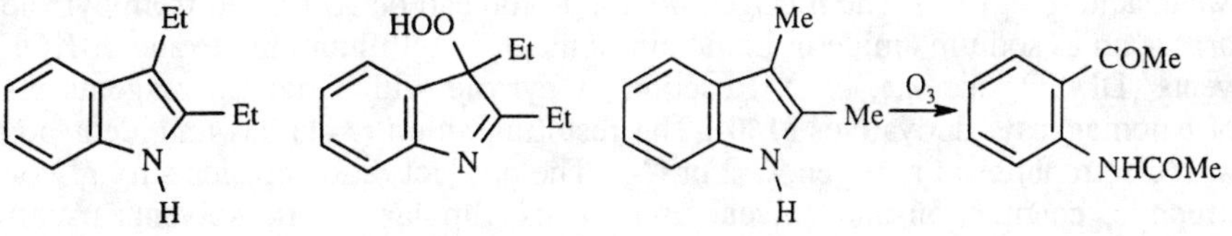

(130)   (131)   (132)   (133)

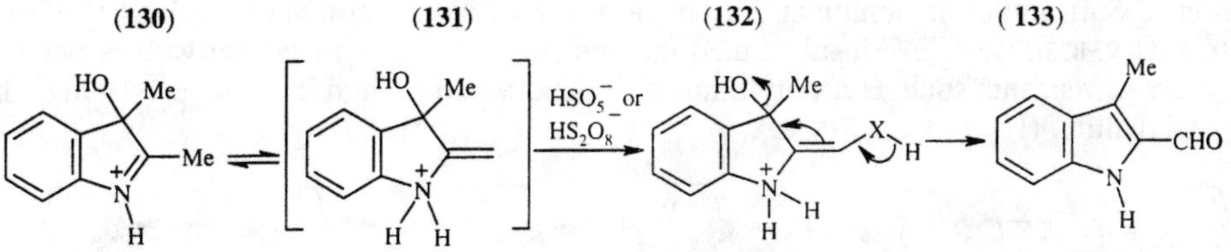

X = O or OSO$_3$

(134)   (135)   (136)

Scheme 26

Singlet oxygenation of 3-substituted indoles in the presence of alcohols followed by treatment with sodium borohydride gives 2-alkoxy-3-hydroxyindolines in high yields. Further reaction with a nucleophile and a Lewis acid forms the basis of a synthesis of 2-substituted indoles (Scheme 27) ⟨82CC977⟩. This represents an alternative approach to C-C bond formation at the 2-position of indoles to that involving the reaction of 2-lithioindoles with electrophiles. Isoindoles and indolizines are also preferentially oxidized in the five-membered ring to give phthalic acid and picolinic acid derivatives (137), respectively.

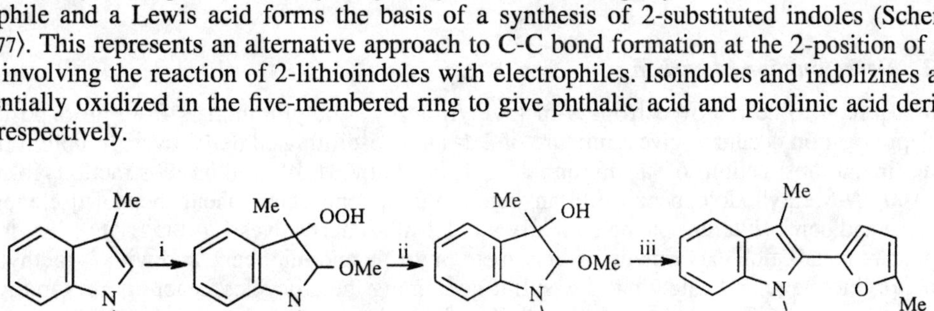

i, $^1O_2$, MeOH; ii, NaBH$_4$; iii, 2-methylfuran, CF$_3$SO$_3$SiMe$_3$

Scheme 27

(137)   (138)   (139)

Thiophenes are reasonably stable to atmospheric oxidation. Ozone attacks the C=C bonds, *e.g.* benzothiophene (138) yields *o*-mercaptobenzaldehyde (139).

3.3.1.6 Reactions with Nucleophiles

Pyrrole, furan and thiophene quite generally react with nucleophilic reagents by proton transfer, and not by substitution or addition. However, after discussion of proton loss from nitrogen and from carbon atoms we consider two major exceptions: (i) the cationic species formed by reaction with electrophiles are highly susceptible to nucleophilic attack and (ii) vicarious nucleophilic substitution and related reactions in rings containing strongly electron-withdrawing substituents.

Five-membered heterocycles also capable to ring-opening and recyclisation processes by nucleophile action. Well known examples include the interconversions of pyrrole, furan, thiophene and selenophene rings: *e.g.* furan and ammonia over a suitable catalyst can give pyrrole ⟨82KGS418⟩, ⟨80DOK1144⟩.

3.3.1.6.1 Deprotonation at nitrogen

Pyrrole is very much less basic than secondary amines but much more acidic. Pyrrole is, however, still a very weak acid (pK_a 17.5). The nitrogen-bound proton can be abstracted from pyrrole by the use of strong bases such as sodium amide in liquid ammonia, *n*-butyllithium in hexane or KOH in dipolar aprotic solvents (DMSO, acetone, *etc.*). Reaction of pyrrole with Grignard reagents results in the formation of halomagnesyl derivatives (140). The resulting anion (141) has ambident properties and may react with electrophiles at nitrogen, C-2 or C-3. The product ratio depends strongly on the nature of the electrophile, counter ion and solvent used. Thus, dipolar aprotic solvents usually favor *N*-substitution. *C*-Substitution is dominant only in case of reactive electrophiles, such as allyl or benzyl halides ⟨62JA43, 73JCS(P1)499⟩. *N*-Unsubstituted indoles similarly form metal derivatives when treated with a variety of reagents such as *n*-butyllithium, sodamide, potassium hydroxide ⟨81S461⟩, Grignard reagents and thallium(I) ethoxide ⟨81S389⟩.

(140)   (141a)   (141b)   (141c)

The reaction of the pyrrole and indole anions with electrophiles are discussed in Section 3.3.1.3.1.

3.3.1.6.2 Deprotonation at carbon

The first proton to be removed from *N*-methylpyrrole by *n*-butyllithium is from an α-position; a second deprotonation occurs to give a mixture of 2,4- and 2,5-dilithiated derivatives. In both furan and thiophene initial abstraction of a proton at C-2 is followed by proton abstraction from C-5 ⟨77JCS(P1)887⟩. *N*-Methylindole, benzo[*b*]furan, benzo[*b*]thiophene, selenophene, benzo[*b*]selenophene, tellurophene and benzo[*b*]tellurophene similarly yield 2-lithio derivatives ⟨77AHC(21)119⟩.

Competitive metallation experiments with *N*-methylpyrrole and thiophene and with *N*-methylindole and benzo[*b*]thiophene indicate that the sulfur-containing heterocycles react more rapidly with *n*-butyllithium in ether. The comparative reactivity of thiophene and furan with butyllithium depends on the metallation conditions. In hexane, furan reacts more rapidly than thiophene but in ether, in the presence of tetramethylethylenediamine (TMEDA), the order of reactivity is reversed ⟨77JCS(P1)887⟩.

Directive effects on lithiation have also been studied. The regiospecific β-metallation of *N*-methylpyrrole derivatives and 2-substituted furans has been effected by employing the directive effect of the oxazolino group ⟨82JCS(P1)1343⟩, but most 2-substituted furans and thiophenes are metallated in the 5-position. 2-Lithio-3-bromofuran forms on treatment of 3-bromofuran with lithium diisopropylamide (LDA) at − 80°C in THF ⟨77HCA2085⟩.

3-Methylthiophene is metallated in the 5-position whereas 3-methoxy-, 3-methylthio-, 3-carboxy- and 3-bromo-thiophenes are metallated in the 2-position ⟨80TL5051⟩. A major effort has been directed towards identification of 2-substituents which would specifically direct metallation to the 3-CH on the thiophene nucleus ⟨90CRV879⟩. The following demonstrated either fully or partially this ability: oxazoline, imidazoline, secondary carboxamide, carboxylate, sulfonamide and α-aminoalkoxide. The ultimate objective is to secure a high-yielding, regioselective route for the synthesis of 2,3-disubstituted thiophenes. Thus, treatment of the analogous imidazoline (**142**) ⟨91T9901⟩ with 2.5 equivalents of *n*-BuLi in THF at − 78°C for 2 hours gave regiospecific (98%) lithiation at position 3. In contrast to this, treatment of (**142**) with LDA (2.2 equiv.) in THF at − 78°C led only to the 5-lithio derivative (78%).

(**142**)   (**143**)   (**144**)   R = H or Ar

The oxazolinyl group directs metallation to the adjacent position. Thus, whereas a 2-furyloxazoline undergoes Friedel-Crafts reactions at the 4-position, metallation allows the introduction of the same functional groups at the 3-position ⟨92T149⟩.

The reactions of the lithiated derivatives are discussed in Section 3.3.3.8.

S-Arylbenzo[*b*]thiophenium ions (**143**) undergo ring opening by cleavage of the S − C_2 bond when treated with NaOMe in MeOH ⟨92CL1357⟩. With 3-unsubstituted compounds, the primary process is abstraction of the proton attached to C-3; subsequent cleavage of the S-C_2 bond results in the formation of acetylenes (**144**) in quantitative yield.

3.3.1.6.3 *Reactions of cationic species with nucleophiles*

Protonation of pyrrole, furan and thiophene derivatives generates reactive electrophilic intermediates which participate in polymerization, rearrangement and ring-opening reactions. Pyrrole itself gives a mixture of polymers (pyrrole red) on treatment with mineral acid and a trimer (**146**) under carefully controlled conditions. Trimer formation involves attack on the neutral pyrrole molecule by the less thermodynamically favored, but more reactive, β-protonated pyrrole (**145**). The trimer (**147**) formed on treatment of thiophene with phosphoric acid also involves the generation of an α-protonated species.

(**145**)   (**146**)

(**147**)

Indolizine gives a stable pyridinium ion and does not polymerize in the presence of acid. Indole undergoes acid-catalyzed dimerization; the 3*H*-indolium ion acts as an electrophile and attacks an unprotonated molecule to give the dimer (**148**). Protonation of the dimer in turn gives an electrophilic species from which a trimeric product can be derived ⟨77CPB3122⟩. *N*-Methylisoindole undergoes acid-catalyzed polymerization, indicating that protonation at C-1 gives a reactive electrophilic intermediate.

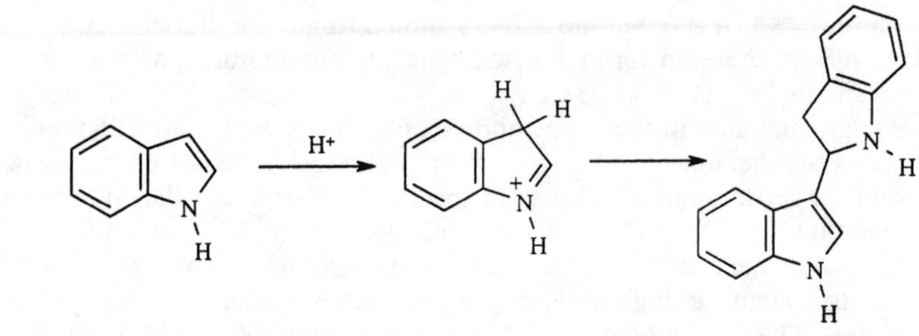

(148)

The chemical consequences of β-protonation are illustrated further by the ring-opening reactions of furans with methanolic hydrogen chloride and of *N*-substituted pyrroles with hydroxylamine hydrochloride (Scheme 28) ⟨82CC800⟩. Proton catalyzed ring opening followed by recyclization is involved in the interconversion of furan and selenophene derivatives using either aqueous hydrolysis followed by reaction with hydrogen selenide or directly using hydrogen selenide ⟨83KGS219⟩. These reactions require protonation at the β-position.

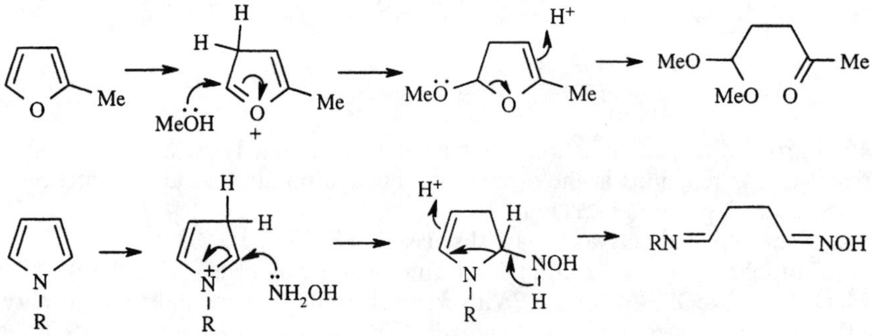

Scheme 28

The so-called ionic method for hydrogenating thiophenes ⟨78T1703⟩ is a further illustration of the chemical consequences of protonation. Protonation of the thiophene ring renders the ring susceptible to hydride ion attack, conveniently derived from triethylsilane (Scheme 29).

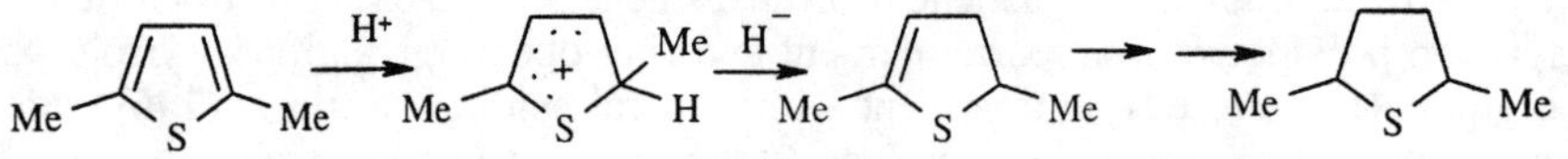

Scheme 29

Indolines are produced in good yield from 1-benzenesulfonylindoles by reduction with sodium cyanoborohydride in TFA at 0°C (Equation 5) ⟨89TL6833⟩. If acyl groups are present at C-2 or C-3 in the substrate, they are reduced to alkyl groups. Indole is also reduced to 2,3-dihydroindole by sodium cyanoborohydride and acetic acid or triethylamineborane and hydrochloric acid. An alternative method for preparing indolines involves treatment of indoles with formic acid (or a mixture of formic acid and ammonium formate) and a palladium catalyst ⟨82S785⟩. Reduction of the heterocyclic ring under acidic conditions probably involves initial β-protonation followed by reaction with hydride ion.

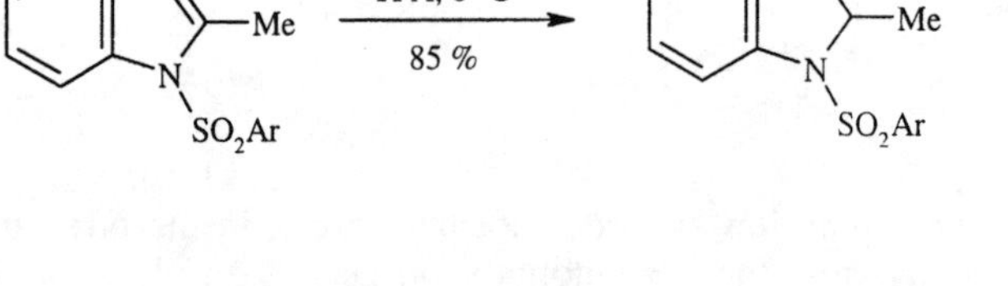

(5)

3.3.1.6.4 *Vicarious nucleophilic substitution and related reactions*

The vicarious nucleophilic substitution of hydrogen in nitrothiophenes is of great synthetic potential ⟨87ACR282, 91S103⟩. Carbanions bearing leaving groups (L) at the carbanion center can initially add to nitrothiophenes forming σ[H] adducts; subsequent elimination of HL, followed by protonation would lead to products which can formally be regarded as having been obtained by substitution of Nu for H. The carbanions are usually generated *in situ* by the action of base on the corresponding CH acids (Scheme 30). With 2-nitrothiophenes, usually H–3 of the thiophene is replaced, if the carbanion is CHLY, but with tertiary carbanions, some 5-substitution can also occur. 3-Nitrothiophene gives only the 2-substituted product. The base employed is usually KOBu[t], KOH, NaOH or NaH; the preferred solvents are DMSO, liquid ammonia or DMF.

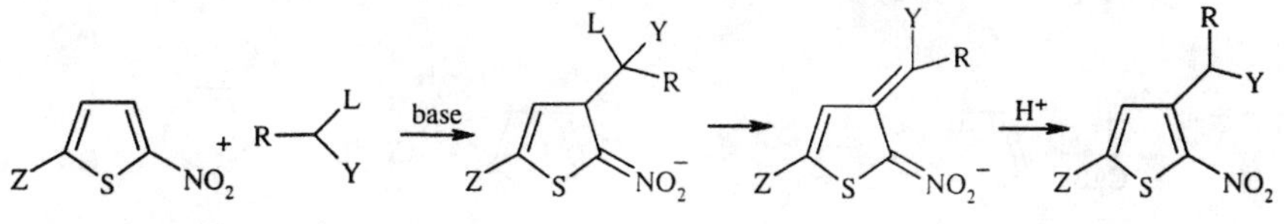

Z = H, Br, I, CN; L = Cl, OAr, SPh

Y = SO₂Ph, CN, CO₂R' *etc.*

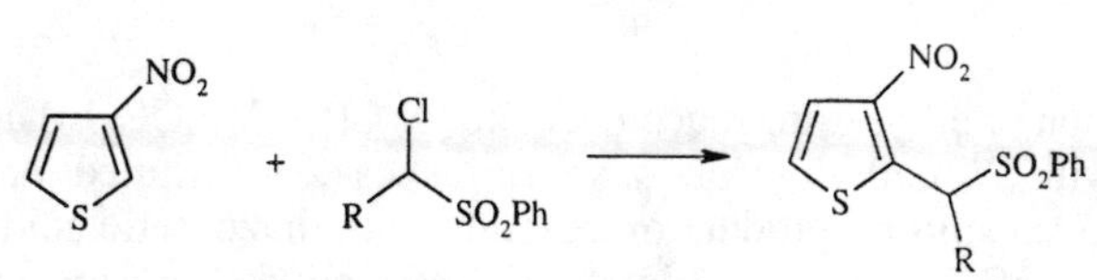

$$Z = H, Br, I, CN; \quad L = Cl, OAr, SPh$$

Scheme 30

In the thiophene ⟨75JHC327⟩ and benzothiophene ⟨78JOC4379⟩ series, nucleophilic substitution of hydride ion (Scheme 31) becomes possible due to the ability of 3d-orbitals of the sulfur atom to stabilize negative charge arising in the anionic σ-complex.

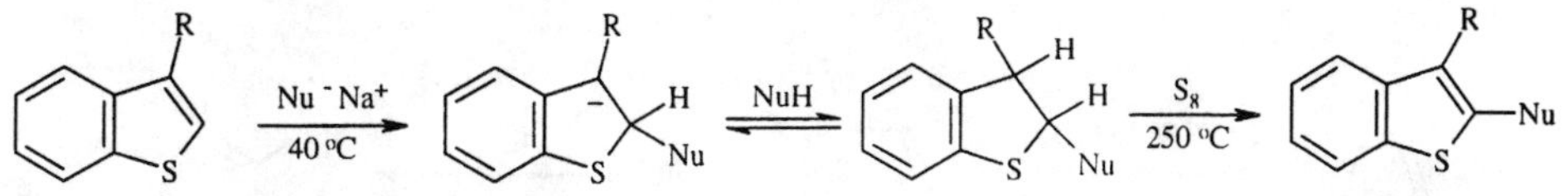

Nu = Et₂N, piperidino, morpholino

R = H, Me

Scheme 31

3.3.1.6.5 *Nucleophilic attack on sulfur*

Convincing evidence has been presented for the intermediacy of an organosulfur ate complex in which the sulfur is linked only to three carbon atoms ⟨92JA7937⟩. The strategy adopted was to synthesize a rigid molecule with ideal geometry favoring formation of the ate complex. The deuterium-labeled bromo compound (149) was converted to the lithio derivative (150). This resulted in equilibration of the isotope label in less than 1 min at −78°C in THF. This means that (150) must have been converted to (150') *via* the ate complex (151). This Li/S exchange at 0°C was 10⁹ times faster than the intermolecular Li/S exchange of Ph₂S with PhLi.

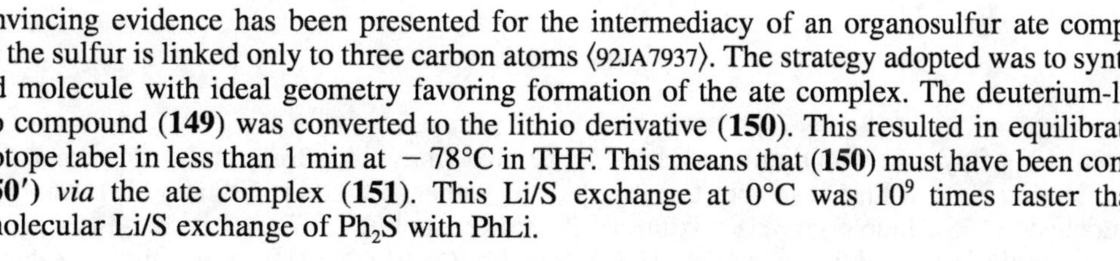

(149)　　　　　(150)　　　　　(151)　　　　　(150')

3.3.1.7 Reactions with Radicals and Electron-deficient Species; Reactions at Surfaces

3.3.1.7.1 *Carbenes and nitrenes*

Furan and thiophene undergo addition reactions with carbenes. Thus, cyclopropane derivatives are obtained from these heterocycles on copper(I) bromide-catalyzed reaction with diazomethane and light-promoted reaction with diazoacetic acid ester (Scheme 32). The copper-catalyzed reaction of pyrrole with diazoacetic acid ester, however, gives a 2-substituted product (152).

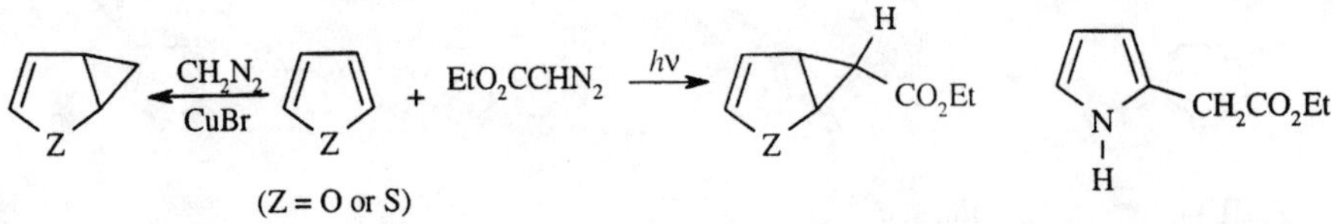

(Z = O or S)

(152)

Scheme 32

Copper-promoted reaction of ethyl 2-diazoacetoacetate [$N_2C(COMe)CO_2Et$] and dimethyl diazomalonate [$N_2C(CO_2Me)_2$] with *N*-methylpyrrole also furnishes 2-substituted derivatives ⟨82JOC3000⟩. Indole similarly yields a 3-substituted product on reaction with diazoacetic acid ester. In the rhodium acetate-catalyzed reaction of 2,5-dichlorothiophene with ethyl 2-diazoacetoacetate, the initial 2,3-cycloadduct (153) fragments with sulfur extrusion and subsequent rearrangement yields the dichlorophenol (154) ⟨81T743⟩. A mixture of cyclopropanes is obtained, however, from the copper(I) bromide-catalyzed reaction of 1-methoxycarbonylpyrrole with diazoacetic acid ester. This is thought to be indicative of the reduced aromaticity of a pyrrole substituted on nitrogen with an electron-withdrawing substituent (Scheme 33).

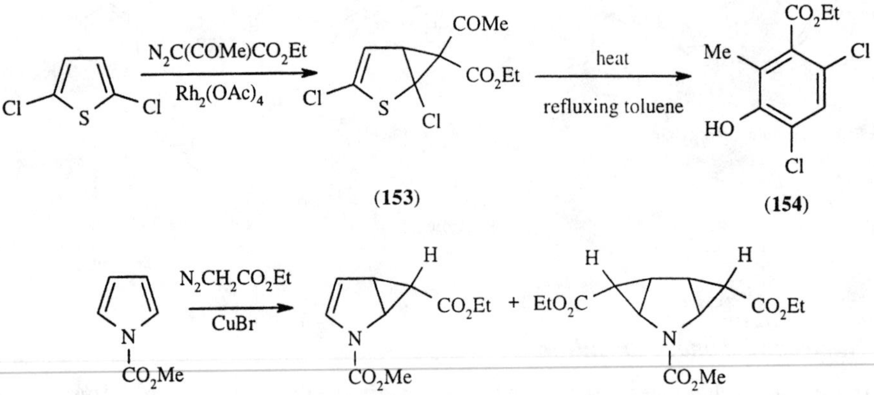

(153)   (154)

Scheme 33

The reaction of pyrrole with dichlorocarbene, generated from chloroform and strong base, gives a bicyclic intermediate which can be transformed to either 3-chloropyridine (155) or pyrrole-2-carbaldehyde (156). Indole gives a mixture of 3-chloroquinoline (157) and indole-3-carbaldehyde (158). The optimum conditions involve phase transfer ⟨76S249, 76S798⟩. Benzofuran reacts with dichlorocarbene in hexane solution to give the benzopyran (159), whereas benzothiophene fails to react.

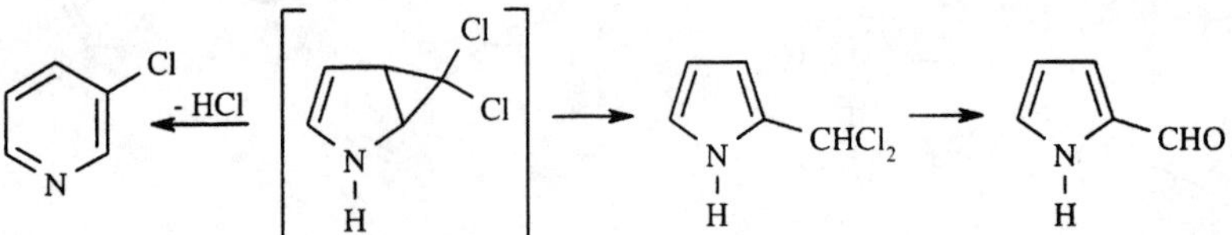

(155)   (156)

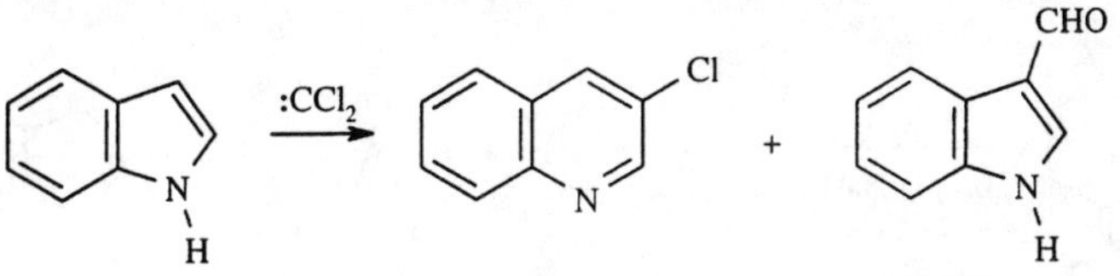

(157)   (158)

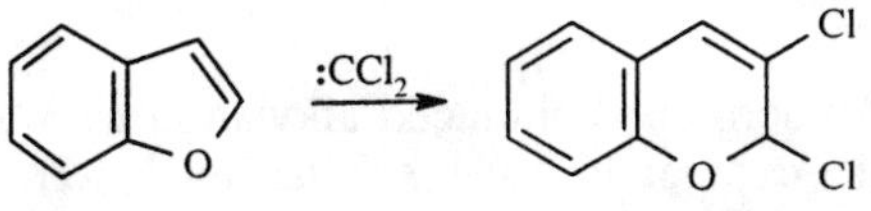

(159)

Furan undergoes 1,4-addition with ethoxycarbonylnitrene to give (**161**; Z = O) which rearranges to the pyrrolinone (**160**). The corresponding reaction with pyrrole gives a mixture of (**162a**) and (**162b**) ⟨64TL2185⟩. Ethoxycarbonylnitrene attacks thiophene and its simple alkyl derivatives at the α-position; subsequent ring opening and reclosure with loss of sulfur leads to pyrroles.

The formation of S,C- and S,N-ylides of thiophenes is discussed in Section 3.3.1.3.2.ii.

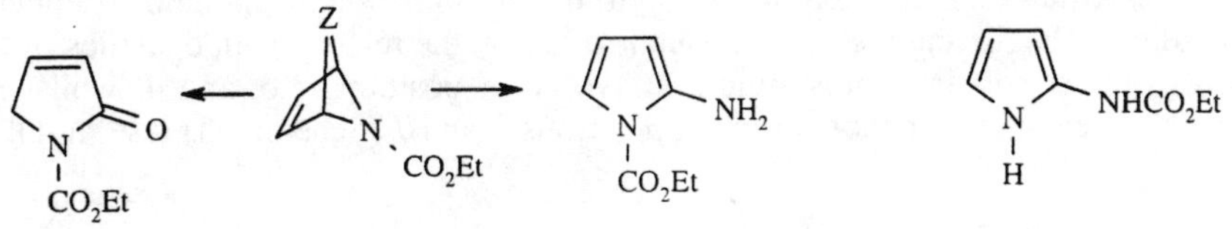

(160)   (161)   (162a)   (162b)

3.3.1.7.2 *Free radical attack*

Pyrroles, furans and thiophenes react preferentially with free radicals at the 2-position. Thus, reaction of pyrrole with benzyl radicals gives 2-benzylpyrrole. With triphenylmethyl radicals, pyrrole behaves like butadiene giving the adduct (**163**). *N*-Methylpyrrole undergoes free radical benzoyloxylation with dibenzoyl peroxide to give the 2-benzoyloxypyrrole (**164**) and 2,5-dibenzoyloxypyrrole (**165**). Furan, however, is converted in good yield to a mixture of *cis* and *trans* addition products analogous in structure to (**163**).

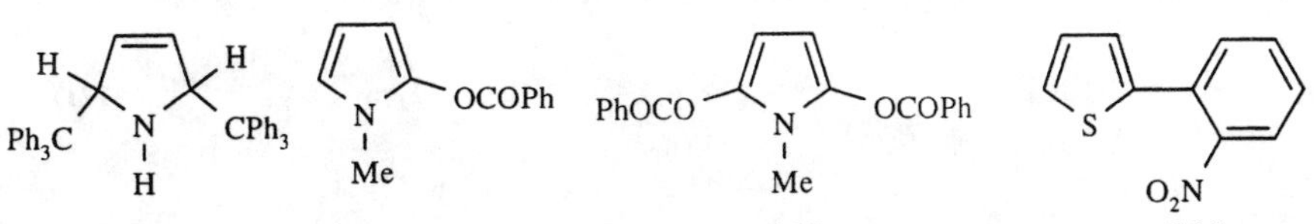

(163)   (164)   (165)   (166)

Arylation of *N*-substituted pyrroles, thiophenes and furans occurs preferentially in the 2-position, *e.g.* the *o*-nitrophenylation of thiophene by phase transfer catalysis yields (**166**) ⟨77TL1871⟩.

Thiophene reacts with phenyl radicals approximately three times as fast as benzene. Intramolecular radical attack on furan and thiophene rings occurs when oxime derivatives of type (**167**) are treated with persulfate ⟨81JCS(P1)984⟩. It has been found that intramolecular homolytic alkylation occurs with equal facility at the 2- and 3-positions of the thiophene nucleus whereas intermolecular homolytic substitution occurs mainly at position 2.

The homolytic substitution of thiophene by electrophilic carbon radicals provides a good method for the synthesis of (2-thienyl)acetic and (2-thienyl)propionic acids. The electrophilic radical, $\cdot CH_2CO_2Et$, generated from ICH_2CO_2Et, H_2O_2 and catalytic Fe^{2+} in DMSO, reacts with thiophene to form (2-thienyl)acetic ester in 62% yield ⟨92JOC6817⟩.

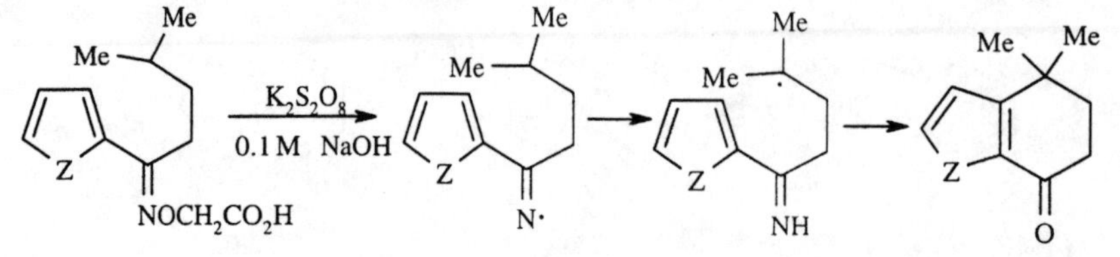

(**167**) Z = O or S

Pyrroles, furans and thiophenes undergo photoinduced alkylation with diarylalkenes provided that the alkene and the heteroaromatic compound have similar oxidation potentials, indicating that alkylation can occur by a non-ionic mechanism (Scheme 34) ⟨81JA5570⟩.

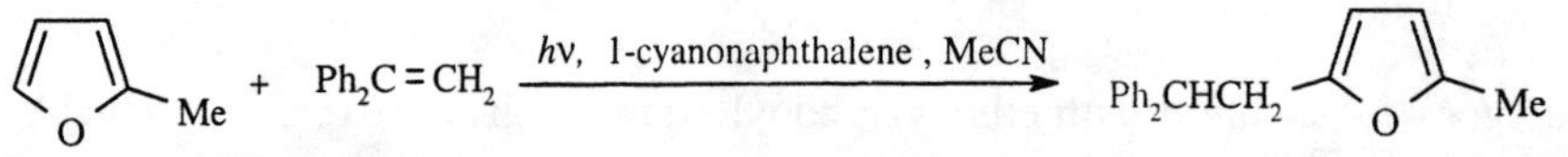

Scheme 34

Radical reactions are facilitated by the fact that pyrroles and indoles can form reasonably stable radical cations in some cases. For instance, photoarylation of indole by 2-iodopyridine is controlled by a photochemical electron-transfer reaction leading to the combination of the indole radical cation and the 2-pyridyl radical. The direction of attack is controlled by the relative spin densities of the possible radical cations. In polar solvents, substitution is favored at positions 3,6, and 4, whilst in nonpolar solvents, there is a preference for substitution at positions 2 and 7 (Scheme 35) ⟨88CPB940⟩.

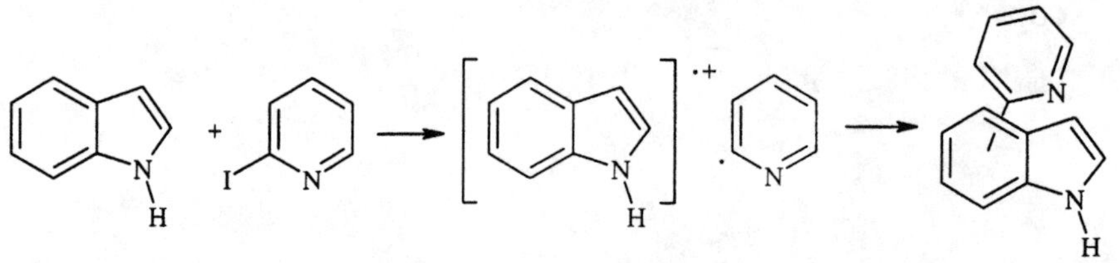

Scheme 35

1-Aroylpyrroles dimerize on treatment with palladium(II) salts; thus, oxidation of 1-benzoylpyrrole (**168**) with palladium acetate in acetic acid gives the 2,2′-bipyrrole (**169**). The ring-closed compound (**170**) is formed as a by-product ⟨81CC254⟩.

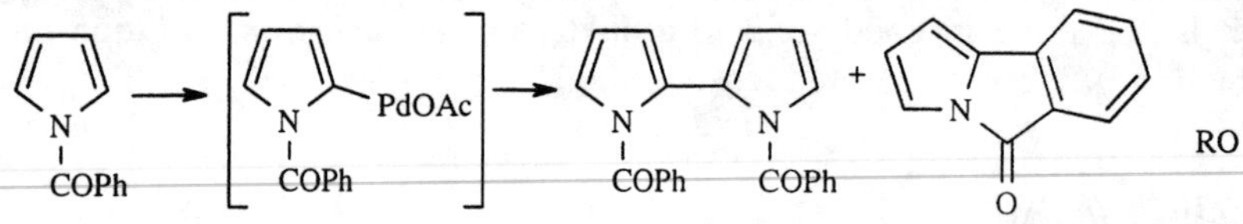

(**168**)   (**169**)   (**170**) 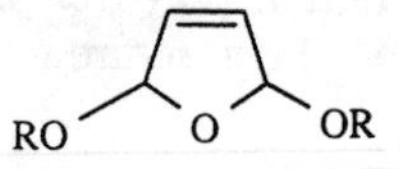   (**171**)

3.3.1.7.3 Electrochemical reactions

Electrolytic oxidation of furan in alcoholic solution gives the corresponding 2,5-dialkoxy-2,5-dihydrofuran (**171**).

3.3.1.7.4 Reactions with free electrons

Pyrroles are not reduced by sodium in liquid ammonia, but the Birch reduction of 2-furoic acid with lithium in liquid ammonia gives the 2,5-dihydro derivative in 90% yield ⟨78OPP94⟩. Sodium–liquid ammonia–methanol reduction of thiophene gives a mixture of Δ^2- and Δ^3- dihydrothiophenes together with butenethiols. Reductive metallation of 1,3-diphenylisobenzofuran results in stereoselective formation of the *cis*-1,3-dihydro derivative (Scheme 36) ⟨80JOC3982⟩.

Regioselective cleavage of dibenzofuran derivatives has been achieved with lithium metal, as exemplified by the preparation of 3-methyl-2-phenylphenol (**172**) ⟨80S634⟩.

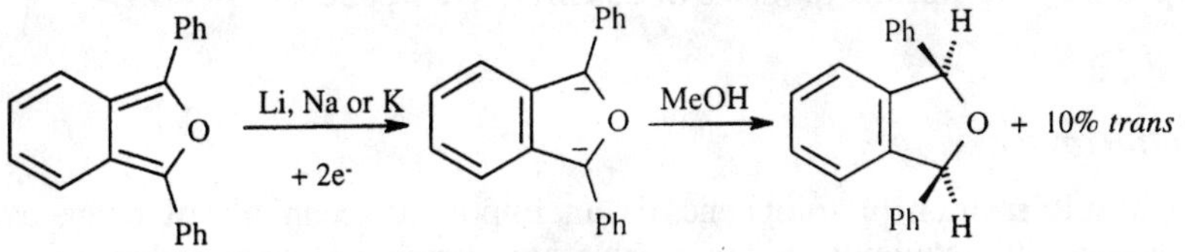

Scheme 36

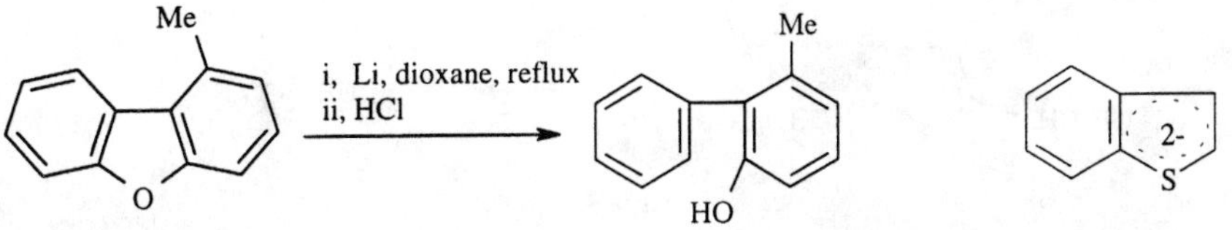

(**172**)   (**173**)

Benzo[*b*]thiophene dianion (**173**) has been prepared by reduction of benzo[*b*]thiophene with sodium metal at $-78°C$ in [$^2H_8$]THF. The ^1H and ^{13}C NMR spectra of the purple solution obtained prove that it is the dianion and not a radical anion. This is the first example of a sulfur-containing $(4n)\pi$ polycyclic dianion. The dianion undergoes oxidation to benzo[*b*]thiophene with oxygen ⟨85CC1033⟩.

3.3.1.7.5 *Catalytic hydrogenation*

Catalytic reduction of pyrroles gives successively Δ^3-pyrroles and pyrrolidines. Tetrahydrofurans are formed by the catalytic reduction of furans with Raney nickel and hydrogen; ring cleavage products may also be formed, *e.g.* (**174**) → (**175**) → $Me(CH_2)_2CH(OH)CH_2OH + Me(CH_2)_4OH$.

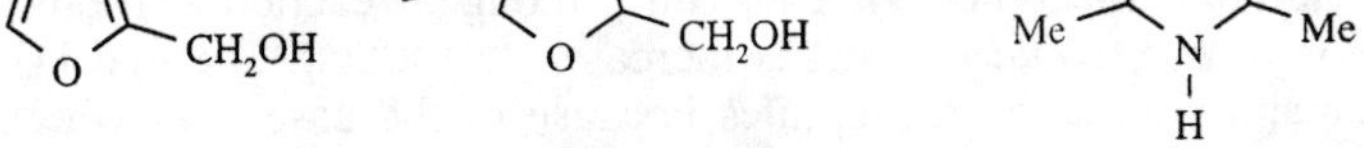

(**174**)   (**175**)   (**176**)

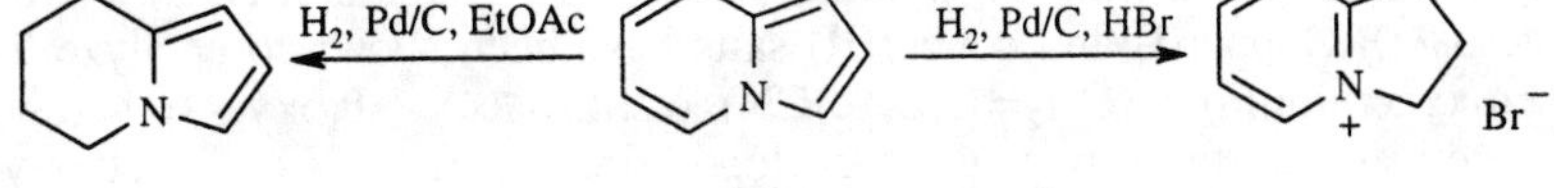

Scheme 37

Vigorous catalytic reduction of indole (H$_2$/Pd/AcOH/HCl, 80°C) results in the formation of *cis*-octahydroindole. Catalytic reduction of isoindoles occurs in the pyrrole ring. Reduction of indolizine with hydrogen and a platinum catalyst gives an octahydro derivative. With a palladium catalyst in neutral solution, reduction occurs in the pyridine ring but in the presence of acid, reduction occurs in the five-membered ring (Scheme 37).

3.3.1.7.6 *Reduction by dissolving metals*

Δ^3-Pyrrolines, *e.g.* (**176**), are formed on the reduction of pyrroles and simple alkylpyrroles with zinc and acid. These are derived from the corresponding α-protonated species ⟨66JA1335⟩.

Indole-2-carboxylic esters undergo reduction with magnesium in methanol to give the related indoline ⟨86TL2409⟩.

Dissolving metals reduce the heterocyclic ring of isoindoles.

Reduction of thiophene and 2-ethylthiophene to the corresponding 2,5-dihydrothiophenes can be carried out with zinc and trifluoroacetic acid. The mechanism is again thought to involve protonation of the thiophene ring followed by transfer of two electrons from zinc and a second proton from the acid ⟨80CC766⟩. The reduction of thiophenes with alkali metals in liquid ammonia can lead to 2,5-dihydrothiophenes, ring-opened products or desulfurized products ⟨85HC(44/1)457⟩.

3.3.1.7.7   *Desulfurization*

Raney nickel desulfurization of thiophenes is an important technique of chain extension. Ring fission is accompanied by saturation of the ring carbon atoms and chain extension by four carbon atoms is effected. This method has been widely used to prepare alkanes, ketones and carboxylic acids and their derivatives ⟨B-74MI30200⟩. An illustrative example is given in Scheme 38; see also CHEC 1.16 and CHEC-II(2)522.

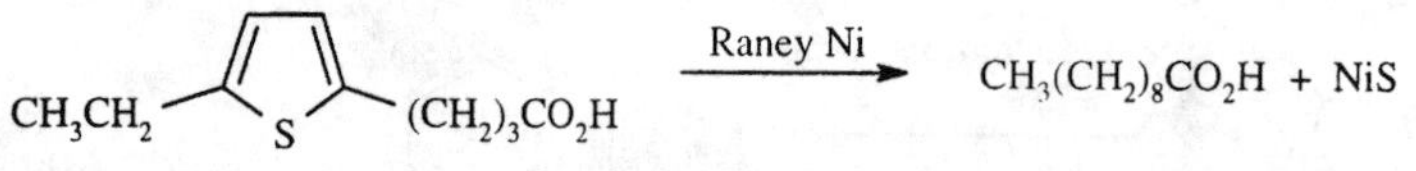

$$CH_3CH_2 \underset{S}{\bigotimes} (CH_2)_3CO_2H \xrightarrow{\text{Raney Ni}} CH_3(CH_2)_8CO_2H + NiS$$

Scheme 38

Nickel boride (prepared by adding $NaBH_4$ to nickel chloride in methanol − THF) is an efficient, non-pyrophoric reagent for the desulfurization of benzo[*b*]thiophene and dibenzothiophene ⟨90CC819, 92JOC1986⟩. The reaction proceeds under very mild conditions and is probably mediated by nickel hydride.

3.3.1.8   Reactions with Cyclic Transition States

3.3.1.8.1   *Heterocycles as inner ring dienes*

Furan has much greater reactivity in cycloaddition reactions compared to pyrrole and thiophene; the latter is the least reactive diene. *N*-Substituted pyrroles often show enhanced diene character compared with the parent heterocycle.

Furan reacts as a diene with powerful dienophiles like maleic anhydride, maleimide and benzyne to give Diels-Alder adducts. The kinetically favored products are the *endo* adducts but the *exo* adducts are thermodynamically preferred (Scheme 39). Thus, on prolonged reaction at room temperature, or on heating, the proportion of *exo* to *endo* adduct is increased ⟨78JOC518⟩. Thermal Diels-Alder reactions of furans are problematic with some dienophiles because of the ease with which they take part in competitive aromatic substitution reactions. Lewis acid catalysis often helps to avoid that problem. The reaction of furan with less reactive dienophiles such as acrylonitrile and methyl acrylate is greatly accelerated by zinc(II) iodide ⟨82TL5299⟩ The reaction of methyl acrylate with furan proceeds in good yield when catalyzed by boron trifluoride etherate ⟨84BCJ3339⟩, and was used in a synthesis of (±) avenaciolide ⟨86BCJ3881⟩. Copper(I) and copper(II) salts have been shown to catalyze the reaction of furan with α-acetoxyacrylonitrile ($CH_2 = C(OAc)CN$) ⟨82HCA1700⟩. Alkoxy ⟨81CC221⟩ and silyloxy groups ⟨80TL3423⟩ activate the furan nucleus to cycloaddition reactions, *e.g.* 3-methoxyfuran readily undergoes [4 + 2] cycloaddition with electron-deficient dienophiles with regio-and stereo-control to give *endo* adducts (Scheme 40).

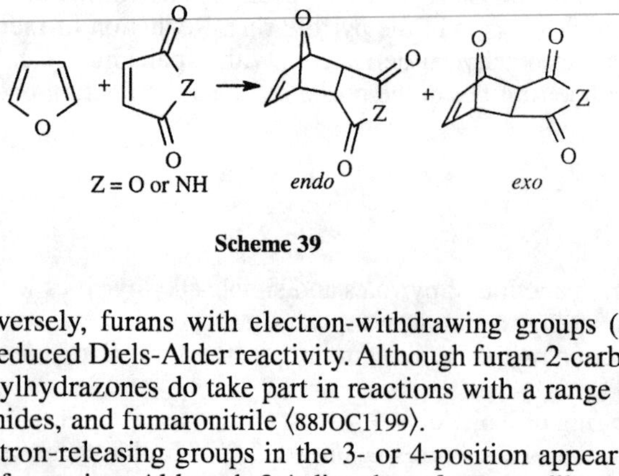

$Z = O$ or NH *endo*   *exo* $Y = CHO, COMe, CO_2Me$ or CN

Scheme 39

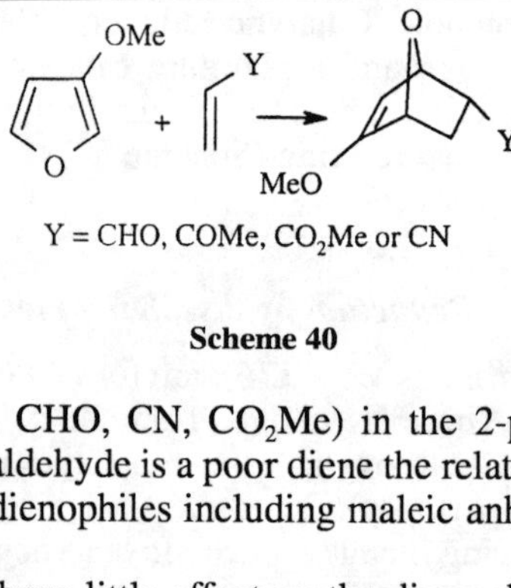

Scheme 40

Conversely, furans with electron-withdrawing groups (*e.g.* CHO, CN, CO_2Me) in the 2-position show reduced Diels-Alder reactivity. Although furan-2-carboxaldehyde is a poor diene the related *N*,*N*-dimethylhydrazones do take part in reactions with a range of dienophiles including maleic anhydride, maleimides, and fumaronitrile ⟨88JOC1199⟩.

Electron-releasing groups in the 3- or 4-position appear to have little effect on the diene character of the furan ring. Although 3,4-dimethoxyfuran readily undergoes [4 + 2] cycloaddition with maleic anhydride or methylmaleic anhydride, high pressure is required to carry out cycloaddition with dimethylmaleic anhydride (Scheme 41) ⟨82HCA1021⟩.

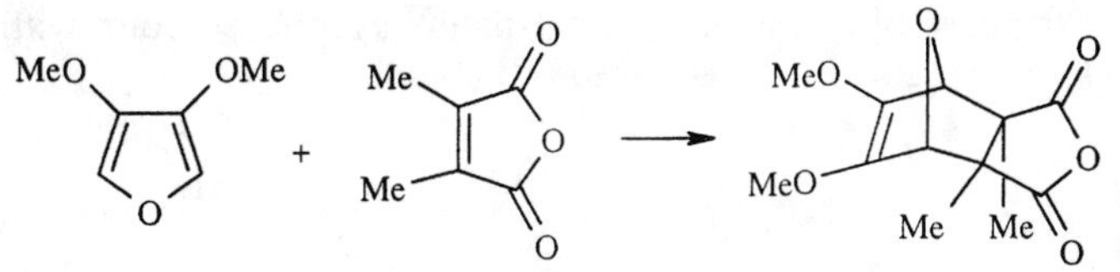

Scheme 41

A disadvantage of traditional Diels-Alder methodology is the ease with which the retro-reaction occurs in some cases, particularly when the reactions have to be conducted at high temperatures. The transition state in the Diels-Alder reaction occupies a smaller volume than the co-reactants and so rates of reactions are significantly increased by methods that produce compression. This can be achieved either by using high pressures ⟨83JOC3830, 83SC537⟩, including examples when 1,4-benzoquinone functions as the dienophile ⟨83HCA218, 83HCA222⟩, or by carrying out the reactions in water. The high pressure technique has been used in a large scale synthesis of cantharidin ⟨85JOC2576⟩, (+)-jatropholones A and B ⟨86JA3040⟩ and, as exemplified in Scheme 42, as a key step in an approach to the synthesis of furano heliangolides ⟨92JOC4512⟩. Diels-Alder reactions of 2-acetoxy- and 2-methylthiofuran proceed smoothly at room temperature and 15 Kbar ⟨87H(26)2347⟩.

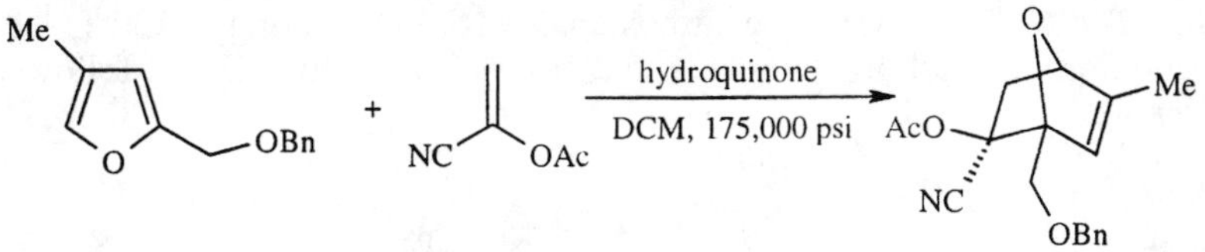

Scheme 42

The development of intramolecular Diels-Alder (IMDA) reactions for the synthesis of natural products ⟨80CR63⟩ and the facility with which substituted furans can be assembled have been reviewed ⟨84OR(32)1, 86CRV795, 87CSR187, 90SL186⟩. Thus, IMDA product from the propargyl sulfone shown in Scheme 43 proceeds *via* the allene and was converted subsequently into the benzenoid sulfone under base catalysis ⟨92CC735⟩.

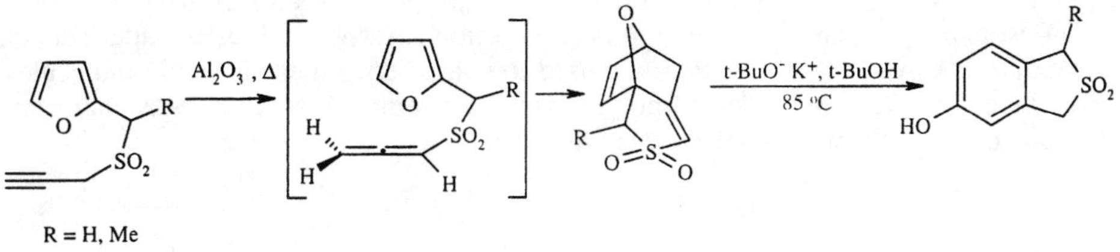

R = H, Me

Scheme 43

Although IMDA reactions are entropically less disfavored than the intermolecular versions they are nonetheless not as simple as it might appear at first. The well known "Alder endo rule" and its modern frontier molecular orbital theoretical interpretation involving secondary orbital interactions, together with steric considerations, serve to explain the kinetic preference for the *endo*-product and the thermodynamic preference for the *exo*-product in intermolecular Diels-Alder reactions. For the IMDA reaction an additional parameter, the ability of the tether that connects the diene to the dienophile to fold into the conformation required in the transition state, has to be considered. A model has been developed that is based on explicit starting material and product conformations using a combination of a conformational search program (WIZARD) and the energy calculating program MM2 ⟨92JA10738⟩.

Pyrroles and indoles can give a wide variety of cycloaddition reactions and this area has seen vigorous activity since 1990 because of the potential of methodology for the synthesis of complex indole alkaloids. The reactions of pyrroles with dienophiles generally follow two different pathways involving either a [4+2] cycloaddition or a Michael-type addition to a free α-position of the pyrrole ring. Pyrrole itself gives a complex mixture of products with maleic anhydride or maleic acid.

The reactions of pyrroles with dimethyl acetylenedicarboxylate (DMAD) have been extensively investigated. In the presence of a proton donor the Michael adducts (**177**) and (**178**) are formed. However, under aprotic conditions the reversible formation of the 1:1 Diels-Alder adduct (**179**) is an important reaction. In the case of the adduct from 1-methylpyrrole, reaction with a further molecule of DMAD can occur to give a dihydroindole (**180**) ⟨82H(19)1915⟩.

E = CO₂Me

(**177**)   (**178**)   (**179**)   (**180**)   (**181**)

N-Amino- and *N*-substituted amino-pyrroles readily undergo Diels-Alder additions and add to activated alkynes at room temperature.

Thiophene fails to undergo cycloaddition reactions with common dienophiles under normal conditions. However, when thiophene is heated under pressure with maleic anhydride, the *exo* adduct (**181**) is formed in moderate yield ⟨78JOC1471⟩ (see also review in ⟨85HC(44/1)671⟩).

Thiophenes can function either as a diene or as a dienophile in an intramolecular Diels-Alder reaction ⟨90CC405⟩. The *N*-(2-thienyl)allene carboxamide (**182**) on heating at 130°C leads to (**183**) by a [4 + 2] cycloaddition in which the thiophene functions as a 4π component, followed by the usual extrusion of sulfur.

(**182**)   (**183**)

Benzo[*b*]furans and indoles do not take part in Diels-Alder reactions. By contrast, the benzo[*c*]-fused heterocycles function as highly reactive dienes in [4 + 2] cycloaddition reactions. Thus, benzo[*c*]furan, isoindole (benzo[*c*]pyrrole) and benzo[*c*]thiophene all yield Diels-Alder adducts (**184**) with maleic anhydride. Adducts of this type are used to characterize these unstable molecules and in a similar way benzo[*c*]selenophene, which polymerizes on attempted isolation, was characterized by formation of an adduct with tetracyanoethylene ⟨76JA867⟩.

(Z = NH, O or S)   (**184**)   (**185**)

Benzo[*c*]furan, generated *in situ* in boiling xylene in the presence of dimethylmaleic anhydride, gives mainly the *exo* adduct; furan itself fails to react with this dienophile ⟨82JOC4011⟩. 1,3-Diphenylbenzo[*c*]furan is also a reactive diene but the corresponding 1,3-dimesityl derivative is inert to several dienophiles, even under forcing conditions ⟨82CC766⟩.

Furans react readily with benzynes, *e.g.* 2-acetoxyfuran yields (**185**). *N*-Methylpyrrole also reacts normally across the 2,5-positions, but pyrrole itself yields 2-phenylpyrrole,

Benzyne, generated from diphenyliodonium 2-carboxylate, reacts with various thiophenes by addition to the sulfur and β-carbon to give, after loss of an acetylene moiety, benzo[*b*]thiophenes in low (< 4%) yield (Scheme 44) ⟨81CC124⟩.

The photosensitized reaction of pyrrole and oxygen yields 5-hydroxy-Δ³-pyrrolin-2-one, probably by way of an intermediate cyclic peroxide (Scheme 45) ⟨76JA802⟩.

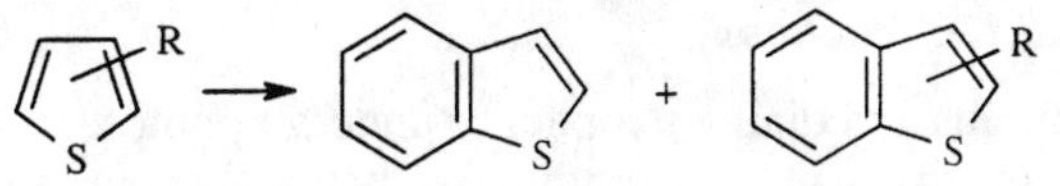

Scheme 44

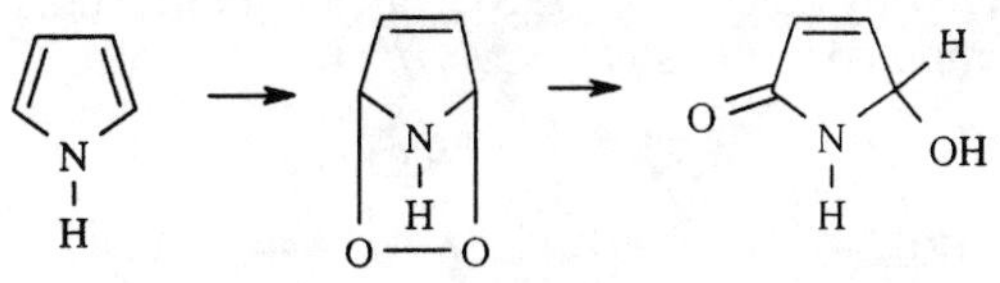

Scheme 45

Furans also form cyclic peroxides on reaction with singlet oxygen ⟨81CC720, 80JCS(P1)1955⟩; these undergo some interesting rearrangements as shown by the formation of the 2-aroyl enol esters (186) from the peroxides derived from 2-arylfurans ⟨82S736⟩.

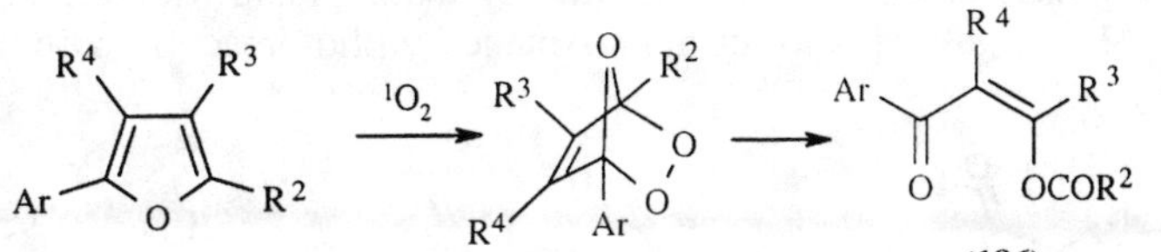

(186)

3.3.1.8.2 Heterocycles as dienophiles

Many five-membered, aromatic heterocycles act as dienophiles in Diels-Alder reactions, both with normal and inverse electron demands. Reactions with normal electron demand occurs only for compounds with electron-withdrawing groups in the heterocyclic ring. An old example is the formation (though only 10%) of a 2:1 adduct (187) on thermal reaction of 1,3-butadiene with furfural. Much better results can be obtained if the electron-withdrawing group is in the β-position. Thus, such derivatives of furan interact with isoprene producing with a good yield a mixture of isomers (188a) and (188b) though regioselectivity is low. In pyrroles and indoles, more than one electron-withdrawing group is required. Thiophenes, in analogy with their behavior as dienes, are poor dienophiles. 2-Thiophenecarboxaldehyde remained unchanged on being heated with 12 equivalents of isoprene at 195°C for 72 hours, while 3-thiophenecarboxaldehyde underwent cycloaddition in less than 6% yield ⟨88JA7188⟩.

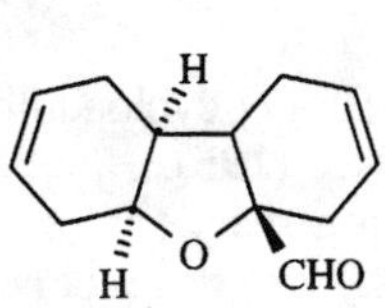

(187)

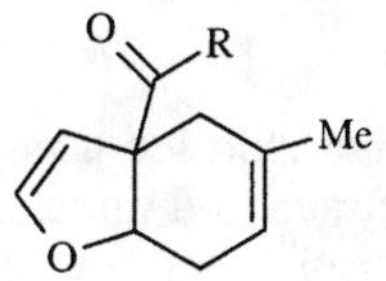

(188a)   R = H, OMe 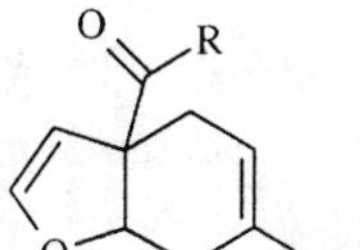   (188b)

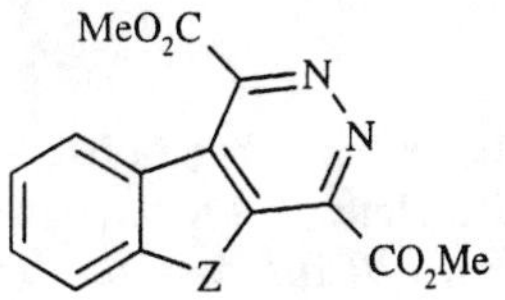

(189)   Z = NR, S

Pyrroles, indoles and benzo[*b*]thiophene act as good dienophiles in inverse electron demand Diels-Alder reactions with 1,2-diazines, 1,2,4-triazines and *symm*-tetrazines. This is examplified by the formation of compounds (189) in excellent yields on interaction of indoles and benzo[*c*]thiophene with dimethyl 1,2,4,5-tetrazine-3,6-dicarboxylate ⟨87JOC4610; 90JOC3257⟩. There are also many examples of such intramolecular reactions, *e.g.* (190 → 191).

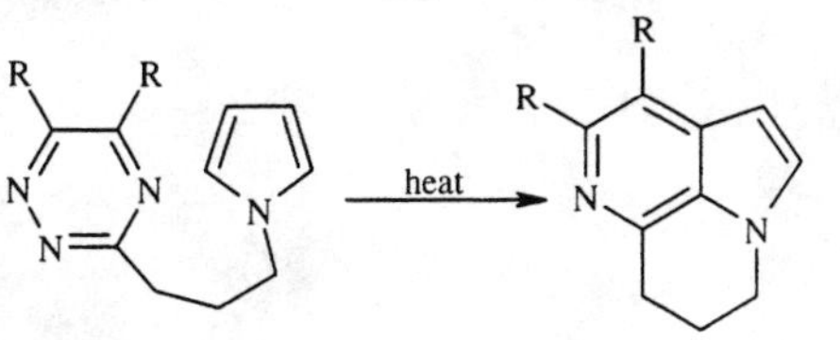

R = H, Me, CO₂Et

(190)   (191)

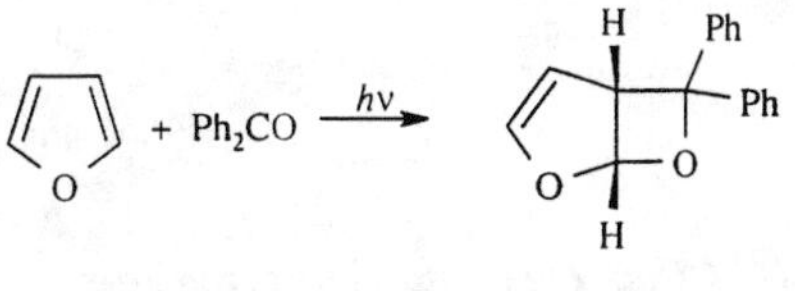

(192)

3.3.1.8.3 [2 + 2] *Cycloaddition reactions*

N-Substituted pyrroles, furans and dialkylthiophenes undergo photosensitized [2 + 2] cycloadditions with carbonyl compounds to give oxetanes. Furan and benzophenone give the oxetane (192). The photochemical reaction of pyrroles with aliphatic aldehydes and ketones results in the regiospecific formation of 3-(1-hydroxyalkyl)pyrroles (*e.g.* 193), *via* an intermediate oxetane which undergoes rearrangement under the reaction conditions ⟨79JOC2949⟩.

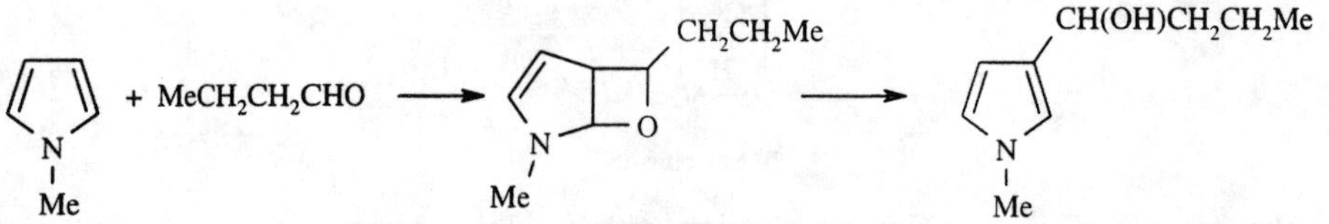

(193)

The photochemically induced [2 + 2] cycloaddition of selenophene with dimethylmaleic anhydride gives a 1:1 adduct (194), but attempts to form an oxetane by photoreaction with benzophenone failed ⟨80JHC1151⟩.

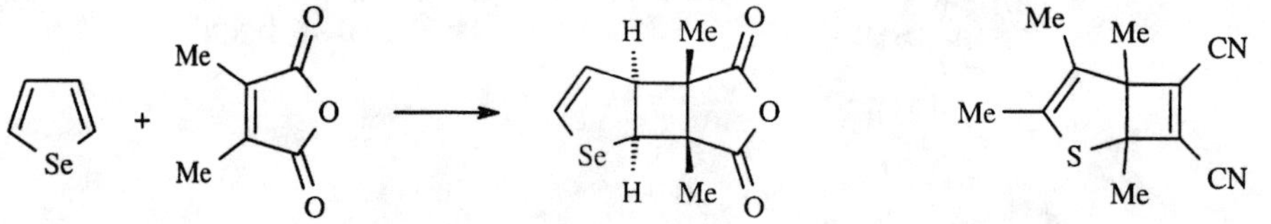

(194)   (195)

3-Aminothiophenes and 3-aminobenzo[*b*]thiophene undergo thermal [2 + 2] cycloaddition reactions with activated alkynes. The reactions are solvent dependent; thus, in non-polar solvents at − 30°C, 3-pyrrolidinothiophene adds to DMAD to give a [2 + 2] cycloadduct which is ultimately converted into a phthalic ester. In methanol, however, a tricyclic product is formed (Scheme 46) ⟨81JOC424⟩.

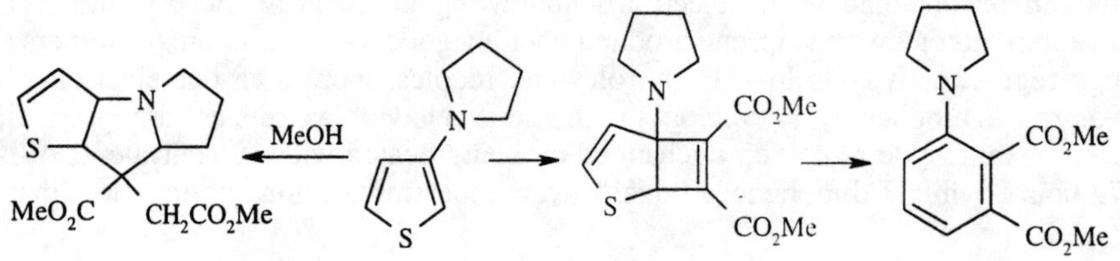

Scheme 46

Thiophenes have been observed to undergo aluminum chloride-catalyzed [2 + 2] cycloaddion with dicyanoacetylene ⟨82JOC967, 82JOC972⟩. Thus, 2,3,4,5-tetramethyl thiophene gives (195).

The benzo[*b*]-fused systems participate in a number of [2 + 2] cycloaddition reactions ⟨81JOC3939, 81TL521⟩. The photocycloaddition products of benzo[*b*]thiophenes and DMAD are dependent on the irradiation wavelength; at 330 nm (196) is formed, while at 360 nm the rearranged product (197) is produced.

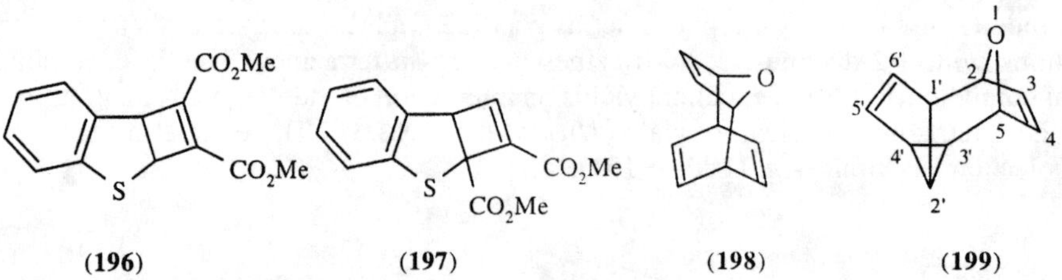

(196)   (197)   (198)   (199)

3.3.1.8.4 *Other cycloaddition reactions*

Photolysis of a mixture of furan and benzene gives mainly the [4 + 4] cycloadduct (198); a substantial amount of the adduct (199) derived by addition of carbons 2 and 5 of furan and 1 and 3 of benzene is also obtained ⟨81JOC2674⟩. Both phenylacetylene and benzonitrile undergo regio- and stereospecific 2,6,2′,5′-photocycloaddition to furan, illustrated in Scheme 47 for the benzonitrile reaction, to give the [4 + 3] cycloadducts ⟨85CC1057, 90JCS(P1)931⟩.

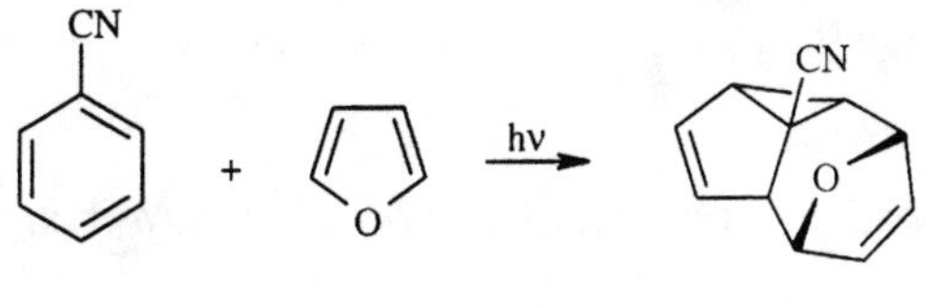

Scheme 47 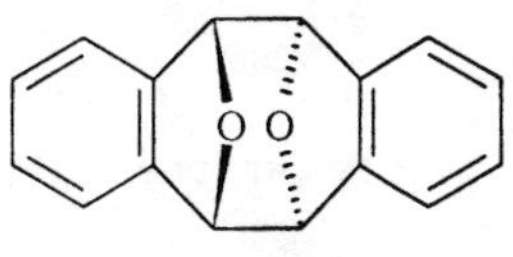   **(200)**

N-Methylisoindole and isobenzofuran give [8 + 8] photodimers. The [8 + 8] dimer (**200**) obtained from isobenzofuran at − 60°C has *anti* stereochemistry ⟨78HCA444, 82CC1195⟩.

Cycloaddition reactions of *C*-vinyl derivatives of five-membered heterocycles are discussed in Section 3.3.3.3.2.

3.3.2 REACTIVITY OF NON-AROMATIC COMPOUNDS

Before turning to the dihydro and tetrahydro derivatives of the fundamental ring systems, we deal with two special classes. The pyrrolenines and the thiophene sulfones both contain two double bonds in the heterocyclic ring, but in each case the conjugation does not include all the ring atoms. Finally we consider hydroxy derivatives, most of which exist predominantly as the non-aromatic carbonyl tautomers.

3.3.2.1 Pyrrolenines and Indolenines

Pyrrolenines and indolenines are much stronger bases than their aromatic analogues. Pyrrolenines readily undergo further alkylation to give quaternary salts (Scheme 48) and form stable hydrochlorides.

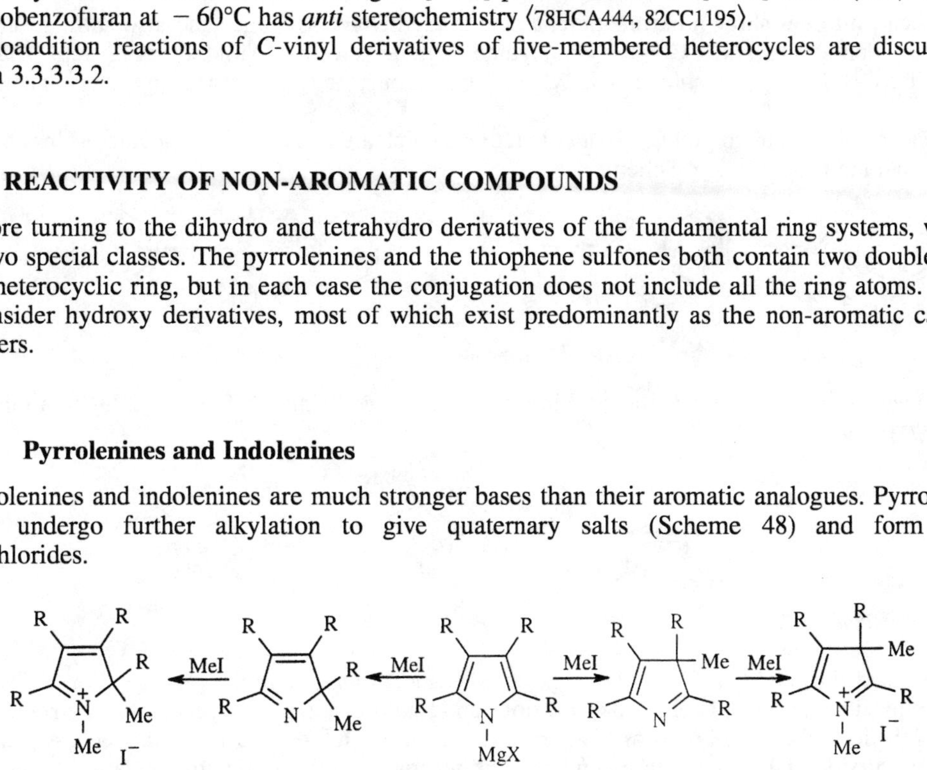

Scheme 48

The derived quaternary salts (*e.g.* **201**) give anhydro compounds (*e.g.* **202**) on treatment with alkali. β-Pentamethylpyrrolenine (2,3,3,4,5-pentamethylpyrrole) undergoes quantitative conversion to the α-isomer (2,2,3,4,5-pentamethylpyrrole) either on heating (> 200°C) or in 1M HCl at room temperature ⟨71CC1093⟩. Indolenines also undergoes an acid-catalyzed rearrangement (*e.g.* **203** → **204**), known as the Plancher rearrangement.

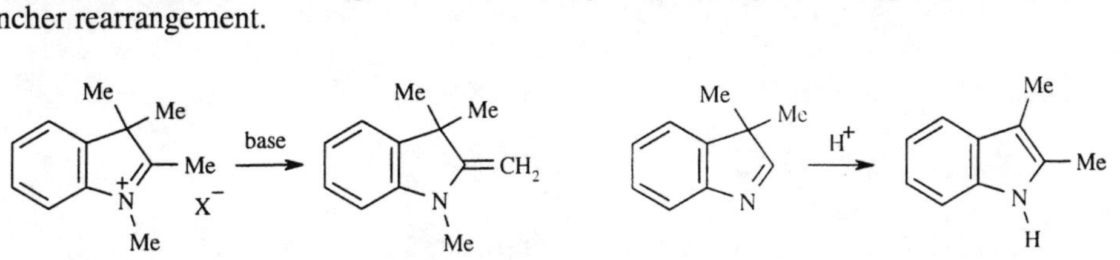

(201)   **(202)**   **(203)**   **(204)**

Reduction of indolenines with sodium and ethanol gives indolines. The pentachloro-α-pyrrolenine (**205**) is in equilibrium with small but finite amounts of the isomeric β-pyrrolenine form (3*H*-pyrrole; **206**), since it forms cycloadduct (**207**) with styrene ⟨80JOC435⟩. Pentachloropyrrole acts as a dienophile in its reaction with cyclopentadiene *via* its ene moiety ⟨81JOC3036⟩.

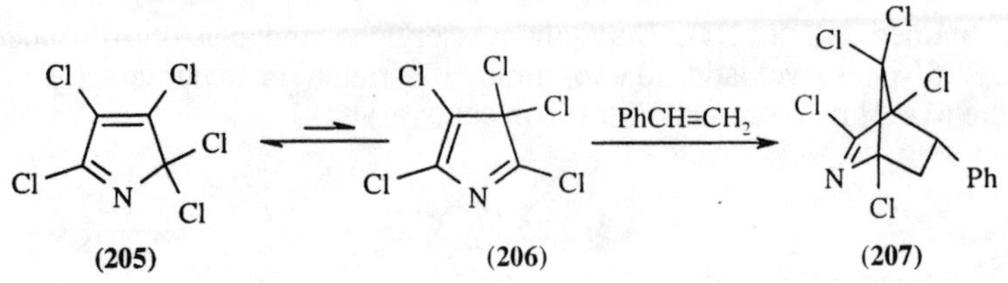

(205)   **(206)**   **(207)**

2*H*-Pyrrole 1-oxides undergo 1,3-dipolar cycloaddition with DMAD and with *N*-phenyl-maleimide ⟨80TL1833⟩.

3.3.2.2 Thiophene Sulfones and Sulfoxides

Thiophene sulfones show no aromatic character, they behave as dienes and also show reactions of compounds containing a C=C bond conjugated with an electron-withdrawing group. Thiophene sulfone itself is highly unstable, but alkyl and aryl groups and fused benzene rings increase the stability.

Thiophene sulfones undergo Diels-Alder reactions which are followed by spontaneous loss of sulfur dioxide from the products, *e.g.* Scheme 49.

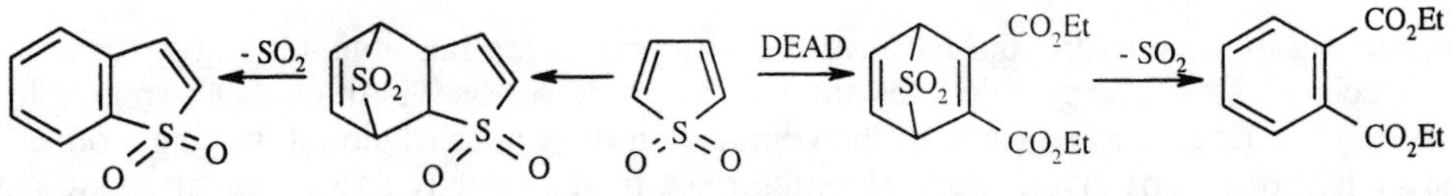

Scheme 49

An azulene synthesis involves the addition of 6-(*N*,*N*-dimethylamino)fulvene (**208**) to a thiophene sulfone ⟨77TL639, 77JA4199⟩.

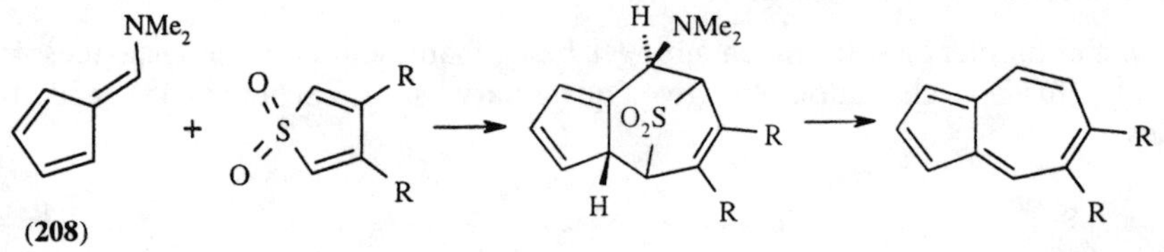

(208)

Halogenated thiophene sulfones (1,1-dioxides) are more stable than the parent sulfone. They have been employed as dienes in Diels-Alder reactions and found to add to a large variety of alkenic bonds, including the formal double bonds of *N*-methylpyrrole, furan and thiophene. The adducts subsequently lose sulfur dioxide and in some cases undergo further rearrangement and aromatization (Scheme 50) ⟨80JOC856, 80JOC867⟩. Reducing agents (*e.g.* Zn/HCl) convert thiophene sulfones to thiophenes; contrast the resistance to reduction of normal sulfones.

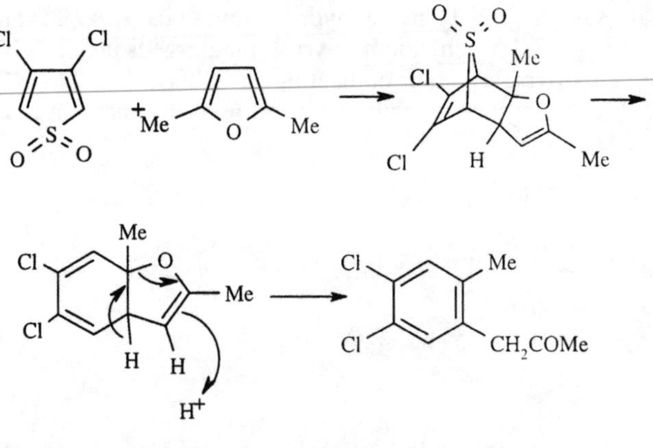

Scheme 50

Benzo[*b*]thiophene sulfone (**210**) reacts as a vinyl sulfone and forms adducts (**209**) and (**211**) when treated with mercury(II) acetate in methanol and with cyclopentadiene, respectively.

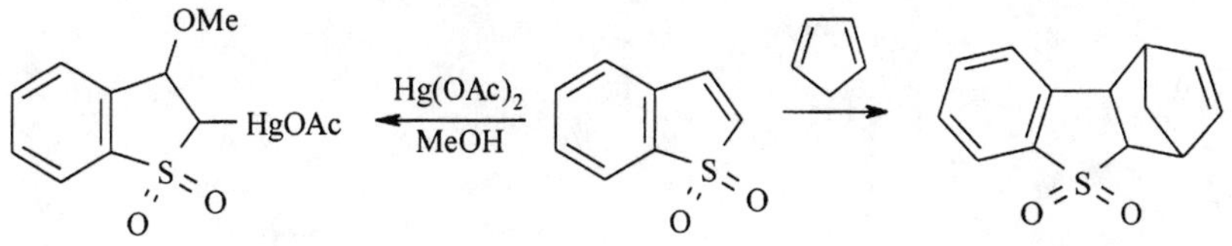

(209)   (210)   (211)

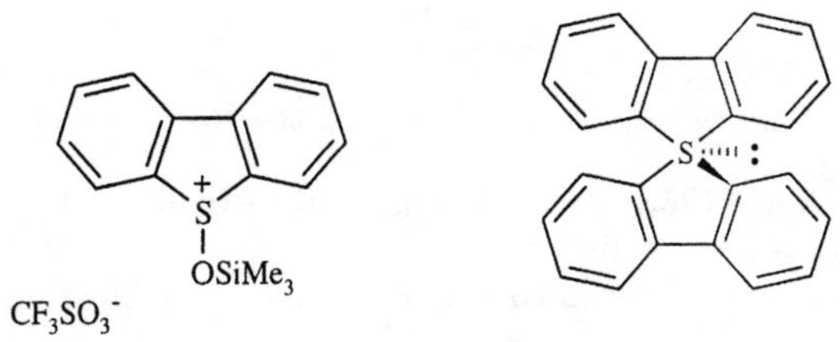

CF$_3$SO$_3^-$
(212)   (213)

The *O*-trimethylsilyl derivative (212) of dibenzothiophene *S*-oxide on treatment at $-78°C$ with 2,2′-dilithiobiphenyl gave the first stable tetracoordinated sulfur compound with four C-S bonds (213) in 96% yield ⟨92CC1141⟩.

3.3.2.3 Dihydro Derivatives

There are two possible types of dihydro-furans and -thiophenes (*cf.* 214, 215) and examples of both are known. There are three possible classes of dihydropyrroles: *N*-unsubstituted 2-pyrrolines (216) are in tautomeric equilibrium with the corresponding 1-pyrrolines (217).

The reaction of these dihydro compounds can be divided into three categories: a pronounced tendency to aromatize, behavior analogous to aliphatic compounds of similar functionally, and other reactions. These are considered in turn.

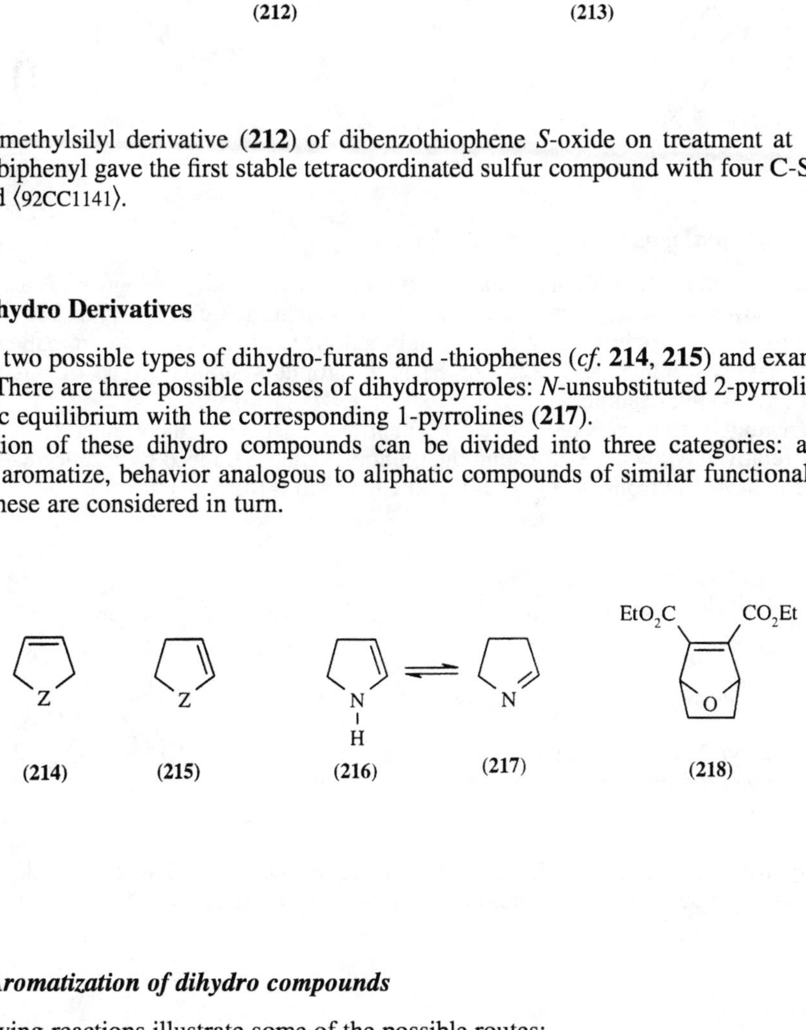

(214)   (215)   (216)   (217)   (218)

3.3.2.3.1 *Aromatization of dihydro compounds*

The following reactions illustrate some of the possible routes:

(a) Retro-Diels-Alder reaction: (218) on pyrolysis → ethyl 3,4-furandicarboxylate and ethylene. Similarly, azanorbornadienes with bulky or electron-withdrawing substituents undergo retro Diels-Alder extrusion of acetylene to give 3,4-disubstituted pyrroles. The adduct (219) from *N*-*t*-butoxycarbonylpyrrole and hexafluorobutyne with 2,4,6-trimethylbenzonitrile oxide gives a product mixture which undergoes smooth fragmentation to give mainly the 3,4-bis(trifluoromethyl)pyrrole (220) together with some *N*-*t*-butoxycarbonylpyrrole (221) ⟨82S313⟩. Alternatively, selective hydrogenation of adducts of type (219) at the less hindered double bond, followed by thermal extrusion of ethylene, provides a convenient route to 3,4-bis(trifluoromethyl)pyrroles ⟨82JOC4778⟩ (see CHEC 3.11.2.7.1). The cycloadduct from 2-acetoxyfuran and benzyne has been used in an analogous manner in the preparation of 1-acetoxybenzo[*c*]furan (222) ⟨81AJC1223⟩.

Ar = 2,4,6-trimethylphenyl

i, 2,4,6-trimethylbenzonitrile oxide; ii, heat, 160 °C

(b) Dehydrogenation: indoline (**223**; Z = NH) + chloranil → indole.
(c) Loss of acetic acid: (**224**) → 2-nitrofuran.
(d) Disproportionation: Δ^3-pyrroline heated over Pt → pyrrole + pyrrolidine.

3.3.2.3.2 *Behavior analogous to aliphatic analogues*

In many reactions, the dihydro compounds resemble their aliphatic analogues. Thus, when Z is nitrogen, (**225**) behaves as a benzylamine, (**223**; Z = NH) as an aromatic amine, and (**226**) as a Schiff's base. Similar comparisons apply when Z is oxygen or sulfur; (**223**; Z = O) is an aromatic ether, (**225**; Z = O) is a dibenzyl-type ether, and (**223**; Z = S) is an aromatic sulfide. Some of this behavior is illustrated by the following examples.

1-Pyrrolines readily form trimers of type (**227**). The trimer dissociates in boiling THF to 1-pyrroline; trimerization is relatively slow at − 78°C and the monomer can be trapped by reaction with acylating reagents to give *N*-acyl-2-pyrrolines, *e.g.* (**228**) with ClCO₂Me ⟨81JOC4791⟩.

N-Methoxycarbonyl-2-pyrroline undergoes Vilsmeier formylation and Friedel-Crafts acylation in the 3-position to give products of type (**229**) ⟨82TL1201⟩. At pH 7, two molecules of 2,3-dihydropyrrole add together to give (**230**), thus examplifying the dual characteristics of 2,3-dihydropyrroles as imines and enamines.

Both 2,3-dihydrofuran and 2,3-dihydrothiophene are converted into a [2 + 2] cycloadduct (**231**) on treatment with tetracyanoethylene under mild conditions ⟨80AG(E)831, 80AG(E)832⟩.

3.3.2.3.3 *Other reactions*

2,3-Dihydrothiophene and its 1-oxide undergo [2 + 3] cycloaddition with DMAD to give unstable sulfonium ylides (*e.g.* **232**). The latter undergoes ring expansion to give the thiepin 1-oxide (**233**) ⟨80AG(E)833⟩.

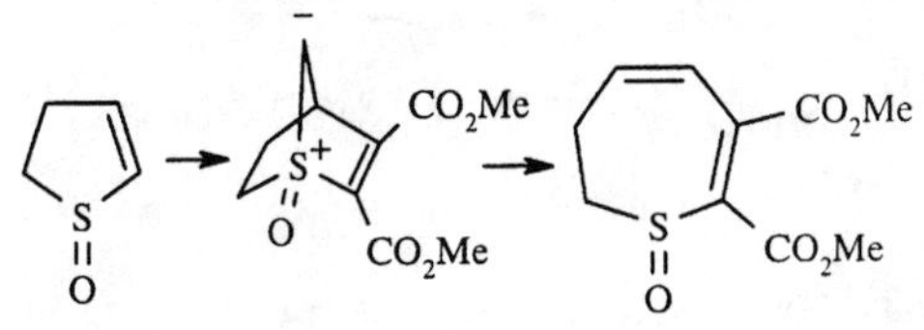

(232)   (233) 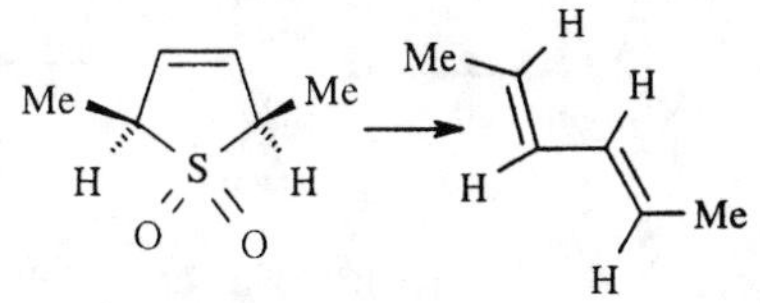   (234)   (235)

The thermal cheletropic extrusion of sulfur dioxide from both *cis* and *trans* isomers of 2,5-dihydrothiophene 1,1-dioxides is highly stereospecific. For example, *cis* 2,5-dimethyl-2,5-dihydrothiophene 1,1-dioxide (234) yield (*E,E*)-hexa-2,4-diene (235) and sulfur dioxide ⟨75JA3666, 75JA3673⟩.

The benzyne adducts prepared from *N*-methylpyrrole (and *N*-methylisoindole) are deaminated conveniently by dichlorocarbene generated under phase-transfer conditions ⟨81JOC1025⟩ to give a convenient route to substituted naphthalenes (236) (and anthracenes) (Scheme 51).

N-Amino- and *N*-substituted amino-pyrroles readily undergo Diels-Alder reactions and add to activated alkynes at room temperature. Loss of *N*-aminonitrene from the resulting adducts yields benzenoid derivatives (Scheme 52) ⟨81S753, 81TL1767⟩.

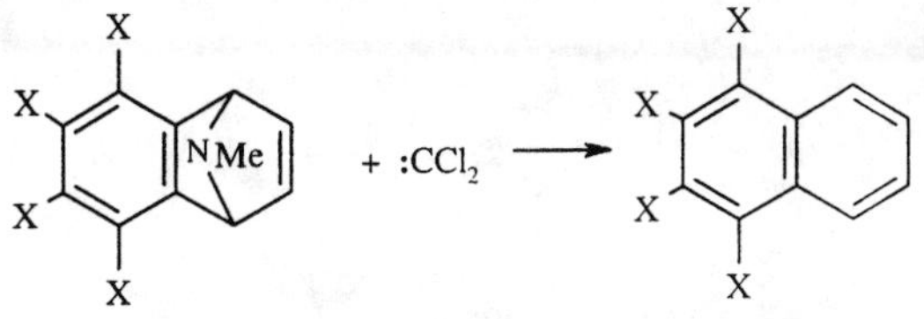

(X = Cl or F)   (236)

Scheme 51

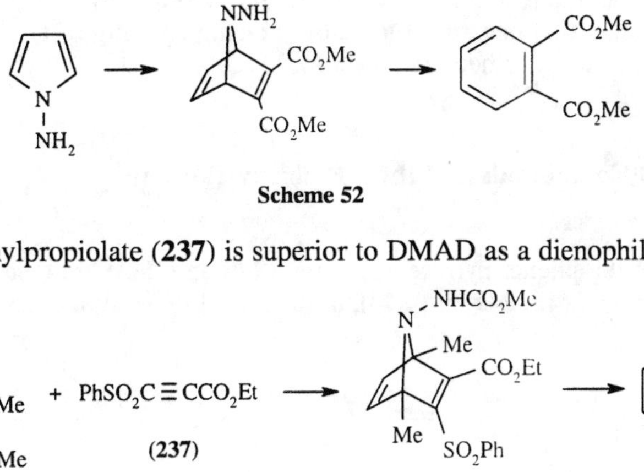

Scheme 52

Ethyl β-phenylsulfonylpropiolate (237) is superior to DMAD as a dienophile (Scheme 53).

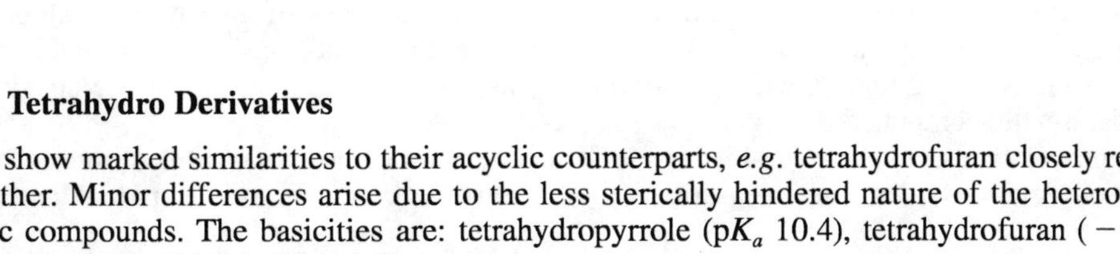

Scheme 53

3.3.2.4 Tetrahydro Derivatives

These show marked similarities to their acyclic counterparts, *e.g.* tetrahydrofuran closely resembles diethyl ether. Minor differences arise due to the less sterically hindered nature of the heteroatoms in the cyclic compounds. The basicities are: tetrahydropyrrole (pK_a 10.4), tetrahydrofuran (-2.1) and tetrahydrothiophene (thiolane) (-4.5).

Pyrrolidine readily forms an *N*-nitroso derivative. This can be lithiated in the 2-position, and subsequent reaction with electrophiles and deprotection yields 2-substituted pyrrolidines ⟨78OS(58)113⟩, as illustrated in Scheme 54. The transformation of tetrahydrothiophene 1,1-dioxide (238) into its 2-bromo derivative (239) is similar in principle. This involves deprotonation with ethylmagnesium bromide followed by electrophilic attack by bromine. Sodium ethoxide treatment of (239) gives buta-1,3-diene in 74% yield ⟨80LA1540⟩.

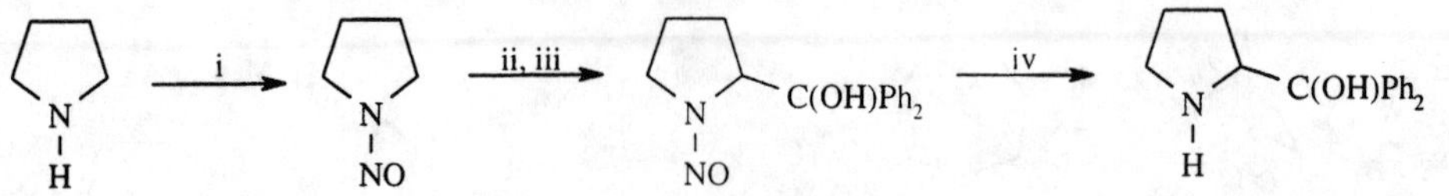

i , EtONO; ii , LDA , THF , -78°C; iii , Ph₂CO; iv, Raney Ni , H₂

Scheme 54

Although tetrahydrofuran is commonly used as a solvent in organometallic chemistry, it does undergo reaction with *n*-butyllithium. Proton-lithium exchange at an α-position is followed by cleavage to ethylene and the enolate anion of acetaldehyde. The half-life of α-lithio-THF is 10 min at 35°C (Scheme 55) ⟨72JOC560⟩.

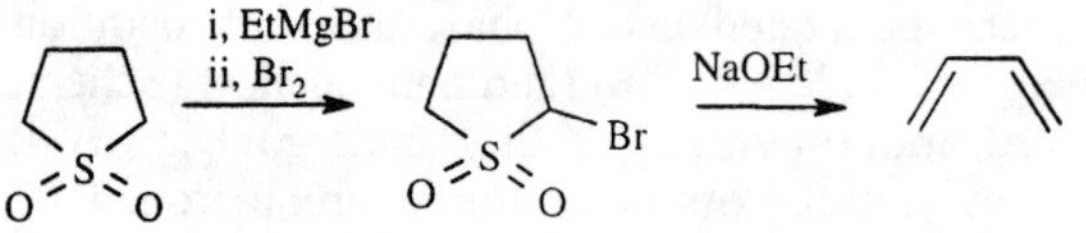

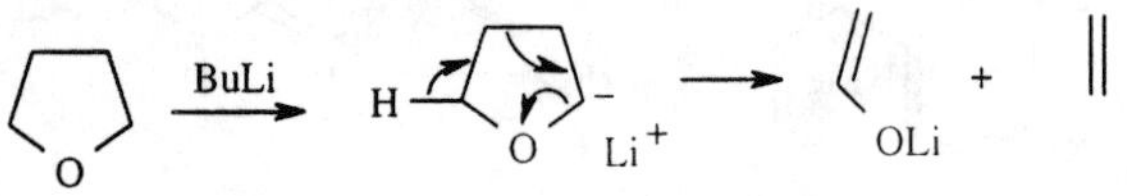

Scheme 55

Ring opening of tetrahydrofuran derivatives has been studied using chlorotrimethylsilane and sodium iodide ⟨87BSF505⟩. 2-Methyltetrahydrofuran is opened predominantly to give the primary iodide, 5-iodopentan-2-ol but the reaction involving 3-methyltetrahydrofuran is less selective. Lithium 4,4′-di-*t*-butylbiphenilide has also been used to cause the ring opening of tetrahydrofuran at − 80°C in the presence of boron trifluoride etherate ⟨91JA1866⟩.

3.3.2.5 Ring Carbonyl Compounds and their Hydroxy Tautomers

3.3.2.5.1 *Survey of structures*

Hydroxy derivatives of thiophene, pyrrole and furan (**240** and **243**) are tautomeric with alternative non-aromatic carbonyl forms (**241, 242** and **244**), as discussed in Section 2.3.5.2.

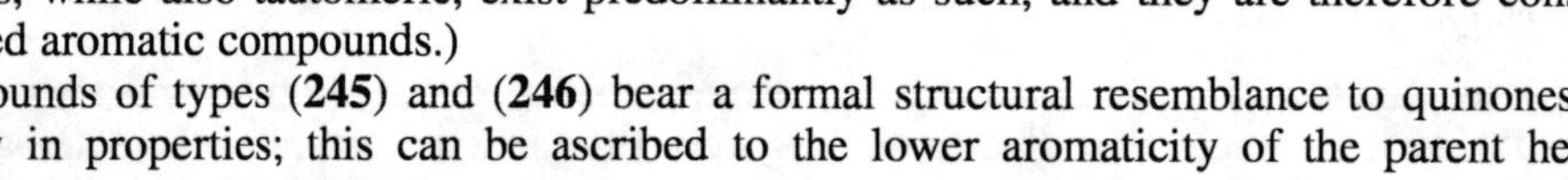

In the majority of cases, the equilibrium lies predominantly in favor of the carbonyl tautomer, and for this reason these compounds are considered in the present section. (Most amino and mercapto analogues, while also tautomeric, exist predominantly as such, and they are therefore considered as substituted aromatic compounds.)

Compounds of types (**245**) and (**246**) bear a formal structural resemblance to quinones but little similarity in properties; this can be ascribed to the lower aromaticity of the parent heterocyclic systems.

3.3.2.5.2 *Interconversion and reactivity of tautomeric forms*

Interconversion of the hydroxy and carbonyl forms of these heterocycles proceeds through an anion (as **247**) or a cation (as **248**) just as the enol (**249**) and keto forms (**252**) of acetone are interconverted through the ions (**250**) or (**251**). Reactions of the various species derived from the heterocyclic compounds are analogous to those of the corresponding species from acetone: hydroxy forms react with electrophilic reagents (**253**) and carbonyl forms with nucleophilic reagents (**254**). In addition, either form can lose a proton (**255**, **256**) to give an anion which reacts very readily (more so than the parent heterocycles) with electrophilic reagents on either oxygen (**257**) or carbon (**258**).

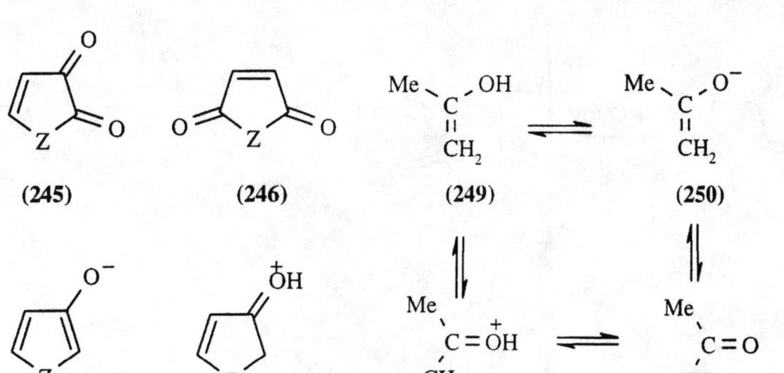

(245)   (246)   (249)   (250)

(247)   (248)   (251)   (252)

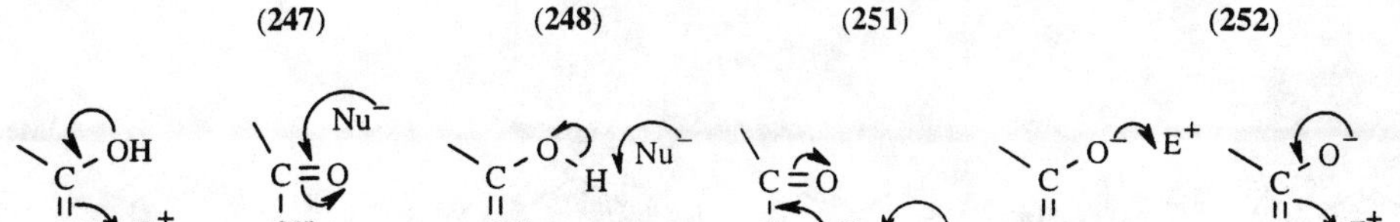

(253)   (254)   (255)   (256)   (257)   (258)

3.3.2.5.3 *Reactions of hydroxy compounds with electrophiles*

Electrophilic substitution reactions at low pH values probably involve the hydroxy form:

(a) Nitrosation (NaNO$_2$-H$_2$O-HCl) to give tautomeric products, *e.g.* (**261**) → (**260**)⇌(**259**), Z = NH, O, S.

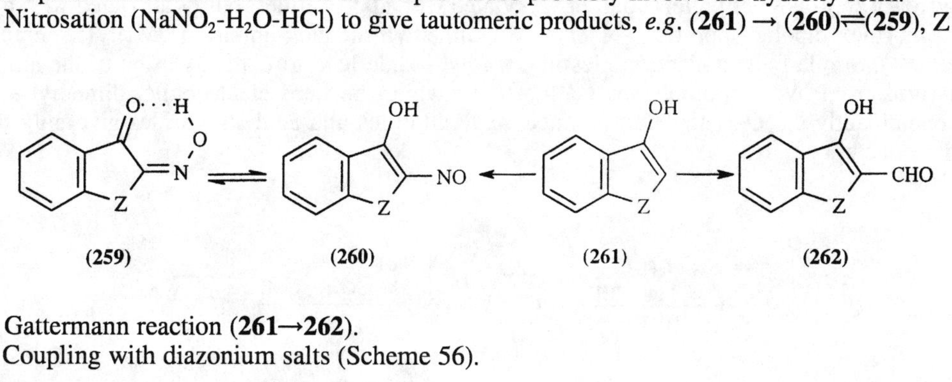

(259)   (260)   (261)   (262)

(b) Gattermann reaction (**261**→**262**).
(c) Coupling with diazonium salts (Scheme 56).

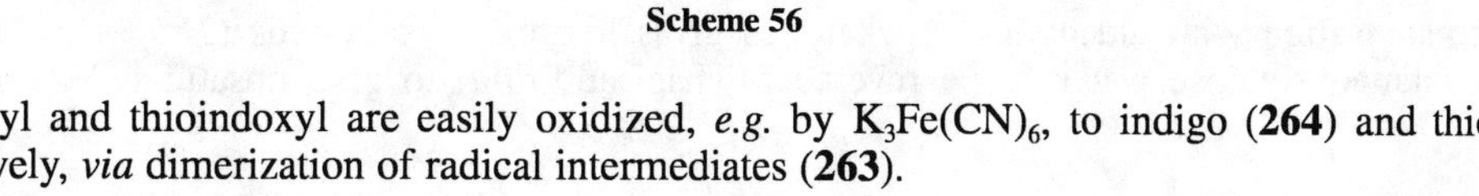

Scheme 56

Indoxyl and thioindoxyl are easily oxidized, *e.g.* by K$_3$Fe(CN)$_6$, to indigo (**264**) and thioindigo, respectively, *via* dimerization of radical intermediates (**263**).

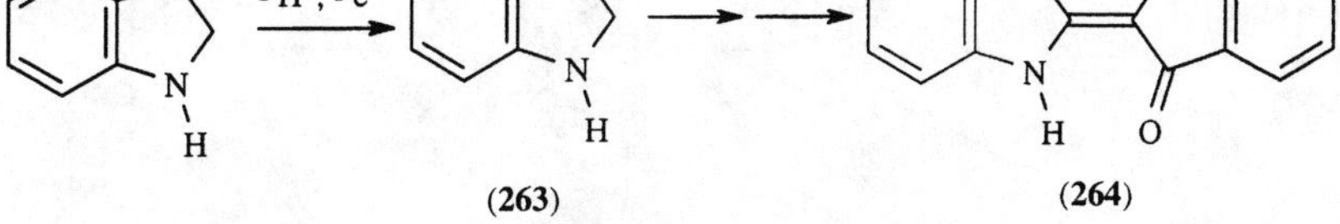

(263)   (264)

3.3.2.5.4　*Reactions of anions with electrophiles*

These heterocyclic compounds undergo many reactions which are similar to those of the acetone enolate. Thus, Claisen condensation and *o*-sulfonylation are exemplified by Scheme 57.

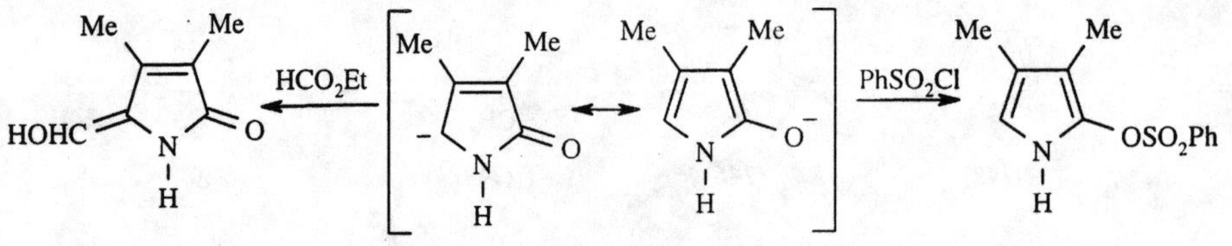

Scheme 57

The indoxyl (3-hydroxyindole) anion undergoes carbon or oxygen alkylation (Scheme 58).

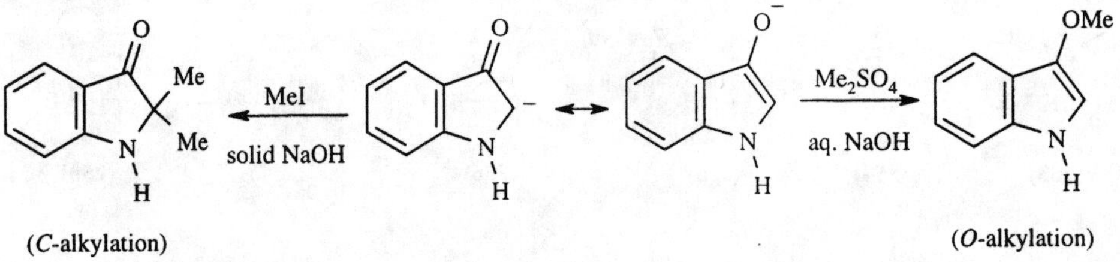

Scheme 58

1*H*-Pyrrol-3(2*H*)-ones are preferentially *C*-alkylated with alkyl halides, but *O*-alkylated with methyl tosylate in the dipolar aprotic solvent *N,N*-dimethylimidazolidinone (DMI) (Scheme 59) ⟨91JCS(P1)3245⟩. Similarly, soft electrophiles like methyl iodide lead to *C*-methylation of the ambident anion derived from hydroxythiophenes ⟨86HC(44/3)1⟩, while the hard electrophile, dimethyl sulfate, gives predominantly the *O*-methylated product. Acyl chlorides and acid anhydrides give only the *O*-acylated products.

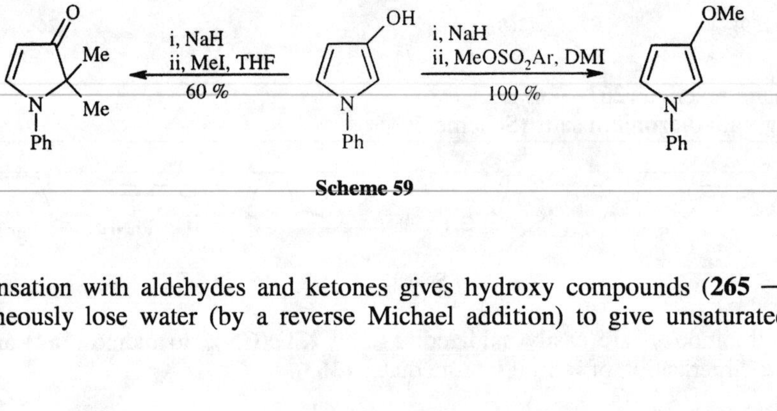

Scheme 59

Aldol condensation with aldehydes and ketones gives hydroxy compounds (**265** → **267**) which usually spontaneously lose water (by a reverse Michael addition) to give unsaturated compounds (**268**).

The following exemplify reactions of the aldol type. 3-Hydroxythiophene with benzaldehyde forms (**269**). Anions derived from oxindole (**270**; Z = NH) and indoxyl (**271**) react with isatin (**272**) to give isoindigo (**273**) and indirubin (**274**).

(**269**)   (**270**)   (**271**)   (**272**)

(**273**)   (**274**)

3.3.2.5.5 Reactions of carbonyl compounds with nucleophiles

Nucleophilic reagents attack the carbonyl carbon atom; the subsequent course of this reaction parallels that in aliphatic chemistry. If the carbonyl group and the heteroatom are adjacent, the ring is usually opened. If they are not adjacent, a carbonyl addition compound results, which often eliminates water spontaneously. The reactions of carbonyl groups in both environments are discussed.

(i) Carbonyl adjacent to heteroatom

Ring opening by nucleophilic reagents necessitates group Z gaining a negative charge, the ease of which depends on the heteroatom (S > O >> NH) and on the ring type (*e.g.* **275** > **276** > **277**). Succinic (**275**; Z = O), maleic and phthalic anhydrides and imides behave similarly to the acyclic acid anhydrides and imides.

(**275**)   (**276**)   (**277**)   (**278**)

The ring opening of phthalimides (**278**; Z = NR) by hydrazine to give a primary amino compound and 1,4-phthalazinedione (**278**; Z = NH-NH) (Ing-Manske reaction) is important in the modified Gabriel synthesis. 2-Coumaranone, its *S*-analogue (**276**; Z = O, S) and the diones (**279**; Z = O, S) react reversibly with hydroxide and alkoxide ions to give salts (as **280**) and esters (as **281**) of the ring-opened acid. The corresponding reactions with indoxyl (**282**; Z = NH) are much more difficult, but in the case of isatin (**272**), the second carbonyl group facilitates ring fission, *e.g.* treatment of (**272**) with sodium hydroxide gives sodium isatinate.

(**279**)   (**280**)   (**281**)   (**282**)   (**283**)

(ii) Carbonyl not adjacent to heteroatom

Carbonyl groups not adjacent to a heteroatom are less stabilized by resonance and react with the relatively weakly nucleophilic 'ketonic' reagents. If carbonyl groups of both types are present, as in

(**279**; Z = O, NH), then the carbonyl group not adjacent to the heteroatom is preferentially attacked. Thus, isatin and indoxyl and their *O*- and *S*-analogues (**279**, **282**) react with hydroxylamine, hydrazine, phenylhydrazine, semicarbazide, *etc.*, to give oximes, hydrazones, phenylhydrazones, semicarbazones, *etc.* (**282** → **283**; **279** → **284**).

The reactive 3-carbonyl group in compounds of type (**279**) undergoes aldol condensation with active methylene compounds; such reactions of isatin with indoxyl, oxindole (Section 3.3.2.5.4) and with thiophenes (Section 3.3.1.5.7.ii) have already been mentioned. These compounds also react with Grignard reagents and phosphorus halides as expected, *e.g.* isatin (**279**; Z = NH) with MeMgBr and PCl₃ yields (**285**) and (**286**), respectively.

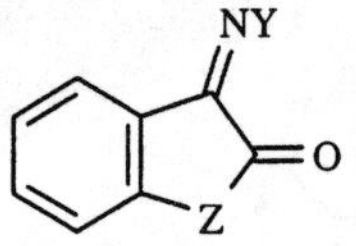

| (**284**) | (**285**) | (**286**) | (**287**) | (**288**) |

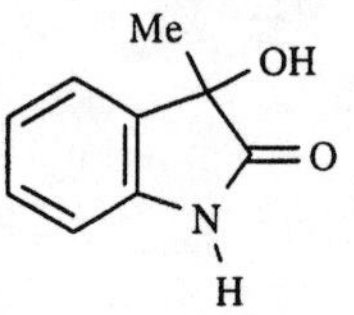

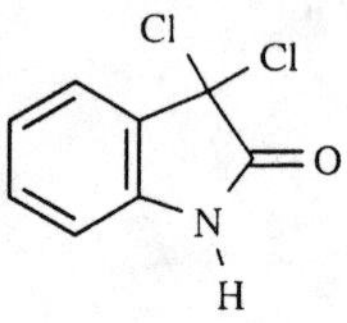

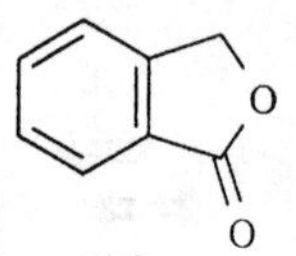

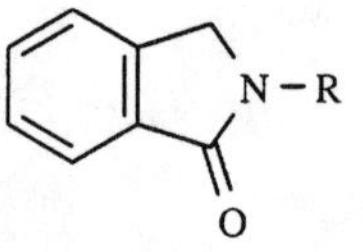

3.3.2.5.6 Reductions of carbonyl and hydroxy compounds

In cyclic anhydrides and imides, one carbonyl group is usually easily reduced; thus, phthalic anhydride with H_2/Ni gives phthalide (**287**), and phthalimides with Zn/HCl yield phthalamides (**288**). Indoxyl and its *O*- and *S*-analogues can be reduced (Zn-HOAc) to indole, *etc.*

3.3.3 REACTIVITY OF SUBSTITUENTS

3.3.3.1 General Survey of Reactivity

3.3.3.1.1 Reaction types

In general, substituents attached to furan, thiophene and pyrrole ring carbon atoms (we consider separately substituents attached to pyrrole nitrogen or thiophene sulfur) react similarly to those on benzenoid nuclei, but there are some important differences:

(a) Some reactions requiring vigorous conditions which succeed in the benzene series fail because the heterocyclic rings are susceptible to attack by electrophilic reagents; see Section 3.3.1.4.

(b) Hydroxy groups attached directly to the heterocyclic nuclei usually exist largely, or entirely, in an alternative, non-aromatic tautomeric form (Section 2.3.5.2); their reactions show little resemblance to those in phenols, and have been considered with the non-aromatic compounds in Section 3.3.2.5. Amino derivatives, although highly reactive, generally exist in the amino form (see Section 3.3.3.4.2).

(c) Benzyl and allyl halides are more reactive than other alkyl halides because the halogen is labilized by electronic shifts of the type shown below; these shifts are enhanced in thienyl- and especially in pyrryl- and furyl-methyl halides.

$$CH_2 = CH - CH_2 - Cl$$

(d) Hydroxymethyl and aminomethyl groups on the heterocyclic compounds are activated in a manner similar to, although less marked than, the chloromethyl derivatives.

The validity of the Hammett relationship log $K/K_0 = \rho\sigma$, has been extensively investigated for five-membered heteroaromatic compounds and their benzo analogues. The ratio ρ(heterocycle) / ρ(benzene) is closest to unity for thiophene. Judged from work on the polarographic reduction of nitro compounds, the ability to transmit electronic effects is HC = CH ≈ S < O < NH.

3.3.3.1.2 Nucleophilic Substitution of Substituents

The tremendous difference in reactivity towards electrophiles, that distinguishes π-deficient and π-excessive heterocycles, is considerably diminished for nucleophilic substitution reactions in which ring substituents are replaced. The reasons for this are the lesser aromaticity of five-membered heterocycles and a rather high polarizability of the pyrrole-like heteroatoms, which depending on the nature of reagent may function as an electron donor or electron acceptor. It is not therefore surprising that five-membered heterocycles show higher than benzene activity not only in reactions with electrophiles but also in reactions of their substituents in nucleophilic substitutions, though of course they remain much less active than azines. Table 3 gives a picture of the relative activity of benzene and heterocyclic derivatives in substitution reactions of a nitro group for *p*-tolylthio- and piperidino groups (Equation 6) (see also Section 3.3.3.7.1).

$$\underset{\text{O}_2\text{N}}{\overset{}{\bigwedge}}\text{X}\overset{}{\bigwedge}\text{NO}_2 \quad \xrightarrow[\text{25 °C}]{\text{Nu}^-} \quad \underset{\text{O}_2\text{N}}{\overset{}{\bigwedge}}\text{X}\overset{}{\bigwedge}\text{Nu} \quad + \text{ NO}_2^- \qquad \textbf{(6)}$$

For both nucleophiles, 2,5-dinitrofuran is the most active substrate, the thiophene derivative follows. On the other hand, the relative reactivity of 1-methyl-2,5-dinitropyrrole and 1,4-dinitrobenzene depends on the nature of the nucleophile. For the $4\text{-MeC}_6\text{H}_4\text{S}^-$ anion, the former is more active by about two powers of ten, but in the piperidinolysis reaction the 1,4-benzene is superior. These phenomena appear to be caused by differences in the polarizability of both substrate and nucleophiles. *p*-Tolylthiolate anion is a softer nucleophile in comparison with piperidine and the pyrrole system is certainly more polarizable than the benzene molecule. Therefore soft-soft interaction of 1-methyl-2,5-dinitropyrrole with $4\text{-MeC}_6\text{H}_4\text{S}^-$ and hard-hard interaction of 1,4-dinitrobenzene with piperidine should occur easier than interactions between reagents with opposite types of softness and hardness.

Table 3 Relative reactivity of 2,5-dinitro derivatives of five-membered heterocycles and 1,4-dinitrobenzene towards some nucleophiles[a]

X	Nucleophile	
	$4\text{-MeC}_6\text{H}_4\text{S}^-$ (in MeOH)	Piperidine (in MeCN)
NMe	1	1
O	$1.7 \cdot 10^3$	$2.4 \cdot 10^6$
S	$1.6 \cdot 10^2$	$4.4 \cdot 10^3$
1,4-dinitrobenzene	$8.8 \cdot 10^{-3}$	9.6

[a] ⟨77JOC3550, 76JOC2824⟩

3.3.3.2   Fused Benzene Rings

3.3.3.2.1   *Electrophilic attack*

Most common reactions of benzene rings involve attack by electrophilic reagents; but since thiophene, pyrrole and furan are more readily attacked than benzene, reaction in fused-ring compounds should occur at a free position in the heterocyclic ring in preference to the benzene ring. This generalization may become invalid if (a) there is a strongly deactivating substituent (*e.g.* CHO, CO$_2$Et, NO$_2$) in the heterocyclic ring or (b) a strongly activating substituent (*e.g.* NH$_2$, OH) is present in the benzene ring. (*e.g.* Bromination of several 4,6-dimethoxyindoles has been shown to occur at C-7 ⟨92T7601⟩.) When both positions of the heterocyclic ring are substituted, substitution in the benzene ring is generally observed, though many examples of substituent displacement (*ipso* substitution) are known. Indole conjugated acids are nitrated in the benzene ring (*e.g.* **289** → **290** → **291**). 3,3-Dialkyl-3*H*-indolium salts similarly nitrate at the 5-position. The reactivity towards electrophiles in benzene ring of benzo[*b*]thiophene is 6 > 5 > 4 > 7.

(**289**)   (**290**)   (**291**)

Carbazole (**292**), dibenzofuran (**293**; Z = O) and dibenzothiophene (**293**; Z = S) behave as diphenylamine, diphenyl ether and diphenyl sulfide, respectively, in their substitution reactions and thus electrophilic substitution occurs at the positions *para* to the heterocyclic atom, as exemplified for:

(a) Dibenzofuran: bromination (Br_2-CS_2), sulfonation ($ClSO_3H$) and formylation (HCN-HCl-$AlCl_3$).

The positional reactivity of dibenzofuran in electrophilic substitution reactions depends on the electrophile. Reaction occurs mostly at the 2- and 3-positions but the ratio of the two products varies ⟨91JOC4671⟩. The reaction of cyanogen bromide catalyzed by aluminum chloride gives an 80% yield of the 2-substituted product together with 15% of the 3-cyano-derivative ⟨92ACS312⟩. Oxidative acetoxylation of dibenzofuran occurs predominantly at the 3-position ($\sim 60\%$) together with attack at the 1-position ($\sim 30\%$) ⟨92ACS802⟩. In this latter reaction, the attack by acetate is on the dibenzofuranium radical cation.

(b) Dibenzothiophene: nitration (HNO_3-AcOH) and bromination (Br_2-CS_2).

(c) Carbazole: acylation (RCOCl-$AlCl_3$), halogenation ($SOCl_2$ or Br_2-CS_2) and sulfonation (H_2SO_4).

(d) Dibenzoselenophene and dibenzotellurophene (**293**; Z = Se, Te) undergo nitration in the 2 position.

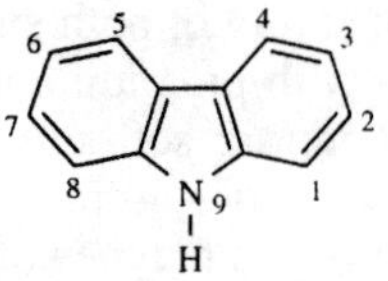

(292) 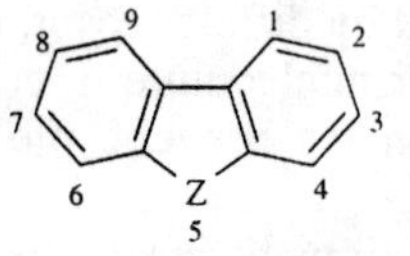   (293) Z = O, S, Se or Te

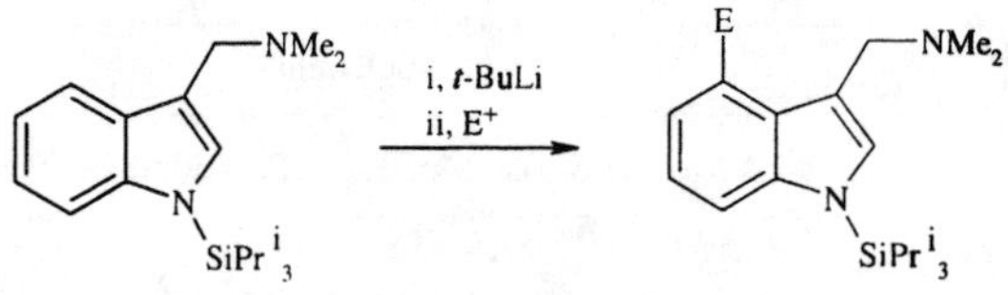

Scheme 60

3.3.3.2.2 Nucleophilic attack

The possibility of activating the benzene ring of the indole nucleus to nucleophilic substitution has been realized by formation of chromium tricarbonyl complexes. For example, the complex from *N*-methylindole (**294**) undergoes nucleophilic substitution with 2-lithio-1,3-dithiane to give a product (**295**) which can be transformed into 1-methylindole-7-carbaldehyde (**296**) ⟨78CC1076⟩.

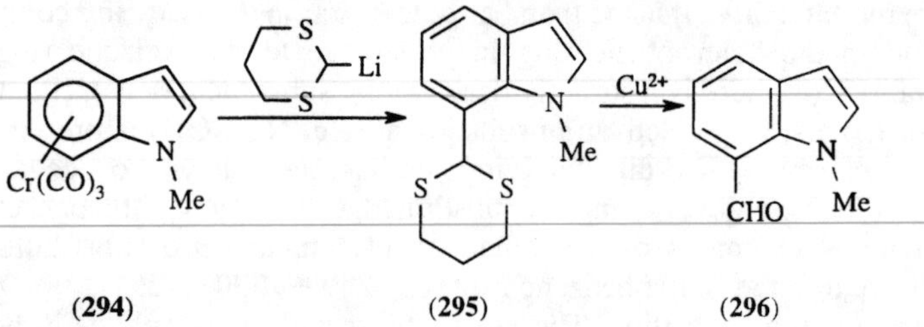

(294)   (295)   (296) 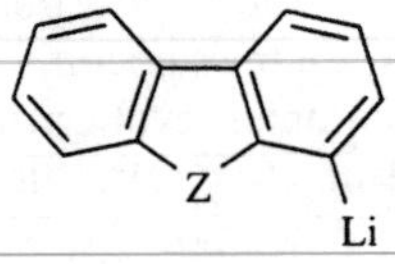   (297)

Lithiation can be diverted away from C-2 of indole by the use of a bulky N-substituent. Although 1-methylgramine is cleanly lithiated at C-2, 1-(triisopropylsilyl)gramine is lithiated selectively at C-4 and can lead to useful 4-substituted indoles: electrophiles include 1,2-dibromoethane, DMF, and diphenyl sulfide (Scheme 60) ⟨93H(36)29⟩.

Lithiation can also be induced in the conventional way from halo compounds, for example, the sodium salt of 5-bromoindole can be converted to the 5-lithio derivative by treatment with *t*-butyllithium, and in turn leads to various 5-substituted indoles ⟨92H(34)1169⟩.

Thallation of 3-acylindoles gives the 4-thallated products, which can be converted to both the 4-nitro and 4-azido derivatives in copper(II)-promoted processes ⟨89H(29)643⟩. The nitro compound is formed by heating the organothallium intermediate with sodium nitrite and copper sulfate in DMF at 100°C. This methodology has been used in a total synthesis of indolactam-V ⟨90T6623⟩.

Thallation also provides a route to 4-iodoindoles from the related 3-acyl compounds. Here the intermediate thallium compound can be treated with iodine and copper(I) iodide in DMF to effect the transformation ⟨84H(22)797, 93JCS(P1)2561⟩.

Alkyllithium compounds metallate dibenzofuran, dibenzothiophene and *N*-alkylcarbazoles (in increasing order of difficulty) to form compounds of type (**297**); substitution occurs *ortho* to the heteroatom as expected from benzene chemistry.

3.3.3.2.3 Reactions with electrons

Selective reduction of indole in the benzene ring can be achieved by treatment with lithium in liquid ammonia, which gives a mixture of the 4,7-dihydro and 4,5,6,7-tetrahydro derivatives.

Birch reduction of indole with lithium metal in THF in the presence of trimethylsilyl chloride followed by oxidation with *p*-benzoquinone gave 1,4-bis(trimethylsilyl)indole (**298**). This is readily converted in two steps into 1-acetyl-4-trimethylsilylindole. Friedel-Crafts acylation of the latter compound in the presence of aluminum chloride yields the corresponding 4-acylindole (**299**) ⟨82CC636⟩.

For the reduction of indolizine see Section 3.3.1.7.5.

i, Li, THF, Me₃SiCl; ii, *p*-benzoquinone; iii, aqueous MeOH; iv, NaH, AcCl;
v, RCOCl, CH₂Cl₂, AlCl₃, r.t.

3.3.3.2.4 Reactions of substituents on benzene rings

Substituents on fused benzene rings undergo the usual reactions expected in the benzene series. The orientation in the diazo coupling reactions with hydroxy compounds (**300**) and (**301**) indicates that there is little 'bond fixation' in dibenzofuran, a distinct contrast to naphthalene.

4-Alkylaminoindoles (**302**) rearrange to 4-amino-l-alkylindoles (**303**) when heated with *p*-toluenesulfonic acid hydrate ⟨82CC1356⟩.

3.3.3.3 Other C-Linked Substituents

3.3.3.3.1 Alkyl groups

The reactivity of alkyl groups on five-membered rings with one heteroatom is similar to that of alkyl groups on benzenoid rings. Because of the high reactivity of the heterocyclic nuclei, specific reactions of the alkyl groups may be difficult to carry out. However, oxidation of alkyl to carboxyl can be achieved (*e.g.* **305** → **306**) ⟨65JOC1453⟩ and selective bromination of alkyl groups (*e.g.* **305** → **304**) has been reported ⟨76SC475⟩. Further bromination of (**304**) yields the dibromomethyl derivative. The practical application of these reactions may require either nuclear deactivation by substitution of electron-withdrawing groups or, in the case of halogenation, that all the nuclear carbon atoms carry substituents.

(304)   (305)   (306)

3.3.3.3.2 *Vinyl groups*

2- And 3-vinyl derivatives of pyrrole, furan, thiophene and their benzologues behave as outer-inner dienes and react with π-electron-deficient alkenes and alkynes to produce the corresponding Diels-Alder adducts. Thus, 3-vinylpyrrole (**307**) in combination with methyl propiolate gives dihydroindoles (**308**) or (**309**) ⟨80JOC4515, 90TL4613, 91H(32)1199⟩.

(307)   (308)   (309)   (310)

By analogy, 2-vinylthiophene reacts with tetrabromocyclopropene by [4 + 2] cycloaddition; subsequent loss of HBr with concominant opening of the cyclopropane ring leads to the benzo[*b*]thiophene derivative (**310**) in 32% yield ⟨90TL4581⟩.

The intramolecular Diels-Alder methodology has been used in approaches to naturally occuring furan derivatives ⟨93JCS(P1)2395⟩. For example, thermolysis of (**311**) gave the product (**312**) in almost quantitative yield by cyclization followed by isomerization of the exocyclic double bond to rearomatize the furan residue. The reaction proved to be highly stereoselective.

(311)   (312)

The "fulgides" are derivatives of bismethylenesuccinic anhydride. As a part of a study on heterocyclic fulgides ⟨85JCS(P1)957⟩, irradiation of the (*E*)-fulgide (**313**) at 366 nm results in the formation of (**314**) which has a deep red color. The reverse ring opening can be brought about by irradiating with white light.

The 2-thienylfulgide (**315**) is similarly photochromic, leading to (**316**) on irradiation at 366 nm.

(313)   (314)   (315)   (316)

In an interesting application of this result, a photoswitchable unit has been incorporated by covalent linkage into a protein molecule ⟨92JA3150⟩. This has resulted in the ability to photoregulate the binding properties of the protein by effecting minor perturbations in its conformations.

1,2-Diarylethenes containing heterocyclic rings have the advantage that they can undergo reversible photocyclization, but at the same time are thermally stable. The closed-ring molecules absorb light at significantly long wavelengths.

In a significant improvement, the central double bond of the diarylethene has been incorporated into a perfluorocyclopentene ring. The diarylethene is thus locked in the *cis*-configuration, thereby preventing fatigue due to *cis-trans* isomerization. The fluorine atoms give added thermal stability to the products ⟨92CC206⟩. In hexane solution (**317**) has λ_{max} at 258 nm. On irradiation with light at 313 nm, cyclization takes place to produce (**318**) with λ_{max} at 517 nm (Equation 7). At photostationary conditions, the ratio (**317**) : (**318**) is 55:45. The reverse reaction could be brought about by irradiating with light of wavelength $\geq$ 500 nm. The coloration (cyclization)-decoloration (ring-opening) cycle could be repeated 14000 times without significant loss of performance (90%).

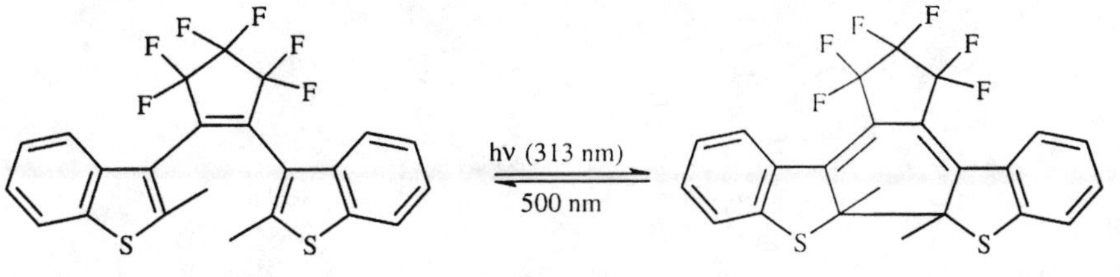

(7)

(**317**)   (**318**)

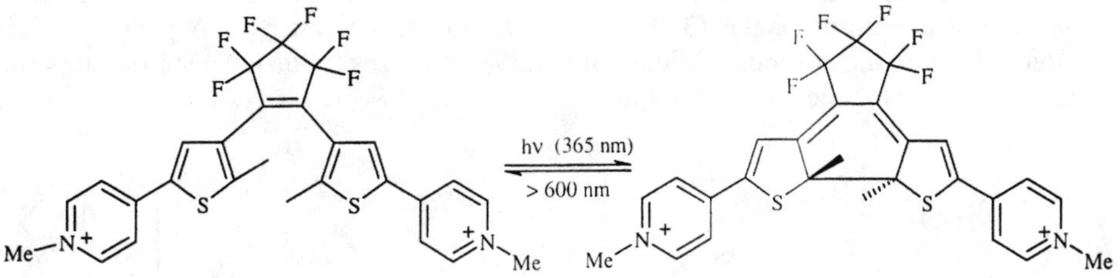

(**319**)   (**320**)

These ideas have been extended in an approach to the design of light-triggered electrical and optical switching devices ⟨93CC1439⟩. The basic concept was to utilize the photochromic property of dithienylethenes as a switch; the crucial advance here was the appropriate choice of substituents on the thiophene rings. In one type of molecule (**319**), the central dithienylperfluorocyclopentene unit has two pyridinium rings attached in such a way that in the cyclized form (**320**) they are at the termini of an extended conjugated system. Irradiation of (**319**) with light of wavelength 365 nm resulted in almost complete conversion to (**320**). The difference in λ_{max} of (**319**) and (**320**) was of the order of 300 nm. The ring opening of (**320**) to (**319**) could be carried out by irradiation with light of wavelength > 600 nm.

3.3.3.3.3 Substituted alkyl groups: general

As discussed previously (Section 3.3.3.1), halomethyl, hydroxymethyl and aminomethyl groups show enhanced reactivity toward nucleophilic attack because of the ease with which the halogen, hydroxy or amino group is lost in (**321**). Both side-chain (**322**) and nuclear substitution products (**323**) have been obtained (Scheme 61). These two possibilities are exemplified by the reaction of furfuryl chloride with sodium cyanide to give (**322a**) and (**323a**).

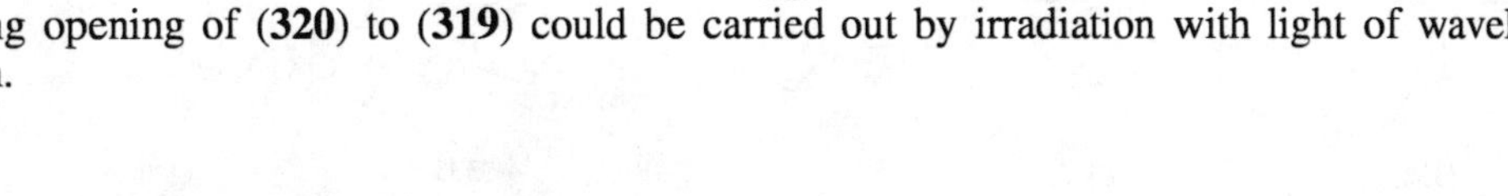

(**321**)   (**322**)   (**323**)   (**322a**)   (**323a**)

Scheme 61

Many examples of Diels–Alder reactions have been carried out on heterocyclic α,β-quinodi-methanes derived in a variety of ways from the corresponding precursors containing substituted alkyl groups. For example in the case of pyrrole derivatives, this methodology leads to the synthesis of indoles and particularly those substituted in the benzene ring. The thieno[3,4-*b*]pyrrole-1,1-dioxide (**324**) reacts with DMAD to give the cycloadduct (**326**), presumably by trapping a pyrrole-2,3-quinodimethane (**325**). The cycloadduct can then be dehydrogenated to the indole (**327**) ⟨92CC549⟩.

The fluoride ion-induced fragmentation of (**328**) is a mild and efficient way of generating of 2,3-dimethylene-2,3-dihydrothiophene (**329**) ⟨88TL2689⟩. In the presence of *N*-phenyl maleimide it gives the adduct (**330**). In the absence of dienophile (**329**) dimerizes to unstable spiro-compound (**331**) acting at the same time as diene and dienophile.

The 3,4-dimethylenethiophene (**333**) has been generated from diazene (**332**) by either thermolysis or photolysis (310–380 nm). The purple-colored biradical (**333**) undergoes intramolecular dimerization to (**334**) at −78°C (the formation of intermolecular dimers is possible at higher temperatures). However, (**333**) is quiet stable up to 160 K at frozen glassy solution in 2-methyltetrahydrofuran.

The biradical (**333**) reacts with oxygen to form the peroxide (**335**). It can also be trapped by electron-poor alkenes: the yields of cycloadducts are almost quantitative. Thus, acrylonitrile gives (**336**).

2,3-Dimethylene-2,3-dihydrofuran (**338**) formed, *e.g.*, from (**337**) has been studied in detail and shown to dimerize into (**339**) in high yield ⟨81JA6691⟩ at temperatures above −30°C ⟨94JOC2594, 94JOC2613⟩. Photolysis in argon matrix at 16 K convert (**338**) into the isomeric dihydrocyclobutafuran ⟨87AG(E)471⟩.

In the majority of the reactions of 3,4-dimethylene-3,4-dihydrofuran with dienophiles, the first reaction involves a conventional Diels-Alder reaction of the furan ring so that the final product, *e.g.* (**340**), may incorporate two molecules of the dienophile ⟨93H(35)57⟩ (Scheme 61a).

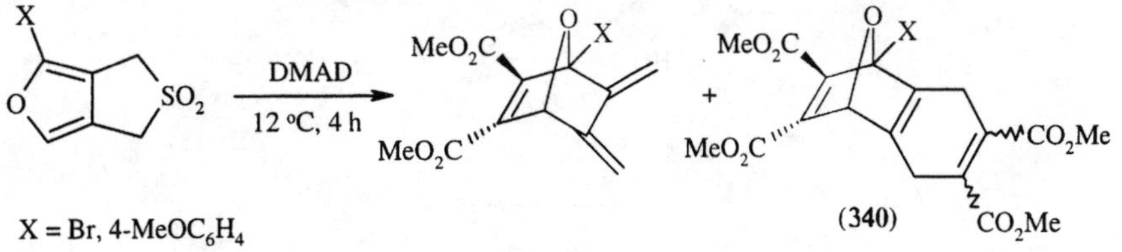

X = Br, 4-MeOC$_6$H$_4$

(**340**)

Scheme 61a

Reduction of tri-(2-thienyl)methyl carbenium tetrafluoroborate by zinc in DME at 65°C produces, *via* the radical (**341**), two dimers (**342**) and (**343**). The former is the kinetically favored product, which changes to the thermodynamically preferred (**343**) on thermal equilibration. This obviously takes place by dissociation into the radical (**341**) ⟨90TL2627⟩.

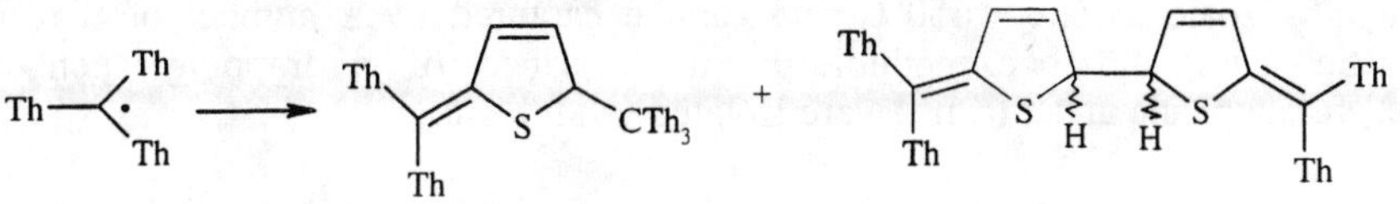

(**341**)   (**342**) Th = 2-Thienyl   (**343**)

3.3.3.3.4 Halomethyl

The furfuryl halides (*cf.* **344**; Z = O) are exceedingly reactive; they are usually not isolated but are used in solution as intermediates because of their instability. The halogen may be replaced directly by amino or alkoxy groups, but with potassium cyanide the S$'_N$ product (**323a**) is also formed. 2,5-Dichloromethylfuran is a precursor for highly conductive high molecular weight poly(2,5-furylene vinylene) ⟨87CC1113⟩.

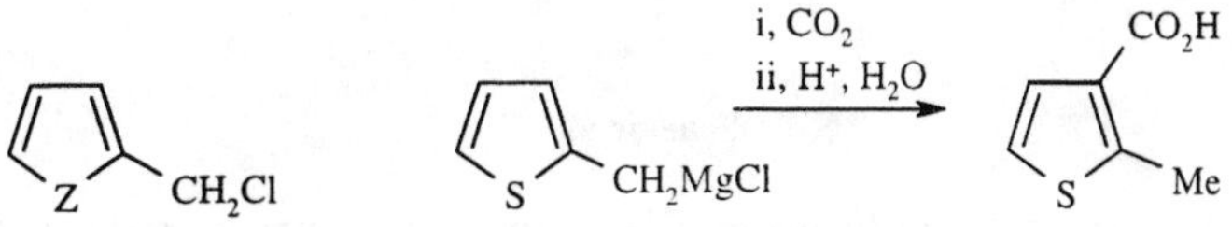

(**344**)   (**345**)   (**346**)

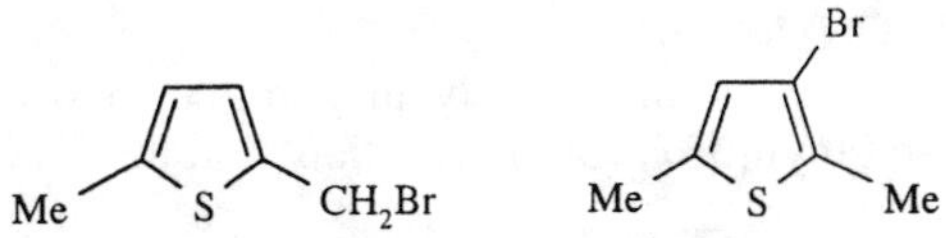

(**347**)   (**348**)

2-Chloromethylthiophene shows similar reactivity to benzyl chloride, in that it is readily converted into 2-cyanomethylthiophene, thiophene-2-carbaldehyde (by treatment with hexamethylenetetramine) and a Grignard reagent (**345**). The latter reacts with electrophiles to give 2-methyl-3-substituted thiophenes (*e.g.* **346**).

2-Bromomethyl-5-methylthiophene gives normal displacement products with amines but it is isomerized on attempted reaction with copper(I) cyanide (**347** → **348**) ⟨48MI30200⟩.

3.3.3.3.5 Hydroxymethyl

Whereas 2-hydroxymethylthiophene reacts normally with hydrogen halides to give 2-halomethylthiophenes, reaction of 2-hydroxymethylfuran (2-furfuryl alcohol) with hydrochloric acid results in formation of laevulinic acid (**350**) *via* the S$'_N$ intermediate (**349**). The conversion of 2-furanacrylic acid

(**351**) into an ester of γ-oxopimelic acid (**352**) by ethanolic hydrochloric acid is a related reaction involving an analogous intermediate.

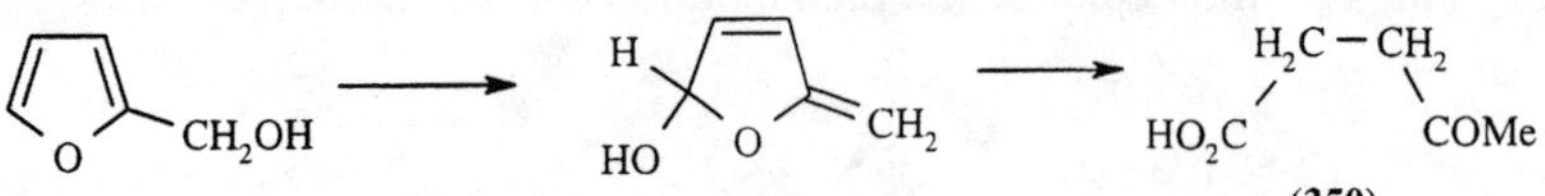

(349)

(350)

(351)

(352)

Reduction of 2-hydroxymethylpyrroles with lithium aluminum hydride or diborane yields the corresponding 2-methylpyrroles.

2,5-Dimethylenefuran has been generated by the flash vacuum pyrolysis of 5-methyl-2-hydroxymethyl benzoate at 630 − 650°C and can be captured by a number of reagents such as bromine, acetic acid, and benzenethiol. In the absence of a trapping agent the dimer, 2,2′,5,5′-[2,2]-furanophane and a trimer were isolated ⟨92MI 206–04⟩.

3.3.3.3.6 Aminomethyl

Aminomethylindoles are particularly important synthetic intermediates. 3-Dimethylaminomethyl-indole (gramine) (**353**) and especially its quaternary salts readily undergo displacement reactions with nucleophiles (Scheme 62).

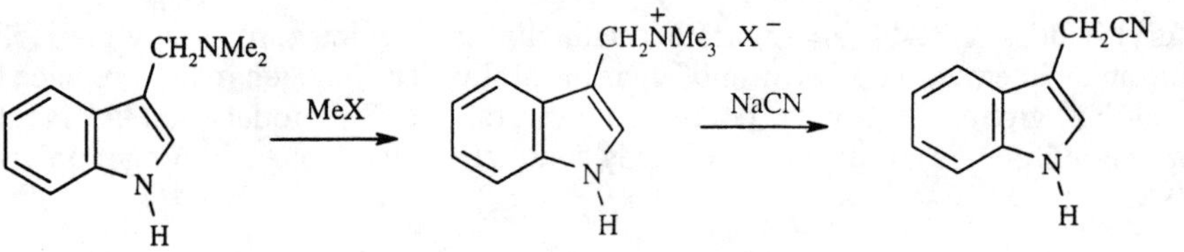

(353)

Scheme 62

(a) Potassium cyanide gives 3-indoleacetonitrile which can in turn be reduced to tryptamine (**354**; Y = CH$_2$NH$_2$), or hydrolyzed to 3-indoleacetic acid (**354**; Y == CO$_2$H).

(b) Diethyl acetamidomalonate gives (**355**), which can be hydrolyzed to tryptophan.

(c) Nitroethane forms (**354**; Y = CHMeNO$_2$).

1-Methylgramine (**356**) generally reacts analogously to gramine, but with potassium cyanide it yields a mixture of the 2-cyano-3-methyl (**357**) (by S$'_N$ reaction) and the 3-cyanomethyl derivatives.

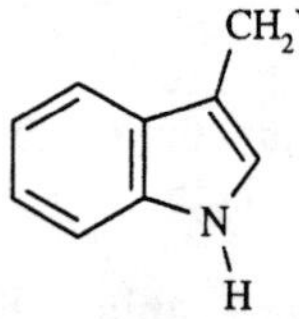

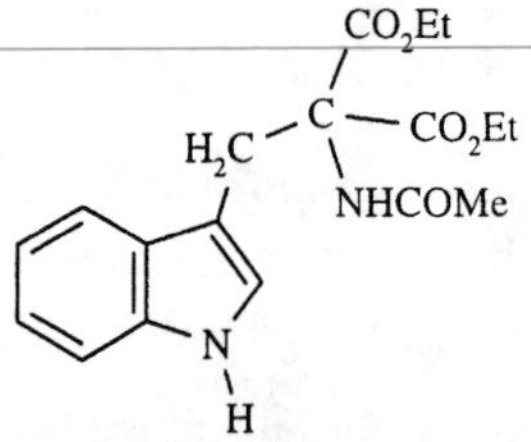

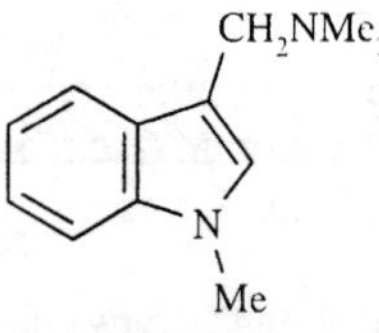

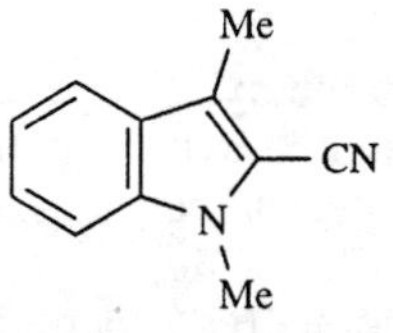

(354)

(355)

(356)

(357)

Indole-2,3-quinodimethanes, generated from 2-methylgramine, undergo intermolecular cyclo-addition reactions with dienophiles similar to that of (**325**) ⟨82T2745⟩.

The dialkylamino group of a 2-dialkylaminomethyl pyrrole can be displaced by various nucleophiles ⟨73S703, B-77MI 202–01, 85CCA627, 90T1791⟩. In a typical example displacement of trimethylamine from the quaternary salt derived from the Mannich base of a 1-arylpyrrole ⟨94S164⟩ gives an azide which can be reduced to give the aminomethylpyrrole (Scheme 63).

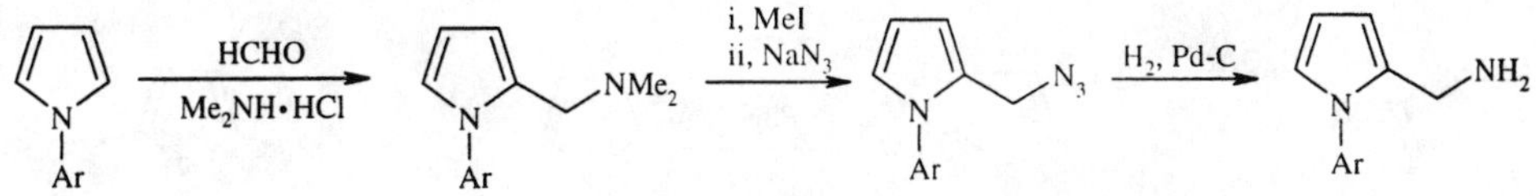

Scheme 63

3.3.3.3.7 Carboxylic acids, esters and anhydrides

Carboxylic acids show most of the standard reactions of benzoic acid. Amides, esters, hydrazides, azides and nitriles can be prepared by standard methods. Thiophenes form stable acid chlorides, furans unstable ones, and *N*-unsubstituted pyrroles do not form them.

The acid dissociation constants of some representative carboxylic acids are given in Table 4.

Table 4 pK_a Values of Pyrrole-, Furan-, Thiophene-, Selenophene- and Tellurophene-carboxylic Acids

Acid	pK_a (H$_2$O, 25°C)
Pyrrole-2-carboxylic acid	4.4
Pyrrole-3-carboxylic acid	5.0
Furan-2-carboxylic acid	3.15
Furan-3-carboxylic acid	4.0
Thiophene-2-carboxylic acid	3.5
Thiophene-3-carboxylic acid	4.1
Selenophene-2-carboxylic acid	3.6
Tellurophene-2-carboxylic acid	4.0
Benzoic acid	4.2

Pyrrole-3-carboxylic acid (**358**) is appreciably weaker than benzoic acid and this is attributed to the stabilization of the undissociated acid by electron release from nitrogen. The 2-carboxylic acids of furan, thiophene, selenophene and tellurophene are all stronger acids than benzoic acid ⟨77AHC(21)119⟩.

pK_a Measurements of 4-*X*-thiophene-2-carboxylic acids and 2-*X*-thiophene-4-carboxylic acids ⟨93JCR(S)300⟩ show that the transmission of substituent effects in the *meta*-like position is quite similar to that observed in *meta*-substituted benzoic acids. The thiophenic σ values are very similar to the benzenic σ values. In both *meta*- and *para*-like disubstituted thiophenes, the hetero ring is able to transmit the electronic effects of the substituents as in the benzene ring, *i.e.*, through the heterocycle as a whole and not *via* the heteroatom.

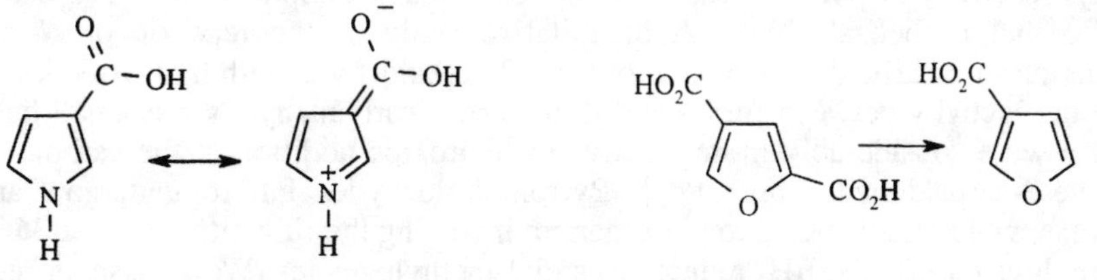

(358a)   (358b)   Scheme 64

Pyrrole-2-carboxylic acid easily loses the carboxylic group thermally. Pyrrole-3-carboxylic acid and furan-2- and -3-carboxylic acids also readily decarboxylate on heating to about 200°C. Thiophene-carboxylic acids require higher temperatures or a copper-quinoline catalyst. In furans, 2-carboxylic acid groups are lost more readily than 3-carboxylic acid groups (Scheme 64).

Decarboxylation often takes place during electrophilic substitution of the nucleus, for example:

(a) Thiophene-2-carboxylic acid and mercuric acetate give tetraacetoxymercurithiophene (**359**; *Z* = S).

(b) 2-Furoic acid and acetyl nitrate give 2-nitrofuran.

(c) Reaction of the anion of pyrrole-2-carboxylic acid with benzenediazonium ion results in the displacement of carbon dioxide rather than hydrogen (Scheme 65).

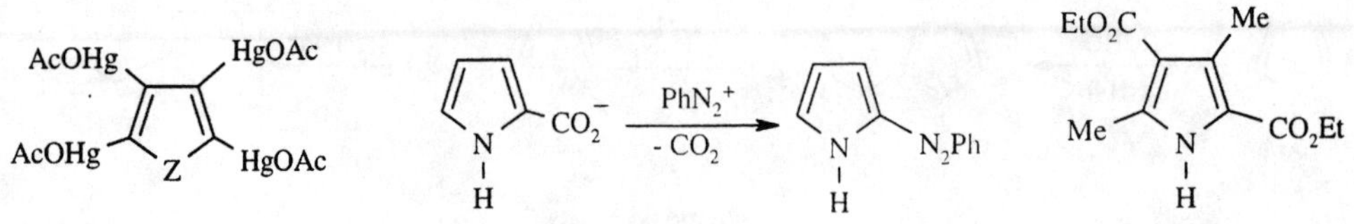

(359)   (360)

Scheme 65

In the pyrrole series, ester groups α to nitrogen are more readily hydrolyzed by alkali, but those in a β position more readily by acid. Thus, in compounds such as diethyl 2,4-dimethylpyrrole-3,5-dicarboxylate (**360**) either ethoxycarbonyl group may be selectively hydrolyzed and, if desired, subsequently eliminated by decarboxylation.

Flash vacuum pyrolysis of 2-methoxycarbonylpyrrole (**361**) gives the ketene (**362**), characterized by IR absorption at 2110 cm^{-1}. On warming to -100 to $-90°C$ the dimer (**363**) is formed ⟨82CC360⟩. Flash vacuum pyrolysis of indole-2-carboxylic acid (**364**) results in loss of water and the formation of a ketene (**365**) showing absorption at 2106 cm^{-1} ⟨82CC360⟩.

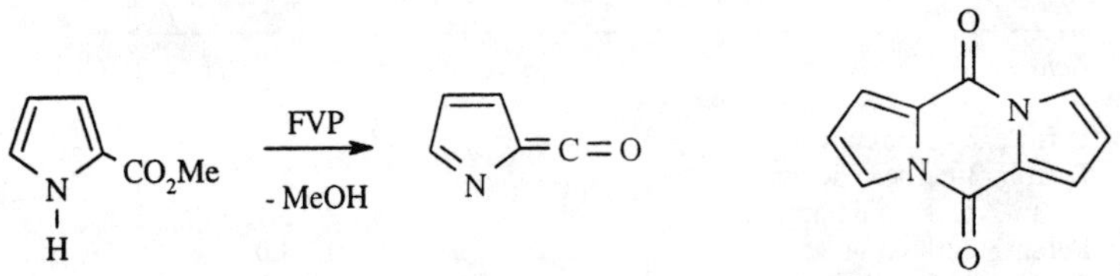

(361)   (362)   (363)

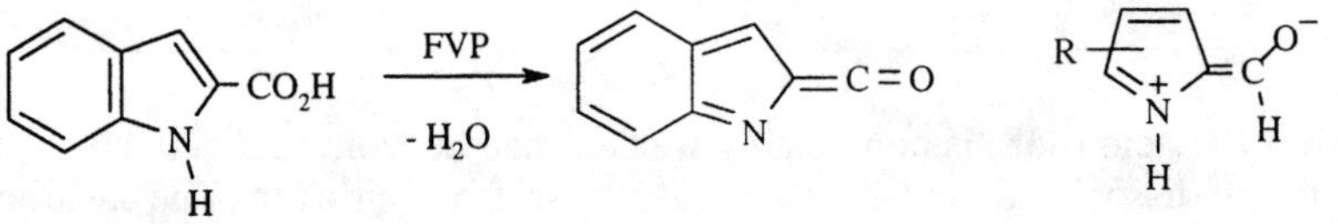

(364)   (365)   (366)

The anhydride of thiophene-2,3-dicarboxylic acid is a precursor of 2,3-didehydrothiophene which is trapped as [4+2] and [2+2] cycloaddition products with dienes ⟨81T4151⟩.

3.3.3.3.8 Acyl groups

The carbonyl reactivity of pyrrole-, furan-, thiophene- and selenophene-2- and -3-carbaldehydes is very similar to that of benzaldehyde. A quantitative study of the reaction of N-methylpyrrole-2-carbaldehyde, furan-2-carbaldehyde and thiophene-2-carbaldehyde with hydroxide ions showed that the difference in reactivity between furan- and thiophene-2-carbaldehydes was small but that both of these aldehydes were considerably more reactive to hydroxide addition at the carbonyl carbon than N-methylpyrrole-2-carbaldehyde ⟨76JOC1952⟩. Pyrrole-2-aldehydes fail to undergo Cannizzaro and benzoin reactions, which is attributed to mesomerism involving the ring nitrogen (see **366**). They yield 2-hydroxymethylpyrroles (by NaBH$_4$ reduction) and 2-methylpyrroles (Wolff-Kishner reduction). The IR spectrum of the hydrochloride of 2-formylpyrrole indicates that protonation occurs mainly at the carbonyl oxygen atom and only to a limited extent at C-5.

The Huang-Minlon reduction of 3-formylfuran surprisingly gave 3-methylene-2,3-dihydrofuran. The product undergoes ene reactions with a number of electron depleted alkenes and provides a route to functionalize the 3-position in furan as shown in Scheme 66 ⟨93TL5221⟩.

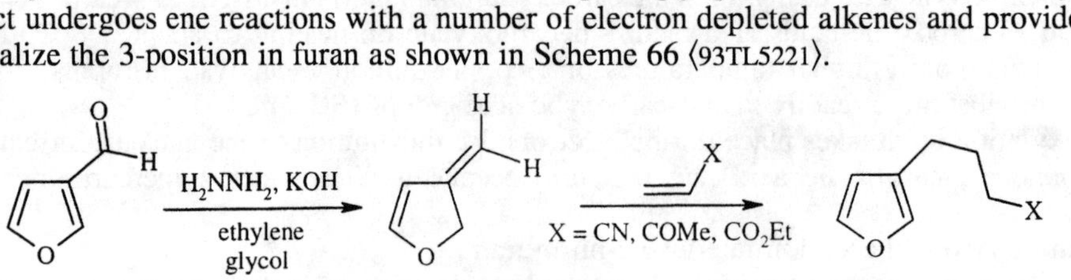

Scheme 66

Acyl-pyrroles, -furans and -thiophenes in general have a similar pattern of reactivity to benzenoid ketones. Acyl groups in 2,5-disubstituted derivatives are sometimes displaced during the course of electrophilic substitution reactions. *N*-Alkyl-2-acylpyrroles are converted by strong anhydrous acid to *N*-alkyl-3-acylpyrroles. Similar treatment of *N*-unsubstituted 2- or 3-acylpyrroles yields an equilibrium mixture of 2- and 3-acylpyrroles; pyrrolecarbaldehydes also afford isomeric mixtures ⟨81JOC839⟩. The probable mechanism of these rearrangements is shown in Scheme 67. A similar mechanism has been proposed for the isomerization of acetylindoles.

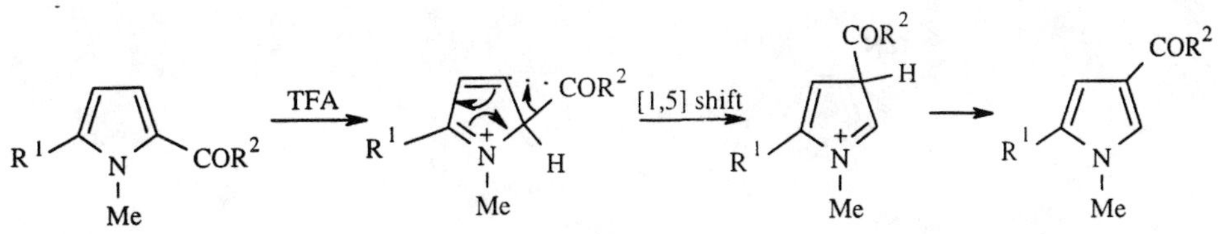

Scheme 67

Diborane reduction of pyrrole and indole ketones affords the corresponding alkyl-pyrroles and -indoles ⟨68T1145⟩.

Carbenes of type (**367**), generated by thermal decomposition of the appropriate tosylhydrazone salts, undergo ring opening more readily when the ring heteroatom is oxygen than when it is sulfur ⟨78JA7927⟩.

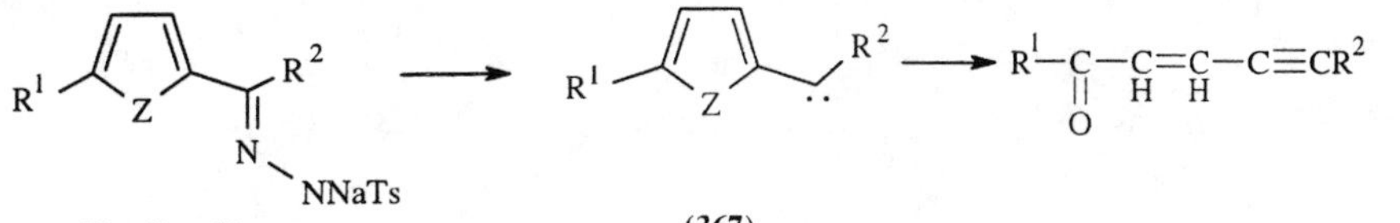

(Z = O or S)   (367)

3.3.3.4 *N*-Linked Substituents

3.3.3.4.1 *Nitro*

Both 2- and 3-nitrothiophenes are reduced by tin and hydrochloric acid to the corresponding aminothiophenes. Reduction of 2,5-dibromo-3,4-dinitrothiophene has given 3,4-diaminothiophene as a stable crystalline solid ⟨83BSF(2)153, 86BSF259, 86BSF267⟩. 2-Acetamidofurans are prepared by the reduction of 2-nitrofurans in the presence of acetic anhydride. 2-Substituted 5-nitrofurans have been reduced to the 5-aminofurans by an electrochemical method ⟨91TL631⟩. Work showed that although catalytic reduction gives 2-aminofurans in low yields they can be trapped ⟨82JOC2483⟩ using ethyl ethoxymethylenecyanoacetate or ethoxymethylenemalononitrile ⟨93JHC113⟩. Benzofuranone (**369**) and not 2-aminobenzofuran is obtained from tin and hydrochloric acid reduction of 2-nitrobenzo[*b*]furan (**368**).

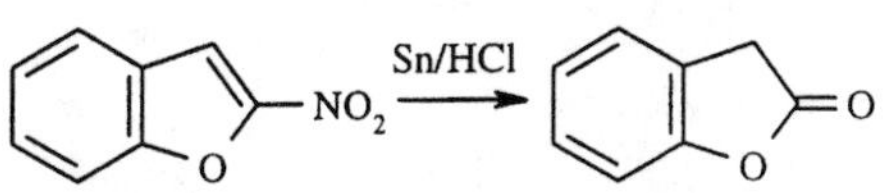

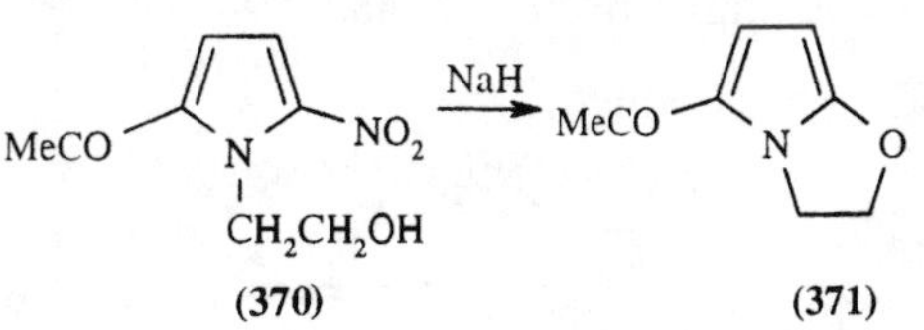

(368)   (369)   (370)   (371)

Although in general the π-excessive nature of the heterocyclic rings under discussion reduces their reactivity to nucleophilic substitution, a number of interesting reactions have been reported in which the leaving group is a nitro group (see also Section 3.3.3.1.2). An example of intramolecular nucleophilic displacement of a pyrrole nitro group is provided by the base-induced cyclization of 2-acetyl-l-(2-hydroxyethyl)-5-nitropyrrole (**370** → **371**) ⟨71JCS(C)2554⟩.

1-Methyl-3,4-dinitropyrrole (**373**) with methanolic sodium methoxide yields product (**374**) which on treatment with trifluoroacetic acid gives the 2-methoxypyrrole (**375**) ⟨78CC564⟩.

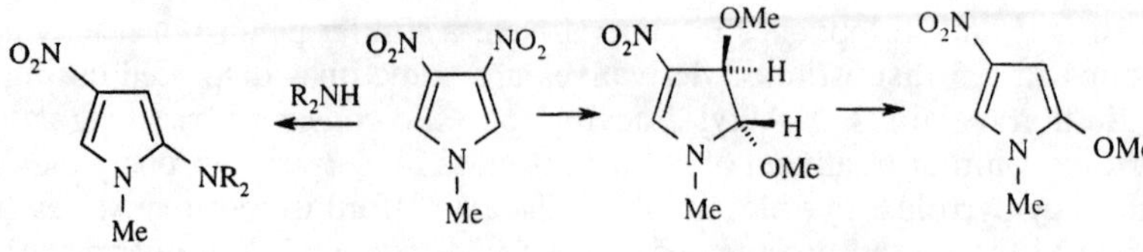

(372)   (373)   (374)   (375)

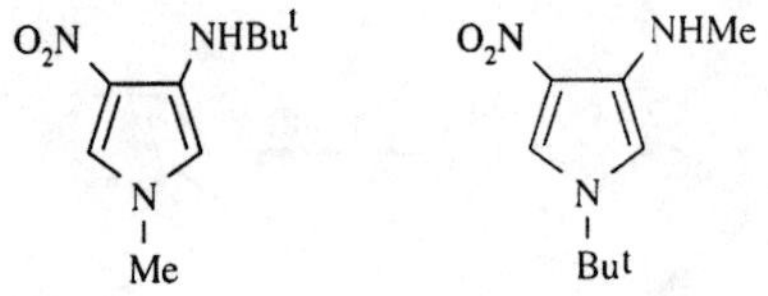

(376)   (377)

With secondary amines such as piperidine or dimethylamine the formal products (**372**) of *cine* substitution are obtained; with primary amines (*e.g.* *t*-butylamine), in addition to the displacement product (**376**), a rearranged product (**377**) is obtained in which the nitrogen-bearing methyl becomes exocyclic ⟨80CC23⟩. Earlier studies on the reaction of 2-nitrothiophene with secondary amines showed that ring-opened products (**378**) are obtained ⟨74JCS(P1)2357⟩.

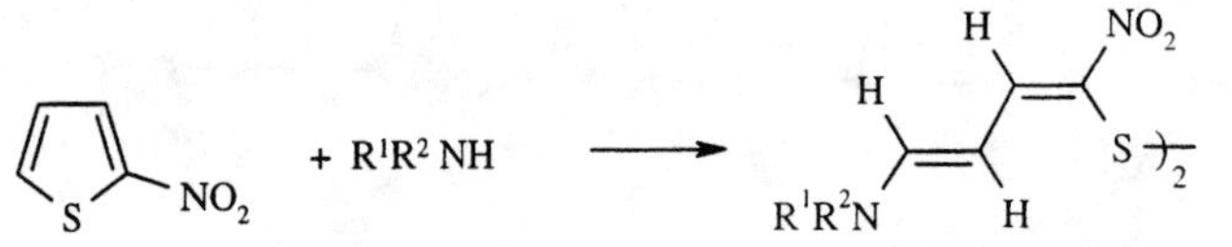

(378)

3.3.3.4.2 *Amino*

The free bases are much less stable than aniline, particularly 2-amino-pyrroles and -furans which are very easily oxidized or hydrolyzed. 2-Aminofurans substituted with electron-withdrawing groups (*e.g.* NO$_2$) are known and 3-amino-2-methylfuran is a relatively stable amine which can be acylated and diazotized. 2-Aminothiophene can be diazotized and the resulting diazonium salt coupled with β-naphthol. 2,3-Diaminothiophene has been prepared and isolated as the hydrobromide. The free base is not stable ⟨85JCR(S)296⟩.

Both 3-amino- and 3,4-diaminothiophene exhibit enaminic character. On acid-catalyzed deuteration, deuterium enters the thiophene α-positions. However, a detailed potentiometric as well as ^{1}H NMR study of the protonation of 3,4-diaminothiophene has revealed that both monoprotonation as well as diprotonation occurs on the nitrogen atoms (Scheme 68) ⟨93JOC4696⟩. The pK_a values in 50% DMSO − water mixture are 3.96 and 0.98.

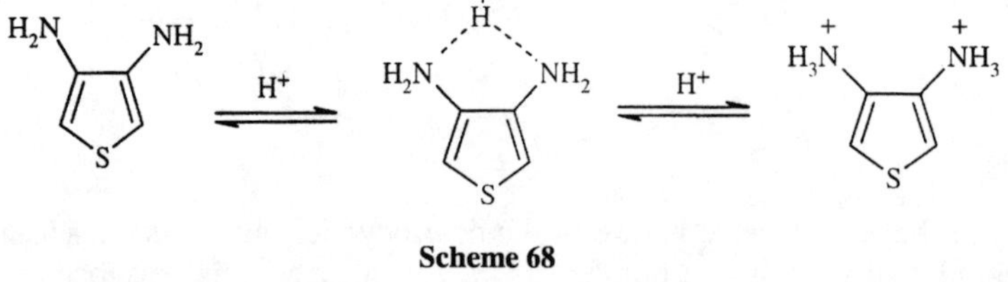

Scheme 68

Most of these compounds exist predominantly in the amino form. However, there are exceptions. 3-Amino-1-tritylpyrrole (**379**) appears to exist in solution exclusively in the imino-Δ^4-pyrroline form (**380**) ⟨83JCS(P1)93⟩. 2-Aminoindole (**381**) exists mainly as the 3H-tautomer (**382**), which is stabilized by amidine resonance.

(379)   (380)   (381)   (382a)   (382b)

The amino compounds form salts and acylamino derivatives which are considerably more stable.

Interesting ring-opening reactions can be initiated by proton loss from the amino group. Thus, 2-amino-3-ethoxycarbonylthiophenes (**383**) with ethanolic sodium ethoxide give cyanothiolenones (**384**) ⟨75JPR86⟩. In a similar sequence 2-amino-3-acetylfurans (**385**) are converted into 3-cyano-2-methylpyrroles (**386**) by aqueous ammonia ⟨78JOC3821⟩.

(383)   (384)   (385)   (386)

3.3.3.4.3 *Azides*

Five-membered ring heterocyclic azides are readily reduced (H_2S, $LiAlH_4$) to the corresponding amines ⟨82JOC3177⟩ (*cf.* Scheme 63). On thermolysis they lose nitrogen, thereby generating nitrenes. For example, thermolysis of 3-azido-2-vinyl derivatives in xylene at 120–130°C yields [3,2-*b*]-fused pyrroles (**387**); nitrene insertion into an *ortho*-disposed imino function similarly yields [3,2-*b*]-fused pyrazoles (**388**) ⟨76ACS(B)391, 77CS(12)1⟩.

(Z = O, S or Se)   (Z = O, S or Se)

(387)   (388)

3.3.3.5 *O*-Linked Substituents

Hydroxy compounds are considered in Section 3.3.2.5, with their non-aromatic carbonyl tautomers.

3.3.3.6 *S*-Linked Substituents

Thiophene-2-sulfonic acid is a strong acid, similar to benzenesulfonic acid. It forms a sulfonyl chloride with phosphoryl chloride which on reduction with zinc yields thiophene-2-sulfinic acid.

Pyrrolethiols, readily obtained from the corresponding thiocyanates by reduction or treatment with alkali, rapidly oxidize to the corresponding disulfides. They are converted into thioethers by reaction with alkyl halides in the presence of base. Pyrrole-, furan- and thiophene-thiols exist predominantly as such rather than in tautomeric thione forms.

The acid-catalyzed rearrangement of 1-methyl-3-phenylthioindole to 1-methyl-2-thiophenylindole proceeds by disproportionation to the 2,3-disulfide and 1-methylindole (Scheme 69) ⟨89CC63, 92JOC2694⟩.

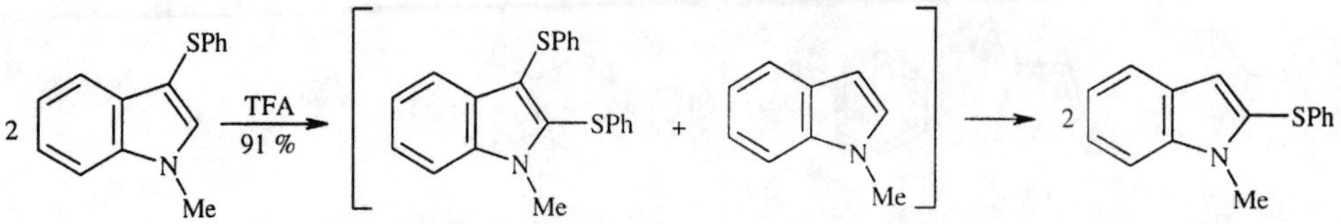

Scheme 69

3.3.3.7 Halo Groups

3.3.3.7.1 Nucleophilic displacement

Halopyrroles do not readily undergo nucleophilic displacement.

Halogen-substituted furans and thiophenes are also relatively inert, although their reactivity is greater than that of the corresponding aryl halides. Kinetic data are available for the nucleophilic displacement of halogen from 2-halofurans with piperidine. 2-Chlorofuran has about the same reactivity as bromobenzene and 2-chloro- and 2-bromo-thiophene have about a tenfold greater rate of reaction than the corresponding benzene compounds ⟨57JOC133⟩. As in the benzene series, the introduction of powerfully electron-withdrawing groups, such as nitro, carboxy or ester groups, greatly facilitates nucleophilic substitution. Halothiophenes which contain a nitro group react very much faster with nucleophilic reagents than the corresponding benzene derivatives, as shown by the rate data in Table 5.

As in benzenoid chemistry, numerous nucleophilic displacement reactions are found to be copper catalyzed. Illustrative of these reactions is the displacement of bromide from 3-bromothiophene-2-carboxylic acid and 3-bromothiophene-4-carboxylic acid by active methylene compounds (*e.g.* $AcCH_2CO_2Et$) in the presence of copper and sodium ethoxide (Scheme 70) ⟨75JCS(P1)1390⟩. Analogously, 2-methoxythiophene could be prepared in 83% yield by refluxing 2-bromothiophene in methanol containing excess sodium methoxide, along with copper(I) bromide as catalyst. For the analogous preparation of 3-methoxythiophene, addition of a polar co-solvent (*e.g.* 1-methyl-2-pyrrolidone) was found to be beneficial ⟨90SC213, 92T3633⟩. In the case of halothiophenes $S_{RN}1$ mechanism has been disclosed ⟨78OPP225⟩.

Table 5 Relative Pseudo-first-order Rates of Displacement of Bromonitrothiophenes and Bromonitrobenzenes with Piperidine at 25°C [a]

Compound	Rate
m-Bromonitrobenzene	1
p-Bromonitrobenzene	1.85×10^2
o-Bromonitrobenzene	1.62×10^3
5-Bromo-2-nitrothiophene	2.84×10^4
2-Bromo-3-nitrothiophene	6.32×10^5
5-Bromo-3-nitrothiophene	Very fast
4-Bromo-2-nitrothiophene	1.36×10^3
4-Bromo-3-nitrothiophene	Very fast

[a]⟨57NKK954⟩

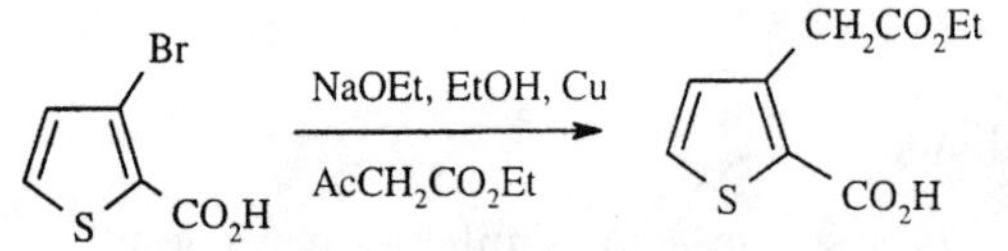

Scheme 70

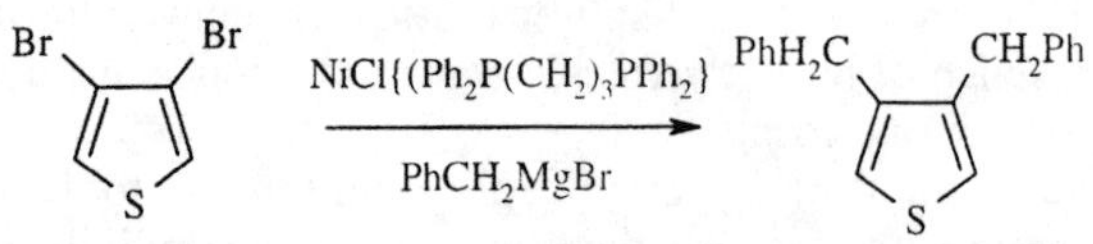

Scheme 71

3-Bromothiophenes give cross-coupled products by reaction with Grignard reagents in the presence of a nickel catalyst (Scheme 71) ⟨80TL4017⟩.

3.3.3.7.2   *Reductive dehalogenation*

Halogen can be removed by catalytic hydrogenation and so it is possible to use halogen as a blocking group in pyrrole chemistry. In the thiophene and selenophene series, α-halogens are preferentially removed by reduction with zinc and acetic acid, as illustrated by the preparation of 3-bromothiophene (**390**) from 2,3,5-tribromothiophene (**389**) ⟨81SC25⟩.

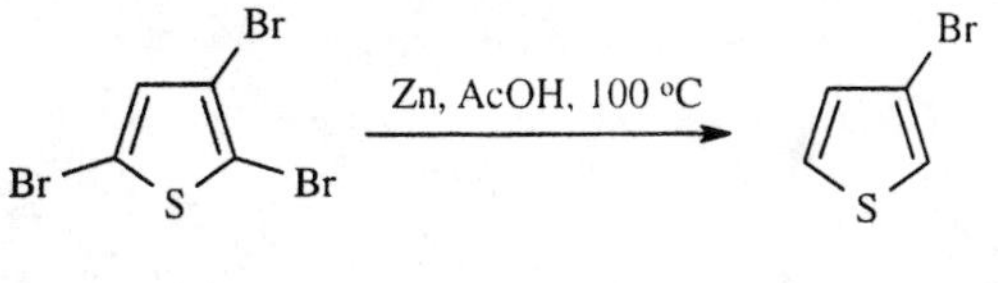

(**389**)   (**390**)

3.3.3.7.3   *Rearrangement*

α-Halothiophenes undergo rearrangement reactions in strongly basic media, resulting in the formation of thermodynamically more stable products. Based on this 'halogen dance' of thienyl halides ⟨84CHEC-(4)741⟩, methods have been developed for the large-scale synthesis of 3-bromo- and 3,4-dibromothiophene. Treatment of 2-bromothiophene with excess sodamide in liquid ammonia, and subsequent quenching with solid NH₄Cl, gives 3-bromothiophene in 73% yield ⟨90SC1697⟩. The corresponding reaction with potassium amide yields 3-aminothiophene ⟨71JOC2690⟩. The mechanism of the rearrangement is given in Scheme 72. The second step is a disproportionation between the metallated molecule and a second molecule of 2-bromothiophene to give 2,3-dibromothiophene which can then form the 3-bromo compound.

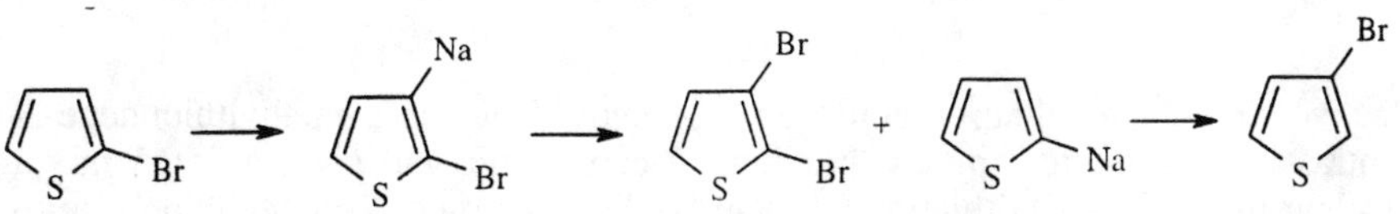

Scheme 72

3.3.3.7.4   *Formation of Grignard reagents*

Grignard reagents can be prepared from 2-bromothiophene and 2-iodofuran; these Grignard reagents show normal reactivity. 3-Iodothiophenes also react with magnesium but 3-bromothiophene only reacts by the entrainment method. The 3-bromo compound, however, reacts smoothly with butyllithium at − 70°C to give 3-thienyllithium. If the reaction is carried out at room temperature, 3-thienyllithium acts as a lithiating agent and an equilibrium mixture of thiophene, 2-lithiothiophene and 3-bromo-2-lithiothiophene is formed. 3-Lithiofurans can similarly be obtained from 3-halofurans and butyllithium.

3.3.3.8   Metallo Groups

3.3.3.8.1   *General*

Although a limited range of Grignard reagents is available, the most widely used group is undoubtedly the lithio group introduced by direct lithiation (see Section 3.3.1.6.2). The ready formation of the lithio derivatives of pyrroles, furans and thiophenes and their benzo-fused derivatives has had a most important impact on the chemistry of these heterocyclic systems. Reaction of the lithiated heterocycles with an extremely wide range of electrophiles leads to heterocyclic derivatives with carbon, nitrogen, oxygen, sulfur and halogen linked substituents.

The dianions derived from furan- and thiophene-carboxylic acids by deprotonation with LDA have been reacted with various electrophiles (Scheme 73). The furan dianions reacted efficiently with aldehydes and ketones but not so efficiently with alkyl halides or epoxides. The thiophene dianions reacted with allyl bromide, a reaction which failed in the case of the dianions derived from furancarboxylic acids, and are therefore judged to be the softer nucleophiles ⟨81JCS(P1)1125, 80TL5051⟩.

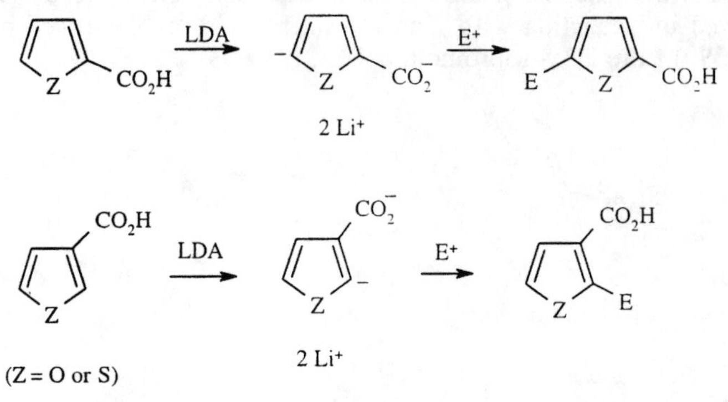

Scheme 73

The dianions of methylated thiophenecarboxylic acids (*e.g.* **391**) are also readily generated by reaction with LDA; they undergo preparatively useful reactions with a range of carbon electrophiles ⟨80JOC4528⟩.

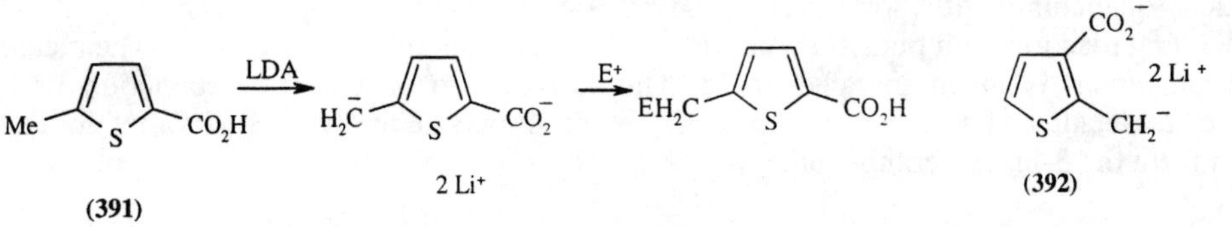

Regioselective side-chain lithiation could also be carried out on 2-methylthiophene-3-carboxylic acid. By contrast, the isomeric 3-methylthiophene-2-carboxylic acid gives a 35:65 mixture of side-chain and nuclear (at position 5) lithiation ⟨80JOC4528⟩. Dilithiation of 2-methylthiophene-3-carboxylic acid with 2 equivalents LDA in THF at $-30°C$ gave only the dilithium derivative (**392**) (80%), which could be reacted with electrophiles at the methylene carbon ⟨89BCJ2725⟩.

Treatment of tri-(2-thienyl)methane with Bu^{n}Li in the presence of TMEDA in THF at $-78°C$ gives exclusively the tri-(2-thienyl)methyllithium (**393**) (96%) without any nuclear lithiation ⟨92CL703⟩. This lithiation is faster than that of triphenylmethane. Treatment of (**393**) with primary alkyl halides leads to alkylation at the carbanion center, forming (**394**). However, secondary alkyl halides give mixture of (**394**) and (**395**).

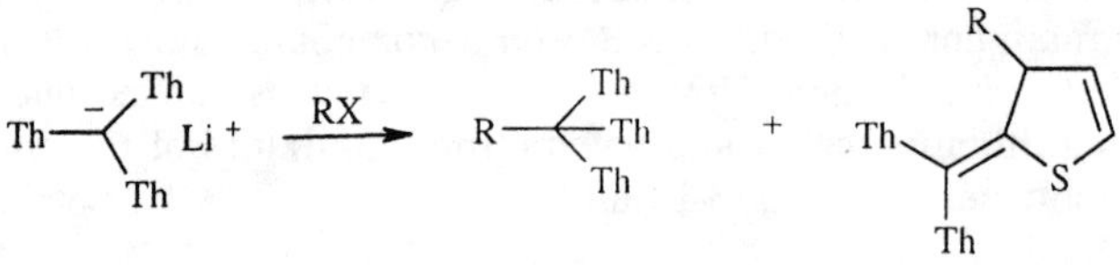

Th = 2-thienyl

To exploit the reactions of the *C*-lithio derivatives of *N*-unsubstituted pyrroles and indoles, protecting groups such as *t*-butoxycarbonyl, *t*-butylcarbamoyl, benzenesulfonyl, dimethylamino and dimethylaminomethyl have been used ⟨81JOC157⟩. This is illustrated by the scheme for preparing *C*-acylated pyrroles (**396**) ⟨81JOC3760⟩. Another useful process involves the *N*-lithiation, carbonation, and C-2 lithiation of indoles, which leads for example to 2-haloindoles in excellent yields ⟨92JOC2495⟩.

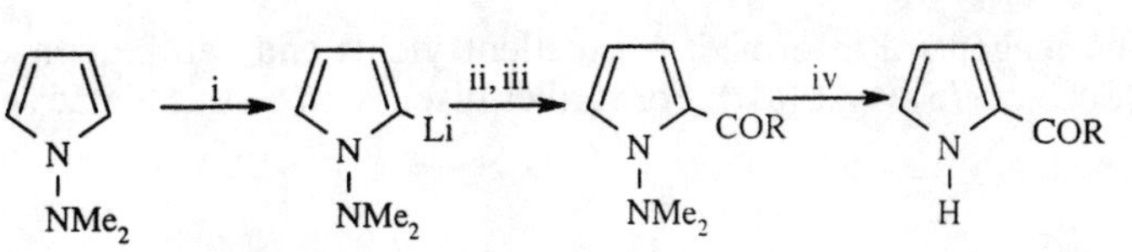

i, BuLi; ii, RCHO; iii, MnO$_2$; iv, Cr$_2$(OAc)$_4$·2H$_2$O   (396)   (397) Z = NR, O or S

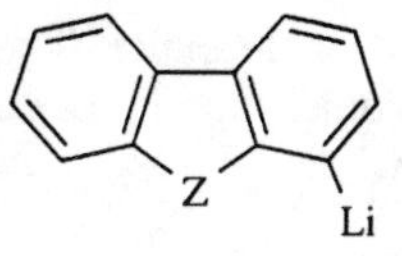

The reactions of the lithio derivatives of benzo[*b*]-fused systems indole, benzo[*b*]furan and benzo[*b*]thiophene are similarly diverse. Since indole and benzo[*b*]thiophene undergo electrophilic substitution mainly in the 3-position, the ready availability of 2-lithio derivatives by deprotonation with *n*-butyllithium is particularly significant and makes available a wide range of otherwise inaccessible compounds. The ready availability of 3-iodoselenophene and hence of 3-lithioselenophene ⟨73CHE845⟩ provides a convenient route to 3-substituted selenophenes. 2-Lithiotellurophenes are especially important precursors of tellurophene derivatives because of the restricted range of electrophilic substitution reactions which are possible on tellurophenes ⟨77AHC(21)119⟩.

Two cautions regarding the use of lithio derivatives need to be given: the possible incursion of rearrangement and ring-opening reactions ⟨78CHE353⟩. For example, the 3-lithio derivative of 1-benzenesulfonylindole, generated from the 3-iodo compound by low temperature (− 100°C) treatment with *t*-butyllithium, rearranges on warming to room temperature to the thermodynamically more stable 2-lithio species ⟨82JOC757⟩. For synthetic applications of ring-opening reactions see Section 3.3.3.8.7.

The lithio derivatives of the dibenzo heterocycles (397) are also preparatively useful since electrophiles attack these systems *para* to the heteroatom. For a review on the lithio heterocycles see ⟨79OR(26)1⟩.

3.3.3.8.2 *Formation of C—C bonds*

The reaction of lithio derivatives with appropriate electrophiles has been utilized in the preparation of alkyl, aryl, acyl and carboxylic acid derivatives. Representative example of these conversions are given in Scheme 74. Noteworthy is the two-step method of alkylation involving reaction with trialkylborane followed by treatment with iodine ⟨78JOC4684⟩.

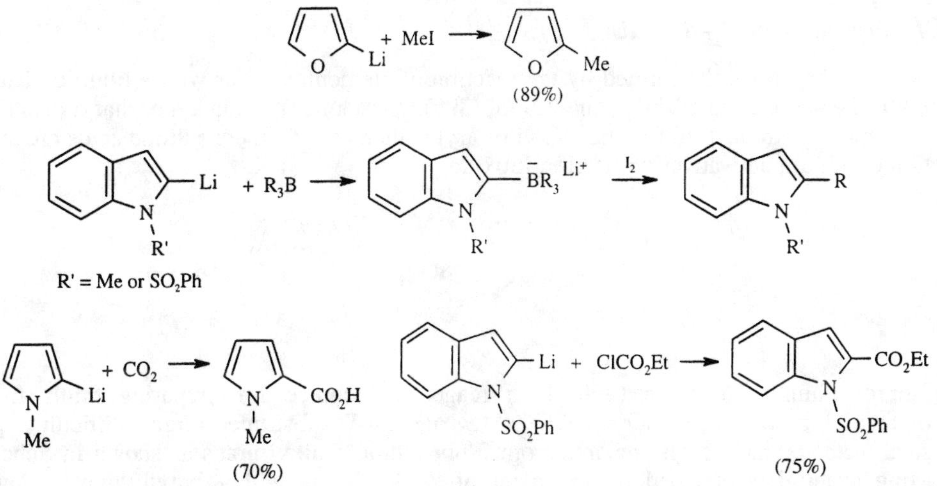

Scheme 74

Other carbon electrophiles which are frequently employed include aldehydes, ketones, esters, nitriles and amides of the type RCONMe$_2$. An indirect method of acylation involves the initial reaction of a lithio compound with an aldehyde followed by oxidation of the resulting secondary alcohol to the corresponding acyl derivative.

Heterocyclic derivatives of a range of metals other than lithium have received considerable attention, especially as precursors for coupling reactions. These derivatives can be prepared either directly from halo compounds or from the lithio compounds. Thus, direct formation of the pyrrolylzinc compounds can be effected under very mild conditions by treatment of an iodide with a zinc-silver couple deposited on graphite. The zinc reagents are formed in excellent yields and can be converted into acylated or allylated products (Scheme 75) ⟨94TL1047⟩. For further discussion on this theme see Section 3.3.3.8.8.

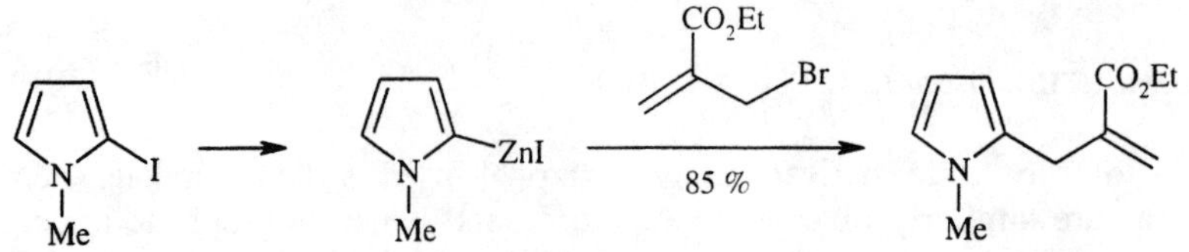

Scheme 75

3.3.3.8.3 Formation of C—O bonds

This can be achieved by an indirect method. The lithio derivative is first reacted with a borate ester. Sequential acid hydrolysis and oxidation yields the corresponding hydroxy derivative. This procedure is illustrated by the conversion of 2-lithiobenzo[*b*]thiophene to 2-hydroxybenzo[*b*]thiophene, which exists predominantly in the 2(3*H*)-one tautomeric form (**398**) ⟨70JCS(C)1926⟩.

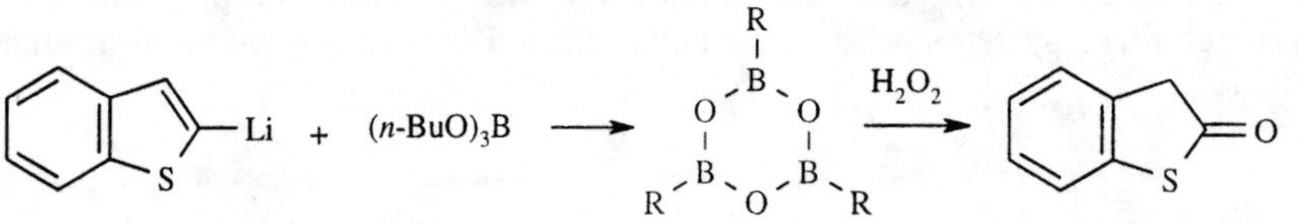

R = 2-benzo[*b*]thienyl

3.3.3.8.4 Formation of C—S bonds

Carbon-sulfur bonds can be formed by the reaction of elemental sulfur with a lithio derivative, as illustrated by the preparation of thiophene-2-thiol (**399**) ⟨70OS(50)104⟩. If dialkyl or diaryl disulfides are used as reagents to introduce sulfur, then alkyl or aryl sulfides are formed; sulfinic acids are available by reaction of lithium derivatives with sulfur dioxide.

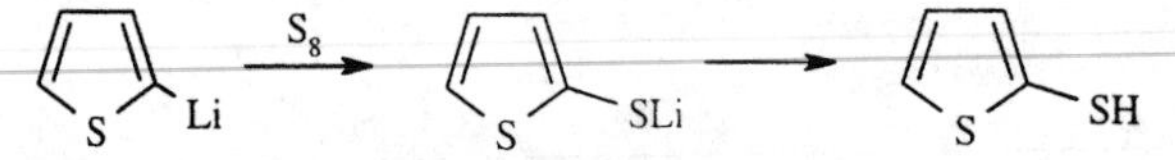

Tetraisopropylthiuram disulfide (**400**) is a reagent of choice for preparing thiols from the corresponding lithio derivatives (Scheme 76) ⟨82TL2001⟩. 2,4-Disubstituted furans, difficult to prepare by classical methods, have been prepared from 2-phenylthio-5-alkylfurans as shown in Scheme 77. The starting material is obtained by treatment of 2-alkylfurans with *n*-butyllithium followed by diphenyl disulfide ⟨81JOC2473⟩. The practicality of this approach thus illustrates the potential of the phenylthio group as a protecting group.

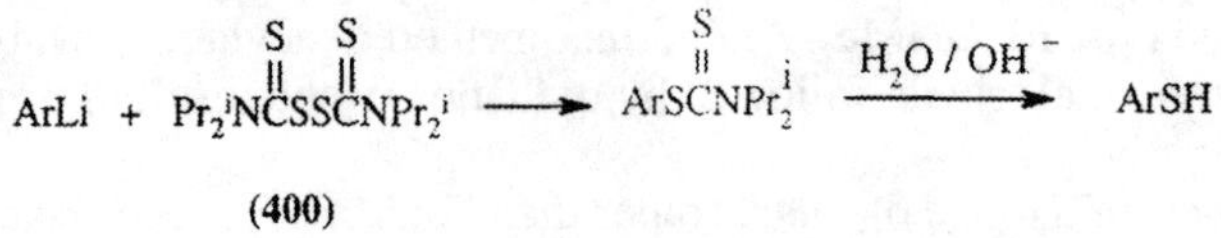

(**400**)

Ar = 2-furyl, 2-thienyl, 2-benzo[*b*]thienyl or 4-dibenzothienyl

Scheme 76

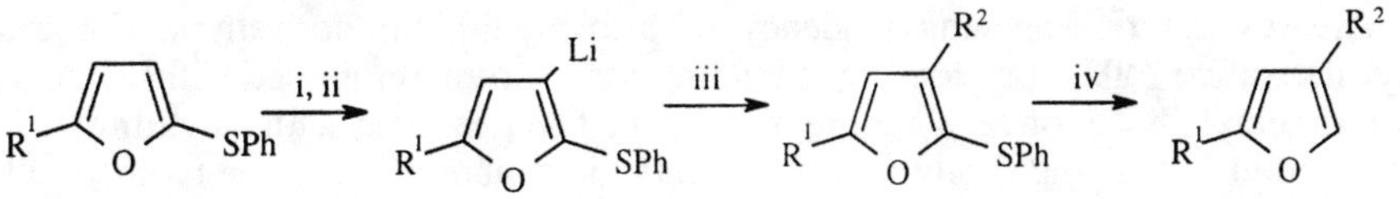

i , Br$_2$; ii, Bu^t Li; iii, alkyl iodide, aldehyde, carbon dioxide or trimethylsilyl chloride; iv, Raney Ni

Scheme 77

2-Thienyl- and 3-thienyl-(tributyl)stannanes react with chlorosulfonyl isocyanate at 20°C to give the corresponding sulfonyl isocyanates by *ipso*-substitution (Scheme 78). The yields are > 90% ⟨93JOC7022⟩.

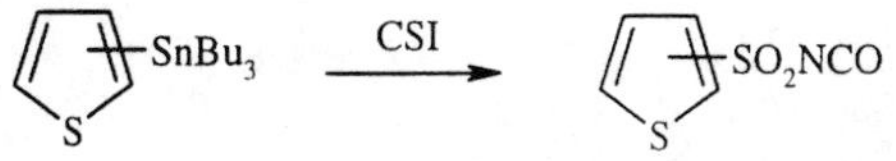

Scheme 78

3.3.3.8.5 Formation of C—N bonds

Azides are formed by the reaction of lithio derivatives with *p*-toluenesulfonyl azide ⟨69JOC3430, 82JOC3177; see also 82TL699⟩, and these in turn can be converted into the corresponding amino compounds by a variety of reductive procedures. Nitro compounds are available by a novel reversal of the general pattern of reaction with electrophiles. This approach requires the initial conversion of the lithio compound into an iodonium salt followed by reaction with nitrite ion. This is illustrated by the preparation of 3-nitrothiophene (Scheme 79) ⟨72CS(2)245⟩. Other nucleophiles, such as thiocyanate ion, which yields the 3-thiocyanate, can be employed. The preparative significance of these reactions is again that products not accessible by electrophilic substitution can be obtained.

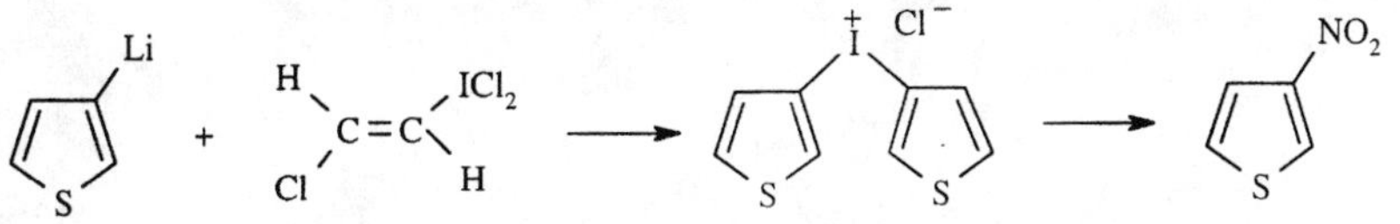

Scheme 79

3.3.3.8.6 Formation of C-halogen bonds

Synthetic procedures are available for the preparation of fluoro, chloro, bromo and iodo compounds from the corresponding lithio derivatives. Perchloryl fluoride (FClO$_3$), *N*-chlorosuccinimide, bromine and iodine are examples of reagents which can be used to introduce fluorine, chlorine, bromine and iodine, respectively.

3.3.3.8.7 Ring-opening reactions

The ring-opening reactions of lithiated derivatives have been reviewed comprehensively. A well-known example of this latter possibility is the ring-opening of 3-lithiobenzo[*b*]furan (**401**) to the lithium salt of 2-ethynylphenol (**402**) ⟨78CHE353⟩.

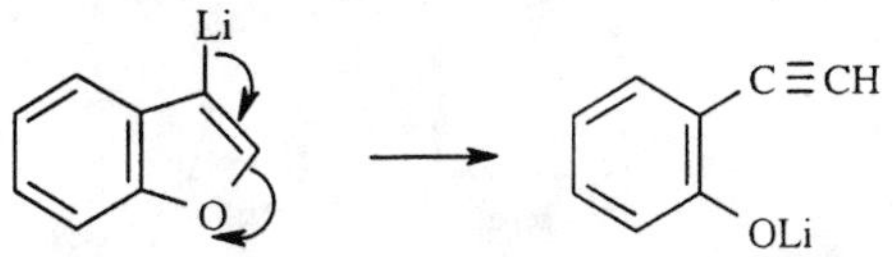

(**401**)   (**402**)

The tendency for the 3-lithio derivatives of furans and thiophenes to undergo ring opening has been exploited for the synthesis of polyunsaturated acyclic compounds. A trimethylsilyl group in the

2-position increases the ring-opening tendency of 3-thienyllithium derivatives. For example, the trimethylsilyl derivative (**403**), prepared by lithiating the 3-bromothiophene with LDA followed by reaction with trimethylsilyl chloride, smoothly ring opened on treatment with butyllithium. Subsequent reaction with methyl iodide and desilylation with potassium fluoride gave the terminal alkyne (**404**) ⟨82JOC374⟩. This sequence also shows that *ortho*-halolithiothiophenes are significantly more stable than the corresponding benzenoid derivatives which are used as benzyne precursors. The preparation of 3-bromothiophene-2-carbaldehyde (**405**) also illustrates this point.

3-Lithio-2,5-dimethylselenophene shows a much greater tendency to undergo ring opening than 3-lithio-2,5-dimethylthiophene ⟨77JHC1085⟩.

i, BuLi, hexane, ether, -20 °C, 30 min; ii, MeI; iii, KF·2H$_2$O, DMF

3.3.3.8.8   Palladium- and nickel-catalyzed cross-coupling reactions

A dominant theme since the mid-1980s in the chemistry of five-membered rings with one heretoatom has been the application of transition metal catalysis, especially the use of Pd or Ni as catalysts for bond formation (for review see ⟨92S413⟩).

The reaction of heterocyclic lithium derivatives with organic halides to form a C-C bond has been discussed in Section 3.3.3.8.2. This cannot, however, be extended to aryl, alkenyl or heteroaryl halides in which the halogen is attached to an *sp*^{2} carbon. Such cross-coupling can be successfully achieved by nickel or palladium-catalyzed reaction of the unsaturated organohalide with a suitable heterocyclic metal derivative. The metal is usually zinc, magnesium, boron or tin; occasionally lithium, mercury, copper, and silicon derivatives of thiophene have also found application in such reactions. In addition to this type, the Pd-catalyzed reaction of halogenated heterocycles with suitable alkenes and alkynes, usually referred to as the Heck reaction, is also discussed in this section.

2-Thienylmagnesium or zinc derivatives can be coupled with vinyl halides, bromo- or iodo-benzene, or ethynyl bromide under Ni or Pd catalysis (Equation 8). The reaction can be extended to the synthesis of β-styrylthiophene; in this case the double-bond configuration is retained in the product.

(8)

M = MgBr or ZnCl

R = -CH=CH$_2$, -Ph, -C≡C—C$_6$H$_{13}$, -CH=CH-Ph

Di- and poly-thienyls can be prepared by cross-coupling of 2- and 3-thienylboronic acids in the presence of Pd^0 catalyst. Another similar method is the cross-coupling of iodothiophenes with stannylthiophenes (Scheme 80).

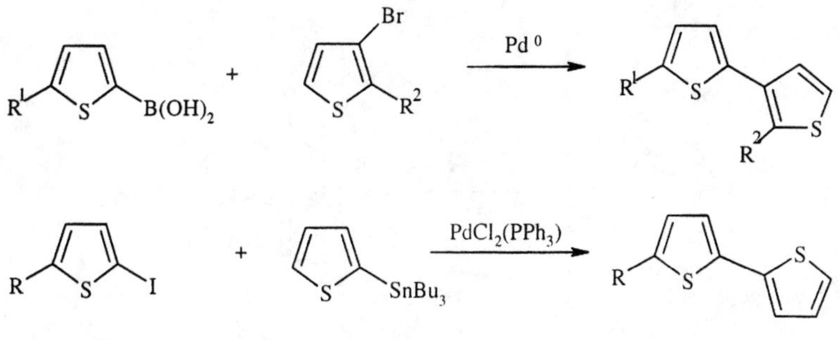

Scheme 80

An interesting application of acetylene coupling is the preparation of the per(acetylenated)thiophene (Equation 9) ⟨88JOC2489⟩.

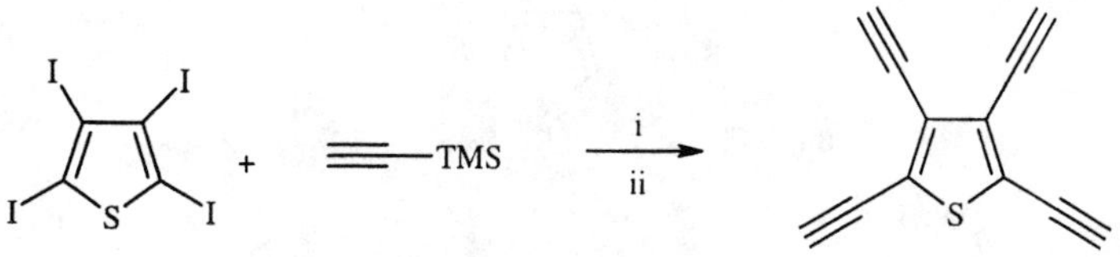

(9)

i, $Pr_2{}^iNH$, $(PhCN)_2PdCl_2$, Ph_3P, CuI

ii, KOH, MeOH

Much work has been directed towards the synthesis of thiophene oligomers and polymers. This is due to the current interest in research on conducting polymers and molecular electronics ⟨92CRV711⟩. Two main approaches have been used for making such polymers: (i) chemical (*e.g.* $FeCl_3$) or electrochemical oxidation of monomeric thiophenes and (ii) transition metal-catalyzed cross-coupling reactions.

Palladium-catalyzed coupling reactions of the Heck type have in many instances involved indole and pyrrole derivatives. Although the mechanisms are complex, organopalladium species are implicated ⟨84H(22)1493⟩. Vinylation of *N*-substituted-3-iodoindoles with amidoacrylate groups provides a useful functionalization of indoles (Scheme 81) ⟨90JOM(391)C23⟩. Yields are improved in intramolecular reactions, *e.g.* (**406 → 407**) and (**408 → 409**) ⟨92H(34)219, 91CPB2830⟩.

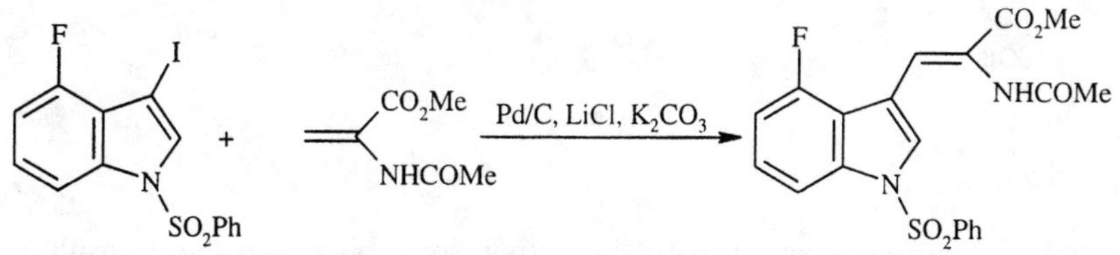

Scheme 81

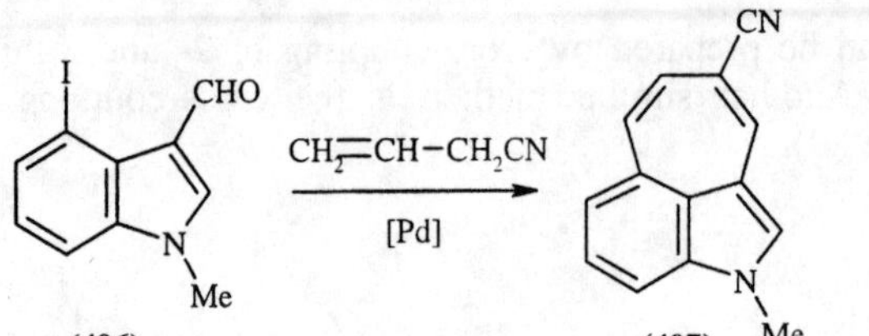

(406)   (407)

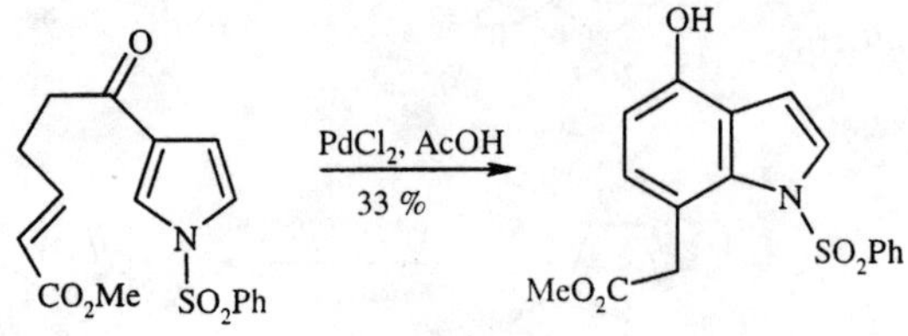

(408)   (409)

Pyrrolyl- and indolyl-stannanes and -boronic acids, which can be prepared from the corresponding organolithium derivatives, have received increasing use in palladium-catalyzed coupling reactions with aryl halides (Scheme 82) ⟨91S613, 92JOC1653⟩.

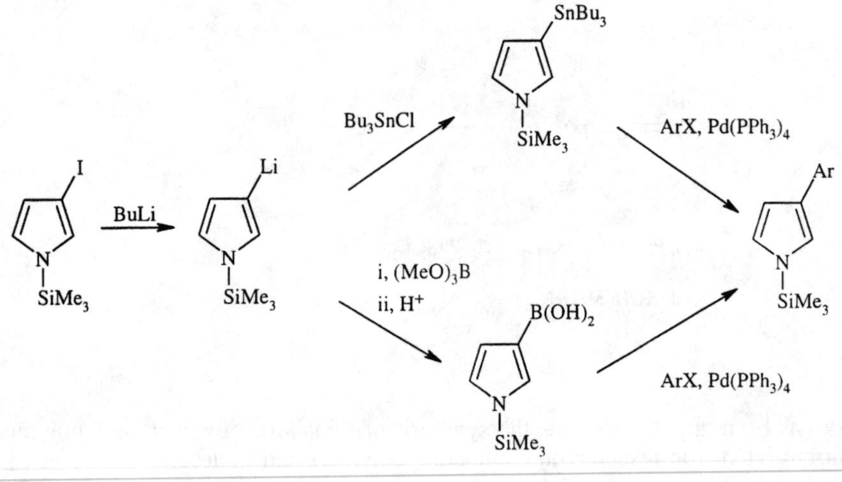

Scheme 82

The arylation of pyrroles can be effected by treatment with palladium acetate and an arene (Scheme 83) ⟨81CC254⟩.

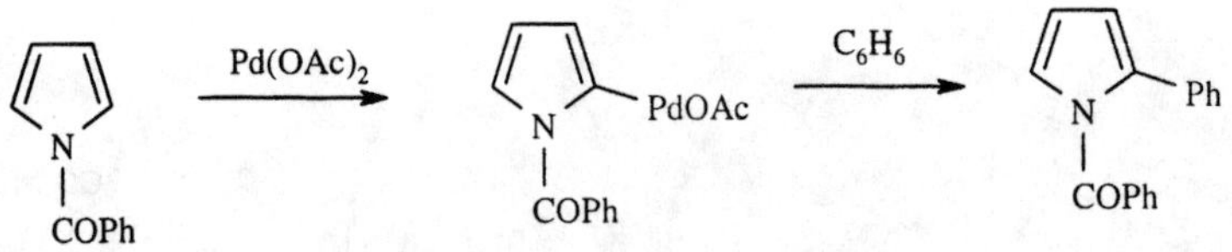

Scheme 83

The *N*-protected pyrrole (410) can be palladiated, but not lithiated, in the 3-position to give the stable complex (411); this is readily converted into the 3-methoxycarbonylpyrrole (412)

⟨82JOM(234)123⟩. The use of palladium derivatives thus further increases the range of transformations made possible through the intermediacy of metallo groups.

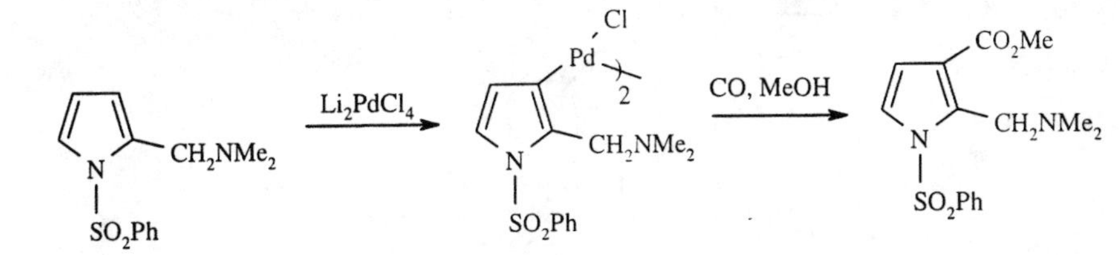

3.3.3.8.9 Mercury derivatives

The classical uses of organomercurials include the replacement of the mercuri group (R-HgCl or R-HgOAc) by hydrogen or halogen. Chloromercurated derivatives of furan, thiophene and selenophene can be acylated with acyl halides ⟨69JOM(17)P21⟩; the range of application of organomercurials seems likely to grow since they have been shown to undergo transmetallation by a variety of transition metal reagents, particularly palladium salts, thus increasing their synthetic potential ⟨82TL713⟩.

The mercuric derivatives can be used instead of the magnesium derivatives in the cross-coupling reaction (Scheme 84).

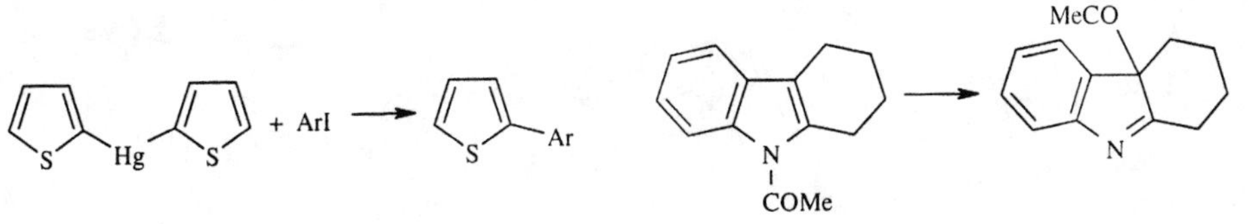

Scheme 84

3.3.3.9 Substituents Attached to the Pyrrole Nitrogen Atom

The thermal reactions of pyrroles include the rearrangement of *N*-substituted pyrroles to *C*-substituted derivatives (Scheme 85). The rearrangement of *N*-acylpyrroles has also been reported to occur in the vapor phase on irradiation.

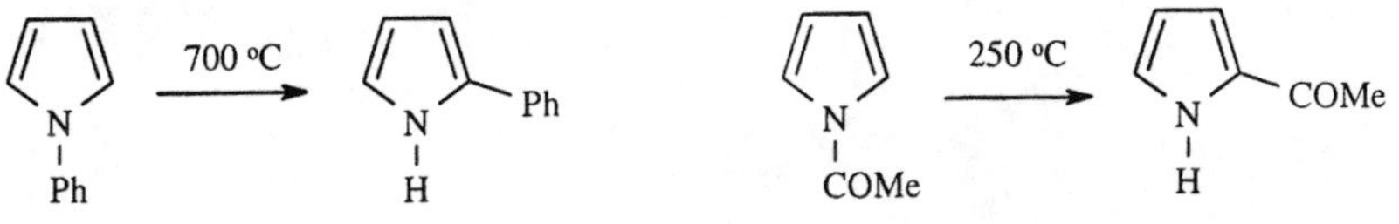

Scheme 85

Photoisomerization of 1-acylindoles yields 3-acylindolenines, as exemplified by the conversion of compound (**413**) into compound (**414**) ⟨81JA6990⟩.

Thermal rearrangement of *N*-chloropyrrole in methanol yields 2-chloropyrrole whereas acid-catalyzed rearrangement gives a mixture of 2- and 3-chloropyrrole and some 2,5-dichloropyrrole ⟨82JOC1008⟩.

Reactions of the *N*-vinyl group in *N*-vinylpyrroles have been reviewed in ⟨B-84MI 202–01⟩.

Both pyrrole and indole can be substituted cleanly on nitrogen by the use of 1-(benzotriazol-1-ylmethyl) derivatives. Reaction of these compounds with Grignard reagents gives substitution products. For example, phenylmagnesium bromide reacts with the indolyl derivative (**415**) to give 1-benzylindole ⟨93T2829⟩. The process can be extended by a sequence of lithiation and alkylation prior to Grignard attack to give branched alkyl products (Scheme 86).

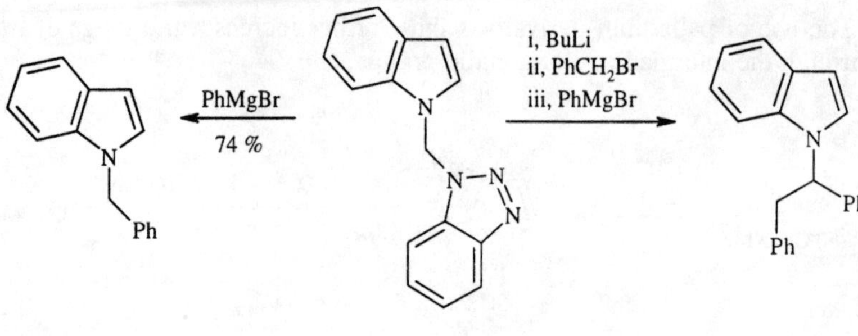

(415)

Scheme 86

3.3.3.10 Substituents Attached to the Thiophene Sulfur Atom

On heating the sulfonium ylide (**416**; R = H) the isomeric bis(methoxycarbonyl)methyl-thiophene (**417**) is formed ⟨78CC85⟩. Thermolysis of the ylide (**416**; R = Cl) yields the thienofuran (**418**) ⟨79CC336⟩. When heated in the presence of copper or rhodium catalysts, (**416**; R = Cl) undergoes cleavage of the carbon-sulfur bond resulting in the formation of carbenoid intermediates which can trapped with activated aromatic substrates or alkenes to yield the corresponding arylmalonates or cyclopropanes, respectively ⟨78CC83, 79CC50⟩.

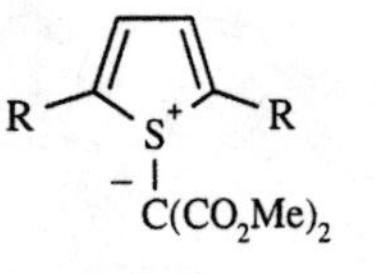

(416) 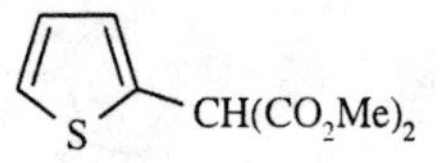   (417) 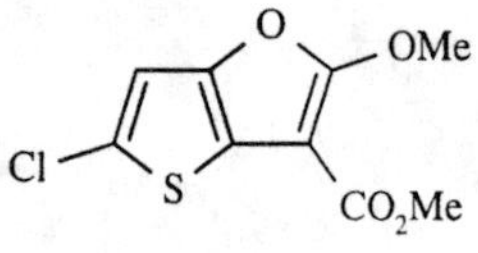   (418)

3.4

Reactivity of Five-membered Rings with Two or More Heteroatoms

3.4.1 REACTIONS AT HETEROAROMATIC RINGS

3.4.1.1 General Survey of Reactivity

In this initial section the reactivities of the major types of azole aromatic rings are briefly considered in comparison with those which would be expected on the basis of electronic theory, and the reactions of these heteroaromatic systems are compared among themselves and with similar reactions of aliphatic and benzenoid compounds. Later in this chapter all these reactions are reconsidered in more detail. It is postulated that the reactions of azoles can only be rationalized and understood with reference to the complex tautomeric and acid-base equilibria shown by these systems. Tautomeric equilibria are discussed in Chapter 2.4. Acid-base equilibria are considered in Section 3.4.1.3 of the present chapter.

3.4.1.1.1 Reactivity of neutral azoles

Replacing a CH group of benzene with a nitrogen atom gives pyridine (**1**); replacing a CH=CH group of benzene with NH, O or S gives pyrrole, furan or thiophene (**3**), respectively.

    (1)   (2)   (3)   (4)   (5)

The azoles (**4**) and (**5**) may be considered to be derived from benzene by two successive steps, one of each of these types. Hence, the chemistry of five-membered aromatic rings with two or more heteroatoms shows similarities to both that of the five- and that of the six-membered aromatic rings containing one heteroatom. Thus, electrophilic reagents attack lone electron pairs on multiply bonded nitrogen atoms of azoles (*cf.* pyridine) (see Section 3.4.1.3), but they do not commonly attack electron pairs on heterocyclic nitrogen atoms in NR groups or on heterocyclic oxygen or sulfur atoms (*cf.* pyrrole, furan, thiophene) (for example, see Section 3.4.1.5.1).

The carbon atoms of azole rings can be attacked by nucleophilic (Section 3.4.1.6), electrophilic (Section 3.4.1.4) and free radical reagents (Section 3.4.1.9.2). Some system, for example the thiazole, imidazole and pyrazole nuclei, show a high degree of aromatic character and usually 'revert to type' if the aromatic sextet is involved in a reaction. Others such as the isoxazole and oxazole nuclei are less aromatic, and hence more prone to addition reactions.

Electron donation from pyrrole-like nitrogen, or to a lesser extent from analogous sulfur or oxygen atoms, helps electrophilic attack at azole carbon atoms, but as the number of heteroatoms in the ring increases, the tendency toward electrophilic attack at both C and N decreases rapidly.

Just as electron displacement toward the nitrogen atoms allows nucleophilic reagents to attack pyridines at the α-position, similar displacements toward the nitrogens of the azoles also facilitate nucleophilic attack at carbon. As in similar reactions with pyridine, formation of the initial adduct involves dearomatization of the ring. The subsequent fate of the adduct depends in part on the degree of aromaticity. Those derived from highly aromatic azoles tend to rearomatize, whereas those of lower aromaticity can take alternative reaction paths. For most neutral azoles, nucleophilic attack at a ring carbon atoms is only possible with very strong nucleophiles.

Where azoles contain ring NH groups, this group is acidic and nucleophiles can remove a proton. Nucleophilic species can also remove ring-hydrogen atoms, particularly those which are α to a sulfur or oxygen atom, as in base-catalyzed hydrogen exchange and metallation reactions (Section 3.4.1.8).

3.4.1.1.2   Azolium salts

All neutral azoles possess a positively charged azolium counterpart. In addition, as discussed in Chapter 2.4, certain 'olylium' species exist which have no neutral counterparts, for example dithiolylium salts.

Azolium systems show much lower reactivity than the corresponding neutal azoles toward electrophiles at ring carbon. Even if an azolium ion contains an additional unquaternized pyridine-like nitrogen, this nitrogen is hardly basic in character. By contrast, azolium cations show a great reactivity toward nucleophiles: at ring carbon atoms, at the hydrogen of ring CH and NH groups, and even at ring sulfur atoms.

In all these azoliums, oxolyliums and thiolyliums, the positive charge facilitates attack by nucleophilic reagents at ring carbon atoms α or γ to the charged heteroatom (Section 3.4.1.6). Hydroxide, alkoxide, sulfide, cyanide and borohydride ions, certain carbanions, amines and organometallic compounds react under mild conditions, usually at a position α to the quaternary center as in (6), to give initial non-aromatic adducts (7) which can be isolated in certain cases but undergo further reaction with alacrity. The most important of these subsequent reactions include:

(i) oxidation, *e.g.* the formation of cyanine dyes, *e.g.* in the thiazole series (Section 3.4.1.6.5 ii);

(ii) ring opening with subsequent closure, *e.g.* the reactions of oxazoliums with amines (Section 3.4.1.6.2);

(iii) ring opening without subsequent closure, *e.g.* the reactions of oxazoliums with hydroxide ion (Section 3.4.1.6.5. ii)

(6)   (7)

Ring hydrogen atoms can be abstracted from the α-carbon atoms of azolium ions by strong bases, as demonstrated in base-catalyzed hydrogen exchange (Section 3.4.1.8.3).

3.4.1.1.3   Azole anions

Azole anions are derived from imidazoles, pyrazoles, triazoles or tetrazoles by proton loss from a ring NH group. In contrast to the neutral azoles, azole anions show enhanced reactivity toward electrophiles, both at the nitrogen (Section 3.4.1.3.6) and carbon atoms (Section 3.4.1.4.1.i). They are correspondingly unreactive toward nucleophiles.

3.4.1.1.4   Azolinones, azolinethiones, azolinimines

These compounds are usually written in the unionized form as in (8; Z = NH, NR, O, S). Canonical forms of types (9) or (10) are important, *i.e.* these compounds can also be considered as betaines formally derived from azolium ions. Many compounds of this type are tautomeric and such tautomerism is discussed in Section 2.4.5.2.

(8)   (9)   (10)

Reactions of these compounds follow logically from the expected electron displacements in the molecules. Their very varied chemical reactivity includes four main possibilities for heterolytic reactions: electrophilic attack at a ring carbon atom β to a ring heteroatom (*e.g.* Section 3.4.1.4.2), or at a carbonyl oxygen atom (Section 3.4.3.7), a thiocarbonyl sulfur atom (Section 3.4.3.8.2) or an imine nitrogen atom (Section 3.4.3.5.5). Nucleophilic attack to remove hydrogen from an NH group (Section 3.4.1.3.6), or at a ring carbon atom, or to a ring heteroatom (Section 3.4.3.12.3) also needs to be considered.

The mode of attack of electrophilic reagents (E$^+$) at ring carbon atoms is β to the heteroatoms as shown, for example, in (11) and (12); the intermediates usually revert to type by proton loss. Halogenation takes place more readily than it does in benzene (Section 3.4.1.4.5). Nitration and sulfonation also occur; however, in the strongly acidic environment the compounds required are present mainly as less reactive hydroxyazolium ions, *e.g.* (13).

(11)   (12)   (13)

The reactions of electrophilic reagents at a carbonyl oxygen atom, a thiocarbonyl sulfur, and an imino nitrogen atom are considered as reactions of substituents (see Sections 3.4.3.7, 3.4.3.8.2, 3.4.3.5.4 and 3.4.3.5.5).

The removal of a hydrogen atom from a heterocyclic nitrogen atom of azolones by nucleophiles acting as bases, *e.g.* (14)→(15), gives mesomeric anions, *e.g.* (15)↔(16)↔(17), which react exceedingly readily with electrophilic reagents, typically:

    (i) at nitrogen, *e.g.* with alkyl halides (Section 3.4.1.3.10);

    (ii) at oxygen, *e.g.* with acylating reagents (Section 3.4.3.7);

    (iii) at the β-carbon atoms, *e.g.* with halogens (Section 3.4.1.4.5), and in the Reimer-Tiemann reaction (Section 3.4.1.4.6).

(14)   (15)   (16)   (17)

3.4.1.1.5 N-Oxides, N-imides, N-ylides of azoles

Azole *N*-oxides, *N*-imides and *N*-ylides are formally betaines derived from *N*-hydroxy-, *N*-amino- and *N*-alkyl-azolium compounds. Whereas *N*-oxides (Section 3.4.3.12.7) are usually stable as such, in most cases the *N*-imides (Section 3.4.3.12.5) and *N*-ylides (Section 3.4.3.12.3) are found as salts, which deprotonate readily only if the exocyclic nitrogen or carbon atom carries strongly electron-withdrawing groups.

The reactivity of these compounds is somewhat similar to that of the azolonium ions, particularly when the cationic species is involved. However, although the typical reaction is with nucleophiles, the intermediate (20) can lose the *N*-oxide group to give the simple α-substituted azole (21). Benzimidazole 3-oxides are readily converted into 2-chlorobenzimidazoles in this way.

(18)   (19)   (20)   (21)

3.4.1.2 Thermal and Photochemical Reactions Formally Involving No Other Species

We consider here fragmentations and rearrangements which involve only the azole molecule itself, without the vital involvement of any substituent or other molecule. Many fragmentations of azoles can be summarized by the transformation (22) → (23) + (24), where (23) represents a stable fragment, particularly N_2, but also CO_2, N_2O, COS or HCN.

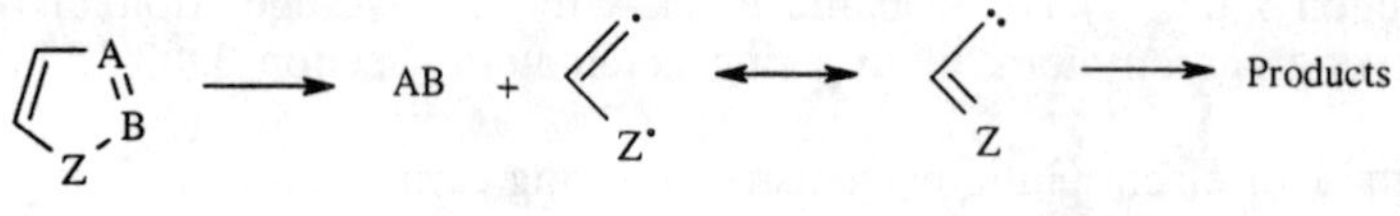

(22)   (23)   (24a)   (24b)

3.4.1.2.1 Thermal fragmentation

Thermal and photochemical fragmentation are often related to the mass spectroscopic breakdown of azole molecules (see Section 2.4.3.8), the latter often providing an indication of the behavior of a given molecule under such stimuli. Such fragmentations are facilitated in the polynitrogenous azoles, and azoles containing several nitrogen atoms undergo ring fission with loss of nitrogen. This is particularly noticeable when two adjacent pyridine-like nitrogen atoms are present. Thermolysis of (25) to give N_2 and a 1,3-dipole, such as Ph-C ≡ N-O, is a useful and general reaction. Azoles containing only two heteroatoms, such as the pyrazole and thiazole systems, are thermally very stable.

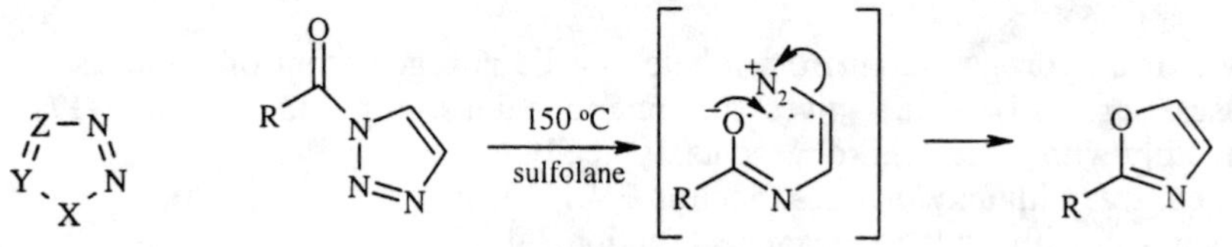

(25)

Scheme 1

Simple triazoles are thermally stable to *ca.* 300°C. However, triazole carboxamides, when heated to 150°C in sulfolane, rearrange with the elimination of nitrogen to give 2-substituted oxazoles. The reaction is general and it is useful for the synthesis of oxazoles with diverse 2-substitutents in excellent yields even for bulky substituents (Scheme 1). This reaction does not occur photochemically ⟨92TL1033⟩.

Thermal or photochemical extrusion of nitrogen from 1-arylbenzotriazoles (26) leads to the formation of carbazoles (29) (Scheme 2). The mechanism is believed to involve cyclization of a diradical (27b) or an iminocarbene (27c) to the 4a*H*-carbazole (28) followed by an aromatizing hydrogen shift.

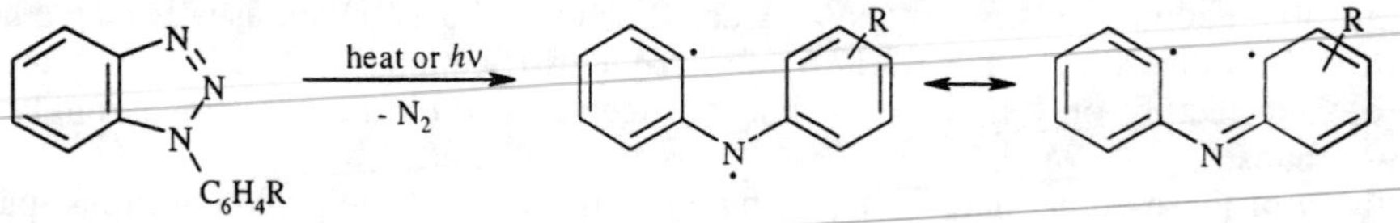

(26)   (27a)   (27b)

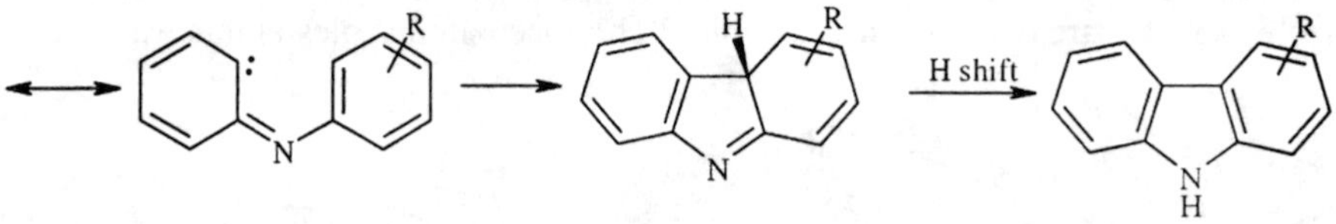

(27c)   (28)   (29)

Scheme 2

1-Vinylbenzotriazoles give indoles on flash pyrolysis at 600°C/10^{-2} Torr. However, depending on the vinyl substituents, side reactions leading to *N*-phenylketenimines or benzonitrile are also observed ⟨90JCS(P1)485⟩.

1,2,5-Oxadiazoles undergo thermal and photochemical ring cleavage at the O(1)-N(2) and C(3)-C(4) bonds to yield nitrile and nitrile oxide fragments, and products derived therefrom. Thus, diphenylfurazan (30, X = O) decomposes under flash vacuum pyrolysis conditions (600°C, 10^{-3} mm

Hg) affording benzonitrile and benzonitrile oxide in nearly quantitative yields ⟨81TH405-01⟩. Benzofurazans are thermally more stable but can be cleaved photolytically.

Although unsubstituted 1,2,5-thiadiazole is stable on heating at 220°C, 3,4-diphenyl-1,2,5-thiadiazole 1,1-dioxide (**30**, X = SO₂) decomposes into benzonitrile and sulfur dioxide at 250°C ⟨68AHC(9)107⟩.

Thermal reactions of 1,4,2-dioxazoles, 1,4,2-oxathiazoles and 1,4,2-dithiazoles are summarized in Scheme 3. The reactive intermediates generated in these thermolyses can often be trapped, *e.g.* the nitrile sulfide dipole with DMAD.

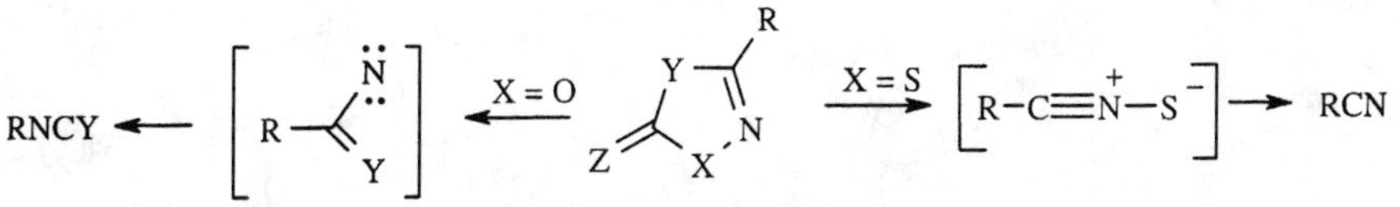

Scheme 3

A well-known reaction of tetrazoles is the thermolysis of 2,5-disubstituted tetrazoles to give nitrilimines by loss of N₂; an example is shown in Scheme 4. The reaction is used not only for its synthetic potential but also as a source of nitrilimines for direct observation and exploration of their behavior. The first direct detection of a thermally generated nitrilimine was the species (**31**) produced by flash vacuum pyrolysis of 5-phenyl-2-trimethylsilyltetrazole ⟨85AG(E)56⟩. The parent nitrilimine, HCNNH, has been generated by the photolysis of tetrazole ⟨96LA1041⟩.

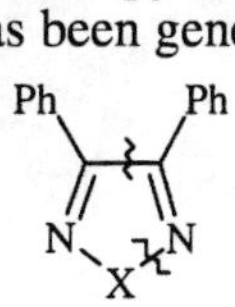

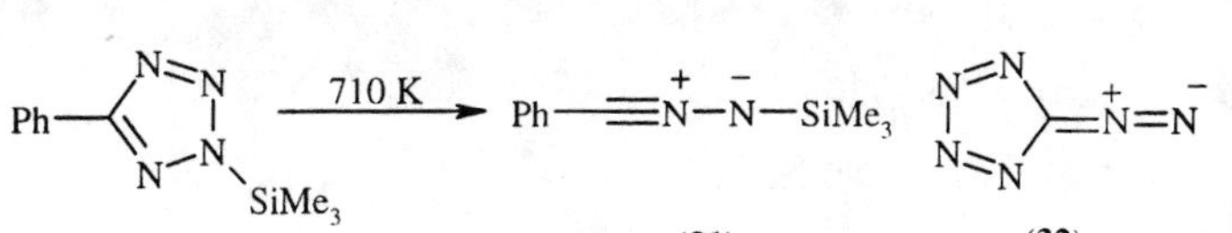

Scheme 4

Thermal degradation of ¹³C labeled 5-diazotetrazole (**32**) provides a source of atomic carbon for the investigation of its carbene reaction ⟨79JA1301⟩.

The flash vacuum pyrolysis of tetrazolo[1,5-*a*]pyridine (**33**) and its benzologues have been investigated in detail ⟨80JA6159, 82BSB997, 86AG(E)480, 92JCS(P1)1062⟩. All cases first gave nitrenes which rapidly underwent subsequent ring insertion, affording cyclic carbodiimides, *e.g.* (**34**) → (**35**).

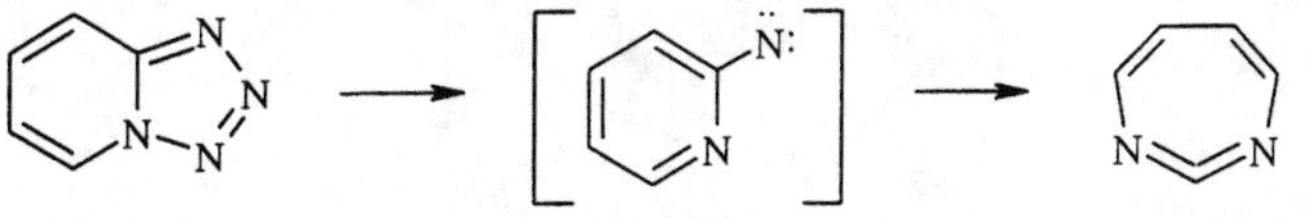

Other classes of heterocycles undergo thermolytic fragmentation to give imidolynitrenes. As typified by the thermolysis of 1,5-diphenyltetrazole (**36**), the intermediates (**37**) can either cyclize on to aromatic rings to form benzimidazoles (**38**) or undergo a Wolff-type rearrangement to carbodiimides (**39**) ⟨81AHC(28)231⟩. Compounds (**40**) and (**41**) thermolyze to give mainly the carbodiimides (**39**) ⟨79JA3976⟩. Pentazoles (**42**) spontaneously form azides, usually below 20°C.

1,2,3,4-Thiatriazoles readily decompose thermally into nitrogen, sulfur and an organic fragment, usually a cyanide, *e.g.* (**43**) → (Buⁱ OCN + ¹⁵N ¹⁴N + S) ⟨76AHC(20)145⟩.

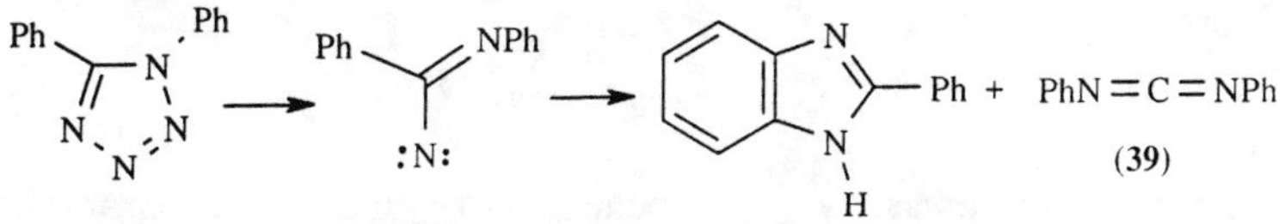

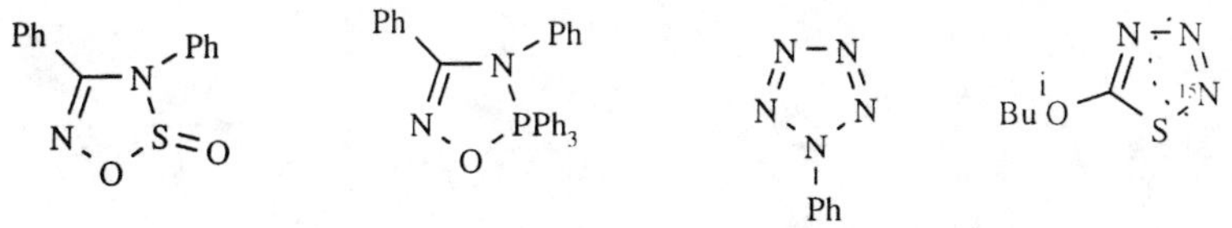

3.4.1.2.2 Photochemical fragmentation

Irradiation is very effective in promoting extrusion of nitrogen from triazole and benzotriazoles. For example, it is well-known that the photolysis of 1-arylbenzotriazoles afford a high yield of the corresponding carbazoles (see Scheme 2) ⟨81AHC(28)231⟩.

Photolysis of 1,2,3-thiadiazole (**44**) gives thiirene (**45**) which can be trapped by an alkyne ⟨70AHC(11)1⟩. 4,5-Diphenyl-1,2,3-thiadiazole (**46**) is photolyzed at low temperatures to the thiobenzoyl-phenylcarbene triplet (**47**). Diphenylthioketene (**48**) is formed on warming ⟨81AHC(28)231⟩.

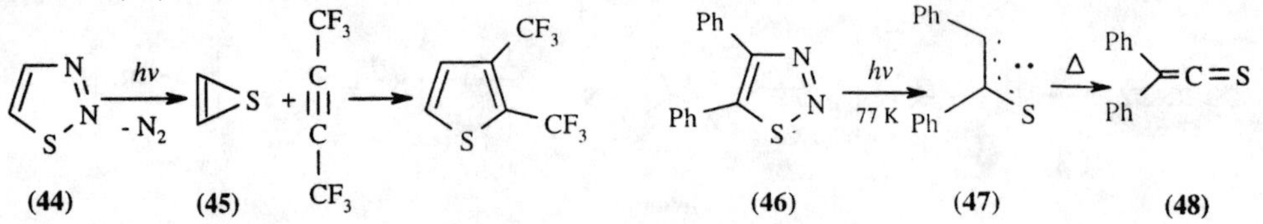

3-Phenylthiazirine (**49**) can be isolated as an intermediate in the photolysis of 5-phenyl-1,2,3,4-thiatriazole and also from other five-membered ring heterocycles capable of losing stable fragments; see Scheme 5 ⟨81AHC(28)231⟩. Photolysis of 5-phenylthiatriazole in the presence of cyclohexene yields cyclohexene episulfide ⟨60CB2353⟩ by trapping the sulfur atom.

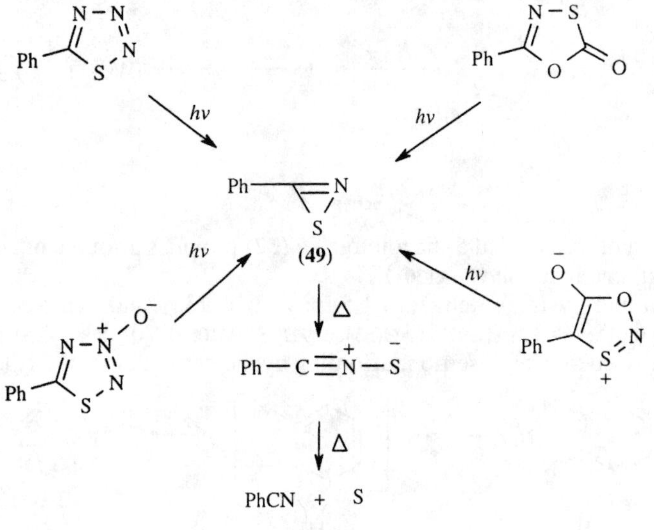

Scheme 5

Diaziridine derivatives (**51**) can be obtained from tetrazoles of type (**50**).

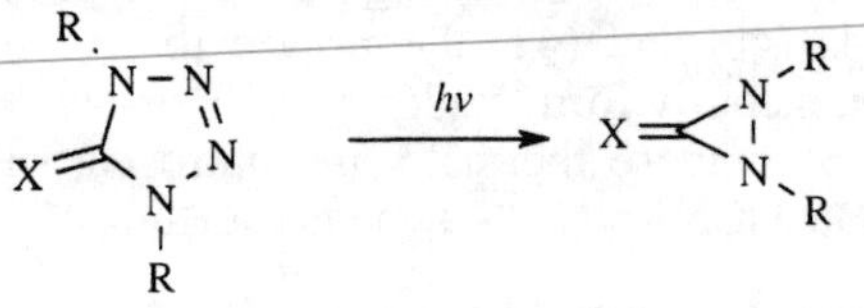

Mesoionic compounds undergo a variety of photochemical fragmentations. Examples are shown in which CO_2 or PhNCO is extruded (Schemes 6 and 7, respectively) ⟨76AHC(19)1⟩.

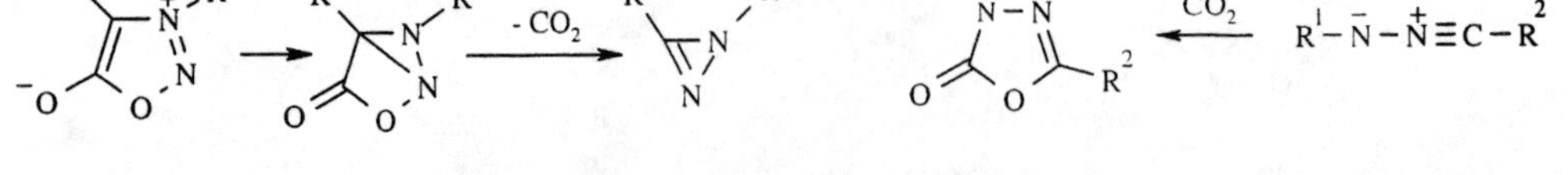

Scheme 6

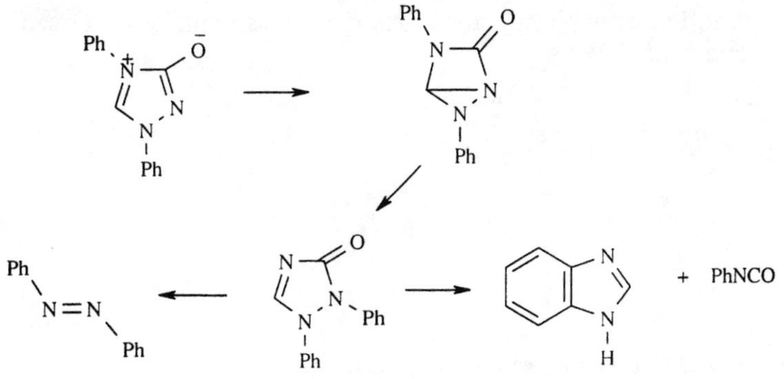

Scheme 7

3.4.1.2.3 Equilibria with open-chain compounds

Azoles of types (**53**) and (**55**) are isomeric with the open-chain compounds (**52**) and (**54**), respectively. Rearrangement between the two pairs is rapid, and the thermodynamically stable isomer is encounted. Thus diazoketones (**52**; X=O) exist as such, but diazothioketones (**52**; X=S) spontaneously ring-close to thiadiazoles (**53**). 1,2,3-Triazoles generally exist as such unless the nitrogen carries a strong electron-withdrawing substituent. Thus 1-cyano- and 1-arenesulfonyl-1,2,3-triazoles (**56**) undergo easy reversible ring-opening to diazo-imine tautomers ⟨74AHC(16)33⟩.

Stable for X = O
Can be stable for X = NR

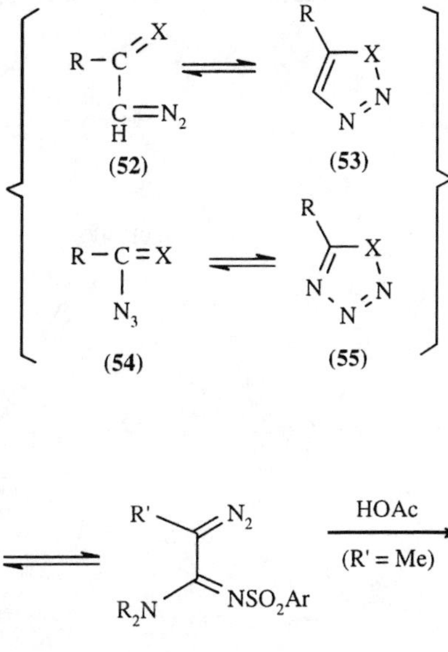

(**52**) (**53**) (**54**) (**55**)

Stable for X = S
Can be stable for X = NR

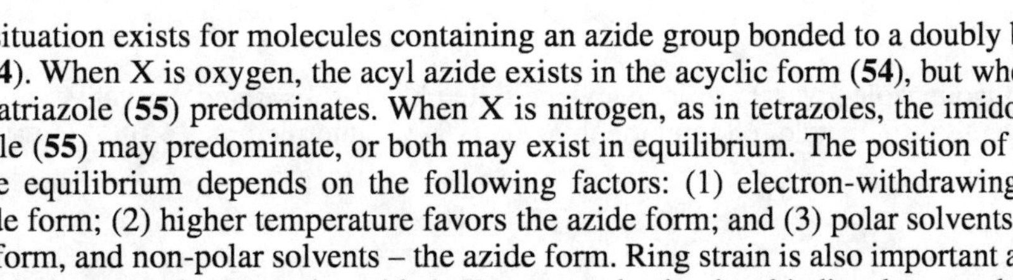

(**56**)

A similar situation exists for molecules containing an azide group bonded to a doubly bound carbon atom as in (**54**). When X is oxygen, the acyl azide exists in the acyclic form (**54**), but when X is sulfur the cyclic thiatriazole (**55**) predominates. When X is nitrogen, as in tetrazoles, the imidoyl azide (**54**) or the tetrazole (**55**) may predominate, or both may exist in equilibrium. The position of the tetrazole-imidoyl azide equilibrium depends on the following factors: (1) electron-withdrawing substituents favor the azide form; (2) higher temperature favors the azide form; and (3) polar solvents tend to favor the tetrazole form, and non-polar solvents – the azide form. Ring strain is also important and two fused five-membered rings are in general avoided. For example, in the thiadiazolotetrazole equilibrium (**57**)⇌(**58**), the system exists in the bicyclic form in the solid state and in the azide form in carbon tetrachloride solution ⟨77AHC(21)323⟩. Fusion with six-membered rings generally is more favorable to a bicyclic tetrazole form. For example, pyridine fusion gives essentially all tetrazole (**59**) ⟨69TL2595⟩ (see Sections 2.2.5.4 and 2.4.5.3.2).

(**57**)   (**58**)   (**59**)

3.4.1.2.4 *Rearrangement to other heterocyclic species*

Many examples are known of rearrangement of azoles involving scrambling of the ring atoms to give a new isomeric azole molecule. Different mechanisms are involved.

For isoxazoles the first step is the fission of the weak N—O bond to give the diradical (**60**) which is in equilibrium with the vinylnitrene (**61**). Recyclization now gives the substituted 2*H*-azirine (**62**) which *via* the carbonyl-stabilized nitrile ylide (**63**) can give the oxazole (**64**). In some cases the 2*H*-azirine, which is formed both photochemically and thermally, has been isolated, in other cases it is transformed quickly into the oxazole ⟨79AHC(25)147⟩.

MNDO results suggest that the activation energies are similar for the gas phase thermal isomerization of isoxazole to oxazole *via* either a nitrile ylide or a keteneimine, through an azirine intermediate. The first step is rate limiting, which is in good agreement with experimental results ⟨90JPO611⟩.

The photorearrangement of pyrazoles to imidazoles is probably analogous, proceeding *via* iminoylazirines (isomerization enthalpy ≈ 42 kJ mol^{-1}) ⟨82AHC(30)239⟩; indazoles similarly rearrange to benzimidazoles ⟨67HCA2244⟩. 3-Pyrazolin-5-ones (**65**) are photochemically converted into imidazolones (**66**) and open-chain products (**67**) ⟨70AHC(11)1⟩. The 1,2- and 1,4-disubstituted imidazoles are interconverted photochemically.

(**60**)   (**61**)

(**64**)   (**63**)   (**62**)

(**65**)   (**66**)   (**67**)

Irradiation of isothiazole gives thiazole in low yield. In phenyl-substituted derivatives an equilibrium is set up between the isothiazole (**68**) and the thiazole (**70**) *via* intermediate (**69**) ⟨72AHC(14)1⟩.

(**68**)   (**69**)   (**70**)

Iminobenzodithioles (**71**) and benzisothiazolethiones (**72**) thermally equilibrate ⟨72AHC(14)43⟩.

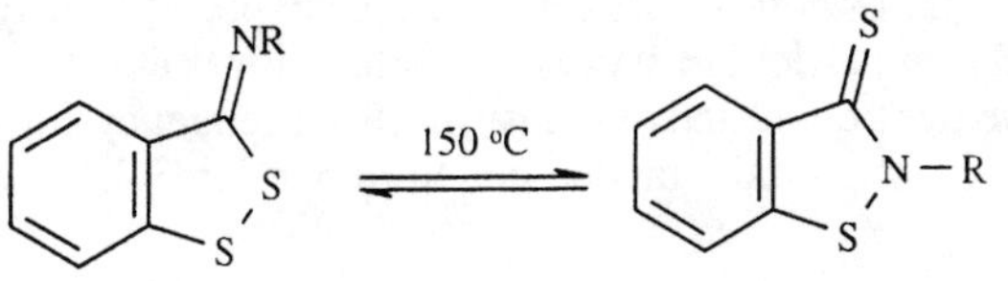

Various 3-heteroallyl-substituted furazans undergo rearrangements in which the oxadiazole is converted into a new five-membered heterocycle bearing a hydroxyiminoalkyl group (Scheme 8) ⟨81AHC(29)141, 90KGS1443⟩. The role of X=Y—Z can be filled by, for example, C=N-O, C=N-N, N=C-N and N=C-S, yielding 3-hydroxyiminoalkyl-1,2,5-oxadiazoles, -1,2,3-triazoles, -1,2,4-triazoles, and -1,2,4-thiadiazoles, respectively.

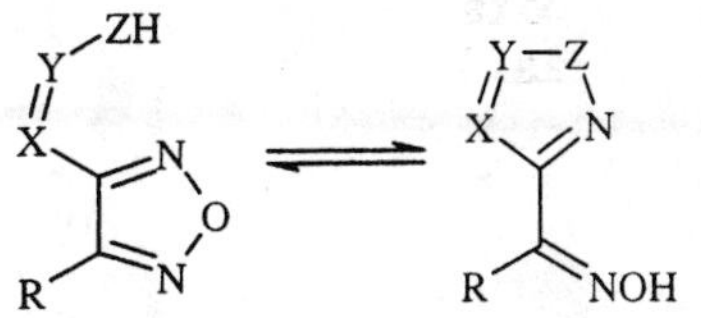

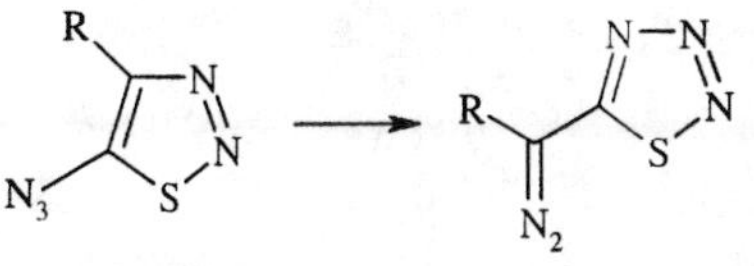

R=CO$_2$Et, PhCO, Ph, *p*-MeOC$_6$H$_4$, H

Scheme 8

Scheme 9

1,2,3-Thiadiazoles rearrange to variously substituted 1,2,3-thiadiazoles ⟨83CC588⟩. Many 5-azido-1,2,3-thiadiazoles rearrange to 1,2,3,4-thiatriazoles (Scheme 9) ⟨88BSB163⟩.

Mesoionic compounds of the type designated ⟨76AHC(19)1⟩ as "A" are capable of isomerism. In one case in the 1,2,4-triazole series, isomerism of the pair (**73**) ⇌ (**74**) has been demonstrated ⟨67TL4261⟩.

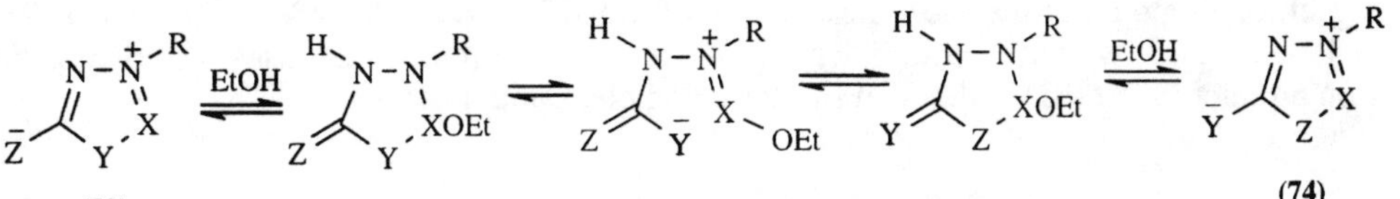

3.4.1.2.5 Polymerization

Imidazoles and pyrazoles with free NH groups form hydrogen-bonded dimers and oligomers ⟨66AHC(6)347⟩ (see Section 2.4.3.5).

3.4.1.3 Electrophilic Attack at Nitrogen

3.4.1.3.1 Introduction

Reactions of this type can be related to the chemistry of simple tertiary aliphatic amines. Thus the lone pair of electrons on the nitrogen atom in trimethylamine reacts under mild conditions with the following types of electrophilic reagents:

  (i)   proton acids give salts;
  (ii) Lewis acids give coordination compounds;
  (iii) transition metal ions give complex formation;
  (iv) reactive halides give quaternary salts;
  (v)   halogens give adducts;
  (vi) certain oxidizing agents give amine oxides.

The analogous reactions of pyridines with these electrophilic reagents at the lone pair on the nitrogen atom are well known. All neutral azoles contain a pyridine-like nitrogen atom and therefore similar reactions with electrophiles at this nitrogen would be expected. However, the tendency for such reactions varies considerably; in particular, successive heteroatom substitutions markedly decrease the ease of reaction. One convenient quantitative measure of the tendency for such reactions to occur is found in the basicity of these compounds; this is treated in Section 3.4.1.3.5 and 3.4.1.3.7.

3.4.1.3.2   Reaction sequence

In azoles containing at least two annular nitrogen atoms, one of which is an NH group and the other a multiply bonded nitrogen atom, electrophilic attack occurs at the latter nitrogen. Such an attack is frequently followed by proton loss from the NH group, *e.g.* (75) → (76). If the electrophilic reagent is a proton, this reaction sequence simply means tautomer interconversion (see Section 2.4.5.1.1), but in other cases leads to the product.

Since the electrophilic reagent attacks the multiply bonded nitrogen atom, as shown for (77) and (78), the orientation of the reaction product is related to the tautomeric structure of the starting material. However, any conclusion regarding tautomeric equilibria from chemical reactivity can be misleading since a minor component can react preferentially and then be continually replenished by isomerization of the major component.

In addition to reaction sequences of type (75) → (76), electrophilic reagents can attack at either one of the ring nitrogen atoms in the mesomeric anions formed by proton loss (*e.g.* 79 → 80 or 81; see Section 3.4.1.3.6). Here we have an ambident anion, and for unsymmetrical cases the composition of the reaction product (80) + (81) is dictated by steric and electronic factors.

3.4.1.3.3   Orientation in azole rings containing three or four heteroatoms

Such compounds contain two or three pyridine-like heteroatoms. For the symmetrical systems (82) and (83), no ambiguity occurs, but for systems (84)-(87) there are at least two alternative reaction sites. It appears that reaction takes place at the nitrogen atom furthest away from the pyrrole-like heteroatom, as shown in (84)-(86) where evidence is available from reactions with alkylating reagents (Section 3.4.1.3.9).

Similar ambiguities arise in the reactions of azole anions. At least as regards alkylation reactions in the 1,2,3-triazole series (**88**), the product appears to depend on the reagent used. In the 1,2,4-triazole series (**89**) a single product is formed, whereas tetrazole (**90**) gives mixtures.

(**87**)   (**88**)   (**89**)   (**90**)

3.4.1.3.4 *Effect of azole ring structure and of substituents*

The ease of attack by an electrophilic reagent at the nitrogen atom of any azole is proportional to ΔE between the ground state and transition state energies. However, ground state structure largely controls the variation in these differences, which hence depend on the electron density on the basic nitrogen atom and the degree of steric hindrance. The number, orientation and type of heteroatoms are very important in determining electron density. Additional pyridine-like nitrogen atoms always reduce the electron density at another pyridine-like nitrogen (compare the reduced basicities of diazines relative to pyridine). Unshared electron pairs on two pyridine-like nitrogen atoms can interact, but the effect on reactivity appears to be small ⟨78AHC(22)71⟩. In the case of pyrrole-like nitrogen, oxygen and sulfur there are two mutually opposed effects: base-strengthening mesomeric electron donation and base-weakening inductive electron withdrawal. The latter is particularly strong for heteroatoms in the opposition and in fact for oxygen and sulfur always dominates over the base-strengthening effect.

The effects of substituents may be rationalized as follows:

(i) Strongly electron-withdrawing substituents (*e.g.* NO_2, COR, CHO) make these reactions more difficult by decreasing the electron density on the nitrogen atom(s). The effect is largely inductive and therefore is particularly strong from the α-position.

(ii) Strongly electron-donating substituents (*e.g.* NH_2, OR) facilitate electrophilic attack by increasing the electron density on the nitrogen. This is caused by the mesomeric effect and is therefore strongest from the α- and γ-positions.

(iii) Fused benzene rings, aryl and alkyl groups, and other groups with relatively weak electronic effects have a relatively small electronic influence.

The foregoing electronic effects are illustrated by the pK_a values given in Section 3.4.1.3.5. Reactions other than proton addition are hindered by all types of α-substituents. However, steric hindrance is less in these five-membered ring heterocycles than that in pyridines because the angle subtended between the nitrogen lone pair and the α-substituent is significantly greater in the five-membered ring compounds and thus the substituent is held further away from the lone pair.

3.4.1.3.5 *Proton acids on neutral azoles: basicity of azoles*

For a general account of the basicity and acidity of azoles see ⟨87AHC(41)187⟩. Gas-phase pK_a values are discussed in Section 3.4.1.3.7.

Mesomeric shifts of the types shown in structures (**91**) and (**92**) increase the electron density on the nitrogen atom and facilitate reaction with electrophilic reagents. However, the heteroatom Z also has an adverse inductive effect; the pK_a of NH_2OH is 6.0 and that of N_2H_4 is 8.0, both considerably lower than that of NH_3 which is 9.5.

(**91**)   (**92**)   (**93**)

The basicities of the parent azole systems in water are shown in Table 1. When both heteroatoms are nitrogen, the mesomeric effect predominates when the heteroatoms are in the 1,3-positions, whereas the inductive effect predominates when they are in the 1,2-positions. The predominance of the mesomeric effect is illustrated by the pK_a value of imidazole (**91**; Z=NH), which is 7.0, whereas that of pyrazole (**92**; Z=NH) is 2.5 (*cf.* pyridine, 5.2). An *N*-methyl group is base-strengthening in imidazole, but base-weakening in pyrazole, probably because of steric hindrance to hydration. When the second heteroatom is oxygen or sulfur the inductive, base-weakening effect increases; the pK_a of thiazole (**91**; Z=S) is 3.5 and that of isoxazole (**92**; Z=O) is 1.3.

The most basic sites of 2-methyl- and 1-methyltetrazole were calculated at the 6-31G level to be N-4 in both cases with protonation energies of 220.5 kcal mol^{-1} and 224.1 kcal mol^{-1} respectively. The experimental pK_a values of the conjugate acids are 2-methyltetrazole, –3.25, and 1-methyltetrazole, –3.00.

Table 1 pK_a Values for Proton Addition[a]

Ring systems	X=NH	X=NMe	X=O	X=S
(ring structure)	2.52	2.06	–2.97	–0.51
(ring structure)	6.95	7.33	0.8	2.53
(ring structure)	1.17	1.25	—	—
(ring structure)	2.45	3.20	—	—
(ring structure)	(1.17)	< 1	–4.9[b]	–4.9
(ring structure)	1.31	0.42	–4.7	—
(ring structure)	—	2.02	–2.20	–0.05
(ring structure)	5.53	5.57	–0.13	1.2

[a] ⟨63PMH(1)1, 71PMH(3)1, B-76MI40200, B-76MI40201⟩.
[b] For methylphenylfurazan.

Substituents are expected to alter the electron density at the multiply bonded nitrogen atom, and therefore the basicity, in a manner similar to that found in the pyridine series. The rather limited data available appear to bear out these assumptions. The additional ring nitrogen atoms in triazoles, oxadiazoles, etc. are quite strongly base-weakening; this is as expected since diazines are weaker bases than pyridine. As regards *C*-substituents, their effects on the pK_a of the parent compounds are as follows:

(i) Methyl groups are weakly base-strengthening due to their mesomeric and inductive electron donor effect: thus in the methylthiazoles the base strengths decrease in the order 2 > 4 > 5.

(ii) Phenyl groups are weak resonance donors, but inductive acceptors. Phenyl groups are therefore expected to reduce the basicity of azoles.

(iii) Amino groups are strong resonance electron donors and hence base-strengthening, particularly if directly conjugated with the basic center.

(iv) Methoxy groups are resonance donors but inductive acceptors. The inductive effect would be expected to be dominant for azoles.

(v) Halogen atoms are inductive acceptors (and weak resonance donors); they are expected to cause a marked decrease in basicity, especially from α-positions.

(vi) Fused benzene rings usually have considerably base-weakening effect; *cf.* the pK_a values of imidazole and benzimidazole ⟨84JHC269⟩. Substituents on the benzene ring in benzazoles should have little effect on the basicity.

Annular nitrogen atoms can form hydrogen bonds, and if the azole contains an NH group, association occurs. Imidazole (**93**) shows a cryoscopic molecular weight in benzene 20 times that expected. Its boiling point is 256°C, which is higher than that of 1-methylimidazole (198°C).

Hydrogen bond basicity is of much relevance to the problem of drug design. Hydrogen bond basicity was shown to correlate with the location of the electrostatic potential local minimum along the axis of the nitrogen lone pair in a series of heterocycles ⟨94JCS(P2)199⟩. The experimental and calculated basicities for oxazole, 2,4,5-trimethyloxazole, and pyridine are shown in Table 2.

Table 2 Hydrogen bond basicities (log K_b)[a]

Compound	Experimental	Calculated
Oxazole	1.67	1.91
2,4,5-Trimethyloxazole	2.65	2.59
Pyridine	2.52	2.51

[a] ⟨89JCS(P2)1355⟩.

3.4.1.3.6 Proton acids on azole anions: acidity of azoles

The acidities of the five parent compounds are compared with that of pyrrole in Table 3. The acidity of the ring system increases as the number of nitrogens increases, the acidity of pyrrole increasing by approximately 2, 4.5 and 5 pK_a units for each successive addition of a nitrogen atom. 1,2,3-Triazole is slightly more acidic than 1,2,4-triazole, but the effect on NH acidity of nitrogen orientation is much less than the effect of the total number of nitrogens ⟨71PMH(3)1⟩.

Table 3 pK_a. Values of Azoles for Proton Loss[a]

Nitrogen positions	pK_a	Nitrogen positions	pK_a
1	16.5	1,2,3	9.26
1,2	14.21	1,2,4	10.04
1,3	14.44	1,2,3,4	4.89

[a] ⟨B-76MI40200, B-76MI40201⟩.

The following data show the NH-acidity (water, 20°C) of some benzazoles. The benzene ring significantly increases the acidity:

Indazole	$pK_a = 13.8$;
Benzimidazole	$pK_a = 12.9$
Purine	$pK_a = 8.93$
Benzotriazole	$pK_a = 8.57$;

Ring substituents can have a considerable effect on the acidity of the system. In the 1,2,4-triazole series a 3-amino group decreases the acidity to 11.1, a 3-methyl group to 10.7, whereas a 3-phenyl group increases the acidity to 9.6, and 3,5-dichloro substitution to 5.2 ⟨71PMH(3)1⟩.

3.4.1.3.7 Basicity and acidity in gas phase

Understanding the behavior of organic bases in solution requires some knowledge of their gas phase (intrinsic) basicities (proton affinities (PA)). These can be determined by ICR methods or by variable-temperature pulsed high-pressure mass spectrometry. Both methods afford basicities (termed thermodynamic *vs.* kinetic basicity), which have been compared in ⟨91JOC179⟩.

Some quoted PA values (kJ mol^{-1}) include: imidazole, 935; 2-methylimidazole, 954; 1-methylimidazole, 950; 4-methylimidazole, 946 ⟨84JOC4379, 90JA1303⟩, 2,4,5-trimethylimidazole, 975; pyrazole, ≈890 ⟨83JCS(P2)1869, 90JA1303⟩, oxazole, 892 ⟨91JA4448⟩ (*cf.* pyridine, 952). The azolium ions have been found to be more sensitive to methyl substituent effects than their conjugate bases. Indeed, the effect of a methyl group on PA comes primarily (> 70%) from interactions in the charged form. C-Methyl groups vicinal to the basic center (N-3) confer extra stabilization because of methyl hydrogen-lone pair interactions, but an N-methyl group at N-1 has a different influence due to partial loss of hyperconjugation. Plots of gas phase basicity against aqueous phase basicity gives different straight lines for NH- and 1-methyl-imidazoles ⟨90JA1303⟩.

Theoretical studies of the basicity of pyrazoles, using the semiempirical approximations as well as the STO-3G and 4-31G methods ⟨83JCS(P2)1869, 84JA6552, 90JA1303, 90JST(205)367⟩ have enhanced the understanding of the differences in basicity between the gas phase and the aqueous solution ⟨84JOC4379⟩. To rationalize the relative gas-phase and solution basicity and acidity of pyrazole, it is necessary to take into account the lone pair/lone pair repulsion in pyrazolate anion (6.5 kcal mol^{-1}), the adjacent NH/lone pair attraction in pyrazole (1.0 kcal mol^{-1}) and the NH$^+$/NH$^+$ repulsion in the pyrazolium cation (6.5 kcal mol^{-1}). Solvation by water, and to a lesser extent by DMSO, modifies these values to the point that the position of the equilibria can be reversed ⟨86JA3237⟩.

The acidity and basicity in the gas phase and in aqueous solution of pyrazole, 1-methylpyrazole, indazole, 1-methyl-, and 2-methylindazole have been measured ⟨88JA4105⟩. From these data it is possible to determine the annulation effect on going from an azole to the corresponding benzazole on the gas phase acidity (6–8 kcal mol^{-1} increase) and on the gas phase basicity (2 kcal mol^{-1} increase). Similarly, it was shown that benzimidazole (PA 954–962 kJ mol^{-1}) is protonated about 40 times faster than imidazole (PA 934 kJ mol^{-1}). The reason for the difference between gas phase and solution basicities is mainly a function of the polarizability of the annulated ring system; this effect disappears in aqueous solution by dispersion of the positive charge through hydrogen bonds ⟨83AG(E)323⟩ (see also ⟨83H(20)1713, 84JHC269⟩).

Although imidazole is a stronger acid than pyrazole in the gas phase (by 2.6 pK_a units) and in DMSO (by 1.3 pK_a units), pyrazole is a slightly stronger acid (by 0.2 pK units) in aqueous solution. This can be explained in terms of the equilibrium shown in Equation (1). In the gas phase electrostatic repulsion between adjacent lone pairs of electrons shifts the equilibrium to the left ⟨86JA3237, 88CHE469⟩.

(1)

Calculated protonation energies of azoles may be used as measures of basicity where the protonation energy is calculated as the difference between the energy of the azole molecule and the most stable protonated species. *Ab initio* calculations at 6-31G*//6-31G level gave the following protonation energies for the azole series ⟨86JPC5604⟩: imidazole, 210.1 (318) kcal mol^{-1}; pyrazole, 227.0 (312); 4H-1,2,4-triazole, 231.8 (307); 1H-1,2,4-triazole, 225.1 (300); 2H-tetrazole, 209.6 (294); 1H-tetrazole, 213.4 (293). Data followed the reverse order of the σ ionization potentials and suggested that the protonation energy of an azole is related to the energy of ionizing an electron from the σ-framework of the molecule. Tetrazoles are the weakest bases among the azoles. Protonation of both 1H- and 2H-tetrazole is predicted to occur at N-4 ⟨86JPC5597⟩.

3.4.1.3.8   *Metal ions* ⟨99AHC(72)1⟩

(i)   Simple complexes

Many examples are known of complexes between metal cations and both neutral azoles and azole anions. Overlap between the *d*-orbitals of the metal atom and the azole π-orbitals is believed to increase the stability of many of these complexes.

Despite the weak basicity of isoxazoles, complexes of the parent methyl and phenyl derivatives with numerous metal ions such as copper, zinc, cobalt, *etc.* have been described ⟨79AHC(25)147⟩. Many transition metal cations form complexes with imidazoles; the coordination number is four to six ⟨70AHC(12)103⟩. The chemistry of pyrazole complexes has been especially well studied and coordination compounds are known with thiazoles and 1,2,4-triazoles. Tetrazole anions also form good ligands for heavy metals ⟨77AHC(21)323⟩.

Isothiazoles react with hexacarbonyls M(CO)$_6$ to give *N*-coordinated M(CO)$_5$ derivatives.

(ii)   Chelate complexes

This is a field in rapid development ⟨90CCR227⟩.

Chelate rings can be formed by azoles containing α-substituents such as *o*-hydroxyphenyl, carbonyl or CH=NH groups. An important bidentate chelating agent is histidine (**94**), and many pyrazoles with substituent groups are known which form bis and tris complexes with many metals, *e.g.* (**95**). Similarly, 2- and 4-α-pyridylthiazoles are bidentate chelating agents. Complex formation of this type has analytical applications; thus 1,3,4-thiadiazole-2,5-dithione has been used as a spot test for bismuth and other metals ⟨58MI40200⟩.

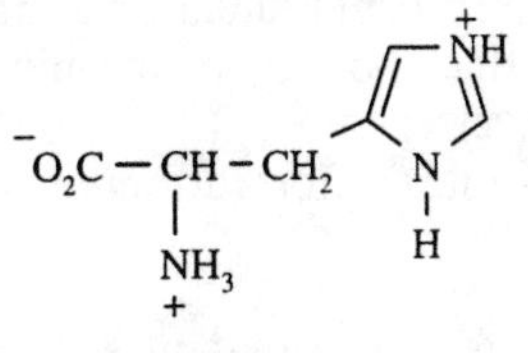

(94) 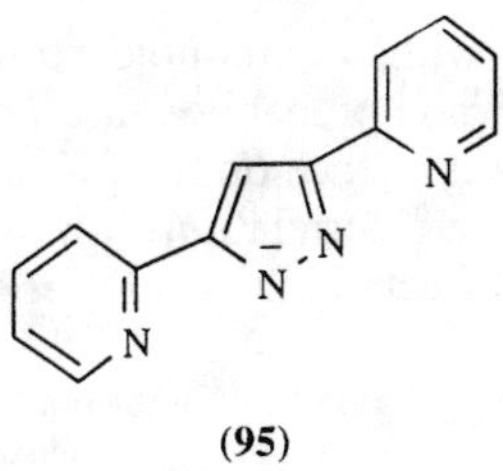 (95)

The extraordinary success of polypyrazolylborate ligands (scorpionates) was summarized ⟨93CRV943⟩.

3.4.1.3.9 *Alkyl halides and related compounds: azoles without a free NH group*

Pyrazoles and imidazoles carrying a substituent on nitrogen, as well as oxazoles, thiazoles, *etc.*, are converted by alkyl halides into quaternary salts. This is illustrated by the preparation of thiamine (**98**) from components (**96**) and (**97**).

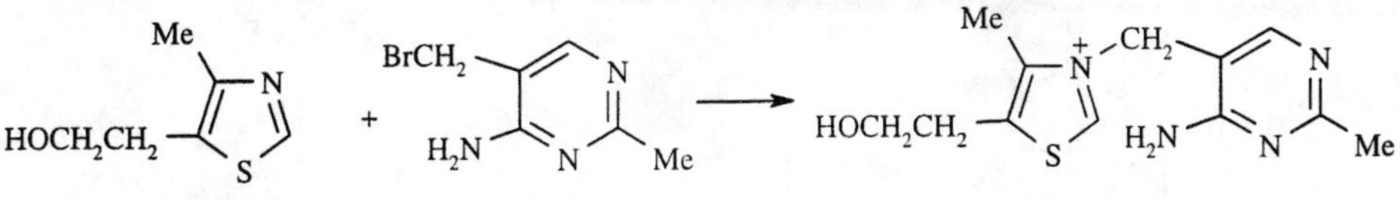

(96) (97) (98)

Azoles having heteroatoms in the 1,3-orientation are more reactive than those in which the arrangement is 1,2. However, the magnitude of the factor varies. Thus oxazole is 68 times more reactive than isoxazole, whereas benzoxazole quaternizes 26 times faster than does 1,2-benzisoxazole ⟨78AHC(22)71⟩.

These reactions are of the S_N2 type and are sensitive to steric effects of substituents in the azole ring. However, these steric effects are significantly less than, for example, in the analogous pyridine derivatives because the angle subtended by the nitrogen lone pair and an α-substituent is about 70° in an azole as opposed to 60° in pyridine. Thus the rate constant for methylation of 2-*t*-butylthiazole by methyl iodide is only 40 times less than that for the corresponding 2-methyl compound. By comparison, in the pyridine series the retardation factor is over 2000 in the same solvent (nitrobenzene). As in six-membered rings, the kinetic consequences of the steric effects of *o*-alkyl groups are related to the E_s parameter. However, buttressing groups can cause special effects as has been investigated extensively in the thiazole series.

The quaternization of 1-substituted imidazoles is usually a facile reaction unless steric factors intervene, or strongly electron-attracting groups are present ⟨82JHC253, 87AHC(42)1, 88AHC(43)174, 93SC209⟩. Thus, 1-acylimidazoles can only be alkylated at N-3 with powerful alkylating agents such as methyl fluorosulfonate or trialkyloxonium fluoroborates ⟨81AG(E)612⟩. Trimethyloxonium fluoroborate would not methylate 1-dimethylaminosulfonylimidazole ⟨89JCS(P1)1139⟩. Regiospecific synthesis of 3-substituted *L*-histidines by alkylation of *N*-butoxycarbonyl-1-phenacyl-*L*-histidine methyl ester at N-3, followed by reductive removal of the phenacyl group has been elaborated (Scheme 10) ⟨87JOC3591⟩.

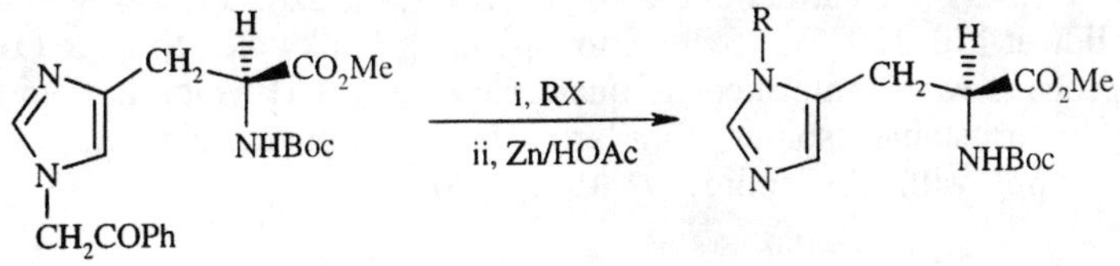

Scheme 10

Annulation of a five-membered aza ring to a benzo ring generally leads to rate retardation in *N*-quaternization reactions similar in magnitude to that for six-membered rings. Exceptions are known: 2,1-benzisoxazole undergoes *N*-methylation faster than isoxazole, and in 2,1,3-benzoxadiazole and 2,1-benzisothiazole the rates are little changed from the corresponding monocyclic rings; however, here we are dealing with *o*-quinonoid structures. The more usual situation is rate retardation by a moderate amount. This is probably caused not by steric effects, but by electronic effects, as is shown by the corresponding influence on the pK_a, values ⟨78AHC(22)71⟩.

Satisfactory Brönsted correlations for α-substituted azoles offer further evidence of the lesser importance of steric effects in the azole series ⟨78AHC(22)71⟩.

Table 4 Heteroatom and Benzo-fusion Effects on Relative Rate Constants for *N*-Methylation[a]

	k_{rel}		
Heterocycle	O	S	NMe
	1	6.9	120
	1	$\begin{pmatrix}15\\1\end{pmatrix}$	$\begin{pmatrix}912\\61\end{pmatrix}$
	1	$\begin{pmatrix}20\\1\end{pmatrix}$	$\begin{pmatrix}56\\2.8\end{pmatrix}$
	1	$\begin{pmatrix}3.6\\1\end{pmatrix}$	$\begin{pmatrix}33\\9.2\end{pmatrix}$
	1	$\begin{pmatrix}9.3\\1\end{pmatrix}$	$\begin{pmatrix}708\\76\end{pmatrix}$
	—	2.8	1

[a] ⟨78AHC(22)71⟩

For both azole and benzazole rings the introduction of further heteroatoms into the ring affects the ease of quaternization. In series with the same number and orientation of heteroatoms, rate constants increase in the order X = O < S < NMe (*cf.* Table 4) ⟨78AHC(22)71⟩. The quaternization of triazoles, thiadiazoles and tetrazoles requires stronger reagents and conditions; methyl fluorosulfonate is sometimes used ⟨78AHC(22)71⟩. The 1- or 2-substituted 1,2,3-triazoles are difficult to alkylate, but methyl fluorosulfonate succeeds ⟨71ACS2087⟩.

Oxadiazoles are difficult to alkylate. However, *N*-methylfurazinium salts are formed on heating furazans with dimethyl sulfate ⟨74AJC1917, 95JCS(P1)1083⟩; the reaction is approximately 7 and 62 times slower, respectively, than the corresponding methylations of 1,2,5-thiadiazole and isoxazole. The *N*-ethyl salts of furazan itself and 3-phenylfurazan have been prepared using triethyloxonium tetrafluoroborate ⟨64JA1863⟩.

1,2,3-Thiadiazoles are quaternized to give 3- or mixtures of 2- and 3-alkyl quaternary salts. In 5-amino-1,2,4-thiadiazole, quaternization takes place at the 4-position (**99**) ⟨64AHC(3)1⟩. 1-Substituted 1,2,4-triazoles are quaternized in the 4-position, and 4-substituted 1,2,4-triazoles are quaternized in the 1- or the 2-position ⟨64AHC(3)⟩. Treatment of 4-alkylated 1,2,4-triazoles (**100**) with a catalytic amount of the alkylating halide at 150–180°C gave the corresponding 1-alkylated triazole (**102**) ⟨90CL347⟩. The reaction seems to proceed *via* an intermediate quaternary salt (**101**) which is then dealkylated to the thermodynamically more stable isomer. Similarly, 1-alkyl-5-phenyltetrazoles are converted into 2-alkyl isomers on heating with alkyl iodide ⟨77AHC(21)233⟩.

(99)　　　　　(100)　　　　　(101)　　　　　(102)

5-Substituted 1,2,3,4-thiatriazoles (**103**) are alkylated only under very forcing conditions with triethyloxonium fluoroborate, but then give the expected products (**104**) ⟨75JOC431⟩.

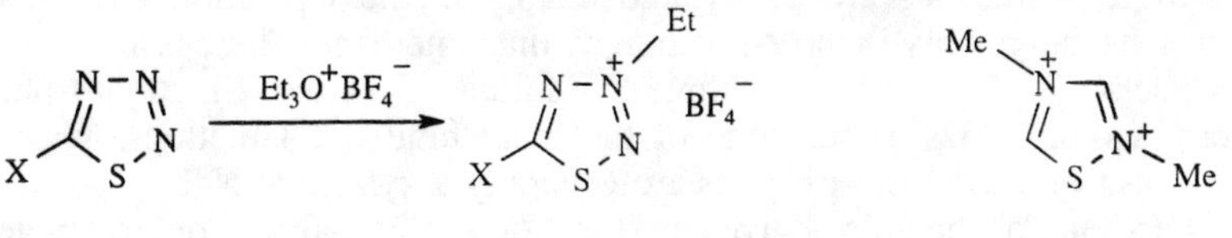

(**103**) X = SAlk   (**104**) X = SAlk   (**105**)

1,2,4-Thiadiazole with $Me_3O^+BF_4^-$ gives the diquaternary salt (**105**); diquaternary salts are also known in the 1,2,4-triazole series.

3.4.1.3.10 Alkyl halides and related compounds: compounds with a free NH group

Pyrazoles and imidazoles with free NH groups are readily alkylated, *e.g.* by MeI or Me_2SO_4. A useful procedure is to use the sodium salt of the azole in liquid ammonia ⟨80AHC(27)241⟩. Recently, dipolar aprotic solvents, especially acetone and dimethylsulfoxide, have found wide application since they strongly enhance the nucleophilicity of N-anions ⟨81S124⟩. Alkylation of NH-azoles under neutral conditions is preferable for alkylating agents which are unstable towards strong bases ($MeOCH_2Cl$, $PhCOCH_2Br$, Ph_3C-Cl, *etc.*). In some instances, *e.g.* in preparation of 1-tritylimidazole, the use of the silver salt of the azole in an inert solvent is also recommended ⟨59CB92⟩. There are very few examples of alkylation of azoles with *tert*-butyl halides in aprotic media ⟨84JCS(P1)481⟩. The yield of 1-*tert*-butylimidazole was low because of a strong E2-elimination process.

Unsymmetrical imidazoles and pyrazoles usually give a mixture of products, the composition of which may depend on the reaction conditions. Thus, the ethoxycarbonylpyrazole (**107**) gives predominantly the isomeric N-methyl derivatives (**106**) and (**108**) under the conditions indicated. The difference in orientation can be related to the stabilization of the tautomeric structure (**107**) by hydrogen bonding (possibly intramolecular), which means that alkylation of the free base gives (**108**). The isomer (**106**) is formed *via* the anion. Benzylation of (**107**) gives mainly the analogue of (**108**) ⟨80JHC137⟩.

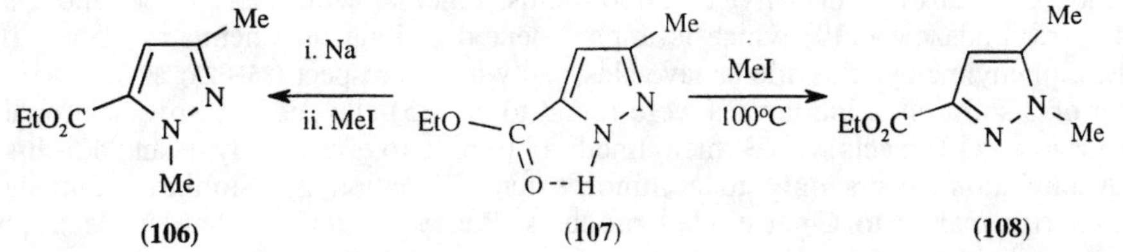

(**106**)   (**107**)   (**108**)

The differential effects of steric hindrance and tautomeric content in the imidazole series are illustrated in Scheme 11 ⟨80AHC(27)241⟩.

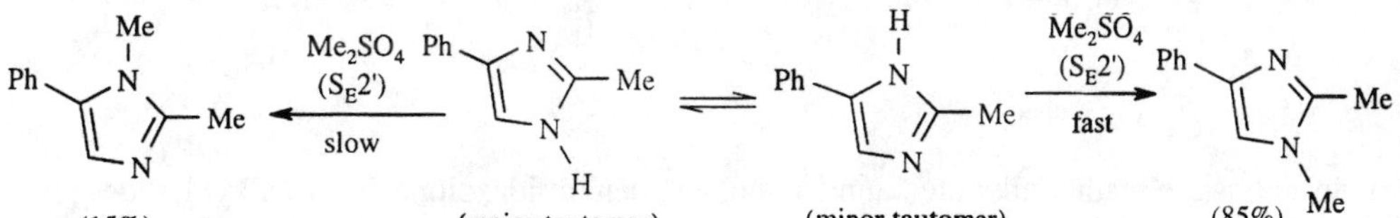

(15%)   (major tautomer)   (minor tautomer)   (85%)

Scheme 11

N-Alkylation of 1,2,3-triazoles and benzotriazoles is readily achieved using (i) alkyl halides, dialkyl sulfates, diazoalkanes, and *p*-tosylates or (ii) the Mannich reaction ⟨84CHEC-(5)697, 91RTC369⟩. When alkyl halides are used sodium alkoxide, sodium hydride ⟨93T10205⟩ or sodium hydroxide ⟨92LA843⟩ is usually employed as the base. The N-alkylation of benzotriazole with alkyl halide proceeds efficiently using powdered NaOH as the base in DMF. The highest yields (80–100%) of the alkylated benzotriazoles are obtained when a four-fold excess of NaOH is employed ⟨91RTC369⟩. N-Alkylbenzotriazoles have been prepared from benzotriazole and alkyl halides using phase-transfer catalysts, *e.g.* KOH/benzene/tetrabutyl-ammonium salts ⟨85H(23)2895⟩, KOH/benzene/polyethylene glycol ⟨90JCS(P2)2059⟩.

N-Unsubstituted 1,2,3-triazoles are methylated mainly in the 1-position with methyl iodide and silver or thallium salts, but mainly in the 2-position by diazomethane. There is also some steric control. For example, 4-phenyl-1,2,3-triazole with dimethyl sulfate gives the 2-methyl-4-phenyl (38%) and 1-methyl-4-phenyl isomers (62%), but none of the more hindered 1-methyl-5-phenyl-1,2,3-triazole ⟨74AHC(16)33⟩. *N*-Unsubstituted 1,2,4-triazoles are generally alkylated at N-1.

Alkylation of tetrazoles as the anions gives mixtures of 1- and 2-alkyl isomers. In general, electron-donating substituents in the 5-position slightly favor alkylation of the 1-position and electron-withdrawing 5-substituents slightly favor the 2-position. Organic media generally favor N-2 alkylation of 5-aryltetrazoles while polar solvents favor attack at N-1.

Direct alkylation of 5-substituted tetrazoles with *t*-butyl alcohol in presence of conc. H_2SO_4 gave high yields of 2-*N*-*t*-butyltetrazoles with a small amount of 1-*N*-*t*-butylation in the case of 5-methyltetrazoles ⟨90KGS1574⟩. This approach has been used to alkylate a wide range of 5-substituted tetrazoles, (R^5 = Alk, Ar, CH=CH$_2$, CF$_3$, NH$_2$) with substituted alcohols as source of the cations R$^+$ ⟨88ZOR2221, 90KGS1643⟩.

Azoles containing an acidic NH-group, *e.g.* 3,5-dimethylpyrazole, react with various alcohols in the presence of a catalytic amount of ruthenium-, rhodium-, and iridium-trialkylphosphite complexes to afford the corresponding *N*-alkyl derivatives with excellent yields ⟨92CL575⟩. Regioselective *N*-alkylation was achieved using alkenes and sulfuric acid ⟨89JHC3⟩.

Polyhalogenoalkanes and dihalogenoethanes have been intensively used to prepare, under PTC conditions, poly(pyrazol-1-yl)alkanes (number of pyrazoles = 2,3,4) and 1,2-di(pyrazol-1-yl)ethanes ⟨84OPP299, 86H(24)2233⟩. The reaction of both nitrogen atoms with a double alkylating agent has been extended to the reaction with *cis*-1,4-dichlorobutene to afford salt (**109**) ⟨89KGS497⟩.

(**109**)   (**110**)   (**111**)

The reaction of indazole with trityl chloride yields, together with expected 1- and 2-substituted derivatives, 3-tritylindazole (**110**) which is unprecedented in indazole chemistry ⟨85BSB421⟩. Benzyl chloride and diphenylmethyl chloride behave classically in this respect ⟨85H(23)2895⟩.

A number of 2-substituted imidazoles were found to be 4(5)-alkylated by soft electrophiles. Thus, 2-phenylimidazole (**112**) reacts with 3-thienylmethyl bromide to give mainly 4- and 4,5-di-substituted products; *N*-alkylation occurs only to a minor extent (Equation 2). Similarly, 2-methoxybenzyl chloride gives rise mainly to *C*-substituted products. Benzyl bromide, a harder electrophile, gives largely *N*-benzyl derivatives of (**112**).

(**112**)   (R=3-thienyl)   45 %   26 %   4 %

Azolone anions are readily alkylated at nitrogen, *e.g.* benzimidazolone-2 with alkyl halides gives the 1,3-dialkyl derivatives (**111**).

N-Arylation of azoles is achieved even with non-activated halobenzenes under modified Ulmann conditions ⟨63ZOB1005, 64ZOB1317, 66KGS143⟩, for a review see ⟨70Rec.Chem.Progr.(31)43⟩. Benzyne also reacts with imidazoles to give *N*-arylimidazoles ⟨70AHC(12)103⟩. 1-(4-Nitrophenyl)azoles are obtained in good yields by direct arylation of the corresponding azole with *p*-fluoronitrobenzene using PTC without solvent ⟨93SC1947⟩. Novel hexa(pyrazol-1-yl)benzenes are easily prepared from hexa-fluorobenzene and pyrazole anions ⟨93H(35)415⟩.

N-Unsubstituted pyrazoles and imidazoles add to unsaturated compounds in Michael reactions; for example, acetylenecarboxylic esters and acrylonitrile readily form the expected addition products. Styrene oxide gives rise, for example, to 1-styrylimidazoles ⟨76JCS(P1)545⟩. Benzimidazole reacts with formaldehyde and secondary amines in the Mannich reaction to give 1-aminomethyl products.

Some Michael additions of unsaturated reagents to imidazoles involve the neutral species of the heterocycles, but others may be reactions of the anion. Fluoride ion catalyzed addition of 2-methyl-4-nitroimidazole (**113**) to a suitable Michael acceptor gives almost quantitative yields of the 1-substituted 4-nitroisomers (**114**) ⟨90JOC3702, 91JCR(S)350⟩.

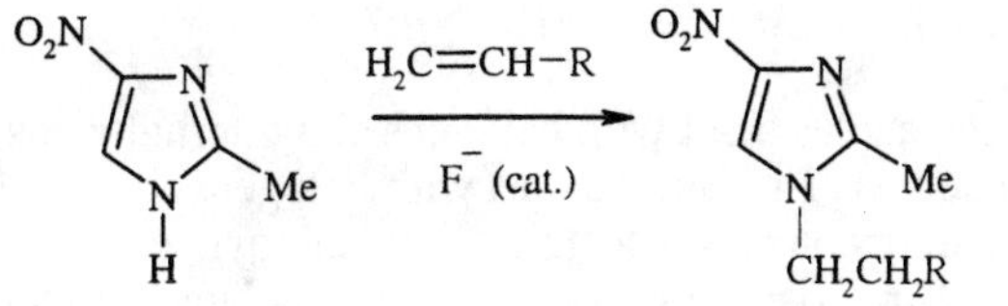

(**113**)   (**114**) 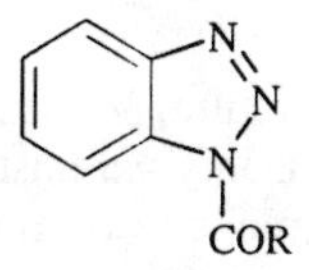   (**115**)

3.4.1.3.11 Acyl halides and related compounds

Azoles containing a free NH group react comparatively readily with acyl halides. *N*-Acyl-pyrazoles, -imidazoles, etc. can be prepared by reaction sequences of either type (**75**)→(**76**) or type (**79**)→(**80**) or (**81**). Such reactions have been carried out with benzoyl halides, sulfonyl halides, isocyanates, isothiocyanates and chloroformates. Reactions occur under Schotten–Baumann conditions or in inert solvents.

When two isomeric products could result, only the thermodynamically stable one is usually obtained because the acylation reactions are reversible and the products interconvert readily. Thus, benzotriazole forms 1-acyl derivatives (**115**) which preserve the 'Kékulé resonance' of the benzene ring and are therefore more stable than the isomeric 2-acyl derivatives. Acylation of pyrazoles also usually gives the more stable isomer as the sole product ⟨66AHC(6)347⟩. The imidazole-catalyzed hydrolysis of esters can be classified as an electrophilic attack on the multiply bonded imidazole nitrogen.

Since *N*-acylation is a reversible process, it has allowed the regiospecific alkylation of, for example, imidazoles to give the sterically less favored derivative. This principle is illustrated in Scheme 12 ⟨80AHC(27)241⟩.

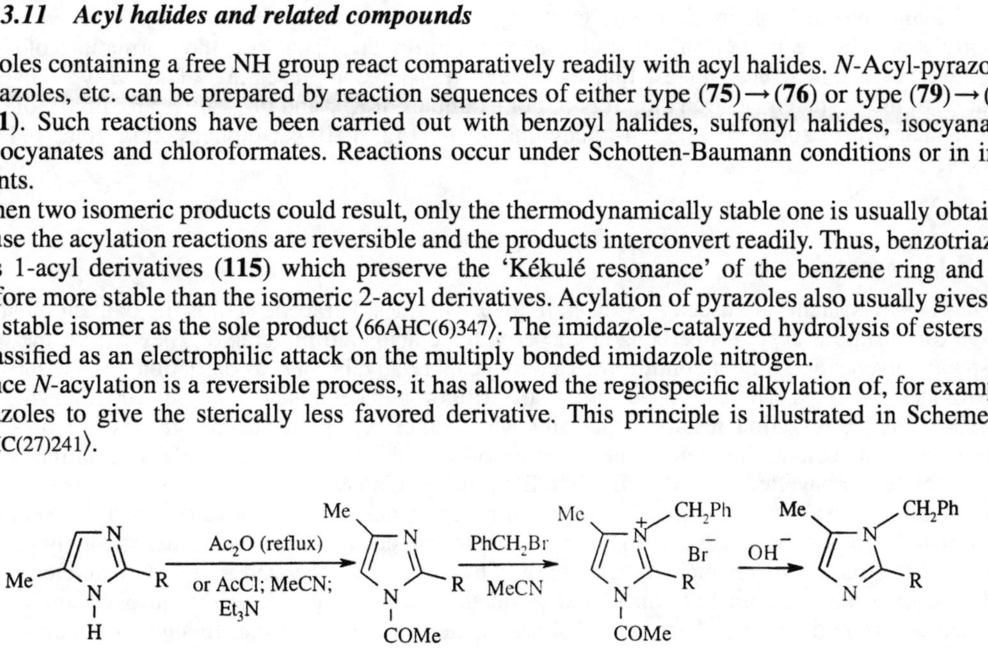

Scheme 12

1,2,3-Triazoles are acylated with acyl halides, usually initially at the 1-position, but the acyl group may migrate to the 2-position on heating or on treatment with base. Thus, acetylation with acetyl chloride often gives 1-acetyl derivatives, which rearrange to the 2-isomers above 120°C ⟨74AHC(16)33⟩.

The more general preparations of *N*-acyltriazoles and *N*-acylbenzotriazoles have utilized acid chlorides in the reactions with 2-trimethylsilyl-1,2,3-triazole ⟨92TL1033⟩, 1-(trimethylsilyl)benzotriazole (**116**) ⟨80JOM(188)141⟩ or 1-(tributylstannyl)benzotriazole ⟨77JOM(137)185⟩. Thus (**116**) reacts with phosgene or sulfuryl chloride to give 1,1'-carbonyl-(**117a**) and 1,1'-sulfonyl-dibenzotriazole (**117b**). Treatment of (**117a**) with alkanols affords 1-alkoxycarbonylbenzotriazoles (**118**). Both (**117a**) and (**117b**) are effective in the dehydration of aldoximes and amides to nitriles under mild conditions ⟨93OPP315⟩.

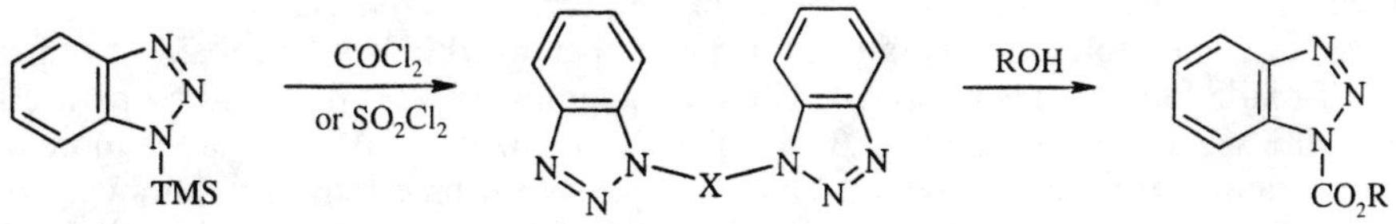

(**116**)   (**117a**) X=CO   (**118**)
          (**117b**) X = SO$_2$

Whether tetrazoles are acylated in the 1- or 2-position depends on the 5-substituent. 2-Acyltetrazoles are unstable (see Section 3.4.3.12.4) ⟨77AHC(21)323⟩; 1-alkylsulfonyltriazoles are also unstable (see Section 3.4.1.2.3).

3.4.1.3.12 Halogens

At room temperature, *N*-unsubstituted azoles react with halogens and interhalogens (*e.g.* ICl) to give *N*-haloazoles, probably *via* unstable adducts. Thus imidazoles with halogens form *N*-halo compounds, which easily rearrange to form *C*-haloimidazoles ⟨70AHC(12)103⟩. *N*-Halopyrazoles are unstable and act as halogenating agents. *N*-Halo-1,2,4-triazoles are more easily isolated, especially when the 3,5-positions are substituted.

Benzotriazole can be chlorinated at the N(1) position by NaOCl ⟨85H(23)2225⟩. 1-Chlorobenzotriazole – a rather stable crystalline compound that has found application in organic synthesis as a selective chlorinating reagent and mild oxidant. *N*-Fluorination of benzotriazole with cesium fluoroxysulfate gives 1-fluorobenzotriazole in 25% yield ⟨91T7447⟩.

Correlations between pK_a values and the equilibrium constants for the formation of iodine complexes with imidazoles suggest that the charge transfer complexes are of the *n*-type involving donation of the unshared electron pair at N-3. For examples of K_{CT}^{298} and pK_a values are: imidazole, 202, 6.95; 1-methylimidazole, 333, 7.33; 4-phenylimidazole, 152, 6.10; 4,5-diphenylimidazole, 141, 5.90 ⟨83BSB923⟩.

3.4.1.3.13 Peracids

Azaaromatic systems are usually oxidized to their *N*-oxides. Peracetic acid is the oxidant most used though for unstable substrates perbenzoic or perphthalic acids are preferable. They permit the use of non-polar solvents and milder conditions. Heterocycles relatively inert to oxidation can be converted to *N*-oxides by the more active performic or trifluoroperacetic acids. *m*-Chloroperbenzoic acid (MCPBA) also gives good results, especially when other easily oxidizable groups are present in heterocyclic molecule and therefore the question of selectivity is important. Thus, 2-aminopyridine was successfully converted with MCPBA into 2-aminopyridine *N*-oxide.

N-Oxidation proceeds on the free base form of heterocycles, in strongly acidic media the substrate exists mainly in the protonated form and this reaction does not occur. Under strongly alkaline conditions the oxidation is hampered because of conversion of the peracid into an unreactive anion.

N-Oxidation may be formally considered as quaternization of pyridine-like nitrogen atom by the HO^+ cation, formed by heterolysis of O-O bond in the peracid molecule. Indeed, common features exist between N-alkylation and N-oxidation: both reactions are second order (first order at each reagent). The reaction constant, ρ, for oxidation of 3- and 4-substituted pyridines by $PhCO_3H$ in aqueous dioxane is –2.35, close to the ρ value for N-alkylation ⟨Chem. of the Heterocyclic N-oxides, N-Y, Acad. Press, 1971⟩.

Despite the rather high basicity of imidazole and benzimidazole attempts to prepare their *N*-oxides by direct oxidation were unsuccessful. Peracids destroy imidazole, depending on the reaction conditions, to give oxamide or urea and ammonia. Oxidation of benzimidazole leads to formation of imidazole-4,5-dicarboxylic acid in low yield. The attempted conversion of oxazoles into *N*-oxides fails and leads to ring opening. Oxidation of other azoles also rarely gives the corresponding *N*-oxides (see below). One can assume that the different ease of N-oxidation of azines and azoles follows the different nature of their highest occupied molecular orbital. In case of the azines, the HOMO is of the *n*-type ⟨74PMH(6)54⟩, therefore the non-bonded electron pair of the nitrogen atom is most available for coordination with an electrophile. By contrast the HOMO of azoles is believed to be occupied by π-electrons ⟨73T2173⟩. In this case an oxidant most likely first removes from substrate a single π-electron, thus forming a radical-cation which undergoes subsequent reactions, *e.g.* nucleophilic attack of peroxide anion with further formation of azolones or ring-cleaved derivatives.

Nevertheless, in some azoles the energies of *n*- and upper π-orbitals are probably comparable and in such cases N-oxide formation is observed. Thus, 1-methylpyrazole is oxidized by peracetic acid to the 2-oxide in 10% yield. 1-Substituted 1,2,3-triazoles are oxidized by MCPBA at the more basic N(3) to give the corresponding triazole *N*-oxides. The yield is lower if an electron-withdrawing substituent is present at the C(4) or C(5) position ⟨87ACS(B)724⟩. Reaction of 3-methylbenzisoxazole (**119**) with sodium hypochlorite or lead tetraacetate gave the 2-oxide (**120**) in 70 and 90% yields, respectively ⟨87H(26)2921⟩.

The oxidation of thiazoles by peroxy acids leads to the corresponding *N*-oxides. Peracetic, MCPBA, permaleic, and trifluoroperacetic acid have been employed for this reaction. Chemical yields range from 4% to 50%, the more basic thiazoles producing higher yields. Thus, thiazole, 2,4-dimethyl- and 4,5-dimethylthiazoles, and 2-phenylthiazole can be oxidized in moderate to good yields. However, neither 4-chloro-2-phenylthiazole nor 5-chloro-2-phenylthiazole could be oxidized. 3-Oxides were also obtained by oxidation of 1,2,3-thiadiazoles and 5-phenylthiatriazole (**121** → **122**) ⟨75T1783⟩.

Oxidation of sulfur-containing azoles quite often leads to the formation of sulfones and sulfoxides. Thus, 3-alkyl-1,2-benzisothiazoles (**123**) with MCPBA give the oxaziridines (**124**), and the use of a chiral 3-alkyl substituent leads to pure diastereomers ⟨90JOC1254, 91JOC809⟩. Reaction of 1,2,3-benzo-thiadiazole with 30% hydrogen peroxide in a mixture of acetic acid and methanol for 45 days afforded (**125**) in 60% yield ⟨90CJC1950⟩. Oxidation of 1,2,3-benzothiadiazole with a variety of other oxidizing agents (MCPBA, 30% hydrogen peroxide, hydrogen peroxide in methylene chloride-acetic acid mixtures, *etc.*) was unsuccessful.

3.4.1.3.14 Aminating agents

Amination at an azole ring nitrogen is known for *N*-unsubstituted azoles ⟨92AHC(53)85⟩. Thus, 1,2,3-triazole ⟨92ZOR1320⟩ and 4,5-diphenyl-1,2,3-triazole ⟨88M1041⟩ with hydroxylamine-*O*-sulfonic acid give 1-amino (**126**) and 2-amino (**127**) derivatives in ratio ~ 4:1 and 1:3, respectively. ⟨74AHC(16)33⟩. Pyrazole affords (**128**) and indazole and tetrazole give comparable amounts of the 1- and 2-amino derivatives.

Azoles without a free NH group are also aminated (usually with *O*-mesitylsulfonylhydroxylamine or *O*-picrylhydroxylamine), giving *N*-aminoazolium salts, *e.g.* (**129**) ⟨94JCS(P1)841⟩.

The diazonium ions of type (**130**) located at a β-position to a nitrogen unsubstituted tetrazole cyclize onto the N-1 position giving the fused tetrazolotriazines (**131**) ⟨89LA83⟩. Similarly, compound (**132**) is formed from 2-(*o*-aminophenyl)benzimidazole ⟨66JHC289⟩.

3.4.1.3.15   Other Lewis acids

Azoles can form stable compounds in which metallic and metalloid atoms are linked to nitrogen. For example, pyrazoles and imidazoles *N*-substituted by B, Si, P, Ga, Ge, Sn and Hg groups are made in this way. Imidazoles and 1,2,3-triazoles with a free NH group can be *N*-trimethylsilylated and *N*-cyanated (with cyanogen bromide or 1-cyano-4-dimethylaminopyridinium bromide ⟨88S470; 91RRC573⟩ (see also Section 3.4.3.12.4)). Imidazoles of low basicity can be *N*-nitrated.

Transition-metal derivatives of triazoles and benzotriazoles have been reviewed ⟨99AHC(72)1⟩. The generally preferred coordination positions appear to be N(3) for 1,2,3-triazoles and N(1) for triazolate anions. When 1-(trimethylsilyl)-1,2,3-triazole is treated with halodiorganylboranes R_2BX (X = Br, Cl), a cyclic tetramer (**133**) is obtained exclusively ⟨89IC4022, 91IC784⟩.

(133)

3.4.1.4   Electrophilic Attack at Carbon

3.4.1.4.1   Reactivity and orientation

(i) Ease of reaction

Replacing a CH = CH group in benzene with a heteroatom (Z) increases the susceptibility of the ring carbon atoms to electrophilic attack noticeably when Z is S, more when Z is O, and very markedly when Z is NH *(cf.* Chapter 2.4). Replacing one CH group in benzene with a nitrogen atom decreases the ease of electrophilic attack at the remaining carbon atoms *(cf.* Chapter 2.4); replacement of two CH groups with nitrogen atoms decreases it further *(cf.* Chapter 2.4). Such deactivation is very strong in nitration, sulfonation and Friedel Crafts reactions, which proceed in strongly acidic media, i.e. under conditions in which the nitrogen atom is largely protonated (or complexed). The effect of a protonated nitrogen atom is considerably greater than, for example, the two nitro groups in *m*-dinitrobenzene. The deactivating effect is less pronounced in reactions conducted under neutral or weakly acidic conditions, where a large proportion of unprotonated free base exists, i.e. as in halogenation and mercuration reactions.

In azole chemistry the total effect of the several heteroatoms in one ring approximates the superposition of their separate effects. It is found that pyrazole, imidazole and isoxazole undergo nitration and sulfonation about as readily as nitrobenzene; thiazole and isothiazole react less readily *(cf.* equal to *m*-dinitrobenzene), and oxadiazoles, thiadiazoles, triazoles, *etc.* with great difficulty. In each case, halogenation is easier than the corresponding nitration or sulfonation. Strong electron-donor substituents help the reaction.

Pyrazoles and imidazoles exist partly as anions (*e.g.* **134** and **135**) in neutral and basic solution. Under these conditions they react with electrophilic reagents almost as readily as phenol, undergoing diazo coupling, nitrosation and Mannich reactions (note the increased reactivity of pyrrole anions over the neutral pyrrole species).

1,2,3-Triazoles can be activated towards electrophilic attack at carbon by the introduction of an *N*-oxide group. N-Oxidation gives rise to better activation of the 5-position than the 4-position.

For general review on electrophilic substitution at carbon atom see ⟨88BSB573; 92H(33)1129; 93AHC(57)291, 96KGS1535⟩.

(ii) Orientation

A multiply bonded nitrogen atom deactivates carbon atoms α or τ to it toward electrophilic attack; thus initial substitution in 1,2- and 1,3-dihetero compounds should be as shown in structures (**136**) and (**137**). Pyrazoles (**136**; Z = NH), isoxazoles (**136**; Z = O), isothiazoles (**136**; Z = S), imidazoles (**137**; Z = NH, tautomerism can make the 4- and 5-positions equivalent) and thiazoles (**137**; Z = S) do indeed undergo electrophilic substitution as expected. Little is known of the electrophilic substitution reactions of oxazoles (**137**; Z = O) and compounds containing three or more heteroatoms in one ring. Deactivation of the 4-position in 1,3-dihetero compounds (**137**) is less effective because of considerable double bond fixation (*cf.* Sections 2.4.3.2.1 and 3.4.3.1.7), and if the 5-position of imidazoles or thiazoles is blocked, substitution can occur in the 4-position (**138**).

(**134**)   (**135**)   (**136**)   (**137**)   (**138**)   (**139**)

The above considerations do not necessarily apply to reactions of electrophilic reagents with pyrazole and imidazole anions (**134**, **135**). The imidazole anion is sometimes (diazo coupling, halogenation, deuterium exchange) substituted in the 2-position (**139**) and the indazole anion in the 3-position (*cf.* Section 3.4.1.4.5).

The Hammett σ^+ constant for the 4(5)-position of imidazole is around –1; for C-2 it is of the order –0.8 ⟨86CHE587⟩. The electrophilic substitutions which do occur at the 2-position invariably involve preformation of an anion at that position. The 2-proton, which should be the least active in a conventional $S_E Ar$ sense, turns out to be the most labile over a wide pH range, and there is a marked rate acceleration on going from imidazole to imidazolium cation. Any negative charge generated at C-2 is stabilized by the adjacent pyrrole-type nitrogen (see Section 3.4.1.8.2).

In condensed heteroaromatic systems with a bridge-head pyrrolic nitrogen atom, π-electron density is always shifted from the electron-rich six-membered ring (formally contains 7 π-electrons) towards the five-membered ring (formally has 6 π-electrons). As a result electrophiles are directed to carbon atoms of the latter. Thus, imidazo[1,2-*a*]pyridines (**140**) unsubstituted at C-3 almost always react with electrophiles at that position.

As predicted by theoretical calculations, the site of electrophilic attack on pyrazolo[3,4-*c*]pyridines (**141**) is C-3 ⟨73JCS(P1)2901, 89AP(322)885⟩.

(**140**)   (**141**)

(iii) Effect of substituents

Just as in benzene, substituents can strongly activate (*e.g.* NH$_2$, NMe$_2$, OMe), strongly deactivate (*e.g.* NO$_2$, SO$_3$H, CO$_2$Et) or have relatively little effect on (*e.g.* Me, Cl) the ring toward further substitution. Further electrophilic substitution generally will not take place on an azole which carries a strong electron-donor group or is strongly activated, as it is in the azolone form. However, these considerations can be affected by basicity considerations: thus a strongly deactivating group can also increase the amount of more reactive neutral molecule.

When the preferred substitution position (*cf.* **136** and **137**) is occupied, activating substituents can facilitate substitution in other positions (*cf.* examples in Sections 3.4.1.4.2 and 3.4.1.4.5); *ipso* attack can also occur if the substituent is itself easily displaced.

In benz- and phenyl-azolones, electrophilic substitution often occurs in the benzene ring; such reactions are considered as reactions of substituents (see Section 3.4.3.2.1 and 3.4.3.4.1).

3.4.1.4.2 Nitration

Nitration of monocyclic compounds is summarized in Table 5. Substitution occurs in the expected positions. The reaction conditions required are more vigorous than those needed for benzene, but less than those for pyridine. Ring nitration of oxazoles is rare, but (**142**) has been obtained in this way ⟨74AHC(17)99⟩.

Table 5 Nitration and Sulfonation of Azoles

Heterocycle	Position substituted	Reaction conditions	
		Sulfonation	Nitration
Pyrazole	4	H_2SO_4/SO_3, 100°C	$HNO_3/H_2SO_4/SO_3$, 100°C
Imidazole	4($\equiv$ 5)	H_2SO_4/SO_3, 160°C	HNO_3/H_2SO_4, 160°C
3-Methylisoxazole	4	HSO_3Cl, 100°C	$HNO_3/H_2SO_4/SO_3$, 70°C
Isothiazole	4	H_2SO_4/SO_3, 150°C[a]	HNO_3/H_2SO_4, 230°C
Thiazole	5	$H_2SO_4/SO_3/Hg$, 250°C	—
4-Methylthiazole	5	H_2SO_4/SO_3, 200°C	$HNO_3/H_2SO_4/SO_3$, 160°C
2,5-Dimethylthiazole	4	H_2SO_4/SO_3, 200°C	$HNO_3/H_2SO_4/SO_3$, 160°C

[a] ⟨66GEP1208303⟩.

(142)

Substituents are sometimes displaced: thus chloroimidazoles are nitrated normally, but iodo analogues suffer nitro-deiodination.

Conversion of 1-methylimidazole into its 2-nitro derivative by successive lithiation and reaction with N_2O_4 ⟨74URP437763⟩ has led to further successful 2-nitrations of 1-substituted imidazoles using N_2O_4, propyl nitrate, and tetranitromethane as quenching agents ⟨80USP4199592, 82JHC253, 89JCS(P1)95, 94H(38)2487⟩.

Pyrazoles can undergo nitration at several positions: 4-bromo-1-methylpyrazole yields the 3,5-dinitro product. 1-Methylpyrazole 2-oxide yields the 5-nitro derivative.

As expected, nitration is facilitated by activating groups such as an amino group; for example, nitration of (143) occurs at about 20°C (HNO_3/H_2SO_4). Sydnones (144) are nitrated readily. The pyrazolinone (145) is nitrated as indicated, and 1,2,4-triazolinones have also been ring nitrated.

(143)   (144)   (145)

3.4.1.4.3 Sulfonation

Sulfonation conditions are given in Table 5. The orientation is as expected ⟨90AHC(47)139⟩. Azolinones react as readily as the corresponding azoles; sulfonation of (145) occurs at the positions indicated (H_2SO_4/SO_3 at 100°C). Imidazoles are readily chlorosulfonated at C-4.

3.4.1.4.4 Acid-catalyzed hydrogen exchange

Acid-catalyzed hydrogen exchange is used as a measure of the comparative reactivity of different aromatic rings (see Table 6). These reactions take place on the neutral molecules or, at high acidities, on the cations. At the preferred positions the neutral isoxazole, isothiazole and pyrazole rings are all considerably more reactive than benzene. Although the 4-position of isothiazole is somewhat less reactive than the 4-position in thiophene, a similar situation does not exist with isoxazole-furan ring systems.

Table 6   Reactivities Toward Acid-catalyzed Deuterodeprotonation[a]

Heterocycle	log (partial rate factor)[b] at the ring positions			
	2	3	4	5
Isoxazole	—	—	4.3	—
Isothiazole	—	—	3.6	—
Furan	8.2	ca. 4.5	ca. 4.5	8.2
Thiophene	8.6	5.0	5.0	8.6
1-Methylpyrazole	—	5.6	9.8	5.6

[a] ⟨79AHC(25)147⟩. [b] Relative to position 1 of benzene = 1.

Imidazoles, because of their high basicity, are very unreactive unless electron-withdrawing substituents are present. The deuteration of oxazole using CF_3CO_2D/D_2O was reported to occur exclusively at the 2 position ⟨90IJC(B)562⟩.

3.4.1.4.5   Halogenation

Imidazoles (see review ⟨93AHC(57)291⟩) and pyrazoles containing an unsubstituted NH group are easily chlorinated (Cl_2/H_2O or N-chlorosuccinimide/$CHCl_3$), brominated ($Br_2/CHCl_3$; $KOBr/H_2O$), and iodinated (I_2/HIO_3). Bromine-DMF-K_2CO_3, which does not affect other acid- or base-sensitive substituents, has been recommended as a general brominating agent for imidazoles. Substitution generally occurs first at the 4-position, but further reaction at other available nuclear positions takes place readily, especially in the imidazole series. When halogenation of the nucleus involves electrophilic attack on anions of type (**135**), the 4-position of imidazole is again initially substituted (earlier work suggested a different orientation). The benzimidazole anion is iodinated at the 2-position; other halogenation generally occurs in the benzene ring.

Imidazoles and halogens form charge transfer complexes ⟨85JCS(P2)531, 88JA2586⟩, and also rather unstable N-halogeno derivatives which rearrange into C-halogeno products. N-Halogenoimidazoles can be intermediates in C-halogenation processes.

In 1-substituted imidazoles C-5 is slightly more reactive than C-4. Partial rate factors are discussed in ⟨90AHC(47)165⟩.

Even apparently deactivated compounds can be quite easily brominated at vacant ring positions, for example 4,5-dicyanoimidazole ⟨91JA6178⟩, 2-methyl-4-nitroimidazole ⟨86JHC913⟩, 2-nitro- ⟨89JCS(P1)95⟩ and 1-methyl-2-nitroimidazole ⟨93CJC427⟩.

Methods are available for the preparation of 2,4,5-triiodoimidazole ⟨83JCS(P1)735⟩, 4,5-diiodoimidazole ⟨91JOC4296, 91JOC5739⟩, and, by reduction, 4-iodoimidazole ⟨83JCS(P1)735, 91JOC5739⟩. Access via lithiated imidazoles is also an attractive route to 2-iodo ⟨90JHC673⟩, 5-iodo ⟨83JCS(P1)271, 90S761⟩, and 2,4-diiodoimidazoles ⟨85H(23)417⟩.

Bromodecarboxylation is known (Scheme 13) ⟨82CHE539⟩.

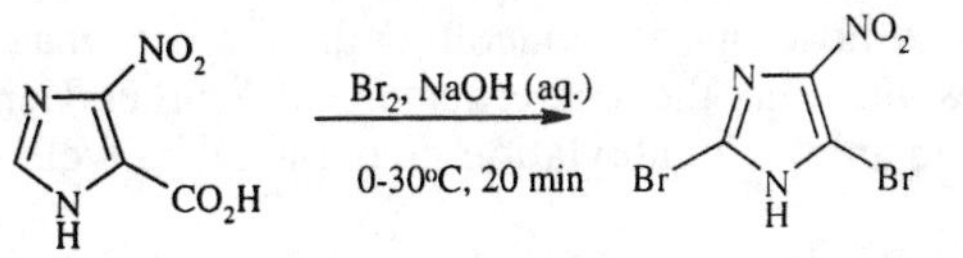

Scheme 13 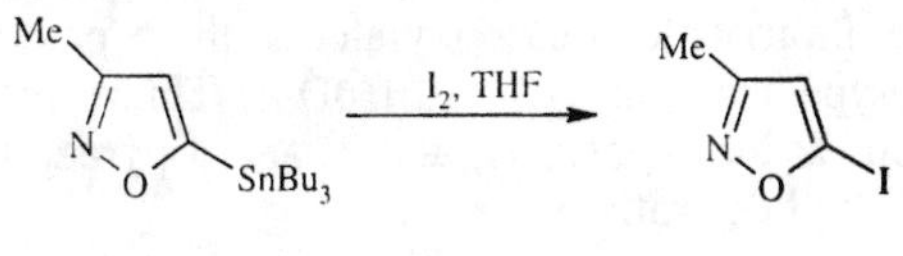   **Scheme 14**

Isoxazoles can be halogenated in the 4-position ⟨63AHC(2)365⟩. Ring bromination of oxazoles with bromine or NBS occurs preferentially at the 5-position and, if this is occupied, at the 4-position ⟨74AHC(17)99; 93AHC(57)292⟩. Aminooxazoles are readily halogenated. Iodinations of isoxazoles are accomplished by using iodine in nitric acid, but only if the 4-position is free. Unlike NCS and NBS, NIS is not a successful iodinating agent for isoxazole. 3-Methyl-5-tributylstannylisoxazole reacted with iodine in THF to give the 5-iodo derivative (Scheme 14) ⟨91T5111⟩.

Isothiazoles with electron-releasing substituents such as amino, hydroxy, or alkoxy in the 3- or 5-position are brominated in high yield in the 4-position. Alkylisothiazoles give lower yields, but 3-methylisothiazole-5-carboxylic acid has been brominated in 76% yield ⟨72AHC(14)1⟩. Again, thiazoles with an electron-releasing substituent in the 2- or 4-position are brominated at the 5-position ⟨79HC(34-1)5⟩.

1,2,3-Triazoles are brominated at the 4- or 5-positions, but only if there is no *N*-substituent ⟨74AHC(16)33⟩. This also applies to 1,2,4-triazoles. *N*-Halo derivatives are frequently isolated as intermediates ⟨81HC(37)289⟩.

3-Amino-1,2,5-thiadiazole is chlorinated or brominated at the 4-position at 20°C in acetic acid. 3-Methyl-1,2,5-thiadiazole can also be chlorinated in the 4-position ⟨68AHC(9)107⟩. Bromination of 2-amino-1,3,4-thiadiazole succeeds in the 5-position ⟨65ACS2434⟩.

1-Substituted tetrazoles undergo electrophilic iodination to give the 5-iodo-derivatives (**146**, R = Alk, Ar, CH$_2$ = CH-) in yields of 55–75% on treatment with I$_2$ in acidic permanganate solution ⟨88KGS1699, 92MI 417-04⟩.

Azolidinones are very easily halogenated, *e.g.* (**147**) gives (**148**) ⟨79AHC(25)83⟩; similar reactions occur in the isothiazolidinone series.

(146)   (147)   (148)

3.4.1.4.6 *Acylation, formylation and alkylation*

Although in general azoles do not undergo Friedel-Crafts type alkylation or acylation, several isolated reactions of this general type are known. 3-Phenylsydnone (**149**) undergoes Friedel-Crafts acetylation and Vilsmeier formylation at the 4-position, and the 5-alkylation of thiazoles by carbonium ions is known. Heating *N*-substituted pyrazoles with benzoyl chloride at 200°C gives quite high yields of 4-benzoylpyrazoles, even in the absence of catalysts. Benzylation of *N*-substituted pyrazoles proceeds similarly in the 4-position ⟨66AHC(6)347⟩. Vilsmeier formylation of pyrazoles at the 4-position is also well-known.

(149)   (150)   (151)

Heating 2-phenyl-4-benzoyl-1,3,4-thiadiazolium chloride (**150**) at 200°C causes the benzoyl group to move to the 2-position as in (**151**) ⟨68AHC(9)165⟩, probably *via* deprotonation.

For the *C*-acylation of imidazoles *via* deprotonation, see Section 3.4.1.8.4.

4-Aminotriazole is carboxylated at the 5-position by heating with aqueous sodium bicarbonate in a Kolbe-type reaction ⟨71JCS(C)1501⟩. 2-Thiazolinones undergo the Gattermann and Reimer-Tiemann reactions at the 4-position, and 3- and 4-pyrazolinone anions on alkylation give 4-alkyl as well as *O*- and *N*-alkyl derivatives.

Treatment of pyrazole with 1-bromoadamantane affords either 4-(1-adamantyl)pyrazole ⟨94TL183, 94H(37)1623⟩ or 3(5)-(1-adamantyl)pyrazole ⟨94CL2079⟩ depending on the conditions (heating in a sealed tube or heating in a microwave oven). (Other cases of *C*-alkylation of azoles are in Section 3.4.1.3.10).

3.4.1.4.7 *Mercuration*

While there appears to have been no general study of the mercuration of azoles, the reaction seems to proceed readily in several systems. Thus pyrazoles are 4-chloromercurated by HgCl$_2$. Mercuration of 1-phenylpyrazole and 1-phenyl-3,4,5-trimethylpyrazole affords organomercury derivatives (**152**) and (**153**) ⟨93AJC1323⟩, the structures have been established using ^{199}Hg NMR spectroscopy.

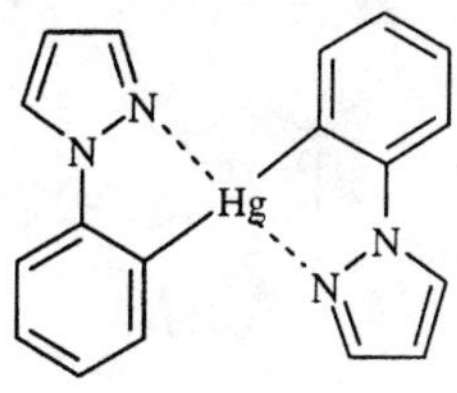

(152) 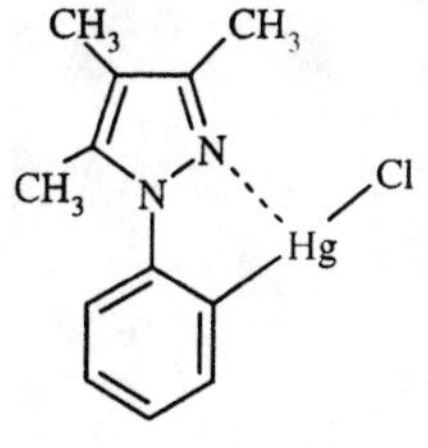   (153)

Oxazoles are mercurated in acetic acid: the ring positions react 5 > 4 > 2 ⟨74AHC(17)99⟩. Thiazoles react under the same conditions and show the same order of ring position reactivities. Isoxazoles can be easily mercurated in the 4-position with mercury (II) acetate ⟨63AHC(2)365⟩. 3-Arylsydnones are mercurated at the 4-position.

1,2,4-Oxadiazoles are rather inert against electrophilic attack. However, electrophilic mercuration of 5-unsubstituted oxadiazoles is possible. 5-Halooxadiazoles are prepared from the 5-mercurio compounds (Scheme 15) ⟨64HCA838⟩; only the 5-iodo derivatives are obtained in good yields.

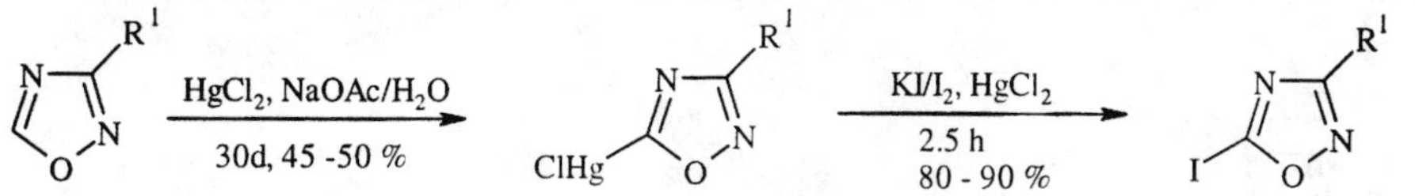

$R^1 = Me, Ph$

Scheme 15

Treatment of tetrazolium salts (154) with mercuric acetate gives the bistetrazolium mercury salts (155) which represent essentially a metal-carbene trapping of the tetrazolium ylide species. Replacement of the mercury atom of (155) is readily achieved with halogens to give the 5-halotetrazolium compounds (156) ⟨93CB1149⟩. These electrophilic reactions at the tetrazole C-5 arise from the lability of the 5-CH proton and they show the necessity of generating carbanionic character at C-5 before electrophilic attack can occur on this strongly π-deficient azole.

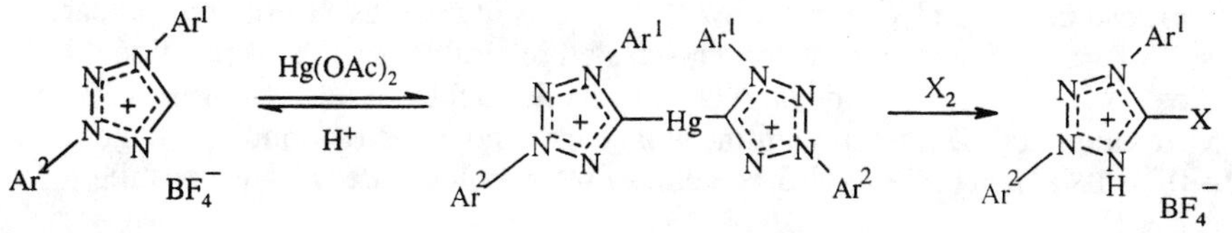

(154)   (155)   (156) X = Cl, Br, I

3.4.1.4.8 Diazo coupling

Diazo coupling is expected to occur only with highly reactive systems, and experiment bears this out. Diazonium ions couple with the anions of *N*-unsubstituted imidazoles at the 2-position (*e.g.* 157 yields 158) and with indazoles (159) in the 3-position. In general, other azoles react only when they contain an amino, hydroxy, or potential hydroxy group, *e.g.* the 4-hydroxypyrazole (160), the triazolinone (161) and the thiazolidinedione (162) (all these reactions occur on the corresponding anions).

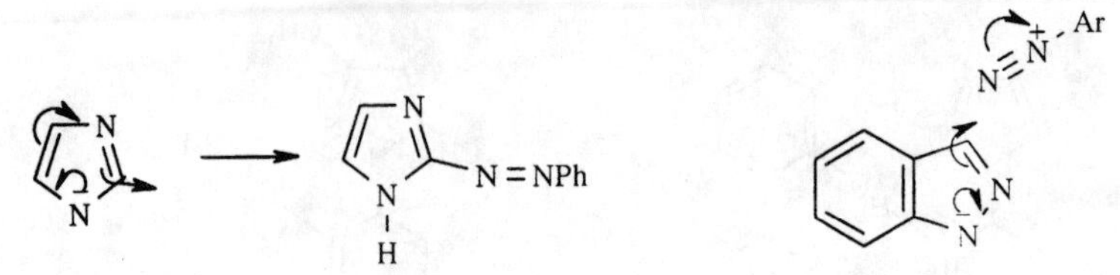

(157)   (158)   (159)

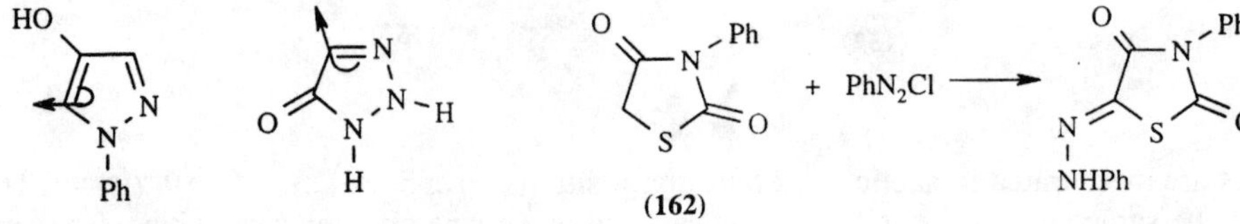

(160)   (161)   (162)

N-Substituted imidazoles do not undergo azocoupling with diazonium salts. However, the diazonium salt formed by diazotization of 1-*o*-aminophenylimidazole (163) easily undergoes intramolecular azocoupling reaction at position 5 to give compound (164). When position 5 is occupied, as in 1-*o*-aminophenyltetrahydrobenzimidazole, azocoupling may occur at position 2 of the imidazole ring though with greater difficulty; in this case compound (165) was obtained ⟨67CI(L)1454, 69KGS916⟩.

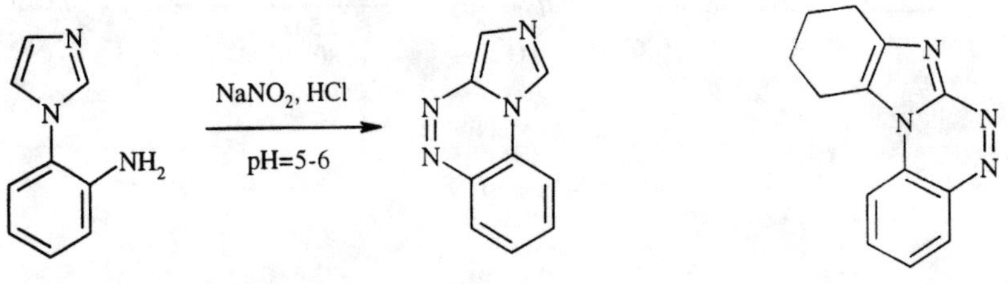

(163)   (164)   (165)

3.4.1.4.9 Nitrosation

Under alkaline conditions, alkyl nitrites nitrosate imidazoles which possess a free NH group in the 4-position ⟨70AHC(12)103⟩. Nitrosation of 3,5-dimethylpyrazoles gives the 4-diazonium salt by further reaction of the nitroso compound with more NO^+. 4-Nitrosopyrazoles (166) were prepared and used as spin traps. 5-Pyrazolinones are often nitrosated readily at the 4-position. Imidazo[1,2-*a*]benzimidazoles are nitrosated with $NaNO_2$ in acetic acid giving 3-nitrosoderivatives (167) ⟨70KGS977⟩. Nitrosation of 2-(dimethylamino)thiazoles under acid conditions gives 5-nitroso-derivatives (168) ⟨88JCS(P1)2209⟩. 3-Alkyl-5-acetamidoisothiazoles undergo 4-nitrosation.

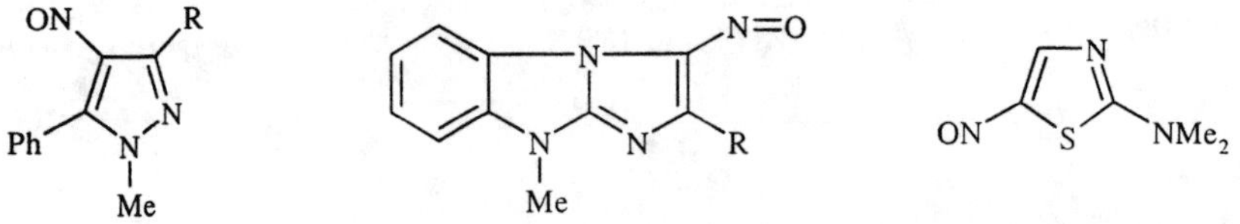

(166) R = Me, Ph   (167) R = Me, Ph   (168)

3.4.1.4.10 Reactions with aldehydes and ketones

Bis(pyrazole) (169) was prepared by the reaction of 3,5-dimethylpyrazole with formaldehyde (a 4-CH_2OH derivative is the probable intermediate) ⟨84ICA99⟩.

Imidazoles are hydroxymethylated by CH_2O at the 4-position; 1-substituted imidazoles react at the 2-position. Isoxazoles can be chloromethylated in the 4-position ⟨63AHC(2)365⟩.

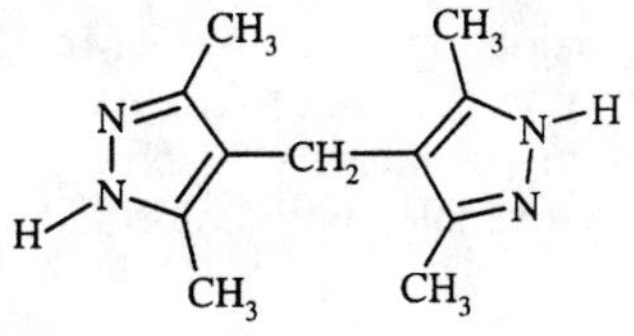

(169) 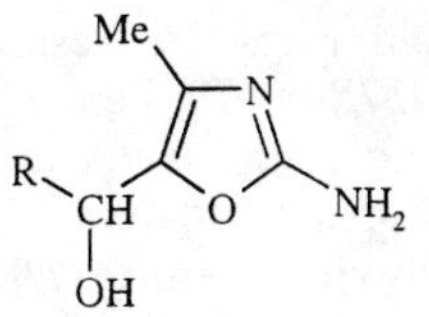　　(170) 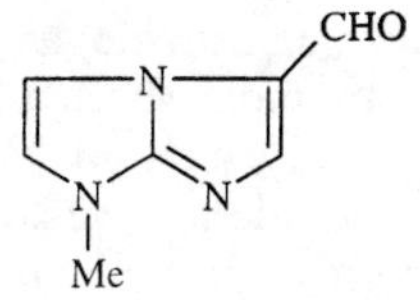　　(171)

2-Amino-4-methyloxazole reacts with aldehydes to give oxazole-5-hydroxymethyl derivatives (**170**) ⟨88JHC815⟩. Oxazoles without strong electron donor groups are inert to aldehydes. Some π-excessive heterocycles with a bridge-head nitrogen atom can be effectively formylated by chloral ⟨80KGS528⟩. Thus, 1-methylimidazo[1,2-*a*]imidazole gives aldehyde (**171**). The reaction seems to proceed in accordance with Equation 3.

$$\text{Het-H} + \text{CCl}_3\text{CHO} \longrightarrow \text{HetCH(OH)CCl}_3 \xrightarrow{\ ^{-}\text{OH}\ } \text{Het-CHO} + \text{CHCl}_3 \qquad (3)$$

Aldehydes and ketones react with azolinones. The reaction between aldehydes and 2-phenyl-5-oxazolinone (**172**; Y = H$_2$), formed *in situ* from PhCONHCH$_2$CO$_2$H and Ac$_2$O, gives azlactones (**172**; Y = RCH). Similar reactions are given by 4-thiazolidinones, *e.g.* (**173**) gives (**174**) ⟨79AHC(25)83⟩, and 4-imidazolinones. In pyrazolin-5-ones the 4-position is sufficiently activated for condensation to occur with ketones in acidic media (Scheme 16) ⟨66AHC(6)347⟩.

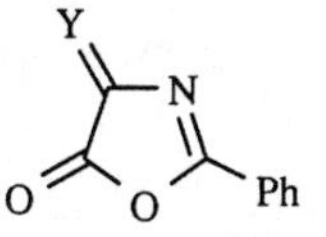

(172) 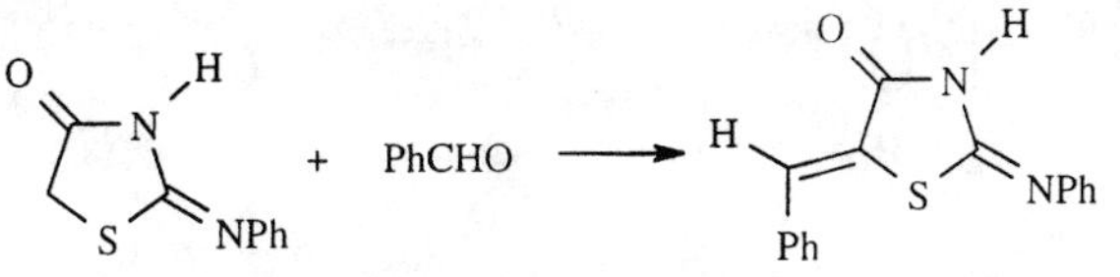　　(173)　+ PhCHO →　(174)

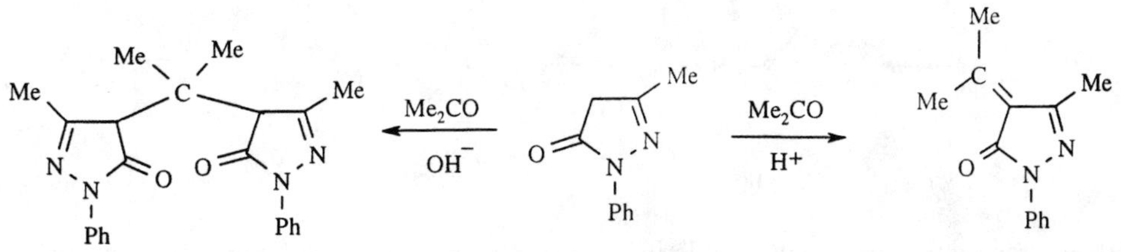

Scheme 16

3.4.1.4.11 Oxidation

The pyrazole ring is generally stable to oxidation and side chains are oxidized to carbonyl groups ⟨66AHC(6)347⟩. 1-Aryl-3-methylpyrazoles (**175**) react with ozone to yield 1,3,4-oxadiazolinones (**176**) ⟨66AHC(7)183⟩.

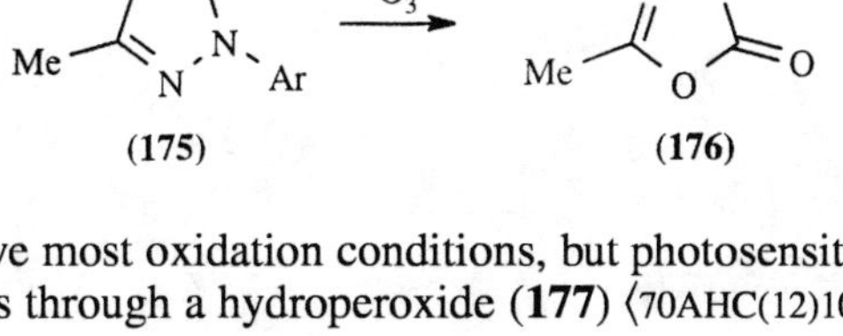

(175)　　(176)

Imidazole rings also survive most oxidation conditions, but photosensitized oxidation of imidazoles can give diaroylbenzamidines through a hydroperoxide (**177**) ⟨70AHC(12)103⟩.

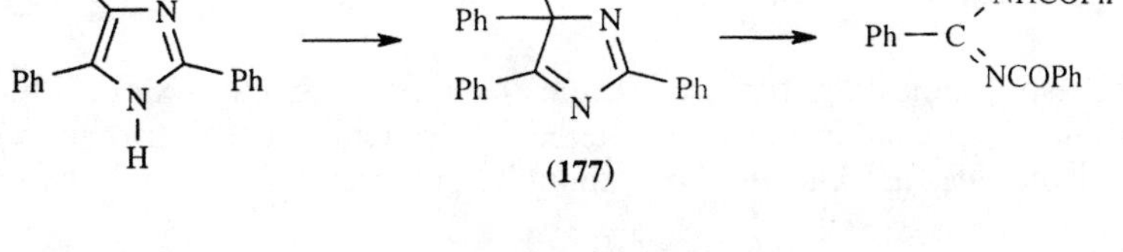

(177)

The thiazole, triazole and tetrazole rings are resistant to oxidation (*e.g.* by $KMnO_4$, CrO_3). Isoxazoles are more susceptible (*e.g.* **178** → **179**, benzil α-monoxime benzoate). The oxazole ring is relatively readily cleaved by oxidizing agents such as potassium permanganate, chromic acid or hydrogen peroxide to give acids or amides. Oxidation of 4,5-diaryloxazoles with chlorine or bromine gives the corresponding benzils in high yield ⟨74AHC(17)99⟩.

Oxidation of azole anions can give neutral azole radicals which could, in principle, be π (**180**) or σ (**181**) in nature. ESR spectra indicate the π-structure (**182**; hyperfine splitting in G) for imidazolyl radicals, but both π and σ-character have been observed for pyrazolyl radicals. Tetrazolyl radicals (**183** ↔ **184**) are also well-known ⟨79AHC(25)205⟩. Oxidation of 2,4,5-triarylimidazole anions with bromine gives 1,1'-diimidazolyls (**185**) which are in equilibrium with the dissociated free radical (**186**) ⟨70AHC(12)103⟩.

Oxidative dimerization of the lithium derivatives of 1-substituted 4,5-dicyanoimidazoles give (**187**) ⟨91JA6178⟩.

3.4.1.5   Attack at Sulfur

3.4.1.5.1   *Electrophilic attack*

A rare example of this type of reaction is the formation of (**189**) in the rhodium catalyzed reaction of di-*t*-butylthioketene with diethyl diazomalonate. This involves the oxathiole (**188**) as an intermediate, which undergoes electrophilic attack by the carbenoid to give (**189**). An X-ray structure was determined ⟨89TL1249⟩.

Certain thiazoles, isothiazoles and benzisothiazoles have been directly oxidized to sulfoxides and sulfones. 4,5-Diphenyl-1,2,3-thiadiazole is converted by peracid into the trioxide (**190**). Although 1,2,5-thiadiazole 1,1-dioxides are known, they cannot be prepared in good yield by direct oxidation, which usually gives sulfate ion analogous to the results obtained with 1,2,4- and 1,3,4-thiadiazoles ⟨68AHC(9)107⟩ (see also Section 3.4.1.3.13).

(190)   Scheme 17

When a hydroxyazole can tautomerize to a non-aromatic structure, oxidation at an annular sulfur atom becomes easy, *e.g.* as in Scheme 17 ⟨79AHC(25)83⟩.

Thiazoles are desulfurized by Raney nickel, a reaction probably initiated by coordination of the sulfur at Ni. The products are generally anions and carbonyl compounds (see Section 3.4.1.9.5).

3.4.1.5.2 Nucleophilic attack

Isothiazoles and isothiazolium cations are attacked by carbanions at sulfur and on recyclization can give thiophenes, as illustrated by (191) → (192). 2-Alkyl-3-isothiazolinones (*e.g.* 193) are also vulnerable to nucleophilic attack at sulfur ⟨72AHC(14)1⟩.

(191)   (192)   (193)

Nucleophilic attack at sulfur is implicated in many reactions of 1,2,4-thiadiazoles; generally, 'soft' nucleophilic attack at sulfur, *cf.* (194) → (195). *n*-Butyllithium with 4,5-diphenyl-1,2,3-thiadiazole yields PhC ≡ CPh, probably by initial nucleophilic attack at sulfur.

(194)   (195)   Scheme 18

Grignard reagents commonly react at S(2) of 1,2-dithiole-3-ones with ring opening (Scheme 18) ⟨73LA247, 74LA1261, 82AHC(31)63⟩.

3.4.1.6 Nucleophilic Attack at Carbon

Because of the increased importance of inductive electron withdrawal, nucleophilic attack on uncharged azole rings generally occurs under milder conditions than those required for analogous reactions with pyridines or pyridones. Azolium rings are very easily attacked by nucleophilic reagents; reactions similar to those of pyridinium and pyrylium compounds are known; azolium rings open particularly readily.

Nucleophilic attack on the ring carbon atoms of azoles occurs readily with oxazole and aza analogues. Such reactions are generally facilitated by additional ring heteroatoms and by electron-attracting substituents, and hindered by electron-donating substituents. A fused benzene ring aids nucleophilic attack on azoles, azolium ions and azolones; this may be rationalized because the loss of aromaticity involved in the formation of the initial adduct is less than that in monocyclic compounds. The orientation of attack is generally between two heteroatoms, as in (196).

(196)

3.4.1.6.1 *Hydroxide ion and other O-nucleophiles*

(i) *Neutral azoles*

Uncharged azoles not containing oxygen or sulfur are often resistant to attack by hydroxide ions at temperatures up to 100°C and above. However, neutral azoles react with hydroxide ions under extreme conditions. A remarkable reaction of this type is the direct hydroxylation of 1-substituted benzimidazoles, naphtho[1,2-*d*]- and naphtho[2,3-*d*]imidazoles and other fused imidazole systems with powdered anhydrous (molten) potassium or sodium hydroxides ⟨78RCR1933⟩. The reaction proceeds at temperature 230–250°C and leads to formation of the corresponding *N*-monosubstituted imidazolone in high yield (*e.g.* (**197**)→(**198**)). Interestingly, imidazo[4,5-*b*]- and imidazo[4,5-*c*]pyridines are hydroxylated exclusively in the imidazole ring, giving compounds (**199**)-(**201**). Non-fused imidazoles can not usually be hydroxylated; however, 1-methyl-4,5-diphenylimidazole under drasic conditions (300°C) gave imidazolone (**202**) (30%). The ease of hydroxylation and the orientation of the nucleophilic attack has been shown to be well predicted by the value of the largest atomic positive charge. The reaction can be considered to be the oxygen analogue of the well-known Chichibabin amination reaction (see Section 3.4.1.6.2). Indeed, in both cases a process is accompanied by evolution of hydrogen gas that can be rationalized in accordance with Scheme 19.

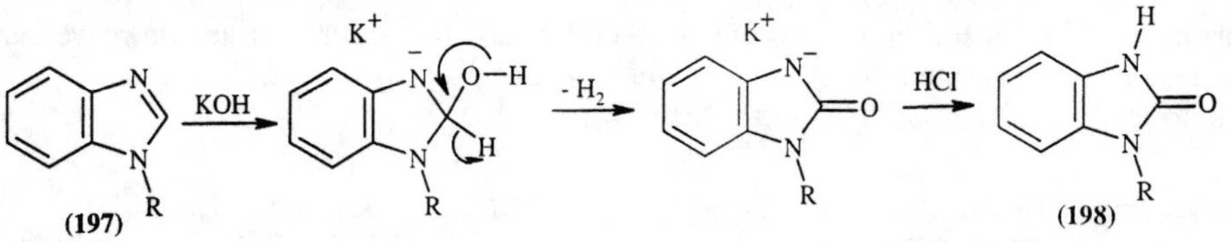

Scheme 19

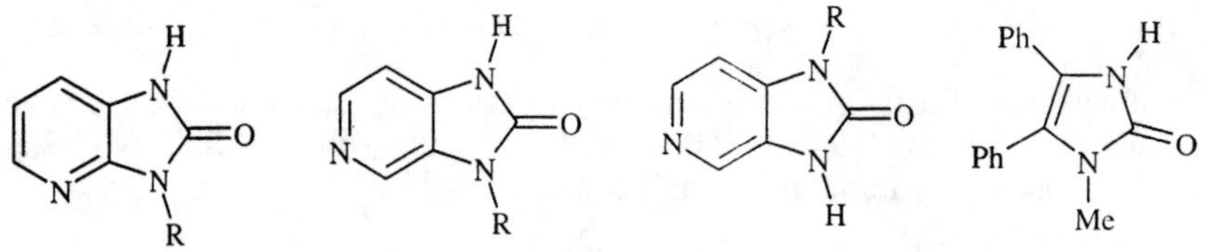

(**199**)   (**200**)   (**201**)   (**202**)

Imidazoles and benzimidazoles (**203**) react with acid chloride and alkali to give compounds of type (**205**), but these are reactions of the cation (**204**). 1,2,4-Triazoles and tetrazoles similarly undergo ring opening.

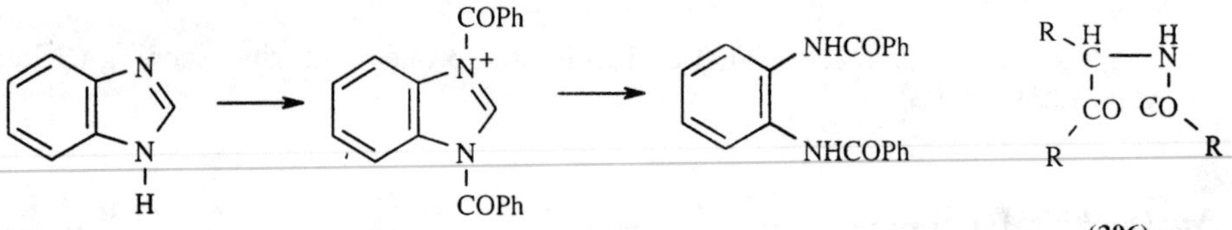

(**203**)   (**204**)   (**205**)   (**206**)

Isoxazoles are also rather stable to nucleophilic attack by OH⁻ at carbon. For reactions with base at a ring hydrogen atom, leading, for example, to ring opening isoxazoles, see Section 3.4.1.8.1.

Oxazoles give acylamino ketone (**206**) by acid-catalyzed ring scission, although they are somewhat more stable than furans. The oxazole ring is also moderately stable to alkali ⟨74AHC(17)99⟩; as expected, reaction with hydroxide ions is facilitated by electron-withdrawing substituents and fused benzene rings.

Oxadiazoles are easily cleaved. 2,5-Dialkyl-1,3,4-oxadiazoles (**207**) in aqueous solution with acid or base give hydrazides (if suitable substituents are present, further reaction can occur; see Section 3.4.3.5.1). 3-Methyl-1,2,4-oxadiazole (**208**) is easily hydrolyzed to acetamidoxime ⟨61CI(L)292⟩.

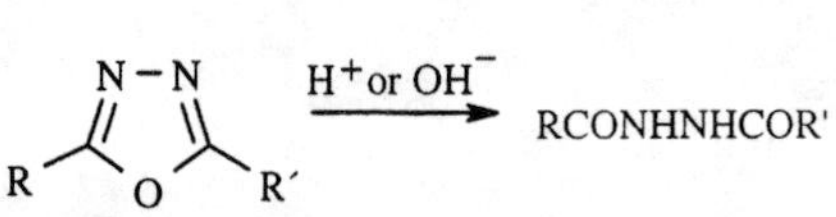

(**207**) 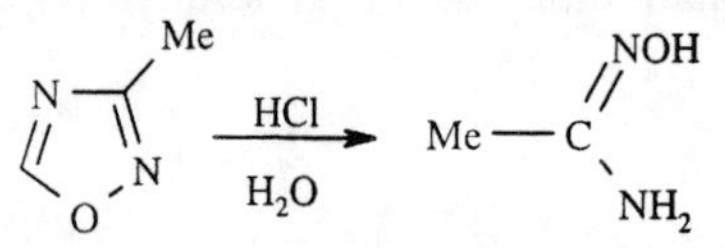   (**208**)

Isothiazoles and thiazoles are rather stable towards nucleophilic attack. Both 5- and 6-nitrobenzothiazole (**209**) add methoxide at C-2; the initial adduct undergoes ring opening to give (**210**) ⟨79AHC(25)1⟩. Unsubstituted 1,2,4-thiadiazole is sensitive to alkali. Substituents stabilize the ring somewhat, but ring-opening reactions are still common, *e.g.* (**211**) → (**212**). 5-Alkylamino-1,2,3,4-thiatriazoles are cleaved by alkali to azide ion and an isothiocyanate; in addition, a Dimroth rearrangement occurs to give a mercaptotetrazole ⟨64AHC(3)263⟩.

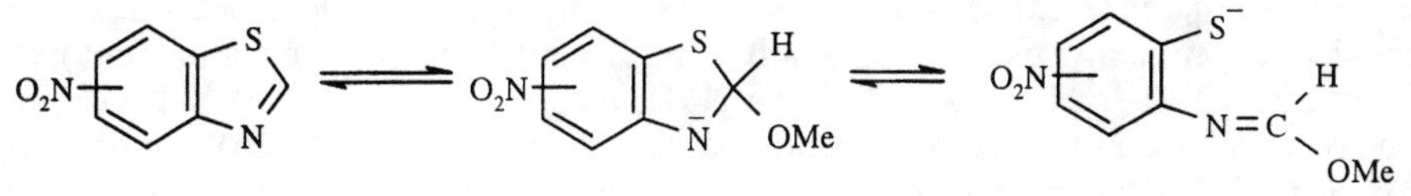

(**209**)   (**210**)

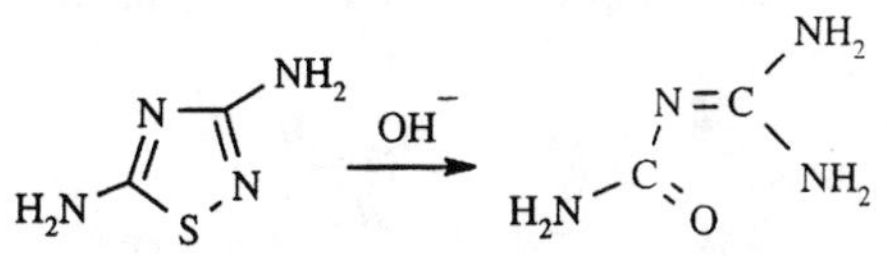

(**211**)   (**212**) 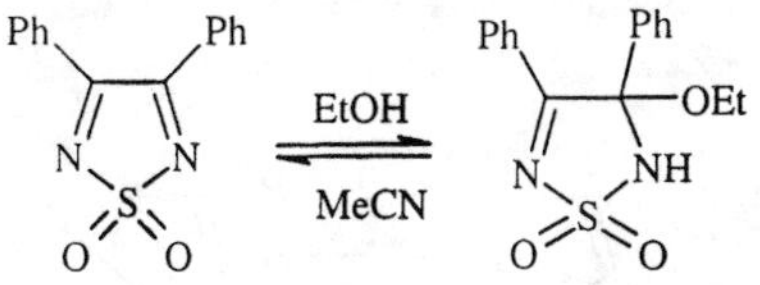   (**213**)

$\lambda_{max} = 328$ nm $\lambda_{max} = 267$ nm

UV spectroscopy has been used to clarify the strong interactions of 1,2,5-thiadiazole dioxides with protic solvents. Based on shifts in the maximum from 328 nm to 267 nm, as well as a doubling in the extinction coefficient, it is likely that the adduct (**213**) is formed in alcoholic media ⟨93JPO341⟩.

(ii) Azolium ions

Azolium ions (**214**) react reversibly with hydroxide ions to form a small proportion of the pseudo base (**215**). The term 'pseudo' is used to designate bases that react with acids measurably slowly, not instantaneously as is normal for acid-base reactions. Fused benzene rings reduce the loss of resonance energy when the hetero ring loses its aromaticity, and hence pseudo bases are formed even more readily by benzothiazolium cations than by thiazolium ions. Pseudo bases carrying the hydroxy group in the α-position are usually formed preferentially. As expected, pseudo base formation for S-containing azoliums is easiest with the dithiolylium and least easy with thiazolium ⟨79AHC(25)1⟩.

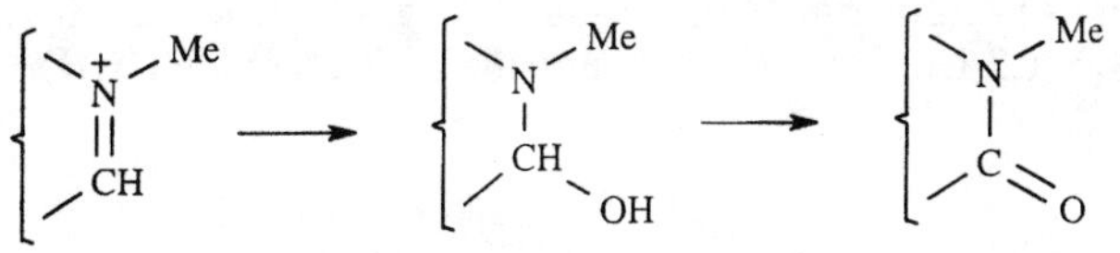

(**214**)   (**215**)

pK_{R^+} values for pseudo base formation are defined by equation (4),

$$K_{R^+} = [H^+] \, [QOH]/[Q^+] \tag{4}$$

where Q^+ and QOH denote the azolium cation and pseudo base, respectively. Some pK_{R^+} values are given in Scheme 20.

5.8   10.8   2.2

Scheme 20 pK_{R^+} values for azolium cations

Some pseudo bases are stable. 1,3-Dithiolylium adds alkoxide ions at the 2-position to give stable adducts which regenerate the starting salts with acids ⟨80AHC(27)151⟩. Pseudo bases can also lose water to give an ether (*e.g.* **216** → **217**).

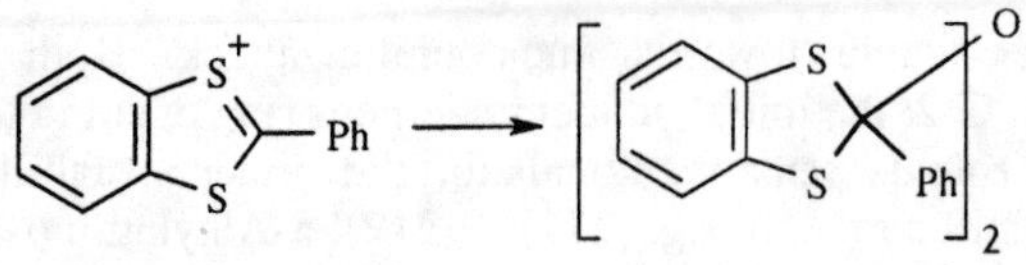

(216)	(217) 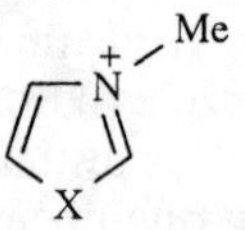	(218)

Oxidation to an azolone is an expected reaction for a pseudo base, but little appears to be known of such reactions. Most commonly, pseudo bases suffer ring fission. Estimated rates of ring opening of (218) are in the ratio 10^9: $10^{4.5}$: 1 for X = O, S and NMe, respectively ⟨79AHC(25)1⟩. Thiazolium salts (219) consume two equivalents of OH⁻ on titration because the pseudo bases (220) ring open to (221), which form anions (222). Quaternized oxazoles (223) are readily attacked by hydroxide to give open-chain products such as (224) ⟨74AHC(17)99⟩, and quaternized 1,3,4-oxadiazoles behave similarly. Quaternary isothiazoles (*e.g.* 225) are cleaved by hydroxide ⟨72AHC(14)1⟩, as are 1,2,4-thiadiazolium salts (226 → 227).

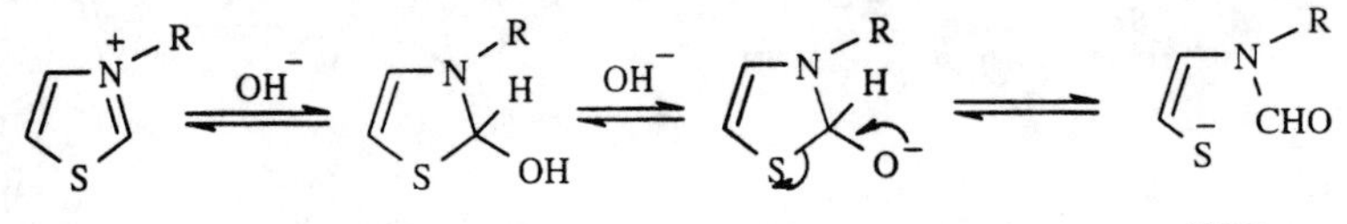

(219)	(220)	(221)	(222)

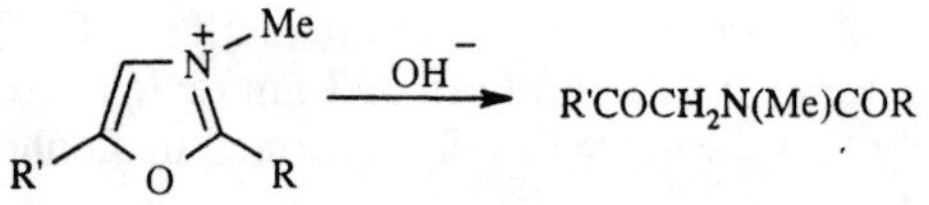

(223)	(224)

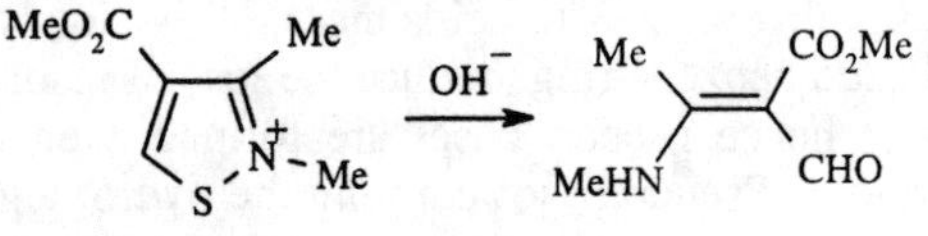

(225) 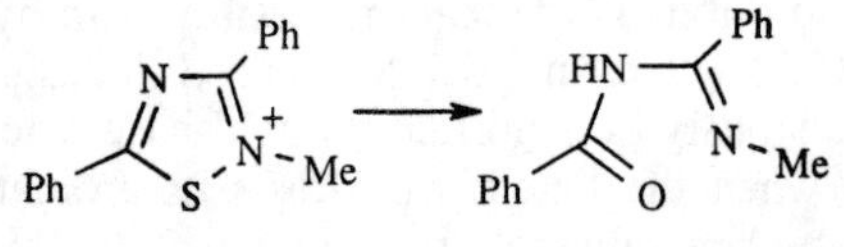	(226)	(227)

1,2-Dithiolylium ions undergo ring opening and degradation with hydroxide (Scheme 21) ⟨66AHC(7)39, 80AHC(27)151⟩. 1,2-Dimethylpyrazolium is degraded to MeNHNHMe.

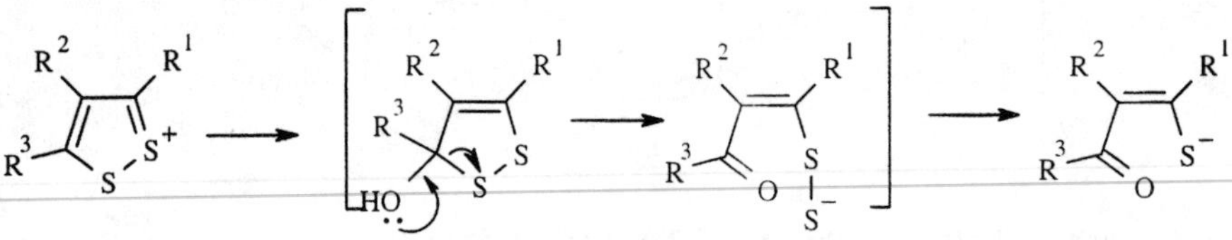

Scheme 21

Oxidative-reductive disproportionation is a rather typical property of some pseudo bases. Thus, 1,3-dimethyl-2-hydroxybenzimidazoline (229), which exists in the solid state in the open-chain form (228), on heating at 165–185°C, is converted to mixture of 1,3-dimethylbenzimidazolone (230) (49%) and 1,3-dimethylbenzimidazoline (231) (46%) ⟨85KGS1694⟩. Evidently, the process proceeds *via* an equilibrium amount of (229) undergoing hydride transfer.

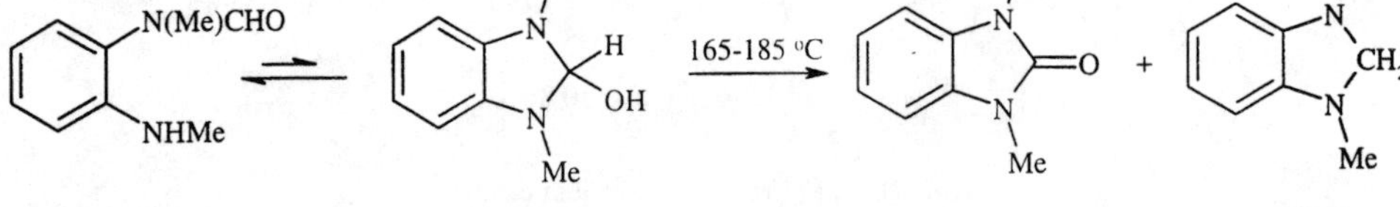

(228)	(229)	(230)	(231)

Reclosure to form a new heterocyclic or homocyclic ring occur in azolium ions carrying suitable substituents; these reactions are considered under the appropriate substituents.

Anthranils are readily cleaved by nitrous acid, presumably by attack of water on *N*-nitroso cations. The first product that can be observed is the nitrosohydroxylamino compound (232), which becomes reduced to the diazonium salt (233) ⟨67AHC(8)277⟩.

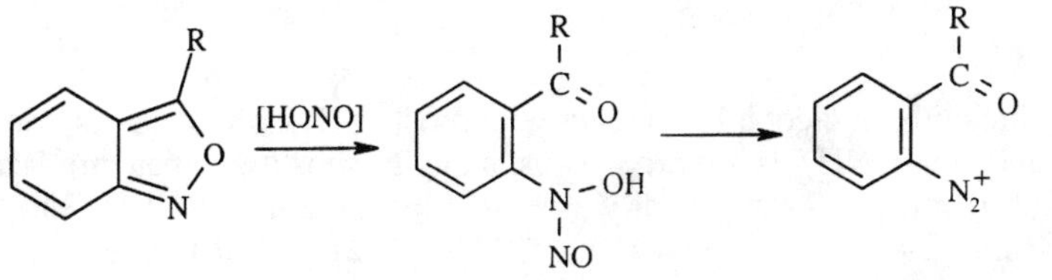

(232) (233)

(iii) Azolinones

Although imidazolinones are usually resistant to hydrolysis, oxazolinone rings are often easily opened. In acid-catalyzed reactions of this type, water converts azlactones (234) into α-acylamino-α,β-unsaturated acids (235) ⟨77AHC(21)175⟩. 1,3,4-Oxadiazolinones are readily opened by hot water to give hydrazine carboxylic acids which undergo decarboxylation.

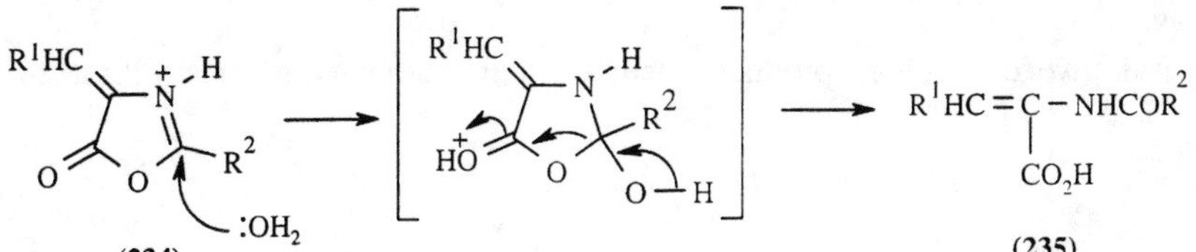

(234) (235)

3.4.1.6.2 Amines and amide ions

(i) Azoles

Oxygen-containing rings can be opened by amines; frequently this is followed by reclosure of the intermediate to form a new heterocycle. Thus, isoxazoles containing electron-withdrawing substituents give pyrazoles with hydrazine, *e.g.* (236; Z = O) → (236; Z = NH), and (237)→(238) ⟨66AHC(6)347⟩. In the benzo series a rather different reaction can occur: 6-nitroanthranil (239) is converted by amines into 2-*R*-indazoles (240) ⟨61JOC3714⟩. Oxazoles heated at 180°C with formamide are transformed into imidazoles ⟨74AHC(17)99⟩. With 2,4-dinitrophenylhydrazine (DNP) in HCl, oxazoles form hydrazones by ring fission (Scheme 22) ⟨74AHC(17)99⟩. Benzoxazole with hydroxylamine gives 2-aminobenzoxazole by elimination of water from the initial adduct. 1,3,4-Oxadiazoles react readily with ammonia and primary amines to give 1,2,4-triazoles.

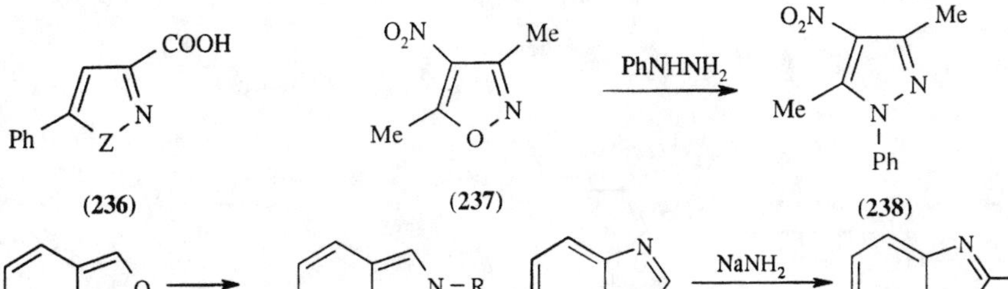

(236) (237) (238)

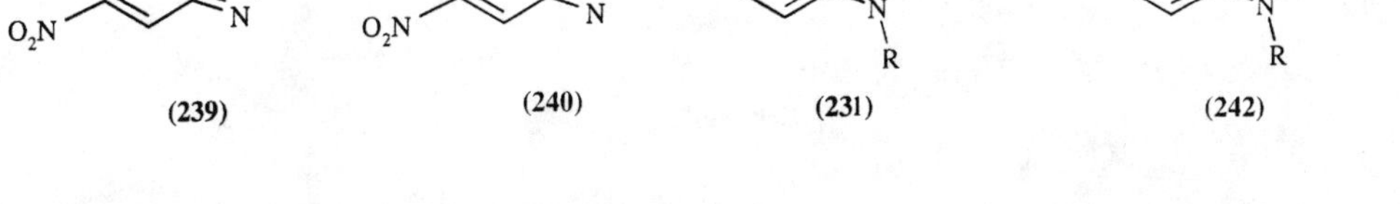

(239) (240) (231) (242)

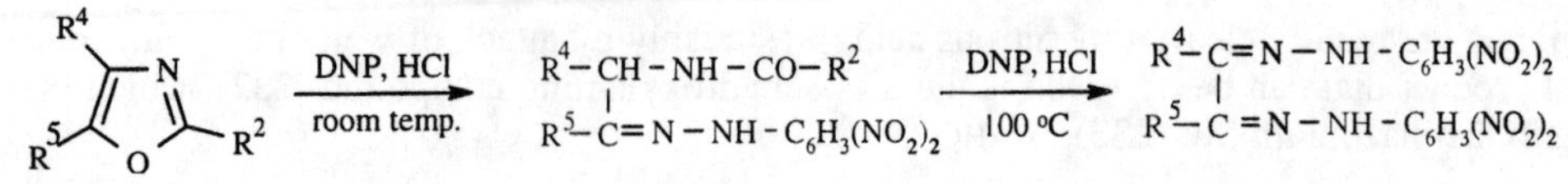

Scheme 22

Amines are insufficiently nucleophilic to react with most azoles which do not contain a ring oxygen and the stronger nucleophile NH_2^- is required. Thus, many N-substituted benzimidazoles, naphtho[1,2-*d*]- and naphtho[2,3-*d*]-imidazoles are readily aminated by sodamide at 110–130°C, forming in good yield the corresponding 2-amino derivatives, *e.g.* (**241**)→(**242**) ⟨78RCR1042, 88AHC2⟩. Thiazoles can be aminated in the 2-position by $NaNH_2$ at 150°C. Non-condensed imidazoles, as well as condensed imidazoles with a free NH-group, do not undergo Chichibabin amination. 9-Substituted purines by the action of KNH_2 in liquid ammonia at –80°C undergo two types of reaction: (1) ionization of the C_8-H bond and (2) addition of amide ion to the C-8 atom with subsequent opening of the imidazole ring ⟨83JOC850⟩. *C*-Nitroimidazoles are aminated by alkaline NH_2OH.

(ii) Azolium ions

Most azolium ions are sufficiently reactive to be attacked by amines. Sometimes the initial adducts are stable: ammonia and primary and secondary amines add to 1,3-dithiolylium salts at the 2-position to give compounds of the types NT_3, RNT_2 and R_2NT, respectively, where T = the 1,3-thiol-2-yl group ⟨80AHC(27)151⟩.

Some azoliums give open-chain products: primary and secondary amines with 1,2-dithiolyliums generally give (**243**) ⟨80AHC(27)151⟩.

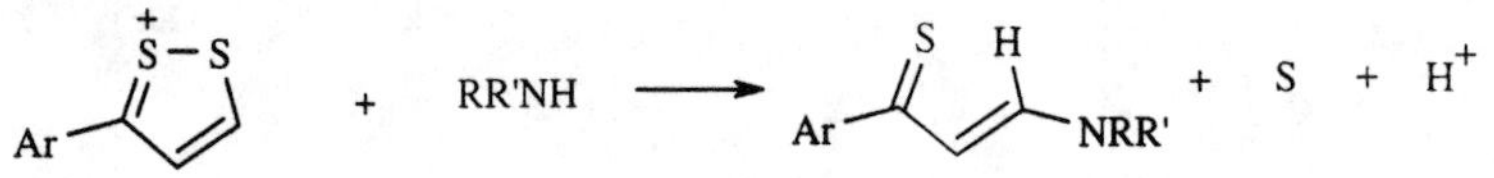

(**243**)

In other cases, reclosure to a new ring occurs: 1,2-dithiolylium ions with ammonia give isothiazoles according to the mechanism shown in Scheme 23 ⟨80AHC(27)151⟩. Treatment of quaternary isothiazoles with hydrazine or phenylhydrazine gives pyrazoles (Scheme 24) ⟨72AHC(14)1⟩, and 1,2,4-thiadiazoliums similarly yield 1,2,4-thiazoles. Oxazolium ions react with ammonium acetate in acetic acid to give the corresponding imidazoles (Scheme 25) ⟨74AHC(17)99⟩.

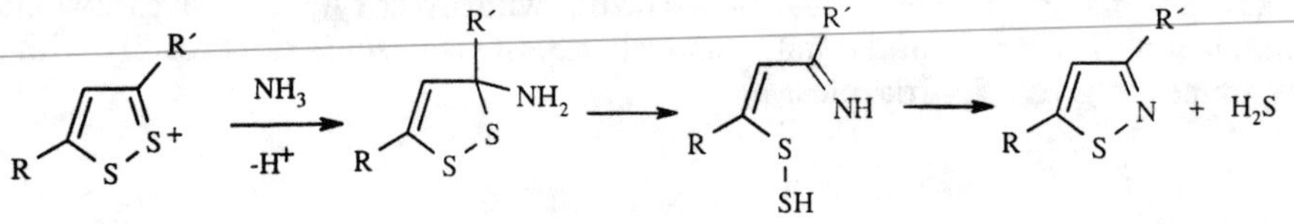

Scheme 23

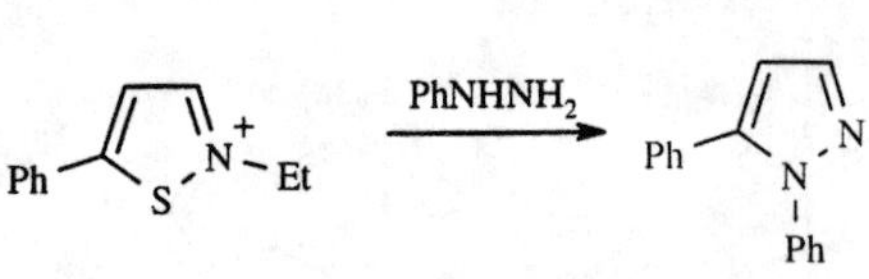

Scheme 24

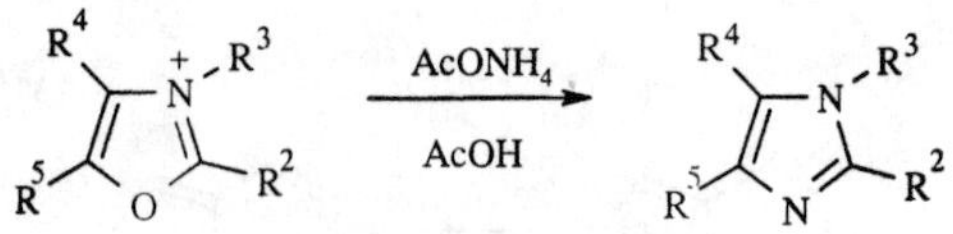

Scheme 25

Scheme 24

3,5-Diaryl-1,2,4-dithiazolium salts (**244**) are highly reactive 3,5-dielectrophiles and readily interact with nucleophiles to give a broad variety of heterocyclic compounds (Scheme 26) ⟨82MI 413-03, 82NKK1518, 85H(23)997⟩.

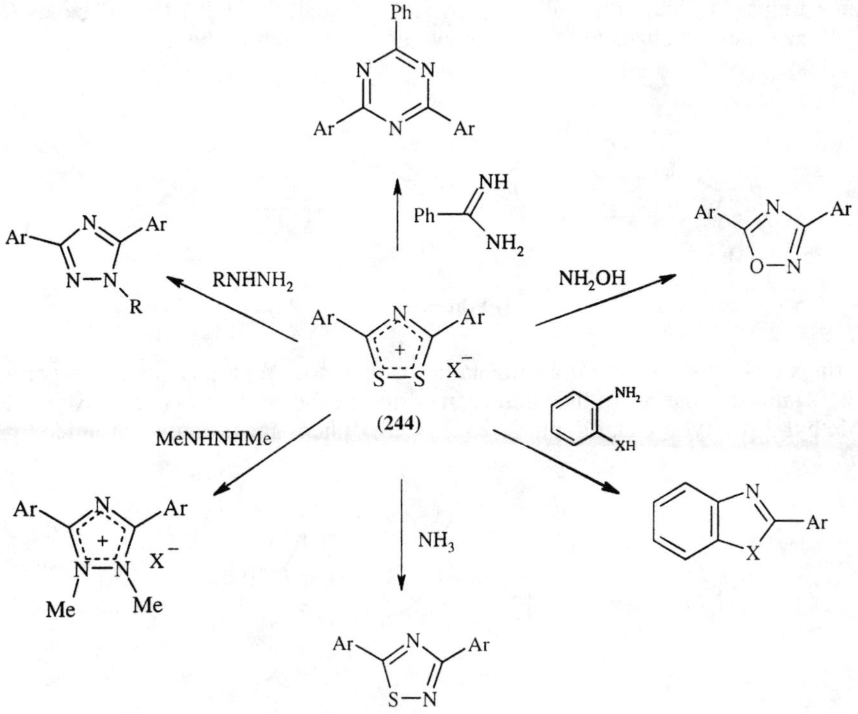

Scheme 26

Where an *N*-methoxy group is present, as in 1-methyl-3-methoxybenzimidazolium cation, elimination of MeOH can give an aromatic product.

3.4.1.6.3 S-Nucleophiles

Data on reactions of sulfur nucleophiles with azoles are sparse. Oxazoles are transformed in low yield into the corresponding thiazoles over alumina with H$_2$S at 350°C ⟨74AHC(17)99⟩. Sulfur nucleophiles such as SH$^-$ or RS$^-$ add to 1,3-dithiolylium salts at the 2-position ⟨80AHC(27)151⟩.

There are a number of examples in which 1-alkoxybenzimidazoles, 1-alkoxy-3-methylbenzimidazolium salts, and 1-alkylbenzimidazole 3-oxides react with anionic sulfur species to give 2-substitution with simultaneous deoxygenation ⟨93CHE127⟩.

3.4.1.6.4 Halide ions

Chloride ions are comparatively weak nucleophiles and do not react with azoles. In general, there is also no interaction of halide ions with azolium compounds.

Benzimidazole 3-oxides, *e.g.* (**245**), react with phosphorus oxychloride or sulfuryl chloride to form the corresponding 2-chlorobenzimidazoles. The reaction sequence involves first formation of a nucleophilic complex (**246**), then attack of chloride ions on the complex, followed by re-aromatization involving loss of the *N*-oxide oxygen (**247**→**248**).

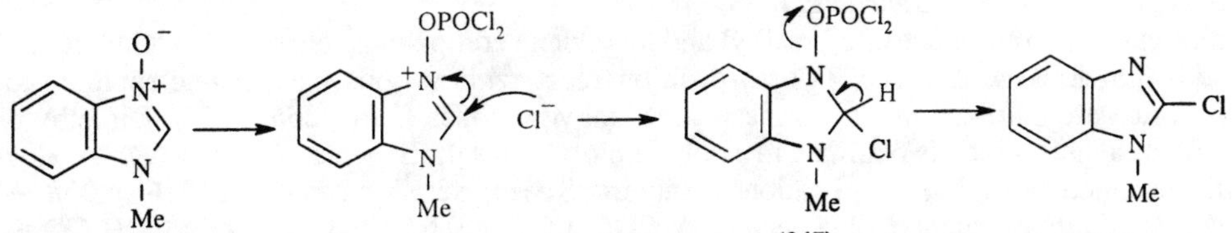

3.4.1.6.5   *Carbanions*

(i) *Organometallic compounds*

In contrast to pyridine chemistry, the range of nucleophilic alkylations that can be effected on neutral azoles is quite limited. Lithium reagents can add at the 5-position of 1,2,4-oxadiazoles (Scheme 27) ⟨70CJC2006⟩. Benzazoles are attacked by organometallic compounds at the C=N α-position unless it is blocked.

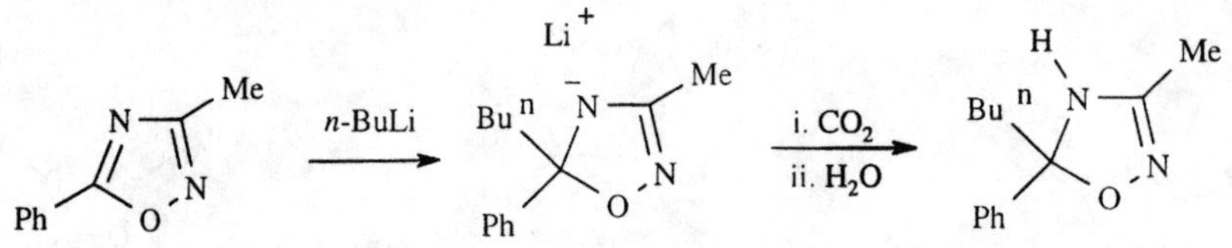

Scheme 27

Azolium rings react readily with organometallic compounds. With a Grignard reagent, conversion **(249)**→**(250)** is known in the benzothiazolium series, and 1,3-benzodithiolyliums give products of type **(251)**. 4-Methyl-2-phenyl-5-oxazolinone **(252)** with phenylmagnesium bromide gives **(253)** ⟨65AHC(4)75⟩.

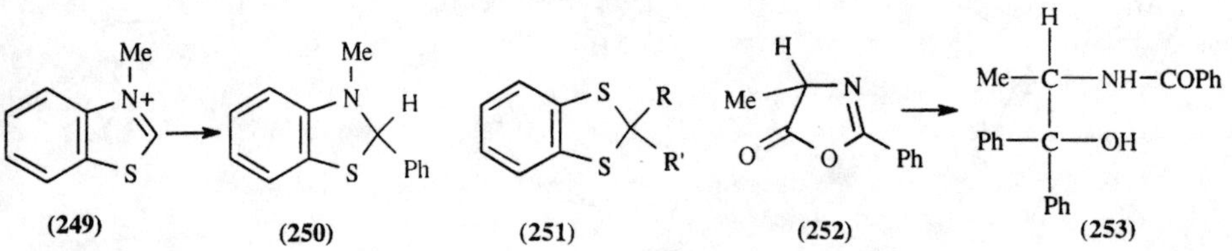

Dihydro allyl adducts like **(254)** are obtained by reaction of thiazoles with allyltributyl tin in the presence of alkyl chloroformates acting as activators of the thiazole ring (Scheme 28) ⟨94JOC1319⟩. This reaction most likely takes place *via* the intermediate azolium salt. Under these conditions even organolithium compounds can add to thiazoles ⟨84TL3633⟩. Similarly, direct ethynylation of thiazole and benzothiazole can be achieved by reaction with bis(tributylstannyl)acetylene (Scheme 29) ⟨94SL557⟩.

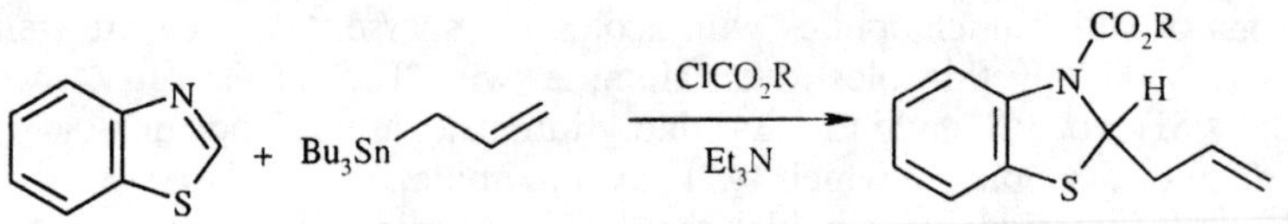

Scheme 28

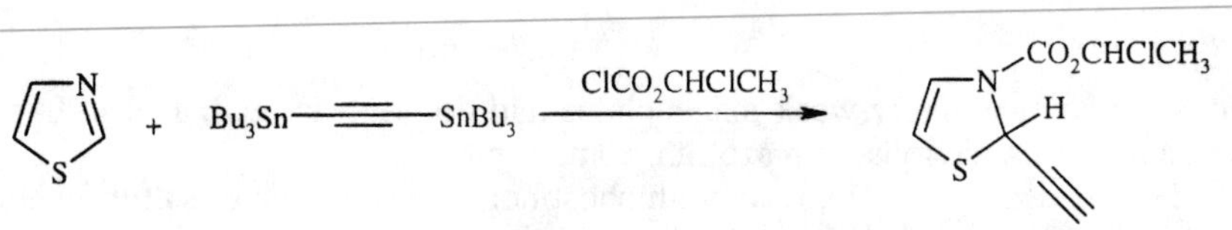

Scheme 29

(ii) *Activated methyl and methylene carbanions*

The mesomeric anions of activated methyl and methylene compounds react with azolium ions. Thus, 1,2-dithiolylium ions with a free 3- or 5-position react with various carbon nucleophiles to give products which are oxidized *in situ* to mesomeric anhydro-bases **(255→256→257)**. Dimethylaniline gives an intermediate which is oxidized to a new dithiolylium salt (Scheme 30) ⟨66AHC(7)39⟩. However, in 3,5-disubstituted 1,2-dithiolylium cations an alternative ring scission can occur **(256 → 258 → 259)** ⟨80AHC(27)151⟩. In this sequence, **(255)** can be ArCOCH$_2$CS$_2$Me, NCCH$_2$CSNH or NCCH$_2$CO$_2$Et. The conversion in the 1,2,4-thiadiazole series of **(260)** into **(261)** is analogous.

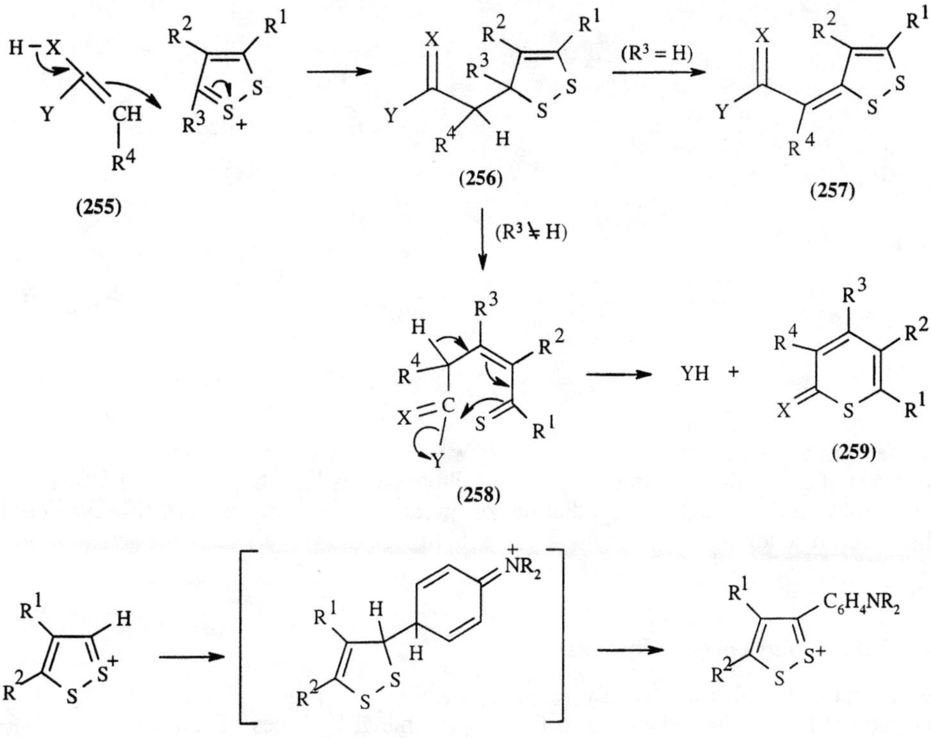

Scheme 30

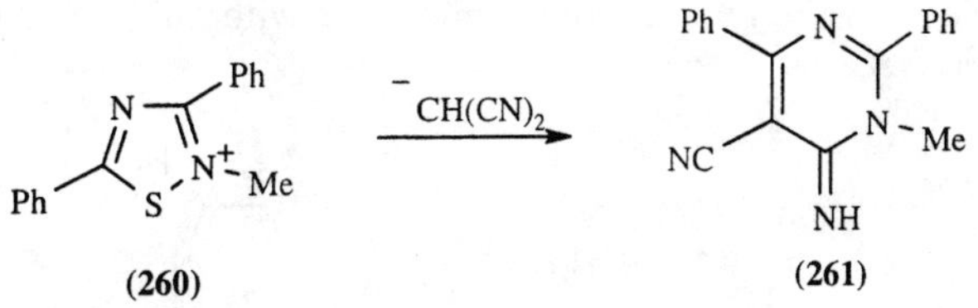

Anhydro bases can attack the α-position, *e.g.* of thiazolium cations, with the formation of adducts capable of oxidation to cyanine dyes, *e.g.* Scheme 31 (see Section 3.4.3.3.4).

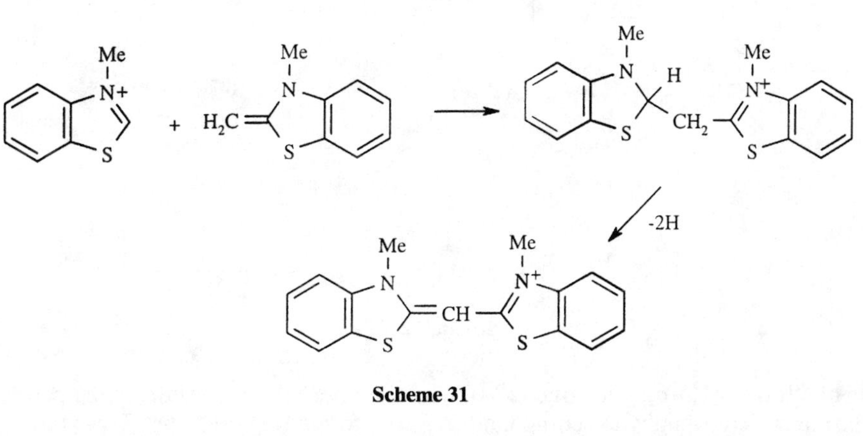

Scheme 31

Active methylene compounds can add to 1,3-dithiolylium ions to give 2-substituted 1,2-dihydro-1,3-dithioles (**263**). Again, addition is often followed by oxidation (to **264**). Alternatively, further addition can occur (to **265**) ⟨80AHC(27)151⟩. In this reaction, (**262**) can be $CH_2(CN)_2$, $CH_2(COMe)_2$ or even MeCOMe. Somewhat similar reactions are shown by 1,3-diarylimidazolium ions.

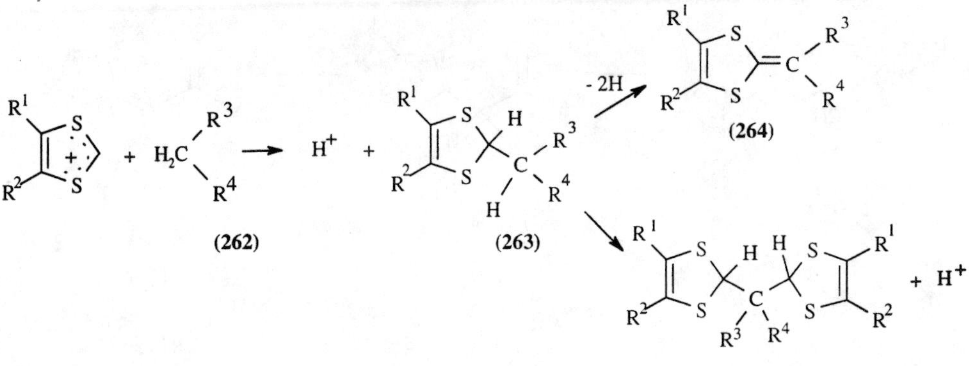

(262)   (263)   (264)

(265)

(iii) Cyanide ions

2-Cyanations of benzimidazolium or benzimidazole N-oxides have been reported ⟨93CHE127⟩.

Cyanopyrazoles are formed by irradiation of pyrazoles in the presence of cyanide ions in photosubstitution reactions.

3.4.1.6.6 Reduction by complex hydrides

Oxygen-containing azoles are readily reduced, usually with ring scission. Only acyclic products have been reported from the reductions with complex metal hydrides of oxazoles (*e.g.* **266→267**), isoxazoles (*e.g.* **268→269**), benzoxazoles (*e.g.* **270→271**) and benzoxazolinones (*e.g.* **272, 273→271**). Reductions of 1,2,4-oxadiazoles always involve ring scission. Lithium aluminum hydride breaks the C-O bond in the ring (Scheme 32) ⟨76AHC(20)65⟩.

Nitrogen azoles are less easily reduced: benzimidazole with lithium aluminum hydride gives 2,3-dihydrobenzimidazole ⟨52CB390⟩.

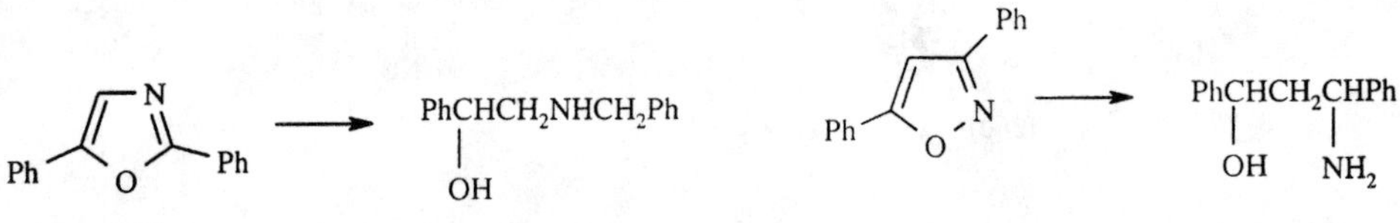

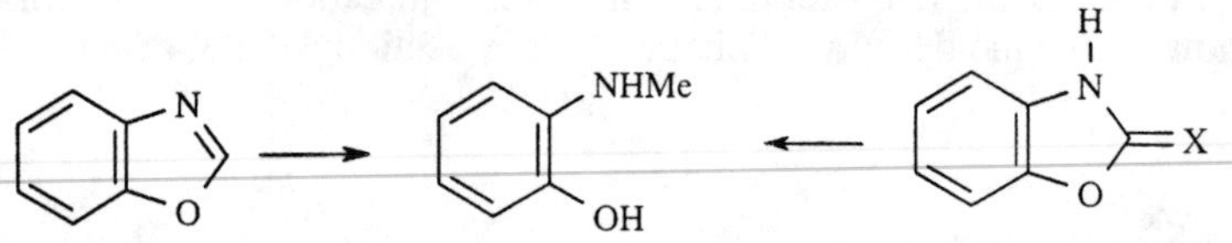

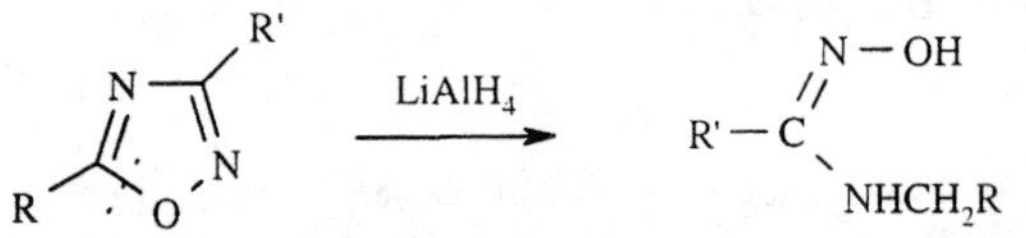

Scheme 32

Reduction of tetrazolo[1,5-*b*]pyridazine (**274**) by sodium borohydride results in the partial reduction of the pyridazine moiety (tetrahydro compound **275** was formed) ⟨76JHC835⟩. The catalytic reduction of the substituted tetrazolo[1,5-*b*]pyrimidine (**276**) afforded the diaminopyrimidine derivative (**277**) ⟨85MIP155606⟩.

Two effective tin reagents, triethylammonium triphenyltin and dibuthyltin dihydride have been offered for reduction of fused tetrazoloazines to aminoazines ⟨87TL5941⟩. The reaction was interpreted to proceed *via* ring opening to an azide and subsequent reduction.

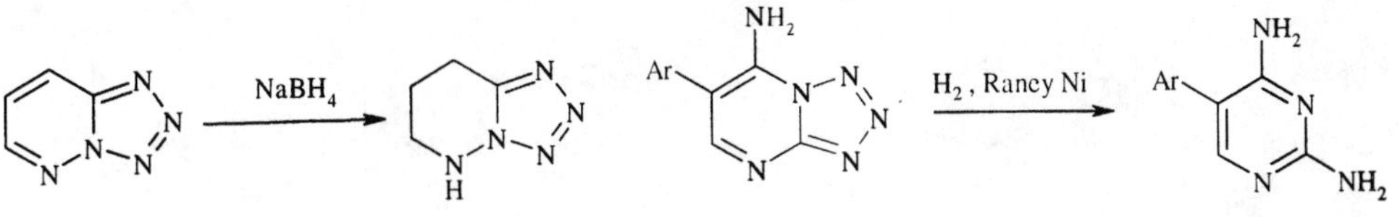

Catonic rings are readily reduced by complex hydrides under relatively mild conditions. Thus, isoxazolium salts with sodium borohydride give the 2,5-dihydro derivatives (**278**) in ethanol, but yield the 2,3-dihydro compound (**279**) in $MeCN/H_2O$ ⟨74CPB70⟩. Thiazolyl ions are reduced to 1,2-dihydrothiazoles by lithium aluminum hydride and to tetrahydrothiazoles by sodium borohydride. The tetrahydro compound is probably formed *via* (**280**), which results from proton addition to the dihydro derivatives (**281**) containing an enamine function. 1,3-Dithiolylium salts easily add hydride ion from sodium borohydride (Scheme 33) ⟨80AHC(27)151⟩.

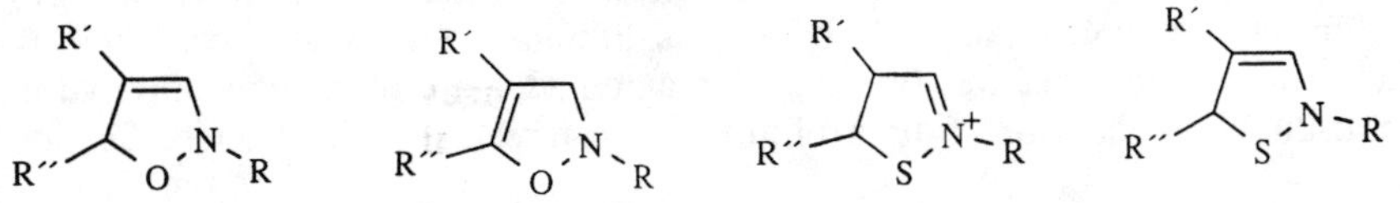

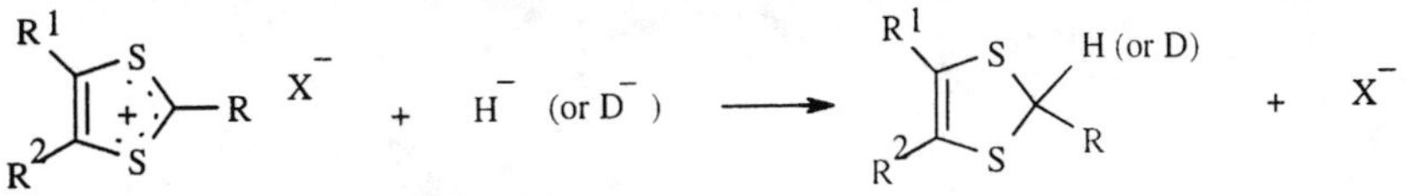

$R = H$, SMe, NR_2, aryl, cyclohexyl

Scheme 33

3.4.1.6.7 *Phosphorus nucleophiles*

Trialkyl- and triaryl-phosphines react with 1,3-benzodithiolylium ions to give a phosphonium salt which is deprotonated by *n*-butyllithium to give (**282**) ⟨76TL3695⟩.

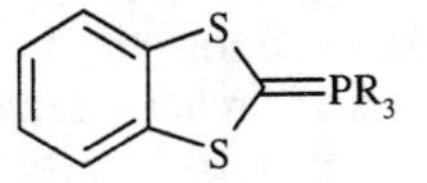

(**282**)

3.4.1.7 Nucleophilic Attack at Nitrogen Heteroatom

1,2,4-Oxadiazoles (and also other 1-oxa-2-azoles) undergo intramolecular nucleophilic displacements of O(1) of the general type (**283 → 284**) ⟨81AHC(29)141, 86JST(136)215, 93AHC(56)49, 94H(37)2057⟩ by the attack on N(2) of oxygen, sulfur, nitrogen, and carbon nucleophiles Z forming the third atom of a side chain in the 3-position of the heterocycle.

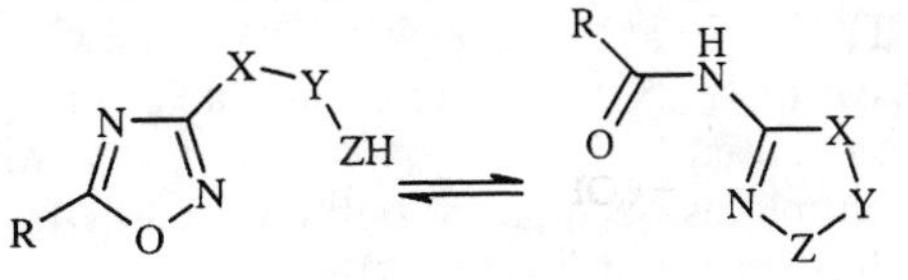

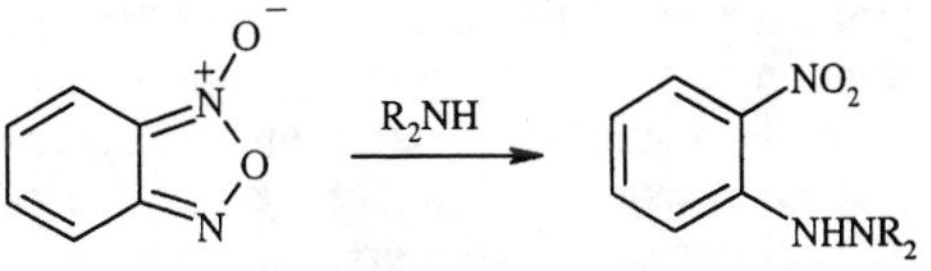

(**283**)   (**284**)   **Scheme 34**

XYZ = CCN, CCO, CNN, CNO, NCC, NCN, NCO, NCS, NNN
X-Y = double or single bond

Nucleophilic attack at N(3) takes place in the benzofuroxan series. For example, the reaction with secondary amines leads to *o*-nitroarylhydrazines (Scheme 34) ⟨88JCS(P1)145⟩.

There are a few examples in which pyridine-like nitrogen atoms of the benzotriazole ring appear to undergo nucleophilic attack by organometallic reagents. Thus, the N(3) atom is attacked by Grignard reagents in (benzotriazol-1-yl)methylethers (**285**) (Scheme 35) ⟨89JOC6022⟩ and in (benzotriazol-1-yl)methyl ammonium salts to give *o*-phenylenediamines (**286**) in 10–40% yield ⟨89JHC1579⟩ (see Section 3.2.1.7).

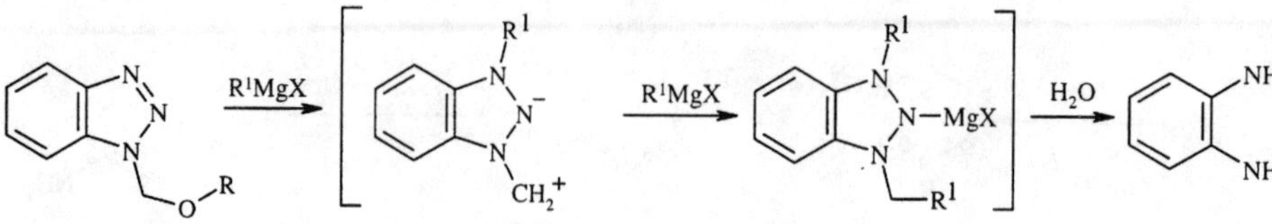

Scheme 35

3.4.1.8 Nucleophilic Attack at Hydrogen Attached to Ring Carbon or Ring Nitrogen

Hydrogens attached to ring carbon atoms of neutral azoles, and especially azolium ions, are acidic and can be removed as protons by bases. Reaction follows the orientations shown in (**287**) and (**288**). The anions from neutral azoles can be stabilized as lithium derivatives, except in isoxazoles and indazoles where ring cleavage occurs. Typically, the anion adds a proton again and hydrogen isotope exchange can result. The zwitterions from azolium rings can react as carbenes (see Section 3.4.1.8.5).

(**287**)   (**288**)

3.4.1.8.1 Metallation at a ring carbon atom

Neutral azoles are readily *C*-lithiated by *n*-butyllithium provided they do not contain a free NH group (Table 7). Derivatives with two heteroatoms in the 1,3-orientation undergo lithiation preferentially at the 2-position; other compounds are lithiated at the 5-position.

N-Unsubstituted pyrazole is readily converted into 3(5)-substituted derivatives in a one-pot sequence, using formaldehyde both for *N*-protection and to mediate the lithiation at the 5-position ⟨89T4253⟩; the transient *N*-(hydroxymethyl) unit is readily removed during workup. Direct lithiation of 1-substituted pyrazoles provides an efficient entry to 1,5-disubstituted pyrazoles ⟨91JHC1849, 92SL327⟩.

Provided that the annular nitrogen is substituted, imidazoles can be lithiated at ring carbon atoms with reactivity order C-2 > C-5 > C-4 ⟨93RTC123, 94H(38)2487⟩. Suitable N-protecting groups have been developed considering the ease of attachment and removal, steric, and electronic properties ⟨84JCS(P1)481, 88JOC1107, 88JOC5685, 89JCS(P1)1139⟩. Whereas butyllithium is frequently the reagent of choice for making lithioimidazoles, LDA has advantages when there are substituents susceptible to nucleophilic attack.

Careful choice of reaction conditions also allows 2-lithiation of imidazoles even in the presence of groups susceptible to attack by the reagent. At −100°C in THF 4,5-dicyano-1-methylimidazole forms the 2-lithio compound; at −80°C butyllithium attacks one of the nitrile functions ⟨92JHC1091⟩. 1-Alkoxyimidazoles are lithiated by *n*-butyllithium (THF, −78°C) at the 2-position ⟨98JOC12⟩.

Polylithiation at both the 2- and 5-positions gives access to 1,2,5-trisubstituted imidazoles in yields dependent on the ability of the 2-anion to further deprotonate at C-5 ⟨83JCR(S)196⟩. As the 5-anion is the more reactive, sequential quenching by different electrophiles becomes possible ⟨84JCS(P1)481⟩.

1-Substituted benzimidazoles are readily lithiated in the 2-position ⟨88KGS147⟩.

Attempted metallation of isoxazoles usually causes ring opening *via* proton loss at the 3- or 5-position (Section 3.4.1.8.6); however, if both of these positions are substituted, normal lithiation occurs at the 4-position (Scheme 36). 2-Lithiooxazoles tend to exist in tautomeric equilibrium with their open chain form. The same is true for 1-*R*-3-lithioindazoles and 1-*R*-5-lithiotriazoles at room temperature.

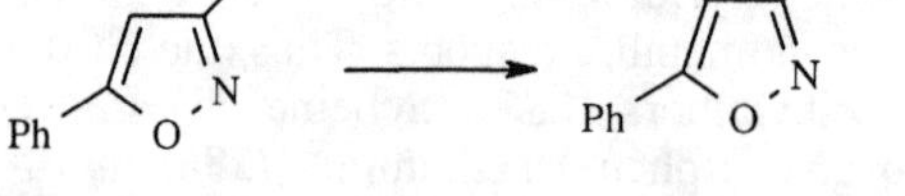

Scheme 36

The treatment of thiazole with *n*-butyl- or phenyllithium leads to exclusive deprotonation at C-2. When the 2-position is blocked, deprotonation occurs selectively at C-5. However, if the substituent at C-2 is an alkyl group, the kinetic acidities of the protons at the α-position and at the 5-position are similar. The reaction of 2,4-dimethylthiazole with butyllithium at –78°C yields the 5-lithio derivative (**289**) as the major product but if the reaction is carried out at higher temperature the thermodynamically more stable 2-lithiomethyl derivative (**290**) is obtained (Scheme 37). The metallation at these two positions is also dependent on the strength and bulk of the base employed ⟨74JOC1192⟩: lithium diisopropylamide is preferred for selective deprotonations at the 5-position.

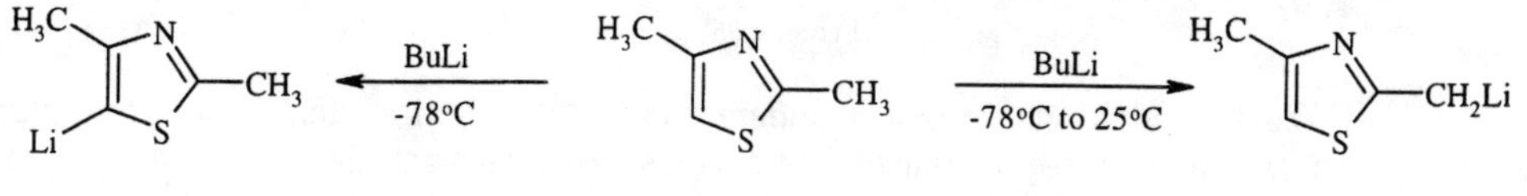

(**289**) (**290**)

Scheme 37

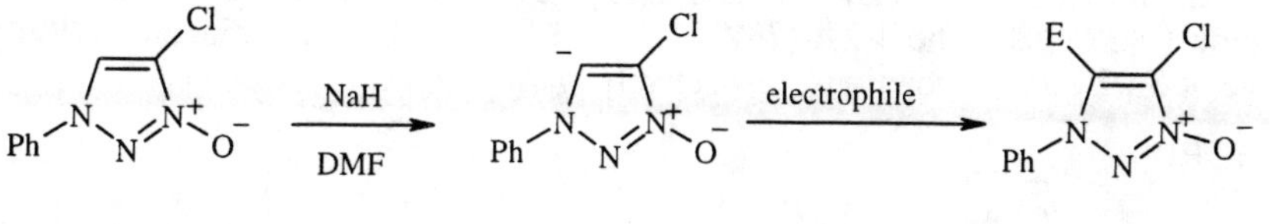

(**291**) (**292**) (**293**)

1-Substituted 1,2,3-triazoles, 1,2,4-triazoles and tetrazoles are metallated by *n*-butyllithium at low temperature at the 5-position. At room temperature 5-lithium derivatives tend to undergo ring opening. No direct lithiation of a 2-substituted 1,2,3-triazole has been reported.

3-Phenyl-5-chlorotriazole 1-oxides (**291**) are deprotonated with NaH in DMF. The resulting anions (**292**) react with carbon, silicon, and sulfur electrophiles to give substituted products (**293**) in good yield ⟨87ACS(B)724⟩. Similarly, deprotonation of 1-substituted pyrazole 2-oxides followed by addition of dimethyl disulfide affords 3- and 5-methylthio as well as 3,5-dimethylthiopyrazoles ⟨92ACS1096, 88BSB573, 92H(33)1129⟩.

Metallation reactions at C-4 of sydnones have been reviewed ⟨95H(41)1525⟩.

Table 7 Lithiation of Azoles by *n*-Butyllithium

Heterocycle	Position lithiated	Temperature (°C)	Ref.
3-Methyl-1-phenylpyrazole	5	—	94H(37)2087
3,5-Disubstituted isoxazoles	4	–70 to –65	79AHC(25)147, 70CJC1371, 85H(23)571,585
3,5-Disubstituted isothiazoles	5	–70	65AHC(4)75, 72AHC(14)1
1-Substituted imidazoles	2	—	85H(23)417, 88CHE117, 94H(38)2487
Oxazoles	2	'low'	74AHC(17)99
Thiazoles	2	–60	—
2-Substituted thiazoles	5	–100	—
1-Phenyl-1,2,3-triazoles	5	–60 to –20	74AHC(16)33, 71CJC1792, 88BSB573, 93AHC(56)155
1-Phenyl-1,2,4-triazole	5	'low'	85H(23)1645
1-Substituted tetrazoles	5	–60	71CJC2139, 66JA4266

3.4.1.8.2 *Hydrogen exchange at ring carbon in neutral azoles*

Base-catalyzed hydrogen exchange has been summarized for five-membered rings ⟨74AHC(16)1, 74KGS1587⟩. In many reactions of this type the protonated azole is attacked by hydroxide ion to form an ylide in the rate-determining step, *e.g.* for imidazole (Scheme 38) ⟨74AHC(16)1⟩. Deuteration of imidazole is fast at the 2-position and much slower at the 4- and 5-positions. Rates fall off for *N*-unsubstituted imidazoles at high pH values because of the formation of unreactive anions. In the case of 1-methylbenzimidazole the rate of hydrogen exchange in the 2-position is independent of the acidity over a wide range, in agreement with the mechanism shown in Scheme 38 ⟨80AHC(27)241⟩. The rates of base-catalysed H-D exchange for *N*-methyl derivatives of benzimidazole, naphtho[1,2-*d*]- and naphtho[2,3-*d*]-imidazoles and perimidine are all the same within a factor of 1.5 ⟨77KGS1544⟩.

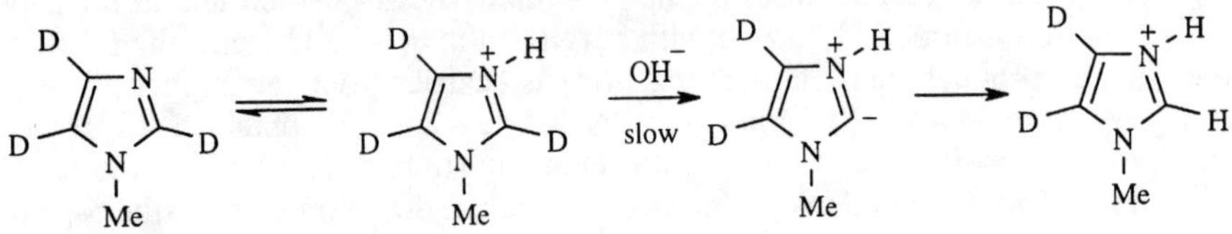

Scheme 38

Under strongly basic conditions oxazoles undergo fast 2-deuteration and slower 5-deuteration ⟨74AHC(17)99⟩. The hydrogen in 5-position of isothiazoles exchanges rapidly under basic conditions ⟨69JHC199⟩. Neutral thiazoles exchange by two competitive mechanisms: at pD 0–11 the conjugated acid exchanges the 2-H *via* the ylide (**294**), whereas at higher pD exchange is at the 2- and 5-positions *via* the carbanion (**295**) and (**296**). The acidities of thiazole (pK_a ~ 28.3) and benzothiazole (pK_a ~ 28.9) have been measured in THF at –60°C by deprotonation at C-2 using lithium tetramethylpiperidide ⟨85CJC3505⟩. The 1,2,4-(**297**), 1,3,4- (**298**) and 1,2,3-thiadiazoles (**299**) all undergo rapid exchange at 5-, 2-, and 5-positions, respectively ⟨74AHC(16)1⟩.

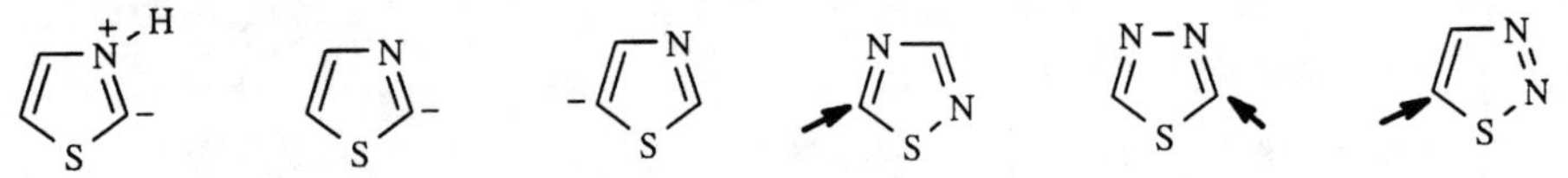

(**294**)   (**295**)   (**296**)   (**297**)   (**298**)   (**299**)

1-Substituted tetrazoles readily exchange the 5-hydrogen for deuterium in aqueous solution. A major rate-enhancing effect is observed with copper(II) or zinc ions due to σ-complexation with the heterocycle. The rate of base-induced proton-deuterium exchange of 1-methyltetrazole is 10^5 times faster than 2-methyltetrazole ⟨77AHC(21)323⟩.

3.4.1.8.3 *Hydrogen exchange at ring carbon in azolium ions and dimerization*

Hydrogen atoms in azolium ions can be removed easily as protons (*e.g.* **300** → **302**); exchange with deuterium occurs in heavy water. The intermediate zwitterion (*e.g.* **301**) can also be written as a carbene. The pK_a values of thiazolium ions range from 16 to 20 ⟨87CRV863, 88B5044⟩.

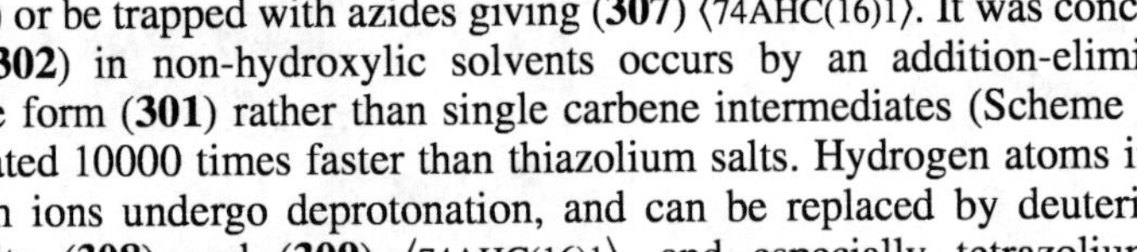

(**300**)   (**301**)   (**302**)

The relative rates of H-isotope exchange in D$_2$O/OD$^-$ for oxazolium, thiazolium and imidazolium are shown in formulae (**303**)-(**305**), respectively ⟨71PMH(4)55⟩. The intermediate carbene, *e.g.* (**302**), can form a dimer (**306**) or be trapped with azides giving (**307**) ⟨74AHC(16)1⟩. It was concluded ⟨91JA985⟩ that dimerization of (**302**) in non-hydroxylic solvents occurs by an addition-elimination mechanism involving the ylide form (**301**) rather than single carbene intermediates (Scheme 39). Thiadiazolium salts are deprotonated 10000 times faster than thiazolium salts. Hydrogen atoms in positions 3 and 5 of 1,2-dithiolylium ions undergo deprotonation, and can be replaced by deuterium ⟨80AHC(27)151⟩. Thiadiazolium salts (**308**) and (**309**) ⟨74AHC(16)1⟩, and especially tetrazolium salts (*e.g.* **310**) ⟨74AHC(16)1⟩, exchange particularly quickly.

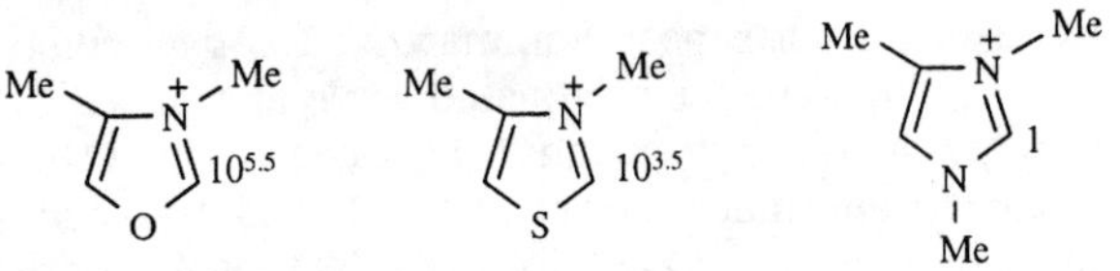

(**303**)   (**304**)   (**305**)

Relative rates of H-isotope exchange in D$_2$O/OD$^-$

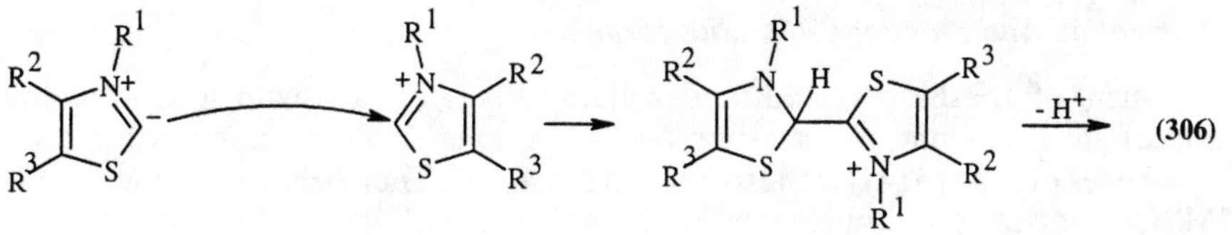

Scheme 39

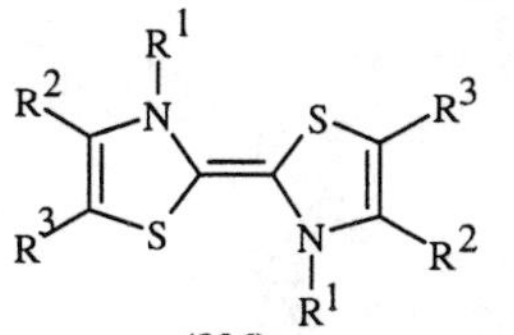

(306) 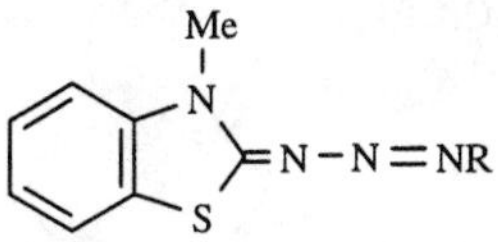   (307) 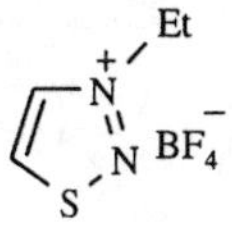   (308) 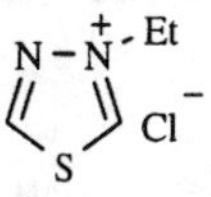   (309) 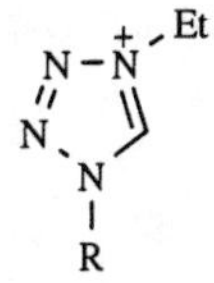   (310)

Base-catalyzed hydrogen exchange occurs at the 3- and 5-positions of 1,2-dimethylpyrazolium salts. 2-Unsubstituted 1,3-dithiolylium salts are easily deprotonated by nucleophilic attack of hydrogen. The intermediate carbene easily undergoes dimerization. Hydrogen exchange can also occur (Scheme 40) ⟨80AHC(27)151⟩.

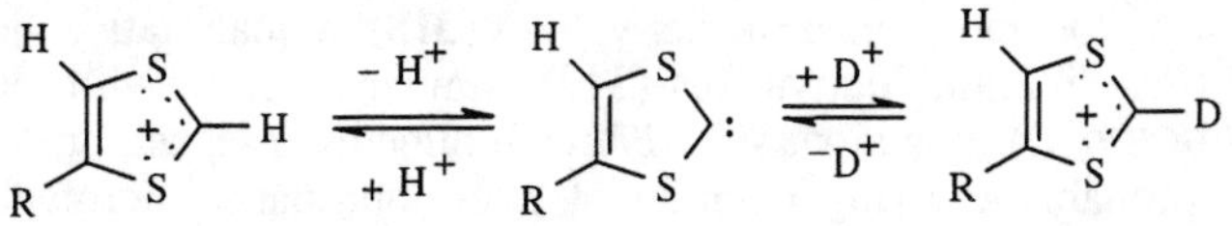

Scheme 40

3.4.1.8.4 C-Substitution via electrophilic attack at N, deprotonation, and rearrangement

1-Substituted imidazoles can be acylated at the 2-position by acid chlorides in the presence of triethylamine. This reaction proceeds by proton loss from the *N*-acylated intermediate (**311**). An analogous reaction with phenyl isocyanate gives (**312**), probably *via* a similar mechanism. Benzimidazoles react similarly ⟨87CHE284⟩, but most pyrazoles do not ⟨80AHC(27)241⟩ (*cf.* Section 3.4.1.4.6).

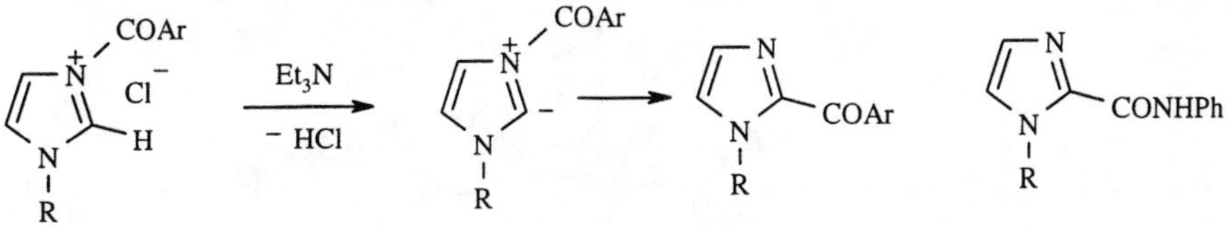

(311)   (312)

Similar imidazolium ylides are implicated in the aroylation of *N*-phenylbenzimidoyl chlorides ⟨92CPB2627⟩, and in reactions of 1-substituted imidazoles with cyanogen bromide to form 2-cyano- or 2-bromo-imidazoles ⟨88S470⟩.

Acid chlorides convert 1-methylbenzimidazol–2-yl-silanes and -stannanes into 2-acylbenzimidazoles. The reaction also works with imidazoles, with the stannanes being more reactive than the silanes. The mechanism is believed to involve initial N-acylation, then loss of the silicon or tin substituent to give a zwitterion, and finally N→C-migration of the acyl group ⟨83JHC1011⟩.

In the case of 2-(trimethylsilyl)thiazole ⟨83TL2901⟩ the acylation of nitrogen and desilylation at C-2 are likely to be concerted giving rise directly to a *N*-acylthiazolium 2-ylide which then evolves to the 2-acylthiazole by a multi-step process (Scheme 41). This hypothesis is supported by a mechanistic study of the reaction of the same silylated thiazole with aldehydes ⟨93JOC3196⟩.

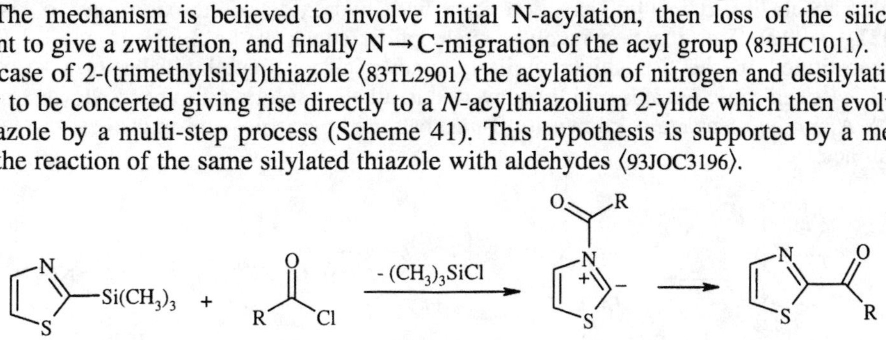

Scheme 41

3.4.1.8.5 *Formation and reactions of stable carbenes*

Careful treatment of 1,3-disubstituted imidazolium salts (*e.g.* **313**) with sodium hydride (in some cases addition of small amount of potassium *tert*-butoxide is also recommended) in tetrahydrofuran gives stable carbenes of type (**314**) (see also Section 2.4.4.2.3). They behave as rather strong bases (pK_a ~ 24 in DMSO) ⟨95CC1267⟩ and nucleophiles. Thus, carbene (**314**) and salt (**313**) form a crystalline bis(carbene)-proton complex (**315**) which is a rare example of an asymmetrical hydrogen bridge between two carbanionic center ⟨95JA572⟩. By the action of BF_3, B_2H_6 or $AlH_3 \cdot NMe_3$ stable mesoionic adducts of type (**316**) are obtained ⟨92JA9724, 93CB2041⟩.

(**313**)	(**314**)	(**315**)	(**316**)

By contrast, 1,2,4-triazole carbene (**317**) displays electrophilic character ⟨95AG1119⟩. Thus, it reacts with alcohols and amines producing triazoline derivatives (**318**) in quantitative yield. Oxygen or sulfur give triazolinone and triazolinthione derivatives (**319**) (similar reaction with tellurium is known for imidazol–2–ylidenes ⟨93CB2047⟩). Reactions of (**317**) with dimethyl maleate or dimethyl fumarate lead to compounds (**320**), probably *via* ring opening of a cyclopropane intermediate with subsequent 1,2-hydrogen shift.

(**318**)	(**317**)	(**319**)
(Nu = OR, morpholino)		(X = O, S)

(**320**)

3.4.1.8.6 *Ring cleavage* via *C-deprotonation*

Isoxazoles unsubstituted in the 3-position react with hydroxide or ethoxide ions to give β-keto nitriles (**321**) → (**322**). This reaction involves nucleophilic attack at the 3-CH group. 1,2-Benzisoxazoles unsubstituted in the 3-position similarly readily give salicylyl nitriles ⟨67AHC(8)277⟩, and 5-phenyl–1,3,4-oxadiazole (**323**) is rapidly converted in alkaline solution into benzoylcyanamide (**324**) ⟨61CI(L)292⟩. A similar cleavage is known for 3-unsubstituted pyrazoles and indazoles; the latter yield *o*-cyanoanilines.

(**321**)	(**322**)	(**323**)	(**324**)

3-Unsubstituted isoxazolium salts (325) lose the 3-proton under very mild conditions, *e.g.* at pH 7 in aqueous solution, to give intermediate acylketenimines (326) which convert carboxylic acids into efficient acylating agents (327) ⟨79AHC(25)147⟩.

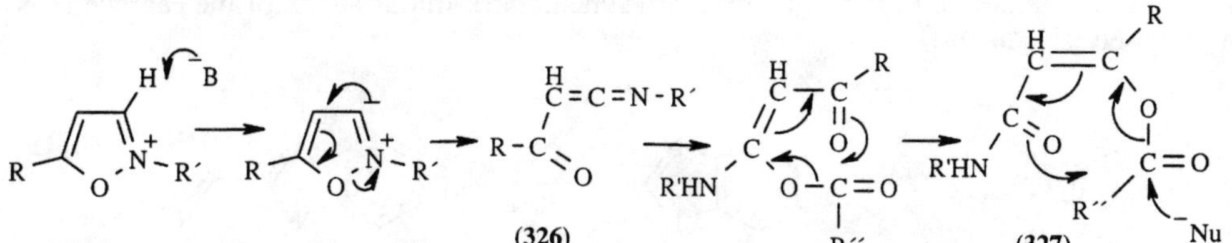

(325)   (326)   (327)

Isoxazoles substituted in the 3-position, but unsubstituted in the 5-position, react under more vigorous conditions to give acids and nitriles (Scheme 42). Anthranils unsubstituted in the 3-position are similarly converted into anthranilic acids by bases (Scheme 43) ⟨67AHC(8)277⟩. Attempted acylation of anthranils gives benzoxazine derivatives *via* a similar ring opening (Scheme 44) ⟨67AHC(8)277⟩.

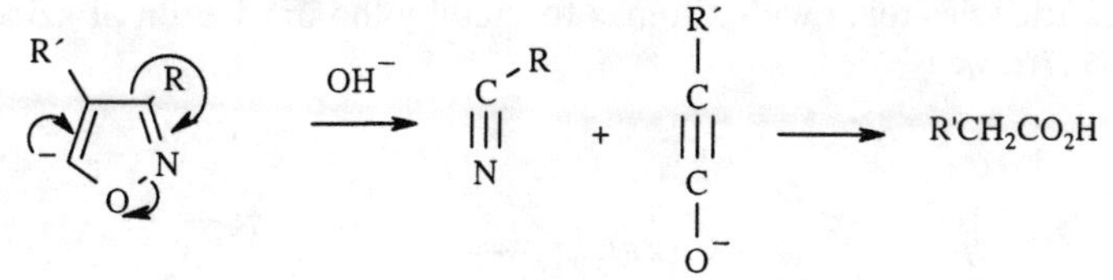

Scheme 42

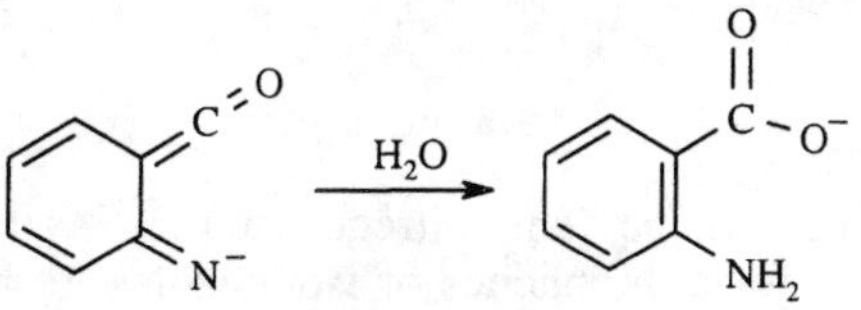

Scheme 43

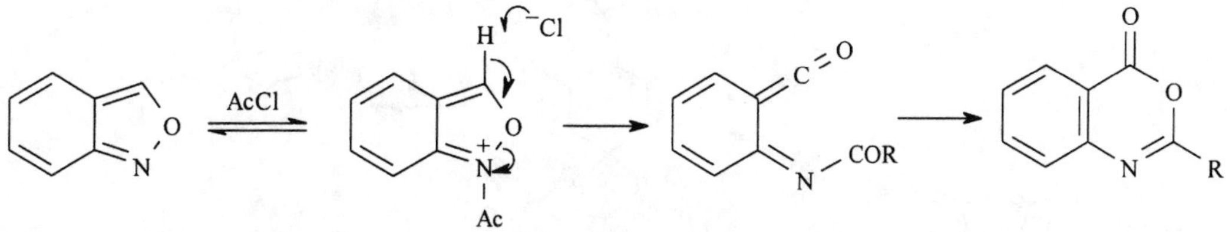

Scheme 44

For ring-opening reactions of *C*-metallated azoles, see Section 3.4.3.10.

3.4.1.8.7 *Proton loss from a ring nitrogen atom*

Pyrazoles, imidazoles, triazoles and tetrazoles are weak NH-acids (Table 3). They form metallic salts (*e.g.* with NaNH$_2$, RMgBr) which are extensively hydrolyzed by water. The anions react very readily with electrophilic reagents on either ring nitrogen or carbon atoms, as discussed in Sections 3.4.1.3 and 3.4.1.4. For example, proton loss from a ring nitrogen gives the highly nucleophilic imidazole anions. This anion can be formed with sodium hydroxide or sodium alkoxide; good results are obtained with sodamide in liquid ammonia ⟨70AHC(12)103⟩.

Azolinones are weak to medium strong acids of pK_a 4–11. They form mesomeric anions which react very readily with electrophilic reagents at the nitrogen, oxygen or carbon atoms, depending on the conditions; see Section 3.4.1.1.4.

3.4.1.9 Reactions with Radicals and Electron-deficient Species; Reactions at Surfaces
3.4.1.9.1 *Carbenes and nitrenes*

Imidazoles react with chloroform at high temperature to form azines by carbene insertion (see CHEC) and trichloromethyl radicals behave similarly ⟨91MI 302–04⟩, but carbenes do not always induce ring expansion. In alkaline medium, chlorodifluoromethane converts benzimidazole and its 2-methyl analogue into the 1-difluoromethyl derivatives (328) (62%, 70%) ⟨82CHE1314⟩. Dichlorocarbene under basic conditions N-alkylates benzimidazole ⟨78CI(L)126⟩, and 1-methylbenzimidazole was shown to couple under the influence of the same reagent (Scheme 45). Initial attack of the carbene at N–3 has been proposed ⟨87H(26)1161⟩.

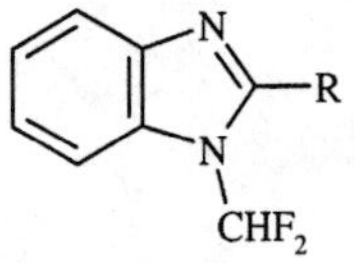

(328) R = H, Me

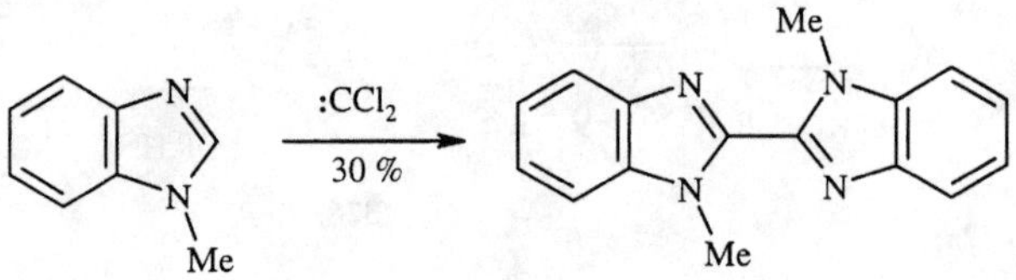

Scheme 45

Pyrazole react with $HCCl_3$ at 550°C (with $:CCl_2$ formation) to give 2-chloropyrimidines in good yields. 1-Alkyl-1,2,4-triazoles react with nitrenes formed by the irradiation of azides to give *N*-imines (Scheme 46) ⟨74AHC(17)213⟩.

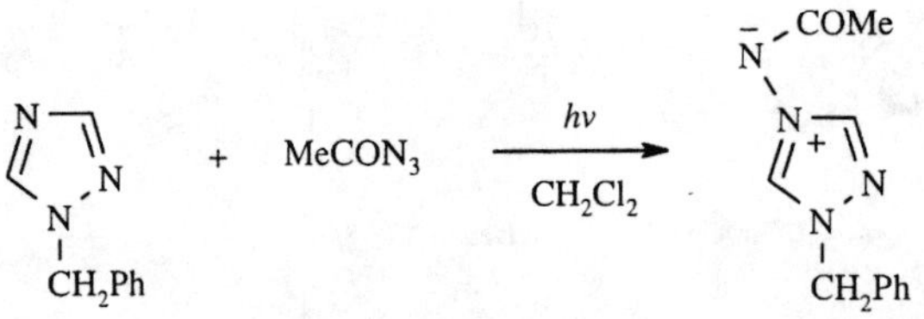

Scheme 46

Most 1,2-dithioles react with carbenes and nitrenes at the S-S bond, leading to insertion, and possible further loss of sulfur to form thiophenes or isothiazoles, respectively (Scheme 47) ⟨74T4113, 76CJC3879⟩.

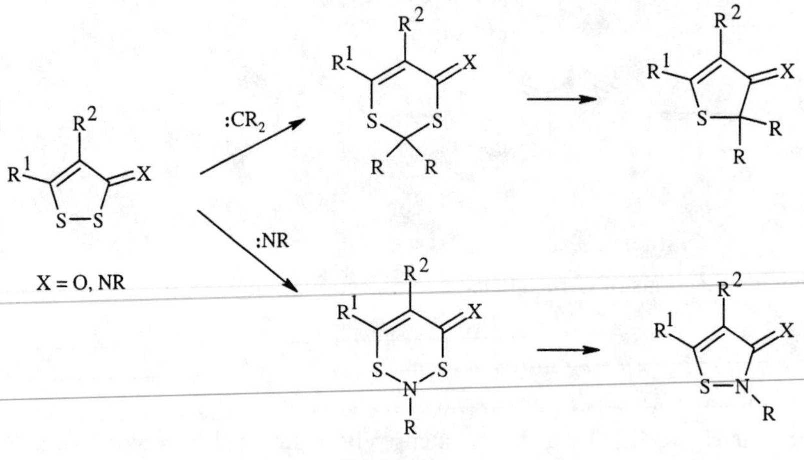

Scheme 47

3.4.1.9.2 *Free radical attack at the ring carbon atoms*

Free radical reactions are still very much less common in azole chemistry than those involving electrophilic or nucleophilic reagents. In some reactions involving free radicals, substituents have little orienting effect; however, rather selective radical reactions are now known.

(i) *Aryl radicals*

Phenyl radicals attack azoles unselectively to form a mixture of phenylated products. Relative rates and partial rate factors are given in Table 8. The phenyl radicals may be prepared from the usual precursors: $PhN(NO)COMe$, $Pb(OCOPh)_2$, $(PhCO_2)_2$ or $PhI(OCOPh)_2$. Substituted phenyl radicals react similarly.

Table 8 Relative Rates and Partial Rate Factors for the Homolytic Phenylation[a] of Five-membered Heterocycles[b]

Heterocycle	Relative rates	Partial rate factors			
		2	3	4	5
Thiazole	1.6	6.2	—	1.0	2.8
2-Methylthiazole	0.6	—	—	1.0	2.4
4-Methylthiazole	1.2	2.9	—	—	4.3
5-Methylthiazole	0.8	3.8	—	1.0	—
Isothiazole	0.95	—	2.7	0.5	2.5
1-Methylpyrazole	0.6	—	0.18	0.03	3.4
1-Methylimidazole	1.2	—	2.7	0.5	2.5

[a] Benzoyl peroxide is the source of the phenyl radicals, except for the first entry, where it is nitrosoacetanilide.
[b] ⟨74AHC(16)123⟩.

(ii) Alkyl radicals

Alkyl radicals produced by oxidative decarboxylation of carboxylic acids are nucleophilic and attack protonated azoles at the most electron-deficient sites. Thus, imidazole and 1-alkylimidazoles are alkylated exclusively at the 2-position ⟨80AHC(27)241⟩. Similarly, thiazoles are attacked in acidic media by methyl and propyl radicals to give 2-substituted derivatives in moderate yields, with smaller amounts of 5-substitution. These reactions have been reviewed ⟨74AHC(16)123⟩; the mechanism involves an intermediate σ-complex.

Similar reactions occur with acyl radicals, for example with the $CONH_2$ radical from formamide ⟨74AHC(16)123⟩.

3.4.1.9.3 Thiation

Numereous derivatives of benzimidazole, naphthoimidazoles and other condensed imidazole systems can be very effectively thiated with elemental sulfur on heating without solvent at 230–260°C. The product of this reaction is the corresponding imidazolin-2-thione formed in excellent yield ⟨67ZOR1518, 95IZV2231⟩. For example, imidazo[4,5-*c*]pyridines (**329**, R^1 = H, Alk, Ar, $C_6H_5CH_2$) gave 1,3-dihydro-2*H*-imidazo[4,5-*c*]pyridine-2-thiones (**330**).

(329) (330) (331) (332)

By analogy, imidazolium salts, *e.g.* (**331**), can be converted with a high yield into the corresponding 2-thione or 2-selenone derivatives (**332**) by heating with sulfur or selenum in DMF in the presence of triethylamine ⟨90CHE1689⟩.

Thiation of N-substituted 1,2,4-triazole has been also achieved ⟨93ZOR1896⟩.

The mechanism of the thiation reaction is unknown, though in the case of quaternary salts one can suppose that it includes formation of the corresponding carbene as a possible intermediate.

3.4.1.9.4 Electrochemical reactions and reactions with free electrons

Neutral rings are reduced by the uptake of an electron to form anion radicals ⟨80AHC(27)31⟩. In isoxazole and oxazole this can be achieved in an argon matrix, but normally ring fission occurs; reduction of 1,2,4-thiadiazoles also usually results in ring cleavage (see also Section 2.4.2.1). Thiazoles containing electron-withdrawing groups, 1,3,4-oxadiazoles, 1,2,5-oxadiazoles and 1,2,5-thiadiazoles on electrochemical reduction yield transient anion radicals which can be characterized by ESR, *e.g.* (**333**) and (**334**). Anion radicals from benzazoles can be more stable, *e.g.* (**335**).

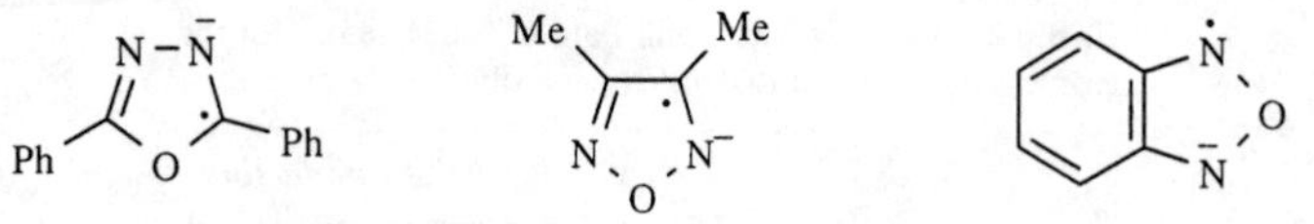

(333)   (334)   (335)

Electrochemical reduction of 1-vinylazoles in acetonitrile or DMF is also a one-electron process in which the generated radical anions dimerize ⟨86CHE253⟩.

Cationic rings are reduced with the uptake of one electron (*e.g.* electrochemically) to give neutral radicals ⟨80AHC(27)31⟩. Examples of radicals which have been detected by ESR are (336) and (337). Such radicals, *e.g.* those from benzothiazolium ions (340), can dimerize (to 342) or undergo further reduction (to 341). 3-Methylbenzothiazolium (339) is reduced in a two-electron wave. The preparative reduction gives a mixture of the dihydro derivative (341) and the dimer (342) ⟨70AHC(12)213⟩. Benzofurazan is reduced polarographically in a six-electron reaction to *o*-phenylenediamine ⟨59MI40200⟩.

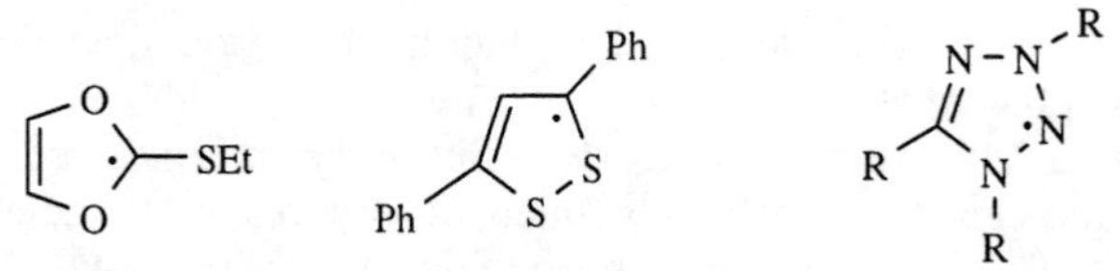

(336)   (337)   (338)

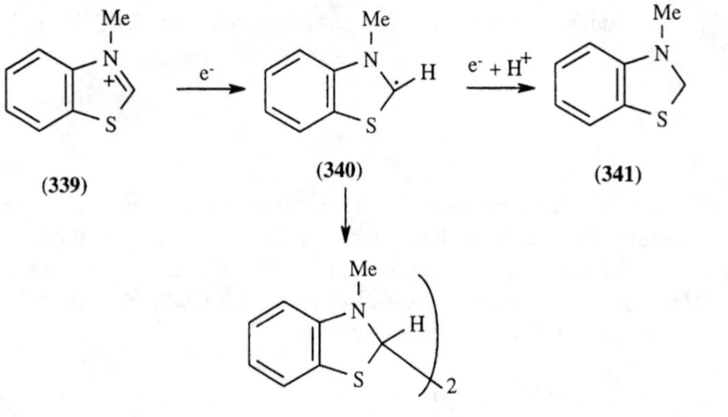

(339)   (340)   (341)

(342)

Calculations indicate that 1,2-dithiolyl radicals have large spin densities at the 3,5-positions and that 1,3-dithiolyl radicals have large spin densities at the 2-position. In agreement, radicals of these types unsubstituted at such positions dimerize very readily; when the position is substituted, the radical is more stable (336 and 337). Reduction of tetrazolium salts gives tetrazolyl radicals (338) which show appreciable spin density on all four ring nitrogen atoms ⟨77AHC(21)323⟩.

1,3,2-Dithiazolium cations (343) are easily reduced to the corresponding stable radicals (344) which in turn can be smoothly oxidized back to the cation ⟨84CC573, 85JCS(D)1405, 87CC66, 90CB881, 92CJC2972⟩. The cations may also be reduced electrochemically; cyclic voltametry shows this to be a reversible process.

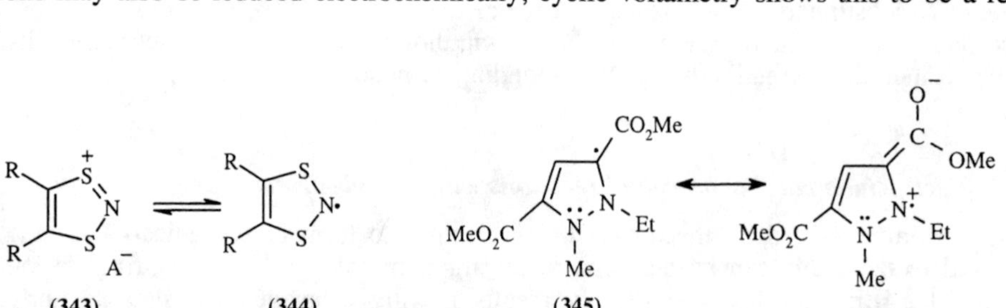

(343)   (344)   (345)   (346)

Electron-withdrawing substitutents stablilize such neutral radicals considerably. Merostabilization is found, for example, in the pyrazolyl derivative (345) ↔ (346) ⟨74JCS(P1)422⟩.

3.4.1.9.5 *Other reactions at surfaces (catalytic hydrogenation and reduction by dissolving metals)*

In general, azoles containing a cyclic oxygen atom are readily reduced, those with cyclic sulfur with more difficulty, and wholly nitrogenous azoles not at all. Pyrazoles are very resistant to catalytic reduction, resisting hydrogenation over nickel at 150°C and 100 atm ⟨66AHC(6)347⟩. So resistant are imidazoles to catalytic hydrogenation that 1-methylimidazole, benzimidazole, and its 1-methyl derivative condense dehydrogenatively in the presence of platinum- or palladium-carbon catalyst and pyridine 1-oxide. Dimerization of unsubstituted imidazole under the same conditions is insignificant (Equation 5) ⟨89CPB1987⟩.

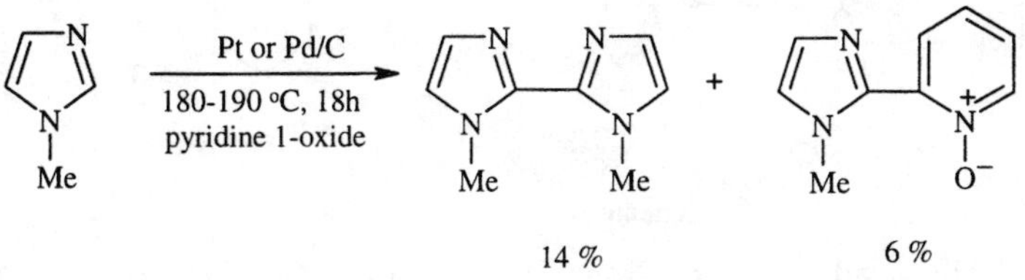

(5)

14 %   6 %

Isoxazoles are readily reduced, usually with concomitant ring fission. Thus, solvent-dependent hydrogenolysis of 3,5-diphenylisoxazole in the presence of palladium on charcoal resulted in open-chain products (**347**) and (**348**) ⟨87T3983⟩.

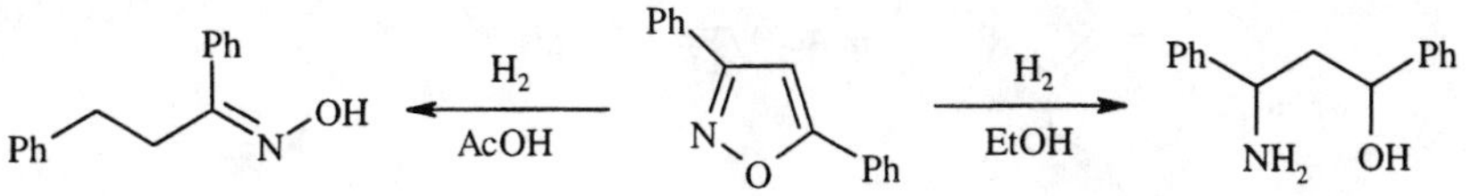

(**347**)   (**348**)

1,2-Benzisoxazoles are easily reduced to various products (Scheme 48) ⟨67AHC(8)277⟩. Chemical or catalytic reduction of oxazoles invariably cleaves the heterocyclic ring (Scheme 49) ⟨74AHC(17)99⟩. For similar reactions of thiazoles, see Section 3.4.1.5.1.

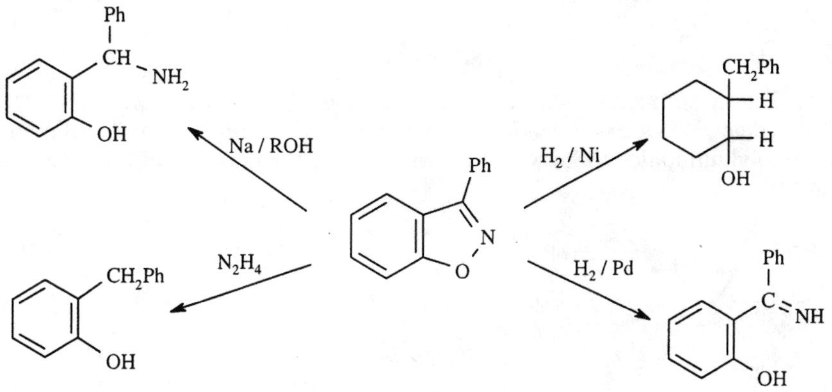

Scheme 48

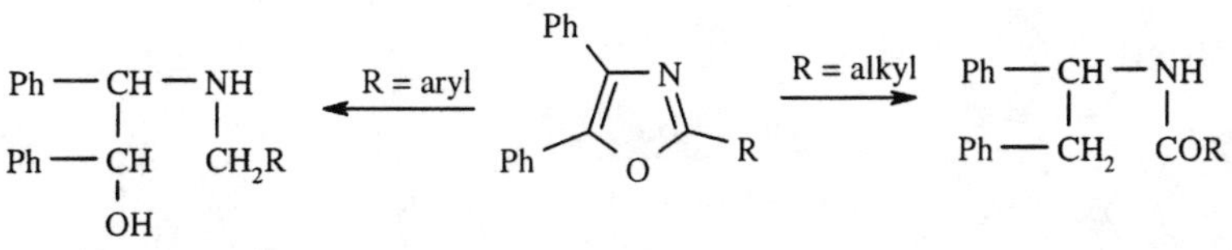

Scheme 49

Catalytic reduction of 1,2,4-oxadiazoles also breaks the N–O bond; *e.g.* (**349**) gives (**350**). Benzofuroxan can be reduced under various conditions to benzofurazan (**351**), the dioxime (**352**) or *o*-phenylenediamine (**353**) ⟨69AHC(10)1⟩. Reduction by copper and hydrochloric acid produced *o*-nitroanilines (Scheme 50) ⟨60AHC(10)1⟩.

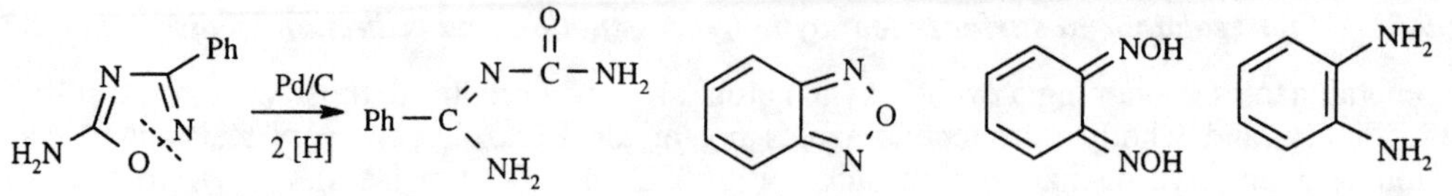

(349) (350) (351) (352) (353)

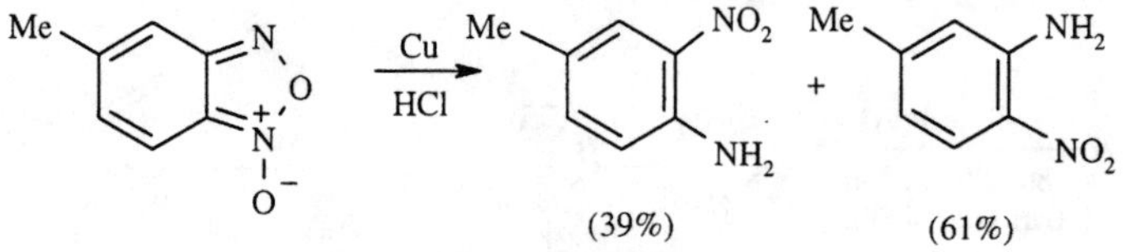

(39%) (61%)

Scheme 50

Isothiazoles are reductively desulfurized by Raney nickel, *e.g.* as in Scheme 51 ⟨72AHC(141)1⟩. 1,2,5-Thiadiazoles are subject to reductive cleavage by zinc in acid, sodium in alcohol, or Raney nickel, *e.g.* Scheme 52 ⟨68AHC(9)107⟩.

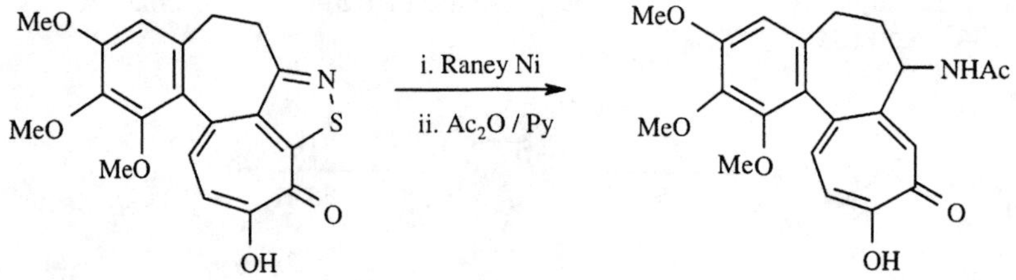

Scheme 51

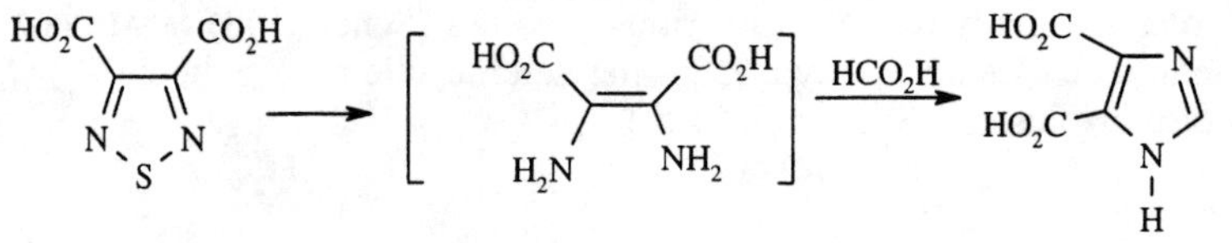

Scheme 52

Reduction of 2-unsubstituted 1,3-dithiolylium salts, *e.g.* (**354**, R = H), with zinc ⟨77JOC2778⟩ or hexacarbonyl-vanadate ⟨75JOC2002⟩ leads to dimerization affording (**355**); the reduction of 2-methylthio-dithiolylium iodide (**356**, R = SMe) with zinc in the presence of bromine, however, gave TTF (**356**) ⟨76ZC317⟩.

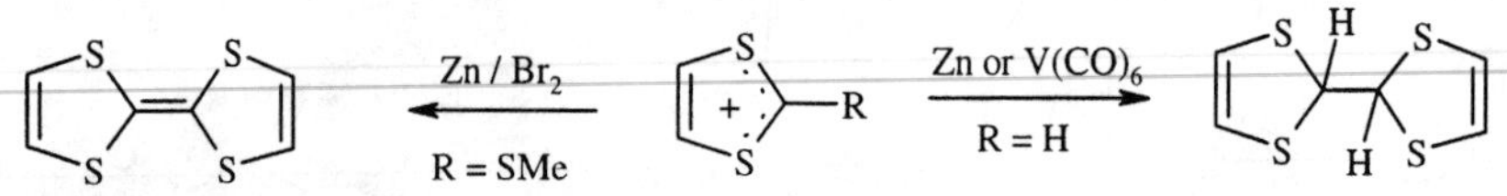

(356) (354) (355)

In the case of condensed heteroaromatic cations reduction can lead to fission of exocyclic N-N bonds (Equation 6):

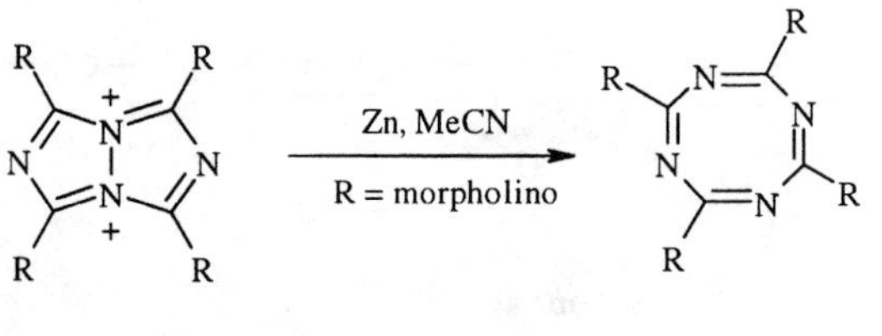

(6)

3.4.1.10 Reactions with Cyclic Transition States

The distinction between Diels-Alder reactions and 1,3-dipolar cycloadditions is semantic for the five-membered rings: Diels-Alder reaction at the F/B positions in (**357**) (four atom fragment) is equivalent to 1,3-dipolar cycloaddition in (**358**) across the three-atom fragment, both providing the four-π-electron component of the cycloaddition. Oxazoles and isoxazoles and their polyaza analogues show reduced aromatic character and will undergo many cycloadditions, whereas fully nitrogenous azoles such as pyrazoles and imidazoles do not, except in certain isolated cases.

(357)   (358)

3.4.1.10.1 *Heterocycles as inner-ring dienes* ⟨96CHEC-II(2)81, 96CHEC-II(2)322, 96CHEC-II(3)277⟩

Diels-Alder reactions of oxazoles afford useful syntheses of pyridines (Scheme 53) ⟨74AHC(17)99⟩. A study of the effect of substituents on the Diels-Alder reactivity of oxazoles has indicated that rates decrease with the following substituents: alkoxy > alkyl > acyl >> phenyl. The failure of 2- and 5-phenyl-substituted oxazoles to react with heterodienophiles is probably due to steric crowding. In certain cases, bicyclic adducts of type (**359**) have been isolated and even studied by an X-ray method ⟨87BCJ432⟩; they can also decompose to yield furans (Scheme 54). With benzyne, generated at 0°C from 1-aminobenzotriazole and lead tetraacetate under dilute conditions, oxazoles form cycloadducts (*e.g.* **360**) in essentially quantitative yield ⟨90JOC929⟩. They can be handled at room temperature and are decomposed at elevated temperatures to isobenzofuran.

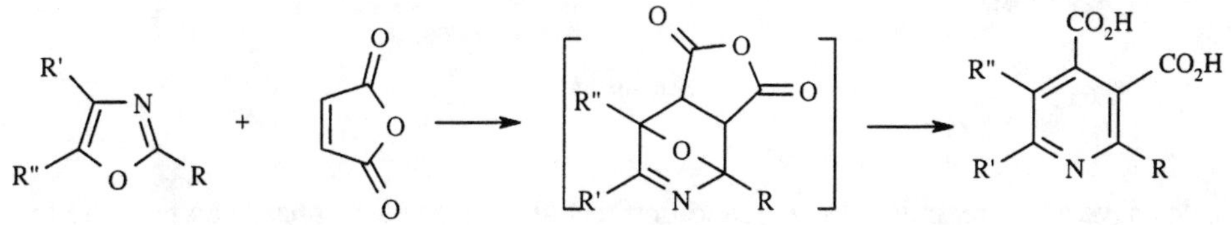

(359)

Scheme 53

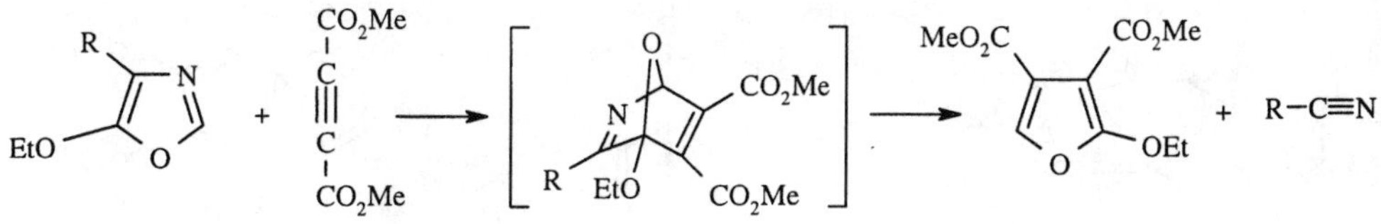

Scheme 54

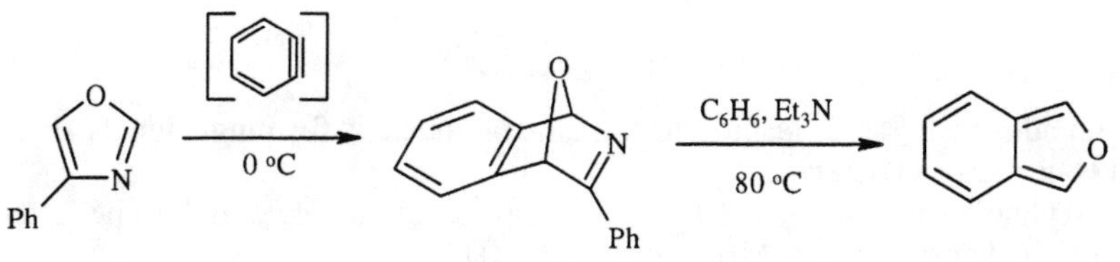

(360)

Reaction of oxazoles with heterodienophiles, including N=N, C=N, C=O, and C=S types, give new heterocycles ⟨89T3535, 88JOC4663⟩. Oxazoles react with singlet oxygen to give bicyclic adducts of type (**361**). The first direct evidence (^{1}H and ^{13}C NMR) for these adducts was obtained by running the reactions at temperatures below –50°C ⟨88TL1007⟩. Between –10°C and RT, the peroxide products began to rearrange or fragment (Scheme 55). The reaction pathway depended on the nature of the substituent at the 2 position of the oxazole. For R^2 = phenyl, triacylamides were formed quantitatively, for R^2 = H, a 1 : 1 mixture of R^4CN and anhydrides was obtained, and for R^3 = methyl, both pathways were observed with the ratio of the former to the latter changing with the number and nature of R^4 and R^5 substituents.

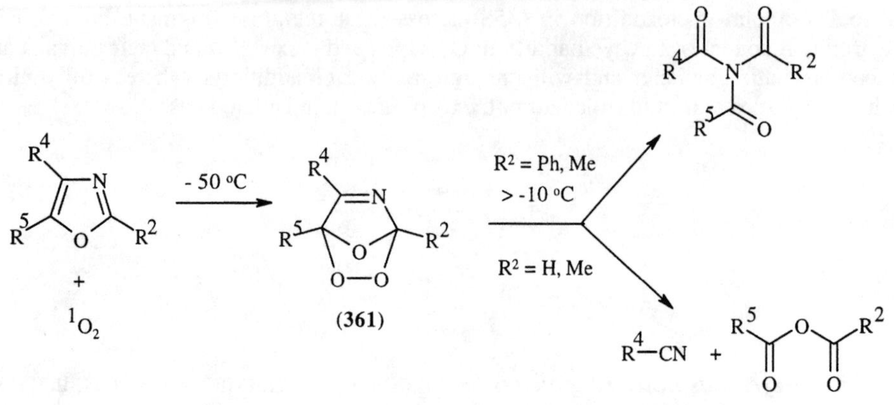

Scheme 55

Isoxazolium salts react with enamines to give pyridinium salts (Scheme 56) ⟨69CPB2209⟩.

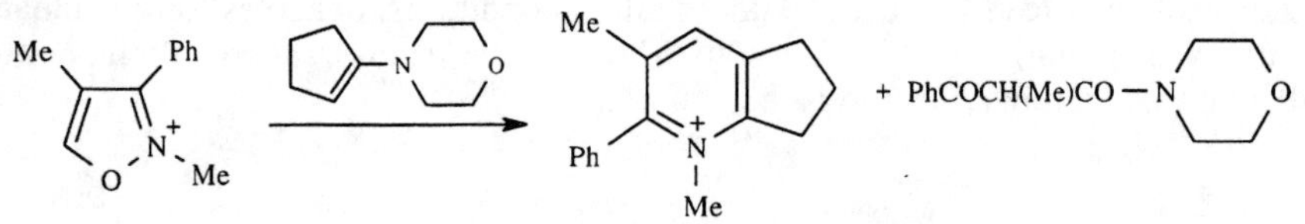

Scheme 56

Thiazoles have a low reactivity in cycloaddition reactions. However, it has been possible to achieve intramolecular Diels-Alder reactions in some cases (*e.g.* **362** → **363**).

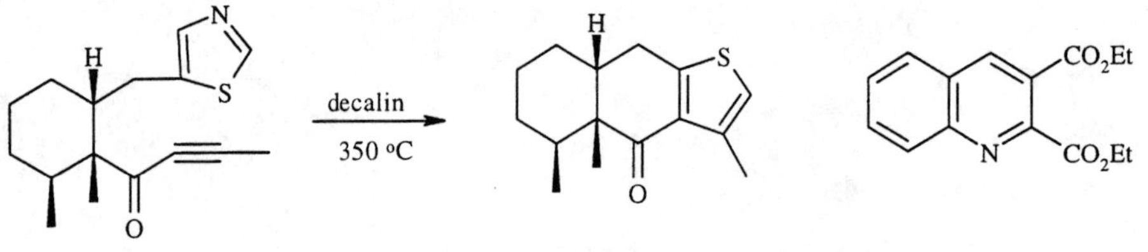

Singlet oxygen adds in a [4 + 2] fashion to the thiazole nucleus forming adducts similar to those of oxazoles (**361**) ⟨83JOC2302, 84TL2935⟩.

Ethynedicarboxylate esters add to 2,1-benzisothiazole to generate quinoline esters (**364**), and benzyne reacts to yield acridine ⟨83H(20)489, 88JCS(P1)2141⟩.

1,3-Dithiolylium-4-olates undergo cycloaddition reactions, *e.g.* as in Scheme 57 ⟨80AHC(27)151⟩. The mesoionic 1,3-dioxolium-4-olates show a similar pattern of reactivity to produce furans ⟨85CC190⟩. Scheme 58 gives an example of cycloaddition in the oxathiazole series.

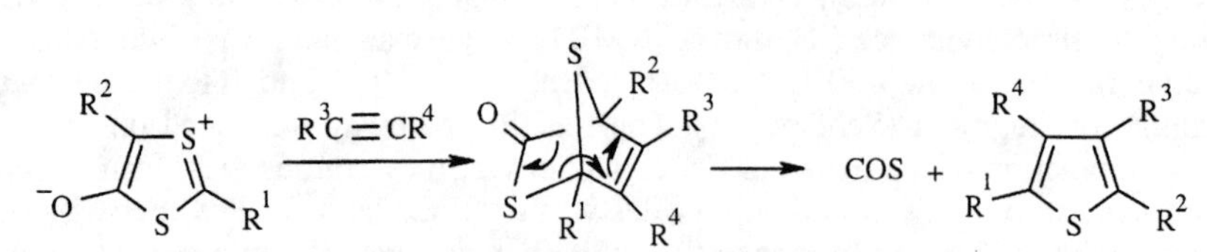

Scheme 57

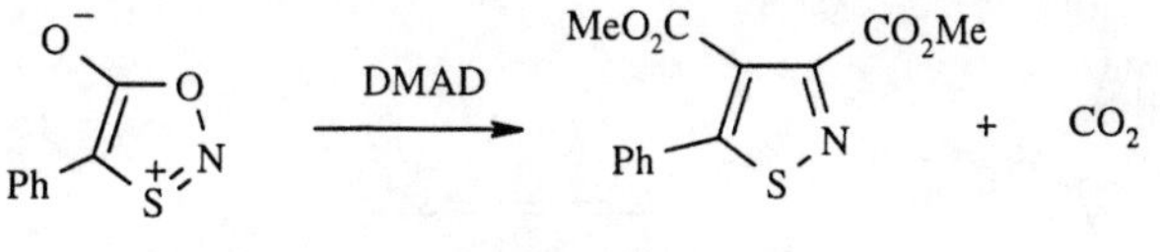

Scheme 58

Enamines and enolate anions react with benzofuroxan to give quinoxaline di-*N*-oxides (Scheme 59) ⟨69AHC(10)1⟩.

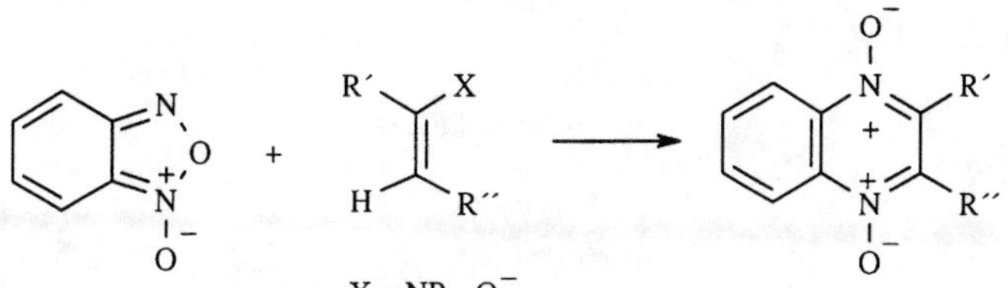

X = NR₂, O⁻

Scheme 59

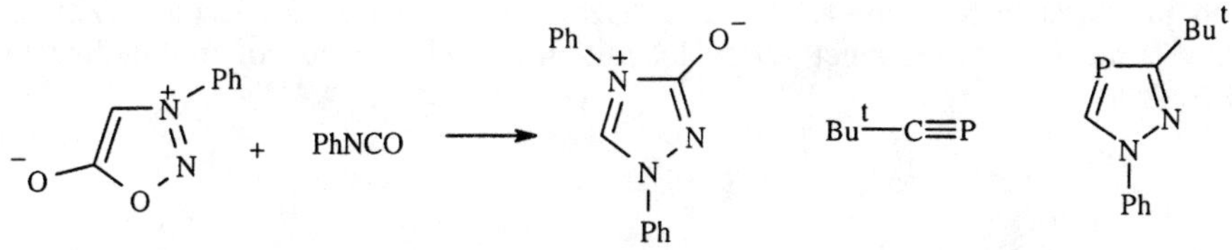

(365)   (366)   (367)   (368)

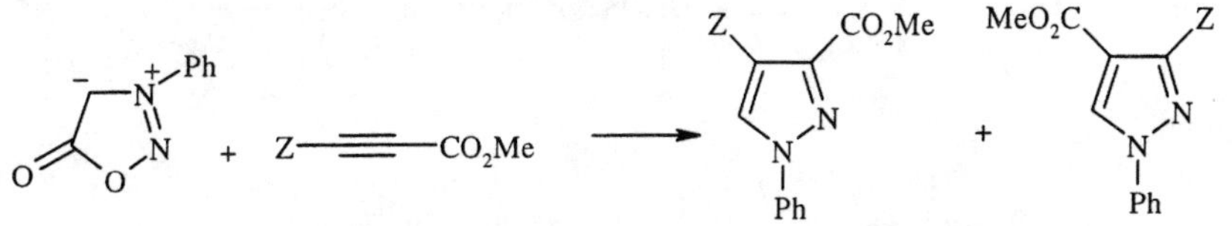

	79	21
Z = CH(OMe)₂, 84 %	79	21
Z = CHO, 93 %	34	66
Z = CH₂OH, 79 %	40	60

Scheme 60

Sydnones can be regarded as cyclic azomethine imines and as such, they undergo thermal cycloaddition reactions with a range of dipolarophiles. Thus, phenyl isocyanates convert (**365**) into 1,2,4-triazoles (**366**) ⟨76AHC(19)1⟩. On photolysis 3,4-diarylsydnones lose carbon dioxide and give nitrile imines, which can also be intercepted by dipolarophiles. Thermal reactions with acetylenic dipolarophiles lead to the formation of pyrazoles (Scheme 60). A preparative disadvantage is that the reactions are rarely completely regioselective with unsymmetrical alkynes ⟨89H(29)967⟩. The phosphaalkyne (**367**) adds to 3-phenylsydnone regioselectively to give the heterocycle (**368**) in high yield ⟨87CB1809⟩.

The 1,2,4-thiadiazolidine (**369**) and the 1,2,4-dithiazolidine (**370**) are interconvertible in the presence of electrophilic nitriles and give the 1,2,4-thiadiazoline-5-ones (**371**) as products (Scheme 61) ⟨91JOC3268⟩. It is suggested that the reaction goes by a consecutive cycloaddition – elimination mechanism *via* hypervalent sulfur intermediates in which the nitrile approaches in the plane of the heterocycle.

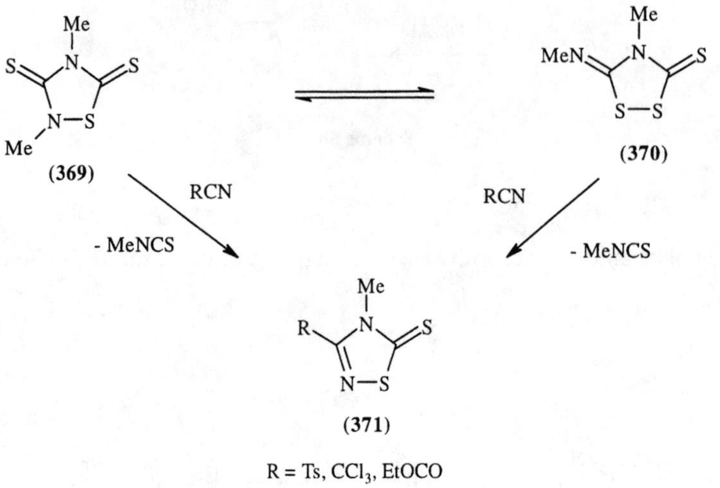

Scheme 61

3.4.1.10.2 *Heterocyclic derivatives as inner-outer ring dienes*

Diazo-substituted 1,2,3-triazoles undergo regiospecific dipolar cycloaddition reactions with electron-rich unsaturated compounds. Thus, 4-diazo-5-phenyl-4*H*-1,2,3-triazole (**372**) reacts with 1-morpholinyl-2-nitroethene by a net 1,7-cycloaddition and elimination of morpholine to give the product (**373**) ⟨87JOC5538⟩.

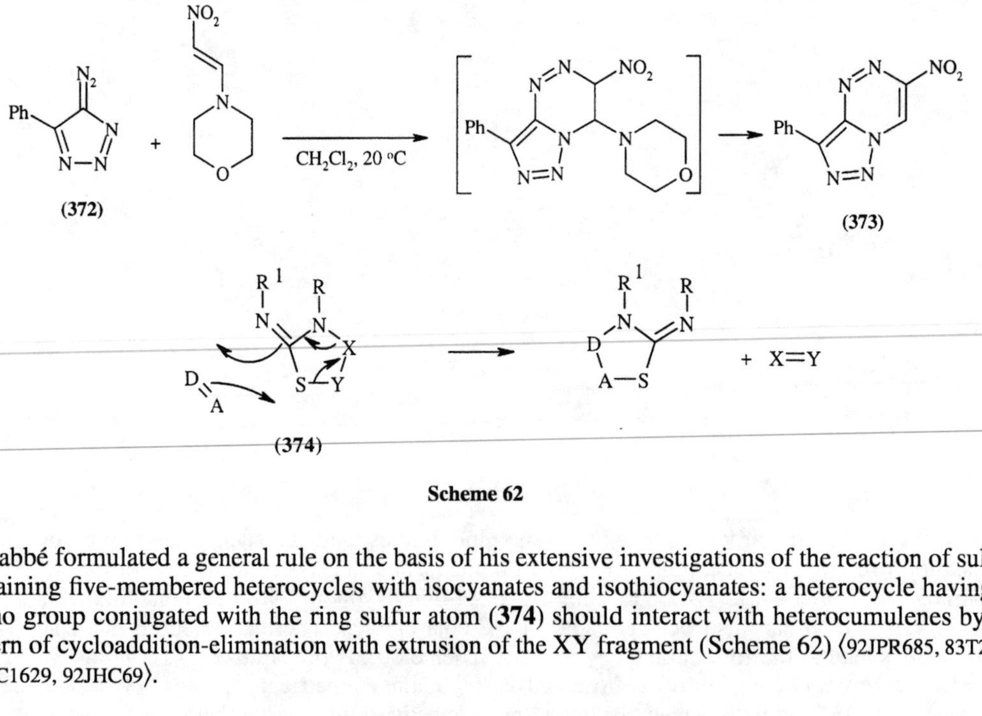

Scheme 62

L'abbé formulated a general rule on the basis of his extensive investigations of the reaction of sulfur-containing five-membered heterocycles with isocyanates and isothiocyanates: a heterocycle having an amino group conjugated with the ring sulfur atom (**374**) should interact with heterocumulenes by the pattern of cycloaddition-elimination with extrusion of the XY fragment (Scheme 62) ⟨92JPR685, 83T2311, 90JHC1629, 92JHC69⟩.

3.4.1.10.3 *Heterocyclic derivatives as outer ring dienes*

Treatment of 4,5-bis(bromomethyl)-3-phenylisoxazole (**375**) with sodium iodide and subsequently with azodicarboxylate gave Diels-Alder cycloadduct (**377**) ⟨91TL4603⟩. Intermediate 1,2-quinodimethyde (**376**) may participate in the reaction as an active 1,3-diene.

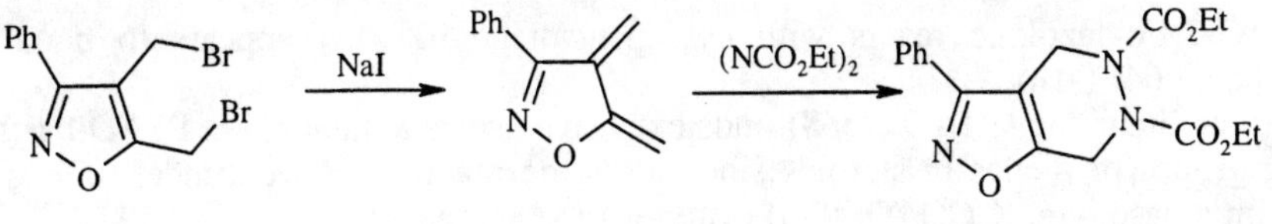

(375)   (376)   (377)

By analogy, flash vacuum pyrolysis of a *p*-chlorobenzoate ester (**378**, Ar = 4-ClC$_6$H$_4$) ⟨90TL1487⟩ produced oxazole-4,5-xylylene (**379**), which could be trapped with SO$_2$ to give adduct (**381**). The unstable intermediate also gave a Diels–Alder adduct (**380**) with methyl acrylate (for the similar reactions see Section 3.3.3.3.3).

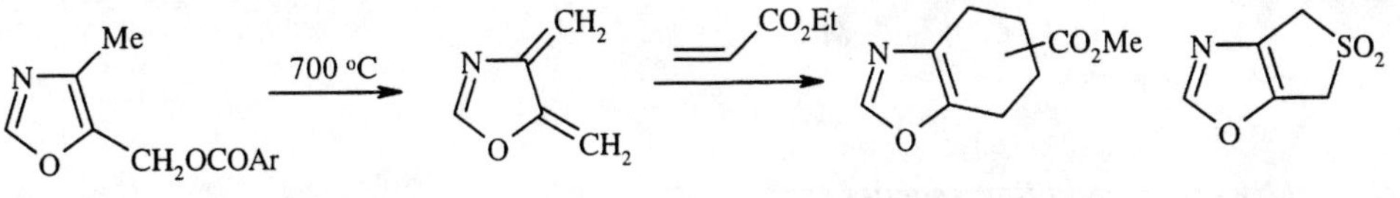

(378)   (379)   (380)   (381)

3.4.1.10.4 Heterocycles as dienophiles

Isoxazoles with electron-withdrawing substituents readily undergo cycloaddition reaction as dienophiles. For example, the nitroisoxazoles (**382**) react with 2,3-dimethylbutadiene to form the adduct (**383**) ⟨90JOC1227⟩. Similar reaction with 4-nitro-2-phenyloxazole (**384**) leads to (**386**) possibly *via* initial adducts (**385**) ⟨93CC978⟩.

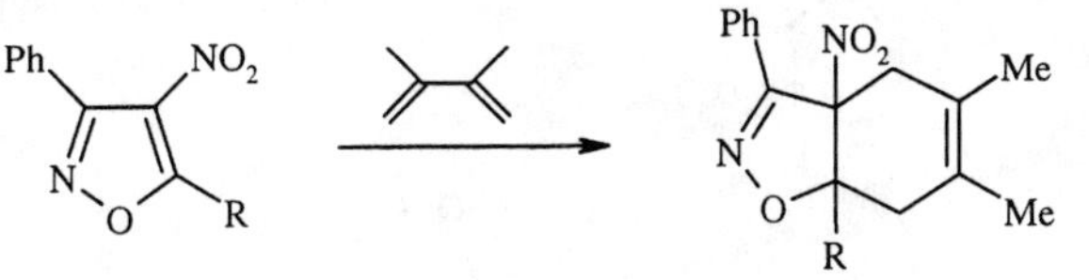

(382)   (383)

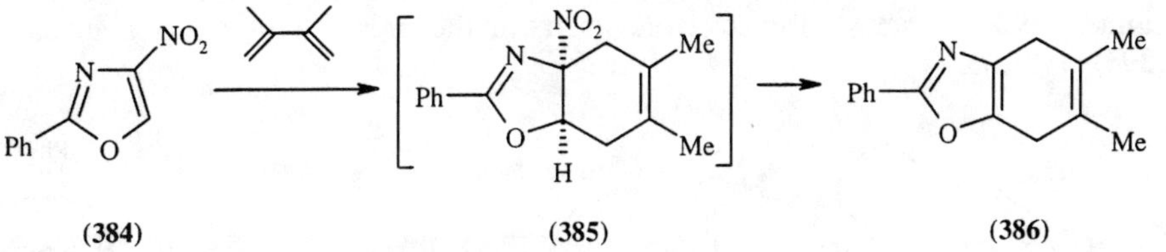

(384)   (385)   (386)

The 4,5 double bond of 2(3H)-oxazolones participates in thermal [4 + 2] cycloaddition reactions. With dialkyl azodicarboxylates the addition occurs at 80°C to give the cycloadducts (**387**) regioselectively and in 83–93% yields (Scheme 63) ⟨92CPB1077⟩. With a chiral substituent attached to N-3, diastereoselectivities as high as 72% have been obtained. Cycloadditions with cyclopentadiene or benzofuran require higher temperatures and longer reaction times, but can yield highly efficient chiral oxazolidin-2-one auxiliaries ⟨94TL721⟩.

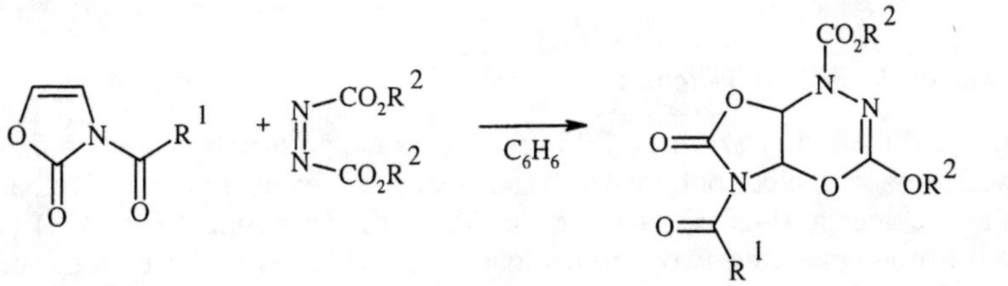

(387)

Scheme 63

1,3-Diacetyl-2-imidazolone reacts with cyclopentadiene in what appears to be a Diels-Alder reaction (Scheme 64) ⟨93JOC3387⟩.

4-Phenyltriazolinedione (PTAD) (**388**) finds extensive use as a dienophile. PTAD has been used as a protecting group for dienes in steroids since it can be readily removed under mild conditions by treatment with a base (*e.g.* K$_2$CO$_2$/DMSO) ⟨85BSF849, 95JOC1828⟩.

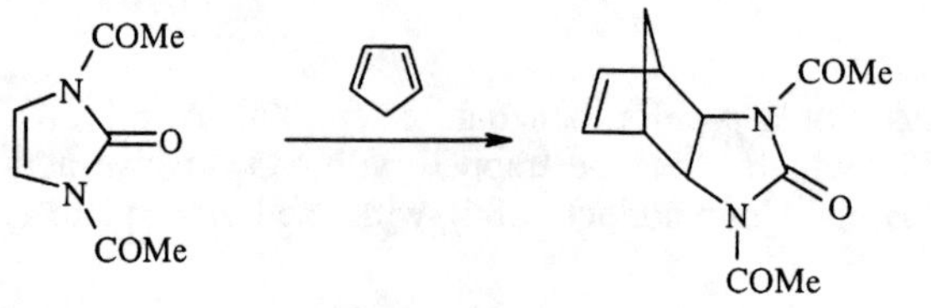

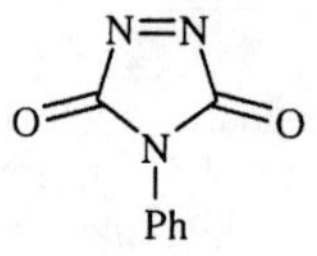

Scheme 64

3.4.1.10.5 [2 + 2]Cycloaddition reactions

Photochemical additions to give four-membered rings are known. Thus, the reactions of imidazoles across the 4,5-bond with benzophenone and acrylonitrile are illustrated by (**389**→**390**) and (**391**→**392**), respectively ⟨80AHC(27)241⟩. Oxazolin-2-one undergoes acetone-photosensitized photochemical addition to ethylene ⟨80CB1884⟩.

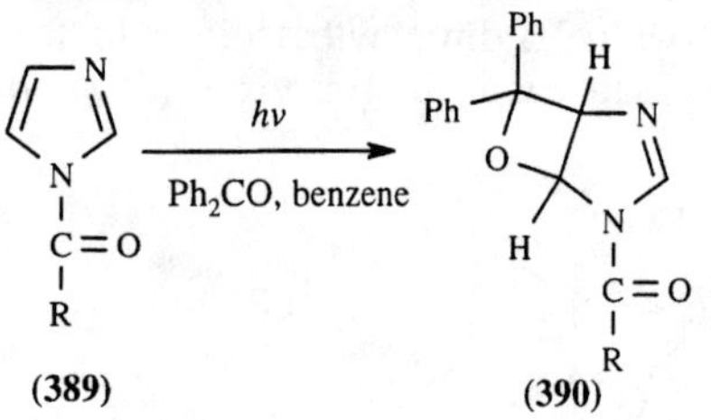

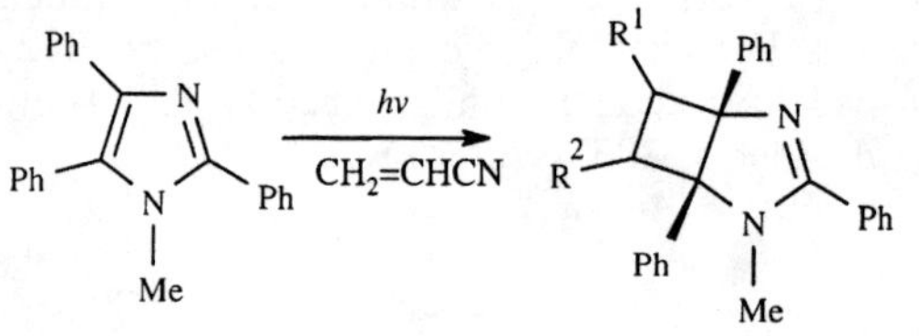

$$R^1 = H, R^2 = CN$$
$$R^1 = CN, R^2 = H$$

The photochemical reaction of 2-phenylbenzoxazole ⟨84CHEC(6)177⟩ forms a head-to-tail dimer, 1,3-diazetidine, (**393**) ⟨87CC578⟩. Benzoxazole itself, in the presence of oxygen, gives a coupled product, (**394**) ⟨74TL375⟩.

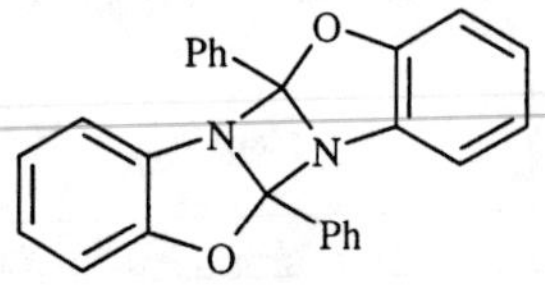

Isothiazoles are reported to yield lactams (**395**) on reaction with diphenyl ketene ⟨85BSB149⟩.

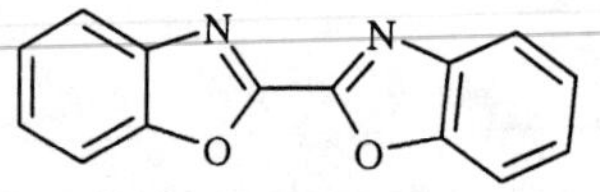

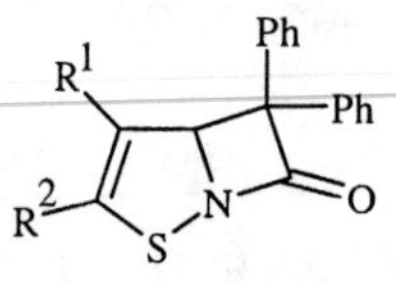

3.4.1.10.6 Other cycloaddition reactions

5-Alkylidene-1,4-dimethyltetrazolines (**396**) behave as electron-rich enamines for inverse-electron demand cycloadditions with electrophilic 4π-systems such as covalent azides. Cycloaddition reactions with alkyl and aryl azides at 0–20°C gave high yields of the spirotriazolines (**398**) ⟨88CB1285⟩. These are formed from a non-concerted two-step cycloaddition which is at the extreme of the 1,3-dipolar cycloaddition mechanistic range. Remarkably, when R^3 was strongly electron-withdrawing, some of the zwitterionic intermediates (**397**) were sufficiently stable for isolation. Heating intermediates (**397**) in CDCl$_3$ gave the 1,2,3,4-tetrazine derivatives with loss of N$_2$.

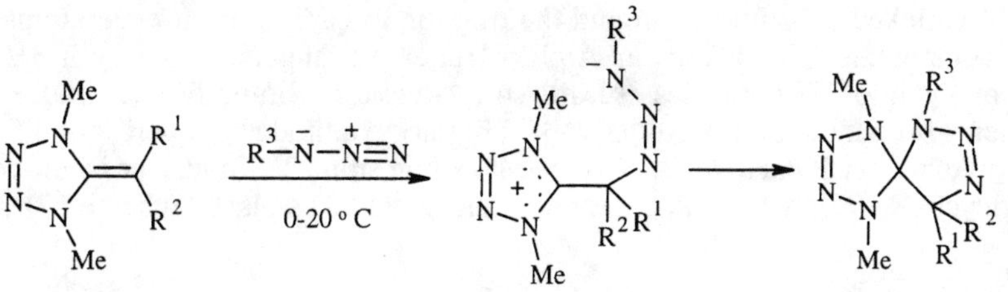

(396)   (397)   (398)

1-Phenyl-1,2,4-triazolium dicyanomethylide (399) undergoes with DMAD 1,3-dipolar cycloaddition ⟨83JCS(P2)1317⟩. The primary product of the reaction (400) can be isolated under carefully controlled conditions. On gentle heating it is transformed into a 1,2,4-triazolo[3,4-*a*]pyridine derivative (401). Many other azolium ylides undergo similar reactions.

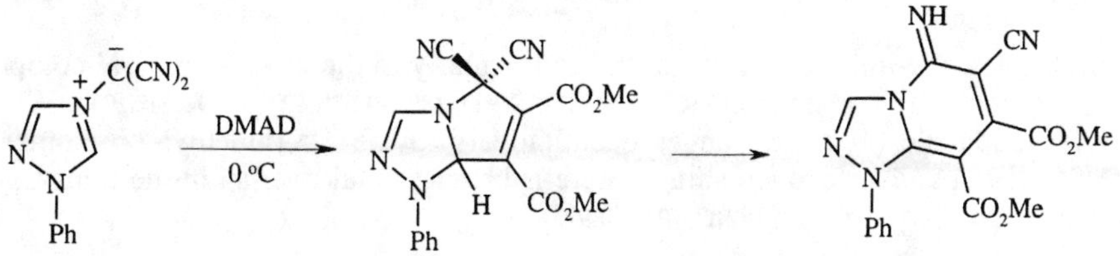

(399)   (400)   (401)

3.4.2 REACTIONS OF NON-AROMATIC COMPOUNDS

Discussion of these compounds is divided into isomers of aromatic compounds, and dihydro and tetrahydro derivatives. The isomers of aromatic azoles are a relatively little-studied class of compounds. Dihydro and tetrahydro derivatives with two heteroatoms are quite well studied, but such compounds become more obscure and elusive as the number of heteroatoms increases. Thus dihydrotriazoles are rare; dihydrotetrazoles and tetrahydro-triazoles and -tetrazoles are unknown unless they contain doubly bonded exocyclic substituents.

S-Oxides of sulfur-containing azoles comprise another class of non-aromatic azoles.

3.4.2.1 Isomers of Aromatic Derivatives

3.4.2.1.1 Compounds not in tautomeric equilibrium with aromatic derivatives

The 3*H*- and 4*H*-pyrazoles and 2*H*- and 4*H*-imidazoles ⟨83AHC(34)2, 83AHC(34)54, 84AHC(35)376, 84AHC(35)414⟩ contain two double bonds in the heterocyclic ring, but in each case the conjugation does not include all the ring atoms; hence the compounds are not aromatic.

The quaternization of 4*H*-imidazoles occurs at the 3-position (Scheme 65) ⟨64AHC(3)1⟩. 4*H*-pyrazoles are also readily monoquaternized.

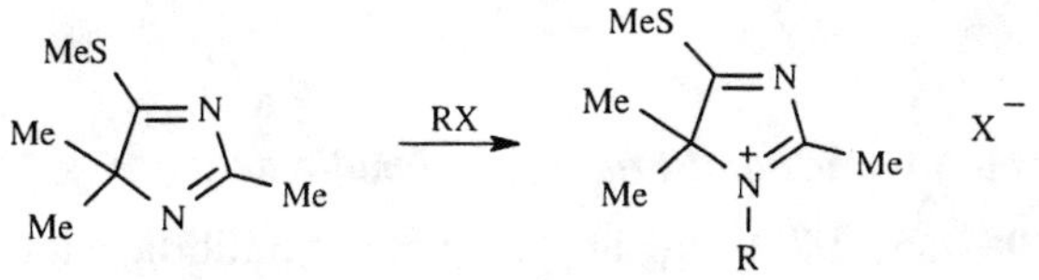

Scheme 65

Dichloropyrazolinones with alkali give alkynoic acids (Scheme 66) ⟨58JA599⟩.

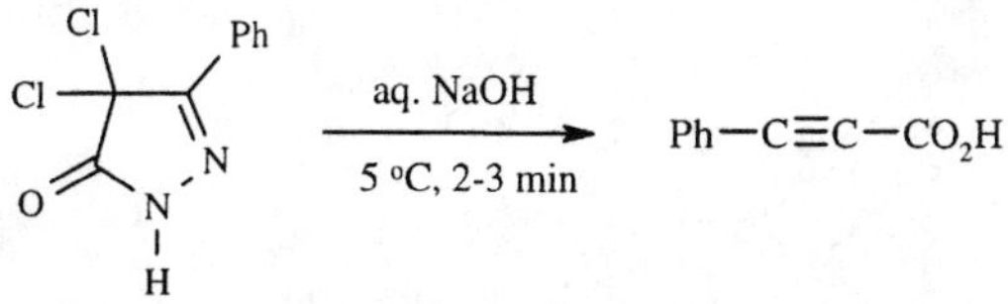

Scheme 66

Migrations of *C*-linked substituents around the ring, on to carbon or nitrogen atoms, are common amongst these compounds. This is the van Alphen-Huttel rearrangement and by it 3*H*-pyrazoles are converted thermally into 1*H*-pyrazoles ⟨83AHC(34)1, 93OPP403⟩. Similarly, 2*H*-imidazoles rearrange thermally to their 1*H*-counterparts ⟨84AHC(35)375⟩. Migratory aptitudes in a series of 4,5-diphenyl-2*H*-imidazoles are in the order PhCH$_2$ > Ph > Et > Me, while spiro-2*H*-imidazoles and -benzimidazoles form bi- and tricyclic species when the C → N rearrangement takes place (Scheme 67) ⟨88JCS(P1)991⟩.

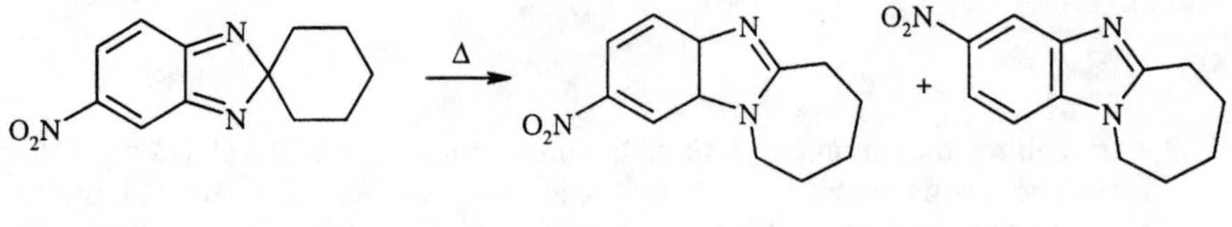

63 : 37

Scheme 67

Like 2*H*-imidazoles, 4*H*-imidazole also rearrange thermally to the more stable 1*H*-compounds, but migration is to carbon in this instance ⟨83CC1082, 87JCS(P1)1389, 91JCS(P1)335, 92PHC107⟩. When heated, 4,4-dimethyl-4*H*-imidazole (**402**) rearranges quantitatively to the 4,5-dimethyl-1*H*-isomer (**403**) by successive (1,5)-methyl and -hydrogen shifts. There is better orbital overlap in the transition state for migration to carbon than to nitrogen ⟨87JCS(P1)1389⟩.

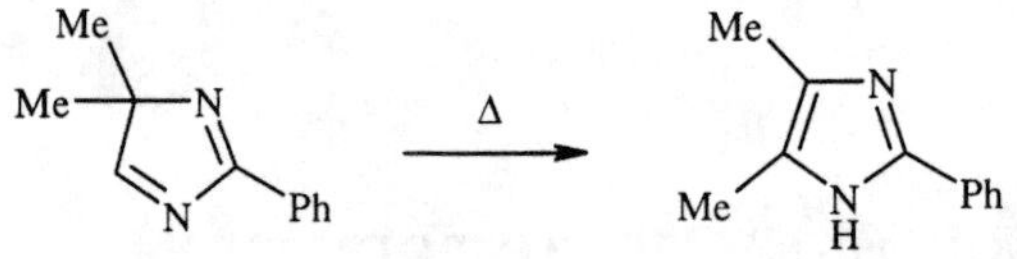

(402)   (403)

3*H*-Pyrazoles are photochemically converted into cyclopropenes, and 3*H*-indazoles react similarly, *e.g.* (**404** → **405**) ⟨70AHC(11)1⟩. If a 3-aryl group is present, an indene can be formed, *e.g.* (**406** → **407**) ⟨83AHC(34)2⟩.

Peracids can oxidize 2*H*-imidazoles to *N*-oxides and *N*,*N*'-dioxides, and sometimes to imidazolinones. Lead dioxide in methanol converts 2*H*-imidazole 1,3-dioxides into stable nitroxide radicals ⟨96CHEC-II(3)146⟩.

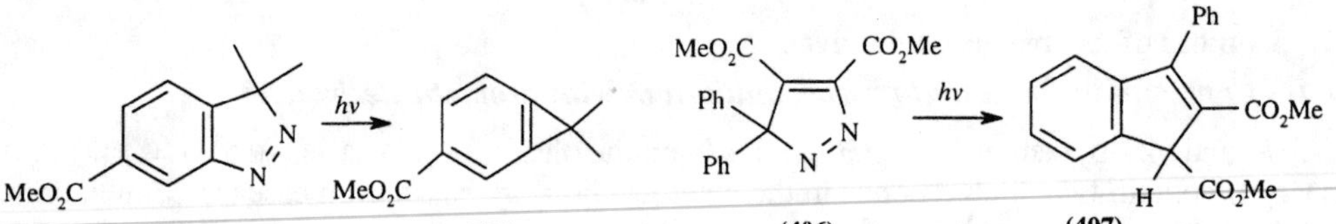

(404)   (405)   (406)   (407)

Addition of nucleophiles to C=N bonds is common in these compounds.

3.4.2.1.2 *Compounds in tautomeric equilibria with aromatic derivatives*

Compounds of type (**408**) and (**409**) are in tautomeric equilibria with 4- or 5-hydroxyazoles. However, the non-aromatic form is sometimes by far the more stable. Thus, oxazolinone derivatives of type (**409**) have been obtained as optically active forms; they undergo racemization at measurable rates with nucleophiles ⟨77AHC(21)175⟩. Reactions of these derivatives are considered under the aromatic tautomer.

(408)   (409)

3.4.2.2 Dihydro Compounds

3.4.2.2.1 Tautomerism

Dihydroazoles can exist in at least three forms (*cf.* Section 2.4.1.3), which in the absence of blocking substituents are tautomeric with each other. The forms in which there is no hydrogen on at least one ring nitrogen normally predominate, because imines are generally more stable than vinylamines in aliphatic chemistry. Thus, for dihydropyrazoles the stability order is Δ^2 (hydrazone) (**410**) > Δ^1 (azo) (**411**) > Δ^3 (enehydrazine) (**412**).

(410)   (411)   (412)

3.4.2.2.2 Aromatization

Δ^4-Imidazolines, Δ^4-oxazolines and Δ^4-thiazolines (**413**), and their benzo derivatives (**414**), are all very easily aromatized (**414→415**), and syntheses which might be expected to yield such dihydro compounds often afford the corresponding aromatic products.

(413)   (414)   (415)   (416)

Dehydrogenation of Δ^2-imidazolines (**416**; Z = NR) gives imidazoles, but requires quite high temperatures and a catalyst such as nickel or platinum. Alternatively, hydrogen acceptors such as sulfur or selenium can be used ⟨70AHC(12)103⟩. Δ^2-Imidazoline derivatives (**417**) are thermally converted into imidazoles (**418**) by a retro-Diels-Alder sequence ⟨93JOC3387⟩.

decalin
185-195 °C
3-4 h
40-79 %

(417)   (418)

Δ^2-Pyrazolines are converted into pyrazoles by oxidation with bromine or Pb(OAc)$_4$ and they can also be dehydrogenated with sulfur. 3,5-Diphenylpyrazoline (**419**) on heating with platinum disproportionates to the pyrazole (**420**) and the pyrazolidine (**421**) ⟨66AHC(6)347⟩.

(419)   (420)   (421)

3-Aminopyrazoline (**422**, R = H) and its 1-phenylderivative (**422**, R = Ph) react with two equivalents of nitrous acid giving in a combined dehydrogenation-diazotization process pyrazole-3-diazonium ions (**424**). Some evidence suggests that the reaction proceeds *via* intermediate red-colored radical-cations (**423**). Indeed, when R in (**422**) is a strongly electron-withdrawing group the formation of the corresponding radical-cation becomes difficult and a rather stable pyrazolin-3-diazonium salts can be isolated. Thus, for 1-phenylsulfonyl-2-pyrazoline-3-diazonium tetrafluoroborate (**425**) the molecular structure has been solved by X-ray crystallography ⟨78ZOR1051, 89ZOR2483⟩.

(**422**)   (**423**)   (**424**)   (**425**)

Pyrazolines have been used as models of intramolecular dyotropy ⟨87T5981, 93JCS(P2)1211⟩. By combining primary deuterium kinetic isotope effects and X-ray crystallography, polycyclic systems, like (**426**), were shown to undergo a double proton transfer to (**427**).

(**426**)   (**427**)   Scheme 68

Phase-transfer catalyzed oxidative dehydrogenation of Δ^2-1,2,3-triazolines by potassium permanganate affords a convenient route to various 1,2,3-triazoles ⟨84CHEC(5)7027, 82JPR857⟩. However, the reaction has limited scope when electron-withdrawing and/or sterically crowded groups are present on the triazoline rings. Nickel peroxide (NiO$_2$) is a selective oxidant for triazolines ⟨90S191⟩. Yields are usually better than those obtained using potassium permanganate (Scheme 68). Steric crowding prevents the bulky permanganate ion from closely approaching the reaction site, whereas the effectiveness of the NiO$_2$ oxidation, possibly *via* an hydroxyl radical, is unaltered ⟨90JOC5891⟩.

The chemistry of Δ^2-1,2,3-triazolines ⟨84AHC(37)217⟩ and Δ^3- and Δ^4-1,2,3-triazolines ⟨84AHC(37)351⟩ has been reviewed.

3.4.2.2.3 Ring contraction

1-Pyrazolines undergo photochemically induced nitrogen elimination and ring contraction to cyclopropanes, *e.g.* (**428**)→(**429**). This is particularly useful for the preparation of strained rings, *e.g.* (**430**)→(**431**) ⟨70AHC(11)1⟩. Δ^2-Pyrazolines unsubstituted in the 1-position lose nitrogen on pyrolysis to give cyclopropanes (*e.g.* **432**; Z = NH → **433**), probably *via* Δ^1-pyrazolines.

(**428**)   (**429**)

(**430**)   (**431**)   (**432**)   (**433**)

Photodecomposition of Δ^2-1,2,3-triazolines gives aziridines. In cyclohexane the *cis* derivative (**434**) gives the *cis* product (**435**), whereas photolysis in benzene in the presence of benzophenone as sensitizer gives the same ratio of *cis*- and *trans*-aziridines from both triazolines which is accounted for in terms of a triplet excited state ⟨70AHC(11)1⟩. Δ^2-Tetrazolines are photolyzed to diaziridines.

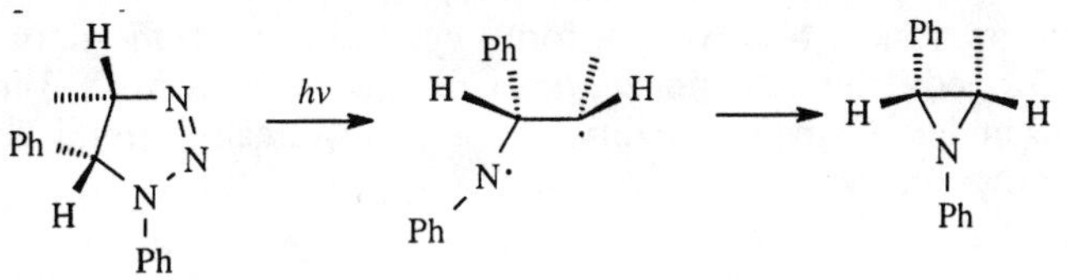

(434) (435)

The hydrolytic decomposition of 1-alkyltriazolines in aqueous buffers ⟨93JOC2097⟩ leads predominantly to 1-alkylaziridines with lesser amounts of 2-(alkylamino)ethanol, alkylamines, and acetaldehyde. The rate of hydrolysis of 1-alkyltriazolines is about twice as fast as that of the analogues acyclic 1,3,3-trialkyltriazenes and varies in the order *t*-butyl > isopropyl > ethyl > butyl > methyl > propyl > benzyl ⟨92JOC6448⟩. The proposed mechanism, involving rate-limiting formation of a 2-(alkylamino)ethyldiazonium ion (436), is shown in Scheme 69.

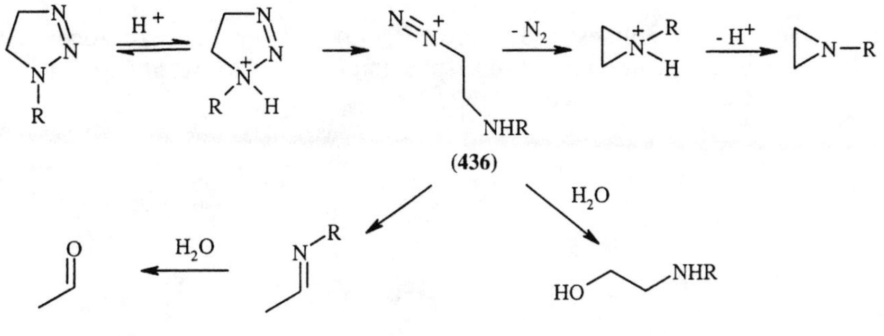

Scheme 69

Fragmentation of Δ^3-1,3,5-thiadiazoline derivatives is summarized in Scheme 70.

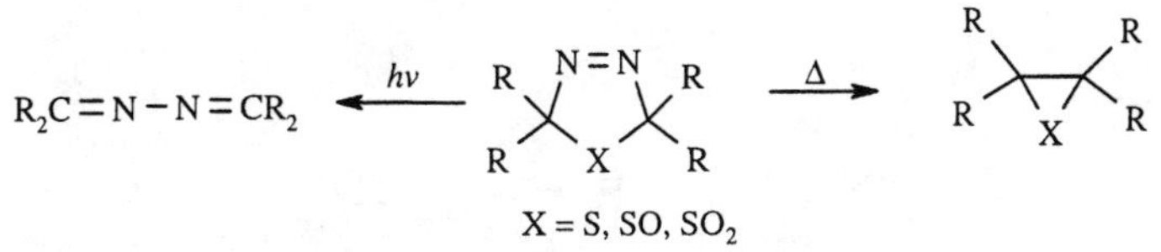

X = S, SO, SO$_2$

Scheme 70

3.4.2.2.4 Other reactions

Dihydro compounds show reactions which parallel those of their aliphatic analogues provided that the aromatization or ring-contraction reactions just discussed do not interfere.

Δ^2-Imidazolines (416; Z=NH) are cyclic amidines and exhibit the characteristic resonance stabilization and high basicity. Δ^2-Oxazolines (416; Z = O) are cyclic imino ethers, and Δ^2-thiazolines (416; Z=S) are imino thioethers; both are consequently easily hydrolyzed by dilute acid.

2-Imidazolines (437) are readily *N*-alkylated or *N*-acylated to form 1-alkyl- (438) or 1-acyl derivatives or 1,3-disubstituted salts (439). Improved alkylations involve pre-generation of the imidazoline anions. With excess alkyl halide the quaternary salt is formed, but suitable 2-substituents allow deprotonation by sodium hydride to give 2-alkylideneimidazolidines (440) ⟨87CB2053⟩.

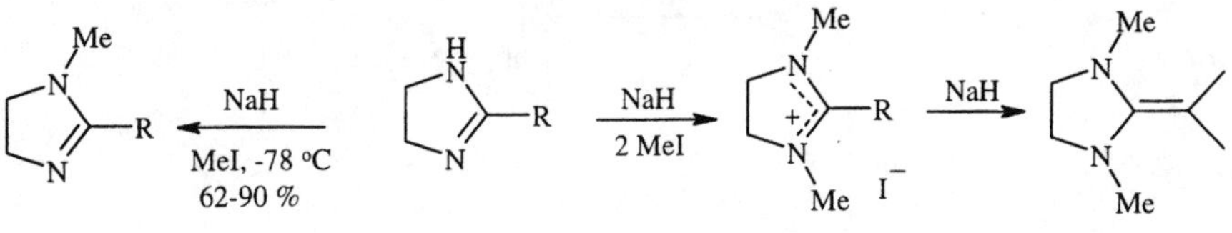

(438)   (437)   (439)   (440)

3-Imidazoline N-oxides have been much studied ⟨84IZV2565, 84TL5809, 85IZV161, 85TL4801, 90CHE643⟩. 1-Hydroxy-3-imidazoline 3-oxides (441) are in tautomeric equilibrium with open chain nitrones. Electron-withdrawing 2-substituents and a donor group in the 4-position favor the cyclic form.

Oxidation gives stable radicals (**442**) which form cycloadducts with dipolarophiles (*e.g.* **443**) ⟨85TL4801⟩. Radicals formed from 3-imidazoline 1-oxides or 1-hydroxy-3-imidazolines can be formylated on nitrogen under Vilsmeier conditions and exocyclically brominated or chlorinated by NBS or NCS on a 4-methyl group.

(441)   (442)   (443)

Δ^2-Pyrazolines and Δ^2-isoxazolines (**432**; Z=NH, O) are cyclic hydrazones and oximes, respectively. 2-Pyrazolines are quaternized at the 1-position (**444**→**445**) ⟨64AHC(3)1⟩. 1,3,4-Oxadiazolines (*e.g.* **446**) are very easily ring-opened ⟨66AHC(7)183⟩.

(444)   (445)   (446)

The so-called homopyrazoles (2,3-diazabicyclo[3.1.0]hex-3-enes) (**447**) exist in equilibrium with 1,4-dihydropyrazines (**448**) ⟨92CB1227⟩.

(447)   (448)

Several stereoselective reducing agents have been used to cleave the N—O bond of isoxazolines. The reductive cleavage method is a well-exploited route to amino alcohols, aziridines, or both, depending upon the reaction conditions ⟨91HC(49)1⟩. LAH is the most selective agent. Ring opening of 3-methyl-4-α-hydroxyisopropyl-2-isoxazoline (**450**) under reductive cleavage conditions with LAH, afforded γ-amino alcohols (**451**) and (**449**) in a 95:5 ratio, respectively. Under catalytic hydrogenolysis conditions with hydrogen / palladium, the stereoselection is reversed to 3:97 of *trans* : *cis*, (**451**) : (**449**) ⟨91HC(49)1⟩.

(449)   (450)   (451)   (452) X = O, S

2-Isoxazoline undergo [2+2] cycloaddition with furan or thiophene to yield the photocycloadducts (**452**) ⟨83CL1357⟩.

4,5-Dihydrooxazoles have been proven to be very useful in organic synthesis. They are inert to a variety of reagents, including Grignard reagents, NaBH₄, and LiAlH₄, and are useful protecting groups for carboxylic acids. There are many methods for the cleavage of 4,5-dihydrooxazoles once they have served their purpose. An effective method for hydrolyzing them back to carboxylic acids employs trifluoromethanesulfonic anhydride (Scheme 71) ⟨92SC13⟩.

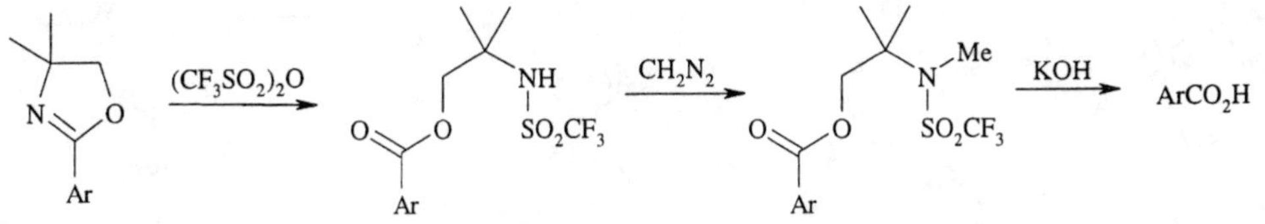

Scheme 71

4,5-Dihydrooxazoles are also effective *ortho* directors for nucleophilic addition to naphthalene and some heteroaromatic derivatives. Tandem additions result if the reaction mixtures are quenched with electrophiles (Scheme 72).

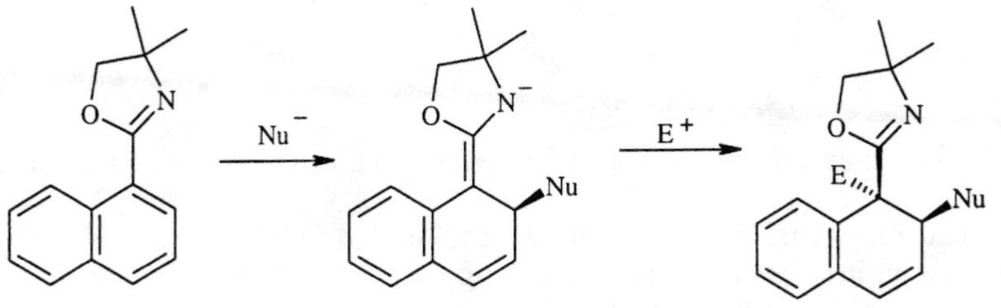

Scheme 72

The thermolysis of 2,1-benzothiazoline 2,2-dioxides generates aza-*o*-xylenes which are trapped *in situ* with maleic acid derivatives to give *cis*-1,2,3,4-tetrahydroquinoline-2,3-dicarboxylic acid products (Scheme 73) ⟨91SL571⟩ (*cf.* **364**).

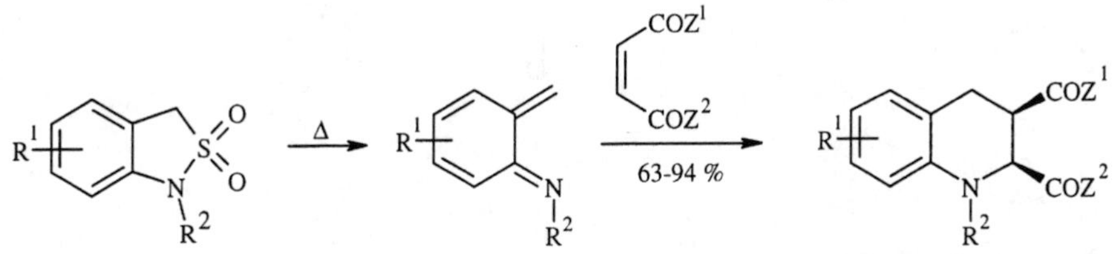

Scheme 73

Reduction of dihydro compounds to the tetrahydro derivatives is sometimes possible. For example, thiazolines are reduced to thiazolidines by aluminum amalgam.

3.4.2.3 Tetrahydro Compounds

3.4.2.3.1 Aromatization

Some tetrahydro azoles can be aromatized, but this is more difficult than in the corresponding dihydro series. Thus, the conversion of pyrazolidines into pyrazoles is accomplished with chloranil. Imidazolidines are aromatized with great difficulty.

3.4.2.3.2 Ring fission

Cleavage of the heterocyclic ring is usually accomplished using degradative procedures which are also applicable in the aliphatic series. Thus a nitrogen-containing ring can be opened by Hofmann exhaustive methylation (*e.g.* **453**→**454**). Pyrazolidines also undergo reactions of the Fischer indole synthesis type (**455**→**456**). The sulfur-containing ring of thiazolidines can be opened by Raney nickel desulfurization.

(453)   (454)   (455)   (456)

Compounds of types (457; R = H) and (458; R = H) are in equilibrium with open-chain forms (459); such tetrahydro compounds are readily hydrolyzed by dilute acid (R ≠ H).

(457)   (458)   (459)

Isoxazolidines sometimes undergo retro 1,3-dipolar cycloaddition to give back alkenes and nitrones ⟨77AHC(21)207⟩.

Thiazolidines undergo hydrolysis to aldehyde and amino thiol under acid or basic aqueous conditions. The reaction involves C—S bond breaking and proceeds by the formation of an iminium thiolate zwitterion intermediate (460) ⟨91JA3071⟩.

The hydrolysis of thiazolidines to aldehydes is conveniently carried out under neutral conditions by the assistance of metal ions such as Hg^{II} or Cu^{II} ⟨93JOC275⟩.

(460)

3.4.2.3.3 Other reactions

These compounds usually show the typical reactions of their aliphatic analogues. 1,3-Dioxolanes (461), tetrahydroimidazoles (457), tetrahydrooxazoles (458) and tetrahydrothiazoles (462) are somewhat less easily ring-cleaved than their acyclic analogues undergo heterolysis (*cf.* previous section), but their properties are otherwise similar.

1-Aryl-5-pyrazolidinones (463) are photochemically ring-contracted to β-lactams ⟨70AHC(11)1⟩.

(461)   (462)   (463)   R = H, R' = Me
                                       R = Me, R' = H

Alkylation of imidazolidines (and their oxo and thio derivatives) is usually carried out in the presence of a strong base such as sodium hydride ⟨86JOC2228, 91JA8546, 91PHA670⟩, potassium carbonate in DMF ⟨89S38⟩, or potassium hydroxide in DMSO ⟨82JOC2787⟩.

Isoxazolidines display the properties consistent with their cyclic hydroxylamine structure. Thus, isoxazolidines behave as nucleophiles and thus can be alkylated or acylated. Isoxazolidines are also strong bases (pK_a 5.05), and undergo reactions such as quaternization, hydrogenolysis, oxidation, thermolysis, photolysis, decomposition by bases, etc. ⟨84CHEC(6)1, 91HC(49)1⟩.

The chemistry of the oxazolidines, like the dihydrooxazoles, is characterized by reactions that take advantage of the opportunity for stereoselectivity. The chiral environment is often provided by the same amino alcohols, such as phenylglycinol or norephedrine, used to prepare dihydrooxazoles.

2-Methoxyoxazolidines (**464**) are useful for the asymmetric formylation of various nucleophiles including silyl enol ethers ⟨92T6011⟩, trimethylsilyl cyanide ⟨93SL921⟩, enamines ⟨90TL4223⟩, and allyl silanes.

The reaction of nucleophiles with 2-alkyl-2-methoxyoxazolidines offers a general method for asymmetric acylation ⟨91T7925⟩.

(464)   (465)   (466)

The sulfur in 1,2-dithiolanes is nucleophilic and is alkylated easily to form 1,2-dithiolanium ions (**465** → **466**) ⟨85JA2807, 85PS(23)169⟩.

3.4.3 REACTIONS OF SUBSTITUENTS

Substituents attached to carbon are considered by classes; substituents linked to ring nitrogen are considered separately because of their differing character.

3.4.3.1 General Survey of Substituents on Carbon

If the reactions of the same substituents on heteroaromatic azoles and on benzene rings are compared, the differences in the reactivities are a measure of the heteroatoms' influence. Such influence by the mesomeric effect is smaller when the substituent is β to a heteroatom than when it is α or γ. The influence by the inductive effect is largest when the substituent is α to a heteroatom.

3.4.3.1.1 *Substituent environment*

The electronic environment of an α-substituent on pyridine (**467**) approaches that of a substituent on the corresponding imino compound (**468**) and is intermediate between those of substituents on benzene (**469**) and substituents attached to carbonyl groups (**470**) (*cf.* discussion in Chapter 3.2). Substituents attached to certain positions in azole rings show similar properties to those of α- and γ-substituents on pyridine. However, the azoles also possess one heteroatom which behaves as an electron source and which tends to oppose the effect of other heteroatom(s).

(467)   (468)   (469)   (470)   (471)

Substituents cannot directly conjugate with β-pyridine-like nitrogen atoms. Azole substituents which are not α or γ to a pyridine-like nitrogen react as they would on a benzene ring. Conjugation with an α-pyridine-like nitrogen is much more effective through formal double bonds; thus the 5-methyl group in 3,5-dimethyl-1,2,4-oxadiazole (**471**) is by far the more reactive as it is so activated by both nitrogen atoms.

In azolium cations, the electron-pull of the positively charged heteroatom is strong, and substituents attached α or γ to positive poles in azolium rings show correspondingly enhanced reactivity.

Azolinones and azole *N*-oxides possess systems which can act either as an electron source or as an electron sink, depending on the requirements of the reaction.

3.4.3.1.2 *The carbonyl analogy*

In aliphatic compounds, reactions of functional groups are often modified very significantly by an adjacent carbonyl group. As would be expected from the discussion in the preceding section, the reactions of certain substituents α and γ to pyridine-like nitrogen atoms in azole rings are similarly influenced. Such effects on substituents can be classified into six groups.

(i) Substituents which can leave as anions are displaced by nucleophilic reagents (472).
(ii) α-Hydrogen atoms are easily lost as protons (473).
(iii) As a consequence of (ii), tautomerism is possible (474 ⇌ 475).
(iv) Carbon dioxide is readily lost from carboxymethyl (476) and carboxyl groups (477).
(v) These effects are transferred through a vinyl group, and nucleophilic reagents will add to vinyl and ethynyl groups (478) (Michael reaction).
(vi) Electrons are withdrawn from aryl groups (479).

Examples of these for both carbonyl and heterocyclic compounds are listed in Table 9.

(472)   (473)   (474)   (475)

(476)   (477)   (478)   (479)

Table 9 Reactivity of Substituents: The Carbonyl Analogy ⟨B-68MI40900⟩

Reaction type	Group	α- or γ-Groups	Compare with
Nucleophilic displacement	Nitro	Are displaced readily	—
	Halogen	Are displaced	Acid chloride
	Alkoxy, Amino	Are displaced when additionally activated	Ester Amide
Proton loss	Hydroxyl	Are acidic	Carboxylic acid
	Amino	Are less basic	Amide
	Alkyl	Become 'active'	Ketone
Tautomerism	Hydroxyl	Exist largely in the oxo form	Carboxylic acid (two equivalent structures)
	Amino	Exist to a small extent only in the imine form	Amide
	Mercapto	Exist largely in the thione form	Thiocarboxylic acid
Decarboxylation	Carboxyl	Decarboxylate at *ca.* 200°C	α-Keto acids
	Carboxymethyl	Decarboxylate at *ca.* 50°C	β-Keto acids
Michael reactions	Vinyl, Ethynyl	Undergo Michael additions readily	α,β-Unsaturated ketones
	β-Hydroxyethyl	Undergo reverse Michael reaction readily (lose H₂O)	β-Hydroxy ketones
Electrophilic attack on phenyl groups	Phenyl	Undergo electrophilic substitution in the *meta* and *para* positions (*ca.* 1: 1)	Phenyl ketones

3.4.3.1.3 Two heteroatoms in the 1,3-positions

The 2-position in imidazoles, thiazoles and oxazoles is electron deficient, and substituents in the 2-position (**480**) generally show the same reactivity as α- or γ-substituents on pyridines. 2-Substituents in azoliums of this type, including 1,3-dithiolyliums, are highly activated.

Substituents in the 4-position of these compounds are also α to a multiply bonded nitrogen atom, but because of bond fixation they are relatively little influenced by this nitrogen atom even when it is quaternized (**481**). This is similar to the situation for 3-substituents in isoquinolines, *cf.* Chapter 3.2. In general, substituents in the 4- and 5-positions of imidazoles, thiazoles and oxazoles show much the same reactivity of the same substituents on benzenoid compounds (but see Section 3.4.3.9.1).

3.4.3.1.4 Two heteroatoms in the 1,2-positions

Substituents on pyrazoles and isoxazoles, regardless of their positions, generally show reactivity closer to that of the same substituent on a benzene ring rather than to that of α- or γ-substituents on pyridine. The (electron-releasing) resonance effect of the 'pyrrole-type' NH group and 'furan-type' oxygen atom appears to be more important than their (electron-withdrawing) inductive effect in pyrazole and isoxazole (**482**). However, some reactions of these types are known (see *e.g.* Section 3.4.3.3.3) and halogen atoms and methyl groups in the 3- and 5-positions of pyrazoles and isoxazoles (**482**) become 'active' if the ring is quaternized (**483**).

Substituents on the isothiazole ring are a little more reactive, especially in 5-position. In cationic rings reactivity is much higher, *e.g.* for substituents in 1,2-dithiolylium salts.

3.4.3.1.5 Three heteroatoms

In the 1,2,4-thiadiazole ring the electron density at the 5-position is markedly lower than at the 3-position, and this affects substituent reactions. 5-Halo derivatives, for example, approach the reactivity of 4-halopyrimidines. The 1,2,4-oxadiazole ring shows a similar difference between the 3- and 5-positions.

Substituents in 1,3,4-thiadiazoles are quite strongly activated, as in the 2-position of pyridine.

In contrast, substituents in 1,2,4-triazoles are usually rather similar in reactivity to those in benzene; although nucleophilic substitution of halogen is somewhat easier, forcing conditions are required.

3.4.3.1.6 Four heteroatoms

Alkyl groups and halogen atoms in tetrazoles are not highly activated unless the ring is quaternized.

3.4.3.1.7 The effect of one substituent on the reactivity of another

The effect of one substituent on the reactivity of another is generally similar to that observed in the corresponding polysubstituted benzenes. However, the partial bond fixation in an azole can lead to differential effects in the mutual interactions of substituents, similar to those found in naphthalene where the benzene ring fusion induces bond fixation. A good example is in the comparison of methyl group reactivity in (**484**); the 5-methyl group condenses with aldehydes easily, the 3-methyl group does not. However, quaternization at nitrogen renders the 3-methyl group reactive.

3.4.3.1.8 Reactions of substituents not directly attached to the heterocyclic ring

In general, substituents removed from the ring by two or more saturated carbon atoms undergo normal aliphatic reactions, and substituents attached directly to fused benzene rings or aryl groups undergo the same reactions as those on normal benzenoid rings.

3.4.3.1.9 Reactions of substituents involving ring transformations

Several classes are known. Dimroth-type rearrangements occur by ring opening and reclosure so that one ring atom changes places with an exocyclic atom. The rearrangement of 5-phenylaminothiatriazole to 1-phenyl-5-mercaptotetrazole in basic solution is reversible (Scheme 74). As the anion, the tetrazole system is the more stable; whereas as the neutral species, the thiatriazole is more stable ⟨76AHC(20)145⟩.

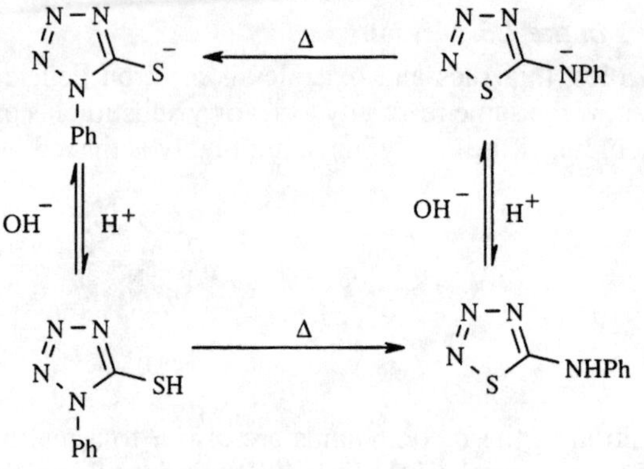

Scheme 74

A different type of rearrangement occurs when suitable side chains are α to a pyridine-like nitrogen atom. In the monocyclic series this can be generalized by Scheme 75. For a given side chain the rate of rearrangement is 1,2,4-oxadiazoles > isoxazoles > 1,2,5-oxadiazoles. Typical side chains include hydrazone, oxime and amidine. Some examples are shown in Table 10 ⟨79AHC(25)147⟩. Similar rearrangements for benzazoles are discussed in Section 3.4.3.2.4.

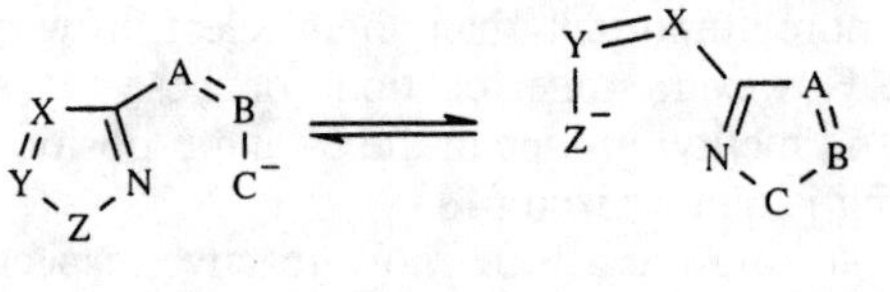

Scheme 75

Table 10 Examples of Rearrangements Involving Three-atom Side
Chains of Azoles[a]

Starting material	Product

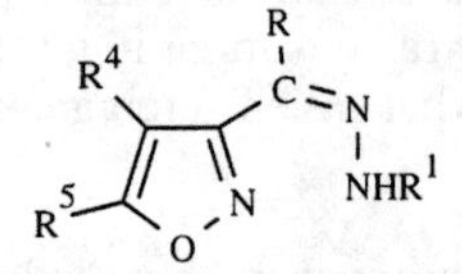

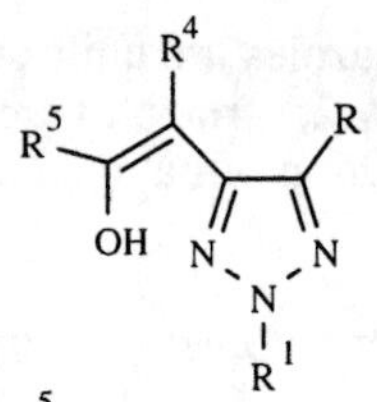

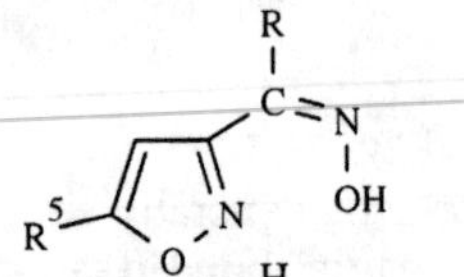

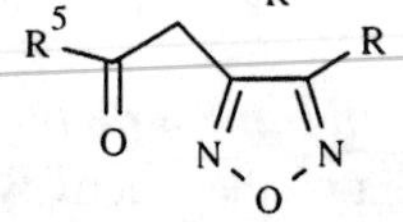

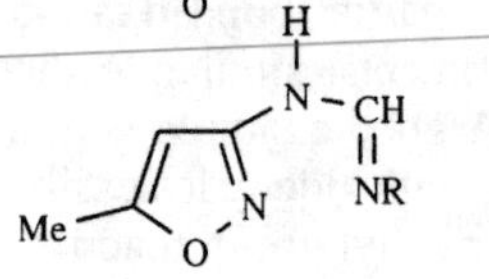

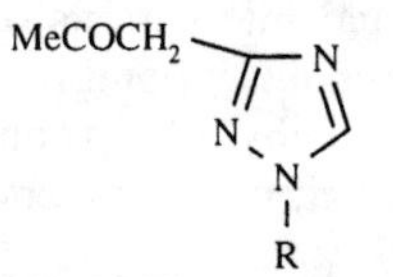

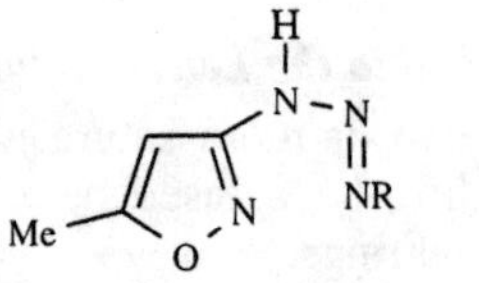

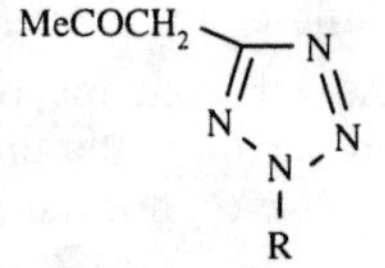

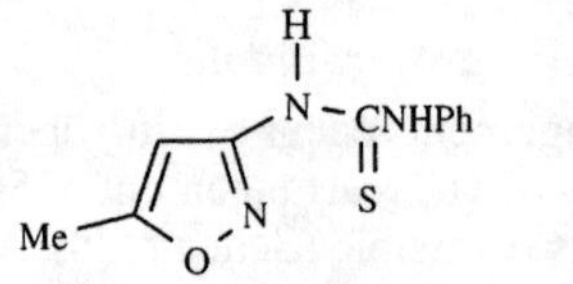

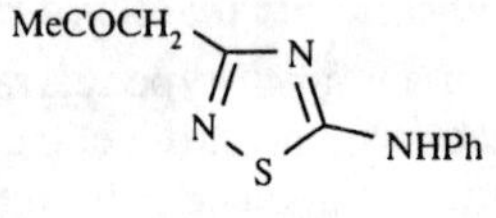

[a] ⟨79AHC(25)147⟩

A somewhat similar type of ring interconversion involving attack on sulfur has been postulated in the 1,2,4-thiadiazole series, *e.g.* (**485** → **486**). Such reactions are common in the 1,2,4-dithiazolium series, *e.g.* (**487** → **488**).

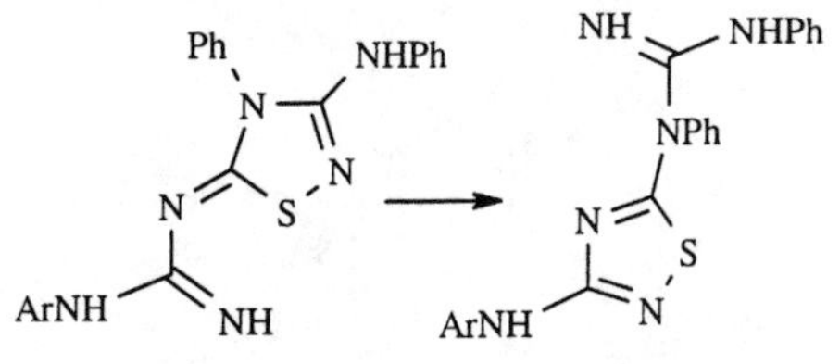

(**485**)　　　　(**486**) 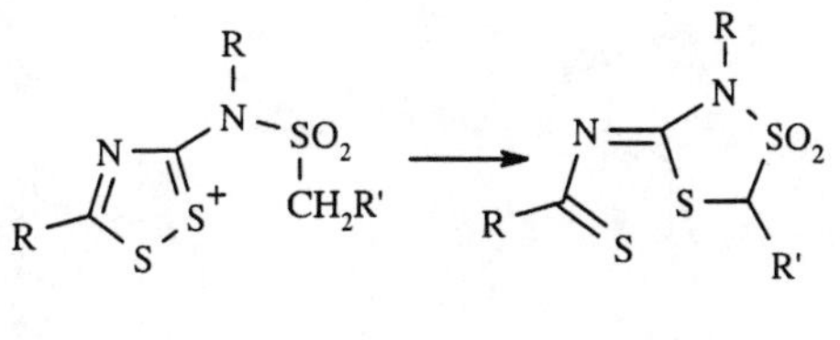　　　　　　(**487**)　　　　　　(**488**)

Cycloadditions including a cyclic S atom and an exocyclic C = X bond are known in the dithiazole series, *e.g.* as shown in Scheme 76 (see also Section 3.4.1.10.2).

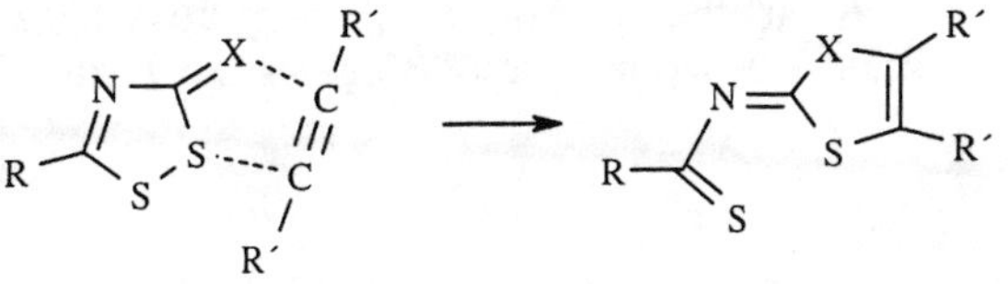

Scheme 76

3.4.3.2　Fused Benzene Rings

3.4.3.2.1　*Electrophilic substitution*

In compounds with a fused benzene ring, electrophilic substitution on carbon usually occurs in the benzenoid ring in preference to the heterocyclic ring. Frequently the orientation of substitution in these compounds parallels that in naphthalene. Conditions are often similar to those used for benzene itself. The actual position attacked varies; compare formulae (**489**)-(**494**) where the orientation is shown for nitration; sulfonation is usually similar.

(**489**) 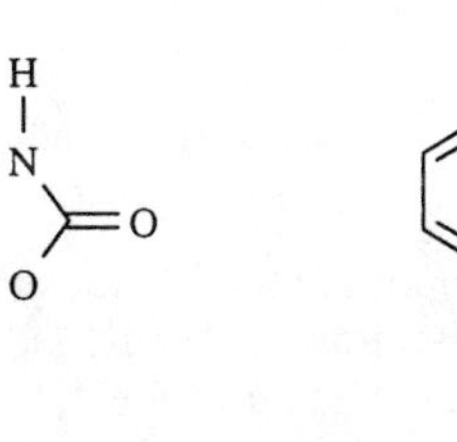　　　　(**490**) 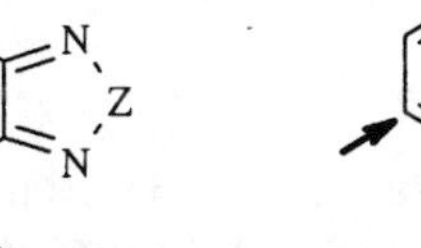　　　　(**491**) 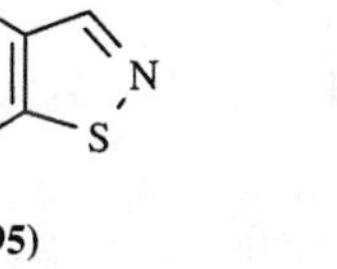　　　　(**492**)

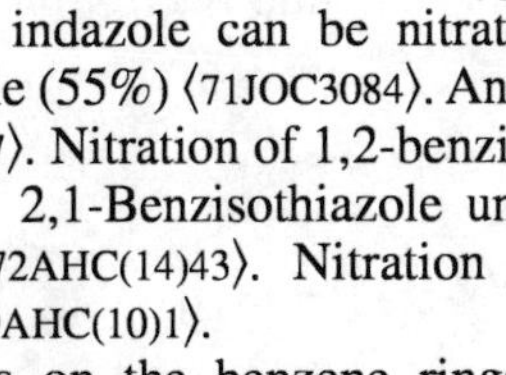

(**493**) 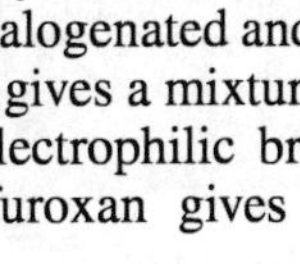　　　　(**494**) 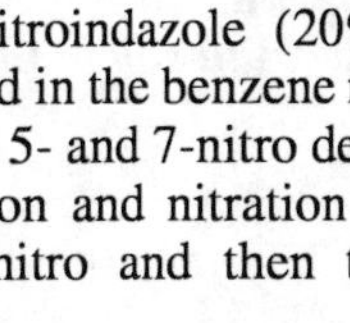　　　　(**495**)

Indazoles show most of the typical benzene electrophilic substitution reactions. However, unsubstituted indazole can be nitrated by acetyl nitrate to 3,5-dinitroindazole (20%) as well as 3-nitroindazole (55%) ⟨71JOC3084⟩. Anthranil is halogenated and nitrated in the benzene ring at position 5 ⟨67AHC(8)277⟩. Nitration of 1,2-benzisothiazole gives a mixture of the 5- and 7-nitro derivatives (**495**) ⟨72AHC(14)43⟩. 2,1-Benzisothiazole undergoes electrophilic bromination and nitration in the 5- and 7-positions ⟨72AHC(14)43⟩. Nitration of benzofuroxan gives the 4-nitro and then the 4,6-dinitro compound ⟨69AHC(10)1⟩.

Substituents on the benzene rings exert their usual influence on the orientation and ease of electrophilic substitution reactions. For example, further nitration (H$_2$SO$_4$/SO$_3$/HNO$_3$) of 4-nitro-benzofuroxan (**496**) gives the 4,6-dinitro derivative, the first nitro group direct *meta* as expected.

Strong electron-donating groups enhance electrophilic substitution and direct *ortho/para*. Thus, dimethylaminobenzofuroxans can be nitrosated and diazo-coupled ⟨69AHC(10)1⟩; bromination of 4-, 5-, 6- and 7-aminobenzothiazoles occurs *ortho* and *para* to the amino group.

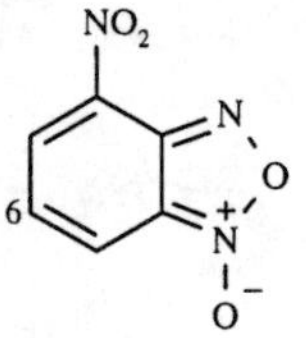

(496) 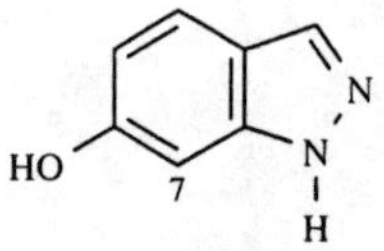   (497) 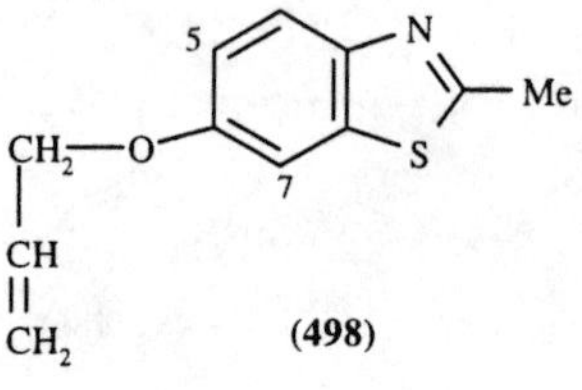   (498)

A heterocyclic ring induces partial double-bond fixation in a fused benzene ring. Hence, for example, diazo coupling occurs at the 7-position of 6-hydroxyindazole (**497**), and Claisen rearrangement of 6-allyloxy-2-methylbenzothiazole (**498**) gives the 7- and 5-allyl products in a ratio of 20:1.

3.4.3.2.2 Oxidative degradation

Vigorous oxidation (*e.g.* with KMnO$_4$) usually degrades fused benzene rings in preference to many azole rings, especially under acidic conditions. Thus, benzimidazoles are oxidized by chromic acid or 30% hydrogen peroxide to imidazole-4,5-dicarboxylic acid ⟨70AHC(12)103⟩, and 2,1,3-benzothiadiazole is oxidized by ozone or potassium permanganate to the dicarboxylic acid (**499**) ⟨68AHC(9)107⟩.

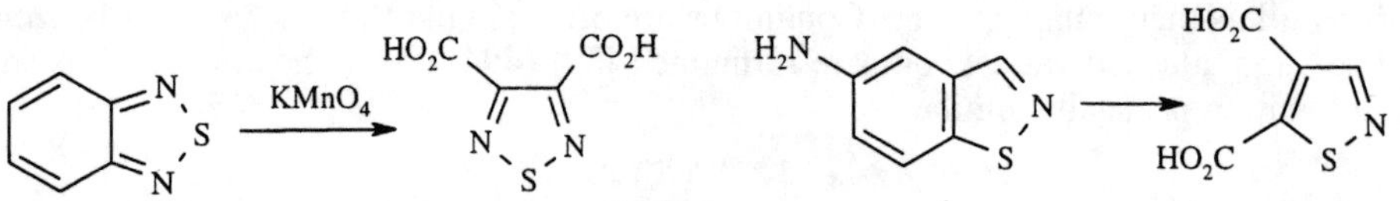

(499)   (500)   (501)

As expected, oxidative degradation of a fused benzene ring is facilitated when it carries electron-donating groups and is hindered by electron-withdrawing substituents. 5-Aminobenzisothiazole (**500**) with potassium permanganate gives the carboxylic acid (**501**) ⟨59JCS3061⟩.

3.4.3.2.3 Nucleophilic attack

Most fused benzene rings are stable toward nucleophilic attack but exceptions are known for highly electron-deficient benzazoles having *o*-quinonoid structures. Thus, sulfur nucleophiles attack 2*H*-benzimidazole-2-spirocyclohexane (**502**) *via* an initial Michael-type 1,4-conjugate addition, followed by a prototropic shift in the adduct (**503a**). When the nucleophile is electron-withdrawing (*e.g.*, phenylsulfonyl) 1,3-dihydro products (**503b**) are isolated. If the nucleophile is electron donating the adducts are oxidized *in situ* to produce (**504**) ⟨96CHEC-II(3)146⟩.

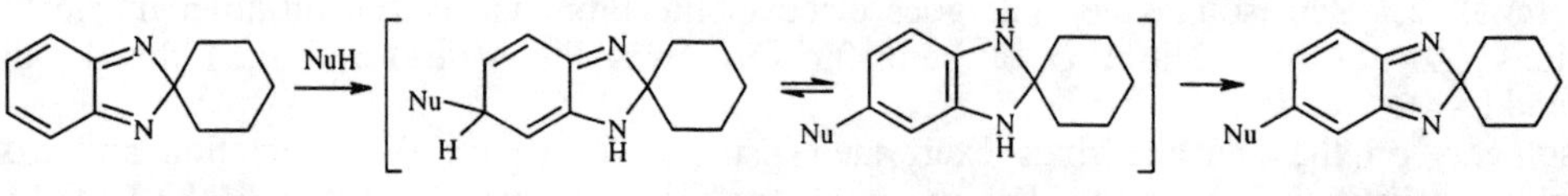

(502)   (503a)   (503b)   (504)

Halogen atoms on benzazole rings can be activated toward nucleophilic displacement by electron-withdrawing groups.

3.4.3.2.4 *Rearrangements*

In the benzazole series, reactions of the type discussed for monocyclic derivatives in Section 3.4.3.1.9 are generalized by Scheme 77 and examples are given in Table 11.

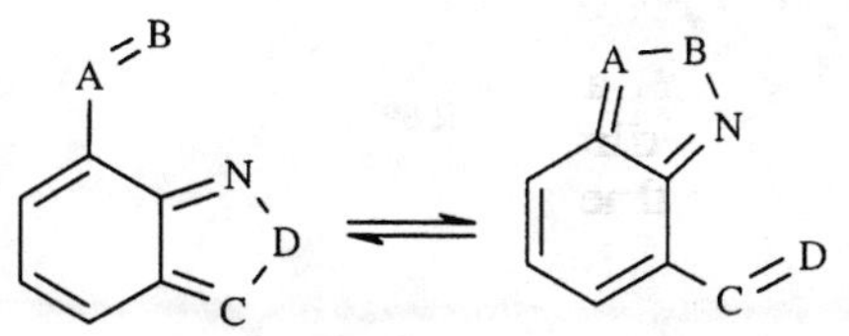

Scheme 77 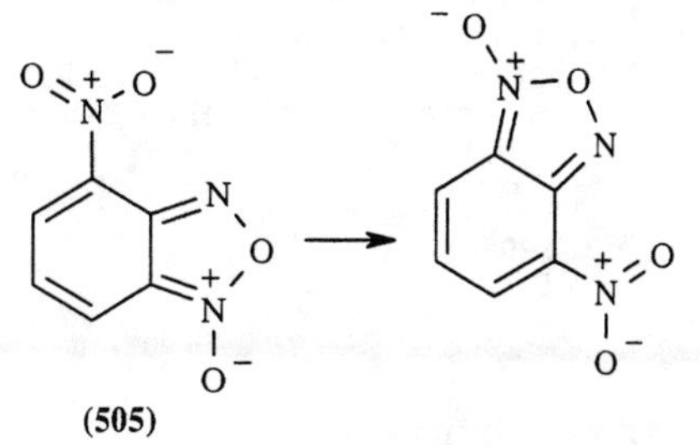   (505)

4-Nitrobenzofuroxan (**505**) undergoes a rearrangement (recognizable as an isomerization in unsymmetrically substituted derivatives) which is an example of this general rearrangement (Scheme 77) ⟨64AG(E)693⟩; see Table 11.

Table 11   Benzazole Rearrangements[a]

Examples of involvement of two-atom side chains

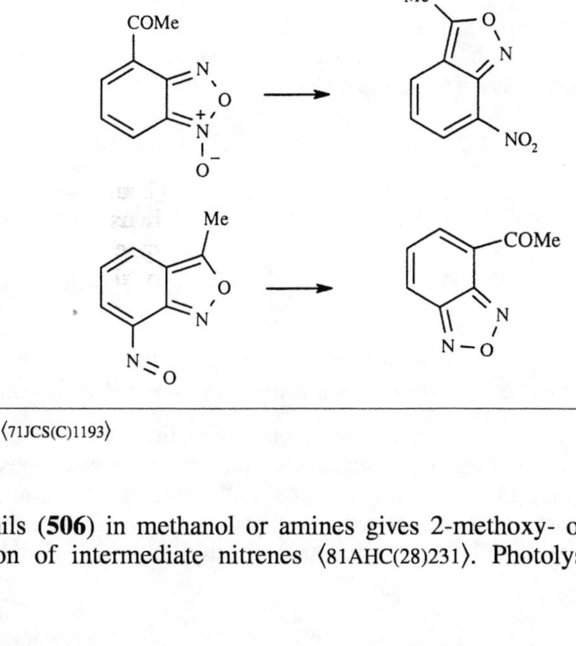

[a] ⟨71JCS(C)1193⟩

Photolysis of anthranils (**506**) in methanol or amines gives 2-methoxy- or 2-amino-3*H*-azepines (**507**) by ring expansion of intermediate nitrenes ⟨81AHC(28)231⟩. Photolysis of 2-alkylindazoles probably also goes through a nitrene intermediate, which either abstracts hydrogen from the solvent to give (509) ⟨81AHC(28)231⟩ or ring expands to yield (508).

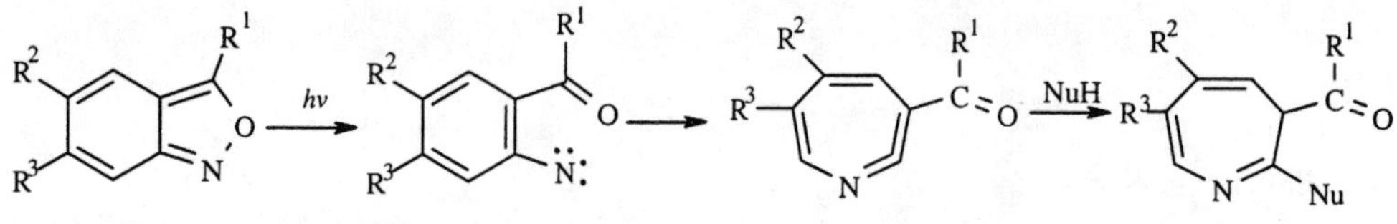

(506)

R¹ = Ph or Me; R², R³ = H or Cl; NuH = MeOH, Et₂NH or PhNH₂

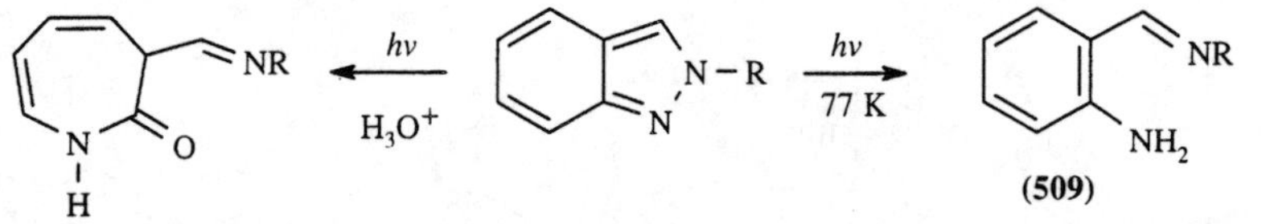

(507)

(508)   (509)

3.4.3.3 Alkyl Groups

3.4.3.3.1 Reactions similar to those of toluene

Alkyl groups attached to heterocyclic systems undergo many of the same reactions as those on benzenoid rings.

(i) Oxidation in solution (KMnO₄, CrO₃, *etc.*) gives the corresponding carboxylic acid or ketone; for example, alkyl groups on pyrazoles are oxidized with permanganate to carboxylic acids ⟨66AHC(6)347⟩, 3-methylisothiazoles are converted by chromium trioxide into the 3-carboxylic acids ⟨72AHC(14)1⟩, and methylthiazoles with SeO₂ give thiazolecarbaldehydes.

(ii) Free radical bromination with *N*-bromosuccinimide often succeeds. Thus, 2,5-disubstituted 4-methyloxazoles on bromination give the 4-bromomethyl compounds ⟨74AHC(17)99⟩, and methyl groups in the 4- and 5-positions of isoxazole (510) and (511) have been brominated with NBS ⟨63AHC(2)365⟩.

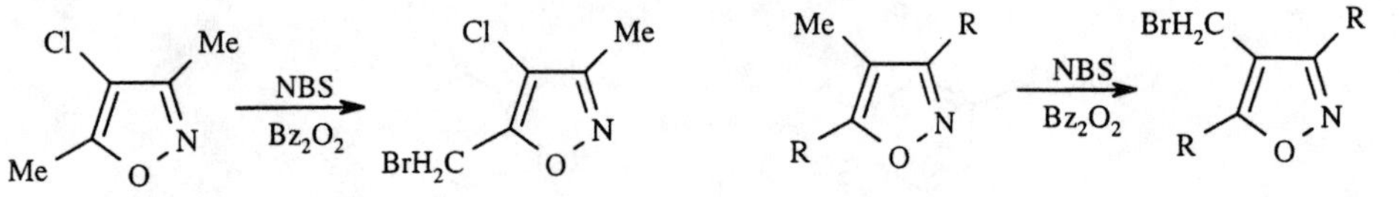

(510)   (511)

(iii) A fused cyclohexeno ring can be converted into a fused benzene ring, *e.g.* (512→513).

(iv) A trichloromethyl group has been converted by antimony trifluoride into a trifluoromethyl group in the 1,2,4-thiadiazole series (Scheme 78).

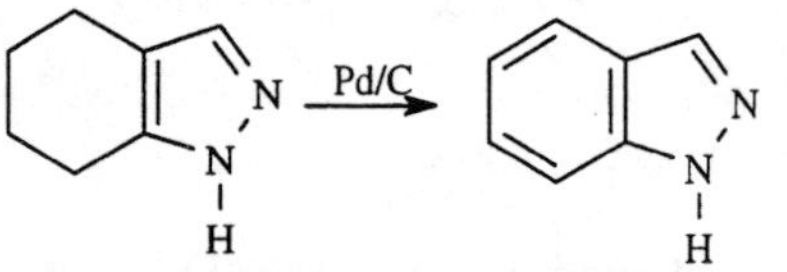

(512)   (513) 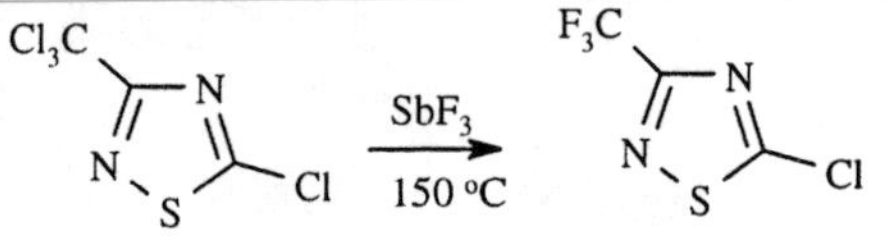   Scheme 78

3.4.3.3.2 Alkylazoles: reactions involving essentially complete anion formation

In addition to the reactions described in the preceding section, alkyl groups in the 2-positions of imidazole, oxazole and thiazole rings show reactions which result from the easy loss of a proton from the carbon atoms of the alkyl group which is adjacent to the ring (see Section 3.4.3.1.2).

Additional nitrogen atoms facilitate such reactions, particularly if they are α or γ to the alkyl group, and, if α act across a formal double bond. Thus, the 5-methyl group in 3,5-dimethyl-1,2,4-oxadiazole is much more reactive than the 3-methyl group in this compound or the methyl groups in 2,5-dimethyl-1,3,4-oxadiazole ⟨76AHC(20)65⟩.

The strongest bases, such as sodamide (NaNH$_2$/NH$_3$, 40°C) or organometallic compounds (BuLi/Et$_2$O, 40°C), convert, for example, 2-methyl-oxazole and -thiazole and 1,2-dimethylimidazole essentially completely into the corresponding anions (*e.g.* **514**), although some ring metallation also occurs (*cf.* Section 3.4.1.8.1). These anions all react readily even with mild electrophilic reagents; thus the original alkyl groups can be substituted in the following ways.

(i) Alkylation, *e.g.* MeI → CH$_2$Me for the formation of (**515**).

(ii) Acylation, *e.g.* the oxadiazole (**516**) undergoes Claisen condensation with ethyl oxalate ⟨76AHC(20)65⟩.

(iii) Carboxylation, *e.g.* CO$_2$ → CH$_2$CO$_2$H in the tetrazole series.

(iv) Reactions with aldehydes, *e.g.* MeCHO → CH$_2$CH(OH)Me in the 1,2-dimethylimidazole series.

3.4.3.3.3 *Reactions of alkylazoles involving traces of reactive anions*

In aqueous or alcoholic solution, certain alkylazoles react with bases to give traces of anions of type (**517**). With suitable electrophilic reagents, these anions undergo reasonably rapid and essentially non-reversible reactions.

(i) A nitroso group gives an imine, as in the probable mechanism of the conversion of (**518**) into (**519**).

(ii) Aliphatic aldehydes can form monoalcohols, *e.g.* (**520**) gives (**521**) ⟨79HC(34-1)5⟩.

(iii) Aromatic aldehydes give styryl derivatives (*e.g.* **522**) by spontaneous dehydration of the intermediate alcohol (*cf.* Section 3.4.3.1.2). 5-Methyl-3-phenyl-1,2,4-oxadiazole (**523**) reacts thus with benzaldehyde in the presence of zinc chloride ⟨76HC(20)65⟩. 3-Nitrobenzaldehyde reacts with 5-methylisothiazole ⟨65AHC(4)75⟩. The 4- and 5-methylthiazoles are unreactive.

(iv) Halogens displace hydrogen atoms, *e.g.* 3,4,5-trimethylpyrazole (**524**) is converted into (**525**) ⟨56LA(598)186⟩.

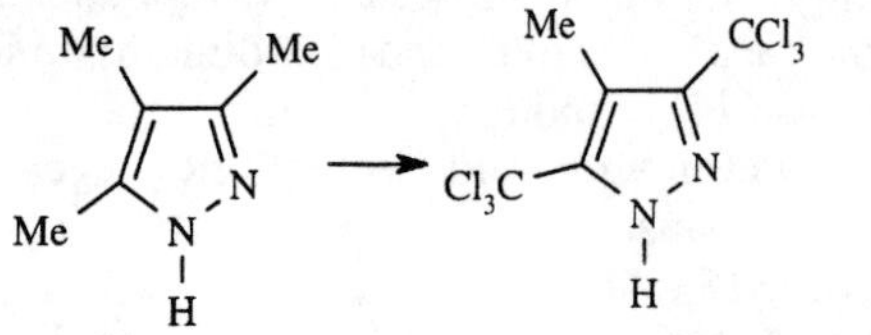

(**524**)   (**525**) 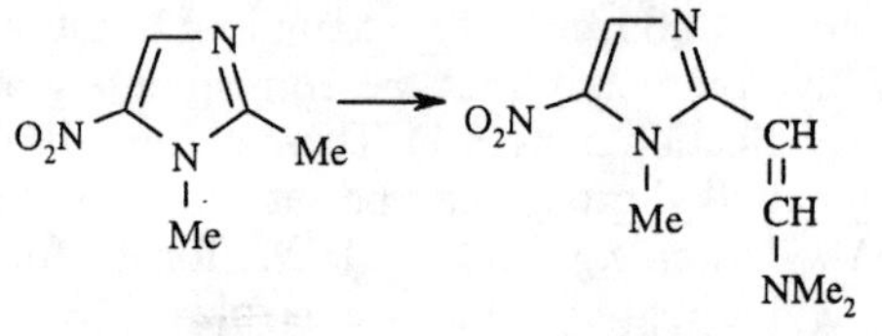   (**526**)   (**527**)

(v) *N,N*-Dimethylformamide acetal gives dimethylaminovinyl derivatives, as in (**526**) → (**527**).

(vi) Pyridine and iodine give pyridylmethyl compounds, *e.g.* (**528**) yields (**529**) ⟨80AHC(27)241⟩.

Reactions of types (i)–(vi) can be catalyzed by alkoxide or hydroxide ions, or amines. Alternatively, an acid catalyst forms a complex of type (**530**) from which proton loss is facilitated.

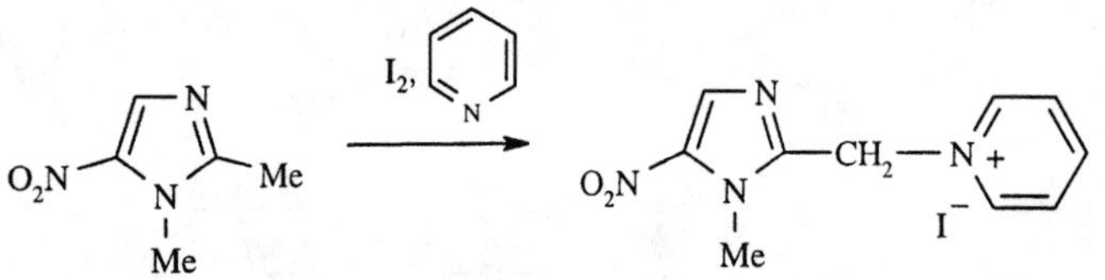

(**528**)   (**529**) 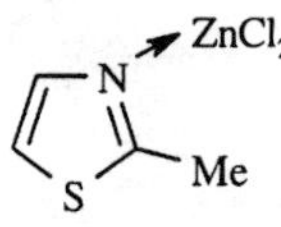   (**530**)

The so-called 'isoxazoline transposition' or Angeli's rearrangement (Scheme 79) involves the conversion of methylfuroxans by treatment with alkoxides or alcoholic alkali hydroxides into the oximes of isoxazolidin-4-ones (**531**) ⟨81G167⟩.

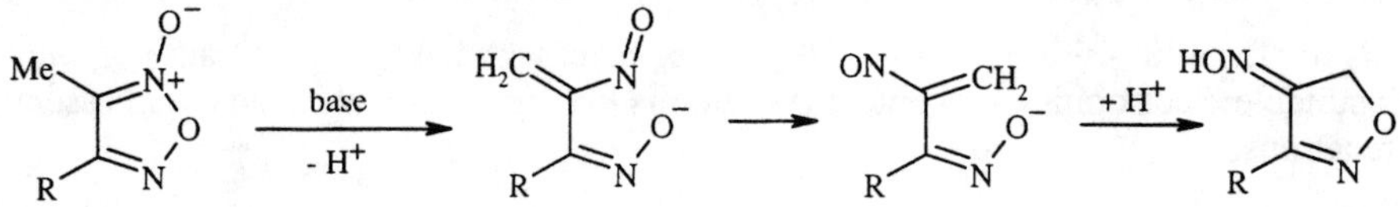

(**531**)

Scheme 79

3.4.3.3.4   C-Alkyl-azoliums, -dithiolyliums, etc.

Proton loss from alkyl groups α or γ to a cationic center in an azolium ring is often easy. The resulting neutral anhydro bases or methides (*cf.* **532**) can sometimes be isolated; they react readily with electrophilic reagents to give products which can often lose another proton to give new resonance-stabilized anhydro bases. Thus, the trithione methides are anhydro bases derived from 3-alkyl-1,2-dithiolylium salts (**533 ⇌ 534**) ⟨66AHC(7)39⟩. These methides are stabilized by electron acceptor substituents such as CN or CO_2R ⟨66AHC(7)39⟩.

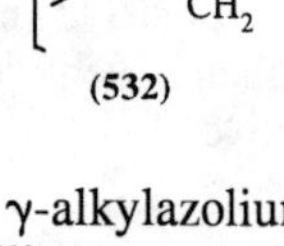

(**532**) 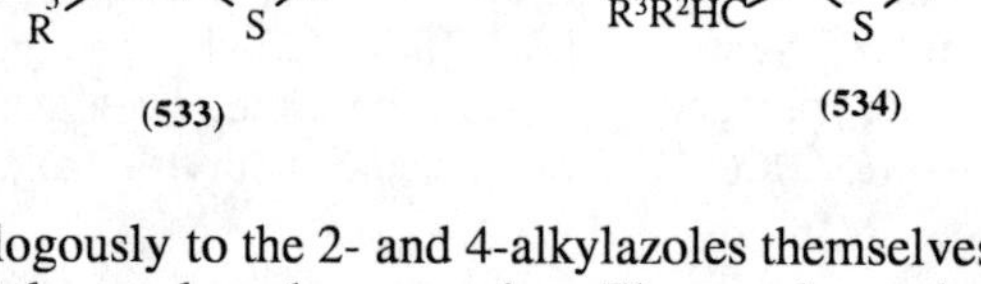   (**533**)   (**534**)

Both α- and γ-alkylazolium ions, analogously to the 2- and 4-alkylazoles themselves, can also react with electrophilic reagents without initial complete deprotonation. They undergo the same types of reactions as the alkylazoles but under milder conditions, and these reactions are often catalyzed by piperidine. Thus, in quaternized pyrazoles 5-methyl groups react with benzaldehyde to give styryl derivatives and can be chlorinated ⟨66AHC(6)347⟩. The methyl groups in quaternized isoxazoles are also reactive, and here piperidine is sufficient as catalyst (Scheme 80) ⟨63AHC(2)365⟩.

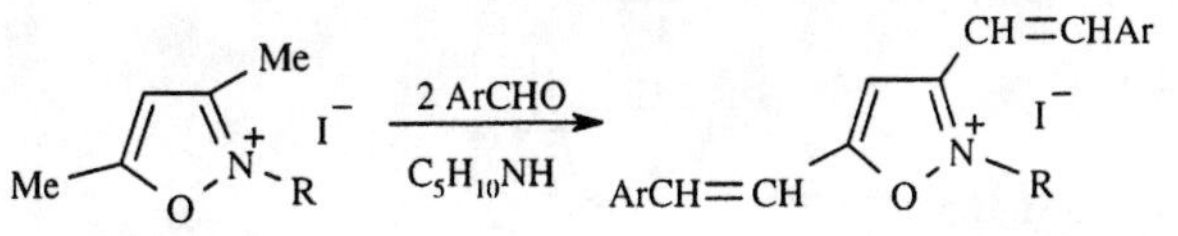

Scheme 80

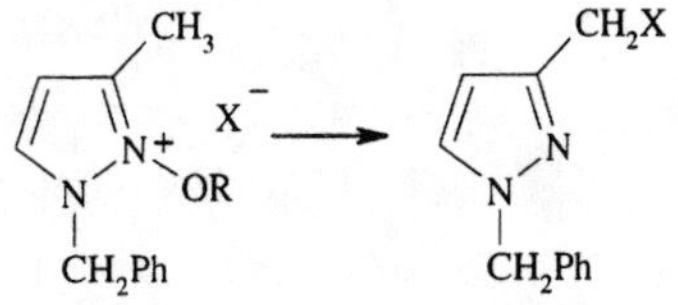

(535)

(536)

In 2-alkoxypyrazoliums (535) methyl groups at positions 3 and 5 are active; in this way pyrazoles (536) were prepared (X = D, I, OMe) ⟨92JCS(P1)2555⟩.

Some weak electrophilic reagents, which are usually inert toward neutral azoles, also react with quaternized azoles. Diazonium salts yield phenylhydrazones (Scheme 81) in a reaction analogous to the Japp–Klingemann transformation of β-keto esters into phenylhydrazones; in the dithiolylium series illustrated the product has bicyclic character. Cyanine dye preparations fall under this heading (see also Section 3.4.1.6.5). Monomethine cyanines are formed by reaction with an iodo quaternary salt, *e.g.* Scheme 82. Tri- and penta-methinecarbocyanines (537; $n = 1$ and 2, respectively) are obtained by the reaction of two molecules of a quaternary salt with one molecule of ethyl orthoformate (537; $n = 1$) or β-ethoxyacrolein acetal (537; $n = 2$), respectively.

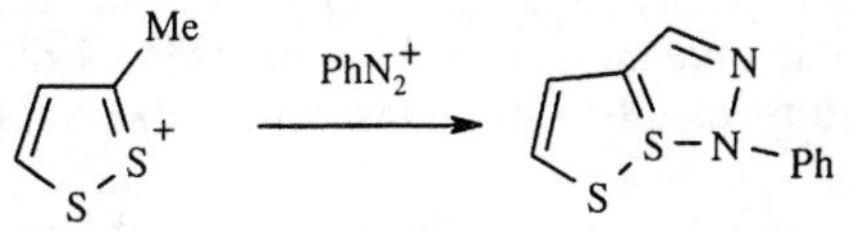

Scheme 81

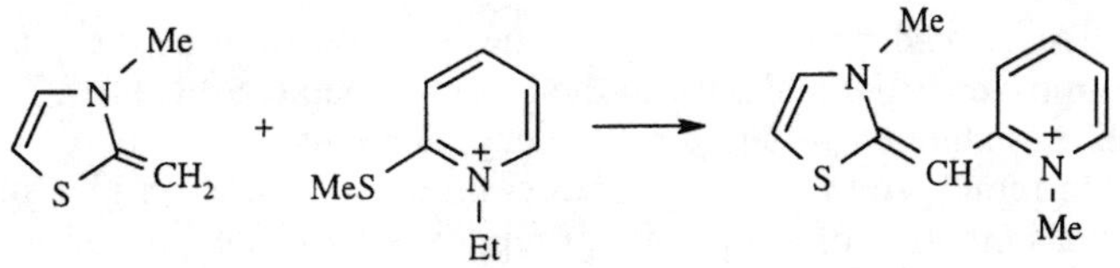

Scheme 82

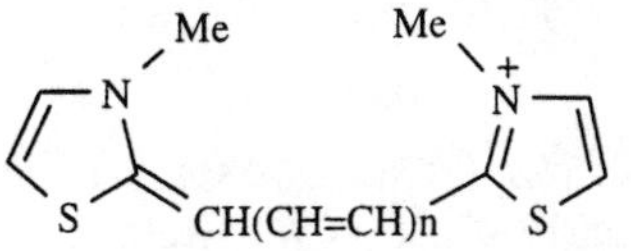

(537)

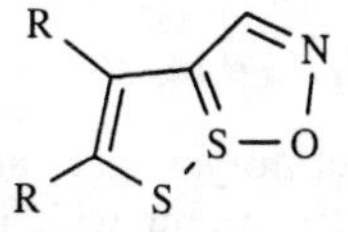

(538)

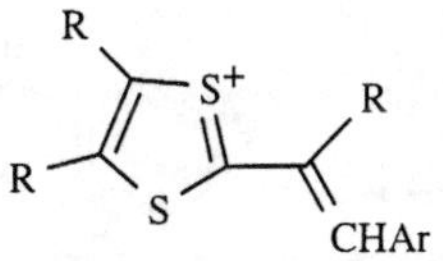

(539)

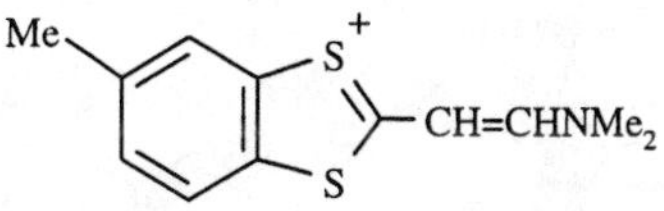

(540)

3-Methyl-1,2-dithiolyliums react with aldehydes to give styryl derivatives, with DMF to give Vilsmeier salts, and on nitrosation form the bicyclic products (538) ⟨80AHC(27)151⟩. 2-Alkyl groups in 1,3-dithiolylium ions also react with aromatic aldehydes to give (539), with DMF to give (540), and with other electrophiles ⟨80AHC(27)151⟩.

In general, methyl groups in the 4- and 5-positions of imidazole, oxazole and thiazole do not undergo such deprotonation-mediated reactions, even when the ring is cationic.

Compounds which can formally be considered as anhydro bases can sometimes react with nucleophiles. Thus, unsaturated azlactones with Grignard reagents give saturated azlactones (Scheme 83) ⟨65AHC(4)75⟩.

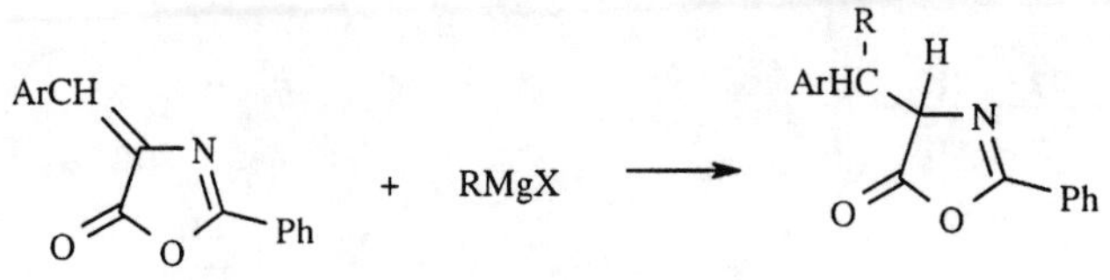

Scheme 83

3.4.3.4 Other *C*-Linked Substituents

3.4.3.4.1 Aryl groups: electrophilic substitution

Electrophilic substitution occurs readily in *C*-aryl groups, often predominantly at the *para* position. Thus nitrations of phenyl-thiazoles, -oxazoles and -imidazoles (HNO$_3$/H$_2$SO$_4$ at 100°C) all yield the corresponding *p*-nitrophenyl derivatives. This is to be contrasted with the situation for α-phenylpyridine, where a mixture of mainly *m*- and *p*-nitrophenyl derivatives is formed. Although in strongly acidic media a *C*-linked aryl group is generally more readily substituted than the ring, the orientation often changes when *C*-phenylazole derivatives are nitrated under less acidic conditions. Thus, 3- and 5-phenylpyrazoles can give under such conditions the 4-nitro derivatives. Such orientation changes have been demonstrated to result from changes in the species undergoing reaction from the azolium ion to the neutral azole.

5-Methyl-3-phenylisoxazole is nitrated as a conjugate acid at the *meta* position but as the free base at the *para* position of the phenyl group ⟨79AHC(25)147⟩. Phenyl groups attached to oxazole rings are nitrated or sulfonated in the *para* position, with relative reactivities of the phenyl groups in the order 5 > 4 > 2 ⟨74AHC(17)99⟩.

3-Phenylisothiazole is nitrated predominantly in the *meta* position of the phenyl group, whereas 4-phenylisothiazole is nitrated *ortho* and *para* in the phenyl group ⟨72AHC(14)1⟩. Nitration of 3-phenyl-1,2,4-oxadiazole gives a mixture of *m*- and *p*-nitrophenyl derivatives ⟨63G1196⟩.

In the 1,2-dithiolylium ion system, 3- and 5-phenyl groups on nitration give mixtures of *para* and *meta* orientation, whereas nitration of a 4-phenyl group gives *para* substitution only ⟨61JA2934⟩.

3.4.3.4.2 Aryl groups: other reactions

3-Arylanthranils (**541**) on thermolysis give acridones (**542**) ⟨81AHC(28)231⟩. 3-Phenylanthranils (**543**) also form acridones (**544**) on treatment with nitrous acid ⟨67AHC(8)277⟩. Related rearrangements are found with 3-heteroarylanthranils (*e.g.* **545** → **546**) ⟨81AHC(28)231⟩.

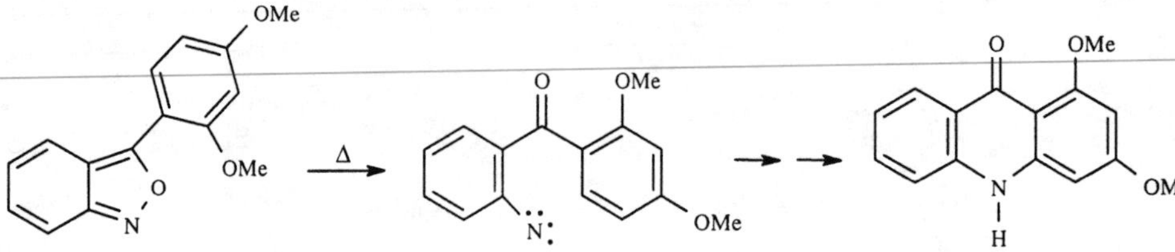

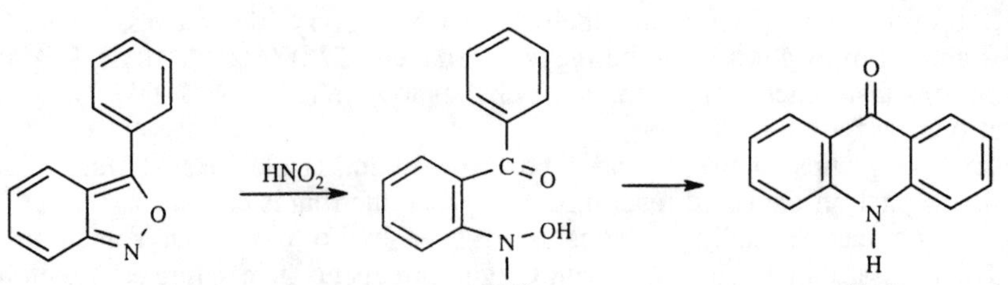

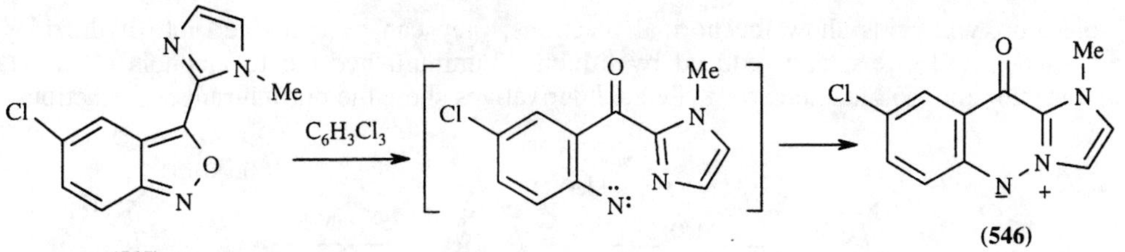

Methyl groups on *C*-linked phenyl attached to oxazoles, isoxazoles and oxadiazoles react with benzylidineaniline to give stilbene derivatives (Scheme 84) ⟨78AHC(23)171⟩.

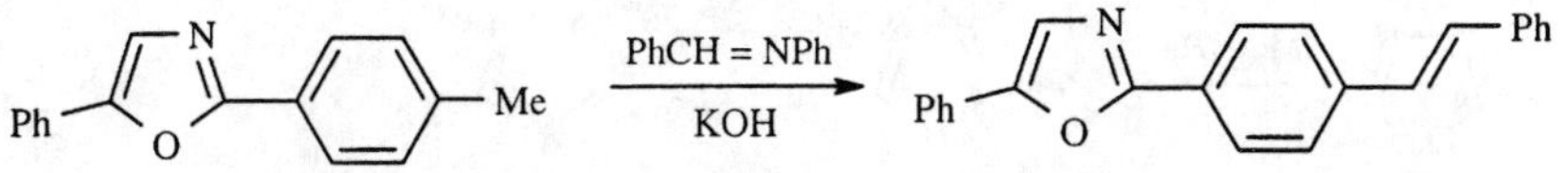

Scheme 84

3.4.3.4.3 Carboxylic acids

Azolecarboxylic acids can be quite strongly acidic. Thus, 1,2,5-thiadiazole-3,4-dicarboxylic acid has first and second pK_a values of 1.6 and 4.1, respectively ⟨68AHC(9)107⟩. The acidic strengths of the oxazolecarboxylic acids are in the order 2 > 5 > 4, in agreement with the electron distribution within the oxazole ring ⟨74AHC(17)99⟩. Azolecarboxylic acids are amino acids and can exist partly in the zwitterionic, or betaine, form *(e.g.* **547**).

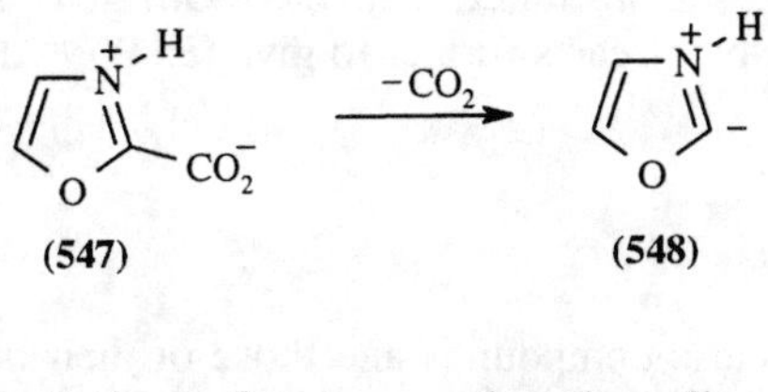

The relatively easy decarboxylation of many azolecarboxylic acids is a result of inductive stabilization of intermediate zwitterions of type (**548**) (*cf.* Section 3.4.1.8.1). Kinetic studies have shown that oxazole-2- and -5-carboxylic acids both are decarboxylated through the zwitterionic tautomers ⟨71JA7045⟩. Thiazole-2-carboxylic acids, and to a lesser extent-5-carboxylic acids, are decarboxylated readily; thiazole-4-carboxylic acids are relatively stable. Isothiazole-5-carboxylic acids are decarboxylated readily, the 3-isomers less so while the 4-isomers require high temperatures. The 1,2,4-, 1,2,5- and 1,3,4-thiadiazolecarboxylic acids are also easily decarboxylated; their stability is increased by electron-donating substituents ⟨68AHC(9)165⟩. Most 1,2,3-triazolecarboxylic acids lose carbon dioxide when heated above their melting points ⟨74AHC(16)33⟩.

Azoleacetic acids with a carboxymethyl group are also decarboxylated readily, *e.g.* all three thiazole isomers, by a mechanism similar to that for the decarboxylation of β-keto acids; *cf.* Section 3.4.3.1.2. The mechanism has been investigated in the oxazole case, (**549**) → (**550**) → (**551**) ⟨72JCS(P2)1077⟩.

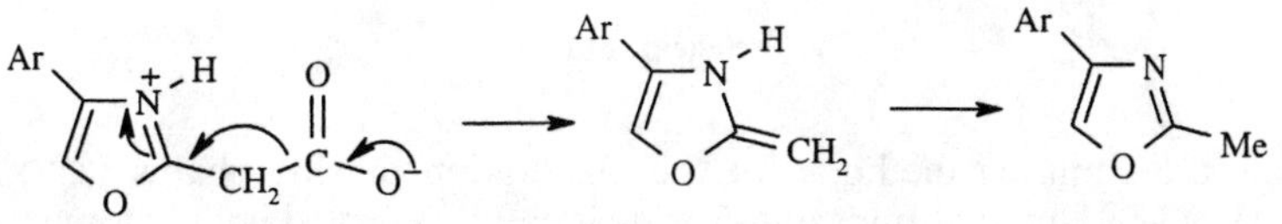

In most other reactions the azolecarboxylic acids and their derivatives behave as expected (*cf* Scheme 85) ⟨37CB2309⟩, although some acid chlorides can be obtained only as hydrochlorides. Thus, imidazolecarboxylic acids show the normal reactions: they can be converted into hydrazides, acid halides, amides and esters, and reduced by lithium aluminum hydride to alcohols ⟨70AHC(12)103⟩. Again, thiazole- and isothiazole-carboxylic acid derivatives show the normal range of reactions.

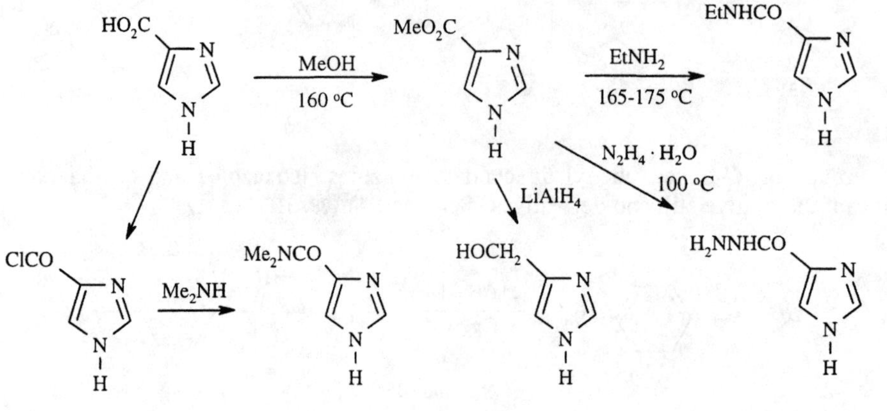

Scheme 85

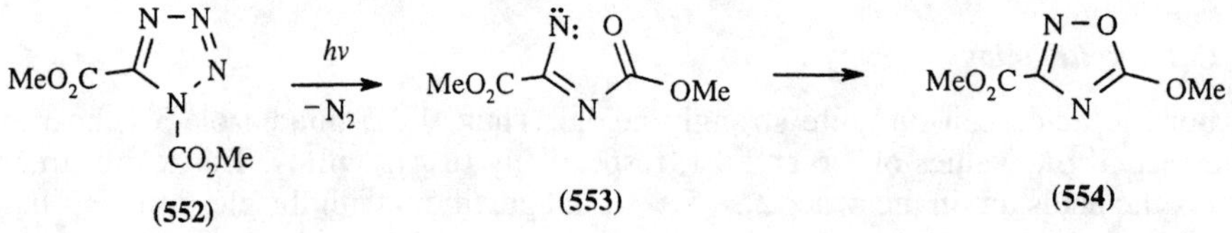

However, in some cases carboxylic acid-derived groups can participate in ring fission-reclosure reactions. Thus, photolysis of 1,5-disubstituted tetrazole (**552**) gives nitrogen and appears to involve the nitrene intermediate (**553**), which reacts further to give (**554**) ⟨77AHC(21)323⟩.

3.4.3.4.4 Aldehydes and ketones

In general, the properties of these compounds and those of their benzenoid analogues are similar. Thus, isothiazole aldehydes and ketones behave normally and form the usual derivatives ⟨72AHC(14)1⟩. Imidazole-2-carbaldehyde exists as a hydrate in aqueous solution. 4-Acetyloxazoles are oxidized to the corresponding acids with sodium hypobromite ⟨74AHC(17)99⟩. Thiazole aldehydes undergo the benzoin and Cannizzaro reactions. However, compounds with aldehyde groups α to an NH group sometimes form dimers, *e.g.* as in the 1,2,4-triazole series (**555**) ⟨70TL943⟩.

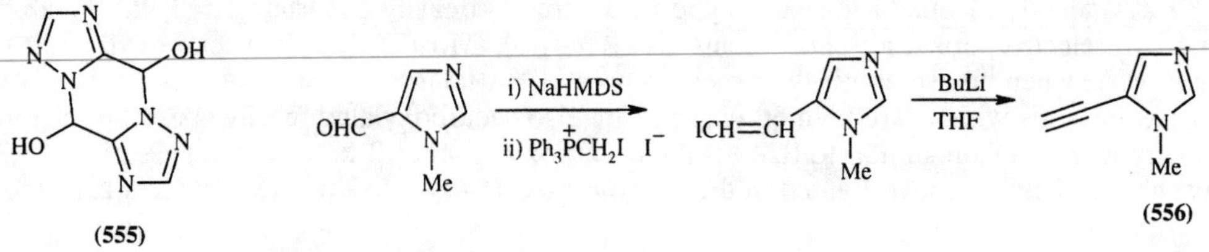

Scheme 86

Wittig reactions are common methods for the elaboration of side chains ⟨83JOC3605, 87CPB4056, 87SC223, 91CI(L)801⟩. When 1-methylimidazole-5-carbaldehyde is heated with iodomethyltriphenylphosphonium iodide, and the iodoethenyl product is treated with *n*-butyllithium in THF, the ultimate product is the highly toxic 5-ethynyl-1-methylimidazole (**556**) (Scheme 86) ⟨91CI(L)801⟩.

Aldehyde groups in *C*-formylazoles can participates in intramolecular cyclizations. Thus, reduction of 1-(2-nitrophenyl)-imidazole-2-carbaldehyde (**557**) with sodium dithionite leads *via* amine (**558**) to imidazo[1,2-*a*]quinoxaline (**519**) with good yield ⟨71KGS570⟩.

The Willgerodt reaction can proceed normally. Thus the 3-acetylpyrazole (**559**) is converted into the morpholide (**560**) ⟨57JCS2356⟩.

Deacylations are known. *C*-Acyl groups in 1,3,4-thiadiazoles are cleaved by sodium ethoxide in ethanol ⟨68AHC(9)165⟩. Imidazole-2-carbaldehyde behaves similarly, yielding imidazole and ethyl formate; this reaction involves an ylide intermediate. 3-Acylisoxazoles (**561**) are attacked by nucleophiles in a reaction which involves ring opening ⟨79AHC(25)147⟩.

Sometimes ring opening and reclosure can occur with participation of a *C*-acyl group. Thus, oxazole derivatives of type (**562**; X = H, Cl or NH$_2$; Y = OH or OEt) rearrange on heating to 255°C by ring opening and recyclization ⟨74AHC(17)99⟩. 3-Acylanthranils (**563**) rearrange to benzoxazinones (**564**) on heating ⟨67AHC(8)277⟩.

3.4.3.4.5 *Vinyl and ethynyl groups*

Vinyl and ethynyl groups attached to an imidazole ring can be catalytically reduced to the saturated (or less unsaturated) species ⟨87SC223, 88JA5571⟩ and cleaved by oxidation ⟨90JHC2189⟩. The corresponding 4-carbaldehyde is formed in 71% yield when 1-methyl-2,5-diphenyl-4-styrylimidazole is oxidized with osmium tetroxide ⟨93JCS(P2)1511⟩. However, they may not react like aliphatic alkenes and alkynes; not all the addition reactions occur normally, Michael additions are known (see CHEC), and the compounds can act as dienophiles in Diels-Alder reactions (Scheme 87) ⟨82CHE835, 88JA5571⟩.

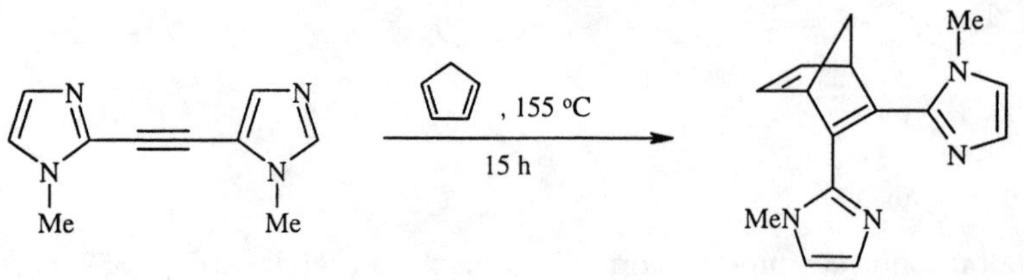

Scheme 87

3.4.3.4.6 Ring fission

Certain α-substituted alkyltetrazoles on pyrolysis yield nitrogen and an alkyne by the mechanism shown in Scheme 88 ⟨77AHC(21)323⟩.

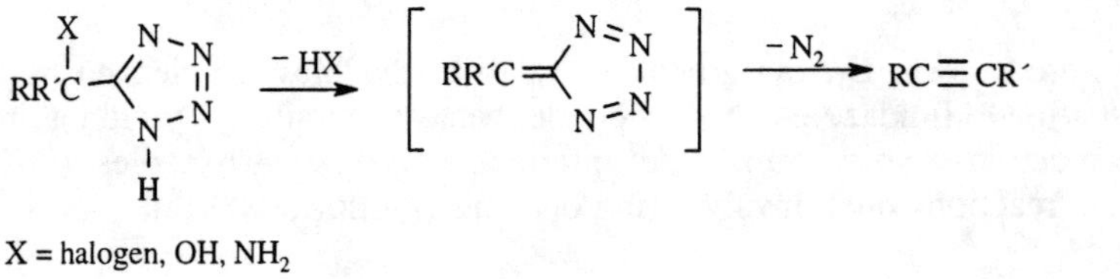

X = halogen, OH, NH$_2$

Scheme 88

When an azole carbene is formed, spontaneous ring fission can occur. The prototypes for these reactions are shown: (**565** → **566**), (**567** → **568**); *cf.* corresponding nitrene reactions (Section 3.4.3.6.2).

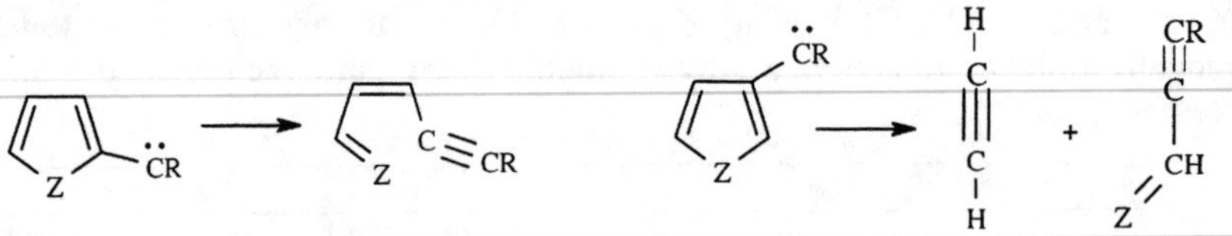

(565)   (566)   (567)   (568)

3.4.3.5 Aminoazoles

3.4.3.5.1 Dimroth rearrangement

The thermal acid- or base-catalyzed interconversion of 5-amino-1-phenyltriazoles (**569**) and 5-anilinotriazoles (**571**) was discovered by Dimroth. It is an example of a general class of heterocyclic rearrangements (**572** ⇌ **573**) now known by the name Dimroth rearrangements ⟨90BSB281, 84JHC627, 84CHEC(5)694, 74AHC(16)33⟩. The original Dimroth rearrangement probably involves a tautomeric diazoimine intermediate (**570**). Electron-attracting and large groups tend to favor the tautomer in which they are on the exocyclic nitrogen. Alkyl groups tend to prefer to reside on the cyclic nitrogen. Many other Dimroth rearrangements of 4- and / or 5-substituted 1,2,3-triazoles are also known ⟨96CHEC-II(4)30⟩.

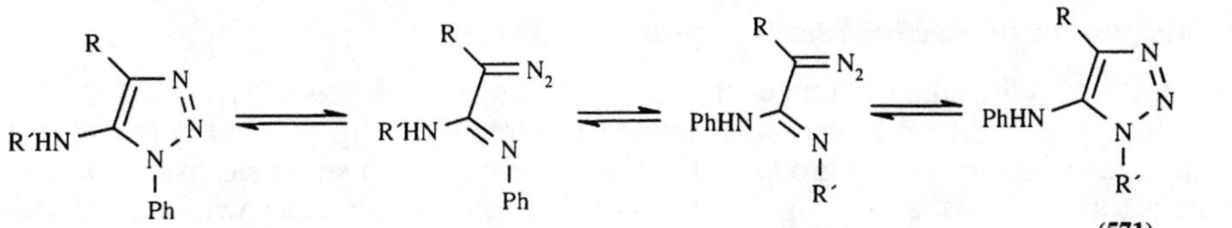

(569)   (570a)   (570b)   (571)

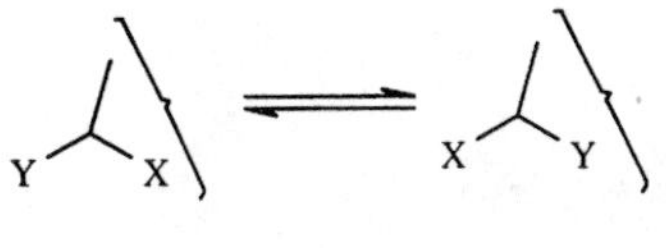

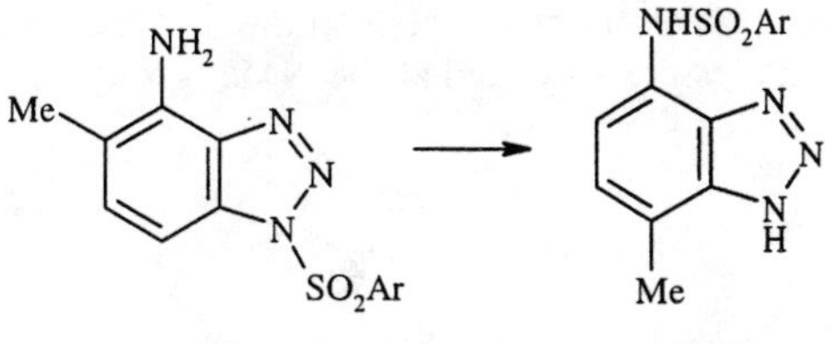

(572)   (573)   (574)   (575)

The benzotriazole ring is more stable and less susceptible to ring opening. Nevertheless, thermal rearrangements of 4-aminobenzotriazoles occur when the N(1) substituent is strongly electron withdrawing (574 → 575) (the reverse rearrangement is not observed). Similar rearrangements are known for 1,2,3-triazolo fused pyrimidines ⟨86BSB679⟩ and triazines ⟨88BSB179⟩.

The thermal behavior of substituted 5-aminotetrazoles involves wide-ranging example of the Dimroth rearrangement together with imidoyl azide-tetrazole ring-chain isomerism ⟨84JHC627⟩. An example with 5-hydrazinotetrazoles is shown in Scheme 89 ⟨88BSB534⟩.

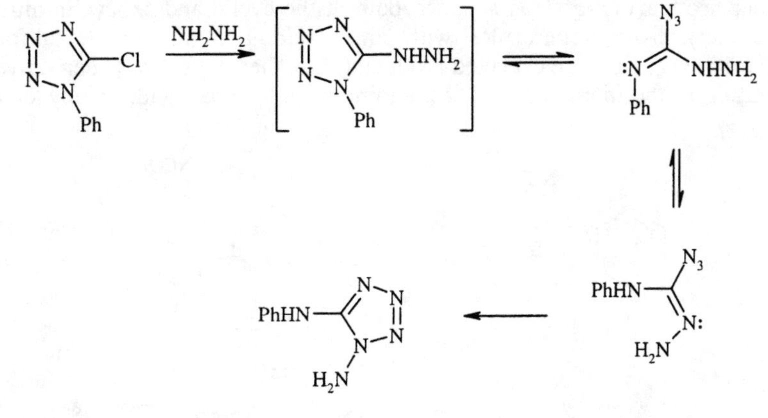

Scheme 89

Many related examples are now known as the general Dimroth rearrangement. For example, 3-ethylamino-1,2-benzisothiazole (576) is in equilibrium in aqueous solution with the 2-ethyl-3-imino isomer (577) ⟨72AHC(14)43⟩. Dimroth rearrangements are known in the 1,2,4-thiadiazoles (578 → 579); see Section 3.4.3.9.1 ⟨68AHC(9)165⟩. For a similar example in the 1,2,3,4-thiatriazole series see Section 3.4.3.1.9.

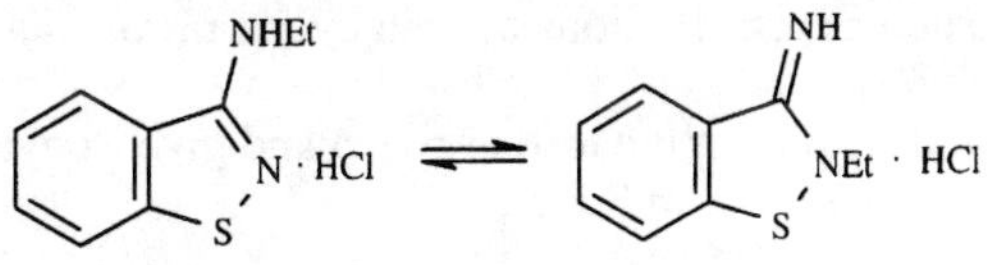

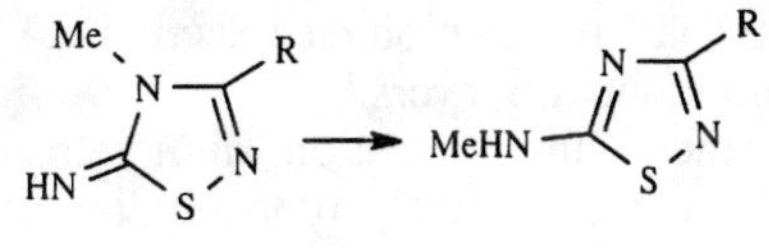

(576)   (577)   (578)   (579)

2-Amino-1,3,4-oxadiazoles (580) ring-open and the products immediately recyclize to triazolinones (581) ⟨66AHC(7)183⟩.

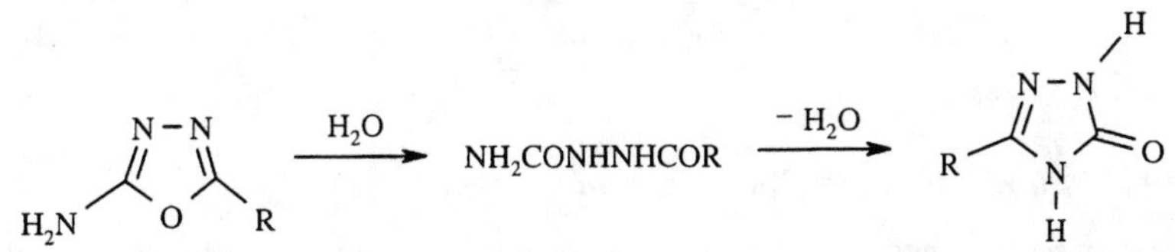

(580)   (581)

3.4.3.5.2　Reactions with electrophiles (except nitrous acid)

In aminoazoles with the amino group α or γ to C=N, canonical forms of type (582b) increase the reactivity of the pyridine-like nitrogen atom toward electrophilic reagents, but decrease that of the amino group. Even when the amino group is β to C=N there is still a small electron flow in the same sense. Consequently, protons, alkylating agents and metal ions usually react with aminoazoles at the annular nitrogen atom (*cf.* Section 3.4.1.3). There are exceptions to this generalization, *e.g.* 4-aminoisothiazole is methylated to the 4-trimethylammonioisothiazole and both 3- and 4-dimethyl-aminopyrazoles are alkylated at the NMe₂ group.

(582a)　　　　　　　(582b)

Other electrophilic reagents form products of reaction at the amino group. This occurs when initial attack at the pyridine-like nitrogen atom forms an unstable product which either dissociates to regenerate the reactants or undergoes rearrangement inter- or intra-molecularly. In reactions of this type, carboxylic and sulfonic acid chlorides and anhydrides give acylamino- and sulfonamidoazoles, respectively. Thus, 3-, 4- and 5-aminothiazoles form acetyl derivatives, sulfonamides and ureas. The 3- and 5-amino-1,2,4-thiadiazoles ⟨65AHC(5)119⟩ can be acylated and sulfonylated; 3-amino-1,2,5- ⟨68AHC(9)107⟩ and 2-amino-1,3,4-thiadiazoles ⟨68AHC(9)165⟩ also behave normally on acylation.

3-Amino-2,1-benzisothiazole (583) is acylated both at the cyclic and exocyclic nitrogen atoms to give (584) ⟨71AJC2405⟩. 5-Aminotetrazoles with nitric acid give nitramines. Sulfonation of the 5-aminopyrazole (585) gives first the expected product, (586), then a disulfonyl derivative (587), which rearranges on heating to the more stable (588). Aminothiazoles react with aldehydes to give Schiff bases.

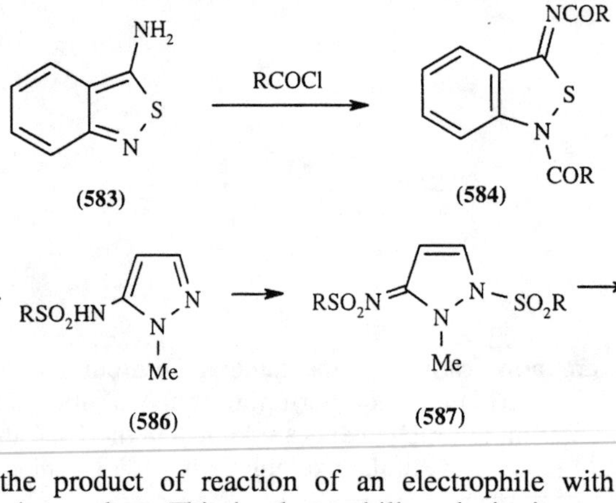

(583)　　　　　　　　　　　　　　(584)

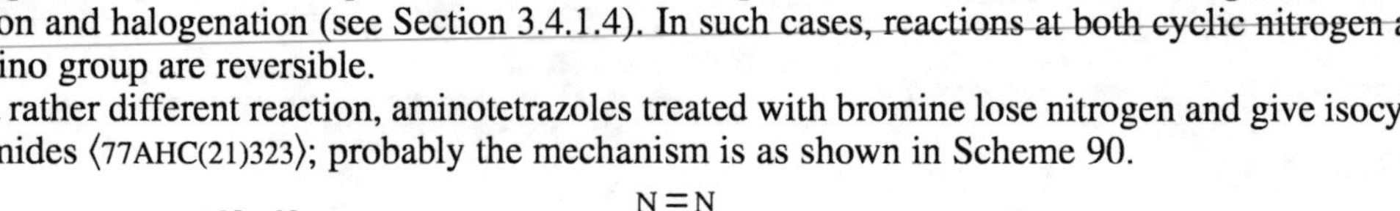

(585)　　　　　(586)　　　　　(587)　　　　　(588)

In still other cases, the product of reaction of an electrophile with an aminoazole is from electrophilic attack at a ring carbon. This is electrophilic substitution and is the general result of nitration and halogenation (see Section 3.4.1.4). In such cases, reactions at both cyclic nitrogen and at an amino group are reversible.

In a rather different reaction, aminotetrazoles treated with bromine lose nitrogen and give isocyanide dibromides ⟨77AHC(21)323⟩; probably the mechanism is as shown in Scheme 90.

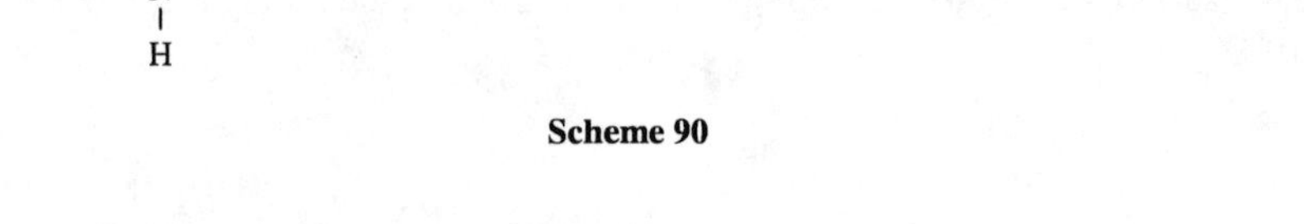

Scheme 90

3.4.3.5.3　Reaction with nitrous acid. Diazotization

Primary amino groups attached to azole rings react normally with nitrous acid to give diazonium compounds *via* primary nitroso compounds. However, the azole series shows two special characteristics: the primary nitroso compounds can be stable enough to be isolated, and diazo anhydrides are formed easily from azoles containing ring NH groups.

(i) Primary nitroso compounds

Attempted diazotization in dilute acid sometimes yields primary nitroso compounds. Reactions of 3- and 5-amino-1,2,4-thiadiazoles with sodium nitrite and acid give primary nitrosamines (*e.g.* **589** → **590**) ⟨65AHC(5)119⟩ which can be related to the secondary nitrosamines (**591**) prepared in the normal way. 1-Substituted 5-aminotetrazoles with nitrous acid give stable primary nitrosamines (**592**). Primary nitrosamines have been isolated in the imidazole series.

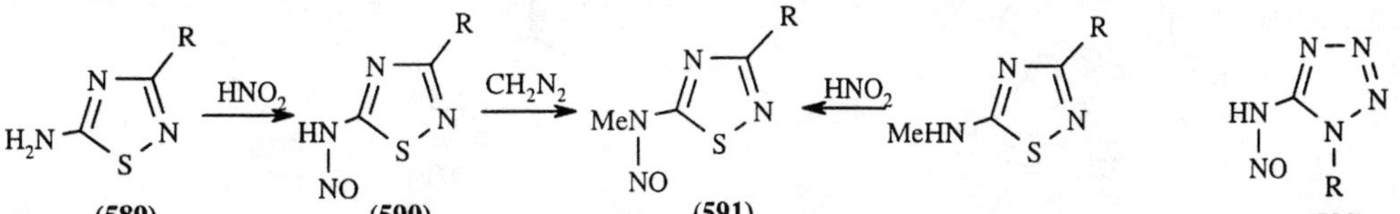

(ii) Diazo anhydrides

Diazotization of aminoazoles with free cyclic NH groups can give diazo anhydrides which show many of the normal reactions of diazoniums ⟨67AHC(8)1⟩. In the pyrazole series these diazo anhydrides (*e.g.* **593**) are particularly stable.

3-Diazopyrazole (**593**) undergoes gas-phase thermal extrusion to form an azirine, probably by the mechanism shown ⟨81AHC(28)231⟩; 4-diazopyrazoles show normal diazonium-type reactions (Schemes 91 and 92) ⟨67AHC(8)1⟩. Analogous diazoimidazoles and diazopurines are known ⟨67AHC(8)1⟩.

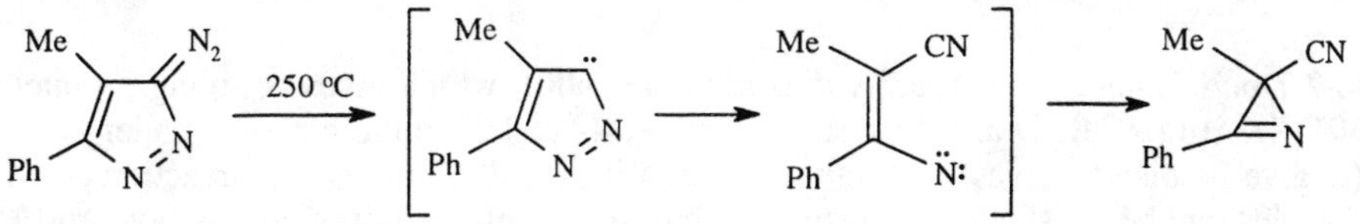

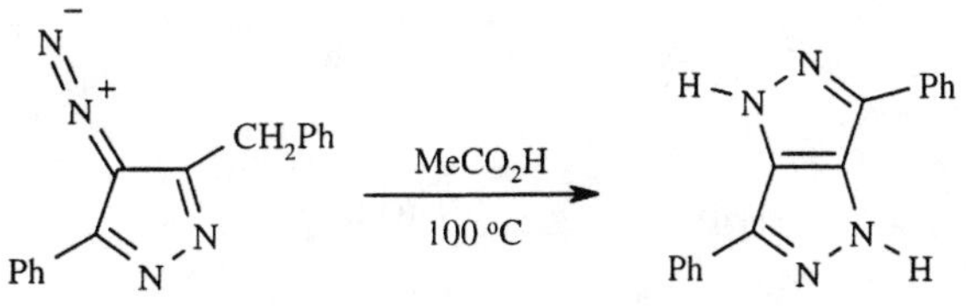

Scheme 91

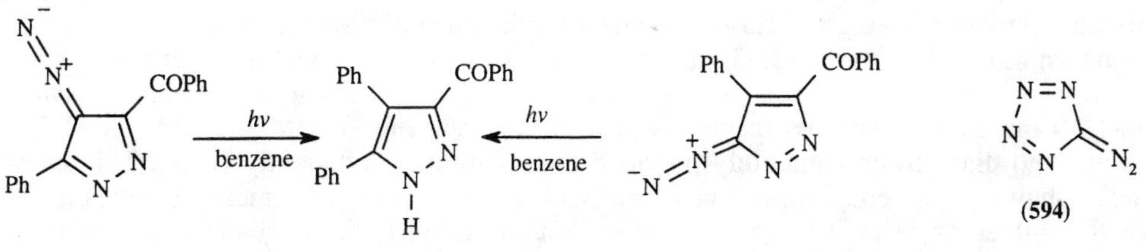

Scheme 92

Diazotetrazole (**594**) has been prepared; on pyrolysis it yields carbon atoms and nitrogen ⟨79JA1303⟩.

(iii) Diazonium salts

Pyridine-2- and -4-diazonium ions are far less stable than benzenediazonium cations. Azole-diazonium salts generally show intermediate stability; provided diazotization is carried out in concentrated acid, many of the usual diazonium reactions succeed. Indeed, azolediazonium salts are often very reactive in coupling reactions.

2-Nitroimidazoles and 2-azidoimidazoles are available *via* the diazonium fluoroborates, and photolytic decomposition of the fluoroborates gives 2-fluoroimidazoles ⟨80AHC(27)241⟩. Conversions of diazonium salts into halogenoimidazoles are quite common ⟨88CPB2730, 93AHC(57)291⟩, with applications to the synthesis of fluoroimidazoles of particular interest.

1-Alkyl-2-aminobenzimidazoles are diazotated only in nitrosylsulfuric acid. Under these extremely acidic conditions, the diazonium salts thus formed exist as highly electrophilic dications. When positions 5 and 6 in such salt (*e.g.* **595**) are occupied, it undergoes azocoupling even with benzene, toluene and xylenes producing azocompounds of type (**597**). If positions 5 or 6 are free, as in salt (**596**), it undergoes self-azocoupling with a molecule of the starting amine to afford a mixture of 5- and 6-azocompounds (**598**). In *N*-arylbenzimidazole-2-diazonium salts, intramolecular azocoupling leading to 1,2,4-benzotriazine derivatives (**599**) is also possible ⟨79KGS867⟩.

(**595**) R = Me
(**596**) R = H

(**597**)

(**598**)

(**599**)

3-Amino-2,1-benzisothiazole is readily diazotized to (**600**), which gives coupling products and the cyanide (**601**) ⟨72AHC(14)43⟩. Diazonium salts from 3-, 4- and 5-aminothiazoles undergo Sandmeyer reactions (to give haloisothiazoles), reductive deaminations and Gomberg-Hey reactions ⟨72AHC(14)1⟩. 5-Aminooxazoles can be satisfactorily diazotized, but the 2-amino compounds cannot ⟨74AHC(17)99⟩.

(**600**)   (**601**)   (**602**)   (**603**)

The 4- and 5-amino-1,2,3-triazoles are diazotizable, *e.g.* the diazonium salt from 4-aminotriazole-5-carboxamide with potassium iodide gives the 4-iodo derivative, and that from 4-amino-1,5-diphenyltriazole gives 1,5-diphenyltriazole in ethanol ⟨74AHC(16)33⟩.

In strong acid the 1,2,4-thiadiazole-3- and -5-diazonium salts have been prepared; the 5-derivatives are very reactive in coupling reactions and undergo Sandmeyer reactions. Diazonium salts from 3-amino-1,2,4-thiadiazoles are less reactive with coupling reagents ⟨65AHC(5)119⟩. Amino-1,3,4-thiadiazoles undergo diazotization smoothly provided the solution is sufficiently acidic. The diazonium salts (**602**) show strong coupling activity and will even couple with mesitylene ⟨68AHC(9)165⟩. 3-Amino-1,2,5-thiadiazole on attempted diazotization forms only the diazoamino compound (**603**) ⟨68AHC(9)107⟩.

3.4.3.5.4 *Deprotonation of aminoazoles*

Canonical forms of type (**582b**) facilitate proton loss from the amino groups; the anions formed react easily with electrophilic reagents, usually preferentially at the exocyclic nitrogen atom. Mono-*N*-anions of aminoazoles are easily formed when the latter are treated with metallic sodium or $NaNH_2$ (KNH_2) in liquid ammonia. They can be further alkylated to afford dialkylamino derivatives in good yield, this reaction being of preparative significance ⟨72CI(L)256⟩ (*cf.* Section 3.2.3.5.4.i). Di-*N*-anions are also obtained when *n*-butyllithium is used as a base. However, the best method to generate dianions appears to be reductive cleavage of hetarylazides with sodium in liquid ammonia ⟨72CI(L)256⟩. There are review articles on the NH-acidity of an amino group, including those in hetarylamines ⟨73RCR65, 83RCR1974⟩.

3.4.3.5.5 Aminoazolium ions/neutral imines

Amino groups on azolium rings can lose a proton to form strongly basic azolinimines, *e.g.* (**604**) yields (**605**). 2-Iminobenzothiazoline with acrylic acid yields (**606**).

(**604**)   (**605**)   (**606**)   (**607**)

In the 1,2-dithiole series such imines, *e.g.* (**607**), are readily isolated; they can be alkylated or protonated ⟨66AHC(7)39⟩.

3.4.3.5.6 Oxidation of aminoazoles

2-Aminobenzimidazoles and 2-aminobenzothiazoles can be oxidizied with sodium hypochlorite to afford the corresponding 2,2'-azocompounds. A rather unusual autooxidation reaction of an heteroaromatic *C*-amino group has been observed on treatment of 1-benzyl-2-aminobenzimidazole (**608**) by excess sodium or potassium in liquid ammonia ⟨67TL2219⟩. The products were 2,2'-azobenzimidazole (**611**) and 2-nitrobenzimidazole (**612**), formed in 60 and 40% yield, respectively. It is supposed that di- (**609**) and tri- (**610**) *N*-anions are generated which are extremely reactive towards oxygen of the air. This is the only known example of such easy autooxidation of an amino group to a nitro group. Modifications of the reaction has been studied ⟨73RCR65⟩.

(**608**)   (**609**)   (**610**)

(**611**)   (**612**)

3.4.3.6 Other *N*-Linked Substituents

3.4.3.6.1 Nitro groups

Nitro groups on azole rings are often smoothly displaced by nucleophiles even more readily than are halogen atoms in the corresponding position. Thus, 2,4,5-trinitroimidazole (**613**) is converted by HCl successively into (**614**) and (**615**) ⟨80AHC(27)241⟩.

(**613**)   (**614**)   (**615**)

Nitro groups are easily reduced, catalytically or chemically, to give amino compounds, *e.g.* 4-nitroisothiazoles give the corresponding 4-amino derivatives ⟨72AHC(14)1⟩. In the pyrazole series, intermediate nitroso compounds can be isolated. Nitrosoimidazoles are also relatively stable.

3.4.3.6.2 Azidoazoles

The most important chemistry of azidoazoles is the fragmentation of derived nitrenes of which the prototypes are (**616** → **617**) and (**618** → **619**). Thus, 5-azido-1,4-diphenyltriazole (**620**) evolves nitrogen at 50°C ⟨70JOC2215⟩. 4-Azido-pyrazoles and-1,2,3-triazoles (**621**) undergo fragmentation with formation of unsaturated nitriles ⟨81AHC(28)231⟩; *cf.* corresponding carbene reactions (Section 3.4.3.4.6).

| (616) | (617) | (618) | (619) |

| (620) | (621) X = N or CR |

Heating 4-azidothiazoles having an imino or aldehyde group at the 5-position affords fused pyrazolo- and isoxazolo[3,4-*d*]thiazoles (Equation (7)). When an alkenyl group is in 5-position of the triazole ring, 4*H*-pyrrolo[2,3-*d*]thiazoles are obtained ⟨92JCS(P1)973⟩.

$$X = O, NR^2 \tag{7}$$

On azido-tetrazole tautomerism in the azole series see Section 2.4.5.3.2.

3.4.3.7 *O*-Linked Substituents

3.4.3.7.1 Tautomeric forms: interconversion and modes of reaction

As discussed in Section 2.4.5.2, hydroxy derivatives of azoles (*e.g.* **622**, **624**, **626**) are tautomeric with either or both of (i) aromatic carbonyl forms (*e.g.* **623**, **627**) (as in pyridones), and (ii) alternative non-aromatic carbonyl forms (*e.g.* **625**, **628**). In the hydroxy 'enolic' form (*e.g.* **622**, **624**, **626**) the reactivity of these compounds toward electrophilic reagents is greater than that of the parent heterocycles; these are analogues of phenol.

| (622) | (623) | (624) | (625) |

| (626) | (627) | (628) |

Interconversion of the hydroxy and carbonyl forms of these heterocycles proceeds through an anion (as **630**) or a cation (as **631**), just as the enol (**633**) and keto forms (**636**) of acetone are interconverted through the ions (**634**) or (**635**). Reactions of the various species derived from the heterocyclic compounds are analogous to those of the corresponding species from acetone: hydroxy forms react with electrophilic reagents (**637**) and carbonyl forms with nucleophilic reagents (**638**). In addition, either form can lose a proton (**639, 640**) to give an anion which reacts very readily with electrophilic reagents on either oxygen (**641**) or carbon (**642**).

The completely conjugated carbonyl forms are usually quite stable and highly aromatic in that after reaction they revert to type. An overall treatment of their reactivity is given in Section 3.4.1.1.4. Electrophilic attack on the oxygen atom of the carbonyl groups, and nucleophilic attack at the carbonyl carbon atom, in reactions which lead to substitution rather than ring opening are discussed in this section. Electrophilic attack at ring carbon (Section 3.4.1.4) and ring nitrogen (Section 3.4.1.3) and nucleophilic attack at ring carbon (Section 3.4.1.6) (other than C=O replacement) are discussed in the sections indicated.

3.4.3.7.2 2-Hydroxyazoles, heteroatoms-1,3

2-Hydroxy-imidazoles, -oxazoles and -thiazoles (**643**; Z = NR, O, S) can isomerize to 2-azolinones (**644a**). These compounds all exist predominantly in the azolinone form and show many reactions similar to those of the pyridones. They are mesomeric with zwitterionic and carbonyl canonical forms (*e.g.* **644a** ↔ **644b**; Z = NR, O, S).

(i) Electrophilic attack on oxygen

2-Azolinones are protonated on oxygen in strongly acidic media. *O*-Alkylation of 2-azolinones can be effected with diazomethane; thiazolinone (**645**) forms (**646**). Frequently *O*- and *N*-alkylation occur together, especially in basic media where proton loss gives an ambident anion (see also Section 3.4.1.3.10).

(ii) Nucleophilic displacement

2-Imidazolinones, 2-oxazolinones and 2-thiazolinones behave as cyclic ureas, thiocarbamates and carbamates, and predictably do not normally react with nucleophilic 'ketonic reagents' such as HCN, RNH_2, $NaHSO_3$, NH_2OH, N_2H_4, PhN_2H_3, or $NH_2CON_2H_3$. Stronger nucleophilic reagents, *i.e.* those of the type that attack amides, generally also react with azolinones. Thus, they can be converted into chloroazoles with $POCl_3$, or PCl_5, *e.g.* (**648**)→(**647**). Similarly, bromoazoles may be prepared using PBr_5. Alkyl substituents on the azole nitrogen atom are usually lost in reactions of this type. Phosphorus pentasulfide converts carbonyl groups into thiocarbonyl groups (*e.g.* **648** → **649**).

(647)   (648)   (649)

3.4.3.7.3   3-Hydroxyazoles, heteroatoms-1,2

Pyrazoles, isoxazoles and isothiazoles with a hydroxy group in the 3-position (**650**; Z = NR, O, S) could isomerize to 3-azolinones (**651**). However, these compounds behave as true hydroxy derivatives and show phenolic properties. They give an intense violet color with iron(III) chloride and form a salt (**652**) with sodium hydroxide which can be *O*-alkylated by alkyl halides (to give **653**; R = alkyl) and acylated by acid chlorides (to give **653**; R = acyl).

(650)   (651)   (652)   (653)

Sometimes compounds which exist predominantly in the hydroxy form give products of *N*-methylation with diazomethane, for example 3-hydroxy-5-phenylisothiazole ⟨63AHC(2)245⟩; of course, the ambident anion (**652**) is an intermediate. 3-Hydroxypyrazoles, under rather severe conditions, can be converted into 3-chloropyrazoles with $POCl_3$ ⟨66AHC(6)347⟩.

3.4.3.7.4   5-Hydroxyazoles, heteroatoms-1,2

5-Hydroxy-isoxazoles and -pyrazoles can tautomerize in both of the ways discussed in Sections 3.4.3.7.3 and 3.4.3.7.5 (**654** ⇌ **655** ⇌ **656**). The hydroxy form is generally the least stable; the alternative azolinone forms coexist in proportions depending on the substituents and the solvent, with non-polar media favoring the CH form (**656**) and polar media the NH form (**655**). The derived ambident anion can react with electrophiles at N, C or O depending on the reagent and conditions.

(654)   (655)   (656)

The hydroxy groups of 5-hydroxypyrazoles are readily replaced by halogens by the action of phosphorus halides.

3.4.3.7.5   4- and 5-Hydroxyazoles, heteroatoms-1,3 and 4-hydroxyazoles, heteroatoms-1,2

The 4- and 5-hydroxy-imidazoles, -oxazoles and -thiazoles (**657**, **659**) and 4-hydroxy-pyrazoles, -isoxazoles and -isothiazoles (**661**) cannot tautomerize to an aromatic carbonyl form. However, tautomerism similar to that which occurs in hydroxy-furans, -thiophenes and -pyrroles is possible (**657** ⇌ **658**; **661** ⇌ **662**; **659** ⇌ **660**), as well as a zwitterionic NH form (*e.g.* **663**). Most 4- and 5-

hydroxy compounds of types (**658**) and (**660**) exist largely in these non-aromatic azolinone forms, although the hydroxy form can be stabilized by chelation (*e.g.* **664**). The derived ambident anions react with electrophilies at O or C. Replacement of the hydroxy group is sometimes possible provided electron-withdrawing groups are present as, for example, in 5-substituted 4-hydroxypyrazoles.

(657)	(658)	(659)	(660)

(661)	(662)	(663)	(664)

4-Hydroxy derivatives of type (**661**) show more phenolic character; thus 4-hydroxyisothiazoles are normally *O*-methylated and *O*-acylated ⟨72AHC(14)1⟩.

4-Chloro substituted thiazoles can be prepared by the reaction of phosphorus oxychloride with the corresponding 4-hydroxythiazoles. This method is also applicable for the preparation of 4-bromo-2-phenylthiazole.

Ring fission occurs readily in many of these compounds. For example, azlactones, *i.e.* 4*H*-oxazolin-5-ones containing an exocyclic C=C bond at the 4-position (**666**), are hydrolyzed to α-benzamido-α,β-unsaturated acids (**667**), further hydrolysis of which gives α-keto acids (**668**), Reduction and subsequent hydrolysis *in situ* of azlactones is used in the synthesis of α-amino acids (*e.g.* **666→665**).

(665)	(666)	(667)	(668)

3.4.3.7.6 *Hydroxy derivatives with three heteroatoms*

These compounds generally exist in carbonyl forms. The oxygen function can be converted into halogen by phosphorus halides. Reactions with electrophiles are quite complex. Thus, urazole (**669**) reacts with diazomethane quickly to yield (**670**), which is more slowly converted into (**671**). Phenylurazole gives (**672**); however, 4-phenylurazole yields (**673**). Oxadiazolinones of type (**674**) can be alkylated at both *O*- and *N*-atoms.

(669)	(670)	(671)

(672)	(673)	(674)

3.4.3.7.7   Alkoxy and aryloxy groups

The alkoxy groups in alkoxyazoles undergo easy dealkylation to the corresponding hydroxyazoles (azolinones) when several nitrogen atoms are present or when they are additionally activated by another substituent. Thus, pyrazolyl ethers are cleaved under vigorous conditions, or more easily if a nitroso group is present. Nucleophilic displacement of alkoxy groups on cationic rings occurs readily, *e.g.*, in quaternary 1,2,3-triazole ethers.

Azoles with alkoxy groups α to nitrogen can rearrange to *N*-alkylazolinones on heating; thus 2-alkoxy-1-methylimidazoles give 3-alkylimidazolin-2-ones and 2-methoxythiazoles behave similarly. *O*-Allyl groups rearrange considerably more readily, *e.g.*, 2-allyloxybenzimidazole gives 1-allyl-2-benzimidazolinone at 180°C (Equation 8) ⟨67AHC(8)143⟩. 5-Allyloxypyrazoles undergo Claisen rearrangement of the allyl group to the 4-position.

(8)

Aryl tetrazolyl ethers (**675**) are reduced by palladium on charcoal to give the arene and the tetrazolinone (**676**) ⟨77AHC(21)323; 97JCS(P2)669⟩; this reaction is used for the removal of a phenolic functionality.

(675)   (676)

3.4.3.8   S-Linked Substituents

3.4.3.8.1   Mercapto compounds: tautomerism

Many mercaptoazoles exist predominantly as thiones. This behavior is analogous to that of the corresponding hydroxyazoles (*cf.* Section 3.4.3.7). Thus oxazoline-, thiazoline- and imidazoline-2-thiones (**677**) all exist as such, as do compounds of type (**678**). However, again analogously to the corresponding hydroxy derivatives, other mercaptoazoles exist as such. 5-Mercaptothiazoles and 5-mercapto-1,2,3-triazoles (**679**), for example, are true SH compounds.

(677)   (678)   (679)

The pattern of reactivity is similar to that discussed for the azolinones is Sections 3.4.1.1.4 and 3.4.3.7.1. A difference is the greater nucleophilicity of sulfur, and thus most reactions of the ambident anion with electrophiles occur at sulfur.

3.4.3.8.2   Thiones

Many azolinethiones show reactions typical of thioamides; in particular, they react with electrophiles at the sulfur atom.

(i) Alkyl halides give alkylthio derivatives, *e.g.* in the imidazoline-2-thione, thiazoline-2-thione and 1-arylpyrazoline-5-thione series.

(ii) Thiones are oxidized, *e.g.*, by iodine, into disulfides. Thus, 5-mercapto-1,2,3,4-thiatriazole is converted into the disulfide (**680**) ⟨64AHC(3)263⟩; similar behavior is known in the tetrazole series.

(**680**)　　　　　(**681**)　　　　　(**682**)

(iii) Thione groups can often be eliminated by oxidation; probably the sulfinic acid is the intermediate. Sometimes the sulfinic acid can be isolated (*e.g.* **681** → **682**), but more often it spontaneously loses SO_2. In this way, thiazoline-2-thiones give thiazoles, 1,2-dithiole-3-thiones (**683**) are converted into 1,2-dithiolylium salts (**684**) and 1,3-dithiole-2-thiones (**685**) into 1,3-dithiolylium salts (**686**) ⟨66AHC(7)39⟩. In the pyrazole series, (**687**) also loses an *N*-methyl group to yield (**688**).

(**683**)　　　　　(**684**)　　　　　(**685**)　　　　　(**686**)

(**687**)　　　　　(**688**)　　　　　(**689**)　　　　　(**690**)

(iv) However, 5-aryl-1,2,4-dithiazoline-3-thiones are oxidized to the 3-ones (**689**→**690**).

(v) Strong oxidation (*e.g.* by $KMnO_4$) forms a sulfonic acid or betaine as, for example, in the pyrazole (**691** → **692**), imidazole, thiazole and tetrazole series.

(**691**)　　　　　(**692**)　　　　　(**693**)　　　　　(**694**)

(vi) Cycloaddition across the C=S bond can lead to spiro derivatives, *e.g.* (**693**) → (**694**).

(vii) Like carbonyl oxygen in azolinones sulfur in azolinethiones can be nucleophilically displaced by a halogen atom. For example, benzimidazoline-2-thiones react with thionyl chloride producing 2-chlorobenzimidazoles.

(viii) On heating 6-aryl-2-methyl-7*H*[1,2,4]triazolo[4,3-*b*][1,2,4]triazole-3(2*H*)-thiones (**695**) undergoes N to S migration of the methyl group, yielding the rearrangement products (**696**) ⟨85BCJ735⟩.

(**695**)　　　　　(**696**)

(ix) Imidazoline-2-thiones are reduced by potassium metal in THF forming stable carbenes (imidazol-2-ylidenes) (**697**) which react very readily with electrophiles ⟨93S561⟩ (Scheme 93). See also Section 3.4.1.8.5.

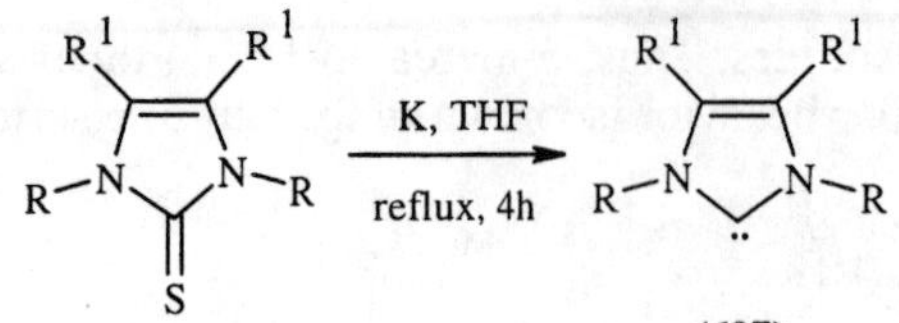

Scheme 93   **Scheme 94**

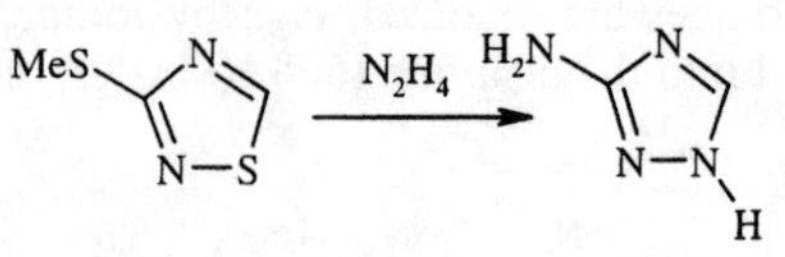

3.4.3.8.3 Alkylthio groups

2-Alkylthiothiazoles rearrange thermally into the 3-alkylthiazoline-2-thiones; in the imidazole series a thermal equilibrium is reached.

Alkylthio groups are oxidized to sulfoxides by H_2O_2 and readily by various oxidizing reagents to sulfones, *e.g.* in the imidazole series. The SR group is replaced by hydrogen with Raney nickel, and dealkylation is possible, *e.g.* of 3-alkylthio-1,2-dithiolyliums to give 1,2-dithiole-3-thiones with various nucleophiles ⟨80AHC(27)15⟩.

Alkylthio groups are sometimes replaced in nucleophilic substitutions, but such reactions are difficult in most neutral azoles. Thus, 3-alkylthio-1,2,4-thiadiazoles resist the action of aniline at 100°C, ammonia at 120°C, molten urea and ammonium acetate. However, hydrazine attacks 3-methylthio-1,2,4-thiadiazole forming 3-amino-1,2,4-triazole (**698**) (Scheme 94) ⟨65AHC(5)119⟩.

Such reactions are easy in cationic derivatives; for example, in the 1,2-dithiolylium series (**699**), substituted cyclopentadienyl ion gives fulvene derivatives (**700**) ⟨66AHC(7)39⟩. 2-Methylthio groups in 1,3-dithiolylium ions are substituted by primary amines or secondary amines ⟨80AHC(27)151⟩, and similar reactions are known for 2-alkylthiothiazoles.

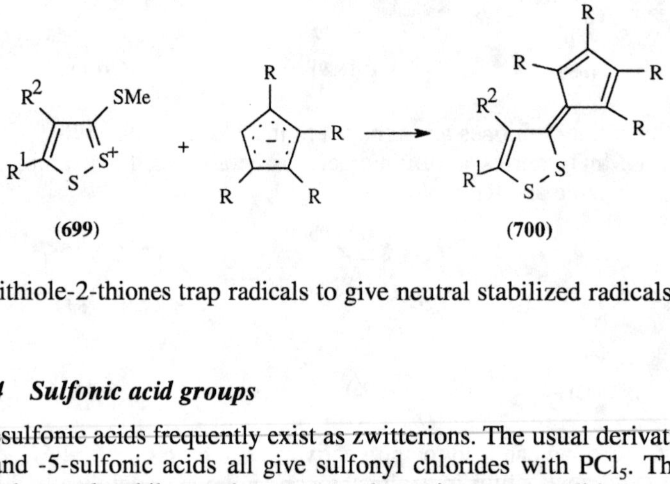

(**699**)   (**700**) 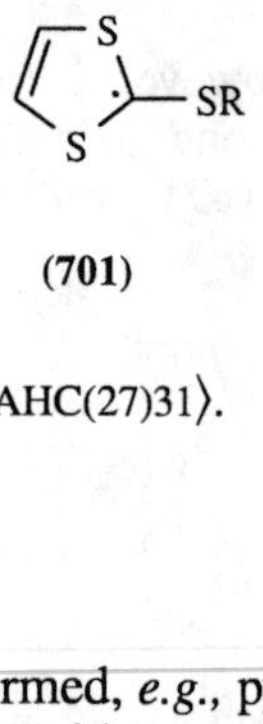   (**701**)

1,3-Dithiole-2-thiones trap radicals to give neutral stabilized radicals (**701**) ⟨80AHC(27)31⟩.

3.4.3.8.4 Sulfonic acid groups

Azolesulfonic acids frequently exist as zwitterions. The usual derivatives are formed, *e.g.,* pyrazole-3-, -4- and -5-sulfonic acids all give sulfonyl chlorides with PCl_5. The sulfonic acid groups can be replaced by nucleophiles under more or less vigorous conditions, *e.g.* by hydroxy in imidazole-4-sulfonic acids at 170°C, and by hydroxy or amino in thiazole-2-sulfonic acids. Benzimidazole-2-sulfonic acids react similarly ⟨88CHE880⟩.

3.4.3.9 Halogen Atoms

3.4.3.9.1 Nucleophilic displacements: neutral azoles

As discussed in Section 3.4.3.1, nucleophilic replacements of halogen atoms are facilitated by mesomeric stabilization in the transition state for some haloazoles, depending on the number and orientation of the ring heteroatoms and halogen. Additional to this, and just as in benzene chemistry, all types of halogen atoms are activated toward nucleophilic displacement by the presence of other electron-withdrawing substituents. Halogen atoms in the 4- and 5-positions of imidazoles, thiazoles and oxazoles and those in all positions of pyrazoles and isoxazoles are normally rather unreactive, but are labilized by an α or γ electron-withdrawing substituent. Reactions of *N*-unsubstituted azoles containing a ring NH group are often difficult because of the formation of unreactive anions under basic conditions.

Halogen atoms in the 2-position of imidazoles, thiazoles and oxazoles can be replaced by the groups NHR, OR, SR, *etc*. The conditions required are more vigorous than those used, for example, for α- and γ-halopyridines, but much less severe than those required for chlorobenzene.

The 4- and 5-haloimidazoles and 4- and 5-halooxazoles are less reactive toward nucleophilic substitution than the 2-halo analogues, but still distinctly more reactive than unactivated phenyl halides. Thus, a bromine atom in the 4- and 5-position of 1-methylimidazole requires lithium piperidide to react, whereas the 2-bromo analogue is converted into 2-piperidinoimidazole by piperidine at 200°C. Sodium isopropoxide replaces the 2-bromo atom of 1-substituted 2,4,5-tribromoimidazole by an isopropoxy group ⟨87JCS(P1)1437⟩. Halogenoimidazoles with a nitro group in the ring are readily subject to halide displacement by thiolate anions with 5-halogeno-4-nitro compounds being three to ten times more reactive than the 4-halogeno-5-nitroimidazoles. Bromide is substituted slightly more readily than iodide, perhaps because of the greater capacity of bromine to help to stabilize a Meisenheimer complex, or there could be steric reasons ⟨82CHE616, 84JHC155, 86JHC1087, 87AJC1415, 87JCS(P1)1437, 91JHC577⟩. Both 2,4,5-tribromo- and 2,4,5-triiodoimidazoles undergo reductive dehalogenation by phenylthiolate ion to the 4,5-dihalogenated products. 5-Halothiazoles react unexpectedly rapidly with methoxide, the 4-halothiazoles less readily.

3-Chloro-5-arylisoxazoles undergo nucleophilic displacement with alkoxide ion. Halogen atoms in the 5-position of the isoxazole nucleus are readily displaced if an activating group is present in the 4-position ⟨63AHC(2)365⟩.

5-Halogens attached to the isothiazole nucleus are more reactive, particularly if there is an electron-withdrawing substituent in the 4-position. However, a 3-halogen atom, even when activated, is less reactive than a halogen in the 5-position, and replacement is often accompanied by ring cleavage, *e.g.* Scheme 95 ⟨72AHC(14)1⟩. 4-Haloisothiazoles are still less reactive, but can react with copper(I) cyanide to give the corresponding nitrile ⟨65AHC(4)107⟩.

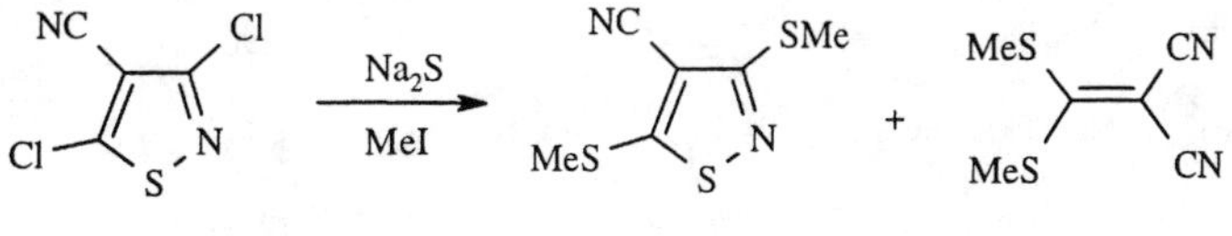

Scheme 95

Halogens attached to the pyrazole nucleus are normally very inert. If there is an electron-withdrawing group in the 4-position, then the halogen atom in the 5-position of a pyrazole ring becomes activated, as for example in Scheme 96 ⟨66AHC(6)347⟩. However, such an electron-withdrawing group in the 4-position only activates the chlorine atom in the 5-position and not one in the 3-position because of the influence of partial bond fixation ⟨66AHC(6)347⟩ (see discussion in Section 3.4.3.1).

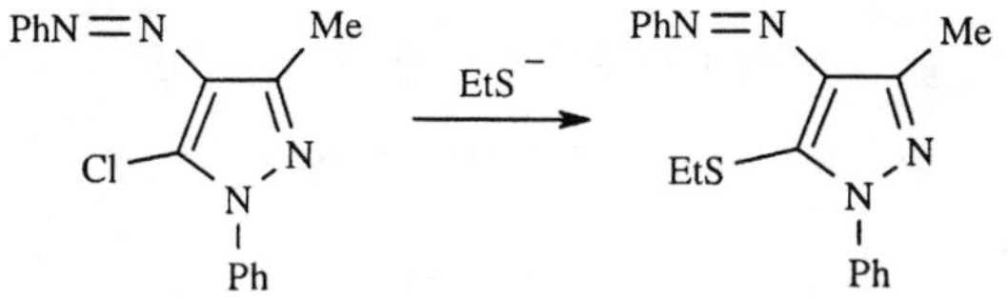

Scheme 96

5-Halo-1-methyl-1,2,3-triazoles undergo substitution reactions with amines, but the 4-halo analogues do not. 5-Chloro-1,4,-diphenyl-1,2,3-triazole with sodium cyanide in DMSO gives the cyano derivative ⟨63JCS2032⟩. 1-Substituted 3-chloro- and 5-chloro-1,2,4-triazoles both react with amines.

5-Chlorine atoms in 1,2,4-oxadiazoles can be replaced by amino, hydroxy or alkoxy groups ⟨76AHC(20)65⟩. 5-Halo-1,2,4-thiadiazoles are also quite reactive: silver fluoride gives the fluorides, in concentrated hydrochloric acid a 5-hydroxy group is introduced, and thiourea reacts, as do various amines. Sodium sulfite gives sulfonic acids, and reactive methylene compounds give the expected substitution products ⟨65AHC(5)119⟩. By contrast, halogens in the 3-position of 1,2,4-thiadiazoles are inert toward most nucleophilic reagents: thus 3-chloro-5-phenyl-1,2,4-thiadiazole resists aminolysis and thiourea; however, a 3-alkoxy group is introduced by sodium alkoxide ⟨65AHC(5)119⟩.

Halogens on the 1,2,5-thiadiazole ring are highly reactive and easily converted into ethers by refluxing with alkoxides ⟨68AHC(9)107⟩. 2-Chloro-1,3,4-thiadiazole and benzylamine give a mixture of (**702**) and (**703**) ⟨68AHC(9)165⟩, the latter resulting from a Dimroth rearrangement (see Section 3.4.3.5.1). With hydrazine, (**704**) is similarly formed ⟨68AHC(9)165⟩.

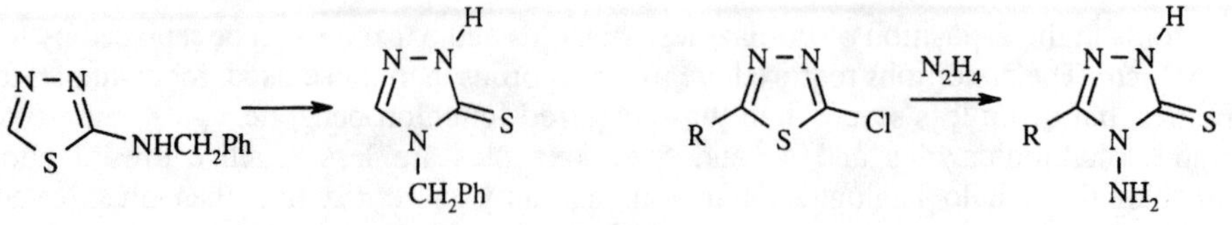

(702)   (703)   (704)

Halogen atoms at the 5-position of tetrazoles are reactive and easily replaced by nucleophiles. 5-Bromo-1-methyltetrazole is significantly more reactive than the 2-methyl isomer ⟨77AHC(21)323⟩.

3.4.3.9.2   Nucleophilic displacements: haloazoliums

Halogen atoms in cationic olium rings are very reactive. The halogen atom in the quaternary salts of 3- and 5-halo-1-phenylpyrazoles is replaced at 80–100°C by hydroxy, alkoxy, thiol, amino or cyano groups ⟨66AHC(6)347⟩. 3-Halo-1,2-dithiolyliums are converted into 1,2-dithiol-3-ones by water and react readily with other nucleophiles ⟨80AHC(27)151⟩.

Quaternization of triazole nitrogen atoms activates towards nucleophilic attack. Thus, the synthetic strategy shown in Scheme 97 can be applied to the preparation of 1-benzyl-4-hydroxy-1,2,3-triazole (**706**) using *p*-methoxybenzyl as a removable activating group. Generally, this synthetic sequence can be used for the preparation of 1-alkyl-4-substituted triazoles by reaction of the intermediate (**705**) with various nucleophiles, followed by removal of the activating group ⟨88BSB573⟩.

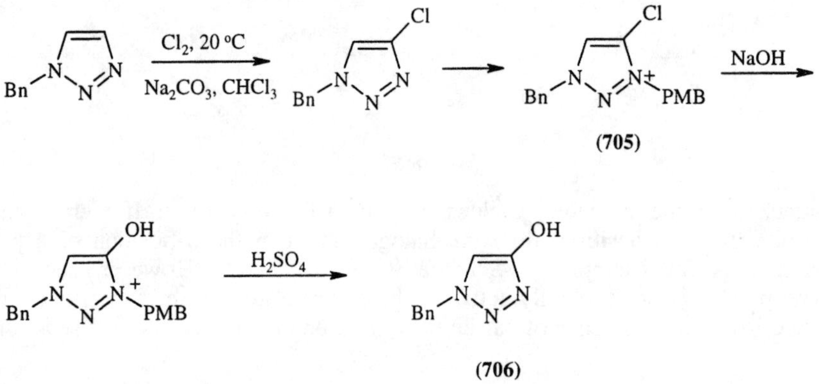

(705)

(706)

Scheme 97   PMB = 4-methoxybenzyl

3.4.3.9.3   Other reactions

Nuclear halogen atoms also show many of the reactions typical of aryl halogens.

(i) They can be replaced with hydrogen atoms by catalytic (Pd, Ni, *etc.*) or chemical reduction (HI or Zn/H_2SO_4). For example, halopyrazoles with HI and red phosphorus at 150°C give pyrazoles ⟨66AHC(6)347⟩, and 5-bromo-1,2,4-thiadiazole is reduced by Raney nickel to the parent heterocycle. 2-Bromothiazole can be reduced electrochemically.

(ii) They give Grignard reagents; however, in the preparation of these it is sometimes necessary to add ethyl bromide to activate the magnesium ('entrainment method'). Pyrazolyl Grignard reagents have been obtained by the entrainment reaction ⟨66AHC(6)347⟩. 4-Iodoisoxazoles give Grignard reagent ⟨63AHC(2)365⟩. The 4- and 5-halooxazoles undergo halogen-metal exchange with *n*-butyllithium to give 4- and 5-lithiooxazoles ⟨74AHC(17)99⟩.

Bromine and iodine groups at any position in *N*-substituted imidazoles undergo lithium exchange. 1-Substituted 2,4,5-tribromoimidazoles exchange in excellent yields at low temperatures in the order 2 > 5 > 4, making selective exchange possible ⟨86TL1635, 87JCS(P1)1445, 87JCS(P1)1453, 91JOC4296, 92JOC3776, 92TL5865⟩. Double bromine-lithium exchange in 1-methyl-2,5-dibromoimidazole gives the 2,5-dianion which selectively forms the aldehyde at C-5 when treated with DMF ⟨85JHC57⟩.

Halothiazoles give Grignard reagents and lithio derivatives. 2- and 5-Bromothiazoles give halogen-lithium exchange on treatment with *n*-butyllithium to afford 2- and 5-lithiothiazoles, respectively. The reaction with 4-bromothiazole under the same conditions leads exclusively to deprotonation at C-2 ⟨88JOC1748⟩, although a number of substituted 4-bromothiazoles react with *n*-butyllithium to give the corresponding 4-lithio derivatives ⟨87S998⟩. The low reactivity of a bromine atom at C-4 is attributed to the effect of the adjacent nitrogen lone pair which destabilizes a developing negative charge at C-4 ⟨92JCS(P1)215⟩.

(iii) Terminal alkynes and alkenes react with 4-bromo and 5-bromooxazoles in the presence of a palladium catalyst to give coupled products in fair to good yields (Scheme 98) ⟨87CPB823⟩.

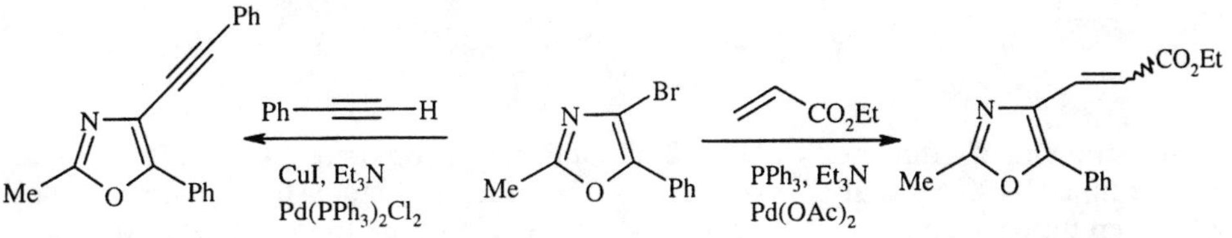

Scheme 98

3.4.3.10 Metals and Metalloid-linked Substituents

Metallated azoles frequently show expected properties, especially if not too many heteroatoms are present. Thus, Grignard reagents prepared from halogen-azoles (see Section 3.4.3.9.3) show normal reactions, as in Scheme 99. 2-Lithioimidazoles react normally, *e.g.* with benzophenone (Scheme 100) ⟨70AHC(12)103⟩. Lithiated imidazoles are not always particularly reactive towards electrophiles and yields may be low ⟨83JA2382, 86TL4095, 88JOC1107, 92JHC1091⟩. The nature of the quenching electrophile is the critical factor. Hard reagents like benzophenone tend to give better yields than the softer methyl iodide; heteroaryl lithiums have hard carbanion centers.

2-Lithioimidazoles are thermodynamically (and possible kinetically) more stable than the 5- or 4-isomers ⟨90JHC673⟩. The adjacent lone pair effect makes 4-lithioimidazoles difficult to prepare, though they are accessible through metal-halogen exchange under certain conditions ⟨81CC1095, 83JCS(P1)279, 83JCS(P1)735, 85JHC57⟩. Before 5-lithioimidazoles can be prepared the 2-position needs to be blocked ⟨81CC1095⟩, but when a 2,5-dilithioimidazole is made it is possible to quench the more reactive 5-anion before the 2-anion ⟨84JCS(P1)481⟩. If the 2-position is unsubstituted a 4- or 5-lithioimidazole (made by metal-halogen exchange) rapidly isomerizes to the 2-isomer. Only very reactive electrophiles are able to functionalize a 5-lithioimidazole when the 2-position is unsubstituted because the reaction must take place at a sufficiently low temperature to prevent transmetallation ⟨91JOC4296⟩.

5-Lithioisothiazoles ⟨72AHC(14)1⟩ and 2-lithiothiazoles undergo many of the expected reactions, including carbonization (with CO_2), formylation (with DMF), alkylation (with alkyl halides), acylation (with acid anhydrides), halogenation (with Br_2 or I_2), *etc.*

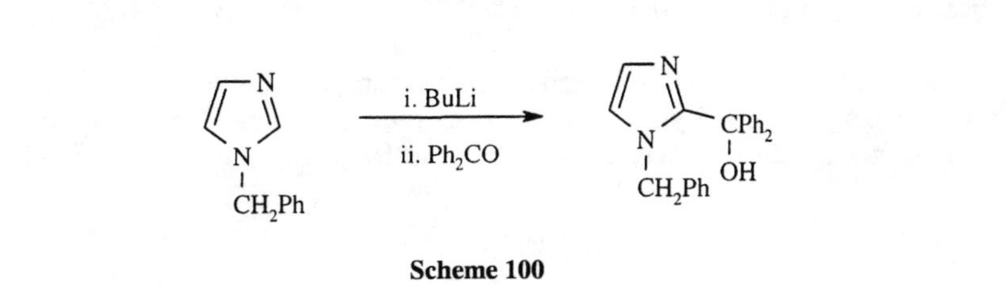

Scheme 99

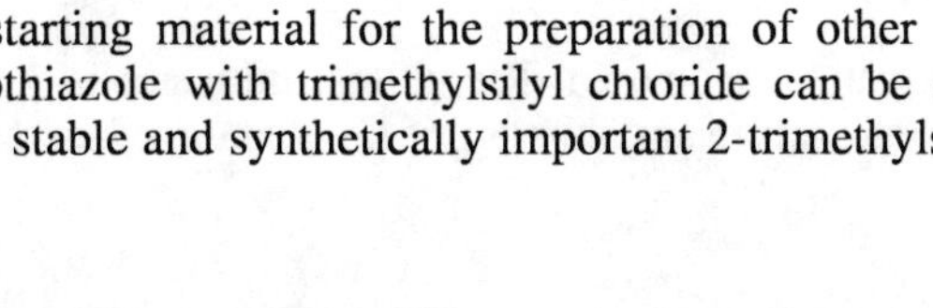

Scheme 100

Lithiothiazoles are the starting material for the preparation of other metallated thiazoles. Thus, transmetallation of 2-lithiothiazole with trimethylsilyl chloride can be carried out in one-pot and multigram scale to give the stable and synthetically important 2-trimethylsilylthiazole (**707**) ⟨84CC655,

95OS(72)21⟩. In the same way, 2- and 4-trimethylstannylthiazoles (**708**) and (**709**) were obtained from the corresponding lithio derivatives and trimethyltin chloride ⟨86S757⟩. 4-Methyl-2-trimethylstannyloxazole reacts with a variety of aryl and heteroaryl halides and tetrakis(triphenylphosphine)palladium to give the 2-substituted products in high yield ⟨87S693⟩.

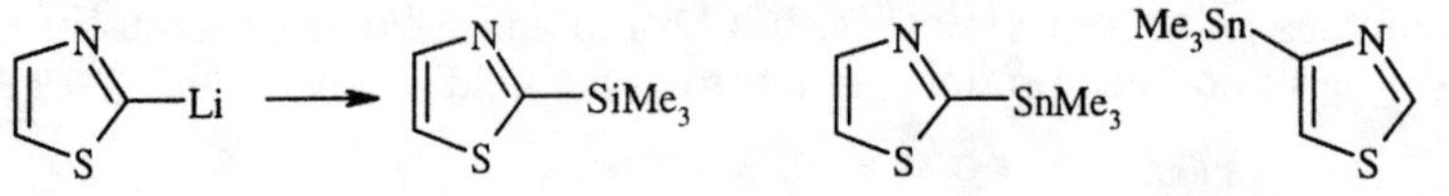

(**707**)   (**708**)   (**709**)

The crystal structure of a dimer of 4-*t*-butyl-2-lithiothiazole incorporating two molecules of diglyme has been determined by X-ray analysis ⟨95AG(E)487⟩. The structure features the lithium atom positioned halfway between nitrogen and C-2, thus providing a carbenoid nature to this metallated thiazole.

Metallation reactions at C-4 of sydnones have been reviewed ⟨95H(41)1525⟩. A variety of electrophiles can be introduced into the 4-position of 3-arylsydnones by way of 4-lithio intermediates ⟨84CHEC(6)365⟩. The method has been used to introduce several sulfur, selenium, and tellurium electrophiles ⟨86BCJ483, 86BCJ487⟩ and to introduce formyl and acetyl substituents. Other organometallic species have been used to introduce vinyl and aryl substituents at C-4, 4-lithio-3-phenylsydnone reacted with copper(I) bromide to give a thermally stable copper species (**710**). This underwent palladium(0) catalyzed coupling to iodobenzenes and to vinyl bromides (Scheme 101) ⟨88JOM(352)C34⟩.

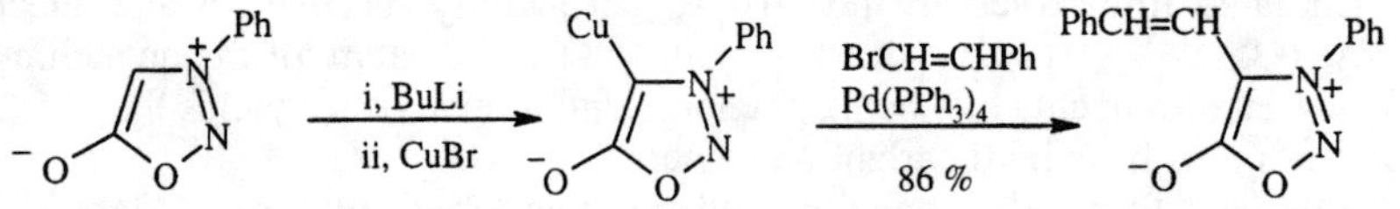

(**710**)

Scheme 101

As the number of nitrogen heteroatoms increases, the stability of lithium azoles decreases: the 5-lithio derivatives of 1,2,3-triazoles (**711**) ring-open spontaneously ⟨74AHC(16)33⟩. 1-Methyltetrazol-5-yllithium decomposes to nitrogen and lithium methylcyanamide above −50°C, although it gives the expected Grignard-like reactions with bromocyanogen, esters, ketones and sulfur at lower temperatures ⟨77AHC(21)323⟩.

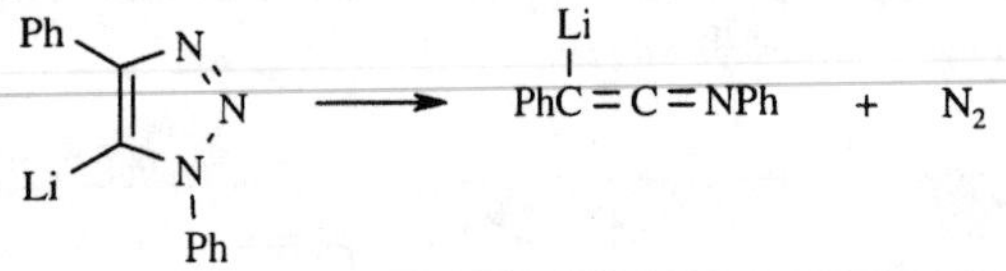

(**711**)

The properties of the lithium derivatives of azoles has been reviewed ⟨85H(23)417, 88CHE117, 94H(33)2487⟩.

Acetoxymercurioxazoles ⟨74AHC(17)99⟩ and acetoxymercurithiazoles with halogens give the corresponding halooxazoles in good yield. 4-Acetoxymercuripyrazoles show many of the reactions of phenylmercury(II) acetate: removal by HCl, conversion to Br by bromine, and to SCH$_2$Ph by (SCN)$_2$/PhCH$_2$Cl.

3.4.3.11   Fused Heterocyclic Rings

A wide variety of such derivatives is known. The mutual influence of heterocyclic rings in them is as expected from their electronic nature and number of heteroatoms. However, some unique reactions arise from the special juxtaposition of the two rings, *e.g.* tetrazolopyridines (**712**) on photolysis yield cyanopyrroles (**713**) ⟨81AHC(28)231⟩. See also Section 3.2.3.6.4.

3.4.3.12 Substituents Attached to Ring Nitrogen Atoms

3.4.3.12.1 N-*Linked azole as a substituent*

It is instructive to consider *N*-substituted azoles in reverse, *i.e.* the azole ring as the substituent linked to some other group. Hammett and Taft σ-constant values for azoles as substituents are given in Table 12. The values show that all *N*-azolyl groups are rather weak net resonance donors, imidazolyl-1 being the strongest. They are all rather strong inductive acceptors, with pyrazole considerably weaker in this respect than imidazole or the triazoles.

Table 12 Hammett and Taft σ-Constant values for Azoles
as Substituents in a Benzene Ring[a]

X	*Azole* Y	Z	σ_I	σ_R°
N	—	—	0.30	–0.06
—	N	—	0.51	–0.15
N	N	—	0.53	–0.10
N	—	N	0.53	–0.12
N	N	—	0.66	–0.10

[a] ⟨81JCR(S)364⟩

N-Linked azole rings behave as good leaving groups, the more so the more nitrogen atoms contained in the ring *(cf.* Section 3.4.3.12.4).

3.4.3.12.2 *Aryl groups*

Electrophilic substitution occurs readily in *N*-phenyl groups, *e.g.*, 1-phenyl-pyrazoles, -imidazoles and -pyrazolinones are all nitrated and halogenated at the *para* position. The aryl group is attacked preferentially when the reactions are carried out in strongly acidic media where the azole ring is protonated.

The azole ring can activate metallation at the *ortho* position of an *N*-phenyl group, as in 1-phenylpyrazoles. If the *N*-aryl group is strongly activated, then it can be removed in nucleophilic substitution reactions in which the azole anion acts as leaving group. Thus 1-(2,4-dinitrophenyl)-pyrazole and -imidazole react with N_2H_4 or NaOMe. Their *N*-picryl derivatives react similarly with $BuNH_2$ or water at a range of pH values ⟨81JA1533, 84JOC3978⟩.

On pyrolysis, 1-arylimidazoles rearrange to 2-arylimidazoles. In other systems pyrolysis causes more deep-seated changes. 1-Arylbenzotriazoles on pyrolysis or photolysis give carbazoles *via* intermediate nitrenes (see Section 3.4.1.2.1). 1-Phenyl-1,2,4-triazole (**714**) is pyrolyzed to isoindole

(715) *via* a carbene intermediate ⟨81AHC(28)231⟩ and another example of the participation of *N*-phenyl groups is found in the formation of benzimidazoles from tetrazoles (see Section 3.4.1.2.1). In the oxadiazolinone series (716), a nitrene intermediate (717) is also probably formed, which then ring closes ⟨70AHC(11)1⟩.

3.4.3.12.3 *Alkyl and alkenyl groups*

N-Alkyl groups in azolium salts can be removed by nucleophilic S_N2 reactions; soft nucleophiles such as PPh_3 and I^- are effective. Sometimes there is competition; for example, in (718) the methyl group is the more readily removed to give mainly *N*-ethylimidazole (719) ⟨80AHC(27)241⟩. This reaction has been studied quite extensively in the imidazolium series. The 1,2- and 1,3-dialkyltriazolium salts undergo nucleophilic displacement on heating ⟨74AHC(16)33⟩, and 2-alkylisothiazolium salts are reconverted into isothiazoles on distillation ⟨72AHC(14)1⟩. Pyrazolium salts similarly give pyrazoles. The benzyl group in *N*-benzylazoles is removed by reduction with sodium in liquid ammonia or catalytic reduction.

N-Alkyl groups in neutral azoles can rearrange thermally to carbon. For example, 2-alkylimidazoles can be prepared in this way in a reaction which is irreversible, uncatalyzed, intramolecular and does not involve radicals ⟨80AHC(27)241⟩.

Methyl and methylene groups attached to an imidazole nitrogen are not particularly "active", but the groups can be metallated ⟨85H(23)417, 85S302, 94H(38)2487⟩. In 1-benzylimidazole the extent of benzylic metallation increases with temperature and with the amount of butyllithium used ⟨84JCS(P1)481⟩. Although 1-phenylthiomethylbenzimidazole is initially lithiated at C-2 at low temperatures, when the temperature is raised rearrangement to the rather unstable methylene-lithiated species occurs. This anion can be trapped by weak electrophiles.

1-Methylbenzotriazole is lithiated readily with *n*-BuLi or LDA at the methyl group and the subsequent reactions with various electrophiles provide 1-substituted benzotriazoles. Other 1-(*n*-alkyl)benzotriazoles also undergo lithiation with LDA ⟨94LA1⟩. 2-Alkylbenzotriazoles behave differently on lithiation. Thus, treatment of 2-methylbenzotriazole with LDA at −78°C produces a coupling product (**720**).

(**720**)

(**721**)

Deprotonation can occur at the α-CH of pyrazole *N*-alkyl groups: treatment of 1-methylpyrazole with *n*-BuLi followed by aldehydes gives products of type (**721**). Such proton loss is facilitated in cationic azido rings, and the ylides so formed sometimes undergo rearrangement. Thus, quaternized 1,2-benzisoxazoles (**722**) lose a proton and then rearrange to 1,3-benzoxazines (**723**) ⟨67AHC(8)277⟩. Quaternized derivatives of benzofuroxan formed *in situ* undergo rearrangement to 1-hydroxybenzimidazole *N*-oxides (**724**) ⟨69AHC(10)1⟩. Reactions of this type are also known for *N*-alkylazolinones.

(**722**)

(**723**)

(**724**)

The benzotriazolate anion is a good leaving group and it can be used in place of a halogen substituent in many reactions. Benzotriazole derivatives are frequently more stable than their chloro or bromo analogues and are easy to prepare. The benzotriazole ring is extremely stable and is opened only under severe conditions, offering advantages over 1,2,3-triazole. Therefore, benzotriazole has become one of the most useful synthetic auxiliary groups.

The benzotriazole moiety of *N*-(α-aminoalkyl)benzotriazoles is readily replaced by hydride upon reduction with sodium borohydride, or with a carbanion by reaction with Grignard or lithium reagents. These are two of the most important reactions of benzotriazole derivatives from which versatile routes have been developed for the synthesis of primary, secondary, and tertiary amines.

1-(Benzotriazol-1-yl)-*N*-triphenylphosphoranylidenemethylamine (**725**), readily available from *N*-(azidomethyl)benzotriazole and PPh$_3$, reacts with Grignard reagents followed by hydrolysis of the intermediates (**726**) to give primary amines ⟨89TL3303⟩. The initial products (**726**) also react with isocyanates, carbon disulfide, aldehydes, epoxides, and alkyl halides to afford a wide variety of carbodiimides, imines, isothiocyanates, aziridines, and secondary amines (Scheme 102) ⟨90S565⟩.

Alkenyl groups in *N*-alkenylazoles act as a weak or moderate π-electron acceptor. Thus, the basic pK_a of 1-vinylimidazole (5.14) is nearly 2 pK units lower than that of imidazole ⟨84JHC171⟩. Examination of ^{15}N NMR spectra of 1-vinylimidazoles and -benzimidazoles shows that the nitrogen attached to the vinyl group resonates in at a higher field than N-3 ⟨91CHE845⟩. In accord with these results 1-vinylimidazoles form less stable complexes with transition metals than 1-alkylimidazoles ⟨89JCR(S)240⟩.

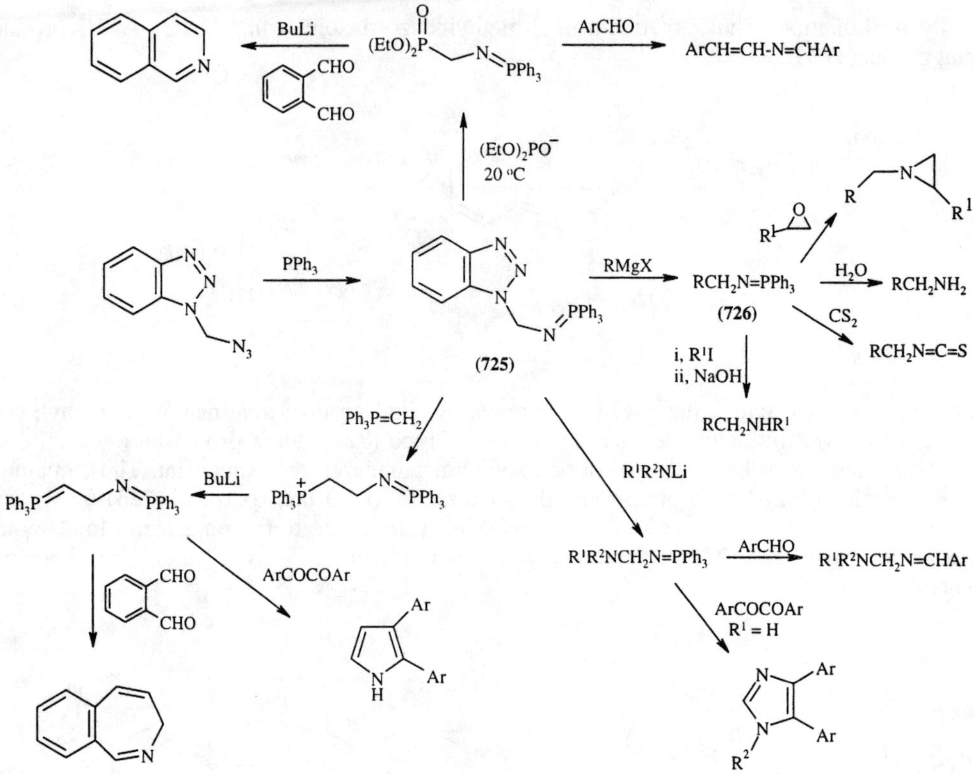

Scheme 102

Additions to *N*-alkenylimidazoles are influenced by conjugation with the heteroring (Equation 9); a vinyl group is not cleaved when 1-vinyl-3-alkyl quaternary salts are thermolyzed and 1-vinylimidazole is not subject to thermal rearrangement ⟨80AHC(27)241, 82CHE1190, 84JHC133⟩.

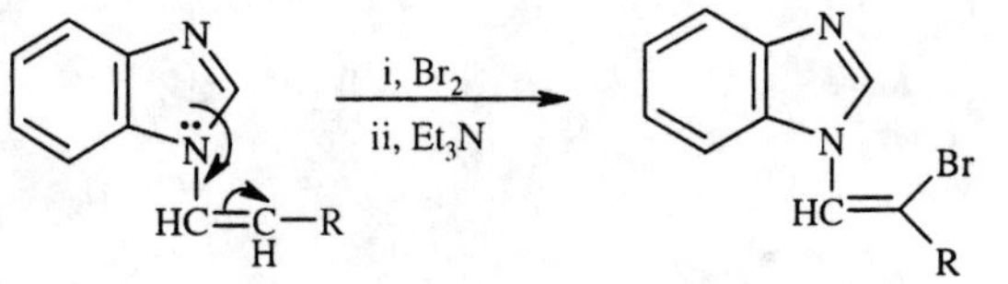

(9)

3.4.3.12.4 *Acyl, carboxy and cyano groups*

An azole ring is quite a good leaving group, far better than NR$_2$. Hence *N*-acylazoles are readily hydrolyzed. Their susceptibility to nucleophilic attack gives rise to the synthetic utility of compounds such as carbonyldiimidazole (**727**) which have been used, for example, in peptide syntheses. *N*-Acylazoles offer mild and neutral equivalents of acid chlorides. The leaving group ability of the azole ring increases with the number of nitrogen atoms it contains. 1-Alkyl-3-acylazolium salts are even more powerful acylating agents ⟨87CC317⟩.

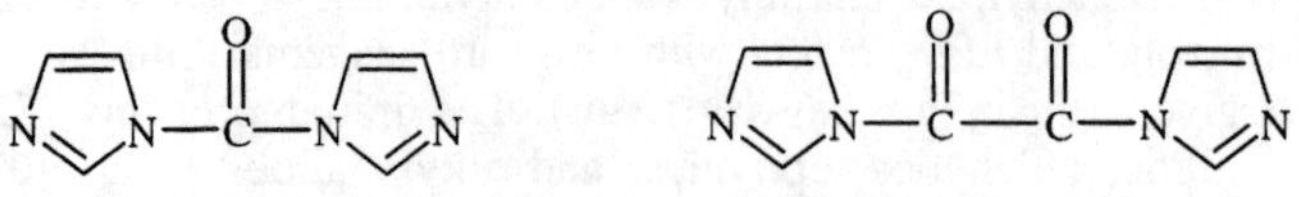

(**727**)   (**728**)

The synthetic utility of azolides has been reviewed in detail ⟨B-86MI 302-02⟩ (see also CHEC), and there are many examples of their use in the synthesis of carboxylic acids, esters, and thioesters ⟨82CPB4242, 83CL1819, 84CPB5044, 85HCA1644, 89H(28)593, 90JHC813, 90SC773, 92JCR(S)196⟩, amides and thioamides ⟨82CPB4242, 85CJC951, 85UKZ860, 89JHC901, 90SC2683, 92JCR(S)196⟩, nitriles ⟨91CPB187⟩, aldehydes and ketones 〈81JOC211, 93H(36)1381, 90MI 302-01, 90SC2683, 93TL5159〉, and phosphorylated imidazoles and benzimidazoles 〈90CHE599〉.

Carbonyldiimidazole (**727**) can be used to convert alcohols into alkyl halides by a one-step reaction using an excess of such reactive halides as allyl bromide or methyl iodide 〈83CPB4189〉, and (**727**) can be made even more reactive by quaternization 〈84CPB5044〉. The disadvantages of (**727**) are its cost and the fact that phosgene must be used in its preparation. An alternative is 1,1'-oxalyldiimidazole (**728**) (made from 1-trimethylsilylimidazole and oxalyl chloride). Transacylation reactions occur very readily with this reagent (Scheme 103) 〈83CL1819, 87CPB4294〉.

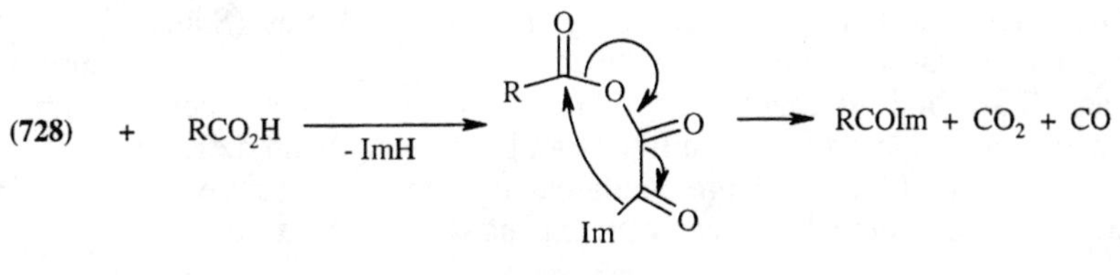

Scheme 103

Acyl derivatives of azoles containing two different environments of nitrogen atoms can rearrange. For example, 1-acyl-1,2,3-triazoles are readily isomerized to the 2*H*-isomers in the presence of triethylamine or other bases; the reaction is intermolecular and probably involves nucleophilic attack by N-2 of one triazole on the carbonyl group attached to another 〈74AHC(16)33〉.

2-Acyltetrazoles may lose nitrogen spontaneously to give oxadiazoles, and thiadiazoles can be prepared similarly from 2-thioacyltetrazoles (Scheme 104) 〈77AHC(21)323〉.

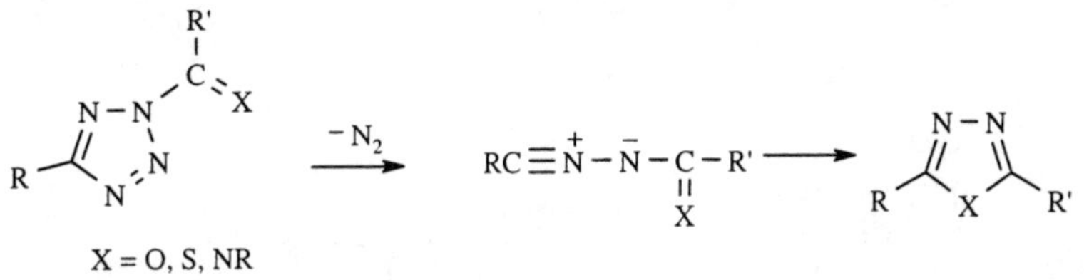

X = O, S, NR

Scheme 104

N-Carboxyazoles, *e.g.* 1-carboxyimidazole, are unstable compounds which readily decarboxylate. However, the ethyl ester of 1-carboxyimidazole is stable enough to be distilled and to form stable salts with acids 〈87H(26)1333〉.

N-Cyanoazoles can be prepared by reactions of azole sodium salts with BrCN 〈95ZOR934〉 or, in the case of 1-cyanobenzotriazole, by reaction of 1-chlorobenzotriazole with NaCN 〈98JOC401〉. These compounds are useful in organic synthesis as mild electrophilic *C*-cyanating reagents or for convenient one-step preparation of carboxamides from carboxylic acids 〈96ZOR903, 98JOC401〉.

3.4.3.12.5　N-*Amino group*

The chemistry of *N*-aminoazoles has been reviewed 〈90AHC(49)117, 92AHC(53)86〉. *N*-Amino groups have little π-interaction with the imidazole ring, and only small inductive effects are apparent when the pK_a values are examined. By contrast, the heteroring has a strong base-weakening effect on an exocyclic amino group. Protonation of 1-aminobenzimidazole, for example, occurs at N-3, the basicity of which is reduced by the amino substituent (pK_a 4.95). As an NH-acid 1-aminobenzimidazole has a pK_a of 28.4 (DMSO, 20°C) (*cf.* pK_a 30.7 for aniline) 〈89KGS221〉 (see also 〈93JCS(P2)1687〉).

N-Amino groups can be alkylated and acylated by way of their anions 〈89CHE180, 92JCS(P1)913〉, and they take part in the usual reactions with carbonyl compounds 〈84CHE1152, 89S843, 93JCR(S)260〉, with carbodiimides 〈88H(27)161〉, and in aza-Wittig reactions 〈88H(27)1935, 94H(37)997〉. 1-Aminobenzimidazoles and their quaternary salts react similarly, and the Schiff bases formed may lead to cyclized products 〈83CHE209, 83CHE314, 86CHE282, 88T7185, 92CHE167〉.

N-Amino groups are replaced by hydrogen on treatment with nitrous acid (*e.g.* **729→730**) 〈80AHC(27)241〉 or phosphorus trichloride (1,2,4-triazole-4-acylimines are converted into triazoles 〈74AHC(17)213〉). *N*-Alkylaminoazoles form stable *N*-nitroso compounds, normally existing as a mixture of *E*- and *Z*-rotamers 〈98T9677〉.

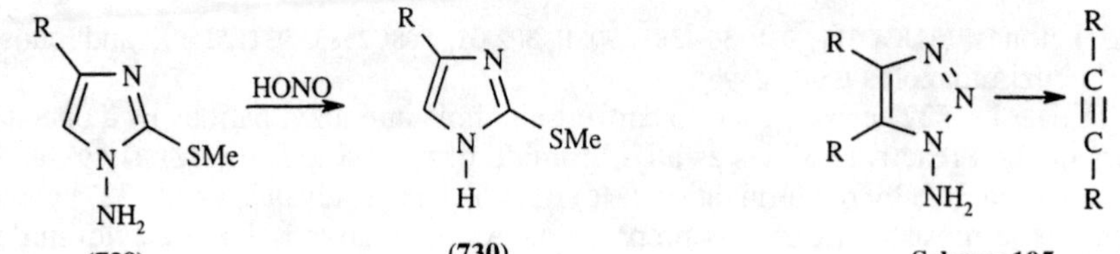

(729)   (730)   Scheme 105

N-Aminoazoles can be oxidized to nitrenes which then fragment or ring expand in various ways. 1-Amino-1,2,3-triazoles lose two moles of nitrogen to give alkynes (Scheme 105) ⟨74AHC(16)33⟩. Oxidation of 1-aminobenzotriazole is one of the most effective methods of generation of benzyne. N-Amino-triazoles (731) and -tetrazoles on oxidation with lead tetraacetate fragment to nitriles ⟨81AHC(28)231⟩; however, the intermediate nitrene can be trapped as an aziridine (732). Similarly, the N-aminopyridazinotriazoles (733) undergo oxidative fragmentation to give open-chain compounds (734) ⟨81AHC(28)231⟩. N-Aminopyrazoles can ring expand to 1,2,3-triazines.

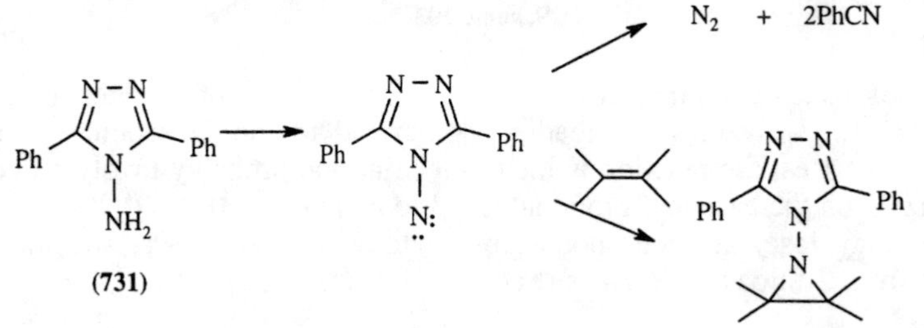

(731)   (732)

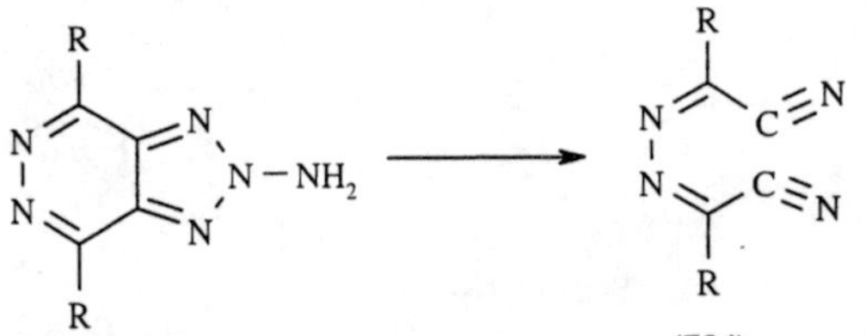

(733)   (734)

Oxidation of 1-aminobenzimidazoles with manganese dioxide or lead tetraacetate can give either 1,1'-azobenzimidazoles (735) or 3-substituted benzo-1,2,4-triazines (736). Electrochemical measurements have shown that the first step in this reaction is removal of an electron from a π-orbital of benzimidazole rather than from the N-amino group. Because the cation radical which is formed must be stabilized by loss of a proton, for (736) to form the 2-substituent must contain an NH or OH group. This is unnecessary for the formation of the azo product (735) which may form *via* a nitrene intermediate.

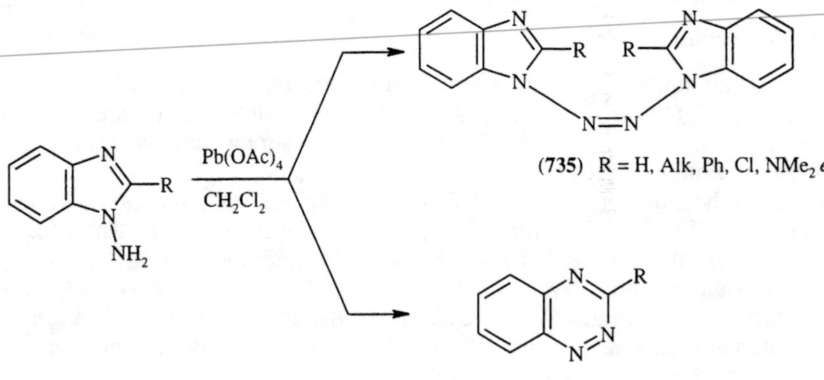

(735) R = H, Alk, Ph, Cl, NMe$_2$ *etc.*

(736) R = NH$_2$, NHR^1, OH

Oxidation of 1,2-diaminobenzimidazole, leading to the formation in high yield of 3-aminobenzo-1,2,4-triazine (739), is thought to proceed through recyclization of an intermediate C-nitrene (737) (possibly *via* diazene intermediate (**738**)) as evidenced by the formation of amine (**739**) with a high efficiency on the thermolysis of 1-amino-2-azidobenzimidazole (**740**) ⟨92MC33⟩. The reaction works well also for other *N*-amino-α-azidoimidazoles ⟨92MC33⟩ and 4-amino-3-azido-1,2,4-triazoles ⟨66TL5369⟩.

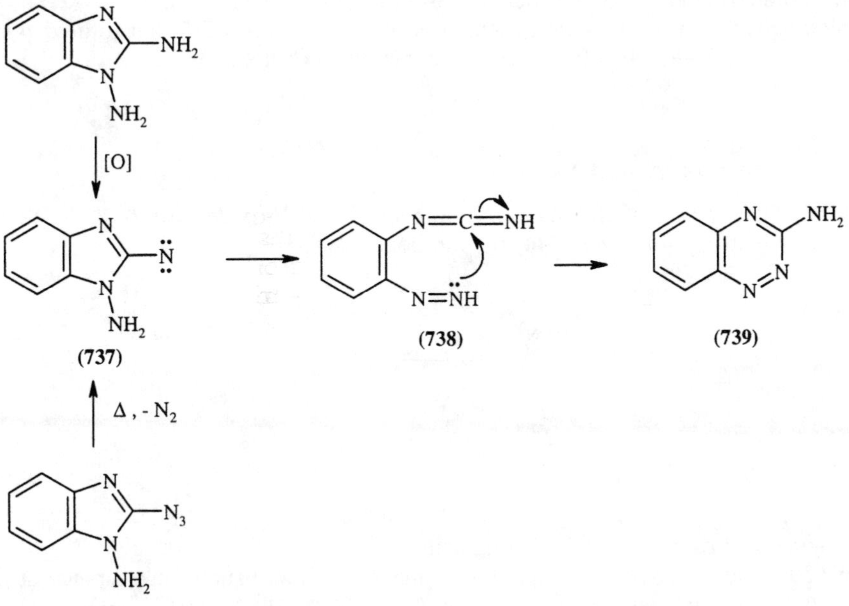

Rearrangement of 1-aminopyrazole and 1-alkylaminopyrazoles into the corresponding 5-amino-pyrazoles has been achieved in 48% aqueous hydrobromic acid; the reaction proceeds by protonation followed by ring opening / ring cyclization in accordance with Scheme 106 ⟨90JCS(P1)809⟩. The 1-alkylaminoindazoles rearrange into 2-alkyl-3-aminoindazoles ⟨92JOC1563⟩.

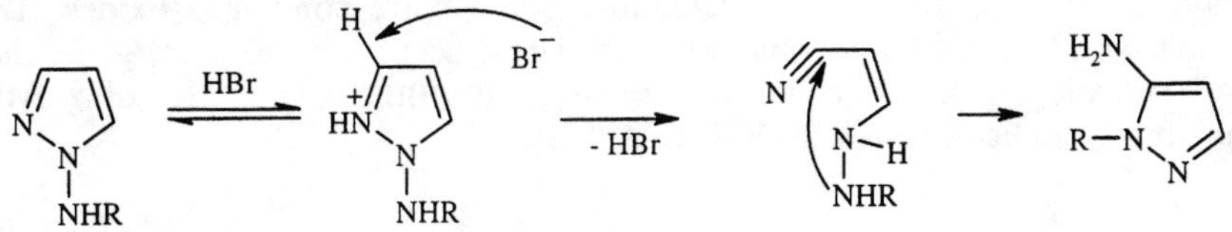

Scheme 106

3.4.3.12.6   N-*Nitro group*

1-Nitropyrazoles rearrange to 4-nitropyrazoles in H_2SO_4 and to 3-nitropyrazoles thermally. Similar rearrangements are known for *N*-nitro-1,2,4-triazoles.

The *N*-nitroimidazoles are stable for a time even in the presence of water, but treatment with concentrated sulfuric acid cleaves the *N*-nitro group, and strong base opens the ring ⟨87PJC913⟩. Much of the interest in such compounds is related to their multi-step complex substitution reactions in which sequential nucleophilic addition of arylamines, ring opening, ring closure, and nitroamide elimination-rearomatization gives 1-aryl-4-nitroimidazoles, *e.g.* (**742**→**741**) ⟨84PJC311, 90PCJ813, 91MI 302-01, 91PJC515, 92AP317, 92PJC943, 93PHC143, 93T3899⟩. This method can also be used to prepare isotopically labeled imidazoles when labeled amino acids are used as the amine ⟨93PHC143⟩.

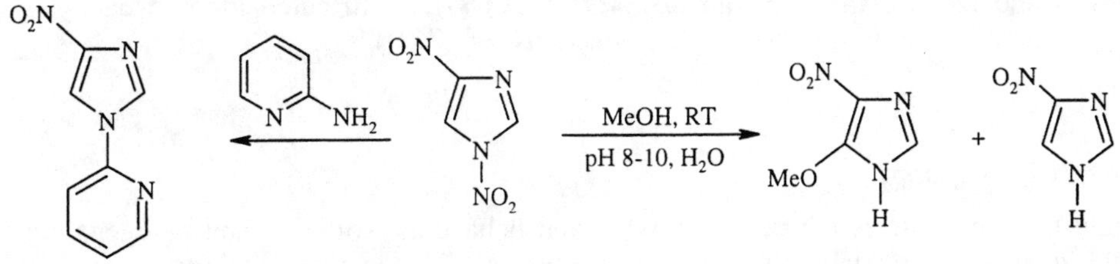

Reaction of 1,4-dinitroimidazole with methanol leads to (743) also *via* "*cine*"-substitution, the reaction being first order with respect to both reactants and also to hydroxide ions.

When heated at 130–140°C in anisole or benzonitrile 1,4-dinitroimidazole (742) forms 4-nitroimidazole (61%) and 2,4-dinitroimidazole (22%). The 2-methyl analogue of (742) gives 2-methyl-4-nitroimidazole (18%) and 2-methyl-4,5-dinitroimidazole (67%).

The complex substitution of 1,4-dinitropyrazoles to afford 4-nitro-3(5)-substituted pyrazoles has been used to prepare the antibiotics formycin and pyrazofurin ⟨91JCS(P1)1077⟩.

3.4.3.12.7 N-*Hydroxy groups and* N-*oxides*

Compounds of this type are often tautomeric: in general, the *N*-oxide form (*e.g.* 745) is favored by polar media, the *N*-hydroxy form (*e.g.* 744) by non-polar media.

(744)   (745)

1-Hydroxypyrazole 2-oxides are quite strong acids.

Alkylation of *N*-oxides, *e.g.*, in basic media by methyl iodide, usually takes place at the oxygen ⟨89S773, 90S795, 91CHE802⟩ as does acylation (by *e.g.* Ac_2O) ⟨89TL4353, 90S795⟩.

Deoxygenation of azole *N*-oxides can be accomplished by catalytic hydrogenation ⟨83JCR(S)76, 89S773⟩, with Raney nickel ⟨89S773⟩, with titanium(III) chloride in methanol ⟨84TL1319⟩, by metal-acid reduction ⟨85CHE1303, 91CHE802⟩, or with phosphorus halides ⟨93CHE127, 98JOC12⟩. When the imidazole *N*-oxides does not have a 2-substituent they readily rearrange to the corresponding 2-imidazolones or 2-benzimidazolones, but 1-hydroxyimidazole 3-oxides will not undergo this rearrangement ⟨93CHE127⟩. 1,2,3-Thiadiazole 3-oxides isomerize on irradiation to the corresponding 2-oxides. The *N*-oxide function in imidazole *N*-oxides, in conjunction with the adjacent carbon atom, gives the molecule properties of a 1,3-dipole, dipolar cycloadditions are known ⟨93CHE127⟩. Oxidation of *N*-hydroxyazoles can give cyclic nitroxyl radicals (*e.g.* 746–748) ⟨79AHC(25)205⟩.

(746)   (747)   (748)   (749)

(750)

The chemistry of imidazole and benzimidazole N-oxides has been reviewed, and methods of preparation have been summarized ⟨88JCS(P1)681, 93CHE127⟩. The oxides are thermally unstable and may be explosive ⟨84CB30, 89S773⟩.

1-Hydroxy-7-azabenzotriazole (749) can be used in peptide coupling reactions, especially with sterically encumbered amines. Although exact mechanistic details have not been established, the faster reaction rates and reduced racemization ⟨93JA4397, 93TL4781⟩ is attributed to base catalysis by the adjacent pyridine nitrogen during the coupling reactions (*cf.* (750)).

3.4.3.12.8 N-*Halo groups*

Generally these derivatives are rather unstable and behave as oxidizing and halogenating agents. 1-Iodoimidazoles are more stable than other analogues. *N*-Chlorobenzotriazole is a useful oxidizing and chlorinating agent ⟨97ZOR285⟩.

3.4.3.12.9 *N-Silicon, phosphorus, sulfur and related groups*

N-Trialkylsilyl- and *N*-trialkylstannylazoles are useful N-protected derivatives for a number of transformations, and suitably volatile for GLC analysis. The groups are thermally stable, but susceptible to hydrolytic and analogous displacements, with the stannyl compounds less readily hydrolyzed than the silyl species.

1-Trimethylsilylimidazole (**751**) is used, with trimethylsilyltriflate as catalyst ⟨84CHE662, 90S104⟩, to make 1-alkylthio- and 1-phenylthio-1-trimethylsilyloxy-alkanes and -cycloalkanes. The use of two equivalents of the thiol prevents the formation of imidazole adducts (Scheme 106a) ⟨90S104⟩.

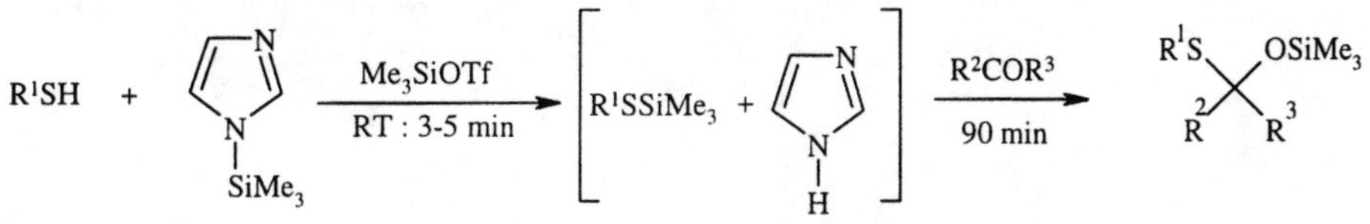

Scheme 106a

Analogues of (**751**) react with derivatives of phosphorus-containing acids to form 1-phosphoryl-imidazoles ⟨90CHE599⟩. The same compound also silylates ketones to give enol silyl ethers and/or siloxyalkylimidazoles (Scheme 107) ⟨87JC271⟩.

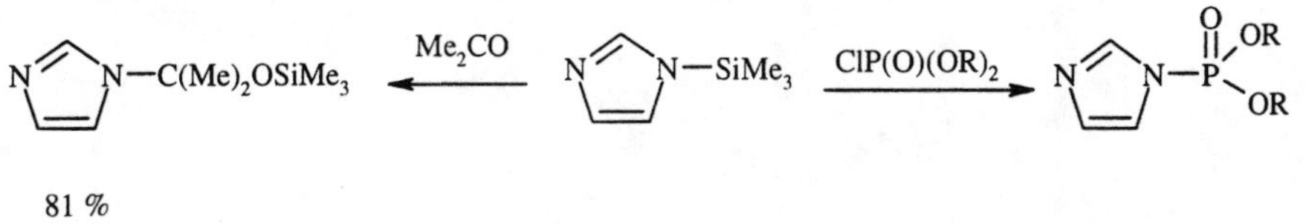

81 %

Scheme 107

N-Phosphorylimidazoles are useful phosphorylating agents ⟨81JA5784, 90CHE599, 94JCS(P1)817⟩.

1,1′-Sulfonyldiimidazoles and 1-chlorosulfinylimidazole have similar applications to the carbonyl analogues in transfer reactions (Scheme 108) ⟨80SC733, 84JHC1489, 85CJC3613, B-86MI 302-02⟩. Treatment of arylsulfonylimidazoles with hydrogen peroxide in alkaline medium gives arylsulfonyl peracids which are useful epoxidizing agents ⟨93SL915⟩. The dimethylaminosulfonyl group is useful for N-protection in lithiation procedures; it can be removed by refluxing with dilute hydrochloric acid ⟨91S1021⟩.

Scheme 108

3.5
Reactivity of Small and Large Rings

3.5.1 GENERAL SURVEY

Chapter 3.5 attempts to give an overview of the reactivity of 'small or large' ring systems treated individually in CHEC and CHEC-II. The great diversity of these systems presents a serious problem of organization. In structuring this chapter on reactivity, the nature of the reaction and the distance of the site of attack from the heteroatoms were used.

3.5.1.1 Neutral Molecules

The reactivity of small (three- and four-membered) heterocyclic rings is dominated by the effects of ring strain, which facilitates all modes of ring opening. Aromaticity is seldom observed, antiaromaticity is present in only a few isolated examples, and thus does not play a general role. Many reactions are initiated by unimolecular ring opening, to give diradicals or ylides, whose reaction products are then observed. Extrusion of stable as well as unstable moieties is assisted by the ring strain. Four-membered systems tend to cleave into two two-membered fragments (each consisting of two former ring atoms and their ligands). Attack on ring carbons concomitant with ring opening is very common, and is usually subject to electrophilic catalysis.

Large heterocycles, *i.e.* those with more than six ring members, often show little effect of ring strain on the reactivities of the neutral molecules. Factors very important for large ring reactivity are unsaturation, especially of polyenic and aromatic character, and the steric accessibility of heteroatoms and functional groups, as well as the possibility of transannular reactions. However, the majority of unsaturated large rings are not aromatic, even where the Hückel rule seems to be obeyed formally.

3.5.1.2 Cations

Onium ions of small and large heterocyclics are usually produced by electrophilic attack on a heteroatom. In most three- and four-membered rings, nucleophilic attack on an adjacent carbon and ring opening follow immediately, stabilizing the molecule. In large rings the onium ion behaves as would its acyclic analogue, except where aromaticity or transannular reactions come into play (each with its electronic and steric pre-conditions). A wide diversity of reactions is observed.

Cations of a different kind may be derived from heterocyclics by removing a leaving group with its bonding electrons, *e.g.* a halide ion from an *N*-halo moiety. Such cations (nitrenium ions) were also assumed to be reactive intermediates in reactions of other *N*-halo heterocycles.

3.5.1.3 Anions

Anions of small heterocycles are little known. They seem to be involved in some elimination reactions of oxetan-2-ones ⟨80JA3620⟩. Anions of large heterocycles often resemble their acyclic counterparts. However, anion formation can adjust the number of electrons in suitable systems so as to make a system conform to the Hückel rule, and render it aromatic if flat geometry can be attained. Anion formation in selected large heterocycles can also initiate transannular reactions (see also Section 3.5.6 below).

3.5.1.4 Radicals

Small-ring radicals with the unpaired electron at the heteroatom or at a carbon adjacent to a heteroatom undergo ring cleavage as the predominant mode of stabilization, as known for oxaziranes ⟨78CJC2985, 77TL4289, 76HCA880, 76JCS(P2)1044⟩, aziridines (CHEC 5.04.3.9), diaziridines (CHEC 5.08.3.2.3), diazirines ⟨79JA837⟩, oxaziridines (CHEC 5.08.3.1.5) and thietanes (CHEC 5.14.3.10.1). The heteroatom is usually retained in the product, except in thiirane cleavages (CHEC 5.06.3.7), where desulfurization occurs. Thietanes, in contrast, are less readily desulfurized (CHEC 5.14.3.10.1). Oxaziridinyls display a variety of reactions, including N-O and N-C cleavage (CHEC 5.08.3.1.5); diaziridinyls behave analogously, with C-N and N-N cleavage (CHEC 5.08.3.2.3).

Large-ring heterocyclic radicals are not particularly well known as a class. Their behavior often resembles that of their alicyclic counterparts, except for transannular reactions, such as the intramolecular cyclization of 1-azacyclononan-1-yl (Scheme 1) ⟨72CJC1167⟩. As is the case with alicyclic ethers, oxepane in the reaction with *t*-butoxy radical suffers abstraction of a hydrogen atom from the 2-position in the first reaction step (Scheme 2) ⟨76TL439⟩.

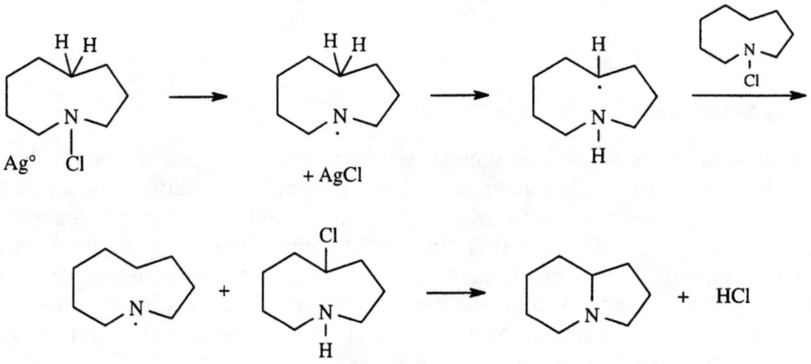

Scheme 1

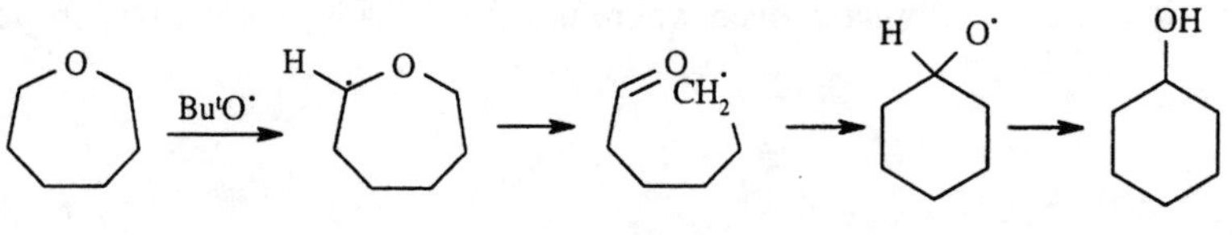

Scheme 2

3.5.2 THERMAL AND PHOTOCHEMICAL REACTIONS, NOT FORMALLY INVOLVING OTHER SPECIES

3.5.2.1 Fragmentation Reactions

Fragmentation reactions are particularly common in small rings. Relief of strain and the gain in stability in forming certain common fragments (such as N_2, CO_2), as felt in the transition state of the rate determining step, are important driving forces. Three-membered rings fragment to give moieties a-b (usually unsaturated) and c. The latter might be a stable molecule, such as CO, but also a carbene or nitrene, atomic sulfur or singlet SO, to name common examples, see Scheme 3.

Extrusion (or elimination) of sulfur from thiiranes and thiirenes is a facile process. Virtually all thiiranes and thiirenes, as well as their oxides and dioxides, undergo thermal extrusion of the sulfur moiety with increasing facility according to the trend $S << SO < SO_2$. The thermolytic desulfurization mechanism is more complex than a simple cheletropic extrusion ⟨85TL2789⟩.

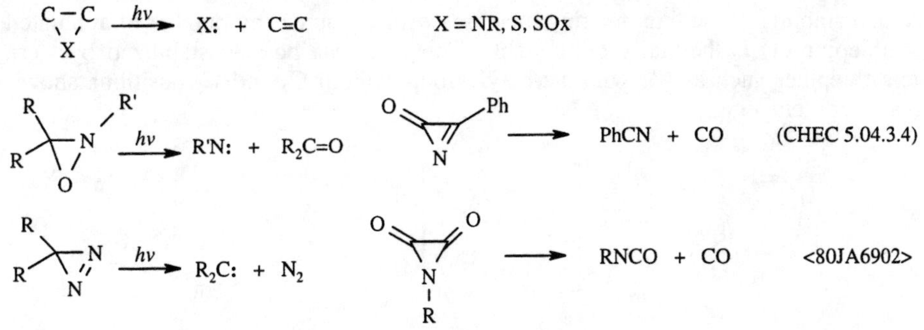

$$C-C \xrightarrow{h\nu} X: \; + \; C=C \qquad\qquad X = NR, S, SOx$$

(with X in ring)

$$R_2C(NR)(O) \xrightarrow{h\nu} R'N: \; + \; R_2C=O$$

$$(O)(Ph)C=N \text{ ring} \longrightarrow PhCN \; + \; CO \qquad (CHEC \; 5.04.3.4)$$

$$R_2C(N)(N) \xrightarrow{h\nu} R_2C: \; + \; N_2$$

$$(O)(O)(NR) \text{ ring} \longrightarrow RNCO \; + \; CO \qquad \langle 80JA6902 \rangle$$

Scheme 3

Four-membered heterocycles most often give fragments containing two ring atoms with the respective ligands. However, [3+1] fragmentation is well known, giving an atom (such as S) or a stable species (such as SO_2) frequently together with a three-membered ring. Examples are found in Schemes 4 and 5. The [2+2] fragmentations are often stereospecific and can be reversible. The reversibility can lead to interconversions as seen in Scheme 6; the RN: moieties of isocyanate and a carbodiimide are exchanged *via* a 3-imino–1,3-diazetidin–2-one ⟨69ACR186⟩.

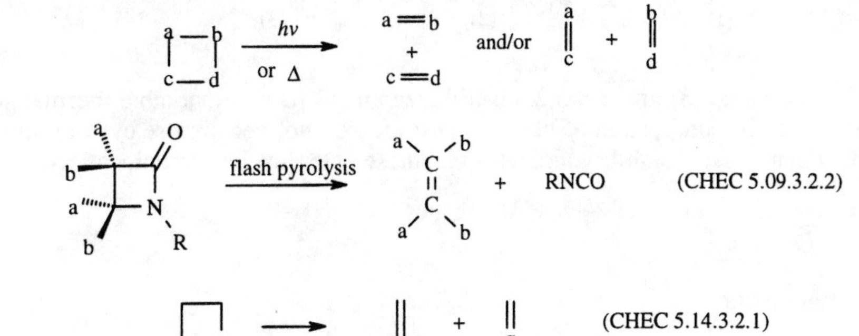

$$\underset{c-d}{a-b} \xrightarrow[\text{or } \Delta]{h\nu} \underset{c=d}{a=b} \; \text{and/or} \; \underset{c}{\overset{a}{\|}} \; + \; \underset{d}{\overset{b}{\|}}$$

$$\text{(azetidinone)} \xrightarrow{\text{flash pyrolysis}} \underset{a}{\overset{a}{}}C=C\underset{b}{\overset{b}{}} \; + \; RNCO \qquad (CHEC \; 5.09.3.2.2)$$

$$\text{(thietane)} \longrightarrow \| \; + \; \underset{S}{\|} \qquad (CHEC \; 5.14.3.2.1)$$

Scheme 4

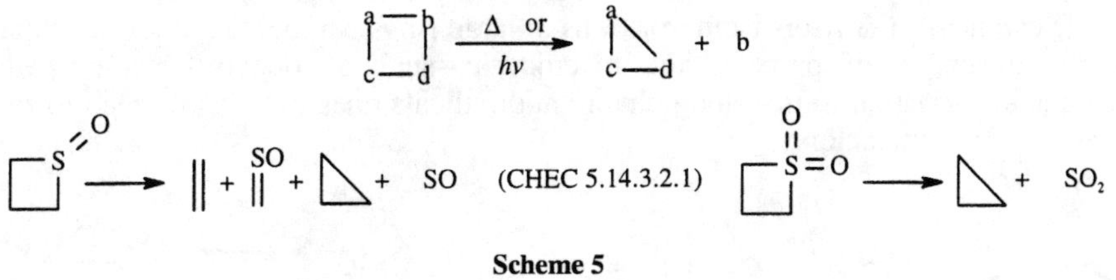

$$\underset{c-d}{a-b} \xrightarrow[h\nu]{\Delta \text{ or}} \underset{c-d}{\overset{a}{\triangle}} \; + \; b$$

$$\text{(thietane S-oxide)} \longrightarrow \| \; + \; \underset{\|}{\overset{SO}{}} \; + \; \triangle \; + \; SO \quad (CHEC \; 5.14.3.2.1)$$

$$\text{(thietane S-dioxide)} \longrightarrow \triangle \; + \; SO_2$$

Scheme 5

$$\begin{array}{c} RN=C=O \\ + \\ R'N=C=NR' \end{array} \rightleftharpoons \text{(diazetidinone)} \rightleftharpoons \underset{NR'}{\overset{NR}{C}} \; + \; \underset{NR'}{\overset{O}{C}}$$

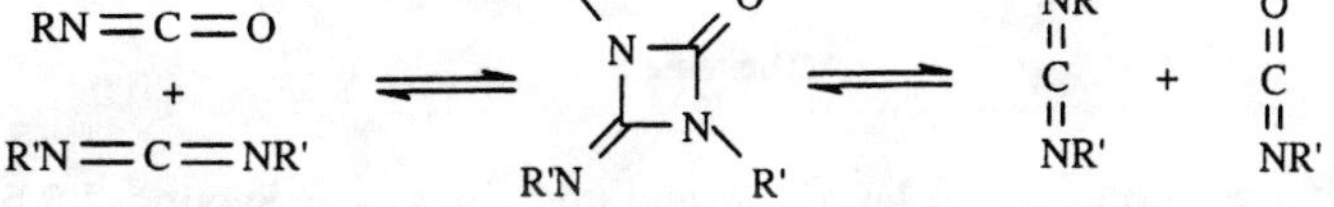

Scheme 6

Nitrogen, CO_2, SO_2, RCN, RNCO and RNCS are particularly common fragments containing two of the original four ring atoms.

The fragmentation of large heterocyclics occurs less readily, since the ring strain is usually less. The most favorable leaving moieties, such as N_2, can of course be extruded easily (often giving diradical species). Thus, 1,2,4-triazepines can lose nitrogen to give pyrroles (Scheme 7; CHEC 5.18.7.2). Most often, fragmentations of large heterocyclics can be classified either as retrocycloadditions, or they are analogous to what would be expected from acyclic counterparts of the large rings. The fragmentations may be orbital symmetry controlled. For example, 2,7-dihydrothiepin 1,1-dioxide loses SO_2 to give *cis*-hexatriene (CHEC 5.17.2.3.2). The orbital control of thermal sulfur extrusions from thiepins has been studied in some detail (CHEC 5.17.2.4.1). In general, simple thiepines thermally extrude sulfur readily, by a valence isomerization to the corresponding benzene episulfide, followed by irreversible loss of sulfur. Because of the thermal instability, the parent thiepine has never been characterized.

The thermal stability of the thiepine ring increases with an increasing number of annulated benzene rings, *e.g.* thiepine (1) is thermally quite stable. Thiepines can be also stabilized by steric effects. Monocyclic thiepines such as (2) with a *t*-butyl group at both C_2 and C_7 positions show no sulfur extrusion at room temperature.

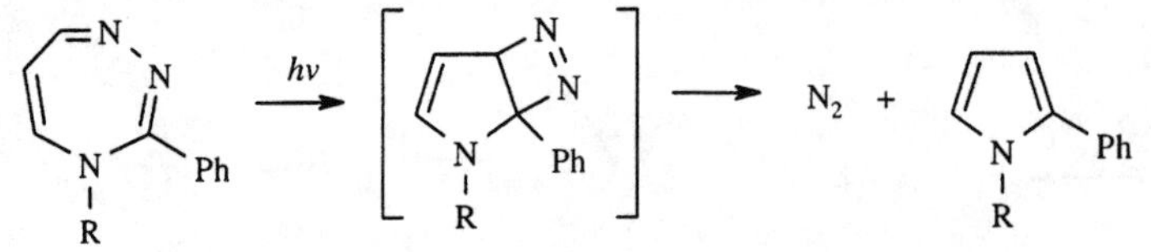

Scheme 7

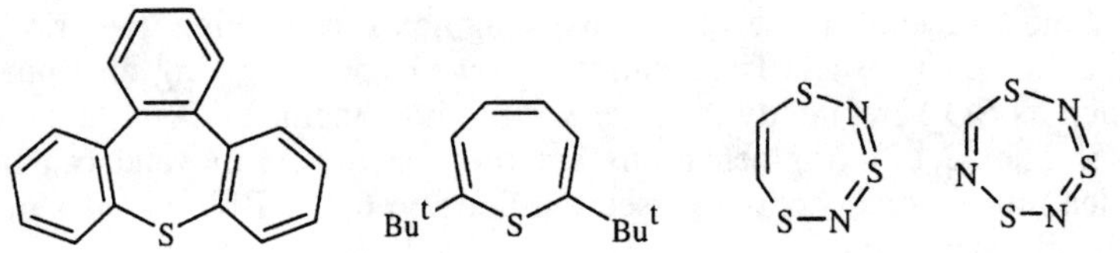

(1)   (2)   (3)   (4)

1,3,5,2,4-Trithiadiazepine (3) and 1,3,5,2,4,6-trithiazepin (4) have remarkable thermal stability for molecules with such a high proportion of heteroatoms; they do not decompose even on the prolonged heating at 180°C. Such unusual stability indicates that these heterocycles are delocalized 10π aromatic systems.

3.5.2.2 Rearrangements

(a) *Cis-trans* isomerizations of substitutes are commonly observed upon heating or irradiating saturated three-membered heterocyclics. The formation of 1,3-diradicals leads to rotation about the single bonds, and the isomerization has been used to probe the bond strengths of the 2,3-bond of such heterocyclics (Scheme 8). *Cis-trans* isomerizations in large rings can be due to retrocycloadditions or the temporary conversion of parts of the heterocycle, such as dehydrogenation-hydrogenation reactions. Diradical formation to give long chain 1,ω-diradicals does not usually lead to recyclization due to unfavorable entropy factors.

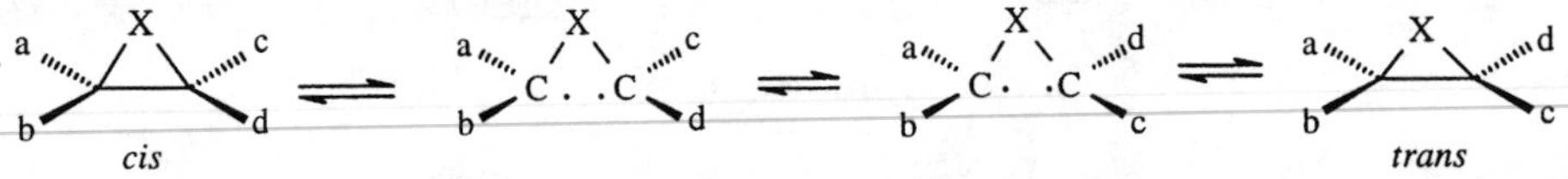

Scheme 8

(b) Hydrogen shifts are common in large, unsaturated rings (see Section 2.5.5.2). Thus, series of 1,5-hydrogen shifts, thermally allowed, connect the 1*H*, 2*H*, 3*H* and 4*H* isomers of *N*-unsubstituted azepines. While the sigmatropic mechanism does allow the interconversion of the isomers, base catalysis has been observed in some cases. Ionic mechanism(s) must therefore be considered (Scheme 9).

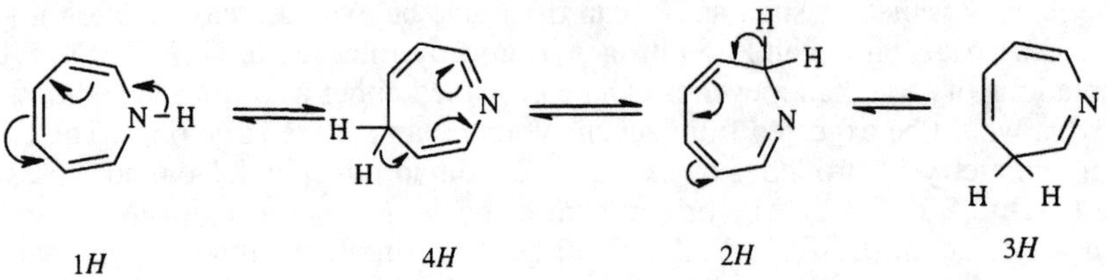

1*H*   4*H*   2*H*   3*H*

Scheme 9

(c) Ring-chain isomerizations are common with small heterocycles, with the ring strain assisting the opening. The reverse reaction is often found where reactive opening products are obtained (see isomerizations). Scheme 10 gives a few examples of irreversible ring openings, and Scheme 11 shows some which are readily or spontaneously reversed.

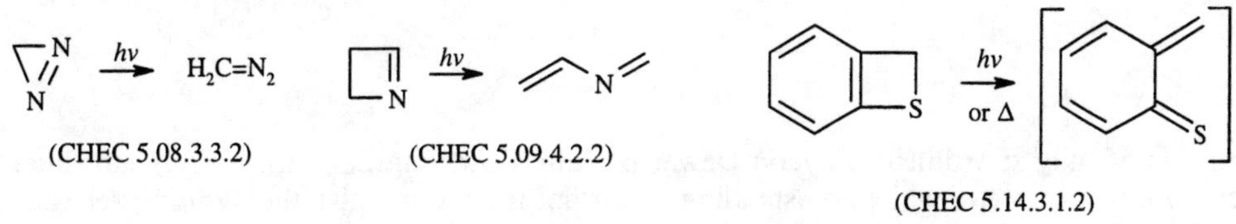

Scheme 10

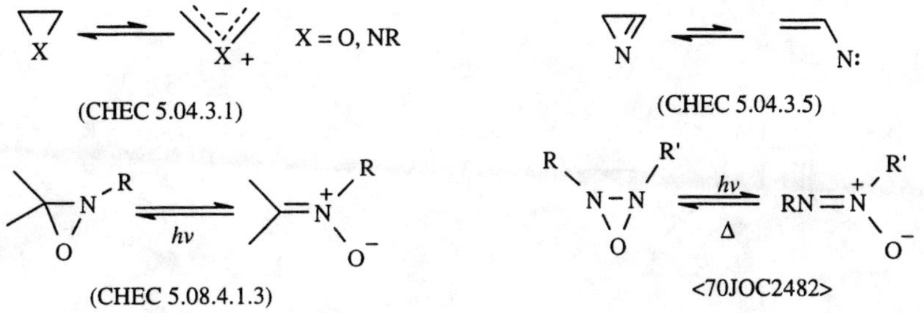

Scheme 11

(d) Ring-ring valence isomerizations occur in small and large rings, with or without changes of ring size. An example of the latter course is the intriguing interconversion of three-membered rings with exocyclic double bonds (Scheme 12) observed with methylene-aziridines ⟨73AG(E)414, 78AG(E)213⟩, -aziridinimines ⟨70AG(E)381⟩ and -diaziridinimines ⟨69AG(E)449⟩.

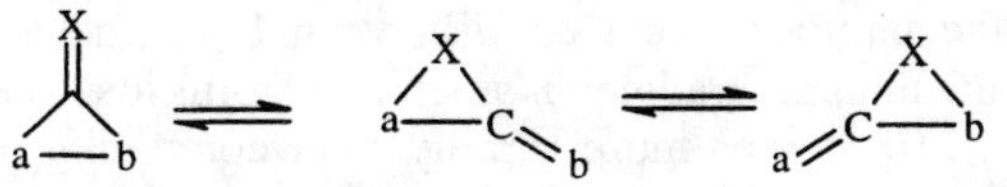

Scheme 12

Ring expansion of small rings is once again favored by ring strain, and many 3 → 5 conversions are known. Four-membered rings can expand to five- or six-membered ones. Examples are given in Scheme 13.

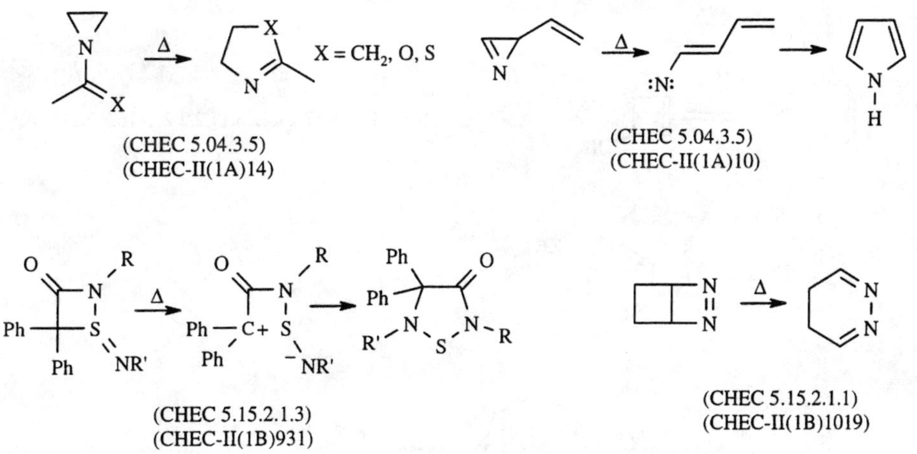

Scheme 13

Thermolysis of the tricyclic azetidines (**5**; X = O, NCO$_2$Et, S or CH$_2$) in solution gives the seven-membered ring heterocycles (**6**) in high yield ⟨83CC941⟩.

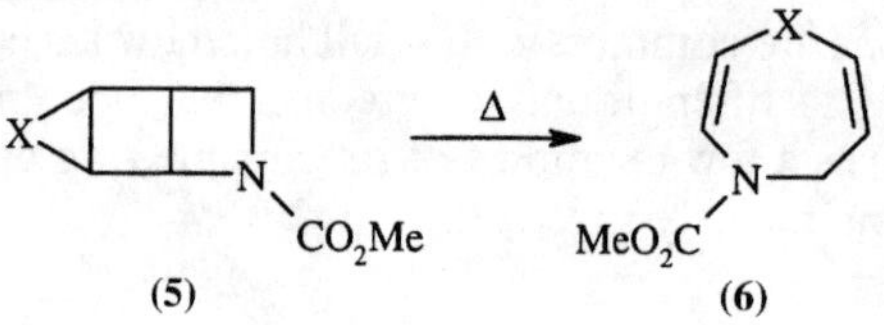

(5)   (6)

Photolysis of fully substituted tricyclic Dewar pyridines carrying 2,3,4-tri(*t*-butyl) substituents (**7**) produces azaprismanes (**8**), which, on standing or further irradiation, give the pyridine derivatives (**9**) ⟨89T3115⟩.

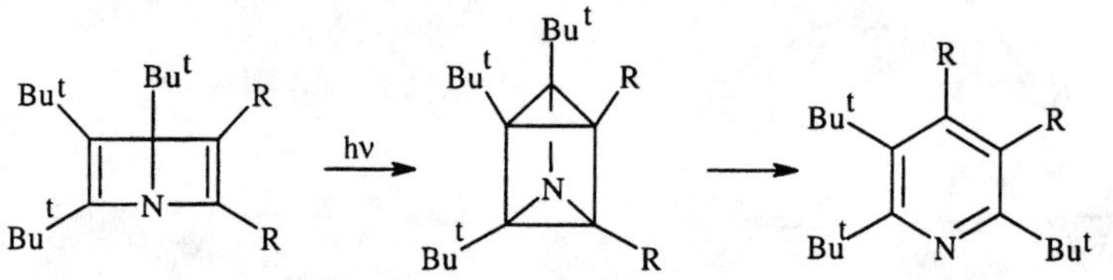

(**7**)   (**8**)   (**9**)

Large rings isomerize to two condensed smaller ones, by transannular reactions of many bond-making mechanisms, and by electrocyclic reactions. Seven-membered, fully unsaturated systems can convert to [3.2.0] and to [4.1.0] isomers. The former conversion is allowed photochemically, the latter thermally. Consequently, the chemistry of azepins, oxepins and thiepins is often governed by the rate and activation barrier (or the photochemical conditions and parameters) prevailing. Thiepins, for example, extrude sulfur *via* the thianorcaradiene isomer, and sulfur loss is likely to occur when a given system isomerizes to the [4.1.0] isomer more rapidly than competing reactions occur through the monocyclic isomer. Depending on the nature of the heteroatom (and the presence of other heteroatoms in the ring) and the substituents, the heterepine – heteranorcaradiene rearrangement can be fast or slow, and one or the other component can be dominant in the equilibrium. Photoinduced rearrangement leads to [3.2.0] systems, which may revert to the seven-membered monocycle thermally-perhaps by homolytic cleavage of the common bond. Both types of bicyclizations are observed in systems containing N, O or S (including SO and SO₂) as heteroatoms. Scheme 14 gives examples.

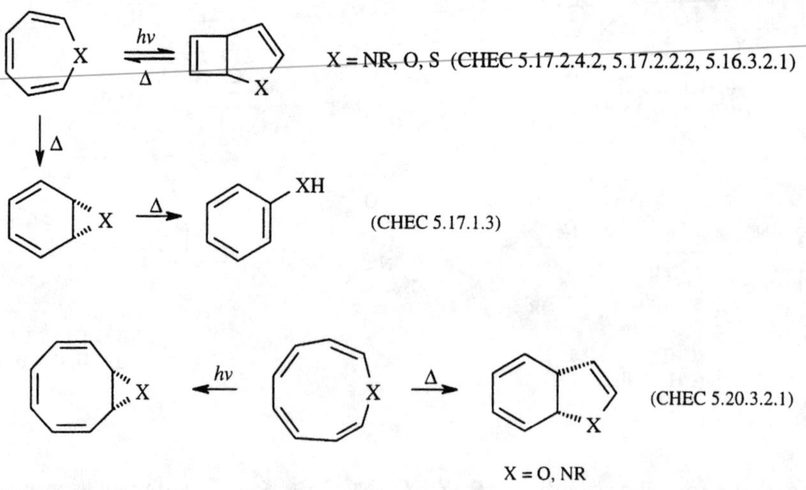

Scheme 14

2*H*-Azocinones are in thermal equilibrium with 8-azabicyclo[4.2.0]octa-3,5-dienones, as measured by NMR (CHEC 5.19.2.3).

Acetoxydithiocin (**10**) is converted photochemically into its dithiabicyclooctadiene valence tautomer (**11**) ⟨78T3631⟩.

(10)   **(11)**

For other examples of valence isomerizations of small and large rings see Section 2.5.5.1.

3.5.3 ELECTROPHILIC ATTACK ON RING HETEROATOMS

3.5.3.1 Protonation

The basicities of saturated heterocycles are similar to those of analogous open chain systems, with the exception of three-membered heterocycles, in which the basicity is markedly reduced. Table 1 gives pK_a values for the equilibria between free and monoprotonated heterocycles. As the ring size increases, the protonated species become more stable and the pK_a values approach those of the open chain analogues. Increasing basicity (thiirane < oxirane < aziridine) prevails in gas phase proton affinities (Table 2) ⟨80JA5151⟩.

The gas phase proton affinities of azetidine and β-lactams were measured by FTICR MS. Azetidine presents a gas phase basicity practically equal to that of *N*-methylethanamine ⟨92JA4728⟩.

In the gas phase, β-lactams are weaker bases than acyclic amides. *Ab initio* calculations show that β-lactams are oxygen bases, but the gap between the oxygen and nitrogen intrinsic basicities is much smaller than in normal amides, a result of redistribution due to hybridization changes ⟨92JA4728⟩.

While lactones are frequently more basic than the corresponding acyclic esters, 2-oxetanone is less basic (~ 6 kcal/mol) than methyl acetate as a consequence of substantial hybridization changes incurred upon cyclization to the four-membered ring ⟨93JA7389⟩. The effect of cyclization of ethers on basicity is the opposite, and oxetane is a stronger base than methyl ethyl ether. Lactones and esters normally undergo protonation on the carbonyl oxygen, favored over the ether oxygen by nearly 18 kcal/mol. Since the 2-oxetanone carbonyl is less basic, the energy difference on protonation at the carbonyl oxygen *versus* protonation at the ether oxygen is substantially reduced to a value of 3 kcal/mol.

Medium-ring saturated monocyclic diamines (**12**) have pK_a values ≥ 12 in aqueous solution ⟨88CC1528⟩, (**12**, n = m = 4) has the highest pK_a^1 resulting from internal hydrogen bonding as suggested by NMR measurements and force-field calculations.

The pK for 10–12-membered ring triazacycloalkanes (**13**) are much higher (pK_a^1 ~ 12.7–13.4) than those of open chain secondary amines, whereas the 13-membered ligand (**13**, n = m = 2, p = 6) has a normal value (pK_a^1 = 10.36).

(12) n, m = 3-5   **(13)**   **(14)**

Tetraazamacrocycles (**14**) generally show two high pK values (pK_a^1 ~ 10.5–11.5; pK_a^2 ~ 9.5–10.5), whereas the other values are distinctly lower (pK_a^3 ~ 1.6–6.9; pK_a^4 ~ 0.8–5.7) (Table 3) ⟨91CRV1721⟩. The second proton can bind to a nitrogen atom *trans* to the first one, whereas the third and forth must

Table 1   Basicities for Some Heterocycles: pKa Values for the Equilibria Between Parent and Monoprotonated Species in Water

Parent species	pKa	CHEC number / ref.	Parent species	pKa	CHEC number / ref.
 (aziridine, N–H)	8.04	63PMH(1)1	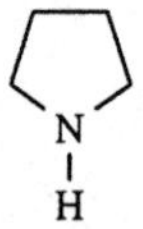 (pyrrolidine, N–H)	11.27	63PMH(1)1
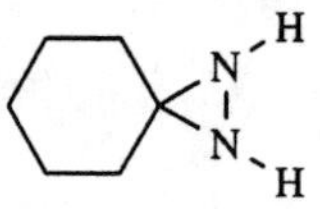 (cyclohexane spiro diaziridine, N–H, N–H)	4.6	5.08.2.3.1	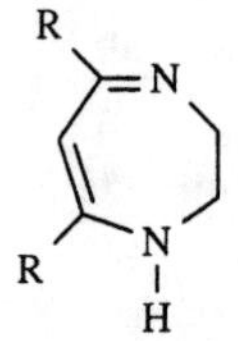 (diazepine, R, R, N–H)	13–14	5.18.2.3
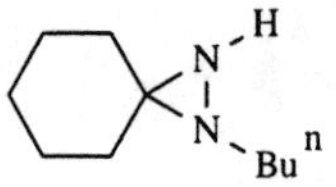 (cyclohexane spiro diaziridine, N–H, N–Bu^n)	6.4	5.08.2.3.1	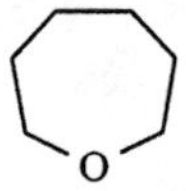 (oxepane)	− 2.02	5.17.2.1.3
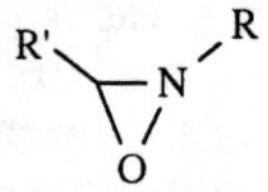 (oxaziridine, R′, R)	0.13 to − 1.81	5.08.3.1.3	 (tetrahydrofuran)	− 2.08	5.17.2.1.3
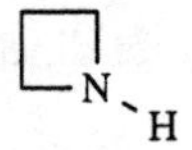 (azetidine, N–H)	11.29	63PMH(1)1			

Table 2   Gas Phase Proton Affinities of Small Heterocycles [a]

Heterocycle	Proton affinity (kJ mol^{-1})	Open chain analogue	Proton affinity (kJ mol^{-1})
 (aziridine, N–H)	902.5	Me-NH-Me	922.6
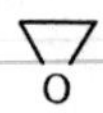 (oxirane, O)	793.3	Me-O-Me	807.9
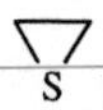 (thiirane, S)	812.9	Me-S-Me	839.7

[a] ⟨80JA5151⟩

bind *cis* to an ammonium group and thus feel the neighboring positive charge more strongly (Scheme 15). As the size of the macrocycle ring increases from 12 to 16 such interactions decrease and the values of the third and fourth pK become larger.

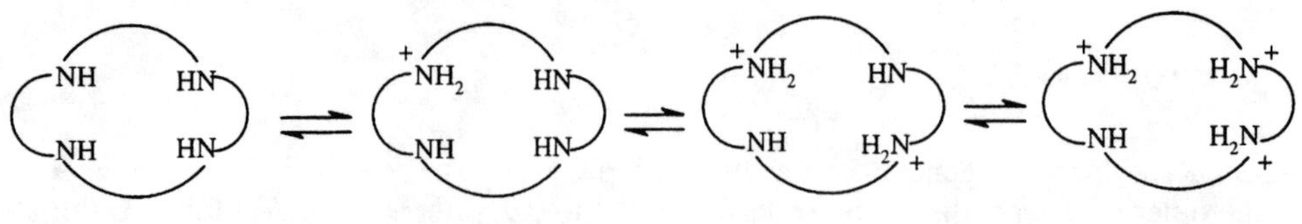

Scheme 15

Table 3 pK_a Values of tetraazacycloalkanes (**14**) (in 0.2 M NaClO$_4$).

Compound	pK_a^1	pK_a^2	pK_a^3	pK_a^4	Temp.(°C)	Ref.
14 (n = m = p = q = 2)	10.51	9.49	1.6	0.8	35	a
14 (n = m = p = 2, q = 3)	10.91	9.91	~1.6	~0.9	35	a
14 (n = p = 2, m = q = 2)	11.50	10.30			25	b
	11.23	10.30	1.5	~0.8	35	a
14 (n = m = 2, p = q = 3)	11.05	9.98	3.5	1.0	25	a
14 (n = m = p = 2, q = 4)	10.57	9.56	4.13	~2	25	c
14 (n = m = p = 3, q = 2)	10.76	9.94	3.6	1.5	25	c

^{a} ⟨80JCS(D)327⟩; ^{b} ⟨84IC4181⟩; ^{c} ⟨83IC2055⟩

Similar observations have been made for larger macrocycles containing five or more nitrogen atoms. These compounds behave as strong bases in their first protonation steps and as weaker bases in the last protonation steps ⟨87IC1243⟩.

3.5.3.2 Complex Formation

Numerous stability constants for metal complexes with triaza-, tetraaza-, and polyazamacrocycles have been measured ⟨91CRV1721⟩. Smaller rings with three and four nitrogen atoms generally form 1:1 species ML (but sometimes also ML$_2$). Larger rings with more nitrogen donors can give protonated species MLH$_n$ and/or binuclear complexes M$_2$L.

Many of these complexes are more stable than those formed with the corresponding open-chain ligands. This "macrocyclic effect" is determined by enthalpic as well as by entropic contributions ⟨74IC2941, 83JCS(D)1189⟩. Still higher stability constants are obtained when side chains with additional donor groups are attached to the macrocycle; thus, tetraacetate (**15**) forms complexes with most metal ions, such as earth-alkali ions, transition-metal ions and lanthanides; these complexes show stabilities among the highest known ⟨82TAL815, 84IC359⟩.

(**15**)   (**16**)   (**17**)

3.5.3.3 Alkylation and Acylation

Alkylation, acylation, *etc.* at the heteroatom lead to substitution of an NH or to onium salts. Three-membered ring onium salts are difficult to isolate, and very weakly nucleophilic counterions must be used, such as BF$_4^-$.

Primary electrophilic attack at N in aziridines is followed by ring transformation. Thus, nitrosation of 2,3-diphenylaziridine (**16**; R = Bu, CH$_2$Ph) in AcOH at 25°C gave the 1-alkyl–1,4-dihydro–3,4-diphenyl–1,2-diazete–2-oxide (**17**) (57%) ⟨91JA2308⟩.

Despite this tendency, examples of alkylation ⟨90JOC1144, 93JOC7848⟩, acylation ⟨87CC153, 93JA6094, 93T4627⟩, sulfonylation ⟨92T6079, 93TL3639⟩, halogenation ⟨92RTC75, 92JCS(P2)1541, 93JA10276⟩, silylation, and phosphorylation of aziridines at nitrogen abound ⟨B-83MI 101–01⟩. Aziridinium salts may be prepared by further alkylation of the aziridine nitrogen ⟨B-83MI 101–01⟩. Aziridines can be alkylated on nitrogen with epoxides producing β-hydroxyamines ⟨92JMC3573⟩. Azirines undergo Michael additions, *e.g.* (**18 → 19**) ⟨92JCR(S)266⟩.

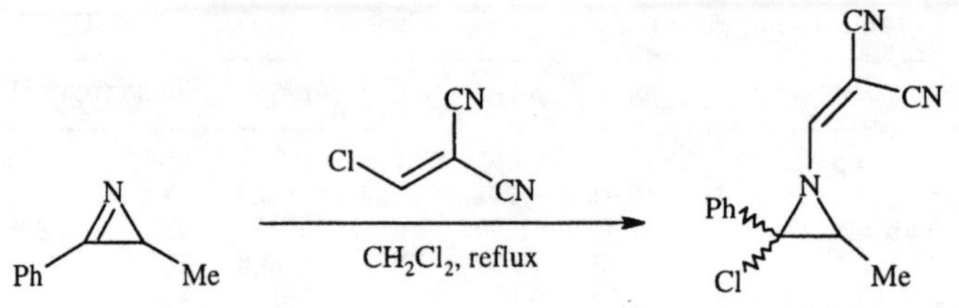

(18)　　　　　　　　　　　**(19)**

Azetidine forms salts and can be acylated (with RCOCl) or nitrosated (with HNO$_2$). Azetidines are readily alkylated at nitrogen by alkyl halides ⟨86JOC5489⟩. The *N*-chlorination of azetidines occurs smoothly with *N*-chlorosuccinimide or *t*-butyl hypochlorite ⟨71JA933, 87CB1971, 88T4447, 89G81⟩.

Nitration of azetidines with acetyl nitrate or, indirectly, *via* dealkylative nitration using HNO$_3$ and acetic anhydride furnished *N*-nitroazetidines **(20)** and **(21)** ⟨90JOC2920, 90SC407⟩. *N*-Perchlorylazetidines are formed by reaction of 1-unsubstituted aziridines with dichlorine heptoxide (Cl$_2$O$_7$) ⟨87ZAAC(547)233, 91TL4985⟩.

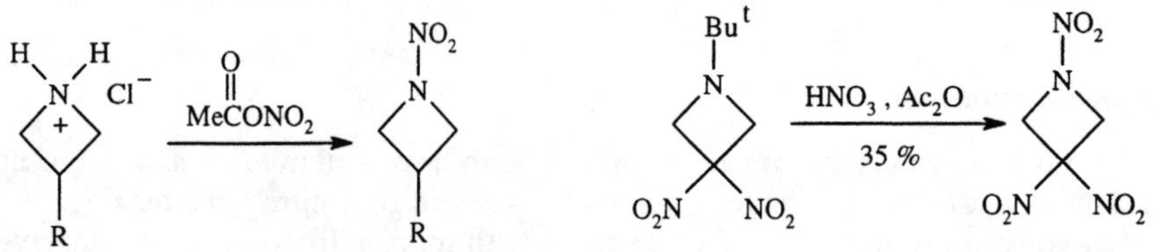

(20)　　　　　　　　　　　**(21)**

In large rings the fate of the onium ions depends mostly on the structure and degree of unsaturation of the particular compound, and the onium salts range from completely stable to highly unstable.

1,4-Dihydro–1,4-diazocine is deprotonated by potassium in liquid ammonia to the dianion which reacts with electrophiles such as methyl iodide, trimethylsilyl chloride, and methyl chloroformate to form the corresponding 1,4-disubstituted derivatives **(22 → 23 → 24)** ⟨79AG(E)967⟩.

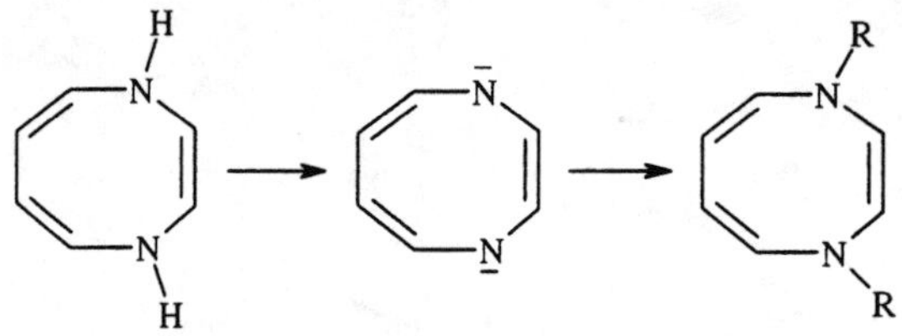

(22)　　　　　　　**(23)**　　　　　　　**(24)**

3.5.4　NUCLEOPHILIC ATTACK ON RING HETEROATOMS

Nucleophilic attack on ring heteroatoms is found most often in two situations: (1) where the heteroatom in question is sulfur, and (2) in small rings with two heteroatoms. For examples of the former, see CHEC 5.14.3.5.2 and CHEC 5.06.3.4. Oxaziridines are attacked on oxygen when bulky ring substituents are present, otherwise the nitrogen is attacked, resulting in nitrogen transfer and the formation of a carbonyl compound (CHEC 5.08.3.1.4), while the ring carbon is altogether inert toward nucleophilic attack.

S-Alkylthiiranium salts, *e.g.* **(25)**, can be desulfurized by fluoride, chloride, bromide, or most conveniently by iodide ions (Equation (1)) ⟨78CC630⟩.

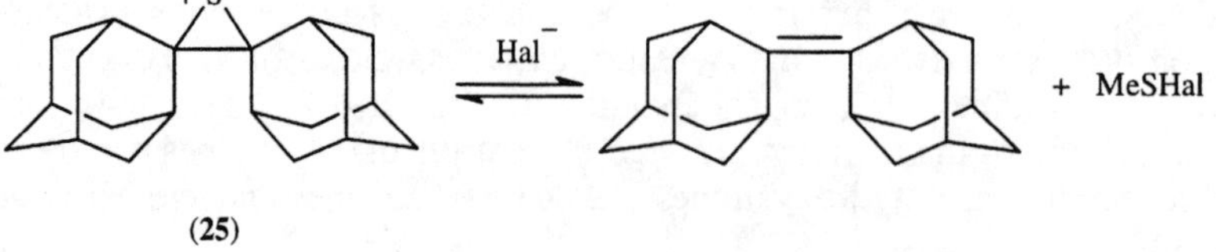

(25)

Stable Wittig reagents react with episulfides to afford the corresponding alkenes (Scheme 16), probably *via* a thiocarbonyl intermediate ⟨87CL357⟩.

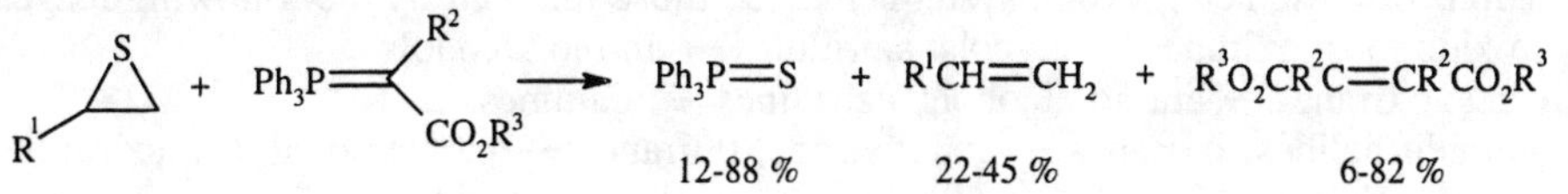

$$R^1 = Me, Et, Ph; \quad R^2 = H, Me; \quad R^3 = Me, Et, Pr^i$$

Scheme 16

Nucleophiles react with *N*-H and *N*-acyl oxaziridines with transfer of NH and *N*-acyl, respectively. Much of this chemistry ⟨91S327⟩ has been carried out with cyclohexanespiro–3′-oxaziridine (**26**): attack at the NH group gives intermediate (**27**) which gives cyclohexanone and amination of the nucleophile. Transfers of NH to *N*-, *O*-, *S*-, and *C*-nucleophiles, enable the syntheses of hydrazines, *N*-amino-peptides, hydroxyamines, sulfenamides, thiooximes, sulfonamides, aziridines, and α-amino acids.

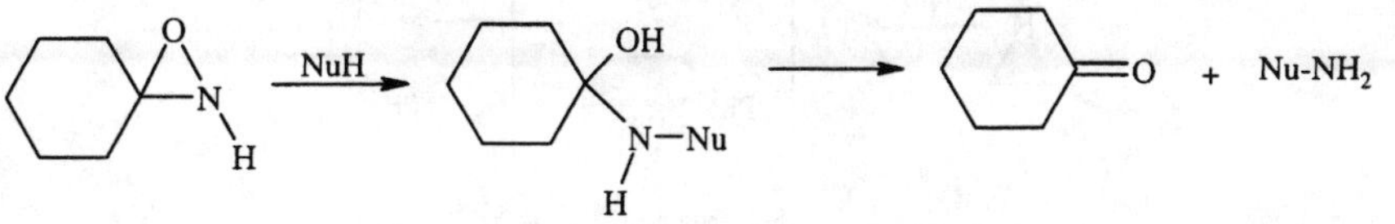

Nucleophiles attack the imide sulfur atom of trithiadiazepines. Trithiadiazepine (**29**, R = H) is converted by potassium *t*-butoxide and iodomethane into 3,4-dimethylthio–1,2,5-thiadiazole (**30**) by a rearrangement in which two of the ring sulfur atoms have become exocyclic; with triphenylphosphine (**29**, R = CO$_2$Me) gives dimethyl 1,2,5-thiadiazole–3,4-dicarboxylate (**28**) ⟨87JCS(P1)217⟩.

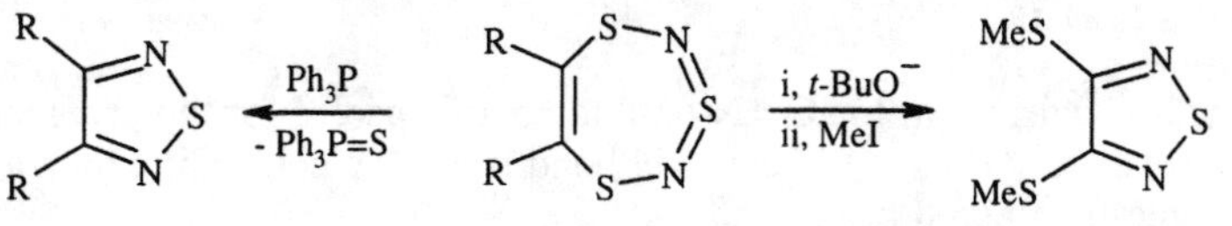

3.5.5 NUCLEOPHILIC ATTACK ON RING CARBON ATOMS

The ring opening of small heterocycles by nucleophilic attack on a carbon adjacent to a heteroatom is exceedingly common. Only in oxaziridines is the ring carbon inert relative to the two ring heteroatoms. In the other three-membered rings, nucleophilic ring opening leads to the corresponding heteroanion or to the XH compound, according as to whether the heteroatom is being protonated before or concurrently with the ring opening, respectively. The reaction can be stereospecific in any of these cases. Alternatively, ring opening can occur before the nucleophilic attack, either after protonation of the heteroatom (to give a carbocation), or due to ylide formation. In the latter cases the reactions become non-stereospecific or partially stereospecific, depending on the timing of the processes involved.

Protonation or Lewis acid complexation of a heteroatom invites nucleophilic attack, including nucleophilic attack by a non-protonated molecule. Oligomerization and polymerization are thus often the results of bringing heterocycles into an acid environment.

3.5.5.1 Reactions of Three-membered Rings

The three-membered rings containing one heteroatom, because of ring strain, are much more reactive than normal ethers, sulfides and amines. Under basic or neutral conditions ring fission takes place preferentially at the least substituted carbon and is accompanied by inversion, *i.e.* S$_N$2 type; under acid conditions these rules do not always apply because of the increasing S_N1 character of the transition state.

The reactions of these heterocyclic systems include those initiated by the following reagents:

(i) Hydroxide ions: oxiranes → glycols, aziridines → amino alcohols.

(ii) Amines: oxiranes → amino alcohols, aziridines → diamines.

(iii) Hydrogen halides: oxiranes → halohydrins, thiiranes → mercaptohalides, aziridines → halo amines.

(iv) Grignard reagents: oxiranes → alcohols, *e.g.* $C_2H_4O + RMgBr \rightarrow RCH_2CH_2OH$.

(v) Catalytic amounts of either a nucleophilic or an electrophilic reagent can induce polymerization; ring fission occurs first, and the ring fission product reacts with additional molecules of the cyclic starting material giving dimers or high polymers, *e.g.* ($\cdots CH_2CH_2\text{-Z-}CH_2CH_2\text{-Z-}CH_2CH_2 \cdots$).

Direct nucleophilic attack at a ring carbon atom of an aziridine is best accomplished when an electron-withdrawing group is present on the nitrogen atom. Examples are shown in Schemes 17 and 18. For reviews see ⟨91COS(6)65, 94AG(E)599⟩.

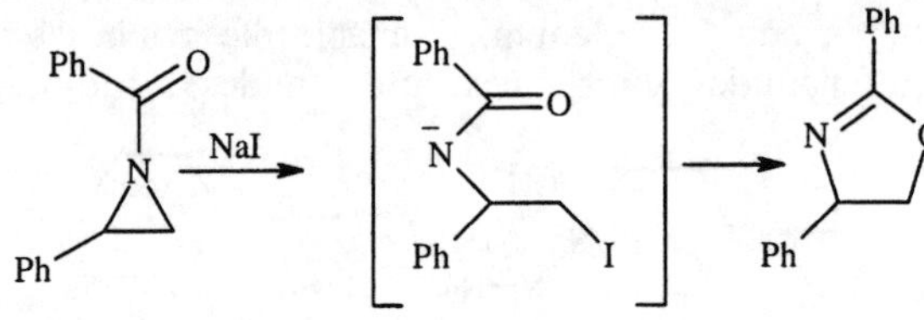

Scheme 17

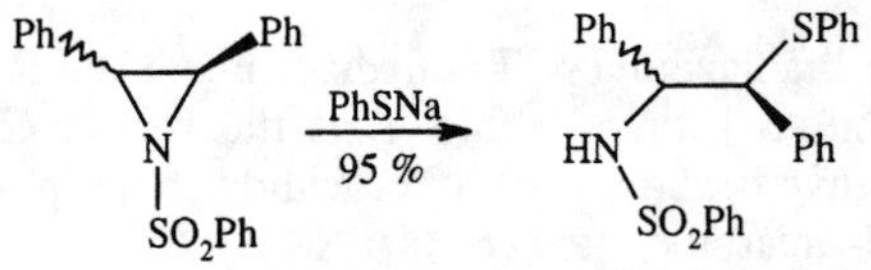

Scheme 18 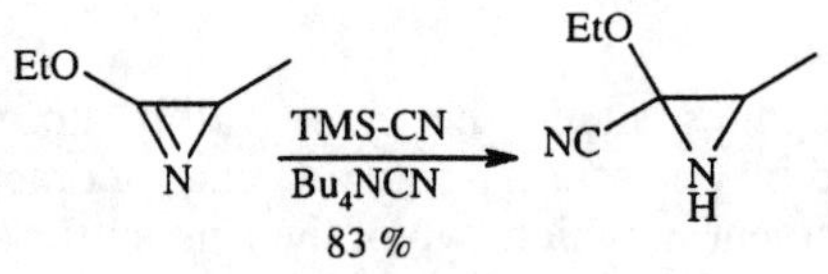   **Scheme 19**

Nucleophilic attack on azirines at the C=N double bond is useful for the preparation of substituted aziridines ⟨B-83MI 101–02, 84CHEC-I(7)47⟩. The C=N bond is more electrophilic than a normal imine due to the strain of the three-membered ring.

Nucleophilic additions to 3-alkoxy ⟨86RTC456⟩ and 3-amino-2*H*-azirines ⟨91AG(E)238⟩ are especially well studied, *e.g.* Scheme 19. Many of these reactions involve assistance by protonation of the nitrogen.

Alkyl-substituted thiiranes (**31**) with 3-chloroallyllithium give pure 2-vinylthietanes (**32**) in good yields ⟨85S1069⟩.

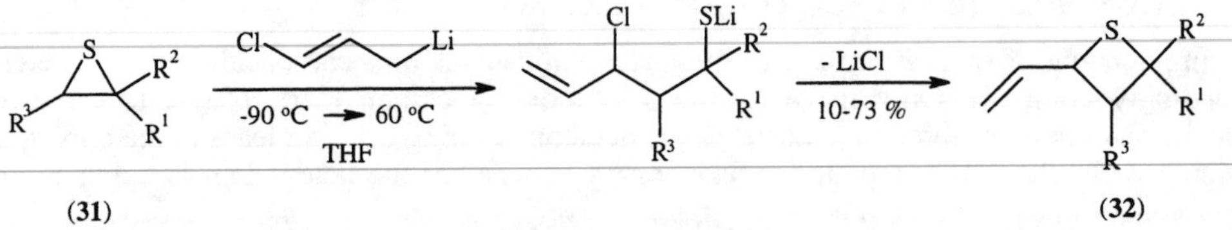

(**31**)   (**32**)

Nucleophilic ring-opening are among the most important reactions of oxiranes ⟨83T2323, 92AG(E)1179, 93JOC1221⟩, driven by (a) the ring strain, (b) the polarization of the C–O bonds in the small ring system, and (c) the basicity of the oxirane oxygen. The stereoselectivity of the ring opening of oxiranes is usually completely *anti* ⟨71G300, B-72MI 103–01⟩.

3.5.5.2 Reactions of Four-membered Rings

The properties of azetidine, oxetane and thietane are intermediate between those of aliphatic amines, ethers and sulfides on the one hand and those of the corresponding three-membered ring systems on the other. Ring fission occurs quite readily. Oxetane reacts with Grignard reagents to give alcohols of type $R(CH_2)_3OH$ and with hydrogen bromide to give 1,3-dibromopropane. Hydrogen halides convert azetidine into γ-halo amines.

3.5.5.3 Reactions of Carbonyl Derivatives of Four-membered Rings

2-Oxetanones (β-lactones) are readily attacked by nucleophilic reagents. Reaction occur by:
(i) Alkyl-oxy fission (**33** → **34**), *e.g.* propiolactone with NaOAc-H$_2$O yields (**34**; Nu = OAc), with MeOH-NaOMe it forms (**34**; Nu = OMe).
(ii) Acyl-oxy fission (**35** → **36**), *e.g.* propiolactone reacts with MeOH-H$^+$ to give (**36**; Nu = OMe).

The reaction of 2-azetidinones (β-lactams) with nucleophilic reagents is accompanied by acyl-nitrogen fission, as is normal for amides (*cf.* **35** → **36**), *e.g.* propiolactam yields β-alanine (NH$_2$CH$_2$CH$_2$CO$_2$H) on hydrolysis.

(33)   (34)   (35)   (36)

Various nucleophilic attacks at C(3) and C(4) of β-lactams have been employed to introduce desired functionalities. Nucleophilic displacement of a halogen at C(3) by nucleophiles, for example, potassium phthalimide, has been performed ⟨93JCS(P1)2357⟩. The α-azidation of 1-hydroxy-2-azetidinones, for example (**37**), with arenesulfonyl azides in the presence of triethylamine affords 3-azido-2-azetidinones (**38**) *via* O-tosylation and S$_N$2′-type of displacement of the tosyloxy group ⟨92JA2741, 93JA548, 93BMC2429⟩.

(37)   (38)

3.5.5.4 Large Rings

Nucleophilic attack on ring atoms of large heterocycles is largely confined to saturated systems, saturated parts of partially unsaturated systems, and to carbonyl functions and the like. These reactions are not fundamentally different from those of the corresponding acyclic systems, except for transannular reactions.

Transannular nucleophilic attack on ring atoms is best known in systems with seven or more ring members. For example, nucleophilic attack by the ring nitrogen on suitably substituted ring carbons in the 3- or 4-position in azepine derivatives has been studied (CHEC 5.16.3.5.2). However, transannular nucleophilic attack can be found already in four-membered heterocycles. The nitrogen of N-*t*-butyl-3-chloroazetidine is, because of the puckered conformation, close enough to the 3-carbon to displace chloride. This, in turn, opens the azabicyclobutanonium ion to give N-*t*-butyl–2-chloromethylaziridine ⟨B-77SH(1)38⟩.

In systems of appropriate geometry, nucleophiles within a side chain may be well connected for attack on ring atoms. For example, an aminomethyl group at the 5-position of a dibenzazepin-2-one was found to attack the carbonyl group (CHEC 5.16.3.5.2). Such reactions should be possible in rings of any size.

Apparent nucleophilic attack on large, fully unsaturated rings may occur by way of attack on a valence tautomer, such as the reaction of oxepin with azide ion. Attack on the oxanorcaradiene valence tautomer leads to ring opening of the three-membered ring, and formation of 5-azido-6-hydroxy-1,3-cyclohexadiene (CHEC 5.17.2.2.4).

3.5.6 NUCLEOPHILIC ATTACK ON PROTONS ATTACHED TO RING ATOMS

The formation of anions by proton abstraction leads to highly diversified types of reactions in small and large heterocycles. However, a few rules do generally apply. In the absence of complicating substituents, nitrogen anions can often be formed from small and large heterocycles, and subsequent alkylation, acylation, *etc.* can be achieved ⟨B-76MI50200, 64HC(19-2)886⟩ (CHEC 5.15.2.3). The N-anions of large rings are usually unexceptional (barring transannular reaction). N-Anions of unsaturated large heterocycles are obtained with difficulty, but they are synthetically useful in the azepine field (CHEC 5.16.3.6).

Proton abstraction from ring carbons of small and large heterocycles often leads to ring opening. Oxiranes, for example, are usually converted to carbenes under conditions under which a carbanion is formed, with a few exceptions such as the synthetically useful 2-triphenylsilyl-2-lithiooxirane (CHEC 5.05.3.5). Likewise, oxaziridines can be opened by nucleophilic attack on a hydrogen at carbon-3 ⟨82JOC419⟩ (CHEC 5.08.3.1.4). Thiirane and thiirene 1,1-dioxides give cyclic carbanions, which are easily ring opened (CHEC 5.06.3.6).

Carbanions at C(2) of the aziridine ring may be generated by deprotonation or by exchange, *e.g.* tin-lithium. A major problem with the deprotonation approach is the necessity of a strong base which may also react by a nucleophilic ring-opening process. *N*-(*t*-Butoxycarbonyl)aziridines may be deprotonated with Bu^{s}Li/TMEDA, as shown in Scheme 20 ⟨94JOC276⟩.

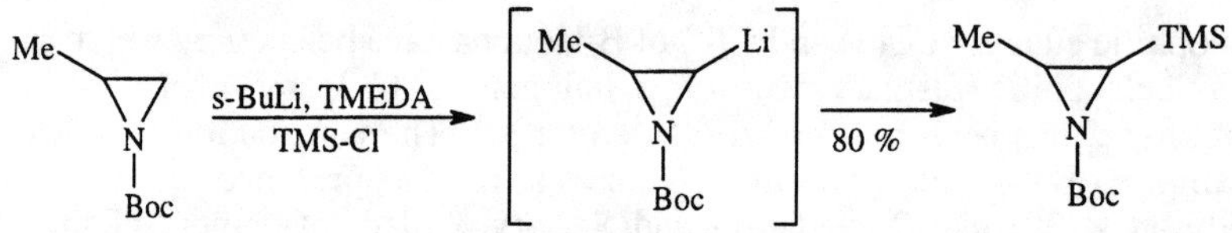

Scheme 20

Various substitutions of hydrogen at position three and four in β-lactams have been performed by electrophilic reagents. The 3-position is activated by the carbonyl function. Alkylation at the 3-position is readily executed *via* the enolate with alkyl halides ⟨90T4733⟩, aldehydes, ketones, carbon dioxide, *etc.* ⟨83LA1152, 83LA1162, 84H(22)1161, 93JOC245⟩. The carboxyl group at C(4) does not interfere as the dilithium salt is alkylated at C(3) with excellent stereocontrol, giving the *trans*-disubstituted lactam (**39**) ⟨90CC720⟩. Hydrolysis leads to unprotected β-alkyl aspartates (**40**) ⟨92SL33⟩.

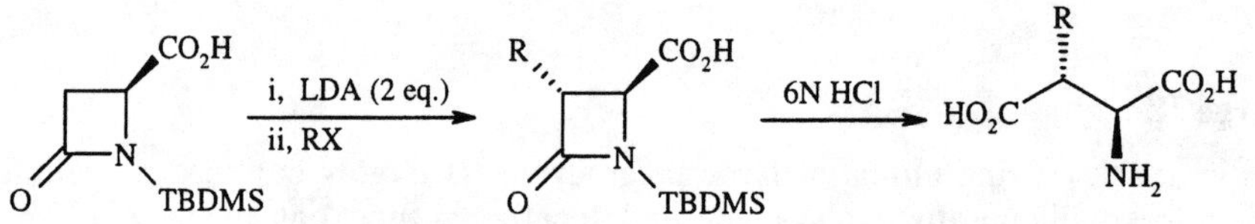

(**39**)   (**40**)

3-Alkyl- and 2,3-dialkyl-thietane 1-oxides, upon deprotonation α to the sulfinyl group, undergo stereospecific ring contraction giving cyclopropylsulfenate anions, which react with methyl iodide to give the corresponding cyclopropyl methyl sulfoxides (Scheme 21). Deprotonation, achieved with lithium cyclohexylisopropylamide, occurs at the α-proton *syn* to the sulfinyl oxygen. Rearrangement occurs stereospecifically with respect to configuration at sulfur, at the migrating residue (retention), and at the migration terminus (inversion) ⟨82CC589⟩.

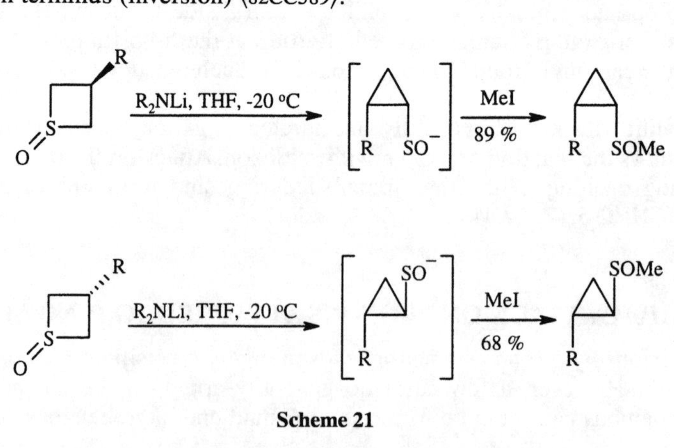

Scheme 21

Large heterocyclics offer a greater variety for ring opening, such as the abstraction of protons from not only the 2- but also the 3-position, leading to ring opening by β-elimination and the formation of ω-unsaturated compounds, such as 6-hydroxy-1-hexene from oxepane (CHEC 5.17.2.1.4).

3.5.7 ATTACK BY RADICALS OR ELECTRON-DEFICIENT SPECIES. OXIDATION AND REDUCTION

3.5.7.1 Reactions with Radicals and Carbenes

Surprisingly little is known about the attack of radicals on small and large heterocycles. Hydrogen abstraction from the heteroatom of small rings leads to ring opening, and in the case of thiiranes to removal of the sulfur (*cf.* Section 3.5.1.4 above). Abstraction of H· exocyclic and α to nitrogen in oxiranes leads to N-O cleavage, and the reaction of vanadium(IV) with the oxygen of 1-oxa-2-azaspiro[2.5]octane gives N-O cleavage and ring expansion to caprolactam (Scheme 22; CHEC 5.08.3.1.5).

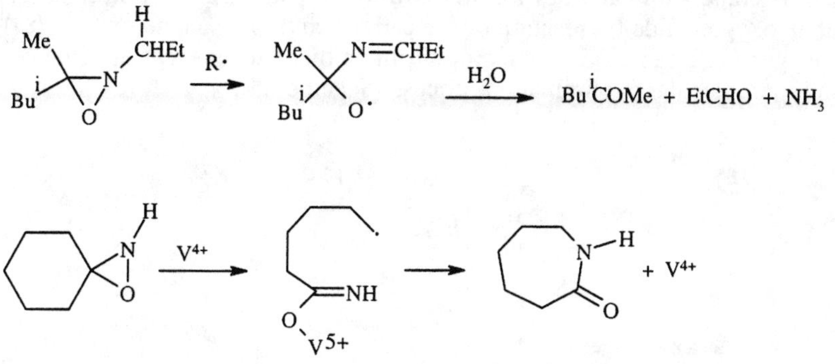

Scheme 22

The formation of a radical on a carbon adjacent to the aziridine ring by cleavage of a carbon-bromine ⟨94SL287⟩ or carbon-selenium bond ⟨90CC434, 93TL4901⟩ with Bu₃SnH/AIBN causes ring-opening in much the same fashion as a cyclopropylcarbonyl radical as shown in Scheme 23.

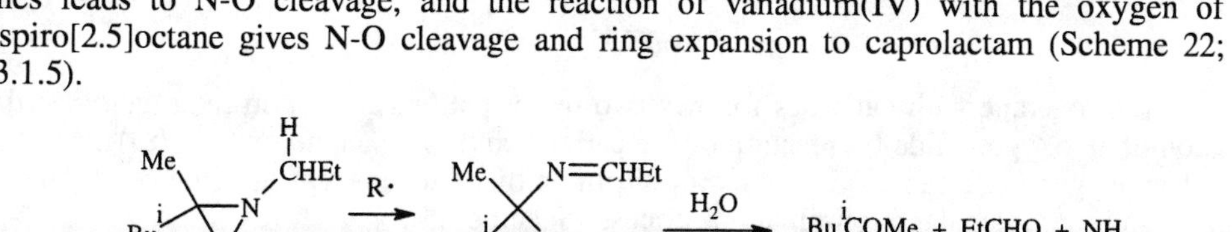

Scheme 23

Thiiranes are desulfurized by radicals (H·, S·), by singlet carbenes and by electrolysis (CHEC 5.06.3.7.1).

Azetidine derivatives, which are less strained, are less sensitive and removal of a hydrogen in the α-position of a substituent on nitrogen does not necessarily lead to ring opening (CHEC 5.09.3.2.5). Thietane rings are opened by radicals attacking on S, while the less strained thiolanes are attacked by hydrogen abstraction at a 2-carbon (CHEC 5.14.3.10.1). In thietane 1,1-dioxides, radicals abstract a hydrogen from the 3-position to give a cyclic radical (CHEC 5.14.3.10.1). Producing radicals exocyclic in the α-position of *N*-substituents of azetidin-2-ones did not result in ring opening (CHEC 5.09.3.2.5).

Saturated large rings may form nitrogen radicals by H abstraction from N, or abstraction may occur in the α- or β-positions in non-nitrogen systems. Oxepane gives the radical in the 2-position, with subsequent cleavage and reclosure of the intermediate carbenoid to cyclohexanol (CHEC 5.17.2.1.5). In unsaturated large systems a variety of reactions, unexceptional in their nature, are found. Some azepines can be brominated by *N*-bromosuccinimide; others decompose under similar conditions (CHEC 5.16.3.7).

Electron-deficient species can attack the unshared electron pairs of heteroatoms, to form ylides, which usually undergo further conversions. Thus, treatment of thiiranes with substituted carbene often gives the corresponding alkenes in good yields by electrophilic attack by carbene on the sulfur atom ⟨75JA2553⟩.

Thiiranes react with benzyne in an efficient synthesis of phenyl vinyl sulfides. The reaction is stereospecific, thus producing *cis*-(phenylthio)stilbene from *cis*-2,3-diphenylthiirane and *trans*-(phenylthio)stilbene from *trans*-2,3-diphenylthiirane (Scheme 24) ⟨84TL2679⟩.

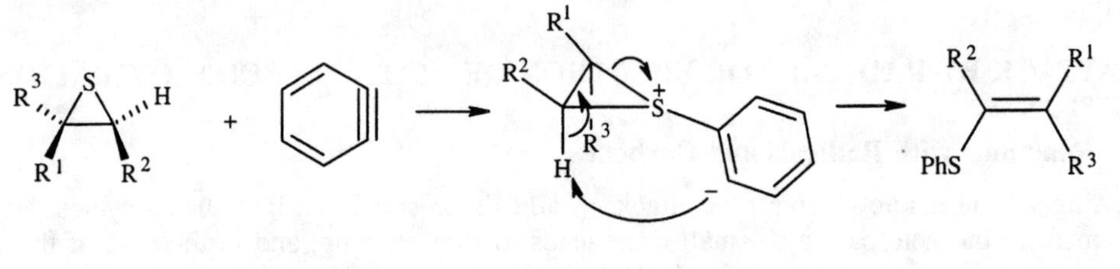

R[1], R[2], R[3] = H, alkyl, aryl

Scheme 24

The reaction of oxetane with carbenes follows two major pathways, carbon-hydrogen insertion or the formation of an oxygen ylide by reaction of the carbene and the oxetane oxygen ⟨85TL193, 85TL197, 91CB1853⟩. The oxygen ylide can produce a tetrahydrofuran by a Stevens rearrangement, or generate an allyl ether by an intramolecular β-elimination process (Scheme 25).

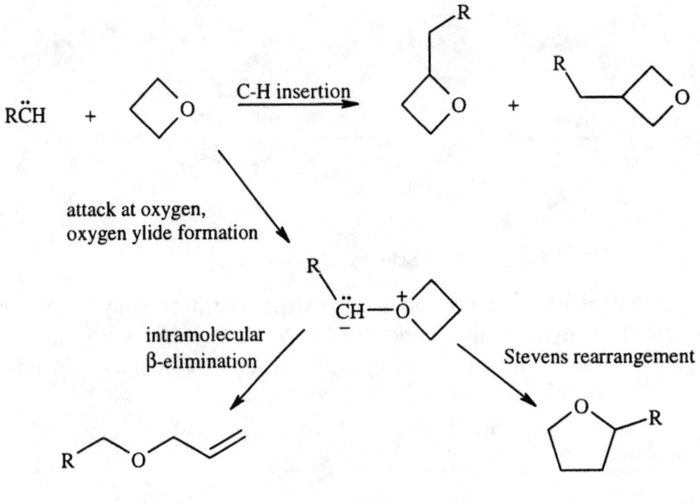

R = H, COOEt, Ph

Scheme 25

Thietane reacts with bis(methoxycarbonyl)carbene to give a S^+-C^- ylide which rearranges to 2,2-bis(methoxycarbonyl)thiolane (CHEC 5.14.3.10.1).

N-Ethoxycarbonylazepine, however, is attacked by dichlorocarbene at the C=C double bonds, with formation of the *trans* tris-homo compound (CHEC 5.16.3.7).

3.5.7.2 Oxidation

Oxidation reactions of small and large heterocycles sometimes have little in common with their open-chain analogues. Oxidations of small heterocyclics include (i) dehydrogenation of ring C-H and N-H bonds leading to the formation of ring C=C, C=N or N=N double bonds, (ii) oxidation of functional groups and (iii) oxidation at the heteroatom, which is typical for sulfur-containing systems. Examples, illustrating items (i) and (ii) are shown in Schemes 26–28 ⟨78TL2469, 83TL825, 93TL6677⟩.

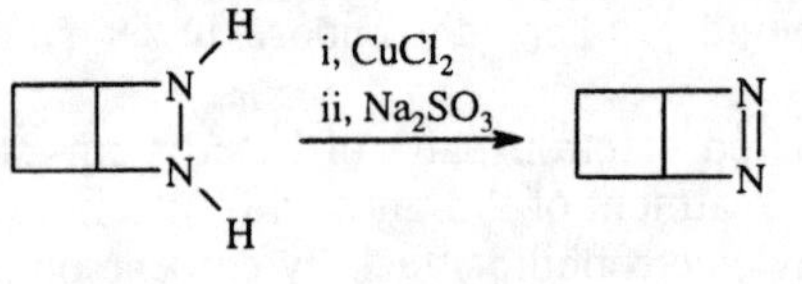

Scheme 26 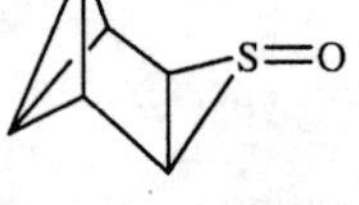   (41)

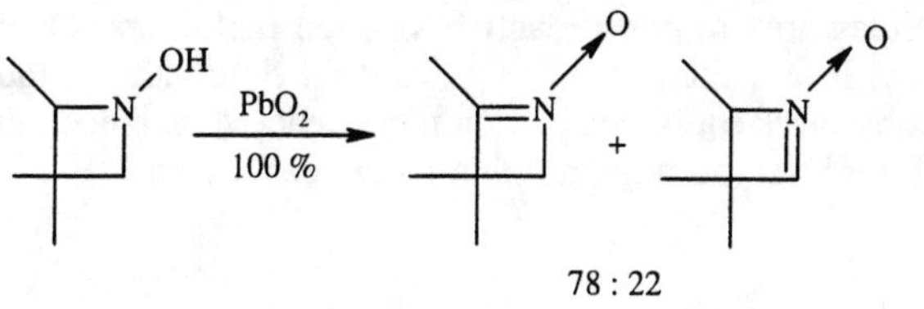

Scheme 27

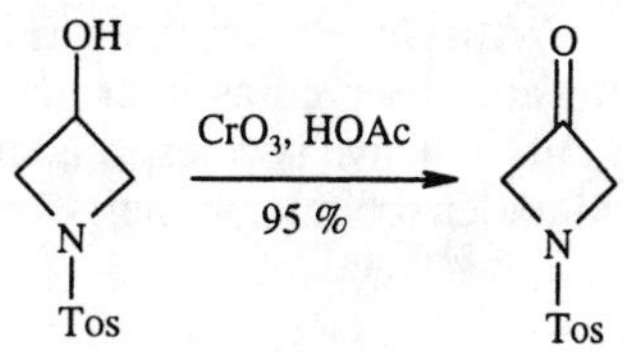

Scheme 28

The isolation of thiirane oxides prepared by oxidation of the corresponding thiiranes is problematic, and the isolation of thiirane dioxides from the direct oxidation of thiiranes or thiirane oxides has yet to be reported. However, highly strained fused-ring thiiranes can be successfully oxidized to thiirane oxides (*e.g.* **41**) with MCPBA or NaIO$_4$ under carefully controlled conditions ⟨82CB3213⟩. Thietane can be oxidized to a sulfone.

3.5.7.3 Reduction

Oxiranes and aziridines are reduced to alcohols and amines, respectively, for example by Ni/H$_2$, Zn-NH$_4$Cl, P-I, Al-Hg, Na/NH$_3$, Li/EtNH$_2$ ⟨71JOC330⟩. Cleavage of aziridines with lithium naphthalenide provides synthetically useful dianions as shown in Scheme 29 ⟨93TL1649, 94JOC3210⟩.

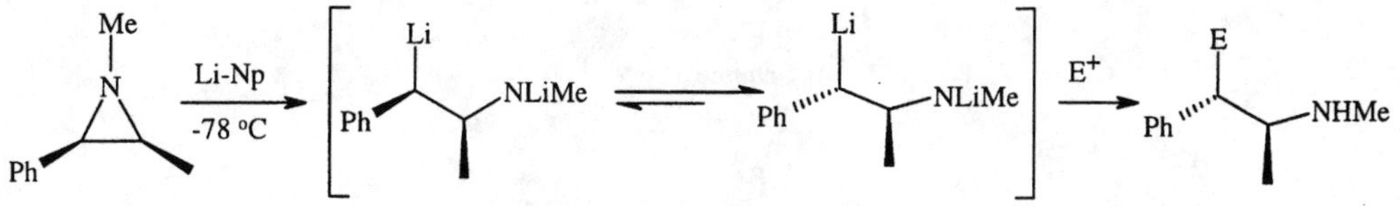

Scheme 29

1,2-Dioxetanes are readily reduced to 1,2-diols (Scheme 30) by lithium aluminum hydride ⟨75CJC1103⟩, thiols ⟨88AG(E)429⟩, and biologically important reductants ⟨89MI 133-02⟩ such as ascorbic acid, tocopherol, dihydronicotinamide adenine dinucleotide (NADH), and riboflavin adenosine diphosphate (FADH$_2$).

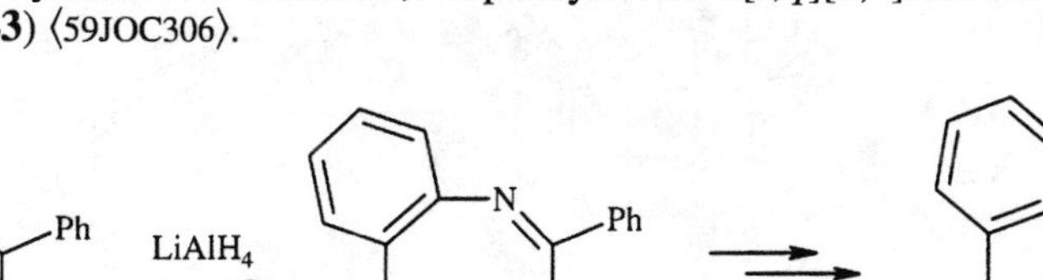

Scheme 30

Lithium aluminum hydride reduction of 6,7-diphenyldibenzo[*e,q*][1,4]diazocine (**42**) affords a ring-contracted product (**43**) ⟨59JOC306⟩.

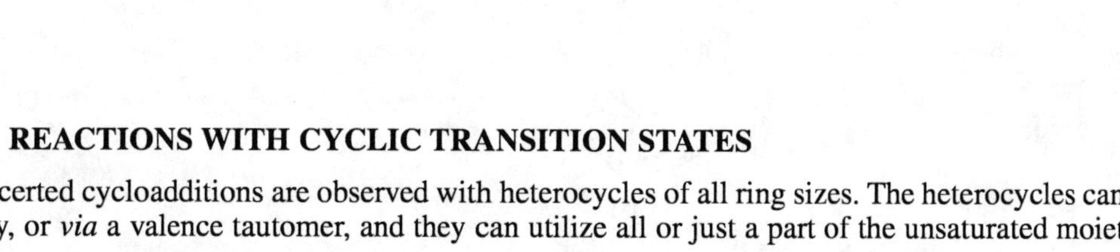

(42)

(43)

3.5.8 REACTIONS WITH CYCLIC TRANSITION STATES

Concerted cycloadditions are observed with heterocycles of all ring sizes. The heterocycles can react directly, or *via* a valence tautomer, and they can utilize all or just a part of the unsaturated moieties in their rings. With three-membered rings, ylides are common reactive valence tautomers. Open chain 4π-systems are observed as intermediates with four-membered rings, and bicyclic valence tautomers are commonly reactive species in additions by large rings. Very often these reactive valence tautomers are formed under orbital symmetry control, by both thermal and photochemical routes.

3.5.8.1 [2 + 4] Cycloadditions

3.5.8.1.1 Heterocycles as dienophiles

Cycloaddition reactions of the C=N bond of azirines are common, *e.g.* Scheme 31 ⟨71AHC(13)45, B-83MI 101-03, 84CHEC(7)47⟩. Azirines can also participate in [4 + 2] cycloadditions with cyclopentadienones, isobenzofurans, triazines, and tetrazines.

The participation of a single double bond of a heterocycle is found in additions of thietes (CHEC 5.14.3.1.1). Azepines and non-aromatic heteronins react in this mode, especially with electron-deficient dienes (Scheme 32; CHEC 5.16.3.8.1).

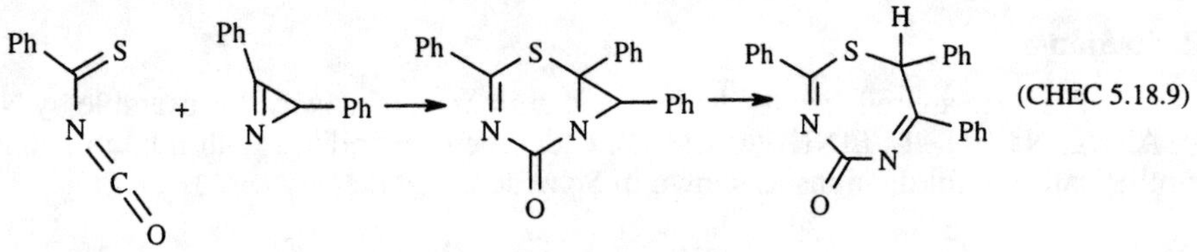

Scheme 31

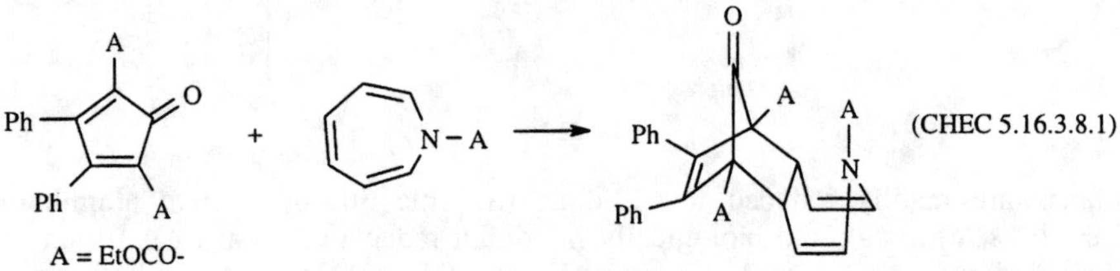

Scheme 32

3.5.8.1.2 Heterocycles as dienes

2,3-Di-*t*-butyl-4-mesitylazete (44) dimerized at 130°C into (45), while further heating at 185°C afforded 1,5-diazocin (46) ⟨91MI 118–03⟩. In the first stage of this reaction azete derivative (44) behaves as diene and dienophile at the same time.

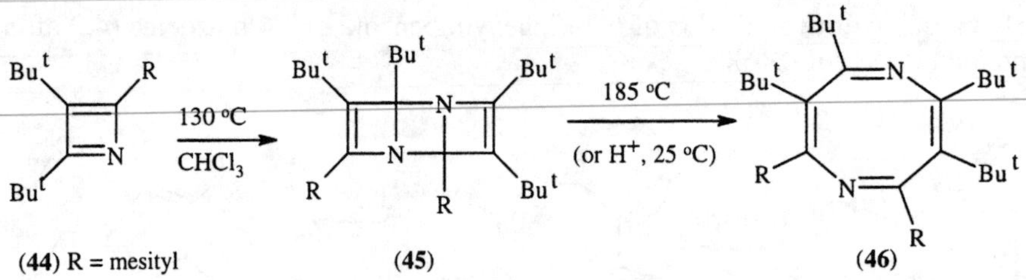

Kinetically stabilized azetes also show a high tendency for cycloaddition with a variety of other reagents. Cycloaddition of (47) with triplet oxygen produced a fully characterized dioxetan adduct (48), which decomposed at 25°C into *t*-butyl cyanide and the α-dione fragment ⟨88AG(E)272⟩.

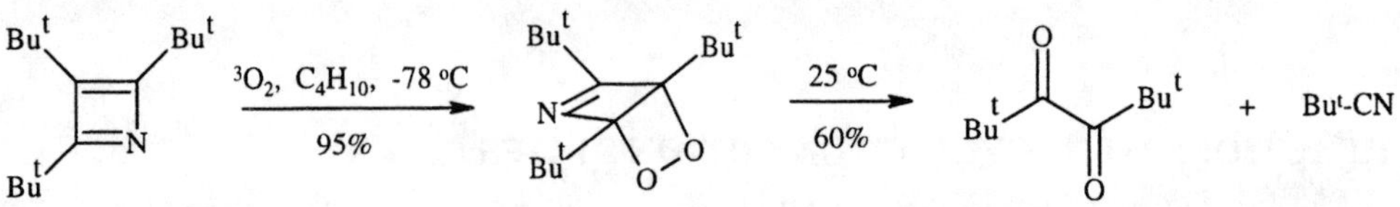

The cycloaddition potential of compound (**47**) is also demonstrated by its reaction with alkynes leading to isolable 1-Dewar pyridines (**50**). The latter were rearranged thermally or by acid catalysis into pyridines (**49**). The photochemical rearrangement of (**50**) led to azaprismane derivative (**51**) ⟨89T3115⟩.

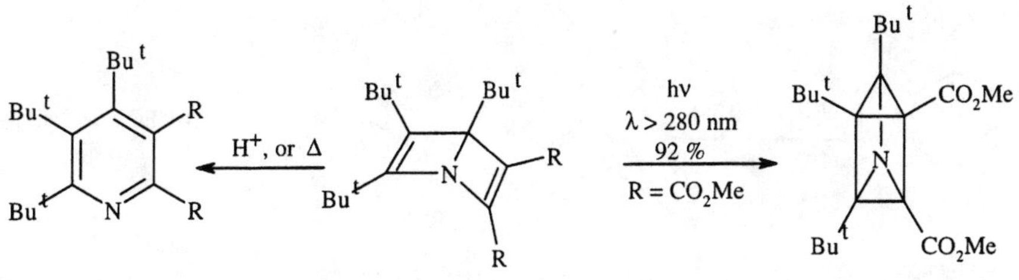

(**49**)   (**50**)   (**51**)

Non-aromatic 1,4-dihydro–1,4-diazocine derivatives (**52**) react readily as dienes with *N*-phenyltriazolinedione, affording the corresponding Diels-Alder adducts ⟨79AG(E)964, 79AG(E)962⟩. 1,4-Oxazocine derivatives behave similarly ⟨83CB2492⟩.

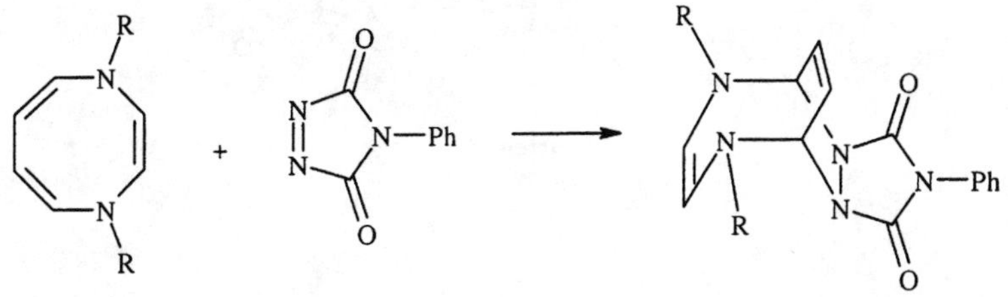

(**52**) R = PhCH$_2$ (**53**)

Diene moieties, reactive in [2 + 4] additions, can be formed from benzazetines by ring opening to azaxylylenes (CHEC 5.09.4.2.3). 3,4-Bis(trifluoromethyl)-1,2-dithietene is in equilibrium with hexafluorobutane-2,3-dithione, which adds alkenes to form 2,3-bis(trifluoromethyl)-1,4-dithiins (Scheme 33; CHEC 5.15.2.4.6).

The four-membered ring of 2*H*-benzo[*b*]thiete (**54**) (see also Section 3.5.2.2.c) can readily undergo retro [2 + 2] ring opening under thermal or photochemical conditions forming an *ortho*-thiobenzoquinone methide (**55**). This highly reactive 8π-electron species can participate in cycloaddition reactions with a variety of dienophiles ⟨93CB775, 94CB955, 94SR23⟩.

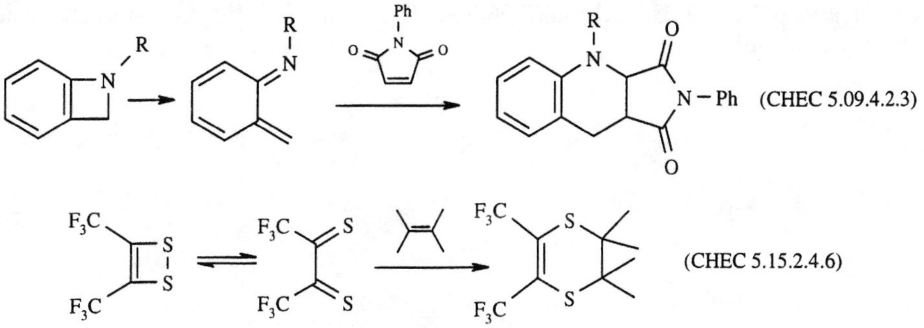

(CHEC 5.09.4.2.3)

(CHEC 5.15.2.4.6)

Scheme 33

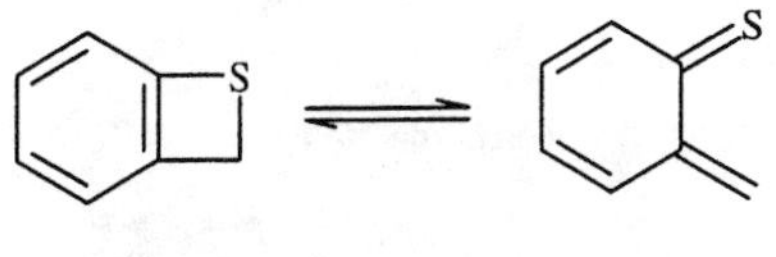

(**54**)   (**55**)

Systems with more than two conjugated double bonds can react by [6π + 2π] processes, which in azepines can compete with the [4π + 2π] reaction (Scheme 34; CHEC 5.16.3.8.1). Oxepins prefer to react as 4π components, through their oxanorcaradiene isomer, in which the 4π-system is nearly planar (Scheme 35; CHEC 5.17.2.2.5). Thiepins behave similarly (CHEC 5.17.2.4.4). Non-aromatic heteronins also react in orbital symmetry-controlled [4+2] and [8+2] cycloadditions (CHEC 5.20.3.2.2).

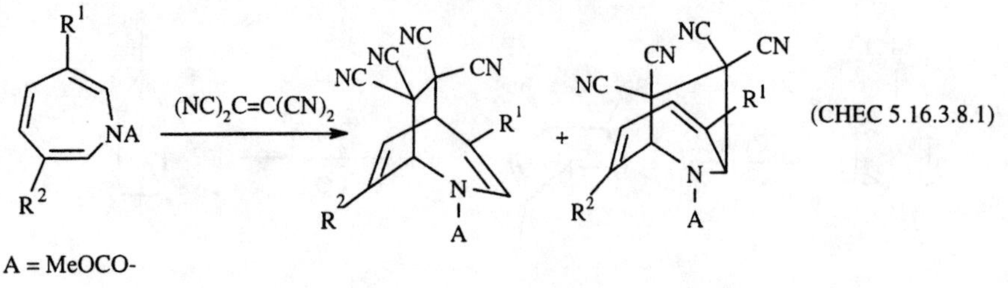

A = MeOCO-

(CHEC 5.16.3.8.1)

Scheme 34

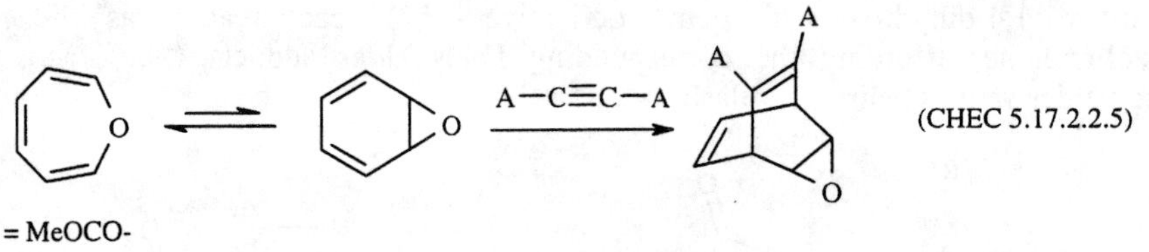

A = MeOCO-

(CHEC 5.17.2.2.5)

Scheme 35

3.5.8.2   1,3-Dipolar Cycloadditions

1,3-Dipolar cycloadditions in which the heterocycle provides the 1,3-dipole are common with three-membered rings, which can provide ylide intermediates, as has been mentioned above. Ylide formation is usually orbital symmetry-controlled and may be achieved by thermolysis or photolysis with the expected stereochemical consequences. Rotation about one of the ylide C-X bonds results in loss of the original stereochemistry. These interconversions often are slow compared to the cycloaddition reactions of the ylides, so that partial stereospecificity is observed. Scheme 36 gives a generalized reaction. Extensive work has been done with aziridines (CHEC 5.04.3.1). Diaziridines behave similarly (CHEC 5.08.3.2.2) ⟨79AHC(24)63⟩. Azirines produce nitrile ylides upon photolysis (Scheme 37; CHEC 5.04.3.2) ⟨91BCJ2757⟩. Examples for oxiranes and thiiranes are found in the corresponding monograph chapters (CHEC 5.05.3.2.1 and 5.06.3.8). The intramolecular version of this reaction is also possible (Scheme 38).

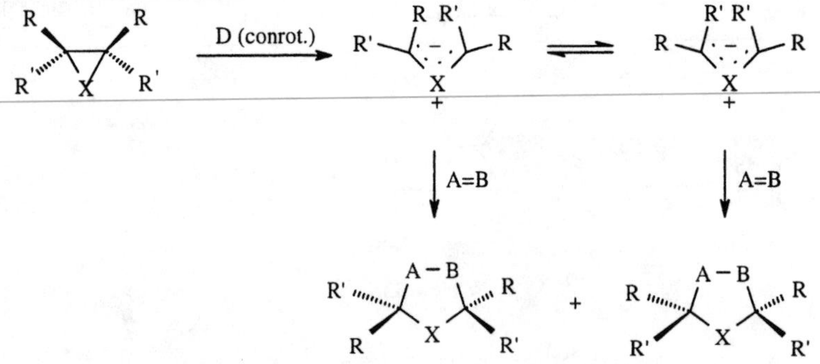

Scheme 36

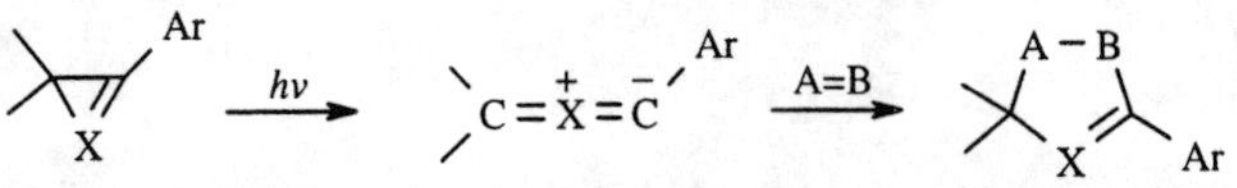

Scheme 37

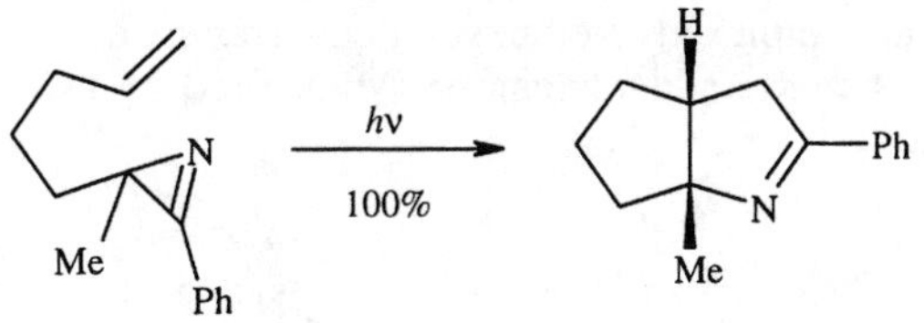

Scheme 38

Heating *N*-lithioaziridines provides 2-azaallyl anions, which undergo concerted cycloaddition reactions with certain alkenes and other anionophiles (Scheme 39) ⟨74AG(E)627, B-88MI 101-02⟩.

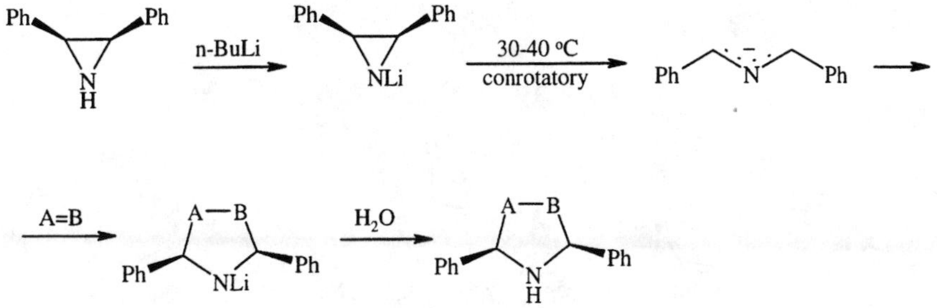

Scheme 39

Heterocyclics of all sizes, as long as they are unsaturated, can serve as dipolarophiles and add to external 1,3-dipoles. Examples involving small rings are not numerous. Thiirene oxides add 1,3-dipoles, such as diazomethane, with subsequent loss of the sulfur moiety (CHEC 5.06.3.8). As one would expect, unsaturated large heterocyclics readily provide the two-atom component for 1,3-dipolar cycloadditions, *e.g.* azepines and thiepins.

3.5.9 REACTIVITY OF TRANSITION METAL COMPLEXES

Metal complexes of heterocyclics display reactivities changed greatly from those of the uncomplexed parent systems. All of the π-electron system(s) of the parent heterocycle can be tied up in the complex formation, or part can be left to do 'alkenic' reactions. The system may be greatly stabilized in the complex, so that reactions, on a heteroatom for example, can be performed which the parent compound itself would not survive. Orbital energy levels may be split and symmetries changed, allowing hitherto forbidden reactions to occur. In short, a multitude of new reaction modes may be made possible by using complexes. Thus, dimerization of azirines with a palladium catalyst may serve as a typical example (Scheme 40) ⟨81JA1289⟩. A variety of other insertion reactions, dimerizations, intramolecular cyclizations, and intermolecular addition reactions of azirines are promoted by transition metals ⟨B-83MI 101-02⟩.

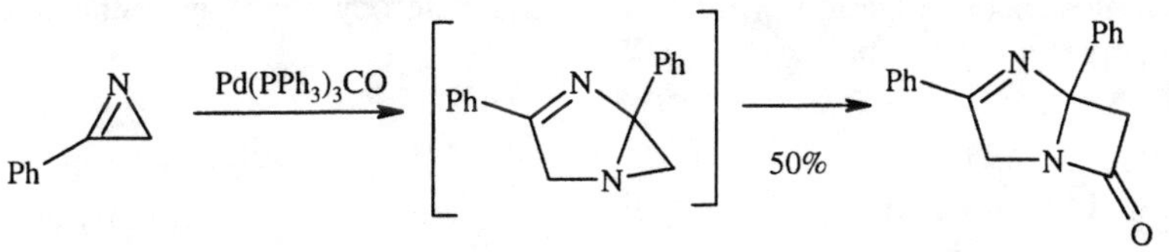

Scheme 40

Since decomplexing is usually possible, many applications await realization, for background see ⟨78JHC1057, 81ACR348, 71JA1123⟩. The chemistry of azepines (CHEC 5.16.3.8.1 and 5.16.3.8.2) and diazepines (CHEC 5.18.2) ⟨78JHC1057, 81ACR348⟩ provides examples.

1*H*-Azepine derivatives form a diene complex with (tricarbonyl)iron, leaving uncomplexed the third of the double bonds. If the 3-position is substituted, two different such complexes are possible, and are in equilibrium, as seen in the ¹H NMR spectrum. An ester group in the 1-position of the complex can be removed by hydrolysis, to give an NH compound which, in contrast to the free 1*H*-azepine, is stable. The 1-position can then be derivatized in the manner usual for amines (Scheme 41). The same (tricarbonyl)iron complex can, by virtue of the uncomplexed 2,3-double bond, serve as the dienophile with 1,2,4,5-tetrazines. The uncomplexed *N*-ethoxycarbonylazepine also adds the tetrazine, but to the 5,6-double bond. Thus, two isomeric adducts can be synthesized by using or not using the complex (Scheme 42; CHEC 5.16.3.8.1).

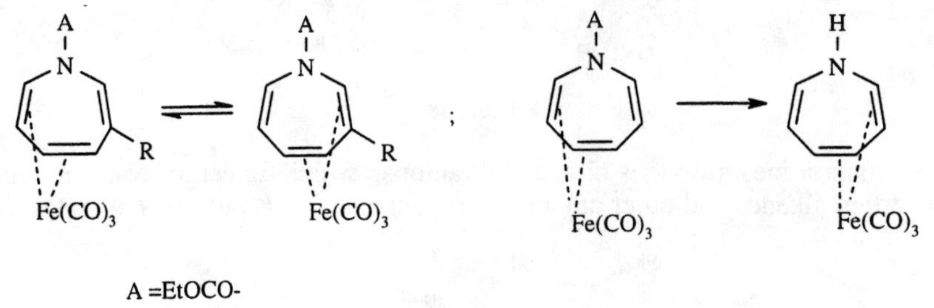

A = EtOCO-

Scheme 41

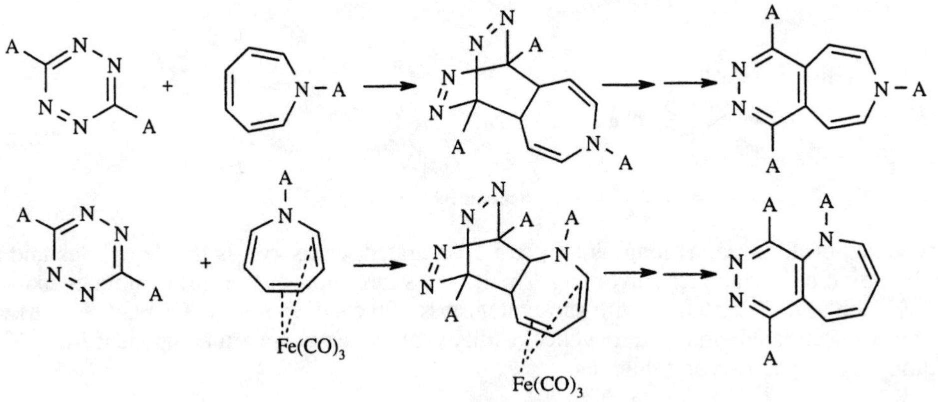

Scheme 42

(Tricarbonyl)iron complexes of 1,2-diazepines do not show the rapid isomerization found in their azepine counterparts (Scheme 41); the iron forms a diene complex with the C=C double bonds in the 4- and 6-positions. The chemistry of the 1,2-diazepine complexes is similar to that of the azepine complexes (CHEC 5.18.2.1) ⟨81ACR348⟩.

Oxepin also forms a diene complex with (tricarbonyl)iron ⟨78JHC1057⟩.

3.5.10 REACTIVITY OF SUBSTITUENTS ATTACHED TO HETEROATOM OR RING CARBON ATOMS

Since small rings are easily opened by various reagents, the development of synthetic procedures involving side-chain functionalization which leaves the heterocyclic ring unchanged is important. Some typical examples are given below.

3.5.10.1 *C*-Linked Substituents

Deprotonation of alkyl groups attached to C(3) of 2*H*-azirines leads to formation of metal-loenamines, which can react with a variety of electrophiles to yield *C*-alkylated products (Scheme 43) ⟨85TL2637⟩.

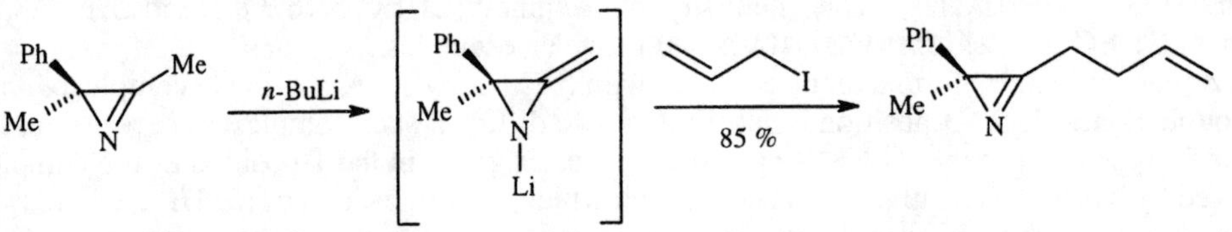

Scheme 43

Optically active aziridines have been prepared in high enantiomeric excess by the enzymatic resolution of *meso* diesters ⟨94AG(E)599⟩. For example, when the *meso*-bis(acetoxymethyl)aziridine (**56**) was subjected to enzymatic hydrolysis with lipase Amano P, the aziridine (**57**) was obtained in 98% *ee* ⟨90TL6663⟩.

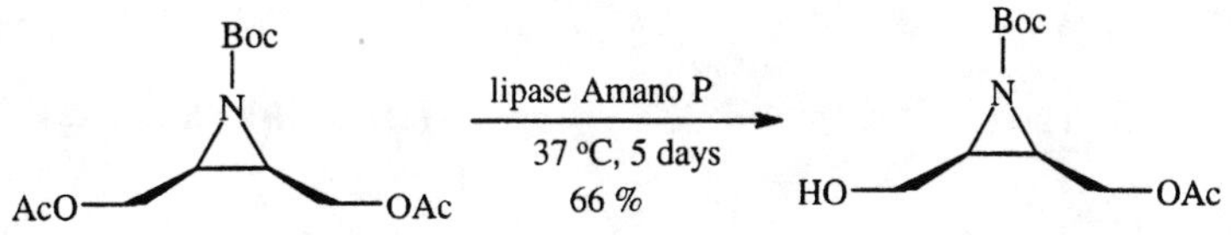

Treatment of certain 2-(hydroxymethyl)aziridines with base can lead to an intramolecular ring-opening reaction to yield an α-amino epoxide, a reaction analogous to the Payne rearrangement of (hydroxymethyl)epoxides. For example, treatment of the aziridine (**58**) with potassium *t*-butoxide provided the epoxide (**59**) in good yield ⟨93TL5127⟩. This type of rearrangement is also believed to occur during the reaction of similar 2-(hydroxymethyl)aziridines with organocuprate reagents ⟨93TL7421⟩.

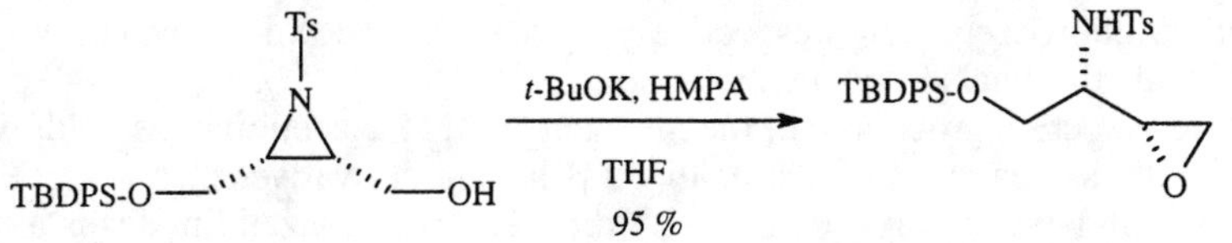

Conventional reactions of oxygen-containing functional groups attached to aziridines, including the addition of hydride and organometallic reagents to ketones and aldehydes, Wittig reactions, carbonyl group reductions, imine and enamine formation from aldehydes and ketones, halide formation from alcohols, and substitution reactions of sulfonates and halides have been summarized ⟨B83MI 101-01⟩. The diastereoselectivity of the addition of organometallic reagents and hydride reagents to carbonyl groups attached to aziridine rings has been studied ⟨57HCA1652, 73BSF2466, 75T997, B-77MI 101-01, 82JCR(S)287, 82JCS(P2)1483⟩. An example is shown in Scheme 44, where chelation-controlled hydride reductions proceeded with high diastereoselectivity ⟨57HCA1652, 82JCS(P2)1483⟩.

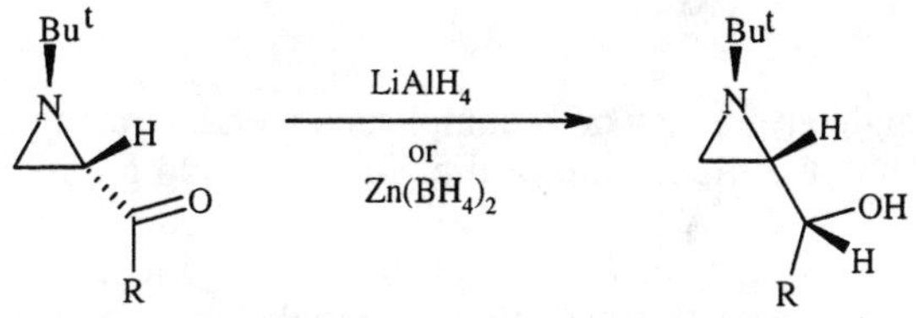

Scheme 44

An example of the aminolysis of an ester is shown in Scheme 45 ⟨91CC538⟩. The reaction proceeds with complete stereoselectivity, and neither the ring nor the *N*-chloro group are affected.

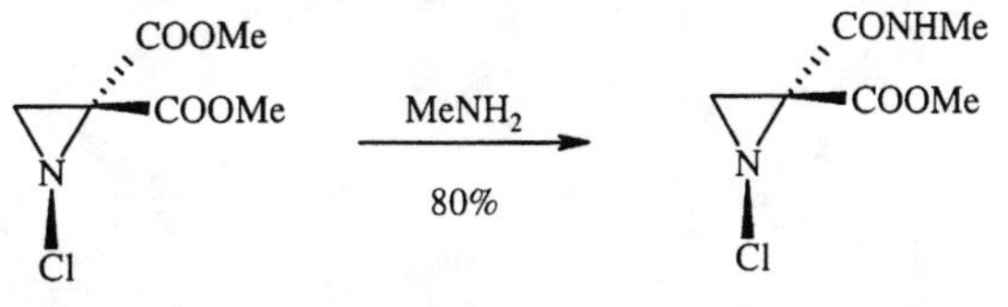

Scheme 45

Michael addition of sodium borohydride or piperidine across 3-methyleneazetidin-2-one (**60**) affords the adducts (**61**) ⟨85T375⟩. 4-(Iodomethyl)azetidin-2-one undergoes nucleophilic substitution by sodium azide in DMF to give the azido compound ⟨88JOC4006⟩.

(60)   (61) R = H, N(CH$_2$)$_5$

3.5.10.2 *N*-Linked Substituents

Deprotection of a nitrogen heteroatom appears to be one of the most widely used reactions of N-linked substituents in small and large heterocycles. The classical removal of suitable substituents at nitrogen has been applied routinely with azetidine derivatives. It has been shown that *N*-debenzylation or *N*-debenzhydrylation works well by hydrogenolysis over a palladium catalyst, with or without added hydrogen chloride ⟨85CC194, 85EUP140437, 88HCA1035, 88SC205, 89EUP299513, 90JMC1561⟩. The *N*-benzyloxycarbonyl and Boc groups are easily removed on azetidine derivatives by means of hydrogenolysis and trifluoroacetic acid, respectively ⟨85CC194, 92T7165⟩. Removal of *N*-tosyl groups can be accomplished by alkali metals in an alcoholic solvent ⟨87URP1323557⟩.

The protection-deprotection procedure of the nitrogen atom of 2-azetidinones holds a prime position in synthetic methodologies leading to functionalized β-lactam derivatives.

The *N*-desilylation of 1-methylsilyl- or 1-(*t*-butyldimethylsilyl)-2-azetidinones is a commonly used process in 2-azetidinone chemistry ⟨91HOU(16B)551⟩. The *t*-butoxycarbonyl group is readily removed with trifluoroacetic acid ⟨86H(24)2539⟩, while the 4-methoxyphenyl protective group is conveniently removed by ceric ammonium nitrate ⟨82JOC2765, 85CJC3613, 93BMC2471, 93TL6325⟩, or anodic oxidation (Equation (2)) ⟨88TL1497, 89USP4834846⟩.

(2)

Removal of an *N*-phthalimido group may be accomplished with hydrazine without destruction of the aziridine ring ⟨70HCA1479, 70JA1784⟩. Reduction of the *N*-phthalimido group with LiAlH$_4$ produces *N*-(isoindolino)aziridines ⟨71JCS(C)988⟩. *N*-Aminoaziridines form hydrazones readily ⟨B-83MI 101-01⟩. Iminophosphoranes bearing an *N*-aziridinyl group have been used to carry out aza-Wittig reactions with carbonyl compounds, resulting in the formation of cumulenes ⟨B-83MI 101-01⟩.

Part 4
Synthesis of Heterocycles

4.1

Overview

4.1.1 AIMS AND ORGANIZATION

The main aim of this part of the book is to provide an introduction to the most efficient ways of making a heterocyclic compound, either by using a known method, or by analogy with existing methods for related compounds. The organization is in accordance with this aim.

The synthesis of a heterocyclic compound can be divided into two parts: ring synthesis, and substituent introduction and modification. The relative importance of the two parts can vary in all proportions, and does vary in a non-random manner for different classes of heterocycles.

The following features generally render the ring synthesis steps of increasing importance relative to substituent modification:

(i) increasing number of heteroatoms;
(ii) increasing number of fused rings;
(iii) decreasing number of endocyclic double bonds.

Substituent modification is based on substituent reactivity as outlined in the reactivity chapters; brief summaries of the scope and limitations of substituent introduction and modification are given for the following important ring systems as they are dealt with:

(i) pyrroles, furans and thiophenes (Section 4.2.3.2);
(ii) pyridines (Section 4.2.4.1);
(iii) azoles (Section 4.3.1.2);
(iv) azines (Section 4.3.1.3);
(v) benzo-fused heterocycles (Section 4.4.1).

In classifying ring syntheses we believe that it is important to group syntheses as follows: (i) those of related classes of compounds, (ii) those from similar precursors and (iii) methods related mechanistically. The system adopted herein attempts to achieve these (not always completely compatible) aims as far as possible.

The synthesis of ring-fused systems is almost always effected sequentially, *i.e.* a bicyclic ring system is formed by the annulation of a second ring on a monocyclic compound (typified by a substituted benzene). (Intramolecular cycloadditions form an important exception to this generalization.) Thus we subdivide ring syntheses first into those leading to monocyclic and those forming polycyclic compounds. This division is not a rigid one. It applies principally to cases where the preformed ring of a bicyclic system is *aromatic:* thus, syntheses of benzimidazoles, for example, are treated separately from those of imidazoles. However, the methods of synthesis of 4,5,6,7-tetrahydrobenzimidazole would show closer analogies to those of 4,5-dimethylimidazole than to those of benzimidazole. Similarly, spiro ring systems are considered with their monocyclic analogues.

Ring-fused systems with ring junction N- or S-atoms are considered separately from their more numerous analogues with only C-atoms at the ring junctions because of considerable differences in the synthetic methods employed.

Mono-, bi- and tri-cyclic systems are further classified firstly according to the number and orientation of their heteroatoms and secondly by the degree of unsaturation in the system. Hence the classification is *not* primarily by ring size, and this enables many related synthetic methods to be discussed together.

In the case of ring-fused systems containing two or more heterocyclic rings, synthetic methods are classified according to the ring being formed. This means that pteridine syntheses are placed in different sections according to routes (1) → (2) or (3) → (2).

501

The system just outlined is that adopted in Chapters 4.2–4.5. The remainder of the present chapter, Sections 4.1.2–4.1.4, now reviews from a mechanistic standpoint the main types of reactions used in the preparation of heterocyclic rings.

4.1.2 RING FORMATION FROM TWO COMPONENTS

4.1.2.1 By Reaction Between Electrophilic and Nucleophilic Carbons

Reactions of this type can occur either between a binucleophile and a bielectrophile, or between two molecules each containing both a nucleophilic and an electrophilic center, *e.g.* $HSCH_2CO_2H$.

Rings of many sizes can be made by this approach. Thus, reaction of a 1,2-binucleophile with a 1,3-bielectrophile leads to a five-membered heterocycle, as would the reaction of a 1,4-binucleophile with a 1,1-bielectrophile.

Table 1 lists some of the common binucleophiles utilized in heterocyclic synthesis, the numerical prefixes referring to the positions of the nucleophilic centers relative to each other. Higher order binucleophiles, *e.g.* 1,5-systems, are analogous.

Table 1   Some Examples of Commonly Encountered Binucleophiles

1,2-Systems	*1,2-Systems*	*1,3-Systems*	*1,3-Systems*	*1,4-Systems*	*1,4-Systems*
H_2NNH_2	H_2NOH	$R\text{-}C(S)NH_2$	$R\text{-}C(=NH)NH_2$	$H_2N(CH_2)_2NH_2$	$R\text{-}C(S)NHNH_2$
$RNHNHR$	$RNHOH$	$R\text{-}C(S)NHR$	$R\text{-}C(=NH)NHR$	$H_2N(CH_2)_2OH$	$R\text{-}C(Se)NHNH_2$
H_2NNR_2	R_2NOH	$R\text{-}C(Se)NH_2$	$H_2N\text{-}C(=NH)NH_2$	$H_2N(CH_2)_2SH$	$R\text{-}C(=NH)NHNH_2$
$RNHNR_2$	$RCH=NNH_2$	$R\text{-}C(Se)NHR$	$H_2N\text{-}C(S)NH_2$		$R\text{-}C(=NH)NHOH$
	$RCH=NOH$		$H_2N\text{-}C(Se)NH_2$		$H_2N\text{-}C(S)NHNH_2$

4.1.2.2 Ring Formation *via* Cycloaddition

Three reactions are of great importance: [2+2] cycloaddition, 1,3-dipolar cycloaddition and Diels-Alder reactions ([4+2] cycloadditions), which lead to four-, five- and six-membered rings, respectively. [3+3] Cycloadditions are known (see Section 4.3.8.2) but are of less importance.

4.1.2.2.1 [2+2] Cycloadditions

Concerted thermal [2+2] cycloadditions forming heterocycles have been reviewed ⟨77AHC(21)245⟩ and cross references to some examples discussed in the present book are given in Table 2.

Table 2   Four-membered Heterocyclic Rings from [2+2]
Cycloaddition Reactions

Heteroatom Position(s)	*Section*	*Precursors with Heteroatom(s)*
1	4.2.1.1.2	C = O, C = S
1,2	4.3.2.2.1	N = N
1,2	4.3.2.2.2	N = O
1,2	4.3.2.2.4	O = O
1,2	4.3.2.2.5	S = O
1,3	4.3.3.1.3	C = S, C = N

The Woodward-Hoffmann rules predict high activation energies for the suprafacial-suprafacial addition of two carbon-carbon double bonds, these may be lowered by polar effects ⟨74AF(3)751⟩. [2+2]

Photocycloadditions are common and usually involve diradical intermediates: photo-excited ketones react with a variety of unsaturated systems (Scheme 1). Both the singlet and the triplet (n, π*) excited states of the ketones will form oxetanes with electron-rich alkenes. With electron-deficient alkenes only the singlet states give oxetanes. Diradicals are the immediate precursors to the oxetanes in all cases, but the diradicals are formed by different mechanisms, depending on the availability of electrons in the two components.

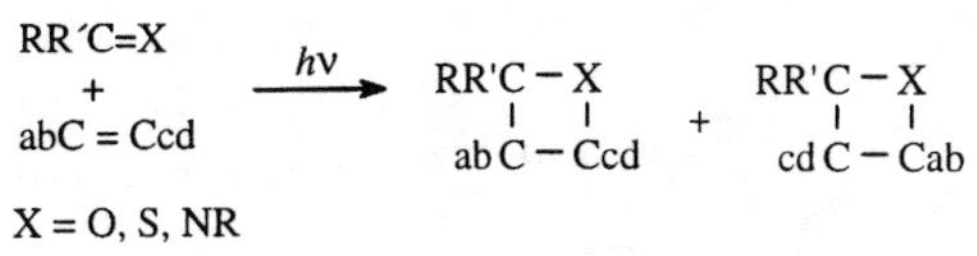

Scheme 1

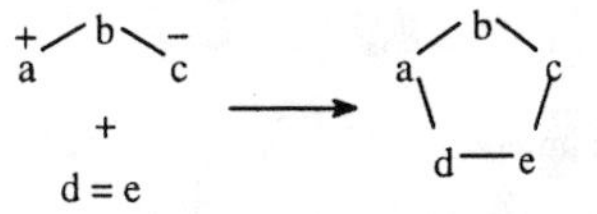

Scheme 2 A 1,3-dipolar cycloaddition reaction

4.1.2.2.2 1,3-Dipolar cycloadditions

This synthesis of a variety of five-membered heterocycles involves the reaction of a neutral 4π-electron-three-atom system, the dipole, with a 2π-electron system, the dipolarophile (Scheme 2). Table 3 illustrates 1,3-dipoles with a double bond and with internal octet stabilization, the propargyl-allenyl anion type.

Table 3 1,3-Dipoles with a Double Bond and Internal Octet Stabilization; Propargyl-allenyl Anion Type

Nitrile ylide	$-\overset{+}{C}{=}N{-}\overset{-}{C}<$ ⟷ $-C{\equiv}\overset{+}{N}{-}\overset{-}{C}<$	*in situ* from	$>C(Cl){=}N{-}CH<$
Nitrile imine	$-\overset{+}{C}{=}N{-}\overset{-}{N}-$ ⟷ $-C{\equiv}\overset{+}{N}{-}\overset{-}{N}-$	*in situ* from	$>C(Cl){=}N{-}NH-$
Nitrile oxide	$-\overset{+}{C}{=}N{-}\overset{-}{O}$ ⟷ $-C{\equiv}\overset{+}{N}{-}\overset{-}{O}$	*in situ* from	$>C(Cl){=}N{-}OH$
Nitrile sulfide	$-\overset{+}{C}{=}N{-}\overset{-}{S}$ ⟷ $-C{\equiv}\overset{+}{N}{-}\overset{-}{S}$	Thermal fragmentation of an oxathiazolone	
Diazoalkane	$>\overset{+}{C}{-}N{=}\overset{-}{N}$ ⟷ $>C{=}\overset{+}{N}{=}\overset{-}{N}$	Usually stable	
Azide	$-\overset{+}{N}{-}N{=}\overset{-}{N}$ ⟷ $-N{=}\overset{+}{N}{=}\overset{-}{N}$	Usually stable	
Nitrous oxide	$\overset{+}{N}{=}N{-}\overset{-}{O}$ ⟷ $N{\equiv}\overset{+}{N}{-}\overset{-}{O}$	Stable	

1,3-Dipoles without a double bond but with internal octet stabilization, the allyl anion type, are shown in Table 4. 1,3-Dipoles without octet stabilization, such as vinylcarbenes and iminonitrenes, are all highly reactive intermediates with only transient existence.

Dipolarophiles utilized in these cycloadditions leading to five-membered heterocycles contain either double or triple bonds between two carbon atoms, a carbon atom and a heteroatom, or two heteroatoms. These are shown in Scheme 3 listed in approximate order of decreasing activity from left to right.

Table 4   1,3-Dipoles without a Double Bond but with Internal Octet Stabilization; Allyl Anion Type

Azomethine ylide	$\overset{+}{C}-N-\overset{-}{C} \longleftrightarrow C=\overset{+}{N}-\overset{-}{C}$	*in situ* from $-\overset{H}{C}-\overset{+}{N}=C\;\;X^-$ or aziridines
Azomethine imine	$\overset{+}{C}-N-\overset{-}{N}- \longleftrightarrow C=\overset{+}{N}-\overset{-}{N}-$	*in situ* from $C=\overset{+}{N}-\overset{N}{N}-H \;\; X^-$
Nitrone	$\overset{-}{C}-\overset{+}{N}=O \longleftrightarrow C=\overset{+}{N}-\overset{-}{O}$	Stable
Azimine	$-\overset{+}{N}-N-\overset{-}{N}- \longleftrightarrow -N=\overset{+}{N}-\overset{-}{N}-$	From heterocycles
Azoxy compound	$-\overset{+}{N}-N-\overset{-}{O} \longleftrightarrow -N=\overset{+}{N}-\overset{-}{O}$	Stable
Nitro compound	$\overset{+}{O}-N-\overset{-}{O} \longleftrightarrow O=\overset{+}{N}-\overset{-}{O}$	Stable
Nitroso imine	$-\overset{+}{N}-O-\overset{-}{N}- \longleftrightarrow -N=\overset{+}{O}-\overset{-}{N}-$	
Nitroso oxide	$-\overset{+}{N}-O-\overset{-}{O} \longleftrightarrow -N=\overset{+}{O}-\overset{-}{O}$	
Carbonyl ylide	$\overset{+}{C}-O-\overset{-}{C} \longleftrightarrow C=\overset{+}{O}-\overset{-}{C}$	From oxiranes or heterocycles
Carbonyl oxide	$\overset{+}{C}-O-\overset{-}{O} \longleftrightarrow C=\overset{+}{O}-\overset{-}{O}$	From carbene + O_2
Carbonyl imine	$\overset{+}{C}-O-\overset{-}{N}- \longleftrightarrow C=\overset{+}{O}-\overset{-}{N}-$	
Ozone	$\overset{+}{O}-O-\overset{-}{O} \longleftrightarrow O=\overset{+}{O}-\overset{-}{O}$	Stable
Thiocarbonyl ylide	$\overset{+}{C}-S-\overset{-}{C} \longleftrightarrow C=\overset{+}{S}-\overset{-}{C}$	From heterocycles
Selenocarbonyl ylide	$\overset{+}{C}-Se-\overset{-}{C} \longleftrightarrow C=\overset{+}{Se}-\overset{-}{C}$	From heterocycles

Alkynic dipolarophiles

NCC≡CCN, CF₃C≡CCF₃, RO₂CC≡CCO₂R, benzyne, R¹COC≡CCOR, HC≡CCO₂R, R²C≡CCO₂R,
R²C≡CR², R²C≡CH (R = Me, Et; R¹ = alkyl, aryl; R² = aryl, heteroaryl)

Alkenic and N=N-containing dipolarophiles

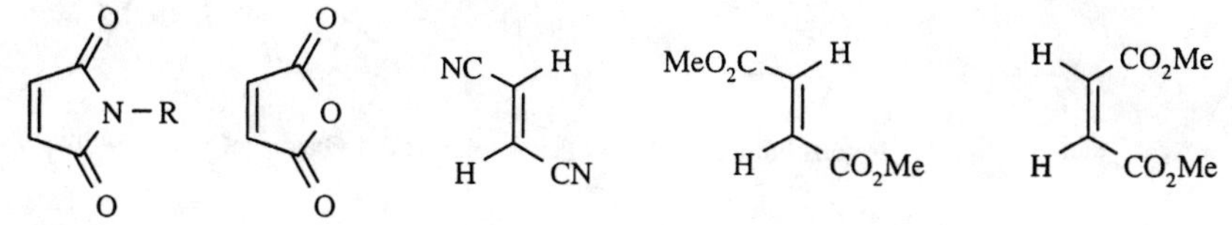

R = Me, Et, Ph

CH₂=CHCN, CH₂=CHCOMe, CH₂=CHCO₂Et, CH₂=CMeCO₂Me, EtO₂CN=NCO₂Et, PhCH=CHNO₂,
PhCCl=CHNO₂, PhCH=CH₂, PhCH=CHPh

Heterocumulenes

R¹CONCO, R¹CONCS, RNCO, RNCS
R¹ = aryl, CCl₃; R = alkyl, aryl

Scheme 3   Frequently used dipolarophiles

Several five-membered ring systems readily available by 1,3-dipolar cycloadditions are shown in Scheme 4. The dotted line indicates how the system was constructed, the line bisecting the two new bonds being formed in the cycloaddition.

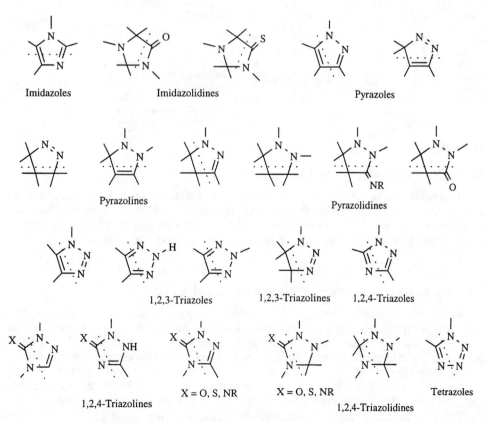

Scheme 4   Some five-membered ring systems available by 1,3-dipolar cycloadditions

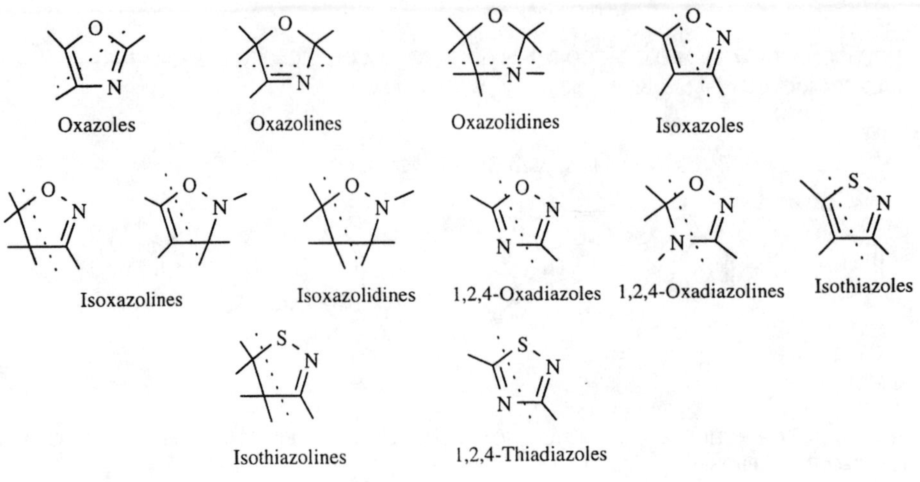

Oxazoles   Oxazolines   Oxazolidines   Isoxazoles

Isoxazolines   Isoxazolidines   1,2,4-Oxadiazoles   1,2,4-Oxadiazolines   Isothiazoles

Isothiazolines   1,2,4-Thiadiazoles

Scheme 4   (Continued)

4.1.2.2.3   Diels-Alder reactions

One or several heteroatoms can be introduced into a six-membered ring from either the diene or the dienophile component of a Diels-Alder reaction. Table 5 gives examples of both these possibilities together with the section where the reaction is covered.

Table 5   Six-membered Heterocyclic Rings Prepared by Diels-Alder Reactions

Type	Heteroatom position(s)	Section	Heterodienes	Heterodienophiles
Monocyclic	1	4.2.2.1	C=C–C=O, C=C–C=N	C=N, C=O, C=S
Monocyclic	1	4.2.4.3.1	1,2,4-Triazines	—
Monocyclic	1,2	4.3.2.4.2	—	N=N, N=O, O=O, N=S
Bicyclic	1	4.4.2.3.4	Ph–N=C	—
Bicyclic	1	4.4.2.3.5	Anthranils	—
Bicyclic	1	4.4.2.4	Benazetes	—
Bicyclic	1,4	4.4.6.1	*o*-Benzoquinone diimine	—
Bicyclic	1,4	4.4.6.2	*o*-Nitrosophenol	—
Bicyclic	1,3,5	4.6.1	C=N–C=S	Azirine

4.1.3   RING CLOSURE OF A SINGLE COMPONENT

In general, syntheses in which a C–Z bond is formed in the last stages are more important for the preparation of monocyclic compounds, whereas those syntheses in which a C–C bond is formed in the last stage are important for the benzo derivatives. The ring closure may be of a chain containing a nucleophile and an electrophile at the ends, or homolytic, or electrocyclic in nature.

4.1.3.1   By Reaction between Electrophilic and Nucleophilic Centers

A set of guidelines predicts the facility of different ring closures ⟨76CC734⟩. For a process involving formation of a five-membered ring by nucleophilic attack of one terminal atom upon the other, the possibilities are shown in Scheme 5. The prefixes *exo* and *endo* indicate whether the breaking bond is exocyclic or endocyclic to the ring being formed. The suffixes refer to tetrahedral, trigonal and digonal carbon atoms, respectively. Of the cases indicated, those termed 5-endo-tet and 5-endo-trig are disfavored when X is a first-row element, in this case nitrogen or oxygen. These restrictions arise from the stereochemical requirements of the respective transition states, but these may not apply when atoms such as sulfur, selenium and tellurium are involved because of the larger atomic radii and bond distances. Thus, a 5-exo-trig ring closure involving sulfur has been observed ⟨76CC736⟩.

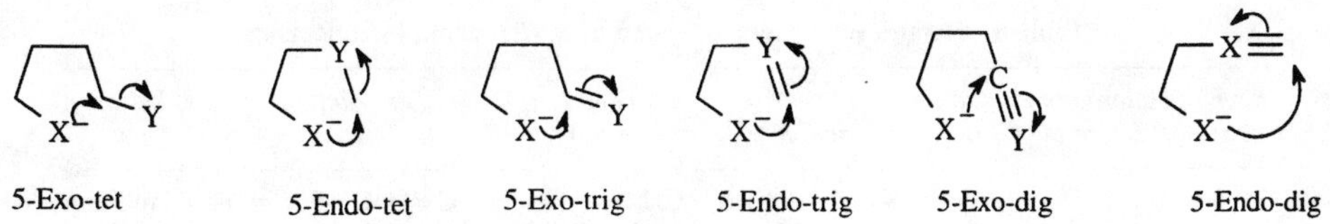

Scheme 5

Application of similar considerations to homolytic ring closures indicates that a 5-exo-trig closure will be preferred to a 6-endo-trig mode (Scheme 6) ⟨80CC482⟩. For cationic closures the 5-endo-trig mode is unknown in contrast to the well-established 6-endo-trig closure (Scheme 7).

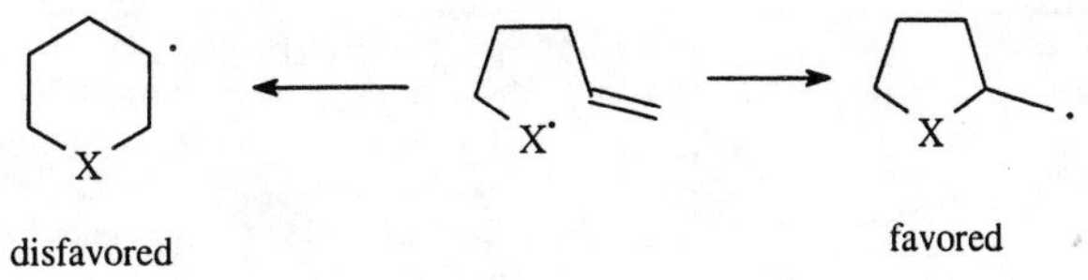

disfavored   favored

Scheme 6

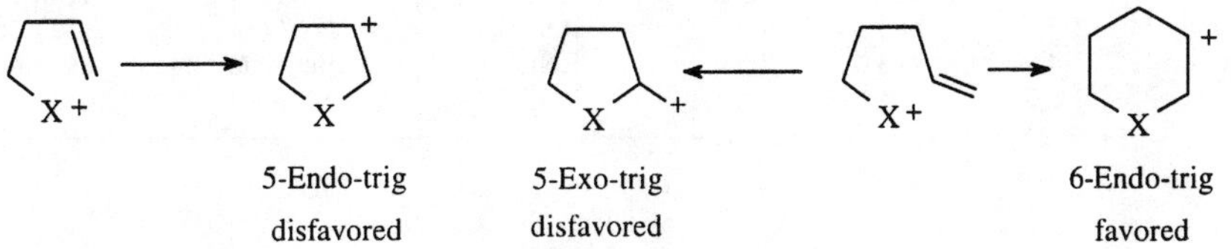

| 5-Endo-trig | 5-Exo-trig | 6-Endo-trig |
| disfavored | disfavored | favored |

Scheme 7

The rules have been extended to rings of other sizes.

4.1.3.2 Electrocyclic Reactions

Electrocyclic ring closures are particularly important in the formation of six-membered rings: many are hetero analogues of the hexatriene – cyclohexadiene transformation (**4 → 5**) and can be considered as Cope rearrangements. As discussed in Section 3.2.1.6.1, they are frequently involved in ring interconversions initiated by nucleophilic attack on a six-membered ring. Further examples are discussed in Sections 4.2.3.6 (preparation of seven-membered rings) and 4.4.8.2.2.ii (formation of bicyclic 6,6 ring systems).

(4)   (5)

Photochemically initiated electrocyclizations can be used to form five-membered rings (*e.g.* Section 4.5.1.1.2.i).

The formation or rupture of ring bridges often involves electrocyclic reactions (*e.g.* Sections 4.4.4.1, 4.4.8.3.4 and 4.6.1).

4.1.3.3 By Radical, Carbene or Nitrene Intermediates

Representative examples of ring syntheses involving carbenoid (Table 6) or nitrenoid (Table 7) intermediates are given. In many cases, the free carbene or nitrene is probably not involved, and the distinction between insertion and addition reactions given in the tables is not always clear cut. Such reactions are particularly useful for the preparation of tricyclic compounds.

Table 6   Heterocyclic Rings Prepared from Carbenoid Intermediates

Ring size	Heteroatom Position (s)	No. of rings	Section	Carbene precursor	Other reaction site	Reaction type
3	1	1	4.2.1.1.1.iii	CH_2N_2	$C=O$, $C=S$, $C=N^+$	Addition
6	1,2	3	4.4.4.4	$R–C \equiv \overset{+}{N}\tilde{N}–R'$	$C=C$	Addition
5	1	3	4.5.1.1.4	RNC	Benzene ring	Insertion
6	1	3	4.5.5	*v*-triazolyl–1	Benzene ring	Insertion

Table 7   Heterocyclic Rings Prepared from Nitrenoid Intermediates

Ring size	Heteroatom Positions	No.of rings	Section	Nitrenoid precursor(s)	Other reaction site	Reaction type
3	1	1	4.2.1.1.1.ii	RN_3, $R_2S^+NH^-$	$C=C$	Addition
3	1,2	1	4.3.2.1	NH_2OSO_3H	$C=N$	Addition
5	1,3	2	4.4.5.1.2	Tetrazole, oxadiazole	Benzene ring	Insertion
5	1	3	4.5.1.1.2.iii	Benzotriazole, pyridotriazole	Benzene ring	Insertion
5	1	3	4.5.1.1.3	RNO_2, RN_3	Benzene ring	Insertion
6	1	3	4.5.1.2	RN_3	CH_3	Insertion
5 (or 6)	1,2,3	3	4.5.5	RN_3	$N=N$	Addition
6	1,2	3	4.5.2	RN_3	$N=O$	Addition
6	1,4	3	4.5.4.2	RNO_2	Benzene ring	Insertion
5	1,2,3	4	4.6.5.1	RN_3	N	Addition

4.1.3.4   By Intramolecular Cycloadditions

An intramolecular cycloaddition reaction results in the simultaneous formation of two new rings. Examples include the formation of an octahydroquinoline (Section 4.4.2.3.4) and a tetra-hydrobenzo[*c*]pyrrole (Section 4.4.3.1) by intramolecular Diels-Alder reactions.

4.1.4   MODIFICATION OF AN EXISTING RING

The conversion of one heterocyclic ring into another is treated as a reaction of the initial compound in the appropriate reactivity section in Chapters 3.2–3.5 (for a review see ⟨B–73MI50300⟩). Here we present a brief overview.

4.1.4.1   Ring Atom Interchange

Such reactions are particularly important in the photochemical isomerization of five- (Sections 3.3.1.2 and 3.4.1.2) and six-membered rings (Section 3.2.1.2); however, they possess relatively little synthetic importance.

Reactions of this type are also implicated in the isomerization of large ring systems through bicyclic ones, *e.g.* the 1,2- to 1,3-diazepine conversion (Scheme 8) (CHEC 5.18.2.2).

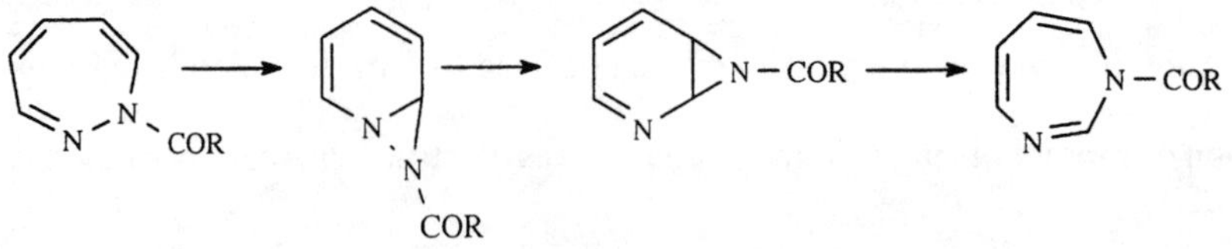

Scheme 8

4.1.4.2   Incorporation of New Ring Atoms: No Change in Ring Size

Important reactions of this type include the replacement of a ring oxygen by a ring sulfur or nitrogen. Table 8 gives some examples and references to the section where they are discussed.

Table 8 Heterocyclic Ring Interconversions: No Change in Ring Size

Product: ring size	Product: heteroatom position(s)	No. of rings	Section	Interconversion(s)
3	1	1	4.2.1.5.2	Oxirane → thiirane; oxirane → aziridine
5	1	1	4.2.3.5.2	Furan → thiophene or pyrrole; oxazole or various mesoionics → furan, thiophene or pyrrole
6	1	1	4.2.3.5.2	α- or γ-Pyrone → α- or γ-pyridone
6	1	1	4.2.4.3.2	Pyrylium → pyridine, pyridinium, thiinium
5	1,2	1	4.3.2.3.4	1,2-Dithiolylium → pyrazole, isothiazole; isoxazole → pyrazole
6	1,3	1	4.3.3.3.4	1,3-Oxazine, 1,3-thiazine → pyrimidine; *s*-triazine → pyrimidine
5	1,2,4	1	4.3.6.1.3	Tetrazole → oxadiazole, thiadiazole, triazole
6	1,3,5	1	4.3.7.3	Pyrimidine, 1,3,5-oxadiazine → *s*-triazine
6	1,3	2	4.4.5.2.3	3,1-Benzoxazine and 3,1-benzothiazine → quinazolone

4.1.4.3 Ring Expansions

Examples treated in this book are summarized in Table 9. One frequently found type consists of ring expansions of cyclic conjugated systems with an exocyclic ylide function, such as pyridine *N*-oxides or *N*-imides; incorporation of the exocyclic half of the ylide function expands the ring by one member.

Table 9 Formation of New Heterocycle by Ring Expansion of Existing Heterocycle

Product: ring size	Product: heteroatom position(s)	No. of rings	Section	Interconversion(s)
4, 5, 6	1	1	4.2.1.5.1	Oxiranes → oxetanes, tetrahydrofurans; azirines → pyrrolidines; tetrahydrofurans → pyrans
5, 6	1	1	4.2.3.5.1	Oxiranes → dihydrofurans; azirine → pyrrole; diketene → pyrones; isooxazoles → pyridones
6	1	1	4.2.4.3.1	Furans → pyridines, pyrones
7	1	1	4.2.4.4	Bicyclic azirines and oxiranes → azepines, oxepins
6	1,2	1	4.3.2.4.4	Pyrrolidines → 1,2-oxazines
7	1,3	1	4.3.3.4.2	Pyridine oxides, 1,3-oxaziniums, pyryliums → 1,3-oxazepines
6	1,4	1	4.3.4.1.4	Aziridine → piperazine, pyrazines; oxiranes → 1,4-dioxanes, 1,4-oxazines; thiiranes – 1,4-dithianes
6	1,2,3	1	4.3.5.3	1,2-Dithioles → 1,2,3-dithiazine
6	1	2	4.4.2.3.5	Indoles, isatins, anthranils → quinolines
7	1	2	4.4.4.4	Quinolines → benzodiazepine
6	1,3	2	4.4.5.2.3	Benzisoxazoles → 1,3-benzoxazines
6	1,4	2	4.4.6.4	Benzofuroxans → quinoxalines
7	1,3,5	2	4.4.8.3.4	Quinoxalines → 3,1,5-benzoxadiazepines
6	1,4	3	4.5.4.1	Benzofuroxans → phenazines
7	1,2	3	4.5.4.4	Acridines → dibenz[*c,f*][1,2]oxazepines

4.1.4.4 Ring Contractions

The isomerization of large rings into bicyclic systems by electrocyclic reaction has been mentioned (Section 4.1.3.2).

Other examples of ring contractions are given in Table 10. They fall into three main classes. The total loss of one or two (or sometimes more) ring members from the heterocycle, concerted with or followed by formation of a new ring, is a versatile synthetic method. Loss of N_2, CO, CO_2, S, SO, SO_2, $H_2C = CH_2$, *etc.* is common.

Table 10   Formation of New Heterocycle by Ring Contraction of Existing
Heterocycle

Product: ring size	Product: heteroatom position(s)	No. of rings in product	Section	Reaction types
3–6	1	1	4.2.1.5.3	Loss of SO_2, C_2H_4; extrusion of C
5,6	1	1	4.2.3.5.3	Extrusion of C
4	1,2	1	4.3.2.2.6	Loss of CO
5	1,2	1	4.3.2.3.4	Loss of CO_2
5	1,2	1	4.3.2.3.4	Extrusion of C_2
5	1,2	1	4.3.6.1.3	Replacement of CO
4	1	2	4.4.2.1.2.ii	Loss of N_2
5	1	2	4.4.2.2.6	Extrusion of C (Wolff, Meerwein, *N*-oxide rearrangements)

The second major reaction type is the extrusion of one or more atoms from the ring to form a new substituent or side chain.

Finally the ring can react with another reagent to exchange two or three ring atoms for one or two provided by the reagent.

4.1.4.5   Ring Closure with Simultaneous Ring Opening

The generalized monocyclic rearrangement has been discussed in Section 3.4.3.1.9; it results in the conversion of one five-membered ring into another.

4.2

Synthesis of Monocyclic Rings with One Heteroatom

4.2.1 RINGS CONTAINING NO ENDOCYCLIC DOUBLE BONDS

For all the saturated monocycles with one heteroatom, intramolecular nucleophilic displacement with the formation of C–Z bond(s) is an important preparative method (Section 4.2.1.2.).

Further significant synthetic routes for the various ring sizes are:
(i) for three-membered rings, electrocyclic addition to double bonds (Section 4.2.1.1);
(ii) for four-membered rings, [2 + 2] photocyclizations;
(iii) for five- and six-membered rings, formation of one C–C bond;
(iv) for six-membered and larger rings, ring expansion of carbocycles (Section 4.2.1.4).

4.2.1.1 From Acyclic Compounds by Concerted Formation of Two Bonds

4.2.1.1.1 *Three-membered rings*

(i) Oxiranes

Oxiranes are formed by direct oxidation of alkenes with oxygen (catalytic) or peracids, *m*-chloroperoxybenzoic acid being commonly used ⟨96CHEC-II(1A)130⟩. This reaction is facilitated by alkylation of the C=C bond of the alkene and hindered by the presence of electron-withdrawing groups. Peracids commoly react stereospecifically – the formation of both bonds overlaps in time, and is followed by loss of carboxylate ion or carboxylic acid. Hydrogen bonding by groups attached to the alkene component steers the peracid to one or the other face of the alkenic double bond, in competition with solvent hydrogen bonding and non-bonding interactions in the activated complex. Experimental evidence exists for a 'butterfly transition state' (**1**) in peroxy acid epoxidations ⟨91JA6281⟩.

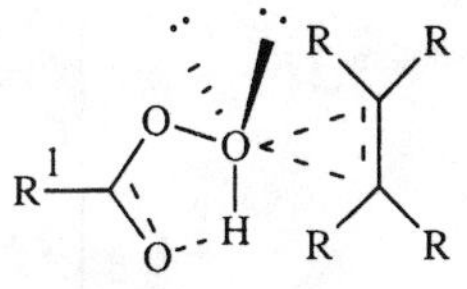

(1)

A powerful synthetic method for enantioselective epoxidation of allylic alcohols uses a mixture of Ti(OPr^i)$_4$, Bu^{t}O$_2$H and (*R,R*)-(+)-diethyl tartrate ⟨80JA5974⟩. The presence of zeolites with a catalytic amount of the Ti(IV)/tartrate complex (Scheme 1) ⟨86JOC1922, 87JA5765⟩ improves the procedure to give high enantiomeric excesses (90–95%). For reviews of the Sharpless asymmetric epoxidation, see ⟨B-85MI 103-03, B-83MI 103-04, B-85MI 103-04, 86CBR38, 87MI 103-04, 89CRV431, 92T2803, B-94MI 103-01⟩.

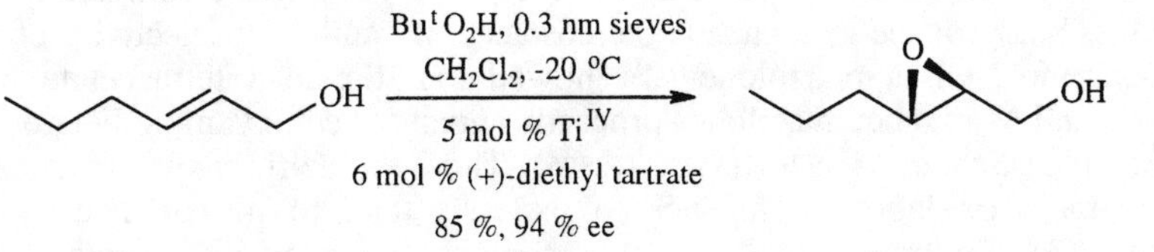

Bu^{t}O$_2$H, 0.3 nm sieves
CH$_2$Cl$_2$, -20 °C
5 mol % Ti^{IV}
6 mol % (+)-diethyl tartrate
85 %, 94 % ee

Scheme 1

511

Electron-deficient C=C double bonds are resistant to electrophilic attack, but are converted into oxiranes by nucleophilic oxidants such as HO_2^-.

Reactions of carbonyl compounds with methylene equivalents are another general way to oxiranes (Equation 1) ⟨96CHEC-II(1A)129⟩. Thus, transfer of methylene from diazomethane to the carbonyl group of 1-fluoro-3-*p*-tolylsulphinylacetone proceeds with high chemo- and enantio-selectivity ⟨92TL5609, 93TL7771, 94T13485⟩. Carbonyl to oxirane transformation occurs when tetrafluorocyclopentadienone is treated with diazomethane ⟨91JOC157⟩.

$$\text{\Large$\searrow$}=O \xrightarrow{\ :\bar{C}H_2\text{-}X\ } \text{\Large$\times$}\!\!\overset{O^-}{\underset{}{}}\!\!X \xrightarrow{\ -X^-\ } \text{\Large$\times$}\!\!\overset{O}{\triangle} \qquad (1)$$

$$X = N_2^+,\ Br,\ {}^+SR_2,\ OS^+Me_2,\ {}^+AsR_3,\ {}^+SeR_2$$

Dimethylsulfonium methylide, $Me_2S=CH_2$, and dimethyloxosulfonium methylide, $Me_2S(O)=CH_2$ (Corey's reagent), are efficient methylene transfer agents, except for hindered or highly enolizable ketones. Generation of $Me_2S(O)=CH_2$ under phase transfer conditions in $CH_2Cl_2/NaOH$ provides a convenient variant of oxirane synthesis with sulfur ylides ⟨85SC749⟩.

(ii) Aziridines

The intermolecular cycloaddition of an electron-deficient species such as a nitrene, a nitrenium ion or a carbene (or their formal equivalents) to the π-bond of an alkene, alkyne, imine, or nitrile is a significant approach to aziridines and azirines (Scheme 2).

Electrophilic nitrogen compounds, such as arenesulfonyloxyamines, can convert alkenes to aziridines without the intervention of free nitrenes ⟨80CC560⟩. The ylide Ph_2S^+-NH^- adds stereospecifically to *E* and *Z* conjugated alkenes, and chiral sulfimides can transfer chirality to the aziridines formed ⟨80T73⟩. These reactions are often named "aziridinations".

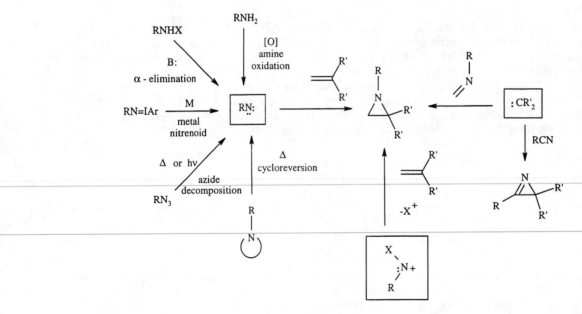

Scheme 2

Aminonitrenes are commonly generated by the lead tetraacetate oxidation of various *N*-aminoheterocycles. Study of the mechanism, stereoselectivity and regioselectivity of aziridination of alkenes with *N*-amino-2-ethylquinazolone (2) showed that the lead tetraacetate oxidation of *N*-aminophthalimide and *N*-aminoquinazolones proceeds *via* an *N*-acetoxyamino heterocycle rather than a nitrene, as shown in Scheme 3 ⟨87CC1362, 89T2875⟩. Thus, the aziridination reaction may be more related to the peracid oxidation of alkenes to oxiranes than to nitrene chemistry. *N*-Acetoxy-aminoquinazolone (3), a nitrogen analogue of a peracid, has been used directly in aziridination reactions.

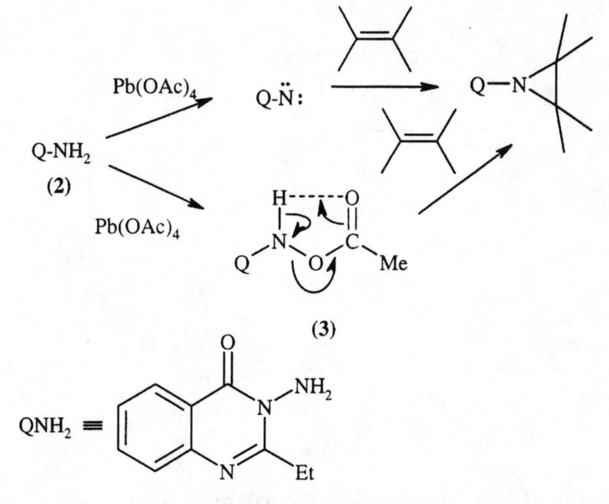

Scheme 3

Sulfur stabilized nitrenes have been implicated in aziridination, *e.g.* (**4**) + (**5**) → (**6**) ⟨94CPB27⟩. Oxynitrenes derived from the oxidation of alkoxyamines are known, but are generally less efficient in aziridination.

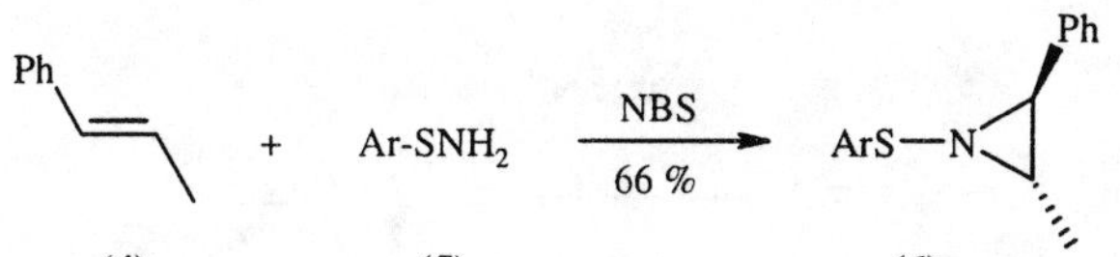

Ar = 2,4-$(O_2N)_2C_6H_3$

Transition metal nitrenoids suitable for the aziridination of alkenes have been developed. The early work involved the reaction of a catalytic amount of an iron or manganese porphyrin with [*N*-(*p*-toluenesulfonyl)imino]phenyliodinane (PhI = NTs) or other nitrogen transfer agents to produce intermediate metal nitrenoids M = NTs which then react with alkenes to produce *N*-tosylaziridines ⟨80CC560, 83JA2073, 88JCS(P2)1517⟩. Copper(I) or copper(II) salts can also catalyze the transfer of a *N*-tosylamino group to alkenes ⟨91JOC6744, 93OM261, 94JA2742⟩. A significant advance is the recent asymmetric aziridinations using chiral copper ligands such as (**7**) (Scheme 4) ⟨91JA726, 91TL7373, 92TL1001, 93JA5326, 93JA5328, 93SL469⟩. The asymmetric induction varies widely with the metal ligand. A stepwise mechanism involving radical intermediates is indicated by non-stereospecific aziridination in some cases ⟨94JA425⟩.

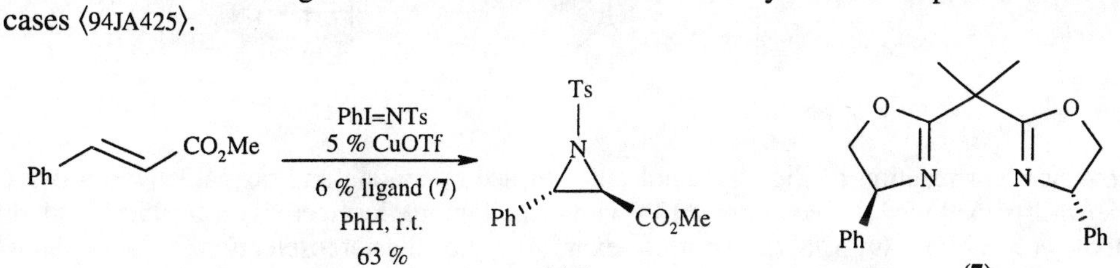

Scheme 4

(iii) Thiiranes

Thiiranes can be made from 'nascent sulfur', such as is obtained by the thermolysis of diethyl tetrasulfide ⟨64HC(19-1)591⟩.

Treatment of bis(trimethylsilyl)sulfide (**8**) with bromine in anhydrous dichloromethane at − 78 °C followed by addition of 1,2-disubstituted alkenes (**9**) to the reaction mixture at the same temperature gave the thiiranes (**10**) in about 30% yields (Scheme 5) ⟨88TL4177⟩. This synthesis is (i) highly selective, since only 1,2-disubstituted alkenes undergo the transformation to thiiranes; and (ii) stereospecific since the stereochemistry of the alkene is retained in the thiirane. The nature of the reagents suggests that the silyl subsituted sulfenyl bromide (**11**), formed by attack of bromine on (**8**), is an intermediate.

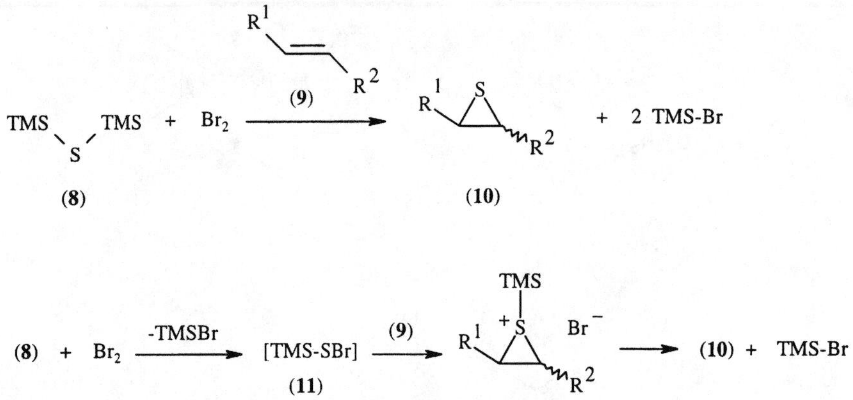

Scheme 5

The reaction of thiocarbonyl compounds with diazoalkenes gives good yields of thiiranes (Scheme 6). The mechanism may be an addition of a carbene across the thiocarbonyl group, especially in the presence of Rh(II) acetate, CuCl, $CuSO_4$, or other metal catalysts ⟨80JOC1481⟩, but involvement of a zwitterion or 1,3,4-thiadiazoline is also possible ⟨75RCR138⟩. This synthetic strategy is widely applied to thiirane derivatives (*e.g.*, allene episulfides and allene episulfide S-oxides) ⟨87TL1787⟩.

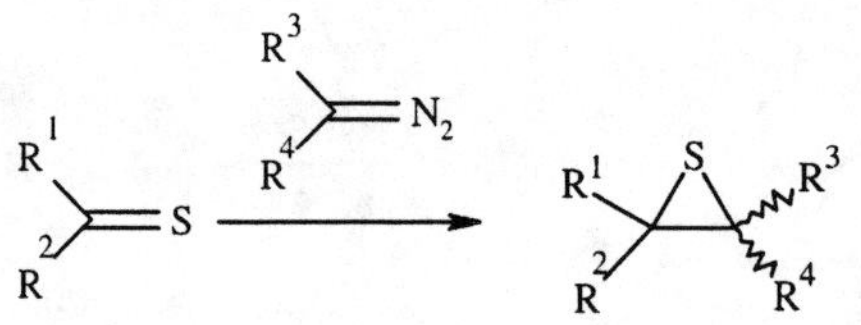

Scheme 6

4.2.1.1.2 Four-membered rings

(i) Electron-deficient alkenes react with singlet $n \rightarrow \pi^*$ excited carbonyl compounds to give oxetanes (*e.g.* **12**), often with high stereospecificity. Isomerization of the alkene in the triplet quenching process can destroy the stereochemical integrity.

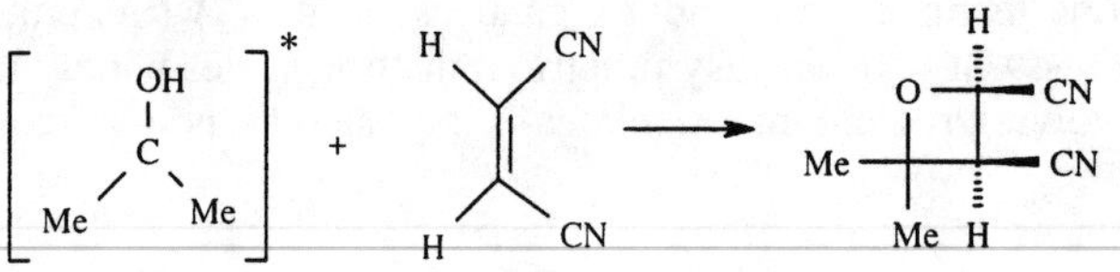

A photochemical reaction of the silyl enol ether of acetophenone and benzaldehyde provided the 2,3-diphenyl-3-trimethylsilyloxyoxetane (**13**) with excellent regioselectivity (> 95:5) and diastereoselectivity (> 95:5) ⟨91TL7037⟩. In this example, the diastereoselection was explained by anti-approach of the two phenyl groups during the carbon-carbon bond forming step from the diradical intermediate (Scheme 7).

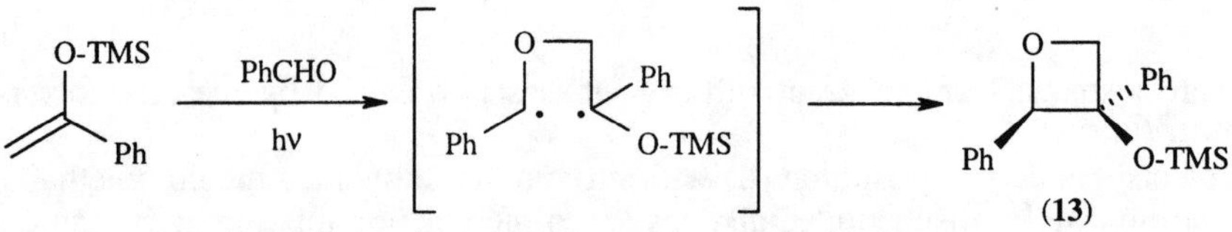

Scheme 7

(ii) Thietanes are produced efficiently from [2+2] photocyclization of thiones to vinyl ethers and other alkenes (see CHEC 5.14.4.1.2, CHEC-II, 1.24.9.2.2). Sulfenes add to electron-rich alkenes to give thietane 1,1-dioxides, *e.g.* (**14**) ⟨77BCJ1179⟩.

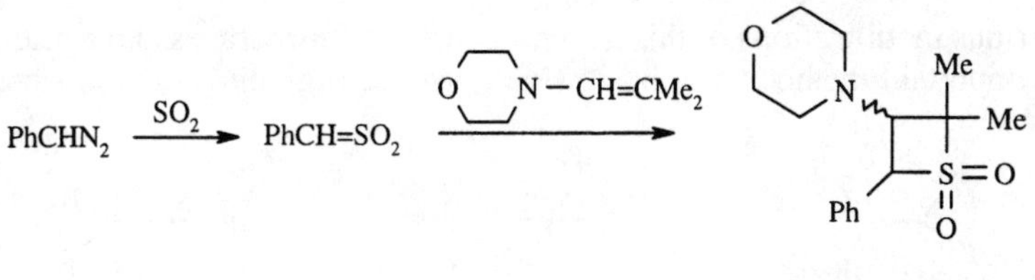

$$PhCHN_2 \xrightarrow{SO_2} PhCH=SO_2 \xrightarrow{}$$

(14) 62%

(iii) 2-Azetidinones (**15**) are obtained from β-haloamines and CO with Pd catalysis (Scheme 8). They also result from the cycloaddition of imines to ketenes, *e.g.* (**17**) → (**18**). The condensation of ketenes with imines to give β-lactams is known as the Staudinger reaction ⟨91COS(5)85, 91T7503⟩. This versatile procedure for the stereocontrolled synthesis of 2-azetidinones involves a [2 + 2] cycloaddition reaction in which the ketene is mostly generated either from acid chlorides and related derivatives in the presence of tertiary amines, thermally ⟨85JOC4231⟩, or photochemically from metal carbenes ⟨88JOC3113⟩.

A number of cycloadditions of imines or imino compounds with a variety of alkenes, including allenes ⟨88HCA1025⟩, vinyl ethers ⟨88TL547⟩, methyl acrylate ⟨86ZOR636⟩, ketene acetals ⟨87JOC365⟩ and electrophilic alkenes ⟨85T1953⟩, afford functionalized azetidines.

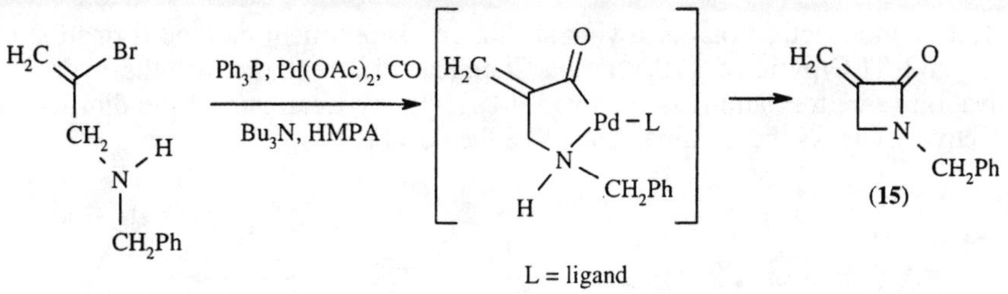

Scheme 8

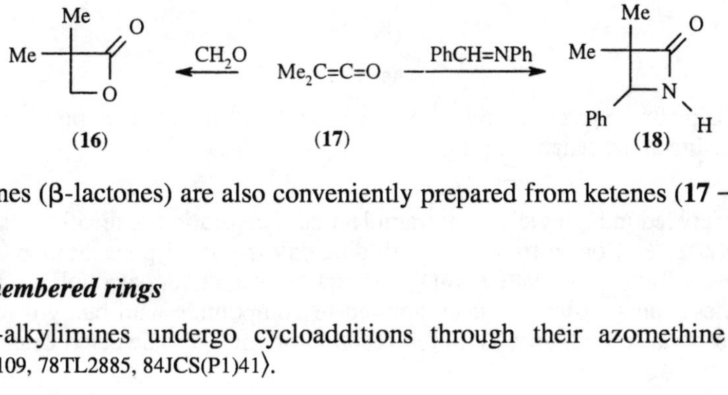

(16)   **(17)**   **(18)**

(iv) 2-Oxetanones (β-lactones) are also conveniently prepared from ketenes (**17** → **16**).

4.2.1.1.3 Five-membered rings

Appropriate *N*-alkylimines undergo cycloadditions through their azomethine ylide tautomers (Scheme 9) ⟨78CC109, 78TL2885, 84JCS(P1)41⟩.

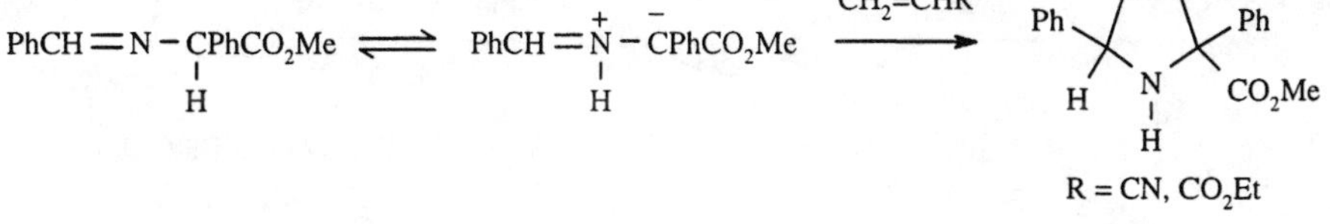

$$PhCH=N-CPhCO_2Me \rightleftharpoons PhCH=\overset{+}{N}-\overset{-}{C}PhCO_2Me \xrightarrow{CH_2=CHR}$$

R = CN, CO$_2$Et

Scheme 9

4.2.1.2 From Acyclic Compounds by Formation of One or Two C-Z Bonds
4.2.1.2.1 Three-membered rings

Oxiranes are formed by the action of alkali on β-hydroxy-halides (**19**; Y = Br, Cl), β-tosylates (**19**; Y = TsO) ⟨84CHEC-(7)115, 85CHE1, 85MI 103-01⟩ and β-hydroxy quaternary ammonium ions (**19**; Y = NMe$_3^+$). Thiiranes are prepared by a similar method, alkali treatment of β-mercaptohalides (**20**). β-Aminohalides and β-aminosulfates (**21**; Y = Br, Cl, SO$_4^-$) give aziridines on heating or on treatment with alkali. Aziridinones with bulky *N*-substituents are synthesized by base-catalyzed cyclization of α-haloamides. Various modifications of this general approach for oxiranes and aziridines are discussed below. The most important methods are those which provide high enantio- and stereo-selectivities of the product formed.

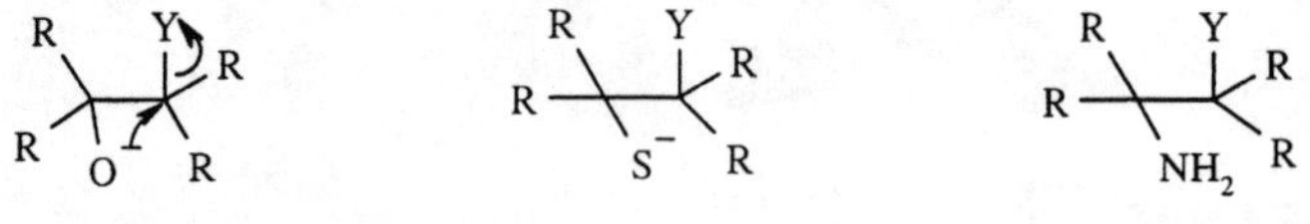

(19)   (20)   (21)

(i) Oxiranes

A three-step protocol ⟨92T10515, 92TL2095⟩ converts chiral diols, efficiently into chiral oxiranes (Scheme 10).

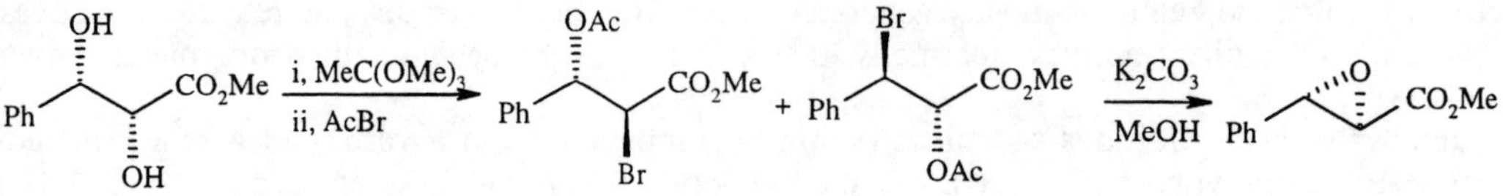

Scheme 10

Similar selective monoactivations of a vicinal diol and subsequent oxirane formation utilize TsCl/NaH ⟨93SC285⟩ and Tf$_2$O/pyridine ⟨93JOC1762⟩. Chemical differentiation of the hydroxy groups by selective activation (*e.g.*, tosylation) is achieved ⟨94TA2485⟩ by treatment of the diol with SOCl$_2$; NaI then regioselectively attacks the terminal carbon (Scheme 11).

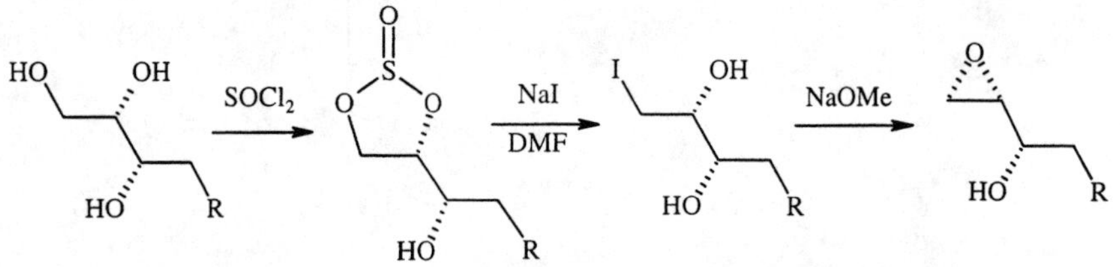

Scheme 11

1.2-Diols are directly converted into oxiranes with Ph$_3$P in the presence of diisopropyl azodicarboxylate (Mitsunobu reaction) ⟨81S1⟩.

(ii) Aziridines

Aziridines are prepared in high yield by intramolecular dehydration of an alkanolamine either in the gas phase ⟨92JAP(K)04217659⟩ or on treatment with diethoxy triphenyl phosphorane ⟨86JOC95⟩ or under Mitsunobu conditions ⟨86H(24)2099, 94H(37)461⟩. Conversion of an aminoalcohol into the corresponding tosylate or methanesulfonate ester has been applied to compounds with bulky or electron-attracting groups on nitrogen because such groups inhibit reaction on nitrogen. Selected examples are shown in Scheme 12 ⟨88SC1391, 93BSF459⟩.

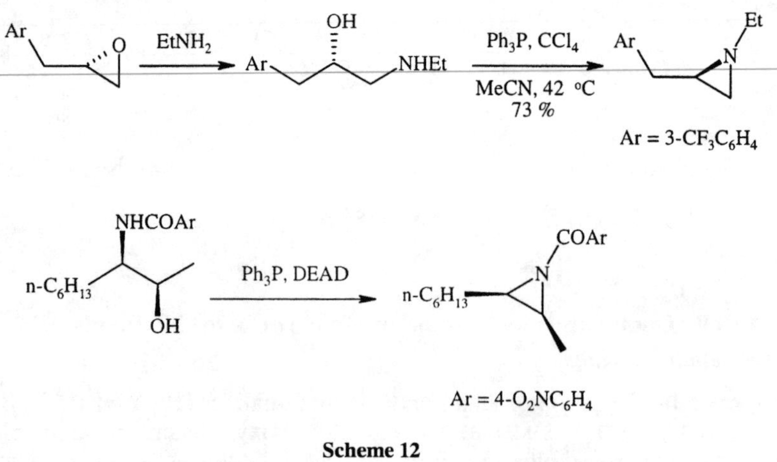

Scheme 12

β-Aminoalcohols may be generated from β-azidoalcohols, which are commonly made from the opening of epoxides with azide ion as shown in Scheme 13 ⟨91TL2533⟩. After conversion of the alcohol into a leading group, the azide may be reduced to an amine which cyclizes to the *N*-unsubstituted aziridine ⟨81LA1633, 83CAR(113)C6, 91TL2533⟩.

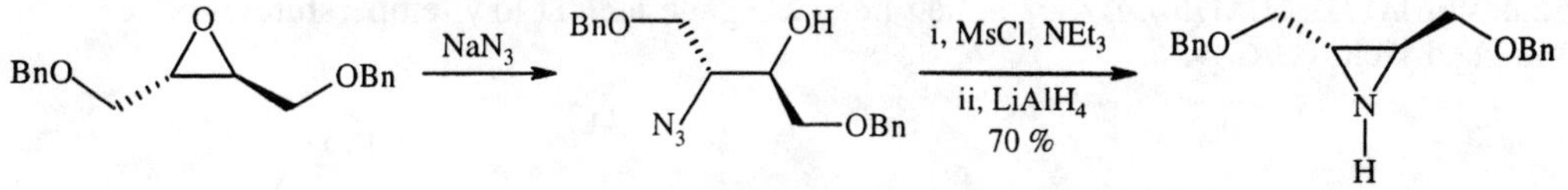

Scheme 13

The cyclization of β-haloamines is the Gabriel synthesis of aziridines ⟨B-83MI 101-01, 84CHEC-(7)47⟩. Imines with halogen-containing side chains on the nitrogen atom have been reductively cyclized *via* the β-haloamines, as shown in Scheme 14 ⟨94SL287⟩.

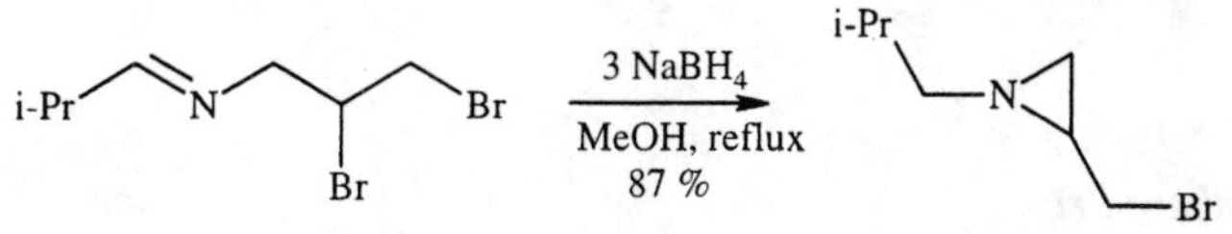

Scheme 14

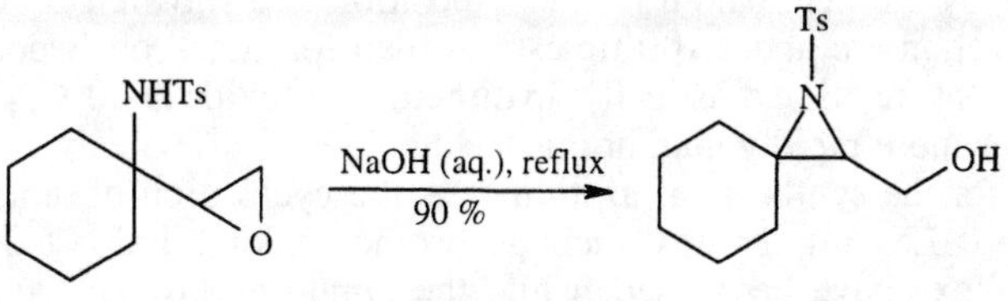

Scheme 15

Aziridines may be prepared by the intramolecular nucleophilic opening of epoxides by amines and their derivatives ⟨84CAR(130)103, 92T5639, 92TL487, 92TL5351⟩ (Scheme 15).

Three groups of polar processes to form aziridines are shown in Scheme 16. In every case, each of the two reactants must be capable of acting formally as either a bis-nucleophile or a bis-electrophile, or they must each have both nucleophilic and electrophilic character. In the aza-Darzens route ⟨B-83MI 101-01, 84CHEC-(7)47⟩, the imine acts as an electrophile at carbon and later as a nucleophile at nitrogen, while the α-haloenolate acts initially as a nucleophile at carbon and later as an electrophile at the same carbon. The roles of the two components are reversed for the polar aziridination route, which is related to the epoxidation reaction. In the last route, the 1,2-dihalide or α-haloenone acts formally as a bis-electrophile while the amine acts as a bis-nucleophile.

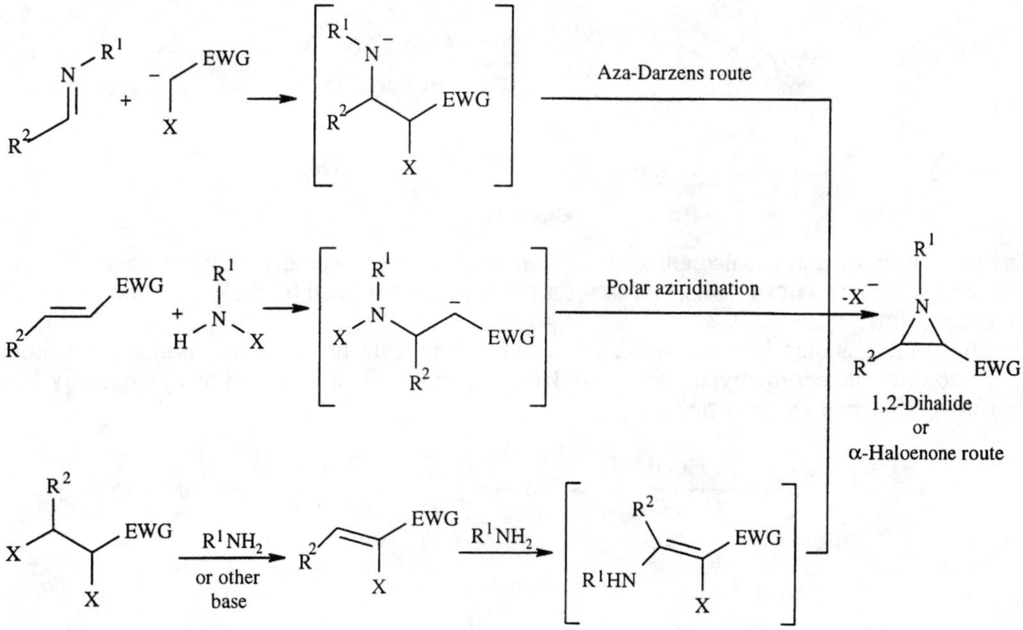

Scheme 16

The electrocyclic ring-opening of aziridines to azomethine ylides is well known, the reverse reaction is a route to aziridines ⟨B-83MI 101-01⟩, and is implicit in the *cis-trans* equilibration of substituted aziridines ⟨67JA1753, 71JA1779⟩. Azomethine ylide (**22**) generated at low temperature, produce aziridine (**23**) in good yield ⟨87JOC3470⟩.

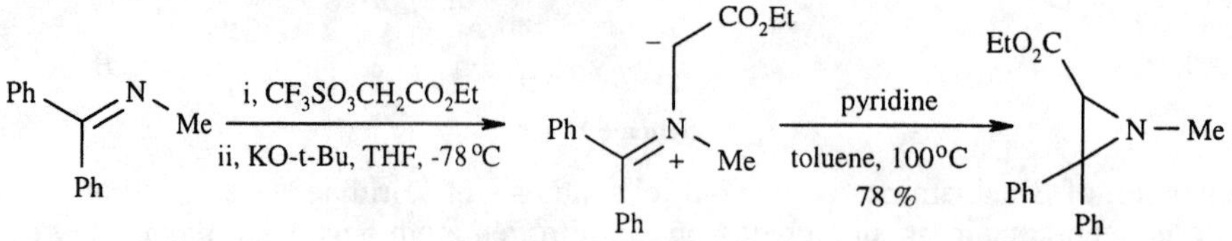

4.2.1.2.2   Four-membered rings

Bond formation α to the heteroatom is a common path to azetidines, oxetanes, thietanes and many oxo derivatives. Attack by the heteroatom on the γ-position is disfavored by the entropy factor; the reaction rates for cyclization are about 100 times less than for attack on β-positions. However, *exo*-unsaturation can change this relation drastically: in dimethyl sulfoxide at 50°C, β-bromopropionate ion cyclizes about 250 times more rapidly than bromoacetate.

A powerful method for the synthesis of azetidines is the cyclization of amines carrying a leaving group at the γ-position. Leaving groups such as bromo, chloro, iodo, tosyloxy, mesyloxy, and trifluoromethanesulfonyloxy have been used, while the amino group can have substituents such as alkyl, aryl, or tosyl ⟨85EUP161722, 87SC469, 88HCA1025, 92MI 118-02, 93JHC1197⟩. For example, the combination of tosylamides and the tosyloxy leaving group gives good results in the stereospecific synthesis of azetidines (**24**) (Scheme 17) ⟨92TL7921, 93TL6677, 94T265⟩.

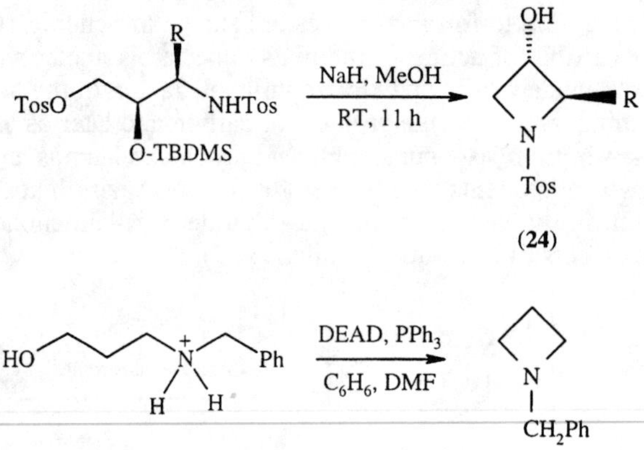

Scheme 17

The cyclization of γ-aminoalcohols or γ-azidoalcohols can produce azetidines *via* activation with phosphines ⟨81JOC3562, 84JCS(P1)2415, 85TL2809, 92OPP209⟩, as exemplified for the conversion (**25**) → (**26**) ⟨83CC682, 84JCS(P1)2415⟩.

Unsubstituted azetidine is synthesized on the kilogram scale by the condensation of 1-bromo-3-chloropropane with benzhydrylamine to give 1-benzhydrylazetidine followed by hydrogenolysis and basic workup (Scheme 18) ⟨88SC205⟩.

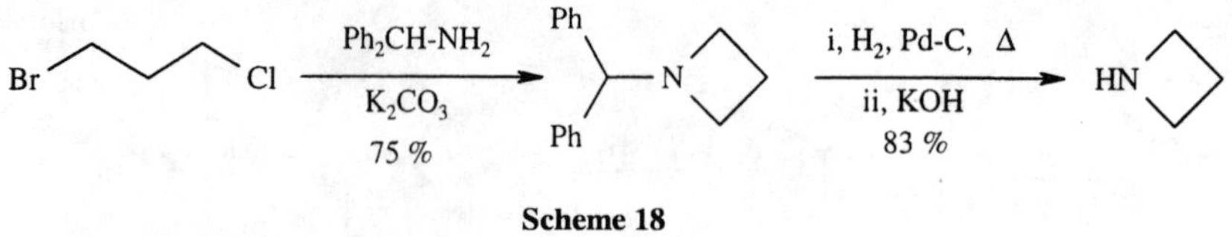

Scheme 18

β-Chloroimines (**27**) are suitable substrates for the synthesis of azetidines; the addition of nucleophiles such as cyanide, hydride, or organolithium reagents gives an adduct, such as γ-chloroamine or γ-chloroamide, which undergoes ring closure in a rate determining step to give azetidines (**28**) ⟨85CC715, 88T3653, 93S89⟩.

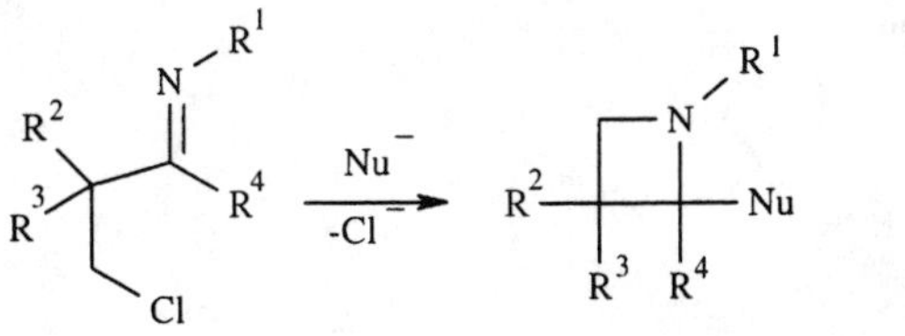

(**27**)   (**28**) Nu = CN, H, Alk, Ar

The numerous syntheses of β-lactams include N–C(2) and N–C(4) categories of bond formations.

The intramolecular condensation of β-amino acids is accomplished by numerous activating agents, including triphenylphosphine/di(2-pyridyl)disulfide ⟨76AG(E)94, 86TL2149, 88T4173, 93CC1153, 93CC1660, 93JCS(P1)2247⟩, triphenylphosphine/CCl$_4$ or NBS ⟨88SC247⟩, triphenylphosphine/CBr$_4$ or Br$_2$, or I$_2$ ⟨80M1117⟩, triphenylphosphine/hexachloroethane ⟨91MI 118-02⟩, diarylphosphinic acid chloride ⟨88CC1242⟩, nitropyridylphosphates ⟨93MI 118-05⟩, phenyl phosphorodichloridate ⟨91JOC2244⟩, bis(5-nitro-2-pyridyl)-2,2,2-trichloroethyl phosphate ⟨87TL2735⟩, benzotriazoloxy-tris[dimethylamino]-phosphonium hexafluorophosphate ⟨88MI 118-03⟩, N,N-dimethyl(chlorosulfinyl)oxyformaldiminium chloride (Scheme 19) ⟨93BMC2419⟩, mesyl chloride/tetrabutylammonium hydrogensulfate ⟨91HCA1213⟩, 2-chloro-1-methylpyridinium chloride ⟨84CL1465⟩, carbodiimides ⟨84JA6414, 86T2685, 87CPB4672, 88T613, 88T619, 89JOC4649⟩, triphenylphosphine/diethyl azodicarboxylate ⟨83CL175⟩, and tosyl chloride/DABCO ⟨84TL3849⟩.

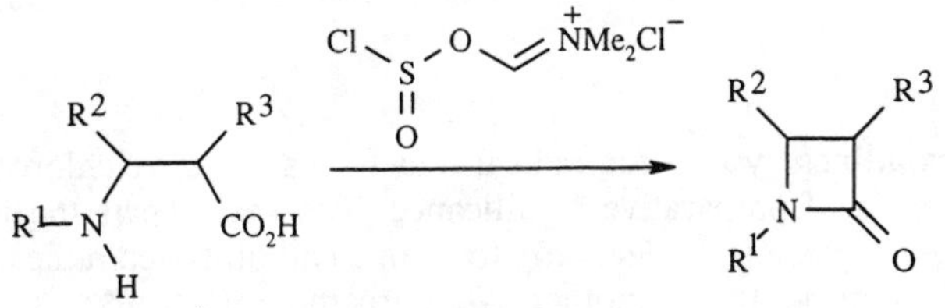

Scheme 19

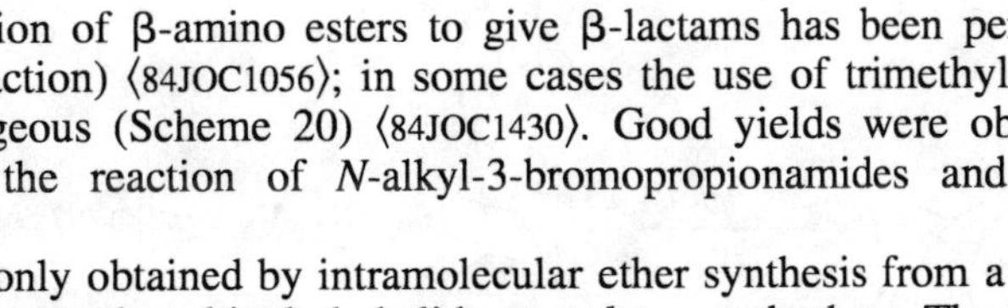

Scheme 20

The cyclocondensation of β-amino esters to give β-lactams has been performed with Grignard reagents (Breckpot reaction) ⟨84JOC1056⟩; in some cases the use of trimethylsilyl esters and amines proved to be advantageous (Scheme 20) ⟨84JOC1430⟩. Good yields were obtained by using phase transfer catalysis in the reaction of N-alkyl-3-bromopropionamides and potassium hydroxide ⟨80H(14)467⟩.

Oxetanes are commonly obtained by intramolecular ether synthesis from a suitably functionalized alcohol. Leaving groups employed include halides, tosylates, and others. The base can range from an alkoxide to a non-nucleophilic amine.

Mildly basic to neutral conditions for the ring closure of 1,3-halohydrins include tetraphenylantimony methoxide as an effective non-basic reagent for oxetane synthesis from 1,3-bromohydrins ⟨90S106⟩. The salts of β-halo acids cyclize in ionizing media to oxetan-2-ones, as do β-diazonium carboxylates ⟨64HC(19-2)787⟩. Thietanes are obtained analogously.

4.2.1.2.3 Five-membered rings

Pyrrolidine (**30**; Z = NH) and thiolane (**30**; Z = S) can be prepared from tetramethylene dibromide (**29**), and tetrahydrofuran (**30**; Z = O) is obtained from the diol (**31**).

γ-Hydroxy and γ-thiol acids (**32**; Z=O, S) usually cyclize spontaneously to give lactones and thiolactones (**33**). γ-Amino acids (**32**; Z=NH) require heating to effect lactam formation (**33**; Z=NH).

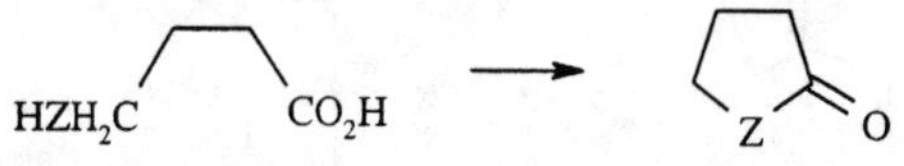

Pyrrolidines may also be prepared by Mannich reactions, *e.g.* the formation of tropinone by reaction (**34**) → (**35**); reactions of this type are involved in alkaloid biogenesis.

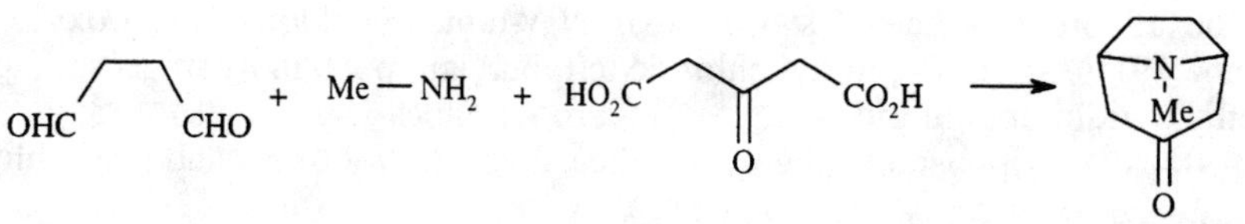

The synthesis of pyrrolidines by the free radical transformation of *N*-chloroamines, the Hofmann-Loeffler-Freytag reaction, is of preparative significance. The key step is the formation of a radical cation which abstracts hydrogen intramolecularly to form a carbon-based radical (Scheme 21(a)). This species then abstracts chlorine from another *N*-chloroamine ⟨60JA1657, 50JA2118⟩. The observed positional selectivity for hydrogen abstraction is a consequence of the preferred adoption of a six-membered transition state. A typical conversion achieved is indicated in Scheme 21(b).

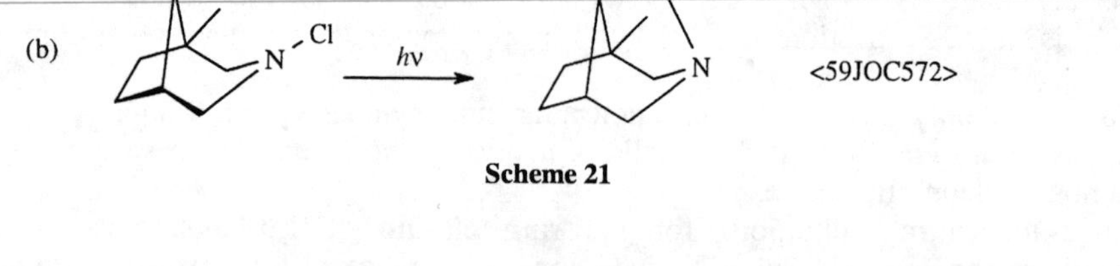

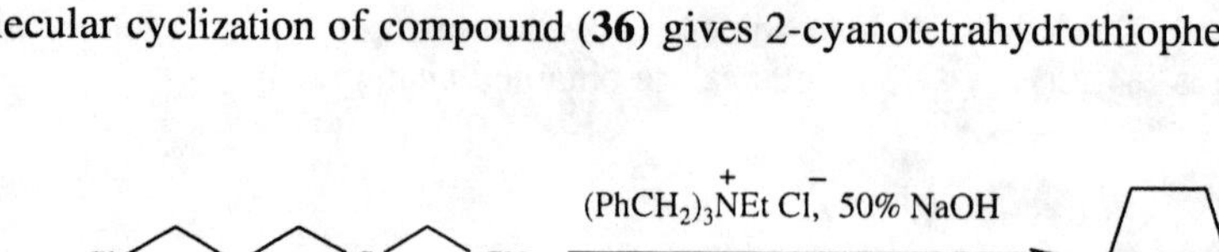

Scheme 21

Intramolecular cyclization of compound (**36**) gives 2-cyanotetrahydrothiophene (**37**) ⟨87S452⟩.

The parent thiocarbonyl ylide (**39**), generated by cesium fluoride-induced desilylation of (**38**), is trapped with alkenes bearing electron-withdrawing substituents and with DMAD to give tetra-hydrothiophenes (**40**) (58–86%) and dihydrothiophene (**41**) (56%), respectively ⟨86CC1073⟩. With alkenes the original stereochemistry of the alkenes is retained.

4.2.1.2.4 Six-membered rings

These methods parallel the synthesis just described for the five-membered rings. As indicated in structures (**42**)–(**49**), standard reactions of aliphatic chemistry can be extended to the preparation of piperidines, tetrahydropyrans and pentamethylene sulfides (**44**; Z = N, O, S); glutarimides, glutaric anhydrides and glutaric thioanhydrides (**46**; Z = N, O, S); and δ-lactams, δ-lactones and δ-thiolactones (**49**; Z = N, O, S).

The traditional route to piperidines by reaction between a 1,5-dihalopentane and ammonia or an amine is illustrated by the synthesis of 4-hydroxypiperidines (**50**) ⟨88JCS(P1)2881⟩.

$$ClCH_2CH_2CH(OH)CH_2CH_2Cl \quad + \quad ArNH_2 \quad \xrightarrow[\substack{DMF,\ 100\ °C \\ 48-82\ \%}]{K_2CO_3,\ NaI}$$

(**50**)

Many examples of the synthesis of tetrahydropyrans are based on the cyclization of 1,5-diols and compounds which can provide a similar electrophilic site for ring closure. Thus, pentan–1,5-diols can be quantitatively cyclized to the pyran in the presence of BuSnCl₃ ⟨88G483⟩.

4.2.1.2.5 Larger rings

Similar methods can also be used for seven-membered and larger rings; however, high dilution techniques must often be utilized. Oxepane, for example, is obtained by the dehydration of hexane 1,6-diol ⟨80BCJ3031⟩.

The synthesis of thiepanes by double nucleophilic reaction in dipolar aprotic solvents improves the yield even under at normal dilution ⟨79SC857⟩. Li$_2$S generated *in situ* from hexamethyldisilathiane and methyllithium was convenient for carbon-sulfur bond formation (Equation (2)) ⟨85JOC4969⟩.

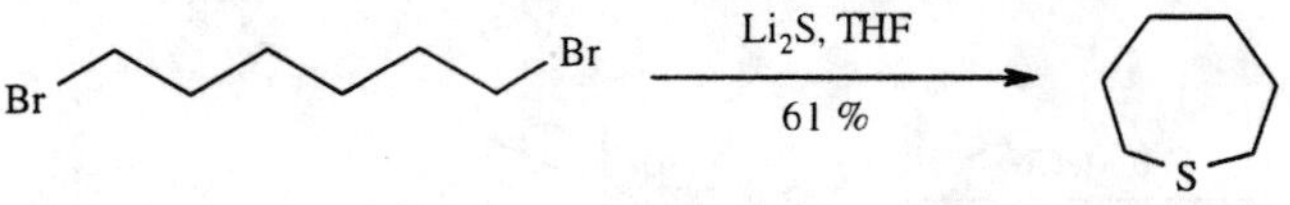

$$\text{Br} \diagup\!\diagdown\!\diagup\!\diagdown\!\diagup \text{Br} \xrightarrow[61\%]{\text{Li}_2\text{S, THF}} \bigcirc_{\text{S}} \qquad (2)$$

4.2.1.3 From Acyclic Compounds by Formation of One C–C Bond

Many of the standard methods of C–C bond formation in aliphatic systems can be extended to heterocyclic systems, *e.g.* the Dieckmann reaction (*cf.* **51** → **52**) and alkylation of active methylene compounds (*e.g.* **53** → **54**).

Of the syntheses involving C-C bond formation, those in which the C(3)–C(4) bond is formed (**55** → **56**) are in general more important than those involving C(2)–C(3) bond formation (**57** → **56**). However, carbenoid insertions of type (**58**) have been used to make β-lactams ⟨72JA1634⟩.

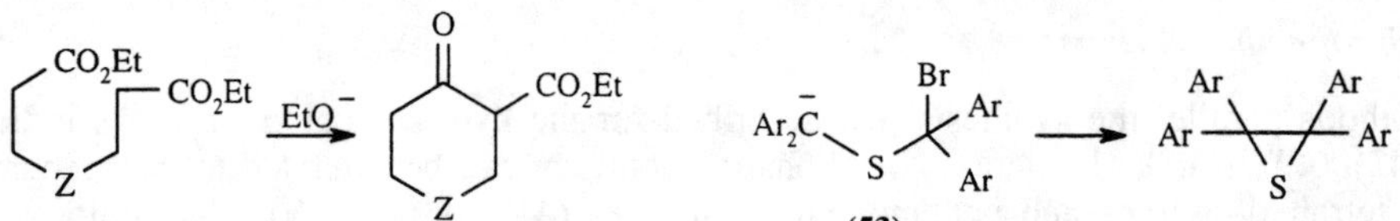

An example of the application of the Dieckmann reaction to the preparation of 3-thiepanone is shown in Scheme 22 ⟨52JA917⟩.

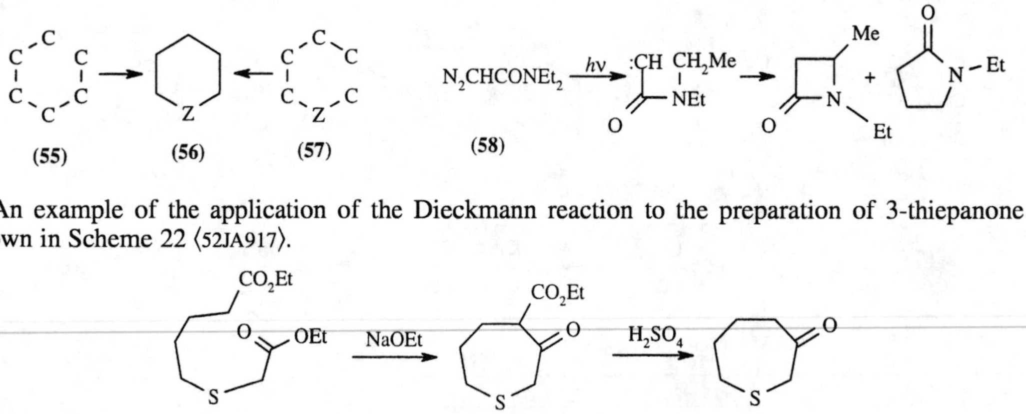

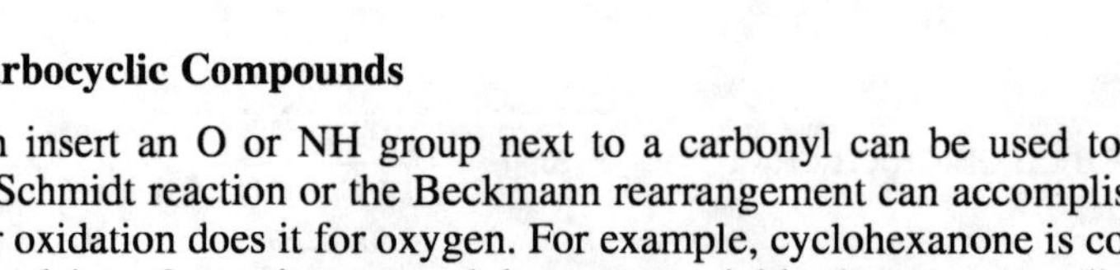

Scheme 22

4.2.1.4 From Carbocyclic Compounds

Reactions which insert an O or NH group next to a carbonyl can be used to form heterocycles (Scheme 23). The Schmidt reaction or the Beckmann rearrangement can accomplish this for nitrogen, the Baeyer-Villiger oxidation does it for oxygen. For example, cyclohexanone is converted in this way into 2-azepinone and into 2-oxepinone; cycloheptanone yields the corresponding eight-membered heterocycles.

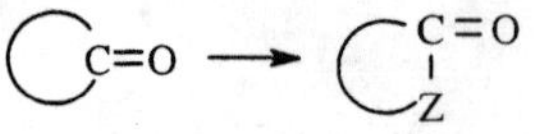

$$\bigcirc\!\!\!\text{C}=\text{O} \longrightarrow \bigcirc\!\!\!\begin{array}{c}\text{C}=\text{O}\\|\\\text{Z}\end{array} \qquad \begin{array}{l}\text{Z} = \text{O by RCO}_3\text{H}\\[6pt]\text{Z} = \text{NH by HN}_3,\ \text{H}^+\end{array}$$

Scheme 23

While these rearrangements are used most often to prepare large rings, the expansion of cyclopropanone to azetidinones is also practical (Scheme 24; CHEC 5.09.3.3.3).

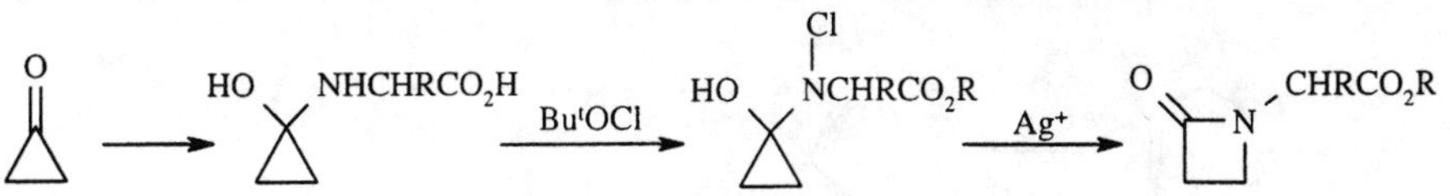

Scheme 24

4.2.1.5 From Other Heterocyclic Compounds

Such syntheses are also considered as reactions of the corresponding starting heterocycles in the relevant reactivity chapter.

4.2.1.5.1 *Reactions involving ring expansion*

The facility with which oxiranes may be prepared and the ease with which they undergo ring opening with nucleophiles or electrophiles make them useful synthons.

Examples of ring-opening reactions with carbanions leading to five-membered heterocyclic ring formation are shown in (**59**) → (**60**) ⟨50JA4368⟩ and (**61**) → (**62**) ⟨66CB2556⟩. 3,4-Epoxyalcohols give hydroxymethyloxetanes (*e.g.* **63**) by ring expansion ⟨80BCJ2895⟩. γ-Hydroxyepoxides on treatment with acids yield tetrahydrofurans (*e.g.* **64** → **65**) ⟨91JA9369⟩, as in the total syntheses of solamin and reticulatacin ⟨94JCS(P1)1975⟩.

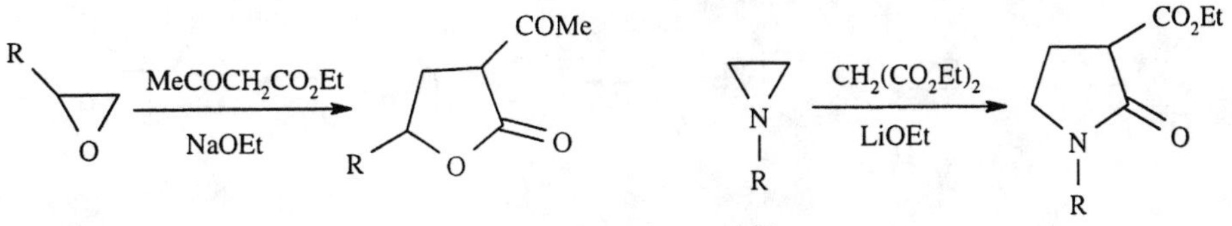

(**59**) R = Me, Ph (**60**) (**61**) (**62**)

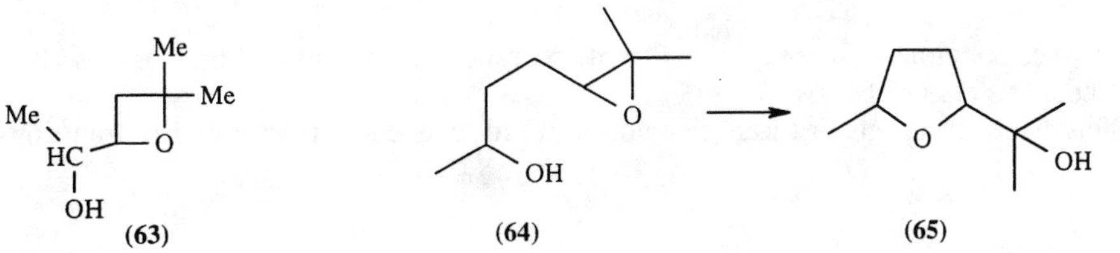

(**63**) (**64**) (**65**)

Certain oxiranes undergo ring opening to carbonyl ylides, and the addition to alkenes, leading to formation of tetrahydrofurans, is stereospecific (Scheme 25).

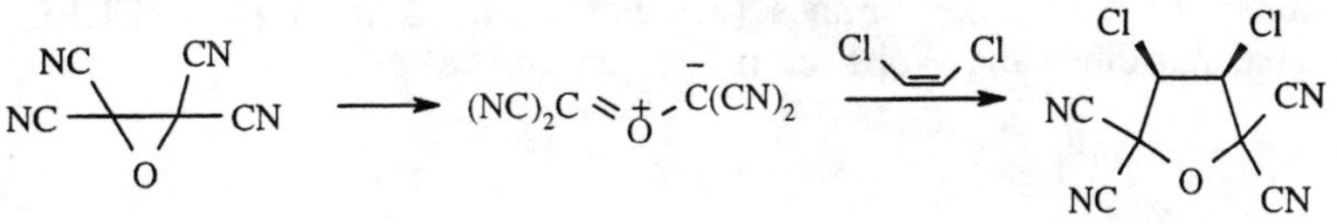

Scheme 25

Azirines can give pyrrolidine derivatives (Scheme 26) ⟨67BCJ2936⟩.

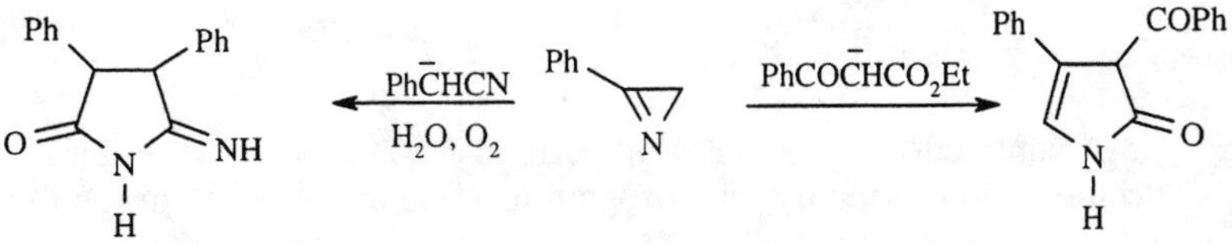

Scheme 26

2-Hydroxymethyltetrahydrofuran is converted into 5,6-dihydro–4*H*-pyran and the acid catalyzed rearrangement of other 2-substituted tetrahydrofurans forms 3-hydroxytetrahydropyrans (Scheme 27). Thiepan-4-one results from ring expansion of tetrahydrothiopyran-4-one (Scheme 28).

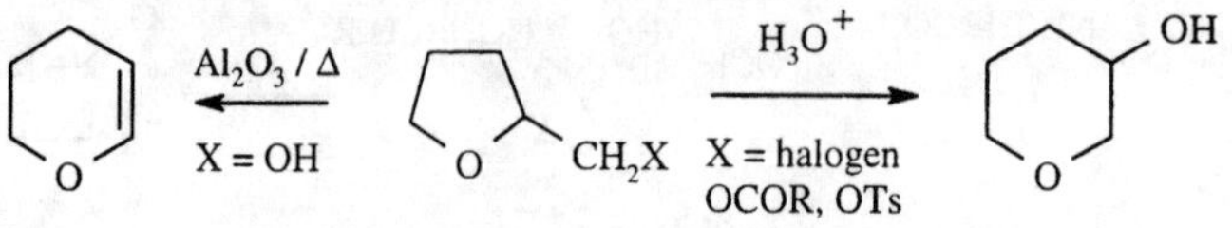

Scheme 27

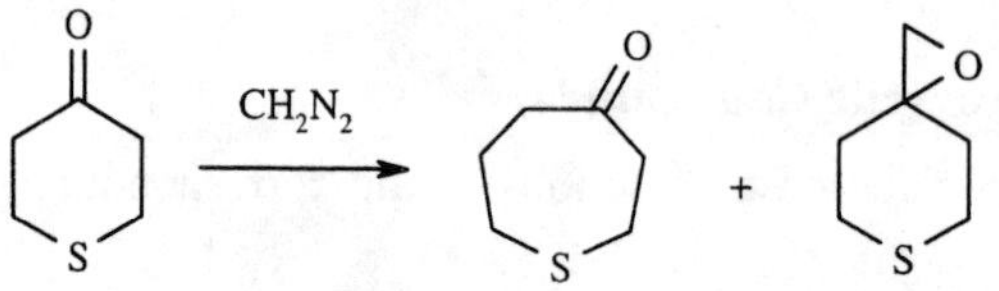

Scheme 28

4.2.1.5.2 Reactions without change in ring size

A classic prototype of these reactions is the conversion of oxiranes into thiiranes by thiocyanate ion (Scheme 29; CHEC 5.06.4.3). Inversion at both ring carbons makes the reaction stereospecific with respect to the *E/Z* relation of the substituents on the oxirane carbons.

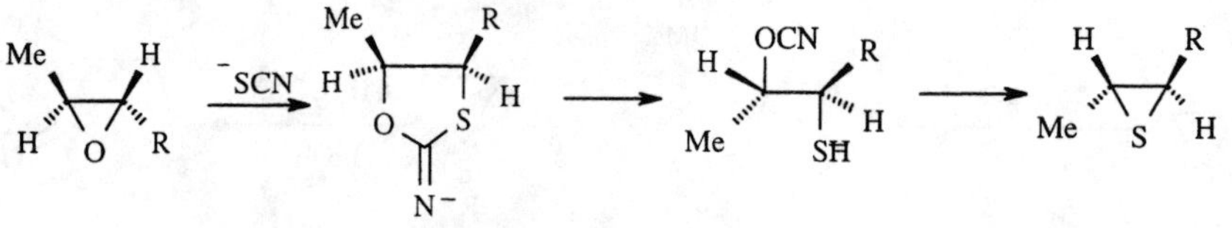

Scheme 29

Aziridines are obtained in one step from oxiranes with iminophosphoranes ⟨76CB814⟩ or phosphoramidate esters ⟨76TL4003⟩.

Many Diels-Alder reactions of furan give addition products containing dihydrofuran rings (Section 3.3.1.8).

4.2.1.5.3 Ring contraction

The loss of CO, S, SO, SO_2 and N_2 by thermolysis or photolysis has been used to make three- and four-membered rings; for example, thiiranes (**67**) are obtained from (**66**) (CHEC 5.06.4.4). Δ^2-1,2,3-Triazolines give aziridines and Wolff rearrangement of (**68**) gives (**69**).

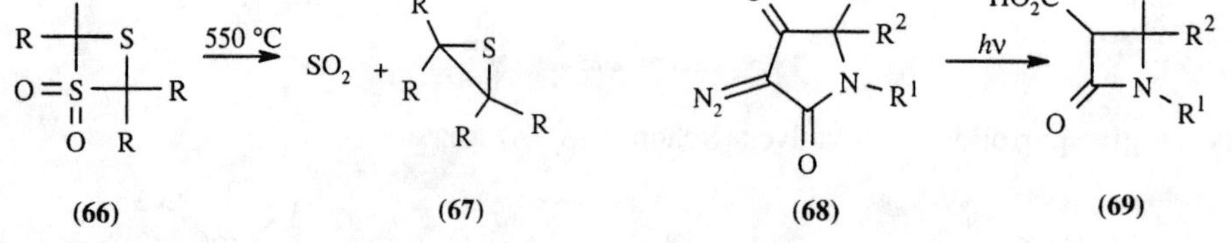

The solvolytic ring contraction of 3-chloropiperidines (**70**) to pyrrolidines proceeds *via* intramolecular displacement of chloride by the nitrogen, forming an aziridinium ion (**71**); subsequent ring opening by a nucleophile gives (**72**) ⟨69AG(E)962⟩. Abnormal Clemmensen reduction can also give a ring contraction as in the conversion (**73**) → (**74**).

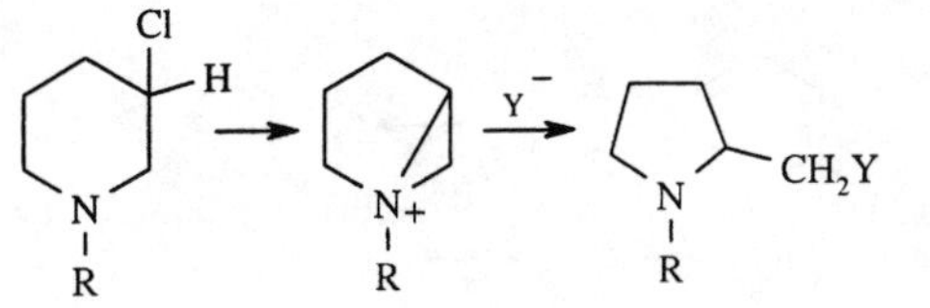

(70)   (71)   (72) Y = R^1NH, OH, CN 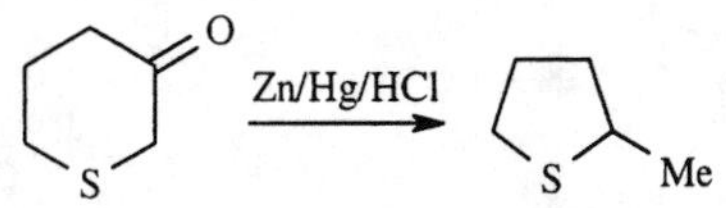   (73)   (74)

Many photo-induced ring contractions are known. An illustrative example is the isomerization of tetrahydrothiin-4-one into a zwitterion before loss of ethylene (Scheme 30) ⟨70JOC584⟩; pyrazolidinones give β-lactams (Section 3.4.2.3.3).

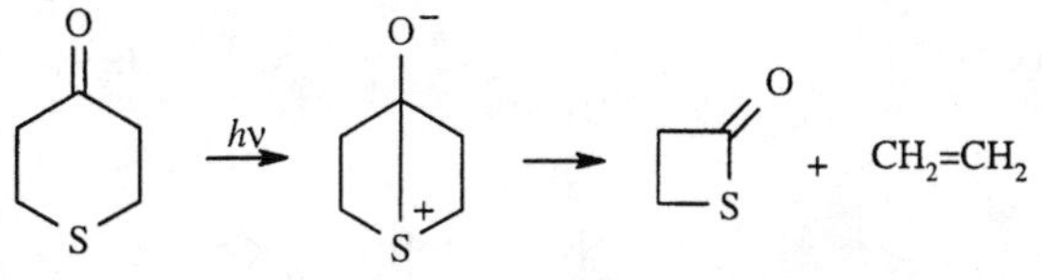

Scheme 30

The photo-rearrangements of certain pyridines and azines into azaprismanes and azabenzvalenes are discussed in Section 3.2.1.2.3.

4.2.2 RINGS CONTAINING ONE ENDOCYCLIC DOUBLE BOND

Most monoheterocycles with one cyclic double bond have been prepared by C–Z bond formation in which the Z atom acts as the nucleophile. However, for six-membered rings of this type, Diels-Alder reactions are especially important. Three-membered rings are also atypical: azirines are often made by C-N bond formation from precursors in which N is electrophilic or has nitrene character, while oxirenes are but fleeting intermediates (CHEC 5.05.6.3).

4.2.2.1 From Acyclic Compounds by Concerted Formation of Two Bonds

(i) Thiete 1,1-dioxides (*e.g.* **75**) result from sulfenes and ynamines ⟨73S534⟩, and cycloaddition of *N*-sulfonylimines (**76**) with alkynyl ethers afforded functionalized 2-azetines (**77**) ⟨91ZOB1389⟩.

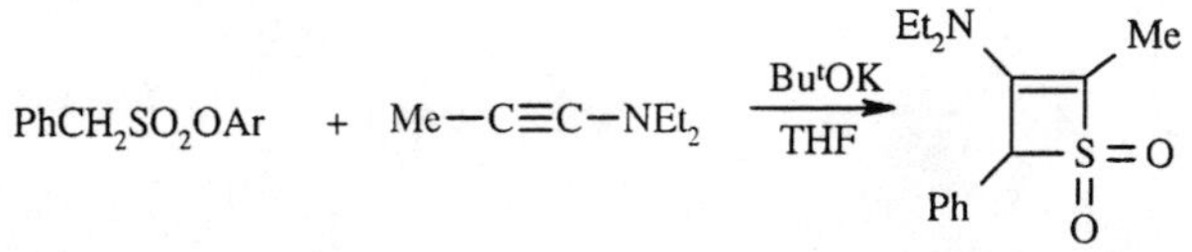

(75)

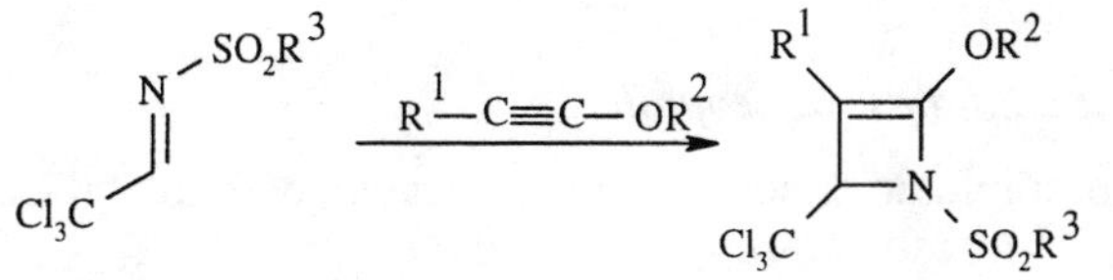

(76)   (77)

(ii) By far the most important reactions of this type are [4 + 2] heterocyclization analogues of the Diels-Alder reaction which present a versatile route to six-membered rings. The heteroatom can originate from the dienophile (*e.g.* Schemes 31–33) or from the diene (*e.g.* Schemes 34 and 35). Whereas α,β-unsaturated carbonyl compounds react best with electron-rich alkenes (Scheme 34), enaminothiones prefer electron-deficient dienophiles (Scheme 35).

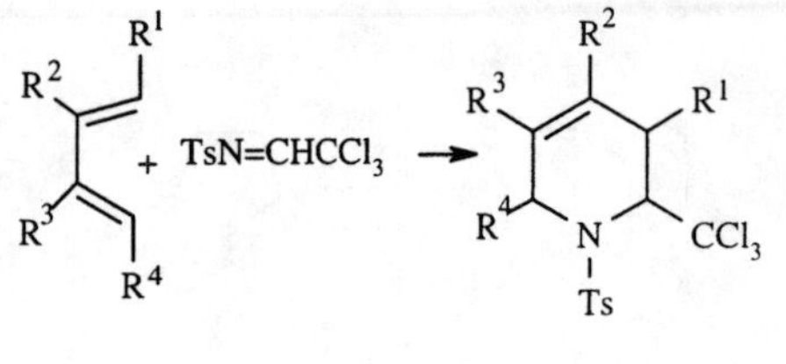

Scheme 31 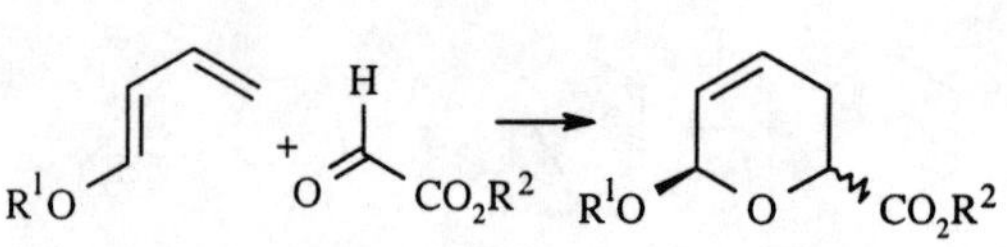   **Scheme 32**

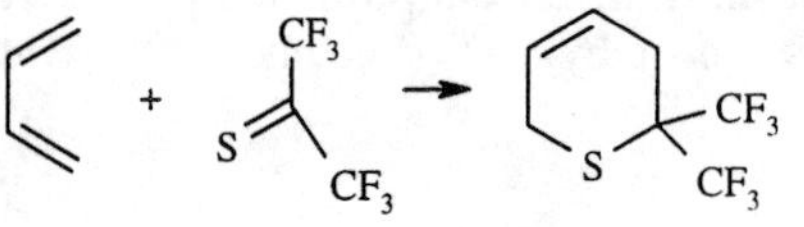

Scheme 33 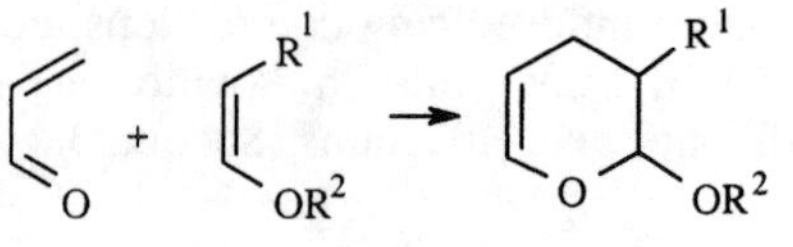   **Scheme 34**

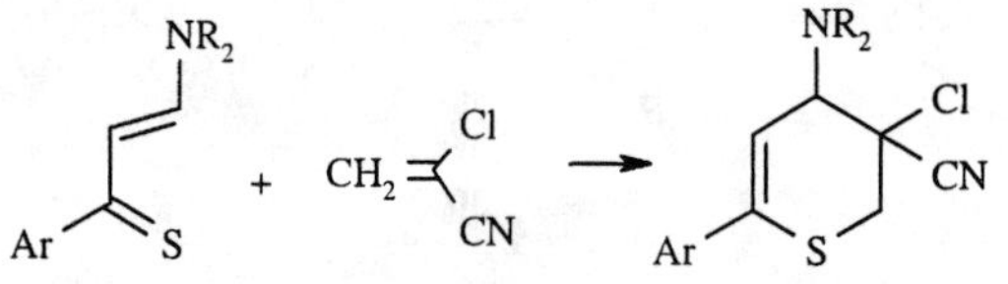

Scheme 35

[4 + 2] Cycloadditions where the 2π component is an imine or iminium species have been much used for the synthesis of reduced pyridines and pyridones ⟨B-87MI 505-01⟩, see also a general review of imine cycloadditions ⟨87H(26)777⟩.

Chiral catalysts, such as the boron derivative **78**, can give piperidinones (**79**) with good enantiomeric excesses ⟨93T1749⟩.

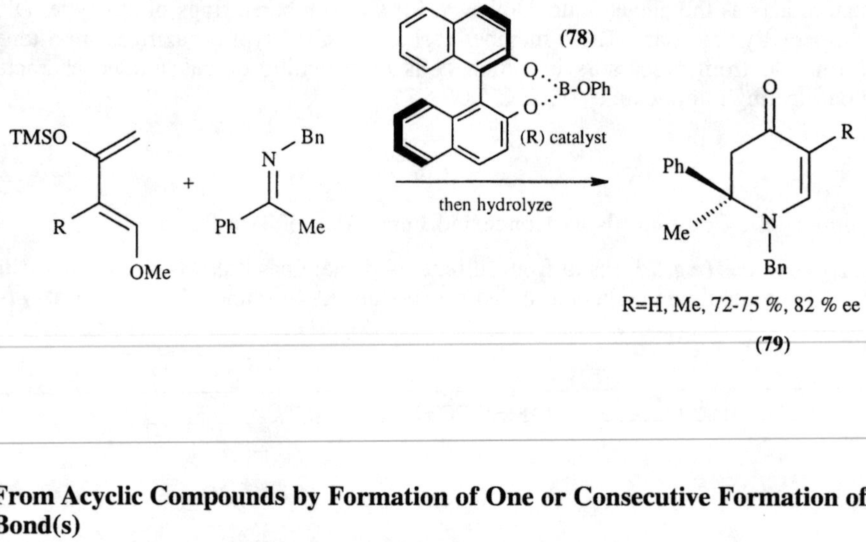

R=H, Me, 72-75 %, 82 % ee

(**79**)

4.2.2.2   From Acyclic Compounds by Formation of One or Consecutive Formation of Two C–Z Bond(s)

4.2.2.2.1   Z Atom component acting as nucleophile

The appropriate reactions of aliphatic chemistry are applicable, and take place particularly readily for the formation of five- and six-membered rings. For example:

(i) 4- and 5-Oxo primary amines yield Δ^1-pyrrolines and Δ^1-tetrahydropyridines, respectively (**80**; $n = 1, 2$).

(ii) Cyclic enol ethers and their thio analogues are formed from keto alcohols and thiols: $Ac(CH_2)_3OH$ gives the dihydrofuran (**81**) on distillation; *cf.* (**82**) → (**83**) → (**84**) for the preparation of Δ^2-dihydro-pyrans and -thiins (Z = O or S).

(iii) Unsaturated lactones are prepared from keto acids, *e.g.* (**85**) → (**86**).

2,3-Dihydrofurans can be prepared efficiently by a Michael addition of β-ketoesters to alkenyl sulfoxides followed by a Pummerer rearrangement (Scheme 36) ⟨92JCS(P1)945⟩.

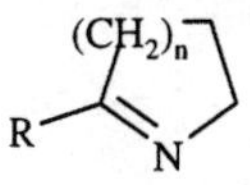

(80) 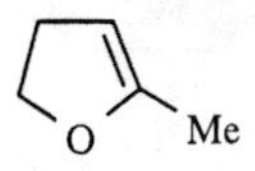 (81) 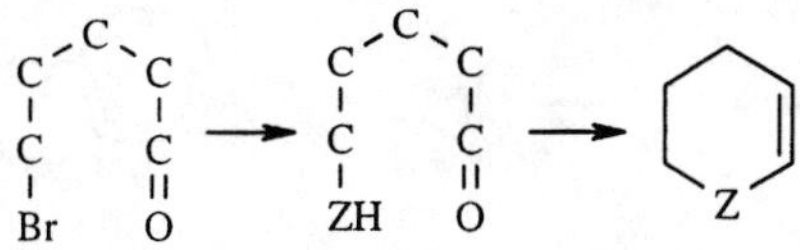 (82) (83) (84)

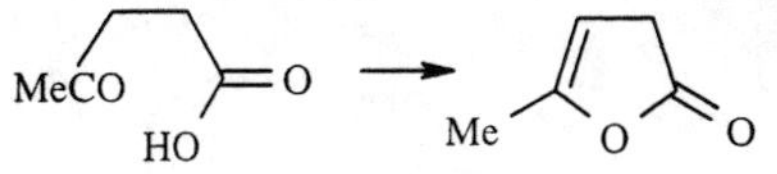

(85) (86)

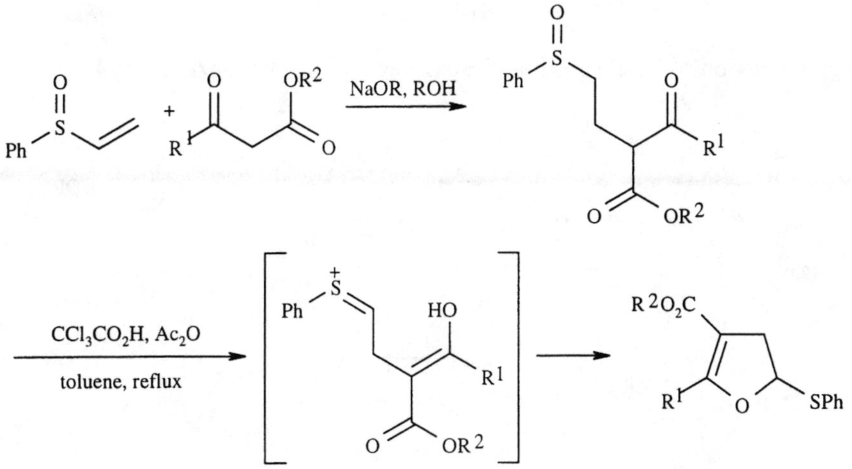

Scheme 36

4.2.2.2.2 Z Atom component acting as electrophile

1-Azirines are prepared by base-catalyzed cycloelimination of imine derivatives, *e.g.* as isolable intermediates in the Neber rearrangement (**87** → **88**) ⟨77JA1514⟩.

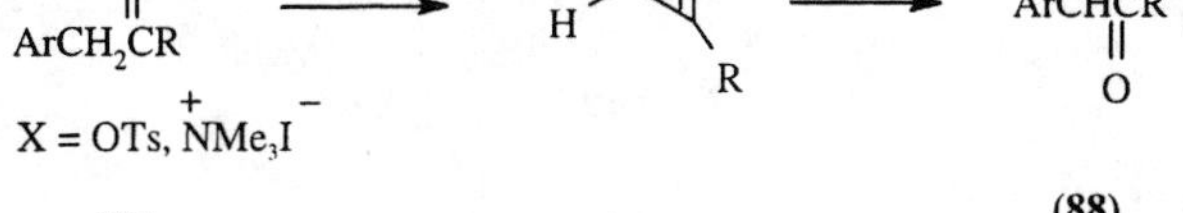

(87) (88)

1-Azirines are also made by carbene addition to nitriles (**89** → **90**) and by thermal or photochemical ⟨68JA2869⟩ elimination of N_2 from vinyl azides (*e.g.* **91** → **92**). Vinyl azides are prepared by the Hassner reaction ⟨68JOC2686, 71ACR9⟩, where iodine azide is first added to an alkene and the resultant β-iodoazide is dehydrohalogenated with base (Scheme 37) ⟨86RTC456⟩.

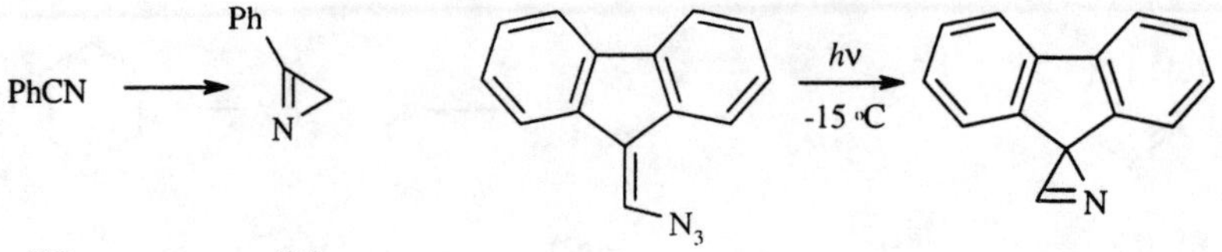

(89) (90) (91) (92)

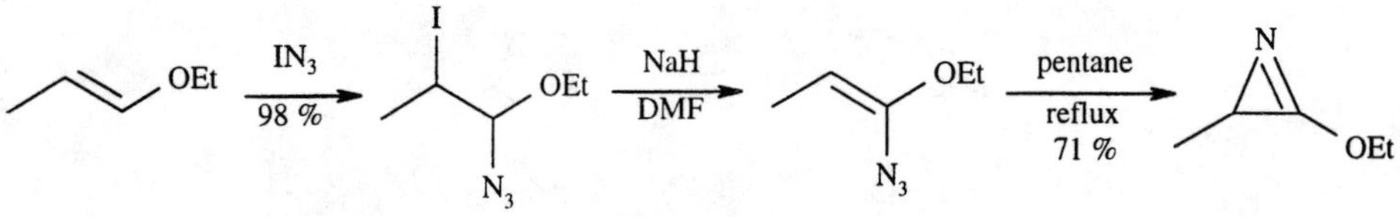

Scheme 37

Themal ring contraction of 3,5-disubstituted isoxazoles (**93**) gives 1-azirines (**94**).

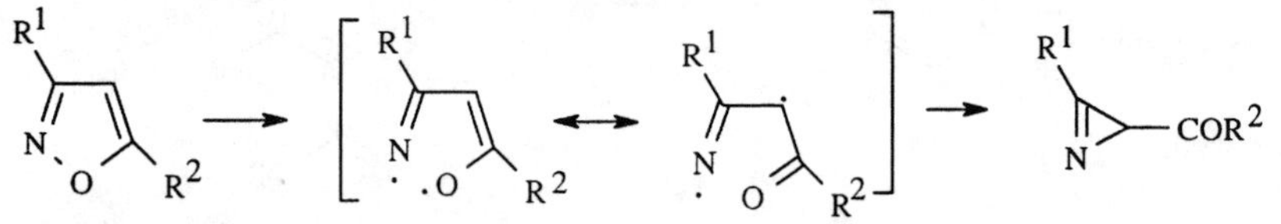

(**93**) (**94**)

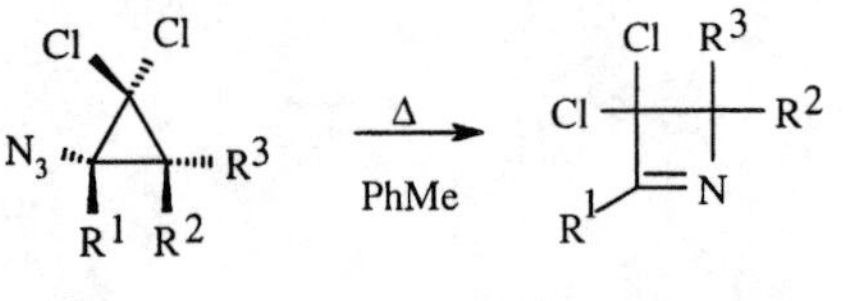

(**95**) (**96**) 50-75%

4.2.2.3 From Carbocycles

Most of the methods of Section 4.2.1.4 apply.

Thermal ring expansion of cyclopropyl azides provides a general route (**95** → **96**) to alkyl- and aryl-azetines ⟨79CB3914⟩.

4.2.2.4 From Heterocycles

The photolytic ring opening of 2*H*-azirines yields nitrile ylides which can be trapped as pyrrolines (**97**) ⟨73JA1945⟩.

Nitrogen extrusion has been used to make fragile molecules: 2-thiirene (**99**) has been obtained by matrix photolysis of 1,2,3-thiadiazole (**98**), and azirines from 4*H*-triazoles (**100** → **101**).

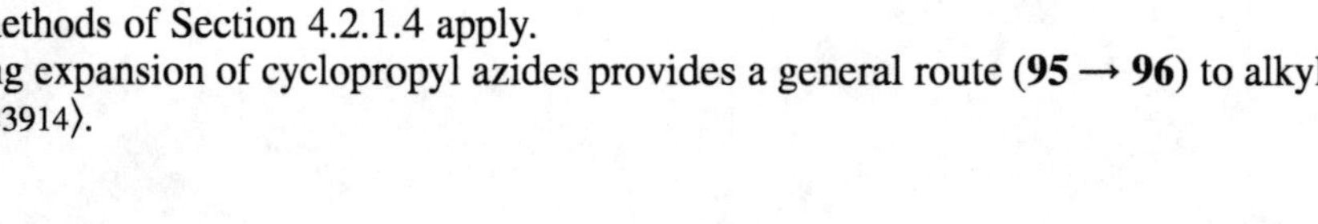

R^3 = CN, CO_2Me (**97**) (**98**) (**99**)

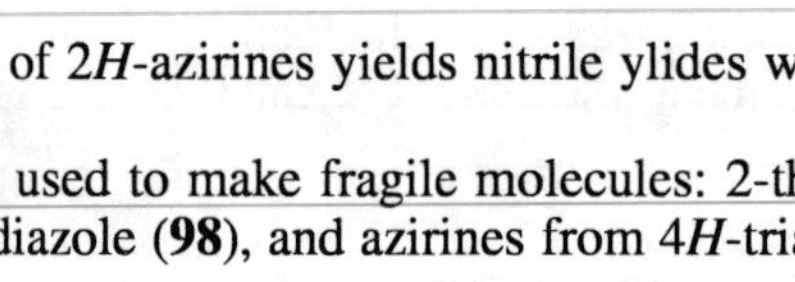

A = ester 85-92%

(**100**) (**101**)

The photo-isomerization of certain pyridines to Dewar pyridines is described in Section 3.2.1.2.2, and the formation of bridged ring 6*H*-1,2-dihydro-3-pyridones in Section 3.2.1.10.4.

4.2.3 RINGS CONTAINING TWO ENDOCYCLIC DOUBLE BONDS

4.2.3.1 Overview

The only four-membered heterocycles with two endocyclic double bonds are the highly reactive azetes; approaches to their synthesis are discussed in CHEC 5.09.5.3.1 and CHEC-II 1.18.5.6.

Very importantly, this section also includes the pyrroles, furans and thiophenes. We deal with their preparation by cyclization methods of type (**102a,b**; Section 4.2.3.3.1), (**103**; Section 4.2.3.3.2), (**104**; Section 4.2.3.3.3) and (**105**; Section 4.2.3.3.4), successively, but precede this by a discussion of their preparation by substituent introduction or modification.

(**102a**)　　(**102b**)　　　(**103**)　　　(**104**)　　　(**105**)

The most important methods of synthesis for these ring systems and the sections in which they are considered are:

(i) Pyrroles: Paal-Knorr (4.2.3.3. 1.i), Knorr (4.2.3.3.3.i), Hantzsch (4.2.3.3.3.ii).
(ii) Furans: Paal-Knorr (4.2.3.3.1.i), Feist-Benary (4.2.3.3.3.ii).
(iii) Thiophenes: Paal-Knorr (4.2.3.3.1.i), Hinsberg (4.2.3.3.4).

We then consider methods for the synthesis of pyrans, dihydropyridines and their oxo derivatives, and finally methods for compounds with larger ring sizes.

4.2.3.2 Synthesis of Pyrroles, Furans and Thiophenes by Substituent Introduction or Modification

For detailed discussion of this topic for pyrroles (CHEC 3.06.7, CHEC-II 2.03.3), furans (CHEC 3.12.6, CHEC-II 2.06.2 and 2.06.3) and thiophenes (CHEC 3.15.9, CHEC-II 2.10.2 and 2.10.4) see the chapters quoted. Some important routes to substituted derivatives are summarized in Table 1.

All these rings undergo easy electrophilic substitution at the 2-position. Particularly for pyrrole and furan the high reactivity often leads to low yields and sometimes it is useful to incorporate deactivating substituents such as CO_2H which can later be removed.

N-Substituted pyrroles, furans and thiophenes can be 2-lithiated, and these lithio derivatives are important synthetic intermediates (Section 3.3.3.8). 2-Mercuri and 2-palladio derivatives are also important (Sections 3.3.3.8.8 and 3.3.3.8.9).

The preparation of β-substituted derivatives is more difficult and different methods have been used in the various series. 2-Tritylpyrrole undergoes electrophilic substitution selectively at the 3-position (Section 3.3.1.4.3) and the trityl group can then be removed. Again, in the pyrrole series the selective hydrolysis of α-CO_2Et in alkali and of β-CO_2Et in acid, followed by decarboxylation, allows the introduction of β-substituents into compounds such as 2,4-dialkylpyrrole-3,5-dicarboxylic esters to afford 3-substituted 2,4-dialkylpyrroles (Section 3.3.3.3.7).

2,4-Disubstituted furans can be prepared by the 3-lithiation of 2-phenylthio-5-alkylfurans, followed by reaction with an electrophile and then desulfurization with Raney nickel (Section 3.3.3.8.4). 3-Furylmercuri acetate can be obtained from furan-2-carboxylic acid (CHEC 3.11.3.9) and transformed to other 3-substituted furans *via* the lithio compound.

3-Lithiothiophene is a key to the synthesis of many 3-substituted thiophenes. It is prepared from 3-bromothiophene, itself obtained from 2,3,4-tribromothiophene by selective Zn reduction (CHEC 3.14.3.8), or by rearrangement of 2-bromothiophene with $NaNH_2$–NH_3 (CHEC 3.15.9.6.3).

4.2.3.3 Synthesis of Pyrroles, Furans and Thiophenes from Acyclic Precursors

4.2.3.3.1 From C₃Z or C₄ units

Three prominent types of reactions fall in this classification: cyclizations by condensation, metal-mediated cyclizations and nitrenoid insertion reactions.

Table 1   Routes to Substituted Pyrroles, Furans and Thiophenes

Substituent	Method[a]	See Section
Aldehyde	S (Vilsmeier-Haack)	3.3.1.5.5
Alkyl	R, Li	3.3.3.8.2
	S (pyrrole anion)	3.3.1.3.1.ii, 3.3.1.5.6
Acyl	S (Friedel Crafts)	3.3.1.5.5
	S (pyrrole anion)	3.3.1.3.1.ii
Amino	Reduce NO$_2$, NO or N$_3$	3.3.3.4.1, 3.3.3.4.3
	S (aminodehalogenation)	3.3.3.7.1
Alkenyl	Pd	3.3.3.8.8
Alkenyl	Pd	3.3.3.8.8
Aryl	S	3.3.1.7.2
	Pd	3.3.3.8.8
Azo	S (pyrrole anion)	3.3.1.5.9
Carboxylic acid	R, Li–CO$_2$,	3.3.3.8.2
	S (pyrrole anion)	3.3.1.3.1.ii
Chloromethyl	S	3.3.1.5.7.iii
Halogen	S	3.3.1.5.4
	Li	3.3.3.8.6
	Hg	3.3.3.8.9
Hydroxy (alcoholic)	S	3.3.1.5.7.i
	Li	3.3.3.8.3
Nitro	S	3.3.1.5.2
	Li	3.3.3.8.5
Sulfonic acid	S	3.3.1.5.3
Ring carbonyl	R, S	3.3.1.5.4, 3.3.1.5.11
	Li	3.3.8.3
Substituted methyl	From CH$_2$Cl	3.3.3.3.3
	From CH$_2$NMe$_2$	3.3.3.3.6

[a]R, preparation usually by ring synthesis; S, preparation by substitution;
Li, *via* lithio derivative; Pd, *via* palladio derivative; Hg, *via* mercuration.

(i) The versatile Paal-Knorr synthesis is the most important preparative method for furans and thiophenes; it is also extensively used for pyrroles. The common starting materials are 1,4-diketones (*e.g.* **106**) or synthetical equivalents which yield:
   (a) pyrroles (**107**; Z = NH or NR) by reaction with NH$_3$ or RNH$_2$;

(106)   (107)

A variety of 1-substituted pyrroles can be prepared using 2,5-dimethoxytetrahydrofuran as a succinaldehyde equivalent ⟨52ACS667, 78JMC962⟩. Even non-nucleophilic amines such as 2,6-dinitroaniline and benzenesulfonamide can be converted to pyrroles in good yield ⟨73SC303, 92HCA2608⟩.
   (b) furans (**107**; Z = O) on treatment with an acidic catalyst such as sulfuric acid ⟨89SC2101, 91JHC225⟩, hydrochloric acid, polyphosphoric acid ⟨84S593⟩, *p*-toluenesulfonic acid, oxalic acid ⟨52CB457⟩, Amberlyst 15 ⟨73S209⟩, acetic anhydride/zinc(II)bromide.
   The 1,4-diketone may be used in protected form, *e.g.* as a ketal. 3,4-Diethoxycarbonylfuran (**109**) is obtained from the acetal (**108**) with sulfuric acid in 68% yield ⟨54JOC1671⟩. 2-Aminofurans are also obtained by cyclization of δ-ketonitriles ⟨90MI 207-01⟩.

(108)   (109)

1,4-Dihydroxyalkynes (**110**) may be considered as masked 1,4-dicarbonyl compounds. Their isomerization–dehydration to furans (**111**) can be accomplished by Pd catalysts ⟨93CC764⟩. Alkynyl ketones of type (**112**) and (**113**) can be cyclized by Pd(0) and Pd(II) catalysts ⟨86TL4893, 87S1022,

91JOC5816⟩. The intermediate (**114**) can either be hydrolyzed or trapped with electrophiles like allyl chloride to give 3-substituted alkenylfurans (**115**).

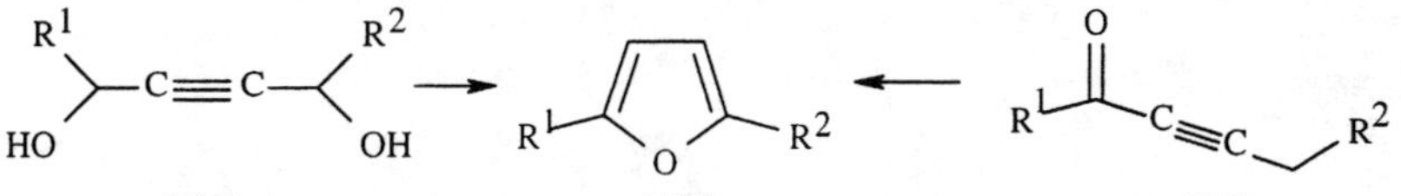

(**110**)   (**111**)   (**112**)

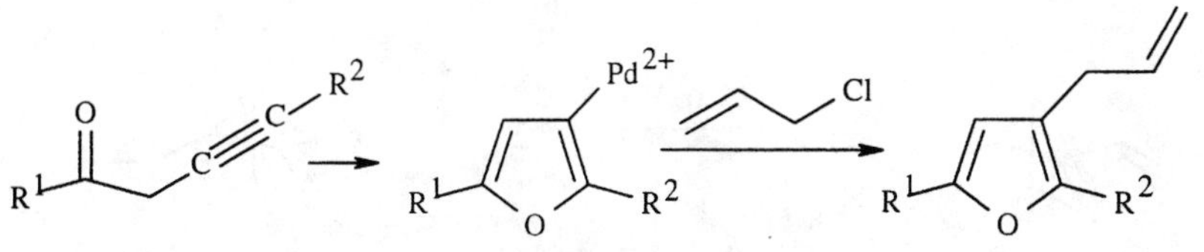

(**113**)   (**114**)   (**115**)

(c) thiophenes (**107**; $Z = S$) by distilling with P_4S_{10}, by sulfuration with a combination of hydrogen sulfide and an acid catalyst ⟨83H(20)1941, 91MI 211-02, 92H(34)2263⟩, or using bis(tributyltin) sulfide in the presence of boron trichloride ⟨92JOC1722⟩ at room temperature to give (**107**; $Z = S$) in high yields.

2-Substituted, 2,3-disubstituted, and 2,3-annulated thiophenes can be prepared by reactions of ketone enolates with carbonodithioic acid O-ethyl S-(2-oxoethyl)ester. Hydrolysis of the resulting aldols, intramolecular addition of thiol group to the carbonyl group, and elimination of two molecules of water lead to the thiophenes (**116**) (Scheme 38) ⟨92HCA907⟩.

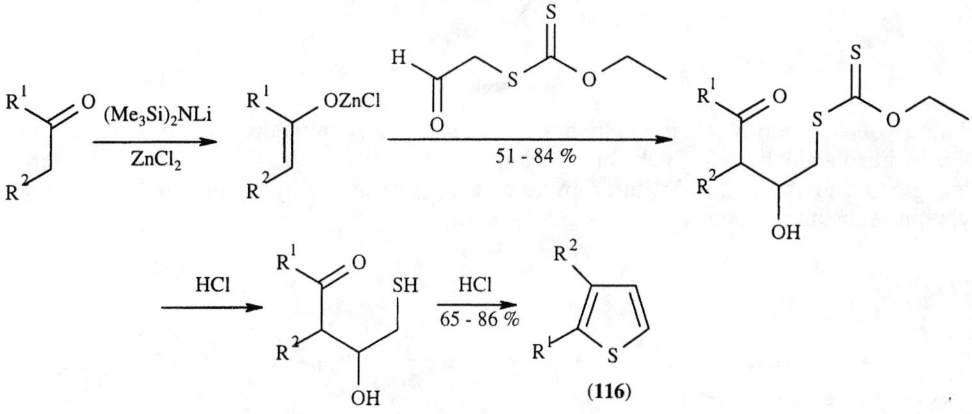

Scheme 38

(ii) 1,2,3,4-Tetrahydroxy compounds are used in further reactions of the same general type. Thus, hexoses (**117**; $R = CH_2OH$) and pentoses (**117**; $R = H$) give 5-hydroxymethylfurfural (**118**; $R = CH_2OH$) and furfural (**118**; $R = H$), respectively. In the presence of hydrochloric acid (or other chlorinating agents), 5-chloromethylfuran-2-carbaldehyde is obtained ⟨82CL617, 92S541⟩.

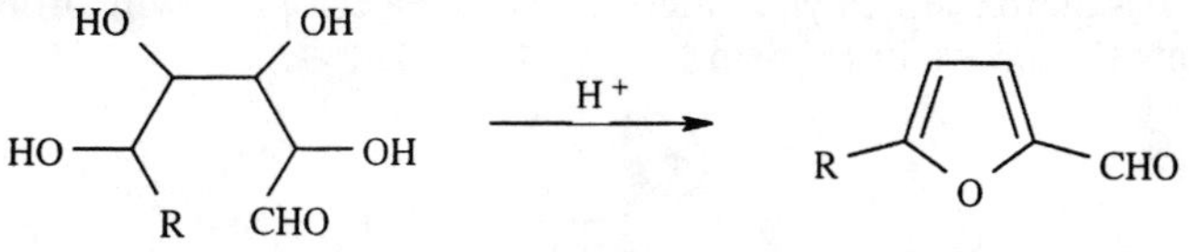

(**117**)   (**118**)

(iii) A variety of other cyclizations of the C_4 system to pyrroles include the following.

Substituted succinonitriles give 3-substituted pyrroles on reduction with DIBAL-H ⟨84TL1659⟩ *via* diimine intermediates which undergo cyclization and aromatization (Scheme 39). This reaction was used in the synthesis of unusual pyrroles (*e.g.* **119**) which were used to prepare sterically shielded porphyrins ⟨94AG(E)889⟩.

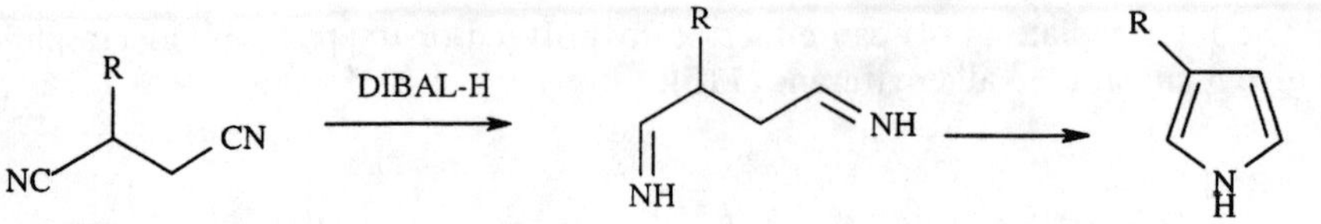

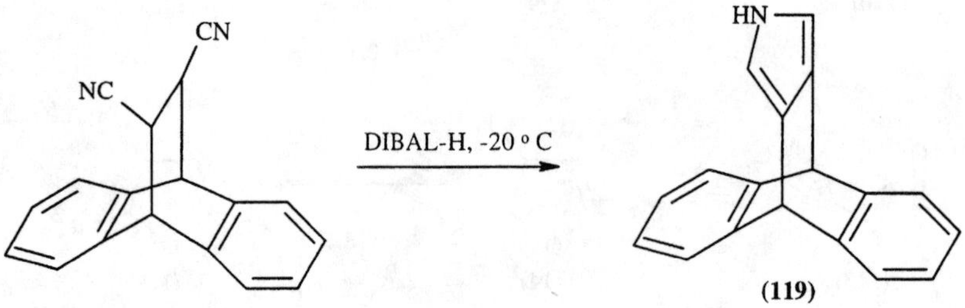

Scheme 39

γ-Nitroketones are reduced to pyrroles by diphenyl disulfide and triphenylphosphine ⟨86JCS(P1)2243⟩. Since γ-nitroketones are accessible by Michael additions of nitroalkanes to enones, this is a potentially versatile method (Scheme 40).

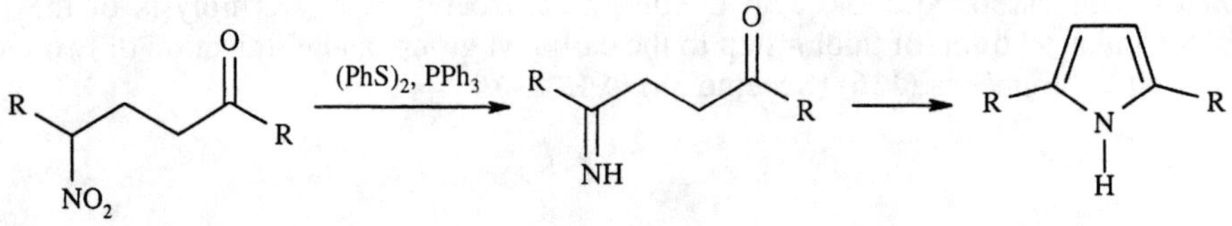

Scheme 40

Alkylation of α-lithio-*N*,*N*-dimethylhydrazones with 2-(iodomethyl)-1,3-dioxolane provides masked γ-hydrazonoaldehydes. Cyclization occurs after acid-catalyzed removal of the dioxolane protecting group and the *N*-unsubstituted pyrrole can be obtained by hydrogenolysis of the *N*,*N*-dimethylamino substituent (Scheme 41) ⟨88JHC1135⟩.

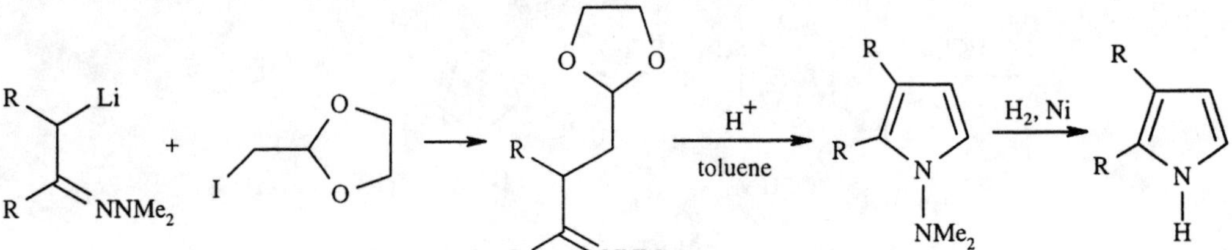

Scheme 41

Pyrrole and indole rings can also be constructed by intramolecular addition of nitrogen to a multiple bond activated by metal ion complexation. Thus, 1-aminomethyl-1-alkynyl carbinols (obtained by reduction of cyanohydrins of acetylenic ketones) are cyclized to pyrroles by palladium(II) salts. In this reaction the palladium(II)-complexed alkyne functions as the electrophile with aromatization involving elimination of palladium(II) and water (Scheme 42) ⟨81TL4277⟩.

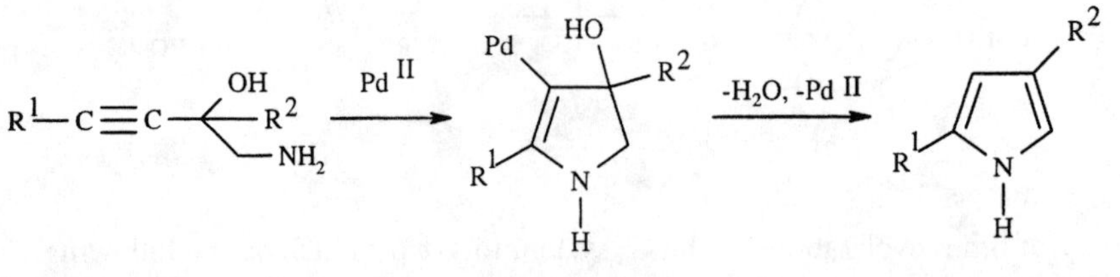

Scheme 42

Another pyrrole cyclization involves nucleophilic addition of amines to electrophilic 1,3-dienes, *e.g.* 2,3-diphenylsulfonyl-1,3-butadiene and 2-acetyl-3-phenylsulfonyl-1,3-butadiene ⟨88TL3041, 91S171⟩.

Elimination of phenylsulfinate anion, induced by base, introduces one double bond. The second double bond can be introduced by DDQ oxidation, as in Scheme 43.

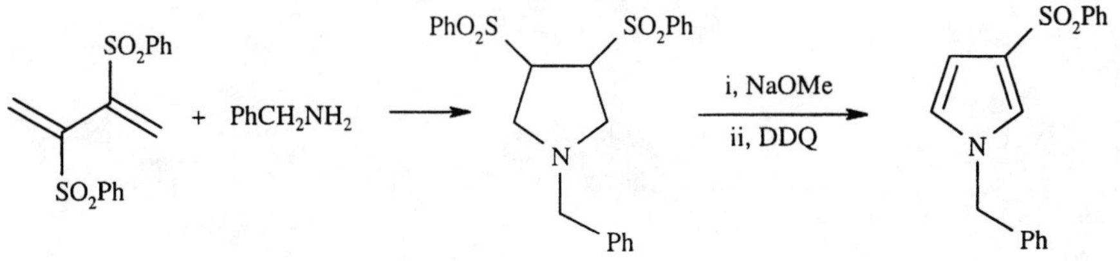

Scheme 43

(iv) Other cyclizations of a C_4 system to give thiophene include pyrolysis of butane with sulfur; this reaction is probably the source of thiophene in coal-tar benzene.

Cyclization reactions effected by intramolecular attack of a heteroatom on a nitrile group provide a useful source of 2-amino heterocycles, *e.g.* (**120**) → (**121**), and numerous syntheses employ this strategy (*vide infra*). Thus, the reaction of α,β-unsaturated nitriles (**122**) with elemental sulfur in basic media, the Gewald synthesis, provides a most convenient route to 2-aminothiophenes (**123**), many of which are a useful as intermediates in the preparation of dyestuffs and pharmaceuticals.

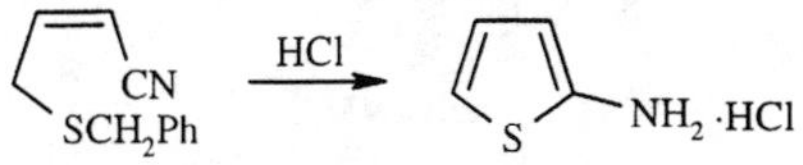

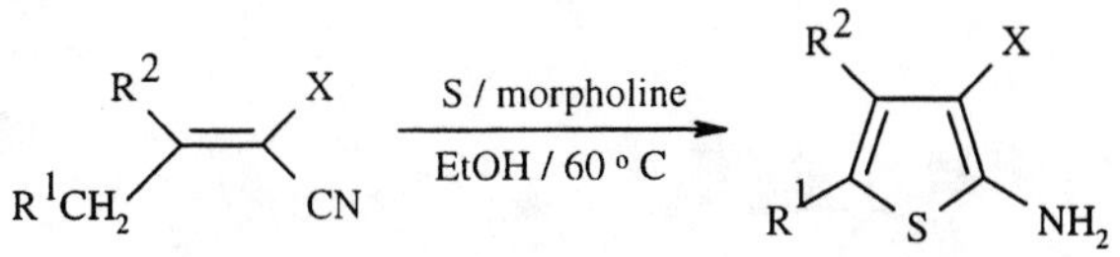

In the previous examples, the sulfur atom acted as a nucleophile. Electron-deficient sulfur species such as sulfenyl ion and its equivalents (*e.g.* disulfide/Lewis acid complexes, sulfenic acids, sulfenyl halides, sulfonium ions, sulfines, *etc.*), can also serve as an electrophile. Oxidative ring closure of enethiols (α-thioketocarboxylic acid) (**124**), which proceeds *via* disulfides, produces thiophenes (**125**) in good yields ⟨86EUP158380, 88JHC367⟩.

(v) Triple bonds are susceptible to nucleophilic addition which can take place in an intra- (Scheme 44) or inter-molecular fashion (*e.g.* **126** → **127**).

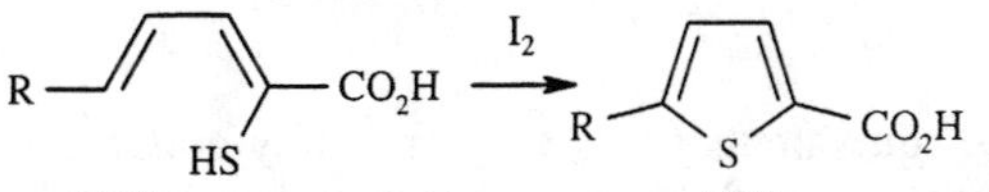

R = Ph, 4-ClC$_6$H$_4$, 4-MeOC$_6$H$_4$, 2-thienyl, 2-chloro-3-thienyl, 2,5-dichloro-3-thienyl

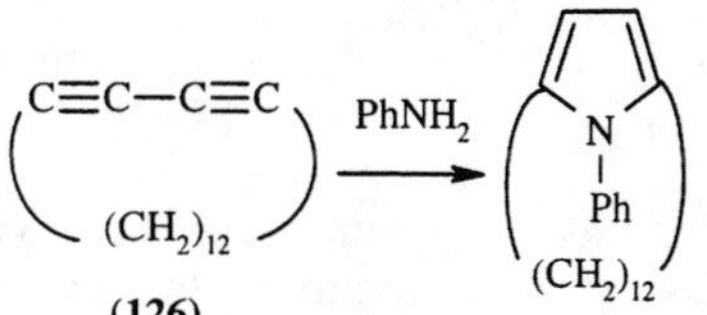

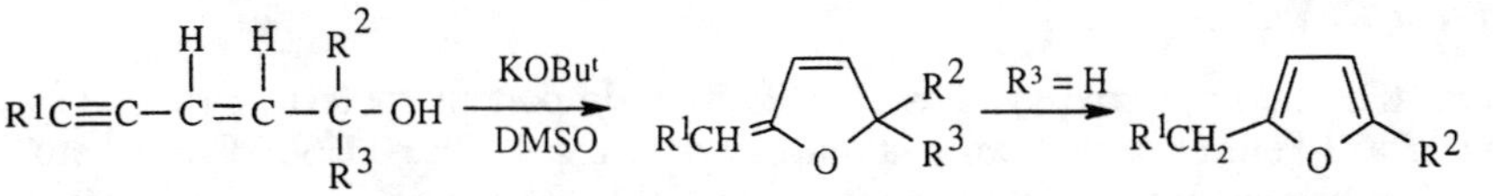

Scheme 44

4.2.3.3.2 *From C₃ZC or C₃ and CZ units*

Many variations of this route involve the formation of the 2,3-bond of a thiophene by the 2-carbon atom acting as a nucleophilic center. Examples are shown in Scheme 45.

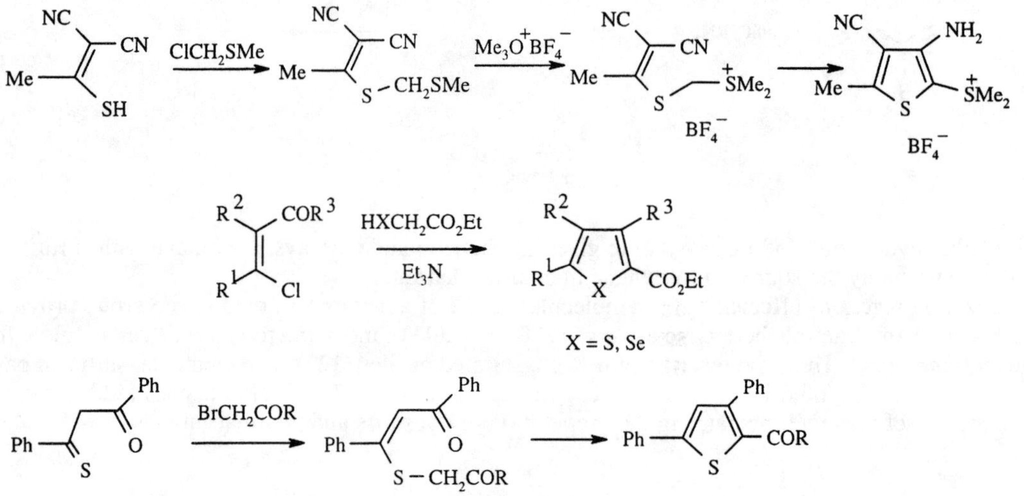

Scheme 45

Electrocyclization of a suitable carbanion can also occur as illustrated in Scheme 46 ⟨78AG(E)676⟩.

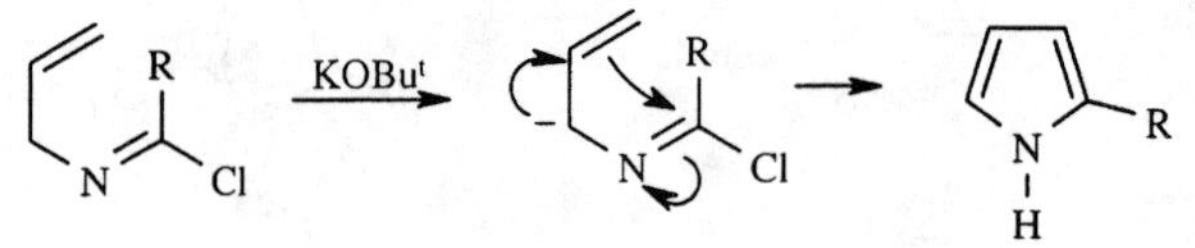

Scheme 46

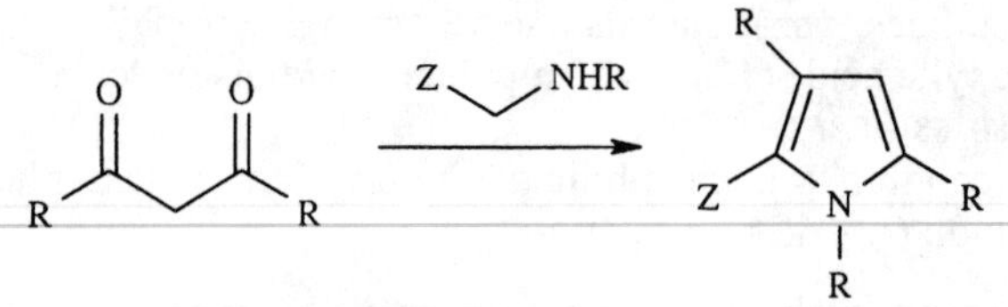

Scheme 47

1,3-Dicarbonyl compounds and their synthetic equivalents can give pyrroles by condensation with amines having an α-electron-withdrawing substituent such as an ester or a ketone (Scheme 47) ⟨82S157⟩. Such condensations can produce two isomers when the dicarbonyl component is unsymmetrical.

4.2.3.3.3 *From C₂ and ZCC units*

(i) The versatile Knorr pyrrole synthesis is the most important route to pyrroles: it involves the condensation of a β-keto ester with an α-amino ketone, *e.g.* (**128 → 129**). The β-keto ester can be replaced by a β-diketone; simple ketones give poor yields. The amino ketone is frequently prepared *in situ* by nitrosation and reduction (*e.g.* with Zn–AcOH) of a second molecule of the β-keto ester.

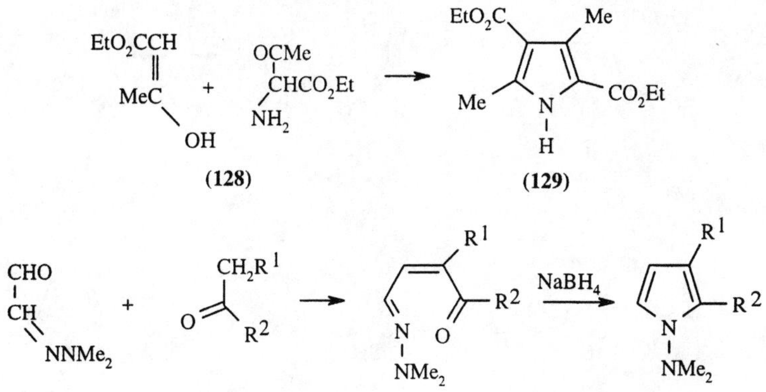

Scheme 48

α-Phenylazo ketones may be used in place of α-oximino ketones ⟨57CB79⟩. Other variations include the use of α-aminoaldimines ⟨78S590⟩, and glyoxal mono-*N,N*-dimethylhydrazone which yields 2,3-disubstituted pyrroles ⟨77CB491⟩ (Scheme 48). α-Aminonitriles afford 3-aminopyrroles ⟨74AG(E)807⟩.

While normal Knorr conditions involve initial N–C bond formation, the overall reaction can also be accomplished under conditions which begin with C–C bond formation. Condensation of the dimethyl ketal of acetamidoacetone with silyl enol ethers give adducts which cyclize to pyrroles (Scheme 49) ⟨88S381⟩.

The Mukaiyama reaction with the dimethoxy acetal of azidoacetaldehyde gives 2,3-disubstituted pyrroles (Scheme 50) ⟨90AG(E)777⟩.

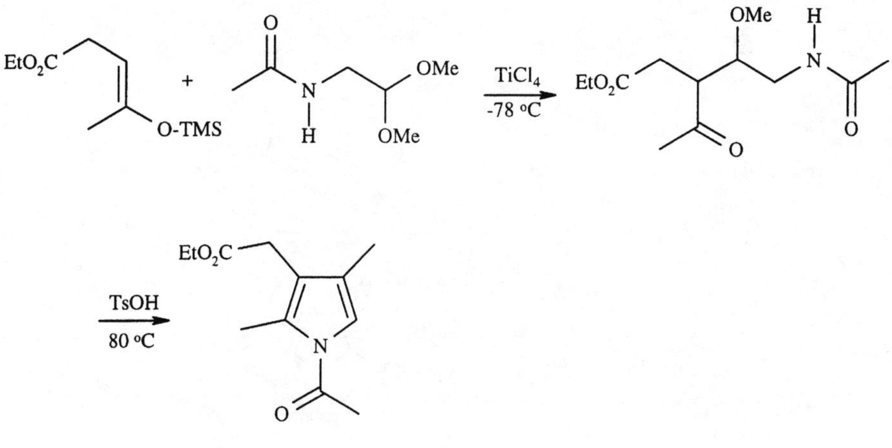

Scheme 49

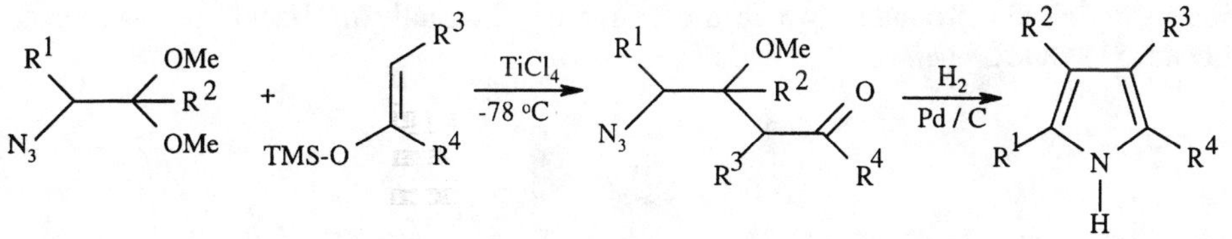

Scheme 50

(ii) α-Halo ketones react with enamines to form pyrroles, and with β-keto esters to give furans. The orientation in the Hantzsch pyrrole synthesis (*e.g.* **130 → 131**) differs from that in the Feist-Benary furan synthesis (*e.g.* **132 → 133**). The employment of activated cyanomethylene derivatives (**134**) for condensation with α-amino ketones provides 2-aminopyrroles (**135**). In comparable fashion, the condensation of α-mercapto carbonyl compounds with malononitrile derivatives provides 2-amino-thiophenes (**136→137**) ⟨79LA328⟩. Similarly, 2-aminofuran-3-nitriles can be obtained by base-catalyzed condensation of acyloins with malononitrile ⟨66CB1002, 80CCC1581, 83CCC3140⟩.

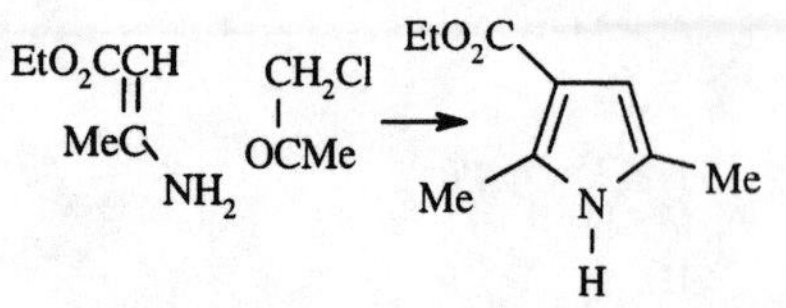

(130)   (131) 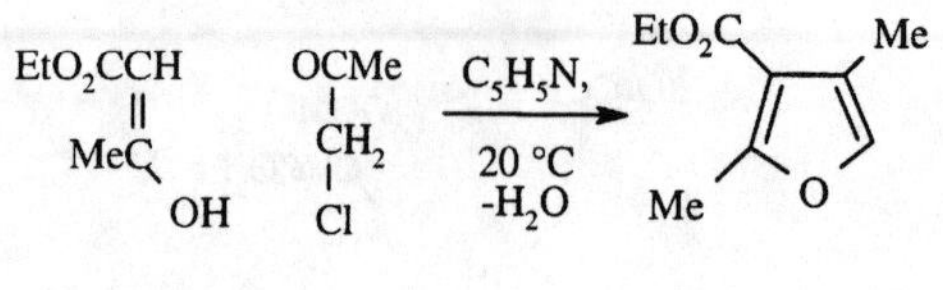   (132)   (133)

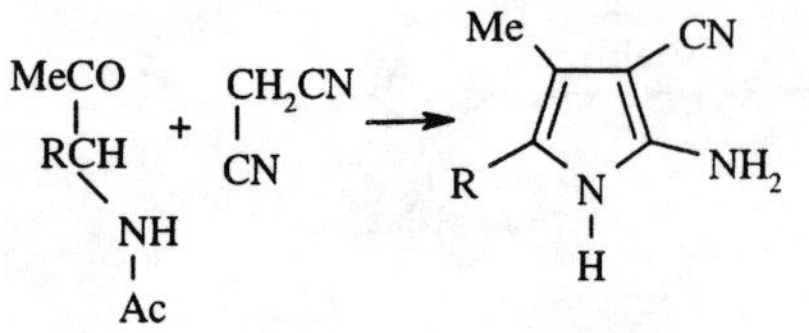

(134)   (135) 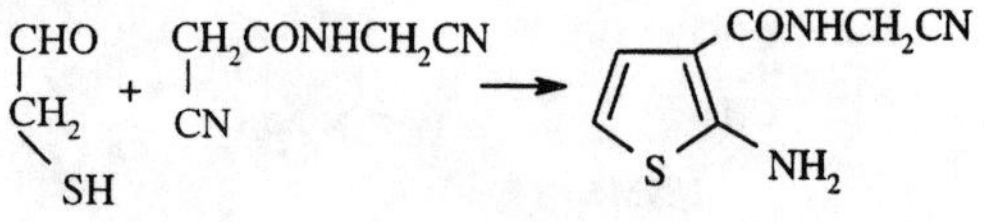   (136)   (137)

The synthesis of furans from α-hydroxycarbonyl compounds frequently utilizes aldoses or ketoses as readily available sources of this functional grouping (**138 → 139**); the resulting polyhydroxyalkyl side chain can be removed easily by oxidative degradation ⟨56MI30300⟩.

Reactions of the Feist-Benary type have been applied to the synthesis of thiophenes (**140 → 141**) ⟨75ZC100⟩. The use of α-halocarbonyl halides provides an entrée to 3-furanones (**142 → 143**) ⟨73RTC731⟩.

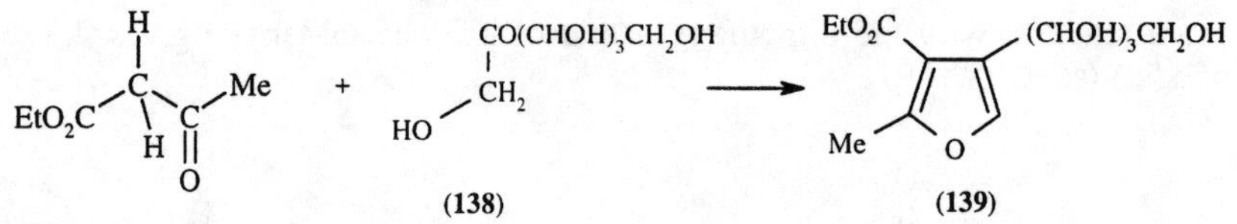

(138)   (139)

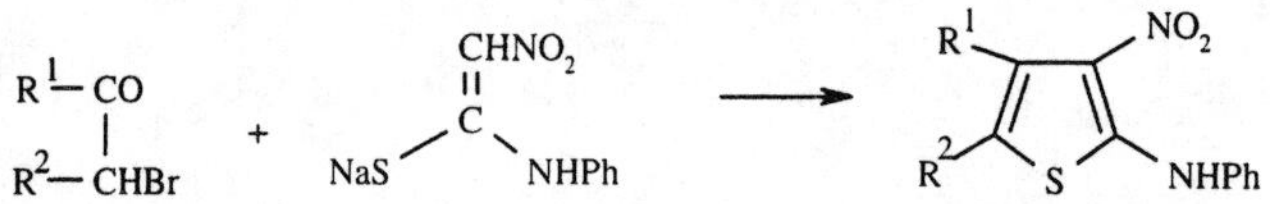

(140)   (141)

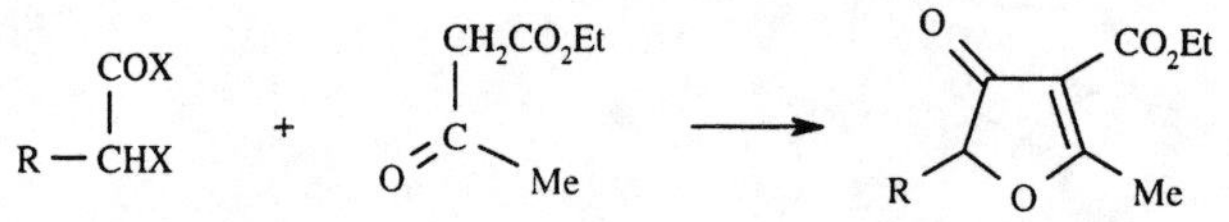

(142)   (143)

Knoevenagel condensation of aldehydes with 1,3-dicarbonyl compounds and subsequent NBS bromination yields allylic bromides, which are converted thermally to 3-acyl- or 3-alkoxycarbonyl furans (Scheme 51) ⟨78JOC4596⟩.

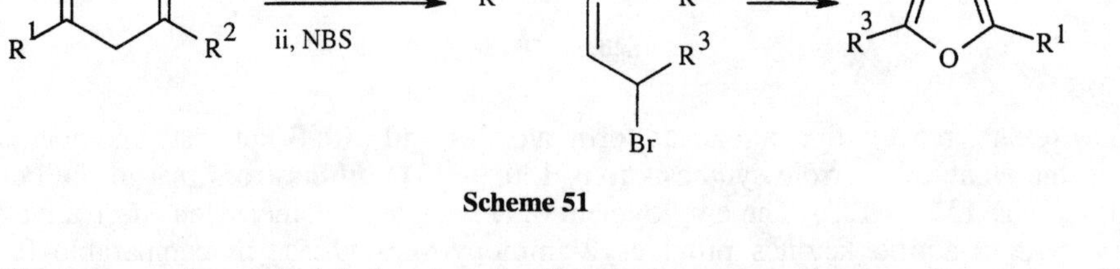

Scheme 51

(iii)   Additions to conjugated triple bonds as in dimethyl acetylenedicarboxylate occur with facility. Thus, base-catalyzed addition of benzoin to DMAD provides a route to furans (**144 → 145**; X = O), and pyrroles result from an analogous addition of α-amino ketones (**144 → 145**; X = NH) ⟨64JA107⟩.

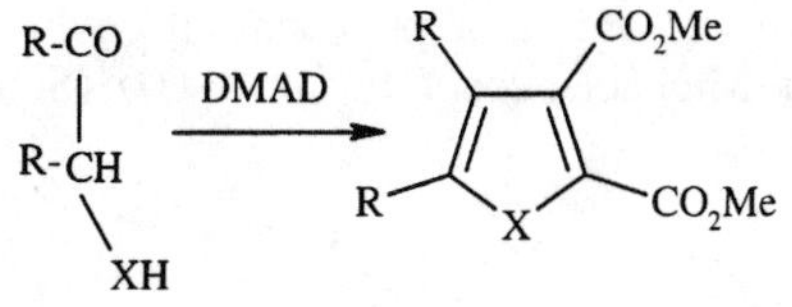

(144) (145) 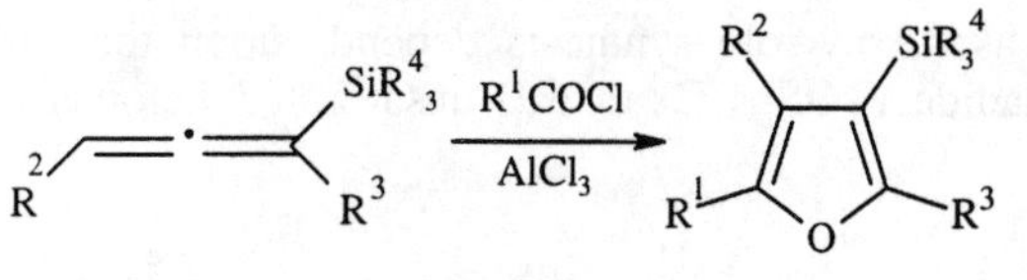 **Scheme 52**

Regioselective [3 + 2] annulation has been reported for the preparation of a wide variety of furans. Reaction of acid chlorides with trialkylsilyl-substituted allenes under the influence of $AlCl_3$ (CH_2Cl_2, $-20°C$) yields the desired compounds (Scheme 52). The intramolecular variant of this method has also been described ⟨89JA4407⟩.

Trofimov's pyrrole synthesis reacts ketoximes with alkynes under strongly basic conditions (Scheme 53) ⟨90AHC178, 94H(37)1193⟩. A key sigmatropic rearrangement of an *O*-vinyloxime is followed by a typical imine-carbonyl condensation. Frequently an *N*-vinylpyrrole formed from the initial pyrrole is a major by-product ⟨92CHE510⟩.

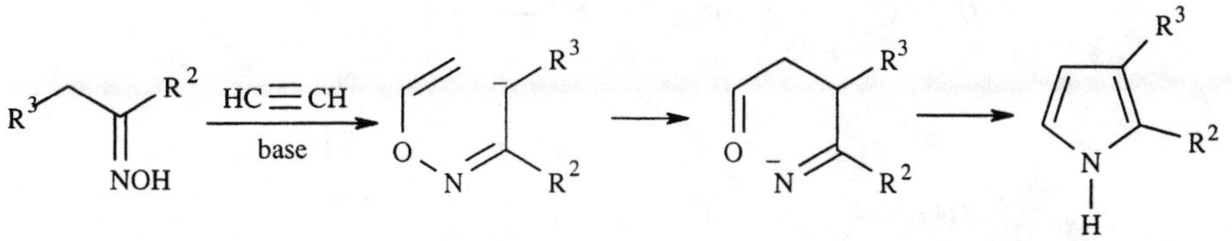

Scheme 53

(iv) The Piloty–Robinson pyrrole synthesis (Scheme 54) is the monocyclic equivalent of the Fischer indole synthesis. The conversion of ketazines into pyrroles under strongly acidic conditions proceeds through a [3,3] sigmatropic rearrangement of the tautomeric divinylhydrazine ⟨74JOC2575⟩. The cyclization of *N*-substituted azines provides 1-substituted pyrroles ⟨71LA(744)81⟩. The reaction can be conducted under relatively mild conditions by converting the azine into the *N,N'*-dibenzoyl derivative prior to thermal ($\sim 140°C$) rearrangement ⟨82CC624⟩.

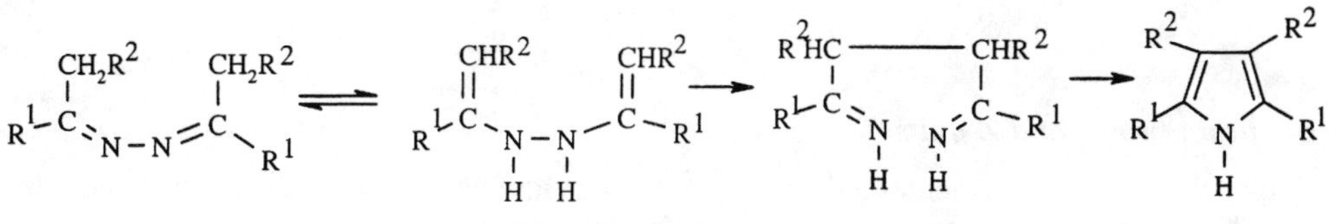

Scheme 54

4.2.3.3.4 From C_2 and CZC units

Ring syntheses based on the combination of $C_\beta C_\beta$, and $C_\alpha X C_\alpha$, fragments depend upon the latter reacting either as a carbanion or as a 1,3-dipole. An example of the former is the Hinsberg thiophene synthesis in which ethyl thiodiglycolate is condensed with a 1,2-dicarbonyl compound under basic conditions (Scheme 55) ⟨79JHC1147⟩. Other carbanion stabilizing groups such as cyano can be used in place of the ethoxycarbonyl group. The same basic method has been used for the synthesis of furans ⟨40RTC423⟩, selenophenes ⟨40RTC423⟩ and pyrroles ⟨65JOC859⟩.

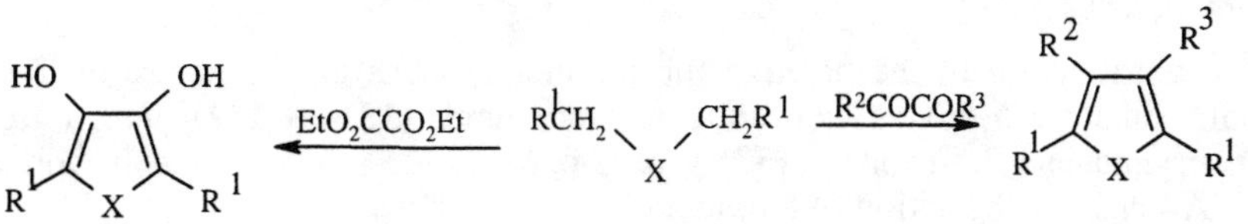

$R^1 = CO_2Et, CN$
$X = O, S, Se, NR$

Scheme 55

A useful pyrrole synthesis depends upon the addition of the anion of *p*-toluenesulfonylmethyl isocyanide (TOSMIC) to α,β-unsaturated ketones or other Michael acceptors ⟨72TL5337⟩ (Scheme 56).

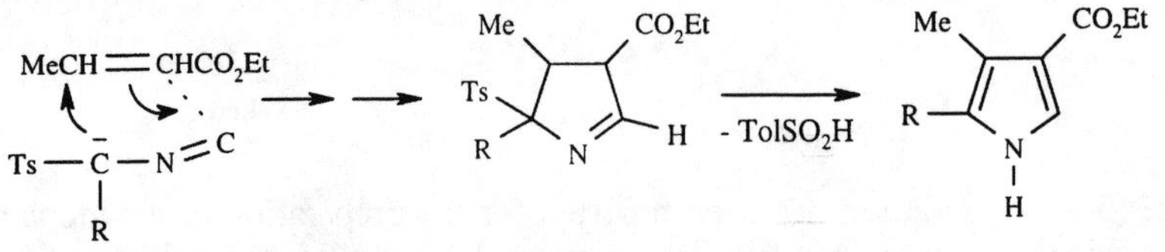

Scheme 56

Other important pyrrole syntheses of this type are cycloadditions involving mesoionic oxazolium 5-oxides, azomethine ylides or isonitriles, *e.g.* (**145** + **146** → **147**) or (**148** + **149** → **150**).

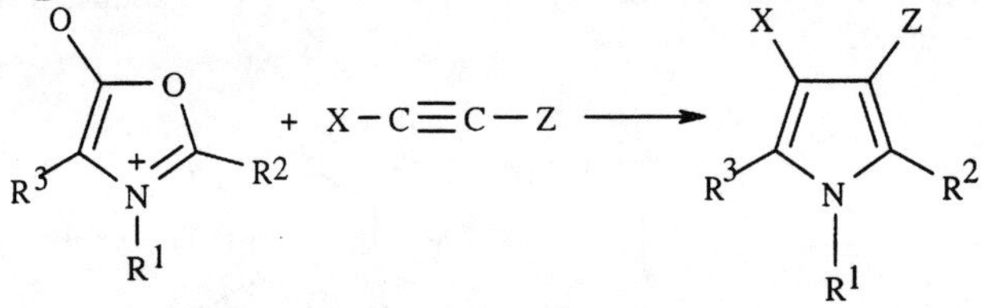

(**145**)   (**146**)   (**147**)

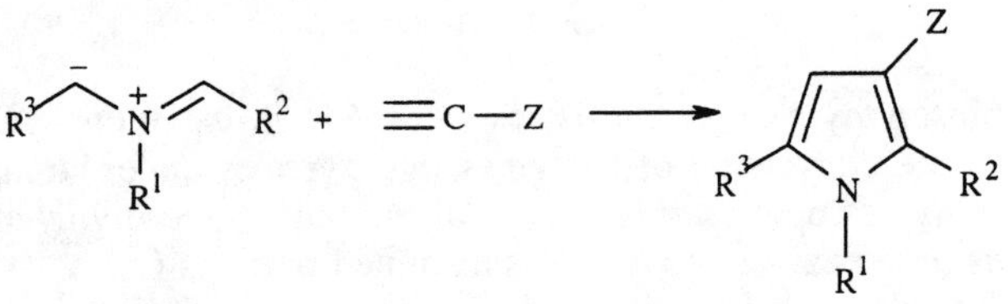

(**148**)   (**149**)   (**150**)

4.2.3.3.5 *From two C$_2$ and Z units*

Diarylethynes react with elemental sulfur to give tetraarylthiophenes in good yields (Equation (3)) ⟨93SUL273⟩. Although the reaction requires rather forcing conditions, it is a practical method, *e.g.* diphenylethyne and elemental sulfur at 220–240°C affords tetraphenylthiophene. The method is also applicable to tetraarylselenophenes.

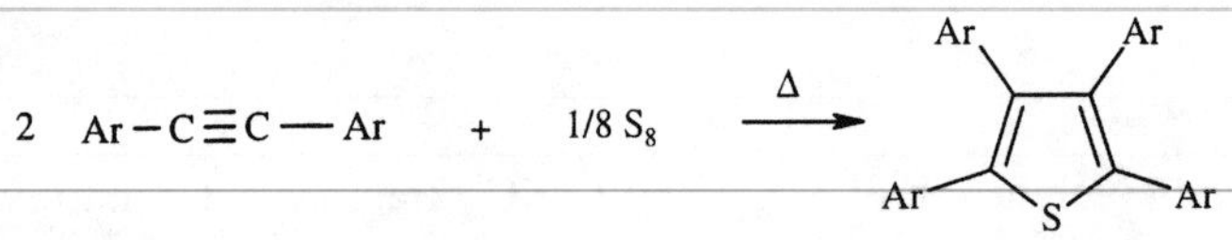

$$2 \quad Ar-C\equiv C-Ar \quad + \quad 1/8\ S_8 \quad \xrightarrow{\Delta} \quad \text{(3)}$$

4.2.3.4 Synthesis of Pyrans, Dihydropyridines and their Thio and Oxo Derivatives from Acyclic Precursors

4.2.3.4.1 *From C$_5$ units*

The degree of unsaturation of the heterocyclic product depends on the nature of the five-carbon starting material: pentane-1,5-diones yield dihydro compounds (**151** → **152**) (which are sometimes oxidized *in situ*); pentane-1,3,5-triones (**153**) give γ-pyrones (**154**; Z = O) by dehydration, and γ-pyridones (**154**; Z = NH) by the action of ammonia.

1,5-Dialdehydes afford 2,6-unsubstituted 4*H*-pyrans while 1,5-ketoaldehydes give the 2-substituted analogues ⟨89AP(322)617⟩ and 1,5-diketones yield the 2,6-disubstituted pyrans. Examples in the latter category include all substitution patterns ⟨92CHE266⟩, fused derivatives ⟨86H(24)1631, 92CHE266⟩, and spiro-linked 4*H*-pyrans ⟨92JCS(P2)1301⟩. The dehydrating reagents used include PPA ⟨89AP(322)617⟩, *p*-

toluenesulfonic acid ⟨89CCC1854⟩, and a mixture of acetic anhydride and boron trifluoride etherate ⟨86H(24)1631, 92CHE266⟩.

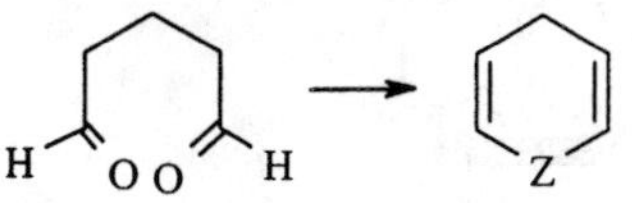

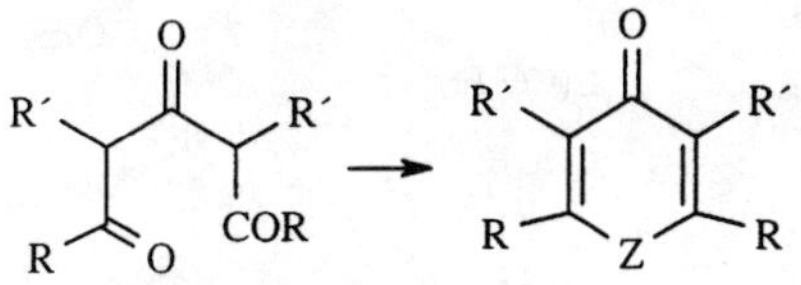

(151)   (152)   (153)   (154)

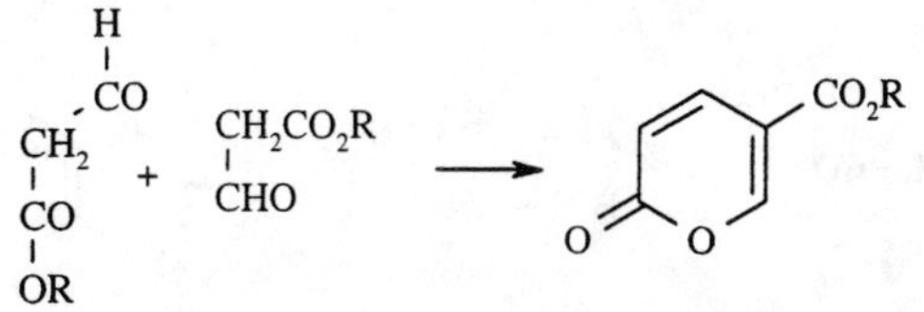

(155)   (156)

Pent-2-ene-1,5-diones may be formed *in situ* (usually by an aldol-type reaction) and subsequently cyclized. Thus, malic acid (*i.e.* hydroxysuccinic acid) with sulfuric acid gives carboxyacetaldehyde (155) which cyclizes spontaneously to coumalic acid (156).

Several syntheses of 2*H*-pyrans are based on the preparation of the acyclic precursors (157b) in the hope that the dienone ⇌ 2*H*-pyran equilibrium will favor the heterocycle (157a). Often the product will contain both valence isomers. Such dienone precursors can be obtained by Knoevenagel condensation of 1,3-dicarbonyl compounds with α,β-unsaturated aldehydes (Scheme 57). Simple 1,3-diketones yield the 2*H*-pyran directly ⟨88IZV1815⟩ and cyclohexan-1,3-diones afford fused pyrans ⟨82S683, 84JHC913, 87JOC1972⟩.

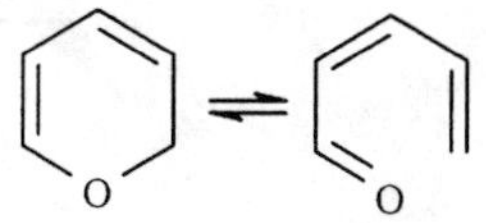

(157a)   (157b)

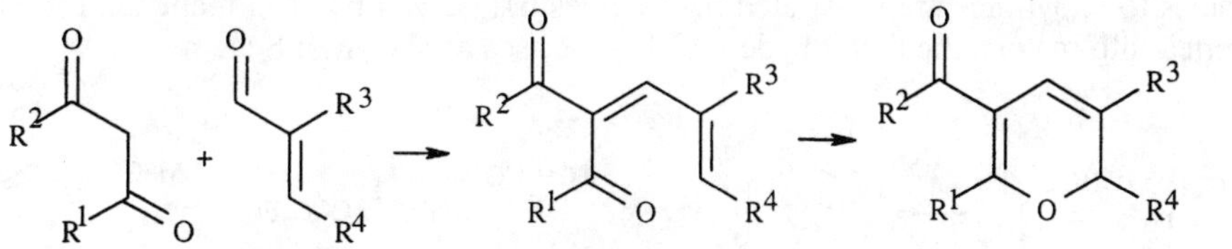

Scheme 57

4.2.3.4.2 With C–C bond formation

(i) The Hantzsch pyridine synthesis affords 1,4-dihydropyridines, although spontaneous oxidation to pyridines often occurs. In its simplest form it involves the condensation of two molecules of a β-keto ester with an aldehyde and ammonia (*cf.* 158 → 159). Compounds resulting from the condensation of ammonia with one of the carbonyl components can be used in the Hantzsch synthesis. Thus, β-aminocrotonic ester (160) can replace the ammonia and one mole of acetoacetic ester in (158). The mechanism of the Hantzsch synthesis has been clarified by ^{13}C and ^{15}N NMR spectroscopy ⟨87T5171⟩.

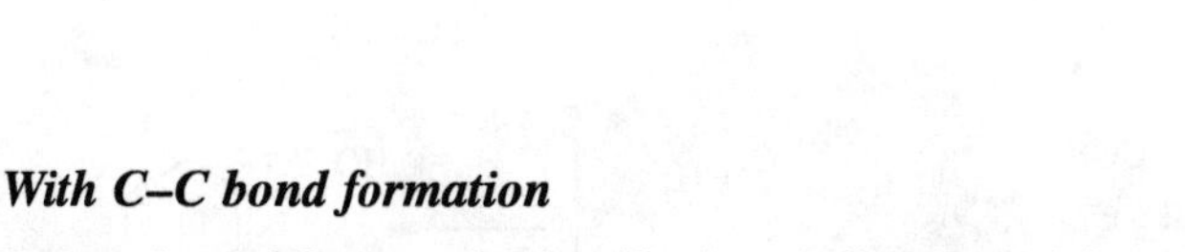

(158)   (159)   (160)

(ii) Related preparations of 2-pyridones involve condensation of enamines and enamides with 1,3-bielectrophiles, *e.g.* Schemes 58 and 59.

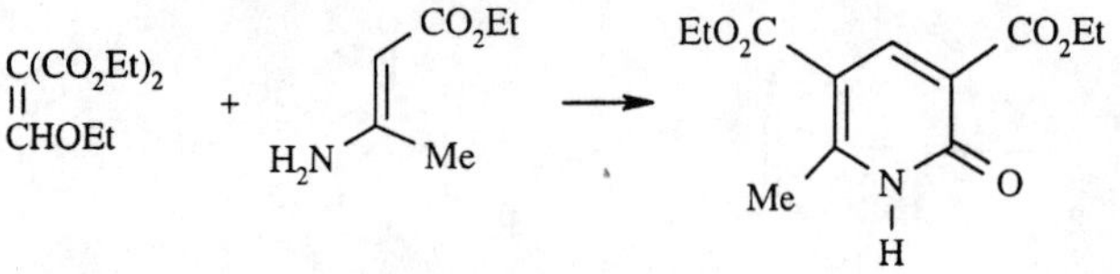

Scheme 58

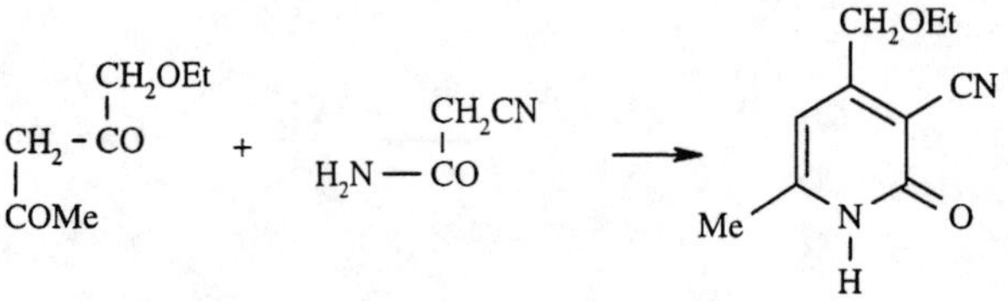

Scheme 59

Numerous pyridines have been synthesized using cyanoacetamide (or cyanothioacetamide), malononitrile, malonamide and closely related compounds, which provide the N–C(2)–C(3) fragment; for reviews see ⟨86H(24)2023, 87H(26)205, 93CR1991⟩. Principal co-reagents are α,β-unsaturated ketones, arylidene malononitriles (review ⟨83H(20)519⟩), and β-diketones or β-ketoaldehydes.

(iii) 4-Pyridones can be made from β-diketone dianions and nitriles (Scheme 60).

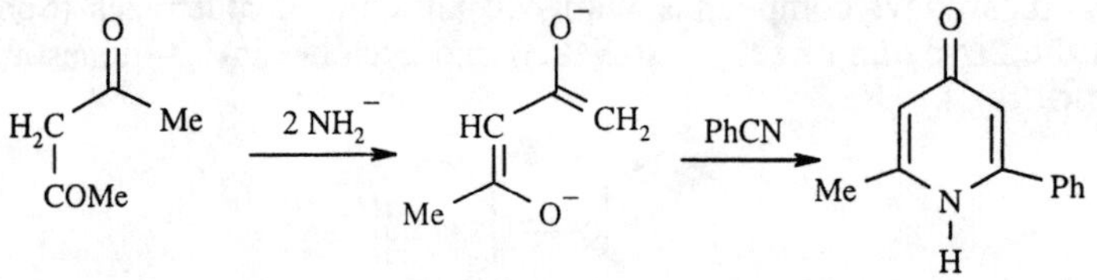

Scheme 60

(iv) Routes to α-pyrones are illustrated in Schemes 61, 62 and 63. The malic acid (Scheme 63) is first converted into malonic hemialdehyde which dimerises as shown in Scheme 61.

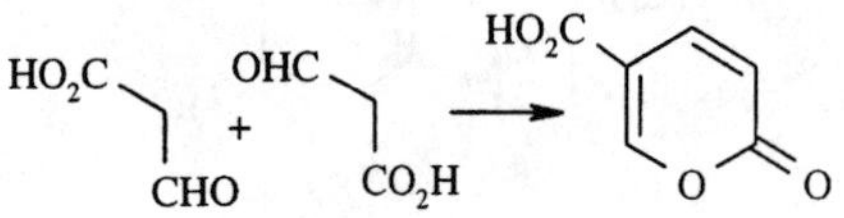

Scheme 61 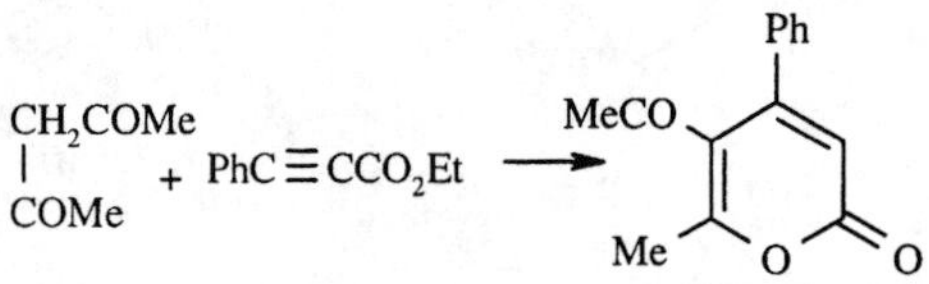　　　　　　　　　　　　　　　　**Scheme 62**

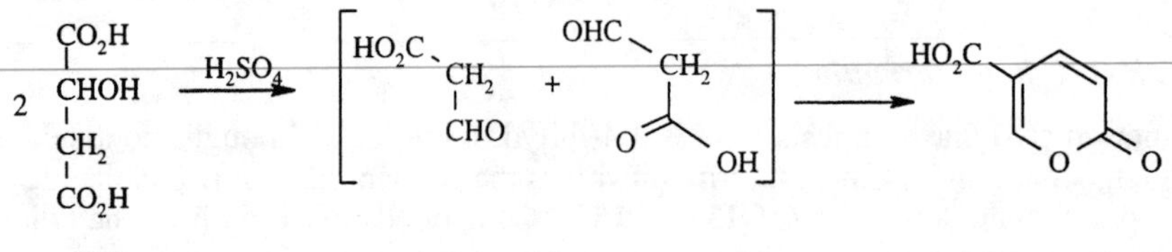

Scheme 63

4.2.3.5 Synthesis of Four-, Five- and Six-membered Rings from Carbocyclic or Heterocyclic Precursors

4.2.3.5.1 With ring expansion

Heating 3-azidocyclopropenes (**161**) neat at 125°C results in a concerted N_2 expulsion and 1,2-C shift to afford the corresponding 2,3,4-trisubstituted azetes (**162**) ⟨86AG(E)842⟩ (see Section 2.5.5.1 on valence isomer equilibrium of **162**).

R = *tert*-butyl, mesityl, 1-adamantyl

| (161) | (162a) | (162b) |

The isomerization of vinyl- or ethynyl-oxiranes is a frequently exploited source of dihydrofurans or furans, as illustrated by (163) → (164) ⟨73JA250⟩ and (165) → (166) ⟨69JCS(C)12⟩. Vinylazirine similarly gives pyrrole (Section 3.5.2.2).

| (163) | (164) | (165) | (166) |

Base-catalyzed dimerization of diketene (167) efficiently yields dehydroacetic acid (168); treatment of diketene with aqueous triethylamine gives 2,6-dimethyl-4-pyrone (169).

| (167) | (168) | (169) | (170) |

Diketene reacts as a masked form of acetoacetic ester with a variety of mono- and di-nucleophiles. Aromatic and heteroaromatic amines give acetoacetanilides which with more diketene cyclize to dioxopyridines, *e.g.* (170) is formed from 4-aminopyridine.

Ring opening of the cycloadducts (172) from oxazoles (171) and dienophiles (XCH=CHY) gives dihydropyridines and frequently pyridines (172 → 173 → 174); entities such as XR^3, HR^3 or XOH can also be lost in the aromatization of intermediate (173).

| (171) | (172) | (173) | (174) |

| (175) | (176) |

Isoxazolium salts with active methylene compounds can undergo ring expansion: thus (175) with $MeCOCH_2CO_2Et$ gives the 4-pyridone (176).

See the sections quoted for ring expansion of 2,3-dihydrothiophenes to thiophene derivatives (3.3.2.3.3) and 1,2-dithiolium cations to thiinthiones (3.4.1.6.5.ii).

4.2.3.5.2 No change in ring size

Furans, thiophenes and pyrroles have all been obtained by addition of alkyne dienophiles to a variety of other five-membered heterocycles, as illustrated in Scheme 64 (see also Sections 3.4.1.9.1 and 4.2.3.3.4). As the alkyne moiety provides carbons 3 and 4 of the resulting heterocycle, this synthetic approach provides an attractive way of introducing carbonyl-containing substituents at these positions, especially as many of the heterocyclic substrates are readily available. Even furans can be converted into other furan derivatives by this method (Scheme 65) ⟨85JHC1233, 87BSF131⟩.

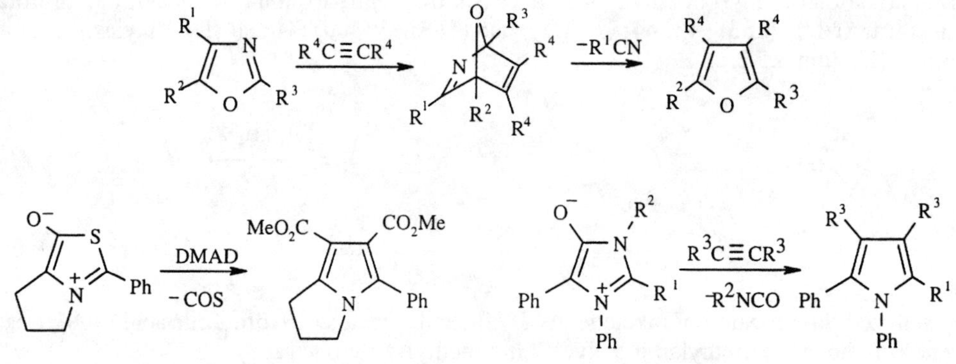

Scheme 64

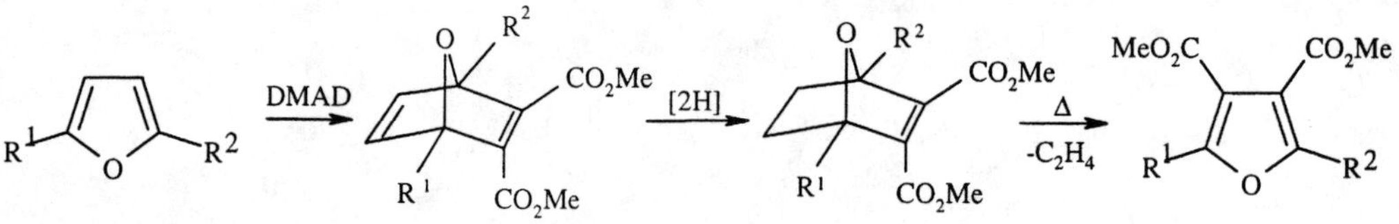

Scheme 65

The conversion of furans into the corresponding thiophenes or pyrroles entails passage of the furan with hydrogen sulfide or ammonia over an alumina-based catalyst at elevated temperatures ⟨B-73MI50300⟩. Conversions of furans into thiophenes are also achieved with hydrogen sulfide and hydrogen chloride at 80–100°C ⟨80DOK(255)1144⟩. Selenophenes can be obtained similarly.

Aminolysis of α- and γ-pyrones is important for the preparation of 2- and 4-pyridones, *e.g.* (**177**) → (**178**) and (**179**) → (**180**) (see also Section 3.2.1.6.4).

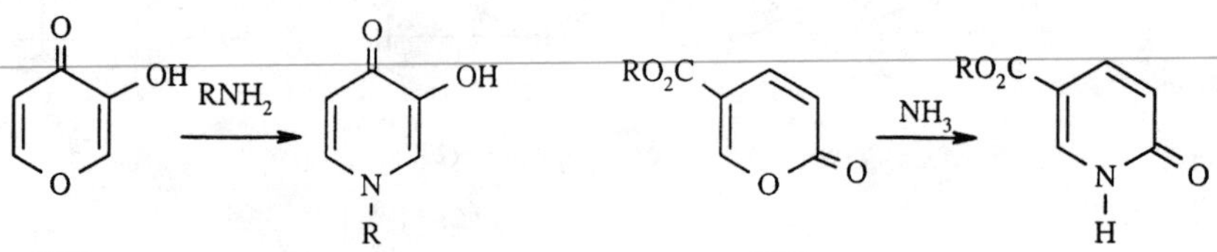

(**177**)   (**178**)   (**179**)   (**180**)

4.2.3.5.3 With ring contraction

The photolytic Wolff ring contraction of diazopyridones (**181**) is a synthesis of pyrrole-2-carboxylic acids *via* carbene (**182**) and ketene (**183**) intermediates ⟨76S754⟩.

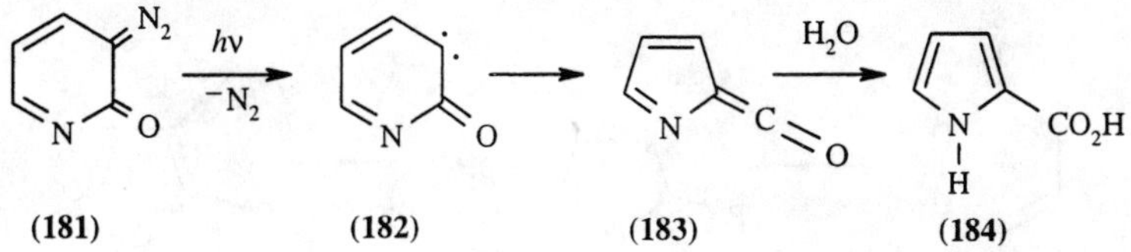

(181) (182) (183) (184)

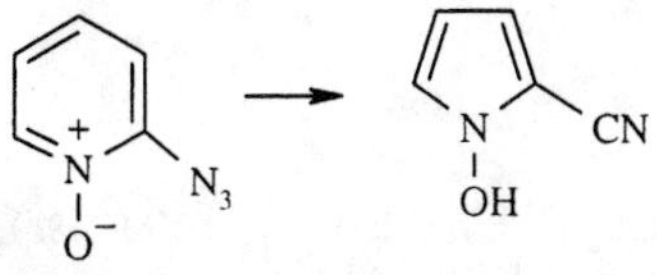

(185) (186)

The thermolysis of 2-azidopyridine *N*-oxides leads to *N*-hydroxy-2-cyanopyrroles (**185** → **186**) ⟨73JOC173⟩ (see also Section 3.4.3.11).

Photolysis of pyridine 1-oxides (**187**) affords a versatile route to a variety of pyrroles (**188**).

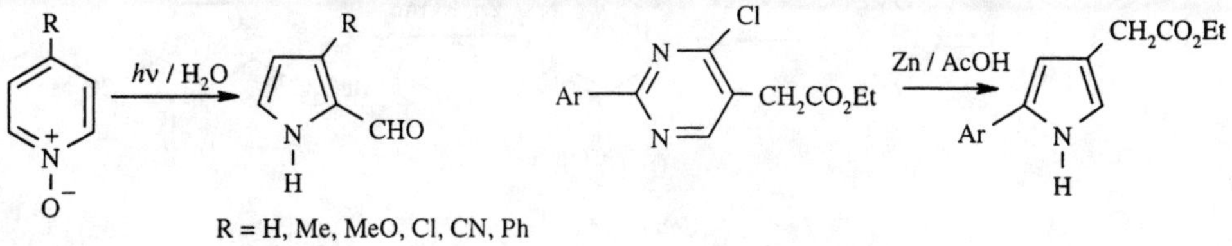

R = H, Me, MeO, Cl, CN, Ph

(187) (188) (189) (190)

Reduction of appropriate 2-arylpyrimidines with zinc in acetic acid provides pyrrolylacetic acids (**189** → **190**) ⟨70JCS(C)1658⟩.

3-Bromopyrones, which are available from pyrones with bromine, on treatment with base yield furancarboxylic acids (Scheme 66) ⟨62JCS5297, 68JCS(C)769, 73JCS(P1)1130, 75H(12)129, 80JOC1524⟩.

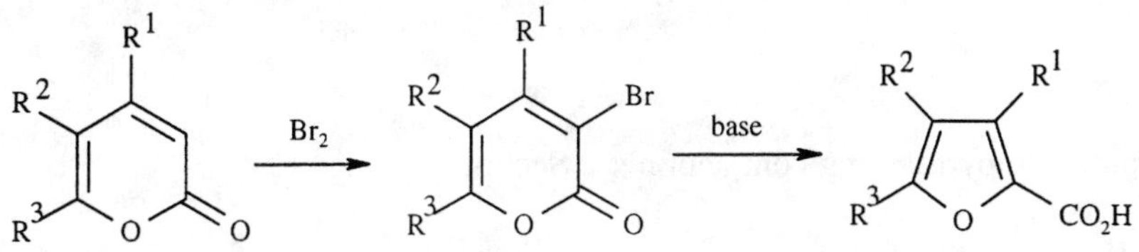

Scheme 66

Related reactions include preparations of furans from pyrylium cations (Section 3.2.1.6.3.v).

1,4-Dithiins ⟨56JA850⟩ and 1,2-dithiins ⟨67AG(E)698⟩ readily lose sulfur on heating, yielding the corresponding thiophene (**191** → **192**; **193** → **194**).

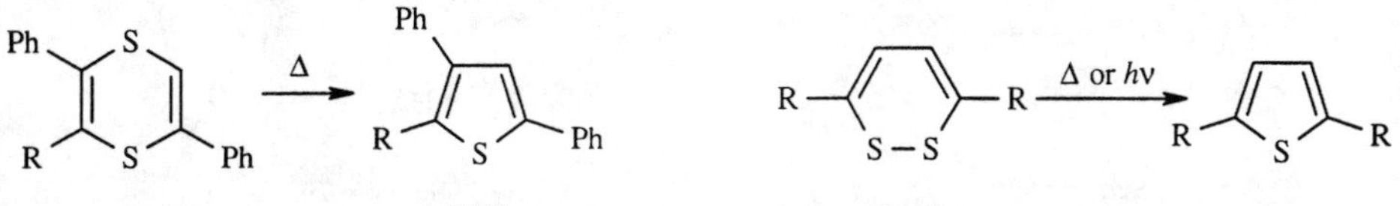

(191) (192) (193) (194)

3,6-Dihydro-1,2-dioxines (**195**) (available from dienes and O_2) can be transformed into furans (**197**) by reduction and rearrangement (into **196**) and subsequent elimination of water, usually by treatment with base. In some cases Co(II) and Fe(II) catalysts have been used ⟨76TL4363, 77JOC1900, 77SC119, 82ACS(B)31, 89JOC3475⟩.

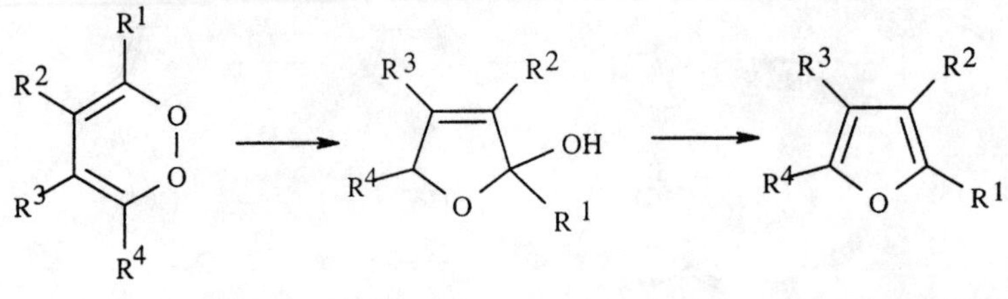

(195)   (196)   (197)

Pyrroles are obtained by reduction of 1,2-diazines ⟨80JMC481⟩. This reaction has been used in conjunction with inverse electron demand Diels-Alder reactions to prepare 3,4-disubstituted pyrrole-2,5-dicarboxylic acid derivatives(Scheme 67). Silyl enol ethers or enamines can also serve as the electron-rich dienophiles; thus, silyl ethers of ester enolates give 3-methoxypyrroles ⟨84JOC4405⟩.

The use of an alkyne as the dienophile allows the synthesis of 3-alkylpyrrole-2,5-dicarboxylate esters ⟨88JOC1405⟩.

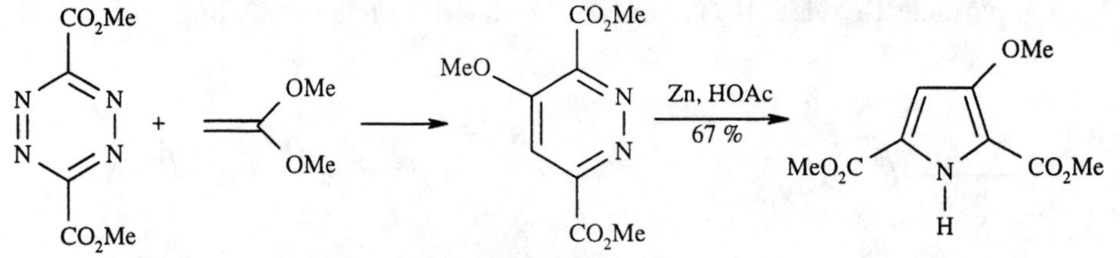

Scheme 67

Similar transformations are based upon 1,2-oxathiin 2,2-dioxides (δ-sultones), which can be converted subsequently into furans ⟨51RTC35⟩, thiophenes ⟨60LA(630)120⟩ (Scheme 68) or pyrroles ⟨61LA(646)32⟩.

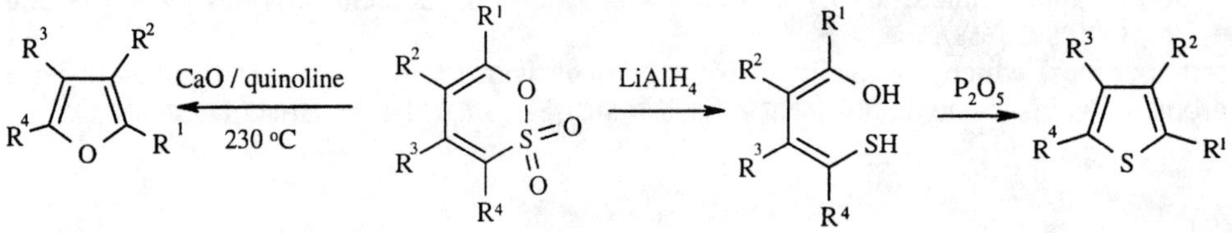

Scheme 68

For a triazepine into pyrrole ring contraction see Section 3.5.2.1.

4.2.3.6 Synthesis of Seven- and Eight-membered Rings

cis-2-Vinylcyclopropanecarboxaldehyde and 2,5-dihydrooxepin interconvert at 50°C (Scheme 69) ⟨69JA2813⟩ and *cis*-1,2-divinyloxirane thermally yields 4,5-dihydrooxepin (Scheme 70) ⟨63JOC1383⟩. Reversible 1,6-addition of SO_2 to *cis*-hexatriene affords 2,7-dihydrothiepin 1,1-dioxide (Scheme 71) ⟨67JA1281⟩.

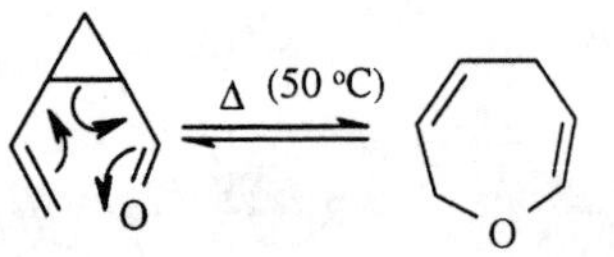

Scheme 69 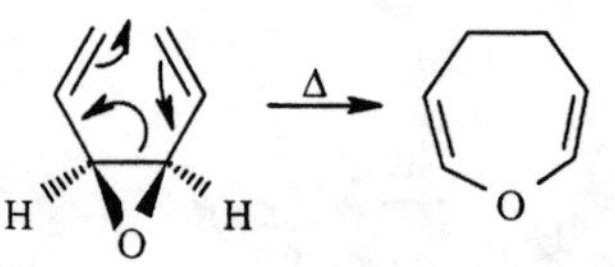   **Scheme 70**

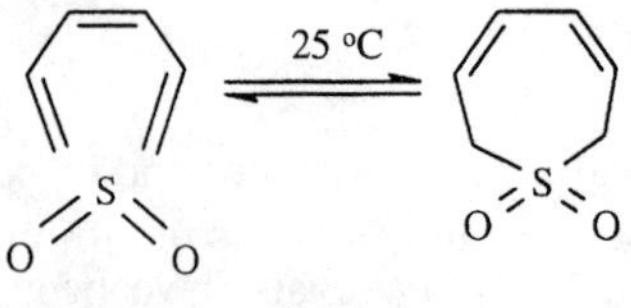

Scheme 71

4.2.4 RINGS CONTAINING THREE ENDOCYCLIC DOUBLE BONDS

4.2.4.1 Synthetic Methods for Substituted Pyridines

There are several significant ring syntheses for pyridines, of which the most important is the Hantzsch synthesis of dihydropyridines (Section 4.2.3.4.2). However, the majority of substituted pyridines are prepared from pyridine itself or from a simple alkyl derivative.

Frequently used methods for introducing substituents into the various positions of the pyridine nucleus include the following:

(i) *Substituents in the 2-position* are often introduced *via* the Chichibabin reaction, which gives 2-aminopyridines (Section 3.2.1.6.4). These can be converted into 2-halopyridines and 2-pyridones (Section 3.2.3.5.3); all are versatile intermediates.

(ii) *Substituents in the 4-position* are most frequently introduced by further transformations of the readily available 4-nitropyridine 1-oxides (Section 3.2.1.4.4).

(iii) *Substituents in the 3-position.* Pyridines can be halogenated (Section 3.2.1.4.7) and sulfonated (Section 3.2.1.4.5) in the 3-position. Yields are better if an activating substituent (which can subsequently be removed) is present in the 2-position and in this case nitration is also feasible. The resulting 3-nitro- and 3-halo-pyridines and pyridine-3-sulfonic acids can be converted into other compounds by the usual methods of benzenoid chemistry.

Methods for obtaining various types of functionally substituted pyridine are summarized in Table 2.

Table 2 Routes to Substituted Pyridines: Cross References to Relevant Sections

Substituent group	Direct introduction of substituent (R)[a]	Obtainable indirectly by or starting from
Acyloxy	—	Hydroxy compounds (3.2.3.7.3), N-oxides (3.2.3.12.5)
Aldehyde	—	Oxidation (3.2.3.3.1), halo compounds (3.2.3.4.4)
Alkoxy	—	Nitro (3.2.3.6.1), hydroxy (3.2.3.7.3) and halo compounds (3.2.3.10.6)
Alkyl	3.2.1.6.8, R	—
Alkylthio	—	Halo (3.2.3.10.6) and
	—	thiocarbonyl compounds (3.2.3.9.2, 3.2.3.9.3)
Amino	3.2.1.6.4, R	Halo (3.2.3.10.6), amido (3.2.3.4.2) and nitro compounds (3.2.3.6.1)
Aryl	3.2.1.6.8, R	—
Arylamino	—	Amines (3.2.3.5.2)
Azo	—	Nitro (3.2.3.6.1) and amino compounds (3.2.3.5.2)
Carboxyl, alkoxycarbonyl, *etc.*	R	Oxidation (3.2.3.3.1, 3.2.3.4.3), halo compounds (3.2.3.10.6)
Cyano	—	Carboxylic acids (3.2.3.4.2), sulfonic acids (3.2.3.9.4), halo compounds (3.2.3.10.6), vinyl compounds (3.2.3.4.5), Reissert compounds (3.2.1.6.8)
Halo	3.2.1.4.7, 3.2.1.6.7	N-Oxides (3.2.1.6.7), nitro compounds (3.2.3.6.1), pyrones and pyridones (3.2.3.7.4), amines (3.2.3.5.3)
Hydrazino	—	Nitramines (3.2.3.6.2), halo (3.2.3.10.6) and nitro compounds (3.2.3.6.1)
Hydroxy (alcoholic)	—	Alkyl (3.2.3.3.1) and vinyl compounds (3.2.3.4.5), carboxylic acids (3.2.3.4.2)
Hydroxy (phenolic)	—	Halo compounds (3.2.3.10.6), sulfonic acids (3.2.3.9.4)
Imino	3.2.1.6.4	Amines (3.2.3.5)
Keto	R	Alkyl compounds (3.2.3.3.3), esters (3.2.3.4.2)
Mercapto	—	Halo compounds (3.2.3.10.6)
Nitro	3.2.1.4.4	Amines (3.2.3.5.2)
Nitroso	3.2.1.4.10	—
Sulfonic acid	3.2.1.4.5	Halo (3.2.3.10.6), thiocarbonyl (3.2.3.9.2) and vinyl compounds (3.2.3.4.5)
Thiocarbonyl	—	Halo compounds (3.2.3.10.6), pyrones and pyridones (3.2.3.7.3)
Vinyl	3.2.3.10.2	Alkyl (3.2.3.3.4) and hydroxy compounds (3.2.3.4.4)

[a]R in this column signifies that compounds containing these substituents are commonly prepared by ring synthesis.

4.2.4.2   Synthesis of Six-membered Rings from Acyclic Compounds

4.2.4.2.1   *From or via pentane-1,5-diones*

As discussed in Section 4.2.3.4.1, pentane-1,5-diones (**200**) can undergo ring closure to give a pyran (**199**) or, in the presence of ammonia, a dihydropyridine (**201**). Oxidative aromatization of these products occurs so easily that it frequently takes place prior to isolation, giving a pyrylium salt (**198**) or a pyridine (**202**). The Hantzsch dihydropyridine synthesis is described in Section 4.2.3.4.2.

(198)   (199)   (200)   (201)   (202)

(203)   (204)   (205)   (206)

The pentane-1,5-dione is usually formed *in situ* by aldol- or Michael-type reactions (**203** → **204** → **200**). Thus, acetaldehyde (**203**; R = H, R′ = Me) and ammonia give 4-picoline and 3-ethyl-4-methylpyridine by formation of the intermediate (**205**), condensation with another molecule of MeCHO, and subsequent dehydrogenation. The same reaction also yields 2-picoline and 5-ethyl-2-methylpyridine *via* the intermediate (**206**). Such reactions are used industrially.

4.2.4.2.2   *From pent-2-ene-1,5-diones*

Ring closure of glutaconic dialdehyde (**208**) with ammonia gives pyridine, hydroxylamine forms pyridine 1-oxide, while primary amines yield pyridinium (**207**) and acids afford pyrylium cations (**209**). Substituted glutaconic dialdehydes and related diketones react similarly. If one of the carbonyl groups is incorporated in a carboxyl group or a modified carboxyl group, α-pyridones and α-pyrones are formed. Pent-2-ene-1,5-diones or nitrogen analogues can be built up from components and subsequently cyclized, *e.g.* NH$_3$ + (**210**) → (**211**), and Me$_2$CO + 2(CO$_2$Et$_2$)$_2$ → (**212**). The conjugate addition of the potassium enolate of methyl ketones to α-oxoketene dithioacetals gives pent-2-ene-1,5-diones, ring closure with ammonium acetate in hot acetic acid then affords excellent yields of 2,6-disubstituted 4-(methylthio)pyridines.

(207)   (208)   (209)

(210)   (211)   (212)

4.2.4.2.3 Other methods

[4 + 2] Cycloaddition reactions with participation of azadienes are of increasing importance in pyridine synthesis (see reviews ⟨87AHC246, B-87MI 505-01, 91COS(5)470⟩). Although a dihydropyridine is an intermediate, spontaneous conversion into a pyridine occurs when a suitable leaving group is present, *e.g.* (213) → (214) (the process is accelerated by sonication) ⟨94T10047⟩.

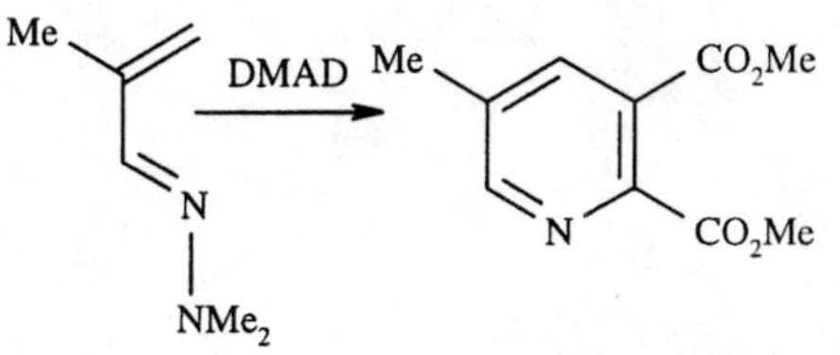

(213)   (214) 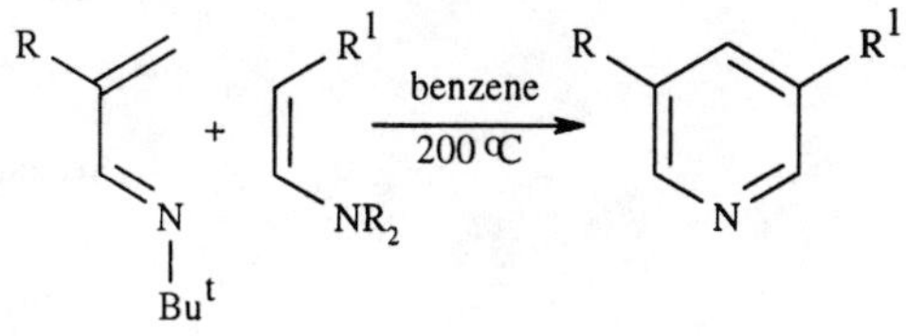   (215)

Similarly, 1-azabutadienes with an N-alkyl substituent react with enamines to give a range of 3,5-disubstituted pyridines (215) ⟨84JOC2691⟩. 2-Azabutadienes are popular choices for ($4\pi_S + 2\pi_S$) cycloaddition reactions; *ab initio* calculations are available ⟨92JOC3031, 92JOC6736⟩. Electron-rich 2-azabutadienes react with alkynes to give pyridines, *e.g.* (216) ⟨82JA1428⟩.

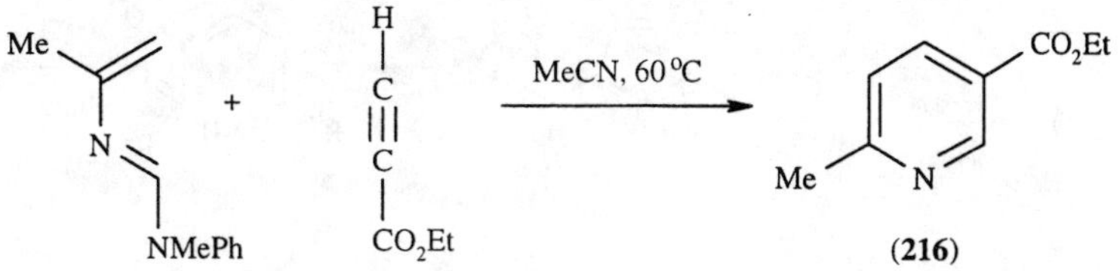

(216)

The cobalt(I)-catalyzed condensation of nitriles (1 mole) or isocyanates (1 mole) with alkynes (2 moles) gives pyridines and 2-pyridones, often in excellent yield (Scheme 72).

The oxime acetate (217) readily undergoes electrocyclization to the dihydropyridine (218), from which pyridine is derived by loss of acetic acid.

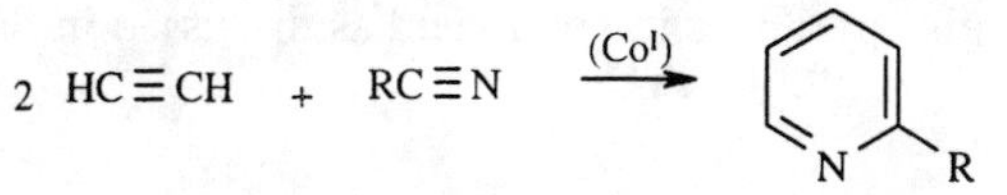

Scheme 72

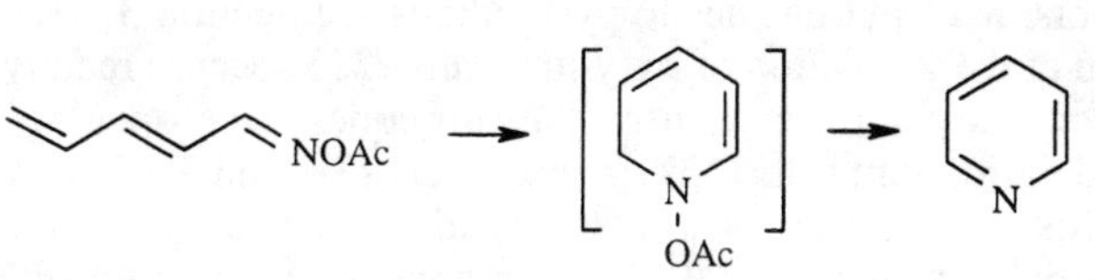

(217)   (218)

Cycloadducts from cyclopentadienones and nitriles spontaneously lose CO to form a pyridine (Scheme 73).

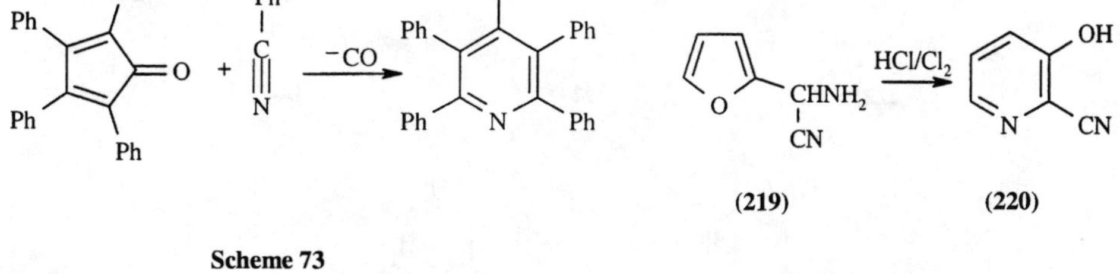

(219)   (220)

Scheme 73

Pyridines are obtained when cyclopentadienes react with phthalimidonitrene, and the resulting bicycles (221) are opened by sodium hydride (Scheme 74) ⟨84H(22)1369⟩.

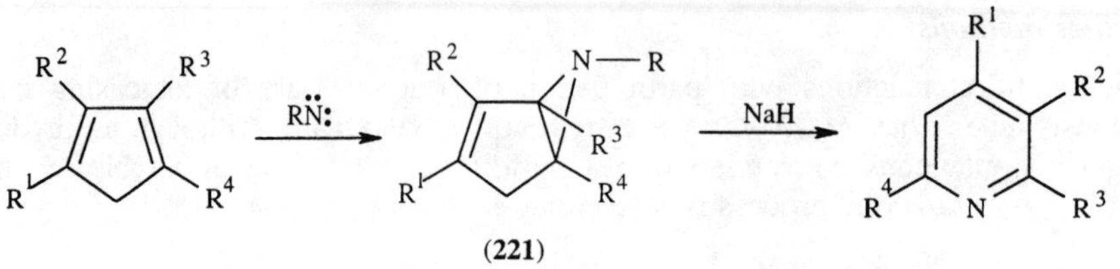

(221)

R = phthalimido

Scheme 74

4.2.4.3 Synthesis of Six-membered Rings from Other Heterocycles

4.2.4.3.1 From five-membered rings

Aminomethylfurans are converted into 3-hydroxypyridines by acid and an oxidizing agent, *e.g.* (219) → (220). 2-Hydroxymethylfurans with chlorine in aqueous methanol give 3-hydroxy-4-pyrones. 3-Hydroxypyridines can conveniently be prepared by reaction of 2-acylfurans with ammonia (Scheme 75). Pyrrole and dichlorocarbene give some 3-chloropyridine (Section 3.3.1.7.1).

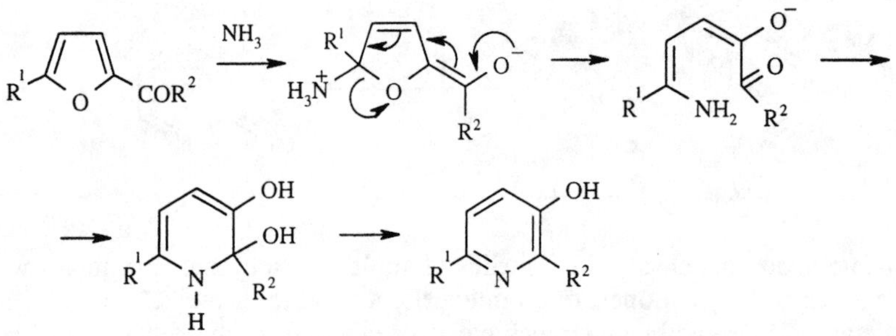

Scheme 75

Oxazoles and dienophiles give pyridines in good yield as discussed in Section 3.4.1.10.1.

4.2.4.3.2 From other six-membered rings

For the thermal conversion of pyridazines to pyrimidines see Section 3.2.1.2.2.

Nucleophilic addition at the 2-position of pyrylium salts (223) occurs readily under mild conditions and when ammonia or primary amines are used the subsequent ring-opening/ring-closure sequences give pyridines (224) and pyridinium salts (222), respectively (Section 3.2.1.6.4.iii). The process is most useful for the synthesis of 2,4,6-trisubstituted pyridine derivatives. Thiinium salts (226) are conveniently prepared from pyrylium salts (225) by treatment with sodium sulfide (Section 3.2.1.6.5). Thiinium salts (226) react with ammonia and amines similarly to their pyrylium analogues.

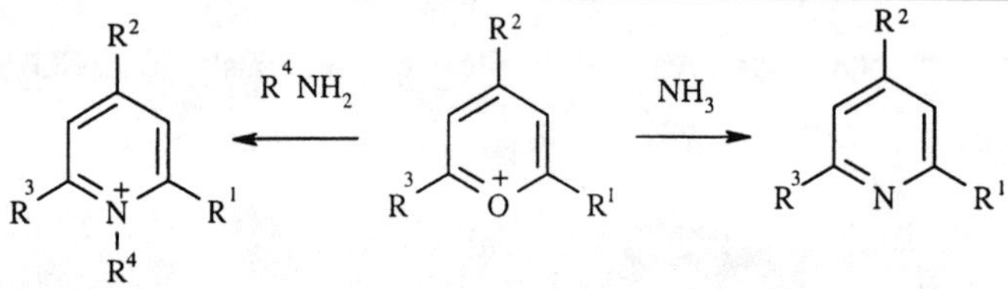

(222)   (223)   (224)

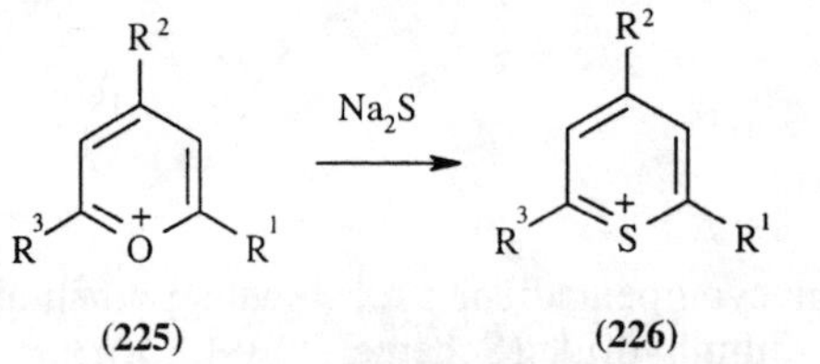

(225)   (226)

'Inverse electron demand' Diels-Alder/retro-Diels-Alder-type reactions, of di- and especially poly-azines with electron-rich dienophiles, interconvert six-membered rings. 1,2,4-Triazines react with enamines and enol ethers to give pyridines (Scheme 76) 〈CHEC-II(5)242〉.

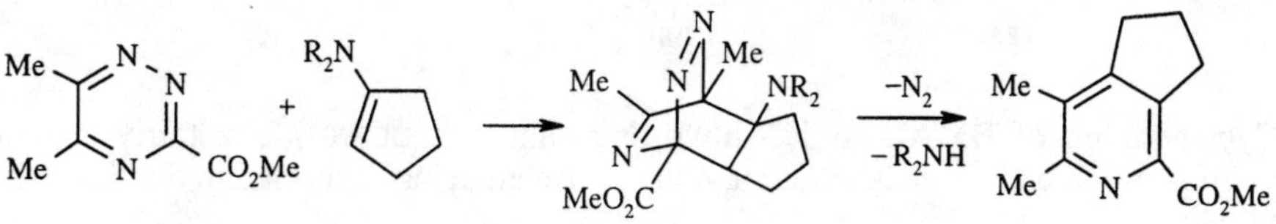

Scheme 76

4.2.4.4 Synthesis of Seven-membered and Larger Rings

1*H*-Azepines (**228**) result from spontaneous valence-bond isomerization of azanorcaradienes (**227**) which are themselves made by reaction of arenes with nitrenes (see CHEC 5.16.4.1.3).

Oxepins are prepared by an analogous method (**230** → **231**); the starting material is made from (**229**) 〈64AG(E)510〉.

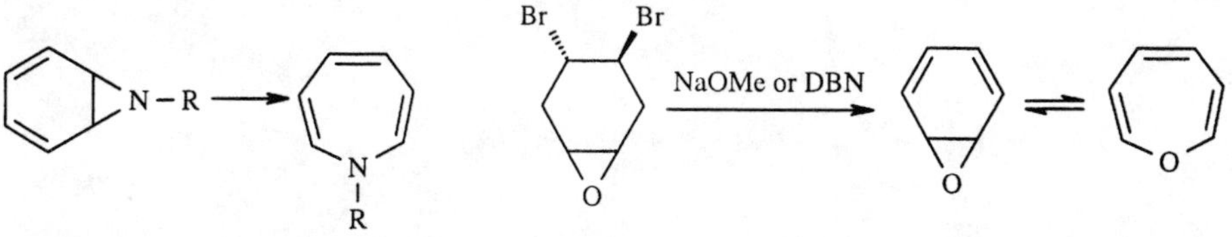

(**227**)   (**228**)   (**229**)   (**230**)   (**231**)

Photolysis of PhN$_3$ in an argon matrix gives 1-aza-1,2,4,6-cycloheptatetraene (**232**) 〈79RTC334〉; in aniline the photoproduct is the 3*H*-azepine (**233**) 〈81AHC(28)231〉.

For related preparations of azepines from pyridines (Section 3.2.3.4.4.iii) and indazole derivatives (Section 3.4.3.2.4) see the sections indicated.

Ring expansion of thiinium salts has been used for the preparation of various thiepins: intermediate (**234**) is prepared using N$_2$C(Li)CO$_2$Et, and decomposes *via* the carbene (**235**) to afford (**236**) (see CHEC 5.17.3.4.2.ii).

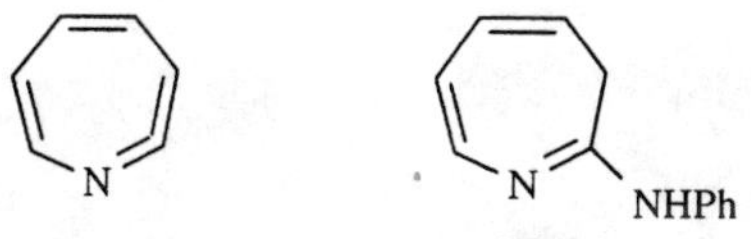

(**232**)   (**233**)

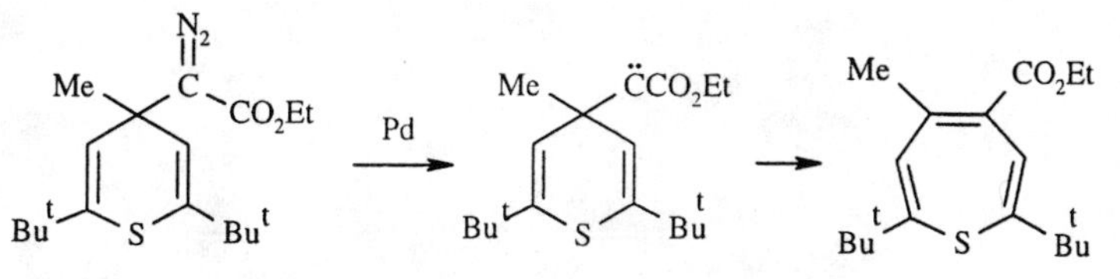

(**234**)   (**235**)   (**236**)

cis-Hexatriene (**237**) reacts with sulfur dioxide to yield 2,7-dihydrothiepine 1,1-dioxide (**238**), which with excess bromine gives 3,4-dibromo-2,3,4,7-tetrahydrothiepine 1,1-dioxide. Treatment of the dibromide with triethylamine gives thiepine 1,1-dioxide (**239**).

      (237)   (238)   (239)

    [2 + 2] Cycloaddition of DMAD to 1,2-dihydropyridines ⟨77JOC2903⟩ is a fairly general route to 1,2-dihydroazocines which proceeds *via* a bicyclic intermediate as described in Section 3.2.2.3.8.

4.3

Synthesis of Monocyclic Rings with Two or More Heteroatoms

4.3.1 SUBSTITUENT INTRODUCTION AND MODIFICATION

4.3.1.1 Overview

In this chapter, synthetic methods are classified by the ring that is being formed. We deal successively with rings containing two, three and four or more heteroatoms. Within each category, classification is first by the relative orientation of the heteroatoms, and then by the ring size.

However, we commence with two brief sections dealing with preparative methods involving substituent introduction and modification for the two large and important classes of monocyclic compounds: those with five and six members and with various heteroatoms.

4.3.1.2 Substituent Introduction and Modification in Azoles

The azoles encompass a very wide range of reactivity (see Chapter 3.4) and possibilities of substituent introduction and modification are very varied. Scheme 1 gives a classification in terms of reactivity type for each position in the various rings.

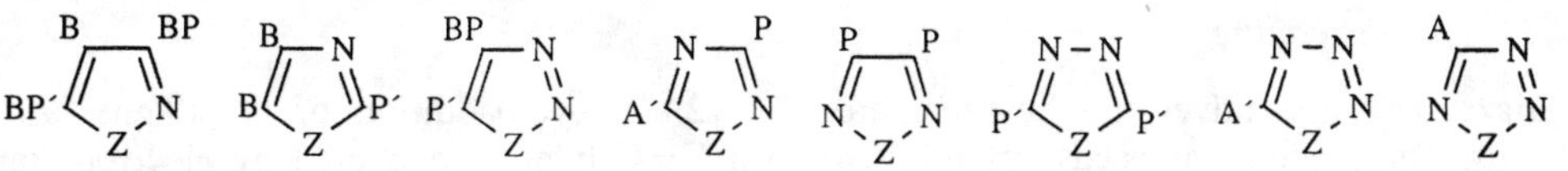

Scheme 1   Character of azole ring positions (Z = NR, O or S)

(i) Positions marked B show reactivity comparable to that of ring carbons in benzene. Substituents can be introduced by electrophilic substitution reactions (Section 3.4.1.4) and show reactivity similar to those of the analogous benzene. Thus, amino groups can be diazotized (Section 3.4.3.5.3) and halogens are unreactive.

(ii) Positions marked P show reactivity similar to those of the 2- and 4-positions in pyridine. The substituents can be introduced by very strong nucleophiles (Section 3.4.1.6); OH compounds exist in the oxo form and can be converted (Section 3.4.3.7) into chloro compounds which are reactive (Section 3.4.3.9.1). Alkyl groups can be deprotonated into anions which undergo many substitution reactions (Section 3.4.3.3).

(iii) Positions marked BP have reactivity intermediate between benzene and pyridine and can be compared to the 3-position of pyridine.

(iv) Positions marked A resemble azines and have the reactivity of the 2-position of pyrimidine, *i.e.* comparable to P but more marked.

(v) Positions marked with a prime (B′, P′, A′) additionally can be lithiated and the lithium replaced in a wide variety of synthetically interesting ways.

4.3.1.3 Substituent Introduction and Modification in Azines

As explained in Chapter 3.2, the reactivity of six-membered rings containing two heteroatoms bears the same relationship to six-membered rings containing one heteroatom as do the latter to benzene. Hence many of the methods listed for the preparation of pyridines by substituent introduction and modification in Table 1 of Section 4.2.4.1 are also applicable to the preparation of analogous azines.

4.3.2 TWO HETEROATOMS IN THE 1,2-POSITIONS

4.3.2.1 Three-membered Rings

(i) Oxaziridines (**9**) are formed by oxidation of Schiff's bases $R_2C = NR'$; $MeCO_3H$, MCPBA or Oxone (potassium peroxymonosulfate) are the usual oxidants. For reviews see ⟨84CHEC-(7)195, 85MI 112–01, 89T5703, 91S327, 92CRV919⟩.

Oxaziridines unsubstituted at nitrogen (**9**, $R' = H$) (in general used as aminating reagents, see Sections 3.5.4 and 3.5.6), are prepared by treatment of carbonyl compounds with chloramine or hydroxylamine-*O*-sulfonic acid in aqueous media ⟨91S327⟩. This method is, however, limited to certain carbonyl compounds with cyclohexanone, followed by butanone, benzaldehyde, and trichloroacetaldehyde giving the best results.

(ii) Diaziridines (**10**) are prepared by the reaction $R_2CO + R'NH_2 + NH_2OSO_3H \rightarrow$ (**10**).

(iii) Diazirines (**11**) are produced by the oxidation of *N*-unsubstituted diaziridines: (**10**; $R' = H$) $+ Ag_2O \rightarrow$ (**11**).

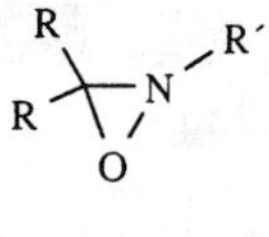

(**9**) 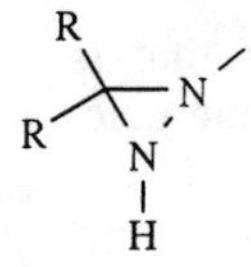 (**10**)  (**11**) 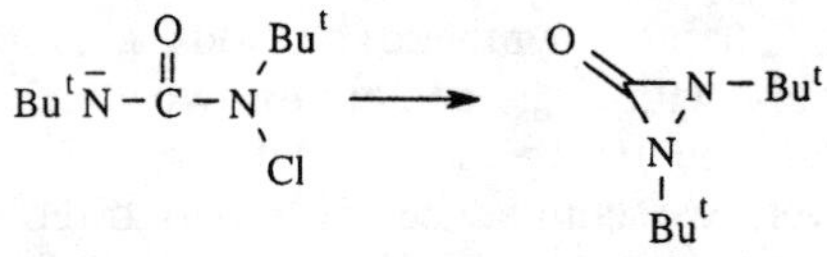 (**12**) (**13**)

(iv) Diaziridinones can be obtained by nucleophilic displacement (**12** → **13**); bulky substituents are required to obtain stable products.

4.3.2.2 Four-membered Rings

4.3.2.2.1 1,2-Diazetidines

1,2-Diazetidine derivatives are best prepared by [2 + 2] cycloaddition of an alkene with azo compounds. The reaction proceeds smoothly when a C=C bond is activated by electron-donating groups such as OR or NR_2 and an azo group by electron-withdrawing functionality (COR, COOR, $CONH_2$, *etc.*) (Scheme 2) ⟨83HC(42)443⟩. The only alternative route to 1,2-diazetidine derivatives is by direct double alkylation of substituted hydrazines with 1,2-dihaloalkanes ⟨78JA2806, 78JOC2785⟩.

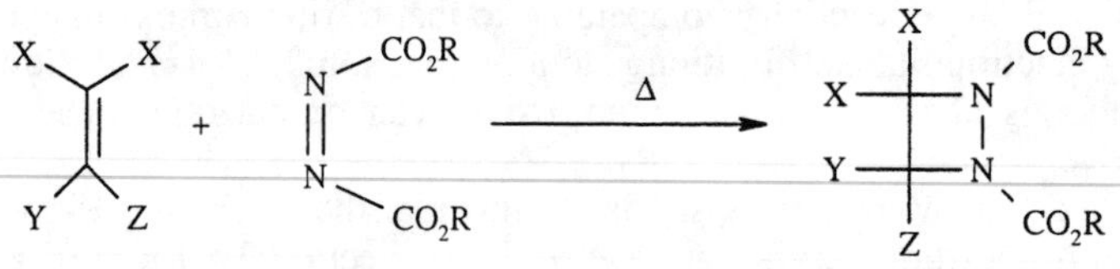

Scheme 2

1,2-Diazetidinones are prepared by the thermal [2 + 2] cycloaddition of ketene with diazo compounds (Scheme 3) ⟨83HC(42)443⟩. Use of ketenimines instead of ketene furnished imino derivatives ⟨67JHC155⟩.

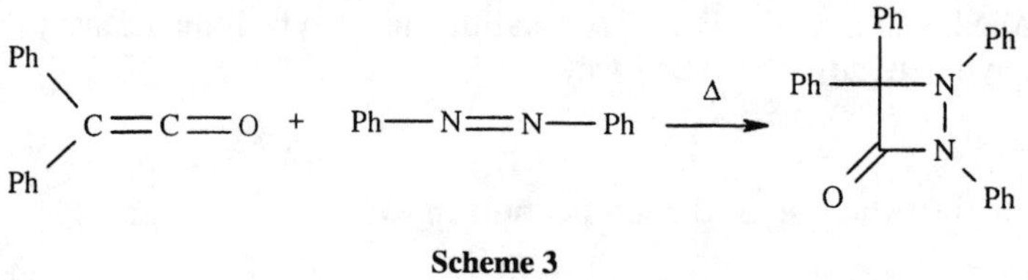

Scheme 3

A large number of diazetidinones have been synthesized by an intramolecular cyclization of haloacetylhydrazones with suitable bases (Scheme 4). A photochemical Wolff rearrangement of

4-diazo-pyrazolidine-3,5-dione in presence of some nucleophiles (H_2O, EtOH) leads to mixture of isomeric diazetidinones (**14**) and (**15**) in comparable quantities ⟨87JCS(P1)899⟩.

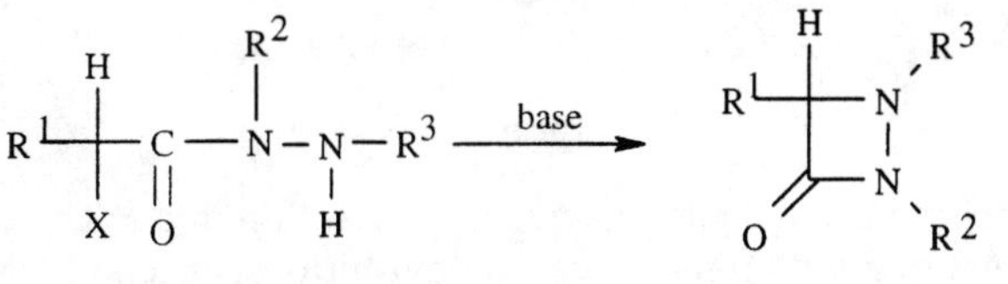

Scheme 4

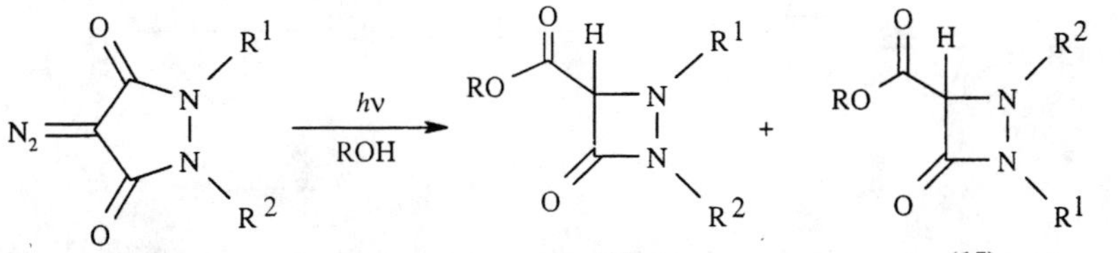

(**14**)   (**15**)

Cyclobutadiene adds ⟨79AJC2659⟩ to azodicarboxylic acid to give fused diazetines (Scheme 5).

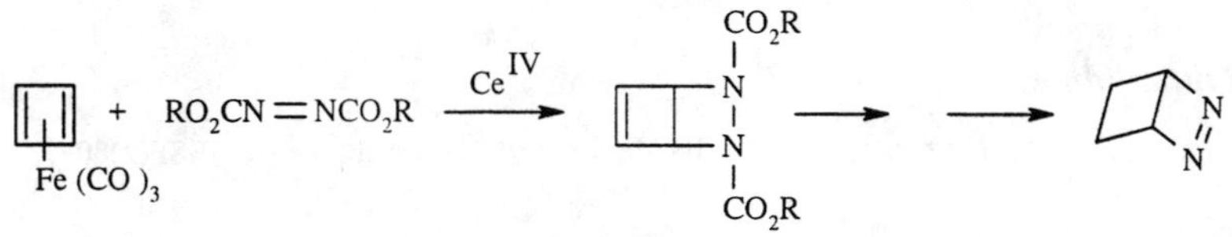

Scheme 5

Oxidative closure of β-dihydroxylamines leads to diazetine dioxides which can be further reduced in two steps to Δ^1-1,2-diazetines (Scheme 6) ⟨75JOC1409⟩.

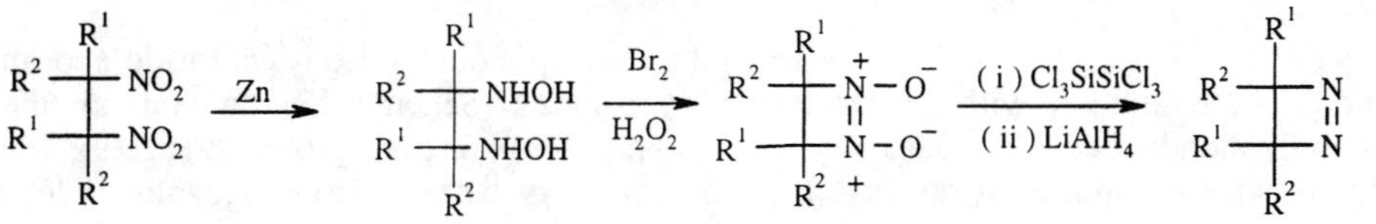

$R^1 = R^2 = Me$

$R^1 = Me, \quad R^2 = Et$

Scheme 6

4.3.2.2.2 1,2-Oxazetidines

Oxazetidines have been obtained by [2 + 2] addition of nitroso compounds to appropriate alkenes. Trifluoronitrosomethane reacts with polyhalogenated ethylenes ⟨69JCS(C)2119⟩, with styrene ⟨65JGU855⟩ or with allenes ⟨73JCS(P1)1561⟩ to give oxazetidines (*e.g.* Scheme 7).

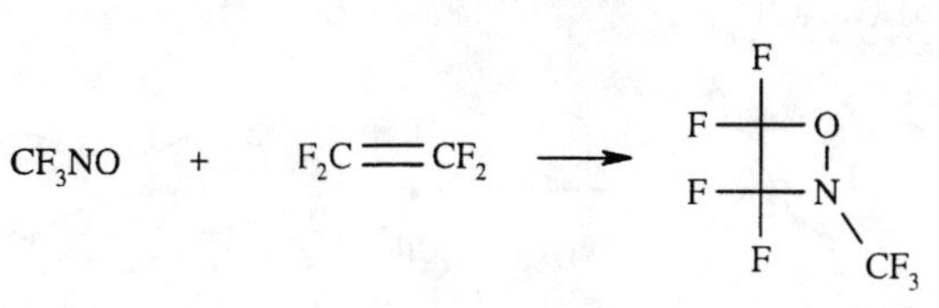

Scheme 7

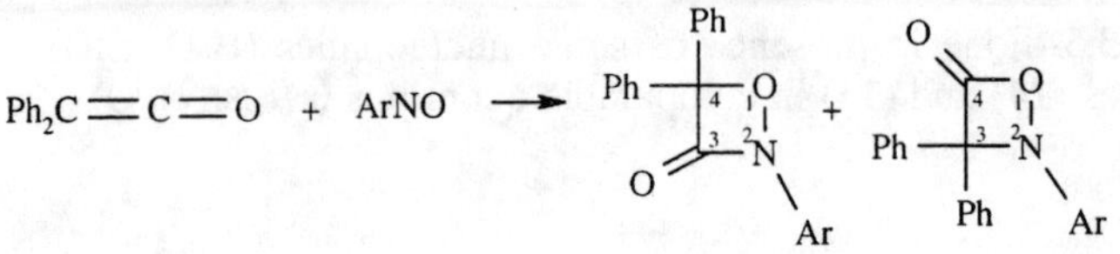

Scheme 8

The addition of nitrosobenzenes to diphenylketene gives two products. The 4-one predominates with *p*-dimethylamino- and the 3-one with *p*-methoxycarbonyl-nitrosobenzene (Scheme 8) ⟨74JOC2552⟩.

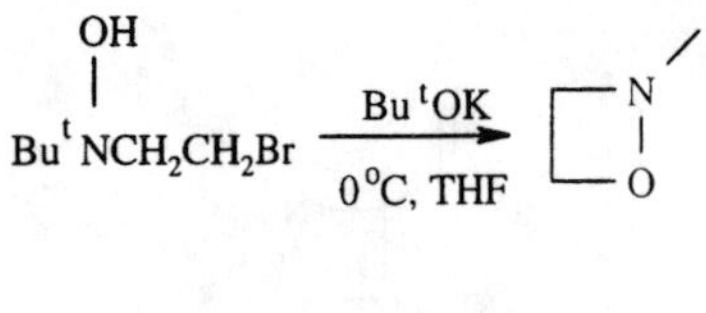

Scheme 9 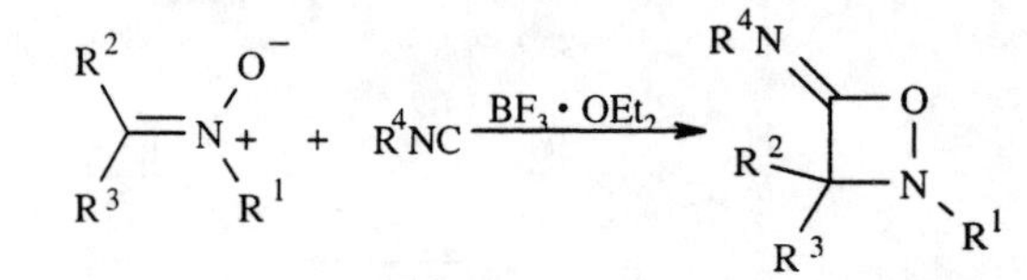   **Scheme 10**

β-Haloalkylhydroxylamines can be converted into oxazetidines (Scheme 9) ⟨71JA4082⟩ and similar cyclizations give *N*-substituted 4,4-diaryl-1,2-oxazetidin-3-ones ⟨68JOC3619⟩.

Nitrones add to isocyanides to afford 4-imino-1,2-oxazetidines (Scheme 10) ⟨85S1083⟩; for a review see ⟨93JHC579⟩.

4.3.2.2.3 1,2-Thiazetidines

1,2-Thiazetidines are made by nucleophilic displacement (Scheme 11) ⟨75BSF(2)807⟩.

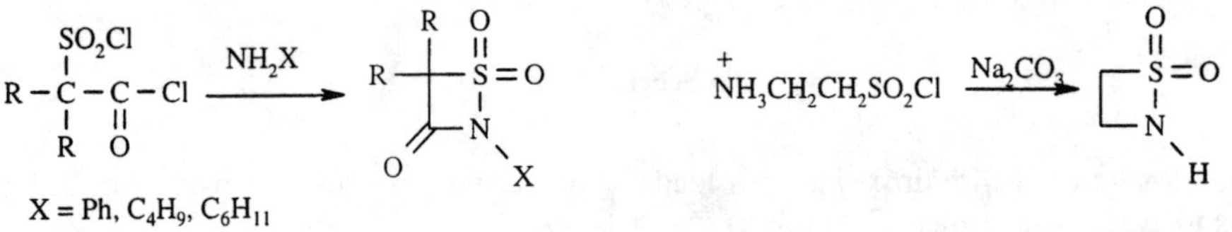

X = Ph, C$_4$H$_9$, C$_6$H$_{11}$

Scheme 11

Sulfenes (RCH=SO$_2$), generated *in situ* from the corresponding sulfonyl chloride and an organic base, undergo cycloaddition with imines to give β-sultams (Scheme 12). In general, the sulfene requires stabilization by benzoyl ⟨70BCJ3543⟩, phenyl ⟨75BCJ480⟩, or ester group ⟨89TL2869⟩; whereas the cycloaddition with an imine works better if the imine is derived from benzaldehyde, avoiding isomerization to an enamine.

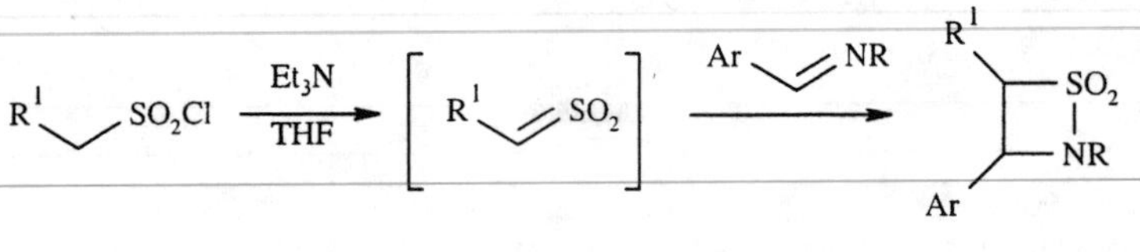

Scheme 12

4.3.2.2.4 1,2-Dioxetanes

The addition of singlet oxygen to alkenes gives dioxetanes: bisadamantylidene forms an unusually stable dioxetane (Scheme 13) ⟨75JA7110⟩.

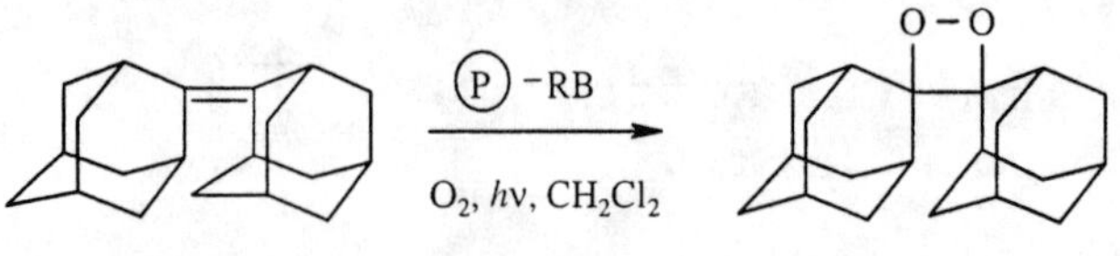

Scheme 13

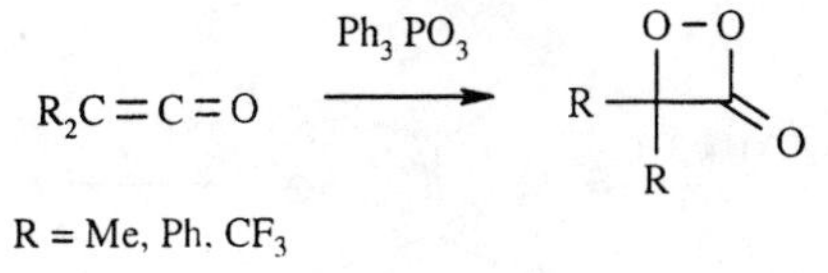

$R_2C{=}C{=}O \xrightarrow{Ph_3PO_3}$ [structure]

R = Me, Ph. CF$_3$

Scheme 14

Ketenes have been converted into 1,2-dioxetan-3-ones by triphenylphosphine ozonide (Scheme 14) ⟨77JA5836, 80CC898⟩. Dioxetanes and dioxetanones have been prepared ⟨80CJC2089⟩, *e.g.* by cyclization of the β-bromohydroperoxide of 2-methyl-2-butene (Scheme 15).

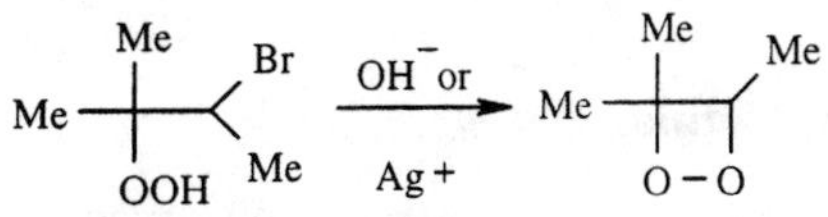

Scheme 15

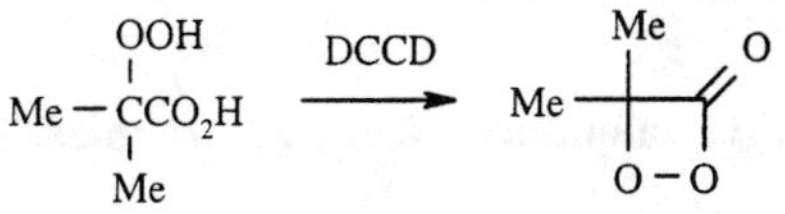

Scheme 16

α-Peroxylactones or 1,2-dioxetan-3-ones are prepared from α-hydroperoxy acids which are cyclized with dicyclohexylcarbodiimide (Scheme 16) ⟨77JA5768⟩.

4.3.2.2.5 1,2-Oxathietanes

Polyhalogenated alkene addition to sulfur dioxide ⟨60JA6181, 66JCS(C)1171⟩ gives 1,2-oxathietane 2-oxide adducts (Scheme 17) ⟨78BAU142, 79BAU106⟩.

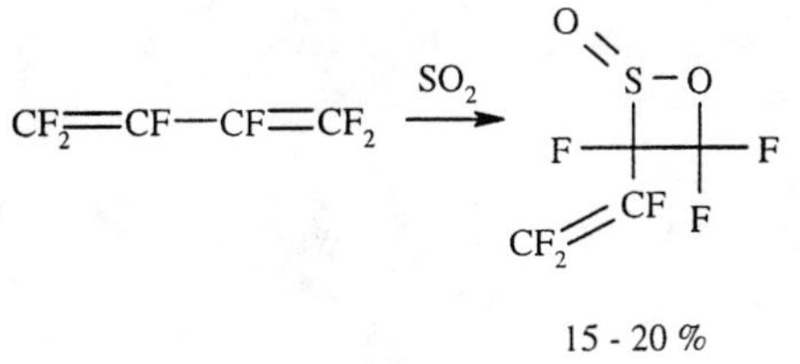

$CF_2{=}CF{-}CF{=}CF_2 \xrightarrow{SO_2}$ [structure]

15 - 20 %

Scheme 17

$Me_3C{-}CPh_2(OH) \xrightarrow{H^+} CH_2{=}C(Me){-}CPh_2(Me) \xrightarrow{SO_3}$ [structure]

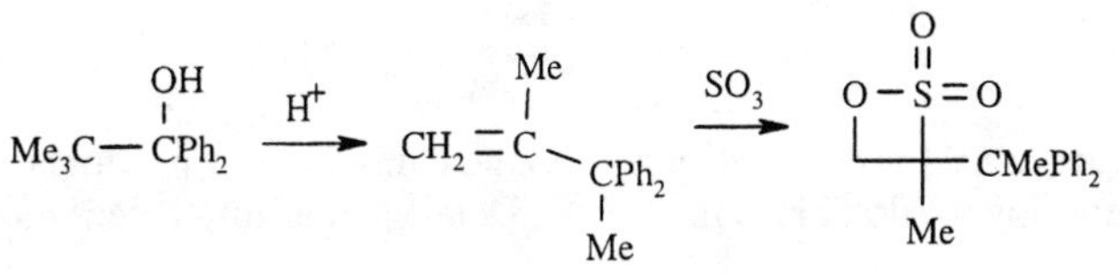

Scheme 18

Acid-catalyzed rearrangement of 1,1-diphenyl-2,2-dimethylpropanol in presence of sulfur trioxide forms a stable 1,2-oxathietane 2,2-dioxide (Scheme 18) ⟨77JCS(P1)247⟩.

4.3.2.2.6 1,2-Dithietanes

Photochemically induced extrusions of carbon monoxide generate 1,2-dithietanes in equilibrium with their valence bond tautomers, dithiones (Scheme 19) ⟨74JA3502⟩.

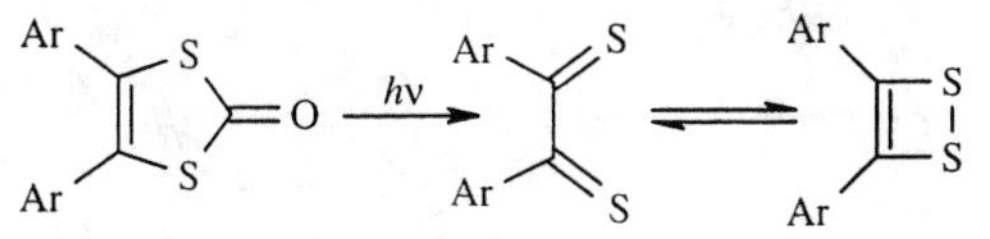

Scheme 19

The intermediacy of 1,2-dithietanes has been proposed in several transformations ⟨B-83MI 135-02, 85JOC1550, 87JA926⟩. An isolable, and well-characterized 1,2-dithietane system is the 1,1-dioxide (**16**) ⟨80JA2490⟩ synthesized by [2 + 3] thermal cycloaddition (Scheme 20).

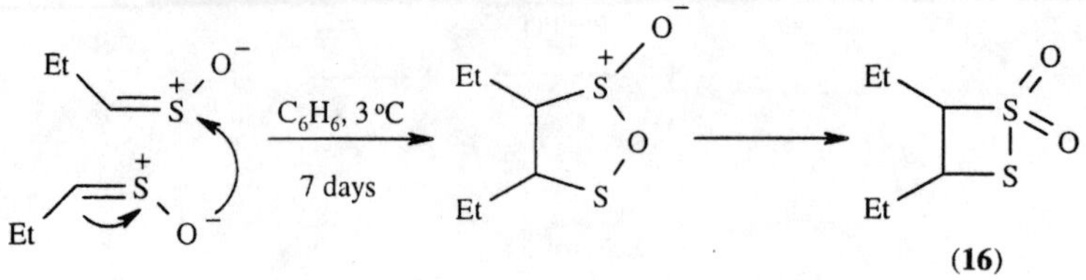

Scheme 20

4.3.2.3 Five-membered Rings: Pyrazoles, Isoxazoles, Isothiazoles, *etc*.

Five-membered rings with two adjacent heteroatoms are most frequently made using a hydroxylamine or hydrazine derivative. However, dipolar cycloadditions are also significant. Methods forming a Z—Z bond are important particularly for sulfur-containing derivatives.

4.3.2.3.1 Synthesis from hydrazine, hydroxylamine and hydrogen disulfide derivatives

(i) Pyrazoles and isoxazoles from 1,3-diketones

The standard syntheses for pyrazoles (**17**) and isoxazoles (**19**) involve the reactions of β-dicarbonyl compounds (**18**) with hydrazines and hydroxylamine, respectively. These reactions take place under mild conditions and are of very wide applicability; the substituents Y can be H, R, Ar, CN, CO$_2$Et, *etc*.

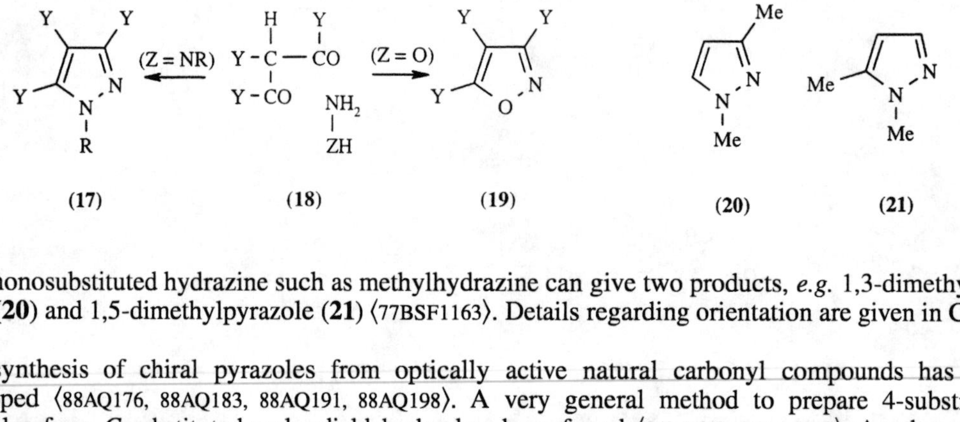

A monosubstituted hydrazine such as methylhydrazine can give two products, *e.g.* 1,3-dimethylpyrazole (**20**) and 1,5-dimethylpyrazole (**21**) ⟨77BSF1163⟩. Details regarding orientation are given in CHEC 4.04.

A synthesis of chiral pyrazoles from optically active natural carbonyl compounds has been developed ⟨88AQ176, 88AQ183, 88AQ191, 88AQ198⟩. A very general method to prepare 4-substituted pyrazoles from *C*-substituted malondialdehydes has been found ⟨84LA649, 85LA1969⟩. Another useful precursor is triformylmethane, from which it is easy to prepare 4-formylpyrazoles ⟨90BSF660⟩.

Reaction of a 1,3-diketone with hydroxylamine gives, *via* the isolable monoxime (**22**) and 5-hydroxydihydroisoxazole (**23**), the isoxazole (**24**). Unsymmetrical 1,3-diketones result in both possible isomers (**24**) and (**25**), but the ratio of the isomeric products can be controlled by the right combination of the 1,3-dicarbonyl component and the reaction conditions used, as described in CHEC 4.16.

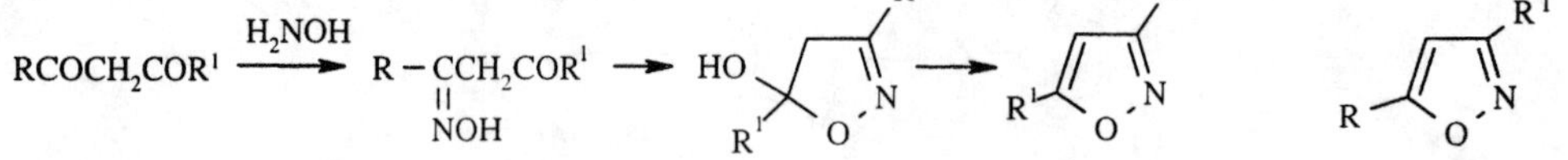

The scope of the possible modifications to the 1,3-dioxo component is illustrated by the following examples. The β-oxoketene dithioacetal (**26**) with a monosubstituted hydrazine gives the pyrazole (**27**)

⟨76BCJ398⟩. The iminium salt (28) with monosubstituted hydrazines gives the 3,5-bis(dimethylamino)-pyrazole (29) ⟨68T4217, 69T3453⟩. β-Ketocyanides yield amino-pyrazoles or -isoxazoles (*e.g.* 30 → 31).

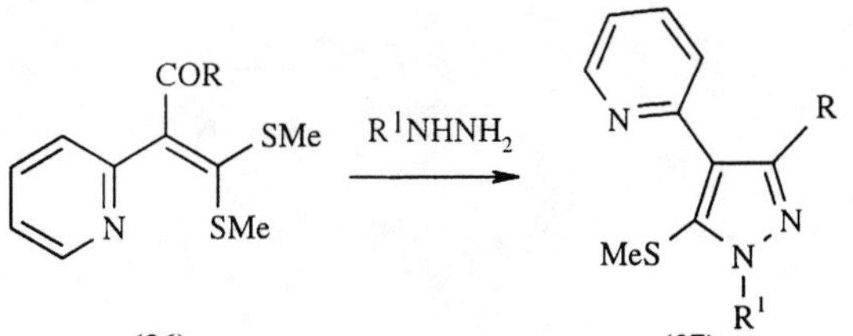

(26) (27)

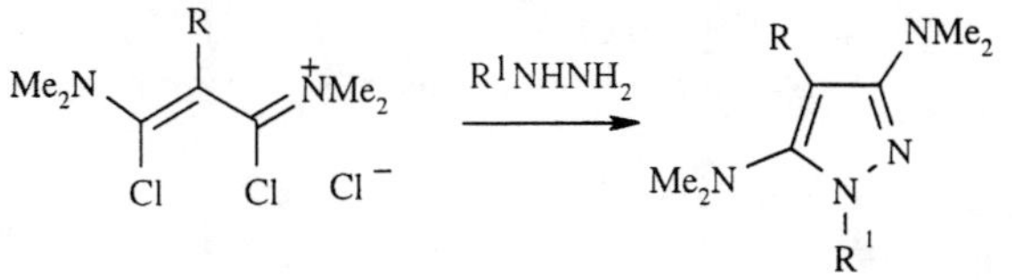

(28) (29)

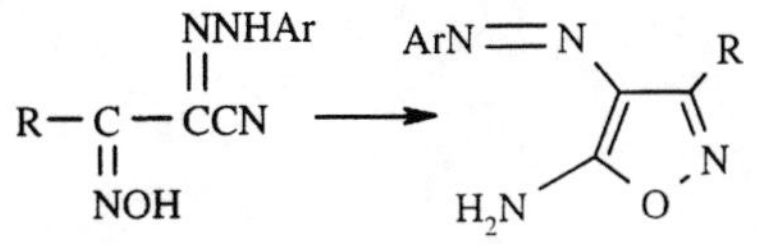

(30) (31) 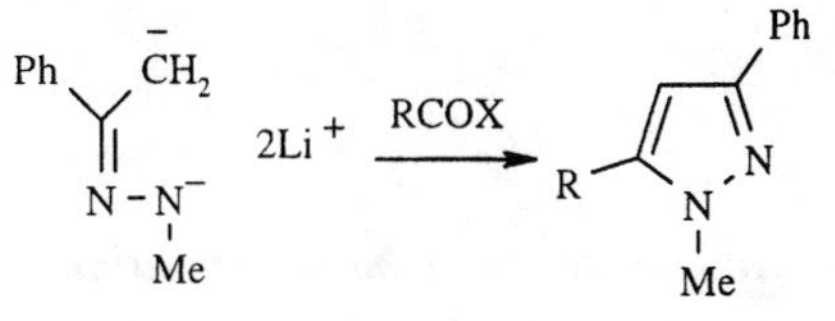 (32) (33)

The dianion (32) of acetophenone methylhydrazone reacts with acid chlorides to give pyrazoles (33; see also CHEC 4.04).

(ii) Pyrazolinones and isoxazolinones

Pyrazolinones and isoxazolinones are prepared from β-keto esters and hydrazine or hydroxylamine by reactions such as (34 → 35) similar to those in (*i*) above. Diketene behaves as a masked β-keto ester. Acetylenecarboxylic esters can be used in place of β-keto esters to give pyrazolinones such as (36) and (37) and the corresponding isoxazolinones. β-Chloro-α,β-unsaturated acid chlorides react similarly (*cf.* 38 → 39).

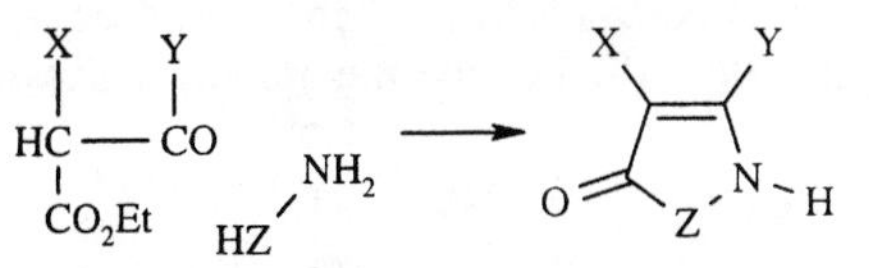

(34) (35) 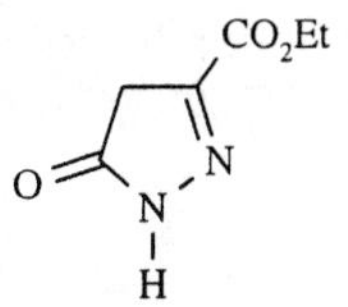 (36) 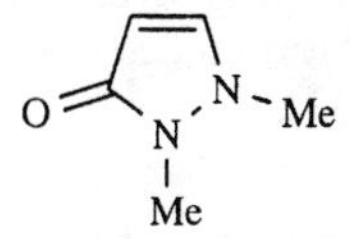 (37)

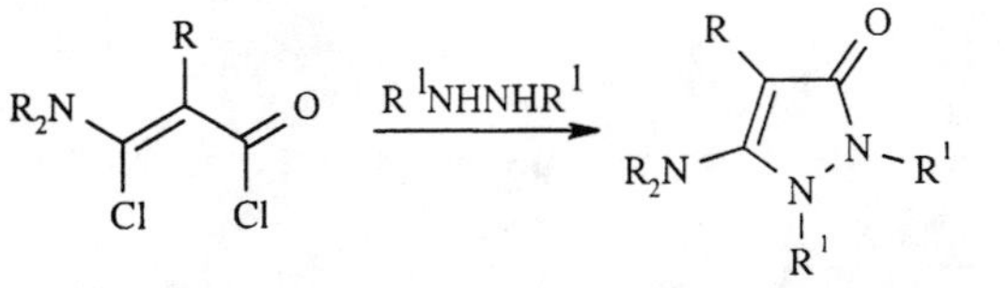

(38) (39)

(iii) Pyrazolines, isoxazolines, pyrazolidines, isoxazolidines and 1,2-dithiolanes

α,β-Unsaturated ketones form pyrazolines and isoxazolines (40 → 41) and the intermediate hydrazones and oximes are often isolated. α,β-Unsaturated ketoximes such as (42) undergo oxidative ring closure with bis(triphenylphosphine)palladium dichloride to give 3,5-diphenylisoxazole (43) ⟨73TL5075⟩.

Tetrahydro compounds (**44**) can be obtained from 1,3-dibromides (with N_2H_4, NH_2OH, S_2^{2-}, *etc.*).

(**40**)   (**41**)

(**42**)   (**43**)   (**44**)

4.3.2.3.2 Synthesis by Z—Z bond formation

The synthesis of *N*-substituted-3,4-trimethylenepyrazoles (**45**) from iminophosphoranes and 2-azido-1-cyclopentene-1-carboxaldehyde has been described (R = Ph, CH_2Ph, CH_2CO_2Et) ⟨88S742⟩.

(**45**)

The β-acylvinyl azides (**46**) lose N_2 forming the isoxazole (**47**) in an anchimerically assisted concerted reaction ⟨75AG(E)775, 78H(9)1207⟩. The N,S-ylid (**48**) on heating undergoes cyclization to isoxazoles (**49**).

(**46**)   (**47**)

(**48**)   (**49**)

Isothiazoles (**51**) may be obtained by the oxidative cyclization of β-thioxoimines (**50**) ⟨88AP863, 89AJC1291, 89JCS(P1)1241, 89JHC1575⟩ (see also CHEC 4.17).

A general route to 1,2-dithiolanes (**53**) involves oxidation of the 1,3-dithiol (**52**) with hydrogen peroxide at 75°C in acetic acid containing potassium iodide. β-Dithio compounds such as (**54**) undergo oxidation in acid to the 1,2-dithiolylium salts (**55**) ⟨76JCS(D)455⟩.

(50)   (51)   (52)   (53)

(54)   (55)

4.3.2.3.3 Other methods from acyclic precursors

Many N-unsubstituted pyrazoles can be obtained from *N*-allyl-*N*-nitrosamides ⟨90H(30)789⟩, *e.g.* (**56**) → (**57**). Reactions of 1-azido-3-chloropropane with Grignard reagents afford intermediates (**58**) that cyclize into azimines of Δ¹-pyrazolines (**59**) ⟨92TL4683⟩.

(56)   (57)   (58)   (59)

Alkynes add nitrile oxides and diazoalkanes to give isoxazoles (**60**) and pyrazoles (**61**), respectively, in 1,3-dipolar cycloadditions. If an alkene is used instead of an alkyne the non-aromatic analogues (**62**; Z = NH, O) result ⟨94AHC(60)261⟩; yields are best when the alkene contains an electron-withdrawing substituent.

(60)   (61)   (62)   (63)   (64)   (65)

In a convenient route to Δ²-isoxazoline *N*-oxides (**65**), the ylide (**64**) adds to the nitrostyrene (**63**) in the presence of copper(I) salts ⟨76JOC4933, 91HC(49)1⟩.

Dimethylsulfonium phenacylide (**67**) undergoes *C*-alkylation with α-chloronitroso compounds (**66**): intermediates (**68**) cyclize to the isoxazolines (**69**) ⟨72T3845⟩.

(66)   (67)   (68)   (69)

Isoxazolidines result from 1,3-dipolar cycloadditions of nitrone or nitrone esters and alkenes (see Equation (1)) ⟨95PHC179⟩.

$$\tag{1}$$

4.3.2.3.4 From other heterocycles

Diaziridinones react with bifunctional carbanions by ring enlargement to give pyrazolinones ⟨92JOC7359⟩; thus (70) with malononitrile affords 1,2-bis-*t*-butyl-3-aminopyrazolin-5-one (71).

Y = CN, CONH₂, CO₂Me

3-Unsubstituted isoxazolium salts (72) react with hydrazines to yield 3-aminopyrazoles (73; X = H, NO₂, Hal, COMe, CO₂Et, R³ = H, Me, Ph, CONH₂) ⟨88S203⟩.

1,3-Dipolar cycloaddition of sydnones with 1-aryl-3,3,3-trifluoropropynes is a general synthesis of 4-trifluoromethylpyrazoles ⟨93JHC365⟩.

1,2-Dithiolylium salts may be converted into pyrazoles, pyrazolium salts and isothiazoles (see Section 3.4.1.6.2.ii). For example, 3-phenyl-1,2-thiolylium salt (74) with hydrazine, methylhydrazine or phenylhydrazine yielded the corresponding pyrazoles (75). 3,4-Dimethyl-1,2-dithiolylium perchlorate (76) with ammonia gave 4,5-dimethylisothiazole (77).

Isoxazoles (78) in the presence of base undergo ring opening to α-ketonitriles. In the presence of hydrazines, 5-aminopyrazoles (79) are obtained.

The anhydro-5-hydroxy-1,3,2-oxathiazolylium hydroxide system (**83**) and DMAD thermally yielded an intermediate 1:1 cycloadduct (**84**) which lost CO_2 forming dimethyl 2-phenylisothiazole-3,4-dicarboxylate (**85**) ⟨72CB196⟩. Irradiation of (**83**) in neat DMAD formed the valence tautomer (**82**) which lost CO_2 to give the nitrile sulfide dipole (**81**) captured by DMAD to form dimethyl 3-phenylisothiazole-4,5-dicarboxylate (**80**) ⟨75JA6197⟩ (see also CHEC 4.17).

(**80**)   (**81**)   (**82**)   (**83**)

(**85**)   (**84**)

Many other heteroring transformations lead to isothiazoles ⟨CHEC-II(3)366⟩.

With hydroxylamine, the pyrylium salt (**86**) undergoes ring opening to an intermediate 1,5-enedione oxime (**87**); conjugate addition of the α,β-unsaturated ketone gives (**88**), which in the presence of acid forms the isoxazole (**89**).

Reaction of (**86**) with a hydrazine resulted in the pyrazoline (**90**). A similar transformation of chromone is described in Section 3.2.1.6.4.iv.

(**86**)   (**87**)   (**88**)   (**89**)   (**90**)

4.3.2.4 Six-membered Rings: Pyridazines, 1,2-Oxazines, *etc.*

The most important synthetic methods involve condensation of hydrazine, hydroxylamine or hydrogen peroxide with a 1,4-oxygenated carbon chain, and these procedures are particularly useful for the preparation of pyridazines and 1,2-oxazines. Other methods include Diels-Alder reactions of a diene with an azo or nitroso compound.

4.3.2.4.1 Synthesis from hydrazine or hydroxylamine derivatives

1,4-Dicarbonyl compounds with a double bond in the 2,3-position condense with hydrazine to give pyridazines (*e.g.* **91** → **92**). If one of the carbonyl groups in the starting material is part of a carboxyl group or a potential carboxyl group, then reactions with hydrazines or hydroxylamine lead to pyridazinones or 1,2-oxazinones (*e.g.* **93** → **94**; Z = NH, NPh, O). Similarly a cyano group leads to an amino or imino product.

(**91**)   (**92**)   (**93**)   (**94**)

Saturated 1,4-dicarbonyl compounds give 1,4-dihydro-pyridazines or -pyridazinones, *etc.*, which are easily oxidized. 1,2-Diketone monohydrazones and esters containing a reactive CH_2 group give 3-pyridazinones (**95** → **96**) ⟨54HCA1467⟩.

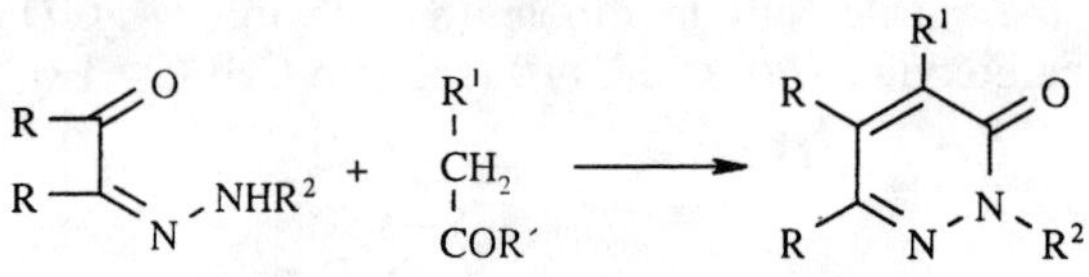

(95)   (96)

4.3.2.4.2 *By cycloaddition reactions*

Pyridazine-3,4-dicarboxylic acid can be conveniently obtained as shown on Scheme 21 ⟨90JHC579⟩. The key step of this conversion is a [4 + 2] cycloaddition of the diazadiene (**98**) (generated *in situ* from the precursor (**97**)) and ethyl vinyl ether.

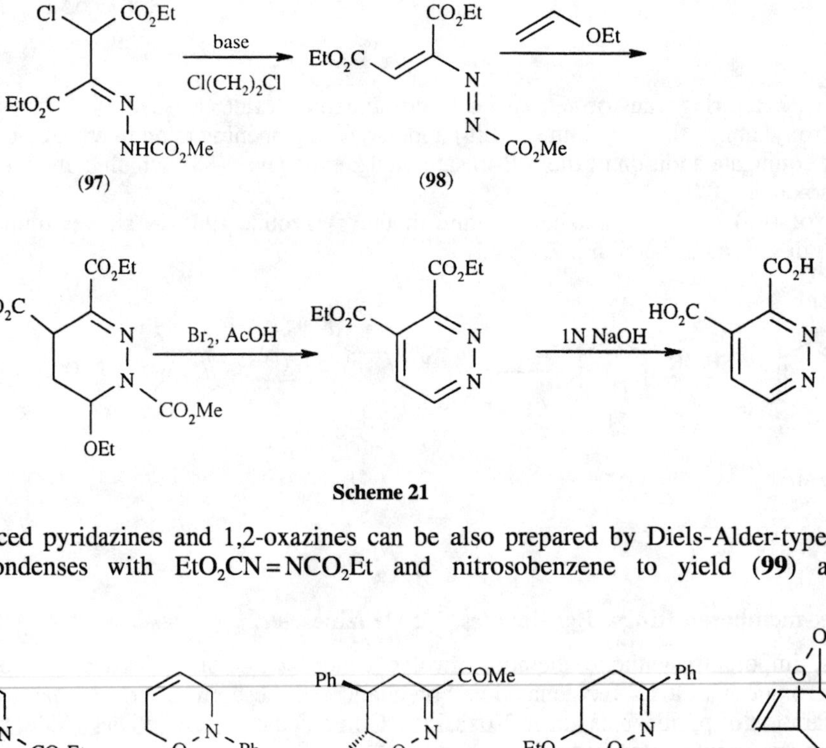

(97)   (98)

Scheme 21

Other reduced pyridazines and 1,2-oxazines can be also prepared by Diels-Alder-type reactions. Butadiene condenses with $EtO_2CN=NCO_2Et$ and nitrosobenzene to yield (**99**) and (**100**), respectively.

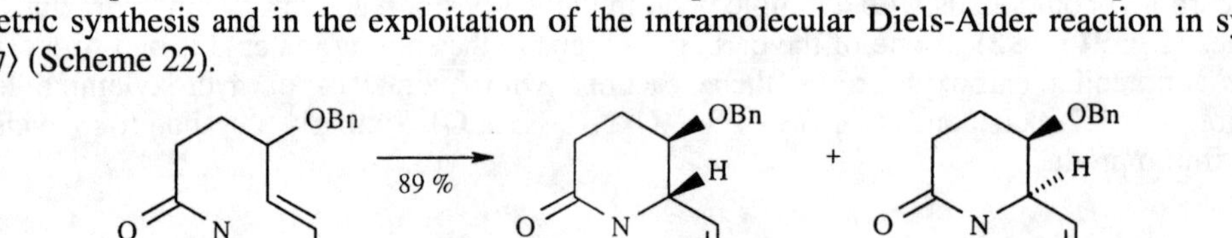

(99)   (100)   (101)   (102)   (103)

The most important advances have been in the development of chiral nitroso dienophiles useful for asymmetric synthesis and in the exploitation of the intramolecular Diels-Alder reaction in synthesis ⟨94S1107⟩ (Scheme 22).

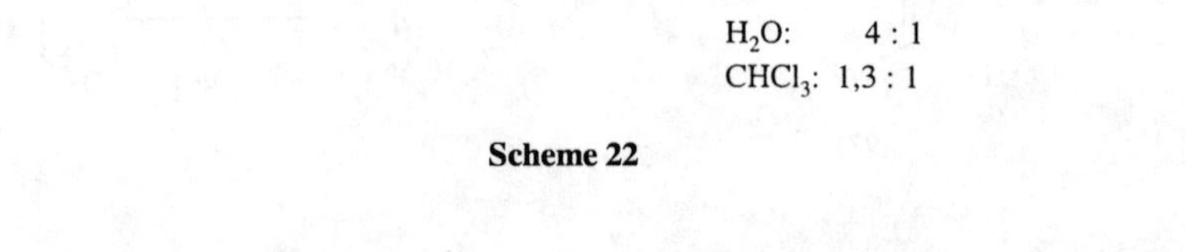

H_2O: 4 : 1
$CHCl_3$: 1,3 : 1

Scheme 22

5,6-Dihydro-4H-1,2-oxazines are prepared by the cycloaddition of nitrosoalkenes and alkenes: $CH_2=C(NO)COMe$ with *trans*-stilbene gives (**101**) ⟨78CC847⟩.

Tetrahydro-1,2-oxazines are accessible by cycloaddition of nitrosoalkenes with electron-rich alkenes, *e.g.* $EtOCH=CH_2+CH_2=CPh—N=O \rightarrow$ (**102**).

Cyclohexadiene and singlet oxygen form (**103**). Most of the important synthetic approaches for 1,2-thiazines use a fragment in which the S-N bond already exists. Syntheses *via* [4 + 2] cycloadditions constitute a major part of the synthetic repertoire for this class of compounds ⟨CHEC-II(6)370⟩. Thus, pentadiene with TsNSO gives (**104**) and perfluoro-1,2-thiazine (**105**) is prepared by the cycloaddition of thiazyl fluoride (FSN) and perfluoro-1,3-butadiene $CF_2=CFCF=CF_2$ ⟨79CC35⟩.

Transient thionitroso-arenes and -alkanes, [RN=S], can be trapped with conjugated 1,3-dienes, yielding the corresponding dihydrothiazines. For example, thionitrosoalkanes have been generated by the thermal decomposition of N-trimethylsilyl-N-chlorosulfenylalkylamines. The thermolysis of (**106**) in the presence of 2,3-dimethylbutadiene affords a 65% yield of the dihydrothiazine (**107**) ⟨90JOU1799⟩. These thionitroso alkanes show good regioselectivity with unsymmetrical dienes.

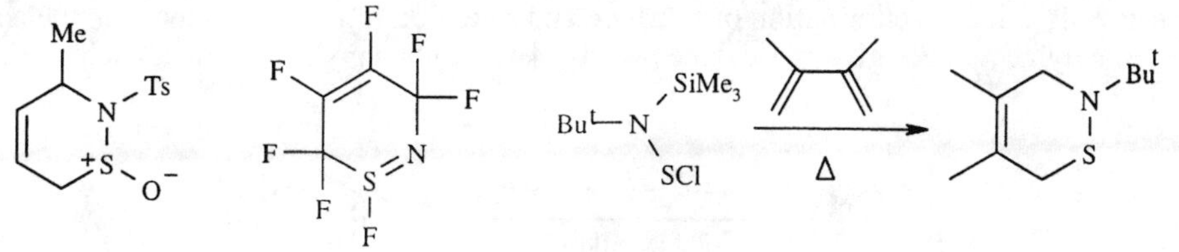

(**104**)   (**105**)   (**106**)   (**107**)

4.3.2.4.3 Other methods from acyclic precursors

The pyridazine dioxide derivative (**108**) was made by intramolecular nitroso compound dimerization as shown (Scheme 23). 1,2-Oxathiin 2,2-dioxides are obtained by the addition of sulfuric acid to α,β-unsaturated ketones, *e.g.* (**109**) → (**110**) ⟨66HC(21-2)774⟩. 1,2-Dithiins are synthesized from conjugated diynes using benzyl thiol: reductive debenzylation of intermediate (**111**) by sodium in liquid ammonia at − 70°C gives, after aerial oxidation, the 1,2-dithiin (**112**) ⟨67AG(E)698⟩.

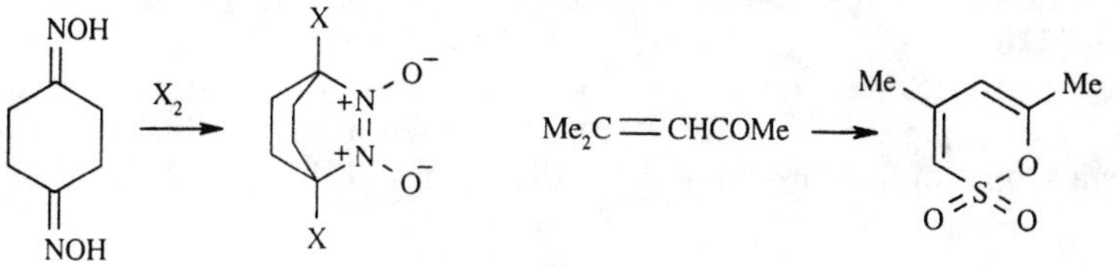

(**108**)   X = Cl or Br   (**109**)   (**110**)

Scheme 23

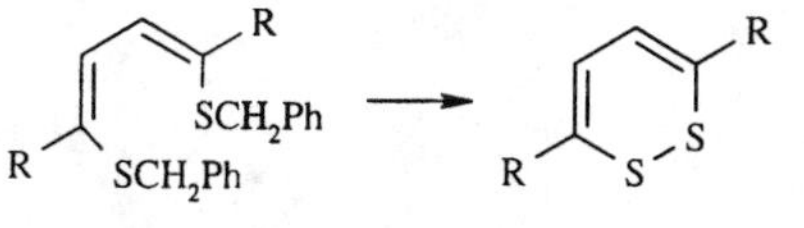

(**111**)   (**112**)

4.3.2.4.4 From other heterocycles

The Meisenheimer-type rearrangement of 1-substituted pyrrolidine 1-oxides gives tetrahydro-2H-1,2-oxazines (**113** → **114**).

1,4-Dihydropyridazines (**117**) result from Diels-Alder addition of *s*-tetrazines (**115**) with electron-rich alkenes (*e.g.* **116**). Frequently the products aromatize, as in (**117**) → (**118**) (see also Section 3.2.1.10.2.iv).

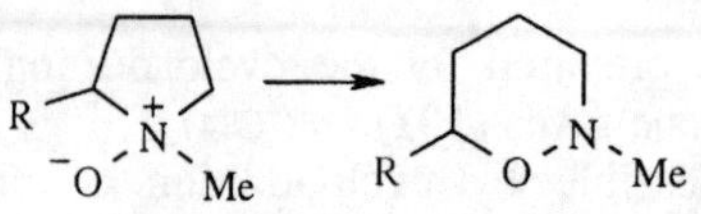

(113)   **(114)**

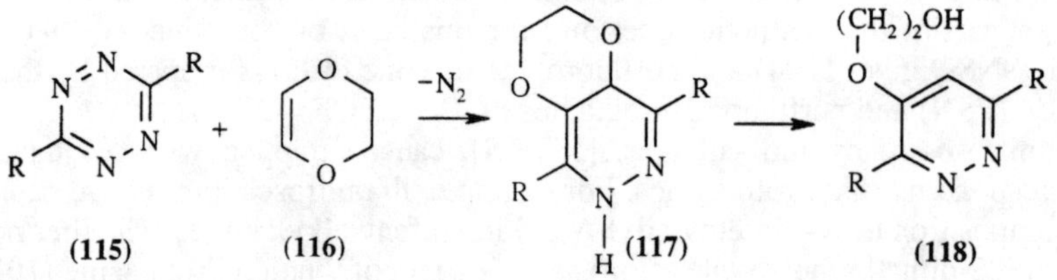

(115)   **(116)**   **(117)**   **(118)**

Inverse electron demand Diels-Alder reactions of 1,2,4,5-tetrazines with alkynes produce pyridazines directly with the elimination of nitrogen and retention of the substituents on the acetylene; thus, tributylstannylacetylenes give pyridazines with a 4-tributylstannyl substituent (Scheme 24).

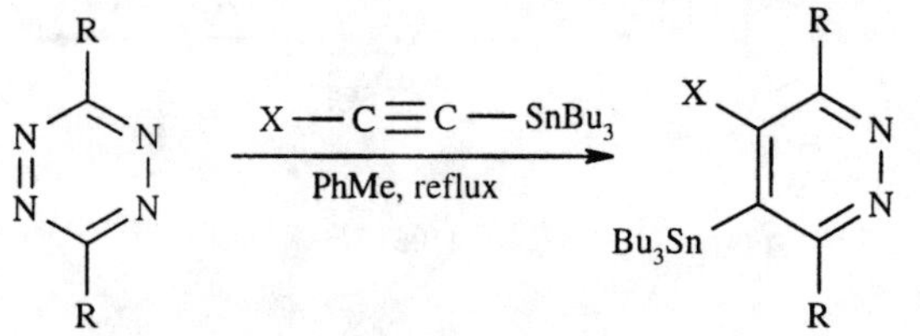

R=Ph, Me, CO_2Me; X=H, Bu, OEt, Ph, Bu_3Sn

Scheme 24

4.3.2.5 Seven-membered Rings

4.3.2.5.1 1,2-Diazepines

1,5-Dihalides and 1,5-ditosylates have been used for the preparation of fully saturated monocyclic systems, *e.g.* (**119** → **120**).

5-Halo-aldehydes and -ketones react with a wide range of substituted hydrazines to give 4,5,6,7-tetrahydro-1,2-diazepines (**121**) ⟨76H(4)1509⟩. The reaction of 1,5-diketones with hydrazine has been much used as a source of 5,6-dihydro-4*H*-1,2-diazepines (**122**) ⟨67AHC(8)21⟩.

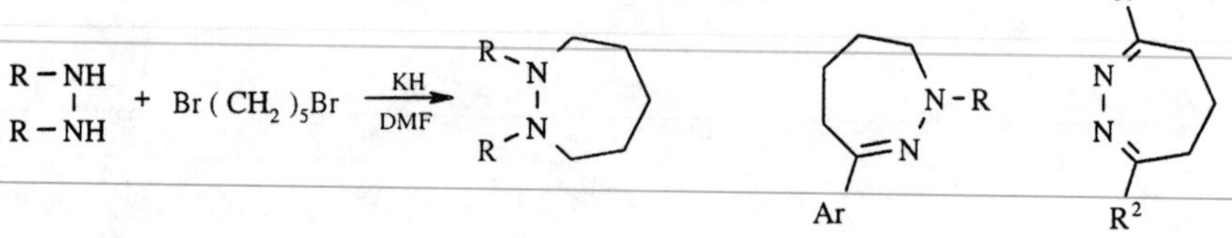

(119)   R = CO_2Et , COPh   **(120)**   **(121)**   **(122)**

Arylhydrazones (**123**) are cyclized with $SOCl_2$ or PCl_3 to 2-aryl-2,3,4,5-tetrahydro-1,2-diazepin-3-ones (**124**) ⟨84CR(B)929⟩.

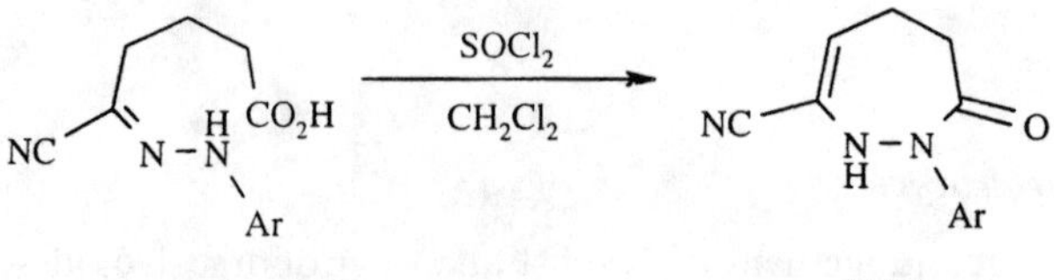

(123)   **(124)**

Reaction of tosylhydrazide with $\alpha,\beta,\gamma,\delta$-unsaturated ketones, followed by dehydrotoluenesulfonylation affords 3*H*-1,2-diazepines (Scheme 25) ⟨84JCS(P1)1581⟩. Tautomeric products (**125**) and (**126**) are in dynamic equilibrium at room temperature but could be separated by high-pressure liquid chromatography at 0°C.

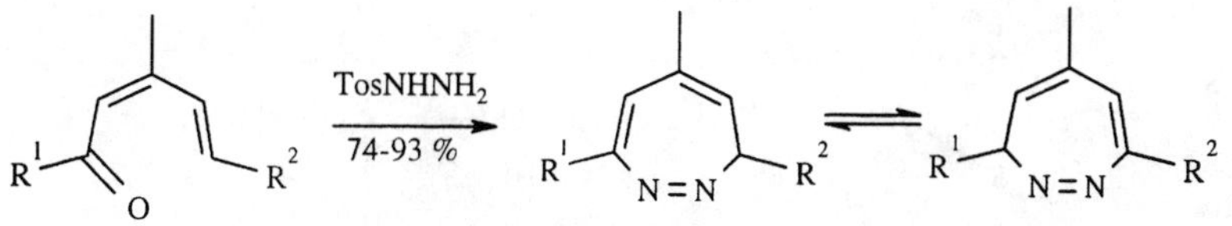

(125) **(126)**

R^1 = Me, Et, CH_2CH_2Ph, $CHMe_2$; R^2=H, Me

Scheme 25

The dihydroxy dihydrazones (**127**) with $AlCl_3$ cyclize to dihydrodiazepines (**128**) in 80–90% yield ⟨87CC582⟩.

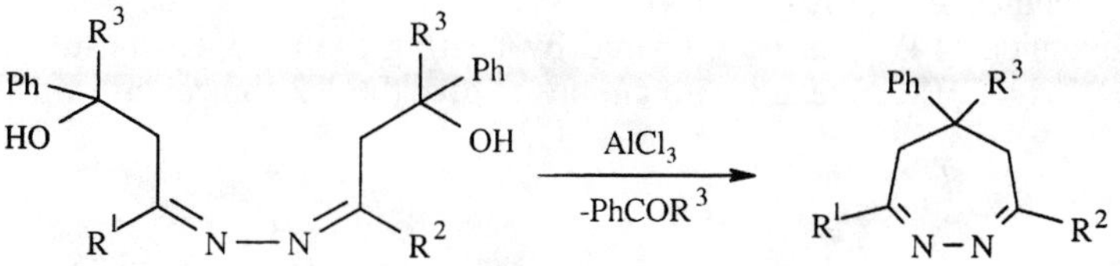

(127) **(128)**

1,7-Electrocyclization of dienyldiazoalkanes, *e.g.* (**129**), provides a general route to 3*H*-1,2-diazepines, *e.g.* (**131**). In the example shown the eight-π-electron cyclization is followed by a 1,5-sigmatropic hydrogen shift in (**130**) ⟨83CC1003⟩. Such hydrogen shifts are rapid at room temperature and the isomer ratio reflects thermodynamic stability.

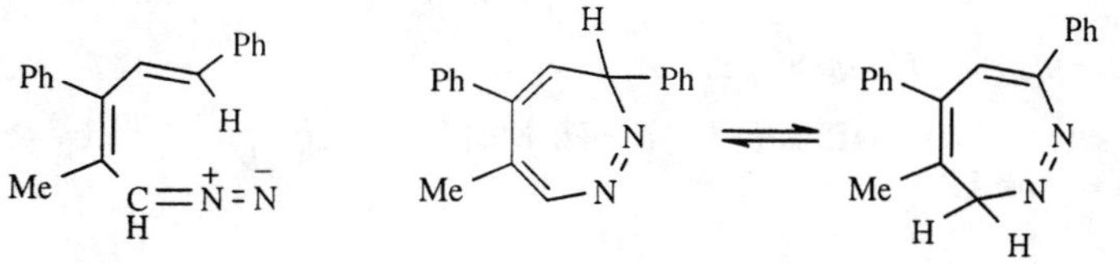

(129) **(130)** **(131)**

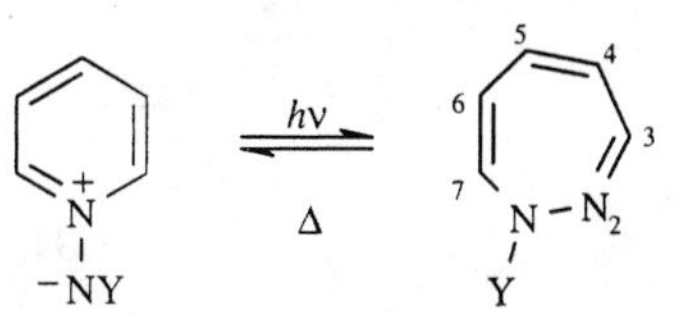

(132) **(133)**

Photochemical conversion of pyridine *N*-imides, *e.g.* (**132**), gives 1*H*-1,2-diazepines (**133**) ⟨81ACR348⟩ (see Section 3.2.3.12.4.iii).

Cycloaddition of diazoalkanes to diazabicyclo[2.2.0]hexanes (**134**) and subsequent extrusion of nitrogen affords diazatricyclo[3.2.0.0^{2,4}]heptanes (**135**) that are easily valence-isomerized to novel dihydrodiazepines (**136**) ⟨84TL297⟩.

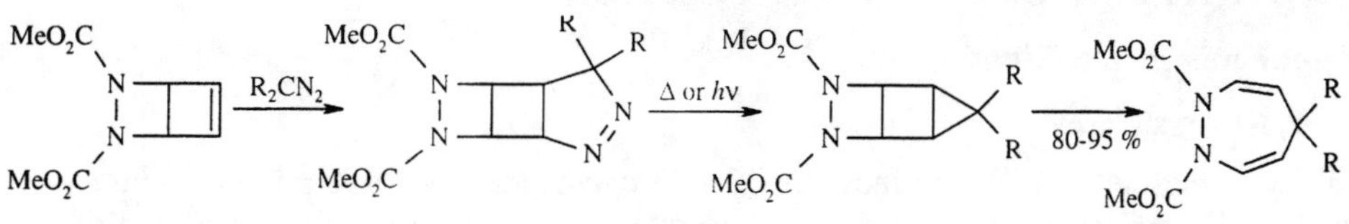

(134) **(135)** **(136)**

R=H, Me, Ph

Reactions of hydrazine and methylhydrazine with pyrylium or thiinium salts, *e.g.* (**137**), provide major routes to 4*H*-1,2-diazepines (**138**) and 1*H*-1,2-diazepines (**139**) ⟨76H(4)1509, 80CJC494⟩.

(**137**)   (**138**)   (**139**)

4.3.2.5.2 1,2-Oxazepines and 1,2-thiazepines

The Meisenheimer rearrangement of *tert*-amine *N*-oxides has been applied to the synthesis of monocyclic 1,2-oxazepines, *e.g.* (**140**) ⟨82H(19)173⟩.

The cyclic sulfonamide (**141**) can be prepared by heating 5-aminopentanesulfonyl chloride. The ketone HC≡CC(O)CH=CHSCN reacts with amines to give the 1,2-thiazepin-5-one system (e.g. **142**) ⟨61CB1606⟩.

(**140**)   (**141**)   (**142**)

4.3.2.5.3 1,2-Dioxepins and 1,2-dithiepins

The monocyclic saturated peroxides (**144**; R = H, Me) have been prepared by the treatment of the hydroperoxides (**143**) with LTA ⟨81S633⟩.

PhCH(R)(CH$_2$)$_3$CMe$_2$OOH $\xrightarrow{\text{LTA}}$

(**143**)   (**144**)   (**145**)

Simple 1,2-dithiepanes are prepared by the oxidation of α,ω-alkanedithiols using hydrogen peroxide, iodine or oxygen. Another useful general route involves the reaction of 1,5-dibromo compounds with sodium disulfide or tetrathiol metalates (MS$_4^{2-}$) of molybdenum and tungsten; thus, the reaction of tetrathiotungstate with 1,5-dibromopentane afforded 1,2-dithiepane (**145**) in 50% yield ⟨92JOC1899⟩.

4.3.3 TWO HETEROATOMS IN THE 1,3-POSITIONS

4.3.3.1 Four-membered Rings

4.3.3.1.1 1,3-Diazetidines

General syntheses of 1,3-diazetidines are [2+2] cycloadditions of C=N-containing substrates (Scheme 26). Dimerization of imines and isocyanates furnishes substituted 1,3-diazetidines. Cycloaddition of isocyanates to carbodiimides and isocyanates with imines are alternative syntheses, as is cycloaddition of iminophosphoranes with carbodiimides and isocyanates ⟨86JCS(P1)2037, 90JCS(P1)1859⟩.

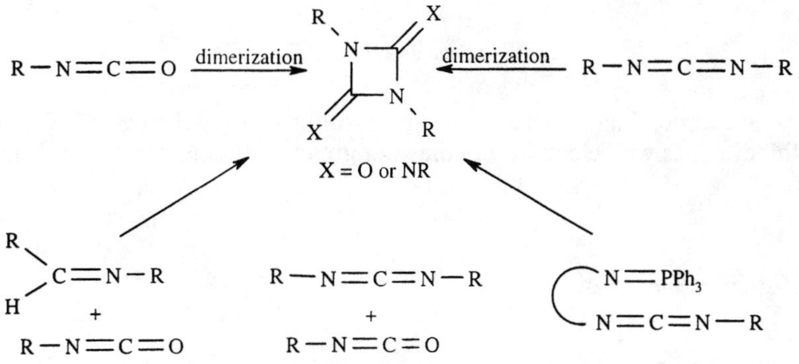

Scheme 26

4.3.3.1.2 1,3-Oxazetidines

[2+2] Cycloadditions are the most common method to generate 1,3-oxazetidin-2-ones ⟨68JOC3088⟩, *e.g.* (**146**) + (**147**) → (**148**).

1,3-Oxazetidin-2-ones (**151**) have been prepared by the cyclization of N-substituted carbamic acid derivatives (**150**), formed *in situ* from isocyanates (**149**) ⟨80S571⟩.

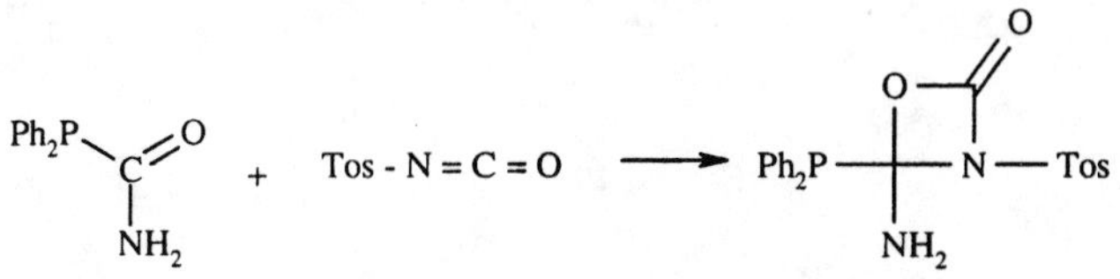

(**146**) (**147**) (**148**)

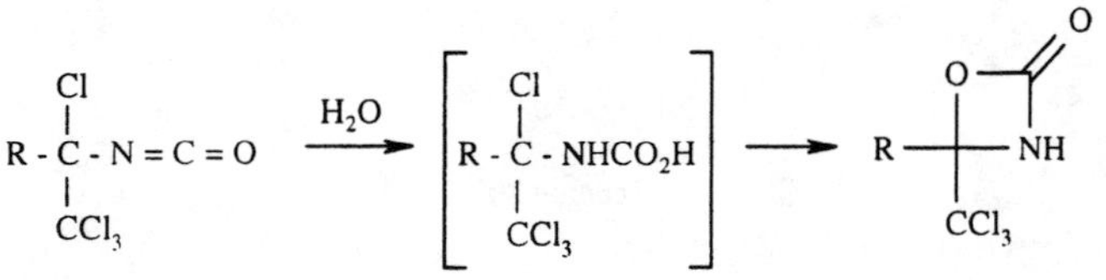

(**149**) (**150**) (**151**)

4.3.3.1.3 1,3-Thiazetidines

1,3-Thiazetidine rings are obtained by cycloaddition of thioamides with a carbon bearing two displaceable geminal halogens. Thus, *N*-benzoyl-*N'*-arylthioureas (**152**) with diiodomethane give the corresponding 2-benzoylimino-1,3-thiazetidine (**153**) in high yield ⟨91JHC177⟩.

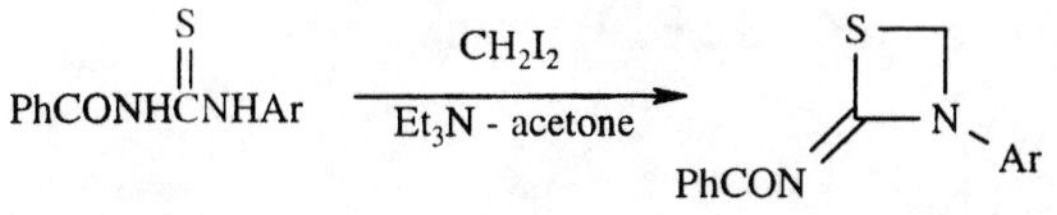

(**152**) (**153**)

Thioketene dimers crack at high temperatures into monomers, which then undergo [2+2] cycloadditions (Scheme 27) ⟨70JOC3470⟩.

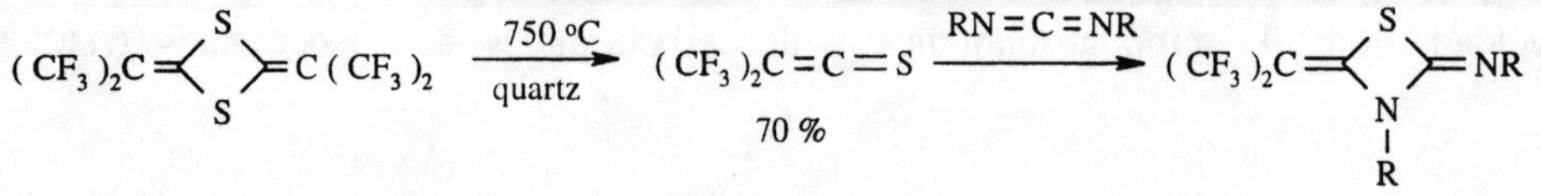

$$(CF_3)_2C \diamondsuit C (CF_3)_2 \xrightarrow[\text{quartz}]{750 \,^\circ\text{C}} (CF_3)_2C = C = S \xrightarrow{RN=C=NR} (CF_3)_2C \diamondsuit NR$$

70 %

Scheme 27

Isothiocyanates add carbodiimides across the carbon-sulfur bond (Scheme 28). The reaction works best for aryl isothiocyanates with electron-withdrawing groups ⟨75JCS(P2)1475⟩.

$$C_6H_{11}N = C = NC_6H_{11} \quad + \quad ArNCS \quad \longrightarrow$$

Scheme 28

4.3.3.1.4 1,3-Dithietanes

Active methylene compounds with carbon disulfide and base form reactive salts which undergo [3 + 1] additions to a variety of alkylating agents, even *gem*-dihaloethylenes (Scheme 29) ⟨77CC207, 80S907⟩. The salts can be oxidized to symmetrical 'desaurin' derivatives (Scheme 30) ⟨62CB2861⟩.

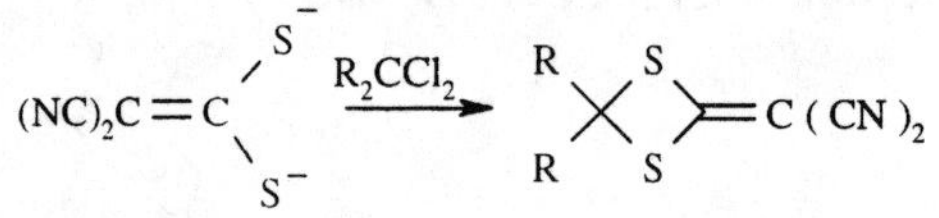

$$(NC)_2C = C \diamondsuit \xrightarrow{R_2CCl_2} \diamondsuit = C (CN)_2$$

Scheme 29

$$X_2CH_2 \xrightarrow[\text{CS}_2]{\text{base}} X_2C = C \diamondsuit \xrightarrow{[O]} X_2C \diamondsuit CX_2$$

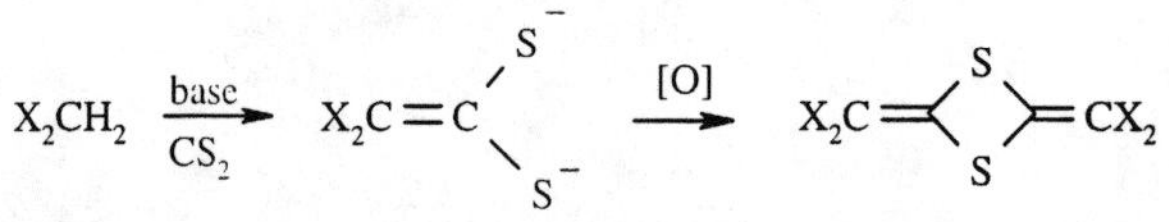

X = CO$_2$R or CN

Scheme 30

"Head-to-tail" dimerization of thiocarbonyl compounds to 1,3-dithietanes (Equation (2)) occurs thermally ⟨77JOC2345, 782289⟩ and photochemically ⟨73BCJ2253, 79CC899, 81JOC3911⟩, or catalysed by base ⟨65JOC1375, 73T2759⟩, a sulfonic acid ⟨70CJC3530⟩, or a complex of MnO ⟨82JOM(224)C31⟩. Diaryl thioketones are resistant to dimerization.

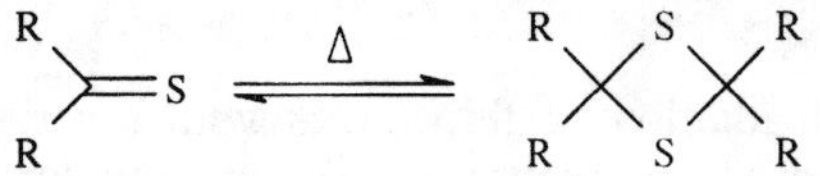

$$\underset{R}{\overset{R}{>}}=S \quad \underset{\longleftarrow}{\overset{\Delta}{\rightleftharpoons}} \quad \diamondsuit \qquad (+\text{ trimers + polymers}) \qquad (2)$$

Some thiones dimerize to 1,3-dithietanes ⟨75BSF(2)1670⟩. Thioketenes dimerize spontaneously; α-keto thioketenes give 'desaurins' (*cf.* Scheme 31) ⟨76BAU1913⟩.

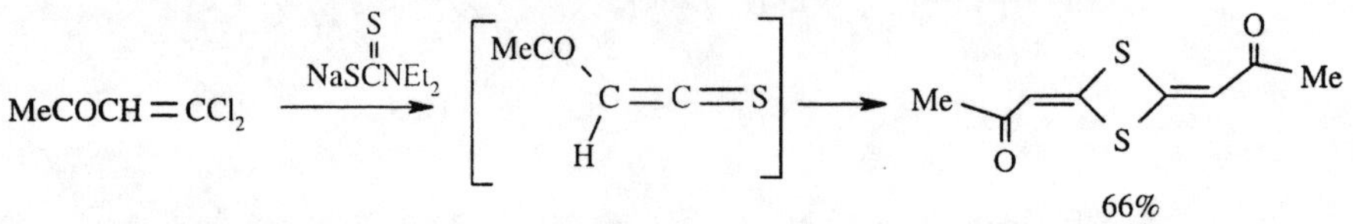

$$MeCOCH = CCl_2 \xrightarrow{NaSCNEt_2} \left[\underset{H}{\overset{MeCO}{>}}C = C = S \right] \longrightarrow$$

66%

Scheme 31

The parent compound 1,3-dithietane was successfully prepared according to Equation (3) ⟨76JA5715, 82JA3119⟩.

(3)

4.3.3.2 Five-membered Rings: Imidazoles, Oxazoles, Thiazoles, Dithiolium Salts and Derivatives

4.3.3.2.1 Overview

We consider successively the synthesis of fully conjugated derivatives by ring closure of type (**154**; Section 4.3.3.2.2), (**155**; Section 4.3.3.2.3) and (**156**; Section 4.3.3.2.4). This is followed by a consideration of methods involving C—C bond formation and/or 1,3-dipolar cycloadditions (Section 4.3.3.2.5), and syntheses of oxo-containing and reduced rings from acyclic precursors (Section 4.3.3.2.6). Finally transformations from other heterocycles are described (Section 4.3.3.2.7).

(154)   (155)   (156)

4.3.3.2.2 Synthesis from C_2 + ZCZ' components

α-Halo ketones react with amides (100°C, no solvent), thioamides (reflux in EtOH) and amidines to give oxazoles, thiazoles and imidazoles (**157** → **158**; Z = O, S, NH), respectively. This is the most important thiazole synthesis, and both the thioamide and the halo ketone components can be varied widely (see CHEC 4.19). Intermediates of type (**159**) can sometimes be isolated.

Guanidines with α-halo ketones form 2-aminoimidazoles. α-Hydroxy ketones also react with amidines to form imidazoles, and a variety of substituents can be introduced into the imidazole nucleus (CHEC 4.08).

The reaction of α-halo ketones with primary amides is appropriate for oxazoles containing one or more aryl groups. Formamide may be used resulting in a free 2-position in the oxazole. Ureas form 2-aminooxazoles (*cf.* CHEC 4.18).

α-Halo ketones react with thioacids to form 1,3-dithiolylium salts (**160**) which are also obtained from α-dimercaptoethylenes (**161** → **162**; *cf.* CHEC 4.32).

(157)   (158)   (159)

(160)   (161)   (162)

α-Haloacyl halides are used for the synthesis of mesoionic rings. The secondary thioamide (**163**) with α-bromophenylacetyl chloride (**164**) gave the 4-oxidothiazolium hydroxide (**165**). Similarly substituted amidines and dithionic acids with the same reagents formed the corresponding imidazolium (**166**) and dithiolylium (**167**) mesoionic systems ⟨77JOC1633, 77JOC1639⟩.

(**163**)   (**164**)   (**165**)   (**166**)

(**167**)   (**168**)   (**169**)

Replacement of the α-halo ketone by a cyano analogue can lead to amino derivatives. The benzenesulfonyl ester of mandelonitrile (**168**) with a primary thioamide, $RCSNH_2$ gives the 4-aminothiazole (**169**).

4.3.3.2.3   Synthesis of imidazoles, oxazoles and thiazoles from acylamino ketones

α-Acylamino ketones on heating with ammonium acetate are converted into imidazoles. 2,4,5-Triarylimidazoles (**171**) were prepared in this way from (**170**) ⟨73CB2415⟩, and the reaction is capable of numerous variations.

Oxazoles may be similarly prepared in good yields. Thus, 5-ethoxy-4-methyloxazole (**173**) was obtained by treating ethyl 2-formamide propionate (**172**) with phosphorus pentoxide in chloroform at 55°C ⟨72JCS(P1)909,914⟩. Known collectively as the Robinson-Gabriel synthesis, these cyclodehydrations can be effected by sulfuric acid or anhydrous hydrogen fluoride (*cf.* CHEC 4.18).

(**170**)   (**171**)   (**172**)   (**173**)

(**174**)   (**175**)   (**176**)

α-Acylamino ketones also provide a convenient synthesis of thiazoles on treatment with phosphorus pentasulfide (Gabriel's method). Substituents are usually restricted to alkyl, aryl and alkoxy derivatives. Thus, the α-acylamino ketone (**170**) with P_4S_{10} gave the thiazole (**174**).

In a related reaction α-amino ketones (**175**) with iminoesters RC(NH)OMe give imidazoles (**175** → **176**).

4.3.3.2.4   Other syntheses of imidazoles, oxazoles, thiazoles, dithiolyliums and oxathiolyliums by cyclization of C₂ZCZ′ components

Reactions of this type include the following:
(i) α-Acylamino Schiffs bases (**177**) with phosphoryl chloride give the 1-substituted imidazoles (**178**) ⟨78LA1916⟩.

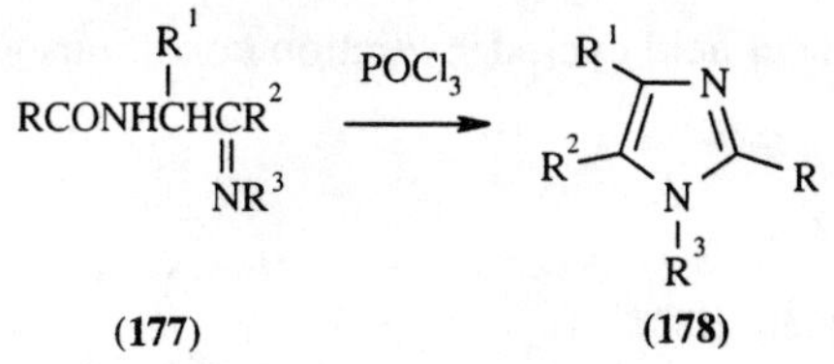

(177) → (178)

Oxidative cyclizations of Schiff bases (179) of 1,2-diaminomaleonitrile are conveniently accomplished using NCS under basic conditions ⟨84S1057⟩.

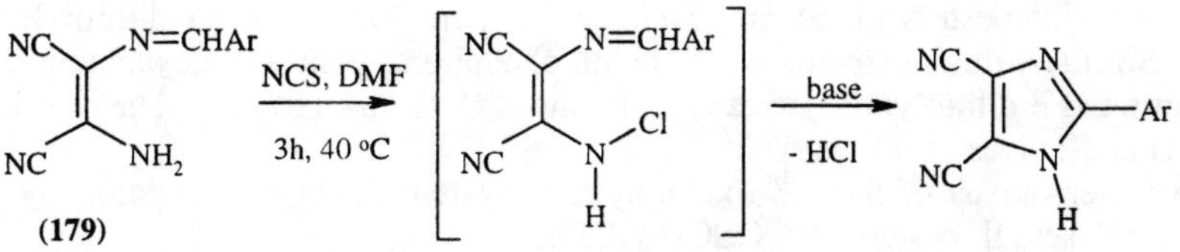

(179)

Reactions of 1,2-diaminoalkenes with alcohols, aldehydes, or carboxylic acids and derivatives at high temperatures in the presence of a dehydrogenating agent are common approaches to the synthesis of 2-substituted imidazoles ⟨89KFZ1246⟩. In the absence of the dehydrogenating agent the products are imidazolines.

(ii) The dimethylsulfonium ylide and aroyl isocyanate addition product (180) on heating cyclized to the 4-hydroxyoxazole (181) ⟨73T1983⟩.

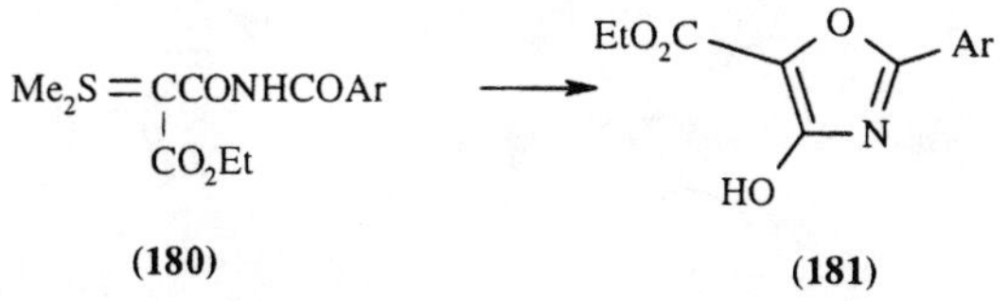

$$Me_2S = CCONHCOAr$$
$$|$$
$$CO_2Et$$

(180)   (181)

The Cornforth synthesis of oxazoles is useful for the preparation of 4-methoxycarbonyl derivatives, which are found in many natural products ⟨92CC1240⟩. In this method, a glycine imidate is reacted with a strong base and methyl formate to give an enolate that cyclizes upon treatment with acetic acid (Scheme 32).

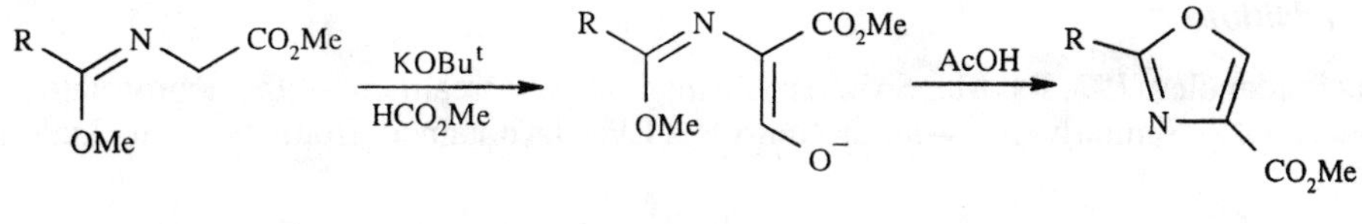

Scheme 32

Another synthesis of 2,5-di- and 2,4,5-trisubstituted oxazoles begins with β-(acyloxy)vinyl azides (182) ⟨89JOC431⟩. An intramolecular aza-Wittig reaction ensues upon treatment of the azides with phosphorus(III) reagents (Scheme 33). Yields of oxazoles, including furyl- and pyridyl-substituted oxazoles, which are difficult to obtain by routes that use strong dehydrating conditions, range from 48–93% when P(OEt)₃ is used.

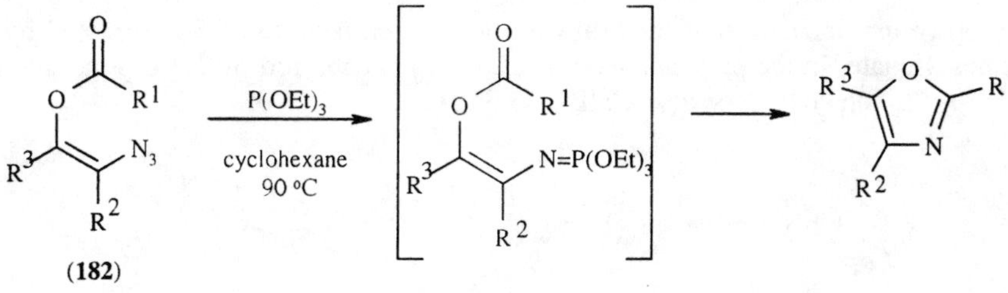

(182)

Scheme 33

(iii) Ring closure of (183) under acid cyclodehydration conditions gave the mesoionic 5-oxidothiazolium system (184).

(183)   (184)

(iv) α-Oxoalkyl dithioesters (185) are cyclized by perchloric acid to dithiolylium salts (186) ⟨80AHC(27)151⟩. Similarly dithiocarbamate (187) with phosphorus pentasulfide and tetrafluoroboric acid gives the 2-amino-1,3-dithiolylium tetrafluoroborate (188) ⟨69CPB1924⟩. The dithiol (189) with $Cl_2C=X$ (X = O or S) gives (190) ⟨76S489⟩.

(v) 2,5-Diaryl derivatives of the 1,3-oxathiolylium system (191) are prepared by acid-catalyzed cyclization of the β-keto thioesters, $ArCOSCH_2COPh$.

(185)   (186)   (187)   (188)

(189)   (190)   (191)

4.3.3.2.5 Synthesis of imidazoles, oxazoles and thiazoles by C—C bond formation or 1,3-dipolar addition

4-Aminoimidazoles (193; R = Me, SMe) are formed on base treatment of the appropriate precursors (192) ⟨75HCA2192⟩. Similarly the 4-aminothiazole (195) is obtained from the cyanoamidine (194) ⟨73JPR497⟩.

(192)   (193)   (194)   (195)

Decomposition of the diazoacetic ester (196) to the keto carbene (197) is promoted by copper(II) trifluoromethanesulfonate. In the presence of nitriles, (197) is captured by 1,3-dipolar addition giving the oxazole (198) ⟨75JOM(88)115⟩ (see also CHEC 4.03.8.1).

(196)   (197)   (198)

4.3.3.2.6 Synthesis of azolinones and reduced rings from acyclic precursors

(i) β-Hydroxy-, β-amino- and β-mercapto-acylamines ($HZCH_2CH_2NHCOR$; $Z = O$, NH, S) cyclize to give Δ^2-oxazolines, Δ^2-imidazolines and Δ^2-thiazolines (**199**).

(ii) 1,2-Difunctional ethanes (**201**; Z, $Z' = O$, S, NH) react with carbonyl chloride and carbonate esters to give 2-oxazolidinones and analogous derivatives (**201 → 200**).

(iii) The 1,3-dithiol-2-one (**203**) is obtained by ring closure of $RCSCHR^1SCSOR$ ⟨76S489⟩ and treatment of $RCOCH = CR^1NHR^2$ with thiocyanogen effects cyclization to (**204**).

(iv) 1,2-Difunctional ethanes (**201**; Z, $Z' = O$, S, NH) react with aldehydes and ketones to form oxazolidines, *e.g.* (**201 → 202**). Such reactions are used extensively to protect *cis*-hydroxy groups (*e.g.* sugars + acetone → isopropylidene sugars) and carbonyl groups (*e.g.* steroidal ketones + ethylene glycol → ethylene ketals).

(199)   (200)   (201)   (202)

(203)   (204)

(v) α-Acylaminocarboxylic acids are converted into 5(4H)-oxazolinones by acid anhydrides (**205 → 206**). In an extension of this reaction, N-acyl derivatives of glycine (**207**) react with aldehydes with concomitant cyclization to give azlactones (**208**); this is the basis of the Erlenmeyer synthesis of amino acids. Treatment of $ROCSNHCHR^1CO_2SiMe_3$ with PBr_3 affords the thiazolidine-2,5-dione (**209**) ⟨71CB3146⟩.

(205)   (206)   (207)   (208)   (209)

(vi) Nitrogen, oxygen and sulfur nucleophiles can add to unsaturated carbon-carbon systems. This synthetic approach is illustrated by the reaction of the propargylamine (**212**) with carbon disulfide. The intermediate dithiocarbamic acid (**213**) cyclizes to the thiazole (**214**) ⟨49JCS786⟩. The NH group of (**212**) adds isocyanates to give ureas (**211**) which are converted by sodium methoxide into 4-methylene-2-imidazolinones (**210**) ⟨63JOC991⟩. N-Allyl-amides, -urethanes, -ureas and -thioureas undergo intramolecular cyclization in 60–96% sulfuric acid to give 2-oxazolines and 2-thiazolines as illustrated by the conversion of N-2-phenylallylacetamide (**215**) into 2,5-dimethyl-5-phenyl-2-oxazoline (**216**) ⟨70JOC3768⟩ (see also CHEC 4.19).

(210)   (211)   (212)   (213)   (214)

(vii) Reaction of RNHCSSH with oxalyl chloride gives the thiazolidine-4,5-dione (**217**) (see CHEC 4.19), and the same reagent with N-alkylbenzamidine (**218**) at 100–140°C formed the 1-alkyl-2-phenylimidazole-4,5-dione (**219**; see CHEC 4.08). Iminochlorides of oxalic acid react with N,N-disubstituted thioureas to give the 2-dialkylaminothiazolidine-4,5-dione bis-imides (**220**).

Phenyliminooxalic acid dichloride likewise yielded thiazolidine derivatives on reaction with thioureas ⟨71KGS471⟩.

(215)   (216)   (217)

(218)   (219)   (220)

4.3.3.2.7 Synthesis from heterocycles

Aziridines (221; Z = O, S, NR) undergo a facile ring opening and subsequent closure on heating with sodium iodide in acetonitrile to give, for example, the oxazoline (222; Z = O). This aziridine ring opening reaction is a particularly attractive route to imidazolines and ring fused imidazolines (222; Z = NR) ⟨62JOC2943⟩ (see Section 3.5.2.2).

Aziridines also undergo ring enlargement on treatment with thiocyanic acid: *cis-* and *trans-*2,3-dimethylaziridines (223) thus gave *trans-* and *cis-*2-amino-4,5-dimethyl-2-thiazolines (224) stereospecifically ⟨72JOC4401⟩.

(221)   (222)   (223)   (224)

Thermolysis of the aziridine (225) with diphenylketene gave the pyrrolidone (226; minor product) and the oxazolidine (227; major product).

(225)   (226)   (227) X = CPh$_2$
                                                                     (228) X = H , CHPh$_2$

The preferential addition to the C=O bond is explained in terms of steric effects ⟨72CC199⟩. Similar addition to diphenylacetaldehyde takes place with the same orientation to give the oxazolidine (228).

The thermal reaction of 2-phenyl-1-azirine (229) with carbon disulfide followed by methylation gave 2-methylthio-5-phenylthiazole (230).

(229)   (230)

Photoisomerization of pyrazoles, isoxazoles and isothiazoles into imidazoles, oxazoles and thiazoles, respectively, is described in Section 3.4.1.2.4.

Catalytic reduction of iminoethers derived from 4-aminoisoxazoles ⟨85TL3423, 87JOC2714⟩ or 4-amino-5(4*H*)-isoxazolones ⟨91S127⟩ leads to α-(acylamino)enaminones (**231**) which cyclize in the presence of bases to form 4-acylimidazoles (Scheme 34). Treatment of the enaminone with a primary amine incorporates a substituted amino function in the β-position with concomitant expulsion of ammonia and allows access to 1-substituted and 1,2-disubstituted 4-acylimidazoles in 75–92% yields ⟨87JOC2714⟩.

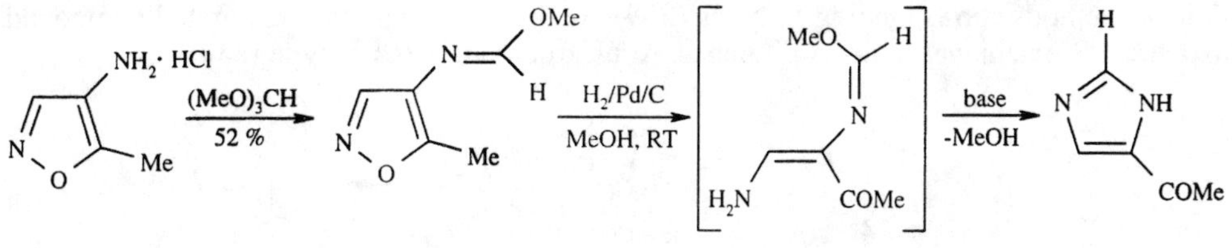

Scheme 34

The oxazole (**232**) heated with formamide gave the imidazole (**233**); oxazolium cations undergo similar conversions. Primary amines convert oxazole-4-carboxylic acids (**234**) at 150°C into imidazoles (**235**) with accompanying decarboxylation ⟨53CB88⟩ (see CHEC 4.07 and 4.18).

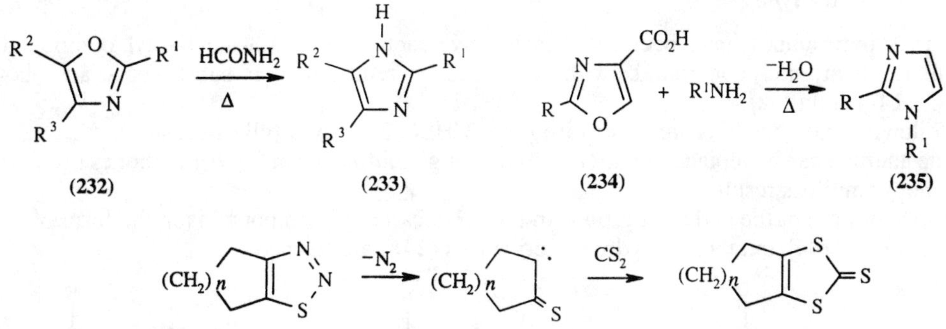

A widely applicable route to substituted imidazoles is photochemical degradation of 1-alkenyltetrazoles (Scheme 34a) ⟨82CC714, 83CC1082⟩. Evolution of N₂ gives *N*-vinylimidoyl nitrenes which cyclize to the 4*H*-imidazoles in good to moderate yields ⟨87JCS(P1)1389, 91JCS(P1)335⟩. When R³ = H, a 1,5-shift in 4*H*-imidazoles results in aromatic imidazoles ⟨84JCS(P1)1933, 85JCS(P1)741⟩. When the vinyl double bond in the starting tetrazole is part of a benzene ring the products are benzimidazoles and carbodiimides ⟨84CHEC-(5)791, 85JCS(P1)1471⟩.

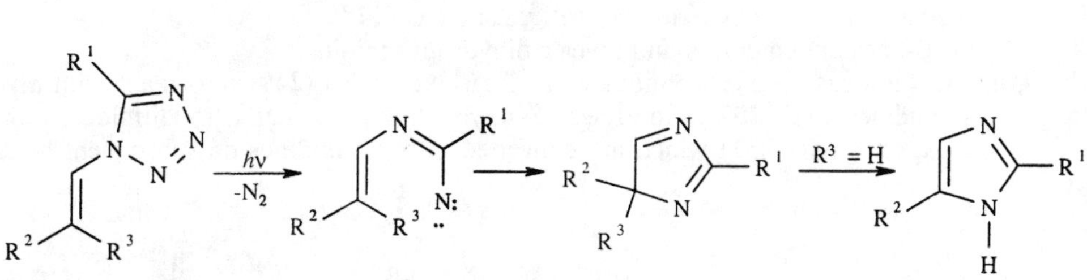

Scheme 34a

Thermolysis of the 1,2,3-thiadiazoles (**236**) in the presence of carbon disulfide leads *via* the thiocarbonyl carbene (**237**) to the ring-fused 1,3-dithiole-2-thione (**238**) ⟨76JOC730⟩.

Photochemical transformations of 2-azidopyrazines into imidazoles are well known. The reaction appears to be more versatile and gives higher yields under thermolysis ⟨83JHC1277, 90JHC711⟩ (see Section 3.2.3.6.4).

4.3.3.3 Six-membered Rings

There are two major routes to six-membered rings, (**239**) and (**240**). For the preparation of pyrimidines, methods corresponding to types (**239**) are the most important. Saturated compounds, *i.e.* 1,3-dioxanes, 1,3-oxathianes and 1,3-dithianes, result from syntheses of type (**240**).

(**239**)   (**240**)   (**241**)   (**242**)

R, R' = H, Me, Ph, CO₂Et;
Y = H, Me, Ph, Br, NO, NO₂
Y' = H, Me, Ph, OMe, SMe, NH₂
Basic catalyst, *e.g.* NaOEt-EtOH

4.3.3.3.1 C₃ + ZCZ type

Numerous pyrimidines have been synthesized by reaction of a 1,3-dicarbonyl compound, or a potential 1,3-dicarbonyl compound, with an amidine; representative substituents are shown in structures (**241**) and (**242**).

The following modifications are noteworthy (see CHEC 2.13 for a full discussion).

(i) The amidine can be replaced by urea, thiourea or guanidine when 2-pyrimidinones (**243**), thiones or 2-aminopyrimidines result.

(ii) If one or both of the carbonyl groups in the 1,3-dicarbonyl compound is in the form of an ester, 4-pyrimidinones (**244**) and their 6-hydroxy derivatives (**245**) are formed.

(**243**)   (**244**)   (**245**)   (**246**)   (**247**)   (**248**)

(iii) Replacement of one or both of the carbonyl groups by a cyano group leads to 4-amino- (**246**) or 4,6-diamino-pyrimidines.

(iv) If the central carbon atom of the carbonyl compound is tetrasubstituted, non-aromatic derivatives are produced, *e.g.* Et₂C(CO₂Et)₂ reacts with urea to yield veronal (**247**).

(v) Use of *N*-methoxyurea gives *N*-oxide derivatives such as (**248**).

(vi) Use of an α,β-unsaturated compound gives a dihydropyrimidine.

Amidines, ureas, thioureas, *S*-alkylisothioureas and carbodiimides (**249**) also react with diketene (**250**) to give pyrimidines (*e.g.* **252**). Amidines, *S*-alkylisothioureas and carbodiimides, however, initially form 1,3-oxazines (*e.g.* **251**) which are converted into pyrimidines on subsequent treatment with acid or base.

(**249**)   (**250**)   (**251**)   (**252**)

β-Chlorovinyl ketones (**253**) and the thioamide (**254**) in the presence of perchloric acid give intermediate thioimidium salts (**255**), which cyclize to yield 1,3-thiazinium salts (**256**) ⟨82T937⟩.

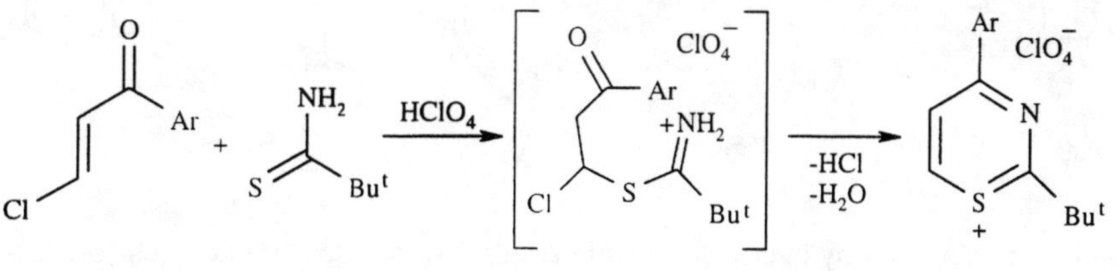

(**253**)   (**254**)   (**255**)   (**256**)

4.3.3.3.2   *ZC₃Z + C (5 + 1) and (6 + 0) Cyclizations*

The method ZC$_3$Z + C is used for the preparation of reduced pyrimidines, oxazines and thiazines as well as for dioxanes, dithianes and oxathianes as mentioned above (*e.g.* **258** → **257**, **259**; Z, Z' = NH, O, S). The Prins reaction yields 1,3-dioxanes ⟨77S661⟩; it involves the acid-catalyzed condensation of alkenes with aldehydes with 1,3-diols as intermediates.

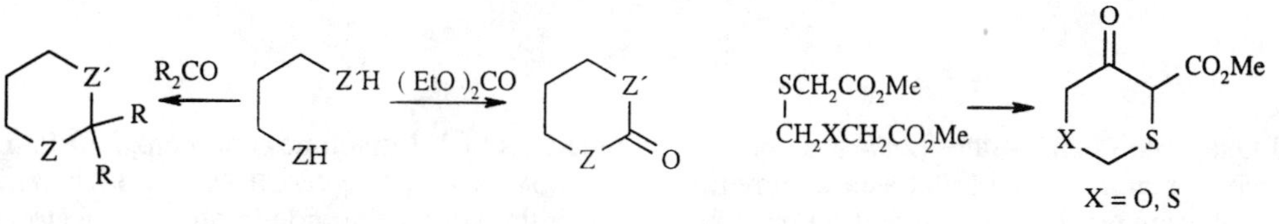

(**257**)   (**258**)   (**259**)   (**260**)   (**261**)

Dieckmann cyclization has been used for the synthesis of non-aromatic heterocycles of this type (**260** → **261**).

Routes to 4*H*-1,3-oxazines and -thiazines involve the cyclization of amides or thioamides with acidic reagents (Scheme 35) ⟨78AHC(23)1⟩. 1,3-Oxazin-2-ones can be made by the thermolysis of carbonyl azides (**262**; Scheme 36) ⟨79CC719⟩.

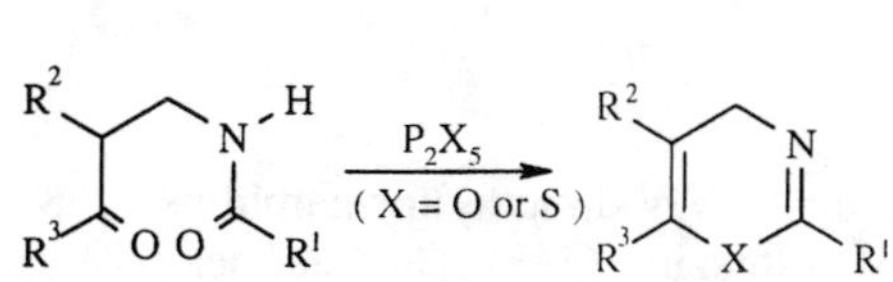

Scheme 35   Scheme 36

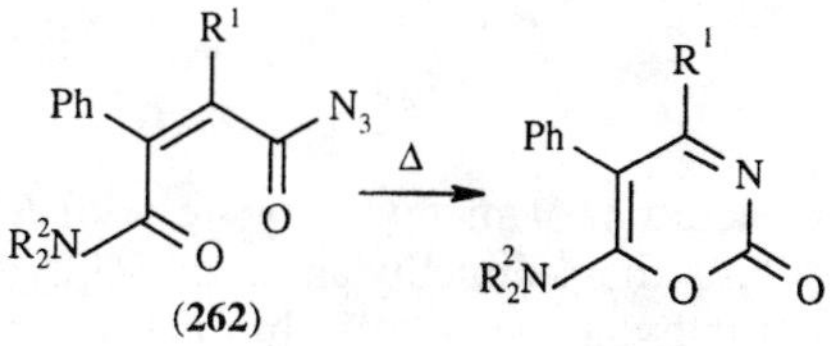

1,3-Oxazin-6-ones (**264**) are made by heating β-acylamino esters (**263**; R^1 = Ar, R^3 = OMe) ⟨74AG(E)533⟩.

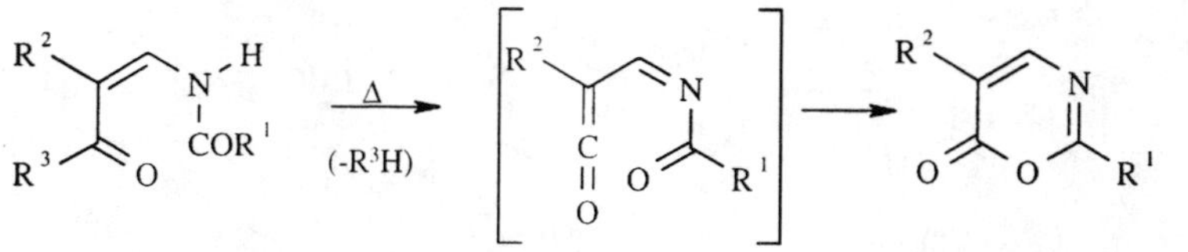

(**263**)   (**264**)

4.3.3.3.3   *[4 + 2] Cyclizations*

1,3-Dioxins (**266**) are obtained from the acid-catalyzed condensation of diketene (**265**) with ketones R^{1}R^{2}C=O.

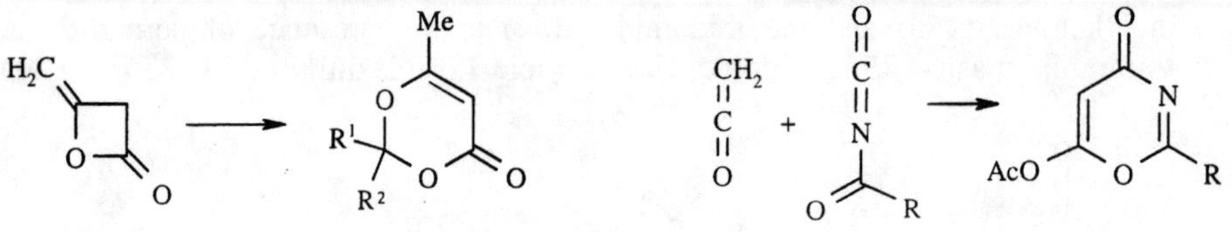

(265)					(266)					Scheme 37

Oxazin-4-ones are obtained by cycloadditions between isocyanates and ketenes (Scheme 37).

Routes to 1,3-oxazinium salts consist of 1,4-cycloadditions either between α,β-unsaturated β-chlorocarbonyl compounds and nitriles or between *N*-acylimidoyl chlorides and alkynes. Tin(IV) chloride is an effective catalyst for both reactions (*cf.* Scheme 38).

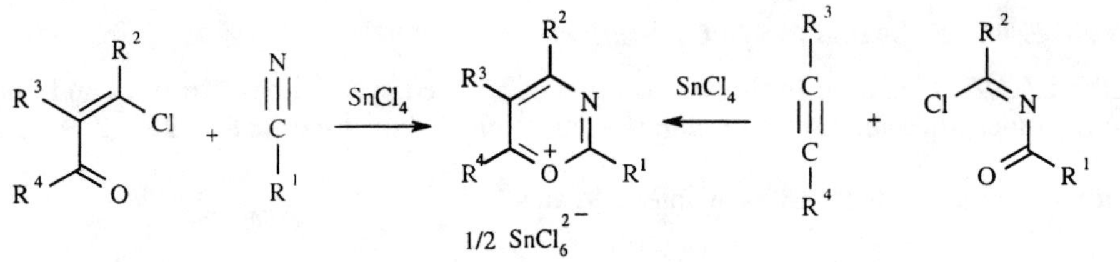

Scheme 38

1,3-Oxazinium perchlorates (**267**) are obtained by reactions of 1,3-diketones and benzonitrile in the presence of perchloric acid and acetic anhydride ⟨88ZOR1561, 88ZOR2232, 91ZOR1986⟩. 1,3-Thiazinium perchlorates are synthesized by treating oxazinium salts with hydrogen sulfide in absolute acetonitrile and then with perchloric acid ⟨72S333⟩.

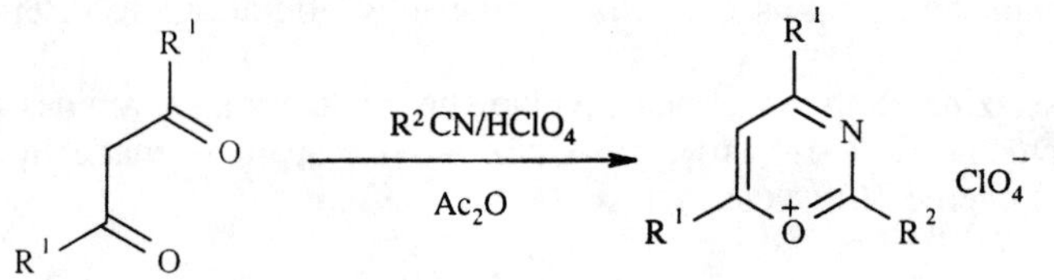

(267)

The cycloaddition of methyl acrylate with *N'*-thiobenzoyl-*N,N*-dimethylformamidine (**268**) under pressure affords the 4-dimethylamino-5,6-dihydro-4*H*-1,3-thiazine (**269**). On treatment with methyl iodide and triethylamine, (**269**) eliminates dimethylamine to give 6*H*-1,3-thiazine (**270**) which is obtained directly through the cycloaddition of methyl acrylate and the methiodide salt of the formamidine ⟨85JOC1545, 87SC1971⟩. The same methodology has been extended to other 6*H*-1,3-thiazines by varying the formamidine ⟨85SL139, 88SL205⟩, or electron-deficient cyclic dienophile.

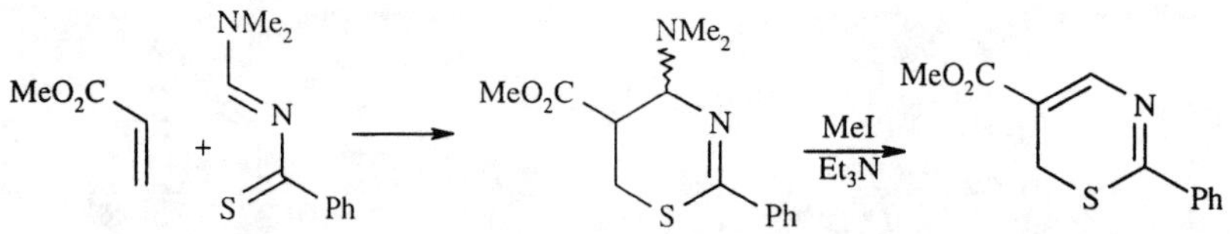

(268)					(269)					(270)

4.3.3.3.4 *Syntheses from heterocycles*

Kinetically stabilized azetes (**271**) cycloadd acceptor-substituted nitriles. The initially formed Dewar pyrimidine (**272**) is subsequently isomerized to the pyrimidine (**273**) ⟨90SL401⟩.

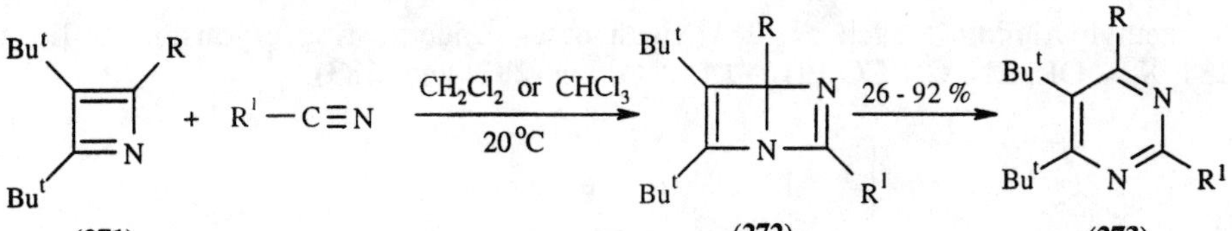

(271) **(272)** **(273)**

R = Bu^t, Mes
R = CO_2Et, CF_3, CN, *p*-Ts, C_6H_4-4-CF_3

2-Trimethylsilyloxyfuran (**274**) reacts with ethoxycarbonylnitrene to give 3-ethoxycarbonyl-3,4-dihydro-1,3-oxazin-2-one (**276**) *via* aziridine intermediate (**275**) ⟨89TL5025⟩.

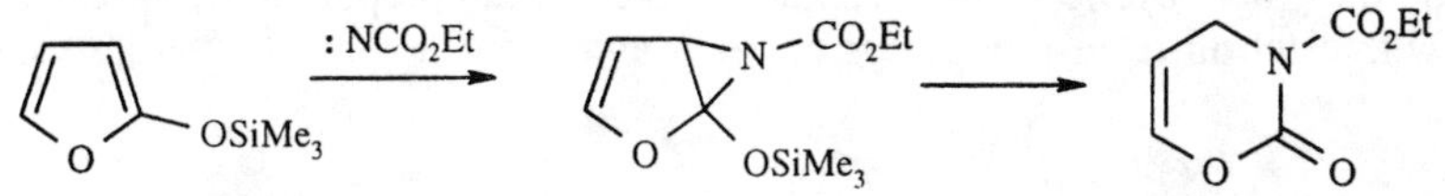

(274) **(275)** **(276)**

1,3-Oxazines and 1,3-thiazines are converted into pyrimidine derivatives by ammonia. For the conversion of 1,2,4-thiadiazoles into pyrimidines see Section 3.4.1.6.5.i.

4,6-Diaryl-1,2,3,5-oxathiadiazines (**277**; from sulfur trioxide and aryl isocyanates) with β-diketones yield pyrimidines (**278**). *s*-Triazine reacts with RCH_2CN to give 4-aminopyrimidines (**279**; see Section 3.2.1.6.1 for a similar reaction), and with electron-rich alkenes and alkynes to yield pyrimidines such as (**280**) from EtC≡CMe (Section 3.2.1.10.2).

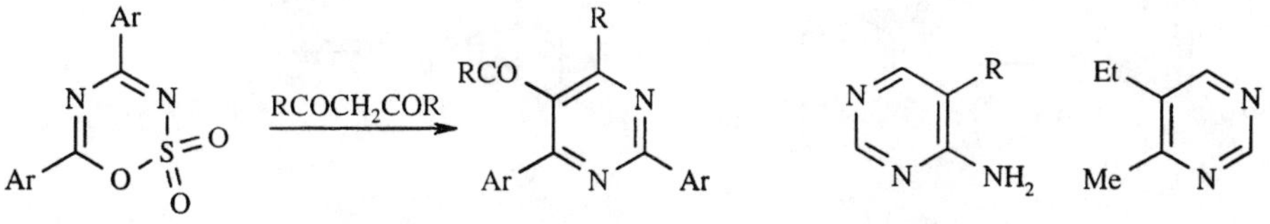

(277) **(278)** **(279)** **(280)**

1,3-Oxazin-6-ones are converted by amines into 4-pyrimidinones (Scheme 39) and the ANRORC reaction (*e.g.* Scheme 40) can be used to prepare pyrimidines from 2-bromopyridines (*cf.* Section 3.2.3.10.5).

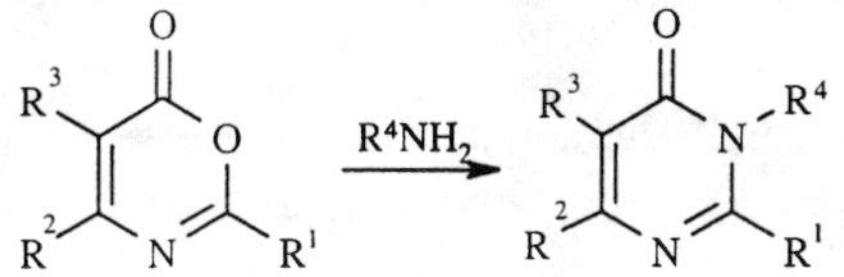

Scheme 39

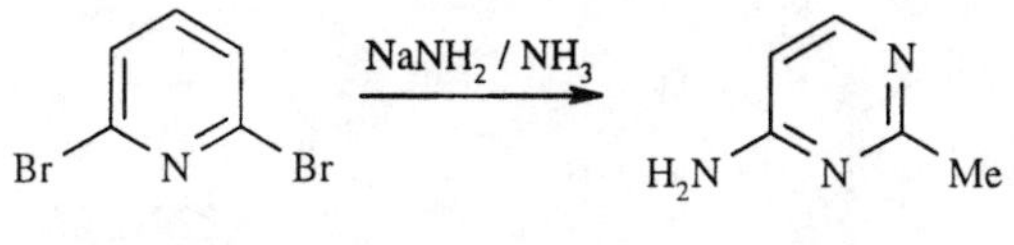

Scheme 40

4.3.3.4 Seven-membered Rings

4.3.3.4.1 1,3-Diazepines

The condensation of 1,4-diamines with a variety of carboxylic acid derivatives, *e.g.* imidate esters, orthoformic esters, *N*-ethoxycarbonylthioamides ⟨77JOC2530⟩, nitriles and ethoxyacetylene, produces the cyclic amidine linkage —N=C(R)NH— ⟨67AHC(8)21, p. 40⟩. Cyclic ureas, —NHC(O)NH—, have been similarly produced using carbonyl chloride, *N,N'*-carbonyldiimidazole, carbon monoxide, thiocarbonyl chloride or carbon disulfide ⟨67AHC(8)21, p. 38⟩.

4-Chloromethylpyrimidines such as (**281**), with bases, undergo ring expansion to 1,3-diazepin-2-ones (**284**; X = OR, CN, CH(CO₂Et)₂) ⟨77CJC895⟩ *via* (**282**) and (**283**).

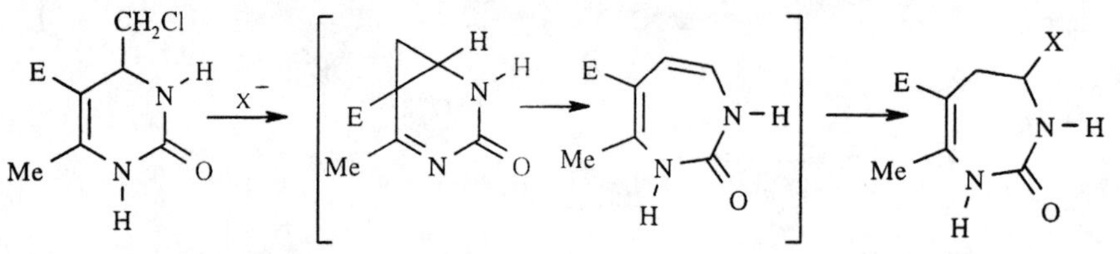

(**281**)   (**282**)   (**283**)   (**284**) E = CO₂R

Fully unsaturated monocyclic 1,3-diazepines have been prepared by thermolysis of 1*H*-*1,2*-diazepines with electron-releasing groups in the 4- and 6-positions ⟨79JOC2683, 80CC444⟩.

4.3.3.4.2 *1,3-Oxazepines and 1,3-thiazepines*

The photochemical rearrangement of aromatic *N*-oxides, *e.g.* (**285**), gives the fully unsaturated 1,3-oxazepine system, *e.g.* (**288**), *via* an oxaziridine intermediate (**286**) which rearranges by a 1,5-sigmatropic shift to (**287**) converted to the product by disrotatory ring opening ⟨76H(4)1391⟩. Similar oxaziridine intermediates appear to participate in thermolytic rearrangement of 1,4-oxazepine derivatives into 1,3-oxazepines ⟨86CC1188, 87CPB3166⟩.

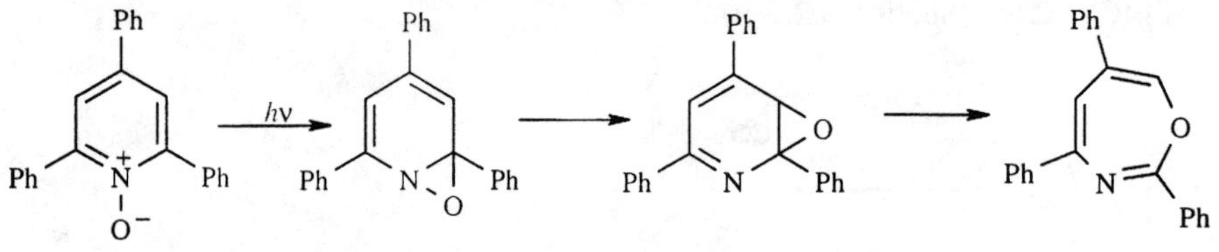

(**285**)   (**286**)   (**287**)   (**288**)

Monocyclic 1,3-oxazepines (**290**) can be prepared by reaction of aliphatic diazo compounds with 1,3-oxazinium perchlorates (**289**) ⟨74S187⟩. Tetra- and penta-phenyl-1,3-oxazepines (**292**; R = H or Ph) have been obtained by the reaction of azide ion with pyrylium salts (**291**) ⟨78H(11)331⟩.

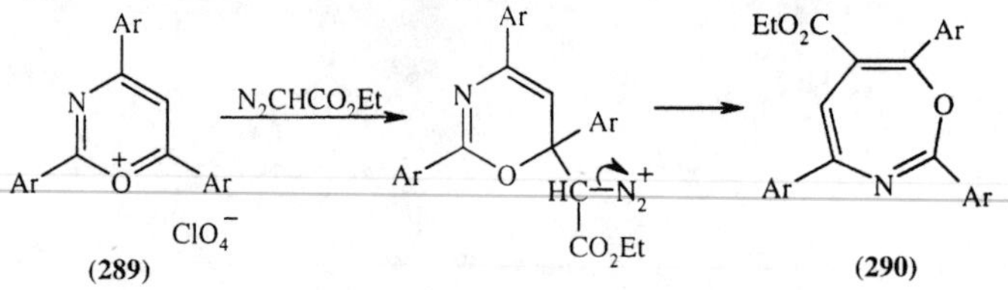

(**289**)   (**290**)

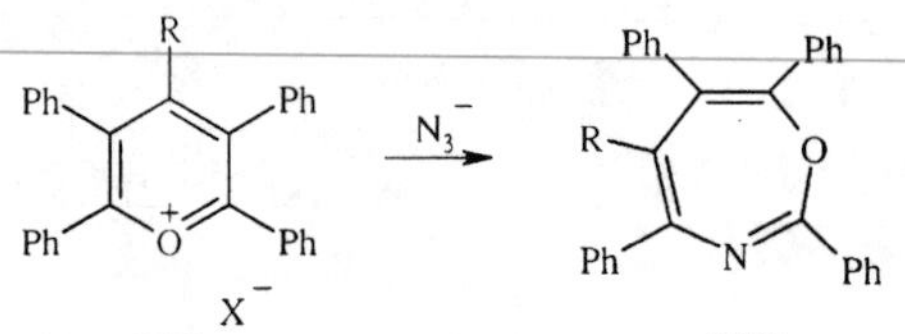

(**291**)   (**292**)

2-Aryl-1,3-oxazepines (**294**) can be made in generally very good yields by photolysis of the bicyclic derivatives (**293**) ⟨84NKK158⟩.

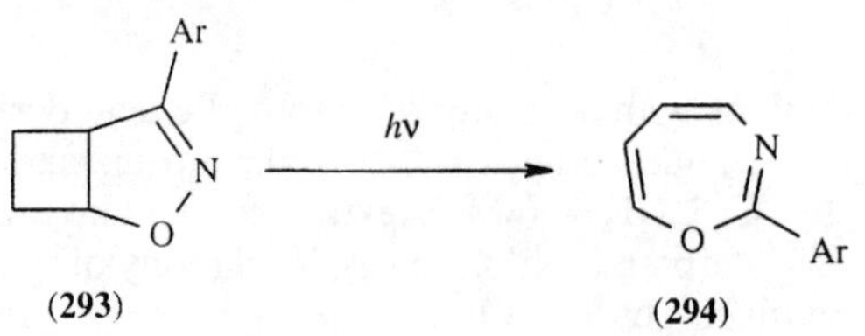

(**293**)   (**294**)

2-Vinylaziridine (**295**) with phenyl isothiocyanate or 4-chlorothiobenzoyl thioglycollate gives (**297**), presumably *via* the intermediate (**296**) ⟨71JOC3076⟩.

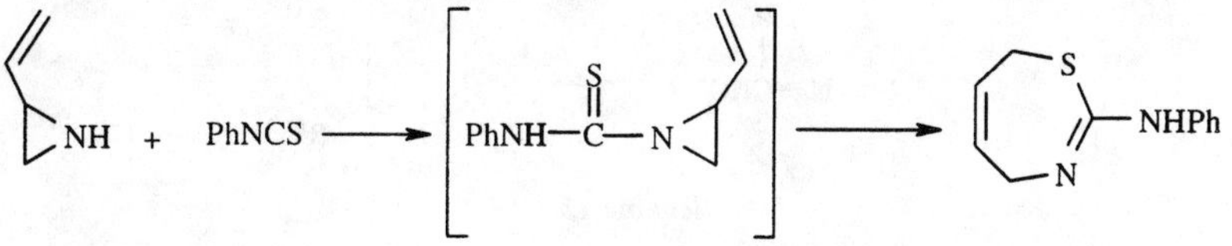

(**295**)   (**296**)   (**297**)

4.3.3.4.3 *1,3-Dioxepins and 1,3-dithiepins*

4,7-Dihydro-1,3-dioxepins (**298**) are prepared by the reaction of *cis*-butene-1,4-diols with aldehydes, and a similar route gave the dithia derivative (**299**) which was converted into the more unsaturated compound (**301**) *via* (**300**) ⟨76TL1251⟩.

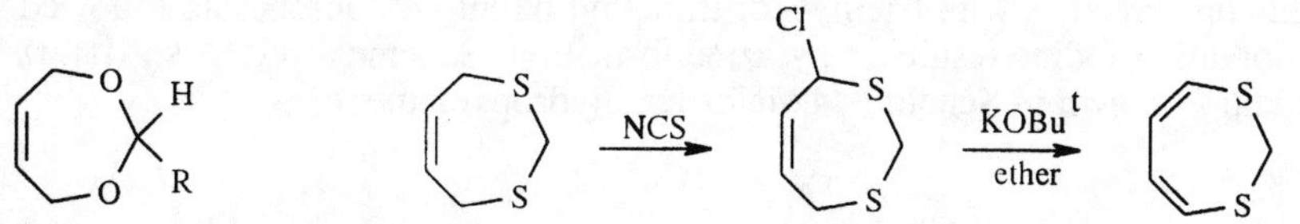

(**298**)   (**299**)   (**300**)   (**301**)

Fully saturated 1,3-dithiepins can be prepared by the Lewis acid-catalyzed exchange reactions of 1,4-butanedithiol with an appropriate diethyl acetal. 1,3-Dithiepan-2-one is obtained *via* the reaction of 1,4-dichlorobutane with alkali trithiocarbonates ⟨72HC(26)598, p. 616⟩.

4.3.4 TWO HETEROATOMS IN THE 1,4-POSITIONS

4.3.4.1 Six-membered Rings

4.3.4.1.1 *Pyrazines from acyclic compounds*

An important preparation of pyrazines (**303**) is from α-amino ketones $RCOCH_2NH_2$ or their monooximes which spontaneously condense to give 2,5-dihydropyrazines (**302**). The α-amino ketones are often prepared *in situ* by reduction of isonitroso ketones, and the dihydropyrazines are usually oxidized to pyrazines before isolation (*cf.* Section 3.2.2.3.3). Catalytic reduction of α-azido ketones also leads to 2,5-dihydropyrazines ⟨80OPP265⟩. Similarly, α-nitro ketones may be reduced to the α-amino ketones which dimerize spontaneously ⟨69USP3453278⟩.

The condensation of an α-diketone or its synthetic equivalent (*e.g.* aroyl cyanide ⟨87S914⟩) with a 1,2-diaminoalkane gives 2,3-dihydropyrazines (Scheme 41), which like their 2,5-analogues can be oxidized by air, or better by MnO_2 in ethanolic KOH, to pyrazines ⟨78MI21400⟩.

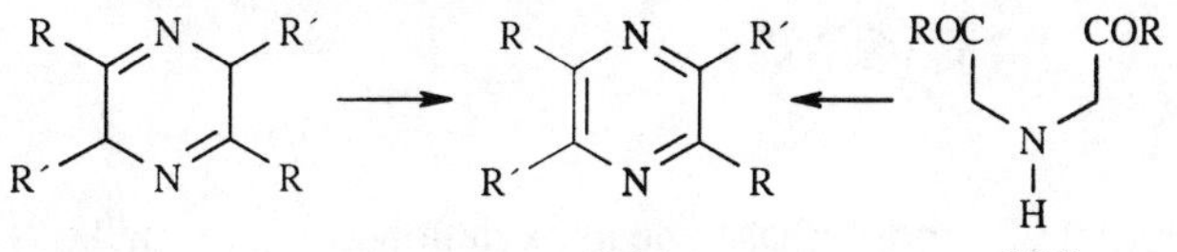

(**302**)   (**303**)   (**304**)

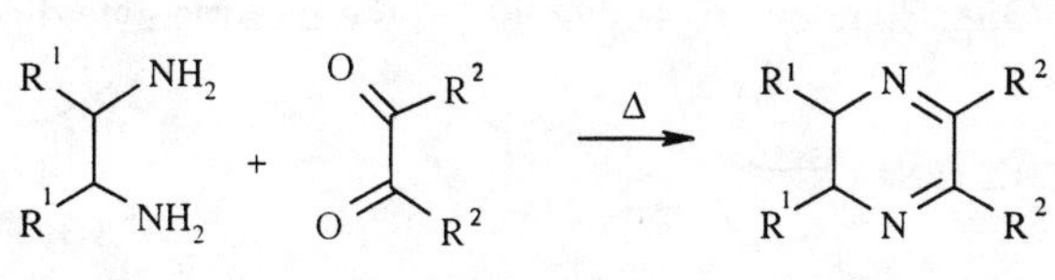

Scheme 41

Pyrazines are obtained by oxidative ring closure of bis(acylmethyl)amines (**304**) with ammonia.

Conversion of α-azido ketones to pyrazines by treatment with triphenylphosphine in benzene (Scheme 42) proceeds in moderate to good yields ⟨69LA(727)231⟩.

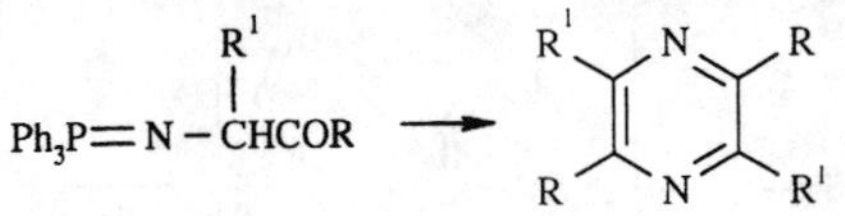

$$Ph_3P = N - CHCOR \longrightarrow$$

Scheme 42

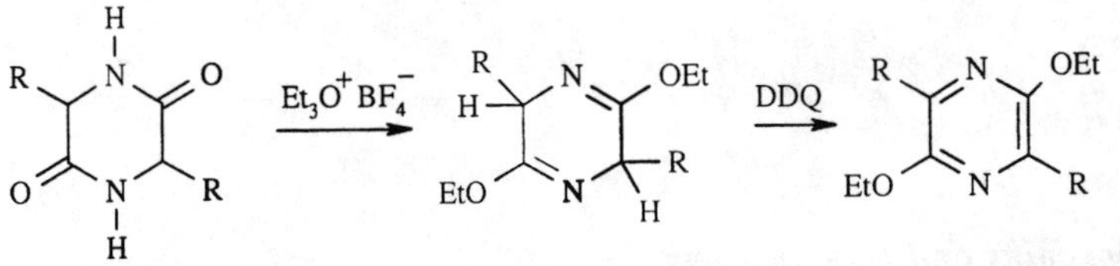

(305)

Scheme 43

Cyclodimerization of α-amino acids [RCH(NH$_2$)CO$_2$H] gives 2,5-dioxopiperazines (305). Treatment of 2,5-dioxopiperazines with triethyl- or trimethyl-oxonium fluoroborate followed by oxidation with DDQ, chloranil or iodine results in pyrazine formation (Scheme 43) ⟨71JCS(P1)2494⟩.

The cycloaddition shown in Scheme 44 yields tetrahydropyrazinones.

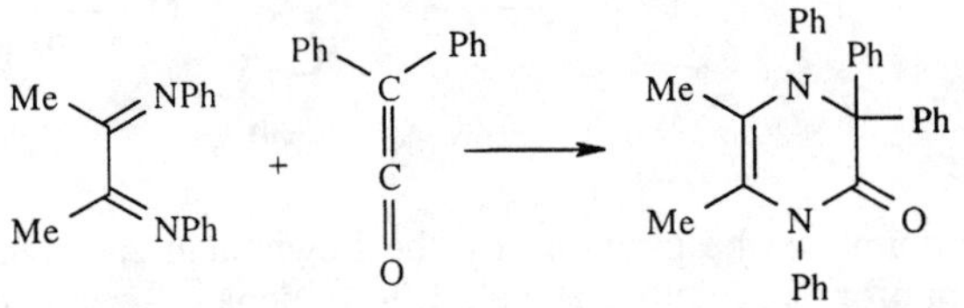

Scheme 44

4.3.4.1.2 *1,4-Dioxins, 1,4-dithiins, 1,4-oxazines and 1,4-thiazines*

A two-step synthesis of 1,4-dioxin from dioxane is *via* 2,3,5,6-tetrachloro-1,4-dioxane which is dechlorinated using magnesium and iodine ⟨39JA3020⟩. The route to 2,5-dimethyl-1,4-dioxin uses 2,5-diiodomethyl-1,4-dioxane (diepiiodohydrin) as precursor (Scheme 45) ⟨57JA6219⟩.

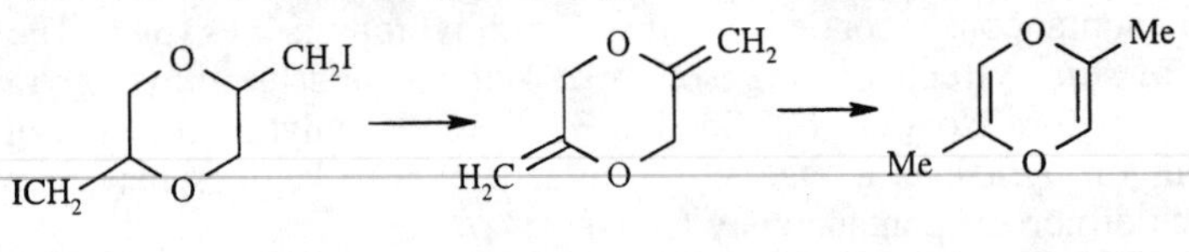

Scheme 45

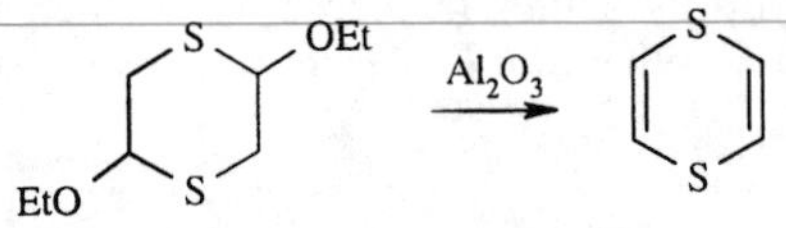

(306)   (307)

Dithiins are prepared by vapor phase dealkoxylation of 2,5-dialkoxy-1,4-dithianes (from HSCH$_2$CH(OEt)$_2$ + H$^+$) over alumina at 260–265°C (306 → 307). 2,5-Diaryl derivatives can be prepared from Bunte salts (308; Scheme 46) with acid *via* the intermediate diol (309).

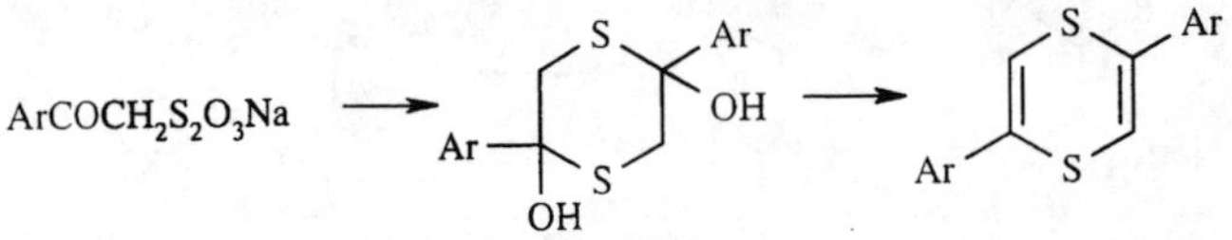

(308)   (309)

Scheme 46

The construction of the 1,4-dithiin ring system *via* Bunte salt formation is severely limited. This difficulty is, however, overcome by using β,β′-dioxodialkyl sulfides and phosphorus pentasulfide or Lawesson's reagent. Such treatment of the sulfides (**310**) (readily obtainable from α-haloketones and sodium sulfide) in refluxing toluene or benzene affords a variety of 1,4-dithiins (**311**) in good yields ⟨84H(22)1527, 89H(29)391, 91PS(57)227⟩. Another approach is based on cyclization between a two-atom and a four-atom fragment. This involves treating alkynes (**312**) with nickel bisdiphenyldithiolene (**313**) in refluxing chlorobenzene in the presence of pyridine. Pyridine appears to be essential in order to avoid further transformation of 1,4-dithiins (**311**) to thiophenes ⟨87SC1683⟩.

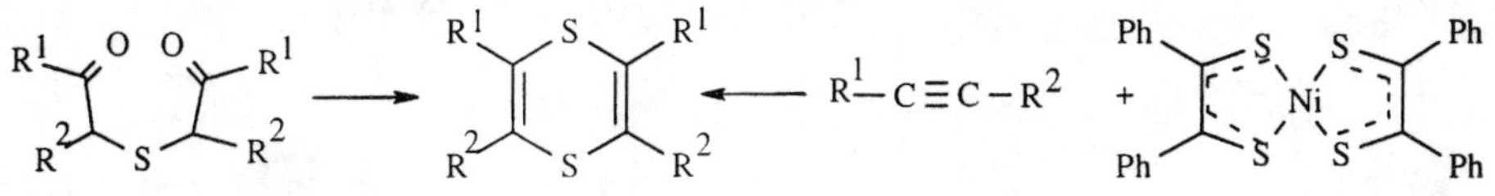

(310) (311) (312) (313)

Addition of sodium sulfide to alkynyl sulfides, $S(C\equiv CR)_2$, also yields 2,6-disubstituted 1,4-dithiins ⟨75RTC163⟩.

2*H*-1,4-Thiazines are formed when the sulfides (**314**) are reacted with ammonia (Scheme 47) ⟨67JMC591⟩. The 3,4-dihydrothiazine-3-thione (**315**) is converted by methyl iodide into the thiazinium salt (**316**), deprotonation of which yields a 4*H*-thiazine (**317**) ⟨69JHC247⟩. 1,4-Thiazine 1,1-dioxides are formed by the cyclodehydration of diacyl sulfones and ammonia (Scheme 48) ⟨72OS(52)135⟩.

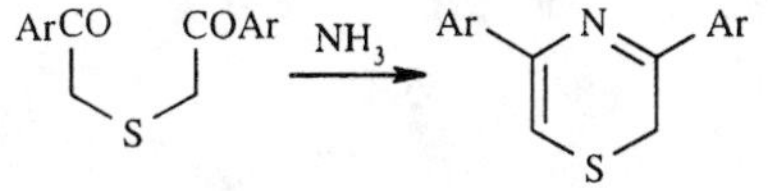

(314)

Scheme 47

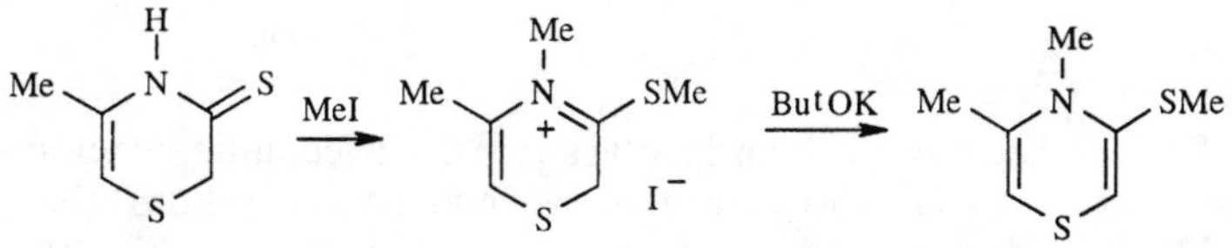

(315) (316) (317)

Dihydro-1,4-oxazines (**319**) are available through the treatment of acetals (**318**) with phosphorus pentoxide in pyridine ⟨79SC631⟩.

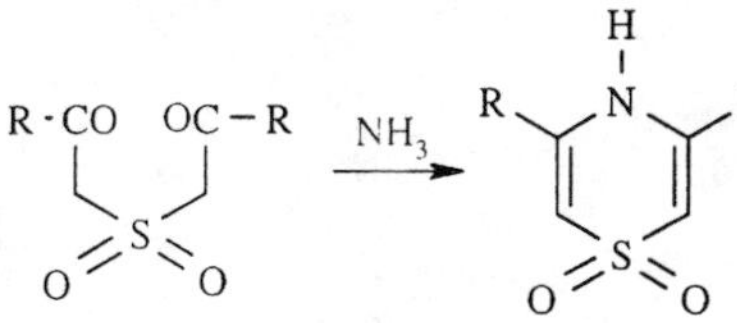

Scheme 48

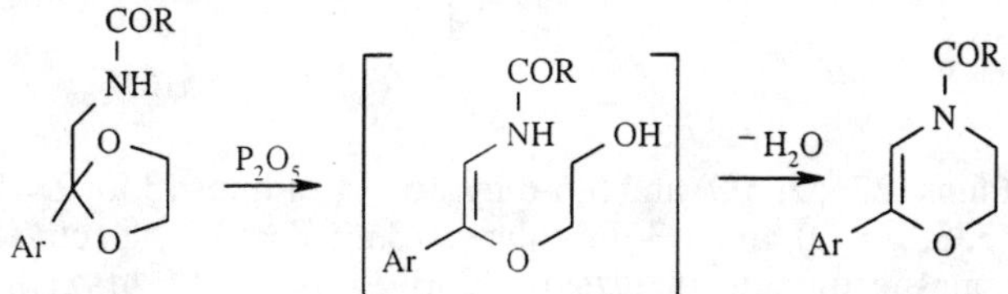

(318) (319)

4.3.4.1.3   Non-aromatic rings from acyclic compounds

Piperazines, dioxanes and dithianes can be prepared as shown ($320, 322 \rightarrow 321$; Z = NH, O, S) from fragments CCZ + CCZ or $C_2 + C_2 + Z + Z$.

(320)   (321)   (322)

4.3.4.1.4   From heterocyclic precursors

1,4-Dithiins are obtained upon photolysis of thiadiazoles, *e.g.* (323) $\rightarrow$ (324) ⟨81JA486⟩. Photolysis of the dithiocarbonate (325) also gives (324; 82%) ⟨73ZC424⟩.

Aziridine is converted into piperazine on NH_3 treatment; 1-substituted aziridines give 1,4-disubstituted piperazines when reacted with Grignard reagents. Azirines (326) with Group VI metal carbonyls give pyrazines (327) and dihydropyrazines (328).

(323)   (324)   (325)

(326)   (327)   (328)

A general type of [3 + 3] heterocyclization involves initial nucleophilic attack on the electrophilic three-membered heterocycle by a 1,3-electrophile-nucleophile. Aziridines (330) with either α-mercapto ketones (329) or with a mixture of a ketone and sulfur give 5,6-dihydro-1,4-thiazines (330 $\rightarrow$ 331 $\rightarrow$ 332). Azirines (333) can be used for the preparation of pyrazinones (334) from α-amino esters $R^2CH(NH_2)CO_2Et$ and of 1,4-oxazinones from α-hydroxy esters ⟨83TL1153⟩.

(329)   (330)   (331)   (332)

(333)   (334)

5,6-Dihydro-1,4-oxathiins (336; Z = O) and 5,6-dihydro-1,4-dithiins (336; Z = S) are easily obtained from 1,3-oxathiolanes (335; Z = O) and 1,3-dithiolanes (335; Z = S), respectively, by treatment with bromine ⟨91S223⟩, *N*-bromosuccinimide ⟨94T7265⟩, chlorine ⟨87JOC5374, 91S223⟩, or sulfuryl chloride ⟨88JCR(M)1401⟩.

(335)   (336)   (337)   (338)   (339)

1,4-Dioxanes (**337**) are produced in excellent yields from oxiranes and dilute sulfuric acid. 1,4-Dioxanes (**337**) are also conveniently obtained by acid-catalyzed condensation of oxiranes with glycols, while use of ethanolamine gives morpholine (**338**). Base-catalyzed reaction of oxiranes with α-amino acids and esters gives tetrahydro-1,4-oxazin-2-ones, *e.g.* propene oxide + $RNHCH_2CO_2H \rightarrow$ (**339**). 1,4-Dithianes have been prepared by the dimerization of thiiranes either in the vapor phase or in the presence of acid catalysts.

4.3.4.2 Seven-membered Rings

4.3.4.2.1 1,4-Diazepines

Ethylenediamine (**340**) and its *N*-substituted analogues with 1,3-dialdehydes or -diketones give 2,3-dihydro-1,4-diazepines, *e.g.* (**341**) ⟨78H(11)S50⟩. 1,4-Diazepin-5-ones, *e.g.* (**342**), can be readily prepared by the reactions of 1,2-diamines with β-keto esters ⟨67AHC(8)21, p. 57⟩. Similarly reactions with malonic acids and esters give 1,4-diazepine-5,7-diones (**343**) ⟨67AHC(8)21, pp. 55, 69, 68CRV747, p. 781⟩.

(340)   MeCOCH₂COMe   (341)   (342)   (343)

The use of vinamidinium salts (**344**) as 1,3-dicarbonyl equivalents in the synthesis of 1,4-diazepinium salts (**345**) may offer advantages ⟨78H(11)550, 78JCS(P1)1453⟩. They have been used to prepare 2,3-dihydro-1,4-diazepines not readily available from the corresponding 1,3-dicarbonyl compounds ⟨93AHC(56)1⟩.

(344)   (345)

R¹, R² = H. alkyl, Ph
R³, R⁴, R⁵ = H, alkyl

Bis-anils of 1,2-diaminocyclopropanes (**346**) undergo a thermal Cope rearrangement followed by hydrogen migration to give the 2,3-diaryldihydrodiazepines (**347**) ⟨78H(11)552⟩.

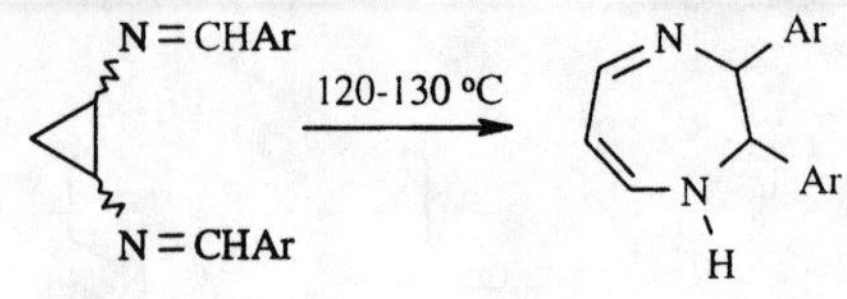

120-130 °C

(346)   (347)

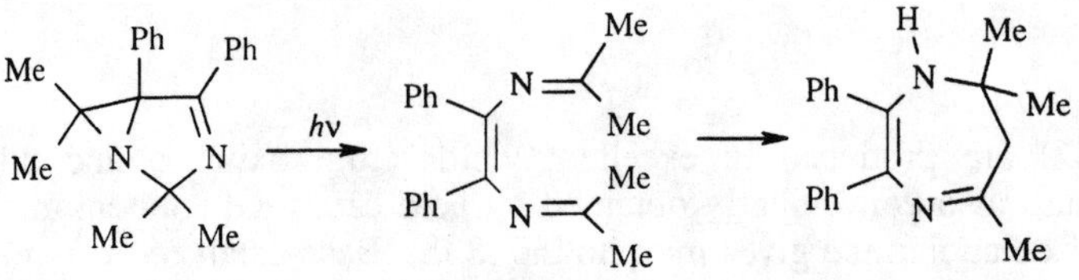

(348)   (349)

The enediimine (348) underwent thermal transformation to (349) ⟨72CC1116⟩.

4-Azidopyridines under photolysis or thermolysis can be converted into derivatives of 6*H*-1,4-diazepine ⟨84CPB4694⟩ (*cf.* Section 3.2.3.6.4).

4.3.4.2.2 *1,4-Oxazepines and 1,4-thiazepines*

The monocyclic systems (351) and (353) have been obtained by the dehydrative cyclization of (350) and (352), respectively.

The use of α-haloacyl halides as the C—C fragment leads to 1,4-oxazepin-3-ones, *e.g.* (354) ⟨77USP4010166⟩.

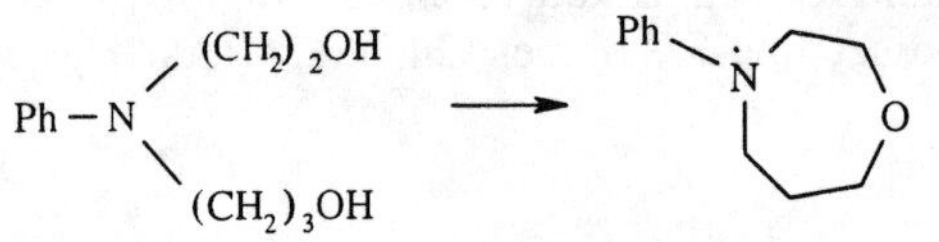

(350)   (351)

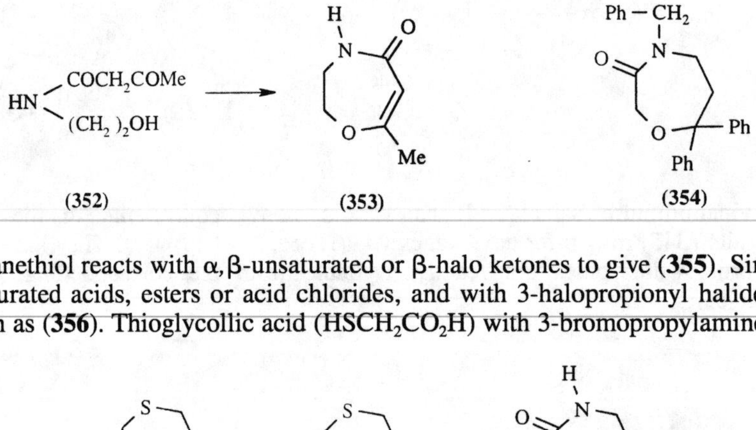

(352)   (353)   (354)

2-Aminoethanethiol reacts with α,β-unsaturated or β-halo ketones to give (355). Similarly, reaction with α,β-unsaturated acids, esters or acid chlorides, and with 3-halopropionyl halides, yields 5-oxo derivatives such as (356). Thioglycollic acid (HSCH$_2$CO$_2$H) with 3-bromopropylamine gives (357).

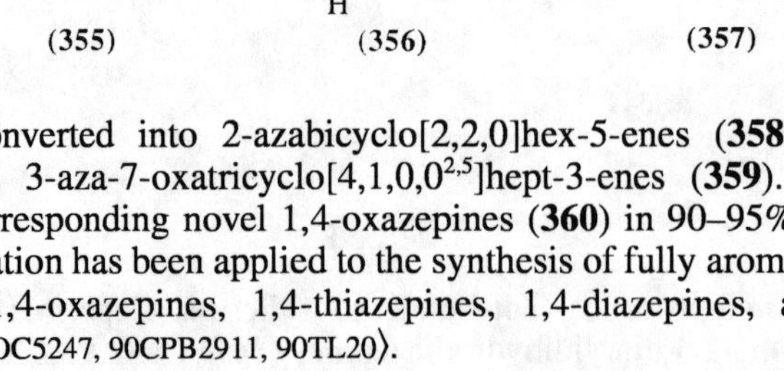

(355)   (356)   (357)

Pyridine is easily converted into 2-azabicyclo[2,2,0]hex-5-enes (358), which can be further transformed into useful 3-aza-7-oxatricyclo[4,1,0,0^{2,5}]hept-3-enes (359). Irradiation of (359) in acetonitrile gives the corresponding novel 1,4-oxazepines (360) in 90–95% yield (Scheme 49). This type of valence isomerization has been applied to the synthesis of fully aromatized 1,4-epines with two heteroatoms, such as 1,4-oxazepines, 1,4-thiazepines, 1,4-diazepines, and azepines ⟨85CPB4572, 86CC1188, 87H(26)3085, 87JOC5247, 90CPB2911, 90TL20⟩.

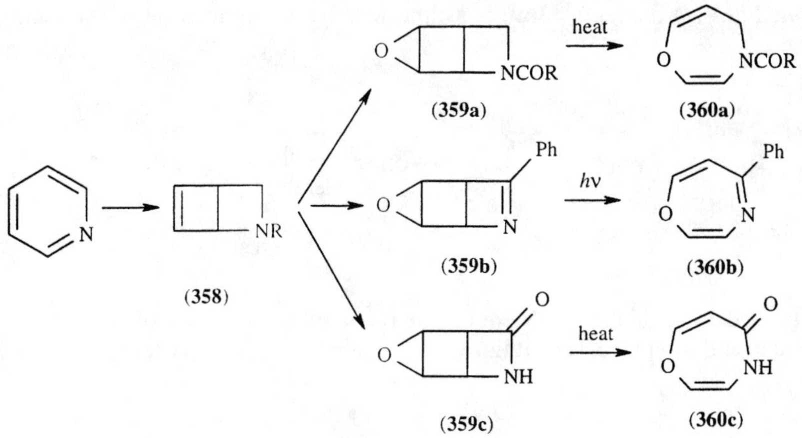

Scheme 49

4.3.4.2.3 1,4-Dioxepins and 1,4-dithiepins

One general method for the preparation of 6,7-dihydro-1,4-dioxepins of type (**362**) involves ring expansion; thus, treatment of 2-(methoxymethyl)-1,3-dioxane (**361**) with dodecylbenzenesulfonic acid at 250°C and simultaneous distillation gave (**362**) with 69% conversion and 84% selectivity ⟨90GEP(O)3823603⟩.

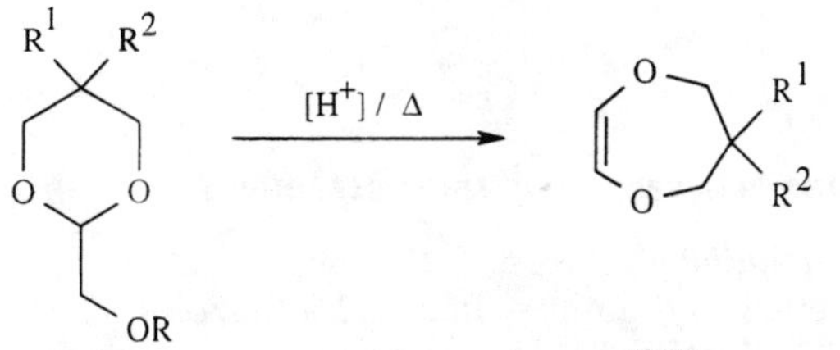

(**361**) (**362**)

The 2,3-dihydro-5*H*-1,4-dioxepins (**363**) and (**365**) can be obtained from 1,4-dioxin-halocarbene adducts (**364**) ⟨77ZC331⟩. Saturated rings of type (**366**) have been prepared by the treatment of cyclic acetals of ethane-1,2-diol with vinyl ethers in the presence of boron trifluoride. 1,4-Dioxepan-5-one (**367**) results from the reaction of bromoform and silver nitrate with aqueous dioxane ⟨60AG415⟩.

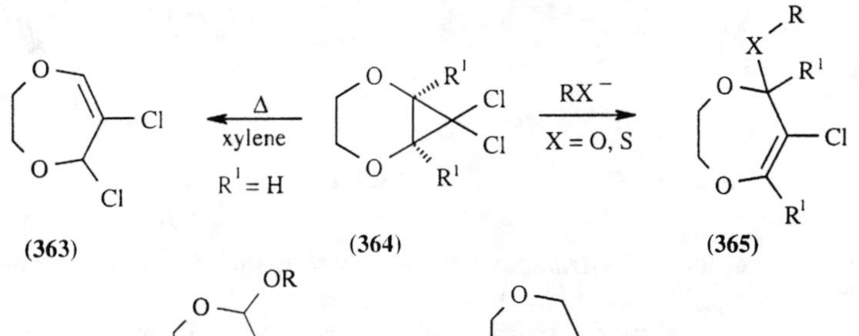

(**363**) (**364**) (**365**)

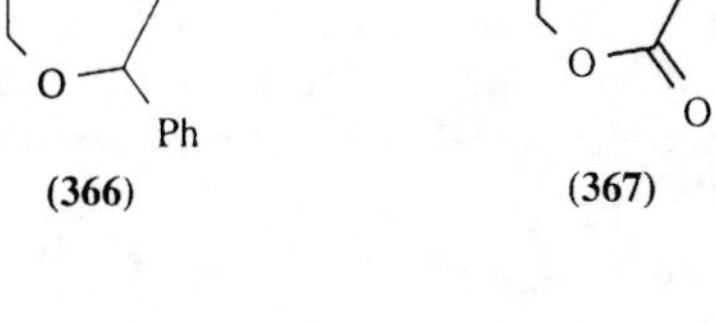

(**366**) (**367**)

The unsaturated 5*H*-1,4-dithiepin (**369**) is synthesized by an elimination of the methyl ether (**368**) ⟨75TL1895⟩.

(368)   (369)

The fully saturated 1,4-dithiepane system can be prepared by reactions of propane-1,3-dithiol with 1,2-dibromoethane, and of ethane-1,2-dithiol with 1,3-dihalopropanes ⟨72HC(26)598, p. 619⟩.

4.3.5 THREE HETEROATOMS IN THE 1,2,3-POSITIONS

4.3.5.1 Three- and Four-membered Rings

Valence isomerization is used in the formation of oxadiaziridines (**370**) and triaziridines ⟨70JOC2482, 80CC1197⟩.

(370)   (371)   (372)

4.3.5.2 Five-membered Rings

4.3.5.2.1 Formation of a bond between two of the heteroatoms

(i) 1,2,3-Triazoles and 1,2,3-thiadiazoles

Diazo ketones are converted by amines into 1,2,3-triazoles and by hydrogen sulfide into 1,2,3-thiadiazoles (**371** → **372**; Z = NR, S). The intramolecular cyclization of suitable precursors is a most useful method for the preparation of the 1,2,3-triazole ring, including *N*-amino- and *N*-imino-triazoles and triazole *N*-oxides.

Oxidative processes leading to N—N bond formation convert the bis-hydrazone (**373**) into the 1-amino-1,2,3-triazole derivative (**374**) [Hg$_2$(OAc)$_2$ or MnO$_2$] ⟨67TL3295, 71JPR882, 71ZC179⟩. The osazone (**375**) with potassium ferricyanide gives the zwitterionic 1,2,3-triazole (**376**) ⟨74T445⟩ (see also ⟨84CRV249, 84JHC1169, 1653, 92AHC(53)85⟩).

(373)   (374)   (375)   (376)

Two alternative routes lead to 2-alkyltriazole 1-oxides from α-dicarbonyl compounds (Scheme 50, routes A and B). Unsymmetrical dicarbonyl compounds frequently, but not invariably, give rise to two isomeric hydrazones and two isomeric oximes and hence two isomeric 1,2,3-triazole 1-oxides ⟨81JCS(P1)503⟩. 2-Phenyltriazole 1-oxide is obtained from glyoxal *via* route A. However, 2-methyl-triazole 1-oxide is prepared from glyoxal by route B in a one-pot process under neutral conditions. 2-Benzyltriazole 1-oxide is obtained similarly. 2,5-Dimethyltriazole 1-oxides are accessible through both routes ⟨86ACS(B)262⟩.

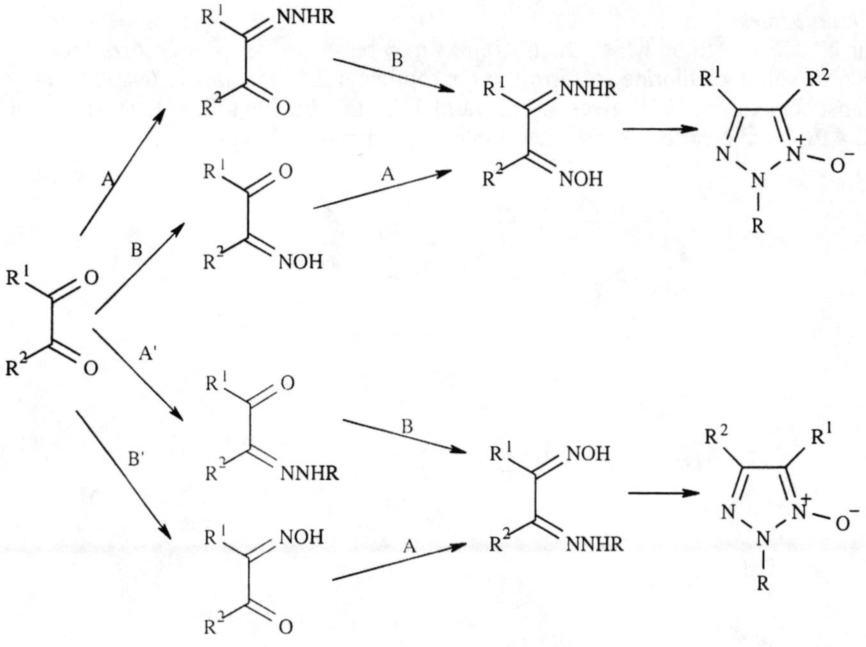

Reagents: A, RNHNH₂; B, NH₂OH

Scheme 50

For a review of triazole syntheses, see ⟨89CHE714, 94H(37)1951⟩.

(ii) 1,2,5-Oxadiazoles

α-Dioximes (**378**) can be cyclized to furazans (**377**) and furoxans (**379**). β-Nitrovinyl azides (**380**) spontaneously cyclize to the furoxan (**381**) with loss of N_2 ⟨75AG(E)775⟩.

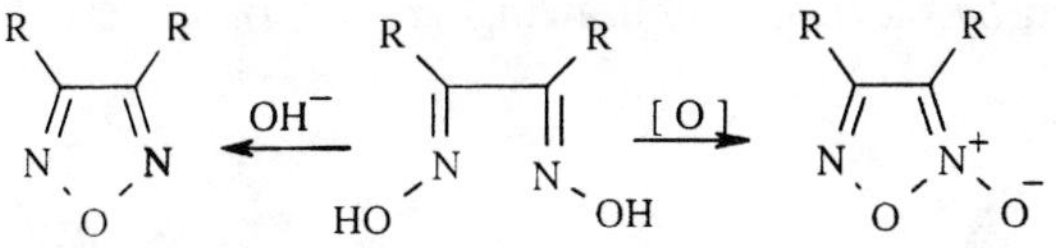

(**377**)   (**378**)   (**379**)

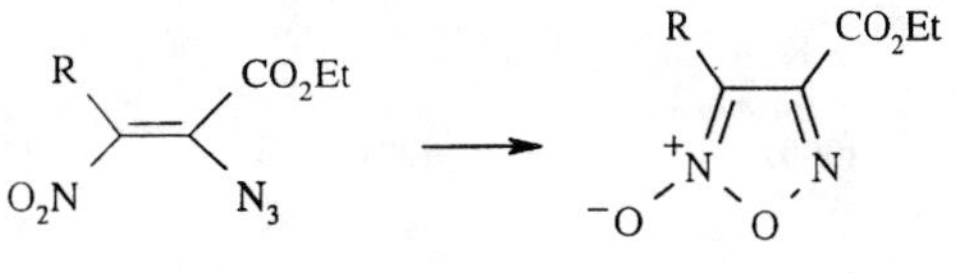

(**380**)   (**381**)

(iii) 1,2,3-Oxathiazoles

1,2,3-Oxathiazole *S*-oxides are prepared by the reaction of thionyl chloride with suitable 1,2-disubstituted ethanes: the 2-aminoethanol (**382**) gives (**383**) (CHEC 4.34).

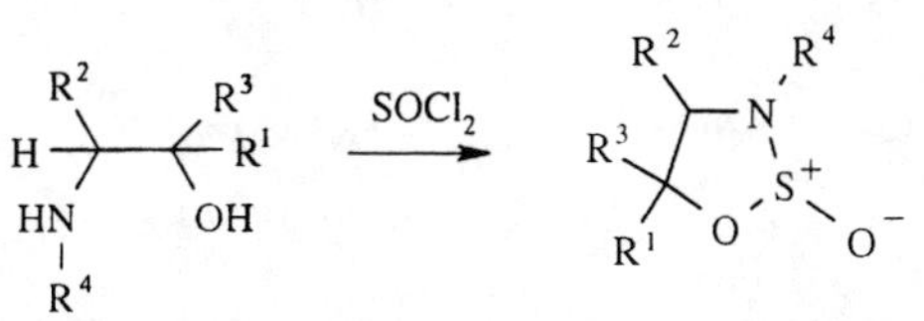

(**382**)   (**383**)

(iv) 1,2,5-Thiadiazoles

An acyclic NCCN system in which the N–C links may be *sp*, *sp²* or *sp³* hybridized reacts with sulfur monochloride or sulfur dichloride to form the appropriate 1,2,5-thiadiazole ⟨68AHC(9)107, 67JOC2823⟩. Thus, diiminosuccinonitrile (**384**) gives 3,4-dicyano-1,2,5-thiadiazole (**385**) ⟨72JOC4136⟩. Tables 16 and 17 in CHEC 4.26 list the various substituted systems prepared in this fashion.

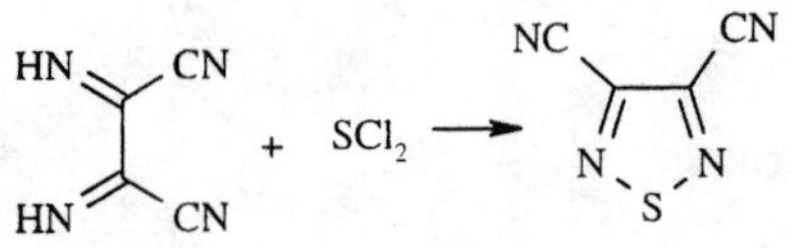

(384)   (385)

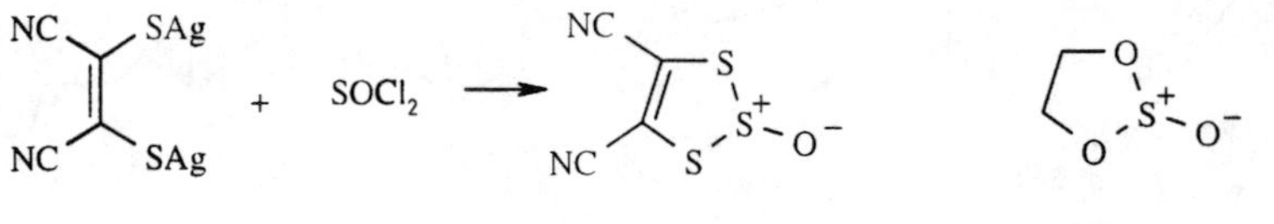

(386)   (387)   (388)

(v) 1,2,3-Trithioles and 1,3,2-dioxathiolanes

4,5-Dicyano-1,2,3-trithiole 2-oxide (**387**) is prepared from the silver salt of 2,3-dimercapto-maleonitrile (**386**) and thionyl chloride ⟨66HC(21-2)1⟩. Similarly, ethylene glycol and SOCl₂ give 1,3,2-dioxathiolane 2-oxide (**388**), the parent saturated five-membered cyclic sulfite (see CHEC 4.33).

4.3.5.2.2   Other methods

(i) By dipolar cycloadditions

Alkynes react with alkyl and aryl azides to give 1,2,3-triazoles (**389** → **390**). Suitable phosphoranes behave similarly; thus (**391**) with cyanazide N₃CN provides the 1,2,3-triazole-1-carbonitrile (**392**). Alkenes which are activated by electron-withdrawing groups, or are strained, give 1,2,3-triazolines (**393**) with azides.

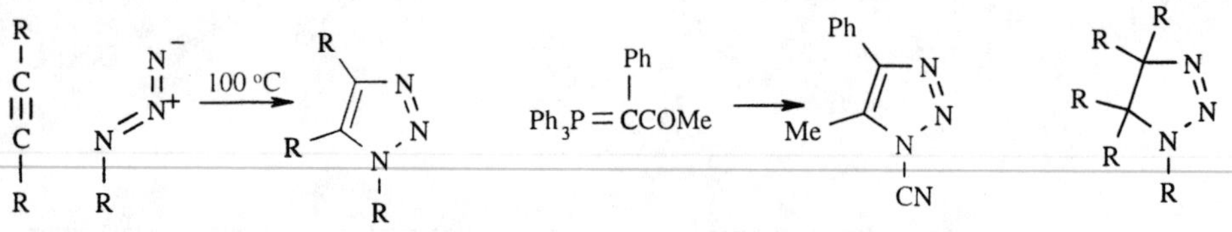

(389)   (390)   (391)   (392)   (393)

(ii) By cyclodehydration

Reaction of the *N*-nitrosoglycine (**394**) with acetic anhydride gives the anhydro-5-hydroxy-1,2,3-oxadiazolium hydroxide (**395**).

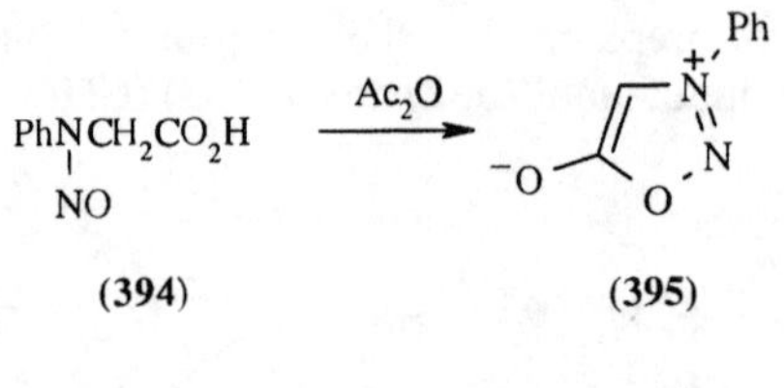

(394)   (395)

(iii)   From other heterocycles

See Sections 3.5.2.2 (ring expansion), 3.4.3.1.9 (monocyclic rearrangement) and 3.5.4 (ring contraction) for further preparations by the types of reaction indicated.

4.3.5.3 Six-membered Rings

The rearrangement of cyclopropenyl azides (**396**) is used for the synthesis of monocyclic 1,2,3-triazines (**397**) ⟨73JOC3149, 79CB1514⟩. However, the most general method is the oxidation of *N*-aminopyrazoles (**398**) with lead tetraacetate or variety of other oxidants ⟨92AHC(53)85⟩.

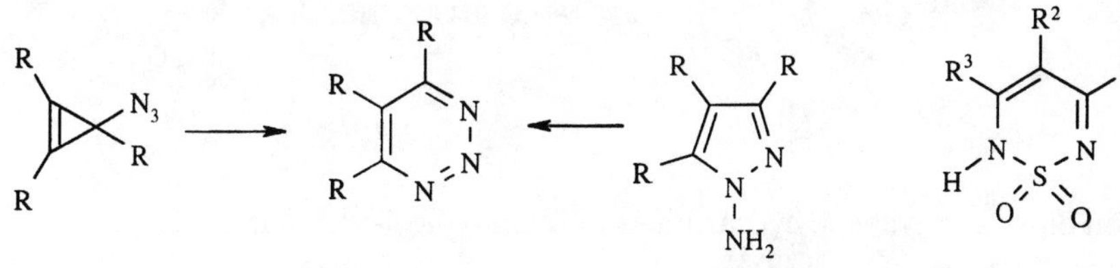

(396)   (397)   (398)   (399)

The 1,2,6-thiadiazine (**399**) is prepared from the β-diketone $R^1COCHR^2COR^3$ and sulfamide $NH_2SO_2NH_2$.

Thermal decomposition of ethyl azidoformate in the presence of 4-phenyl-1,2-dithiole-3-thione leads to 1,2,3-dithiazine (**400**; Scheme 51) ⟨76CJC3879⟩.

Cyclization of (**401**) in CCl_4 — NEt_3 gives a 1,3,2-dioxazine (**402**) ⟨80IZV2669⟩.

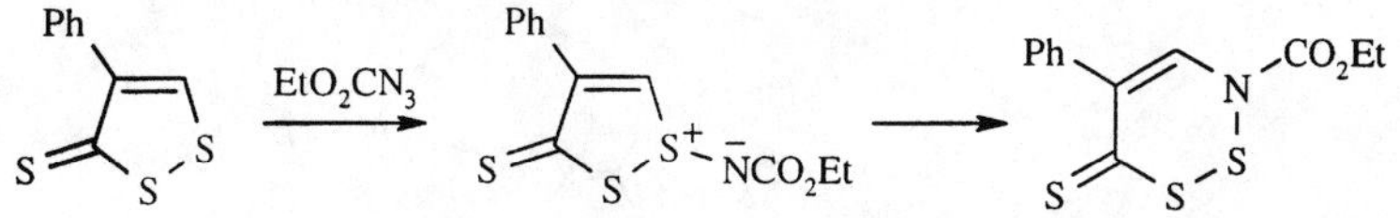

(400)

Scheme 51

(401)   (402)

4.3.6 THREE HETEROATOMS IN THE 1,2,4-POSITIONS

4.3.6.1 Five-membered Rings

4.3.6.1.1 *From acyclic intermediates containing the preformed Z—Z' bond*

(i) Diacylhydrazines (**403**) yield 1,3,4-oxadiazoles (**404**; Z = O) on heating or on treatment with $SOCl_2$, 1,3,4-thiadiazoles (**404**; Z = S) with P_2S_5 and 1,2,4-triazoles (**404**; Z = NR') with primary amines. 1-Substituted 1,2-diacylhydrazines are cyclized by strong acid to 2,3,5-trisubstituted 1,3,4-oxadiazolium salts (**405**) ⟨70JCS(C)1397⟩. Cyclization of acyl thiosemicarbazides (**406**) with sulfuric acid or phosphorus halides gave 5-substituted 2-amino-1,3,4-thiadiazoles (**407**) ⟨80JHC607⟩.

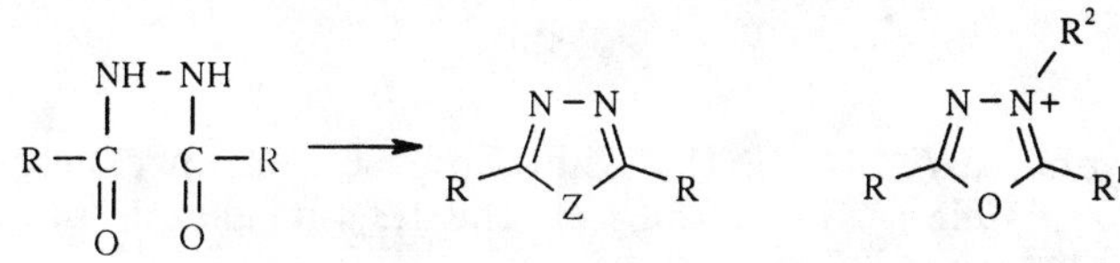

(403)   (404)   (405)

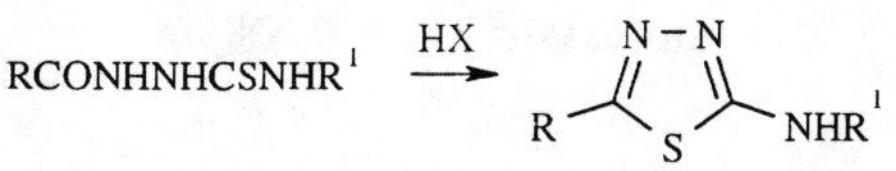

$$RCONHNHCSNHR^1 \xrightarrow{HX}$$

(406)   (407)

(ii) Similar ring closures can be carried out oxidatively. Lead tetraacetate causes hydrazone cyclization at a carbonyl oxygen atom as in the conversion of (**408**) into the 1,3,4-oxadiazolyl ether (**409**) ⟨76NKK782⟩. The semicarbazones (**410**) yield the 2-amino-5-benzoyl-1,3,4-oxadiazoles (**411**) ⟨72AC(R)11, 76MI40300⟩.

(**408**)   (**409**)   (**410**)   (**411**)

The conversion of the benzylidene hydrazidines (**412**) into the 4-arylamino-1,2,4-triazoles (**413**) was effected with mercury(II) oxide ⟨77BCJ953⟩.

(**412**)   (**413**)

(**414**)   (**415**)   (**416**)

The direction of ring closure can often be influenced by the conditions. The substituted thiosemicarbazone (**415**) with $Al_2O_3/CHCl_3$ formed the 1,2,4-triazoline-3-thione (**416**) but MnO_2/C_6H_6 afforded the thiadiazoline (**414**) ⟨70JCS(C)63⟩.

(iii) Amidoximes and amidrazones (**417**; Z = O, NH) react with acid chlorides, *etc.*, to give 1,2,4-oxadiazoles and 1,2,4-triazoles (**418**; Z = O, NH). In a related reaction, conversion of an amidine into its *N*-chloro derivative (**419**) with sodium hypochlorite and addition of potassium thiocyanate yield 1,2,4-thiadiazoles (**421**); the intermediate (**420**) undergoes spontaneous ring closure.

(**417**)   (**418**)

(**419**)   (**420**)   (**421**)

(iv) Reaction of a hydrazide ($RCONHNH_2$) with phosgeneimonium chloride (**422**) led to the 2-dimethylamino-1,3,4-oxadiazole (**423**) ⟨75AG(E)806⟩. The 1,3,4-thiadiazole (**426**) was made by an analogous reaction from (**424**) and (**425**).

(**422**)   (**423**)   (**424**)   (**425**)   (**426**)

This approach is suitable for the formation of mesoionic 1,3,4-thiadiazoles. The thiohydrazide (**427**) with phosgene, thiophosgene or an isocyanide dichloride leads to ready ring closure and formation of the mesoionic 1,3,4-thiadiazoles (**428**; X = O, S, NR, respectively) ⟨B-79MI40300⟩.

The imino dichloride (**430**) with the *N*-hydroxythioamide (**429**) gives the 1,3,5-oxathiazole (**431**) ⟨71AP763⟩.

(v) Dimethyl *N*-ethoxycarbonylthiocarbonimidate (**433**) with a monosubstituted hydrazine gives the 1,2,4-triazolinone (**432**), and with hydroxylamine the 1,2,4-oxadiazolinone (**434**) ⟨73JCS(P1)2644⟩.

4.3.6.1.2 *From acyclic intermediates by formation of the Z—Z' bond*

1,2,4-Thiadiazoles are conveniently prepared from thioamides or analogous substrates by oxidative dimerization which can be effected by halogens, hydrogen peroxide, sulfur halides, *etc.*; *cf.* the conversion of the thioamide (**435**) into (**436**; Hector's base) by hydrogen peroxide ⟨65AHC(5)119⟩. Commencing with the thiourea (**437**) gives the alternatively substituted product (**438**) ⟨72ZC130, 82AHC(32)285⟩.

The reaction of trichloromethanesulfenyl chloride (**440**) with amidines (**439**) and mild base is a general preparation for 5-chloro-1,2,4-thiadiazoles (**441**) ⟨65AHC(5)119⟩. Iminochloromethanesulfenyl chlorides (**442**; from RNCS + Cl₂) react with amidines such as PhC(NH)NHR′ to give 1,2,4-thiadiazolines (**443**) ⟨71T4117⟩. Chlorocarbonylsulfenyl chloride (**444**) (prepared from trichloromethanesulfenyl chloride and sulfuric acid) reacts with ureas, thioureas and guanidines to give 1,2,4-thiadiazolidine derivatives (445) ⟨70AG(E)54, 73CB3391⟩.

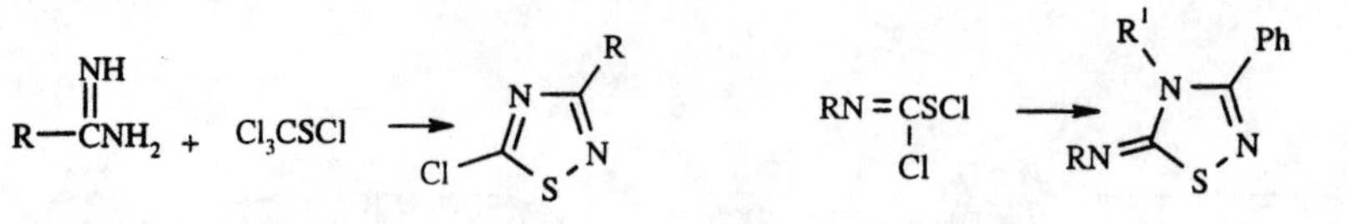

(439)   (440)   (441)   (442)   (443) R =Ar, ArCO

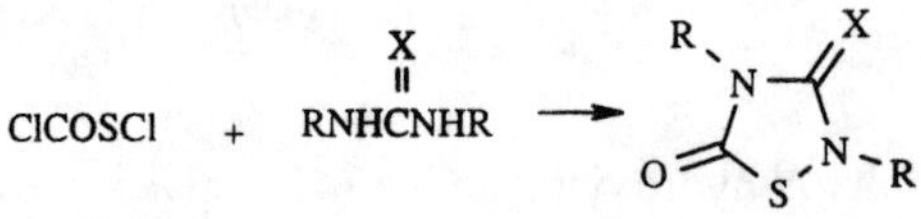

(444)   (445)

The reaction of *S,S*-disubstituted sulfur diimides (446) with oxalyl chloride (447) in the presence of triethylamine gives 1,2,5-thiadiazole-3,5-dione (448) ⟨72LA(759)107⟩.

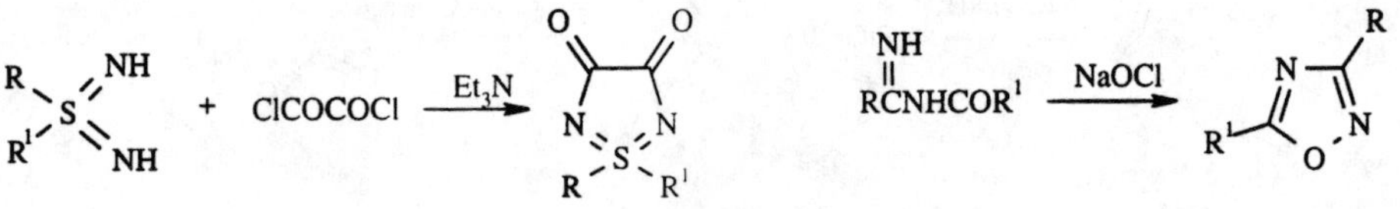

(446)   (447)   (448)   (449)   (450)

N—O bond formation by oxidative procedures has found less application. However, the 1,2,4-oxadiazole system (450) can be prepared by the action of sodium hypochlorite on *N*-acylamidines (449) ⟨76S268⟩.

4.3.6.1.3   *From heterocycles*

(i) By dipolar cycloadditions

1,3-Dipolar additions have been extensively used for the synthesis of 1,2,4-triazoles. The nitrilimine (452) reacts with the tetrazole (451; Z=NH) to give the triazole (453). The reaction of (452) with the tetrazole (451; Z=O) forms the tetrazole (454), which on treatment with base affords the triazolinone (455) ⟨93JCR(S)306⟩.

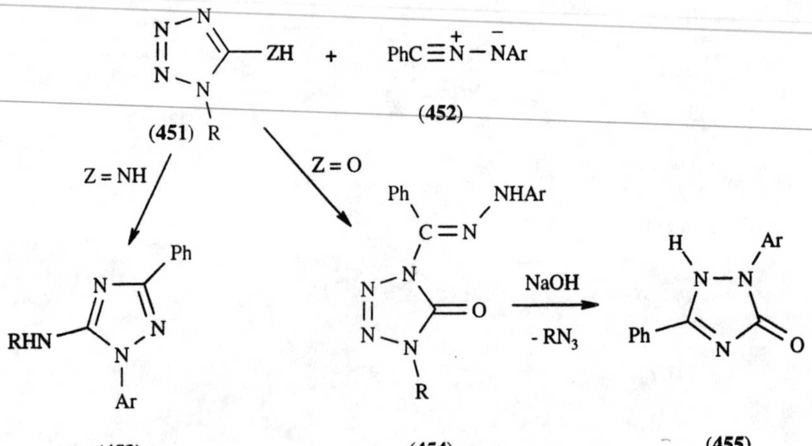

(451) R   (452)

(453)   (454)   (455)

The 2,5-disubstituted tetrazole (456; X=O, S, NR) undergoes thermal ring opening to the 1,5-dipolar species (457) which readily loses N_2 to give the conjugated nitrilimine (458 ↔ 459).

1,5-Dipolar cyclization of (**459**) leads to oxadiazoles, thiadiazoles and triazoles (**460**) ⟨70CB1918, 65CB2966⟩ (see also CHEC 4.13).

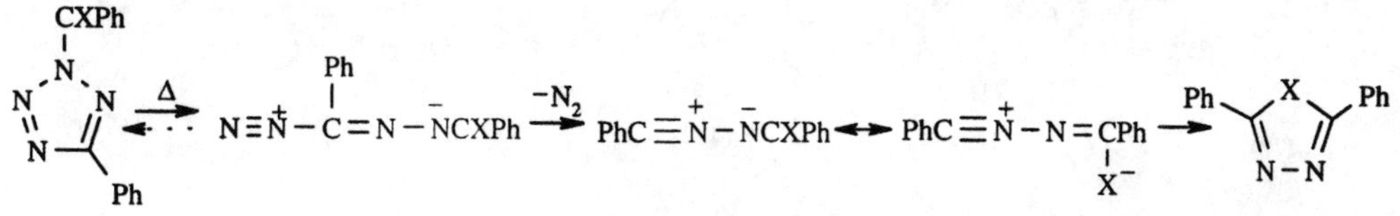

(**456**) (**457**) (**458**) (**459**) (**460**)

X = O, S, NR

N-Phenylsydnone (**461**) on irradiation gives 4-phenyl-1,2,4-oxadiazolin-5-one (**465**). When labeled carbon dioxide is passed through the solution during irradiation, it is incorporated into (**465**) ⟨66TL4043, 71JOC1589⟩. These results were rationalized in terms of an initial valence isomerization to (**462**) and loss of CO₂ to give (**463**), which underwent ring opening to the 1,3-dipole (**464**). This dipole subsequently reacted with carbon dioxide to give (**465**). For trapping with PhNCO see Section 3.4.1.9.1.

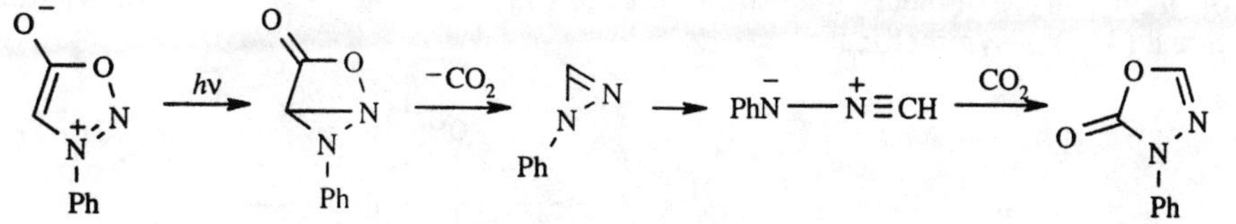

(**461**) (**462**) (**463**) (**464**) (**465**)

Intramolecular trapping of the intermediate (**467**) from the photolysis of methyl tetrazole-1,5-dicarboxylate (**466**) gave methyl 5-methoxy-1,2,4-oxadiazole-3-carboxylate (**468**) (see also Section 3.4.3.12.4).

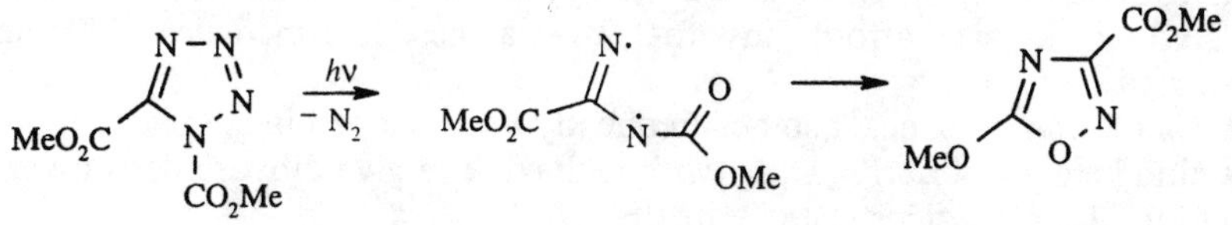

(**466**) (**467**) (**468**)

The 1,3-dipolar cycloaddition reaction between nitrile sulfides (**470**), produced by the thermal decomposition of oxathiazolones (**469**), and nitriles is a general method for the synthesis of 3,5-disubstituted 1,2,4-thiadiazoles (**471**) of unambiguous structure ⟨84CHEC-(6)463⟩. More recently trichloroacetonitrile ⟨86PS(26)151⟩ and tosyl cyanide ⟨93JHC357⟩ have been used as acceptor molecules to give the 5-trichloromethyl and 5-tosyl derivatives in high and moderate yields, respectively. Displacement of the tosyl group by a range of nucleophiles leads to a wide variety of 5-substituted analogues.

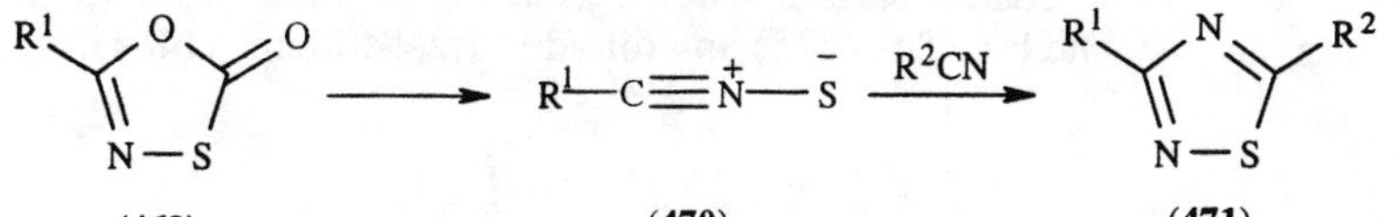

(**469**) (**470**) (**471**)

R^1 = aryl, alkyl, CO₂Et; R^2 = aryl, CO₂Et, CH₂CO₂Et, CCl₃, SO₂Tol

(ii) By ring contraction

4-Pyrimidones (**472**) react with hydrazine hydrate at 130–140°C to give the 1,2,4-triazole (**473**) ⟨83H(20)1243⟩. 1,3,5-Oxadiazinium salts (**475**) with hydroxylamine give the 1,2,4-oxadiazoles (**474**) and with hydrazines form the 1,2,4-triazole derivatives (**476**). The substituents in these cationic species are usually aryl, restricting the appeal of these ring interconversions ⟨65CB334, 67CB3736⟩.

(472)   (473)   (474)   (475)   (476)

(iii) By the 'monocyclic rearrangement'

Section 3.4.3.1.9 gives details of ring interconversions involving three-atom side chain displacements at both N and S ring atoms.

4.3.6.2 Six-membered Rings

4.3.6.2.1 1,2,4- Triazines

The reaction of 1,2-dicarbonyl compounds (477) with amidrazones (478) is the best method for the synthesis of alkyl-, aryl- or hetaryl-substituted 1,2,4-triazines (481) ⟨78HC(33)189⟩. Mixtures result unless the dione (477) is symmetrical.

(477)   (478)   (479)   (480)   (481)

Various extensions are possible (see CHEC 2.19 for full details). Use of aminoguanidines, semicarbazide and thiosemicarbazide gives respectively the 3-amino-1,2,4-triazine, and the 3-one and 3-thione derivatives. Use of α-keto esters and α-keto cyanides gives 5-ones and 5-amino derivatives, respectively. α-Hydroxy ketones afford dihydro-1,2,4-triazines. Intermediates (479) and (480) can sometimes be isolated.

For the synthesis of 1,2,4-triazines, another frequently used method is the reaction of α-acylamino and α-thioacylamino ketones (482; X = O, S) with hydrazine to give dihydro derivatives (483) which can be oxidized to the 1,2,4-triazines ⟨78HC(33)189, p. 197⟩.

(482)   (483)

1,2,4,5-Tetrazines (484) undergo Diels-Alder reactions with C-N multiple bonds. Imidates ⟨69JHC497⟩ or, that is better, thioimidates ⟨83JOC621, 83TL4511, 85JA5745⟩ thus afford 1,2,4-triazines (487) which are formed *via* intermediate bicycles (485) and dihydro-1,2,4-triazines (486).

(484)   (485)   (486)   (487)   (488)

On reaction of tetrazines (484) with cyanamides, the bicyclic intermediates (488) lead directly to 3-amino-1,2,4-triazines (487; $R^1 = NR_2$) by elimination of nitrogen ⟨79CZ230⟩.

α-Hydrazono oximes (**489**) with ortho esters give triazine 4-oxides (**490**) ⟨77LA1713⟩.

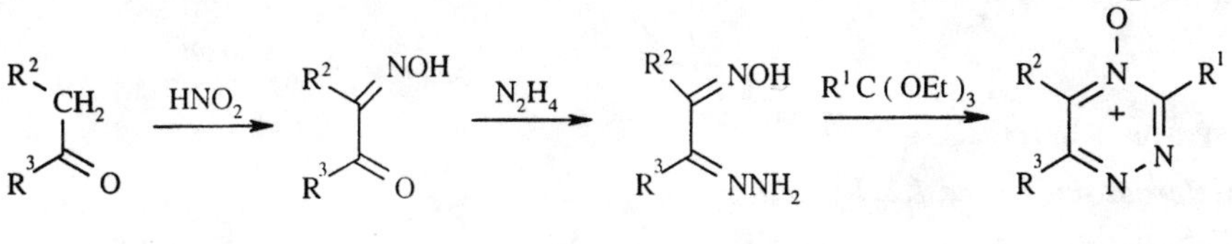

(**489**)   (**490**)

4.3.6.2.2 Rings containing O or S atoms

α-(Ethoxycarbonyl)methylsulfonylurea (**492**; $X = SO_2$) cyclizes to the 1,2,4-thiadiazine (**491**) on treatment with base ⟨59JA5655⟩, and a similar cyclization occurs with the oxygen analogue (**492**; $X = O$) to give (**493**) ⟨79JHC161⟩.

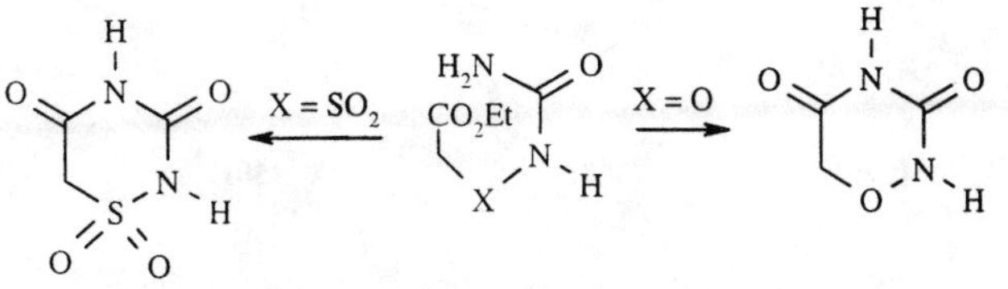

(**491**)   (**492**)   (**493**)

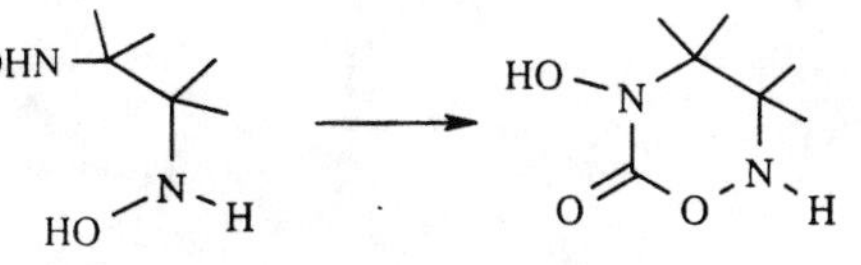

(**494**)   (**495**)

The 1,2-bishydroxylamine (**494**) with phosgene gives the 5-hydroxy-1,2,5-oxadiazine (**495**) ⟨80AP35⟩. Acid-catalyzed dehydration of *N*-(2-hydroxyethyl)-*N*′-acylhydrazines (**496**) is a general route to 4,5-dihydro-1,3,4-oxadiazines (**497**) ⟨64JOC668⟩.

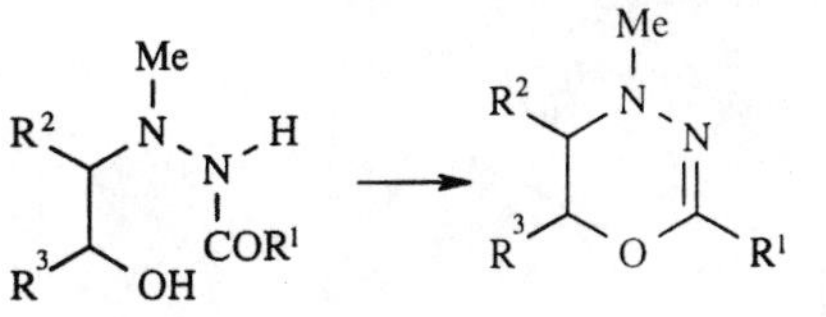

(**496**)   (**497**) 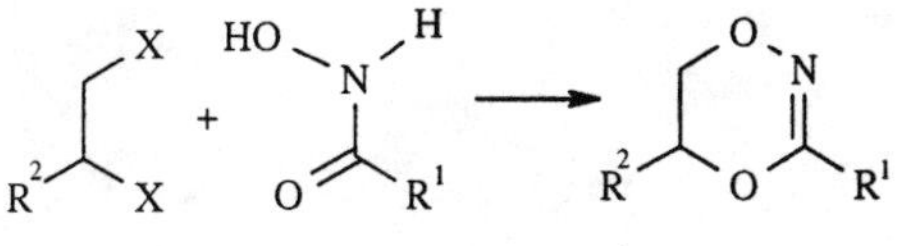   Scheme 52

1,4,2-Dioxazines are prepared by di-*O*-alkylation of hydroxamic acids (Scheme 52) using 1,2-dihalides or 1,2-dimesylates ⟨71JOC284, 75NKKIO41⟩.

Condensation of α-halo ketones with thiohydrazides (Scheme 53) is a general route to 1,3,4-thiadiazines ⟨76JPR(318)971⟩.

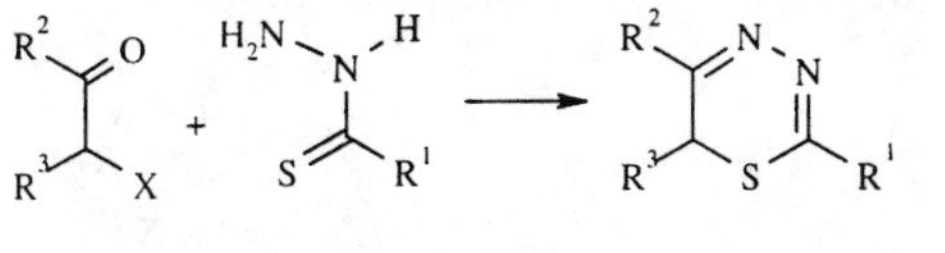

Scheme 53

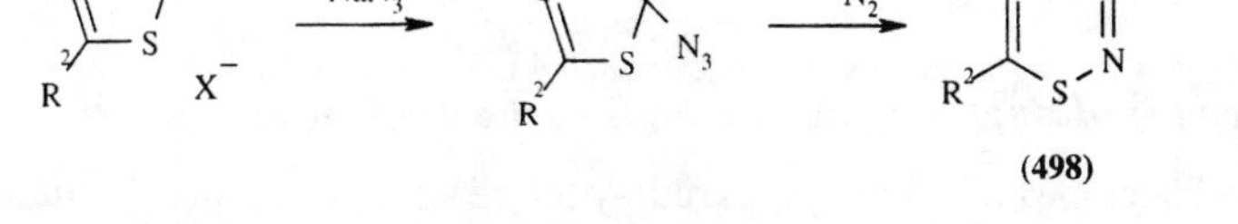

(**498**)

Scheme 54

Preparation of 1,4,2-dithiazines (**498**) involves the ring expansion of 1,3-dithiolium salts with azide (Scheme 54) ⟨76JPR(318)127⟩.

4.3.6.3   Seven-membered Rings

4.3.6.3.1   *Heteroatoms in the 1,2,4-positions*

Fully unsaturated 2*H*-1,2,4-triazepines (**501**) are formed by the cycloaddition of 1-azirines (**499**) to 1,2,4,5-tetrazines (**500**). The initial product rearranges by a 1,5-hydrogen shift to give (**501**) ⟨74TL2303⟩.

The 4*H*-1,2,4-triazepine system (**503**) can be prepared *via* a photochemical rearrangement of 3,4,7-triaza-2,4-norcaradienes (**502**) ⟨76TL2459⟩.

(**499**)   (**500**)   (**501**)

(**502**)   (**503**)

The β-functionalized isocyanate (**504**; X = O) reacts with methylhydrazine to give (**505**) ⟨77S756⟩, and (**504**; X = S) reacts with both alkylhydrazines and hydrazine itself to give the (**506**) ring system.

3-Oxo-1,2,4-oxadiazepines (**507**) have been prepared by the reaction of hydroxyurea with some α,β-unsaturated ketones ⟨75TL2979⟩. *syn*-ω-(Benzylamino)propiophenone oximes (**508**) react with phosgene to give the 7-oxo-1,2,6-oxadiazepines (**509**) ⟨75CB3387⟩, and with formaldehyde to give (**510**) ⟨80CB3373⟩.

(**504**)   (**505**) X = O   (**507**)
                        (**506**) X = S

(**508**)   (**509**) X = O
                        (**510**) X = H$_2$

4.3.6.3.2   *Seven-membered rings with heteroatoms in the 1,2,5-positions*

Hexahydro-1,4,5-oxadiazepines (**511**) are prepared by the reaction of monosubstituted hydrazines with O(CH$_2$CH$_2$Cl)$_2$ ⟨75MI51803⟩.

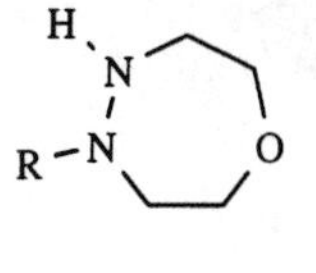

The sulfides or sulfones (513) react with hydrazine to give (512) or (514) depending on R^1 ⟨70JHC431, 72T2307⟩.

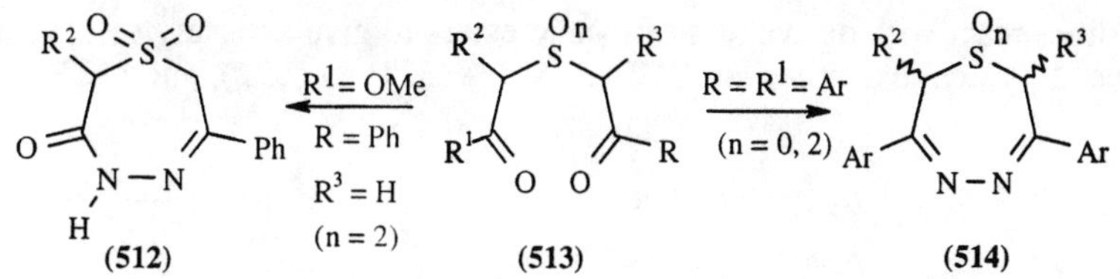

4.3.7 THREE HETEROATOMS IN THE 1,3,5-POSITIONS

4.3.7.1 *s*-Triazines

Biguanides [$H_2NC(=NH)NHC(=NH)NH_2$] react with lactones, amides, ortho esters, esters, acid anhydrides and acid chlorides to produce a wide range of 6-substituted 2,4-diamino-1,3,5-triazines (515) ⟨59HC(13)1⟩. Biguanides with carbodiimides, isothiocyanates and ketones give corresponding melamines, thiones and dihydro derivatives, respectively.

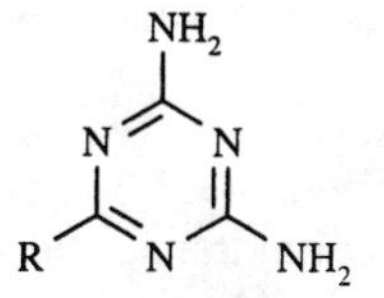

The condensation of *N*-cyanoamidines (517) with chloromethyliminium salts (516), prepared *in situ* from amides $R^1CONR^2R^3 + PCl_3$, provides an efficient, convenient route to many 1,3,5-triazines ⟨80S841, 81AJC623⟩. For R^2 and $R^3 =$ alkyl, the chloro-1,3,5-triazines (518; $R^5 =$ Cl) are produced. However, for $R^2 =$ aryl, the amino derivative (518; $R^5 = NR^2R^3$) may become the major product.

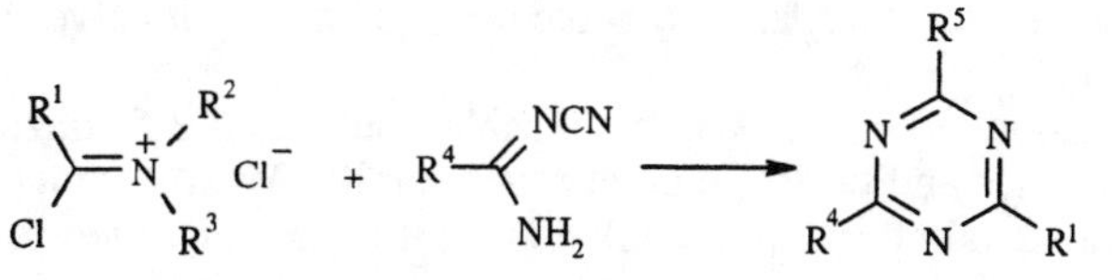

The condensation of *N*-cyanoamidines with thiocarbamate esters provides routes to 1,3,5-triazine thioethers (Scheme 55), and dimethylamino-1,3,5-triazines (Scheme 56). Melamine (520) can be prepared by fusing dicyanamide (519) ⟨40MI22000⟩.

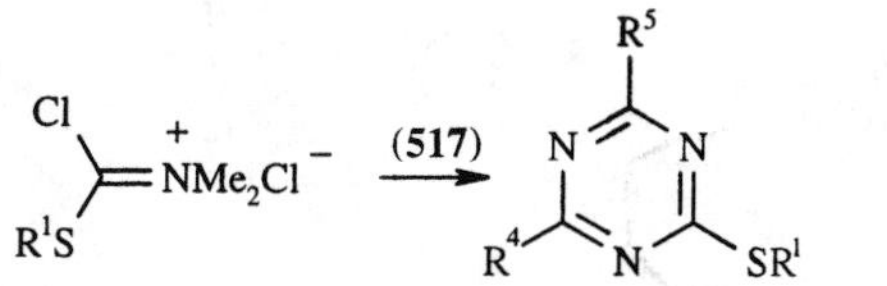

Scheme 55

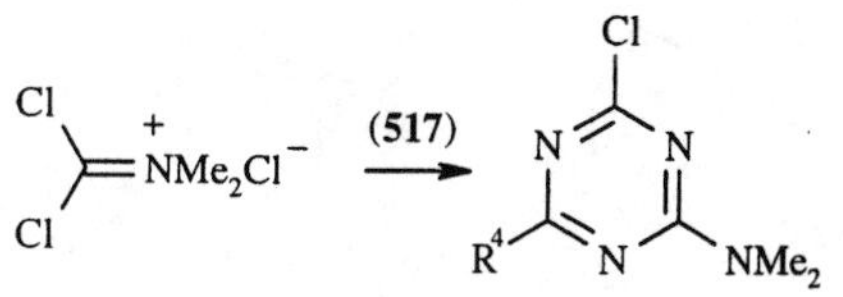

Scheme 56

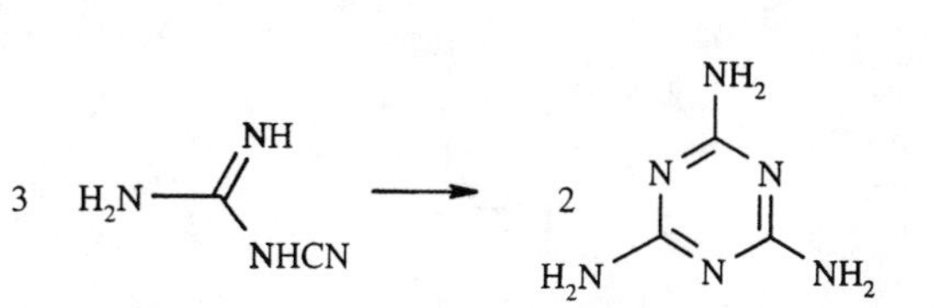

Amidines react with derivatives of cyanic esters to give a variety of 1,3,5-triazines with an aryloxy substituent. Typical examples are $RC(NH)NH_2 + (521) \rightarrow (522)$, and $(523) \rightarrow (524)$ ⟨72AG(E)949⟩.

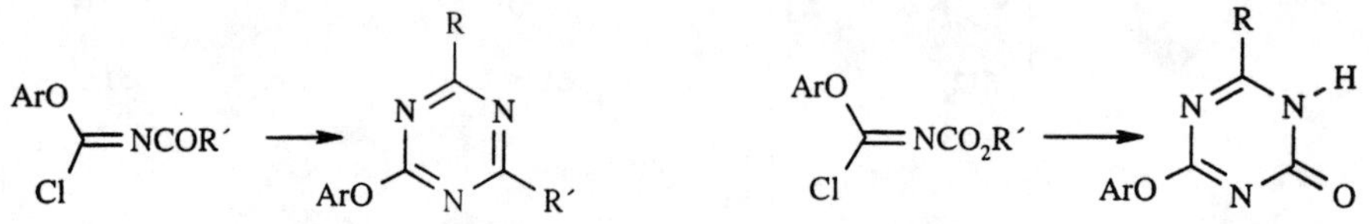

Cyclotrimerization of nitriles is the best-known route to 1,3,5-triazines (for a detailed discussion see CHEC 2.18). The reaction is of value for preparing the symmetrical derivatives only. Nevertheless, many important triazines, such as cyanuric chloride, are made in this way. Other cyclotrimerization reactions are useful; thus, an easy route to 1,3,5-triazine involves heating ammonium acetate with ethyl orthoformate.

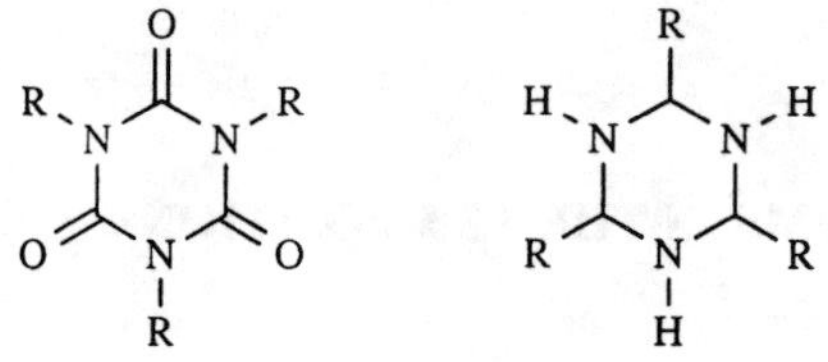

Trimerization of imidates is a valuable route to 1,3,5-triazines. Imidates can be considered as activated nitriles and cyclotrimerize more readily. Most symmetrical 2,4,6-trialkyl-1,3,5-triazines are easily formed, although large alkyl substituents may give rise to steric hindrance ⟨61JOC2778⟩. Symmetrical isocyanurates (525) are readily available from isocyanates, RNCO; catalysts include tertiary amines, phosphines and sodium methoxide. Aldehydes RCHO and ammonia give hexahydro-1,3,5-triazines (526), known as 'aldehyde ammonias' ⟨73JOC3288⟩.

4.3.7.2 Compounds Containing O or S Atoms

1,3,5-Trioxanes (527; Z = O) are trimers of aldehydes or ketones formed by acid-catalyzed condensations of the monomers. 1,3,5-Trithiane (527; Z = S) is prepared by passing H_2S through formaldehyde and hydrochloric acid ⟨43OSC(2)610⟩. Passage of H_2S through acetaldehyde in dilute hydrochloric acid affords not only the 1,3,5-trithiane but also the 1,3,5-oxadithiane and 1,3,5-dioxathiane analogues ⟨66HC(21-2)633⟩.

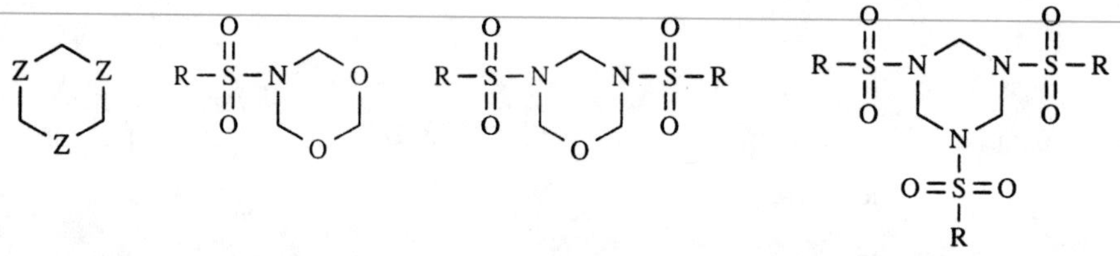

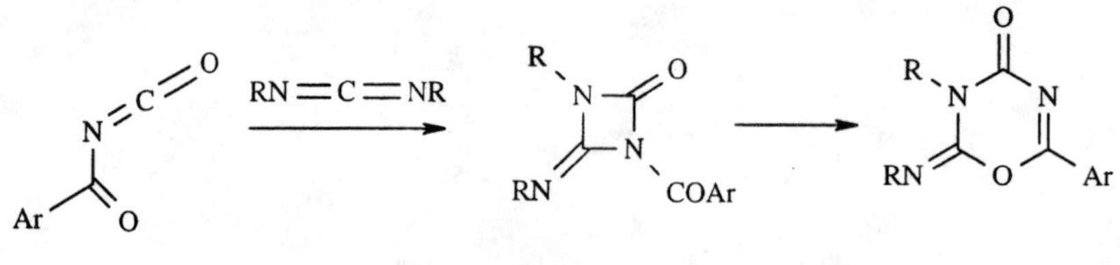

Scheme 57

Sulfonamides RSO_2NH_2 react with formaldehyde to give *N*-sulfonyl-1,3,5-dioxazines (528), -1.3,5-oxadiazines (529) or -1,3,5-triazines (530) according to the relative quantities of reagents used ⟨75JCS(P1)772⟩. Aroyl isocyanates with carbodiimides give imino-1,3,5-oxadiazinones ⟨79BSF(2)499⟩ *via* the [2+2] adduct which rearranges (Scheme 57) ⟨79BSF(2)499⟩.

Treatment of *N,N'*-disubstituted ureas with methanal followed by cyclodehydration of the resulting *N,N'*-bis(hydroxymethyl)ureas (531) with P_2O_5 furnishes 3,5-dialkyl-5,6-dihydro-2*H*-1,3,5-oxadiazin-4(3*H*)-ones (532) in high yields ⟨85JAP(K)6067471, 85JAP(K)6067472, 85JAP(K)60104075, 85JAP(K)60104076⟩.

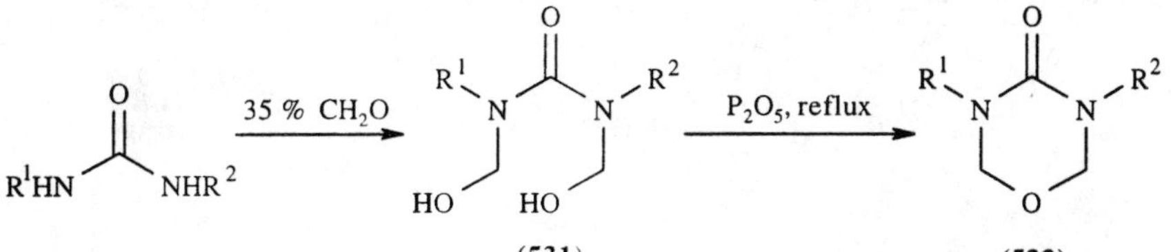

Trimerization of methyl isocyanate (MeNCO) in the presence of tri-*n*-butylphosphine gives a 1,3,5-oxadiazine (533) ⟨73CR(C)(277)795⟩. In the presence of carbon dioxide, the reaction leads to the 1,3,5-oxadiazinimine (534) derived from CO_2 (1 mol) and the isocyanate (2 mol) ⟨74BSF1497⟩.

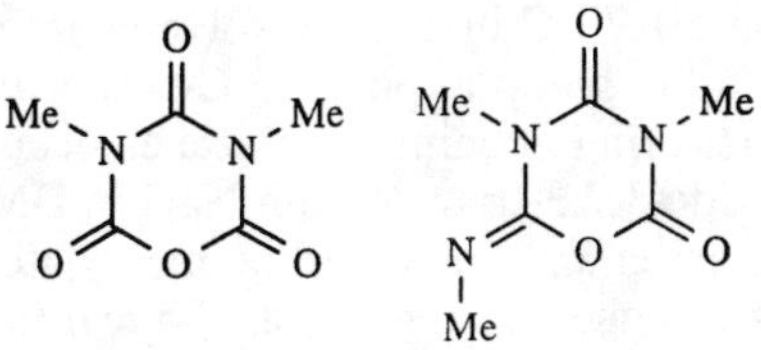

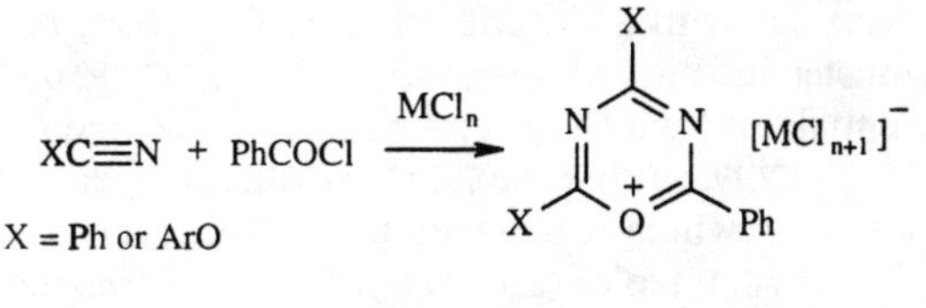

Scheme 58

Benzoyl chloride condenses with benzonitrile or aryl cyanates in the presence of aluminum trichloride to give 1,3,5-oxadiazinium salts (Scheme 58) ⟨67CB3736⟩.

4.3.7.3 Synthesis from Heterocyclic Precursors

2-Substituted 4- or 5-halopyrimidines with sodamide in liquid ammonia give 4-methyl-1,3,5-triazines (Scheme 59); the reaction is general and yields are good; *cf.* the related conversion of 6-substituted 2-bromopyridines into pyrimidines (Section 4.3.3.3.3). 4-Amino-5-nitrosopyrimidines are converted into 1,3,5-triazines by acetic anhydride or phosphorus oxychloride (Scheme 60). 1,3,5-Triazines can be obtained from 1,3,5-oxadiazinium cations (Section 3.2.1.6.1.iii).

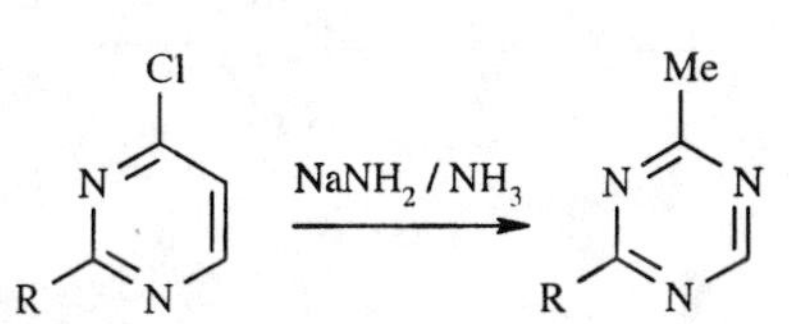

Scheme 59

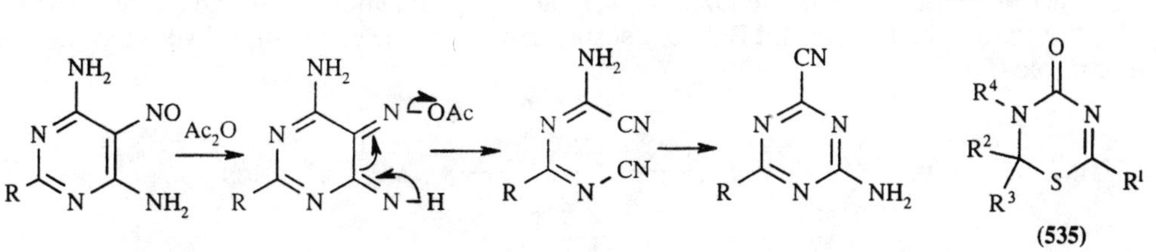

Scheme 60

1,3,5-Thiadiazines (535) are obtained by the cycloaddition of thioaryl isocyanates (R^1CSNCO) with imines $R^2R^3C=NR^4$.

4.3.7.4 Seven-membered Rings

The 1,3,5-thiadiazepine (538) has been prepared by the thermal rearrangement of the [4+2] cycloadduct (537) of the azirine (536) and thiobenzoyl isocyanate ⟨74JOC3763⟩.

(536)   (537)   (538)

4.3.8 FOUR OR MORE HETEROATOMS

4.3.8.1 Five-membered Rings

Most transformation of other azoles into tetrazoles involve azido substituents although the rearrangement of substituted 5-amino-1,2,3,4-thiatriazoles is an exception (see below). 5-Azido-1,2,3-triazoles with a CO_2Me group at C-4 (539) rearranged at 50–70°C in organic solvents into the 5-diazoester substituted tetrazoles (540) ⟨98JCS(P2)785⟩. The CO_2Me group at position C-5 appears to be essential: and with aryl groups at C-5, nitrogen evolution and ring opening to triazenes occurred ⟨85T4621, 88T3617, 89T749, 90BSB213, 90BSB281⟩. Treatment of 5-azido-1,2,3-triazole with NaH in DMSO gave an anion which rearranged to the 5-diazomethyltetrazole anion and this could be methylated to give *N*-methyltetrazoles ⟨88CC1144⟩. The 1,2,3-triazoles (541) with a phenylazo group at C-4 rearranged into 2-phenyl-5-carboxamido tetrazoles (542) on recrystallization from acetic acid ⟨58ACS1236⟩.

(539)   (540)   (541)   (542) R = Me, Et

The thermal conversion of 1-substituted 5-aminotetrazole (543) into a 5-substituted aminotetrazole (545) is rationalized by a 1,5-dipolar cyclization of (544).

(543)   (544)   (545)

Treatment of 5-substituted amino-1,2,3,4-thiatriazoles (546) with aqueous base results in competitive rearrangement to 5-thiotetrazoles (547) and fragmentation to isothiocyanates (548) and azide ion ⟨67JOC3580⟩. The use of solid $Ba(OH)_2$ suspended in toluene containing HMPA has given 80% yield of tetrazole (547).

(546)   (547)   (548)

2-Alkyl-2-chlorosulfonylcarbamoyl chlorides (549) with methylhydrazine yield 1,2,3,5-thia-triazolidines (550) ⟨77JCR(S)238⟩ (CHEC 4.28).

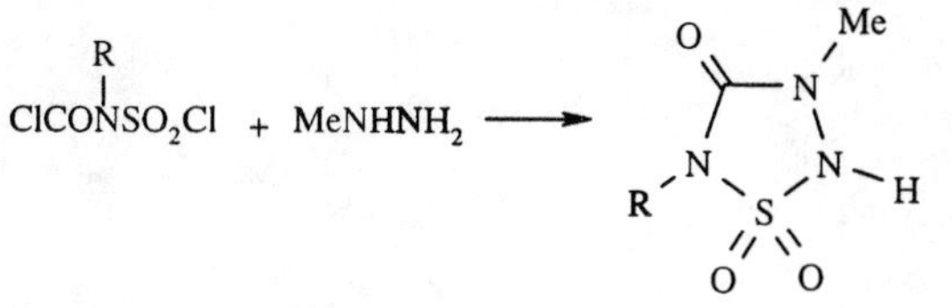

(549)   (550)   (551)

Explosive 1-arylpentazoles (**551**; Ar = Ph, 4-Me$_2$NC$_6$H$_4$, *etc.*) can be prepared by reaction of aryldiazonium salts with azide anion ⟨85AG515, 98JCS(P2)2243⟩.

4.3.8.2 Six-membered Rings

Dihydro-1,2,4,5-tetrazines (**552**) can be prepared from PhCN and N$_2$H$_4$.

1,4,2,5-Dioxadiazines (**553**) are available by [3 + 3] cycloaddition of PhCH=NR$^+$—O$^-$ and R'C≡N$^+$—O$^-$. 1,4,3,5-Oxathiadiazines (**554**) result from hydration of O$_2$S(NCO)$_2$.

Photolysis of 2,4,5-triphenyl-1,2,3-triazole 1-oxide gives the triphenyl-1,3,4,5-oxatriazine shown in Scheme 61 ⟨80AJC2447⟩.

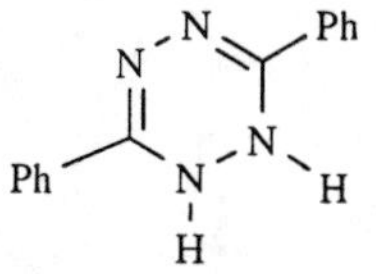

(552) 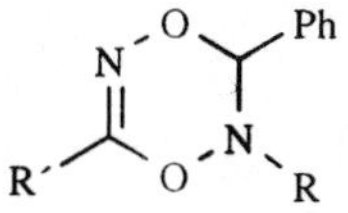   (553) 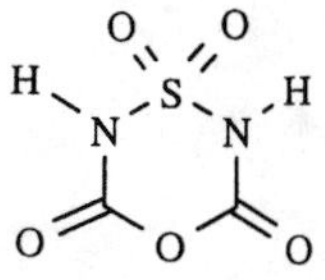   (554)

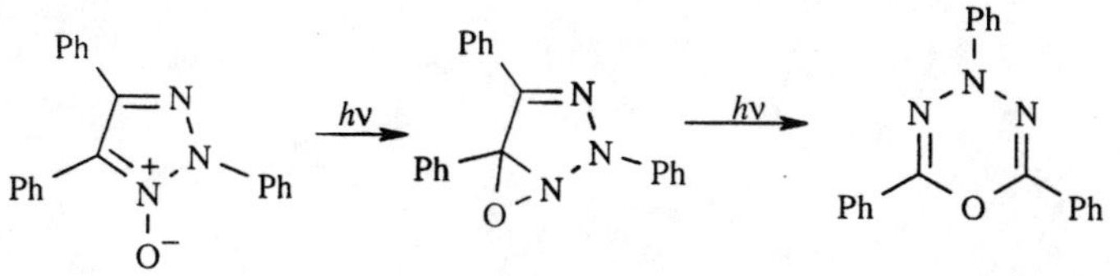

Scheme 61

Of the thiatriazines, the 1,2,4,6-system has been the most studied. There has been particular interest in *N*-alkyl-1,2,4,6-thiatriazine 1,1-dioxides because of their herbicidal, fungicidal, and histamine H$_2$-antagonist activity ⟨86TL123, 87S170⟩. The preparation of 1-chloro-1,2,4,6-thiatriazines (**576**) by reaction of imidoylamidines (**575**) with an excess of sulfur dichloride was reported ⟨80JOU1303⟩.

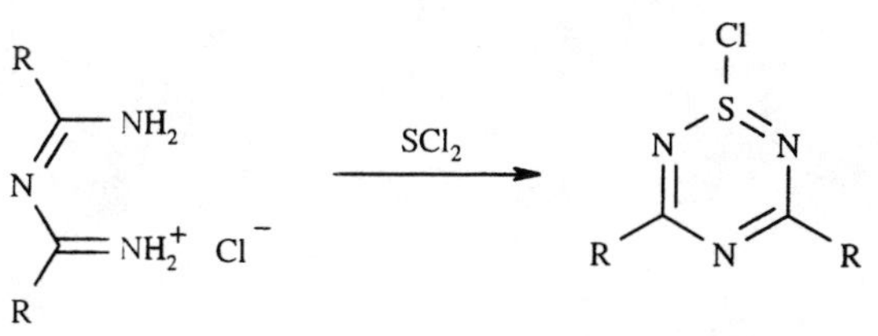

(575)   (576)

A convenient and versatile reaction leading to aryl derivatives of these compounds involves heating an arylamidine with S$_3$N$_3$Cl$_3$ in refluxing acetonitrile ⟨85JA1346⟩. An older reaction which leads to 1,2,4,6-thiatriazines is that of *N*-halo-amidines with thiolates, or with *N*-sulfenylamidines. This reaction provides a wide range of 1-, 3-, and 5-substituted products ⟨54CB1079, 62CB147⟩.

Sulfonyl diisocyanate gives the thiatriazine (**577**) on treatment with ammonia ⟨58CB1200⟩. Ketones and 86% hydrogen peroxide give 3,3,6,6-tetrasubstituted 1,2,4,5-tetroxanes (**578**) ⟨80JCR(S)35⟩.

(577)   (578)   (579)   (580)

Symmetrical 1,2,4,5-dioxadiazines are formed by [3 + 3] dimerization of arene nitrile oxides in the presence of pyridine ⟨74JCS(P1)1951⟩. The sulfonylsulfur diimide (579) cyclizes to the 1,3,2,4,6-dithiatriazine (580) on heating in DMF ⟨71AG(E)264⟩.

4.4

Synthesis of Bicyclic Ring Systems without Ring Junction Heteroatoms

Material in this chapter is arranged firstly by the number of heteroatoms in the heterocyclic ring, secondly by the orientation of the heteroatoms to the fused ring (usually benzenoid), and finally by the size of the heterocyclic ring.

4.4.1 SYNTHESIS BY SUBSTITUENT INTRODUCTION AND MODIFICATION

Most benzo-fused heterocyclic systems are constructed from a substituted benzene by synthesis of the heterocyclic ring. Similarly most bicyclic heterocycles with heteroatoms in both rings commence with a monoheterocycle and build on the second heterocycle. However, substituent modification and, to a lesser extent, substituent introduction are also important, particularly in the later stages of a synthesis, and we now survey available methods for this.

4.4.1.1 In the Heterocyclic Ring

The reactivity of heterocyclic rings is modified but not radically changed by benzo- or hetero-ring fusion. We therefore refer readers to the appropriate sections dealing with the analogous monocyclic rings:
(i) pyrroles, furans and thiophenes (Section 4.2.3.2);
(ii) pyridines (Section 4.2.4.1);
(iii) azoles (Section 4.3.1.2);
(iv) azines (Section 4.3.1.3).
Benzo-ring fusion tends to facilitate reaction: it reduces the loss of resonance energy in non-aromatic transition states and intermediates.

4.4.1.2 In the Benzene Ring

Again, the reactivity of a benzene ring, although modified, is not radically changed by hetero-ring fusion. The reactivity sections of this book have dealt with the reactions of fused benzene rings and we refer now to those sections.

Since thiophene, pyrrole and furan are more readily attacked by electrophiles than is benzene, in the corresponding benzo-fused heterocycles attack generally, but not invariably, occurs in the heterocyclic ring (Section 3.3.3.2.1).

Conversely in benzopyridines and benzazines, electrophilic attack usually occurs in the benzene ring (Section 3.2.3.2).

Benzazoles occupy an intermediate position, but in most cases electrophilic attack occurs in the benzene ring (Section 3.3.3.2.1).

4.4.2 ONE HETEROATOM ADJACENT TO RING JUNCTION

4.4.2.1 Three- and Four-membered Rings

4.4.2.1.1 Three-membered rings

Fused-ring three-membered heterocycles fall into three classes:

605

(i) In the majority of cases the fused ring is saturated or partially saturated: the synthetic methods utilized for the corresponding non-fused systems apply.

(ii) Compounds of the 'benzene oxide' and 'benzene imide' types are tautomeric with, and generally exist predominantly as, the seven-membered oxepin and azepine rings (see Sections 2.5.5.1 and 3.5.2.2). For benzene polyepoxides and related derivatives see CHEC 5.07.6.

(iii) True fused-ring oxirenes and thiirenes are known only as intermediates (see CHEC 5.07.8).

4.4.2.1.2 Four-membered rings

(i) Azetidines, azetidinones. Synthetic routes to penicillins (CHEC 5.11; CHEC-II, 1.20.6), cephalosporins (CHEC 5.10; CHEC-II, 1.19.5) and other fused azetidines (CHEC 5.12; CHEC-II, 1.21.9) are described in the chapters quoted.

Photolytically induced valence bond isomerism of heterocycles is a useful method: 2-amino-5-chloropyridine gives (1) ⟨61JA2967⟩, and (2) → (3) illustrates the conversion of a seven-membered ring into a 4,5-fused derivative ⟨71JOC1934⟩. Benzazetidines are available by ring contraction (4 → 5) ⟨80CC471⟩.

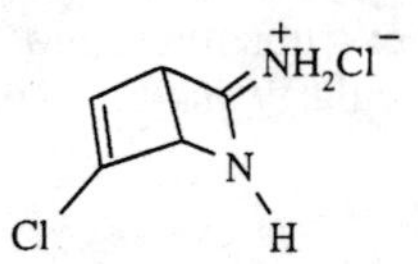

(1) 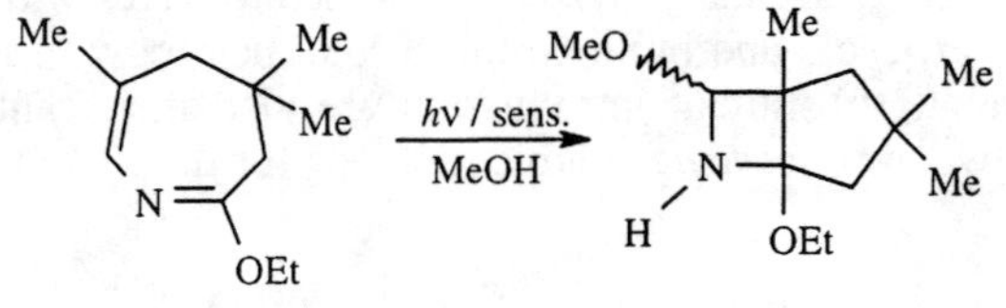 (2) (3)

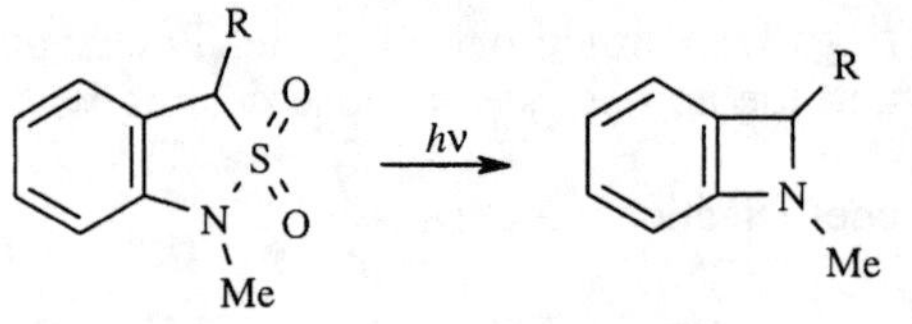

(4) (5)

(ii) Benzazetes are obtained by thermolysis of 1,2,3-benzotriazines (6 → 7) ⟨75JCS(P1)45⟩. The [2 + 2] photocycloaddition of dimethyl acetylenedicarboxylate to 2-pyridone gives (8) ⟨88JHC731⟩ by a thermal rearrangement of the initially formed [2 + 2] cycloaddition product.

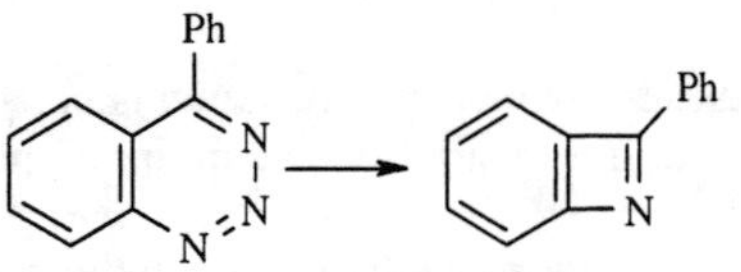

(6) (7) 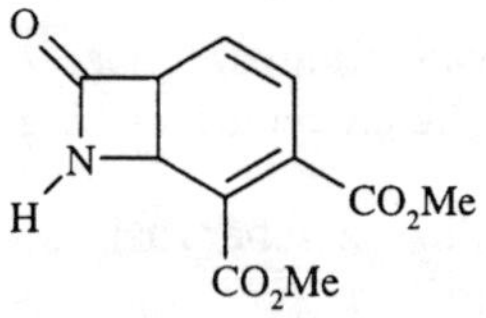 (8) 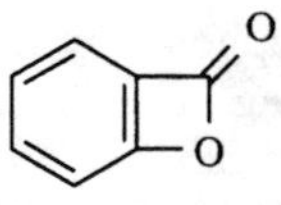 (9)

(iii) Benzoxetan-2-one (9) has been prepared in an argon matrix by CO_2 loss from phthaloyl peroxide ⟨73JA4061⟩.

(iv) The preparation of fused thietane derivatives is discussed in CHEC 5.14.4.2 and CHEC-II, 1.25.

(v) See Sections 3.3.1.8.3 and 3.4.1.10.5 for the preparation of fused oxetanes by [2 + 2] cycloadditions.

4.4.2.2 Five-membered Rings

4.4.2.2.1 Survey of syntheses for indoles, benzofurans and benzothiophenes

$$(10) \qquad (11) \qquad (12) \qquad (13)$$

We deal successively with methods to construct the Z-C(2) **(10)**, the ring-C **(11)**, the C(2)-C(3) **(12)** and the ring-Z bonds **(13)**, and methods from other heterocycles. Table 1 gives an overview of the most important methods for preparing these compounds.

Table 1   Important Ring Syntheses for Indoles, Benzo[*b*]furans, Benzo[*b*]thiophenes and their Derivatives

Ring	Synthesis type	Synthesis name	Section
Indoles	**(10)**	Nenitzescu	4.4.2.2.2.iii
	(10)	Reissert	4.4.2.2.2.iv
	(11)[a]	Fischer	4.4.2.2.3.i
	(11)	Bischler	4.4.2.2.3.iii
	(11)	Gassman	4.4.2.2.3.iv
	(12)	Madelung	4.4.2.2.4.i
	(11)	Heck	4.4.2.2.3.v
Benzo[*b*]furans and benzo[*b*]thiophenes	**(10)**	—	4.4.2.2.2.iv, v, vi
	(11)	—	4.4.2.2.3.ii, iii, iv
	(12)	—	4.4.2.2.4.ii
Indolines and analogues	**(10)**	—	4.4.2.2.2.i
Indoxyls	**(11)**	—	4.4.2.2.3.vi
Oxindoles	**(10)**	—	4.4.2.2.2.ii
	(11)	Brunner	4.4.2.2.3. vii
Indolenines	**(11)**[a]	Fischer	4.4.2.2.3.i
Isatins	**(11)**	—	4.4.2.2.3.viii

[a]Classified under **(11)** from the point of view of precursors although mechanistically should strictly be **(10)** for the indole ring-forming step.

4.4.2.2.2 Ring closure by formation of Z-C (2) bond

(i) Indolines (**15**; Z=NH) and their *S*- and *O*-analogues are prepared from α-substituted β-phenylethyl bromides (**14**) which cyclize spontaneously, on heating or on treatment with alkali.

The intramolecular addition of the heteroatom to a suitably disposed double bond is the basis of a variety of ring syntheses. The cyclization of *o*-allylphenols to 2,3-dihydrobenzofurans (**16** → **17**; X=O) frequently accompanies the Claisen rearrangement of allyl aryl ethers, and is promoted by acid catalysis ⟨68JCS(C)1837⟩. 2,3-Dihydrobenzothiophenes ⟨66JOC413⟩ and 2,3-dihydrobenzoselenophenes ⟨67ZOR597⟩ have been obtained through analogous rearrangements (**16** → **17**; X=S, Se). *o*-Allylanilines can be converted into indolines (**16** → **17**; X=NH) ⟨61JA3319⟩. Cyclization of *o*-(2-chloroallyl)phenols leads directly to 2-methylbenzofurans (**18** → **19**) ⟨76JCS(P1)1⟩.

$$(14) \qquad (15) \qquad (16) \qquad (17)$$

$$(18) \qquad (19)$$

The employment of non-protic electrophiles for the foregoing type of cyclization, as illustrated by (16; X = O) + (MeS)$_2$SMe$^+$SbCl$_6^-$ → (20), leaves a useful point of departure for further transformations ⟨81JCS(P1)3106⟩.

(ii) Oxindoles (22; Z = NH) and the corresponding *S*- and *O*-heterocycles are formed by spontaneous cyclization of acids of type (21).

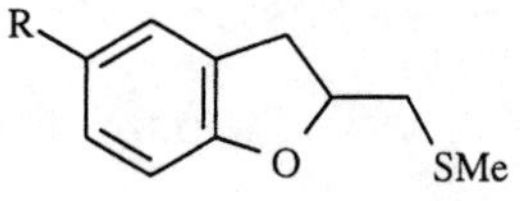

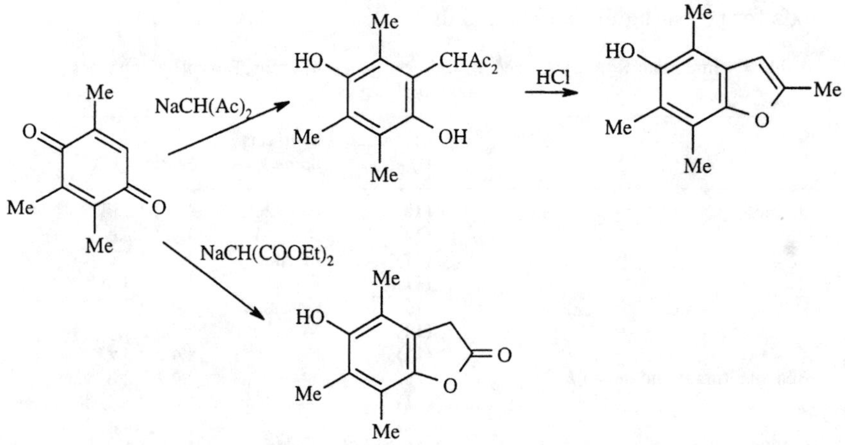

Scheme 1

(iii) The addition of α-keto carbanions to *p*-quinones having at least one unoccupied nuclear position provides syntheses of benzo[*b*]furans and α-benzo[*b*]furanones (Scheme 1) ⟨40JA133⟩. In the Nenitzescu indole synthesis, a quinone is reacted with a β-aminocrotonate (Scheme 2) ⟨51JCS2029, 53JCS1262, 66CI(L)117, 70JA3740⟩.

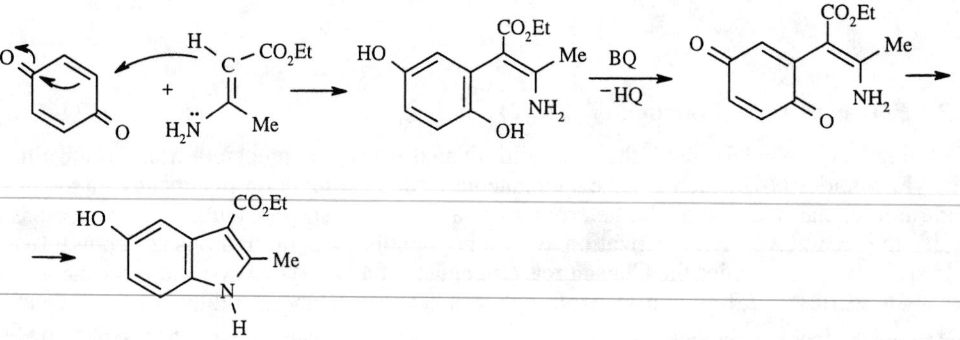

Scheme 2

(iv) Fully aromatic derivatives result from the cyclizations of compounds of type (23) or of equivalents which can be constructed in a variety of ways, and are often not isolated.

Amino groups in compounds of type (23, Z = NH) are frequently derived from nitro groups. In the Reissert indole synthesis, *o*-nitrotoluene undergoes Claisen condensation with oxalic ester to yield the pyruvic ester (24). When this is reduced with Zn-AcOH the corresponding amino derivative spontaneously cyclizes to the 2-ethoxycarbonylindole (25) ⟨63OS(43)40⟩.

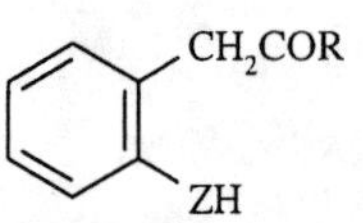

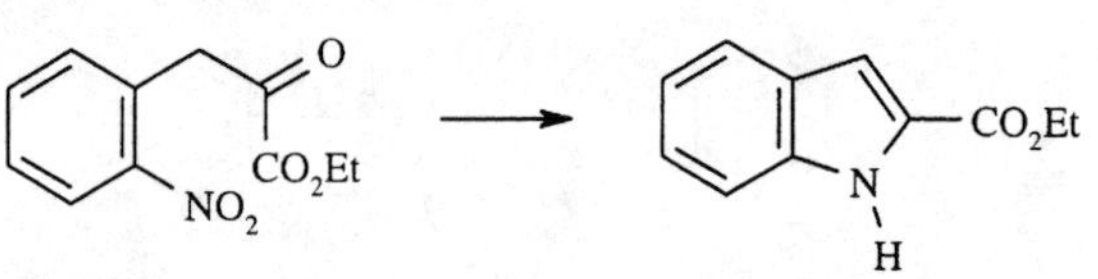

2-Methyl- and 2-aryl-substituted benzo[*b*]thiophenes (**26**) can be synthesized starting from *o*-toluenethiol in three steps, which involves the intramolecular addition of thiols, produced by acid hydrolysis, to the carbonyl group followed by dehydration as the final step (Scheme 3) ⟨91JHC173⟩.

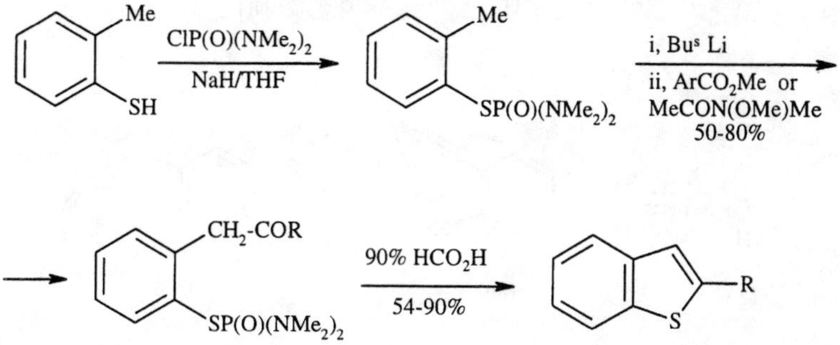

Scheme 3

Reduction of the *o*-nitrophenylnitroethylene (**27** → **28**) ⟨70JOC1248⟩ and the transformation shown in Scheme 4 (R = Me, MeO, CO$_2$Me) ⟨81H(16)1119⟩ are related reactions. Compounds of type (**27**) can be obtained through palladium-catalyzed coupling ⟨86CPB2362⟩.

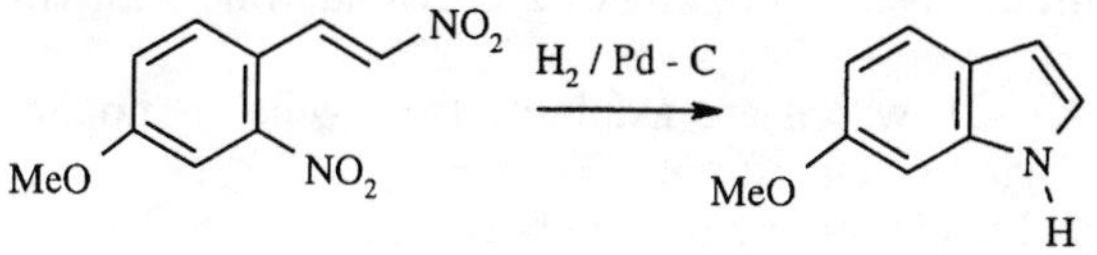

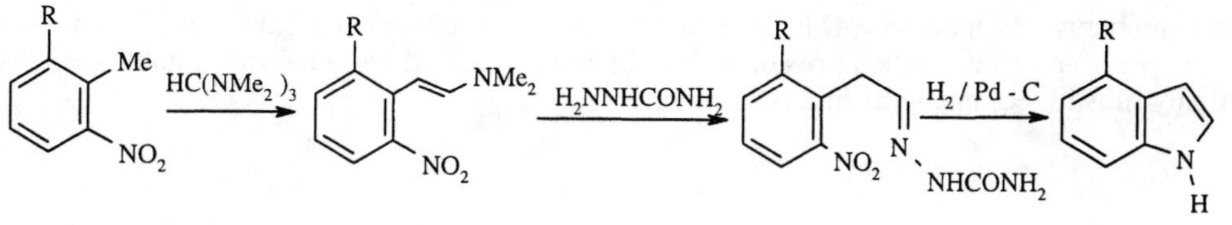

Scheme 4

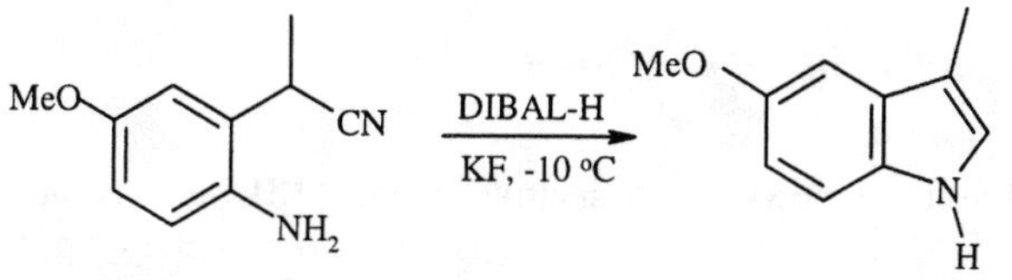

Indoles result from *o*-nitro- and *o*-amino-phenylacetonitriles by reduction of the cyano group, *e.g.* (**29**) → (**30**) ⟨91T7195, 94SC639⟩.

The cyclization of *o*-allyl- or vinyl-phenols into benzo[*b*]furans can be effected by palladium catalysts (Schemes 5 and 6) ⟨73TL739, 75BCJ1533, 75G1151, 76BCJ3662, 77S122, 78CC687, 81JA2318, 85JOC1282⟩.

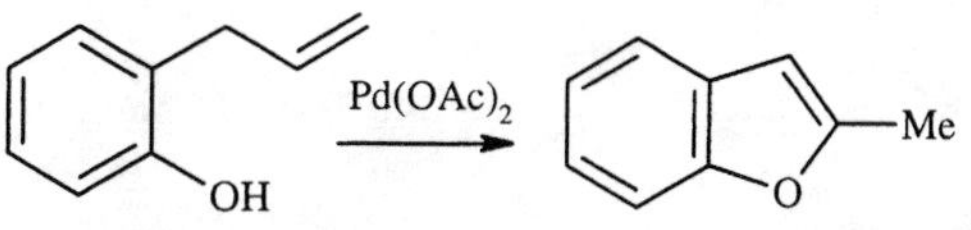

Scheme 5

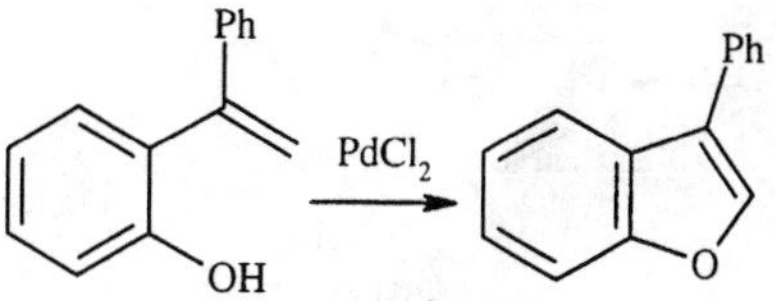

Scheme 6

(v) The addition of dimethylsulfoxonium methylide to carbonyl groups is the basis of a benzo[*b*]furan synthesis from *o*-acylphenols, *e.g.* (**31**) → (**32**) ⟨66TL683⟩. In an analogous synthesis of indoles, *o*-acylanilines react with dimethylsulfonium methylide (**33** → **34**) ⟨70G652⟩.

(vi) The reaction between copper (I) acetylides and *o*-halo-phenols or -anilines provides a general and convenient route to 2-substituted benzofurans or indoles (**35** → **37**; X=O, NH) ⟨66JOC4071, 84JA4218, 85TL5963⟩. The practicalities of this method have been enhanced by improved methods for the synthesis of the required starting materials. *o*-Acetamidophenylalkynes (**36**, X=NCOMe) can be directly cyclized to indoles using PdCl$_2$ as a catalyst. The alkynes can be prepared by copper-catalyzed coupling of terminal alkynes with o-bromoacetanilide ⟨89JOC5856⟩. The required arylalkynes can also be prepared by coupling of aryl triflates with tri-*n*-butylstannylalkynes ⟨89TL2581⟩. Another route to the starting materials involves palladium-catalyzed coupling of 2-ethynylaniline with vinyl or aryl triflates ⟨89TL2581⟩.

Heating *o*-alkynylphenols (**36**, X=O), which are available from *o*-halogenophenols and alkynes, without or with palladium catalysts, yields benzofurans (**37**, X=O).

For ring closure of an *o*-substituted azide on to a double bond see Section 3.3.3.4.3.

4.4.2.2.3   Ring closure by formation of ring—C bond

(i) The Fischer indole synthesis

The Fischer acid-catalyzed conversion of an *N*-arylhydrazone (**38**) into an indole is probably the most versatile method for making indoles; certainly it is the most widely applied ⟨B-82MI 203–01, 88CHE709, 93OPP609⟩. The mechanism involves a [3,3] sigmatropic Claisen-type rearrangement of an enehydrazine tautomer (**39**) to give intermediate (**40**) which spontaneously cyclizes by loss of ammonia, probably *via* indoline (**41**), to an indole (**42**). For unsymmetrical ketones, two isomeric indoles are possible and the general result is that the indole derived from the more stable (usually the more highly substituted) enehydrazine is formed.

The Fischer cyclization is very versatile in terms of the functional groups which can be tolerated at positions 2 and 3, and in the aromatic ring. The reaction is somewhat retarded by strong electron-withdrawing groups but even nitroaryl hydrazones can be cyclized under appropriate conditions ⟨73CHE31, 84CPB2126, 86JMC2415, 88CHE154, 92JMC4086⟩. A variety of protonic and Lewis acids can be used to effect the cyclization. Strong acids such as hydrochloric or sulfuric acid in water, methanol, ethanol or glacial acetic acid are frequently used ⟨86JMC2415, 92JMC4823⟩. Phosphorus trichloride in benzene is mild and effective ⟨81CC563⟩. Polyphosphoric acid is also effective ⟨89TL2099⟩. Zinc chloride is the most commonly used Lewis acid ⟨93JMC2908⟩. A number of solid catalysts for high temperature reaction have been developed and these presumably function as Lewis acids. These include alumina, $Al_2O_3 - MgF_2$, and $AlPO_4$ ⟨83BSB715, 88CHE1191⟩.

Recent detailed studies of the effect of acid strength and solvent isotope effects have provided evidence that the reaction in strongly acidic solution may proceed through an intermediate in which the aromatic ring is protonated ⟨91JOC3001, 93JOC228⟩. This would be expected to substantially lower the activation energy of the sigmatropic rearrangement step. The interpretation of the effect of acid strength is that the less substituted (but kinetically preferred) enehydrazine is formed by C-deprotonation and rapidly rearranges (Scheme 7).

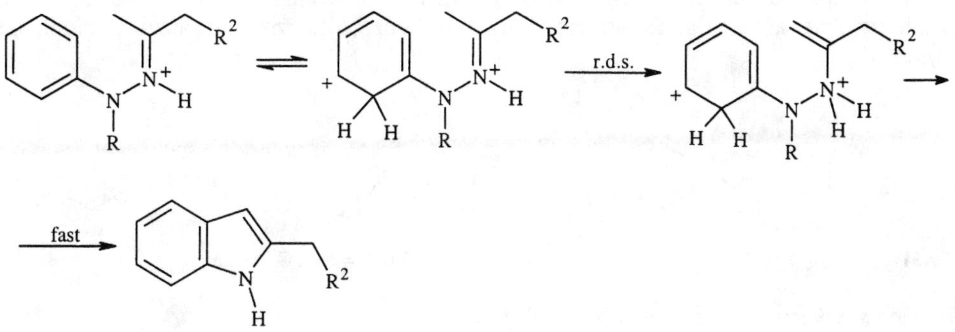

Scheme 7

Electron-releasing *meta*-substituents in the aryl ring favor 6- over 4-substituted indoles, whereas the opposite is usually true for electron-withdrawing substituents ⟨57JCS3175⟩. Formal 1,4-methyl group migrations have been observed in the cyclization of mesitylhydrazones (**43** → **44**) ⟨80JA4772⟩. The preferred direction of cyclization of arylhydrazones of unsymmetrical ketones varies with the acid catalyst ⟨69JCS(B)446⟩: high acidity and temperature favor cyclization to the less substituted position ⟨78JOC3384⟩. However, the arylhydrazone of a ketone $R^1CH_2COCHR_2^2$ often yields the 3*H*-indole (**45**) in preference to the 1*H*-indole (**46**) ⟨55JCS2519⟩. If formation of an endocyclic double bond is precluded, then a methylene-indoline may result (**47** → **48**).

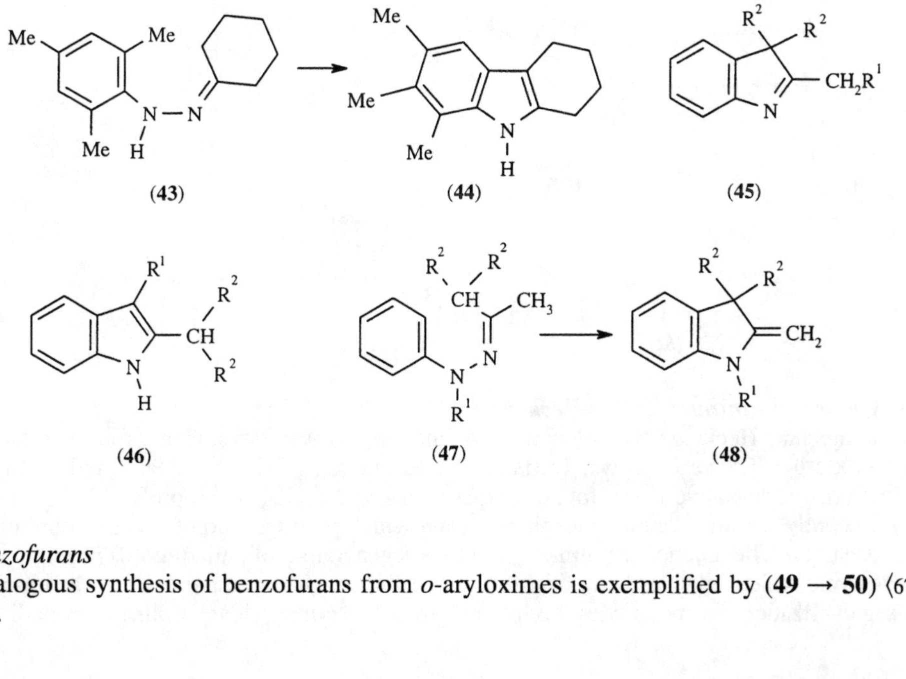

(ii) Benzofurans

An analogous synthesis of benzofurans from *o*-aryloximes is exemplified by (**49** → **50**) ⟨67TL2867, 73KGS31⟩.

(49)   **(50)**

(iii) Bischler indole synthesis and related methods for O- and S-analogues

The Bischler synthesis of indoles from α-arylamino ketones is acid catalyzed (**51** → **52**) ⟨1892CB2860, 1893CB1336⟩. α-Aryloxy and α-arylthio ketones can be cyclized similarly to benzo[*b*]furans ⟨1900LA(312)237⟩ and benzo[*b*]thiophenes ⟨49RTC509, 70JCS(C)2621⟩ respectively. Concurrent migrations of groups occur from position 3 to position 2, especially under vigorous conditions.

To obtain compounds unsubstituted at positions 2 and 3, cyclization of the acetals (**53**) using polyphosphoric acid catalysis gives indoles ⟨81JOC778⟩, benzo[*b*]furans, and benzo[*b*]thiophenes (**54**) ⟨71T1253⟩.

(51) X = NR, O, S   **(52)**   **(53)** X = NR , O , S , Se   **(54)**

(iv) Gassman synthesis of indoles and benzo[b]furans

The Gassman synthesis of indoles depends on a sigmatropic rearrangement to generate the ring-carbon bond. An *N*-chloroaniline react with a β-keto sulfide to form a sulfonium salt (**55**); this is deprotonated to an ylide (**56**) which then rearranges and cyclizes (**56** → **57** → **58**); desulfurization then gives an indole ⟨79JA5512⟩. Likewise, an α-ethoxycarbonyl sulfide in place of the β-keto sulfide leads *via* (**60**) to an oxindole (**61**) or isatin (**59**) ⟨80JCR(S)347⟩. The synthetic approach has been extended to benzo[*b*]furans (**62** → **63**) ⟨75SC325⟩.

(55)   **(56)**   **(57)**   **(58)**

(59)   **(60)**   **(61)**

(62)   **(63)**

(v) Indole synthesis by intramolecular Heck reaction

The intramolecular Heck reaction of *N*-allyl-*o*-haloanilines was developed as an effective indole synthesis ⟨80JOC2709⟩. The reaction was initially carried out using Pd(OAc)$_2$, Ph$_3$P and triethylamine. An examination of alternative bases found CH$_3$CO$_2$Na and Na$_2$CO$_3$ to be preferable in some cases ⟨87TL5291⟩. Recently, a water-soluble phosphine ligand which permits mild cyclization conditions was introduced ⟨92SL715⟩. The reaction presumably requires regeneration of palladium(0) *in situ* to maintain the catalytic cycle which involves oxidative addition at the aryl halogen bond. The reaction proceeds by 5-*exo*-trig cyclization *via* indolinylmethylpalladium and *exo*-methyleneindoline intermediates. The latter are sometimes isolated ⟨93JCS(P1)1941⟩. The reaction is applicable to both ring- and *N*-substituted indoles (Scheme 8) ⟨91JOC3048, 92TL8011⟩.

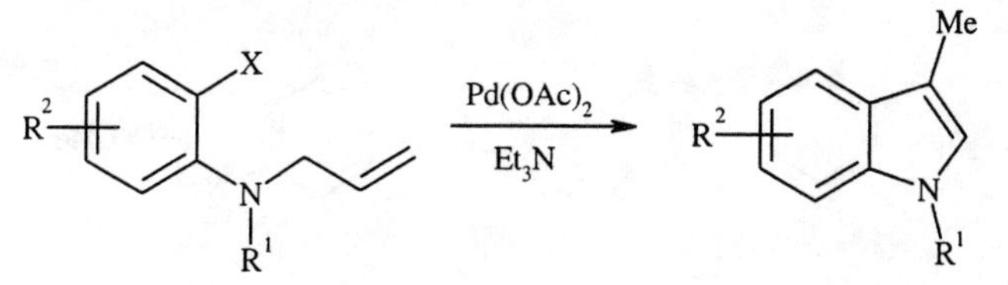

X = Br, I;   R¹ = H, SO₂Me *etc.*;   R² = H, OMe, NO₂ *etc.*

Scheme 8

(vi) Indoxyls and their analogues

Indoxyls and their oxygen and sulfur analogues are prepared by the cyclization of anilino-, phenoxy- and phenylthio-acetic acids, respectively (**64** → **65**), with NaNH₂ (for Z = NH), P₂O₅ (for Z = O) and H₂SO₄ (for Z = S).

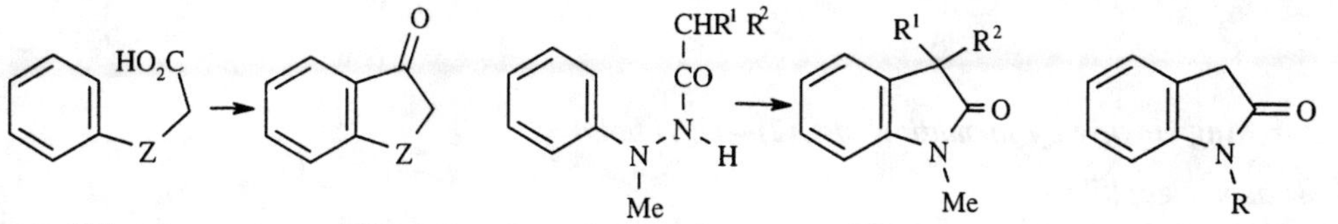

(**64**)   (**65**)   (**66**)   (**67**)   (**68**)

(vii) Oxindoles

Under conditions (CaO, 200°C) different to those used to convert hydrazones in the Fischer indole synthesis, phenylhydrazides (**66**) give oxindoles (**67**) (Brunner synthesis). Treatment of *N*-chloroacetyl derivatives of primary or secondary arylamines with aluminum chloride also provides oxindoles (PhNRCOCH₂Cl → **68**) ⟨54JCS1697⟩.

(viii) Isatins

Intramolecular acylation is involved in both isatin syntheses shown: (**69** → **70** → **71**) and (**72** → **71**) ⟨75AHC(18)1⟩.

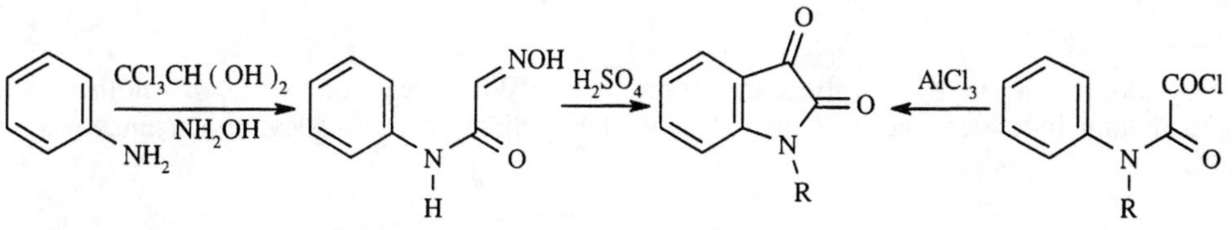

(**69**)   (**70**)   (**71**)   (**72**)

(ix) Reaction via arynes

The intramolecular addition of a carbanion to an aryne has been applied to the synthesis of indoles (**73** → **74** → **75**) ⟨75CC745⟩ and oxindoles (**76** → **77** → **78**) ⟨80JA3646⟩.

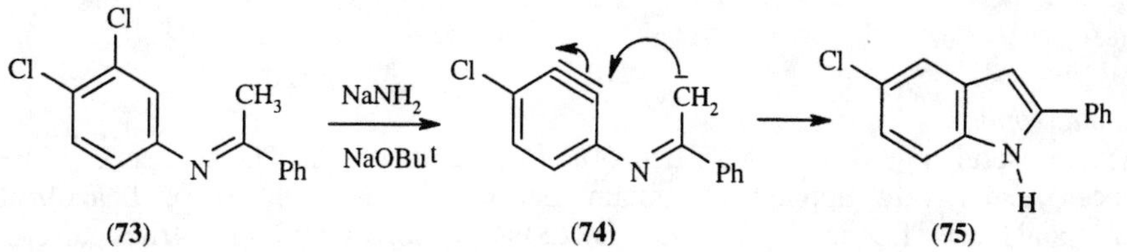

(**73**)   (**74**)   (**75**)

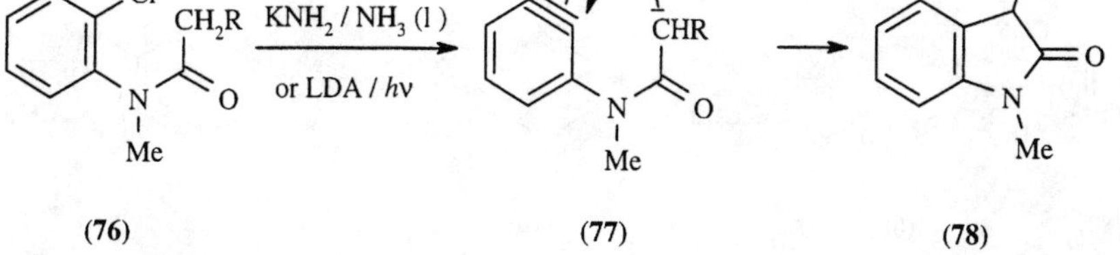

(**76**)   (**77**)   (**78**)

(x) Photochemically mediated cyclizations

These have been used to produce indolines ⟨80T1757⟩ and 2,3-dihydrobenzofurans ⟨78JA2150⟩ (Scheme 9).

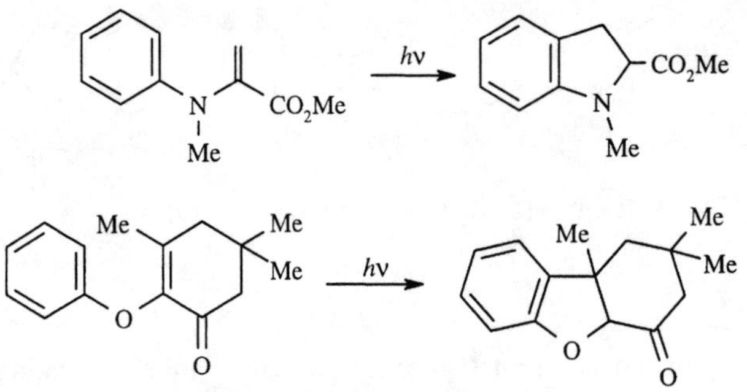

Scheme 9

4.4.2.2.4 *Ring closure by formation of C(2)—C(3) bond*

(i) C(3) as nucleophile

The Madelung synthesis of indoles (**79** → **80**) from *N*-acyl-*o*-toluidines originally necessitated heating with sodamide at 250°C; however, the stronger bases *n*-butyllithium or LDA cause reaction at 20°C ⟨81JOC4511⟩. Milder conditions can also be employed if the methyl group is activated as in (**81** → **82**) ⟨68JA7008⟩.

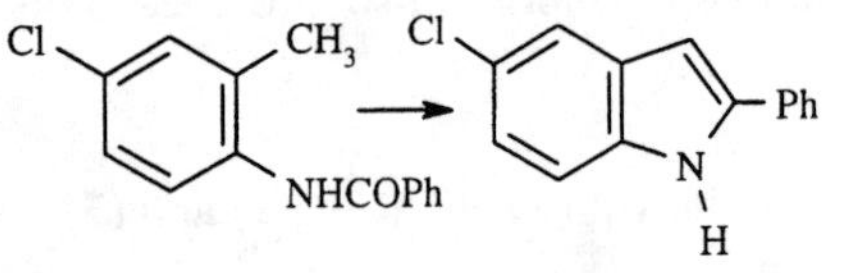

(79) **(80)**

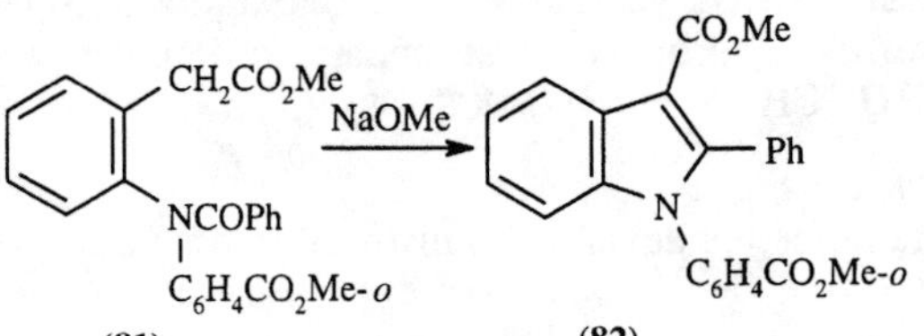

(81) **(82)**

The ring closure (**83** → **84**) involves an intramolecular Wittig reaction ⟨81CC14⟩. Another variation on the Madelung indole synthesis is provided by the cyclization of *o*-isocyanobenzenes (**85** → **86**) ⟨77JA3532⟩.

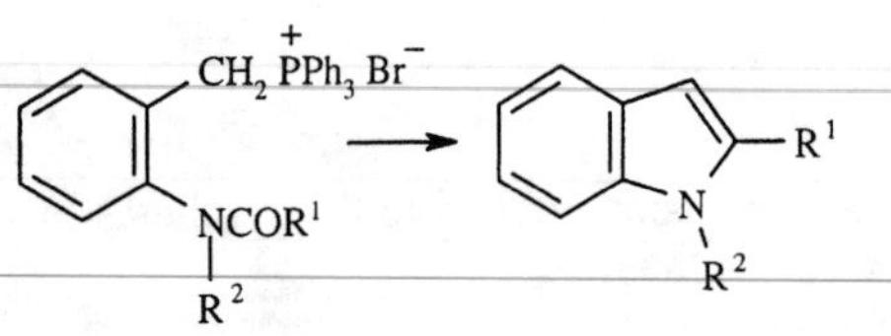

(83) **(84)**

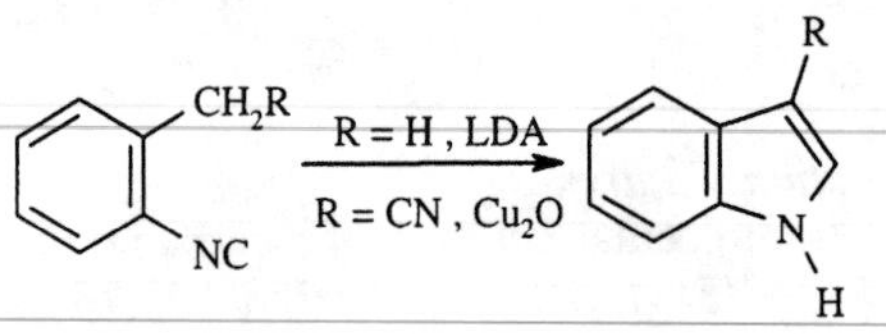

(85) **(86)**

(ii) C(2) as nucleophile

These include useful ring syntheses for benzo[*b*]-fused compounds. The sequence (**87**) → (**88**) → (**89**) has been extensively applied to obtain benzo[*b*]furans ⟨48JCS2254⟩, benzo[*b*]thiophenes ⟨31LA(488)259⟩ and, less frequently, indoles ⟨27JCS1937⟩. Corresponding nitriles afford 3-amino derivatives, *e.g.* (**90**) → 3-amino-2-aroylbenzofurans.

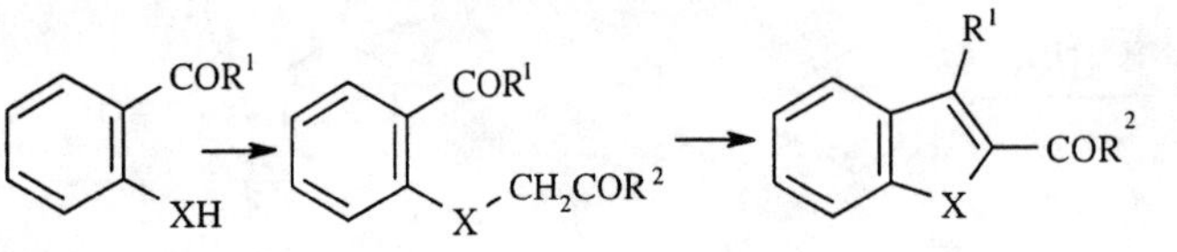

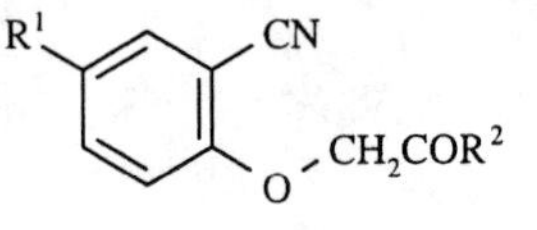

(87) **(88)** **(89)** **(90)**

Halomethylcarbonyl compounds provide one-carbon components. For example, phenacyl bromide with *o*-acylanilines leads to indoles (**92**) ⟨72JOC3622⟩ by intramolecular aldol condensation of intermediate (**91**).

(**91**)   (**92**)

4.4.2.2.5 Ring closure by formation of ring—Z bond

This is the least common of the four types of ring closure. It is illustrated by the transformations (**93**) → (**94**) ⟨72JHC879⟩ and (**95**) → (**96**) ⟨75CR(C)(281)793⟩. The latter illustrates the use of nitrene intermediates.

(**93**)   (**94**)   (**95**)   (**96**)

Both these examples deal with annulations to heterocyclic rings, but analogous transformations can lead to benzothiophenes and indoles, respectively. Conversely, annulations to heterocyclic rings can also utilize the methods enumerated in the previous sections.

4.4.2.2.6 From other heterocycles

Preparations from six-membered rings by reactions analogous to those known for acyclic compounds are illustrated by the Wolff rearrangement (**97** → **98**) and analogues of the Meerwein rearrangement (**99** → **100**) ⟨72JCS(P1)787⟩ and (**101** → **102**) ⟨75JHC981⟩.

(**97**)   (**98**)   (**99**)   (**100**)

(**101**)   (**102**)

Quinoline 1-oxides can be rearranged photochemically into indoles (**103** → **104** → **105**) or *N*-hydroxyindoles (**106** → **107**). Cinnolines are reductively ring-contracted into indoles (Section 3.2.1.6.9.ii).

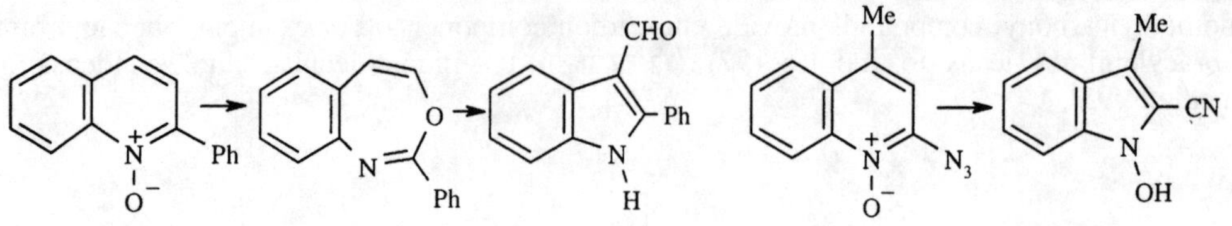

(103)   **(104)**   **(105)**   **(106)**   **(107)**

The carbocyclic ring of substituted indolines can be constructed by intramolecular Diels-Alder reactions of 2-(*N*-pent–3-ynylamino)pyridazines (Scheme 10) ⟨84JOC2240⟩.

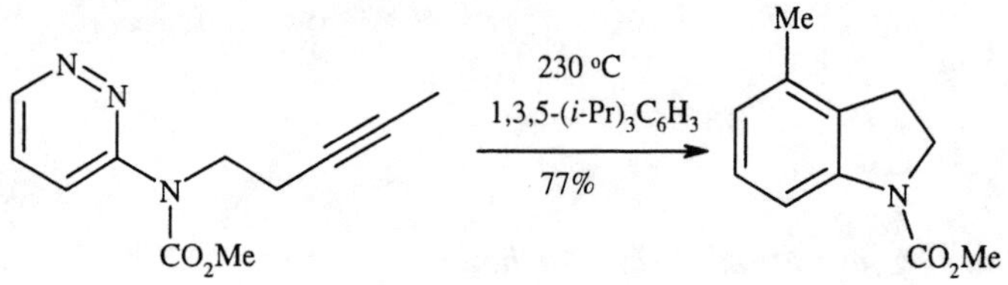

$$230\ ^\circ C$$
$$1,3,5\text{-}(i\text{-Pr})_3C_6H_3$$
$$77\%$$

Scheme 10

4.4.2.3 Six-membered Rings

4.4.2.3.1 *Survey of synthetic methods for quinolines, benzo[b]pyrans and their derivatives*

(108)   **(109)**

The important methods involve ring closure of *o*-substituted anilines and phenols (type **108**) and cyclization of *o*-unsubstituted aniline, *etc.*, derivatives (type **109**). Additionally, cycloadditions and transformations from other heterocycles are considered. Table 2 gives an overview of the important methods for preparation of derivatives of these types.

Table 2   Important Ring Syntheses for Quinolines, Benzo[*b*]pyrans, Benzo[*b*]thiins and their Derivatives

Ring	Synthesis type	Synthesis name(s)	Section
Quinolines	(108)	Friedländer, Pfitzinger	4.4.2.3.2.i (*a, b*)
	(109)	Skraup, Doebner-von Miller, Baeyer, Riehm	4.4.2.3.3.ii
Quinolones	(108)	Camps	4.4.2.3.2.ii
	(109)	β-Keto ester	4.4.2.3.3.i (*a*)
Tetrahydroquinolines	(109)	—	4.4.2.3.3.iii (*a*)
Benzo[*b*]pyryliums	(108)	—	4.4.2.3.2. i (*c*)
Coumarin, and chromones	(108)	Kostanecki-Robinson	4.4.2.3.2.i (*d*)
Coumarins	(109)	von Pechmann	4.4.2.3.3. i (*b*)
Chromones	(109)	Simonis	4.4.2.3.3.i (*b*)
Chromans and tetrahydrobenzothiins	(109)	—	4.4.2.3.3.iii (*a*)

4.4.2.3.2 *Ring closure of o-substituted anilines or phenols*

(*i*)   *From or* via *o-substituted cinnamoyl derivatives*

ortho-Substituted benzenes of type (**111**; Z = O, S, NH) can undergo ring closure (**111** → **110**, **112**). Amines of type (**111**; Z = NH), which usually cyclize spontaneously, are often prepared *in situ* by reduction of nitro compounds, *e.g. o*-nitrocinnamic acid with $(NH_4)_2S$ gives 2-quinolone.

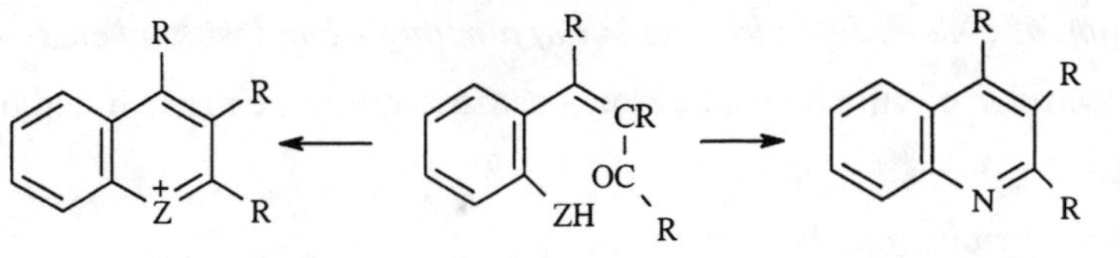

(110)   (111)   (112)

Important reactions that involve an aldol reaction to form the intermediate (111) *in situ* include:

(a) The Friedländer synthesis of quinolines from *o*-amino benzaldehydes and ketones (*e.g.* **113** → **114**).

(b) The Pfitzinger synthesis of quinoline–4-carboxylic acids from a ketone and isatinic acid (obtained *in situ* from isatin), *e.g.* (**115**) yields (**116**).

(c) The preparation of benzopyrylium ions from ketones and *o*-acylphenols (**118** → **117**).

(d) The Kostanecki-Robinson synthesis which can lead to coumarins (**118** → **119**) or chromones (**120** → **121**).

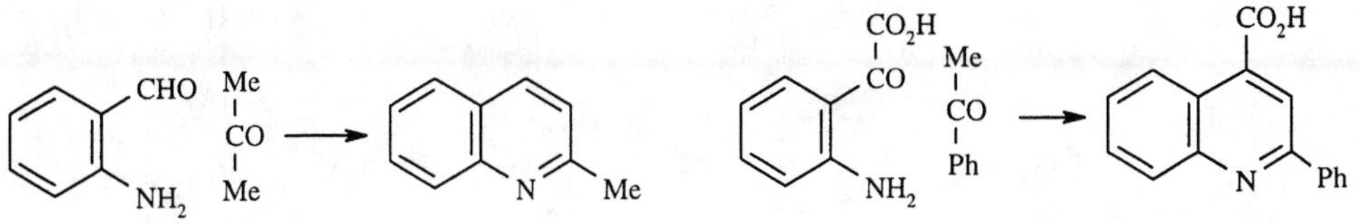

(113)   (114)   (115)   (116)

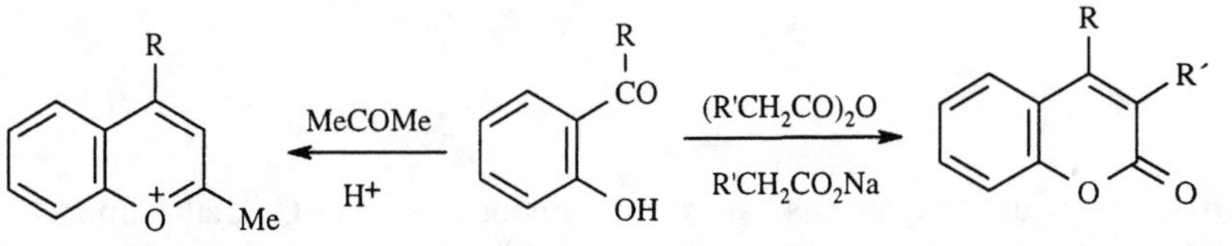

(117)   (118)   (119)

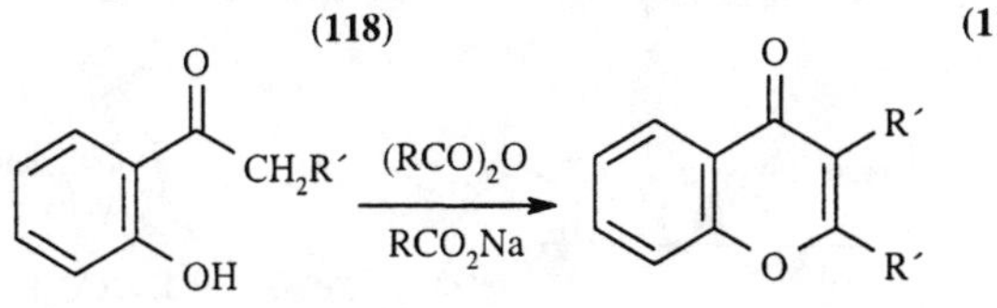

(120)   (121)

(ii) From other o-substituted benzenes

Standard reactions of aliphatic chemistry can be applied; for example, chroman (**123**) can be prepared by ring closure of (**122**). 4-Quinolones result from the Camps reaction (**124** → **125**), and flavonone (**127**) by cyclization of (**126**).

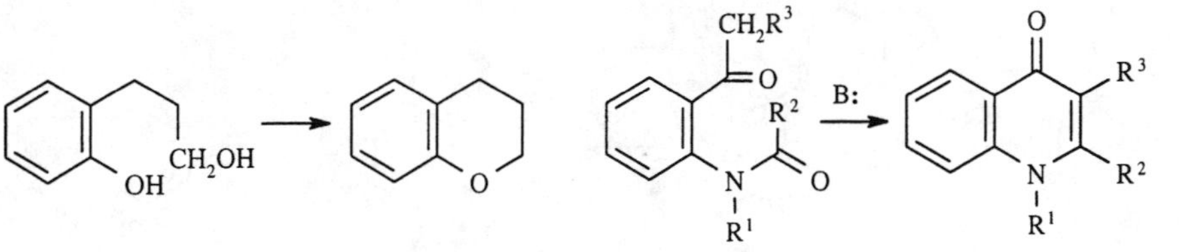

(122)   (123)   (124)   (125)

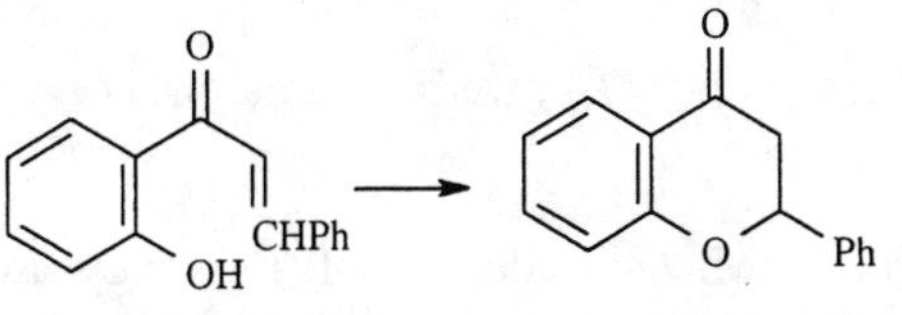

(126)   (127)

4.4.2.3.3 *Formation of a C—C bond by reaction of a multiple bond with a benzene ring*

These reactions involve electrophilic attack on a benzene ring which is activated by the heteroatom (as in **128**).

(i) *Quinolones, benzopyrones and benzothiinones*

(a) Anilines and β-keto esters (**131**) give either Schiff's bases (**132**) at 20°C, or at 100°C the more slowly formed but more stable amides (**130**). Cyclization of the amide (**130**) with H_2SO_4 at 100°C yields the 2-quinolone (**129**), whereas the Schiff's base (**132**) is converted into the 4-quinolone (**133**) at 280°C.

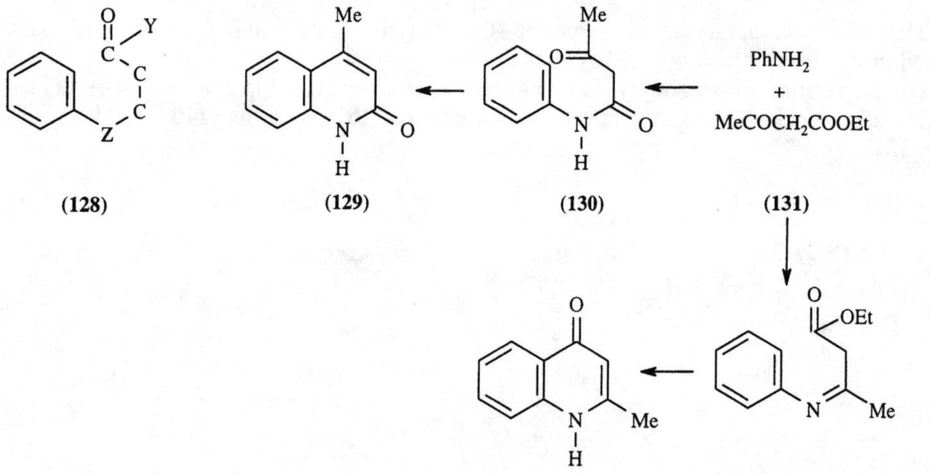

(128)   (129)   (130)   (131)

(133)   (132)

The most frequently used synthesis of 4-quinolinones is the Conrad–Limpach cyclization ⟨77HC(32)139, 77HC(32)146⟩ and its modification due to Gould and Jacobs (Scheme 11) ⟨92T7373⟩.

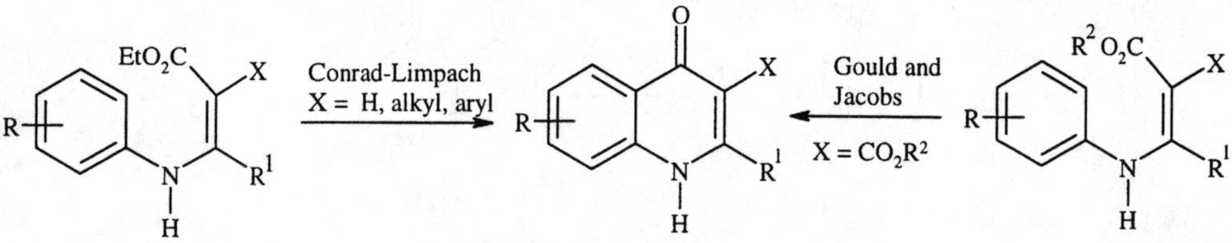

Scheme 11

(b) Phenols and β-keto esters (**135**) give either coumarins (**134**; von Pechmann reaction) or chromones (**136**; Simonis reaction) under the conditions indicated.

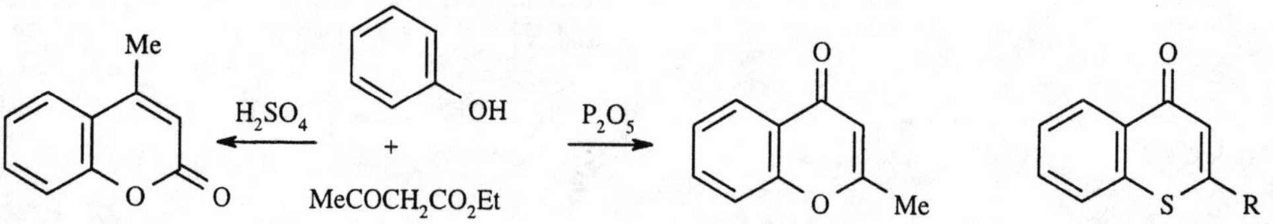

(134)   (135)   (136)   (137)

(c) Thiophenols and β-keto esters (with P_2O_5) give benzothiinones (**137**).

(ii) *Quinolines*

In the reactions listed in Table 3, Michael addition of a primary aromatic amine to an α, β-unsaturated aldehyde or ketone (prepared *in situ*) is followed by cyclization and oxidation of the intermediate dihydroquinoline to a quinoline (**138** → **141**).

Table 3 Formation of Quinolines from Anilines

Name of reaction	Starting materials	Catalyst	Intermediate carbonyl compound	Oxidizing agent
Skraup	Glycerol	H_2SO_4	$CH_2=CHCHO$	As_2O_5, $ArNO_2$,[a] or m-$NO_2C_6H_4SO_3H$
Doebner-von Miller	RCHO and R'CH$_2$CHO	$ZnCl_2 - HCl$	$RCH=CR'CHO$	$ArN=CHR$[b]
Baeyer	RCHO and R'CH$_2$COR	HCl, 20°C	$RCH=CR'COR$	$ArN=CHR$[b]
Riehm	RCOR and R'CH$_2$COR	HCl, 200°C	$R_2C=CR'COR$	None (RH lost from product)[c]

[a]Nitro compound corresponding to the amine. [b]Schiff's base from RCHO and amine.
[c]Dihydroquinolines, *e.g.* (**140**; R = Me), can be isolated.

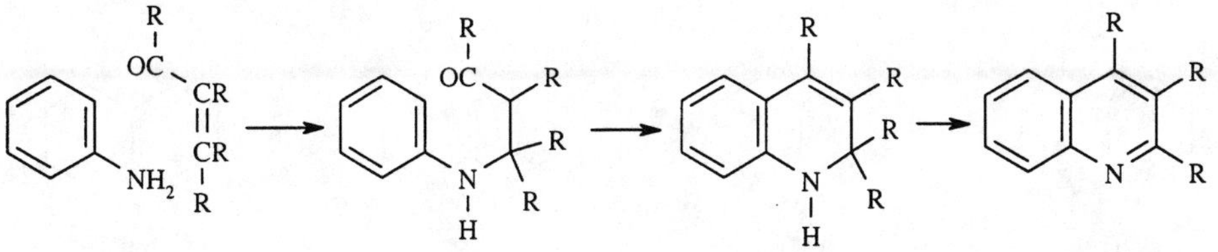

(**138**)   (**139**)   (**140**)   (**141**)

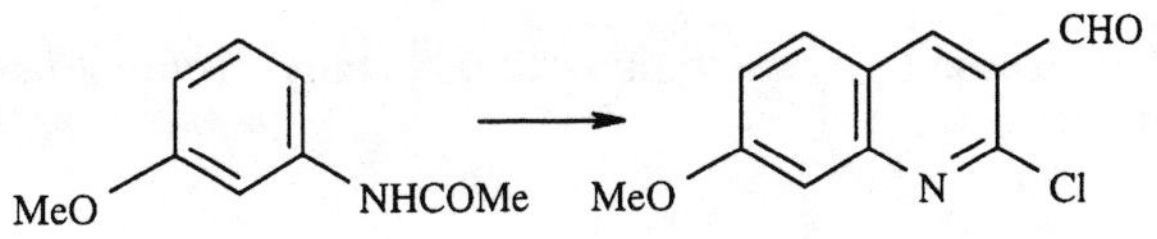

(**142**)   (**143**)

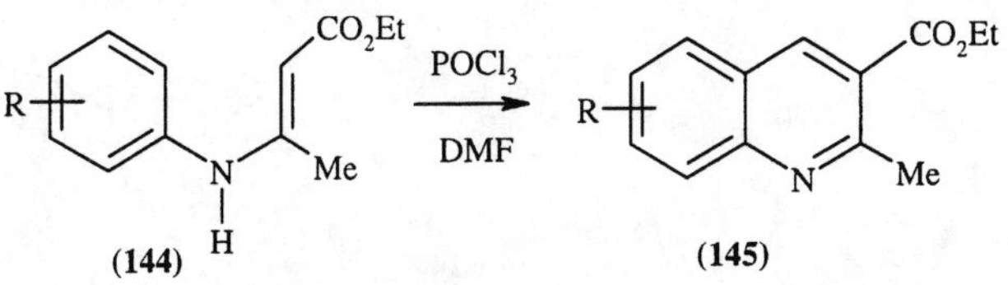

(**144**)   (**145**)

R = OMe or NO$_2$, 60–73 %

Acylanilines react with the Vilsmeier-Haack reagent (POCl$_3$ − HCONMe$_2$) to give quinolines in good yield (*e.g.* **142** → **143**); the reaction has been reviewed ⟨93H(35)539⟩. Reaction of the Vilsmeier reagent with the vinylogous enamides (**144**) gives quinolines (**145**) ⟨83TL517⟩.

Photochemical cyclization of unsaturated imines of type **145a** can give excellent yields of quinolines (Scheme 12) ⟨93TL5321⟩.

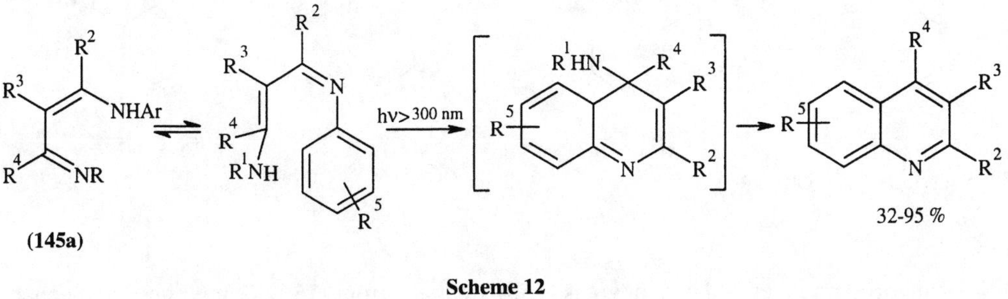

(**145a**)   32–95 %

Scheme 12

The Combes synthesis of quinolines (Scheme 13) has been reviewed ⟨92CHE845⟩.

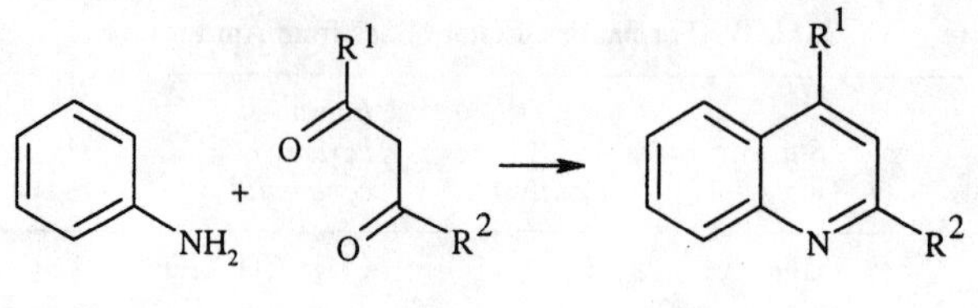

Scheme 13

(iii) Partially saturated rings

Ring closure of compounds of types (**146**), (**148**), (**150**), (**152**) and (**154**) can give a wide variety of partially saturated rings.

(a) Tetrahydroquinolines, chromans and tetrahydrobenzothiins (**147**) result from (**146**; Y = OH, OR, OTs or halogen) and from reactions of type (**148** → **149**).

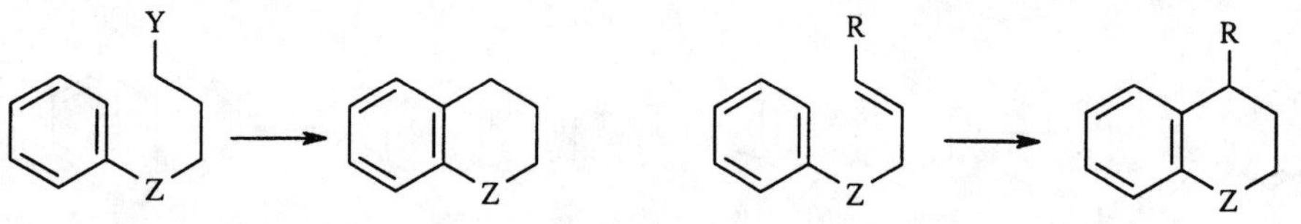

| (146) | (147) | (148) | (149) |

Mercuric acetate-induced cyclization gives a high yield of tetrahydroquinolines from 2-(2-propen-1-yl)anilines (Scheme 14) ⟨91AJC1749⟩.

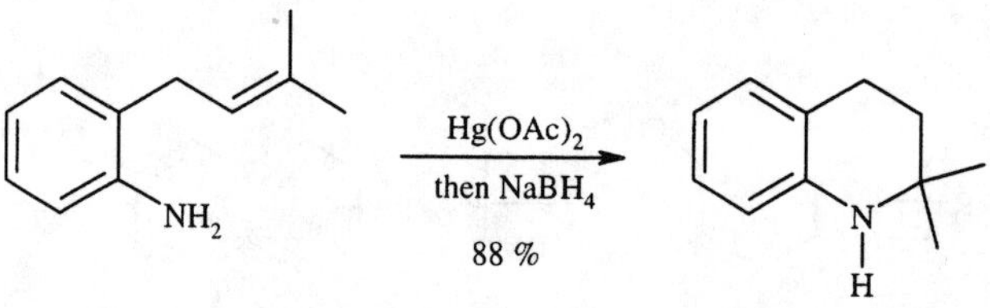

Scheme 14

(b) Dihydro-4-quinolones, chromanones or dihydrobenzothiinones (**151**) are obtained from (**150**; Y = halogen, OH, OR or NR₂).

(c) Imino derivatives (**153**) are formed by the cyclization of nitriles (**152**).

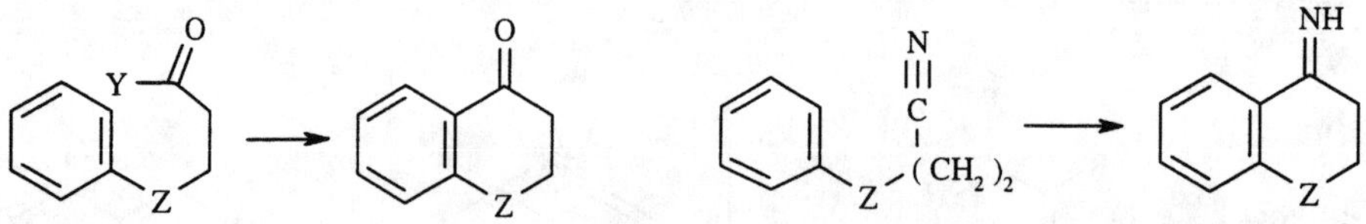

| (150) | (151) | (152) | (153) |

Y = halogen , OH , OR , NR₂

(d) 1,2-Dihydroquinolines and benzopyrans (**155**) are made from (**154**). Sometimes the mechanism is complex. The propargyl ether (**156**) is converted on heating into 2*H*-chromene (**157**) *via* a Claisen rearrangement, a 1,5-hydrogen shift and electrocyclic ring closure (Scheme 15).

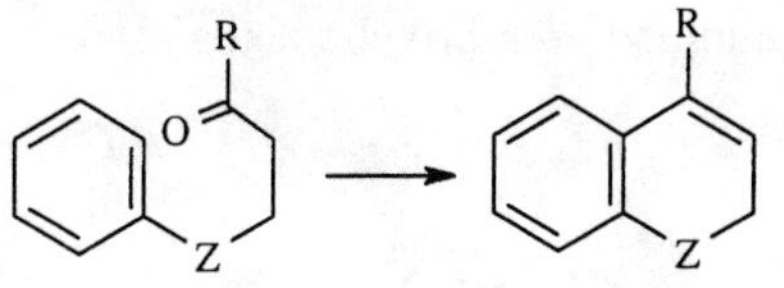

(154)   (155)

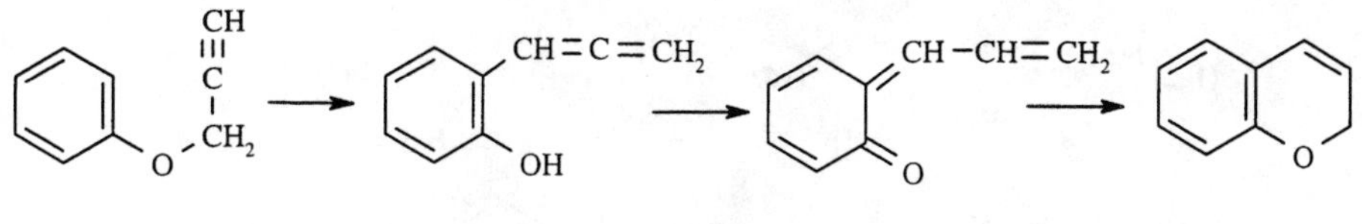

(156)   (157)

Scheme 15

4.4.2.3.4 Synthesis via cycloaddition reactions

Intramolecular Diels-Alder reactions simultaneously form both rings. Precursors can be used which do *not* contain a heteroatom in either the 4π or the 2π component (*e.g.* Scheme 16).

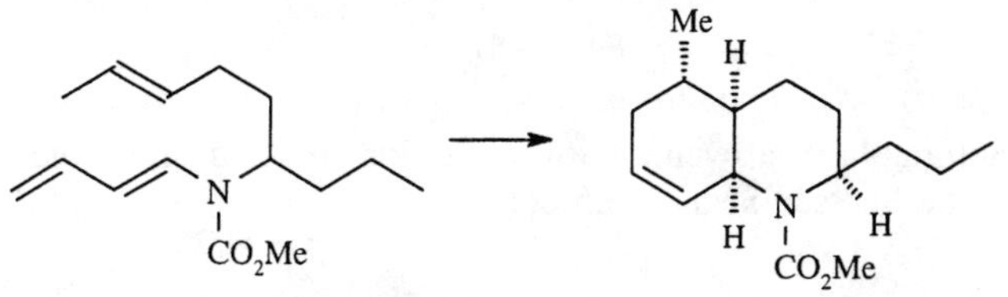

Scheme 16

Intermolecular cycloaddition reactions include that of (**158**) and RCH=CHOEt to give quinoline (**159**). The 2-quinolone (**161**) is obtained from (**160**) and HC≡COEt.

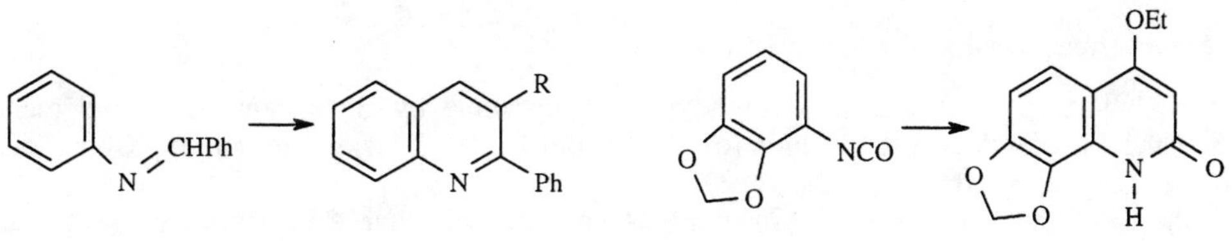

(158)   (159)   (160)   (161)

4.4.2.3.5 Synthesis from heterocycles

Indoles with dibromocarbene are ring-expanded to 3-bromoquinolines (**162**) and benzofurans behave analogously (Section 3.3.1.7.1). Isatins (**163**) with CH_2N_2 form 3-hydroxy-2-quinolones (**164**).

With malonic acid they form 2-quinolone-4-carboxylic acids (166) by acid-catalyzed rearrangements of intermediates (165).

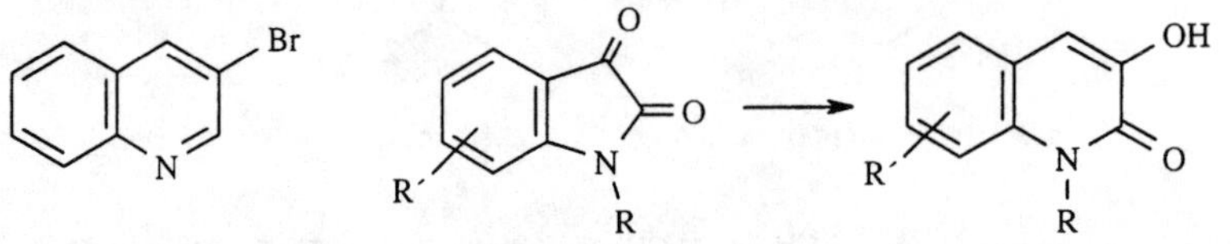

(162)   (163)   (164)

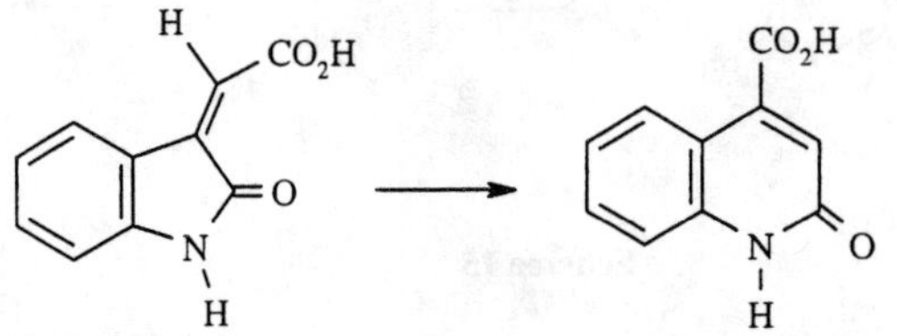

(165)   (166)

Anthranils undergo ring expansion to quinoline derivatives on treatment with various *C*-nucleophiles (see Scheme 17).

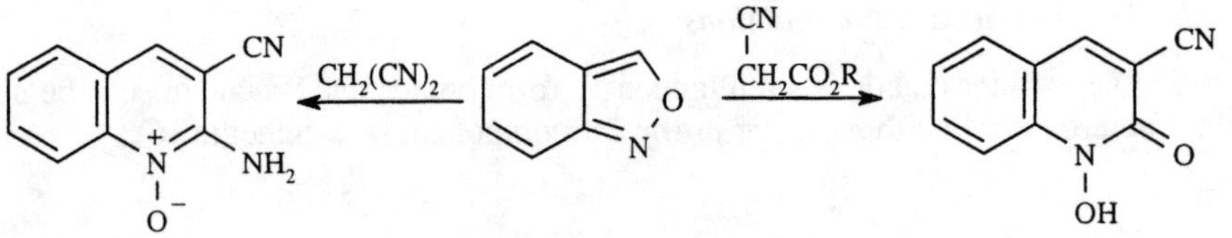

Scheme 17

Alkynes react with anthranils in a cycloaddition reaction to form quinolines (Scheme 18). For a similar reaction of benzazetidine see Section 3.5.8.1.2.

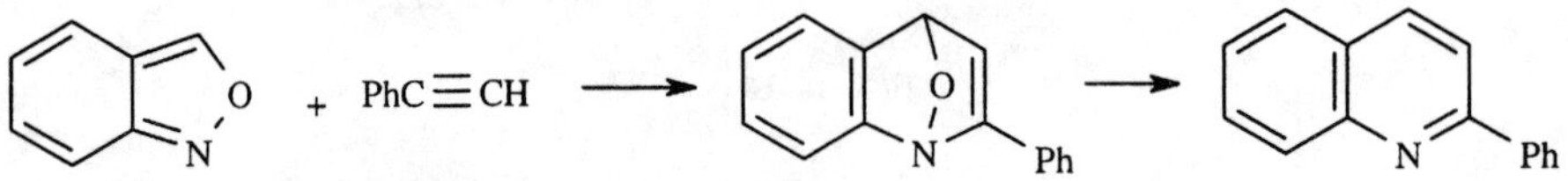

Scheme 18

4.4.2.4 Seven-membered and Larger Rings

Tetrahydro-1-benzazepin-2-ones are formed from α-tetralone by Beckmann or related reactions (CHEC 5.16.4.1.1). Classical ring closures (of the Friedel-Crafts, Dieckmann, *etc.* types) can also be applied to benzazepine synthesis ⟨74AHC(17)45⟩.

The synthesis of the benzoxepin (167) involves an intramolecular Wittig reaction (Scheme 19) ⟨68JOC2591⟩.

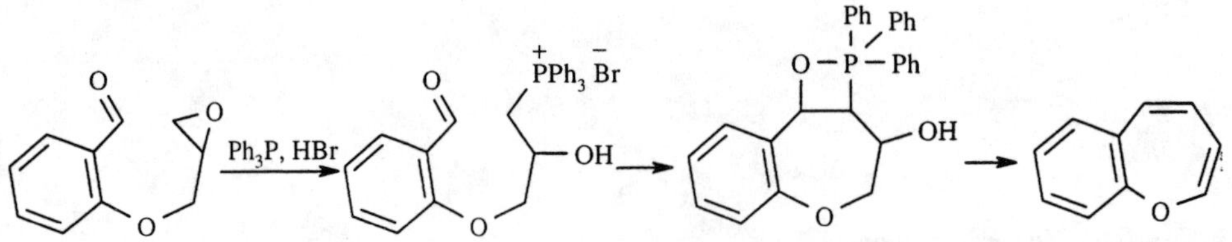

(167)

Scheme 19

Diels-Alder addition of the benzazete (**168**) to tetraphenylcyclopentadienone (**169**) followed by CO loss yields the pentaphenylbenzazocine (**170**) ⟨75JCS(PI)45⟩. Intramolecular Wittig reaction of (**171**) gives some 3,4-dihydro-2*H*-benzoxocin ⟨74JOC3038⟩.

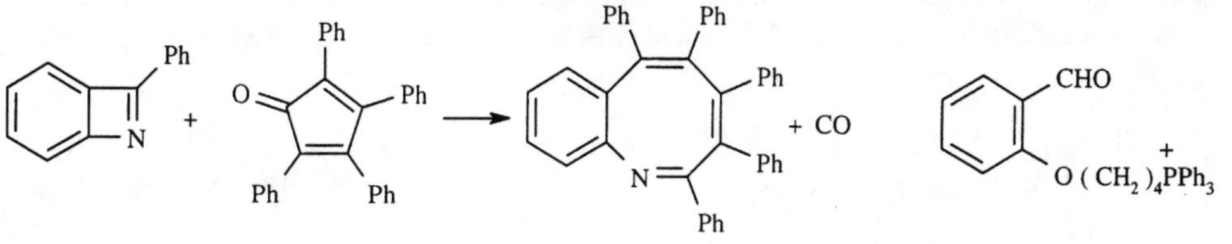

(**168**)   (**169**)   (**170**)   (**171**)

4.4.3   ONE HETEROATOM NOT ADJACENT TO RING JUNCTION

4.4.3.1   Five-membered Rings: Isoindoles and Related Compounds

(i) Compounds of types (**172**) and (**173**) can be prepared from *o*-disubstituted benzenes by standard reactions. Thus, 2-alkyl- ⟨88CZ85⟩ and 2-aryl-2*H*-isoindoles ⟨87CZ155⟩ are prepared from 1,2-bis-(bromomethyl)benzene *via* Polonovsky elimination in dihydroisoindole *N*-oxides in acetic anhydride/triethylamine (Scheme 20). The application of this type of ring formation to ene-1,4-dione systems utilizes a reducing agent as in (**174** → **175**) ⟨65CC272⟩.

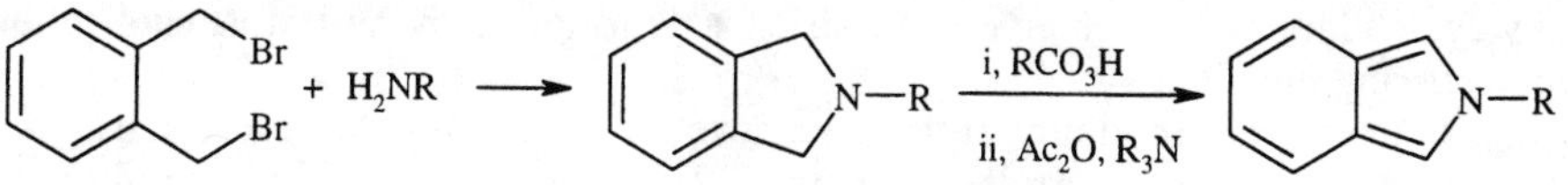

Scheme 20

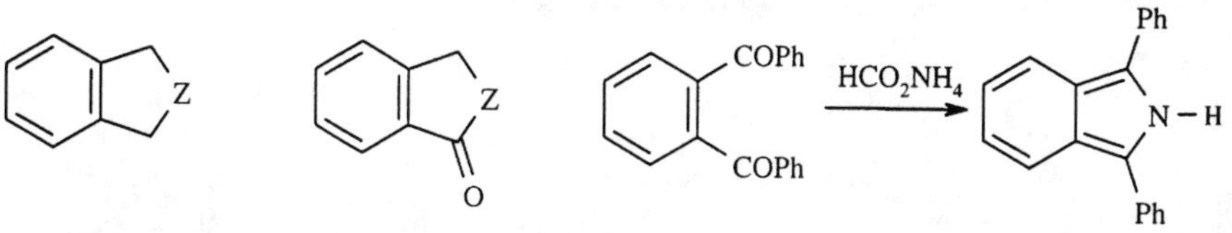

(**172**)   (**173**)   (**174**)   (**175**)

o-Diaroyl- and *o*-bis(trimethylacetyl)-benzenes (**176**) with P$_4$S$_{10}$ in the presence of NaHCO$_3$ afford benzo[*c*]thiophenes (**177**) stabilized by aryl or *t*-butyl groups at the 1- and 3-positions ⟨90S670, 90PS(54)209, see also 93CC172⟩.

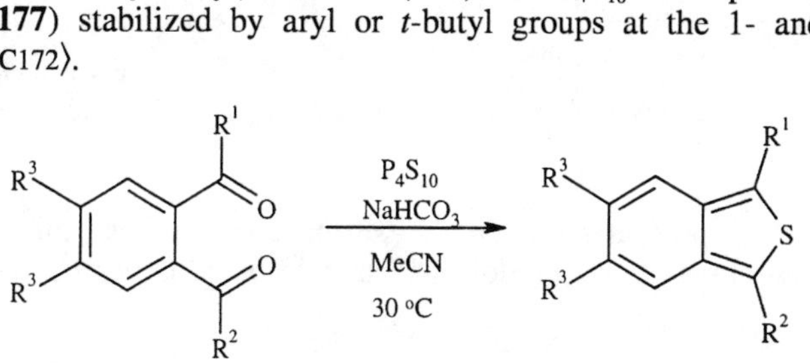

(**176**)   (**177**)

R^1, R^2 = aryl, *t*-Bu   R^3 = H, Me

(ii) A versatile isoindole synthesis (Scheme 21) proceeds through intramolecular carbanion addition to an aryne and subsequent aromatization by base-promoted elimination of hydrogen cyanide ⟨77T2255⟩.

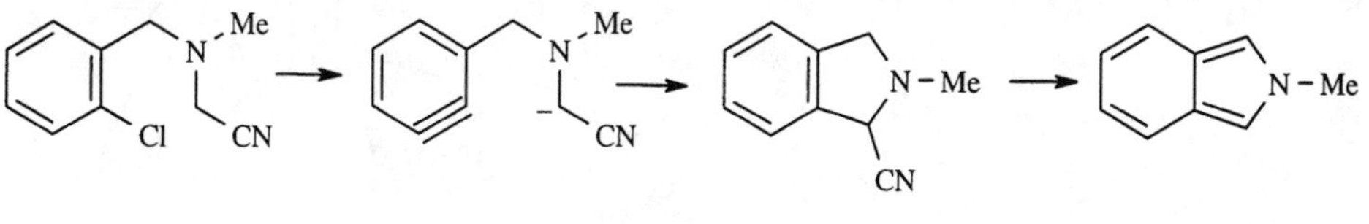

Scheme 21

Synthesis of benzo[c]furans and isoindoles (181) is also possible by the addition of benzyne to the respective monocycles (178), followed by reduction (179 → 180) and pyrolysis. In an alternative procedure, (179) is reacted with 3,6-bis(2-pyridyl)-1,2,4,5-tetrazine, which affords (181) under far less vigorous conditions *via* a retro Diels-Alder reaction of the intermediate (182). 4-Phenyl-1,2,4-triazoles pyrolyze to form isoindoles (Section 3.4.3.12.2).

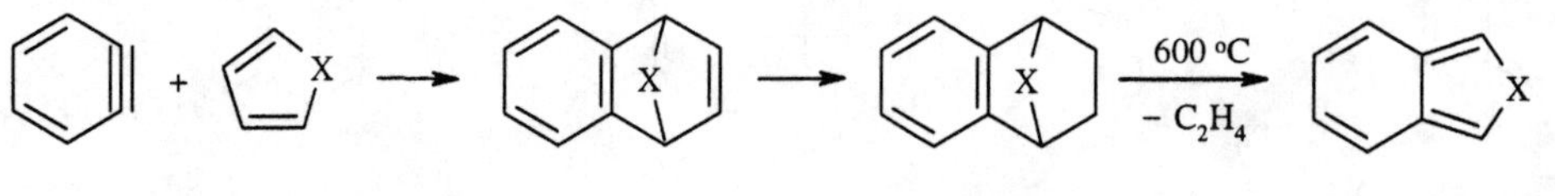

(178)	(179)	(180)	(181) X = O, NR

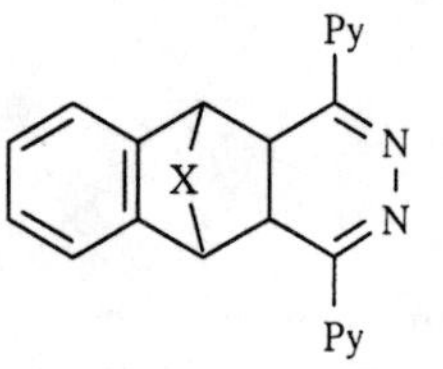

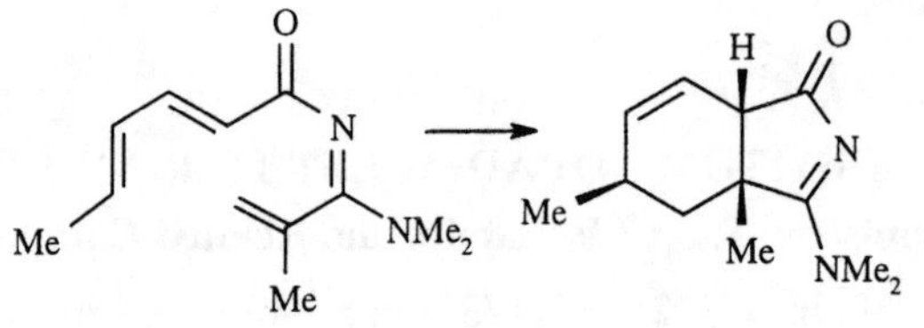

(182)	(183)	(184)

In the intramolecular cycloaddition reaction (183 → 184) ⟨81HCA1515⟩, the stereochemistry is rationalized by cycloaddition through an *exo* transition state followed by reversible enolization to yield the more stable *cis* ring junction.

(iii) The replacement of rhodium from rhodacycles forms condensed furans, thiophenes, selenophenes, tellurophenes and pyrroles (Scheme 22). Replacement of the rhodium by sulfur, selenium or tellurium is effected by direct treatment with the element, by oxygen with *m*-chloroperbenzoic acid, and by nitrogen with nitrosobenzene.

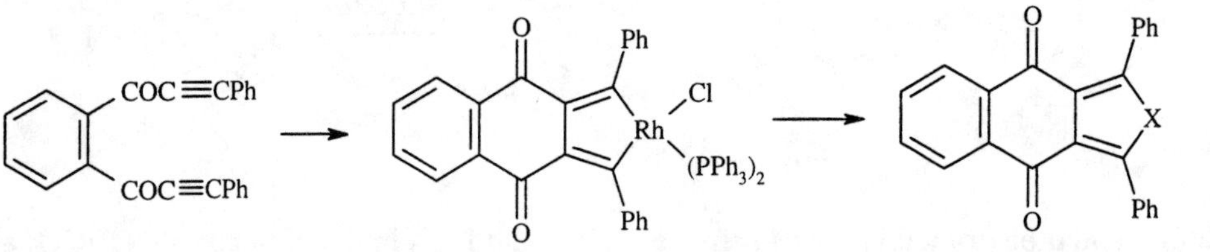

X = O, S, Se ⟨71LA(754)64⟩

X = Te ⟨75S265⟩

Scheme 22

(iv) The isoindole ring is present in the phthalocyanine complexes and one method for their preparation is from 1,3-diiminodihydroisoindole which is prepared by cyclization of *o*-phthalonitrile (Scheme 23) ⟨83JA1539⟩.

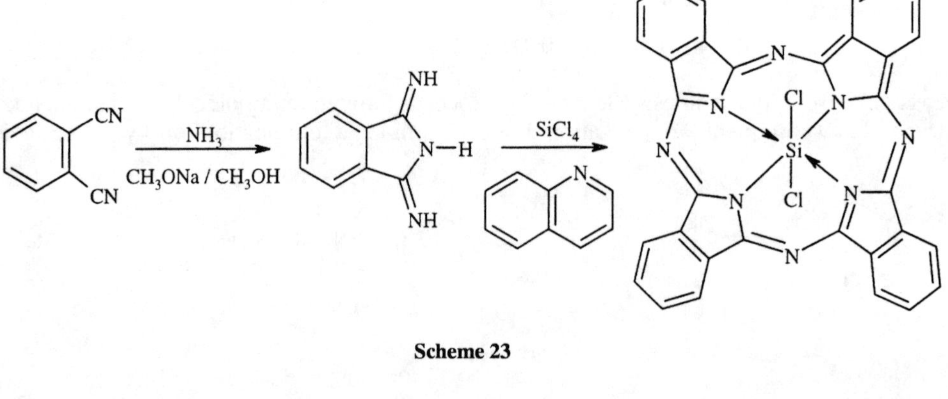

Scheme 23

4.4.3.2 Six-membered Rings

4.4.3.2.1 Overview of ring syntheses of isoquinolines, benzo[c]pyrans and their derivatives

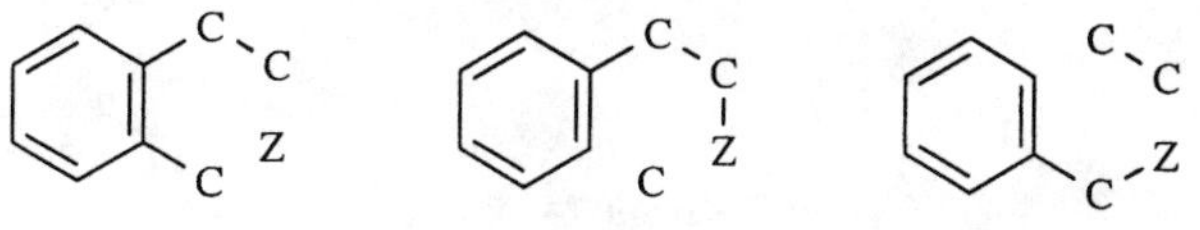

(185)   (186)   (187)

We deal successively with methods of types (185), (186) and (187). Important methods are summarized in Table 4.

Table 4 Important Ring Syntheses for Isoquinolines, Benzo[c]pyrans and Derivatives

Ring	Synthesis type	Synthesis name(s)	Section
Isoquinolines	(185)	—	4.4.3.2.2
	(186)	Pictet-Gams	4.4.3.2.3.ii
	(187)	Pomeranz-Fritsch	4.4.3.2.4
3,4-Dihydroisoquinolines	(186)	Bischler-Napieralski	4.4.3.2.3.i
Tetrahydroisoquinolines	(186)	Pictet-Spengler	4.4.3.2.3.ii
Benzo[c]pyrylium salts	(185)	—	4.4.3.2.2

a

4.4.3.2.2 Ring closure of an o-disubstituted benzenes

Homophthalaldehyde (188) gives isoquinoline, isoquinoline 2-oxide, 3,4-benzopyrylium salts, and 2-alkyl- and 2-aryl-isoquinolinium salts (189) by reaction with NH_3, NH_2OH, H^+ or RNH_2, respectively.

The transformation (190 → 191) exemplifies the use of transition metal reagents; the reaction probably involves aminopalladation of the C=C bond ⟨77JOC1329⟩. 2,3-Dihydro-4(1*H*)isoquinolones are obtained by Dieckmann cyclization of *N*-(*o*-alkoxycarbonylbenzyl)glycine ester derivatives (192 → 193).

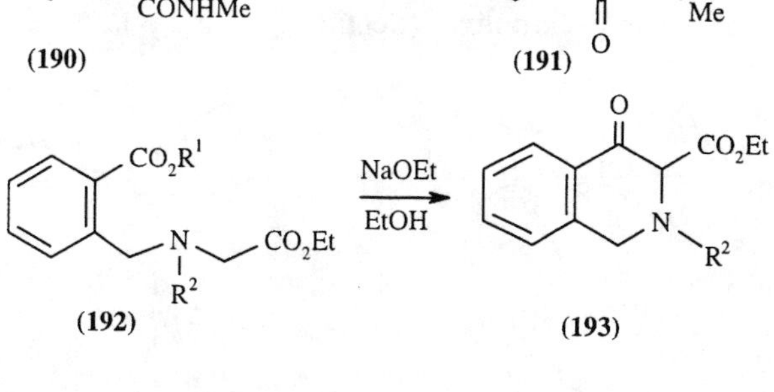

(188)   (189)

(190)   (191)

(192)   (193)

Cyclizations of *o*-formyl or *o*-acylderivatives of β-phenethylamine give 3,4-dihydroisoquinolines ⟨88H(27)2403⟩ (Scheme 24).

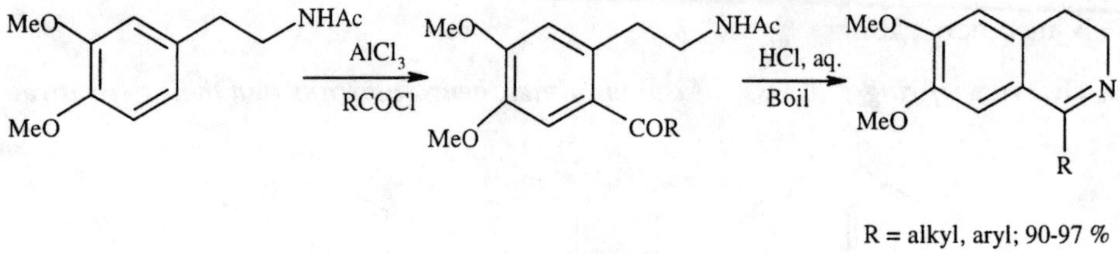

R = alkyl, aryl; 90–97 %

Scheme 24

Nitrogen ylide cyclization gives isoquinolines (**194**) as shown in Scheme 25 ⟨87JCS(P1)921⟩.

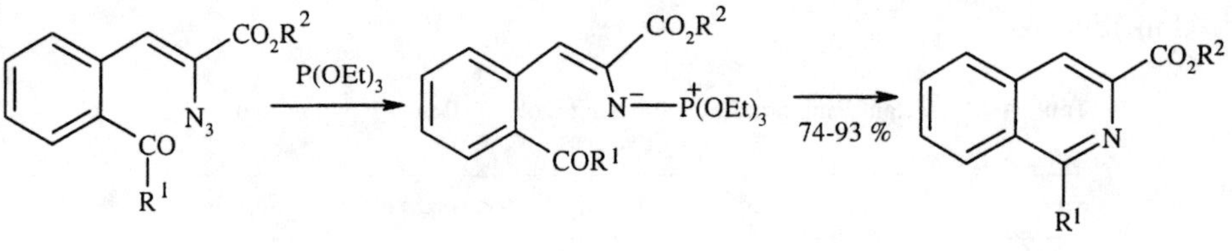

74–93 %

(**194**)

Scheme 25

4.4.3.2.3   *From a β-phenethylamine*

(i) In the Bischler-Napieralski synthesis of 3,4-dihydroisoquinolines (**195**) from *N*-acylated 2-phenethylamines, the amide carbonyl group is condensed with a benzene ring using acid catalysis (*e.g.* P_2O_5, $POCl_3$, $H_3PO_4 - P_2O_5$). Electron-releasing substituents in the *meta* position generally facilitate reaction, but in the *para* position they can inhibit cyclization. *m*-Substituted phenethylamides form mainly 6-substituted dihydroisoquinolines.

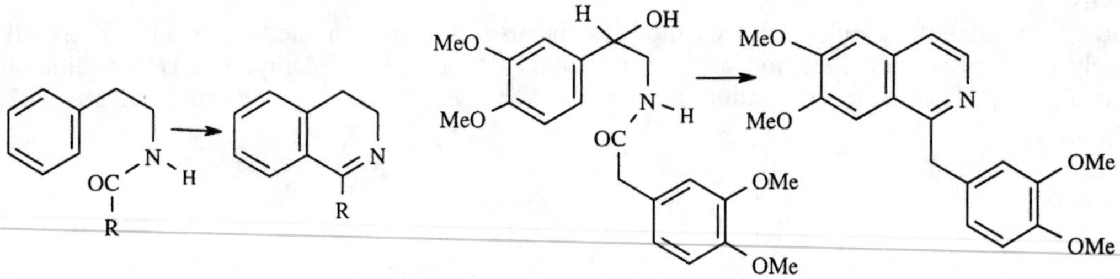

(**195**)   (**196**)   (**197**)

(ii) Pictet-Gams preparation of isoquinolines from *N*-acylated 2-hydroxyphenethylamines, *e.g.* (**196**) → papaverine (**197**), utilizes similar conditions.

(iii) A Mannich-type reaction is used in the Pictet-Spengler synthesis of tetrahydroisoquinolines (**198** → **199**). Indoles similarly give β-carbolines (Section 3.3.1.5.7.iv).

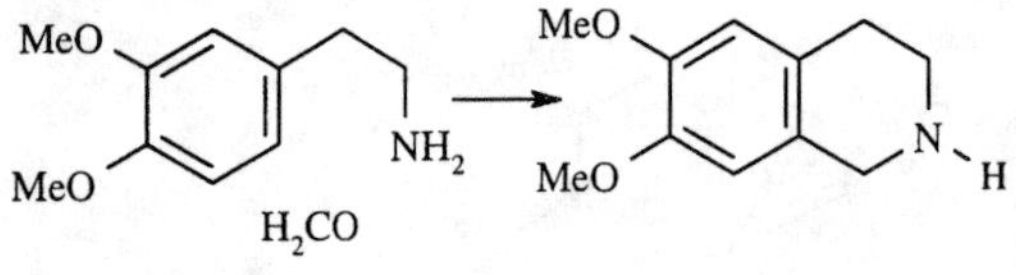

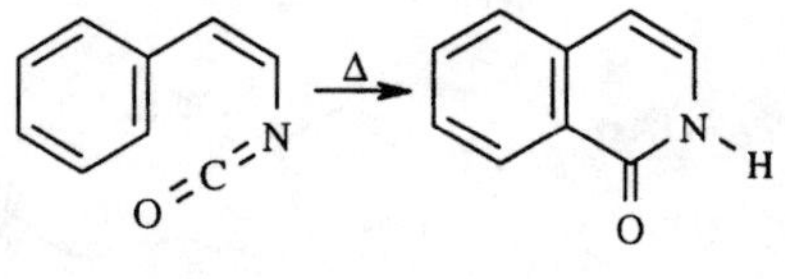

(**198**)   (**199**)   (**200**)   (**201**)

(iv) Styryl isocyanates, readily available by Curtius rearrangement of cinnamoyl azides, undergo thermal or Friedel-Crafts cyclization to 1-isoquinolones (**200** → **201**).

(v) 1-Isoquinolones (**203**) are obtained by a Pd-catalyzed CO insertion reaction of *o*-bromophenylethylamines (**202**).

(**202**)   (**203**)   (**204**)   (**205**)

4.4.3.2.4 *From a benzylimine*

The Pomeranz-Fritsch synthesis of isoquinolines is illustrated by the sequence PhCHO + NH$_2$CH$_2$CH(OEt)$_2$ at 160°C → (**204**) → (**205**). See Section 3.3.1.5.6 for the synthesis of a thienotetrahydropyridine.

4.4.3.3 Seven-membered and Larger Rings

3-Benzoxepin (**207**) can be synthesized as shown in (**206** → **207**) ⟨66CB634⟩. The benzothiepin (**208**) is obtained by condensation of phthalaldehyde with S(CH$_2$CO$_2$Et)$_2$ and following hydrolysis ⟨53JA6332⟩.

(**206**)   (**207**)   (**208**)

Nitrenes insertion is effective for the synthesis of azepines by thermolysis of *o*-styryl azidocinnamates ⟨86JCS(P1)1113⟩. Thus, the vinylazidocinnamate (**209**) yields the 3-benzazepine (**210**) ⟨90JCS(P1)2929⟩.

(**209**)   (**210**)

4.4.4 TWO HETEROATOMS 1,2 TO RING JUNCTION

4.4.4.1 Four-membered Rings

2-Mesityl-2*H*-benzo[*c*]-1,2-thiazetidine 1,1-dioxide is prepared from the corresponding benzothiatriazine by photochemically induced elimination of N$_2$ (Scheme 26) ⟨72LA(763)46⟩.

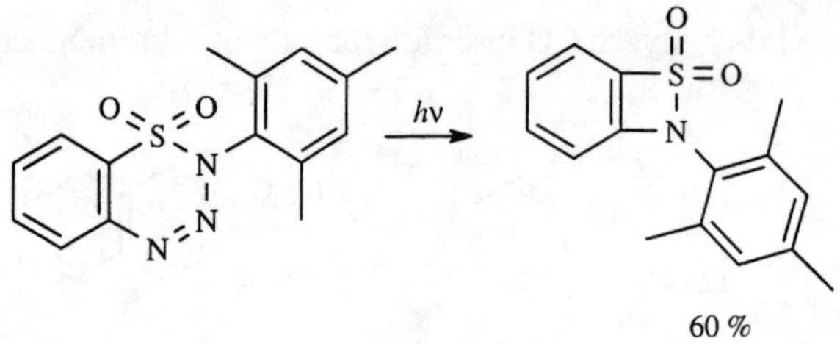

Scheme 26 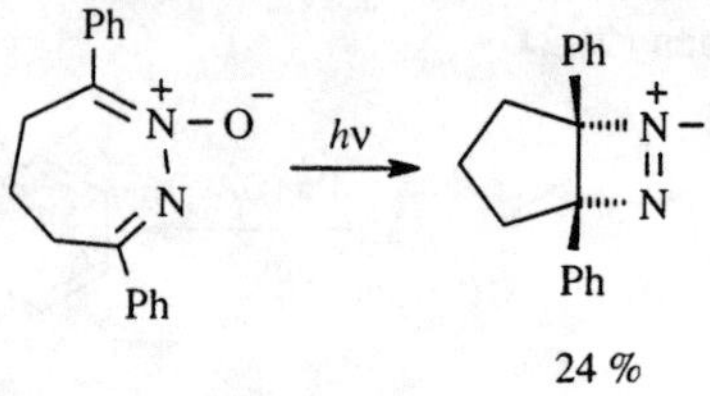   **Scheme 27**

Photoisomerization has been reported to give fused derivatives of 1,2-diazetidines (Scheme 27) ⟨68CC686, 69JA2818⟩.

4.4.4.2 Five-membered Rings

4.4.4.2.1 Indazoles

Indazoles (**212**) are formed by spontaneous cyclization of *o*-acylphenylhydrazines (**211**). Certain *o*-toluenediazonium salts cyclize spontaneously to indazoles (**213** → **214**); yields are good when the methyl group is activated by an *ortho* or *para* electron-withdrawing group. Pyrazoles are prepared similarly (Section 4.3.2.3.3).

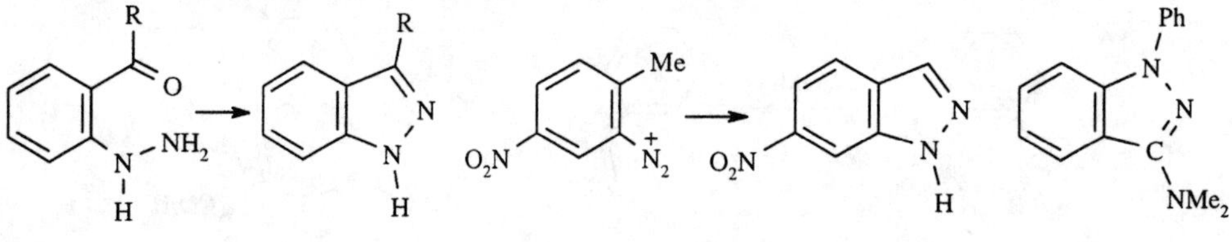

(**211**)   (**212**)   (**213**)   (**214**)   (**215**)

Diphenylhydrazine (Ph$_2$NNH$_2$) is converted into indazole (**215**) by Cl$_2$C=NMe$_2^+$. Oxidative cyclizations are also known: thus, fully methylated 6-(benzylidenehydrazino)uracils (**216**) on photolysis or in hot AcOH give the pyrazolo[3,4-*d*]pyrimidines (**217**) ⟨75BCJ1484, 92KGS219⟩.

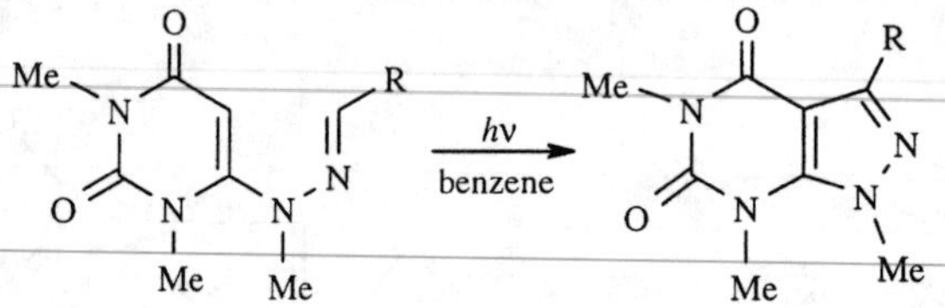

(**216**)   (**217**)

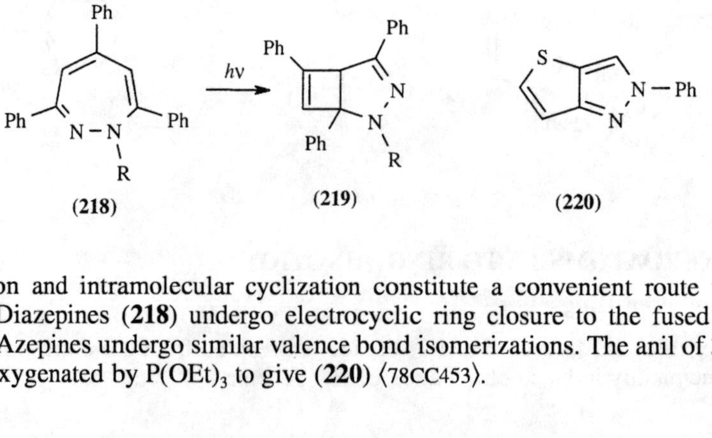

(**218**)   (**219**)   (**220**)

Ring contraction and intramolecular cyclization constitute a convenient route to some bicyclic systems. 1*H*-1,2-Diazepines (**218**) undergo electrocyclic ring closure to the fused pyrazole system (**219**) ⟨71CC1022⟩. Azepines undergo similar valence bond isomerizations. The anil of 3-nitrothiophene-2-aldehyde is deoxygenated by P(OEt)$_3$ to give (**220**) ⟨78CC453⟩.

A new synthesis of 2,3-diaminoindazoles (**222**) involves the treatment of *o*-azidobenzaldimines (**221**) with tertiary phosphine *via* the Staudinger reaction followed by hydrolysis ⟨89TL6237⟩.

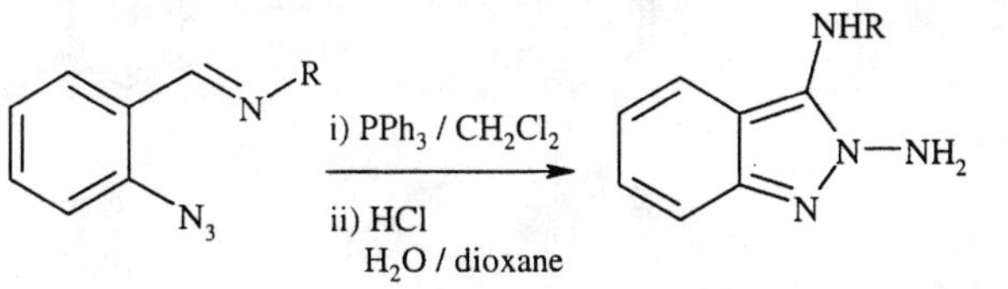

i) PPh$_3$ / CH$_2$Cl$_2$

ii) HCl
H$_2$O / dioxane

(**221**)   (**222**) R = Ph, Pr^n, 4-MeOC$_6$H$_4$

4.4.4.2.2 *Anthranils, benzisothiazoles and saccharins*

(i) Anthranils (**224**) are formed by spontaneous cyclization of *o*-acylphenylhydroxylamines (**223**) (themselves made by reduction of the corresponding nitro compounds). Treatment of the β-amino ketone (**225**) with lead tetraacetate gives the anthranil (**226**).

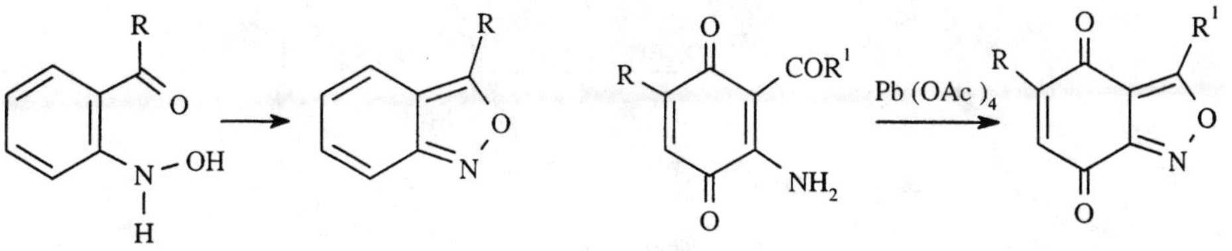

(**223**)   (**224**)   (**225**)   (**226**)

(ii) Benz[*d*]isothiazoles (**228**) are prepared from sulfenyl chlorides (**227**), and their benz[*c*] isomers result from the H$_2$O$_2$ oxidation of *o*-aminothiobenzamides (**229** → **230**). Iminobenzodithioles equilibrate thermally with benzisothiazolethiones (Section 3.4.1.2.4).

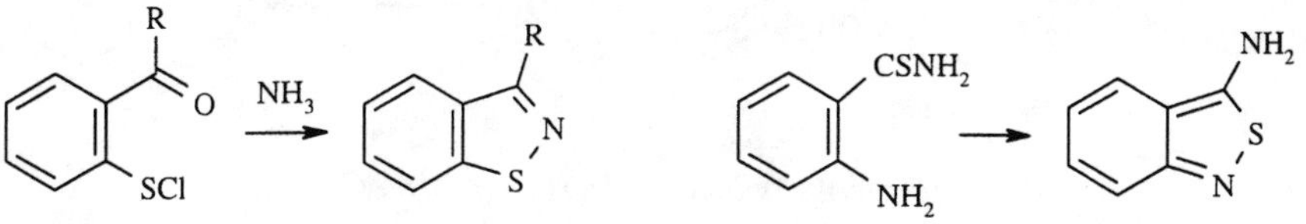

(**227**)   (**228**)   (**229**)   (**230**)

(iii) Saccharins are obtained by KMnO$_4$ oxidation of *o*-methylbenzenesulfonamides (*e.g.* **231** → **232**).

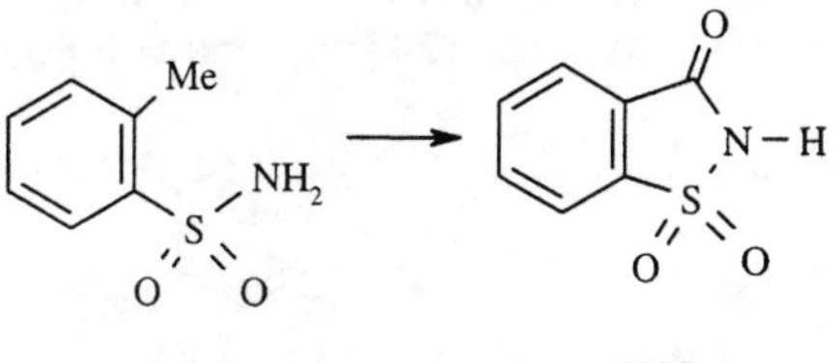

(**231**)   (**232**)

4.4.4.3 Six-membered Rings
4.4.4.3.1 *Cinnolines*

o-Alkenyl-, *o*-alkynyl- and *o*-acyl-diazonium ions cyclize spontaneously to give cinnolines or cinnolones (**233** → **234**; **235**, **237** → **236**). These are the usual preparative methods for these classes of compounds. However, Friedel-Crafts cyclization of mesoxalyl chloride hydrazones (**238**) ⟨61JCS2828⟩ or benzene ring-N bond formation, as in (**239** → **240**) ⟨83S52⟩, can also be used.

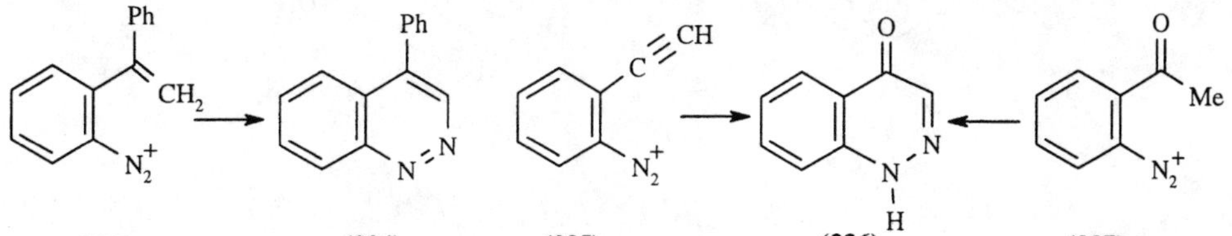

(**233**)   (**234**)   (**235**)   (**236**)   (**237**)

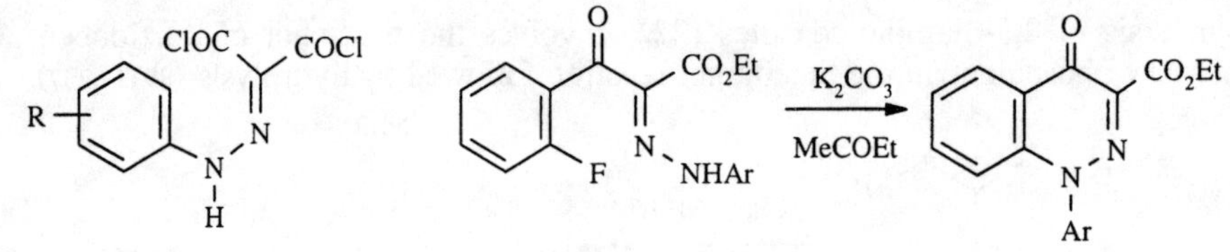

(238)   (239)   (240)

DABCO catalyses the cyclization of the 2-acetylphenylhydrazone of nitroformaldehyde to 4-methyl-3-nitrocinnoline (Scheme 28) ⟨84JOC289⟩.

Iron cyclopentadienyl complexes of 3-(2-chlorophenyl)propanones are cyclized with hydrazine into 1,4-dihydrocinnoline complexes, demetalated with sodium amide to give 3- or 3,4-disubstituted cinnolines (70–80%) (Scheme 29) ⟨88JHC1107⟩.

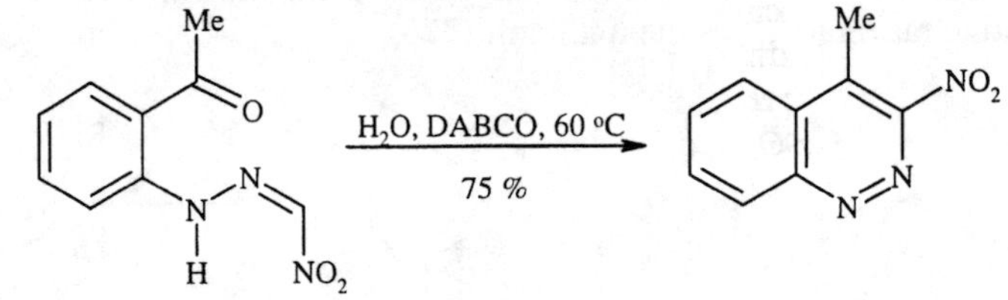

Scheme 28

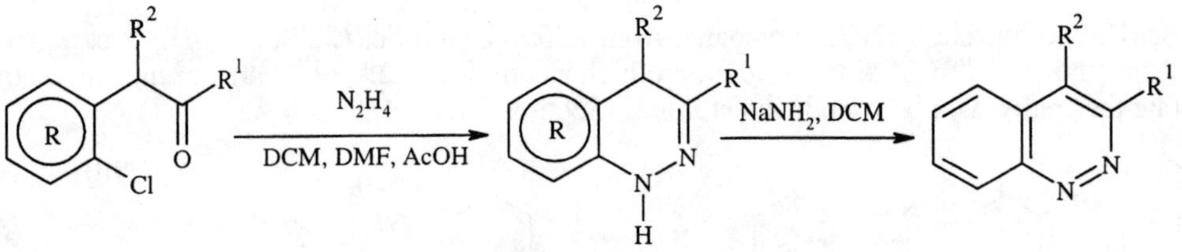

$R = Fe(Cp^*)PF_6$; $R^1 = Me$, Ph; $R^2 = H$, Me

Scheme 29

Preparations of cinnolines by expansion of five-membered heterocycle rings include the oxidation of N-aminooxindoles (**241** → **242**), the treatment of isatogens with ammonia (Scheme 30) and the base-catalyzed conversion of 1-aminodioxindoles (**243**) into cinnolin-3-ones.

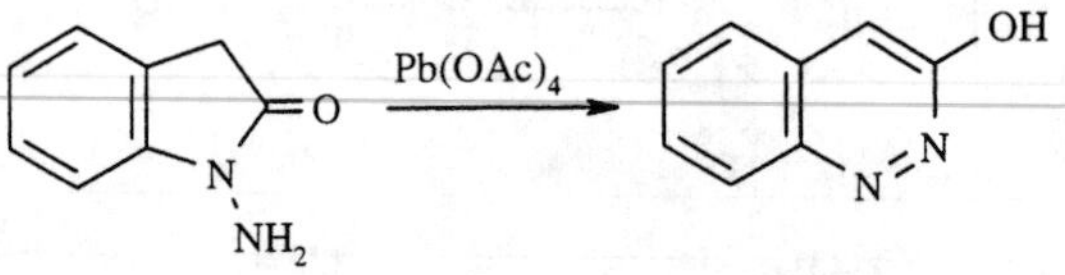

(241)   (242)

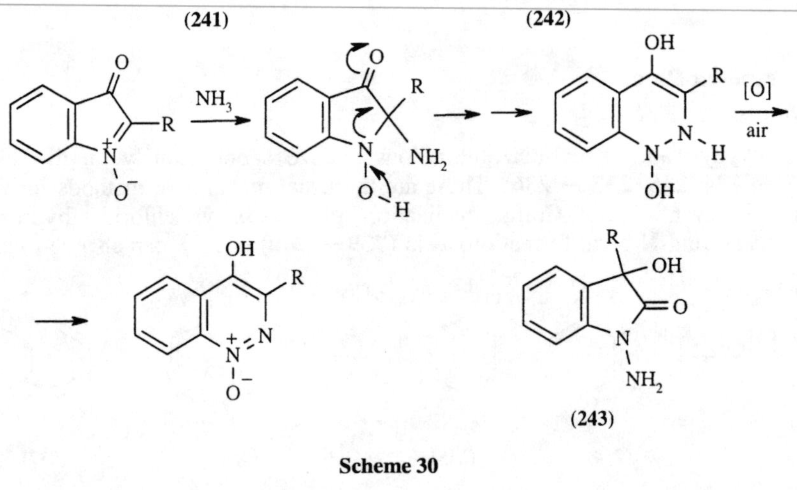

(243)

Scheme 30

4.4.4.3.2 Rings containing O or S atoms

1,2-Benzoxathiin 2,2-dioxides (**245**) are prepared from 2-acylphenols by reaction with a sulfonyl chloride, followed by base-catalyzed cyclization of the sulfonate ester (**244**) ⟨66HC(21-2)792⟩.

| **(244)** | **(245)** | **(246)** | **(247)** |

An example of 1,2-thiazine synthesis in which the creation of the S—N bond is the final step is the thermolysis of azides (**246**) to afford the expected $1\lambda^4$-1,2-thiazines (**247**) ⟨86JCS(P1)483⟩.

The 2,1-benzothiazines (**249**) are the products from *N*-chlorosuccinimide and potassium hydroxide with 2-aminostyrenes (**248**) ⟨79TL3969⟩, and the sulfone (**251**) is obtained by the action of base on the tosylhydrazone (**250**) ⟨66JOC3531⟩.

| **(248)** | **(249)** | **(250)** | **(251)** |

Routes to 3,4-dihydro-2*H*-1,2-benzothiazine dioxides (**253**) include the cyclization of aminosulfonic acids (**252**) or cyanosulfonamides (**254**). 2,4-Dihydro-1*H*-2,1-benzothiazine dioxides (**256**) are normally prepared by thermolysis of the sodium sulfonates (**255**) or amino sulfonamides (**257**) ⟨71CB1880⟩.

| **(252)** | **(253)** | **(254)** |

| **(255)** | **(256)** | **(257)** |

4.4.4.4 Seven-membered Rings

In an intramolecular 1,3-dipolar reaction, the nitrilimines (**259**), generated by the reaction of the hydrazoyl chlorides (**258**) with triethylamine at 80°C, cyclize to give 1*H*-1,2-benzodiazepines (**261**) ⟨79S380⟩. At 20°C using silver carbonate, the cyclopropa[*c*]cinnolines (**260**) are isolated; they rearrange *via* ring expansion and hydrogen migration to give the benzodiazepines (**261**) ⟨81JOC1402⟩.

(258)   (259)   (260)   (261)

The photolysis of quinoline *N*-imides (**262**) (which are in equilibrium with their dimers) ⟨77JOC1856⟩ gives 1*H*-1,2-benzodiazepines (**263**). Reactions of this type have also provided routes to pyrido-, thieno- and furo-1,2-diazepines ⟨79CPB2183, 79H(12)471⟩.

(262)   (263)

(264)   (265)   (266)

N-Tosylsulfimides, *e.g.* (**264**), are converted into 1,2-benzothiazepines (**266**) by triethylamine. In the absence of base the intermediate (**265**) can be isolated ⟨81JCS(P1)1037⟩.

N-Acyl-1,3,4,5-tetrahydro-2,1-benzoxazepines (**268**) are synthesized from the *N*-chloro derivatives (**267**). The reaction proceeds *via* initial *ipso* attack followed by 1,2-carbon migration ⟨84JCS(P1)2255, 87T2577, 90T7247⟩.

(267)   (268) R^1 = H, Me; R^2 = Me, Ph

4.4.5   TWO HETEROATOMS 1,3 TO RING JUNCTION

4.4.5.1   Five-membered Rings

4.4.5.1.1   *Ring closure of o-disubstituted benzene or hetarene*

(i) *o*-Hydroxy-, *o*-mercapto- and *o*-amino-anilides (**269**; Z = O, S, NH) cyclize (*e.g.* by heating at 150°C or refluxing with H$_2$O − HCl) to benzoxazoles, benzothiazoles and benzimidazoles (**270**; Z = O, S, NH), respectively. The anilides are often prepared and cyclized *in situ* by heating the corresponding *o*-substituted anilines with a carboxylic acid, anhydride, acid chloride, ester, nitrile, amidine, *etc. o*-Substituted anilines with the phosgeneiminium chloride Cl$_2$C=NMe$_2^+$Cl$^-$ give benzoxazoles, benzothiazoles and benzimidazoles (**271**) containing a 2-dimethylamino substituent ⟨73AG(E)806⟩.

Similar *ortho*-substituted heterocycles react similarly, providing entry into a large number of ring-fused systems, including purines (Traube synthesis).

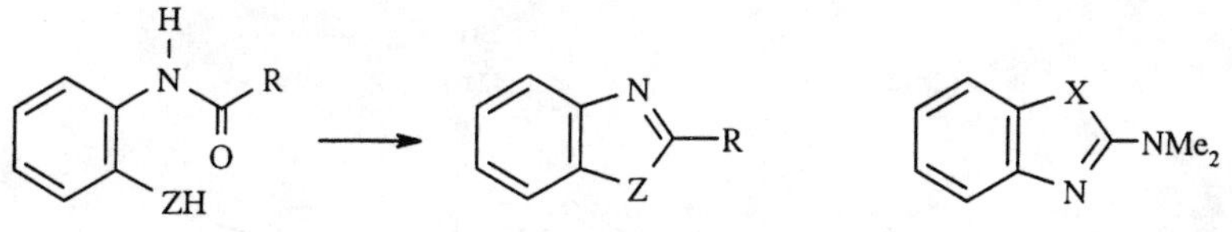

(269) (270) (271) X = O , S , NR

Similar methods lead to ring closure on a variety of other ring systems. The substituted pyrimidinethione (**272**) with polyphosphoric acid formed the thiazolo[5,4-*d*]pyrimidine (**273**) ⟨65JOC1916⟩. Using phosphorus pentasulfide as the thiation agent, reaction with the α-acylamino-carbonyl system contained in (**274**) led to thieno[2,3-*d*]thiazoles (**275**) ⟨70CHE1515⟩, pyrazolo[5,4-*d*]-thiazoles (**276**) ⟨74CHE813⟩ and pyrazolo[4,5-*d*]thiazoles (**277**) ⟨64CHE165⟩.

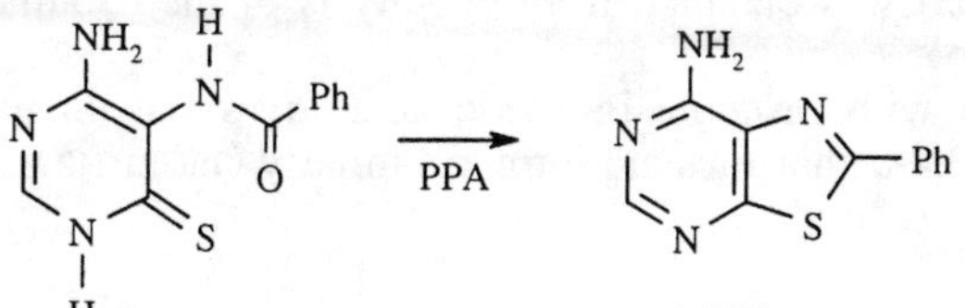

(272) (273)

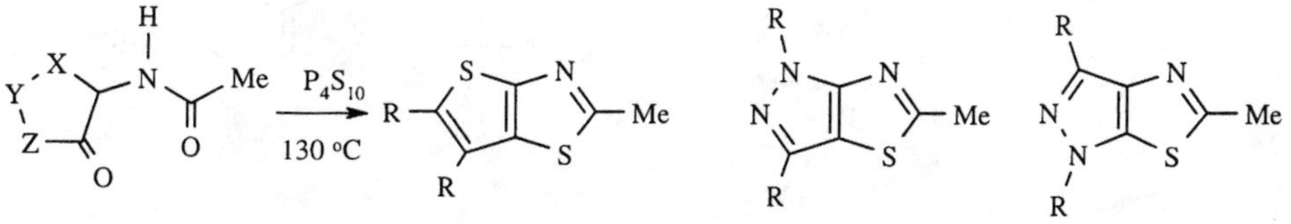

(274) (275) (276) (277)

(ii) Benzoxazolones, benzothiazolones and benzimidazolones (**278**) are prepared by the reaction of carbonic acid derivatives [CO(OEt)$_2$, COCl$_2$ or ClCO$_2$Et] with the corresponding *o*-substituted anilines.

(iii) If an aldehyde, ketone or *gem*-dihalo compound is used in place of the carbonic acid derivative, the corresponding compound with non-aromatic five-membered ring is formed (**279** → **280**). Schiff bases of *ortho*-phenylenediamines can be oxidatively cyclized, undoubtedly *via* tautomeric benzimidazolines, to 2-substituted benzimidazoles by active manganese dioxide, lead tetraacetate, nickel peroxide, or barium permanganate ⟨88SC1537⟩.

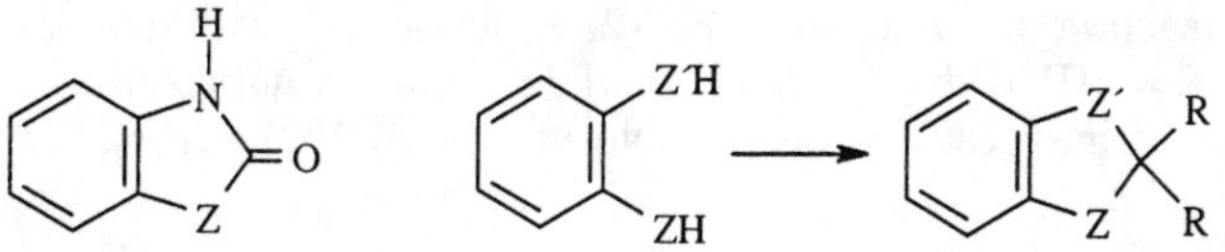

(278) (279) (280)

4.4.5.1.2 Other methods

Oxidative C—S bond formation (I$_2$, Br$_2$ or SOCl$_2$) converts thioanilides PhNHCSR into benzothiazoles (**281**) (the Jacobson-Hugershoff synthesis). Thus, ArNHCSCH(CO$_2$Et)$_2$ with bromine yields the benzothiazole (**282**) ⟨73ZC176⟩.

Nitrene-like intermediates can lead to C—N bond formation and thus to imidazole derivatives. 1,5-Diphenyltetrazole (**283**) fragments to (**284**), which is trapped intramolecularly to form (**285**) (Section 3.4.1.2.1).

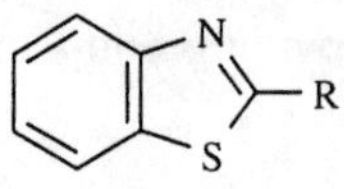

(281) 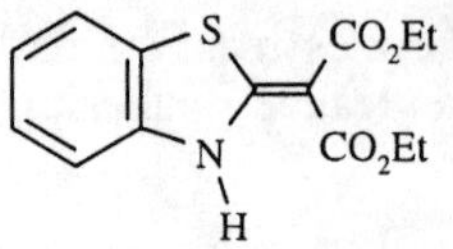 (282)

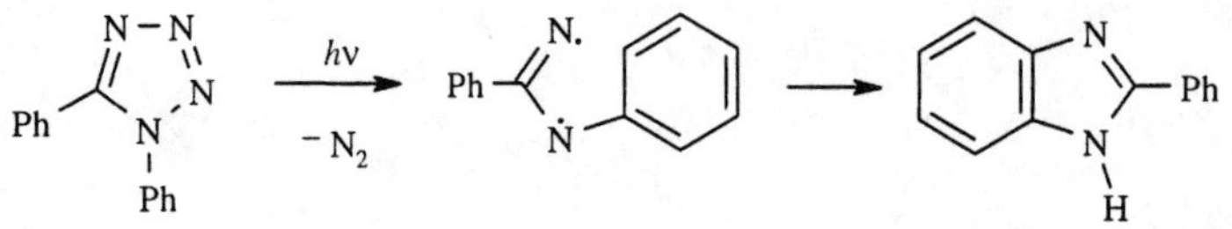

(283) (284) (285)

Photolysis of (**286**, X = C) also gives the nitrene intermediate (**284**) by loss of CO_2, and thus (**285**) (Section 3.4.3.12.2). Some (**285**) was obtained from photolysis of the oxathiadiazole 2-oxide (**286**, X = S) with loss of SO_2 ⟨68TL325⟩.

The 2-thienylthiourea (**287**) with bromine in acetic acid gives the thieno[3,2-*d*]thiazole (**288**) ⟨71AJC1229, 78JHC81⟩. Pyrazolo[3,4-*d*]thiazoles are formed similarly ⟨76GEP2429195⟩.

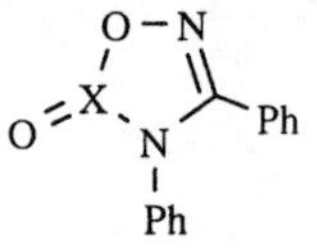

(286) 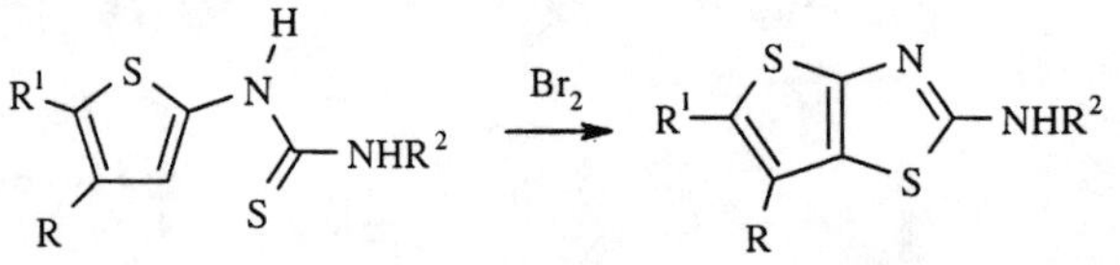 (287) (288)

4.4.5.2 Six-membered Rings

4.4.5.2.1 *Quinazolines and azinopyrimidines by cyclization procedures*

The usual precursor is an appropriately *ortho*-disubstituted benzene. Thus, quinazolines (**290**) can be prepared by the reaction of *o*-acylanilines (**289**; R = alkyl) with amides $RCONH_2$. Heating anthranilic acid (**289**; R = OH) with amides or amidines yields 4-quinazolinones (**291**). The second nitrogen can be introduced into (**289**) before ring closure, as in (**292**) + $HC(OEt)_3 \rightarrow$ (**293**).

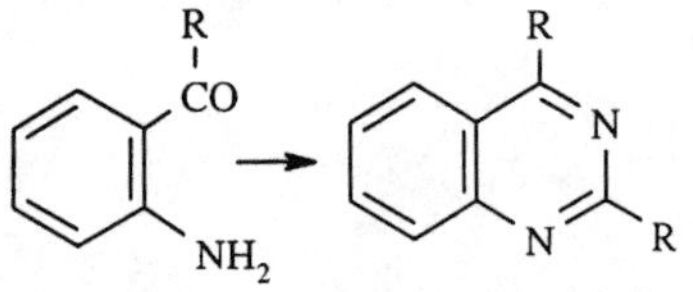

(289) (290) 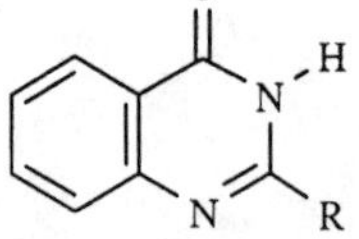 (291) 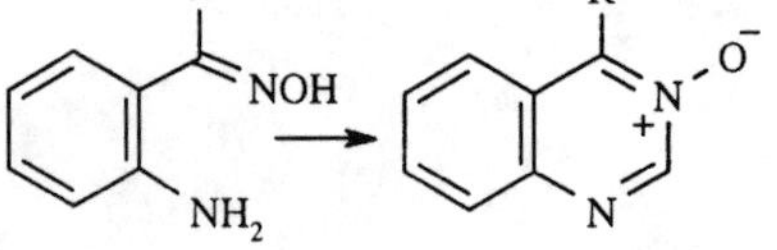 (292) (293)

The synthesis of pyrimidopyrimidines, pyrimidopyridazines and pteridines is illustrated in Scheme 31. Full details are given in the appropriate chapters of CHEC and CHEC-II. In addition to the syntheses that resemble those of quinazolines from anthranilic acid, the high reactivity of the

4-chlorine in a pyrimidine to nucleophiles and that of the 5-position in 6-aminouracil to electrophiles are exploited.

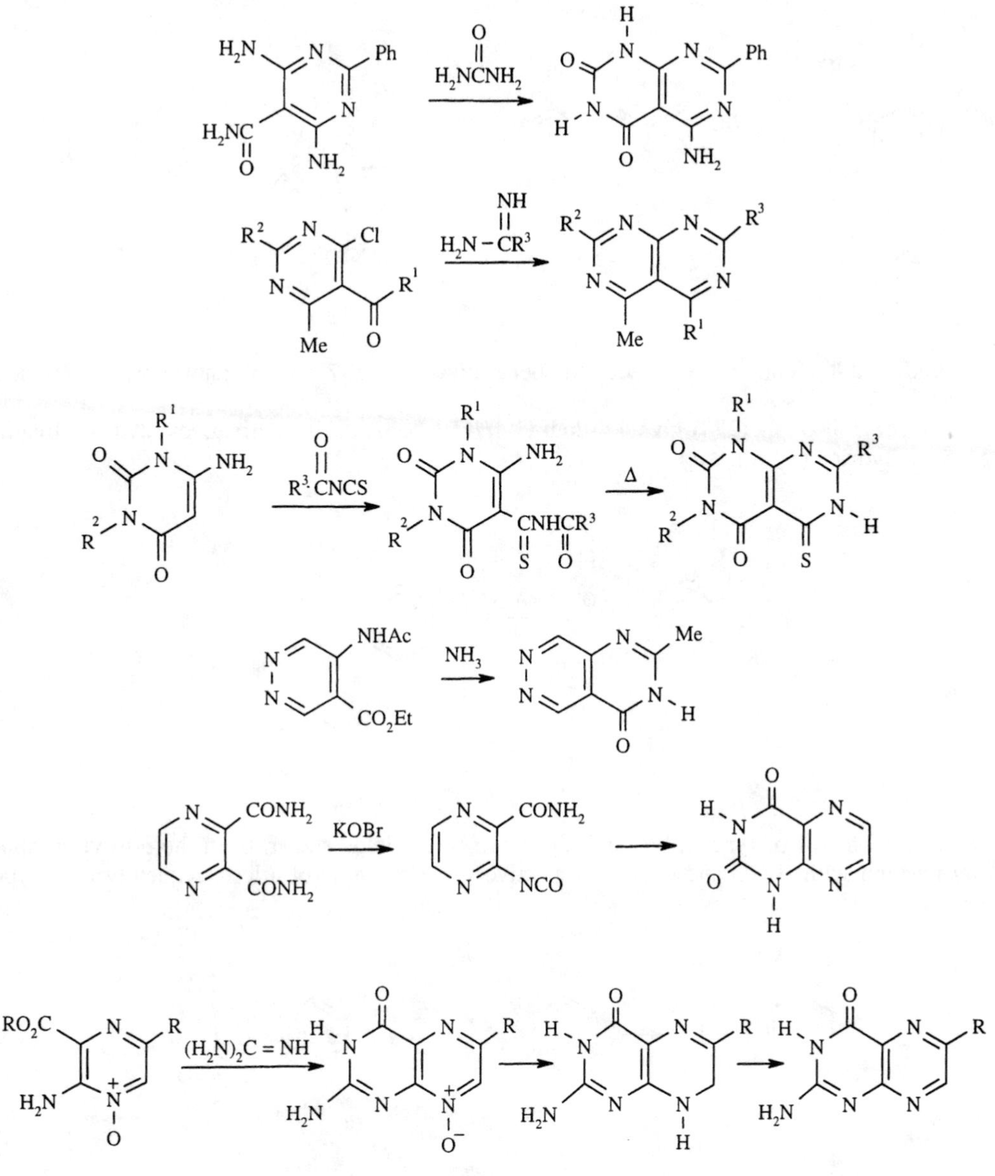

Scheme 31

4.4.5.2.2 *Rings containing O or S atoms*

1,3-Benzoxazin-4-ones are made by the cyclization of *O*-benzoylsalicylamides or reactions between phenyl salicylates and benzamidines ⟨13LA(409)325⟩. The first method has wide applicability, and when 2-acylmercaptobenzamides are used 1,3-benzothiazin-4-ones are obtained (Scheme 32) ⟨67BSF(2)4441⟩.

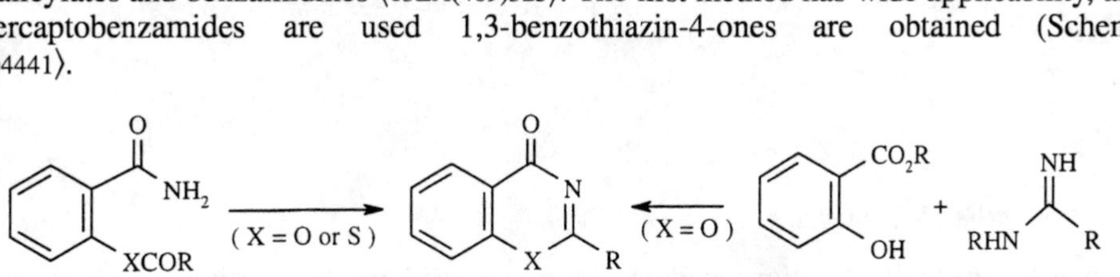

Scheme 32

2H-1,3-Benzothiazines (295) are available through a Bischler-Napieralski-type cyclization of amides (294) ⟨77ACH(92)317⟩.

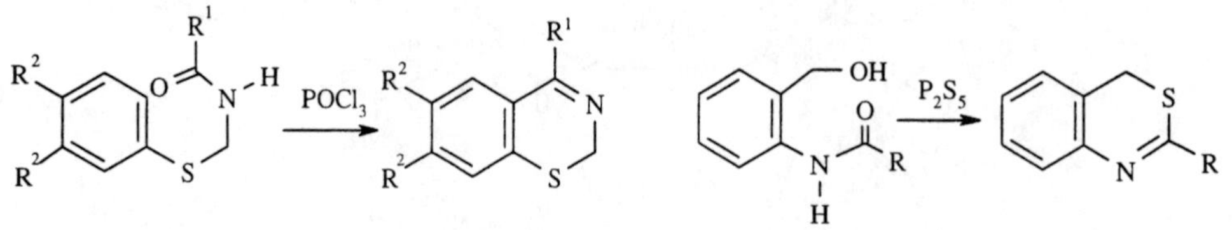

(294)　　　　　　　　(295)　　　　　　　(296)　　　　　　　(297)

The amide (296) can be cyclized to benzothiazine (297) with phosphorus pentasulfide ⟨1894CB3509⟩.

2-Aminobenzyl alcohols (298) give 4,4-dialkyl-4H-3,1-benzoxazine derivatives (299) on treatment with acetic anhydride ⟨1883CB2576⟩.

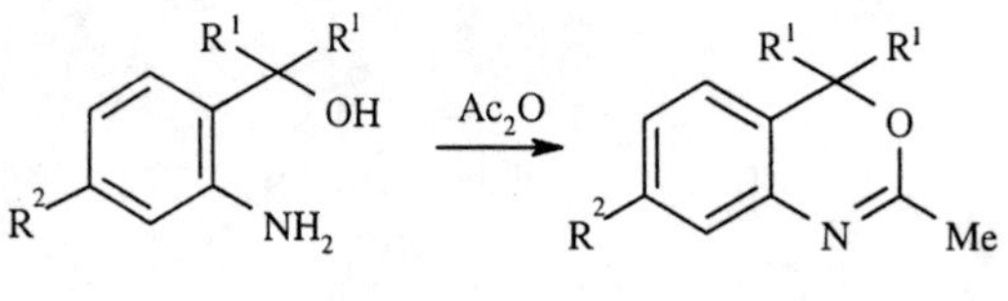

(298)　　　　　　　　　　(299)

Saturated derivatives of types (300) and (302) can be made as shown from the o-tolyl compound (301) by reaction with H₂CO and COCl₂, respectively. An example of a less frequently used type of ring closure is (303) → (304).

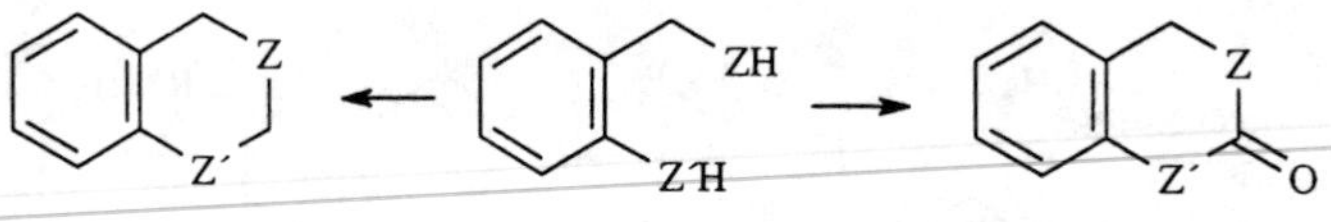

(300)　　　　　　　　(301)　　　　　　　(302)

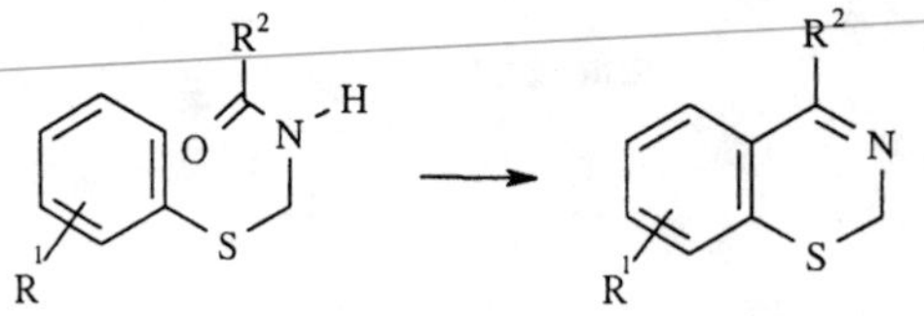

(303)　　　　　　　　(304)

4.4.5.2.3　From other heterocycles

N-Alkylbenz[d]isoxazolium cations undergo base-catalyzed ring expansion to 1,3-benzoxazines (305 → 306 → 307) (Section 3.4.3.12.3). Benzoxazinones are obtained by heating 3-acylanthranils

(Section 3.4.3.4.4) or acylating anthranils (Section 3.4.1.8.6). 1,3-Benzoxazine and 1,3-benzothiazine derivatives can be converted into quinazolone compounds as shown in sequences exemplified by (**308**) → (**309**) and (**310**) → (**311**).

(**305**)   (**306**)   (**307**)

(**308**)   (**309**)   (**310**)   (**311**)

4.4.5.3 Seven-membered Rings

Benzoxazepines are obtained photolytically from quinoline 1-oxides (Section 3.2.3.12.5.iv).

4.4.5.3.1 Seven-membered rings with heteroatoms 1,3 to ring junction

The fully unsaturated 1,3-benzodiazepine (**313**) is formed by a photoreaction of the 1-substituted isoquinoline *N*-imide (**312**) ⟨80CPB2602⟩. The same principle has been applied to prepare thieno-, furo- and pyrrolo-fused 1,3-diazepines ⟨80CC454, 81CPB1539⟩.

(**312**) E = CO$_2$R   (**313**)

1,3-Benzodiazepin-4-ones (**315**) have been synthesized by the Cu$_2$O-catalyzed cyclization of (**314**); the competing route to indoles is disfavored by bulky R groups ⟨79TL1039⟩.

(**314**)   (**315**)   (**316**)   (**317**)

The copper-catalyzed insertion of isocyanides (**316**) into the O—H bond of alcohols gives a high yielding route to 4,5-dihydro-3,1-benzoxazepines (**317**) ⟨78TL2087⟩.

Fully unsaturated 1,3-benzoxazepine system (**319**) can be obtained by intramolecular aza-Wittig reaction ⟨90S455⟩ of the esters (**318**) induced by triethyl phosphite ⟨90S455, 92CC81⟩.

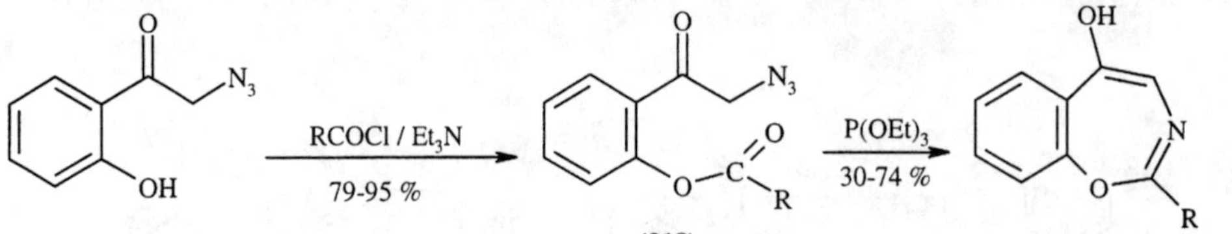

(**318**)   (**319**)

R = H, Me, *cyclo*-Pr, Ph

4.4.5.3.2   *Seven-membered rings with heteroatoms 2,4 to ring junction*

Treatment of (**320**) as shown gives 2,4-benzodiazepin-1-one (**321**) ⟨75JHC903⟩.

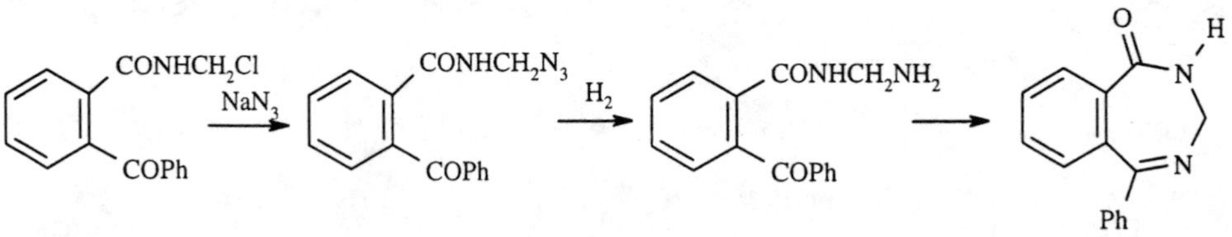

(**320**)   (**321**)

1,4-Dihalo compounds react with thioureas or thioamides ⟨77HCA2872, 75CPB1764, 73RTC20⟩: syntheses of 2,4-benzothiazepines, *e.g.* (**322**) and (**324**), from *o*-xylyl dibromides (**323**) are shown. *o*-Chloromethylbenzoyl halides similarly give 2,4-benzothiazepin-5-ones and 4-bromobutyryl chloride gives the 1,3-benzothiazepin-4-one system.

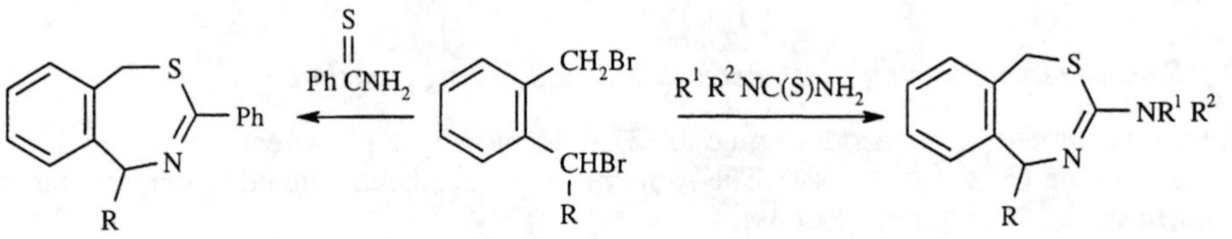

(**322**)   (**323**)   (**324**)

One-pot reactions involving successive nucleophilic attack by oxygen and nitrogen on a 1,1-bis-carbon electrophile are used in the synthesis of 2,4-benzoxazepine systems (**325**) and (**326**) ⟨72JHC1209, 75FES773⟩.

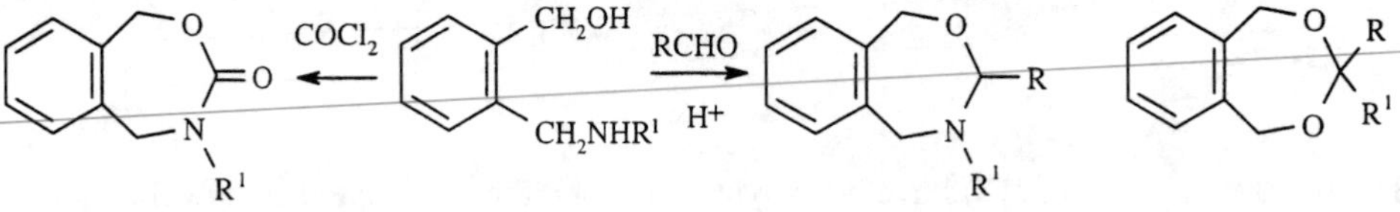

(**325**)   (**326**)   (**327**)

Reaction of 1,2-benzenedimethanol with aldehydes or related compounds gives 3*H*-2,4-benzodiox-epins (**327**).

The analogous 1,5-dihydro-3*H*-2,4-benzodithiepin system can be prepared by the reaction of 1,2-benzenedimethanethiol with either methylene iodide in the presence of base or with aldehydes or ketones in the presence of acid.

4.4.6   TWO HETEROATOMS 1,4 TO RING JUNCTION

4.4.6.1   Quinoxalines and Azinopyrazines

Quinoxalines (**328**) are prepared from *o*-phenylenediamines and α-diketones ⟨86JCR(S)16, 92JOC6749⟩.

Heterocyclic *o*-diamines react analogously, as in the preparation of pteridines (**329** → **330**) and pyrazinopyrazines (**331** → **332**).

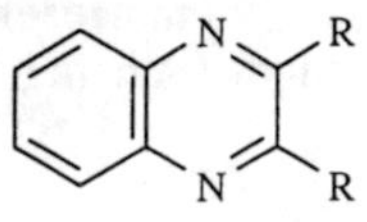

(328) 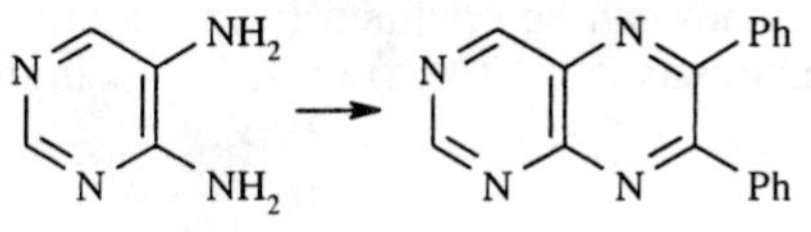   (329)   (330)

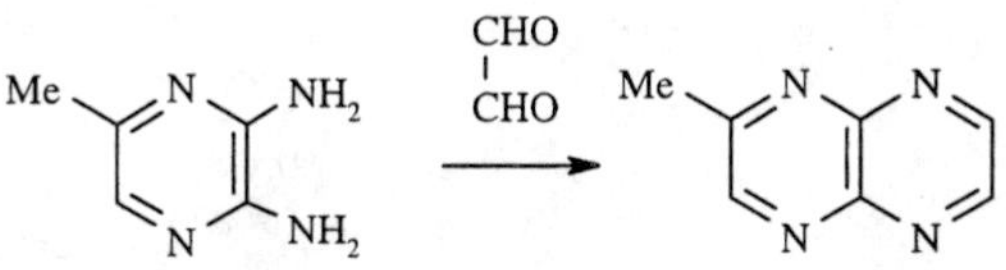

(331)   (332)

Ketomethylene structures react with 6-amino-5-nitrosopyrimidines ⟨49MI21600⟩, providing a general route to 6- and 7-substituted pteridines from ketones and aldehydes (Scheme 33) ⟨54JCS2881, 56JCS213⟩.

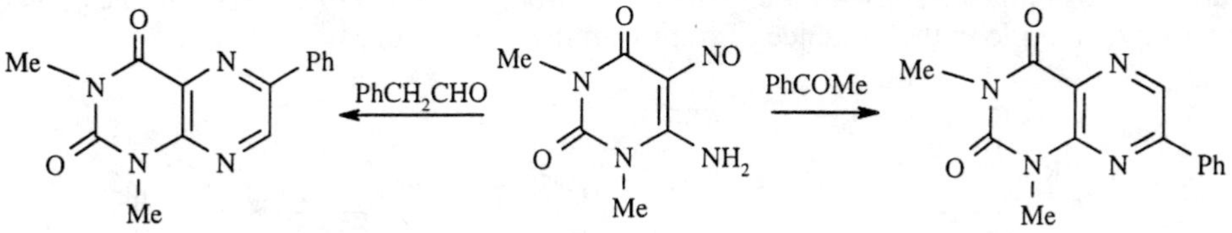

Scheme 33

The cycloaddition reaction (**333**) + (**334**) → (**335**) illustrates the preparation of tetrahydroquinoxalines from *o*-benzoquinone diimines.

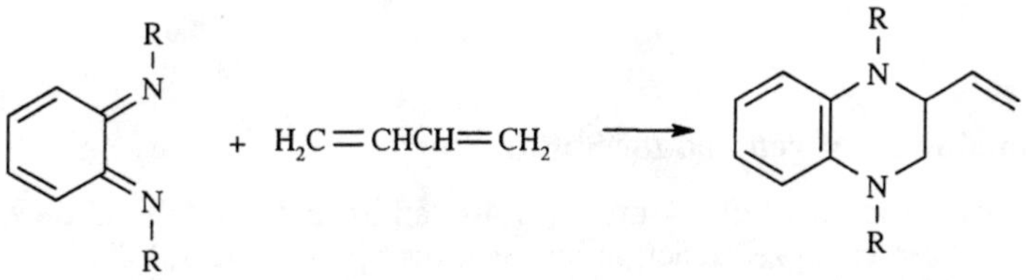

(333)   (334)   (335)

R = COAr , SO₂Ar

4.4.6.2 1,4-Benzoxazines and 1,4-Benzothiazines

2*H*-1,4-Benzoxazines (**337**) result by the cyclization of acetals (**336**) in acid solution ⟨79M257⟩.

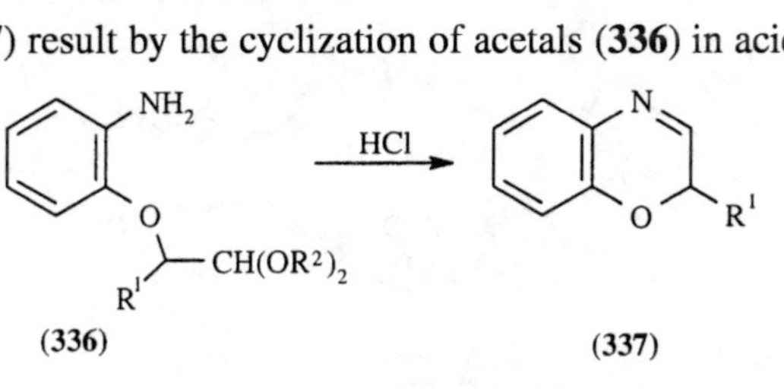

(336)   (337)

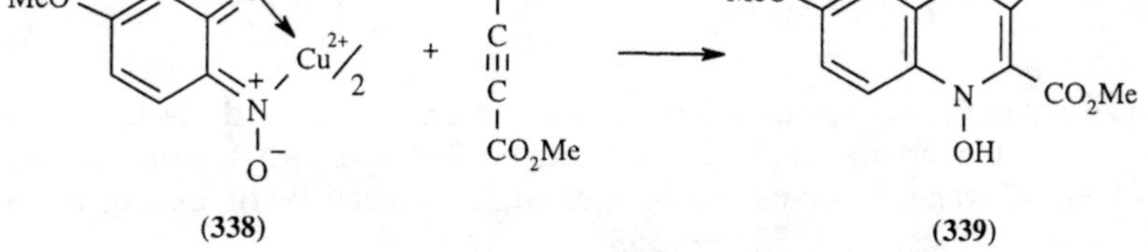

(338)   (339)

4*H*-1,4-Benzoxazines are available by reaction of the copper complexes derived from *o*-nitrosophenols with alkynic dienophiles (**338** → **339**).

2*H*-1,4-Benzothiazines can be obtained from 2-aminothiophenols, by their reactions with either α-bromocarbonyl compounds ⟨70AC(R)383⟩ or active methylene compounds (Scheme 34) ⟨76JCS(P1)1146⟩.

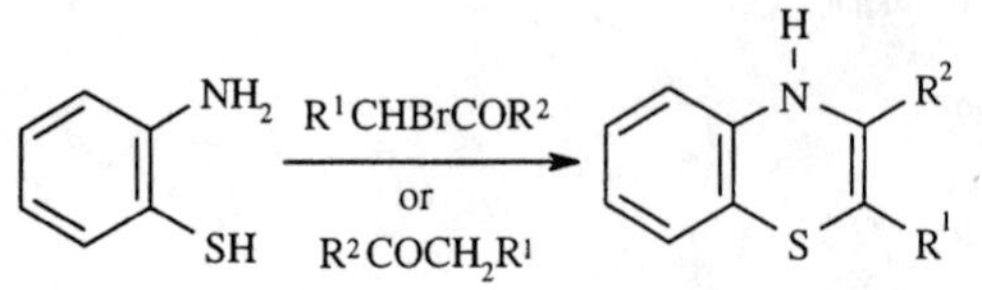

Scheme 34   **Scheme 35**

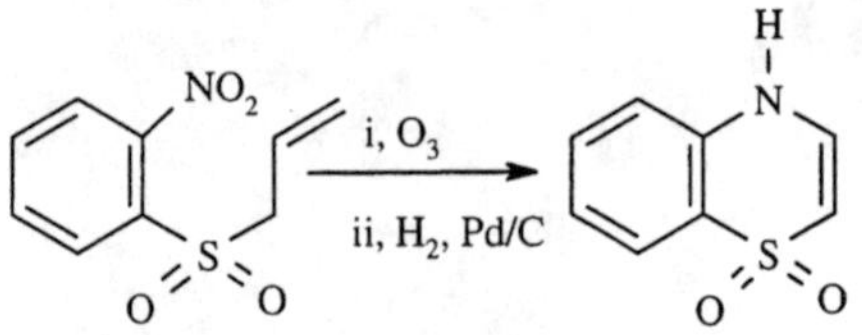

4*H*-1,4-Benzothiazine 1,1-dioxide may be synthesized from the sulfone (**340**) by ozonolysis and hydrogenation of the ozonide over palladium on carbon (Scheme 35) ⟨68TL1041⟩.

Dihydro-1,4-benzoxazines (**342**) are prepared from 2-hydroxyacetanilides (**341**) with 1,2-dibromo-ethane and sodium hydroxide in acetonitrile containing a phase transfer catalyst ⟨79S541⟩. 1,1-Dioxides of dihydro-1,4-benzothiazines (**344**) are generated through the cyclization of imines (**343**) with methanesulfonyl chloride in the presence of triethylamine ⟨79CI(L)26⟩.

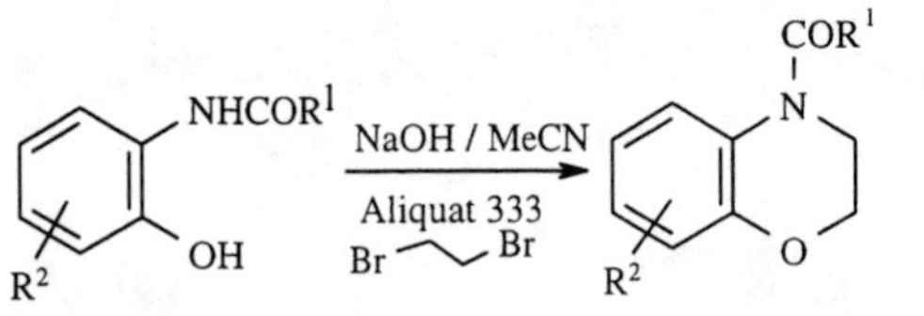

(**341**)   (**342**) 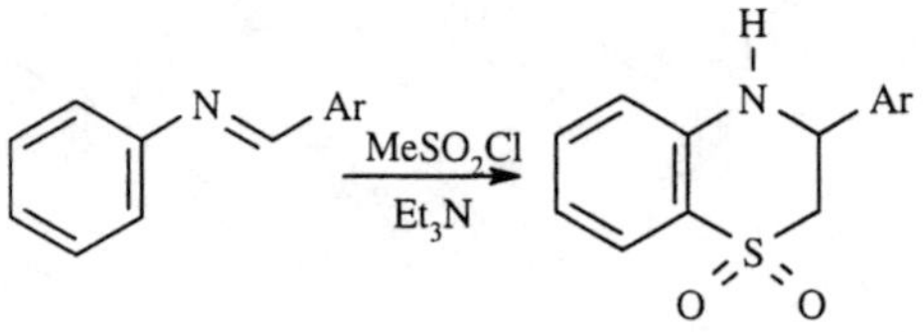   (**343**)   (**344**)

4.4.6.3 Rings Containing Oxygen and /or Sulfur Atoms

The monobenzo-fused derivatives of 1,4-dioxin, 1,4-oxathiin and 1,4-dithiin, (**345**), (**346**) and (**347**), can all be prepared by base-catalyzed reaction between the appropriate 1,2-disubstituted benzene and an α-haloketal *via* an intermediate 2-alkoxy-2,3-dihydro derivative (**348**). The pyrolysis of the acetoxy derivative (**349**) at 450°C gives (**345**; 80%) ⟨67ZC152⟩. 2-Hydroxy-2-phenyl-1,4-benzodioxane, from catechol and phenacyl bromide, is dehydrated to (**345**) by thionyl chloride in pyridine.

In the 1,4-benzoxathiane series again the best yield (76%) is obtained by pyrolysis of the 2-acetoxy derivative (**350**) ⟨66HC(21-2)852⟩.

Elimination of EtOH from 2-ethoxy-1,4-benzodithiane (**351**) readily gives 1,4-benzodithiin (**347**; 75%) ⟨66HC(21-2)1143⟩.

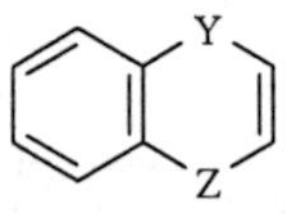

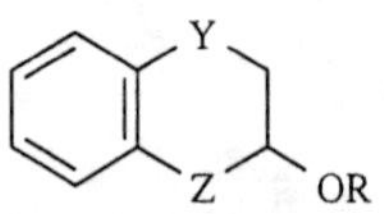

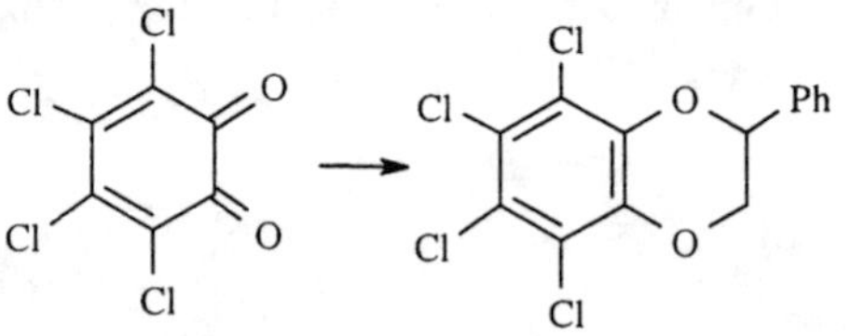

(**345**) Y = Z = O	(**348**) Y , Z = O or S , R = alkyl	(**352**)	(**353**)
(**346**) Y = S , Z = O	(**349**) Y = Z = O , R = COMe		
(**347**) Y = Z = S	(**350**) Y = S , Z = O , R = COMe		
	(**351**) Y = Z = S , R = Et		

Routes to benzo-fused derivatives of 1,4-dioxanes, 1,4-oxathianes and 1,4-dithianes make use of anions or dianions of the appropriate 1,2-disubstituted benzene. An alternative approach to the synthesis of 1,4-benzodioxanes involves Diels-Alder addition reactions of alkenes across the quinone function of 1,2-benzoquinones, *e.g.* (**352**) → (**353**).

4.4.6.4 Synthesis from Heterocyclic Precursors

Initial nucleophilic attack and ring opening are involved in the conversion of benzofuroxans into quinoxaline di-*N*-oxides by treatment with imines, enamines, carbonyl compounds and active methylene compounds (**354** → **355**) (Section 3.4.1.10.1).

(354)   (355)

4.4.6.5 Seven-membered Rings with Two Heteroatoms 1,4 to the Ring Junction

4.4.6.5.1 1,4-Benzodiazepines

Reactions between 1,5-bis nitrogen nucleophiles and 1,2-bis electrophile equivalents have been used in the synthesis of 1,4-benzodiazepines ⟨67AHC(8)21⟩. Chloroacetyl chloride has been much used as the two-carbon fragment in the synthesis of 2-oxo-1,4-benzodiazepine systems, *e.g.* (**356**) and (**357**) ⟨68CRV747⟩.

(356)   (357)

One of the commercially important routes to 1,4-benzodiazepin-2-ones (**361**) from 2-aminobenzophenones (**358**) involves the reaction of (**359**) with ammonia. Intermediate (**360**) can be isolated.

(358)   (359)   (360)   (361)

Quinazoline 3-oxides, *e.g.* (**362**), react with ammonia and primary amines to give 2-amino-1,4-benzodiazepine 4-oxides (**363**) ⟨79JMC1⟩.

(362)   (363)

4.4.6.5.2 1,4- and 4,1-Benzoxazepines, 1,4- and 1,5-benzothiazepines and 1,4-benzodioxepins

o-Aminobenzyl alcohols give 4,1-benzoxazepin-2-ones (364) ⟨65FES323⟩, o-hydroxybenzylamines give 1,4-benzoxazepin-3-ones (365) ⟨66JHC237⟩, and anthranilic acids give 4,1-benzoxazepine-2,5-diones (366). Compounds of type (364) are also obtained by the reaction of o-aminobenzyl alcohols with chloroacetic esters ⟨71JOC305⟩. Compound (367) can be cyclized by thionyl chloride to give the 1,4-benzoxazepin-2-one system (368).

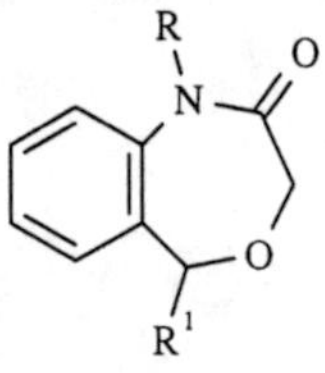

(364) 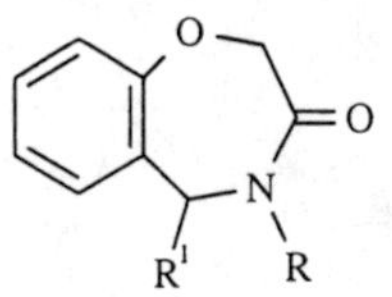　　　(365) 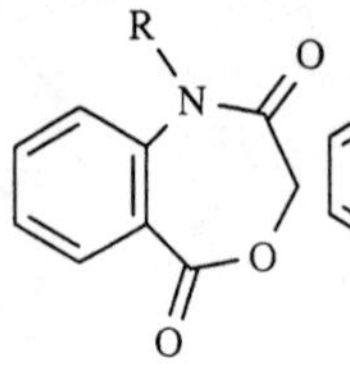　　　(366) 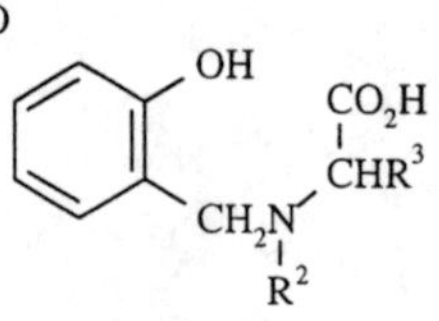　　　(367) 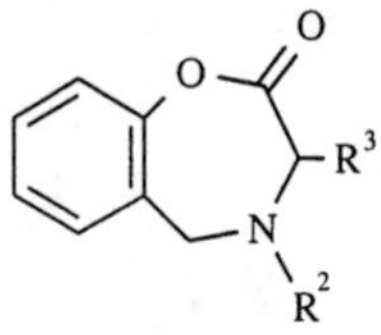　　　(368)

Salicylamide with 2-chlorophenylacetic acid followed by acetyl chloride gives the 3,5-dione (369). Similarly phenacyl bromide gives (370).

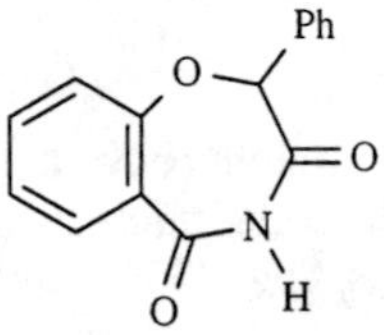

(369) 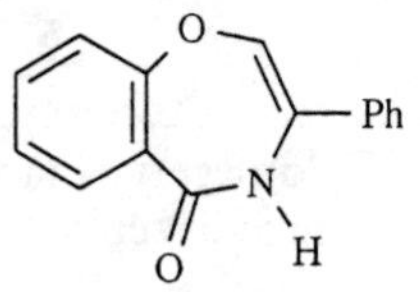　　　(370) 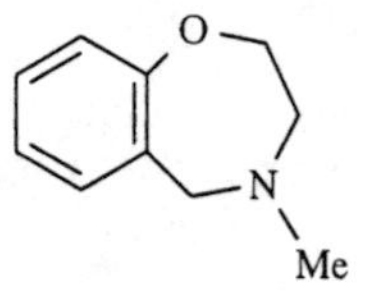　　　(371) 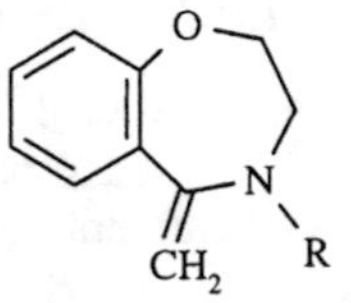　　　(372)

The reaction of methylamine with o-chloromethylphenyl 2-chloroethyl ether gives (371). o-(2-Chloroethoxy)acetophenone reacts with amines to give (372) ⟨63HCA1696⟩ and the analogous benzophenones react with formamide to give analogues of (371) with a 5-phenyl substituent ⟨69JPS1460⟩.

Anthranilic acids and esters ⟨75BSF(2)283⟩ with halohydrins give the 4,1-benzoxazepin-5-one system (373). o-Methylaminobenzamide with ethylene oxide gives (374) which is cleaved to (375) by ammonia ⟨66JOC4268⟩.

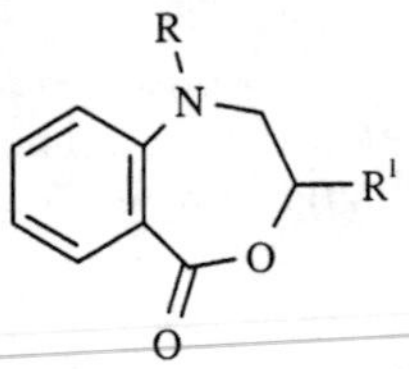

(373) 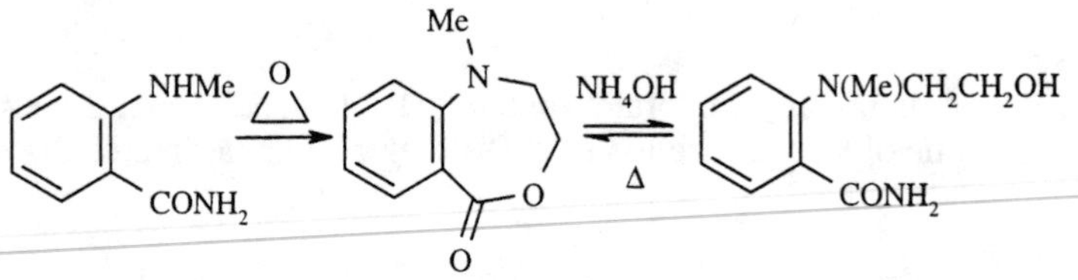　　　(374)　　　(375)

1,4-Benzothiazepines can be constructed from substrates of type (376) and an appropriate C—C—N fragment; thus, 2,3-dihydro-1,4-benzothiazepin-5-one (377) can be prepared by reaction of (376; R = OH) with aziridine. The reaction of 2-mercaptoaryl ketone (376; R = Ph) with 2-bromoethylamine is a two-stage process; intermediate (378) cyclizes in the presence of pyridine to give (379). The reaction of (376; R = OMe) with chloroacetonitrile in the presence of alcohols (R¹OH) gives (380) ⟨74OPP287⟩.

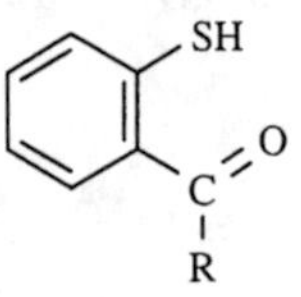

(376) 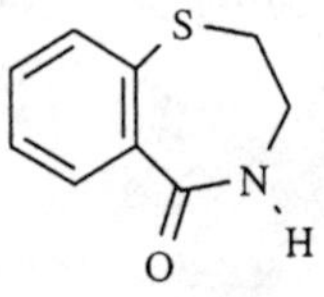　　　(377) 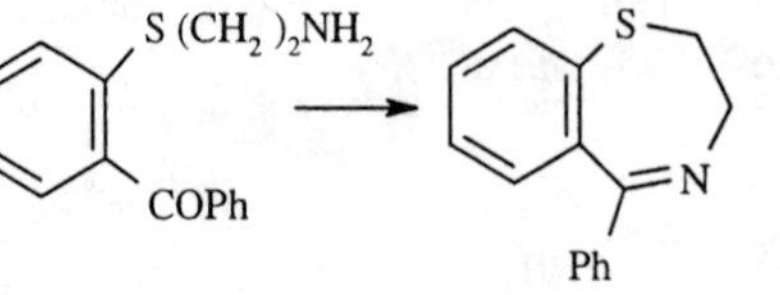　　　(378)　　　(379) 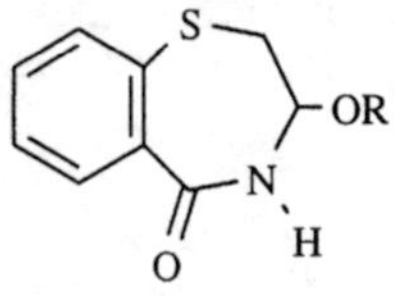　　　(380)

The 1,4- and 1,5-benzothiazepines (382) and (384) have been prepared by the photochemical reactions of the benzisothiazole (381) and the benzothiazole (383), respectively, with ethyl vinyl ether ⟨81TL529, 81TL2081⟩.

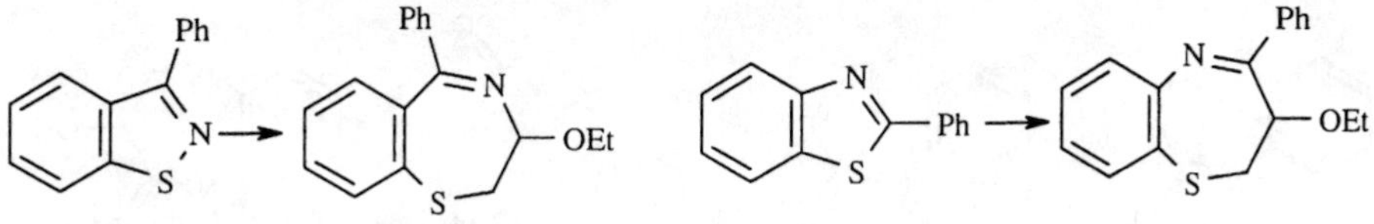

(381) (382) (383) (384)

5H-1,4-Benzodioxepin-ones and -diones have been prepared ⟨72HC(26)319, p. 339⟩: sodium salicylate and 2-chloroethanol give (385) ⟨75BSF(2)277⟩ and the methyl ester of 2-acetyl-6-chloro-3,5-dimethoxy-phenoxyacetic acid with hydrochloric acid gives (386). The dicarbonyl compound (387) is prepared by heating 2-carboxy-5,6-dimethoxyphenoxyacetic acid in acetic anhydride. Dicarbonyl compound (388) is prepared from chloroacetylsalicylic acid.

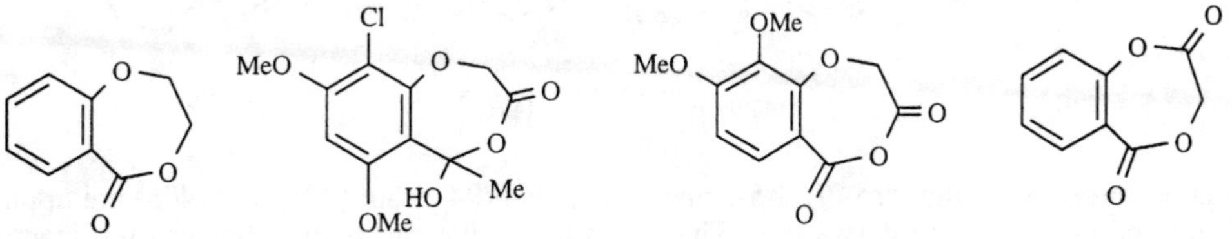

(385) (386) (387) (388)

4.4.6.6 Seven-membered Rings with Two Heteroatoms 1,5 to the Ring Junction

1,5-Benzodiazepine synthesis has been reviewed ⟨B-91MI 906-01⟩. The method generally applied the appropriately o-disubstituted benzene with a 1,3-bis carbon electrophile, e.g. o-phenylenedia-mine + RCOCHR'COR → (389) → (390) for 1,5-benzodiazepines.

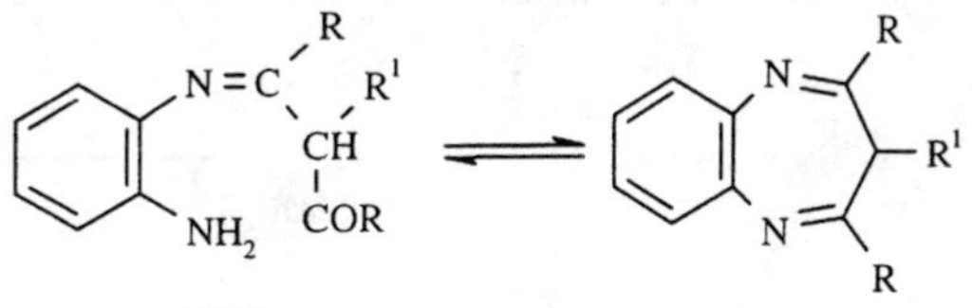

(389) (390)

o-Aminophenols react with a variety of functionalized three-carbon chains: 3-bromo-1-chlor-opropane gives (391) and 3-chloropropionyl chloride gives the analogous 4-oxo derivative. Similarly α,β-unsaturated ketones give (392), β-keto esters give (393) and 1,3-oxazolid-5-ones give (394).

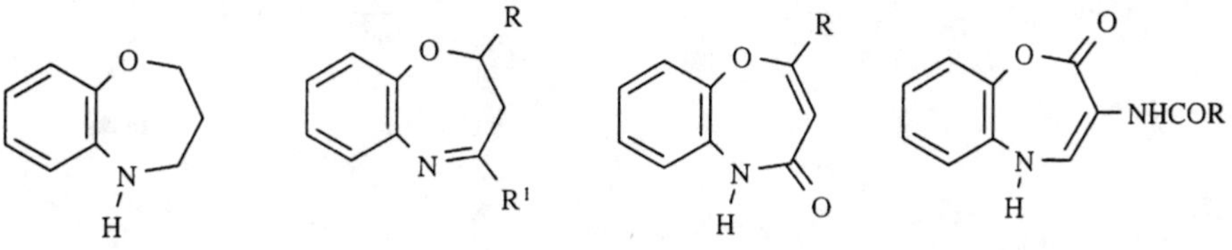

(391) (392) (393) (394)

In a similar fashion, 2-aminothiophenol can be reacted with 1,3-bis-carbon electrophiles to give various types of 1,5-benzothiazepine. Thus, 1,3-diphenylpropynone gives (395), reaction with β-keto esters gives products of type (396), reaction with diketene gives (397), and the reaction with methyl 3-arylglycidates gives (398).

(395)   (396)   (397)   (398)

The thiazolium salt (399) undergoes base-induced ring expansion *via* (400) and (401) to give (402) ⟨80TL2429⟩, a direct parallel of the conversion of analogous oxazolium salts into 1,5-benzoxazepines.

(399)   (400)   (401)   (402)

The major routes to 3,4-dihydro-2*H*-1,5-benzodioxepins (404) from (403) and (405) are applicable to a wide range of substituted derivatives. The 3-oxo derivative can be prepared *via* the reaction of 1,2-dihydroxybenzene with chloroacetonitrile ⟨75CJC2279⟩ or *via* a Dieckmann cyclization ⟨74USP3799892⟩.

(403)   (404)   (405)

2,3-Dihydro-5*H*-1,4-benzodioxepin (408) and halogenated derivatives (407) and (409) are obtained from 1,4-benzodioxin (406) by dihalogencarbene addition, thermolysis, and reduction ⟨83AG(E)64⟩.

(406)   (407)   (408)   (409)

4.4.7 TWO HETEROATOMS 2,3 TO RING JUNCTION

4.4.7.1 Six-membered Rings

o-Diacylbenzenes with hydrazine form phthalazines (410). Monoxo compounds of type (411) result from *o*-acylbenzoic acids and hydrazine or hydroxylamine, and dioxo derivatives (412) from phthalic acid derivatives with N_2H_4, NH_2OH or H_2O_2.

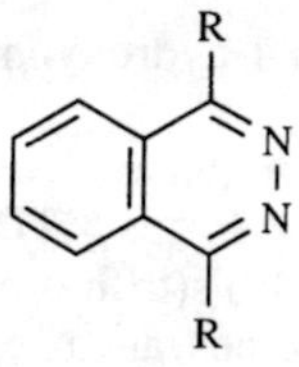

(410) 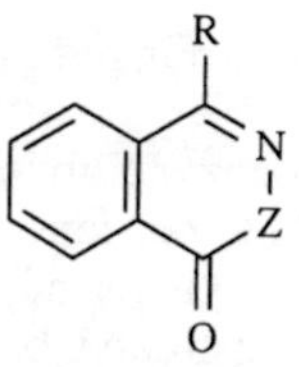 (411) 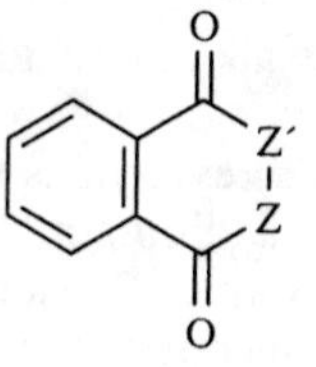 (412) 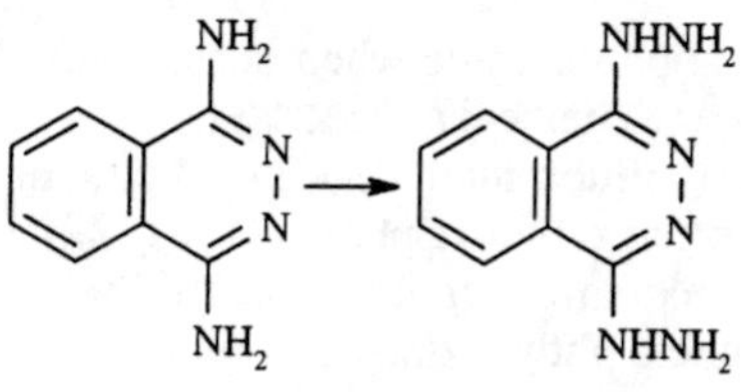 (413) (414)

Phthalodinitrile is converted by methanolic hydrazine into 1,4-diaminophthalazine (413) ⟨68JHC111⟩, and with excess of hydrazine 1,4-dihydrazinophthalazine (414) is obtained, *via* the 1,4-diamino compound ⟨59JOC1205⟩.

Many other [*c*]-fused pyridazines have been prepared similarly. 6-Methylpyridazine-3,4-dicarbaldehyde (415) with hydrazine gives 3-methylpyridazino[4,5-*c*]pyridazine (416) ⟨73JHC1081⟩. 5-Oxo- and 5,8-dioxo-substituted pyridazino[4,5-*c*]pyridazines are prepared from ethyl 3-formylpyridazine-4-carboxylates and diethyl pyridazine-3,4-dicarboxylates, respectively, with hydrazine. Pyridazine-3,4-dicarbonitrile (417) with hydrazine gives the diamino heterocycle (418) ⟨67JHC393⟩.

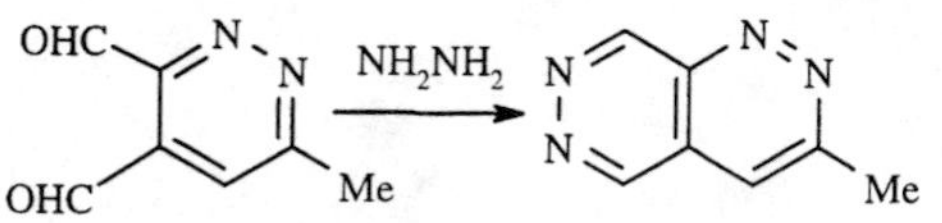

(415) (416) 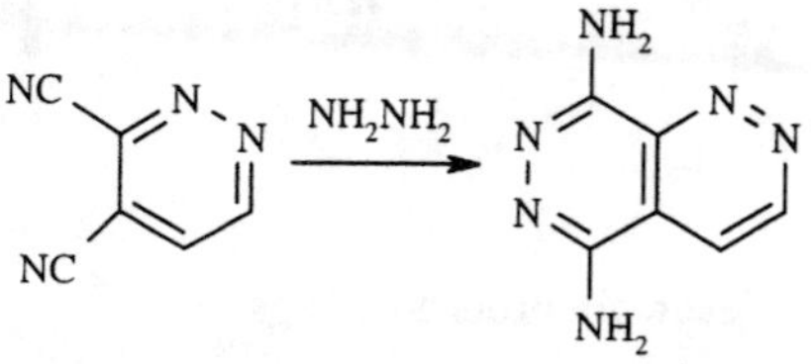 (417) (418)

Pyrazino[2,3-*d*]pyridazine-5,8-dione (420) can be prepared from pyrazine-2,3-dicarboxylic acid anhydride (419). Condensation of hydrazine with ethyl 5-acylpyridazine-4-carboxylates (421) gives pyridazino[4,5-*d*]pyridazin-1(2*H*)-ones (422) ⟨79M365⟩.

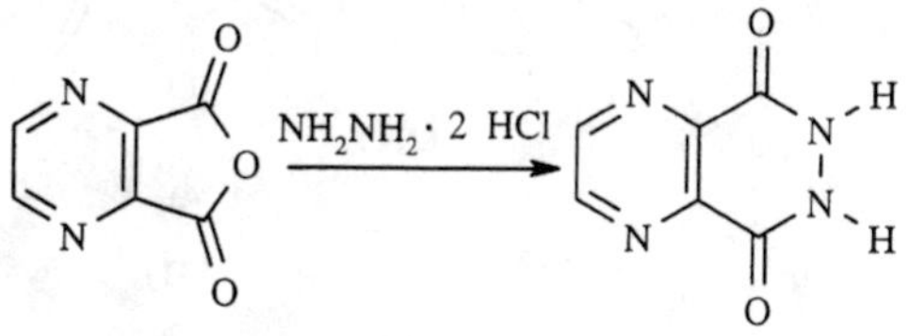

(419) (420) 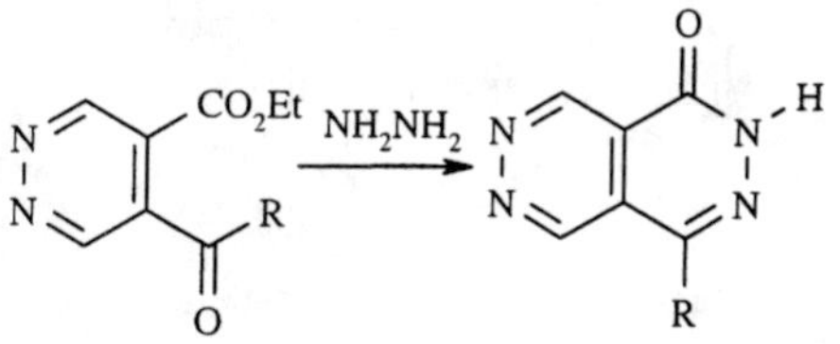 (421) (422)

The 1,2,3-thiadiazolo[4,5-*d*]pyridazines (423) are prepared similarly from the appropriate thiadiazoles ⟨76JHC301⟩.

Tetrahydropyridazines and analogues can be made as indicated in Scheme 36.

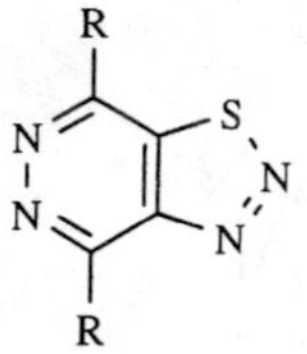

(423) 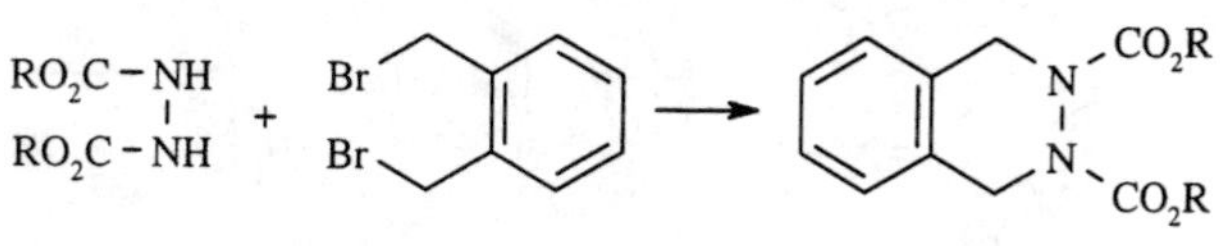 Scheme 36

Aromatic aldazines undergo oxidative cyclization to 1-arylphthalazines (424 → 425) on treatment with aluminum chloride/triethylamine at 180°C.

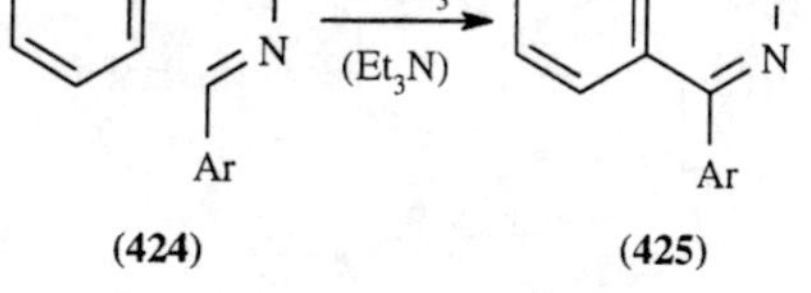

(424) (425) 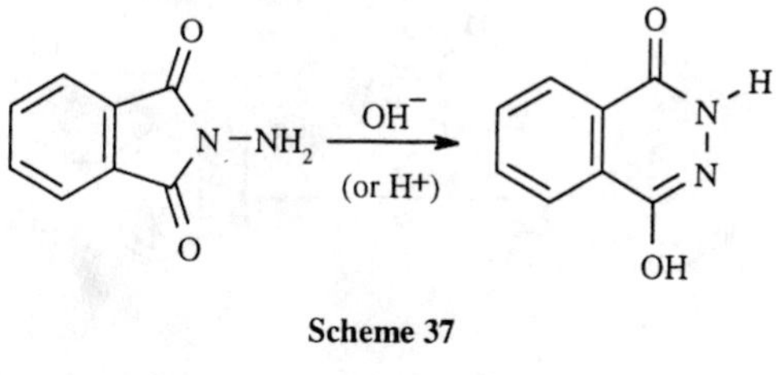 Scheme 37

2-Aminophthalimide when heated with dilute alkali or acid is rearranged into 4-hydroxyphthalazin-1(2*H*)-one (Scheme 37) ⟨55JCS852⟩.

3,6-Bis(trifluoromethyl)-1,2,4,5-tetrazine (**426**) cycloadds to benzyne giving a quantitative yield of 1,4-bis(trifluoromethyl)phthalazine (**427**). Cycloaddition of (**426**) to benzene also forms (**427**) in good yield by oxidation of the initial dihydro derivative. Reasonable yields of 1,4-bis(trifluoromethyl)-6-substituted phthalazines are obtained when electron donor-substituted benzenes are used in this reaction, and the regioselectivity is in agreement with calculations ⟨87AG(E)332⟩.

(**426**)   (**427**)

4.4.7.2 Seven-membered Rings

Reactions of appropriate carbonyl precursors with hydrazine have been used to obtain compounds such as 3,5-dihydro-4*H*-2,3-benzodiazepin-4-ones (**428**) ⟨76H(4)509⟩ and 2,5-dihydro-1*H*-2,3-benzodiazepin-1-ones (**429**) ⟨67AHC(8)21⟩.

(**428**)   (**429**)   (**430**)   (**431**)

Reaction of the isobenzopyrylium cation (**430**) with N_2H_4 gives (**431**) ⟨93KGS1475⟩. 1*H*-2,3-Benzodiazepines (**433**) ⟨73JCS(P1)2543⟩ and analogous thienodiazepines ⟨80JCS(P1)1718⟩ are obtained from electrocyclic reaction of diazo derivatives (**432**).

(**432**)   (**433**)

Dipolar cycloaddition of sydnones with benzocyclobutene yield 3*H*-2,3-benzodiazepines, *e.g.* (**434**).

(**434**)

The 2,3-benzoxazepin-1-one system (436) is prepared by the reaction of benzonitrile oxide with the benzopyranone (435) ⟨80JCS(P1)846⟩. The tetrahydro-3,2-benzothiazepine 3,3-dioxide (438) was prepared from (437) by intramolecular sulfonamidomethylation ⟨76CC470⟩.

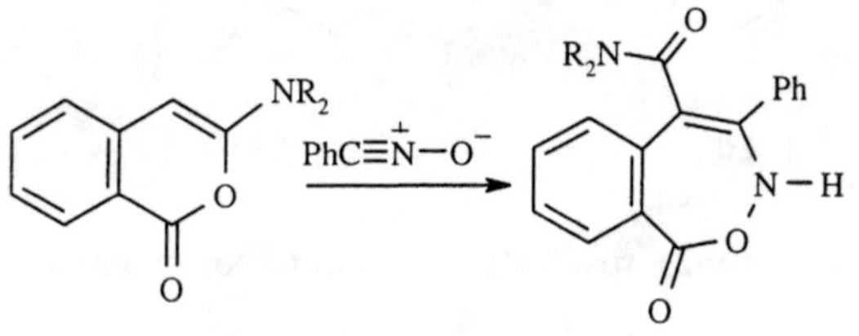

(435)   (436)

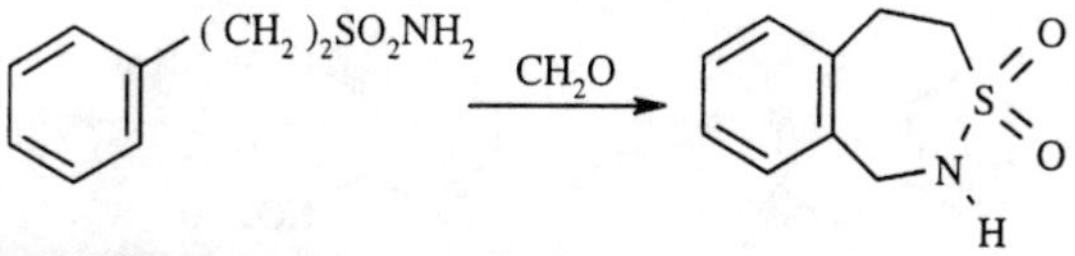

(437)   (438)

4.4.8   THREE OR MORE HETEROATOMS

4.4.8.1   Five-membered Heterocyclic Rings

o-Phenylenediamine is readily converted by HNO_2 into 1,2,3-benzotriazole (439) and by $SOCl_2$ into 2,1,3-benzothiadiazole (440). The 4,5-diaminopyrazole (442) gives the dihydro-pyrazolo[3,4-*c*]-[1,2,5]thiadiazole *S*-oxide (443) with thionyl chloride ⟨68JMC1164⟩ and the aromatic system (441) with sulfur monochloride ⟨81JOC4065⟩. Lead tetraacetate in glacial acetic acid effects closure between the primary amino group and the azo group of (444) giving the triazolo[4,5-*d*]pyrimidine (445) ⟨72CPB605⟩.

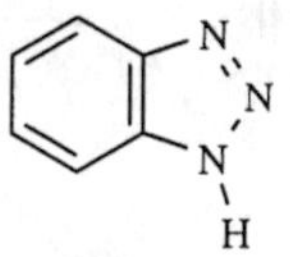

(439) 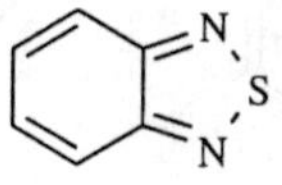   (440)

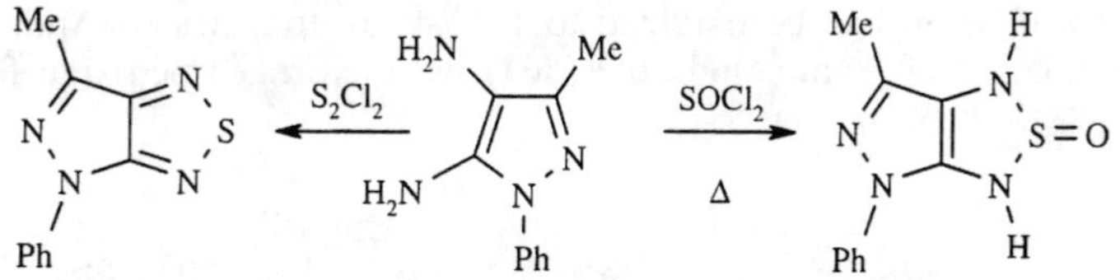

(441)   (442)   (443)

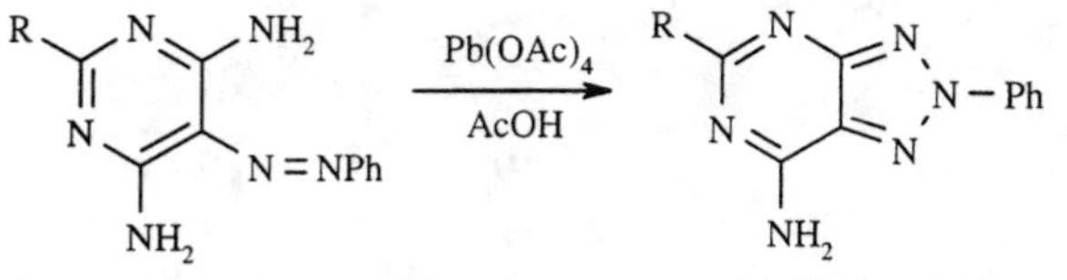

(444)   (445) 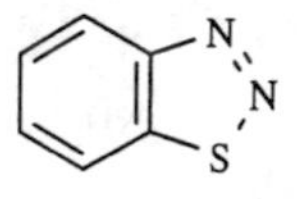   (446)

o-Mercaptoaniline and nitrous acid form 1,2,3-benzothiadiazole (446). Benzofurazans can be obtained by the benzazole rearrangement (Section 3.4.3.2.4).

4.4.8.2   Six-membered Heterocyclic Rings

4.4.8.2.1   *Three heteroatoms in the 1,2,3-positions*

(i) 1,2,3-Benzotriazines are prepared by methods of the type illustrated for (447) → (448) and (449) → (450), which resemble those used for the synthesis of cinnolines (*cf.* Section 4.4.4.3.1).

(447)   (448)   (449)   (450)

Dihydro-1,2,3-benzotriazines (**452**) result from 2-aminobenzylamines (**451**) with nitrous acid ⟨78HC(33)91⟩.

(451)   (452)   (453)   (454)

2-Nitrobenzaldehyde arylhydrazones with a halogen give compounds (**453**), which on treatment with base form 2-aryl-4-oxido-1,2,3-benzotriazinium betaine 1-oxides (**454**; R = Ar) ⟨74JOC2710⟩.

Oxidation of 2-azidophenyl ketone hydrazones (**455**) affords the 2-azidophenyldiazoalkanes (**456**) which can be cyclized thermally to 1,2,3-benzotriazines (**457**) ⟨75JCS(P1)31⟩. Similarly, 2-aminophenyl ketone hydrazones (**458**) give 1,2,3-benzotriazines (**457**) on oxidation with lead tetraacetate ⟨75JCS(P1)31⟩.

(455)   (456)   (457)   (458)

1-Amino- (**459**) and 3-amino-quinazolin-2-ones (**460**) ⟨75JCS(P1)31⟩, as well as 1-amino- and 2-amino-indazoles ⟨92AHC(53)85⟩ can be oxidized to 1,2,3-benzotriazines. *C*-Amino compounds can also give triazines; oxidation of 3-aminoindazole (**461**) with hydrogen peroxide forms 1,2,3-benzo-triazin-4-one (**462**) ⟨1899LA(305)289⟩.

(459)   (460)   (461)   (462)

(ii) 1,2,3-Benzoxathiazine and 2,1,3-benzothiadiazine derivatives are prepared by insertion of the S atom; thus, (**463**) results from *o*-C$_6$H$_4$(OH)CH$_2$NHR with SOCl$_2$, and (**464**) and (**465**) result from reaction of sulfamide, SO$_2$(NH)$_2$, with *o*-acylphenols and *o*-acylanilines, respectively.

(463)   (464)   (465)   (466)   (467)

1,2,3-Benzothiadiazine 1,1-dioxides, which are cyclic sulfonylhydrazides, are prepared by treating appropriate hydrazones with phosphorus pentachloride, *e.g.* (**466**) → (**467**).

4.4.8.2.2 Three heteroatoms in the 1,2,4- or 1,3,4-positions

(i) 1,2,4-Benzotriazines

2-(2-Nitrophenyl)hydrazides (**468**) give 1,2-dihydro-1,2,4-benzotriazines (**469**) with sodium amalgam in ethanol. In most cases the initial dihydro compounds (**469**) are oxidized by potassium ferricyanide to the aromatic 1,2,4-benzotriazines (**470**) ⟨78HC(33)189⟩. Similarly, reduction of the nitrohydrazones (**471**) affords 1,2,4-benzotriazines (**470**) ⟨78HC(33)189, p. 666⟩.

Cyclization of the 2-acylaminoazobenzenes (**472**) leads to 2-aryl-1,2,4-benzotriazinium salts (**473**) ⟨74GEP2241259⟩. 2-Nitrophenylamidines (**474a**) with base afford 1,2,4-benzotriazine 1-oxides (**475a**) ⟨78HC(33)189⟩. Similar treatment of the 2-nitrophenyl-ureas (**474b**), -thioureas (**474c**) and -guanidines (**474d**) yields the 3-hydroxy- (**475b**), 3-mercapto- (**475c**) and 3-amino-1,2,4-benzotriazine 1-oxides (**475d**) or their tautomers ⟨78HC(33)189⟩.

o-Benzoquinone (**476**) with formamidrazone (**477**) gives 1,2,4-benzotriazine (**478**) in low yield ⟨68CB3952⟩. Better yields of (**480**) were obtained from 1,2-naphthoquinone (**479**) using semicarbazide or thiosemicarbazide; compounds (**480**) are apparently formed regiospecifically ⟨78HC(33)189, p. 725⟩. Reaction of benzofuroxan with nitrile oxides affords 1,2,4-benzotriazine 1,2,4-trioxides (**481**) ⟨89CC986⟩.

3-Substituted-1,2,4-benzotriazines can be prepared by cyclization of diarylformazans (**482** → **483**) ⟨82T1793⟩.

1,2-Dihydro-1,2,4-benzotriazines (**485**) are obtained from the reaction of *N*-(2-chlorophenyl)-imidates (**484**) with hydrazines. They can be oxidized to 1-alkyl-1,2,4-benzotriazinium salts (**486**) ⟨67MI21900⟩.

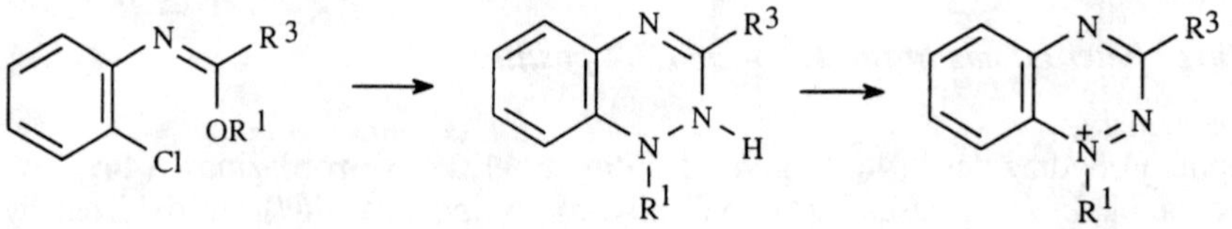

(**484**)   (**485**)   (**486**)

Nitrosation of the hydrazinouracil (**487**) leads to the pyrimidotriazine (**488**). Pyrimidotriazines can also be made by the sequence (**489**) → (**490**).

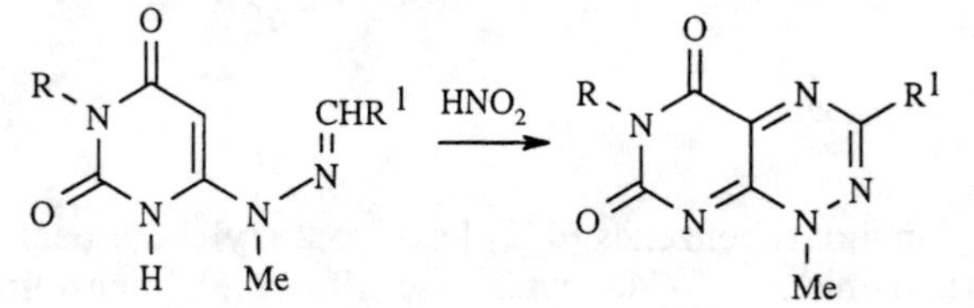

(**487**)   (**488**)

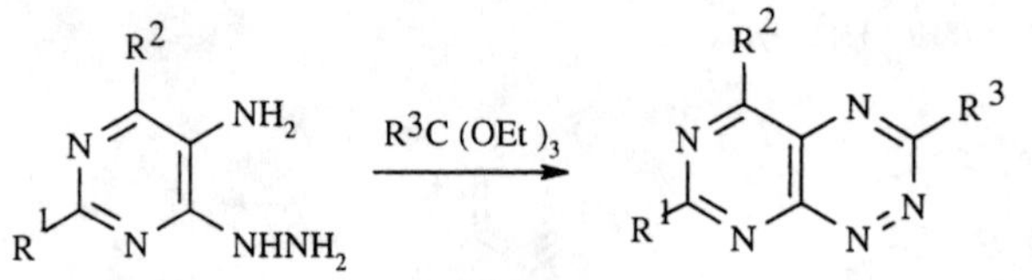

(**489**)   (**490**)

Oxidation of 1,2-diaminobenzimidazoles with Pb(OAc)$_2$ affords 3-amino-1,2,4-benzotriazines with good yield ⟨92AHC(53)85⟩ (see also Section 3.4.3.12.5).

(ii) Six-membered rings containing O or S atoms

1,3,4-Benzoxadiazines (**492**; X = O) result from ring closure of the phenylhydrazide (**491**; X = O, Y = Br or NO$_2$) with displacement of Y ⟨80JOC3677⟩. Routes to 1,2,4-benzoxadiazines include reaction of *N*-aryl-*S*,*S*-dimethylsulfimides with nitrile oxides, and oxidation of *N*-arylamidoximes to give the benzoxadiazines (**493**). Both reactions involve nitrosoimine intermediates (Scheme 38).

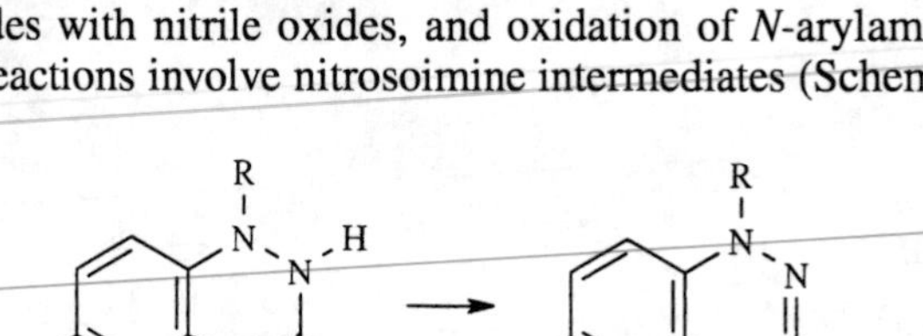

(**491**)   (**492**)

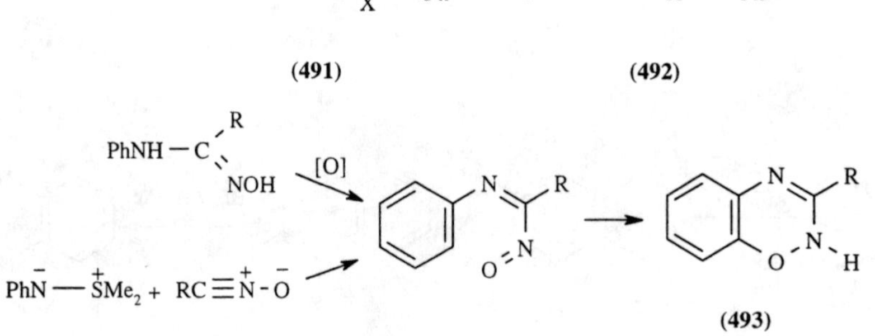

(**493**)

Scheme 38

An electrocyclic reaction is involved in the reversible formation of the oxadiazine ring in (**494**) → (**495**).

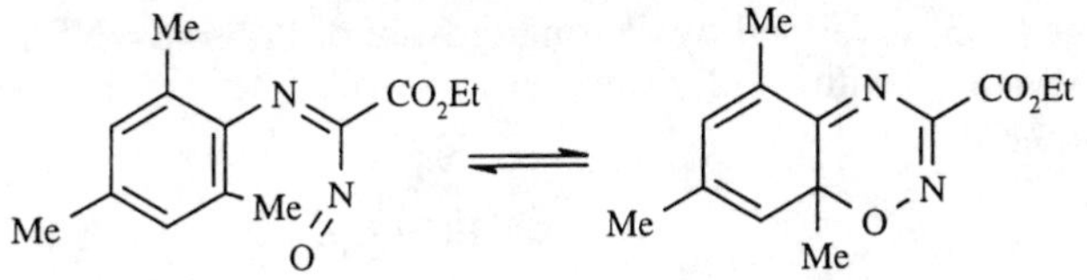

(494) (495)

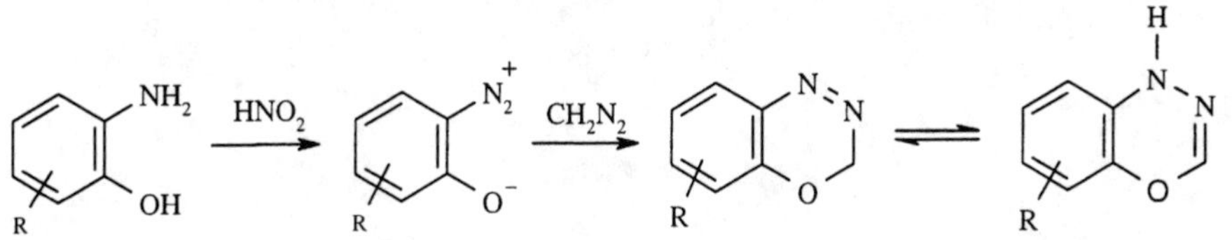

Scheme 39

Diazotization of 2-aminophenols followed by reaction with diazomethane gives 1,3,4-benzox-adiazines (Scheme 39); the initially formed 2H-isomers readily tautomerize to the more stable 4H-isomers ⟨70CB331⟩.

1,3,4-Benzothiadiazines (**492**; X = S) are prepared by cyclization of (**491**; X = S, Y = Br or NO₂) ⟨80JOC3677⟩, and analogous compounds result from cyclization of the chlorohydrazones (**496**) with triethylamine ⟨81JCS(P1)2245⟩. Intramolecular aromatic sulfonation gives 1,2,4-benzothiadiazine 1,1-dioxides (**498**) from (**497**) and aluminum trichloride ⟨79JCS(P1)1043⟩.

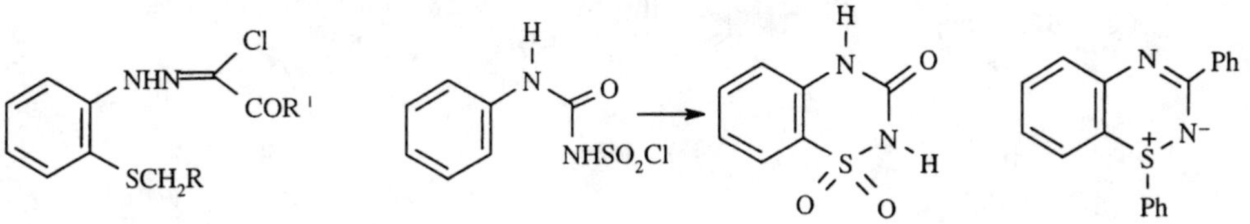

(496) (497) (498) (499)

Treatment of *N*-(2-phenylthio)phenylbenzamidine with *N*-chlorosuccinimide (NCS) leads to the cyclic sulfimide (**499**) ⟨78CC1049⟩.

2-Aminobenzenesulfonamides are ring closed directly to 4H-1,2,4-benzothiadiazine 1,1-dioxides (**500**) with a variety of carboxylic acids, employing polyphosphoric acid trimethylsilyl ester as the cyclization agent ⟨83S851, 85JAP(K)6025984, 90JMC1721⟩.

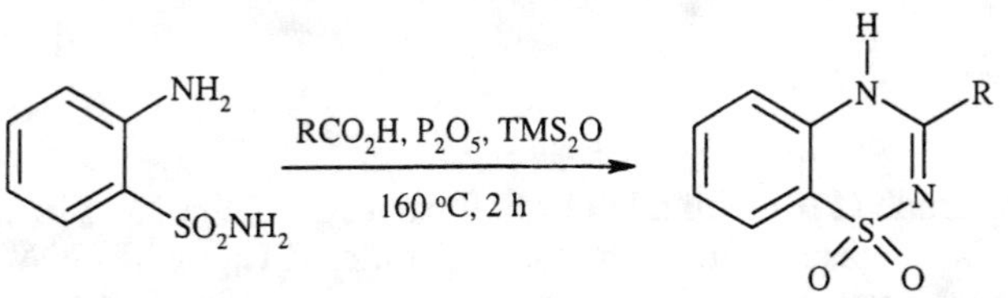

(500)

4.4.8.2.3 Four heteroatoms

The only known representative of 1,2,3,4-tetrazines (**502**) has been obtained with 81% yield on oxidation of 1-amino-5-phenyl-1,2,3-triazolo[4,5-*d*]-1,2,3-triazole (**501**) with lead tetraacetate ⟨88CC1608⟩.

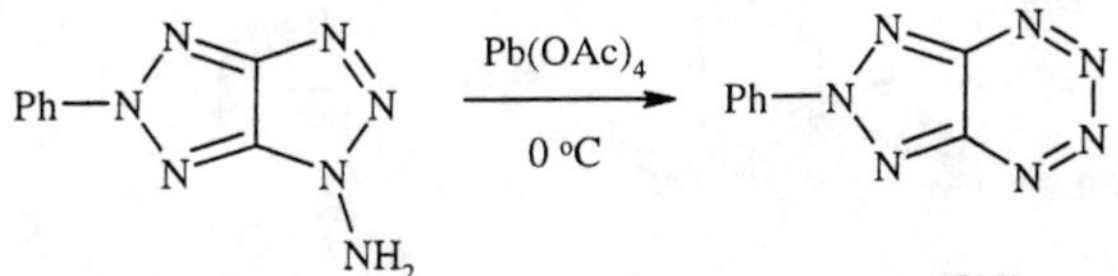

(501) (502)

The benzodithiadiazine (**503**) is formed by the ring closure of PhN=S=NSCl ⟨83CC74⟩.

1,2-Hydrazinedicarboxylates with 1,2-disulfenyl chlorides give 1,4,2,3-benzodithiadiazines (Scheme 40) ⟨71AG(E)408, 74CB771⟩.

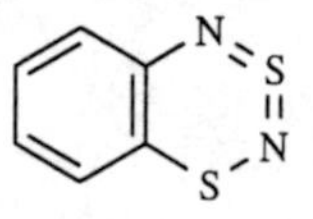

(503)

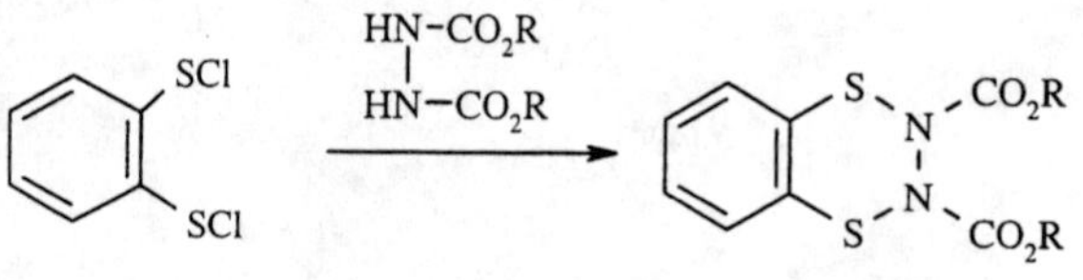

Scheme 40

4.4.8.3 Seven-membered and Larger Rings

4.4.8.3.1 *Heteroatoms 1,2,4 to ring junction*

1*H*-1,2,4-Benzotriazepines (**506**) can be prepared by reaction of hydrazidoyl chlorides (**504**) with cyano compounds *via* the nitrilium salt (**505**) ⟨74T195⟩.

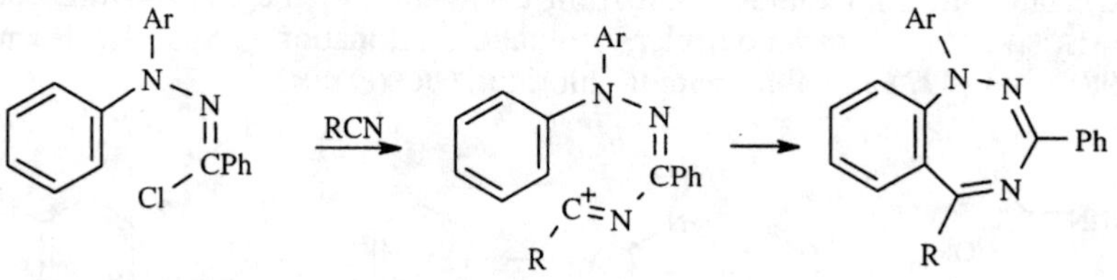

(504)   **(505)**   **(506)**

2-Benzoylarenesulfonyl chlorides (**507**) with amidines, guanidine or *S*-alkylthioureas give arenesulfonyl derivatives (**508**; R = Ph, NH$_2$ or SMe) which cyclize on heating to 1,2,4-benzothiadiazepine 1,1-dioxides (**509**) ⟨68JHC719⟩.

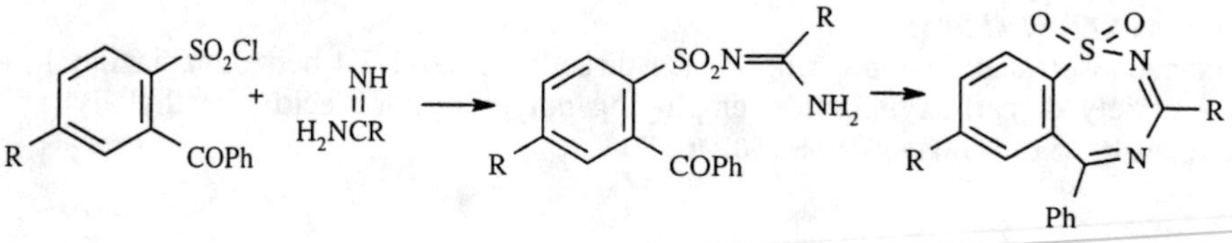

(507)   **(508)**   **(509)**

2-Aminomethylsulfonamide (**510**) with ethyl orthoformate gives the 1,2,4-benzothiadiazepine (**511**) ⟨60JA1594⟩, and (**513**) is prepared from the sulfoximide (**512**) with *N,N'*-carbonyldiimidazole ⟨72CB2575⟩. The cyclic sulfoximides (**513**) are also prepared from (**514**) with hydrazoic acid ⟨72CB2575⟩.

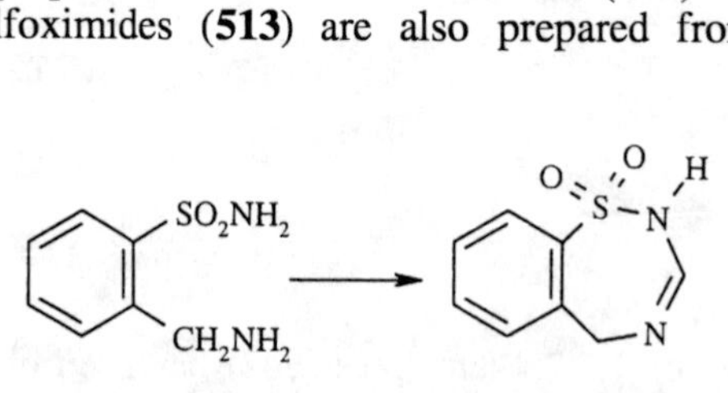

(510)   **(511)**

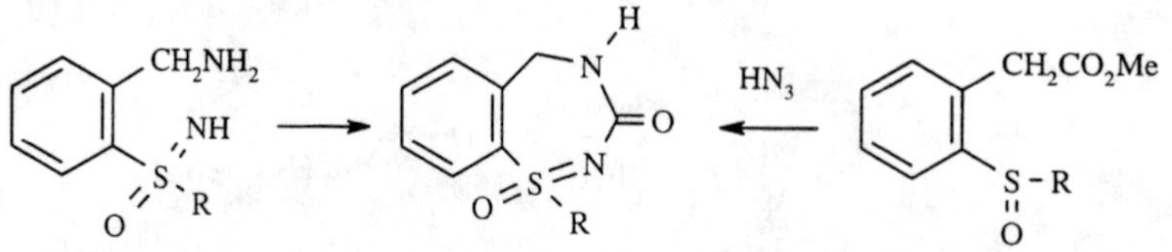

(512)   **(513)**   **(514)**

4.4.8.3.2 Heteroatoms 1,2,5 to ring junction

1H-4,5-Dihydro-1,2,5-benzotriazepin-4-ones (**516**; $R^1 = CO_2Et$ or CN) can be prepared by cyclization of the diazonium salts (**515**).

| (515) | (516) | (517) | (518) |

The 1,2,5-benzothiadiazepine 1,1-dioxide (**518**) and its 3,4-dihydro analogue have been prepared by the reaction of 2-nitrobenzenesulfonyl chloride with φ-aminoacetophenone to give (**517**) followed by reductive cyclization ⟨79JHC835⟩.

4.4.8.3.3 Heteroatoms 1,3,4 to ring junction

The hydrazones or semicarbazones of 2-aminoaryl ketones (**519**) react with paraformaldehyde to give the 2,3-dihydro-1H-1,3,4-benzotriazepine (**520**) ⟨70BCJ135⟩, and with ethyl chloroformate to give the 2-oxo analogue (**521**) ⟨74JPS838⟩. Compounds of type (**521**) can also be prepared by the reaction of 2-aminoaryl ketones with ethoxycarbonylhydrazine.

| (519) | (520) | (521) |

1-[(2-Aminophenyl)methyl]-1-phenylhydrazines (**522**), prepared from phenylhydrazine and 2-nitrobenzyl chloride, readily cyclized with triethyl orthoacetate and with orthopropionate to give the 4,5-dihydro-3H-1,3,4-benzotriazepines (**523**) (*ca.* 60% yield) ⟨85JHC1105⟩. 2-Aminobenzoyl hydrazides have similarly been used in the preparation of 1,3,4-benzotriazepin-5-ones (**524**) by reaction with ortho esters ⟨76JOC2732, 76JOC2736, 67JHC967⟩.

| (522) | R^1 = H, Me; R^2 = Me, Et | (523) | (524) |

Arylazoarylimines (**527**), with a methyl or ethyl group *ortho* to the imine function (prepared by oxidation with manganese dioxide of arylaminohydrazones from anilines (**525**) and chlorohydrazones

(**526**)), underwent thermal intramolecular cyclization in refluxing xylene catalysed by DABCO to give the 1*H*-4,5-dihydro-1,3,4-benzotriazepines (**528**) in 50–87% yields ⟨86JHC1795⟩.

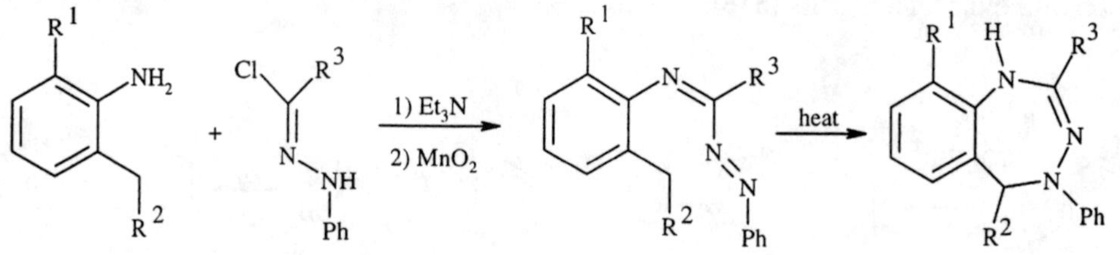

(**525**)   (**526**)   (**527**)   (**528**)

R^1 = Me, Et; R^2 = H, Me; R^3 = H, CN, CO_2Me

Treatment of salicylic hydrazide in toluene with a single carbon insertion unit, such as carboxylic acid anhydride, acid chloride, and orthoester, in the presence of an equimolecular amount of methanesulfonic acid gave the 1,3,4-benzoxadiazepin-5-ones (**530**) (43–68% yield), *via* the *O*-acylation intermediates (**529**) ⟨92S929⟩.

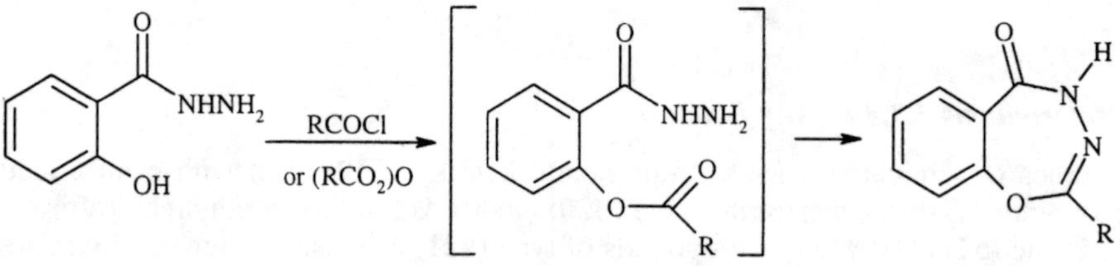

(**529**)   (**530**)

R = H, Me, Pr^i, Ph, CH_2Cl, CO_2Et

2-Mercaptobenzoylhydrazide with formaldehyde gives (**531**) ⟨53JOC1380⟩ and *o*-thiocyanatoaceto-phenone with semicarbazide gives (**532**) ⟨79JCR(S)395⟩.

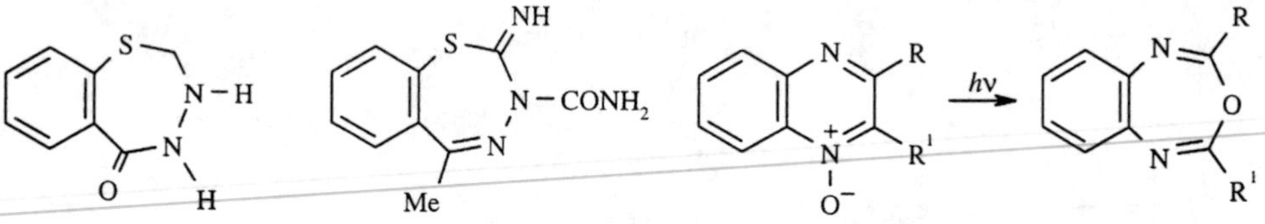

(**531**)   (**532**)   (**533**)   (**534**)

4.4.8.3.4 *Heteroatoms 1,3,5 to ring junction*

The photoisomerizations of quinoxaline 1-oxides (**533**) give the 3,1,5-benzoxadiazepines (**534**) ⟨67TL1873, 68ACS877⟩. The photoisomerization of quinazoline 3-oxides (**535**; R = H or Me) leads, by the rearrangement shown, to the isolable 1,3,5-benzoxadiazepines (**536**).

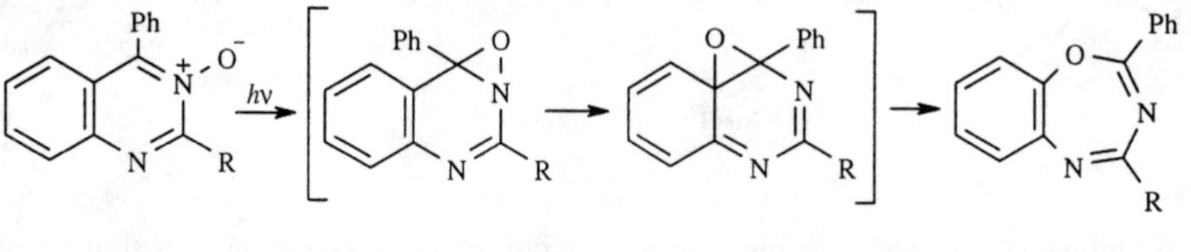

(**535**)   (**536**)

2-Phenylbenzazete (**537**) with nitrile oxides $ArC \equiv N^{+} - O^{-}$ gives labile adducts (**538**) which rearrange to the 1,3,5-benzoxadiazepines (**542**) ⟨75CC740⟩, probably *via* (**539**), (**540**) and (**541**).

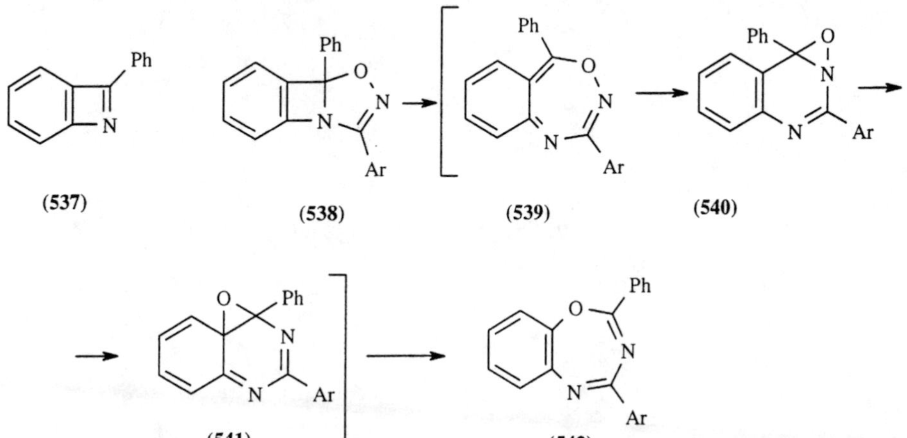

(**537**)   (**538**)   (**539**)   (**540**)

(**541**)   (**542**)

4.4.8.3.5 *Four or more heteroatoms*

The 1,2,4,5-benzotetrazepine (**545**) is produced in good yield by the oxidative ring closure of (**544**) or by the action of zinc and aqueous sodium hydroxide on (**543**) ⟨55CB1284⟩.

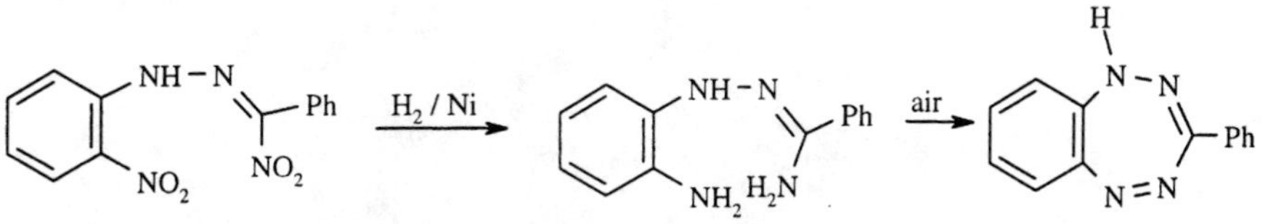

(**543**)   (**544**)   (**545**)

Reaction of benzeno-1,2-bis(sulfenyl) chlorides (**546**) with bis(trimethylsilyl)sulfurdiimide gives benzo-1,3$\lambda^4\delta^2$,5,2,4-trithiadiazepines (**547**) ⟨85CC396, 87JCS(P1)211⟩.

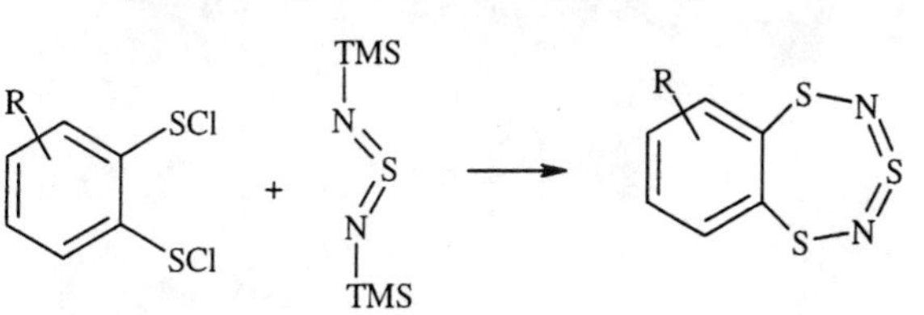

(**546**) R = H, Me   (**547**)

Aromatic and heteroaromatic fused pentathiepins, *e.g.* (**548**), have been prepared by reactions of S_2Cl_2 with vicinal dithiols or their salts. The dithiols can be generated *in situ* (Scheme 41) ⟨80JOC5122, 84JOC1221, 85JA3871⟩.

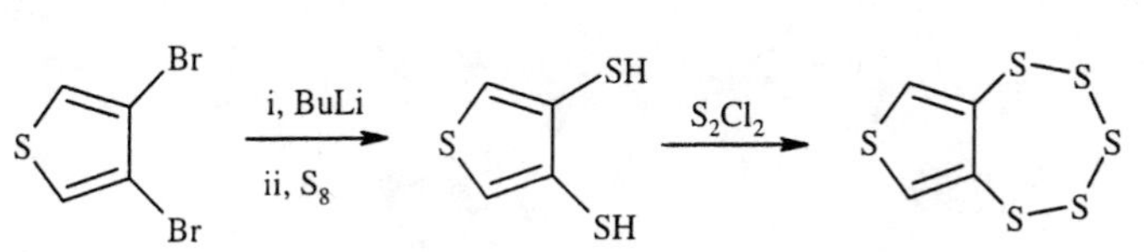

(**548**)

Scheme 41

4.5

Synthesis of Tri- and Poly-cyclic Ring Systems without Ring Junction Heteroatoms

The arrangement of the material in this chapter follows closely that for the bicyclic analogues: it is ordered first by the mutual relationship of the fused rings, second by the number of heteroatoms and finally by the size of the heterocyclic ring. We define 'adjacent' as in (1) and 'non-adjacent' fused rings as in (2). Rings of type (3) form a class of *peri*-annulated heterocyclic systems which are of growing importance.

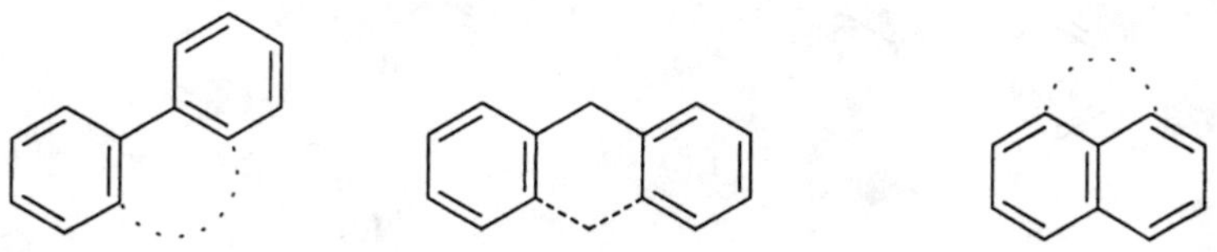

(1)　　　　　　　(2)　　　　　　　(3)

4.5.1　TWO ADJACENT FUSED RINGS, ONE HETEROATOM

4.5.1.1　Five-membered Heterocyclic Ring

4.5.1.1.1　Overview of synthetic methods for carbazoles, dibenzofurans and dibenzothiophenes

(4)　　　　　　　(5)

Most of the important methods involve C—C (4) or C—Z (5) bond formation and these two classes are considered in turn. There are a variety of miscellaneous methods.

4.5.1.1.2　Formation of C—C bond

(i) The photochemically initiated cyclizations of diphenyl ethers to dibenzofurans (6 → 9) ⟨75S532, 75AJC1559⟩, diphenyl sulfides to dibenzothiophenes (7 → 10) ⟨75S532⟩ and diarylamines to carbazoles (8 → 11) ⟨66CC272, 66TL661⟩ normally require the presence of an oxidizing agent such as iodine or oxygen. The oxidative ring closure of diphenyl ethers to dibenzofurans (6 → 9) is also effected by palladium(II) acetate ⟨75JOC1365, 76JCS(P1)1236⟩, as is that of diphenylamines to carbazoles ⟨75JOC1365, 86CPB2672, 92JCS(P1)3439, 94TL1695⟩. *o*-Iododiarylamines on treatment with Pd(OAc)$_2$ also yield carbazoles ⟨84T1919⟩. This approach has been extended to the conversion of *N*-arylenaminones to tetrahydrocarbazoles (12 → 13) ⟨80JOC2938⟩.

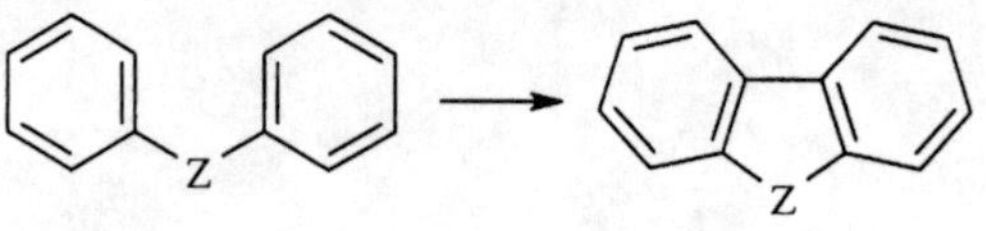

(6) Z = O
(7) Z = S
(8) Z = NH

(9) Z = O
(10) Z = S
(11) Z = NH

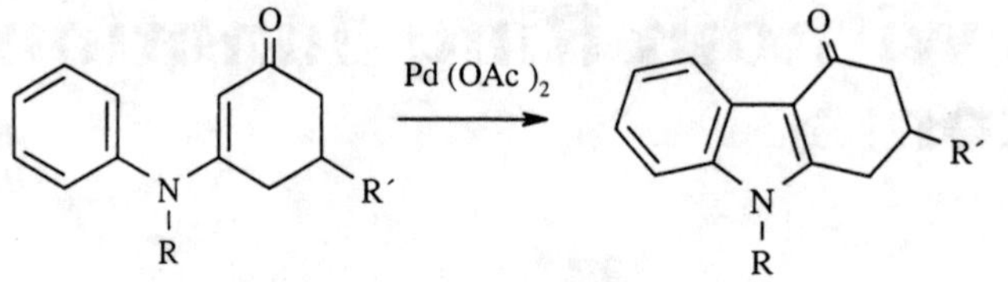

(12)　　　　(13)

Sulfuric acid promotes cyclization of *N,N*-diphenylhydroxylamine (14) to carbazole (11) ⟨13CB3306⟩. The parallel conversions of diphenyl sulfoxide (15) and diphenyl selenoxide to dibenzothiophene (10) ⟨23CB2275⟩ and dibenzoselenophene ⟨39CR(199)531⟩ are effected by sodamide.

(14)　　　　(15)

(ii) Dibenzofurans (9) ⟨81BCJ2374⟩, dibenzothiophenes (10) ⟨36JCS1435⟩ and *N*-substituted carbazoles ⟨52JCS2276⟩ are formed by spontaneous Pschorr-type cyclization of appropriate diazonium salts, thus (16) gives dibenzofuran (9).

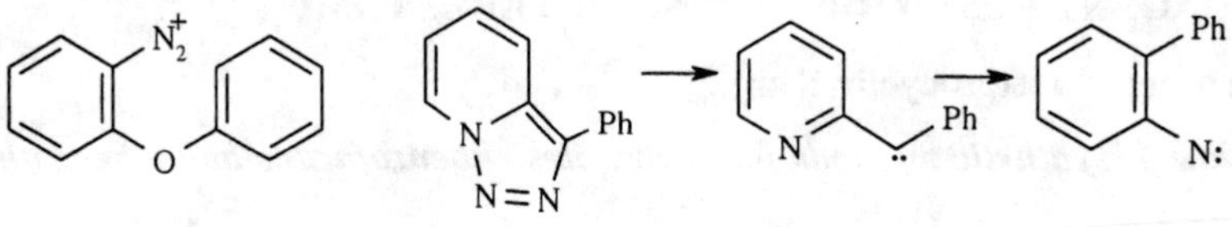

(16)　　　(17)　　　(18)　　　(19)

(iii) Carbazoles are prepared by the Graebe–Ullmann synthesis: 1-arylbenzotriazoles on pyrolysis yield carbazoles ⟨57JCS110⟩, see Section 3.4.1.2.1. Pyrolysis of 3-phenyltriazolo-pyridine (17) is more complex, ultimately providing carbazole (11) *via* (18) and (19) ⟨75JA7467⟩.

(iv) Substituted carbazoles are made by acid-catalyzed condensation of indoles with 2,5-hexanedione (Scheme 1) ⟨67AJC2715, 77JCS(P1)1698, 87CPB425, 90H(31)401⟩.

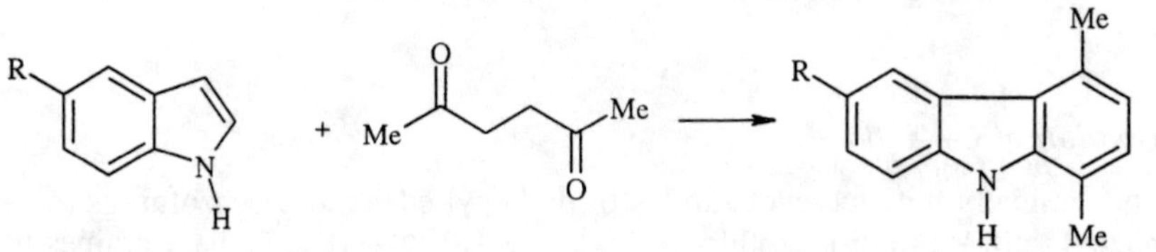

Scheme 1

Indoles with 2-substituents such as methyl or benzyl are cyclized with enones to carbazoles in acetic acid in the presence of Pd/C. This reaction involves electrophilic addition at C-3 and subsequent aromatization (Scheme 2) ⟨88T5215⟩.

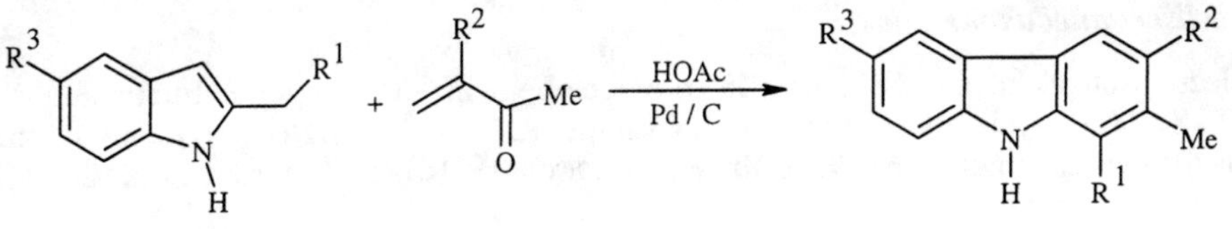

Scheme 2

(v) The Borsche synthesis of tetrahydrocarbazoles, *e.g.* (**20** → **21**), is a special case of the Fischer indole synthesis in which cyclohexanone phenylhydrazones are used as the starting materials.

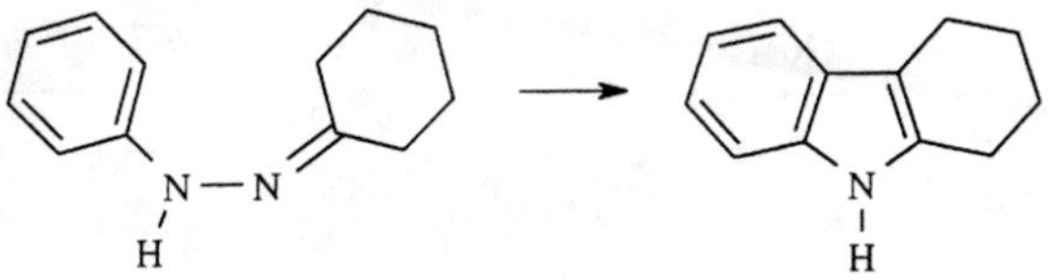

(20)   (21)

4.5.1.1.3   *Formation of C—Z bond*

Electrophilic attack by the *ortho* heteroatom of biphenyl derivatives is involved in the conversions (**22** → **23**) ⟨82JOC3585⟩ and (**24** → **10**) ⟨62JOC4111⟩.

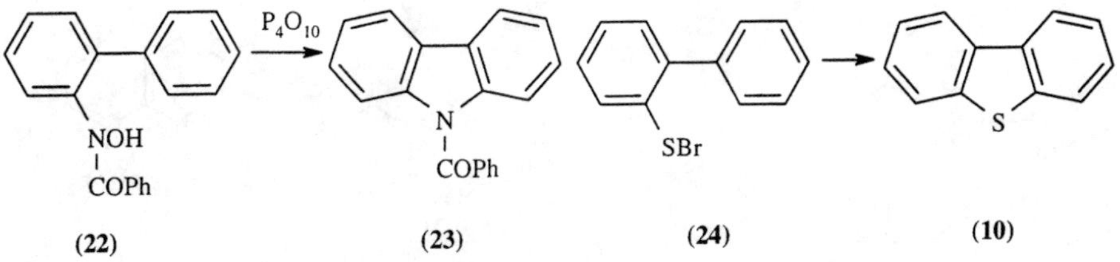

(22)   (23)   (24)   (10)

A variety of pyrrole ring closure reactions are conveniently formulated as proceeding *via* nitrene intermediates, although it is doubtful whether a free nitrene is involved. Pyrolysis of *o*-nitrobiphenyls with iron(II) oxalate ⟨61T(16)80⟩, or reduction under milder conditions with triethyl phosphite ⟨65JCS4831⟩ or tris(trimethylsilyl) phosphite ⟨79TL375⟩, leads to the carbazole, as does the pyrolysis or photolysis of 2-azidobiphenyls (Scheme 3) ⟨75JA6193⟩.

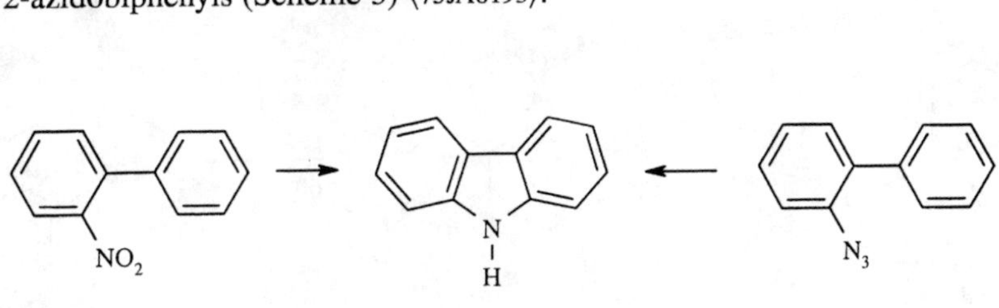

Scheme 3

4.5.1.1.4 *Miscellaneous methods*

Intramolecular attack of the carbenes (25) provides benzo[*b*]cyclohepta[*d*]-furans and -thiophenes (26; X = O, S). Photolysis of 2-biphenyl isocyanide (27 → 28) ⟨72JOC3571⟩ and thermolysis of 2-biphenylsulfonyl diazomethane (29 → 30 → 31) ⟨72CC893⟩ also result in ring expansion.

(25)   (26) X = O, S   (27)   (28)

(29)   (30)   (31)

The extrusion of sulfur from phenothiazine and thianthrene, leading to carbazole ⟨75CJC2293⟩ and dibenzothiophene ⟨37RTC627⟩, respectively, is effected by thermolysis in the presence of copper bronze.

4.5.1.2 Six-membered Rings

Phenanthridines are obtained:
(i) By cyclodehydration of acylated 2-aminobiphenyls (32 → 33).
(ii) By photochemical dehydrogenation of azomethines (34 → 33); dihydro compounds are initially formed and aromatization is effected with either oxygen or iodine. Phenanthridine formation by this method is improved by the presence of BF$_3$ ⟨88TL5213⟩.

(32)   (33)   (34)   (35)   (36)

(iii) On thermolysis in hexadecane at 230°C the azide (35) gives (36) by insertion of the nitrene into the methyl group.
(iv) Radical cyclization of 2-arylidenaminodiphenyls (37) leads to phenanthridines (39), possibly *via* radical intermediates (38) ⟨85S107⟩.

(37)   (38)   (39) X = H, Cl, OMe

(v) Another well known route to phenanthridine and related systems is the intramolecular cyclization of 2-amino-2′-formyldiphenyls obtained *in situ* from the corresponding 2′-lithium derivatives (Scheme 4), *cf.* Section 3.4.3.4.4, conversion (**558 → 519**).

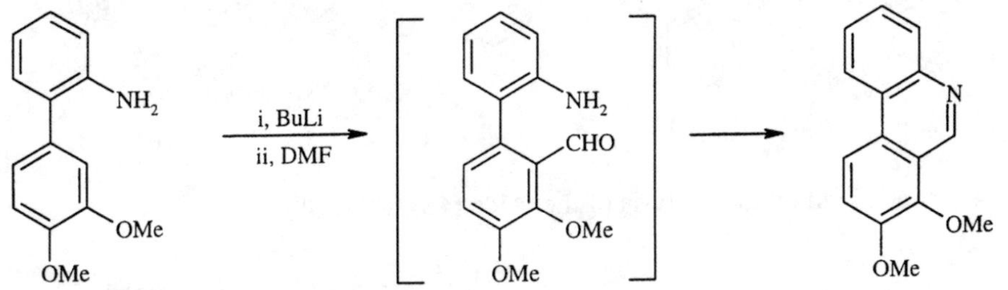

Scheme 4

4.5.2 TWO ADJACENT FUSED RINGS, TWO HETEROATOMS

Benzo[*c*]cinnoline derivatives of different oxidation levels are prepared by *N–N* bond formations from suitable 2,2′-disubstituted biphenyls as outlined in Scheme 5.

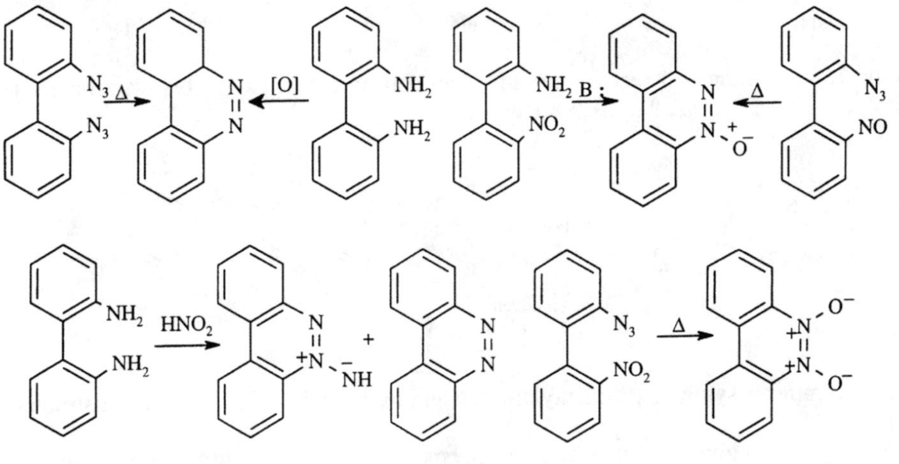

Scheme 5

Controlled reduction of 2,2′-dinitrobiphenyl is achieved with 2, 1, and 0.5 molar equivalents of sodium borohydride in the presence of palladium on carbon and sodium hydroxide to give benzo[*c*]cinnoline, its *N*-oxide, and *N,N′*-dioxide, respectively, in high yields (Scheme 6).

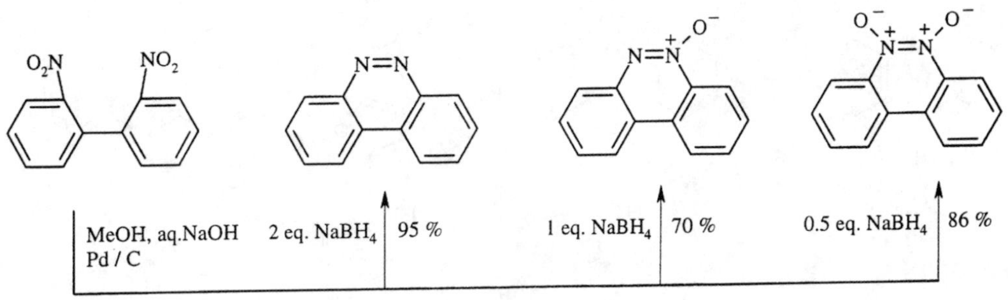

Scheme 6

Benzo[*c*]cinnoline is also available by irradiation of azobenzene.

Dibenzo[1,2]dithiins are prepared by oxidation of appropriate dithiols and related starting materials, *e.g.* (**40**) → (**41**).

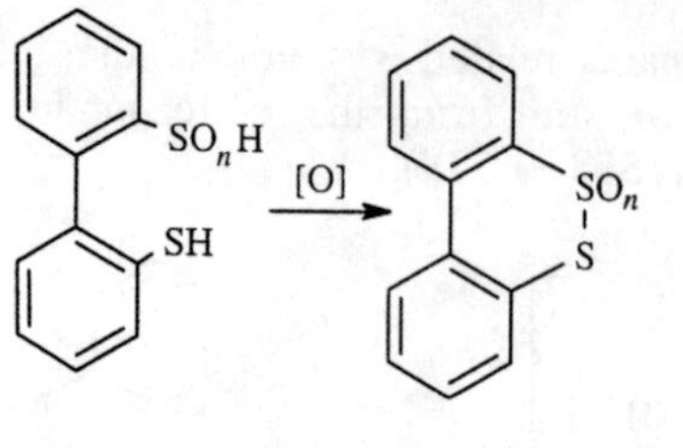

(40)   (41) 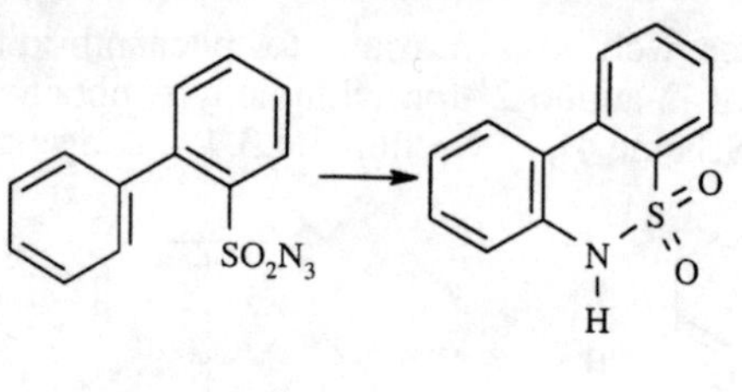   (42)   (43)

Sultone (**43**) is obtained by thermolysis of the azide (**42**) ⟨69JA1219⟩.

4.5.3   TWO NON-ADJACENT FUSED RINGS, ONE HETEROATOM

(i) *o*-Anilinobenzoic acids give 9-chloroacridines (**45** → **44**) and acridones (**45** → **46**) under the conditions indicated.

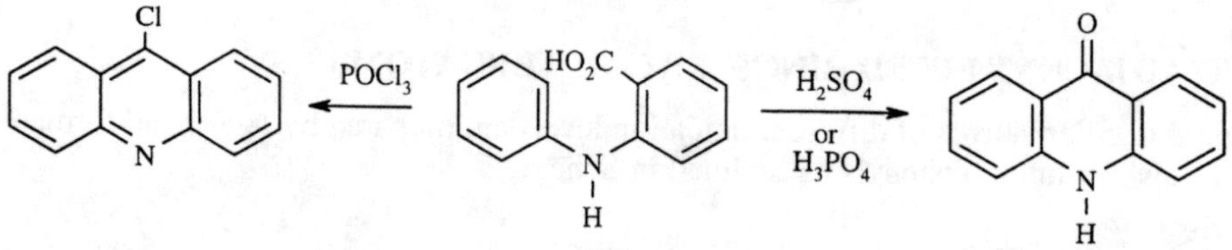

(44)   (45)   (46)

In Scheme 7, an aminouracil reacts with an *o*-halobenzaldehyde as 1,3-bis electrophile to provide a simple preparation of 5-deazaflavins ⟨82CC1085⟩.

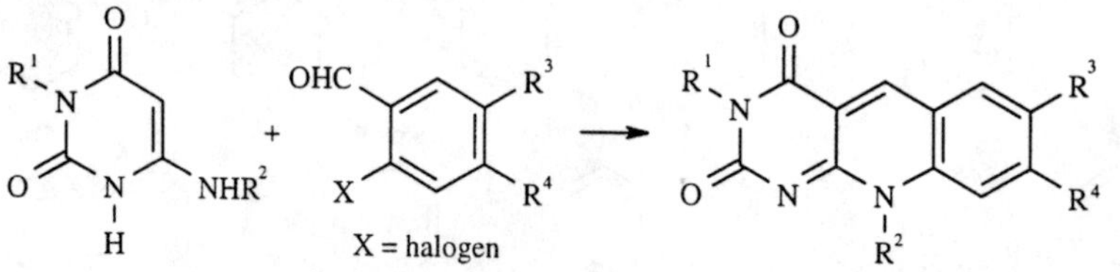

Scheme 7

(ii) In the Bernthsen synthesis, diphenylamines and carboxylic acids form 9-substituted acridines (**47** → **48**)

(iii) 3-Arylanthranils give acridones on thermolysis (**49** → **50**) or on nitrous acid treatment (Section 3.4.3.4.2).

(iv) The generation and trapping of *o*-quinone methides with a dienophile are illustrated in Scheme 8.

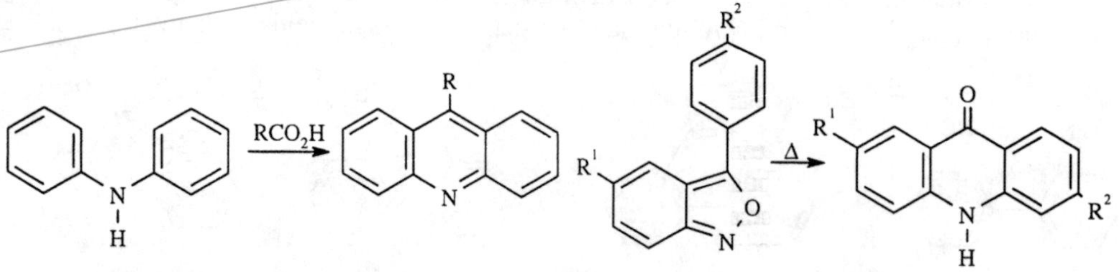

(47)   (48)   (49)   (50)

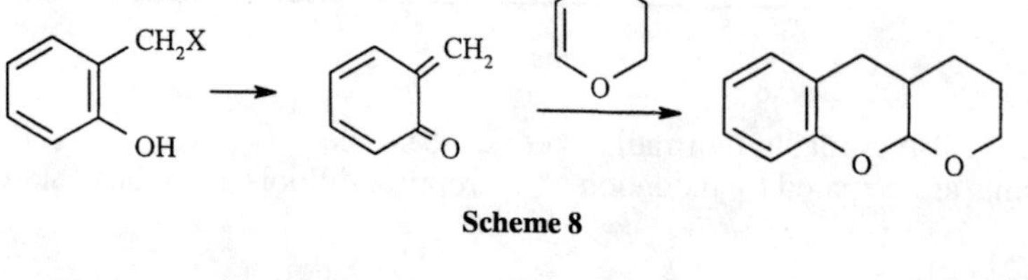

Scheme 8

4.5.4 TWO NON-ADJACENT FUSED RINGS, TWO HETEROATOMS

4.5.4.1 Phenazines

Phenazines can be obtained from *o*-nitrodiphenylamines by reduction or from *o*-aminodiphenyla-mines by oxidative techniques. The preferred method is that of phenazine 9,10-dioxides from benzofuran, thus benzofuroxans, itself with hydroquinone, gives the 2-hydroxy derivative (Scheme 9).

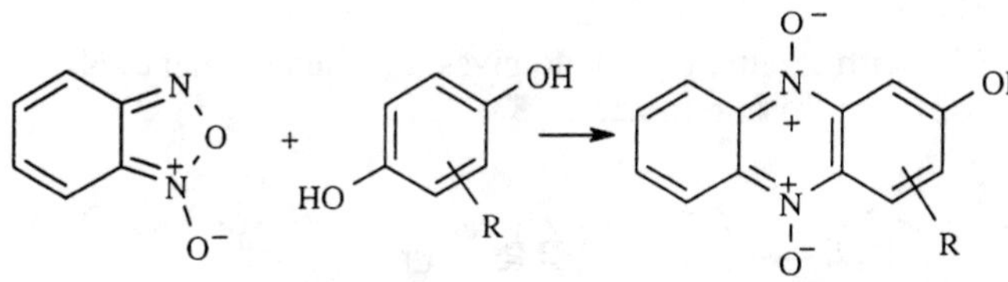

Scheme 9

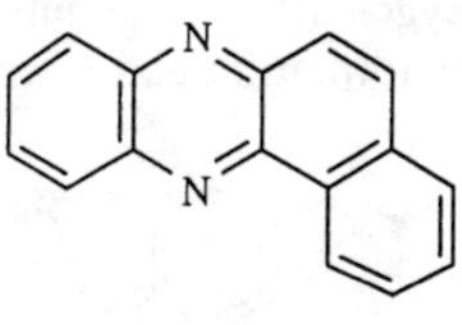

o-Phenylenediamine with *o*-quinones gives phenazines, thus α-naphthoquinone yields (**51**).

The C=C bond of an enamide functions as a nucleophile in the synthesis of the 3,4-dihydrophenazin-1(2*H*)-ones (**52** → **53**) ⟨82S852⟩.

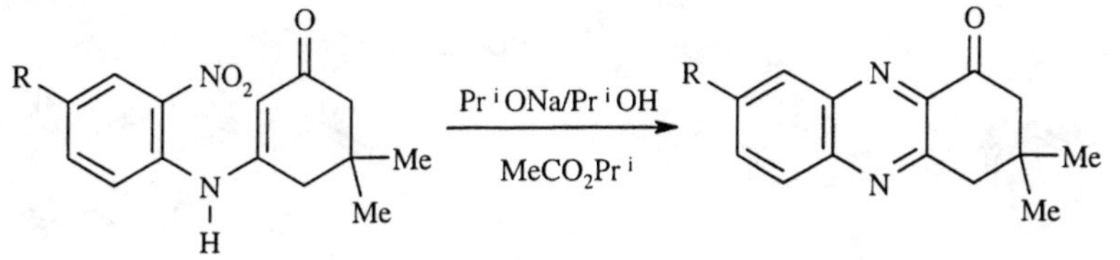

o-Nitroanilinopyridines, *e.g.* (**54**), are cyclized to pyrido[2,3-*b*]- or -[3,4-*b*]-quinoxalines (**55**) by reduction with iron(II) oxalate ⟨74JCS(P1)1965⟩.

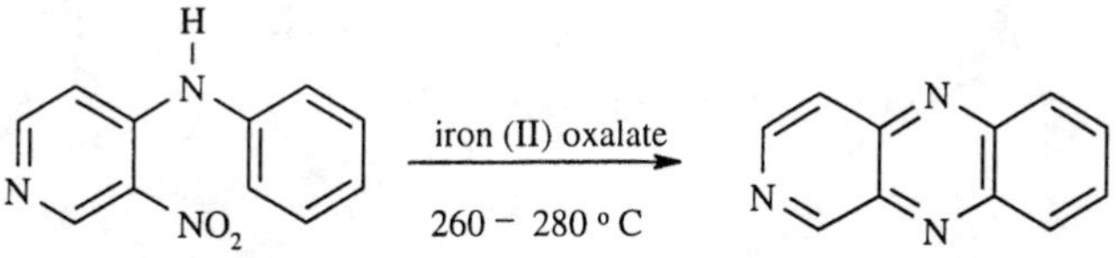

4.5.4.2 Phenoxazines and Phenothiazines

Traditional routes to phenoxazines include the thermolysis of 2-aminophenol and catechol, the latter acting as an acid catalyst, or catechol and ammonia. Phenothiazines are prepared similarly by heating diphenylamines with sulfur (Scheme 10) ⟨B-78MI22701⟩. 2-Hydroxy- (or mercapto-) 2',4'-dini-trodiphenylamines cyclize to phenoxazines (or phenothiazines) in base by elimination of nitrous acid. This reaction is complicated by Smiles-type rearrangement so that mixtures of isomeric products are obtained (Scheme 11).

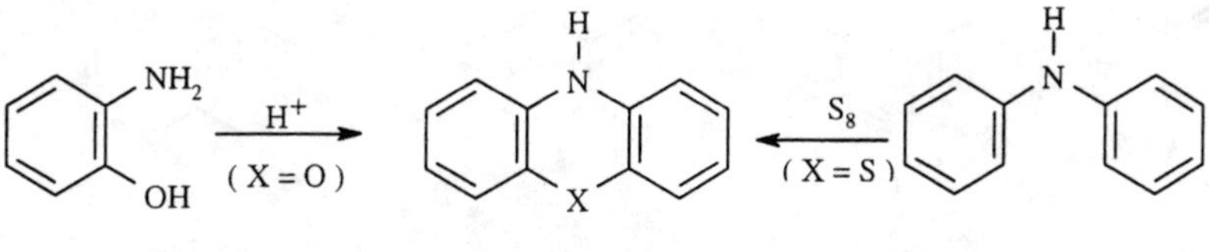

Scheme 10

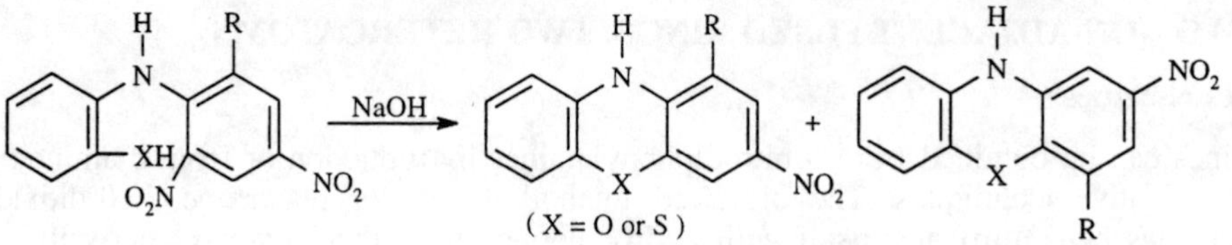

Scheme 11

Deoxygenation of the diaryl sulfide (**56**) with triethyl phosphite gives the phenothiazine (**58**) in good yield *via* formation and rearrangement of the spiro intermediate (**57**).

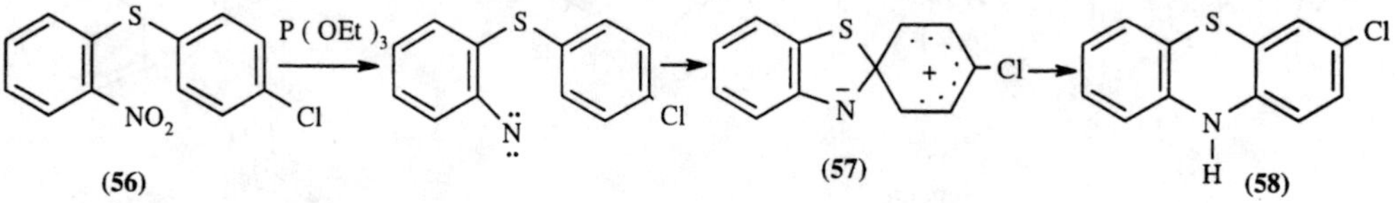

(**56**) (**57**) (**58**)

Phenoxazin-3-ones and phenothiazin-3-ones can be prepared by condensation of 2-aminophenols or -thiols with quinones. Alizarin Green G (**61**), for example, is obtained from (**59**) and (**60**). Similarly, 2-aminothiophenols and 6-chloro-2-methoxy-1,4-benzoquinone (**62**) afford phenothiazin-3-ones (**63**).

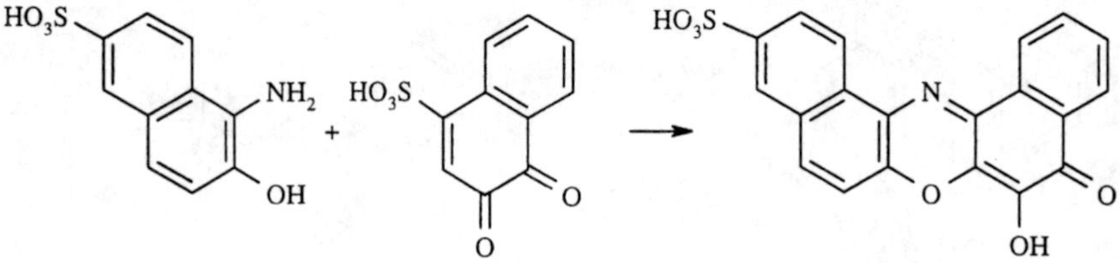

(**59**) (**60**) (**61**)

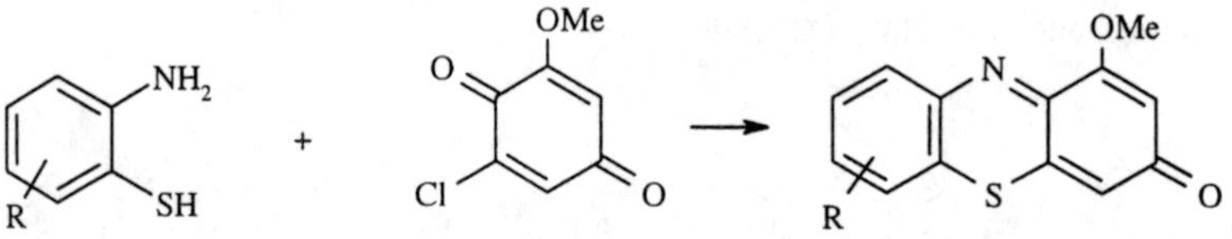

(**62**) (**63**)

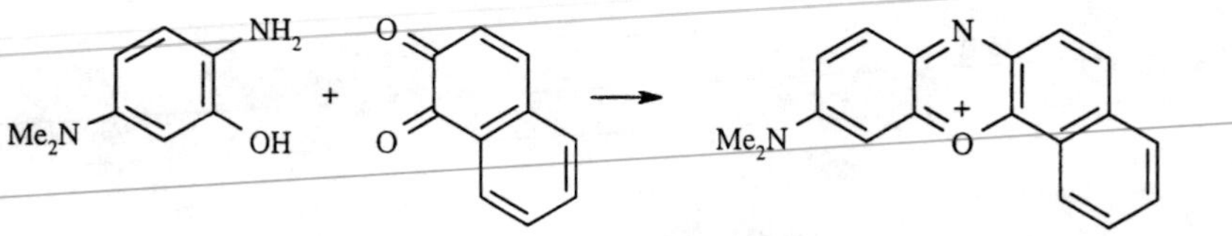

(**64**) (**65**) (**66**)

o-Aminophenols react with quinones to give phenoxazonium salts, *e.g.* (**64**) + (**65**) → (**66**).

The triazaphenothiazine (**69**) is prepared from the 3-aminopyridine-2-thione (**67**) with the dihalopyrimidine (**68**).

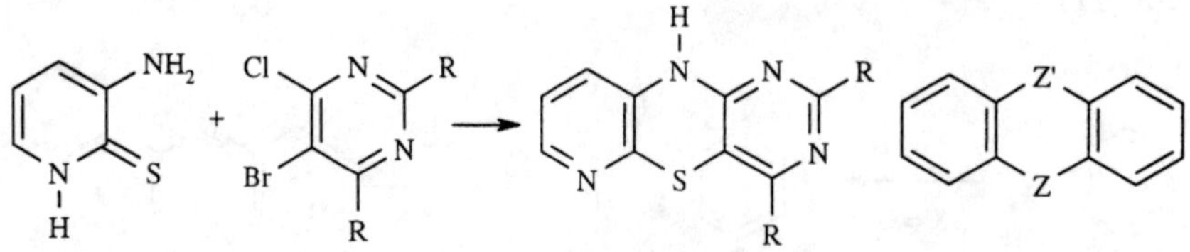

(**67**) (**68**) (**69**) (**70**)

4.5.4.3 Dibenzo[1,4]dioxin, Phenoxathiin and Thianthrene

Dibenzo[*b,e*][1,4]dioxin (**70**; Z, Z' = O) is prepared by heating 2-chlorophenol, potassium carbonate and copper ⟨57JA1439⟩.

A general route to phenoxathiins (**70**; Z = O, Z' = S) utilizes the reaction of diphenyl ethers with sulfur.

One route to thianthrene (**70**; Z, Z' = S) involves reaction of sulfur monochloride with benzene over aluminum chloride ⟨66HC(21-2)1155⟩.

4.5.4.4 Dibenzoxepins and Dibenzothiepins

Unstable dibenz[*c,f*][1,2]oxazepines (**73**; R = CN, Cl) are major products of the UV irradiation of 9-cyano- and 9-chloro-acridine 10-oxides (**71**) in benzene. No oxaziridine tautomer (**72**) was detectable by UV spectroscopy.

(**71**)   (**72**)   (**73**)

Dibenzo[*b,e*][1,4]thiazepines have been prepared by the thermolysis of 2-azido-2′,6′-dimethyldiphenyl sulfide ⟨70CC233⟩.

Thioxanthen-9-one 10,10-dioxides with sodamide undergo ring expansion to dibenzo[*b,f*]-[1,4]thiazepin-11-one 5,5-dioxides ⟨75JHC1211⟩.

4.5.5 *PERI*-ANNULATED HETEROCYCLIC SYSTEMS

Due to proximity effects, *peri*-cyclizations leading to this type of heterocycle have some specificity ⟨66CRV567, 90AHC(51)1⟩. Thus, when five-membered and especially six-membered *peri*-rings are formed, cyclization occurs much more readily than in the case of *ortho*-disubstituted arenes with similar structures. At the same time cyclizations involving the formation of four- or seven-membered *peri*-rings proceed with greater difficulty since they demand considerable bond distortions.

Naphtho[*b,c*]thiete (**75**) was obtained with high yield as a stable compound on photolysis of (**74**) or (**76**). Nitrogen and oxygen analogues of (**75**) are unstable, but can in some instances be trapped.

(**74**)   (**75**)   (**76**)

The cyclization of *peri*-aminonaphthoyl compounds allows the preparation of three main types of benzo[*c,d*]indole derivatives (**77**) – (**79**). 1,2-Dihydrobenzo[*c,d*]indoles (**80**) are unstable and easily autooxidized to (**79**). Naphtho[1,8-*c,d*]dichalcogenols (**81**) were obtained in good yield on treatment of 1,8-dilithionaphthalene with sulfur, selenium or tellurium.

(**77**)   (**78**)   (**79**)   (**80**)   (**81**) (X = S, Se, Te)

Syntheses of perimidine (**82**) and naphtho[1,8-*d,e*]triazine (**83**) derivatives usually start from 1,8-diaminonaphthalene by procedures which are quite similar to the synthesis of benzimidazoles and benzotriazoles from 1,2-diaminobenzene ⟨81RCR1559; 95AQ151⟩. Intramolecular carbenoid and nitrenoid insertions are also quite effective for the preparation of *peri*-condensed heterocycles. Thus, photolysis of 1-naphthyl-1,2,3-triazoles (**84**) leads to benzo[*d,e*]quinolines (**86**), possibly *via* carbene intermediate (**85**) ⟨87JCS(P1)413⟩. Similarly, on photolysis or thermolysis of 8-azido-1-arylazonaphthalenes (**87**) naphtho[1,8-*d,e*]triazine derivatives (**88**) are formed along with *N*-aryliminobenzo[*c,d*]indazoles (**89**) ⟨78JOC2508, 82JOC1996⟩.

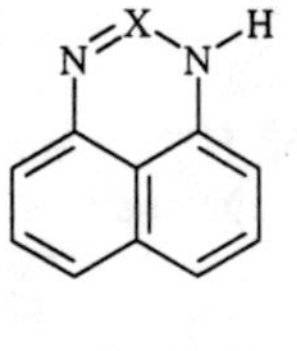

(**82**) (X=CH)
(**83**) (X=N)

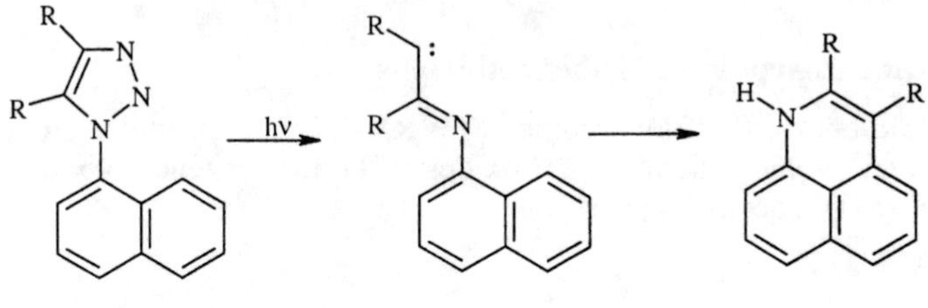

(**84**)

(**85**)

(**86**) R=CO₂Me, CN

2-Dimethylaminonaphtho[*b,c*]oxepinium salts (**91**) have been obtained from *peri*-hydroxynaphthalenes (**90**) in accordance with Scheme 12 ⟨90AHC(51)1⟩.

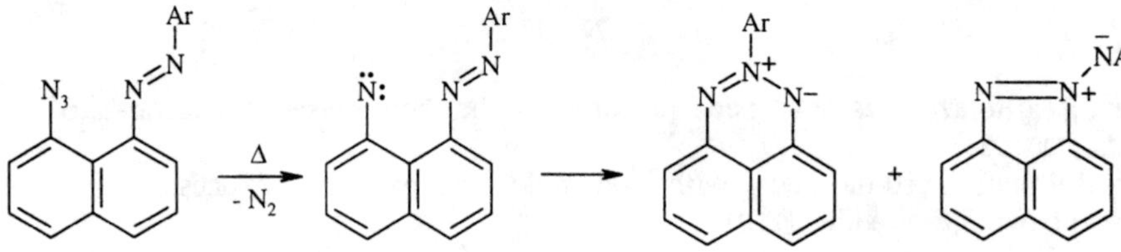

(**87**)

(**88**)

(**89**)

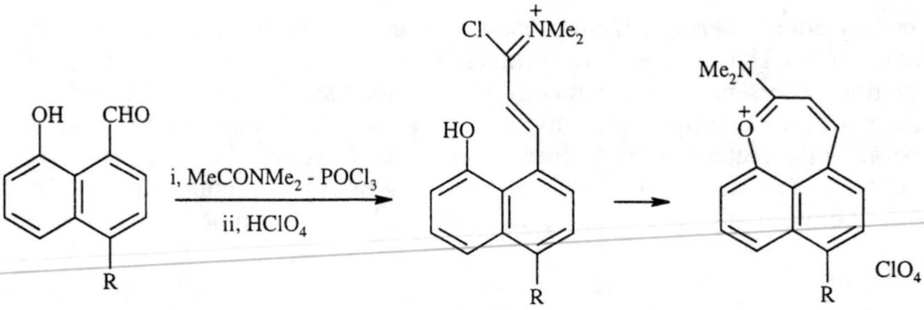

i, MeCONMe₂ - POCl₃

ii, HClO₄

(**90**)

(**91**)

Scheme 12

4.5.6   THREE FUSED RINGS

The tribenz[*b,d,f*] oxepin (**93**; Z=O) is made from the biphenyl-2-yl ether (**92**; Z = O), and the thio analogue (**93**; Z=S) is prepared similarly ⟨65TL299⟩.

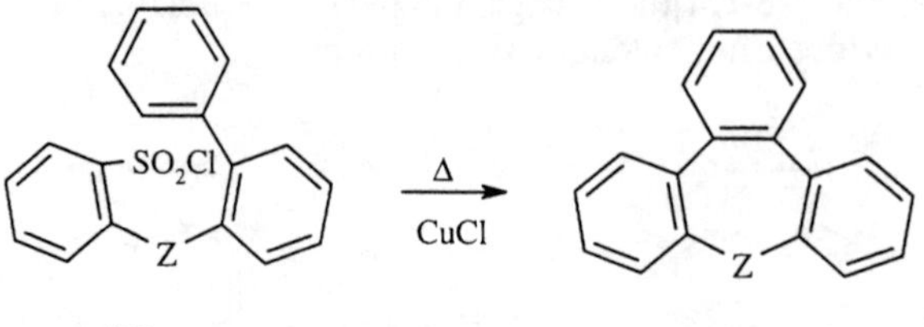

Δ

CuCl

(**92**)

(**93**)

4.6

Synthesis of Fused Ring Systems with Ring Junction Heteroatoms

This chapter considers the formation of rings containing one or more N or S atoms at a ring junction. For nitrogen, this is possible with the retention of a three-coordinate neutral, or four-coordinate positively charged, nitrogen atom. We consider successively the formation of small (three- or four-membered) rings with such a feature, five-membered and then six-membered rings with one ring junction N atom, and finally rings containing two ring junction N atoms.

A very large number of these systems with ring junction heteroatoms exists, and this number is constantly increasing. Only illustrative examples of the preparation of such systems can be given here. The synthetic methods for the formation of this type of heterocycle can be usefully classified as follows: (i) various cyclocondensations between the corresponding heterocyclic derivatives and bifunctional units, (ii) intramolecular cyclizations of electrophilic, nucleophilic or (still rare) radical type, (iii) cycloadditions, (iv) intramolecular oxidative coupling, (v) intramolecular insertions, (vi) cyclization of open-chained predecessors, (vii) various reactions (quite often unusual) which are specific for each type of system. Examples given below illustrate all these cases.

4.6.1 FORMATION OF THREE- OR FOUR-MEMBERED RINGS WITH ONE N ATOM AT A RING JUNCTION

Cycloaddition of azirines can give fused systems as shown in Scheme 1 ⟨72TL1353, 74TL1487⟩.

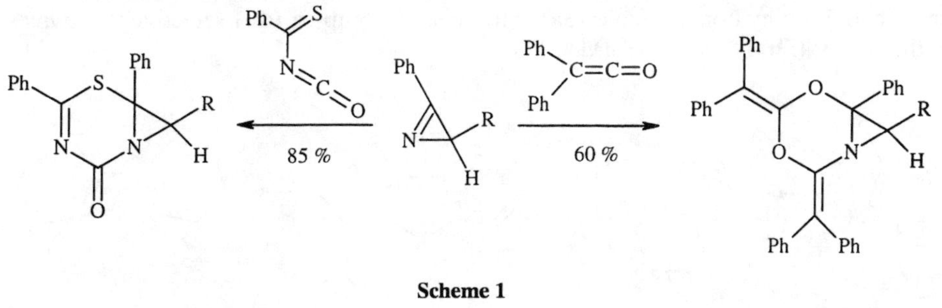

Scheme 1

Different strategies (I, II, III, and IV) for the synthesis of penams, the backbone of penicillins, are schematically outlined in Scheme 2.

Use of the Staudinger reaction is exemplified by the synthesis of substituted penam derivatives (**3**) from 4,5-dihydro-1,3-thiazole (**2**) and either dichloroacetyl chloride (**1**; $R^1 = R^2 = Cl$) or azidoacetyl chloride (**1**; $R^1 = N_3$, $R^2 = H$).

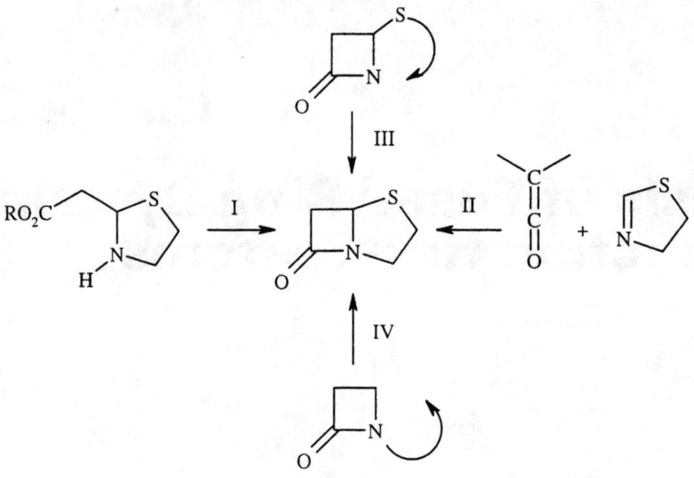

Scheme 2

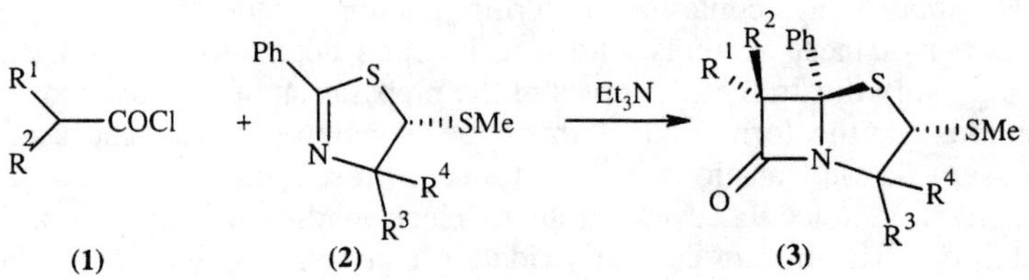

(1)   **(2)**   **(3)**

4.6.2 FORMATION OF A FIVE-MEMBERED RING WITH ONE N ATOM AT A RING JUNCTION

4.6.2.1 No Other Heteroatoms

4.6.2.1.2 [5–5] Systems

Various intramolecular reactions are widely used to prepare this type of heterocycle. Intramolecular acylation has been used frequently ⟨84AHC(37)1⟩. Houben-Hoesch cyclization of 1-β-cyanoethylpyrrole (**4a**) gives 1-oxo-2,3-dihydropyrrolizine (**5a**). The methyl group in (**4b**) selectively activates C-2 so that only the isomer (**5b**) is formed ⟨66JA1305⟩.

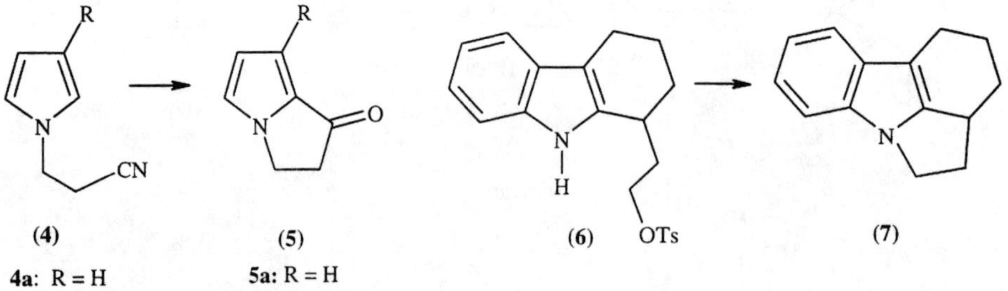

(4)   **(5)**   **(6)**   OTs   **(7)**

4a: R = H   **5a**: R = H
4b: R = Me   **5b**: R = Me

Intramolecular alkylation of suitably substituted pyrroles and indoles gives pyrrolizines in excellent yields: thus (**6**) affords (**7**) ⟨80JCS(P1)97⟩.

2-Acylpyrroles react with allylphosphoranes to give pyrrolizines (**9**). Butadiene (**8**) has been isolated and cyclized subsequently ⟨66JOC2912⟩. However, (**8**; $R^1 = R^2 = H$, $R^3 = CO_2Me$) cyclizes only into 3*a*-azaazulenone ⟨73CB1993⟩.

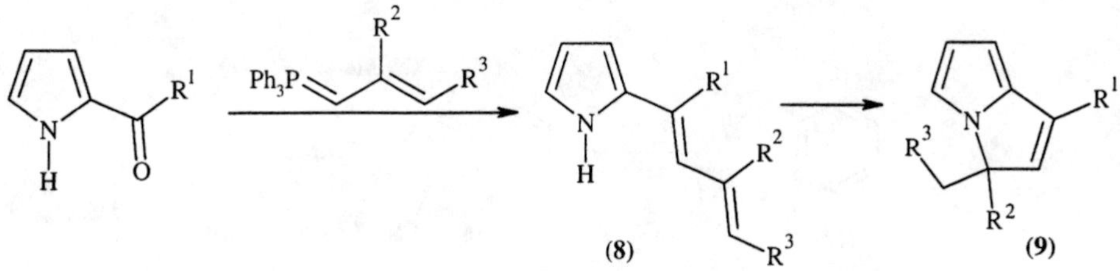

(8)   (9)

2-Formylpyrrole anion and vinyltriphenylphosphonium bromide give (**10**) by an intramolecular Wittig reaction ⟨64JA2744, 66JOC870⟩.

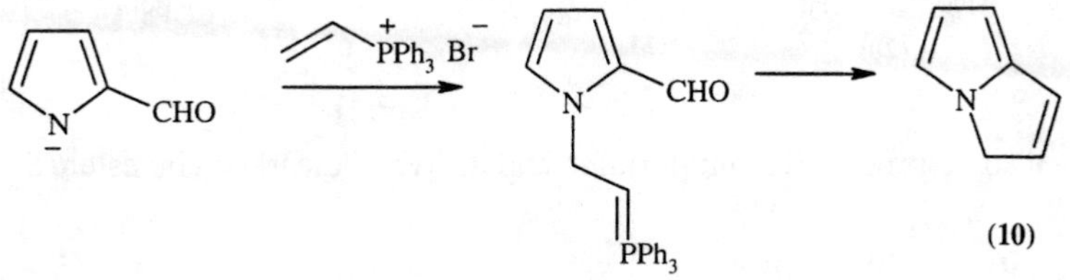

(10)

Thermal or photochemical ring opening of aziridines, when applied to ring-fused aziridines, generate ylides which may be trapped. The ring-fused aziridine (**11**) in refluxing xylene gives the azomethine ylide (**12**), converted by diethyl acetylenedicarboxylate into the 3*H*-pyrrolo-[1,2-*c*]imidazole (**13**).

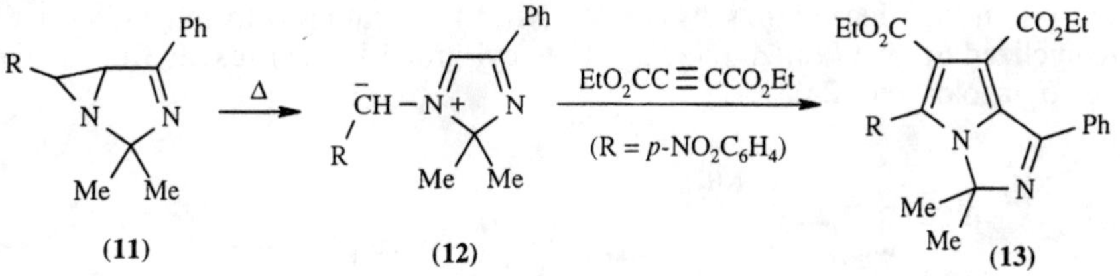

(R = *p*-NO_2C_6H_4)

(11)   (12)   (13)

4.6.2.1.2 *[5–6] Systems*

2-Substituted indolizines (**15**) are formed in Chichibabin reactions of 1-[alkyl(aryl)carbonyl-methyl]-2-alkylpyridinium halides (**14**) ⟨72AJC1003⟩. Various modifications of this method are used for the preparation of many other pyrrolo[1,2-*a*]azines and pyrrolo[1,2-*a*]azoles.

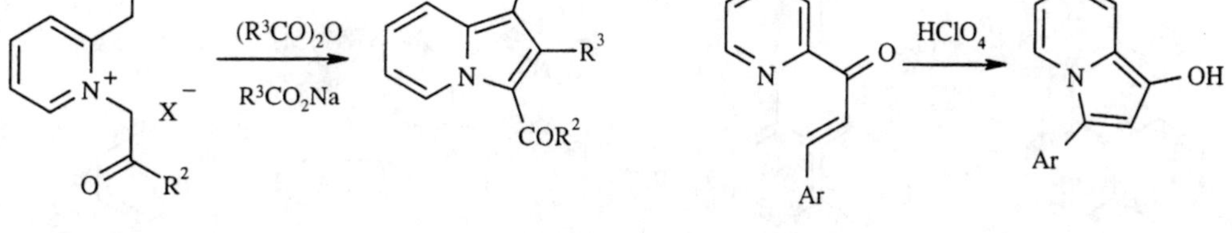

(14)   (15)   (16)   (17)

R^1, R^2 = H, Me, Ph
R^3 = Me, Et, Ph

Indolizines can also be made by intramolecular Michael additions, *e.g.* 2-acetylpyridine + ArCHO → (**16**) → (**17**).

Pyridinium ylides undergo cycloadditions with suitable alkynes, generating indolizines, *e.g.* (**18**) → (**19**) ⟨65JA3651⟩. Diphenylthiirene dioxide (**20**) in its reactions with pyridinium, quinolinium and isoquinolinium phenacylylides behaves as an acetylene equivalent. The pyridinium phenacylylide (**21**) gave the indolizine (**22**) with elimination of SO_2 and loss of hydrogen ⟨73BCJ667⟩.

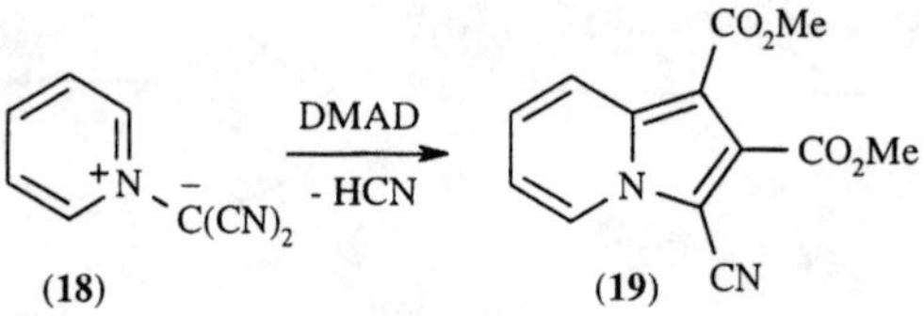

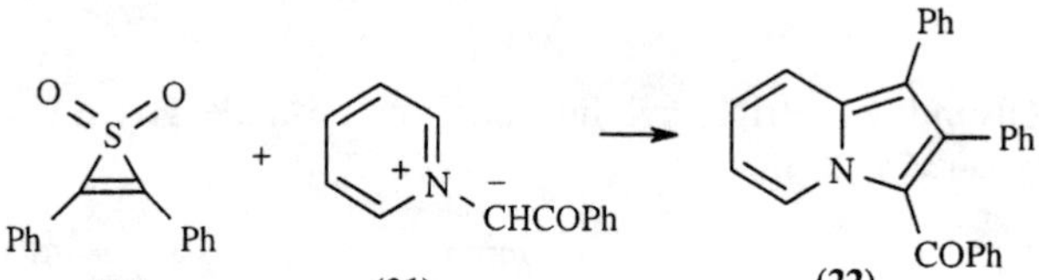

Indolizines can also be produced from pyridine and acetylenedicarboxylic ester (Section 3.2.1.3.7).

4.6.2.2   One Additional Heteroatom

4.6.2.2.1   *Pyrazolo-fused systems*

Cyclization of an *N*-amino group with an α-methyl substituent in *N*-aminoazinium salts or in *N*-aminoazoles is the most frequently used route to pyrazolo-fused systems ⟨92AHC(53)85⟩. Thus, *N*-amino-imidazolium and -thiazolium salts (**23**; Z = NR, S) on heating with anhydrides in the presence of a base, are cyclized to give (**25**; Z = NR, S). The acyl group in (**25**) results from acylation of the initially formed pyrazoloazole (**24**).

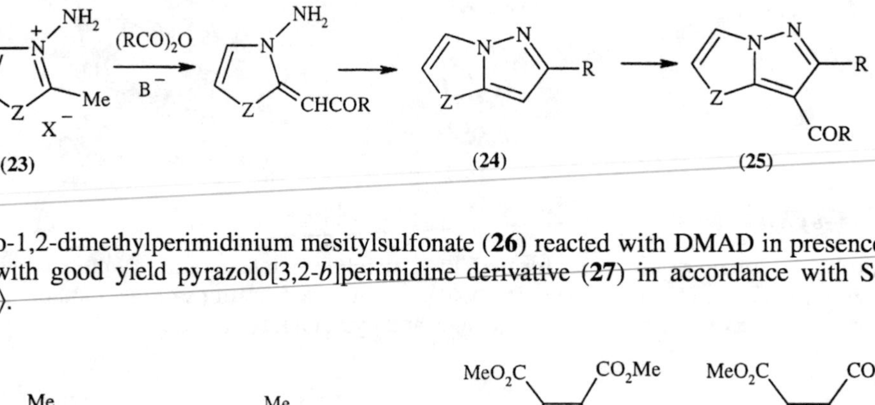

3-Amino-1,2-dimethylperimidinium mesitylsulfonate (**26**) reacted with DMAD in presence of base to afford with good yield pyrazolo[3,2-*b*]perimidine derivative (**27**) in accordance with Scheme 3 ⟨83CPB1378⟩.

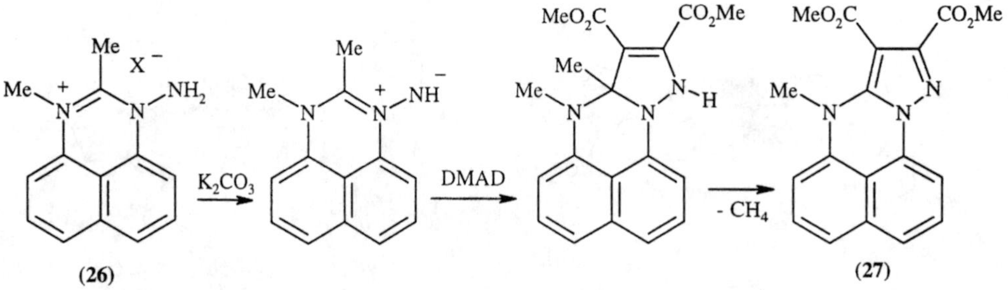

Scheme 3

1-Aminopyridinium iodide with dimethyl chlorofumarate forms a readily aromatized dihydropyrazolo[1,5-*a*]pyridine (**29**) *via* an initially formed 1:1 adduct (**28**); for a similar example see Section

3.2.3.12.4.iii. The 2-substituted pyridines (30) with triethyl phosphite yield the pyrrolopyrazoles (31) ⟨79JOC622⟩.

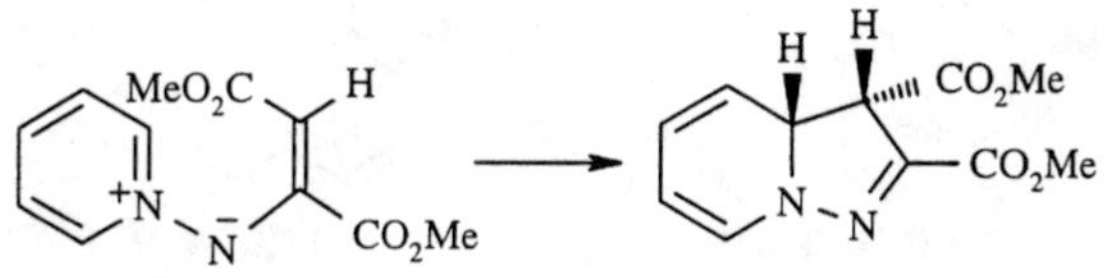

(28)   (29)

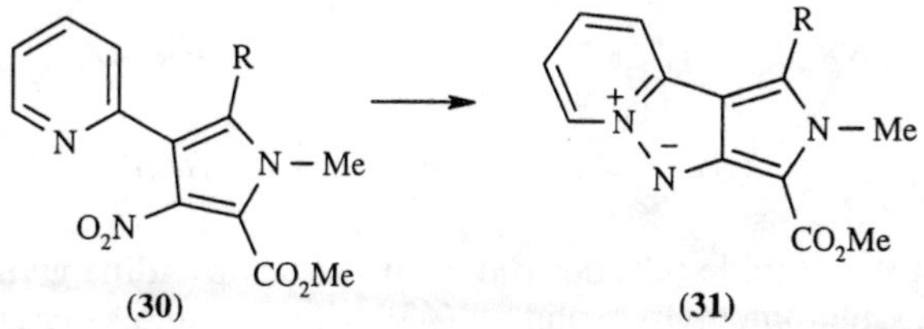

(30)   (31)

4.6.2.2.2 *Imidazo-fused systems*

Most syntheses of imidazo-fused systems involve cyclization of amino substituted heterocycles (or their tautomers), suitably substituted at an adjacent ring position. This provides a convenient route to imidazo[2,1-*b*]thiazoles, imidazo[1,2-*a*]imidazoles (Scheme 4), imidazo[1,2-*b*]pyrazoles, imidazo[1,5-*a*]imidazoles and imidazo[1,5-*c*]thiazoles.

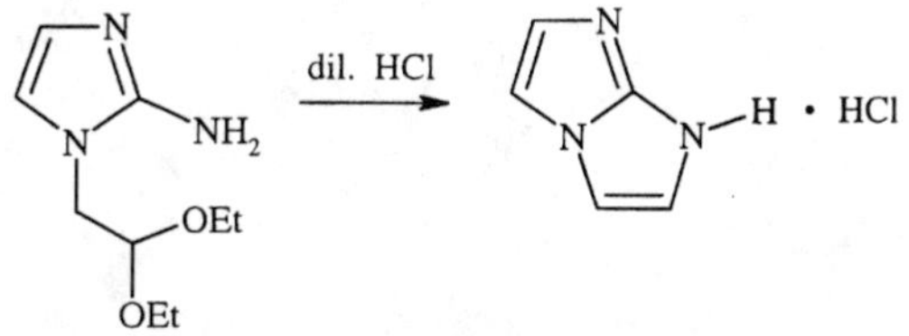

Scheme 4

2-Aminopyridine (32) with PhCOCHBrPh gives (33), and 2-aminobenzothiazole (34) with iminochloride (35) gives (36).

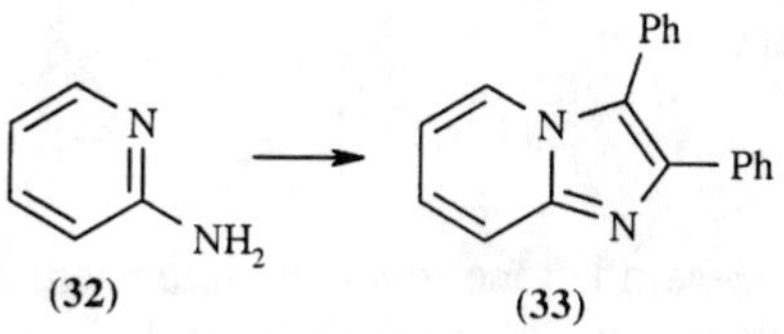

(32)   (33)

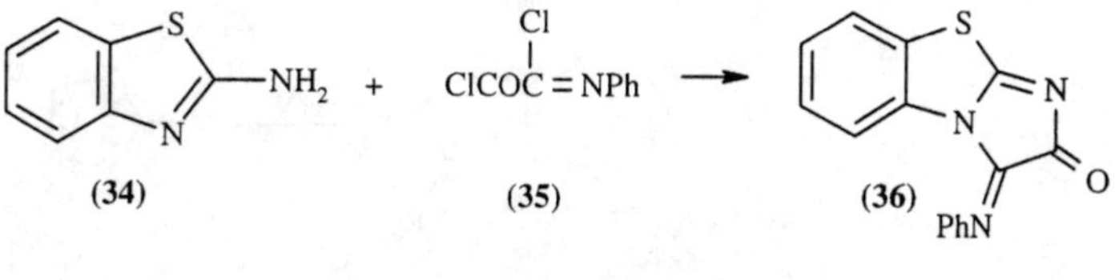

(34)   (35)   (36)

Corresponding reduced rings (*cf.* 38, 40) can be obtained from α-aziridinyl heterocycles (37, 39). Related ring systems are obtained by formation of an alternative C—N bond, *e.g.* the alcohol (41) on heating with sulfuric acid affords racemic tetramisole (42) ⟨68JOC1350⟩.

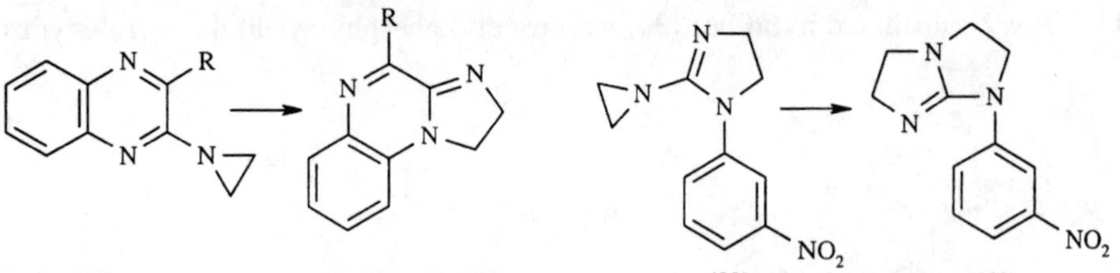

(37)   (38) (R = Cl, OMe)   (39)   (40)

(41)   (42)

With diphenylketene, the sulfimide (43) derived from 2-aminopyridine gives imidazo[1,2-*a*]pyridin-3-one (44) ⟨77H(8)109⟩. 2-Chloromethylpyrrolidine (45) with CS_2 and potassium carbonate forms the dithiocarbamate (46) which undergoes intramolecular alkylation on sulfur to give (47) ⟨63JOC981⟩.

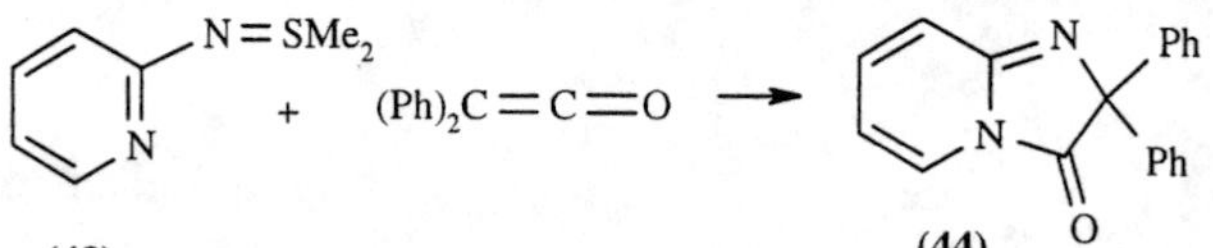

(43)   (44)

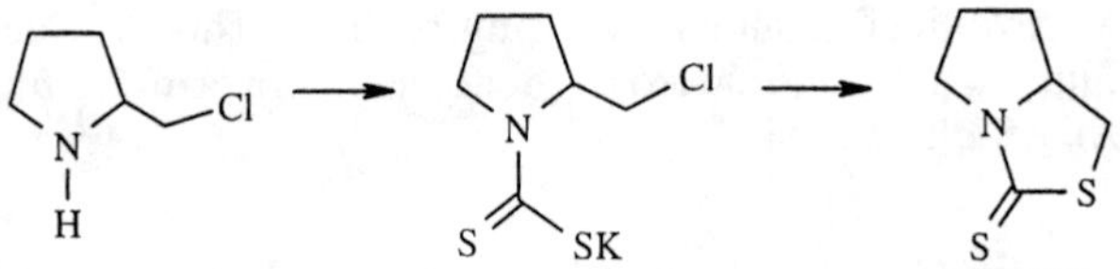

(45)   (46)   (47)

4.6.2.2.3 Thiazolo-fused systems

The usual precursor is a cyclic thioamide. On formation, the *N*-β-chloroethylthiazolidine-2-thione (48) spontaneously ring-closed to the tetrahydrothiazolo[2,3-*b*]thiazolium chloride (49) ⟨71CHE1534⟩. Similar reactions of imidazole analogues of (48) give compounds of type (50).

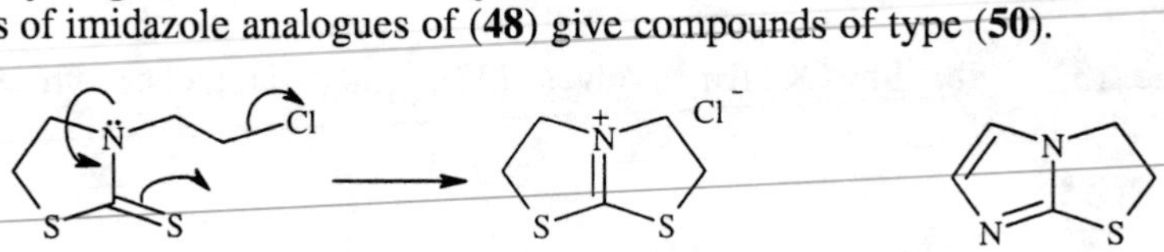

(48)   (49)   (50)

The thiazole-2-thione (51) with an α-halo ketone gives the intermediate (52) which is cyclized by strong acid into the thiazolo[2,3-*b*]thiazolium salt (53) ⟨77HC(30-1)1⟩. A wide variety of [5,5]-fused systems are prepared in this way.

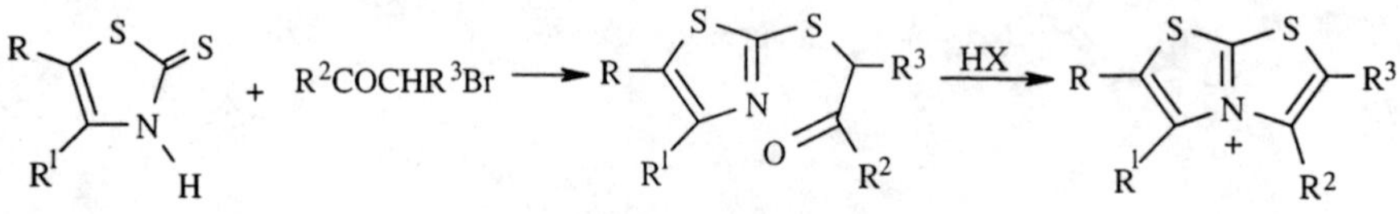

(51)   (52)   (53) X^-

Conversion of (1-methylimidazol-2-yl)thioglycolic acid (54) into the ring-fused mesoionic system (55) requires acetic anhydride ⟨79JOC3803⟩ and is accompanied by an acylation characteristic of reactive mesoionic systems (*cf.* conversion **24** → **25**). The thioether (**56**) with ethanolic sodium ethoxide gives the 3-benzylthiazolo[3,2-*a*]benzimidazole (**57**).

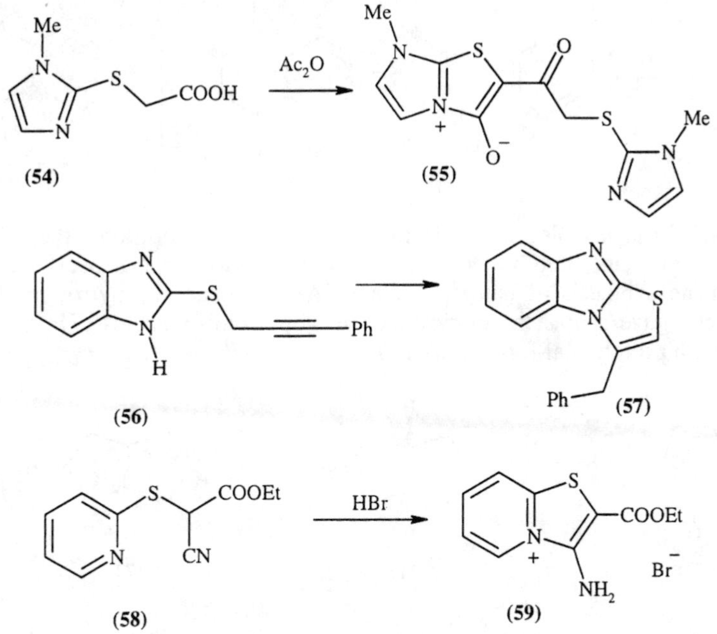

(54) (55)

(56) (57)

(58) (59)

The cyano group is also a precursor for intramolecular cyclization. The thioacetonitrile derivative (**58**) with HBr formed 3-amino-2-ethoxycarbonylthiazolo[3,2-*a*]pyridinium bromide (**59**) ⟨78JCR(S)407⟩. This is a general reaction for such annulation to a variety of five- and six-membered heterocycles.

2(1*H*)-Pyridinethione (**60**) with an α-halo ketone gives the 2-β-oxoalkylthiopyridine (**61**) which is ring-closed with strong acid to the thiazolo[3,2-*a*]pyridinium system (**62**). The 3-hydroxypyridine analogue of (**60**) cyclizes to (**63**). Further variations are described in CHEC 4.10 and 4.29; CHEC-II, 8.10.6.

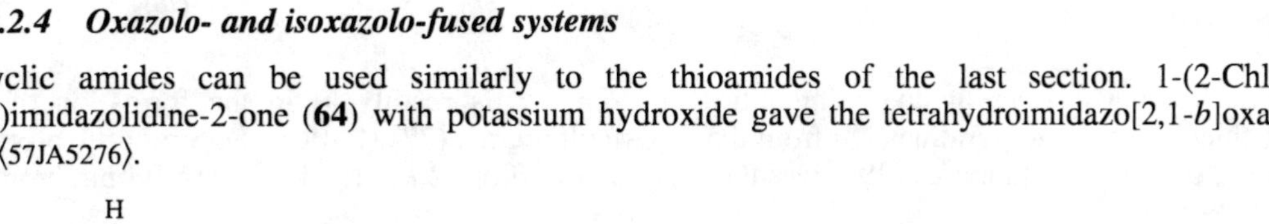

(60) (61) (62) (63)

4.6.2.2.4 *Oxazolo- and isoxazolo-fused systems*

Cyclic amides can be used similarly to the thioamides of the last section. 1-(2-Chloro-ethyl)imidazolidine-2-one (**64**) with potassium hydroxide gave the tetrahydroimidazo[2,1-*b*]oxazole (**65**) ⟨57JA5276⟩.

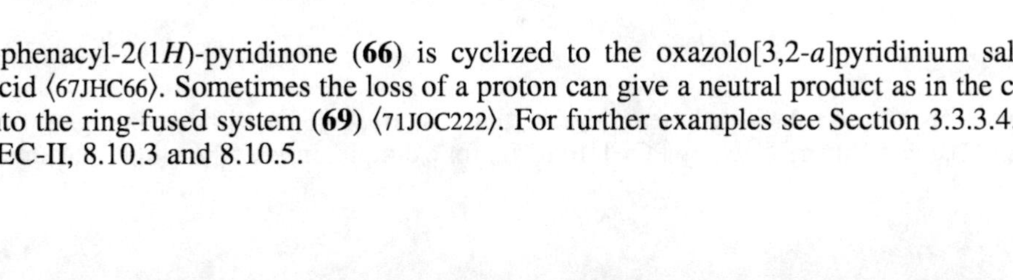

(64) (65) (66) (67)

The 1-phenacyl-2(1*H*)-pyridinone (**66**) is cyclized to the oxazolo[3,2-*a*]pyridinium salt (**67**) by sulfuric acid ⟨67JHC66⟩. Sometimes the loss of a proton can give a neutral product as in the conversion of (**68**) into the ring-fused system (**69**) ⟨71JOC222⟩. For further examples see Section 3.3.3.4.1; CHEC 4.29; CHEC-II, 8.10.3 and 8.10.5.

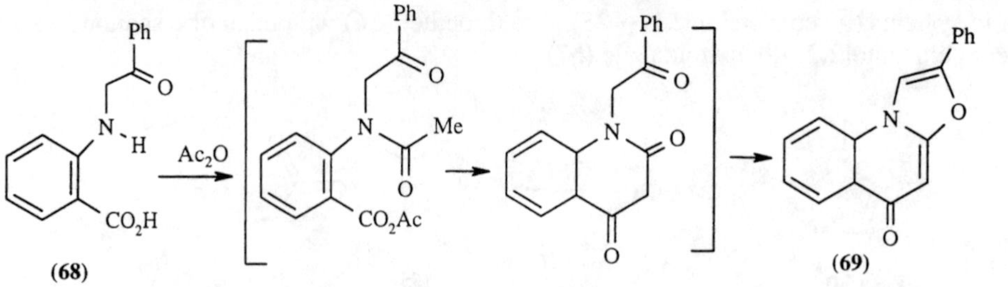

(68)   (69)

Perhydropyrrolo[1,2-b]isoxazoles result from 1,3-dipolar cycloaddition of cyclic nitrones with alkenes. The high regio- and stereoselectivity of this cycloaddition has been used to control the stereochemistry in the synthesis of natural products. As one example, pyrroline N-oxide (70) and 3,4-dimethoxystyrene gave a diastereomeric mixture of pyrroloisoxazoles (71) and (72), in nearly quantitative yield with preferential formation of (71).

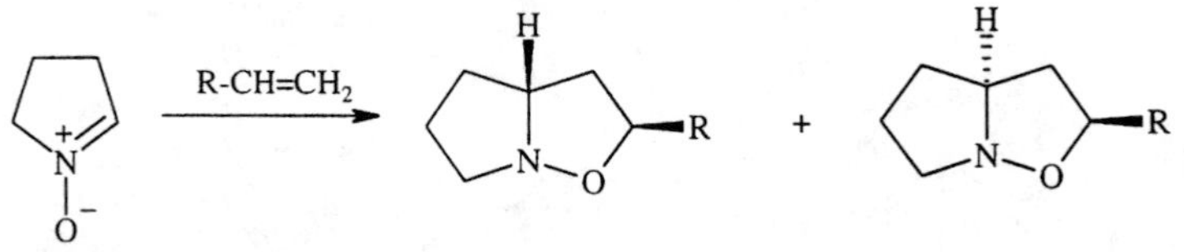

(70)   (71)   (72)

R = 3,4-(MeO)$_2$C$_6$H$_3$

4.6.2.3   Two Other Heteroatoms

4.6.2.3.1   1,2,4-Triazolo[b]-, 1,2,4-thiadiazolo[b]- and 1,3,4-thiadiazolo[b]-fused systems

The 1,2-diamino-4-methylthiazole (73) with phosgeniminium chloride Cl$_2$C=NMe$_2^+$ gives the thiazolo[3,2-b][1,2,4]triazole derivative (74) ⟨73AG(E)806⟩. Reaction of the sulfimide (43) with nitrile oxides RCNO forms [1,2,4]triazolo[1,5-a]pyridine 3-oxides (75) in good yield. This method is applicable to analogous pyrimidines and pyrazines ⟨76JCS(P1)2166, 78BCJ563⟩. Lead tetraacetate oxidation of (76) gives compounds of type (75) but without the N-oxide group.

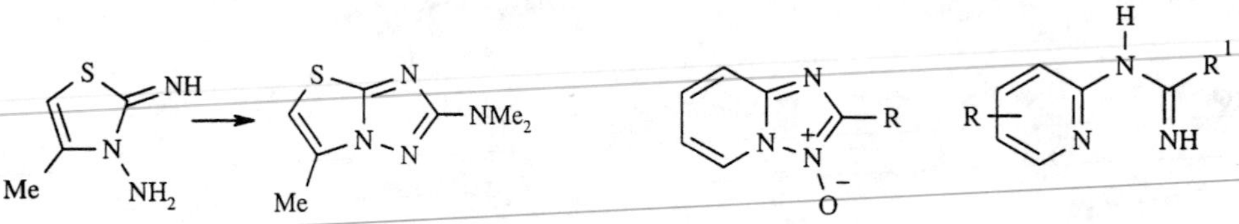

(73)   (74)   (75)   (76)

Oxidative ring closure on to a ring nitrogen atom occurs readily as in the formation of the [1,2,4]thiadiazolo[2,3-a]pyridine (78) from the 2-pyridylthiourea (77) ⟨75JHC1191⟩. Similarly, bromine oxidation of 2-thiazolylthiourea (79) gives the 2-aminothiazolo[3,2-b] [1,2,4]thiadiazolylium bromide (80) ⟨71JPR1148⟩.

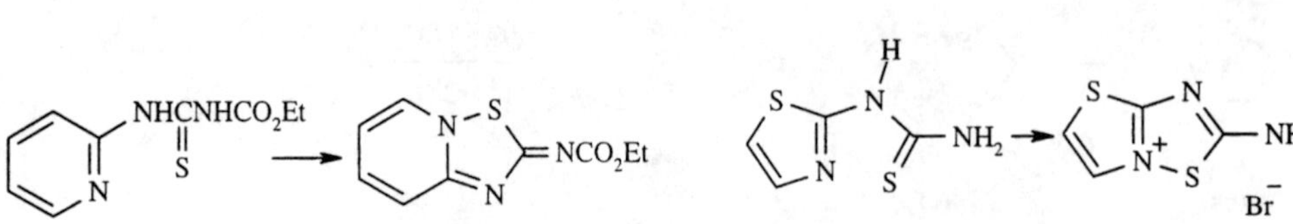

(77)   (78)   (79)   (80)

Reaction of an amino-substituted heterocyclic thiol such as (81) with acylating agents gives compounds (82), which are cyclized by POCl$_3$ to form, for example, imidazo[2,1-b] [1,3,4]thiadiazoles

(83) ⟨review: 97KGS1155⟩. Bromine oxidation of the cyclic thiourea **(84)** forms 2,3,5,6-tetra-hydroimidazo[1,2-*d*][1,2,4]thiadiazol-3-ones **(85)** ⟨73JPR539⟩.

Heating 1-(thiazolyl-2)-5-phenyl-1*H*-tetrazole **(86)** induces extrusion of N₂ and transformation into 2-phenylthiazolo[3,2-*b*][1,2,4]triazole **(87)** ⟨90FES953⟩.

4.6.2.3.2   *1,2,4-Triazolo[c]- and 1,2,4-thiadiazolo[c]-fused systems*

α-Hydrazino nitrogen heterocycles are readily converted into 1,2,4-triazolo[*b*]-fused systems, as exemplified by **(88)** → **(89)** ⟨71JOC10⟩. Oxidation of the 2-pyrazolyl hydrazone **(90)** with Pb(OAc)₄ in CH₂Cl₂ gives, *via* azine **(91)**, the 3*H*-pyrazolo[5,1-*c*]-1,2,4-triazole **(92)** ⟨79TL1567⟩.

The diazopyrazole **(93)** with various phosphorus ylides afforded 3*H*-pyrazolo[5,1-*c*]-1,2,4-triazoles **(94)** by elimination of triphenylphosphine ⟨79TL1567⟩. Reaction of 1-chloroisoquinoline with 5-phenyltetrazole gives 3-phenyl-1,2,4-triazolo[3,4-*a*]isoquinoline **(96)** *via* intermediate **(95)** ⟨70CB1918⟩ (*cf.* conversion **86** → **87**).

Trichloromethanesulfenyl chloride (**98**) converts 2-aminopyridine (**97**) into the intermediate sulfenamide (**99**) which with an aromatic amine cyclizes to (**101**), probably *via* the intermediate (**100**) ⟨75JOC2600⟩.

4.6.2.3.3   *1,2,3-Triazolo*[c]-*fused systems*

An example of synthesis by N—N bond formation is shown in the conversion (**102**) → (**103**). Nickel peroxide as oxidizing agent has advantages over NBS or Pb(OAc)$_4$ ⟨91T2851⟩.

4.6.2.4   Three Other Heteroatoms: Fused Tetrazoles

Compounds of type (**104**) are readily available by intramolecular dehydrogenation of appropriate triazenopyrazoles ⟨82AG(E)698, 87CB1375⟩. Fused tetrazoles with three unsubstituted nitrogens arise by cyclizations of azides: see Sections 2.2.5.4, 2.4.5.3.2, 3.4.1.2.3 and 3.2.3.6.4.

4.6.3   FORMATION OF A SIX-MEMBERED RING WITH ONE N ATOM AT A RING JUNCTION

4.6.3.1   Ring Formation Using a Three-atom Fragment

Six-membered rings are formed by reaction of a three-carbon fragment with an aza heterocycle containing any of (i) an α-methyl, (ii) an α-amino or (iii) a potential α-mercapto group. Examples of

(i) and (ii), respectively, are (**105**) → (**106**) and the reaction of 3-aminobenzisoxazole (**107**) with methyl propiolate to form (**108**) (together with a regioisomer) ⟨72CB794⟩. Heating (**109**) with mercury(I) acetate and acid affords the 4-phenyl-2*H*-thiazino[3,2-*a*]benzimidazole (**110**).

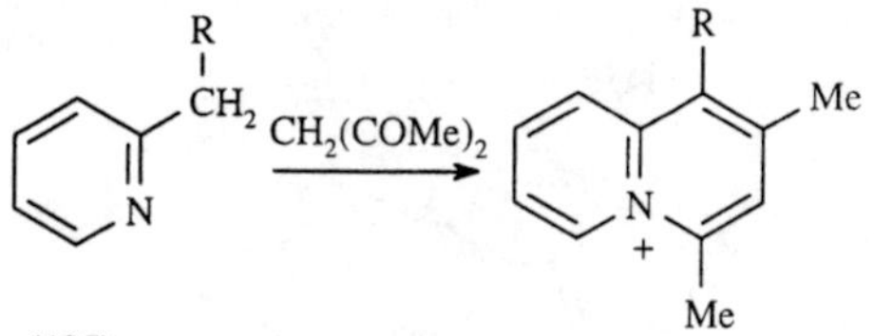

(**105**)   (**106**) 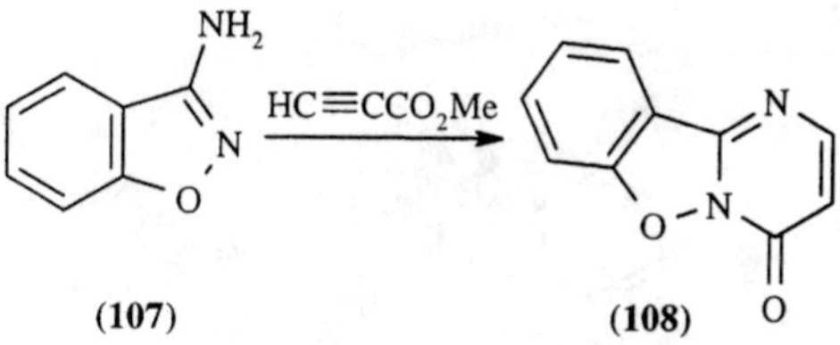   (**107**)   (**108**)

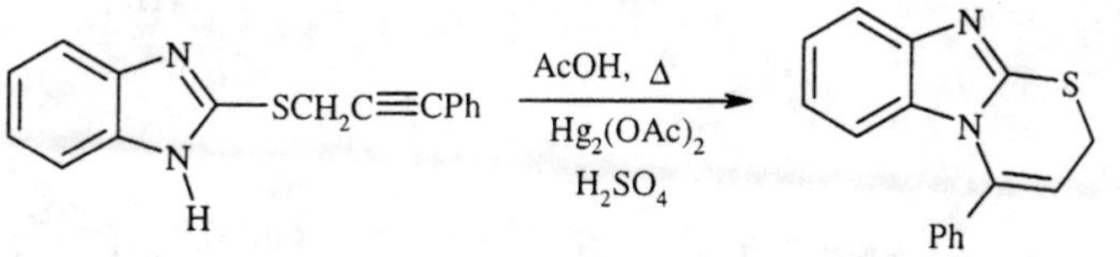

(**109**)   (**110**)

Reactions of this type have been used extensively for the preparation of diazinodiazines, and examples include the formation of compounds (**111**), (**112**), (**113**) and (**114**) by the routes shown.

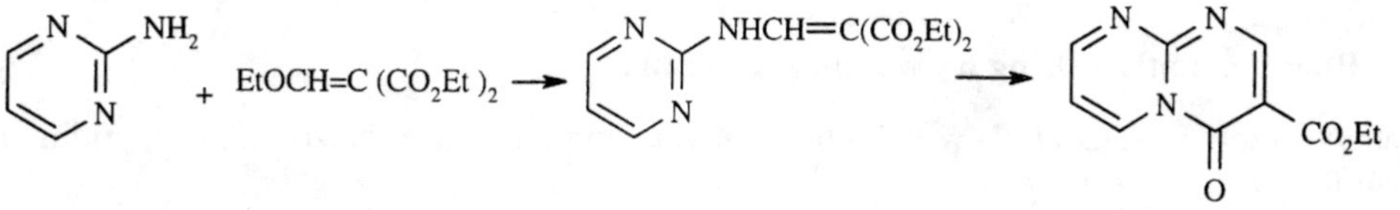

(**111**)

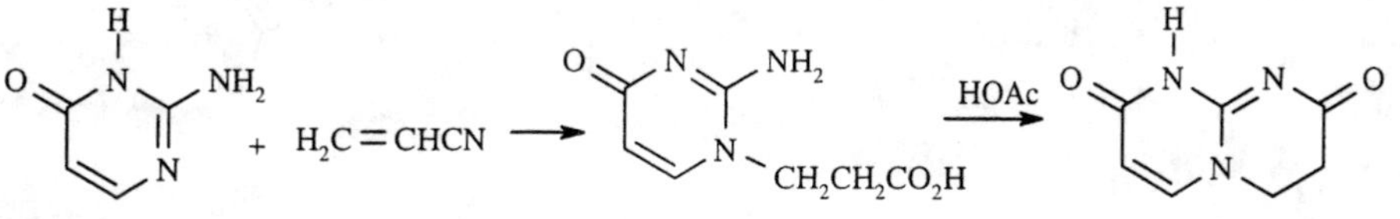

(**112**)

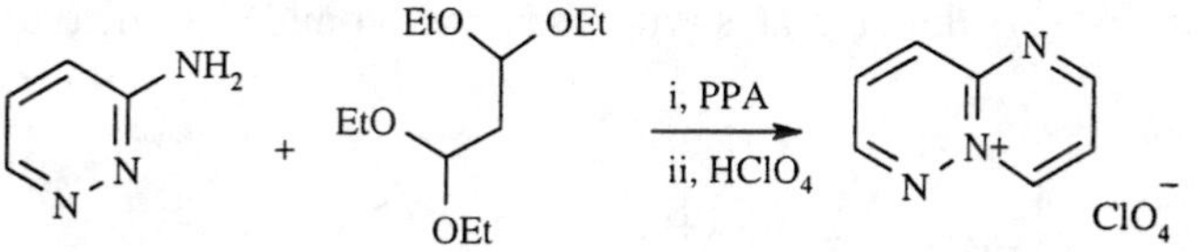

(**113**)

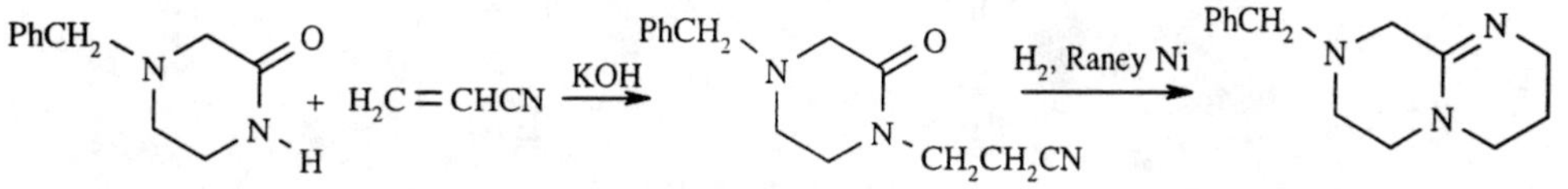

(**114**)

Methods analogous to those described above, but using a CCS or CNC three-atom fragment, are illustrated by the syntheses of compounds (**115**), (**116**) [from 2-aminopyridine and MeC(OMe)=NCN], (**117**) [from 3-amino-5-methyl-1,2,4-triazole and CH(OEt)=NCN], (**118**) [from 2-aminobenzimidazole and cyanoguanidine], and the fused mesomeric 1,3,5-triazine betaine shown in Scheme 5.

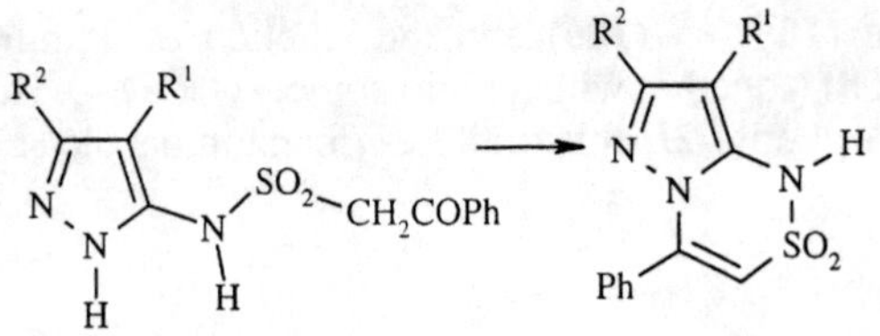

(115)

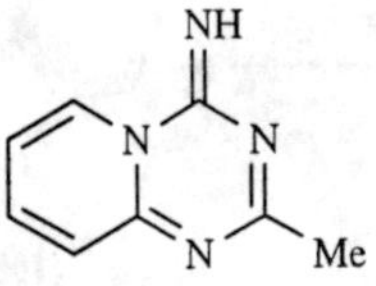

(116) 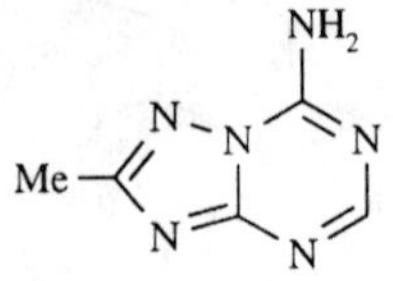   (117) 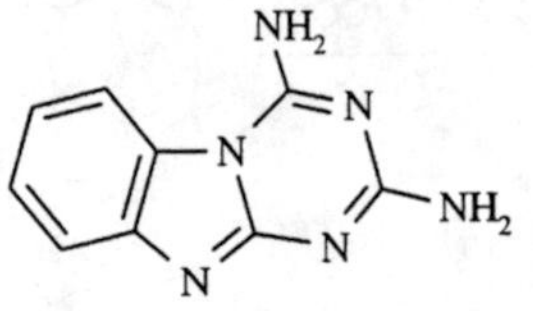   (118)

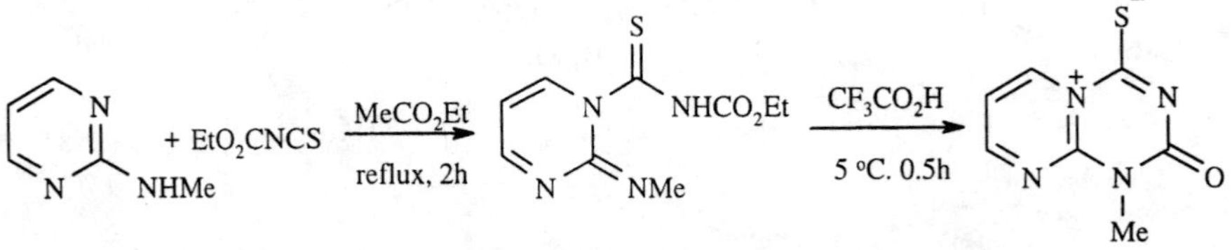

Scheme 5

4.6.3.2   Ring Formation Using a Two-atom Fragment

2-Methylthiazolium salts (**119**) with α-dicarbonyl compounds give thiazolo[2,3-*b*]pyridinium salts (**120**) ⟨86JHC1899⟩.

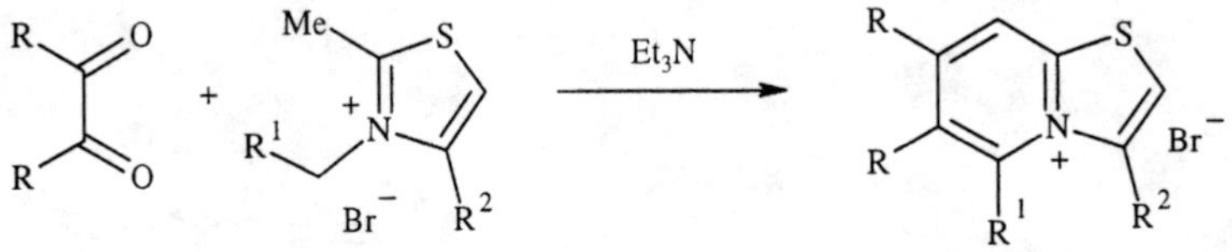

(119)   (120)

N-Amino-4-phenylthiazole-2(3*H*)thione reacts with phenacyl bromides to yield the products (**121**) as bromide salts.

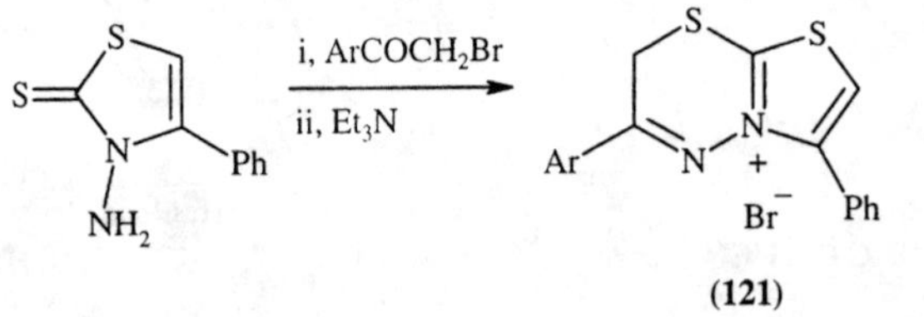

(121)

4.6.3.3   Ring Formation Using a One-atom Fragment

4-Amino-3-hydrazino-1,2,4-triazoline-5-thione (**122**) and aldehydes form unstable products (**123**), which readily air-oxidize to deeply colored tetrazines (**124**). This is a sensitive method for determining aldehydes ⟨70CC1719⟩.

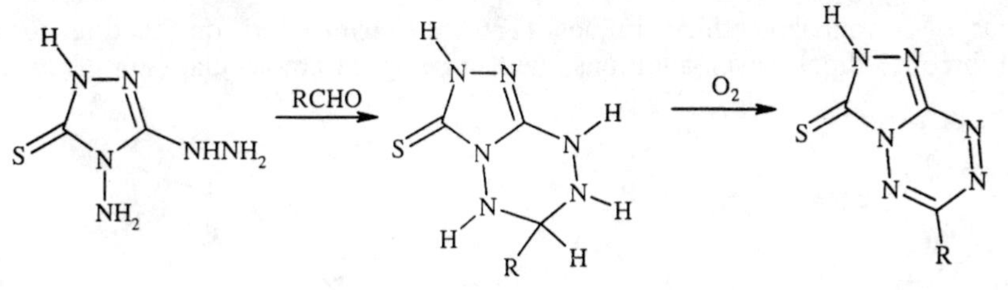

(122)   (123)   (124)

Other syntheses of this type are exemplified by conversions (125) → (126) and (127) → (128) by the action of HCO$_2$Et and of two equivalents of RNCO, respectively.

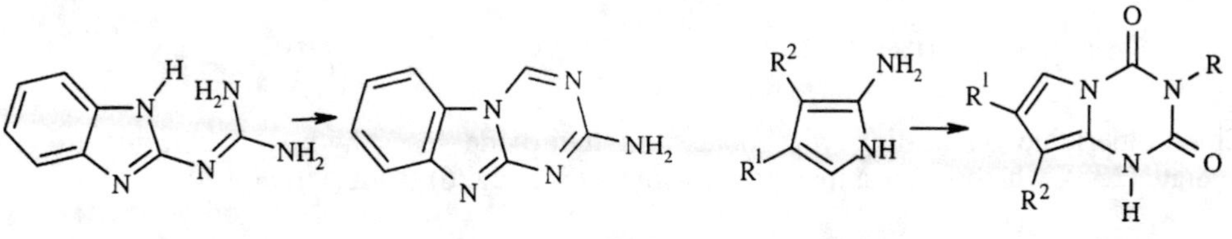

(125)   (126)   (127)   (128)

4.6.3.4 Cycloaddition and Ring-Transformation Reactions

1,4-Dipolar cycloadditions lead to six-membered rings. Rearrangements may be encountered. Thiazole or 2-methylthiazole (129; R = H and Me) with DMAD forms an initial 1,4-dipolar species (130). Reaction of (130) with a second DMAD gives a 1:2 adduct, presumably (131). Ring opening to (132), followed by cyclization in the alternative mode, resulted in (133) ⟨78AHC(23)263⟩ (see also CHEC 4.19). For the similar reactions of pyridine with DMAD and pyridazine with maleic anhydride see Section 3.2.1.3.7.

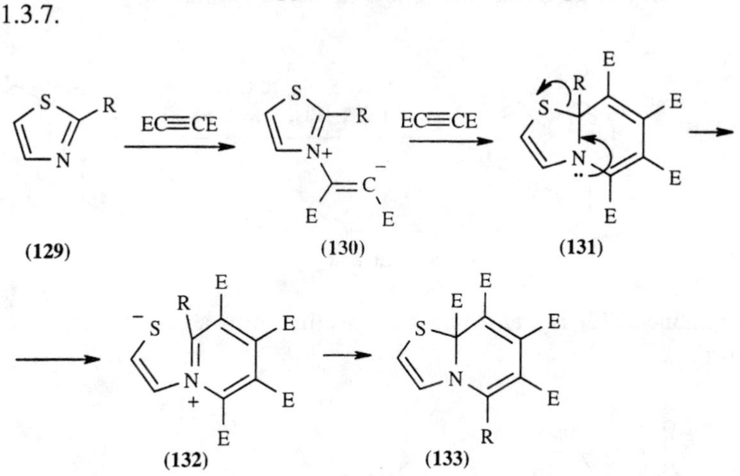

(129)   (130)   (131)

(132)   (133)

E = CO$_2$Me

Diels-Alder reaction of oxazolines and thiazolines (134) with tetrazines yield bicyclic compounds (135) ⟨84AP(317)237⟩.

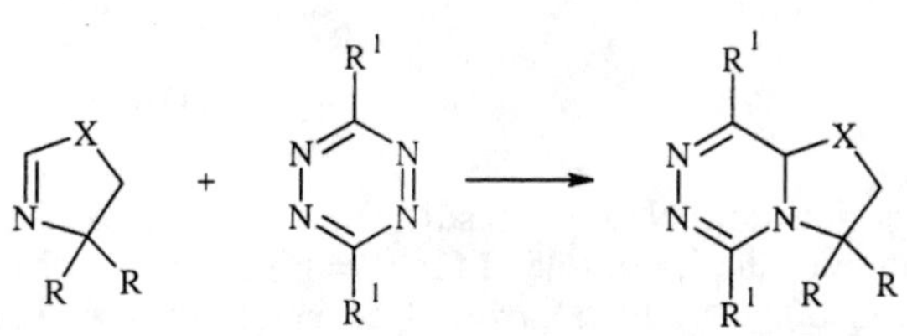

(134) X = O, S   (135)

C,C,N_α-Triaryl-N_β-cyanoazomethine imines (**136**) are trimerized on heating to (**137**) *via* a succession of three 1,3-dipolar cycloadditions, the last being intramolecular ⟨80AG(E)906⟩.

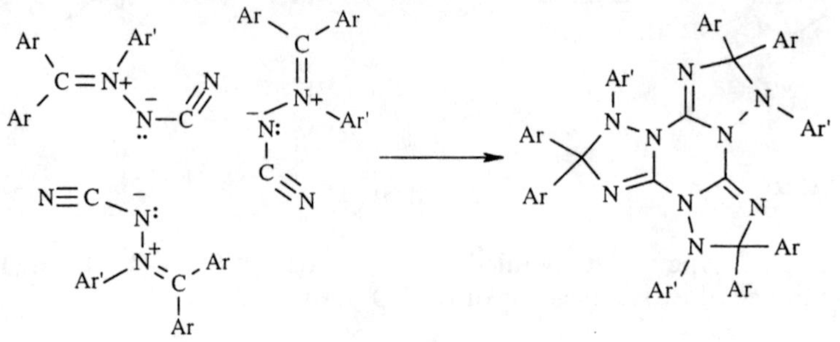

(**136**)   (**137**)

N-Chloropiperidine on heating with potassium hydroxide loses hydrogen chloride to give the intermediate (**138**), which forms a mixture of trimers (Scheme 6) ⟨59HC(13)1, p. 446⟩.

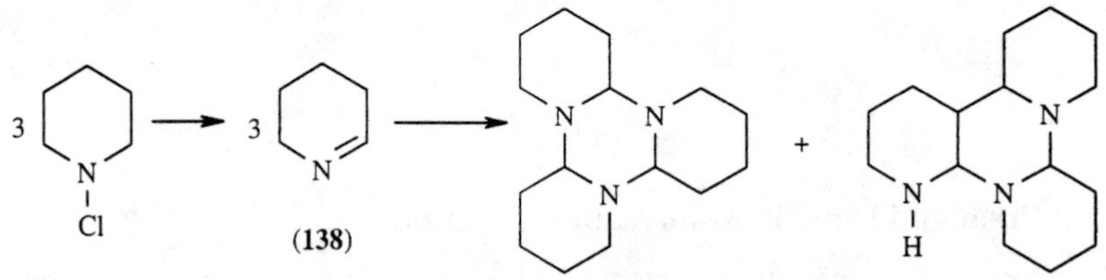

Scheme 6

Mixed 'trimerization' of heterocyclic compounds is a potentially valuable route to fused systems, *e.g.* isoquinoline with 2-chloro-4,5-dihydroimidazole (Scheme 7) ⟨81S154⟩.

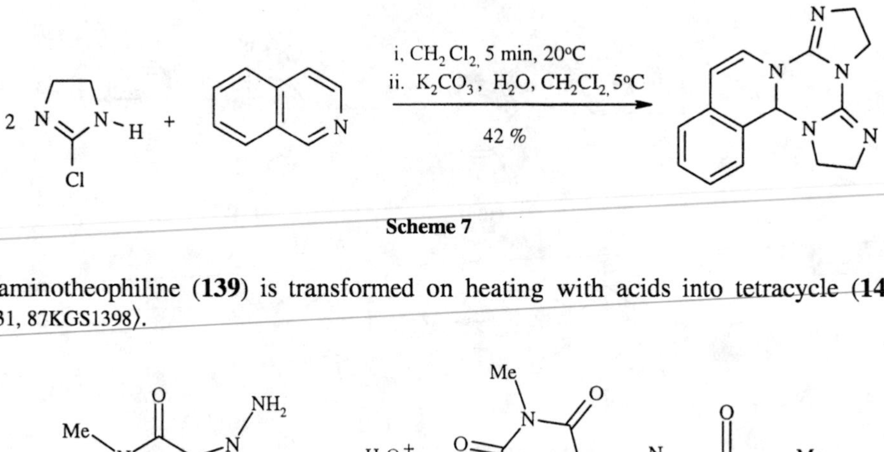

i, CH$_2$Cl$_2$, 5 min, 20°C
ii. K$_2$CO$_3$, H$_2$O, CH$_2$Cl$_2$, 5°C

42 %

Scheme 7

7,8-Diaminotheophiline (**139**) is transformed on heating with acids into tetracycle (**140**) (74%) ⟨87CPB4031, 87KGS1398⟩.

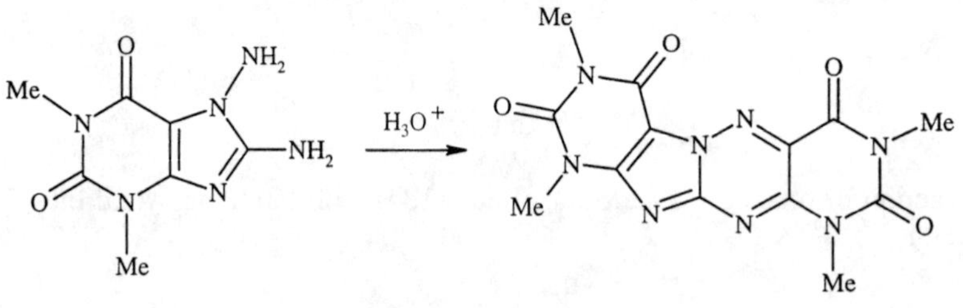

(**139**)   (**140**)

Compounds with a ring junction N atom result from the photochemical cyclization of *cis*-1-styrylimidazoles. Irradiation of the imidazole (**141**; R^1 = H) in methanol in the presence of I$_2$ gave imidazo[2,1-*a*]isoquinoline (**142**) ⟨76JCS(P1)75⟩. *trans*-1-Styrylbenzimidazole similarly formed successively the *cis* isomer and the benzimidazo[2,1-*a*]isoquinoline (**143**). For similar reaction of 1-styrylpyridinium, see Section 3.2.2.3.9.

(141)   (142)   (143)

Intramolecular [4+2] cycloaddition reactions are illustrated by thermolysis of the benzocyclo-butene (144) to the tetracyclic system (145), *via* trapping of a quinodimethane by the nitrile group.

(144)   (145)

For [3+3] cycloaddition of azafulvalene ketenes see Section 3.3.3.3.6.

4.6.3.5 Other Methods

Azolealdehydes with an α-NH group dimerize (Section 3.4.3.4.4). Cyclic N^+—N- links can be formed using nitrene intermediates (Section 3.4.3.4.2).

4.6.4 FORMATION OF A SEVEN-MEMBERED RING WITH ONE N ATOM AT A RING JUNCTION

Ylides (146) are cyclized by 30% aqueous hydrogen peroxide to afford the pyrido[1,2-*a*]azepines (147) ⟨91HCA1095⟩. The pyrido[1,2-*a*]azepinium system (148) is also obtainable by intramolecular free radical cyclization of 5-(1-pyridinio)-pentyl iodide ⟨91T4077⟩.

(146)   (147)   (148)

Pyrrolo[1,2-*a*]azepines occurs in a number of natural products and compounds of biological interest. The pyrroloalkyl epoxide (**149**) is cyclized to the alcohol (**150**) (85% yield) by Ti(OPr^i)$_3$Cl ⟨87JOC819⟩. The allylsilyl substituted *N*-acylimium salts (**151**) cyclize to (**152**) ⟨92H(34)37⟩.

(**149**)   (**150**)   (**151**)   (**152**)

4.6.5   TWO NITROGEN ATOMS AT A RING JUNCTION

4.6.5.1   Five-membered Rings

A wide variety of methods are available: the following examples are intended only to be illustrative. Reaction of pyrazole with (chlorocarbonyl)phenylketene (**153**) forms the zwitterionic pyrazolo[1,2-*a*]pyrazole (**154**) ⟨80JA3971⟩.

(**153**)   (**154**)   (**155**)   (**156**)

The pyrazoline (**155**) cyclized to the pyrazolinium salt (**156**) on heating with thionyl chloride ⟨68JOC3941⟩.

Heating 2,2′-diazidoazobenzene (**157**) at 58°C formed (**158**); at 70°C a second ring closed to give the tetraazapentalene (**159**) ⟨67JA2618⟩.

(**157**)   (**158**)   (**159**)

The substituted azobenzene (**160**) with triethyl phosphite gives the triazapentalene (**162**) *via* the 2-substituted benzotriazole (**161**) which undergoes a second deoxygenative cyclization ⟨74CL951⟩.

(**160**)   (**161**)   (**162**)

The betaine (**163**) reacts with DMAD to give product (**164**) which easily undergoes thermal fragmentation to (**165**) followed by another cycloaddition to form (**166**) ⟨81JA7743⟩.

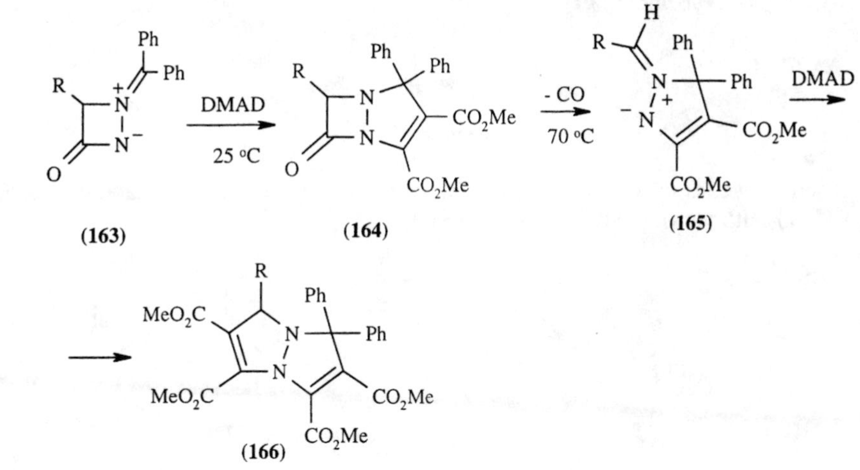

(**163**) (**164**) (**165**) (**166**)

'Criss-cross' cycloaddition of azines also involves two successive 1,3-dipolar cycloadditions (Scheme 8) ⟨76S349⟩.

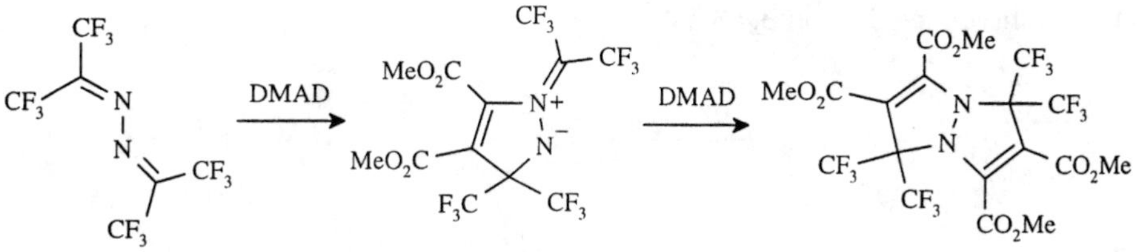

Scheme 8

4.6.5.2 Six-membered Rings

3,6-Pyridazinedione (**167**) readily condenses with butadiene, 2,3-dimethylbutadiene and coumalic acid to give the respective Diels-Alder adducts (**168**).

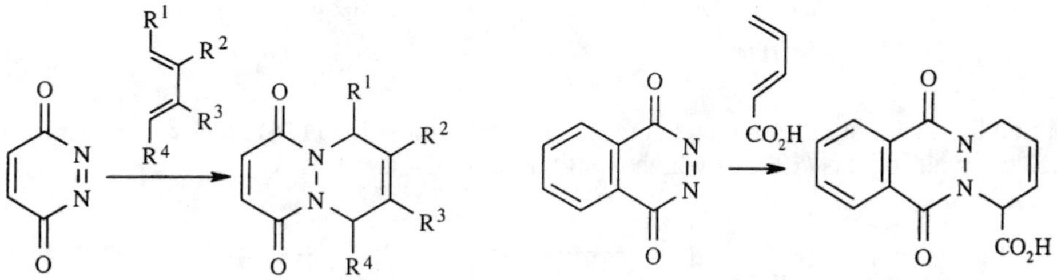

(**167**) (**168**) (**169**) (**170**)

Phthalazine-1,4-dione reacts similarly, *e.g.* (**169**) → (**170**).

4.6.6 SULFUR AT A RING JUNCTION

Preparations of bicyclic sulfonium ions utilize thiacyclohexanes with electron-withdrawing group in the α-position; these can be converted into an anion and subsequently cyclized into the sulfonium ion by reaction with 1,3-dibromopropane or 1,4-dibromobutane (Scheme 9) ⟨90TL3027⟩.

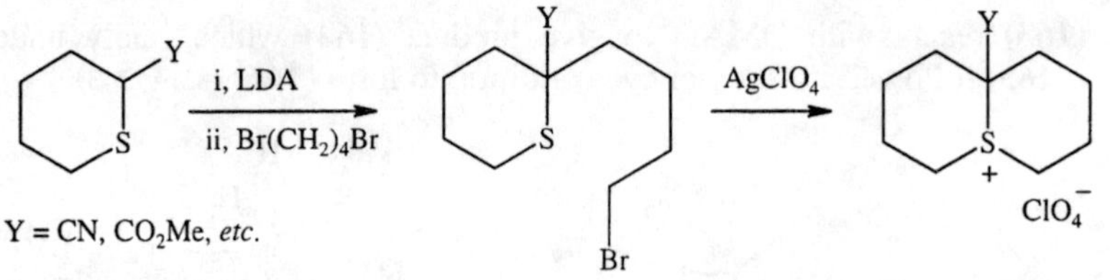

Scheme 9

The benzo-fused bicyclic sulfonium salts (172) can be prepared as cycloadducts of 2-thiana-phthylium ions (171) with various butadienes ⟨91TL5571⟩.

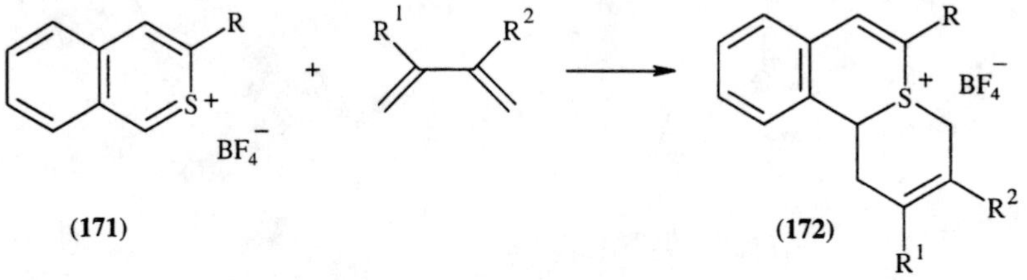

(171)   (172)

The best known compounds are of type (173). Available synthetic methods are discussed in CHEC 4.38.5; the following are illustrative. The method of Scheme 10 has been widely used (CHEC 4.38.5.2.2): X can be S or NMe; Y = OMe, SMe or NMe; Z = O, S, Se or NR.

For related syntheses see Section 3.4.3.3.4.

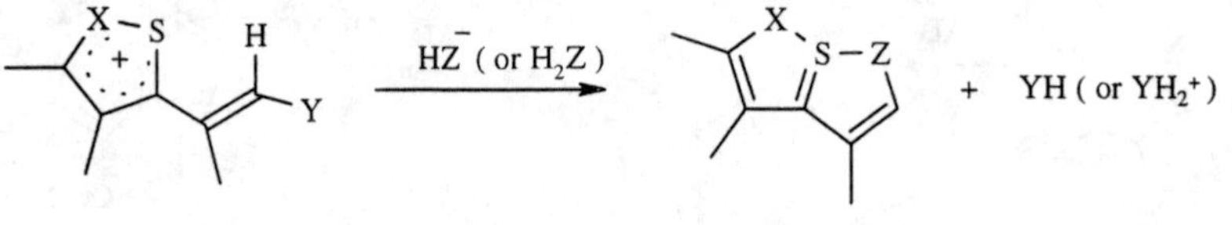

(173)

Scheme 10

1,6,6aλ^4-Trithiapentalenes are obtained from 1,3,5-triketones with P_2S_5, *e.g.* (174) → (175) ⟨71AHC(13)161⟩.

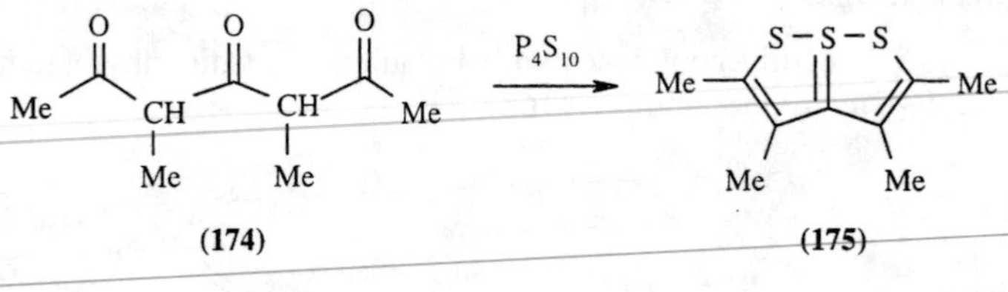

(174)   (175)

1,2-Dithiolylium cations carrying a leaving group at the 3-position (176) react with active methylene compounds to give products (177) in which X = O or S ⟨80AHC(27)151⟩.

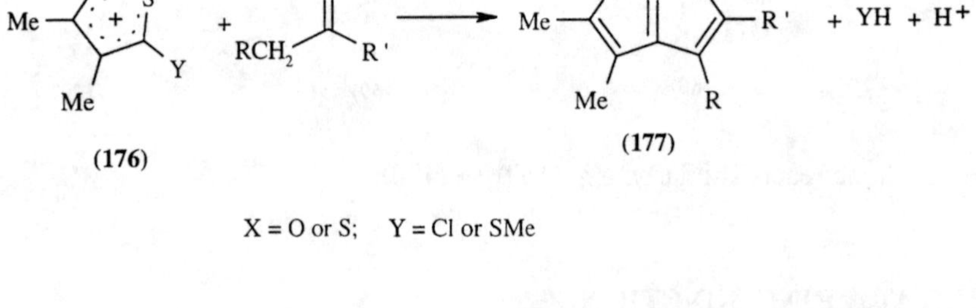

(176)   (177)

X = O or S;   Y = Cl or SMe

Part 5
Appendixes

Appendix A

Introduction to "Comprehensive Heterocyclic Chemistry – II"

1 SCOPE, SIGNIFICANCE, AND AIMS

1.1 Scope

Heterocyclic compounds possess a cyclic structure with two or more different kinds of atoms in the ring. This work is devoted to organic heterocyclic compounds in which the ring contains at least one carbon atom; all atoms other than carbon are considered the heteroatoms. Carbon is still by far the most common ring atom in heterocyclic compounds, but as the number and variety of heteroatoms in the ring increase there is a steady transition to the expanding domain of inorganic heterocyclic systems. Since rings can be of any size, from three-membered upwards, and since the heteroatoms can be drawn in almost any combination from a large number of the elements (though nitrogen, oxygen, and sulfur are still by far the most common), the number of possible heterocyclic systems is almost limitless. An enormous number of heterocyclic compounds is known and this number is increasing very rapidly. The literature of the subject is correspondingly vast and of the three major divisions of organic chemistry, aliphatic, carbocyclic, and heterocyclic, the last is by far the largest. Over 13 million compounds are recorded in *Chemical Abstracts* and approximately half of these are heterocyclic.

1.2 Significance

Heterocyclic compounds are very widely distributed in nature and are essential to life, playing a vital role in the metabolism of all living cells. For example, the following are heterocyclic: the pyrimidine and purine bases of the genetic material DNA; the essential amino acids proline, histidine, and tryptophan; the vitamins and coenzyme precursors thiamine, riboflavine, pyridoxine, folic acid, and biotin; the B_{12} and E families of vitamins; the photosynthesizing pigment chlorophyll; the oxygen-transporting pigment hemoglobin and its breakdown products, the bile pigments; the hormones kinetin, heteroauxin, serotonin, and histamine; and most sugars. There are a vast number of pharmacologically active heterocyclic compounds, many of which are in regular clinical use. Some of these are natural products, for example antibiotics such as penicillin and cephalosporin; alkaloids such as vinblastine, ellipticine, morphine, and reserpine; and cardiac glycosides such as those of digitalis. However, the large majority of pharmaceuticals are synthetic heterocyclics which have found widespread use, *inter alia* as anticancer agents, analeptics, analgesics, hypnotics, and vasopressor modifiers.

Other heterocyclics serve as pesticides, insecticides, herbicides, and rodenticides. Other important practical applications include dyestuffs, copolymers, solvents, photographic sensitizers and developers, and antioxidants and vulcanization accelerators in the rubber industry. Many heterocycles are valuable intermediates in synthesis.

The successful application of heterocyclic compounds in these and many other ways, and their utility in applied chemistry and in more fundamental and theoretical studies, stems from their very complexity; this ensures a virtually limitless series of structurally novel compounds with a wide range of physical, chemical, and biological properties, spanning a broad spectrum of reactivity and stability. Another consequence of the varied chemical reactivity of heterocycles, including the possible destruction of the heterocyclic ring, is their increasing use in the synthesis of specifically functionalized non-heterocyclic structures.

1.3 Aims of CHEC and of the Present Work

All of the above aspects of heterocyclic chemistry are mirrored in the contents of the present work. The scale, scope, and complexity of the subject, already referred to, with its correspondingly complex system of nomenclature, can make its study initially somewhat daunting. One of the main aims both of the first edition of *Comprehensive Heterocyclic Chemistry*, *i.e.*, CHEC, and of the present work is to minimize this problem by presenting a comprehensive account of fundamental heterocyclic chemistry. The emphasis is on basic principles and, as far as possible, on unifying correlations in the structure, chemistry, and synthesis of different heterocyclic systems and the analogous carbocyclic structures. The motivation for this effort was the outstanding biological, practical, and theoretical importance of heterocyclic chemistry, and the absence of an appropriate major modern treatise when CHEC was conceived in 1980.

At the introductory level there are several good textbooks on heterocyclic chemistry, although the subject is scantily treated in most general textbooks of organic chemistry. At the specialist, research level there are two established ongoing series, "Advances in Heterocyclic Chemistry" edited by Katritzky and "The Chemistry of Heterocyclic Compounds" edited by Weissberger and Taylor, devoted to a very detailed consideration of all aspects of heterocyclic compounds, which together comprise some 100 volumes. CHEC was designed to fill the gap between these two levels, *i.e.*, to give an up-to-date overview of the subject as a whole (particularly in the general chapters) appropriate to the needs of teachers and students and others with a general interest in the subject and its applications, and to provide enough detailed information (particularly in the monograph chapters) to answer specific questions, to demonstrate exactly what is known or not known on a given topic, and to direct attention to more detailed reviews and to the original literature. Mainly because of the extensive practical uses of heterocyclic compounds, a large and valuable review literature on all aspects of the subject has grown up over the last few decades. References to all of these reviews are now immediately available: reviews dealing with a specific ring system are reported in the appropriate monograph chapters both of CHEC and of CHEC-II; reviews dealing with any aspect of heterocyclic chemistry which spans more than one ring system are collected together in a logical, readily accessible manner in CHEC, Chapter 1.03.

2   ARRANGEMENT OF THE WORK IN VOLUMES

2.1   Relationship of CHEC-II to CHEC

CHEC, published early in 1984, covered the literature through to 1982. In the intervening years, publication in the field has proceeded apace, and the success of CHEC has encouraged us to bring it up to date with a new edition, CHEC-II. With some exceptions discussed below, CHEC-II concentrates on material published in 1982 and later; to have reissued all the material from CHEC would have produced an overly long work.

CHEC-II keeps very much to the organization of the material used in CHEC with the following major changes:

(a) the "general chapters" are not repeated;

(b) instead of the individual chapters on applications (*e.g.*, as pharmaceuticals, agrochemicals, polymers) which appeared in Volume 1 of CHEC, applications are treated in CHEC-II under the relevant ring system;

(c) chapters summarizing heterocyclic ring systems containing the "less common elements" (*i.e.*, other than C, N, O, and S) appear in their logical places throughout CHEC-II and are not collected in a separate volume;

(d) the volume order is changed to follow a logical pattern of increasing ring size.

2.2   Arrangement of CHEC-II in Volumes

Heterocyclic compounds are normally classified according to the size of the heterocyclic ring and the nature and number of the heteroatoms. The present work is organized on this basis.

Volume 1A covers three- and **Volume 1B** four-membered heterocycles, together with all fused systems containing a three- or four-membered heterocyclic ring.

Volume 2 covers five-membered rings with one heteroatom together with their benzo- and other carbocyclic-fused derivatives.

Volumes 3 and 4 cover five-membered rings with two heteroatoms, or more than two heteroatoms, respectively, each with their fused carbocyclic compounds.

Volumes 5 and 6 cover six-membered rings with one, or more than one, heteroatom respectively, again with fused carbocyclic compounds.

Volumes 7 and 8 cover systems containing at least two fused heterocyclic five- and/or six-membered rings. Volume 7 deals with such biheterocyclic rings without a ring junction heteroatom. Volume 8 deals with those containing ring junction heteroatoms, together with systems with three five- and/or six-membered fused heterocyclic rings and with spiro cyclic systems.

Volume 9 covers seven-membered and larger heterocyclic rings including all fused derivatives (except those with three- or four-membered rings which are included in Volume 1).

Volume 10 contains the Author and Ring Indexes.

Volume 11 contains the Subject Index.

2.3   Arrangement of CHEC in Volumes

Part 1 (Volume 1) deals with (a) the nomenclature and the literature of heterocyclic compounds (Chapters 1.02 and 1.03); (b) various special topics (Chapters 1.04–1.16); and (c) rings containing less common heteroatoms (Chapters 1.17–1.22).

The literature chapter presents an organized collection of references to leading papers, review articles, and books dealing with all aspects of heterocyclic chemistry which span more than one ring system. Other reviews which deal with a specific ring system are reported in the appropriate monograph chapter.

The special topics discussed are (a) the biological aspects of heterocyclic compounds, *i.e.* their biosynthesis, toxicity, metabolism, role in biochemical pathways, and their uses as pharmaceuticals, agrochemicals, and veterinary products; (b) the use of heterocyclic compounds in polymers, dyestuffs and pigments, photographic chemicals, semiconductors, and additives of various kinds; and (c) the use of heterocyclic compounds as intermediates in the synthesis of non-heterocyclic compounds.

The less common heteroatoms are those other than nitrogen, oxygen, and sulfur (and selenium and tellurium which are treated alongside sulfur), *i.e.*, phosphorus, arsenic, antimony, bismuth, the halogens, silicon, germanium, tin, lead, boron, and the transition metals.

Part 2 (Volumes 2 and 3) deals with mono- and polycyclic compounds with one or more six-membered heterocyclic rings, with nitrogen, oxygen, and sulfur as the heteroatoms. Volume 2 contains the six-membered rings with one nitrogen atom (Part 2A) and Volume 3 the six-membered rings with two or more nitrogens and with oxygen and sulfur heteroatoms (Part 2B).

Part 3 (Volume 4) covers five-membered heterocyclic rings with one oxygen, sulfur, or nitrogen as the heteroatom and **Part 4** (Volumes 5 and 6) covers the same sized rings with more than one heteroatom. Rings with two or more nitrogen atoms only (Part 4A) are in Volume 5 and those with nitrogen, oxygen, and sulfur (Part 4B) are in Volume 6.

Part 5 (Volume 7) covers heterocyclic compounds with rings smaller than five and larger than six. (The separate monograph chapters deal with three-membered rings with nitrogen, oxygen, and sulfur, with two heteroatoms, and their fused derivatives, four-membered rings with nitrogen, oxygen, and sulfur, with two or more heteroatoms, and their fused derivatives, cephalosporins, penicillins, seven-membered rings with nitrogen, oxygen, sulfur, and with two or more heteroatoms, eight-membered rings, larger rings, crown ethers and cryptands, and heterocyclophanes.)

Part 6 (Volume 8) contains the Author and Subject Indexes, together with a Ring Index and a Physical Data Index which are described in Section 1.01.7.

3   RATIONALE FOR ARRANGEMENT OF MATERIAL IN EACH VOLUME

3.1   Major Division of Carbocyclic and Heterocyclic Chemistry

Carbocyclic compounds are very usefully divided into (a) saturated (alicyclic) compounds, (b) aromatic compounds, and (c) the intermediate partially unsaturated (alicyclic) compounds. Heterocyclic compounds can be subdivided in exactly the same way, and this is equally useful.

On the whole, the physical and chemical properties of saturated and partially unsaturated alicyclic compounds closely resemble those of the analogous acyclic compounds formally derived by cleavage of the carbon ring at a point remote from any functionality. Relatively small (but often significant) differences in properties arise from conformational effects and from strain effects in small rings. These differences can be striking in properties which are particularly sensitive to molecular shape.

In marked contrast to these are the fully "unsaturated" aromatic compounds, epitomized by benzene, in which the carbocyclic rings formally consist of a conjugated set of alternating single and double bonds. Such systems have a specially stabilized cyclic π-electron system in which all of the bonding molecular orbitals are completely filled and the antibonding orbitals are all empty. The concept of aromatic properties initially associated with six-electron systems, as in benzene, the cyclopentadienyl anion, and the cycloheptatrienyl cation, was later extended to any planar, monocyclic, fully conjugated polyene with a closed shell of $(4n+2)$ π-electrons. Although not easily rigorously defined and quantified, the concept of aromaticity has been of enormous value in the understanding of carbocyclic chemistry. Since it is associated with molecular orbital energies, aromaticity is not particularly sensitive to the nature and number of the ring atoms and the concept is thus of equal generality in heterocyclic chemistry. The main features of carbocyclic aromatic systems are (a) their stability, and hence their ready formation and regeneration after chemical attack, (b) their tendency to undergo substitution reactions which preserve the aromatic system rather than addition reactions which destroy it, (c) the uniformity of the bond lengths in the ring which are not alternating single and double, and (d) their special spectroscopic characteristics, particularly NMR. All of these features reappear, to a greater or lesser extent, in heteroaromatic chemistry.

3.2   Saturated Heterocyclic Compounds

As noted above, the formation of an alicyclic ring from an acyclic compound makes relatively little difference to the properties of the compounds. The same principle applies to the formation of a saturated heterocyclic compound from the corresponding acyclic compound, providing that the environment of the heteroatom is not changed significantly. Thus, saturated cyclic ethers, sulfides, and amines are very similar in physical and chemical properties to the analogous dialkyl ethers, sulfides, and dialkyl- and trialkylamines. Differences arise in small ring compounds where chemical reactivity is enhanced greatly by strain in three-membered rings and to a lesser, but still significant, extent in four-membered rings. Furthermore, properties which depend critically on steric requirements, particularly of lone pairs of electrons on the heteroatom, can be significantly altered. Thus a very good approximation of the properties of tetrahydropyran and piperidine can be obtained from those of ethyl propyl ether and ethylpropylamine, respectively. Piperidine is a typical secondary aliphatic amine of the same base strength as ethylpropylamine. However, it is substantially more reactive as a nucleophile, because of the reduction in steric encumbrance of the nitrogen lone pair caused by ring formation.

3.3   Partially Unsaturated Heterocyclic Compounds

The same general principles also extend to the partially unsaturated rings, although with extra complications expected from the presence of the double bond(s), especially in small rings. If the double bond is conjugated with the heteroatom, then the expected consequences of electron delocalization are observed; thus such oxygen and nitrogen compounds are enol ethers and enamines, respectively, and will show the appropriate modified reactivity. Again, as expected, dihydro heteroaromatic compounds usually oxidize very readily to the aromatic compound.

3.4   Heteroaromatic Compounds

Although the range of heteroaromatic structures has expanded considerably since the 1950s, the central core of the subject is still based on the 6π-electron system. These structures are related to, and formally derived from, benzene by successive replacement of one or two annular CH groups by trivalent or divalent heteroatom groups, respectively; the overall pattern of filled bonding molecular orbitals is retained. Thus replacement of one CH group by O^+, S^+ or N gives the six-membered pyrylium, thiopyrylium, or pyridine systems, and replacement of two adjacent CH groups by O, S, or NH gives the five-membered furan, thiophene, or pyrrole. Multiple replacements are also possible and systems with up to four heteroatoms in five- and six-membered rings are common. The 6π-electron structure is preserved since the trivalent and divalent heteroatoms contribute one and two electrons, respectively, to the aromatic orbitals.

3.5   Characteristics of Heteroatoms in Rings

The replacement of one CH group in benzene by —N= to give pyridine introduces an electron-withdrawing heteroatom into the ring. The electron-withdrawing effect is accentuated when a CH in benzene is replaced by a positively charged atom (NR^+, O^+ or S^+). Thus, the six-membered heteroaromatic rings are electron deficient (π-deficient). The introduction of further heteroatoms into a six-membered ring reinforces these effects. Thus, for example, the chemistry of pyrazine is related to that of pyridine in much the same way as that of pyridine is related to benzene.

The replacement of two CH groups in benzene by a neutral NR, O, or S introduces into the new ring an electron-donating heteroatom. This electron-donor character is accentuated in the pyrrole anion where N^- is introduced. Thus the five-membered rings with one heteroatom are electron rich (π-excessive), and the chemistry of pyrrole, furan, and thiophene is dominated by this effect.

However, five-membered rings containing two or more heteroatoms necessarily possess both a pyridine-like heteroatom *and* a pyrrole-like heteroatom and thus their chemistry shows similarities to both those of the six-membered rings *and* of the five-membered rings with one heteroatom.

3.6 The General Chapters

Three general chapters on structure, reactivity, and synthesis precede the monograph chapters in each of Parts 2 to 5 of CHEC. The purpose is to introduce each family of ring systems and to emphasize the logical correlations within them, and so to help in the understanding of known reactions and in the prediction of new ones. These 12 general chapters thus provide an overview of the whole subject of heterocyclic chemistry and they should be of particular interest to students and teachers. These general chapters also appeared in the 1985 version of the *Handbook of Heterocyclic Chemistry* and it is planned that they will appear in a new revised version of the *Handbook*.

4 ORGANIZATION OF INDIVIDUAL MONOGRAPH CHAPTERS

CHEC-II is essentially composed entirely of monograph chapters. Monograph chapters dealing with single ring systems are divided into the three sections dealing with structure, reactivity, and synthesis (as are the general chapters) and are given a fourth section, where appropriate, on applications and important compounds. Where two or more ring systems are covered in one chapter, they are, where appropriate, treated together in the same sections. The following conventions concerning the treatment of fused rings should be noted.

Fused benzene and other carbocyclic rings are treated as substituents. Thus quinoline, for example, is considered as a substituted pyridine (albeit a very special and important one) and is treated alongside other substituted pyridines in the discussion of its structure, reactivity, and synthesis. Reactions of quinoline at positions 1 to 4 are considered as reactions at ring atoms, whilst reactions at positions 5 to 8 are regarded as reactions of the "substituent". Structures containing two or more *non*-fused heterocyclic rings are treated in the monograph chapter appearing last in the sequence. However, fused heterocyclic ring systems are treated in separate monograph chapters. In CHEC-II fused heterocycles appear in Volumes 7 and 8 unless they contain a small or large heterocyclic ring and are then treated in Volumes 1 and 9 respectively.

The applications sections of monograph chapters provide access to important compounds used in medicine or industry, to industrially important sources of compounds, and to key natural products.

As far as practical, a standard arrangement of the material in the various chapters has been followed in the belief that this will:
 (a) assist readers in retrieving information;
 (b) indicate what is *not known* as well as what is known;
 (c) facilitate comparisons between different ring systems;
 (d) assist authors in the organization of their material.

Whereas the chapters in CHEC were organized in four main sections: structure, reactivity, synthesis, applications; the chapters in CHEC-II are organized in 12 sections, as follows.

4.1 Introduction

If appropriate, this commences with a brief historical piece, and comments on the relationship of the new chapter to the corresponding chapter in CHEC, and also gives general references to reviews of the material. The scope of the chapter is outlined with a survey of the various structural types and nomenclature of the parent, its non-conjugated isomers, partially reduced compounds, oxo compounds and benzo derivatives. Distinction is made here between the structural types possible and those which are known and treated in the chapter.

4.2   Theoretical Methods

Since the publication of CHEC, theoretical methods have become of much greater importance. This section outlines the scope of what has been done with *ab initio* and semiempirical (*e.g.*, AM1) molecular orbital methods, and with molecular mechanics.

4.3   Experimental Structural Methods

Methods covered include X-ray, neutron and electron diffraction, microwave spectroscopy, and their results in terms of molecular dimensions. NMR spectroscopy is treated in some detail as befits its importance, not only proton, but particularly ^{13}C, ^{15}N, and, where appropriate, other nuclei. The section on mass spectrometry briefly covers fragmentation patterns. UV/Fluorescence, IR/Raman, photo-electron spectroscopy, ESR, and dipole moments are covered as appropriate.

4.4   Thermodynamic Aspects

Boiling points and melting points are considered from the point of view of intermolecular forces between the molecules, together with solubilities and chromatographic behavior, both gas and liquid chromatography. The topic of aromaticity and stability in general is covered as befits its importance. Conformations, particularly of the cyclic non-planar compounds, are dealt with. A section on tautomerism covers both prototropic tautomerism (annular and of substituents) and ring-chain tautomerism.

4.5   Reactivity of Fully Conjugated Rings

CHEC-II utilizes this heading in place of "reactivity at heteroaromatic rings" used in CHEC. The change has been made because of the difficulty in defining exactly the term "aromatic". "Cyclically conjugated" is defined here as any ring that does not contain an sp^3-hybridized carbon or nitrogen atom in it. Thus, "fully conjugated" includes all heteroaromatic rings, antiaromatic rings, and some rings that have little aromaticity although they are completely conjugated.

In most cases, the introduction and overview give a general survey of the reactivity and make comparisons with analogous ring systems, as well as referring to the corresponding chapter in CHEC. The following sections are then considered in turn:

(a) thermal and photochemical reactions that are unimolecular;

(b) electrophilic attack at nitrogen; the further detailed organization corresponds to the corresponding chapter in the *Handbook of Heterocyclic Chemistry*;

(c) electrophilic attack at carbon, organized as in the *Handbook of Heterocyclic Chemistry*;

(d) where the ring contains cyclic sulfur, electrophilic attack at sulfur is next covered;

(e) nucleophilic attack at carbon is dealt with according to the nucleophilic atom which is carrying out the attack;

(f) nucleophilic attack at hydrogen attached to carbon (deprotonation) is considered, for both neutral and cationic rings;

(g) reactions with radicals and electron-deficient species (carbenes, nitrenes) and also reactions at surfaces (heterogeneous catalysis) and reductions;

(h) intermolecular cyclic transition state reactions.

4.6   Reactivity of Non-conjugated Rings

The different types of non-conjugated rings are classified as follows.

(a) Isomers of aromatic compounds are dealt with in two classes: those which are not in equilibrium with the corresponding fully conjugated derivative, and then those which are in such an equilibrium. For the latter class, the main discussion should come under the corresponding tautomeric derivative (which will often be a cyclic hydroxy compound).

(b) Dihydro derivatives of various types are considered with emphasis on the role of carbonyl compounds and discussion of the ease of aromatization.

(c) Tetrahydro derivatives are dealt with followed, if appropriate, by hexahydro derivatives.

4.7   Reactivity of Substituents Attached to Ring Carbon Atoms

With this section, the classification in the *Handbook of Heterocyclic Chemistry* has been followed; thus a general survey of the effect of rings on the reactions of substituents is followed by a survey of the effect of rings on reactions of individual substituents in the order: fused benzene rings, *C*-linked, *N*-linked, *O*-linked, *S*-linked substituents, halogens, and metals. Substituents attached to cyclic nitrogen and sulfur are dealt with as described below.

4.8   Reactivity of Substituents Attached to Ring Heteroatoms

Substituents attached to ring nitrogen are discussed in order of the atom linking the substituent to the ring nitrogen atom just as described above. If the ring contains sulfur, then substituents attached to sulfur atoms are next considered. These are often oxygen, but may also be carbon and nitrogen.

4.9   Ring Syntheses Classified by Number of Ring Atoms in Each Component

A similar treatment to that in CHEC is followed.

4.10   Ring Synthesis by Transformation of Another Ring

A similar treatment to that in CHEC is followed; some methods are dealt with by cross-references.

4.11   Synthesis of Particular Classes of Compounds and Critical Comparison of the Various Routes Available

The present overview is intended to cover materials from both CHEC-II and CHEC.

4.12   Important Compounds and Applications

Specialized applications are no longer dealt with in a separate chapter (as was done in CHEC) and these sections cover the important advances since the mid–1980s.

5   THE REFERENCE SYSTEM

The same reference citation system is employed as was used for CHEC. It rapidly becomes familiar with use and has distinct advantages over the more common superscript number method. In this system reference numbers appear neither in the text, in tables, in footnotes, nor at the end of chapters. Instead, each time a reference is cited, there appears in angle brackets a letter code assigned to the journal, preceded by the year (tens and units only, except for non-twentieth century references) and followed by the page number. For example, "It was shown ⟨80TL2727⟩ that . . ." where "80" refers to 1980, "TL" to *Tetrahedron Letters* and "2727" to the page number. For journals which are published in separate parts, or which have more than one volume per year, the appropriate part or volume is indicated, for example ⟨73JCS(P2)1594⟩ refers to *J. Chem. Soc., Perkin Trans. 2*, 1973, p. 1594. A full list of journal codes is reproduced in each volume. Patents have three-letter codes as appropriate, for example ⟨60USP2922790⟩ refers to *US Pat.* 2 922 790 (1960). Books which are frequently referred to are also given a code. Journals and books which are referred to rarely are given a miscellaneous code (MI) starting each year and numbered sequentially 1, 2, 3 etc. Books are indicated by the prefix "B-".

This reference system is considered to be more useful than the conventional superscript number method since it enables the reader to see immediately in which year and in which journal (at least for the more common journals whose letter codes soon become familiar) the work cited was published. The reader is thus able to go directly to the original literature reference without having to consult a bibliography.

All references for chapters in a given volume are collected together in a merged list at the end of that volume (where they are most easily located). There are no separate chapter bibliographies. In the final list, references are given both in code and in full conventional form, with authors' names. They appear in an ordered sequence, numerically by year, then alphabetically by journal code, and then by page number. Cross-references to the text citation are also given in the reference list.

Chemical Abstract references are given when these are likely to help; in particular they are given for all patents, and for less accessible sources such as journals whose language is other than English, French, or German, company reports, obscure books, and theses.

6 THE INDEXES

6.1 Author Index

The Author Index of CHEC-II has over 50,000 names compiled from the references in the text and tables. The style is such, for example

Meyers, A.I., 1, 237, 319 ⟨78TL5179⟩; 3, 423 ⟨81JOC3881⟩,

that the reader can proceed to the text page where the work is cited, directly to the original literature, or to the literature reference at the end of each volume. More details are given at the beginning of the index in Volume 10.

6.2 Ring Index

The Ring Index for CHEC-II is similar in style and organization to the CHEC, Patterson, and *Chemical Abstracts* ring indexes. It gives the formula, full name, and ring numbering of all the parent ring systems. Under these are included all the heterocyclic compounds mentioned in the text, tables, formulae, equations, and schemes (except for trivial cases, *e.g.*, when pyridine is used as a solvent). Compounds are classified according to their fully unsaturated parent compound. Thus substituted, partially saturated, and fully saturated derivatives of, say, pyrrole, are all indexed under pyrrole. Benzo and similar derivatives are indexed under the most unsaturated parent system, for example quinoline, thienofuran, benzoxepin. This provides a useful cross-referencing of the main text material where benzo derivatives are classified as substituted derivatives of the parent monocyclic system (see above).

Index entries are divided into two categories, primary and secondary. Primary index entries are used when a significant part of the text is devoted to a particular ring system. Secondary index entries are used when a heterocyclic system is mentioned in a chapter devoted to another (primary) system. This may be, for example, as a starting material or as a product, or in a comparison of properties.

6.3 Subject Index

The Subject Index of over 80,000 entries has been compiled from keywords, names, and formulas in the text and tables. It covers general classes of compound, specific compounds, general types of reactions, specific and named reactions, spectral and other properties, and other topics in heterocyclic chemistry. More details are again given at the beginning of the index in Volume 11.

Appendix B

Short Contents of "Comprehensive Heterocyclic Chemistry – II"

Volume 1A   Three-membered Rings, with all Fused Systems containing Three-membered Rings

1.01   Aziridines and Azirines: Monocyclic
1.02   Aziridines and Azirines: Fused-ring Derivatives
1.03   Oxiranes and Oxirenes: Monocyclic
1.04   Oxiranes and Oxirenes: Fused-ring Derivatives
1.05   Thiiranes and Thiirenes: Monocyclic
1.06   Thiiranes and Thiirenes: Fused-ring Derivatives
1.07   Three-membered Rings with One Selenium or Tellurium Atom
1.08   Phosphiranes, Phosphirenes, and Heavier Analogues
1.09   Three-membered Rings with One Silicon, Germanium, Tin, or Lead Atom
1.10   Three-membered Rings with One Boron Atom
1.11   Diaziridines and Diazirines
1.12   Oxaziridines and Oxazirines
1.13   Thiaziridines and Thiazirines
1.14   Three-membered Rings with Two Oxygen and/or Sulfur Atoms
1.15   Three-membered Rings with Two Heteroatoms including Selenium or Tellurium
1.16   Three-membered Rings with Two Heteroatoms including Phosphorus to Bismuth
1.17   Three-membered Rings with Two Heteroatoms including Other Elements

Volume 1B   Four-membered Rings, with all Fused Systems containing Four-membered Rings

1.18   Azetidines, Azetines, and Azetes: Monocyclic
1.19   Cephalosporins
1.20   Penicillins
1.21   Other Fused Azetidines, Azetines, and Azetes
1.22   Oxetanes and Oxetenes: Monocyclic
1.23   Oxetanes and Oxetenes: Fused-ring Derivatives
1.24   Thietanes and Thietes: Monocyclic
1.25   Thietanes and Thietes: Fused-ring Derivatives
1.26   Four-membered Rings with One Selenium or Tellurium Atom
1.27   Four-membered Rings with One Phosphorus, Arsenic, Antimony, or Bismuth Atom
1.28   Four-membered Rings with One Silicon, Germanium, Tin, or Lead Atom
1.29   Four-membered Rings with One Boron or Other Atom
1.30   Four-membered Rings with Two Nitrogen Atoms
1.31   Four-membered Rings with One Oxygen and One Nitrogen Atom
1.32   Four-membered Rings with One Sulfur and One Nitrogen Atom
1.33   Four-membered Rings with Two Oxygen Atoms
1.34   Four-membered Rings with One Oxygen and One Sulfur Atom
1.35   Four-membered Rings with Two Sulfur Atoms
1.36   Four-membered Rings with Two or More Heteroatoms including Selenium or Tellurium
1.37   Four-membered Rings with Two Heteroatoms including Phosphorus to Bismuth
1.38   Four-membered Rings with Two Heteroatoms including Silicon to Lead
1.39   Four-membered Rings with Three Heteroatoms

Volume 2   Five-membered Rings with One Heteroatom and Fused Carbocyclic Derivatives

2.01   Pyrroles and their Benzo Derivatives: Structure
2.02   Pyrroles and their Benzo Derivatives: Reactivity
2.03   Pyrroles and their Benzo Derivatives: Synthesis
2.04   Pyrroles and their Benzo Derivatives: Applications
2.05   Furans and their Benzo Derivatives: Structure
2.06   Furans and their Benzo Derivatives: Reactivity
2.07   Furans and their Benzo Derivatives: Synthesis
2.08   Furans and their Benzo Derivatives: Applications
2.09   Thiophenes and their Benzo Derivatives: Structure
2.10   Thiophenes and their Benzo Derivatives: Reactivity
2.11   Thiophenes and their Benzo Derivatives: Synthesis
2.12   Thiophenes and their Benzo Derivatives: Applications
2.13   Selenophenes
2.14   Tellurophenes
2.15   Phospholes
2.16   Arsoles, Stiboles, and Bismoles
2.17   Siloles, Germoles, Stannoles, and Plumboles
2.18   Boroles
2.19   Five-membered Rings with Other Elements

Volume 3   Five-membered Rings with Two Heteroatoms and Fused Carbocyclic Derivatives

3.01   Pyrazoles
3.02   Imidazoles
3.03   Isoxazoles
3.04   Oxazoles
3.05   Isothiazoles
3.06   Thiazoles
3.07   1,2-Selenazoles
3.08   1,3-Selenazoles
3.09   1,2-Dioxoles and 1,2-Oxathioles
3.10   1,3-Dioxoles and 1,3-Oxathioles
3.11   1,2-Dithioles
3.12   1,3-Dithioles
3.13   Two Adjacent Heteroatoms with at least One Selenium or Tellurium
3.14   Two Nonadjacent Heteroatoms with at least One Selenium or Tellurium
3.15   Two Adjacent Heteroatoms with at least One Phosphorus, Arsenic, or Antimony
3.16   Two Nonadjacent Heteroatoms with at least One Phosphorus, Arsenic, or Antimony
3.17   Two Adjacent Heteroatoms with at least One Boron
3.18   Two Nonadjacent Heteroatoms with at least One Boron
3.19   Two Adjacent Heteroatoms with at least One Other Element
3.20   Two Nonadjacent Heteroatoms with at least One Other Element

Volume 4   Five-membered Rings with More than Two Heteroatoms and Fused Carbocyclic Derivatives

4.01   1,2,3-Triazoles
4.02   1,2,4-Triazoles
4.03   1,2,3-Oxadiazoles
4.04   1,2,4-Oxadiazoles
4.05   1,2,5-Oxadiazoles
4.06   1,3,4-Oxadiazoles
4.07   1,2,3-Thiadiazoles
4.08   1,2,4-Thiadiazoles
4.09   1,2,5-Thiadiazoles
4.10   1,3,4-Thiadiazoles
4.11   1,2-Oxa/thia-3-azoles

4.12   1,3-Oxa/thia-2-azoles
4.13   1,2-Oxa/thia-4-azoles
4.14   1,4-Oxa/thia-2-azoles
4.15   Five-membered Rings with Three Oxygen or Sulfur Atoms in 1,2,3-Positions
4.16   Five-membered Rings with Three Oxygen or Sulfur Atoms in 1,2,4-Positions
4.17   Tetrazoles
4.18   Oxatriazoles
4.19   1,2,3,4-Thiatriazoles
4.20   1,2,3,5-Thiatriazoles
4.21   Three or Four Heteroatoms including at least One Selenium or Tellurium
4.22   Three or Four Heteroatoms including at least One Phosphorus
4.23   Three or Four Heteroatoms including at least One Arsenic or Antimony
4.24   Three or Four Heteroatoms including at least One Other Element
4.25   Pentazoles

Volume 5   Six-membered Rings with One Heteroatom and Fused Carbocyclic Derivatives

5.01   Pyridines and their Benzo Derivatives: Structure
5.02   Pyridines and their Benzo Derivatives: Reactivity at the Ring
5.03   Pyridines and their Benzo Derivatives: Reactivity of Substituents
5.04   Pyridines and their Benzo Derivatives: Reactivity of Reduced Compounds
5.05   Pyridines and their Benzo Derivatives: Synthesis
5.06   Pyridines and their Benzo Derivatives: Applications
5.07   Pyrans and their Benzo Derivatives: Structure
5.08   Pyrans and their Benzo Derivatives: Synthesis
5.09   Pyrans and their Benzo Derivatives: Applications
5.10   Thiopyrans and their Benzo Derivatives
5.11   Six-membered Rings with One Selenium or Tellurium Atom
5.12   Six-membered Rings with One Phosphorus Atom
5.13   Six-membered Rings with One Arsenic, Antimony, or Bismuth Atom

Volume 6   Six-membered Rings with Two or More Heteroatoms and Fused Carbocyclic Derivatives

6.01   Pyridazines and their Benzo Derivatives
6.02   Pyrimidines and their Benzo Derivatives
6.03   Pyrazines and their Benzo Derivatives
6.04   1,2-Oxazines and their Benzo Derivatives
6.05   1,3-Oxazines and their Benzo Derivatives
6.06   1,2-Thiazines and their Benzo Derivatives
6.07   1,3-Thiazines and their Benzo Derivatives
6.08   1,3-Dioxins, Oxathiins, Dithiins, and their Benzo Derivatives
6.09   1,4-Dioxins, Oxathiins, Dithiins, and their Benzo Derivatives
6.10   1,2,3-Triazines and their Benzo Derivatives
6.11   1,2,4-Triazines and their Benzo Derivatives
6.12   1,3,5-Triazines and their Benzo Derivatives
6.13   1,2,3-Oxadiazines and 1,2,3-Thiadiazines
6.14   1,2,4-Oxadiazines and 1,2,4-Thiadiazines
6.15   1,2,5-Oxadiazines and 1,2,5-Thiadiazines
6.16   1,2,6-Oxadiazines and 1,2,6-Thiadiazines
6.17   1,3,4-Oxadiazines and 1,3,4-Thiadiazines
6.18   1,3,5-Oxadiazines and 1,3,5-Thiadiazines
6.19   Dioxazines, Oxathiazines, and Dithiazines
6.20   Six-membered Rings with 1,2,4-Oxygen or Sulfur Atoms
6.21   1,2,4,5-Tetrazines
6.22   Other Tetrazines and Pentazines
6.23   Other Six-membered Rings with Four or Five Nitrogen, Oxygen, or Sulfur

6.24 Six-membered Rings with Two or More Heteroatoms with at least One Selenium or Tellurium
6.25 Six-membered Rings with Two or More Heteroatoms with at least One Phosphorus
      Six-membered Rings with Two or More Heteroatoms with at least One Arsenic to Bismuth
6.27 Six-membered Rings with Two or More Heteroatoms with at least One Silicon to Lead
6.28 Six-membered Rings with Two or More Heteroatoms with at least One Boron

Volume 7   Fused Five- and Six-membered Rings without Ring Junction Heteroatoms

7.01 Bicyclic 5–5 Systems: Two Heteroatoms 1:1
7.02 Bicyclic 5–5 Systems: Three Heteroatoms 1:2
7.03 Bicyclic 5–5 Systems: Four Heteroatoms 1:3
7.04 Bicyclic 5–5 Systems: Four Heteroatoms 2:2
7.05 Bicyclic 5–5 Systems: Five Heteroatoms 2:3 and Six Heteroatoms 3:3
7.06 Bicyclic 5–6 Systems: Two Heteroatoms 1:1
7.07 Bicyclic 5–6 Systems: Three Heteroatoms 1:2
7.08 Bicyclic 5–6 Systems: Three Heteroatoms 2:1
7.09 Bicyclic 5–6 Systems: Four Heteroatoms 1:3
7.10 Bicyclic 5–6 Systems: Four Heteroatoms 3:1
7.11 Bicyclic 5–6 Systems: Purines
7.12 Bicyclic 5–6 Systems: Other Four Heteroatoms 2:2
7.13 Bicyclic 5–6 Systems: Five Heteroatoms 2:3 or 3:2
7.14 Bicyclic 5–6 Systems: Six and Seven Heteroatoms
7.15 Bicyclic 6–6 Systems: Two Heteroatoms 1:1
7.16 Bicyclic 6–6 Systems: Three Heteroatoms 1:2
7.17 Bicyclic 6–6 Systems: Four Heteroatoms 1:3
7.18 Bicyclic 6–6 Systems: Pteridines
7.19 Bicyclic 6–6 Systems: Other Four Heteroatoms 2:2
7.20 Bicyclic 6–6 Systems: Five or More Heteroatoms
7.21 Tricyclic Systems: Central Carbocyclic Ring with Fused Five-membered Rings
7.22 Tricyclic Systems: Central Carbocyclic Ring with Fused Five- and Six-membered Rings
7.23 Tricyclic Systems: Central Carbocyclic Ring with Fused Six-membered Rings

Volume 8   Fused Five- and Six-membered Rings with Ring Junction Heteroatoms

8.01 Bicyclic 5–5 Systems with One Ring Junction Nitrogen Atom: No Extra Heteroatom
8.02 Bicyclic 5–5 Systems with One Ring Junction Nitrogen Atom: One Extra Heteroatom 1:0
8.03 Bicyclic 5–5 Systems with One Ring Junction Nitrogen Atom: Two Extra Heteroatoms 2:0
8.04 Bicyclic 5–5 Systems with One Ring Junction Nitrogen Atom: Two Extra Heteroatoms 1:1
8.05 Bicyclic 5–5 Systems with One Ring Junction Nitrogen Atom: Three Extra Heteroatoms 2:1
8.06 Bicyclic 5–5 Systems with One Ring Junction Nitrogen Atom: Three Extra Heteroatoms 3:0
8.07 Bicyclic 5–5 Systems with One Ring Junction Nitrogen Atom: Four Extra Heteroatoms 2:2
8.08 Bicyclic 5–5 Systems with One Ring Junction Nitrogen Atom: Four Extra Heteroatoms 3:1
8.09 Bicyclic 5–6 Systems with One Ring Junction Nitrogen Atom: No Extra Heteroatom
8.10 Bicyclic 5–6 Systems with One Ring Junction Nitrogen Atom: One Extra Heteroatom 1:0
8.11 Bicyclic 5–6 Systems with One Ring Junction Nitrogen Atom: One Extra Heteroatom 0:1
8.12 Bicyclic 5–6 Systems with One Ring Junction Nitrogen Atom: Two Extra Heteroatoms 1:1
8.13 Bicyclic 5–6 Systems with One Ring Junction Nitrogen Atom: Two Extra Heteroatoms 2:0
8.14 Bicyclic 5–6 Systems with One Ring Junction Nitrogen Atom: Two Extra Heteroatoms 0:2
8.15 Bicyclic 5–6 Systems with One Ring Junction Nitrogen Atom: Three Extra Heteroatoms 3:0
8.16 Bicyclic 5–6 Systems with One Ring Junction Nitrogen Atom: Three Extra Heteroatoms 2:1
8.17 Bicyclic 5–6 Systems with One Ring Junction Nitrogen Atom: Three Extra Heteroatoms 1:2
8.18 Bicyclic 5–6 Systems with One Ring Junction Nitrogen Atom: Four Extra Heteroatoms 3:1
8.19 Bicyclic 5–6 Systems with One Ring Junction Nitrogen Atom: Four Extra Heteroatoms 2:2
8.20 Bicyclic 5–6 Systems with One Ring Junction Nitrogen Atom: Four Extra Heteroatoms 1:3
8.21 Bicyclic 5–6 Systems with One Ring Junction Nitrogen Atom: Five Extra Heteroatoms 3:2
8.22 Bicyclic 6–6 Systems with One Ring Junction Nitrogen Atom: No Extra Heteroatom
8.23 Bicyclic 6–6 Systems with One Ring Junction Nitrogen Atom: One Extra Heteroatom 1:0

8.24 Bicyclic 6–6 Systems with One Ring Junction Nitrogen Atom: Two Extra Heteroatoms 2:0
8.25 Bicyclic 6–6 Systems with One Ring Junction Nitrogen Atom: Two Extra Heteroatoms 1:1
8.26 Bicyclic 6–6 Systems with One Ring Junction Nitrogen Atom: Three Extra Heteroatoms 3:0
8.27 Bicyclic 6–6 Systems with One Ring Junction Nitrogen Atom: Three Extra Heteroatoms 2:1
8.28 Bicyclic 6–6 Systems with One Ring Junction Nitrogen Atom: Four Extra Heteroatoms 3:1
8.29 Bicyclic 6–6 Systems with One Ring Junction Nitrogen Atom: Four Extra Heteroatoms 2:2
8.30 Bicyclic 6–6 Systems with One Ring Junction Nitrogen Atom: Five Extra Heteroatoms 3:2
8.31 Bicyclic Systems with Two Ring Junction Nitrogen Atoms
8.32 Bicyclic Systems with Ring Junction Sulfur, Selenium, or Tellurium Atoms
8.33 Bicyclic Systems with Ring Junction Phosphorus, Arsenic, Antimony, or Bismuth Atoms
8.34 Bicyclic Systems with Ring Junction Boron Atoms
8.35 Three Heterocyclic Rings Fused (5:5:5)
8.36 Three Heterocyclic Rings Fused (5:6:5)
8.37 Three Heterocyclic Rings Fused (5:6:6)
8.38 Three Heterocyclic Rings Fused (6:5:6)
8.39 Three Heterocyclic Rings Fused (6:6:6)
8.40 Systems with a Spirocyclic Heteroatom
8.41 Compounds containing a Spiro Phosphorus Atom

Volume 9 Seven-membered and Larger Rings and Fused Derivatives

9.01 Azepines and their Fused-ring Derivatives
9.02 Oxepanes and Oxepines
9.03 Thiepanes and Thiepines
9.04 1,2-Diazepines
9.05 1,3-Diazepines
9.06 1,4-Diazepines
9.07 1,2-Oxazepines and 1,2-Thiazepines
9.08 1,3-Oxazepines and 1,3-Thiazepines
9.09 1,4-Oxazepines and 1,4-Thiazepines
9.10 1,2-Dioxepins, 1,2-Oxathiepines, and 1,2-Dithiepines
9.11 1,3-Dioxepanes, 1,3-Oxathiepanes, and 1,3-Dithiepanes
9.12 1,4-Dioxepanes, 1,4-Oxathiepanes, and 1,4-Dithiepanes
9.13 Seven-membered Rings with Three Heteroatoms 1,2,3
9.14 Seven-membered Rings with Three Heteroatoms 1,2,4
9.15 Seven-membered Rings with Three Heteroatoms 1,2,5
9.16 Seven-membered Rings with Three Heteroatoms 1,3,5
9.17 Seven-membered Rings with Four or More Heteroatoms
9.18 Eight-membered Rings with One Nitrogen Atom
9.19 Eight-membered Rings with One Oxygen Atom
9.20 Eight-membered Rings with One Sulfur Atom
9.21 Eight-membered Rings with Two Heteroatoms 1,2
9.22 Eight-membered Rings with Two Heteroatoms 1,3
9.23 Eight-membered Rings with Two Heteroatoms 1,4
9.24 Eight-membered Rings with Two Heteroatoms 1,5
9.25 Eight-membered Rings with Three Heteroatoms
9.26 Eight-membered Rings with Four or More Heteroatoms
9.27 Nine-membered Rings
9.28 Ten-membered Rings or Larger with One or More Nitrogen Atoms
9.29 Ten-membered Rings or Larger with One or More Oxygen Atoms
9.30 Ten-membered Rings or Larger with One or More Sulfur Atoms
9.31 Ten-membered Rings or Larger with One or More Nitrogen and Oxygen and/or Sulfur Atoms
9.32 Ten-membered Rings or Larger with One or More Oxygen and Sulfur Atoms
9.33 Rings containing Selenium or Tellurium
9.34 Rings containing Phosphorus
9.35 Rings containing Arsenic, Antimony, or Bismuth
9.36 Rings containing Silicon to Lead
9.37 Rings containing Boron
9.38 Rings containing Other Elements

Volume 10   Author and Ring Indexes

Author Index
Ring Index

Volume 11   Subject Index

Appendix C

Short Contents of "Comprehensive Heterocyclic Chemistry"

Volume 1

1.01 Introduction
1.02 Nomenclature of Heterocycles
1.03 Review Literature of Heterocycles
1.04 Biosynthesis of Some Heterocyclic Natural Products
1.05 Toxicity of Heterocycles
1.06 Application as Pharmaceuticals
1.07 Use as Agrochemicals
1.08 Use as Veterinary Products
1.09 Metabolism of Heterocycles
1.10 Importance of Heterocycles in Biochemical Pathways
1.11 Heterocyclic Polymers
1.12 Heterocyclic Dyes and Pigments
1.13 Organic Conductors
1.14 Uses in Photographic and Reprographic Techniques
1.15 Heterocyclic Compounds as Additives
1.16 Use in the Synthesis of Non-heterocycles
1.17 Heterocyclic Rings containing Phosphorus
1.18 Heterocyclic Rings containing Arsenic, Antimony or Bismuth
1.19 Heterocyclic Rings containing Halogens
1.20 Heterocyclic Rings containing Silicon, Germanium, Tin or Lead
1.21 Heterocyclic Rings containing Boron
1.22 Heterocyclic Rings containing a Transition Metal

Volume 2

2.01 Structure of Six-membered Rings
2.02 Reactivity of Six-membered Rings
2.03 Synthesis of Six-membered Rings
2.04 Pyridines and their Benzo Derivatives: (i) Structure
2.05 Pyridines and their Benzo Derivatives: (ii) Reactivity at Ring Atoms
2.06 Pyridines and their Benzo Derivatives: (iii) Reactivity of Substituents
2.07 Pyridines and their Benzo Derivatives: (iv) Reactivity of Non-aromatics
2.08 Pyridines and their Benzo Derivatives: (v) Synthesis
2.09 Pyridines and their Benzo Derivatives: (vi) Applications
2.10 The Quinolizinium Ion and Aza Analogues
2.11 Naphthyridines, Pyridoquinolines, Anthyridines and Similar Compounds

Volume 3

2.12 Pyridazines and their Benzo Derivatives
2.13 Pyrimidines and their Benzo Derivatives
2.14 Pyrazines and their Benzo Derivatives
2.15 Pyridodiazines and their Benzo Derivatives

2.16   Pteridines
2.17   Other Diazinodiazines
2.18   1,2,3-Triazines and their Benzo Derivatives
2.19   1,2,4-Triazines and their Benzo Derivatives
2.20   1,3,5-Triazines
2.21   Tetrazines and Pentazines and their Benzo Derivatives
2.22   Pyrans and Fused Pyrans: (i) Structure
2.23   Pyrans and Fused Pyrans: (ii) Reactivity
2.24   Pyrans and Fused Pyrans: (iii) Synthesis and Applications
2.25   Thiopyrans and Fused Thiopyrans
2.26   Six-membered Rings with More than One Oxygen or Sulfur Atom
2.27   Oxazines, Thiazines and their Benzo Derivatives
2.28   Polyoxa, Polythia and Polyaza Six-membered Ring Systems

Volume 4

3.01   Structure of Five-membered Rings with One Heteroatom
3.02   Reactivity of Five-membered Rings with One Heteroatom
3.03   Synthesis of Five-membered Rings with One Heteroatom
3.04   Pyrroles and their Benzo Derivatives: (i) Structure
3.05   Pyrroles and their Benzo Derivatives: (ii) Reactivity
3.06   Pyrroles and their Benzo Derivatives: (iii) Synthesis and Applications
3.07   Porphyrins, Corrins and Phthalocyanines
3.08   Pyrroles with Fused Six-membered Heterocyclic Rings: (i) *a*-Fused
3.09   Pyrroles with Fused Six-membered Heterocyclic Rings: (ii) *b*- and *c*-Fused
3.10   Furans and their Benzo Derivatives: (i) Structure
3.11   Furans and their Benzo Derivatives: (ii) Reactivity
3.12   Furans and their Benzo Derivatives: (iii) Synthesis and Applications
3.13   Thiophenes and their Benzo Derivatives: (i) Structure
3.14   Thiophenes and their Benzo Derivatives: (ii) Reactivity
3.15   Thiophenes and their Benzo Derivatives: (iii) Synthesis and Applications
3.16   Selenophenes, Tellurophenes and their Benzo Derivatives
3.17   Furans, Thiophenes and Selenophenes with Fused Six-membered Heterocyclic Rings
3.18   Two Fused Five-membered Rings each containing One Heteroatom

Volume 5

4.01   Structure of Five-membered Rings with Several Heteroatoms
4.02   Reactivity of Five-membered Rings with Several Heteroatoms
4.03   Synthesis of Five-membered Rings with Several Heteroatoms
4.04   Pyrazoles and their Benzo Derivatives
4.05   Pyrazoles with Fused Six-membered Heterocyclic Rings
4.06   Imidazoles and their Benzo Derivatives: (i) Structure
4.07   Imidazoles and their Benzo Derivatives: (ii) Reactivity
4.08   Imidazoles and their Benzo Derivatives: (iii) Synthesis and Applications
4.09   Purines
4.10   Other Imidazoles with Fused Six-membered Rings
4.11   1,2,3-Triazoles and their Benzo Derivatives
4.12   1,2,4-Triazoles
4.13   Tetrazoles
4.14   Pentazoles
4.15   Triazoles and Tetrazoles with Fused Six-membered Rings

Volume 6

4.16   Isoxazoles and their Benzo Derivatives
4.17   Isothiazoles and their Benzo Derivatives

4.18   Oxazoles and their Benzo Derivatives
4.19   Thiazoles and their Benzo Derivatives
4.20   Five-membered Selenium-Nitrogen Heterocycles
4.21   1,2,3- and 1,2,4-Oxadiazoles
4.22   1,2,5-Oxadiazoles and their Benzo Derivatives
4.23   1,3,4-Oxadiazoles
4.24   1,2,3-Thiadiazoles and their Benzo Derivatives
4.25   1,2,4-Thiadiazoles
4.26   1,2,5-Thiadiazoles and their Benzo Derivatives
4.27   1,3,4-Thiadiazoles
4.28   Oxatriazoles and Thiatriazoles
4.29   Five-membered Rings (One Oxygen or Sulfur and at least One Nitrogen Atom) Fused with Six-membered Rings (at least One Nitrogen Atom)
4.30   Dioxoles and Oxathioles
4.31   1,2-Dithioles
4.32   1,3-Dithioles
4.33   Five-membered Monocyclic Rings containing Three Oxygen or Sulfur Atoms
4.34   Dioxazoles, Oxathiazoles and Dithiazoles
4.35   Five-membered Rings containing One Selenium or Tellurium Atom and One Other Group VIB Atom and their Benzo Derivatives
4.36   Two Fused Five-membered Heterocyclic Rings: (i) Classical Systems
4.37   Two Fused Five-membered Heterocyclic Rings: (ii) Non-classical Systems
4.38   Two Fused Five-membered Heterocyclic Rings: (iii) $1,6,6a\lambda^4$-Trithiapentalenes and Related Systems

Volume 7

5.01   Structure of Small and Large Rings
5.02   Reactivity of Small and Large Rings
5.03   Synthesis of Small and Large Rings
5.04   Aziridines, Azirines and Fused-ring Derivatives
5.05   Oxiranes and Oxirenes
5.06   Thiiranes and Thiirenes
5.07   Fused-ring Oxiranes, Oxirenes, Thiiranes and Thiirenes
5.08   Three-membered Rings with Two Heteroatoms and Fused-ring Derivatives
5.09   Azetidines, Azetines and Azetes
5.10   Cephalosporins
5.11   Penicillins
5.12   Other Fused-ring Azetidines, Azetines and Azetes
5.13   Oxetanes, Oxetes and Fused-ring Derivatives
5.14   Thietanes, Thietes and Fused-ring Derivatives
5.15   Four-membered Rings with Two or More Heteroatoms and Fused-ring Derivatives
5.16   Azepines
5.17   Oxepanes, Oxepins, Thiepanes and Thiepins
5.18   Seven-membered Rings with Two or More Heteroatoms
5.19   Eight-membered Rings
5.20   Larger Rings except Crown Ethers and Heterophanes
5.21   Crown Ethers and Cryptands
5.22   Heterophanes

Volume 8

Data Index
Author Index
Subject Index
Ring Index

Appendix D
Miscellaneous (MI) References

⟨40MI20100⟩   N. M. Cullinane and W. T. Rees; *Trans. Faraday Soc.*, 1940, **36**, 507.

⟨40MI22000⟩   P. McClellan; *Ind. Eng. Chem.*, 1940, **32**, 1181.

⟨48MI30200⟩   J. Lecocq; *Ann. Chim. (Paris)*, 1948, **3**, 62.

⟨51MI40100⟩   C. C. J. Roothaan; *Rev. Mod. Phys.*, 1951, **23**, 69.

⟨51MI40101⟩   G. G. Hall; *Proc. R. Soc. London, Ser. A*, 1951, **205**, 541.

⟨56MI30300⟩   F. García-González; *Adv. Carbohydrate Chem.*, 1956, **11**, 97.

⟨57MI40100⟩   H. A. Staab, W.Otting and A. Veberle; *Z. Electrochem.*, 1957, **61**, 1000.

⟨57MI40101⟩   D. F. Othmer, P. W. Maurer, C. J. Molinary and R. C. Kowalski; *Ind. Eng. Chem.*, 1957, **49**, 125.

⟨58MI40200⟩   A. K. Majumdar, M. M. Chakrabatty; *Anal. Chim. Acta*, 1958, **19**, 372.

⟨59MI40100⟩   T. Balaban; *Stud. Cercet. Chim.*, 1959, **7**, 257.

⟨59MI40200⟩   R. Schindler, H. Will and L. Holleck; *Z. Electrochem.*, 1959, **63**, 596.

⟨64MI40100⟩   G. Pouzard, L. Pujol, J. Roggero and E. J. Vincent; *J. Chim. Phys. Phys. Chim. Biol*, 1964, 613.

⟨64UP30100⟩   C. W. Bird; unpublished observations, 1964.

⟨65MI40100⟩   F. Eloy; *Fortschr. Chem. Forsch.*, 1965, **4**, 807.

⟨B-67MI20100⟩   H. Budzikiewicz, C. Djerassi and D. H. Williams; 'Mass Spectrometry of Organic Compounds', Holden-Day, San Francisco, 1967.

⟨67MI21900⟩   L. Pallos and P. Benko; *Ind. Chim. Belg.*, 1967, **32**, 1334.

⟨B-68MI40900⟩   A. Albert, 'Heterocyclic Chemistry', Athlone Press, University of London, 1968.

⟨70MI20100⟩   V. M. S. Gil and A. J. L. Pinto; *Mol. Phys.*, 1970, **19**, 573.

⟨70MI30100⟩   N. M. Pozdeev, L. N. Gunderova and A. A. Shapkin; *Opt. Spektrosk.*, 1970, **28**, 254.

⟨B-70MI40100⟩   J. A. Pople and D. L. Beveridge; 'Approximate Molecular Orbital Theory', McGraw-Hill, New York, 1970, pp. 57–84.

⟨B-71MI20100⟩   J. F. Stoddart; 'Stereochemistry of Carbohydrates', Wiley-Interscience, New York, 1971, pp.72–92.

⟨B-71MI30100⟩   H.-I. Dauben, J. D. Wilson and J. L. Laity; in 'Non-benzenoid Aromatics', ed. J. P. Snyder; Academic Press, New York, 1971, vol. 2, p.167.

⟨B-72MI103-01⟩   J. C. Buchanan and H. Z. Gable; in 'Selective Organic Transformations', ed. B. S. Thygarajan, Wiley, New York, 1972, vol. 1, p.1.

⟨B-73MI20100⟩   'Nitrogen NMR', ed. M. Witanowski and G. A. Webb; Plenum Press, London, 1973.

⟨B-73MI40100⟩   M-Witanowski, L. Stefaniak and H. Januszewski; in 'Nitrogen NMR', ed. M Witanowski and G. A. Webb, Plenum, New York, 1973, p.163.

⟨B-73MI50300⟩   H. C. van der Plas; in 'Ring Transformations of Heterocycles', Academic Press, New York, 1973.

⟨B-74MI30200⟩   A. I. Meyers; 'Heterocycles in Organic Synthesis', Wiley, New York, 1974, p.243.

⟨74MI40100⟩   W. A. Lathan, L. A. Curtiss, W. J. Hehre, J. B. Lisle and J. A. Pople; *Prog. Phys. Org. Chem.*, 1974, **11**, 175.

⟨74MI50100⟩   R. Moriarty; *Top. Stereochem.*, 1974, 271.

⟨B-75MI50100⟩   L. J. Bellamy; 'The Infrared Spectra of Complex Molecules', Chapman & Hall, London, 1975.

⟨B-75MI50101⟩   L. J. Bellamy; 'Advances in IR Group Frequencies', Chapman & Hall, London, 1975.

⟨75MI51803⟩   R. Glinka and B. Kotelko; *Acta Pol. Pharm.*, 1975, **32**, 525 (*Chem. Abstr.*, 1976, **85**, 32 970).

⟨B-76MI40200⟩   K. Schofield, M. R. Grimmett and B. R. T. Keene; 'The Azoles', Cambridge University Press, London, 1976, p.60.

⟨B-76MI40201⟩ K. Schofield, M. R. Grimmett and B. R. T. Keene; 'The Azoles', Cambridge University Press, London, 1976, p.281.

⟨76MI40300⟩ G. Werber, F. Buccheri, N. Vivona and M. Gentile; *Chim. Ind. (Milan),* 1976, **58**, 382.

⟨B-76MI50200⟩ R. M. Acheson; 'Introduction to the Chemistry of Heterocyclic Compounds', Wiley, New York, 1976.

⟨B-77MI101-01⟩ W. L. F. Amarego; in 'Stereochemistry of Heterocyclic Compounds, Part 1', Wiley, New York, 1977, p.11.

⟨B-77MI202-01⟩ R. A. Jones and G. P. Bean; 'The Chemistry of Pyrroles', Academic Press, London, 1977.

⟨B-77MI30100⟩ R. A. Jones and G. P. Bean; 'The Chemistry of Pyrroles', Academic Press, London, 1977, p.463.

⟨B-77MI40100⟩ K. F. Freed; in 'Modern Theoretical Chemistry. Semiempirical Methods of Electronic Structure Calculation. Part A: Techniques', ed. G. A. Segal, Plenum, New York, 1977, pp. 201–253.

⟨78MI21400⟩ T. Akiyama, Y. Enomoto and T. Shibamoto; *J. Agric. Food Chem.*, 1978, **26**, 1176.

⟨B-78MI22701⟩ M. Sainsbury; in 'The Chemistry of Carbon Compounds', ed. S. Coffey; Elsevier, Amsterdam, 2nd edn., 1978, vol. IV, p.427.

⟨78MI30100⟩ S. H. Gerson, S. D. Worley, N. Bodor, J. J. Kamiuski and T. W. Flechtner; *J. Electron Spectrosc. Relat. Phenom.*, 1978, **13**, 421.

⟨B-78MI30102⟩ R. J. Abraham and P. Loftus; 'Proton and Carbon-13 NMR Spectroscopy', Heiden, London, 1978.

⟨B-79MI20101⟩ E. L. Eliel and K. M. Pietrusiewicz; in 'Topics in Carbon-13 NMR Spectroscopy', ed. G. C. Levy; Wiley, New York, 1979, vol. 3, pp. 171–282.

⟨B-79MI20102⟩ G. C. Levy and R. L. Lichter; 'Nitrogen-15 Nuclear Magnetic Resonance Spectroscopy', Wiley, New York, 1979.

⟨79MI30101⟩ J. Fabian; *Z. Phys. Chem. (Leipzig)*, 1979, **260**, 81.

⟨79MI30102⟩ A. Kh. Mamleev and N. M. Pozdeev; *Zh. Strukt. Khim.*, 1979, **20**, 1114.

⟨79MI40100⟩ M. Baudet and M. Gelbcke; *Anal. Lett.*, 1979, **12**, 641.

⟨B-79MI40300⟩ Ramsden; in 'Comprehensive Organic Chemistry', eds. D. H. R. Barton and W. D. Ollis, Pergamon Press, Oxford, 1979, vol. 4, p.1207.

⟨B-79MI50100⟩ C. W. Rees; in 'New Trends in Heterocyclic Chemistry', ed. R. B. Mitra, N. R. Ayyangar, V. N. Gogte, R. M. Acheson and N. Cromwell; Elsevier, Amsterdam, 1979, p.365–372.

⟨79MI50102⟩ V. F. Kalasinsky, E. Block, D. E. Powers and W. C. Harris; *Appl. Spectrosc.*, 1979, **33**, 361.

⟨80MI40101⟩ J. B. Collins and A.J.Streitwieser, Jr.; *Comput. Chem.*, 1980, **1**, 81.

⟨81MI40100⟩ J. A. Webb, M. Witanowski and L. Stefaniak; *Annu. Rep. NMR Spectrosc.*, 1981, 118.

⟨B-82MI203-01⟩ B. Robinson; 'The Fisher Indole Synthesis', Wiley, New York, 1982.

⟨82MI413-03⟩ I. Shibuya; *Kagiken Nyusu Kagaku Kogyo Shiryo*, 1982, **17**, 44.

⟨B-82MI505-02⟩ A. T. Balaban *et al*; 'Pyrylium Salts: Synthesis, Reactions and Physical Properties', Academic Press, New York, 1982.

⟨B-83MI101-01⟩ J. A. Deyrup; in 'The Chemistry of Heterocyclic Compounds', ed. A. Hassner, Wiley, New York, 1983, vol. 42, part 1, p.1.

⟨B-83MI101-02⟩ V. Nair; in 'The Chemistry of Heterocyclic Compounds', ed. A. Hassner, Wiley, New York, 1983, vol. 42, part 1, p.215.

⟨B-83MI101-03⟩ H. W. Moore and D. M. Goldish; in 'The Chemistry of Functional Groups', eds. S. Patai and Z. Rappoport, Wiley, New York, 1983, suppl. D, p.321.

⟨B-83MI103-04⟩ K. B. Sharpless; in 'Proceeding of the R. A. Welch Foundation Conference on Chemical Research, XXVII', Houston, TX, 1983, p.59.

⟨B-83MI135-02⟩ V. Zoller; in 'Small Ring Hetrocycles', ed. A. Hassner, Wiley, New York, 1983, vol. 42, chap. 11, part 1, p.596.

⟨83MI718-04⟩ B. Andondonskaja-Renz and H. J. Zeitler; *Anal. Biochem.*, 1983, **133**, 68.

⟨83MI718-09⟩ C. E. Lunte and P. T. Kissinger; *Anal. Biochem.*, 1983, **133**, 377.

⟨B-83MI20100⟩ A. J. Kirby; 'The Anomeric Effect and Related Stereoelectronic Effects at Oxygen', Springer-Verlag, Berlin, 1983.

⟨84MI718-06⟩ C. E. Lunte and P. T. Kissinger; *Anal. Chim. Acta*, 1984, **158**, 33.

⟨B-84MI202-01⟩ B. A. Trofimov and A. I. Mihaleva; 'N-Vinylpyrroles', Nauka, Novosibirsk, 1984.

⟨85MI103-01⟩ M. E. Borredon, M. Delmas and A. Gaset; *Informations Chimie*, 1985, 129.

⟨B-85MI103-03⟩ B. E. Rossiter; in 'Asymmetric Synthesis', ed. J. D. Morrison, Academic Press, New York, 1985, vol. 5, p.193.

⟨B-85MI103-04⟩ M. G. Finn and K. B. Sharpless; in 'Asymmetric Synthesis', ed. J. D. Morrison, Academic Press, New York, 1985, vol. 5, p.247.

⟨B-85MI112-01⟩ M. J. Haddadin and J. P. Freeman; in 'The Chemistry of Heterocyclic Compounds: Small Ring Heterocycles', ed. A. Hassner, Wiley, New York, 1985, vol. 42, p.283.

⟨85MIP155606⟩ K. G. Dave; *Indian Pat. IN 155 606* (1985) (*Chem. Abstr.*, 1986, **105**, 208 912t).

⟨86MI302-02⟩ M. Ogata; in 'Annual Report of Shionogi Res. Laboratories', 1986, N 36, p.1.

⟨86MI927-01⟩ O. Ouamerali and J. Gayoso; *Int. J. Quant. Chem.*, 1986, **29**, 1599.

⟨87MI103-04⟩ T. Katsuki; *J. Synth. Org. Chim., Jpn.,* 1987, **45**, 90.

⟨87MI301-01⟩ J. Elguero, R. Faure and J. Llinares; *Spectros. Lett.*, 1987, **20**, 149.

⟨B-87MI505-01⟩ D. J. Boger and S. M. Weinreb; 'Hetero-Diels-Alder Methodology in Organic Synthesis (Organic Chemistry Series, vol. 47)', Academic Press, London, 1987.

⟨B-88MI101-02⟩ W. H. Pearson; in 'Studies in Natural Products Chemistry', ed. Rahman, Elsevier, Amsterdam, 1988, vol. 1, p.323.

⟨88MI118-03⟩ S. Kim and T. A. Lee; *Bull. Korean Chem. Soc.*, 1988, **9**, 189 (*Chem. Abstr.*, 1988, **109**, 170 150).

⟨89MI133-02⟩ W. Adam, B. Epe, D. Schiffmann, F. Vargas and D. Wild; *Free Rad. Res. Commun.*, 1989, **5**, 253.

⟨89MI502-01⟩ J. S. Cha, M. S. Yoon, K. W. Lee and J. C. Lee; *Bull. Korean Chem. Soc.*, 1989, **10**, 75.

⟨B-89MI718-02⟩ W. Pfleiderer, J. Rehse and J. Schnabel; in 'Pteridines and Biogenic Amines in Neuropsychiatry, Pediatrics, and Immunology', eds. R. A. Levine, S. Milstein, D. M. Kuhn and H. C. Curtius, Lakeshore, Grosse Point, 1989, p.71.

⟨B-90MI201-02⟩ M. P. Sammes; in 'Chemistry of Heterocyclic Compounds: Pyrroles Part 1', Wiley-Interscience, New York, 1990, vol. 48, p.549.

⟨90MI206-01⟩ S. B. Ginderich and P. W. Jennings; *Adv. Oxygenated Processes,* 1990, **2**, 117.

⟨90MI207-01⟩ J. A. Ciller, N. Martin, S. Seoane and J. L. Soto; *Trends Heterocycl. Chem.*, 1990, **1**, 19 (*Chem. Abstr.*, 1994, **120**, 269 954).

⟨90MI302-01⟩ S. Kim and S. Lee; *Bull. Korean Chem. Soc.*, 1990, **11**, 544 (*Chem. Abstr.*, 1991, **114**, 164109).

⟨B-91MI103-01⟩ D. A. Boykin and A. L. Baumstark; in 'Applications of Oxygen-17 NMR Spectroscopy to Structural Problems in Rigid Planar Organic Molecules', 17O NMR Spectroscop. Org. Chem., CRC Press, New York, 1991, p.154.

⟨91MI118-02⟩ B. Chung, C. Paik and C. Nah; *Bull. Korean Chem. Soc.*, 1991, **12**, 456.

⟨91MI118-03⟩ M. Regitz; *Nachrichten aus chemie*, 1991, **39**, 9.

⟨91MI211-02⟩ J. Wang and F. Fan; *Gaodeng Xuexiao Huaxue Xuebao*, 1991, **12**, 1200 (*Chem. Abstr.*, 1992, **116**, 151 465v).

⟨91MI302-01⟩ S. Takasu, H. Takai, T. Takagi and H. Fujiwara; *Anal. Sci.*, 1991, **7** (Suppl. Part 1), 845 (*Chem. Abstr.*, 1992, **116**, 255537).

⟨91MI302-04⟩ M. Ehsan; *Sci. Int. (Lahore)*, 1991, **3**, 217 (*Chem. Abstr.*, 1992, **116**, 106181).

⟨91MI502-01⟩ M. J. Silvester; *Aldrichimica Acta*, 1991, **24**, 31.

⟨B-91MI906-01⟩ R. I. Figer and A. Walser; in 'The Chemistry of Heterocyclic Compounds', ed. E. C. Taylor, Interscience, New York, 1991, vol. 50, p.1.

⟨B-92MI101-01⟩ W. B. Jennigs and D. R. Boyd; in 'Cyclic Organonitrogen Stereodynamics', eds. J. B. Lambert and Y. Takeuchi, VCH Publishers, New York, 1992, p.105.

⟨92MI118-02⟩ N. Verbruggen, M. Van Montagu and E. Messen; *FEBS Letters,* 1992, **308**, 261.

⟨92MI206-04⟩ P. S. Chen and C. H. Chou; *J. Chin. Chim. Soc. (Taipei),* 1992, **39**, 251.

⟨92MI301-01⟩ A. D. Dzhuraev, K. M. Karimkulov, A. G. Makhsumov and N. Amanov; *Fiziol. Akt. Veshch.*, 1992, **24**, 71.

⟨92MI417-04⟩ Yu. V. Grigor'ev and P. N. Gaponik; *Vesti Akad. Navuk Belarusi, Ser. Khim. Navuk,* 1992, 73 (*Chem. Abstr.*, 1992, **116**, 255 555).

⟨92MI502-01⟩ Y. C. Kim, J. K. Kyong, S. K. Kim and D. J. Koo; *J. Korean Chem. Soc.*, 1992, **36**, 180 (*Chem. Abstr.*, 1992, **116**, 234833).

⟨92MI502-03⟩ N. M. Yoon, Y. S. Shon and J. H. Ahn; *Bull. Korean Chem. Soc.*, 1992, **13**, 199.

⟨92MI718-05⟩ R. Klein; *Anal. Biochem.*, 1992, **203**, 134.

⟨92MI718-11⟩ S. D. Wong and J. E. Gready; *Pteridines*, 1992, **3**, 115.

⟨93MI101-02⟩ L. Fourie, K. J. Van der Merwe, P. Swart and S. S. De Kock; *Anal. Chim. Acta*, 1993, **279**, 163.

⟨93MI118-05⟩ Y. H. Lee, C. H. Lee, J. H. Lee and W. S. Choi; *Bull. Korean Chem. Soc.*, 1993, **14**,

415 (*Chem. Abstr.*, 1993, **119**, 270 852).

⟨B-93MI718-06⟩ R. Klein; in 'Chemistry and Biology of Pteridines and Folates', eds. J. E. Ayling, M. G. Nair and C. M. Baugh, Plenum Press, New York, 1993, p.43.

⟨93MI718-08⟩ R. Klein and C. A. Groliere; *Chromatography*, 1993, **36**, 71.

⟨B-94MI103-01⟩ M. B. Smith; 'Organic Synthesis', McGraw-Hill, New York, 1994, p.275.

⟨B-94MI502-06⟩ O. N. Chupakhin, V. N. Charushin and H. C. van der Plas; 'Nucleophilic Aromatic Substitution of Hydrogen', Academic Press, San Diego, 1994.

⟨B-94MI602-01⟩ D. J. Brown, R. F. Evans, W. B. Cowden and M. D. Fenn; 'The Pyrimidines', Wiley, New York, 1994.

⟨94MI718-02⟩ R. Klein, I. Tatischeff, G. Tham and N. Mano; *Chirality*, 1994, **6**, 564.

⟨B-97MI502-07⟩ A. F. Pozharskii, A. T. Soldatenkov and A. R. Katritzky; 'Heterocycles in Life and Society', Wiley, Chichester, 1997, p.63.

Subject Index

As described in the introductory material, the arrangement of this Handbook is such that the physical properties and reactions of particular types of compounds are very systematically treated, and information on them can usually be rapidly assessed by consultation of the detailed contents under the 'Structure' and 'Reactivity' sections, respectively. However, it is not as simple to access the *preparation* of particular heterocyclic compounds for reasons which are discussed in the 'Synthesis' section. A major aim of this index is to provide such information; entries referring to the synthesis of a given compound are therefore denoted with an asterisk.

It has not been possible, nor was it desirable, to index a compound or reaction at every mention in the text. Instead the essential features of compounds have been chosen, and individual compounds indexed specifically only where they are discussed in some detail.

Acridine, 662*
    aromaticity, 46
    as π-acceptor, 24
    basicity, 178
    Bernthsen synthesis of, 662*
    Chichibabin amination, 204
    9-chloro-*, 662
    π-donor character, 24
    10-methyl-9-R-9,10-dihydro-, aromatization, 241
    nitration, 252
    nucleophilic hydroxylation, 198
    oxidation, 194
    10-oxides, 9-chloro-, UV irradiation, 665
    10-oxides, 9-cyano-, UV irradiation, 665
Acridinium ion, N-methyl-, reaction with sodamide, 206
Acridizinium ions
    cycloaddition, 228
    [4 + 4]photocycloaddition, 236
Acridones, 199*, 444*, 662*
    from 3-arylanthranils, 445
Acyl
    migration, 134
    radical, reaction with heterocycles, 225
Acylation, 192, 310
    5-membered rings, one heteroatom, 310–312
        two heteroatoms, 392
    6-membered rings, 192
    small and large rings, 483–484
Adenine, 1-ethyl-, Dimroth rearrangement, 268
Alizarin Green G, 664
Alkali metal ions, crown ether/cryptand selectivity, 147
Alkoxy groups, nucleophilic displacement of, 276
Alloxan, 276
Alloxanic acid, 276
Alloxazine, 25
    crystal structure, 25
Anhydro bases, 259, 442
    reaction with cationic rings, 215
Annular
    elementotropy, 134
    metallotropy, 108
    prototropy, 133, 162
    tautomerism, 85, 132, 163
ANRORC reaction, 284, 579, 601, 646
Anthranils*, 629
    3-acyl-, rearrangement, 447
    3-aryl-, thermolysis, 444, 662
        to acridones, 444
    cleavage by nitrous acid, 401
    cycloaddition reaction, 622
    halogenation and nitration, 437
    6-nitro-, reaction with amines, 401
    photolysis, 439
    ring expansion, 622
    ring opening, 413
Antiaromaticity, 157
Aromaticity, 43–44, 46, 126, 128, 157
    5-membered rings, one heteroatom, 79, 80
        two heteroatoms, 125–128
    6-membered rings, 43
    of 1,2-δ³-diazetine, 158
    small and large rings, 157–159
    structural indices, 45, 126
Aromatization, 241
    of dihydrocompounds, 427

of 10-methyl-9-R-9,10-dihydroacridines, 241
Arsabenzene, 47
    deshielding of α-protons, 47
1-Aza-1,2,4,6-cycloheptatetraene*, 549
Aza-Darzens reaction, 517
Azaphilic addition, 220
Azaprismane, 493
Azathiaphenanthrene, thermal rearrangement, 296
Azepine
    annular prototropy, 162
    cycloaddition reactions, 492
    transition metal complexes, 495
1H-Azepine*, 549
    calculations, 147
    ¹H NMR, 151
1-R-Azepines, non-aromatic, 158
3H-Azepine*, 549
    3-methyl-, ring inversion barriers, 160
Azetes*, 540
    cycloaddition reactions, 492, 578
    stabilized derivatives, 158
    valence bond tautomerism, 161
Azetidine*, 518
    fused*, 606
    N-inversion, 159
    ¹⁵N NMR spectra, 152
    reactions at nitrogen, 484
2-Azetidinones, 515*, 519, 523*
    nucleophilic attack, 487
Azetines*, 525
    alkyl-*, 528
    aryl-*, 528
Azine-carboxylic acids, 263
    equilibrium with betaines, 262
    reaction with bifunctional C-nucleophiles, 214
Azines, 23, 38, 46, 51
    acid-catalyzed hydrogen exchange, 188
    alkoxy, reactions, 276
    alkyl derivatives, proton loss, 256
    N-amination, 184
    amines, nucleophilic displacement of amino group, 268
    amino derivatives, 269
        proton loss, 267
    as π-acceptors and π-donors, 24
    as σ- and π-ligands, 179
    atomic π-charges, 22
    α-azido, equilibrium with fused tetrazoles, 52
    azido derivatives, reactions, 271
    base-catalyzed hydrogen exchange, 221
    basicity, 177–178
    bond lengths, 24
    carboxylic acids, Hammick reaction, 263
    catalytic hydrogenation, 226
    Chichibabin amination, 204
    cross-coupling reactions, 286
    cycloaddition reactions, 227–235
    π-deficiency, 23
    diazonium salts, 266
    dihydro-, antiaromatic anions from, 244
        aromatization, 239
        disproportionation, 240
    gas–liquid chromatography, 43
    halogen, 282, 284, 286
        halogen–metal exchange, 280
        reactions via S_{AE} mechanism, 284

reactions via S_RN1 pathway, 283
halogenation, 190
hydrazino derivatives, 270
hydroxy, structure and reactions, 272
internal bond angles, 24
IR spectra, 37
meta-bridging cyclization, 214
nitration, 186
nitro derivatives, nucleophilic displacement of nitro
  group, 269
  reduction to amines, 270
nitroso derivatives, 272
^{13}C NMR, 29
^{1}H NMR, 27
^{14}N and ^{15}N NMR, 34
nucleophilic attack at ring nitrogen, 220
organometallic, metal–metal exchange, 287
oxidative amination and alkylamination, 205
N-oxides, nitration, 187
  nucleophilic amination, 206
  reaction with dipolarophiles, 233
    with Grignard reagents, 212
  reactivity of, 293
  reduction, 220
  metallation, 221
reaction with carbenes and nitrenes, 223
  with free radicals, 223
  with Lewis acids, 184
  with metals, 225
reduction, 218, 226
ring metallation, 220
substituent in fused benzene ring, 254
substituent reactivity, 252
N-substituents, general reactivity, 288
Ullmann reaction, 282
UV absorption bands, 35
Azinethiones, 275
  acidity, 179
  reactions, 278
Azinium ions
  N-amino-, cyclization to pyrazolo-fused systems, 670
  hydrogen exchange, 222
  reaction with N-nucleophiles, 206
  stabilized ylides, 290
Azinones
  acidity, 179
  anions, reaction with electrophiles, 222
  carbonyl frequencies, 38
  hydrogen exchange, 221
  mass spectra, 40
  nitration, 187
  nucleophilic displacement of carbonyl oxygen, 274
Aziridination reactions, 512
Aziridines*, 512, 516, 524
  electrophilic attack at N, 483
  from nitrenes, 512
  from δ^2-1,2,3-triazolines, 429
  Gabriel synthesis of, 517*
  geometry, 148
  N-inversion, 159
  nitrogen pyramidal inversion, 141
  NMR data, 150
  ^{15}N NMR spectra, 152
  nucleophilic attack, 486
  ring enlargement, 574, 584

ring opening, 669
2-vinyl-, ring expansion with phenyl isothiocyanate,
  581
Azirine, 528
  cycloaddition reactions, 492, 602, 667
  dimerization, 495
  nucleophilic attack on, 486
  ring enlargement, 584
1-Azirine*, 527
  cycloaddition to 1,2,4,5-tetrazines, 598
  NMR data, 150
  2-phenyl-, ring enlargement, 574
1H-Azirine, antiaromaticity, 157
2-Azirine, ab-initio calculations, 146
2H-Azirines, ring opening, 528
Azlactones*, 395, 573
  reaction with Grignard reagents, 443
  ring opening, 401
Azocine, structure, 160
1H-Azocines, anion, ^{1}H NMR, 159
Azoles, 103, 415
  N-acetyl, carbonyl frequencies, 120
  acidity, 379
  N-acyl-, reactions, 468
  acylation, 385
  aldehydes, properties, 446
  alkoxy-, rearrangement, 458
  alkyl-, deprotonation, 441
  N-alkyl-, reactions at N-alkyl group, 466
  N-alkylation, 383
  alkylthio, reactions, 460
  N-amination, 387
  amino-, N-anions, 452
  N-amino-, cyclization to pyrazolo-fused systems, 670
    reactions, 469
  N-anions, ^{1}H NMR, 105
    oxidation, 396
  aromaticity, 125
  C-aryl-, conformation, 131
  N-aryl-, reactions, 465
  N-arylation, 384
  3-atom side chain, rearrangement of, 437
  azido-, fragmentation, 454
  basicity, 378
  benzoannulation, influence on aromaticity of, 127
  carbenes, 128, 448
  carboxylic acids, properties, 445
  catalytic hydrogenation, 417
  chloro-, reactions, 461
  complexes with metal ions, 380
  cross-coupling reactions, 463
  cycloaddition reactions, 418–425
  Dimroth-type rearrangements of derivatives involving
    3-atom side chains, 436
  dipole moments, 103
  π-electron distribution, 95
  germilenes, 128
  halo-, reactions, 460
  N-halo-, 386
    reactions, 461, 472
  halogenation, 391
  homolytic alkylation, 415
    phenylation, 415
  hydroxy-, derivatives, 454
    reactions, 456

tautomerism, 454
N-hydroxy-, reactions, 472
ionization potentials, 96
IR spectroscopy, 117
ketones, properties, 447
lithiation, 409
mass spectrometry, 121
melting and boiling points, 122
mercapto-, tautomerism, 458
mesoionic compounds, isomerism, 375
metallated, 463
molecular geometry, 99
Na$^+$, K^{+M} and Al$^+$ complexes, 98
N-nitro-, reactions, 471
^{13}C NMR spectra, 108
^{1}H NMR spectra, 104
^{15}N NMR, tautomerism, 113
orientation of electrophilic attack, 376
N-oxides, reactions, 472
photoelectron spectroscopy, 120
quaternization, 381
reaction with Lewis acids, 388
    with organometallics, 404
    with peracids, 386
silylenes, 128
solubility, 124
substituent introduction and modification, 551
sulfur-containing, oxidation, 387
    ^{1}H NMR spectral data, 105
thiation, 415
thiones, oxidation, 459
Ullmann reaction, 384
UV spectroscopy, 115
1-vinyl-, electrochemical reduction, 416
Azolesulfonic acids, 460
Azolides, synthetic utility of, 468
Azolidines
    ^{13}C NMR, 112
    ^{1}H NMR, 110
Azolidinones, halogenation, 392
Azolines
    ^{13}C NMR, 112
    ^{1}H NMR, 109
Azolinethiones, reactions, 458
Azolinones
    carbonyl frequencies, 120
    ^{13}C NMR, 111
    ^{1}H NMR, 106
    reaction with aldehydes and ketones, 395
2-Azolinones, reactions with electrophiles, 455
Azolium ions
    C-alkyl-, reaction at alkyl group, 442
    π-deficiency, 95
    halo-, reactions, 462
    ^{1}H NMR, 106
    reaction with carbanions, 404
        with hydroxide, 399
        with N-nucleophiles, 402
Azolone anions, alkylation, 384
Azolyl groups, σ constants, 465
Azonine, 163
1*H*-Azonine, calculation of, 159

Baeyer–Villiger oxidation, 522
Barbituric acid, derivatives, hypnotics, 16

Base-catalysed H–D exchange, 221, 410
Beckmann rearrangement, 522
Benzazepine*, 622, 627
Benzazepinones*, 622
Benzazete*, 606
    Diels–Alder addition, 623
    ring expansion, 623
    2-phenyl-
        as a dienophile, 231
        cycloaddition–ring expansion, 655
Benzazines, electrophilic substitution, 252
Benzazocine, pentaphenyl-, 623
Benzazoles, rearrangement, 439
    aromaticity, 127
    UV spectroscopy, 115
Benzene oxide, calculations, 147
Benzimidazo[2,1-*a*]isoquinoline, 680
Benzimidazole, 575*, 632
    acylation, 411
    1-alkyl-2-amino-, diazotization, 452
    2-amino-, oxidation, 453
    1-amino-2-azido-, thermolysis, 471
    1,1′-azo-, 470
    2,2′-azo-*, 453
    carbene insertion, 413
    1,2-diamino-, oxidation, 470
    2-diazonium salts, N-aryl-, intramolecular azocoupling,
        452
    1-difluoromethyl derivatives*, 414
    H–D exchange, 410
    2-nitro-*, 453
    oxidative degradation, 438
    3-oxides, reaction with phosphorus oxychloride, 403
        with sulfuryl chloride, 403
    reduction, 406
    trans-1-styryl-, cyclization, 680
    1-substituted, hydroxylation, 398
    N-substituted, amination, 402
    2-sulfonic acids, 460
    thiation, 415
    1-trimethylsilyl-, annular elementotropy, 134
2H-Benzimidazole-2-spirocyclohexane, attack by sulfur
        nucleophiles, 438
    Michael-type 1,4-conjugate addition, 438
Benzimidazolones, 633
Benzisothiazole, 5-amino-, oxidative degradation, 438
    oxidation, 396
1,2-Benzisothiazole, nitration, 437
2,1-Benzisothiazole, 3-amino-, acylation, 450
        diazotization, 452
    Diels–Alder reactions, 420
Benz[*d*]isothiazoles*, 629
Benzisoxazole, 3-amino-, reaction with methyl propiolate,
        677
1,2-Benzisoxazole, quaternization, 467
    rearrangement to 1,3-benzoxazines, 467
    reduction, 417
Benz[*d*]isoxazolium, N-alkyl-, ring expansion, 636
Benzo[*c*]cinnoline*, 661
    1,10-dimethyl-, thermal fragmentation, 174
Benzo[*b*]cyclohepta[*d*]furans*, 660
1,3-Benzodiazepine*, 637
1,5-Benzodiazepine, 643
1,4-Benzodiazepines*, 641
1*H*-1,2-Benzodiazepines*, 631–632

1*H*-2,3-Benzodiazepines*, 646
3*H*-2,3-Benzodiazepines*, 646
2,4-Benzodiazepinone*, 638
1,3-Benzodiazepin-4-ones*, 637
1,4-Benzodiazepin-2-ones*, 641
2,3-Benzodiazepin-1-ones, 2,5-dihydro-1*H*-*, 646
2,3-Benzodiazepin-4-ones, 3,5-dihydro-4*H*-, 646
5*H*-1,4-Benzodioxepin, 2,3-dihydro-*, 644
5*H*-1,4-Benzodioxepinones, 643
3*H*-2,4-Benzodioxepins*, 638
2*H*-1,5-Benzodioxepins, 3,4-dihydro-*, 644
1,4-Benzodioxin, reaction with carbenes, 644
Benzo[1,2-*h*:4,3-*h'*]diquinoline, as proton sponge, 177
Benzodithiadiazine*, 652
3*H*-2,4-Benzodithiepin, 1,5-dihydro-*, 638
1,4-Benzodithiin, 2-nitro-, [2+2]photodimerization, 236
1,3-Benzodithiolylium ions, reaction with phosphines, 407
Benzofuran*, 611
    2,3-dihydro-, 607*, 614
    2-methyl-, 608
    13C NMR, 65
    1H NMR, 63
    UV spectra, 69
Benzo[*b*]furan, 608*-609, 612, 614
    acetylation, 311
    deprotonation, 320
    Gassman synthesis of, 612*
    halogenation, 310
    3-lithio-, ring opening, 361
    2-nitro-, reduction, 353
    reactions of lithio derivatives, 359
    reaction with carbenes, 324
Benzo[*c*]furan, 624*
    Diels–Alder reaction, 330
    1H NMR, 64
    UV spectra, 69
Benzofuranone*, 353
Benzofurazan, polarographic reduction, 416
Benzofuroxan
    cycloaddition, 649
    nitration, 438
    4-nitro-, rearrangement, 439
    quaternized, rearrangement to 1-hydroxybenzimidazole
        N-oxides, 467
    reduction, 417
    ring enlargement, 641
    ring opening, 407
Benzo[*c*]indole, UV spectra, 69
Benzo[*c,d*]indole*, 665
    1,2-dihydro-, 665
Benzo[*c*]isothiazole*, 629
    3-(o-aminophenyl)-, 139
Benzopyranone*, 647
Benzopyrans*, 620
Benzo[*b*]pyrans, 616
Benzopyridine, *see also* Quinoline and Isoquinoline
Benzopyrylium ions*, 617
Benzo[*b*]pyrylium ions, 207
2-Benzopyrylium salts, cycloaddition, 227
3,4-Benzopyrylium salts*, 625
Benzo[*d,e*]quinolines*, 666
Benzoselenophene
    2,3-dihydro*, 607
    1H NMR, 63

UV spectra, 69
Benzo[*b*]selenophene, deprotonation, 320
Benzo[*c*]selenophene, 1H NMR, 64
    UV spectra, 69
Benzo[*b*]tellurophene
    deprotonation, 320
    mass spectra, 75
    1H NMR, 63
    UV spectra, 69
1,2,4,5-Benzotetrazepine*, 655
1,2,4-Benzothiadiazepine*, 652
    1,1-dioxides 651–652
1,3,5-Benzothiadiazepine, 654
1,2,5-Benzothiadiazepine 1,1-dioxide, 653
1,3,4-Benzothiadiazine*, 651
2,1,3-Benzothiadiazine*, 648
1,2,3-Benzothiadiazine 1,1-dioxide, 649
1,2,4-Benzothiadiazine 1,1-dioxide*, 651
4*H*-1,2,4-Benzothiadiazine 1,1-dioxide*, 651
1,2,3-Benzothiadiazole*, 647
2,1,3-Benzothiadiazole*, 647
    oxidative degradation, 438
Benzothiatriazine, N2 elimination, 627
Benzothiazepine*, 642
1,4-Benzothiazepine*, 642
1,5-Benzothiazepine*, 643
2,4-Benzothiazepine*, 638
3,2-Benzothiazepine 3,3-dioxide, tetrahydro-*, 647
1,3-Benzothiazepin-4-one*, 638
1,4-Benzothiazepin-5-one, 2,3-dihydro-*, 642
1,3-Benzothiazine, 637
4*H*-1,4-Benzothiazine 1,1-dioxide, 640
2*H*-1,2-Benzothiazine dioxides, 3,4-dihydro-*, 631
2,1-Benzothiazines*, 631
2*H*-1,3-Benzothiazines*, 636
    Bischler–Napieralski cyclization, 636*
2*H*-1,4-Benzothiazines*, 640
Benzothiazole, 632*, 633
    acidity, 410
    6-allyloxy-2-methyl-, Claisen rearrangement, 438
    2-amino-, cyclization, 671
        oxidation, 453
    2-(2-hydroxyphenyl)-, 138
    ring enlargement, 642
2,1-Benzothiazoline 2,2-dioxides, thermolysis, 431
Benzothiazolium ions
    3-chloropropyl-, ring expansion, 644
    3-methyl-, electrochemical reduction, 416
Benzothiazolones, 633
Benzothiepin*, 627
2*H*-Benzo[*b*]thiete, retro [2+2] ring opening, 493
2-Benzothiinium, 4-oxido-, cyclodimerisation, 233
Benzothiinones*, 618
    dihydro-*, 620
Benzothiins, tetrahydro-*, 620
Benzo[*b*]thiins, 616
Benzo[*b*]thiophene, 609*, 612*, 614*
    acetylation, 311
    3-amino-, [2+2]cycloaddition, 332
    as dienophile, 331
    3-*t*-butyl-*, 314
    chloromethylation, 315
    deprotonation, 320
    dianion, 327
    2,3-dihydro-*, 607

1,1-dioxide, aromaticity, 81
halogenation, 310
2-hydroxy-, tautomerism, 87
2-lithio-, conversion to 2-hydroxybenzo[b]thiophene, 360
mass spectra, 75
nitration, 308
^{13}C NMR, 65
^{1}H NMR, 63
nucleophilic substitution of hydrogen, 323
reactions of lithio derivatives, 359
reactivity towards electrophiles, 305, 343
sulfonation, 308
sulfone, 334
UV spectra, 69
ylide formation, 301
Benzo[c]thiophene*, 623
    Diels–Alder reaction, 330
    ^{1}H NMR, 64
    UV spectra, 69
Benzo[b]thiophenium ion, 1,2,3,5-tetramethyl-, 59, 81
Benzo[b]thiophenium ions, S-aryl-, ring opening, 321
1H-1,3,4-Benzotriazepine, 2,3-dihydro-*, 654
1,3,4-Benzotriazepines, 1H-4,5-dihydro-*, 654
1H-1,2,4-Benzotriazepines*, 652
3H-1,3,4-Benzotriazepines, 4,5-dihydro-*, 654
1,2,5-Benzotriazepin-4-ones, 1H-4,5-dihydro-*, 653
1,3,4-Benzotriazepin-5-ones*, 654
1,2,3-Benzotriazine*, 647
    dihydro-*, 648
    hydrolysis, 199
    hydrolytic ring cleavage, 199
    nucleophilic attack at ring nitrogen, 220
    thermal fragmentation, 174
    thermolytic ring contraction, 606
1,2,4-Benzotriazine*, 649
    1,2-dihydro-*, 649*, 650
    1-oxide, 183, 649*
    2-oxides, 183
    1,2,4-trioxide*, 649
1,2,4-Benzotriazines, 3-substituted, 470
1,2,4-Benzotriazinium salts, 2-aryl-*, 649
1,2,3-Benzotriazin-4-one*, 648
    3-amino-, oxidation to nitrenes, 292
    ring–chain tautomerism, 176
Benzotriazolate anion, as a leaving group, 366, 467
Benzotriazole, 133, 647*
    N-alkylation, 383
    4-amino-, thermal rearrangement, 449
    aminoalkyl-, aminoalkylation by, 316
    N-(α-aminoalkyl)-, reactions, 467
    annular prototropy, 113
    1-aryl-, photochemical extrusion of nitrogen from, 370
    N-chlorination, 386
    1-chloro-, 469, 472
    1-cyano-, 469
    N-dialkylaminomethyl-, annular elementotropy, 134
    N-fluorination, 386
    H–H coupling constants, 110
    Jacobsen–Hugershoff synthesis, 633*
    2-methyl-, dimerization, 467
    1-nitro-, ring–chain tautomerism, 138
    transition-metal derivatives, 388
    1-(tributylstannyl)-, reaction with acid chlorides, 385
    1-[tri-(n-butyl)stannyl]-, 108

1-(trimethylsilyl)-, reaction with acid chlorides, 385
1-vinyl-, flash pyrolysis, 370
Benzo-1,3λ^{4}δ^2, 5,2,4-trithiadiazepines*, 655
1,3,5-Benzoxadiazepines, 654*, 655
3,1,5-Benzoxadiazepines*, 654
1,3,4-Benzoxadiazepin-5-ones*, 654
1,2,4-Benzoxadiazines*, 651
1,3,4-Benzoxadiazines*, 651
1,2,3-Benzoxadiazoles, ring–chain tautomerism, 139
1,2,3-Benzoxathiazine*, 648
1,2-Benzoxathiin 2,2-dioxides*, 631
1,3-Benzoxazepine*, 638
2,4-Benzoxazepine*, 638
4,1-Benzoxazepine-2,5-diones*, 642
2,1-Benzoxazepines, N-acyl-1,3,4,5-tetrahydro-*, 632
3,1-Benzoxazepines, 4,5-dihydro-, 637
1,4-Benzoxazepin-2-one*, 642
2,3-Benzoxazepin-1-one*, 647
4,1-Benzoxazepin-5-one*, 642
1,4-Benzoxazepin-3-ones*, 642
4,1-Benzoxazepin-2-ones*, 642
3,1-Benzoxazine, 4,4-dialkyl-4H-*, 636
1,3-Benzoxazines*, 636
1,4-Benzoxazines, dihydro-*, 640
2H-1,4-Benzoxazines*, 639
4H-1,4-Benzoxazines*, 640
1,3-Benzoxazin-4-ones*, 635
Benzoxazole*, 632
    2-phenyl-, [2+2]cycloaddition reactions, 424
    dimerization, 424
Benzoxazolones, 633
Benzoxepin*, 622
3-Benzoxepin*, 627
Benzoxetan-2-one, 606
2H-Benzoxocin, 3,4-dihydro-, 623
Biazoles
    N,N-linked, conformation, 130
2,2′-Bi-1,3-dioxolane, conformation, 129
Biheteroaryls, N,C-linked, conformation, 130
2,2′-Bipyridyl, 282
4,4′-Bipyridyl*, 226
4,4′-Bipyrimidines, 281
2,2′-Bipyrrole*, 326
2,3′-Biquinolyl, 226
3,3′-Biselenenyls, conformation, 82
2,2′-Bis-1H-imidazole, 109
    frozen tautomerism, 109
Bismabenzene, 47
    deshielding of α-protons, 47
Bis(pyrazole), 394
Bis-pyrazolyl-1-methanes, conformation, 130
4,4-Bis(1,3,5-thiadiazinylidene), 275
2,2′-Bithienyl, conformation, 82
Boiling points, 42
    5-membered rings, one heteroatom, 79
        two heteroatoms, 124
    6-membered rings, 41
Borirene, aromaticity, 157
Breckpot reaction, 519
Bromination, *see also* Halogenation
Bunte salts, 583

Carbazole, 58, 370*, 655
    alkali metal salts, solid-state structure, 60
    N-alkyl-, metallation, 345

aromaticity, 80
base strength, 306
electrophilic substitution, 344
Graebe–Ullmann synthesis of, 658*
9-methyl-, N-protonation, 299
^{13}C NMR, 66
^{1}H NMR, 64
^{15}N NMR, 68
N-protonation, 299
reactivity towards electrophiles, 343
N-substituted by Pschorr-type cyclization, 658
tetrahydro-, 657
–, Borsche synthesis, 659
UV spectra, 69
Carbene, stable, 106
Chichibabin amination, 204, 207, 402
Chlorination, *see also* Halogenation
Chlorophylls, 56
Chroman*, 617, 620
Chromanones, 620
2*H*-Chromene, 621
electrocyclic ring opening, 243
Chromones*, 617–618
formation, Simonis reaction, 618*
Kostanecki–Robinson synthesis of, 617*
Mannich reaction, 193
reaction with alkali, 203
with hydroxylamine, 207
Cinnoline*, 629
nitration, 252
reduction, 218
Cinnolones*, 630
Clemmensen reduction, 247
Conformation
azoles, C-aryl, 131
biazoles, N,N-linked, 130
2,2′-bi-1,3-dioxolane, 129
biheteroaryls, N,C-linked, 130
3,3′-biselenenyls, 82
bis-pyrazolyl-1-methane, 130
2,2′-bithienyl, 82
dibenzo[*b,g*]thiazocinium salts, 130
1,3-dioxolane, 129
1,3-dithiolane, 130
furan, 2,3- and 2,5-dihydro-, 85
2-(2-furyl)pyrrole, 82
imidazolidines, 1,3-diacyl-, 130
pyrrole, 2,5-dihydro-, 85
selenophene, tetrahydro-, 85
tetrakis-pyrazolyl-1-methane, 130
2-(2-thienyl)pyrrole, 82
thiophene, 2,3- and 2,5-dihydro-, 85
tetrahydro-, 85
1,2,4-trioxolane, 130
tris-pyrazolyl-1-methane, 130
Core ionization energies, 78
Cotarnine, pseudo-bases, 246
Coumarin*, 617–618
bromination, 185
Kostanecki–Robinson synthesis of, 617*
Michael addition of hydrogen cyanide, 217
reaction with alkali, 203
with malonic ester, 214
Criss-cross cycloaddition, 683
Cross-coupling reactions, 286, 363, 463

Crown ethers, selectivity for alkali metal ions, 147
Cryptands, selectivity for alkali metal ions, 147
Cyanine dyes, 261, 405, 443
Cyanuric acid, tautomeric structure, 52
Cyanuryl chloride, ring cleavage, 206
(3.3.3)Cyclazine, 16
Cycloaddition reactions
3-membered rings, 494
5-membered rings
[2 + 2], 332, 336
[2 + 3], 336
[4 + 2], 346
[4 + 4], 333
6-membered rings
[2 + 4], 227
[4 + 4], 235
8-membered rings, 493
Cyclopropa[*c*]cinnolines*, 631

Desaurins, 568
Dewar pyridines, 173, 480, 493
Dewar pyrimidine, 173, 578
Dialuric acid, 276
2,3-Diazabicyclo[3.1.0]hex-3-enes *see* homopyrazoles
1,4-Diazepine-4-oxides, 642
Diazepines*, 632
1,2-Diazepines*, 564
annular tautomerism, 163
1,3-Diazepines*, 579
1,4-Diazepines*, 585
annular tautomerism, 163
–non-aromatic nature, 158
1*H*-1,2-Diazepines, electrocyclic ring closure, 628
thermolysis, 580
1-tosyl-1*H*-1,2-, conformation, 150
1,3-Diazepin-2-ones, 580
1,2-Diazete, 158
1,3-Diazete, 158
1,2-Diazetidine*, 552
fused derivatives, 628
1,3-Diazetidine*, 566
Diazetidinones, 552
Diazetines*, 553
Diazines, 46
[2 + 4]cycloaddition, 228
N-oxides, ^{17}O NMR, 34
photoelectron spectroscopy, 41
reaction with organometallic compounds, 210
Diazinodiazines*, 677
Diaziridine*, 552
derivatives, from tetrazoles, 372
N-inversion, 159
NMR data, 150
Diaziridinones, 552
reaction with bifunctional carbanions, 560
Diazirine*, 552
NMR data, 150
1,2-Diazocine
1,2-dihydro-, anti-aromatic, 158
valence bond tautomerism, 162
1,4-Diazocine, 1,4-dihydro-, 158
cycloaddition reactions, 493
^{1}H and ^{13}C NMR, 159
structure, 160
N-substitution, 484

Diazonines, bridged, ionization energies of, 156, 157
Diazonium salts, Pschorr-type cyclization, 658
Dibenzo[*e,q*][1,4]diazocine, 6,7-diphenyl-, reduction, 491
Dibenzodioxin, crystal structure, 25
Dibenzo[*b,e*][1,4]dioxin, 25, 665*
Dibenzo[1,2]dithiins*, 661
Dibenzofuran, 657
   aromaticity, 80
   electrophilic substitution, 344
   metallation, 345
   ^{13}C NMR, 66
   ^{1}H NMR, 64
   reactivity towards electrophiles, 343
   UV spectra, 69
Dibenzoselenophene*, 658
   aromaticity, 81
   nitration, 344
   UV spectra, 69
Dibenzotellurophene
   nitration, 344
   UV spectra, 69
Dibenzo[*b,e*][1,4]thiazepines*, 665
Dibenzo[*b,f*][1,4]thiazepin-11-one 5,5-dioxides*, 665
Dibenzo[*b,g*]thiazocinium salts, conformation, 130
Dibenzothiophene, 323, 657*
   aromaticity, 80
   electrophilic substitution, 344
   metallation, 345
   S-methyl cations, 59
   ^{13}C NMR, 66
   ^{1}H NMR, 64
   reactivity towards electrophiles, 343
   'spiro dimer', 335
   UV spectra, 69
   ylide formation, 301
Dibenzotriazole, 1,1'-carbonyl-, reaction with acid
      chlorides, 385
   1,1'-sulfonyl-*, 385
Dibenz[*c,f*][1,2]oxazepines*, 665
Dieckmann cyclization, 577, 625
Diimidazole, 1,1'-carbonyl, 469
   carbonylation with, 652
1,1'-Diimidazolyls, 396
Dimroth rearrangement, 268, 435, 449, 602
   of 5-aminotetrazoles, 140, 449
   of 1-ethyladenine, 268
   of 3-ethylamino-1,2-benzisothiazole, 449
   of 1,2,3-triazoles, 448
1,2,4,5-Dioxadiazines*, 604
1,4,2,5-Dioxadiazines*, 603
Dioxane, 584*, 577
   ring expansion, 587
1,3-Dioxane, 2-(methoxymethyl)-, ring expansion, 587
   Prins reaction synthesis, 577
   reaction as acetal, 247
1,4-Dioxane*, 585
   benzo-fused derivatives, 640
   reaction as bis-ether, 247
Dioxathiaazapentalene, valence bond structures, 139
1,3,5-Dioxathiane*, 600
1,3,2-Dioxathiolane, 590
   2-oxide*, 590
1,3,2-Dioxazine*, 591
1,4,2-Dioxazines*, 597
1,3,5-Dioxazines, N-sulfonyl-*, 601

1,4,2-Dioxazoles, thermal reactions, 371
1,2-Dioxepins*, 566
1,3-Dioxepins*, 581
1,4-Dioxepins*, 587
1,2-Dioxetanes*, 554
Dioxetanones, 555
1,4-Dioxin, 582*
   benzo-fused derivatives*, 640
   electrophilic addition reactions, 236
   [2 + 2]photodimerization, 236
1,3-Dioxins*, 577
1,2-Dioxocine, anti-aromaticity, 158
1,4-Dioxocine, planarity, 158
   reactivity, 158
   valence bond tautomerism, 162
1,2-Dioxolane, ^{17}O NMR, 113
1,3-Dioxolanes, 432
   conformation, 129
   ring-strain energies, 125
1,3-Dioxolium-4-olates, cycloaddition reactions, 420
1,3-Dipolar cycloaddition
   of sydnones, 560
   pyridinium ions, 1-imide, 293
1,4,2,6-Dithiadiazines, extrusion of sulfur, 236
1,2-Dithiane, 40
1,3-Dithiane, 40
Dithianes*, 577, 584
1,4-Dithianes, 640*, 585
1,5-Dithia-2,4,6,8-tetrazocine, structure, 160
   UV spectra, 153
1,3,2,4,6-Dithiatriazine*, 604
1,2,3-Dithiazine*, 591
1,4,2-Dithiazine*, 598
1,4,2-Dithiazoles, thermal reactions, 371
1,2,4-Dithiazolidine, reaction with nitriles, 421
1,2,4-Dithiazolium ions, 3,5-diaryl-, recyclization, 403
1,3,2-Dithiazolium ions, stable radicals, 416
1,2-Dithiepins*, 566
1,3-Dithiepins*, 581
1,4-Dithiepins*, 587
1,2-Dithietanes*, 555
1,3-Dithietanes*, 568
Dithiin, 1,5-diphenyl-, Vilsmeier reaction, 236
1,2-Dithiin, 563*, 582, 584*
1,4-Dithiin*, 640
   electrophilic addition reactions, 236
   extrusion of sulfur, 236
   metallation, 237
   [2 + 2]photodimerization, 236
   reaction at sulfur, 236
1,3-Dithiolane, conformation, 130
   ring enlargement, 584
   ring-strain energies, 125
1,2-Dithiolanes*, 559
   alkylation of sulfur, 433
1,2-Dithiole-3-ones, ring opening, 397
1,2-Dithioles, intramolecular charge transfer, 117
   reaction with carbenes and nitrenes, 414
1,3-Dithiole-2-thione*, 575
1,3-Dithiolylium ions*, 572
   2-alkyl, reactions, 443
   deprotonation, 411
   reduction, 407
   reduction-dimerization, 418
   ring expansion, 598

1,2-Dithiolyl radicals, 416
1,3-Dithiolyl radicals, spin densities, 416
1,2-Dithiolylium ion, C-phenyl, nitration, 444
1,2-Dithiolylium ions, 559*
  3-alkyl-, anhydro bases, 442
  3-alkylthio-, dealkylation, 460
  aromaticity, 126
  conversion to thiazoles, 402
  3-methyl-, reactions at methyl group, 443
  reaction with methylene compounds, 684
  ring opening with hydroxide, 400
1,3-Dithiolylium-4-olates, cycloaddition reactions, 420

Ehrlich color reaction for pyrroles, indoles and furans,
  314

Fervenulin, 289
Flavines, 5-deaza-*, 662
Flavone, nitration, 262
Flavonone*, 617
  Camps reaction, 617*
Flavylium ions, reaction with C-nucleophiles, 214
Fluorene, ^{13}C NMR, 66
Fluorination, *see also* Halogenation
Fulgides, Diels–Alder reactions, 348
  photochromism, 346
Furan, 58, 541–544*, 624
  2-acetamido-, 353*
  2-acetyl-, conformation, 84
  acetylation, 311
  acid-catalyzed hydrogen exchange, 307
  acyl-, carbonyl reactivity, 353
  alkylation, 313
  2-amino-, stability, 354
  aromaticity, 80
  as a diene, 328
  as dienophiles, 331
  base strength, 306
  2-carboxylic acid, conformation, 84
  catalytic reduction, 327
  conformation, 84–85
  conversion to pyrroles, 319
  core ionization energies, 78
  2-cyano-, 61
  cycloaddition, 328
  [2 + 2]cycloaddition, 332
  [4 + 2]cycloaddition, 328
  [4 + 4]cycloadduct with benzene, 332
  deprotonation, 320
  2,5-di-*t*-butyl-*, 313
  dihydro-*, 541
  2,3-dihydro-*, 526
    conformation, 85
    [2 + 2]cycloadduct, 336
  2,5-dihydro-, conformation, 85
  2,5-dimethylene-, 350
  2,3-dimethylene-2,3-dihydro-, Diels–Alder reactions,
    348
  3,4-dimethylene-3,4-dihydro-, Diels–Alder reactions,
    349
  2,5-dinitro-, reactivity towards nucleophiles, 343
  2,4-disubstituted, 360
  π-donation, 58
  Ehrlich color reaction, 314
  electrolytic oxidation, 318

  electrophilic oxidation, 318
  Feist–Benary synthesis, 535*
  3-formyl-, Huang–Minlon reduction, 352
  free radical attack, 325
  Gattermann aldehyde synthesis, 311
  halogenation, 309
  Houben–Hoesch ketone synthesis, 311
  2-hydroxy-, tautomers, 87
  3-hydroxy-, tautomers, 87
  2-hydroxymethyl-, reaction, 349
    ring opening, 349
  interconversion, 320
  2-iodo-, Grignard reagent, 357
  IR spectra, 73
  2-lithio-, reactions, 359
  3-lithio derivatives, 357*, 361
  Mannich reaction, 315, 316
  mass spectra, 73
  2-mercapto-, tautomers, 89
  mercuration, 316
  mercury derivatives, 365
  nitration, 307–308
  ^{13}C NMR, 65
  ^{1}H NMR, 62
  ^{17}O NMR, 67
  Paal–Knorr synthesis, 530
  O-protonated, inversion barrier, 59
  O-protonation, 59
  rates of reaction, 303
  reaction with acetone, 315
    with carbenes, 324
    with singlet oxygen, 331
  reduction, 327
  ring opening, 342
  sulfonation, 308
  synthesis from acyclic precursors, 529
  tetrahydro-, 519*, 523*
    basicities, 337
    conformation, 85
    core ionization energies, 78
    derivatives, ring opening, 338
    ionization potentials, 78
    mass spectra, 75
    ^{1}H NMR, 64
  2-trimethylsilyloxy-, nitrene addition 579
  UV spectra, 68
  vertical ionization energies, 76
  2-vinyl-, [4 + 2]cycloaddition, 346
Furan-carbaldehydes, carbonyl reactivity, 352
Furan-3-carbaldehydes, conformation, 83
Furancarboxylic acids*, 543
  Birch reduction, 326
  dianions, 358
  pK$_a$, 351
  reactions, 351
3-Furanones*, 536
2,5-Furanophane, conformation, 81
Furanopyridophane, conformation, 81
Furazans*, 589
  diphenyl-, flash vacuum pyrolysis, 370
  geometries, 98
  3-heteroallyl-, rearrangements, 375
  rearrangement, 375
Furfural*, 531
  conformation, 83

nitration, 307
  thermal reaction with 1,3-butadiene, 331
Furfuryl halides, 349
Furoxan*, 589
  geometry, 98
  methyl-, Angeli's rearrangement, 442
  open-chain tautomers, 98

Geometry by theoretical methods, 59
Glutaric anhydrides and thioanhydrides, 521
Glutarimides, 521
Gramine(3-dimethylaminomethyl indole), reactions, 344, 350

'Halogen dance', 357
Halogenation
  5-membered rings, 308, 386
  6-membered rings, 188
Hammett & Taft σ constants, 465
Hantzsch synthesis, 539, 535
Hantzsch–Widman names, 12
Heptaazapentalene anion, valence bond structures, 140
Heptaazaphenalene, 16
Hetarynes, 282
Heterocyclic cations, ^{13}C NMR, 30
Heterocyclic α, β-quinodimethanes, Diels–Alder reactions, 348
Heterocyclic quinones, 276
Homopyrazoles, equilibrium with 1,4-dihydropyrazines, 430

Imidazo[1,2-*a*]benzimidazoles, nitrosation, 394
  aromaticity, 127
Imidazo[1,2-*a*]imidazole*, 671
  1-methyl-, formylation, 395
Imidazo[2,1-*a*]isoquinoline*, 680
Imidazole, 569, 575*
  N-acetyl-, as an acylating agent, 310
  N-alkenyl-, reactions, 468
  alkylation, 383
  amino-, tautomeric forms, 137
  1-o-aminophenyl-, diazotization, 394
  N-anions, oxidation, 396
  annular tautomerism, 132
  aromaticity, 126
  2-azido-*, 451
  azocoupling, 394
    with diazonium salts, 394
  carbene insertion, 413
  1-carboxy-, 469
  catalytic reduction, 417
  1-chlorosulfinyl-, reactions, 473
  1-cyano-, 271
  [2 + 2]cycloaddition reactions, 424
  dimerization, 417
  C-ethynyl-, 448
  5-ethynyl-1-methyl-*, 446
  2-fluoro-*, 451
  formation of iodine complexes, 386
  from photorearrangement of pyrazoles, 374
  halogenation, 391
  hydroxy-, tautomeric forms, 137
  hydroxymethylation, 394
  2-lithiation, 408
  lithio-, 464

mercapto-, tautomeric forms, 137
Michael reactions, 384
nitration and sulfonation, 390
2-nitro-*, 451
^{13}C NMR, 110
^{1}H NMR, 104
1,1'-oxalyldi-, 469
phenyl-, nitration, 444
N-phosphoryl-, as a phosphorylating agent, 473
stable carbenes, 128, 412
  ^{1}H NMR, 106
cis-1-styryl-, photochemical cyclization, 680
1-substituted-, acylation, 411
1,1'-sulfonyldi-, reactions, 473
tetrahydro-, 432
1-trimethylsilyl-, reactions, 473
2,4,5-trinitro-, nitro group displacement, 453
C-vinyl-, 448
2*H*-Imidazole, ^{13}C NMR, 111
  ^{1}H NMR, 107
  oxidation, 426
  thermal rearrangement, 426
4*H*-Imidazole, ^{13}C NMR, 111
  ^{1}H NMR, 107
  quaternization, 425
  thermal rearrangement, 426
Imidazole-2-carbaldehyde, deformylation, 447
Imidazolidine-2-one, 1-(2-chloroethyl)-, cyclization, 673
Imidazolidines, 1,3-diacyl-, conformation, 130
  alkylation, 432
  ring–chain tautomerism, 138
2-Imidazoline*, 574
  N-alkylation, 429
3-Imidazoline 1-oxides, radicals, 430
3-Imidazoline 3-oxides, 1-hydroxy-, ring–chain tautomerism, 429
3-Imidazoline N-oxides, 429
3-Imidazolines, 1-hydroxy-, radicals, 430
δ^{2}-Imidazolines*, 573
  aromatization, 427
Imidazoline-2-thiones, conformational analysis, 130
  reduction, 460
Imidazolinones*, 573
Imidazolium ions, thiation, 415
Imidazolone, N-monosubstituted*, 398
2-Imidazolone, 1,3-diacetyl-, Diels–Alder reaction, 424
Imidazo[2,1-*b*]oxazole, tetrahydro-*, 673
Imidazo[1,2-*a*]pyridines, 389
Imidazo[4,5-*b*]pyridines, hydroxylation, 398
Imidazo[4,5-*c*]pyridines, hydroxylation, 398
  thiation, 415
Imidazo[1,2-*a*]pyridin-3-one*, 672
Imidazo[1,2-*a*]quinoxaline*, 446
Imidazo[2,1-*b*][1,3,4]thiadiazoles*, 674
Imidazo[1,2-*d*][1, 2,4]thiadiazol-3-ones, 2,3,5,6-tetrahydro-*, 675
Imidazo[2,1-*b*]thiazoles, 671
Indazole, 628*
  (1-adamantyl)-, interconversion of isomers, 134
  2-alkyl-, photolysis, 439
  2-alkyl-3-amino-*, 471
  3-amino-, oxidation, 648
  azocoupling, 393
  2,3-diamino-*, 629
  6-hydroxy-, diazo coupling, 438

3-lithio-, 408
nitration, 437
reaction with trityl chloride, 384
tautomerism, 133
4,5,6,7-tetrahydro-, aromatization, 440
trityl, interconversion of isomers, 134
Indigo, 57
Indole, 58, 609–615*
1-acyl-, photoisomerization to 3-acylindolenines, 365
3-acyl-, by Vilsmeier–Haack reaction, 311
5-acyl-, formation by Vilsmeier–Haack reaction, 311
4-alkylamino-, rearrangement to 4-amino-1-alkylindoles, 345
C-alkylation, 313
2-amino-, 3H -tautomer, 354
4-amino-1-alkyl-, from 4-alkylaminoindoles, 345
as dienophiles, 331
3-azophenyl-, conversion to 3-nitroindole, 308
base strength, 306
1-benzenesulfonyl-, 3-lithio derivative, 359
reduction, 322
1-(1-benzotriazol-1-ylmethyl)-, reactions, 366
1-benzyl-*, 365
Birch reduction, 345
Bischler synthesis of, 612*
catalytic reduction, 327
5-chloro-, N-nitrosation, 318
condensation with 2,5-hexanedione, 658
cross-coupling reactions, 363
3-cyano-, 312
N-derivatives, metallation at benzene ring, 344
diazo coupling, 316
4,6-dimethoxy-, bromination, 343
3-dimethylaminomethyl-, see Gramine
Ehrlich color reaction, 314
electrolytic oxidation, 318
electrophilic oxidation, 318
Fischer synthesis, 431, 610*, 659*
from flash pyrolysis of 1-vinylbenzotriazoles, 370
from 1-phenyl-1,2,4-triazole, 465
Gassman synthesis*, 612
Grignard reagents, 300
2-halo-*, 358
halogenation, 309
Heck synthesis*, 612
N-hydroxy-*, 615
3-iodo-N-substituted, Heck reaction, 363
C-2 lithiation of, 358
N-lithio-, TMEDA complex, structure, 60
Madelung synthesis*, 614
N-methyl-, deprotonation, 320
Nenitzescu synthesis, 608*
^{15}N NMR, 68
nitration, 308
3-nitro-, from 3-azophenylindole, 308
3-nitroso, 318
^{13}C NMR, 65
^{1}H NMR, 63
N-potassio-, TMEDA complex, structure, 60
reactions of lithio derivatives, 358–359
reaction with carbenes, 324
reduction, 218
Reissert synthesis, 608
ring expansion with carbenes, 621
N-substituted-3-iodo-, 363

sulfonation, 308
UV spectra, 69
Indole-7-carbaldehyde, 1-methyl-*, 344
Indolenines
acid-catalyzed rearrangement, 333
Plancher rearrangement, 333
reduction, 333
Indole-2,3-quinodimethanes, cycloaddition reactions, 350
Indoline*, 322, 607, 613, 616
dehydrogenation, 336
3H-Indolium ion, as an electrophile, 321
3,3-dialkyl-, nitration, 343
Indolizine, 58, 669*
base strength, 306
reduction, 327
Indophenine test, for thiophene, 315
Indoxyl, 339, 613*
Infrared spectroscopy
5-membered rings, one heteroatom, 69
two heteroatoms, 117
6-membered rings, 37
small and large rings, 153–155
Iodination, see also Halogenation
Ionization potentials, 41, 78, 96
Isatin*, 612–613
reactions, 342
ring enlargement, 621
ring fission, 342
Isobenzofuran, [8 + 8] photodimer, 333
Isobenzopyrylium cation, ring expansion, 646
Isocoumarins, conversion to isoquinolones, 207
Isoindole, 58, 623–624*
base strength, 306
N-t-butyl-, diazo coupling, 316, 317
catalytic reduction, 327
Diels–Alder reaction, 330
N-methyl-, [8 + 8] photodimer, 333
^{1}H NMR, 64
acid-catalyzed polymerization, 321
^{1}H NMR, 64
1-phenyl-, acetylation, 311
reduction by dissolving metals, 327
tautomers, 86
Isoquinoline*, 625–626
basicity, 178
Chichibabin amination, 204
1-chloro-, reaction with 5-phenyltetrazole, 675
3,4-dihydro-*, 625
Bischler–Napieralski synthesis, 626
N-imide, photolytic ring expansion, 637
mixed trimer with 2-chloro-4,5-dihydroimidazole, 680
nitration, 252
oxidation, 253
2-oxide, bromination, 192
Picket–Gams preparation*, 626
Pomeranz–Fritsch synthesis, 627
reaction with organometallic compounds, 210
reduction, 218, 226
Reissert compounds, 217
sulfonation, 252
tetrahydro-, Emde reaction, 246, 247
Picket–Spengler synthesis*, 626
1,2,3,4-tetrahydro-, retro-Diels–Alder, 247
Isoquinolinium ion, 1-acyl, reduction, 219
N-alkyl-, oxidative imination, 206

cycloaddition, 228
  2-(2,4-dinitrophenyl)-, reaction with aniline, 207
  1-methyl-, reduction, 219
Isoquinolinium salts, Bradsher reaction, 228
Isoquinolones, from isocoumarins, 207
1-Isoquinolones*, 199, 626
4(1*H*)Isoquinolones, 2,3-dihydro-*, 625
  Dieckmann cyclization, 625*
Isothiazole*, 558
  3-alkyl-5-acetamido-, nitrosation, 394
  amino-, tautomeric forms, 135
  4-amino-, alkylation, 450
  aromaticity, 126
  conversion to thiophenes, 397
  cycloaddition reaction, 424
  4,5-dimethyl-*, 560
  halogenation, 392
  hydroxy-, tautomeric forms, 135
  mercapto-, tautomeric forms, 135
  nitration and sulfonation, 390
  ^{13}C NMR, 111
  ^{1}H NMR, 105
  oxidation, 396
  3-phenyl-, nitration, 444
  4-phenyl-, nitration, 444
  reductive desulfurization, 418
  ring opening, 397, 399
  substituted, tautomeric forms, 135
Isothiazole-3,4-dicarboxylate, 2-phenyl-*, 561
Isothiazole-4,5-dicarboxylate, 3-phenyl*, 561
Isothiazolium ions, recyclization to thiophenes, 397
Isoxazole, 557–559*, 561*
  amino-, tautomeric forms, 135
  aromaticity, 126
  4,5-bis(bromomethyl)-3-phenyl-, Diels–Alder reaction, 422
  chloromethylation, 394
  3,5-diphenyl-*, 557
    hydrogenolysis, 417
  halogenation, 391
  hydroxy-, tautomeric forms, 135
  isomerization to oxazole, 374
  mercapto-, tautomeric forms, 135
  metallation, 408
  5-methyl-3-phenyl-, nitration, 444
  nitration and sulfonation, 390
  ^{17}O NMR, 113
  ^{1}H NMR, 104
  photosensitised oxidation, 396
  reaction with free radicals, 415
  recyclization to pyrazoles, 401
  ring cleavage, 413
  ring contraction, 528
  substituted, tautomeric forms, 135
Isoxazolidines, 432, 557*, 560
  retro 1,3-dipolar cycloaddition, 432
Isoxazoline*, 557,559
2-Isoxazoline, [2 + 2]cycloaddition, 430
  3-methyl-4-α-hydroxyisopropyl-, ring opening, 430
$δ^2$-Isoxazoline, photolysis, 580
$δ^2$-Isoxazoline N-oxides*, 559
Isoxazolinone*, 557

Isoxazolium ions
  reaction with enamines, 420
    with hydrazines, 560
  reduction, 407
Isoxazolo[3,4-*d*]thiazoles*, 454

Japp–Klingemann reaction, 259

Kolbe reaction, 193, 392

β-Lactams*, *see also* 2-Azetidinones, 522
  synthesis, 519
δ-Lactams, 521
Lactones, 520
β-Lactones, 515
δ-Lactones, 521
Large rings
  aromaticity, 158
  basicity, 481
  conformation, 160
  cycloaddition rections, 492
  metal complexes, 483
  radical attack, 489
  reactivity, 475, 477
  rearrangement to small rings, 480
  structure, 141
  tautomerism, 478
  transannular interactions, 161, 477
Lawesson's reagent, 275
Lumazine, 6-chloro-1,3-dimethyl-, reaction with 1,2-diaminoethane, 206

Magnetic susceptibility, 46
Mannich reaction, 193, 315, 520
Mass spectrometry, 73
  5-membered rings, one heteroatom, 73
    two heteroatoms, 121
  6-membered rings, 138–141
  small and large rings, 155, 156
Meerwein rearrangement, 615
Meisenheimer rearrangement, 563, 566
Melamine, 600
Melting points, 42
  5-membered rings, one heteroatom, 79
    two heteroatoms, 124
  6-membered rings, 41
3-Membered rings
  bond length and angles, 150
  geometry, 150
  NMR data, 150
  reactivity, 476, 485, 486
  structure, 141
  synthesis
    bicyclic, no ring junction heteroatoms, 605
    by 2 concerted bond formation, 511
    fused rings with ring junction heteroatom, 667, 668
    monocyclic rings
      one heteroatom, 511–518
      two heteroatoms, 552
      three heteroatoms, 558
4-Membered rings
  geometry, 150
  reactivity, 476, 486, 487
    of carbonyl derivatives, 487
  structure, 141

synthesis
  bicyclic, no ring junction heteroatoms
    with one heteroatom, 605
    with two heteroatoms, 627, 628
  by 2 concerted bond formation, 513
  from carbocyclic, or heterocyclic precursors, 540
  fused rings with ring junction heteroatom, 667, 668
  monocyclic rings with one heteroatom
    no endocyclic double bond, 514, 515
    one endocyclic double bond, 518, 519
    with two heteroatoms, 522
    with three heteroatoms, 558
5-Membered rings
  formylation, 394
  one heteroatom, 79
    aromaticity, 80
    azides, 355
    carbonyl-hydroxy tautomers, 339
    dihydro-, aromatization, 335
    dinitro derivatives, nucleophilic attack, 342
    formation of C–C bonds, 359
      of C–halogen bonds, 361
      of C–N bonds, 361
      of C–O bonds, 359
    Grignard reagents, 357
    mercury derivatives, 365
    molecular geometry, 79
    nitro derivatives, reactions, 353
    oxidation, 311
    pK_a values, 351
    photochemical reactions, 298, 299
    photoelectron spectroscopy, 76
    reaction with carbenes, 323
    reactivity, 297
      directing effects of substitutents, 304–305
      N-nitroso formation, 337
      nucleophilic attack at sulfur, 323
    relative reactivity, 303
    Vertical Ionization Energies, 76
    X-ray diffraction, 60
  reduction by dissolving metals, 327
  structural types, 91
  synthesis
    bicyclic, no ring junction heteroatom
      one heteroatom adjacent to ring junction, 607
        not adjacent to ring junction, 623, 624
      two heteroatoms 1,2 to ring junction, 628
        1,3 to ring junction, 632
      three or more heteroatoms, 647
    fused ring systems
      one N atom at ring junction, 668
      two N atoms at ring junction, 682, 683
      S at ring junction, 683, 684
    monocyclic rings
      1,2-dihetero, 556
      1,3-dihetero, 569
      1,2,3-trihetero, 588
      1,2,4-trihetero, 591
      four or more heteroatoms, 602, 603
    monocyclic rings, one heteroatom
      no endocyclic bonds, 515, 519–521

two endocyclic double bonds, 540
  tri- and poly-cyclic rings, no ring junction
    heteroatoms, 657
two and more heteroatoms
  alkyl-thio, reactions, 460
  as dienophiles, 331
  base-catalyzed hydrogen exchange, 409
  melting and boiling points, 124
  ^{13}C NMR spectra, proton–proton coupling constants, 110
  oxidation, 395
  oxidative degradation, 438
  photochemical reactions, 369, 370
  photoelectron spectroscopy, 121–123
  N-protection, 473
  prototropic tautomerism in the solid state, 132
  pseudo bases, oxidative-reductive disproportionation, 400
  reactivity, general survey, 367
    nucleophilic attack at sulfur, 397
  solubilities, 124
  stable carbenes, 412
  structural types, 55
  X-ray diffraction, 99
6-Membered rings
  calculations, 22
  cyclodimerizations, 232, 235
  geometry from microwave spectra, 26
  heteroaromatic cation, ^{1}H NMR, 27
  nomenclature, 15
  oxidation, 193, 253
    with endocyclic double bonds, ^{1}H NMR, 28
  N-oxides, mass spectra, 38
  photoelectron spectroscopy, 41
  prototropic tautomerism, 47
  radical cations, 237
  reactivity, 169
    aromatic ring, 169
    non-aromatic ring, 235
    nucleophilic attack at ring nitrogen, 220
    nucleophilic displacement of alkoxy groups, 276
    nucleophilic hydroxylation, 198
    photochemical reaction, 172
    radical-cation electrophilic substitution, 254
    substituents, 248
  REPE (resonance energy per electron), 44
  solubility in water, 41
  structural types, 15
  synthesis
    bicyclic rings
      one heteroatom adjacent to ring junction, 616
        not adjacent to ring junction, 625
      two heteroatoms adjacent to ring junction, 629–631, 634–636, 638–640
      –not adjacent to ring junction, 644–646
      three or more heteroatoms, 647
    fused rings with ring junction heteroatoms
      one N atom at ring junction, 683
      two N atoms at ring junction, 676
      S atom at ring junction, 683
    monocyclic rings
      1,2-dihetero, 561
      1,3-dihetero, 576
      1,4-dihetero, 581
      no endocyclic double bonds, 521

one endocylic double bond, 525, 526
two endocyclic double bonds, 540
three endocyclic double bonds, 564
four or more heteroatoms, 603–604
1,2,3-trihetero, 591
1,2,4-trihetero, 596
1,3,5-trihetero, 599
tri- and poly-cyclic rings, no ring junction
heteroatoms, 657
with exocyclic carbonyl or thione groups, NMR data,
28, 32
X-ray diffraction, 24
7-Membered rings
reactivity, 475
structural types, 141
synthesis
bicyclic rings
one heteroatom adjacent to ring junction, 622, 623
not adjacent to ring junction, 627
two heteroatoms 1,2 to ring junction, 631, 632
1,3 to ring junction, 637–638
1,4 to ring junction, 641
1,5 to ring junction, 646–647
2,3 to ring junction, 646–647
2,4 to ring junction, 638
three or more heteroatoms, 652
fused rings, N atom at a ring junction, 681–682
monocyclic rings
1,2-dihetero, 565
1,3-dihetero, 579
1,4-dihetero, 585
1,2,4-trihetero, 598
1,2,5-trihetero, 598–599
1,3,5-trihetero, 602
two endocyclic double bonds, 544
three endocyclic double bonds, 549–550
Mercuration
5-membered rings, 316, 365, 392
6-membered rings, 192
Metallation, 345
5-membered rings, 345, 464
6-membered rings, 280
Metal–metal exchange, 281
Methylene blue, oxidation, 255
Microwave spectroscopy
5-membered rings, one heteroatom, 61
two heteroatoms, 99
6-membered rings, 26
small and large rings, 150
Mills–Nixon effect, 132
MO calculations
theoretical methods, 20
Morpholine, reactivity, 247, 248
Muscone*, 312

Naphtho[1,8-c,d]dichalcogenols*, 665
Naphtho[1,2-d]imidazole
aromaticity, 128
Chichibabin amination, 402
H–D exchange, 410
hydroxylation, 398
Naphtho[2,3-d]imidazole
aromaticity, 128
Chichibabin amination, 402
H–D exchange, 410

hydroxylation, 398
Naphthoimidazoles, thiation, 415
Naphtho[2,3-d]-1,2,3-oxadiazole, ring–chain tautomerism,
139
Naphtho[b,c]oxepinium, 2-dimethylamino-*, 666
Naphtho[b,c]thiete*, 665
Naphtho[2,3-b]thiophene, S-methyl cations, 59
Naphtho[1,8-d,e]triazine*, 666
Neber rearrangement, 527
Nitration
of azines, 186
of arylazines, 262
of benzazines, 252
of 5-membered ring
with one heteroatom, 307–308
with two and more heteroatoms, 444
of pyrons, 187
Nitrenes
from oxidation of N-aminoazoles, 470
from oxidation of N-aminopyridones, 292
use in aziridine formation, 512
Nitrogen deprotection, 498
Nitrophenol, from 5-nitropyrimidin-2-one, 214
Nitrosation
of six-membered rings, 193
of pyrrole, 317
Nuclear magnetic resonance
^{13}C NMR
5-membered rings, one heteroatom, 27, 65–66
two or more heteroatoms, 108–112
6-membered rings, 27
chemical shifts, 27–30, 32–33
coupling constants, 30–32
of sulfur containing azoles, 111
small and large rings, 151
^{1}H NMR
5-membered rings, one heteroatom, 62–64
two or more heteroatoms, 103–107
6-membered rings, 26–27
chemical shifts, 26–27
coupling constants, 27
small and large rings, 150
heteroatom NMR
5-membered rings, one heteroatom, 66–68
two or more heteroatoms, 112, 113
6-membered rings, 33–34, 66–68
small and large rings, 152

1,2,4-Oxadiazepines, 3-oxo-*, 598
1,2,6-Oxadiazepines, 7-oxo-*, 598
1,4,5-Oxadiazepines, hexahydro-*, 598
1,2,5-Oxadiazine, 5-hydroxy-*, 597
1,3,5-Oxadiazine*, 601
N-sulfonyl-*, 601
1,3,4-Oxadiazines, 4,5-dihydro-*, 597
1,3,5-Oxadiazinium salts*, 601
reaction with hydroxylamine, 595
Oxadiazinones*, 601
Oxadiaziridines*, 588
Oxadiazole*, 469, 595
2-amino-1,3,4-, Dimroth rearrangement, 449
aromaticity, 126
ring opening, 399
1,2,3-Oxadiazole, ring–chain tautomerism, 138

1,2,4-Oxadiazole*, 592, 595
  catalytic reduction, 417
  5-methoxy-3-carboxylate, methyl, photolysis, 595*
  ^{1}H NMR, 104
  3-phenyl-, nitration, 444
  recyclization, 407
  reduction, 406
1,2,5-Oxadiazole, 589
  anion radicals, 415
  electrochemical reduction, 415
  ^{13}C NMR, 111
  ^{1}H NMR, 104
  thermal and photochemical ring cleavage, 370
1,3,4-Oxadiazole, 591*
  anion radicals, 415
  electrochemical reduction, 415
  ^{13}C NMR, 111
  ^{1}H NMR, 104
1,3,4-Oxadiazolines, ring-opening, 430
1,2,4-Oxadiazolinone, 594
1,2,4-Oxadiazolin-5-one, 4-phenyl-*, 595
1,2,3-Oxadiazolium hydroxide, anhydro-5-hydroxy-*, 590
1,3,4-Oxadiazolium salts, 591*
1,3,4-Oxadiazolyl ether, 592*
1,3,5-Oxadithiane*, 600
Oxaheterocinyl-anions, ^{1}H NMR, 159
1,4,3,5-Oxathiadiazines*, 603
Oxathianes*, 577
1,4-Oxathianes*, 640
1,3,5-Oxathiazole*, 593
1,2,3-Oxathiazole S-oxides*, 589
1,4,2-Oxathiazoles, thermal reactions, 371
Oxathiazolones, thermal decomposition, 595
1,2-Oxathietanes*, 555
1,4-Oxathiin*, 640
  5,6-dihydro-*, 584
1,2-Oxathiin 2,2-dioxides*, 563
1,3-Oxathiolane, ring-strain energies, 125
  ring enlargement, 584
1,3-Oxathiolylium, 572
1,3,4,5-Oxatriazine, triphenyl-*, 603
1,2-Oxazepines*, 566
  Meisenheimer rearrangement, 566
1,3-Oxazepines*, 580
1,4-Oxazepines*, 586
Oxazetidines*, 553
1,3-Oxazetidines*, 567
Oxazetidinones*, 567
1,2-Oxazines, 561*
  pericyclic reactions, 244
1,3-Oxazines*, 576
1,4-Oxazines*, 582
1,3-Oxazinium salts*, 578
  perchlorates, ring expansion, 580
  reaction with N-nucleophiles, 207
Oxazinones, 561, 577*
  reaction with alkali, 203
1,3-Oxazin-6-ones, photochemical isomerization, 176
  recyclization into 4-pyrimidinones, 579
Oxaziridine*, 552
  cyclohexanespiro-3'-, as reagent for electrophilic
    amination, 485
  N-inversion barrier, 159
  NMR data, 150
  3,3-pentamethylene-, 279

1,4-Oxazocine, planarity, 158
Oxazole, 569
  amino-, tautomeric forms, 137
  aromaticity, 126
  Cornforth synthesis, 571*
  4,5-diaryl-, oxidation, 396
  Diels–Alder reactions, 418
  4,5-dihydro-, cleavage to carboxylic acids, 430
    as directors for nucleophilic addition, 431
  from isomerization of isoxazole, 374
  hydroxy-, tautomeric forms, 137
  5-hydroxymethyl derivatives, 395
  2-lithio-, equilibrium with open chain form, 408
  mercapto-, tautomeric forms, 137
  mercuration, 393
  methyl derivatives, free radical bromination, 440
  nitration, 389
  ^{13}C NMR, 111
  ^{1}H NMR, 104
  phenyl-, nitration, 444
  reaction with free radicals, 415
  recyclization to imidazoles, 401
  reduction, 417
  tetrahydro-, 432
  with diverse 2-substitutents*, 370
  4,5-xylylene, cycloaddition reactions, 422
Oxazolidine, 573, 574*
  2-methoxy-, asymmetric formylation, 433
  5-methoxy-, for the asymmetric formylation of
    nucleophiles, 433
  ring–chain tautomerism, 138
Oxazoline*, 574
  Diels–Alder reaction, 679
  2-furyl-, Friedel–Crafts reactions, 321
2-Oxazolines*, 573
δ^2-Oxazolines, 573
δ^4-Oxazolines, aromatization, 427
Oxazolin-2-one, [2 + 2]cycloaddition reactions, 424
5(4H)-Oxazolinones*, 573
Oxazolium ions, conversion to imidazoles, 403
2(3H)-Oxazolones, [4 + 2]cycloaddition, 424
Oxazolo[3,2-a]pyridinium*, 673
Oxazones, reaction with ammonia and amines, 208
Oxepane, 522*
Oxepin, 549*, 493
  calculations, 147
  conformation, 150
  4,5-dihydro-*, 544
  reaction with azide ion, 487
Oxetane, 514*, 519*
  hydroxymethyl-*, 523
  reaction with carbenes, 490
  X-ray crystal structure, 148
2-Oxetanones*, 515
  nucleophilic attack, 487
4-Oxido-2-benzothiinium, cyclodimerization, 233
Oxindole*, 608, 612, 613
  N-amino-, oxidative ring expansion, 630
  Brunner synthesis of, 613*
Oxirane*, 511, 516
  conversion to thiiranes, 524
  ethynyl-, isomerization to dihydrofurans, 541
  geometry, 148
  NMR data, 150
  ^{17}O NMR data, 152

nucleophilic ring-opening, 486
ring expansion, 523, 585
vinyl-, isomerization to furans, 541
Oxirene, antiaromaticity, 157
Oxonin, geometry of, 159

Penams*, 667
Pentakis(pentafluoroethyl)-1-azaprismane, 174
Pentamethylene sulfides*, 521
Pentathiepins*, 655
Pentazoles, 371, 603*
Perimidine, 23–24, 666*
  anion, alkylation, 223
  as π-acceptor, 24
  2,3-dihydro-aromatization, 241
  1,2-disubstituted-, 201
  H–D exchange, 410
  radical-cation electrophilic substitution, 254
  thiation, 208
  1-R-, Chichibabin amination, 204
    nucleophilic hydroxylation, 198
  2-thiones, 208
Perimidinium ions, 3-amino-1,2-dimethyl-, cycloaddition, 670
  1,3-dialkyl-, 200
Perimidones, 1,3-dialkyl-, 200
α-Peroxylactones, 555
Phenalenes, 1,6-diaza-, radical-cation electrophilic substitution, 254
Phenanthridine*, 660
  Chichibabin amination, 204
  nitration, 252
9-Phenanthridones*, 199
Phenanthro[9,10-d]imidazole, 1-methyl-2-phenyl-*, 242
Phenazine*, 663
  9,10-dioxides*, 663
  oxidation, 253
Phenazin-1(2H)-ones, 3,4-dihydro-*, 663
Phenothiazine, 25, 664*
  N-alkyl-, reaction at sulfur, 237
  crystal structure, 25
  extrusion of sulfur, 660
  radical cations, 237
    electrophilic substitution, 254
Phenothiazinium ion, 25
  crystal structure, 25
Phenothiazin-3-ones*, 664
Phenoxathiin, 25, 665*
  crystal structure, 25
Phenoxazine, 25, 664*
  crystal structure, 25
  radical cations, 237
Phenoxazinium ion, 25
  crystal structure, 25
Phenoxazin-3-ones*, 664
Phenoxazonium salts, 664
Phosphabenzene, 47
  deshielding of α-protons, 47
Phosphole, aromaticity, 80
[8 + 8] Photodimers, 333
Photoswitchable units, 348
Phthalazine*, 644
  1-aryl-*, 646
  1,4-bis(trifluoromethyl)-*, 646
  1,4-diamino-*, 644

1,4-dihydrazino-*, 645
1,4-dione, cycloaddition, 683
  oxidation, 253
Phthalazin-1(2H)-one, 4-hydroxy-*, 646
1(2H)Phthalazinone, 1-benzoyl-, 199*
Phthalide, reaction, 342
Phthalimide, N-amino-, oxidation, 512
  Ing–Manske reaction, 342
  ring opening, 342
Phthalocyanine, 56, 624*
2-Picoline, 256–257
  1-oxide, Claisen condensation, 258
Piperazine*, 584
  2,5-dioxo-*, 582
  reaction as bis-lactam, 247
  reactivity, 247
δ²-Piperideine, 1-methyl-, 246
Piperidine, 521
  N-chloro-, thermal trimerization, 680
  conversion to pyridines, 246
3-Piperidone, Clemmensen reduction, 247
Piperidinones, 526
Polyaza rings, hydrolytic ring cleavage, 199
Proton sponge, 177
Pseudoazulenes, 259
Pseudo-base, 199–200, 240, 399–400
  cotarnine, 246
  equilibrium with open-chain compounds, 246
  formation, pK_{r+} values, 399
  oxidative–reductive disproportionation, 400
Pseudocyanides, 216
Pteridine*, 634, 639
  covalent hydration, 198
  reaction with ammonia, 205
Pterins, tetrahydro-, oxidation, 248
Purine
  aromaticity, 126
  assignment of isomeric structures, 113
  π-deficiency, 95
  9-substituted, reaction with potassium amide, 402
  Traube synthesis, 633
  2,6,8-trichloro-, 274
    halogen exchange, 286
2H-Pyran, 53
  tautomerism, 239
4H-Pyran, 5,6-dihydro-*, 524
2H-Pyranes, electrocyclic ring opening, 243
Pyran-2-one
  bromination, 191
  conversion to 2-pyridones, 207
  cyclodimerization, 234
  Diels–Alder reactions, 228
  [2 + 2] photodimers, 235
  photolysis, 174
Pyran-4-one
  2,6-dimethyl-, cyclodimerization, 235
  photoisomerization, 175
  reaction with ammonia and amines, 207
    with Grignard reagents, 212
α-Pyranones, Diels–Alder reactions, 228
Pyrans, dihydro-, 526*
  from acyclic precursors, 538*
  3-hydroxytetrahydro-*, 524
  tetrahydro-*, 521

Pyrazine*, 581, 584
  azido-, pyrolysis, 271
    photochemical transformations, 576
  chloro-, alkylation and arylation by cross-coupling, 287
  2,3-dicarboxylic acid anhydride, reactions, 645
  dihydro-, aromatization, 581
  1,4-dihydro-, antiaromatic, 44
  from pyridazines, perfluoro-, 174
  N-oxides
    reaction with phosphoryl chloride, 209
    Reissert–Henze reaction, 217
  reduction, 218
Pyrazinium salt, 1-methyl-3-oxido-, 1,3-dipolar
  cycloaddition, 233
Pyrazinones, tetrahydro-*, 582
Pyrazinopyrazines*, 639
Pyrazino[2,3-d]pyridazine-5,8-dione*, 645
Pyrazole*, 557, 559–560
  acylation, 385
  alkylation, 383
  allyloxy, Claisen rearrangement, 458
  5-allyloxy-, Claisen rearrangement, 458
  amino-, tautomeric forms, 135
  3-amino-*, 560
  5-amino-*, 471, 560
    sulfonation, 450
  annular tautomerism, 132
  aromaticity, 126
  carbene insertion, 414
  catalytic reduction, 417
  4-chloromercuration, 392
  4,5-diamino-, cyclization, 647
  diazo-, 451
    cycloaddition, 675
  3-diazo-, thermal extrusion, 451
  4-diazo-, 451
  dihydro-, tautomerism, 427
  dimers, 113
  3,5-dimethyl-, 113
    hydrogen-bonded trimer, 113
    nitrosation, 394
    reaction with formaldehyde, 394
    tautomerism, 132
  dimethylamino-, alkylation, 450
  from thiadiazine 1-oxides, 238
  fused systems*, 628, 670
  halogenation, 391
  hydroxy-, tautomeric forms, 135
  mercapto-, tautomeric forms, 135
  3-methyl-4-nitro-, desmotropy, 132
  5-methyl-4-nitro-, desmotropy, 132
  Michael reactions, 384
  nitration and sulfonation, 390
  4-nitroso-, 394
  ^{13}C NMR, 110–111, 121
  ^{1}H NMR, 104, 107
  oxidation, 395
  1-phenyl-, mercuration, 392
  1-phenyl-3,4,5-trimethyl-, mercuration, 392
  photorearrangement to imidazoles, 374
  3(5),4-polymethylene-, tautomers, 132
  reaction with 1-bromoadamantane, 392
  1-substituted, lithiation, 408
  tetramers, 113
  4-trifluoromethyl-*, 560

  3,4,5-trimethyl-, chlorination, 442
  3,4-trimethylene-, , 558
  Vilsmeier formylation, 392
3H-Pyrazole, the van Alphen–Huttel rearrangement*, 426
  conversion to cyclopropenes, 426
Pyrazolidine-3,5-dione, of 4-diazo-, Wolff rearrangement,
  552
Pyrazolidines, 557
  conversion to pyrazoles, 432
  Fischer indole synthesis type reaction, 431
5-Pyrazolidinones, 1-aryl-, ring contraction, 432
Pyrazoline, 557*, 561*
  3-amino-, dehydrogenation-diazotization, 427
  azimines*, 559
1-Pyrazoline, ring contraction, 428
  azimines*, 559
2-Pyrazoline, quaternization, 430
  aromatization, 427
  3-diazonium tetrafluoroborate, 1-phenylsulfonyl-, 428
  ^{1}H NMR, 106
  protonation site, 113
  pyrolysis to cyclopropanes, 428
Pyrazolines, as models of intramolecular dyotropy, 428
δ^2-Pyrazolines, protonation site, 109
Pyrazolin-5-ones*, 557, 560
  condensation with ketones, 395
  dichloro-, ring opening with alkali, 425
  nitrosation, 394
Pyrazolium, 2-alkoxy-, 443
Pyrazolo[3,2-b]perimidine*, 670
Pyrazolo[1,2-a]pyrazole*, 682
Pyrazolo[1,5-a]pyridine, dihydro-*, 670
Pyrazolo[3,4-c]pyridines, site of electrophilic attack, 389
Pyrazolo[3,4-d]pyrimidines*, 628
Pyrazolo[3,4-c][1,2,5]thiadiazole S-oxide, dihydro-*, 647
Pyrazolo[3,4-d]thiazoles*, 454, 634
Pyrazolo[4,5-d]thiazoles*, 633
Pyrazolo[5,4-d]thiazoles*, 633
3H-Pyrazolo[5,1-c]-1,2,4-triazoles*, 675
Pyrazolyl-1-methanes, conformation, 130
Pyridazine*, 561
  4-azido-, photolysis, 271
  4-carboxylates, ethyl 5-acyl-, reactions, 645
  3,4-dicarbonitrile, reactions, 645
  1,4-dihydro-*, 563
  isomerization to pyrimidine, 174
  1-oxides, reaction with phosphorus oxychloride, 209
  N-oxides, nitration, 187
  2-(N-pent-3-ynylamino)-, cyclization, 616
  perfluoro-, rearrangement to pyrimidines, 174
  tetrahydro-*, 645
3,6-Pyridazinedione, cycloaddition, 683
Pyridazinium, 1-methoxy-, reaction with cyanide, 217
Pyridazinones, 561
Pyridazino[4,5-c]pyridazines*, 645
Pyridazino[4,5-d]pyridazin-1(2H)-ones*, 645
Pyridine*, 418, 541, 545, 546
  alkyl-, photolytic isomerization, 174
  1-alkyl, conversion to 4,4'-bipyridyl diquaternary salts,
    216
  N-alkylation, 180
  2-amino-, 204
    cyclization, 671, 676
    cyclodimerization, 236
  2-amino-5-chloro-, valence bond isomerism, 606

N-arylation, 181
basicity, substituents effect, 178
2-bromo-, 579
Chichibabin amination, 204
2,6-diamino-, 204
  nitrosation, 193
2-(diazomethyl)-, thermolysis, 265
2,6-dichloro-, halogen exchange, 284
dihydro-, 541*, 545*
  electrocyclic ring opening, 243
  from acyclic precursors, 538*
  isomerization, 243
  tautomerism, 239
1,2-dihydro-, as 1,3 dienes, 244
  asymmetric synthesis with Grignard reagents, 211
4-dimethylamino, use in acylation, 181
electrophilic halogenation, 190
fluorination, 209
formation by Diels–Alder reactions, 419
2-formyl-, conformation, 26
from piperidines, 246
from pyrylium cations, 207
hydrogen exchange, 221
2-hydroxy-, prototropic tautomerism, 47
3-hydroxy-, Kolbe reaction, 193
  oxidation, 194
2-hydroxymethyl-3-hydroxy-, 193
N-imides, photochemical ring expansion, 565
introducing substituents, 545
iodine complex, 182
ionization potentials, 41
IR spectra, 37
mercuration, 192
1-methyl-1,2-dihydro-, [2 + 2]cycloaddition, 245
  as 1,3 diene, 245
nitration, 186
^{1}H NMR, 26–27
^{15}N NMR, 34
^{13}C NMR, 30
nucleophilic hydroxylation, 198
oxidation, 194
1-oxide*, 183, 546
  2-alkoxy-, thermal rearrangement to N-alkoxy-2-pyridones, 276
  amino-, diazonium salts, 267
  2-azido-, photolysis and termolysis, 543
  halogenation, 191
  mercuration, 192
  nitration, 187
  2-phenyl-, nitration, 262
  reaction with acid anhydrides, 294
    with enamines, 215
  Reissert–Henze reaction, 217
  ring opening with sodium acetylide, 214
  sulfonation, 188
  thioalkylation, 208
4-(2-oxoalkyl)-, 212
α,γ-phenyl, amine–imine tautomerism, 51, 269
  benzyl- Kröhnke reaction, 290
  2,6-dichloro-, halogen exchange, 284
  1,2-dihydro-, electrophilic substitution, 244
  4-dimethylamino-, use in acylation, 181
  2,4-dioxo-*, 541
  electrophilic halogenation, 189
  Hantzsch synthesis of, 539*

3-hydroxy-, Kolbe reaction, 193
  ionization potentials, 41
  nitration, 262
3-phenyltriazolo-, pyrolysis, 658
photoelectron spectroscopy, 41
reaction with organometallic compounds, 210
  with phosphorus oxychloride, 209
reduction, 217, 226
C-styryl-, 246
substituted, general synthetic methods, 545
sulfonation, 188
sulfur trioxide, 200, 206
  as a sulfonating agent, 295
tetrahydro-*, 526
2-vinyl-, adduct with dimethylamine, 265
Pyridine-2-thione, 3-amino-, 664
  cyclization, 673
  N-(2,4,6-trinitrophenyl)imine, nitration, 188
5-(1-Pyridinio)-pentyl iodide, radical cyclization, 681
Pyridinium ion, 27, 546*
  N-acyl-, 25
    reaction with activated methylene compounds, 212
      with metal enolates, 213
      with organometallic compounds, 211
  N_1-acyl-2-amino-, rearrange to 2-acylamino derivatives, 266
  1-alkoxy-, reaction with cyanide, 217
  1-alkyl-, Kröhnke reaction, 291
    Ladenburg rearrangement, 291
    loss of alkyl groups, 289
  N-alkyl-, oxidative amination, 206
    reaction with organometallic compounds, 211
  1-amino-*, 184
    prototropic equilibrium with N-imides, 291
    reactions, 292
  1-arylthio-, 295
  bromide, 1-cyano, 206
  1-(2,4-dinitrophenyl)-, Zincke reaction, 206
  1-ethoxycarbonyl-, reaction with phosphites, 209
    1-phenyl-, nitration and sulfonation, 291
  N-fluoro-, as fluorinating agents, 182
  H–D exchange, 222
  1-imide, 1,3-dipolar cycloaddition, 293
  N-ω-iodoalkyl-, intramolecular radical cyclization, 224
  1-methyl-, photolysis, 174
    reaction with hydroxide, 199
  nitriles, nucleophilic displacement of CN, 264
  N-nitro-, use as nitrating agent, 293
    tetrafluoroborate, 184
  3-oxido-, as 1,3-dipole, 232
    rearrangement, 175
  1-(4-pyridyl)-, 206
  1-(4′-pyridyl)-, nucleophilic displacement, 268
  reduction, 219
  N-styryl-, 246
  N-substituted, hydrogen exchange, 222
  1-vinyl-, Michael addition, 291
  ylides, cycloadditions, 670
    reaction with carbenes and nitrenes, 223
2(1H)-Pyridinone, 1-phenacyl-, cyclization, 673
Pyrido[1,2-a]azepines*, 681
Pyridone, 33, 268*, 542*
  alkylation, 274
  N-amino-, oxidation to nitrenes, 292
  anion, reaction with electrophiles, 222

diazo-, ring contraction, 542
from aminopyrones, 268
nitration, 187
prototropic tautomerism, 47
reaction with alkali, 202
with N-nucleophiles, 207
Pyrid-2-one*, 199, 207
N-alkoxy-, from 2-alkoxypyridine N-oxides, 276
aromaticity, 45–46
Chichibabin amination, 207
cycloaddition, 606
cyclodimerization, 236
diazo coupling, 193
1-ethyl-, aromaticity, 46
from pyran-2-ones, 207
oxidation, 194
Pyrid-4-one*, 540, 541
aromaticity, 45
Chichibabin amination, 207
Pyrido[2,3-*b*]quinoxalines*, 663
Pyrido[3,4-*b*]quinoxalines*, 663
2-Pyridylcarbene, interconversion with phenylnitrene, 265
2-Pyridylthiourea, ring closure, 674
Pyridynes, formation and reactions, 282, 283
Pyriliums, 3-oxido-, as 1,3-dipoles, 232
rearrangement, 175
Pyrimidine*, 576
alkoxy-, Hilbert–Johnson reaction, 276
rearrangement to N-alkylpyrimidinones, 276
allyloxy-, 277
4-chloromethyl-, ring expansion, 580
covalent hydration, 198
diazo coupling, 193
2,4-dichloro-, halogen exchange, 286
dihydro-, tautomerism, 239
dihydroxylation, 194
5,6-dihydroxylation, 194
2,5-dimethyl-, reaction with benzaldehyde, 258
from pyridazines, perfluoro-, 174
halo-, conversion to triazines, 601
halogen, Heck reaction, 282
hydroxy-, ethers, 274
hydroxymethylation, 193
Mannich reaction, 193
nitriles, nucleophilic displacement of CN, 264
nitrosation, 193
oxidation, 194
N-oxides, 183
Reissert–Henze reaction, 217
4-phenyl-, nitration, 262
reaction with organometallic compounds, 210
reduction, 218
2-stannyl-, 281
2,4,6-trichloro-, halogen exchange, 285
Pyrimidinium ion
N-acyl-, reaction with enol ethers, 213
anhydro-bases, 259
1-methyl-2-amino-, Dimroth rearrangement, 268
reaction with hydroxide, 200
1-trimethylsilyl-, reaction with enol ethers, 213
Pyrimidinones
N-alkyl-, from alkoxypyrimidines, 276
Pyrimid-2-one, 5-nitro, conversion to nitrophenol, 214
amination with HMDS, 276

Pyrimid-4-ones*, 579
reaction with hydrazine hydrate, 595
Diels–Alder reaction, 234
Pyrimidopyridazines*, 634
Pyrimidopyrimidines*, 634
Pyrimido[4,5-*e*]-1,2,4-triazine-6,8-dione, 5,7-dimethyl-, 205
Pyrimidotriazines*, 650
Pyrone, 30, 546
2-amino, Dimroth rearrangement to pyridones, 268
2,6-dimethyl-, reaction with benzaldehyde, 259
nitration, 187
nucleophilic displacement of carbonyl oxygen, 274
2-Pyrone*, 540
mass spectrum, 40
4-Pyrone, 2,6-dimethyl-*, 541
reaction with hydrazine, 207
Pyrrole, 58, 542–544*
2-acetyl-1-(2-hydroxyethyl)-5-nitro-, cyclization, 353
acid-catalyzed hydrogen exchange, 307
acyl-, carbonyl reactivity, 353
2-acyl-, acid-catalyzed rearrangement, 307
cyclization, 669
N-acyl-, preparation, 301
acylation, 310
alkali metal salts, 301
N-alkyl-, preparation, 301
C-alkylation, 313
2-amino-*, 535
stability, 354
N-amino-, Diels–Alder reactions, 337
aminomethylation, 315
3-amino-1-trityl-, tautomerism, 354
aromaticity, 80
1-aroyl-, dimerization with palladium(II), 326
as dienophiles, 331
base strength, 306
1-benzenesulfonyl-, 304, 307
acylation, 311
N-benzenesulfonyl-, 301
N-*t*-butoxycarbonyl-, reaction with hexafluorobutyne, 335
2-carbaldehyde, conformation, 83
Wittig reaction, 669
carbaldehydes, carbonyl reactivity, 352
carboxamide, thermal fragmentation to cyanide, 312
2-carboxylic acid*, 542
pK_a, 351
3-carboxylic acid, pK_a, 351
carboxylic acids, reactions, 351
pK_a, 351
catalytic reduction, 327
N-chlorination, 301
N-chloro-, rearrangement, 365
conformation, 83
core ionization energies, 78
cross-coupling reactions, 364
2-cyano-, 61, 312*
photochemical rearrangement to 3-cyanopyrrole, 298
1-β-cyanoethyl- Houben–Hoesch cyclization, 668
deprotonation at nitrogen, 320
2,4-diacyl-, 311
2-dialkylaminomethyl-, reactions, 350
diazo coupling, 316
2,5-dicarbaldehydes, 311

2,5-dicarboxylic acid, derivatives*, 544
2,5-dihydro-, conformation, 85
Ehrlich color reaction, 314
electrolytic oxidation, 318
electrophilic oxidation, 318
free radical attack, 325
from furans, 319
2-(2-furyl)-, conformation, 82
Gattermann aldehyde synthesis, 311
Grignard reagents, 300
halogenation, 308
Hantzsch synthesis of, 535*
Houben–Hoesch ketone synthesis, 311
2-hydroxy-, tautomers, 87
3-hydroxy-, tautomers, 87
N-hydroxy-, tautomers, 86
N-hydroxy-2-cyano-*, 543
2-hydroxymethyl-, reduction, 350
N- or C-hydroxymethylation, 314
interconversions, 320
IR spectra, 70
Knorr synthesis, 534*
Mannich reaction, 315
mass spectra, 73
mercuration, 316
2-methoxycarbonyl-, flash vacuum pyrolysis, 352
2-methoxycarbonyl-1-methyl-, formylation of, 304
N-methyl-, deprotonation at carbon, 320
1-methyl-2,5-dinitro-, reactivity towards nucleophiles, 343
1-methyl-3,4-dinitro-, reactions, 353
nitration, 307–308
3-nitro-*, 307
nitrosation, 317
^{13}C NMR, 65
^{1}H NMR, 62
^{14}N NMR, 67
Paal–Knorr synthesis, 530
pentaaryl-, rotational barrier, 82
pentachloro-, as a dienophile, 333
    equilibrium, 333
photosensitized reaction with oxygen, 330
Piloty–Robinson synthesis, 537
protonation, 321
2,3-quinodimethane, Diels–Alder reactions, 348
reaction with acetone, 315
    with carbenes, 324
    with dienophiles, 329
reactions of the C-lithio derivatives, 358
reduction by dissolving metals, 327
N-substituted, [2 + 2]cycloadditions, 332
    rearrangement to C-substituted derivatives, 365
1-substituted derivatives, cross-coupling, 364
sulfonation, 308
synthesis, 538
    from acyclic precursors, 529
    Mukaiyama reaction, 535*
tautomers, 85
tetrahydro-, basicities, 337
2-(2-thienyl)-, conformation, 82
3-(trifluoroacetyl)-, conformation, 83
N-(triisopropylsilyl)-, acylation, 310
1-trityl-, 303
Trofimov's synthesis, 537
UV spectra, 68

vertical ionization energies, 76
Vilsmeier–Haack acylation, 310
3-vinyl-, [4 + 2]cycloaddition, 346
N-vinyl-, 365, 537*
2H-Pyrrole, 1-oxides, 1,3-dipolar cycloaddition, 334
Pyrrole-2-carboxylic acids, synthesis by ring contraction
    of diazopyridones*, 542
Pyrrolenines, alkylation, 333
Pyrrolidines*, 515, 519, 523, 524
    basicities, 337
    2-chloromethyl-, cyclization, 672
    core ionization energies, 78
    Hofmann–Loeffler–Freytag reaction, 520*
    ionization potentials, 78
    mass spectra, 75
    ^{1}H NMR, 64
    ^{14}N NMR, 67
    1-oxides, rearrangement, 563
1-Pyrroline*, 526, 528
    trimers, 336
2-Pyrroline, N-methoxycarbonyl-, Friedel–Crafts
    acylation, 336
    N-methoxycarbonyl-, Vilsmeier formylation, 336
δ^3-Pyrroline, disproportionation, 336
Pyrroline N-oxide, cycloaddition, 674
Pyrrolizines*, 668
    ^{1}H NMR spectra, 64
Pyrrolo[1,2-a]azepines*, 682
3H-Pyrrolo[1,2-c]imidazole*, 669
Pyrroloisoxazoles*, 674
Pyrrolo[1,2-b]isoxazoles, perhydro-*, 674
1H-Pyrrol-3(2H)-ones, alkylation, 340
Pyrrolopyrazoles*, 671
1H-Pyrrolo[2,3-b]pyridine, 184
    N-amination, 184
Pyrrolo[3,2-d]pyrimidines, 233
4H-Pyrrolo[2,3-d]thiazoles*, 454
Pyrrolylacetic acids*, 543
Pyrrolylzinc compounds, 359
Pyrylium ion*, 546
    2-amino-, Dimroth rearrangement, 268
    anhydro-bases, 259
    conversion to pyridines, 207
    4-methoxy-2,6-dimethyl-, nucleophilic displacement of
        alkoxy, 276
    ^{1}H NMR, 26, 27
    reaction with azide ion, 580
        with hydrazines, 566
        with hydroxide, 202
        with C-nucleophiles, 214
        with phosphines, 209
        with sulfite, 208
    reduction, 219, 226
    ring opening, 561
    structure, 25

Quinazoline, 634*
    Chichibabin amination, 204
    covalent hydration, 198
    2,4-dichloro-, halogen exchange, 285
    nitration, 252
    2-ones, N-amino-, oxidation, 648
    3-oxides, photoisomerization of, 654
        reaction with nucleophiles, 641

Quinazolone*, 637
 N-amino-, oxidation, 512
 N-amino-2-ethyl-, aziridination of alkenes with, 512
Quinoline*, 616, 619, 621, 622
 3-amino-4-nitro-, 205
 basicity, 178
 Bayer synthesis, 619
 3-bromo-*, 621
 4-carboxylic acids, Pfitzinger synthesis of, 617
 Chichibabin amination, 204
 Combes synthesis*, 619
 2-cyano-, formation by Reissert–Henze reaction, 217
 1,2-dihydro-*, 218, 620*
 Doebner–von Miller synthesis, 619*
 electrophilic chlorination, 190
 Friedländer synthesis, 617*
 N-imides, photolytic ring expansion, 632
 nitration, 252
 oxidation, 253
 1-oxide
  bromination, 192
  nitration, 253
  reaction with phosphorus oxychloride, 209
  Reissert–Henze reaction, 217
  ring reduction rearrangement, 615
 reaction with organometallic compounds, 210
 reduction, 218, 226
 Riehm synthesis, 619
 sulfonation, 252
 tetrahydro-*, 620
Quinolinium ion, N-alkyl-, oxidative amination, 206
 1-benzoyl-, Reissert reaction, 217
 1-methyl-
  reaction with cyanides, 216
   with Grignard reagents, 211
  sodium borohydride reduction, 219
Quinolinones, tautomerism, 50
4-Quinolinones, Conrad–Limpach cyclization, 618*
Quinolizidine, quaternization, 249
Quinolizinium ions, reaction with piperidine, 207
2-Quinolone*, 199, 618, 621
 4-carboxylic acids*, 622
 [2 + 2]cycloaddition, 235
 3-hydroxy-*, 622
 nitration, 253
4-Quinolone, 617*, 618*
 diazo coupling, 193
 dihydro-, 620
 formation by Camps reaction, 617
Quino[7,8-h]quinoline, proton sponge, 177
Quinoxaline*, 638
 di-N-oxides, 421, 641*
 oxidation, 194
 1-oxides, photoisomerizations, 654
 reduction, 218
 tetrahydro-*, 639

Reumycin, 289
Ring–chain isomerizations, 479
Ring–chain tautomerism, 52, 373
 1,2,3-benzoxadiazoles, 139
 imidazolidines, 138
 1,2,3-oxadiazole, 138
 oxazolidines, 138
 1,2,3-thiadiazoles, 139

1,2,3-triazoles, 138
Ring-strain energies
 for 1,3-dioxalane, 125
 for 1,3-dithiolane, 125
 for 1,3-oxathiolane, 125
Rosindoles*, 314
Rotaxane, 53

Saccharins*, 629
Selenadiazoles, aromaticity, 126
Selenophene, 624
 aromaticity, 80
 carbaldehydes, carbonyl reactivity, 352
 2-carboxylic acid, conformation, 84
  pK$_a$, 351
 carboxylic acids, pK$_a$, 351
 chloromethylation, 315
 conformation, 84
 [2 + 2]cycloaddition, 332
 deprotonation, 320
 2-hydroxy-, tautomers, 87
 interconversion, 320
 IR spectra, 70
 3-lithio-, 359
 3-lithio-2,5-dimethyl-, ring opening, 362
 mass spectra, 73
 2-mercapto-, tautomers, 89
 mercury derivatives, 365
 ^{13}C NMR, 65
 ^{1}H NMR, 62
 ^{77}Se NMR, 67
 relative rates of electrophilic substitution, 303
 ring heteroatom attack by halogens, 301
 tetrahydro-,
  conformation, 85
  ionization potential, 78
  mass spectra, 75
  ^{1}H NMR, 64
 UV spectra, 68
 vertical ionization energies, 76
Small and large rings, non-aromaticity of large rings, 158
 photochemical reactions, 476
 photoelectron spectroscopy, 156, 157
 N-protection, 498
 X-ray diffraction, 148
Small heterocyclic systems, spectra, 154
Small rings
 aromaticity and antiaromaticity, 157
 basicity, 481
 carbanions, 488
 cis–trans rearrangement reactions, 478
 cycloaddition reactions, 492
 electrophilic attack, 483
 fragmentation, 477
  reactions, 476
 mass spectrometry, 155
 oxidation, 490
 radical attack, 489
 reactivity, 475
  of C-linked substituents, 496
  of N-linked substituents, 498
  with nucleophiles, 485
 reduction, 491
 ring expansion, 479
 ring–ring valence isomerizations, 479

spectra, 155
Stibabenzene, 47
  deshielding of α-protons, 47
Strychnine, 57
Sulfones, by oxidation of thiazoles, isothizoles,
      benzisothiazoles, 396
Sulfoxides, by oxidation of thiazoles, isothiazoles,
      benzothiazoles, 396
Supramolecular structures, 53, 180
Sydnone
  aromaticity, 126
  3-aryl-, mercuration, 393
    Friedel–Crafts acetylation, 392
    Vilsmeier formylation, 392
  cycloaddition of, 421, 646
  1,3-dipolar cycloaddition, 560
  metallation, 409, 464
  N-phenyl-, irradiation, 595
Synthesis of heterocycles, general principles, 502

Tautomerism
  5-membered rings, one heteroatom, 85
    two or more heteroatoms, 131
  6-membered rings, 47
  small and large rings, 161
  thiol–thione, in pyridines and azines, 51
  valence bond, 52–53, 161
Tellurophene*, 624
  aromaticity, 80
  2-carboxylic acid, conformation, 84
    pK$_a$, 351
  carboxylic acids, pK$_a$, 351
  conformation, 84
  deprotonation, 320
  IR spectra, 70
  2-lithio-, 359
  mass spectra, 73
  ^{13}C NMR, 65
  ^{1}H NMR, 62
  ^{125}Te NMR, 67
  relative rates of electrophilic substitution, 303
  ring heteroatom attack by halogens, 301
  tetrahydro-, ^{1}H NMR, 64
    ionization potential, 78
    vertical ionization energies, 76
Tetraazamacrocycles, selectivity for transition-metal ions,
      147
Tetraazapentalene*, 682
Tetracoordinated sulfur compound, 335
Tetrahydrofuran, see Furan, tetrahydro-
Tetramisole*, 671
1,2,4,5-Tetrazine
  3,6-bis(trifluoromethyl)-, cycloadditions, 646
  [1 + 4]cycloaddition, 232
  Diels–Alder reactions, 231, 564, 596, 598
  dihydro-*, 603
  1,4-dihydro-, stable radical cations, 242
  6-ethyl-3-phenyl-1,6-dihydro-, crystal structure, 25
    homoaromaticity, 25
  hydrolysis, 199
  hydrolytic ring cleavage, 199
  nucleophilic attack at ring nitrogen, 220
  1,2,3,4-tetrahydro-, formation by reduction, 242
  thermal fragmentation, 172
  UV spectra, 35

1,2,3,4-Tetrazines*, 651
Tetrazole*, 602
  acylation, 386
  1-alkenyl-, photochemical degradation, 575
  alkylation, 384
  5-amino-, 602
    bromination, 450
    Dimroth rearrangement, 140, 449, 602
    nitrosamines, 451
  annular tautomerism, 132
  aromaticity, 126
  diazo-, pyrolysis, 451
  5-diazo-, ^{13}C labeled, thermal degradation of, 371
  1,5-dicarboxylate, methyl-, photolytic rearrangement,
      595
  1,5-dimethoxycarbonyl-, photolysis, 446
  1,5-diphenyl-, thermolysis, 371
  2,5-disubstituted- thermal ring opening, 594*
  equilibrium with imidoyl azide, 373
  5-hydrazino-, rearrangement, 449
  metallation, 409
  N-methyl-, base-induced H–D exchange, 410
  ^{13}C NMR, 110
  ^{1}H NMR, 104
  1-phenyl-5-aryloxy-, 458
  2-phenyl-5-carboxamido-, 602
  photolysis, 372
  ring opening, 399
  1-substituted, halogenation, 392
  α-substituted alkyl-, pyrolysis, 448
  thermolysis, 371
  1-(thiazolyl-2)-5-phenyl-1H-, thermolysis, 675
  5-thio-*, 602
Tetrazolines, 5-alkylidene-1,4-dimethyl-, cycloaddition
      reactions, 424
Tetrazolium ions, 5-halo-, 393
  mercuration, 393
Tetrazolo[1,5-b]pyridazine, reduction, 406
Tetrazolo[1,5-a]pyridine, flash vacuum pyrolysis of, 371
Tetrazolo-triazines, reduction, 218
Tetrazolyl radicals, spin density, 416
1,2,4,5-Tetroxanes, 3,3,6,6-tetrasubstituted *, 603
Theophylline, 7,8-diamino, cyclomerization, 680
Thermodynamics
  5-membered rings, one heteroatom, 79
    two or more heteroatoms, 124
  6-membered rings, 41
  small and large rings, 157
Thiabenzenes, reactions of, 237
1,3,5-Thiadiazepine*, 602
1,2,4-Thiadiazine*, 597
1,2,6-Thiadiazine*, 591
  1-oxides, extrusion of sulfur monoxide, 238
1,3,4-Thiadiazines*, 597
1,3,5-Thiadiazines*, 601
Thiadiazole*, 469, 595
  aromaticity, 126
  photodimerization, 584
1,2,3-Thiadiazole*, 588
  5-azido-, rearrangement to 1,2,3,4-thiatriazole, 375
  ^{13}C NMR, 111
  ^{1}H NMR, 105
  photolysis, 528
    to thiirene, 372
  rearrangement, 375

ring–chain tautomerism, 139
thermolysis, 575
1,2,4-Thiadiazole*, 592, 595
 amino-, nitrosamines, 451
 3-amino-, diazonium salts, 452
 3-diazonium salts, 452
 5-diazonium salts, 452
 Dimroth rearrangements, 449
 13C NMR, 111
 1H NMR, 105
 ring opening, 397
 reaction with free radicals, 415
1,2,5-Thiadiazole*, 590
 anion radicals, 415
 3,4-dione*, 594
 1,1-dioxide, 3,4-diphenyl-, thermal fragmentation, 371
 electrochemical reduction, 415
 halogenation, 392
 1H NMR, 105
 reductive cleavage, 418
1,3,4-Thiadiazole*, 591
 amino-, diazonium salts, 452
 2-amino-, bromination, 392
 2-chloro-, Dimroth rearrangement, 462
 13C NMR, 111
 1H NMR, 105
1,2,4-Thiadiazolidine*, 594
 interconversion with 1,2,4-dithiazolidine, 421
δ³-1,3,5-Thiadiazoline, fragmentation, 429
1,3,4-Thiadiazolium chloride, 2-phenyl-4-benzoyl-,
  benzoyl migration, 392
1,2,3-Thiadiazolo[4,5-*d*]pyridazines*, 645
[1,2,4]Thiadiazolo[2,3-*a*]pyridine*, 674
2-Thianaphthylium ions, cycloaddition, 684
Thiane, 40
 strain energies, 47
Thianthrene, 25, 40, 665*
 crystal structure, 25
 extrusion of sulfur, 660
1,2,4,6-Thiatriazines*, 603
Thiatriazole, 5-phenylamino-, Dimroth rearrangement to
  1-phenyl-5-mercaptotetrazole, 435
1,2,3,4-Thiatriazoles
 from 5-azido-1,2,3-thiadiazoles, 375
 5-phenyl-, 13C NMR, 111
 photolysis, 372
 5-substituted amino-, 602
 thermolysis, 371
1,2,3,5-Thiatriazolidines*, 602
1,2-Thiazepines*, 566
1,3-Thiazepines*, 580
1,4-Thiazepines*, 586
1,2-Thiazetidines*, 554
1,3-Thiazetidines*, 567
Thiazines*, 631
1,2-Thiazines*, 563
1,4-Thiazines*, 582
6*H*-1,3-Thiazines*, 578
1,3-Thiazinium ions*, 577–578
 reaction with N-nucleophiles, 207
2*H*-Thiazino[3,2-*a*]benzimidazole, 4-phenyl-, 677
Thiazirine, 3-phenyl-*, 372
Thiazole*, 569, 573
 CH-acidity, 410
 5-alkylation, 392

amino-, diazonium salts, 452
 Sandmeyer reaction, 452
 tautomeric forms, 137
aromaticity, 126
Chichibabin amination, 402
cycloaddition, 679
desulfurization, 397
1,2-diamino-4-methyl-, cyclization, 674
Diels–Alder reactions, 420
2-(dimethylamino)-, nitrosation, 394
hydroxy-, tautomeric forms, 137
lithiation, 409
mercapto-, 458
 tautomeric forms, 137
mercuration, 393
2-methylthio-5-phenyl-*, 574
nitration, 391
13C NMR, 111
1H NMR, 105
oxidation, 396
phenyl-, nitration, 444
ring opening, 399
sulfonation, 391
tetrahydro-, 432
2-thione, cyclization, 672
2(3*H*)thione, N-amino-4-phenyl-, reaction with
  phenacyl bromides, 678
2-(trimethylsilyl)-, acylation, 411
Thiazolidine-2,5-dione*, 573
Thiazolidines, 138
 hydrolysis, 432
 ring–chain tautomerism, 138
2-Thiazolines*, 573, 574
 Diels–Alder reaction, 679
δ⁴-Thiazolines, aromatization, 427
2-Thiazolinones, Gattermann reaction, 392
 Reimer–Tiemann reaction, 392
Thiazolium ions, CH-acidity, 410
 2-methyl-, reaction with α-dicarbonyl compounds, 678
 reduction, 407
Thiazolo[3,2-*a*]benzimidazole, 3-benzyl-*, 673
Thiazolo[2,3-*b*]pyridinium*, 678
Thiazolo[3,2-*a*]pyridinium*, 673
Thiazolo[3,2-*a*]pyridinium bromide, 3-amino-2-
  ethoxycarbonyl-*, 673
Thiazolo[5,4-*d*]pyrimidine*, 633
Thiazolo[3,2-*b*] [1,2,4]thiadiazolylium, 2-amino-*, 674
Thiazolo[2,3-*b*]thiazolium*, 672
 tetrahydro-, 672
Thiazolo[3,2-*b*][1,2,4]triazole*, 674
 2-phenyl-*, 675
2-Thiazolylthiourea, oxidation, 674
Thieno[*c*]cinnolines*, 317
Thienodiazepines*, 646
Thieno[3,2-*b*]pyridine N-oxide, nitration, 187
Thieno[3,4-*b*]pyrrole-1,1-dioxide, Diels–Alder reactions,
  348
Thieno[2,3-*d*]thiazole*, 633
Thieno[3,2-*d*]thiazole*, 634
Thieno[3,2-*b*]thiophene, 615
Thienylboronic acids, cross-coupling, 363
3-Thienyllithium, 357
2-Thienylmagnesium cross-coupling, 362
Thienyl-(tributyl)stannanes, 361
2-Thienylzinc, cross-coupling, 363

Thiepane, 522, 524
Thiepanone*, 522
Thiepan-4-one*, 524
3-Thiepanone, Dieckmann reaction, 522*
Thiepine
  1,1-dioxide*, 549
    2,7-dihydro-*, 544
  thermal stability, 478
4H-Thiepinium ion, structure, 152
Thiepins, 549
Thietane*, 514
  fused derivatives*, 606
  2-vinyl-*, 486
Thietane-2-one*, 525
Thiete 1,1-dioxides*, 525
Thiinium ion, 25, 27, 548*
  reaction with lithium aryls, 195
    with hydrazines, 566
    with O-nucleophiles, 202
Thiiniums, 3-oxido-, as 1,3-dipoles, 232
Thiinones, 30
Thiins, dihydro-*, 526
2H-Thiins, Diels–Alder reactions, 245
Thiirane, 513, 524*
  dimerization, 585
  1,1-dioxide, ^{33}S NMR, 152
  from oxiranes, 524
  NMR data, 150, 152
  1-oxide, ^{33}S NMR, 152
  reaction with benzyne, 490
Thiiranium salts, S-alkyl-, desulfurization, 484
2-Thiirene*, 372, 528
  ab-initio calculations, 146
  antiaromaticity, 157
  dioxide, diphenyl-, cycloaddition, 670
Thioindole, 1-methyl-3-phenyl-, rearrangement, 355
Thiolactones, 520
δ-Thiolactones, 521
Thiolane*, 519, 521
  basicities, 337
1,2-Thiolylium ion, 3-phenyl-, reaction with hydrazine, 560
Thiophene*, 397, 542–544
  acid-catalyzed hydrogen exchange, 307
  acid-catalyzed rearrangement, 307
  acyl-, carbonyl reactivity, 353
  acylation, 311
  alkylation 313
  amino-, 354
  2-amino-, 533, 535*
    diazotization, 354
    Gewald synthesis, 533
  3-amino-, [2+2]cycloaddition, 332
  aromaticity, 80
  as diene/dienophile, 329
  base strength, 306
  bis(trifluoromethyl)-, photolysis, 299
  2-bromo-, Grignard reagent, 357
  3-bromo-, reaction with Grignard reagents, 357
  bromonitro-, halogen reactivity, 357
  carbaldehydes, carbonyl reactivity, 352
  carboxamide, thermal fragmentation to cyanide, 312
  2-carboxylic acid, conformation, 84
    pK$_a$, 351
  3-carboxylic acid, pK$_a$, 351

carboxylic acids, dianions, 358
  pK$_a$, 351
  reactions, 351
2-chloromethyl-, reactivity, 349
chloromethylation, 315
conformation, 83–84
core ionization energies, 78
cross-coupling reactions, 363
3-cyano-, 61
deprotonation, 320
desulfurization, 328
dialkyl-, [2+2]cycloadditions, 332
2,3-diamino-*, 354
3,4-diamino-*, 353
2,5-di-t-butyl-*, 314
2,3-didehydro-, cycloaddition products, 352
2,3-dihydro-, conformation, 85
  [2+2]cycloadduct, 336
  [2+3]cycloaddition, 336
2,5-dihydro-, conformation, 85
  1,1-dioxides, sulfur dioxide extrusion from, 337
3,4-dimethylene-, Diels–Alder reactions, 348
2,3-dimethylene-2,3-dihydro-, Diels–Alder reactions, 348
2,5-dinitro-, reactivity towards nucleophiles, 343
1,1-dioxides, 2,5-dihydro-, extrusion of sulfur dioxide, 337
π-donation, 58
electron density, 58
Feist–Benary synthesis, 536*
free radical attack, 325
from isothiazoles, 397
geometry, 61
halo-, S$_{RN}$1 mechanism, 356
α-halo-, 357
halogenation, 309
Hinsberg synthesis, 537*
hydroxy-, C-methylation, 340
3-hydroxy-, aldol reaction with benzaldehyde, 341
  tautomers, 87
2-hydroxymethyl-, reactions, 349
indophenine test, 315
interconversions, 320
intramolecular Diels–Alder, 330
iodo-, cross-coupling, 363
ionic method for hydrogenating, 322
IR spectra, 70
3-lithio derivatives, ring opening, 362
Mannich reaction, 315,316
mass spectra, 73
2-mercapto-, tautomers, 89
mercuration, 316
mercury derivatives, 365
methoxy*, 356
nitration, 307–308
nitro-, vicarious nucleophilic substitution of hydrogen, 323
2-nitro-, reaction with amines, 354
3-nitro-*, 361
^{13}C NMR, 65
^{1}H NMR, 62
^{33}S NMR, 67
oligomers and polymers*, 363
oxidation by peracid, 302
Paal–Knorr synthesis, 530

per(acetylenated)*, 363
2-phenyl-, rearrangement to 3-phenyl-thiophene, 298
S-protonated, inversion barrier, 59
S-protonation, 59
reaction with aldehydes, 315
    with carbenes, 324
reduction by dissolving metals, 327
relative rates of electrophilic substitution, 303
selective debromination, 357
sulfonation, 308
sulfones, Diels–Alder reactions, 334
2-sulfonic acid, reactions, 355
synthesis from acyclic precursors, 529
tetrahydro-, basicities, 337
    conformation, 85
    core ionization energies, 78
    ionization potential, 78
    mass spectra, 75
    ^{1}H NMR, 64
2-thiol*, 360
    conformation, 83
2,3,5-tribromo-, selective debromination, 357
UV spectra, 68
vertical ionization energies, 76
2-vinyl-, [4+2]cycloaddition, 346
S,N-ylides, 301
Thiophene-2-carbaldehyde, 3-bromo-*, 362
Thiophenium bis(alkoxycarbonyl)methylides, 301
    reactions, 366
Thiophenium ions, 301
2-Thiopyridones, 46
Thiopyran, *see* Thiin
Thiopyrylium, *see* Thiinium ions
Thioxanthen-9-one 10,10-dioxides, ring expansion, 665
Thymine, [2+2]cyclodimerization, 235
Toxoflavins, demethylation, 289
Transition metal complexes, reactions, 495
Transition metal catalyzed cross-coupling reactions, *see*
    Cross-coupling reactions
3,4,7-Triaza-2,4-norcaradienes, photochemical
    rearrangement, 598
Triazapentalene*, 682
Triazaphenothiazine*, 664
Triazenopyrazoles, dehydrogenation, 676
2*H*-1,2,4-Triazepine*, 598
4*H*-1,2,4-Triazepine*, 598
1,2,4-Triazepines, nitrogen elimination, 477
1,2,3-Triazine, 182, 183, 470, 591*
    [2+4]cycloaddition, 229
    hydrolysis, 199
    hydrolytic ring cleavage, 199
    N-imines, reaction with dipolarophiles, 234
    iodo-, Heck reaction, 282
        with enamines, 549
        with organometallic compounds, 210
    thermal fragmentation, 172
1,2,4-Triazine*, 596
    3-amino-, 596
    derivatives, photochemical hydration, 199
    2-oxides, 3-fluoro-, 267
    4-oxides*, 597
    reaction with cyanide ions, 216
1,3,5-Triazine*, 599, 601
    azido-, photolysis, 271
    2,4-diamino-, 599

hexahydro-*, 600
nucleophilic ring cleavage, 215
N-sulfonyl-*, 601
2,4,6-triamino-, 265
tricarbaldehyde, synthetic equivalent, 264
2,4,6-trichloro-, halogen exchange, 285
2,4,6-tris(dimorpholinomethyl)-, 264
2,4,6-tris(trichloromethyl-, 265
1,2,4-Triazine-5(2H)-ones, 3-methylthio-, N-amination,
    184
1,2,3-Triazinium 2-dicyanomethylides, 182
Triaziridines*, 588
Triazole
    4-amino-, Kolbe-type reaction, 392
    5-lithio-, 408
1,2,3-Triazole*, 588*, 590
    acylation, 385
    acyl migration, 134
    alkylation, 383
    amino-, diazotization, 452
    1-amino-*, 588
    4-amino-, carboxylation, 392
    annular tautomerism, 132
    1-arenesulfonyl-, ring–chain tautomerism, 373
    aromaticity, 126
    N-arylimides, rapid exchange of aryl rings, 107
    5-azido-, rearrangement to tetrazoles, 602
    1-benzyl-4-hydroxy-*, 462
    bromination, 392
    1-carbonitrile*, 590
    1-cyano-, 108
    1-cyanosulfonyl-, ring–chain tautomerism, 373
    4-diazo-5-phenyl-4*H*-, 1,7-cycloaddition, 422
    Dimroth rearrangement, 449
    group migrations, 134
    lithiation, 409
    N-(*p*-methoxybenzyl) migration, 134
    microwave spectrum, 102
    1-naphthyl-, photolysis, 666
    ^{13}C NMR, 110
    ^{1}H NMR, 104
    ^{15}N NMR, 113
    NOE discrimination of isomers, 103
    1-oxides*, isomer synthesis, 588
        photolytic ring expansion, 603
    ring–chain tautomerism, 138
    ring opening, 399
    trialkylstannyl migration, 134
    1-trifluoromethylsulfonyl-, open-chain isomer, 108
    2-trimethylsilyl-, reaction with acid chlorides, 385
    trimethylsilyl migration, 134
1,2,4-Triazole*, 591–596
    1-alkyl-, reaction with nitrenes, 414
    alkylation, 384
    annular tautomerism, 132
    aromaticity, 126
    4-arylamino-*, 592
    lithiation, 409
    ^{13}C NMR, 110
    ^{1}H NMR, 104
    ^{15}N NMR, 114
    1-phenyl-, photolysis to indole, 465
    stable carbenes, 412
    N-substituted, thiation, 415
    transition-metal derivatives, 388

1,2,3-Triazoline*, 590

Triazolinedione, 4-phenyl-, (PTAD), as a dienophile, 424

δ^2-1,2,3-Triazolines, dehydrogenation, 428
   1-alkyl-, hydrolytic decomposition, 429
  photodecomposition, 428

1,2,4-Triazoline-3-thione*, 592

1,2,4-Triazoline-5-thione, 4-amino-3-hydrazino-, reaction with aldehydes, 678

1,2,4-Triazolinone*, 594

1,2,4-Triazolium dicyanomethylide, 1-phenyl-, 1,3-dipolar cycloaddition, 425

1,2,4-Triazolo[3,4-*a*]isoquinoline, 3-phenyl-*, 675

1,2,4-Triazolo[1,5-*a*]pyridine 3-oxides*, 674

Triazolo[4,5-*d*]pyrimidine*, 647

1,2,3-Triazolo[4,5-*d*]-1,2,3-triazole, 1-amino-5-phenyl-*, 651

Tribenz[*b,d,f*]oxepin*, 666

Tribenz[*b,d,f*]thiepine, 666

1,3,5-Trioxanes*, 600

1,2,4-Trioxolane, conformation, 130

Trithiadiazepine
  calculations, 147
  nucleophilic attack, 485
  structure, 160
  thermal stability, 478

1,3,5-Trithiane*, 600

1,6,6aλ^4-Trithiapentalenes*, 684

Trithiatriazepine, calculations, 147

1,3,5,2,4,6-Trithiazepine, thermal stability, 478

Tri-(2-thienyl)methane, lithiation, 358

Tri-(2-thienyl)methyl carbenium tetrafluoroborate, reduction, 349

1,2,3-Trithiole, 590
  2-oxide, 4,5-dicyano-*, 590

Types of heteroatoms, 11

Ultraviolet spectroscopy, 35, 68–69
  5-membered rings, one heteroatom, 68
    two or more heteroatoms, 114
  6-membered rings, 35
  small and large rings, 153

Uracil
  addition of sulfite ion, 208
  6-(benzylidenehydrazino)-, cyclization, 628
  [2 + 2]cyclodimerization, 235
  1,3-dimethyl-, oxidation, 194
  hydrazino-, nitrosation, 650
  photohydration, 198
  Reimer–Tiemann reaction, 193

Urazole, reaction with diazomethane, 457

Verdazyls, reduction, 243

Veronal*, 576

Vitamin B12, 56

Xanthone, 201
  reduction, 275

Xanthylium ions, 207
  covalent addition of chloride ion, 209
  reaction with C-nucleophiles, 214
    with hydroxide, 201

Yohimbine, 57